AN ENCYCLOPAEDIA OF THE HISTORY OF TECHNOLOGY

AN ENCYCLOPAEDIA
OF THE
HISTORY OF TECHNOLOGY

EDITED BY

IAN McNEIL

ROUTLEDGE
LONDON AND NEW YORK

First published 1990
by Routledge
First published in paperback 1996
by Routledge
11 New Fetter Lane, London EC4P 4EE

Routledge is an International Thomson Publishing company

© 1990, 1996 Routledge

Set in 10/12pt Ehrhardt, Linotron 202,
by Promenade Graphics, Cheltenham
Printed in Great Britain by
Richard Clay Ltd., Bungay, Suffolk

All rights reserved. No part of this book may be reprinted or reproduced or utilized in any form or by any electronic, mechanical, or other means, now known or hereafter invented, including photocopying and recording, or in any information storage or retrieval system, without permission in writing from the publishers.

British Library Cataloguing in Publication Data
An Encyclopaedia of the history of technology
   1. Technology–History
   I. McNeil, Ian
   609   T15
   ISBN 0-415-14792-1

Library of Congress Cataloging in Publication Data
An Encyclopaedia of the history of technology/edited by Ian McNeil.
1062pp.   4.25cm.
Bibliography
Includes index.
ISBN 0-415-14792-1
1. Technology–History. I. McNeil, Ian.
T15.E53   1989
609–dc20   89-10473

CONTENTS

Preface xiv

Introduction: Basic Tools, Devices and Mechanisms 1
Ian McNeil

The place of technology in history 1
Science and technology 2
The archaeological ages 4
The seven technological ages of man 5
The first age: man, the hunter, masters fire 5
The second age: the farmer, the smith and the wheel 11
The third age: the first machine age 22
The fourth age: intimations of automation 27
The fifth age: the expansion of steam 31
The sixth age: the freedom of internal combustion 37
The seventh age: electrons controlled 40

PART ONE: MATERIALS 45

1. Non-Ferrous Metals 47
   A. S. Darling

   Neolithic origins 47
   Copper 47
   Tin and bronze 57
   Metallurgy in the Roman world 66
   Brass and zinc 72
   The emergence of nickel 96
   The light metals, aluminium and magnesium 102
   Age hardening alloys 121
   Development of high temperature alloys 124

	Powder metallurgy	128
	Sintered carbide cutting tools	135
	Titanium and the newer metals	141
	Niobium	144
2.	**Ferrous Metals** *W. K. V. Gale*	146
	Introduction	146
	Wrought iron: the prehistoric era to AD 1500	147
	Cast iron: 1500–1700	149
	Mineral fuels	153
	Steam power and early industrialization	154
	Steel	159
	The industrial iron age: 1800–1850	160
	The steel age	167
	Modern steelmaking	176
3.	**The Chemical and Allied Industries** *Lance Day*	186
	Introduction	186
	Pottery, ceramics, glass	190
	Textile chemicals	199
	Fuels	206
	Polymers: rubbers, plastics and adhesives	216
	Heavy inorganic chemicals	220

PART TWO: POWER AND ENGINEERING — 227

4.	**Water, Wind and Animal Power** *J. Kenneth Major*	229
	Water power	229
	Water turbines	242
	Wind power	245
	Animal power	260
5.	**Steam and Internal Combustion Engines** *E. F. C. Somerscales*	272
	Introduction	272
	Steam engines	273
	Steam turbines	288
	Internal combustion engines	303

	Gas turbines	329
	External combustion engines	341
	Appendix	342

6. Electricity — 350
Brian Bowers

Static electricity	350
Current electricity	351
Michael Faraday	354
Generators	356
Arc lighting	362
The filament lamp	365
Central power stations	369
Transmission: AC v DC	370
Economics: off-peak loads	372
Measurement	373
Electromagnetic engines	377
Practical electric motors	381
Modern electric motors	384
The steam turbine	385
Electricity today	385

7. Engineering, Methods of Manufacture and Production — 388
A. K. Corry

Introduction	388
Bronze and iron age tools	388
Early machines	390
Measurement	391
General machine tools	392
Mass production	404
Twentieth-century organization of production	412
Welding, electro-forming and lasers	417
Wartime advances	420
Control revolution and electronic metrology	422
Flexible manufacturing systems	424
The automatic factory	427

PART THREE: TRANSPORT — 429

8. Roads, Bridges and Vehicles — 431
Ian McNeil

Road construction	431
Early road transport	438

	Powered road transport: initial experiments	439
	Bicycles	442
	Motor cycles	447
	The motor car	449
	Automotive engines	453
	Trams and trolleybuses	456
	Buses	457
	Electrically operated vehicles	458
	Road transport ancillaries	459
	Road maps	461
	Bridges	462
	Tunnels	467
	Earthmoving and roadbuilding machinery	470
9.	**Inland Waterways** *John Boyes*	474
	The ancient world	474
	The British Isles	475
	France	482
	The Low Countries	489
	Germany	494
	The Rhine	499
	Italy	500
	Greece	501
	Sweden	501
	The Soviet Union	502
	Eastern Europe	504
	Spain and Portugal	505
	The Suez Canal	505
	Japan	506
	Canada	506
	The United States	509
	The Panama Canal	514
	Canal and river craft	514
	The contemporary scene	515
10.	**Ports and Shipping** *A. W. H. Pearsall*	519
	Oar and sail	519
	Steamships	527
	Merchant shipping	529
	Warships	532
	Submarines	537
	Hovercraft and hydrofoils	538
	Lifeboats and lifesaving	539

	Anchors and cables	540
	Lights and buoys	541
	Navigation	543
	Charts and sailing directions	546
	Ports and harbours	547
	Shipbuilding and dockyards	551
	Diving	553
11.	Rail	555
	P. J. G. Ransom	
	Railways before locomotives	555
	The first steam locomotives	558
	The railway at work	562
	The first trunk lines	565
	Railway promotion	567
	Main line motive power and operation	568
	Early railway development in the United States	572
	Continental Europe	574
	Narrow gauge	575
	Japan and China	576
	Pullman and wagons-lit	577
	Larger locomotives	579
	Compounds	581
	Specialized railways	583
	Early electrification	584
	Monorails	586
	The peak years	586
	The aftermath of war	593
	Internal combustion	594
	Late steam	596
	Diesel take-over in the United States	598
	Post-War Europe	598
	British Railways	599
	Freight containers and bulk freight	601
	New high-speed lines	603
	Surviving steam	605
	Railway preservation	606
12.	Aeronautics	609
	J. A. Bagley	
	Early attempts at flight	609
	Balloons	609
	Airships	614
	Heavier-than-air flying machines: the pioneers	617
	Steam power	619

Gliders	621
The Wright brothers	622
European pioneers	624
Military and commercial applications	626
The inter-war years	629
The Second World War	632
Demise of the flying-boat	633
Expansion of civil aviation	634
Introduction of jet propulsion	634
Jet airliners	636
Supersonic commercial aircraft	638
Jet-supported flight	639
Helicopters and rotary wings	640
Convertible and hybrid aircraft	644
Recreational aircraft and gliders	644
Man-powered flight	646

13. **Spaceflight** — 648
    John Griffiths

Black powder rockets	648
Spaceflight pioneers	649
Vengence Weapon Two	649
Post-War research	651
Manned spaceflight and the space race	652
Satellite technology	656
Probes to the moon	658
Probes to the planets	659
Launch vehicles	661

PART FOUR: COMMUNICATION AND CALCULATION — 663

14. **Language, Writing, Printing and Graphic Arts** — 665
    Lance Day

Language	665
Writing	666
The invention of printing	669
The growth of printing	671
Technological innovation in the nineteenth century	674
Colour printing	681
Office printing	682
Optical character recognition	684

15. Information: Timekeeping, Computing, Telecommunications and
    Audiovisual Technologies   686
    Herbert Ohlman

    Introduction   686
    The evolution of information technologies   687
    The timing of inventions   691
    Timekeeping   694
    Counting, calculating and computing   697
    The telegraph   710
    The telephone   717
    The gramophone   720
    Radio and radar   725
    Photography   729
    Facsimile and television   743
    Communications satellites   746
    Information storage today   748

PART FIVE: TECHNOLOGY AND SOCIETY   759

16. Agriculture: The Production and Preservation of Food and Drink   761
    Andrew Patterson

    Introduction   761
    Hunter gatherer to farmer   762
    Arable farming   767
    Sowing   772
    Fertilizers   774
    Pest control   775
    Weed control   777
    Crop rotation   779
    Harvesting   779
    Farm traction   787
    Dairy farming   791
    Poultry farming   796
    Food preservation   796
    Conclusion   801

17. Textiles and Clothing   803
    Richard Hills

    Introduction   803
    Textile fibres   804
    Early textile processes   808
    The Middle Ages   812
    The seventeenth and early eighteenth centuries   818

xi

	The industrial revolution	823
	The nineteenth and twentieth centuries	835
	Velvet, towelling and carpet weaving	845
	Knitting machines	846
	Lace machines	848
	Sewing machines	848
	Man-made finishes and fibres	849
	Clothing manufacture	850
	Sewing	851
	Fastenings	852
	Waterproof clothing and elastic	853
	Cleaning	854
	Footwear	854
18.	**Building and Architecture** *Doreen Yarwood*	855
	Primitive building	855
	Trabeated construction	857
	Masonry	859
	Advanced timber construction	861
	Brick and tile	866
	The arch	872
	The vault	874
	The dome	879
	Cantilever construction	882
	Roofing	883
	Plaster	885
	Concrete and cement	886
	Iron and glass	893
	Modern industrial construction	899
19.	**The Domestic Interior: Technology and the Home** *Doreen Yarwood*	902
	Surfaces, coverings and decoration	902
	Furnishings and furniture	906
	Heating and lighting	911
	Washing, bathing and toilet facilities	917
	Cleaning in the home	922
	Laundering	928
	The kitchen and cooking	934
	Plastics in the home	947

20.	Public Utilities *R. A. Buchanan*	949
	Introduction	949
	Water supply	950
	Power supply	957
	Waste disposal	962
	Roads and postal services	963
	Telegraph and telephone services	964
	Conclusion	966
21.	Weapons and Armour *Charles Messenger*	967
	Prehistoric weapons	967
	The bronze age and classical era	968
	The Dark Ages	971
	The age of chivalry	972
	The introduction of gunpowder	975
	The Renaissance	976
	Seventeenth and eighteenth centuries	978
	The industrial age	983
	The First World War	989
	The Second World War	998
	Weapons today	1006

The Contributors — 1012

Index of Names — 1018

Index of Topics — 1033

PREFACE

Dr Johnson wrote, 'A man may turn over half a library to make one book'. In the present case around a score of writers have turned over about as many libraries to make this Encyclopaedia. The Book of Proverbs states, 'God hath made man upright; but they have sought out many inventions'. Whatever one may think about Charles Darwin's 'Descent of Man', it is a fact that man walked upright, giving him a pair of hands which he could use for manipulation, rather than ambulation, and his cranial capability enabled him to evolve many inventions. This book tells the story of these inventions from stone axe to spacecraft, from cave dwelling to computer. The objective has been to simplify the study of the History of Technology by putting into the hands of the reader, be he or she student or layman, a single volume telling the whole story in twenty-two chapters, each written by an acknowledged expert.

The content and layout of this book are based on an analysis of human needs. From earliest times man has existed in a fundamentally hostile environment and has had to use his wits in the struggle for survival. From the start, this has involved his remarkable power of invention. Other primates, such as chimpanzees, have been known to add one stick of bamboo to another to enable them to reach and hence to enjoy a banana otherwise out of reach. Many species of birds show remarkable ingenuity in the construction of their nests, while insects like the ant, the wasp and the bee display a constructive capacity which could be mistaken for genuine creativity, but these examples are no more than instinctive and isolated responses to a set of circumstances peculiar to the species. Only God knows why man is the only species of animal capable of inventive thought and equipped with the dexterity to make practical use of his ideas.

The Encyclopaedia had its inception during the period when I was Executive Secretary of the Newcomen Society for the Study of the History of Engineering and Technology, and worked from an office within the Science Museum in London's South Kensington. In this position I was able to call upon a host of specialists, many of whom are members of the Society, and some also on the curatorial staff of that excellent institution. Thus while the conception, chapter contents and planning were my responsibility, the execution of the work was dependent on the contributors. I would like to thank them all for keeping to my original plan and layout, only adding topics that I had inadvertently omitted, and for the excellent chapters that they have written.

The final text has benefited enormously from the work of Mrs Betty Palmer who has laboured hard and long to cut out duplications, correct errors and generally shape the disparate typescripts into a uniform and coherent style. I would like to record my thanks to her, as well as to Jonathan Price and Mark Barragry of Routledge for their patience, good humour and encouragement. My gratitude must also go to the proof readers John Bell, George Moore and Jenny Potts and to the indexer, Dr Jean McQueen, whose work has contributed so much to the usefulness of the Encyclopaedia. I would like also here to acknowledge the generosity of the Trustees of the Science Museum for permission to reproduce over 60 of the illustrations contained in this book and thank the staff of the Science Museum Photographic Library for their assistance in tracking down photographs sometimes specified only vaguely.

Lastly I would like to thank my wife for her patience and forbearance. The period of gestation of this book has been longer than the others I have written and has caused a greater amount of paperwork to accumulate around my desk than usual. She has put up with it all with admirable fortitude.

Ian McNeil
Banstead, Surrey

To the memory of
THOMAS NEWCOMEN
who built
the first engine to work
without wind, water or muscle power

INTRODUCTION

BASIC TOOLS, DEVICES AND MECHANISMS

IAN McNEIL

THE PLACE OF TECHNOLOGY IN HISTORY

It is strange that, in the study and teaching of history, so little attention is paid to the history of technology. Political and constitutional history, economic history, naval and military history, social history – all are well represented and adequately stressed. The history of technology is neglected in comparison yet, in a sense, it lies behind them all. What monarchs and statesmen did in the past, how they fought their wars and which side won, was largely dependent on the state of their technology and that of their enemy. Their motivation was more often than not economic, and economic history and the history of technology can surely be considered as twin hand-maidens, the one almost totally dependent on the other. So far as social history is concerned, the lot of the common man, as of his king and his lords, was usually directly related to the state of technology prevailing at any particular time and place, whatever political and economic factors may also have been of influence.

Technology is all around us: we live in a world in which everything that exists can be classified as either a work of nature or a work of man. There is nothing else. We are concerned here with the works of man, which are based on technological and, to some extent, aesthetic factors. It is a sobering thought that every man-made object of practical utility has passed through the process of conception, testing, design, construction, refinement, to be finally brought to a serviceable state suitable for the market. Aesthetics may have entered into the process of development and production at some stage, increasingly so in our present consumer age, although from a glance at some of the products on the market, one might well question the makers' artistic sensibility.

It is even more sobering, however, to try to contemplate a world in which one had absolutely no knowledge of history, of one's own country or of the world at large. It is almost impossible to imagine a citizen of an English-speaking

INTRODUCTION

country being in a state of total ignorance of William the Conqueror, of Henry VIII and his six wives, of Napoleon and the Duke of Wellington, of Lord Nelson, of Abraham Lincoln and Gettysburg, of Kaiser Wilhelm, of Adolf Hitler and Auschwitz. These are the very stuff and characters that make up the pages of conventional history. Yet there are also Johann Gensfleisch zum Gutenberg, Leonardo da Vinci, McAdam and Telford, the Stephensons and the Brunels, Edison and Parsons, Newcomen and Watt, Daimler, Benz and Ford, Barnes Wallis, Whittle, von Braun, Cockcroft, Shockley, Turing and von Neumann and many others. It is interesting to consider which group had the greater influence on the lives of their contemporaries. Even more, which group has had the more long-lasting influence on the man in the street of later generations. It is a matter of regret that space does not allow us, in the present volume, to deal in a biographical manner with these and the many other inventors involved, but only with their works. To do so would require a whole shelf of books, rather than just a single volume.

We might well question the value of studying the history of technology. One answer is much the same as that for history as a whole. By studying the past, one should, with wisdom, be able to observe its successes while perceiving its mistakes. 'Study the past, if you would divine the future,' Confucius is said to have written some 2500 years ago, and even if this is an apocryphal quotation, the precept holds good. In fact it seems self-evident that, in the normal course of events, in the process of invention or of engineering design, the inventor or designer starts his quest with a good look at the present and the past. Inventors, though not necessarily ill-natured, tend to be dissatisfied with things around them. The endeavour to invent arises when their dissatisfaction becomes focused on a single aspect of existing technology. Typically, the inventor seeks a method of improving on past and present practice, and this is the first step in the process of moving forward to a new solution. Thus the history of technology and the history of invention are very much the same.

Why study the history of technology? One could argue that it is a discipline with all the essential elements needed to give a good training to the mind, if such an exercise be considered desirable. Then there is another school of thought; a growing body of people find the study is its own reward. They are willing to pursue it for the pure fun of it. Though many of them may be professionals, they are in fact truly amateurs in the exact sense of the word. Long may they flourish and continue to enjoy the pursuit of knowledge in this field for its own sake.

SCIENCE AND TECHNOLOGY

It is important at the start to distinguish between science and technology, for science as such can have no place in the present volume. Though the dividing line is sometimes imprecise, it undoubtedly exists. In our context, at least, science is the product of minds seeking to reveal the natural laws that govern the world in which we live and, beyond it, the laws that govern the universe. Technology, on the other hand, seeks to find practical ways to use scientific discoveries profitably, ways of turning scientific knowledge into utilitarian processes and devices.

It is quite clear where the line must be drawn in most cases. The steam engine, for instance, the first source of mechanical power and the first heat engine, was to release man from reliance on his own or animal muscles or the fickleness of wind and water. For a short period in the seventeenth century scientists, mostly dilettantes, took a lively interest in the possibility of harnessing the power of steam, but little came of their curiosity. Nor did that of certain less scientific but more practical experimenters such as Sir Samuel Morland, 'Magister Mechanicorum' to King Charles II, Captain Thomas Savery or Denis Papin, the French scientist who invented the pressure cooker and worked for some time at the Royal Society, lead to the crucial breakthrough. Claims may be, and have been made for any one of these to have 'invented' the steam engine but, without question, it was Thomas Newcomen, a Dartmouth 'ironmonger', who devised and built the world's first practical steam engine, which was installed for mine-pumping at Dudley Castle in 1712. There is equally little doubt that Newcomen was a practical man, an artisan with little or no scientific knowledge or any training in scientific matters. Science and scientists had little direct or indirect influence on the early development of the steam engine. The prestigious Royal Society, founded as recently as 1662, did not even honour Newcomen.

The situation was little different when Sir Charles Parsons patented and produced the first practical steam turbine in 1884. True, Parsons had a top-drawer upbringing and education. Sixth son of the Earl of Rosse, he was privately tutored until he went to Trinity College, Dublin, and then to Cambridge University. There, the only pure science that he studied was pure mathematics, before starting an engineering apprenticeship. This was before there was any established School of Engineering at Cambridge, but he did attend such few lectures that were given on Mechanisms and Applied Mechanics. That was all the 'scientific' training given to the man who was to revolutionize both marine propulsion and the electrical supply industry.

But matters are not always so clear-cut. Take horology, for instance, or timekeeping. The men who evolved the first calendars, who observed the difference between the twelve cycles of the moon and the one of the sun, were astronomers, scientists. Admittedly they were working towards the practical solutions of how to predict the seasons, the flooding of the River Nile, the times for sowing and the time for harvest. But they were scientists. Technology entered into the matter only when mechanical timekeeping had arrived, when clock and watchmakers and their predecessors had devised practical instruments to cut up the months into days, the days into hours and the hours into minutes and, later, seconds. These were technologists. They were practical men who made their living by making instruments with which scientists and others could tell the time.

Perhaps the matter may best be summed up by a quotation, supposedly originating from Cape Canaveral or one of the other stations involved in United States NASA Space Programme. One of the engineers is speaking: 'When it works,' he is reported to have said, 'it's a scientific breakthrough. When it doesn't, it's those b—— engineers again.'

Purists, of course, would doubtless dispute the difference between engineering and technology. The latter – the science of the industrial arts, as the *Concise Oxford English Dictionary* puts it – includes engineering but is a much wider concept. Engineering – mechanical, civil, electrical, chemical etc., with further sub-divisions into smaller sectors – is defined in the same work as the 'application of science for the control and use of power, especially by means of mechanics'. It is but a part of technology, although a large and important part.

One further possible source of confusion exists. It is clear that the astronomer, the man who looks through the telescope, is a scientist. On the other hand, the scientific instrument maker, the man who made the telescope, is a technologist. In some cases, like those of Galileo and Sir William Herschel, they may be one and the same man. However, as space is at a premium, we must forgo the telescope as a part of technology and consider it the prerogative of the editor of an Encyclopaedia of the History of Science, just as we would consider the violin and the bassoon as musical topics although the craftsmen who originally made them were undoubtedly technologists.

THE ARCHAEOLOGICAL AGES

The neglect of technology, the near-contempt in which archaeologists and historians seem to hold it, is all the more surprising when one considers that it was one of the former who originated what is now the standard classification of the archaeological ages, and which is based on technological progress. Christian Jurgensen Thomsen, who became Curator of the Danish National Museum in 1816, first started the system that is used world-wide today. He had previously read a work by Vedel Simonsen which stated that the earliest inhabitants of Scandinavia had first made their tools and weapons of wood or stone, then of copper or bronze and finally of iron. This inspired him to arrange his collections by classifying them into the three ages of Stone, Bronze and Iron and, from 1819, visitors to the museum were confronted with this classification. It first appeared in print in 1836, in his guidebook to the museum.

The scheme was by no means universally accepted until, in 1876, François von Pulski, at the International Congress of Archaeology in Budapest, added a Copper Age between the Stone and Bronze Ages and, in 1884, published his book on the Copper Age of Hungary. This added the final seal of approval and thenceforth the world took wholeheartedly to Thomsen's classification. Yet although it was clearly based on the materials from which tools were made, and such tools are the predecessors of industry, industrial archaeology and industrial history are only grudgingly accepted and taught but sparingly in the majority of centres of learning.

One archaeologist who was convinced that we should look upon pre-history primarily as a history of technology was Professor V. Gordon Childe who studied, rather than the rise and fall of civilizations, the rise and fall of technologies – the technologies of hunting and weapon-making, of herding and domesticating animals, of crop-growing and agriculture, of pottery and metal working. Childe held that one should not study the palace revolutions that enabled one pharaoh to displace another, but the technologies that enabled one tribe or nation to overcome another in battle and the technologies that enabled people to produce such a surplus of food in the valley of the Nile or the Tigris or Euphrates that great states could be set up. Of recent years more and more archaeologists have been adopting Professor Childe's approach.

THE SEVEN TECHNOLOGICAL AGES OF MAN

When studying the history of mankind from the point of view of technological development, it is possible to distinguish seven to some extent overlapping ages: 1. the era of nomadic hunter-gatherers, using tools and weapons fashioned from easily available wood, bone or stone and able to induce and control fire; 2. the Metal Ages of the archaeologist, when increasing specialization of tasks encouraged change in social structures; 3. the first Machine Age, that of the first clocks and the printing press, when knowledge began to be standardized and widely disseminated; 4. the beginnings of quantity production when, with the early application of steam power, the factory system began irreversibly to displace craft-based manufacture; 5. the full flowering of the Steam Age, affecting all areas of economic and social life; 6. the rapid spread of the internal combustion engine, which within 50 years had virtually ousted steam as a primary source of power; 7. the present Electrical and Electronic Age, which promises to change human life more swiftly and more radically than any of its predecessors.

THE FIRST AGE: MAN, THE HUNTER, MASTERS FIRE

The history of technology can be said to be older than man himself, for the hominids that preceded *Homo erectus* and *Homo sapiens* were the first to use

Table 1: A summary of the material ages

Age	Approx. dating	Notes
EOLITHIC (Dawn Stone Age)	c.10 million ybp	Origins of tool making
LOWER PALAEOLITHIC (Old Stone Age)	5 to $1\frac{3}{4}$ million ybp	Itinerant hunter tribes. Hand axes widespread
MIDDLE PALAEOLITHIC (Old Stone Age)	400,000 to 35,000 ybp	
UPPER PALAEOLITHIC (Old Stone Age)	35,000 to 12,000 ybp	Origins of blade technology
MESOLITHIC (Middle Stone Age)	12,000 to 7,000 BC	
NEOLITHIC (New Stone Age)	6,000 to 3,000 BC	Agrarian revolution. The beginnings of towns
AENEOLITHIC or CHALCOLITHIC (Bronze Age)	3,000 to 1,500 BC	Writing known in Near East. Earliest copper articles in Egypt c.4000 BC. Tin (hence bronze) discovered in Mesopotamia c.3000 BC
IRON AGE	started c.1500 BC	

Note: ybp indicates years before the present. The dates given are approximate: the same event took place in different countries at different times.

tools. Australopithecenes, typically Taung Man, whose skull was turned up by Dr Louis Leakey and his wife Mary in 1925 in the Olduvai Gorge in Tanzania, was one of the earliest and has been found associated with simple stone tools as well as potentially useful flakes of stone, the by-products of the tool-making process (see Figure 1). Australopithecus, originating probably between two and three million years ago, was the first of man's predecessors to walk upright. This ability was to lead to the whole story of technology, for it made available a pair of forelimbs and hence the ability to grasp sticks or stones and later to fashion them for particular purposes and to sharpen them to a cutting edge. The first of the hominids was Ramepithecus, thought to date back as far as fourteen million years and closely related to the great apes. However, it appears to have taken eleven or twelve million years for the tool-making habit to emerge.

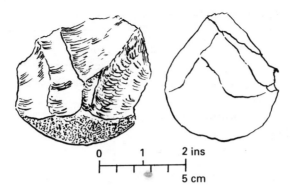

Figure 1: Basalt side-chopper; over 2.5 million years old from the Oldurai Gorge, Tanzania.
After M. D. Leakey.

The ability to fashion stone tools was followed by a further advance otherwise unknown in the animal kingdom. No other species has the ability to make fire. It is one of man's most wonderful accomplishments and one which was to lead to innumerable benefits.

'Making' fire is not the same as 'using' fire: the use of natural sources of fire, such as volcanoes, meteorites, spontaneous chemical combustion or the focusing of the sun's rays through a raindrop, clearly predated the ability to generate fire. In early tribal societies an important function was the tending of a source of fire, started from one of the natural sources and which must, at all costs, be kept alight, fed and nurtured. It is said that, even today, there are Tasmanian and Andamanese tribes who have not mastered the art of making fire but have to borrow it from their neighbours. The first hominid known to have made fire was the *Homo erectus* (originally classed as *Sinanthropus pekinensis*) of Choukoutien in China. Many layers of charcoal have been uncovered there in the caves that they used, indicating intermittent occupation and fire making over a period of many years. This activity dates from about 600,000 BC.

The uses to which fire was put were many and may be summarized as: for warmth, for cooking, for the curing of hides, for protection in scaring off wild animals, and as a focus for the social life of the tribe after darkness had fallen. At a later period it was used also for hollowing out logs to make primitive boats, and in firing pots, bricks and tiles, while the extraction of copper and iron from their ores, the very bases of the metallurgical eras, and the subsequent working of those metals into tools, weapons and ornaments, was entirely dependent on fire. The making of glass objects was also based on the control of fire.

The ability to make fire at will was thus one of the first major advances in the early history of technology. There were two principal methods of doing so, by impacting flint and iron or iron pyrites, and by the generation of heat by the

INTRODUCTION

Figure 2: Fire drills from northern Queensland Australia, Alaska and the Kalahari.

friction of a hard stick, or fire-drill, against a softwood block, or hearth. While the flint (silicon dioxide) method seems the more likely to have occurred by chance and is therefore likely to be the earlier, it does require the addition of dried grass or some other suitable tinder to make a fire. On the other hand, the fire-drill, which would seem to imply a higher degree of intellectual capacity for its conception, provides its own tinder from the friction of the hard, pointed stick on the soft wood of the hearth (see Figure 2). Possibly a later development of the fire-drill was the addition of a doughnut-shaped stone, drilled through its centre, held in place by a tapered peg to the drill, which would act as a flywheel by its own inertia. Some authorities have interpreted this artefact as merely a digging stick. It does seem, however, that the makers of so sophisticated a tool must have gone to an inordinate amount of toil and trouble to bore out the flat, circular, stone to weight a digging stick to which a weight could easily be attached by tying with a thong or cord.

The fire-drill was rotated simply between the two hands of the operator, limiting the number of revolutions it made before its direction of rotation had to be reversed. The stroke could easily be increased from the nine or ten revolutions that would be made by a $\frac{1}{4}$-inch diameter stick between average hands before reversing, by a quite simple addition. This was to wind a piece of cord or thong once round the stick and then to tie the ends to a bent piece of springy wood in the shape of a bow. Thus evolved the bow-drill, used as much for drilling holes as for starting fires and one of the first multi-component machines to be invented. Indeed, some archaeologists have propounded the use of such a drill as a component of an elementary machine tool, in which a weight and lever arm comprised the tool feed, the tool of hollow bone being fed with powdered flint at the cutting edge, the drill being rotated by a bow. Such a machine is purely conjectural, but the bow is known to have turned lathes in the Romano-British period of the Iron Age.

The bow when used as a weapon supposedly invented by the people of Bir-el-Ater in Tunisia in the middle to late Stone Age, was also the first energy-storing device. The energy of the bowman is gradually put into the bow as it is drawn and stored until released instantaneously at the moment of shooting. This was a considerable advance on the spear-thrower, a sling which merely extended the leverage of a man's arm. Bone, for example from a deer's antler, was used as the bow, with an animal sinew as the string, sometimes as a substitute for a suitably flexible piece of wood.

Wood was, of course, a natural and usually easily obtained material which by the nature of its growth would suggest itself to primitive man for many purposes – for digging, for spears and clubs, and for use as a lever in many situations. Bone came into service in slivers for making needles and for digging on a grander scale as in the Neolithic flint mines such as Grime's Grave in Norfolk, where the quantity of flints removed suggests that they must have been a commodity of primitive trade. The shoulder-blade of an ox is flat and splayed out in such a way as to make an ideal natural shovel, while a part of the antler of the red deer would serve as a suitable pick. Similar flint mines have been found in Belgium, Sweden, France and Portugal. In some cases the shafts of such mines are as much as thirteen metres deep and extend at the bottom into galleries where the richest seams are to be found. Trade in these flints, sometimes in the raw state and sometimes shaped into finished tools, was international as well as within their countries of origin. International commerce was thus established, probably several thousand years before the Bronze Age and, it seems likely, long before the introduction of agriculture and settled centres of population.

Bone, ivory and horn found use for making spear-tips, fish-hooks and harpoons, as well as needles. Fish was a valuable addition to the diet of hunting and food-gathering peoples as it was to agricultural communities, and fishing increased as the building of boats became possible. This appears to have occurred about 7000 BC. Even boat-building, however, was much dependent on the mastery of fire to hollow out logs.

The development of tools in the Stone Age

Owing to its density and hardness, stone was probably the most popular material for tools in the Stone Age. Thanks to its durability, it is also the most common material of such tools as have survived the centuries since they were in use. The oldest and, at the same time, the most primitive that are undoubtedly man-made, or made by his predecessors, are the pebble tools found by Richard Leakey in Kenya which have been dated at 2.6 million years old. These include the characteristic core and flake tools; the flakes produced as waste in the process of developing the core were put to good use, as many of them had sharp cutting edges. Characteristically such core pebble tools were only flaked to produce a sharp edge at one end. It is notable that in this, the world's oldest industry, dating back probably some 5 million years, 'tools to make tools' were included, hammer-stones and anvil stones being found in the lowest levels at Olduvai in Tanzania.

So-called hand axes, on the other hand, were bifaced, that is to say, sharpened by flaking all round the periphery. The development of this type of tool is also attributed to peoples in Central Africa, supposedly dating from about half a million years ago. This was a general purpose tool, serving not only as an axe but also for piercing and scraping the hides of animals. Not only pebbles were used in their manufacture: some show signs of having been quarried from the natural rock.

Where long parallel-sided flakes were produced, usually from flint, chert or obsidian, they represent the so-called blade-tool industries. From these basic knife blades a number of variants have been found: gravers, spokeshaves, saw blades, planes and drills have been identified by palaeontologists, although the common man might have some difficulty in distinguishing some of them. All belong to the Upper Palaeolithic period that is, say, from 35,000 to about 13,000 years ago when hunting was still the primary source of food.

The Mesolithic, or Middle Stone Age, lasting approximately from 12,000 to 7000 BC, saw a revolution in the making of stone tools. The techniques of grinding and polishing, already applied to bone and ivory in Middle and Upper Palaeolithic times, began to be used for the surfacing of stone tools. An axe with a smooth surface would be much easier to use for felling trees, though its advantages with some other types of tool do not seem to be so evident. Basalt and epidiorite, finer grained igneous rocks, are more easily ground and polished than flint and it is supposed that the technique probably originated in regions where these rocks were in common use for tool-making. This grinding and polishing was probably at its peak around 6000 to 5000 BC, declining in importance after 3000 BC when copper and then bronze came into use.

The grinding and polishing process generally involved rubbing the tool against a slab of wetted sandstone or similar hard rock, sand being used as an abrasive powder if only a non-friable rock was available as a grinding base. Some axes of the Neolithic and Bronze Ages, probably used for ceremonial purposes, have a very high polish suggestive of a final burnishing with skins and

BASIC TOOLS, DEVICES AND MECHANISMS

a polishing powder. These have generally been found associated with the burials of tribal chieftains.

The production of the basic core and flake tools was a skilled occupation using one of two methods – pressure flaking or percussion flaking. In the former, a tool of bone, stone or even wood was pressed against the core so as to split off a flake and the process was repeated. In percussion flaking a hammer stone was repeatedly struck against the core or against an intermediate bone or wooden tool applied to its edge. Either process requires a high degree of skill, acquired through long practice and much experience and, in the case of the more complex shapes such as barbed and tanged arrowheads, indicates a degree of specialization at an early date.

The adze, roughly contemporary with the hafted axe, similarly developed into a polished tool about 6000 BC in the Middle East with the general adoption of agriculture as a method of food production. The spokeshave is also of this period; of course, it was not at that time used for wheel spokes but more for refining spear shafts, needles, awls, bows and the like, in wood or bone.

It is interesting to note that the impulse to create and the ability to produce images of animals (including men and women) seems to date from late Palaeolithic times at least, that is before about 12,000 BC. Relief carving on cave walls, modelling in clay and powdered bone paste, and cave wall painting were all included in the artistic activities of the Gravettian and Magdelenian cultures that were established in the Dordogne region of France. Black oxide of manganese and the ochres or red and yellow oxides of iron, generally ground to a powder and mixed with some fatty medium, were the colours generally used and probably represent man's first excursions into the world of chemistry as well as that of art. Mammoth, woolly rhinoceros, bison, reindeer, horse, cave lion and bear have all been found in these paintings, mostly of men fighting with bows and arrows, credited to the people of Bir-el-Ater, Tunisia, in the Middle to Late Palaeolithic period. Another rock painting in Spain shows a woman collecting honey with a pot or basket and using a grass rope ladder, another early invention extant in this period.

THE SECOND AGE: THE FARMER, THE SMITH AND THE WHEEL

The change from nomadic hunter to settled agricultural villager did not happen overnight, even over centuries. It must have taken several thousand years. It started some time about 10,000 BC, when a great event took place – the end of the last Ice Age when the melting ice flooded the land and brought to life a host of plants that had lain dormant in seeds. Among these was wild wheat as well as wild goat grass. It was the accidental cross-fertilization of these that led to the much more fruitful bread wheat, probably the first plant to be sown as a crop, which was harvested with a horn-handled sickle with sharpened flints set into the blade with bitumen. At an oasis near the Dead Sea, as at other places, a village grew into a city: this was Jericho, with fortified walls and buildings, at first of reed and mud, then of unbaked clay until baked brick was used between 8000 and 6000 BC. Barley and millet were also grown and harvested (see Chapter 16). As well as the already domesticated dog, the sheep, the goat and the onager, a form of ass, were added to the domestic animals. Pottery was made not only by the old method of smoothing together coils of clay but also on the newly invented potter's wheel. Copper was used for ornaments in Egypt about 4000 BC (see Chapter 1). It came into more general use for making tools 1000 years later, at about the time that tin, which could be alloyed with it to make bronze, was discovered in Mesopotamia. Iron was not discovered as a useful material until about 1500 BC (see Chapter 2).

Social influences of copper and iron

The social implications of copper and iron were very different. Copper, as later bronze, was something of a rarity and consequently expensive when it had been worked into a tool or weapon by someone with the skill and knowledge to do so. It came, too, at a time before there was an establishment, a hierarchy of king and priest and counsellor. Wealth was the only uncommon denominator and wealth could be equated with worldly success in the business that mattered most – success in agriculture. Copper tools, and weapons, thus became available only to the powerful, to those who were already wealthy, and had the effect of increasing their power and multiplying their wealth. It thus tended to create an elitist society in which the majority, who formed the lower ranks, were still confined to grubbing the earth with tools of bone and wood and stone. It was socially divisive, helping only the rich to become richer in creating the small agricultural surpluses that were to be at the foundation of subsequent cultures.

The technology of ironmaking and the forging of tools and weapons from the refined metal was a more complex process, a more specialized business. On the other hand, iron became more widely available than copper or bronze, it was far cheaper and could be made into much better and longer lasting tools and weapons. Thus, once the techniques and the craftsmen became established, iron tools became more generally available to a wider spectrum of the population than had those of copper or bronze. Iron has rightly been called the democratic metal, the metal of the people, for so it was in comparison with its predecessor.

The common ground

On the other hand, the introduction of metallurgy, whether of bronze or iron, and its processes did become the start of a new way of living in which speciali-

BASIC TOOLS, DEVICES AND MECHANISMS

zation and the division of labour were important factors. Metal workers were a class of specialists who needed specialist equipment and who depended for their sustenance on the labours of their fellow men, the farming community for whom they provided the tools. Many ancillary trades, too, were involved, in the quarrying or mining of the ores to be smelted. The construction of furnaces and the manufacture of crucibles were to become other objects of specialization. The human lungs and the blowpipe produced a very limited area of high temperature for smelting, so that the blowpipe was virtually restricted to goldsmiths. Around 3000 BC bellows were developed, at first from the whole skins of sheep, goats or pigs, or from clay or wooden pots topped with a flexible membrane of leather. Copper ore was not to be found in sufficient quantity in the 'fertile crescent' where it was first used. Thus traders and carriers were required using, at first, pack animals and later wheeled vehicles, then riverboats and sea-going ships. All these called for further specialists to produce them. These in turn needed further tools with which to work and added to the number of specialists who had to be fed.

It has been suggested that copper ore was mined at Mount Sinai as early as 5000 BC and 2000 years later in Oman in the south of Arabia. These are respectively some 1400 and 2000km (875–1250 miles) from Mesopotamia, a long haul for the unrefined ore and a powerful incentive towards the invention of the wheel. It was about 3500 BC that the wheel was first added to a primitive sledge at Erech in Sumeria. Strictly speaking, the invention consisted more of the axle to which the wheel was fixed, for previously wheel-like rollers had been placed beneath the sledges, especially when heavy loads were involved.

Some of the heaviest of these were the great obelisks, a characteristic of Egyptian civilization, which were as much as 500 tonnes in weight and 37m (122ft) high when erected. They were quarried in the horizontal position, in one piece, from around 1470 BC. It is believed that the huge blocks were cut out of the parent granite by continual pounding from a round stone of dolerite so as to generate a narrow trench all round; the final undercutting to separate the obelisk must have been an awesome task. Apart from the glory brought to the name of the pharaoh who had ordered the obelisk, it had little use except as the gnomen of a sundial.

The plough was a vital invention for a civilization that was becoming increasingly dependent on agriculture, and a great improvement on the hoe as an instrument for tilling (see Figure 3). An intermediate device, which survived in the Hebrides until the nineteenth century as the caschrom, was the lightweight man-plough and it was not until about 3000 BC that animals started to be harnessed to the plough or the cart. In warfare, horses were at first used to draw two-wheeled chariots: men did not learn to ride horses until about 2000 BC and the saddle is a much later invention which did not become truly effective until

INTRODUCTION

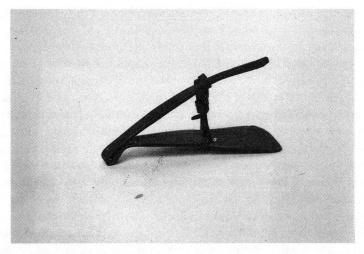

Figure 3: The plough. The development of more effective instruments for tilling the soil lies at the heart of agricultural production. An Egyptian hand digging implement, *c.* 1500 BC. See Chapter 16.

stirrups were added, allowing the horseman a better purchase with sword or spear.

There has always been a need to join one piece of material to another, be it wood, leather or metal, or to join wood to stone as in the hafted axe. Thongs and cords or ropes of fibres woven together served at first, but both the nail and the rivet were known to the coppersmiths of Egypt well before 2500 BC. Such nails were used in sailing ships some five centuries before that date. These again were Egyptian and represent the first known application of a natural source of power other than animal power, the wind. The use of wind power to drive the rotating sails of a windmill does not appear to have occurred for another three millennia or more.

The Iron Age naturally brought with it, among other things, iron nails, but for large timbers, such as are used in ships, where these were to be joined, it was common to use trenails. These were no more than closely fitting dowels and they remained the conventional method of joining ships' and other heavy timbers until iron replaced wood as the basic construction material in the nineteenth century AD. There is some doubt about the woodscrew: examples exist from Roman sites between AD 400 and 450, but in a very corroded state so that it is impossible to be sure of their purpose. All that we know for sure is that the first illustration of a woodscrew is in Agricola's *De Re Metallica*, first published in Venice in 1566. This remarkable and detailed work does not mention that there was anything unusual about the woodscrew, from which it has been assumed that by that time it was an article generally in common use.

Another method of joining wood, known at least by 3000 BC, was the mortise and tenon joint, implying the use by that date of quite sophisticated carpenter's tools. These must have included saws and chisels and a high degree of skill must have been needed to use them effectively.

The first type of hinged tool, as we know it today, with two separate components, was in fact made in one piece, depending on the spring of one part of the metal for the relative movement between the two blades. The Egyptians, perhaps as early as 4500 BC, had shears similar to a tailor's scissors (and they are still made in steel in Japan). With these they could cut silhouettes, and it is suggested that this resulted in the Egyptian convention of making carved or painted representations in profile. The same principle was then applied to making tongs for those employed in coppersmith's work. Previously there was no alternative to holding the hot metal between two stones.

It may appear to the reader that an excessive number of events and inventions are recorded as having taken place in the period 3500–3000 BC. There may be some unintentional distortion of the facts due possibly to the dating of the earliest surviving written records. The first known writing was the cuneiform or wedge-shaped script evolved by the Sumerians about 3400 BC. Before that everything depended on memory and speech, the only method of recording and recalling the past. Egyptian hieroglyphic writing followed within a couple of centuries and by 2000 BC the Egyptians had reduced their system to an alphabet of twenty-four letters. In contrast, the Chinese, in advance in so many other techniques, were not so in writing. It is known that Chinese writing was well established by 1700 BC, but the date of its origin is not known.

Pots and baskets

The development of both tools and weapons increased the demand for containers in which to remove the spoil of excavation or to preserve or to cook the winnings of the hunt. Basketwork is a characteristic of the Neolithic Age and is a development of the weaving of rushes to make floor coverings for mud huts. Such forerunners of the carpet date from some time before 5000 BC. The same weavers learned to work in three dimensions so as to produce baskets in which grain could be stored. By 3000 BC the skill was widespread.

Similarly pottery, in its broadest sense, did not start with the making of pots (see Chapter 3). Long before vessels were made – as early it seems as 25,000 BC – small figures representing human beings were moulded in clay and baked, at first, in the sun. The making of pots by coiling strips of clay in a spiral and then moulding them together is supposed to date from about 7000 BC, as is the moulding of clay to take the form of the inside of a basket. Sun drying would be insufficient to produce a watertight vessel and slow baking in a kiln would be required. The kiln process was first used in Mesopotamia and Persia around 4000 BC and in Egypt within the next thousand years.

The advance of the wheel

The inventor of the potter's wheel wrought, perhaps, better than he knew. A simple turntable mounted on a central pivot was turned by an assistant so that the potter had both hands free with which to manipulate the clay. It originated about 3500 BC in Mesopotamia, not far from Sumeria where, at about the same time at Erech, came that great advance in technology, the addition of an axle to a sledge on to which a wheel could be fixed, thus to form a primitive cart or wagon. Both of these concepts were, of course, preceded by the use of a round flat stone with a hole in the centre to act as a flywheel when spinning thread for weaving. Doubtless the wagon at Erech was an experiment and doubtless the same idea came to others about the same time: technology has a habit of working like that. But to the anonymous experimenter at Erech must go our thanks, for what would we do without wheels and axles today?

Glass

Until the plastics age of the twentieth century, and many might say into and beyond it, glass was the ultimate material for making containers. It resists all substances – except hydrofluoric acid. It exists in nature in the form of obsidian, a volcanic rock of which man-made artefacts have been found, such as arrowheads. The earliest man-made glass is dated at about 4000 BC in Egypt, as a simple glaze on beads. Not until about the seventh century BC are Assyrian examples of small decorated jugs found, made by casting the glass round a clay core which could then be scraped out. A great advance was the blowing iron, allowing larger and thinner vessels to be made. This originated in the first century under the Romans.

Gearing

We know little more of when and where gear wheels originated than we do of the invention of the wheel. Aristotle (*c.* 384 BC) recorded seeing a train of friction wheels set in motion, that is a series of contiguous wheels with smooth peripheries but without teeth. Ctesibius of Alexandria is said, by Vitruvius, to have constructed a water-clock with gears about 150 BC. In this, a primitive rack was mounted on a floating drum and meshed with a circular drum so as to rotate it. This is the earliest reference to toothed gearing, but no mention is made of the materials used.

Gearing, then, developed in two materials – in wood for large installations transmitting power; and in metal, usually bronze or brass initially, for timekeeping and other related astronomical instruments. The earliest surviving 'mathematical gearing', as the latter is known, is probably the Antikythera mechanism now in the National Archaeological Museum, Athens, which takes its name from that of a Greek island off which it was found in a wreck in AD 1900 and is thought to date from the first century BC. It is supposed, from the complexity of this mechanism, of which no less than thirty-one gear wheels survive, that this was far from the first example to be made and that, by that date, there was an established tradition of making mathematical gearing. The teeth on such an early example of gears were filed with a straight tooth profile and a root angle of approximately 90 degrees.

Wooden power transmission gears for mills date at least as far back as Roman times, a comprehensive range of carpenter's tools being available such as lathe, plane, bow drill, saw, chisel, awl, gimlet and rasp. It was customary at that time to bind the ends of shafts with iron hoops as well as to line the pivot sockets of timbers to provide a bearing surface. The blacksmith was equipped with forge hearth, bellows, hammer, tongs and anvil by the time of Vitruvius, who wrote about these matters in about 25 BC.

The traditions established at this time in both carpentry and blacksmithing were to continue well into the nineteenth century with little change in techniques but with great improvements in the products made. Improved tooth profiles, so as to provide true rolling contact between gears, were introduced. Bevel gears were substituted for the earlier 'lanthorn and trundle', the latter being a wheel with pegs, and later shaped teeth, set into its flat surface, the lanthorn resembling a lantern or bamboo birdcage, two discs being connected by a number of pins or dowels equally spaced near to their periphery. The use of wrought iron in clock gears from the Middle Ages developed into the use of cast-iron mill gearing in the 1770s, but mortise gears – wooden teeth let into a cast-iron rim – persisted into the nineteenth century. Hornbeam was the most popular wood until lignum vitae, the wood of the guaiacum tree, a native of South America, began to be imported in the eighteenth century.

Helical gears were developed by Robert Hooke in 1666. There being then no machinery for cutting such teeth, he built them up from a series of laminations which were staggered or progressively displaced when the faces of the teeth were then filed smooth. Ten years later Hooke devised the universal joint named after him and contributed much to later mechanical engineering.

Early machines in Egypt

It was not until about the time of Christ that Hero of Alexandria classified the five basic machines as the lever, the wheel and axle, the wedge, the pulley and the screw, but the first three of these had been in common use since about 3000 BC.

The *shaduf* for irrigation and the balance beam for weighing were applications of the lever. It was, as we have already said, at about this time that written records started with the Sumerian invention of cuneiform script usually inscribed on clay tablets. The same period saw the first attempts at standardization of weights and of linear measures, the span, the palm, the pace, the inch and the cubit all being based on parts of the human body by the Egyptians.

The Egyptians were also the first large-scale builders, largely using huge quantities of slave labour rather than mechanization, or craftsmen working off their tax dues or debts. No pulleys, for instance, were used in raising the thousands of huge blocks of limestone for the pyramids. The great pyramid of Cheops covered over 5.3 hectares (13 acres) and contained some 2.3 million blocks of over 1 tonne each. It was 146m (479ft) high. Though it was a natural progression for a civilization that had the wheel, the Egyptians did not have the pulley. The first depiction of it is in an Assyrian relief dating from the eighth century BC, in which it is clearly shown in use on a building site hauling up a bucket or basket, the workman on the ground grasping a pointed mason's trowel.

In spite of the replacement of the blowpipe by bellows, at first operated by hand but later by the feet, articles of iron were of limited size even towards the end of the Egyptian Empire, that is, until about 600 BC. Drill bits, chisels, rasps, door hinges, edging for ploughshares, bearings, spindles and hoes are typical. Wood and stone were the principal materials of the Egyptians. Cast iron, however, was developed by the Chinese as early as the fourth century BC.

Greece and Rome

The Greeks were great builders but, apart from a few exceptions such as Archimedes, were theoretical scientists rather than practical technologists. Their contributions to sciences such as mathematics and astronomy, were considerable – not to mention philosophy – but they were not great inventors except, perhaps, in the production of mechanical devices to strike the worshipping plebeians in the temples with a sense of awe. Falling weights drove some of these, but more common was the use of hot air or hot water, even steam being brought into service as well as a form of windmill. The Romans, although a far more practical people, invented little of their own but did much to adapt the principles, used by the Greeks only for their temple 'toys', to large-scale practical applications such as could be used 'for the public good'.

The Greek and Roman empires lasted for about a thousand years, from 600 BC to about AD 400, during which period the Chinese made some remarkable advances in technology. They had cast iron as early as 350 BC, some thirteen centuries before it was known in the West; they developed the double-acting box bellows; steel was produced in the second century BC; they invented paper-making about AD 100. Gunpowder, mainly used in fireworks, is another Chinese invention. It is surprising that there seems to have been so little transfer of technology to the West in spite of so many travellers passing along the 'Silk Road'. These, however, were merchants or royal or papal envoys. Technology was, perhaps, above the intellectual level of the merchants and below the notice of the envoys. Marco Polo, for instance, in his *Travels*, recorded by Rustichello of Pisa in 1298–9, records the glories of architecture, customs, weapons and armour, food, gemstones, crops, natural history, governments and rulers but rarely, if ever, records seeing a technological process in twenty years of wanderings in the Middle East, India and China.

One practical invention of the Greeks was the horizontal waterwheel, the predecessor of the turbine, now known more commonly as the Norse mill. This was the first form of power of non-animal origin except, perhaps, the sailing ships used by the Egyptians as early as 2500 BC. Its use spread northwards throughout Europe. Except for the shaft from the wheel to the millstones above and the bearings, which were usually of iron, practically the whole construction was of wood which became the principal material of millwrights for the next thousand years.

The Roman mill, described by Vitruvius in about AD 180, was the first machine in which gears were used to transmit power. This mill had a vertical wheel driving the horizontal upper millstone through lanthorn and trundle gears. However, the Romans were well supplied with slaves and hence not encouraged to invest in labour-saving mechanization. Watermills did not increase greatly in number until the fourth and fifth centuries AD, towards the end of the era of the Roman Empire. By the time of the Domesday Book, completed in AD 1086, this survey was to record over 5600 in use in England alone, mostly used for corn milling but possibly a few for ore crushing and for driving forge hammers.

In building, the Romans used cranes frequently fitted with a treadmill to turn the windlass, the rope running in pulleys. The most powerful of them were of about 6 tonnes' lifting capacity. The stone blocks were lifted by means of a 'lewis', a dovetail cut by a mason in the upper surface into which a wedge-shaped metal anchor was fitted and locked in place by a parallel-sided metal key. The key of the lewis and the wedged anchor could be released when the stone had been positioned, even under water.

As well as the extensive network of roads across the Roman Empire, which included many bridges, a great number of aqueducts were built to supply water to their cities. The construction of river bridges often involved the building of coffer dams of timber piles, sealed with clay, and in the building of these two machines were used: first, the pile driver and, second, the 'snail' or Archimedean screw to drain the water from the completed coffer dam.

Water supply was of the greatest importance, for the Romans in the cities are said to have used more than 270 litres (59 UK gallons, 71 US gallons) per person per day. Rome in the fourth century AD had over 1350 fountains and 850 public baths, while flowing water was also used for flushing the plentiful sewers. Some houses at Pompeii had as many as thirty water taps each. The common use of lead pipes for water distribution, producing lead poisoning and resulting in brain damage, is held by some to have been one of the causes of the decline of the Roman Empire.

The Dark Ages

The last Roman troops left Britain in AD 436 and all contact between Britain and Rome had ended by 450. To a great extent the Roman legacy of roads, bridges and aqueducts died, the relics being allowed to fall into disuse and decay, but a good deal of knowledge was preserved as well as the skills to transmute it into practice. Many of the engineering crafts were kept alive by the monastic orders who became rich on the basis of the products that they made but largely from the water-powered mills that they built and operated.

The Romans, for instance, had superseded the half horsepower (or one donkey power) Greek or Norse mill with the Vitruvian mill which generated up to three horsepower. As civilization could no longer depend on large numbers of slaves, there was a demand for such mechanization as was available and the successors of the Romans continued to build water mills. At the time of the Domesday survey there was an estimated population in England, south of the Severn to the Trent, of nearly $1\frac{1}{2}$ million and hence one watermill for every 250 people, the vast majority being devoted to corn milling. Later, from AD 1000 on, mills were used for beer-making and for fulling cloth, particularly woollen fabrics. Later applications were for forge hammers and bellows, for paper-making, crushing woad and bark for tanning, for grinding pigments and also for making cutlery, for water lifting and irrigation, and for saw mills, lathe drives and wire-drawing. In the twelfth century engineers turned their attention to harnessing wind power and the first post mills resulted. Tidal mills also existed, but had the disadvantage that their working hours varied each day. The first mention of a post mill is in about 1180 in Normandy. Post mills were of necessarily limited size and hence power, but tower mills, which first appeared in the fourteenth century, could double or treble this. They did not come into extensive use until the sixteenth century, particularly then in the Netherlands.

Naturally most inventions of the period were related to agricultural improvements, textile production or building construction, or else for military applications. Agricultural performance was greatly improved with the development of the horse collar which transferred the load from the neck of the horse to the shoulders, thus not interfering with its breathing. Though more costly in food, needing oats, the horse continued to replace the ox and nailed iron shoes took the place of the earlier cord or leather sandals with which horses had been shod. The heavy wheeled plough and the harrow also appeared and helped to increase production. The three-field system of crop production was introduced. In the military arena, stirrups improved the purchase of a horseman allowing him to transfer the momentum of his mount to the weapon in his hand, be it sword, axe or lance.

Some idea of the state of manufactures in the sixth and seventh centuries is given by artefacts found in the excavation of the Sutton Hoo ship in 1939 near the River Deben, not far from Ipswich in Suffolk. There an impression of a ship, its timbers decayed and disintegrated, was found in one of seventeen mounds, many still unopened, although there were signs that others had been disturbed in the sixteenth century. It was, if not a royal burial place, a royal memorial; no body was found, although there were signs of a coffin. A helmet, a sword of gold with garnet fittings, a battle axe, a spear, silver and bronze bowls, a gold buckle or reliquary, intricately worked and patterned, a purse of Merovingian coins, silver-mounted drinking horns, spoons and jewellery were among the buried treasure.

The ship itself had been carvel-built of 2.8cm (1in) planks for sailing or rowing by thirty-six men. It was 24m (79ft) long by 4.4m (14.5ft) beam and some 23 tonnes displacement with an estimated speed of 14kph (7.6 knots) when fully manned. The strakes were mounted on 7.5–12.5cm wide ribs at 90cm centres but 45cm apart near the stern, where a steering oar or rudder had been fixed. A variety of iron fastenings had been used in the construction – rib nails, keel plate spikes, steerage frame bolts, gunwhale spikes and keel scarf nails and thole pins to form rowlocks for the oars. Once in place, these were clenched over, after iron roves or diamond-shaped washers had been placed over the shanks of the various fittings. The long axes of the roves were all placed fore and aft in the vessel. The Sutton Hoo ship thus shows a high degree of practical and artistic craftsmanship in Saxon days. From the coins, it has been dated at about AD 630.

Some five centuries were to elapse, after its emergence in China, for the rudder mounted on a sternpost to be adopted in Europe in the thirteenth century. This was a great improvement on the steering oar for it was of larger area and, being well beneath the water level, it was far less affected by the waves. Simultaneous improvements in the rigging and an increase in hull size meant that longer journeys could be undertaken. The properties of the lodestone had been known since Roman times but it was not until the late twelfth century that the compass began to appear in Europe as an aid to navigation. The first crossing of the Atlantic was made by the Genoese Christopher Columbus, then in the service of Spain, in 1492. He observed the difference between magnetic and true north, a fact already known but now confirmed from a different location.

Considerable advances were made in the development of textile machinery. The rope-driven spinning wheel replaced hand spinning, while the weaving loom developed into its box-shaped frame with roller, suspended reed and shedding mechanism. Water-power was sometimes applied to spinning as well as to the fulling of cloth. These developments came in the thirteenth and fourteenth centuries.

Between AD 1100 and 1400 universities were founded in many European cities, particularly in Italy, signalling the start of a period of higher learning for its own sake. Towards the end of this period, the technique of paper-making, originating in China about AD 100, reached Europe via the Middle East, North Africa and Spain where it had existed since 1100. By 1320 it had reached Germany, paving the way to the printing of books.

Apart from the building of many fortified castles and some notable manor houses, the twelfth and thirteenth centuries were the peak of the construction of Europe's many cathedrals, marvels of architecture, their lofty slenderness seeming to defy the laws of nature. One of the most grandiose and eloquent was begun at Chartres in 1194 and completed in 1260. In the more prosaic field of vernacular architecture, the use of chimneys, which started about 1200, added considerably to the comfort of the occupants. Another improvement was the introduction of window glass on a small scale. Though it was a Roman invention, its use did not become at all common until the seventeenth century. Stained glass, of course, was of earlier date, its use at Augsburg cathedral dating from 1065.

THE THIRD AGE: THE FIRST MACHINE AGE

Timekeeping

The history of timekeeping, at least by mechanical means, is very much the history of scientific instrument making (see also Chapter 15). Although scientists may have conceived the instruments they needed for astronomical observation, a separate trade of craftsmen with the necessary skills in brass and iron working, in grinding optical lenses, in dividing and gear-gutting and many other operations grew up. It is impossible to say whether it was a scientist or a craftsman who was the first to calculate the taper required in the walls of an Egyptian water-clock to ensure a constant rate of flow of the water through the hole at the bottom as the head of water diminished. But water-clocks, together with candle-clocks and sandglasses were the first time measuring devices which could be used in the absence of the sun, so necessary with the obelisk, the shadow stick and the sundial. Once calibrated against a sun timepiece, they could be used to tell the time independently. On the other hand, portable sundials to be carried in the pocket became possible once the compass needle became available from the eleventh century AD. By the time of the Romans, the water-clock had been refined to the state that the escape hole was fashioned from a gemstone to overcome the problem of wear, much as later mechanical clockmakers used jewelled bearings and pallet stones in their escapements.

The sand hourglass had one advantage over the water-clock: it did not freeze up in a cold climate. On the other hand it was subject to moisture absorption until the glassmaker's art became able to seal the hourglasses. Great care was taken to dry the sand before sealing it in the glass. Candle clocks were restricted to the wealthy, owing to their continual cost.

Mechanical clocks, in the West, were made at first for monasteries and other religious houses where prayers had to be said at set hours of the day and night. At first, though weight-driven, they were relatively small alarms to wake the person whose job it was to sound the bell which would summon the monks to prayer. Larger monastic clocks, which sounded a bell that all should hear, still had no dials nor any hands. They originated in the early years of the fourteenth century. When municipal clocks began to be set up for the benefit of the whole population, the same custom prevailed, for the illiterate people would largely be unable to read the numbers on a dial but would easily recognize and count the number of strokes sounded on a bell. The weights that drove the clock were also used to power the striking action and to control the speed of the movement through a 'verge' escapement. Dials and hands were often added to clocks at a later date, as at Wells and Salisbury, first dating from 1386 and 1392. So also were jacks or 'jacques' which, in dramatic fashion, appeared and struck the hours at the appointed times.

The most remarkable clock of the age was that completed by Giovanni di Dondi in 1364 after sixteen years' work. Giovanni, whose father Jacopo is credited with the invention of the dial in 1344, was lecturer in astronomy at Padua University and in medicine at Florence and also personal physician to the Emperor Charles IV. He fortunately left a very full manuscript describing in detail his remarkable clock from which modern replicas have been made (one is in the Smithsonian Institution in Washington and the other in London's Science Museum), for the original has not survived. It had separate dials for the five planets then known and even included a perpetual calendar for the date of Easter driven by a flat-link chain. The whole was driven by a single central weight. All the gears were of brass.

Galileo's observations of the swinging altar lamp in the cathedral of Pisa marked the start of the use of the pendulum as a means of controlling the speed of clocks. Having no watch, he timed the swing of the lamp against his own pulse and established the time of the pendulum's swing, finding that it varied not with its amplitude but according to the length of the pendulum. The Dutch astronomer Christiaan Huygens turned this knowledge to good effect when he built the first pendulum clock in 1656. Within twenty or thirty years the average error of a good clock was reduced from some fifteen minutes to less than the same number of seconds in a day. The pendulum was a great advance but, like the weight drive, still only suitable for fixed and stationary clocks.

The coiled spring drive rather than the falling weight was first used by the Italian architect Filippo Brunelleschi in a clock built around 1410, and the problem of the decrease in the pull of the spring as it unwound was solved soon after that date by the incorporation of a conical spool, the fusee. The verge escapement was replaced by the anchor escapement which greatly reduced the arc of the pendulum, invented by William Clement about 1670. It was Robert Hooke who devised the balance spring to drive the escapement and thereby obviated the use of the pendulum in 1658. This enabled truly portable clocks and watches to be made for the first time. Huygens again was one of the first to make a watch with a balance spring, but this was probably not until 1674, a little later than Hooke's invention. At last, with a main spring and a balance spring, a timepiece could now be made entirely independent of gravity. Accurate clocks that could run at sea were essential to mariners for establishing longitude. Such a clock was made by John Harrison in 1761 and enabled him to win a £10,000 prize offered by the British government. On a nine-week trip to Jamaica, it was only five seconds out, equivalent to 1.25 minutes of longitude.

The very first watches, almost small clocks, were Italian 'orlogetti' but Germany, particularly Nuremburg, became the leading centre of watchmakers early in the sixteenth century. By about 1525 other centres had started up in France, at Paris, Dijon and Blois. German supremacy was soon eclipsed, a disastrous effect of the Thirty Years War which ended in 1648. By this time many French watchmakers who were Huguenots had fled the country, a number settling in Geneva to help found the industry for which Switzerland is still famous. Others settled in London, mainly in Clerkenwell, a great stimulus to the British watch trade.

Clockmakers were at first largely drawn from blacksmiths, gunsmiths and locksmiths and were itinerant craftsmen, for municipal public clocks and monastic clocks had to be built where they were to be installed. Only later, when timepieces became smaller, could the customer take them away from the maker's workshop or could they be delivered to the user complete and working. It was from the same groups of craftsmen that the early scientific instrument makers came, producers of celestial and terrestrial globes, astrolabes, armillary spheres and orreries for the use of astronomers, while, for surveyors and cartographers, chains, pedometers, waywisers, quadrants, circumferentors, theodolites, plane tables and alidades were among the instruments in demand. Together with the Worshipful Company of Clockmakers, a guild founded in 1631, stood the Spectacle Makers Company whose members supplied telescopes for probing the skies and microscopes for looking into the minuscule mysteries of nature. Both these instruments appear to have originated in the Dutch town of Middleburg in the workshops of spectacle makers.

Optics

The telescope originated, so history relates, in the shop of Johannes Lippershey, a spectacle maker of Middleburg, in 1608. Two children playing in this unlikely environment put two lenses in line, one before the other, and found the weathervane on the distant church tower miraculously magnified. Lippershey confirmed this and, mounting the lenses in a tube, started making telescopes commercially. He applied for a patent, but was opposed by claims from other Dutch spectacle makers. The secret was out and, within a year, a Dutch 'perspective' or 'cylinder' was displayed at the Frankfurt fair, was on sale in Paris, seen in Venice and Padua and, by the end of 1609, was being made in London.

These telescopes were made virtually without any understanding of the principles of optics, but needed only a competence in the grinding and polishing of lenses, craftsman's work. The true inventor of the microscope is not known, there being several claimants to the invention. Galileo, by 1614, is reported to have seen 'flies which looked as big as lambs' through a telescope with a lengthened tube, but Zacharias Jansen, one of the spectacle makers of Middleburg, a rival contender with Lippershey for the telescope, is a possible candidate. Early users were certainly Robert Hooke who used his own compound instrument to produce results published in his *Micrographia* in 1665 and Anton van Leeuwenhoek of Delft who was reporting his observations with a simple microscope to the Royal Society by 1678.

The surveyor's quadrant is an instrument of particular importance, for it was the first to which Pierre Vernier's scale was fixed so that an observer could read an angle to an accuracy of one minute of arc. The invention dates from 1631 and the earliest known example was made by Jacob Lusuerg of Rome in 1674. For linear rather than angular measurement the Vernier gauge has been a standard instrument in engineering workshops for many years. It seems that, once they had grasped the principle, the Lusuergs wanted to keep it in the family for it was Dominicus Lusuerg, who lived in Rome from 1694 to 1744, who manufactured a pair of gunner's calipers with a Vernier scale for measuring the bores of cannon and the diameter of cannon balls.

The crank

An important development in the Middle Ages was that of mechanisms for the interconversion of rotary and reciprocating motions. The cam had been known to the Greeks: it was illustrated by Hero of Alexandria. In the early Middle Ages the crank came into use (see Figure 4). First a vertical handle was used to turn the upper stone of a rotary quern, in itself an improvement on the saddle quern for hand-grinding corn. About AD 850 the same simple mechanism was applied to the grindstone for sharpening swords. In the fourteenth century it was used to apply tension to the strings of the crossbow, while it was frequently

Figure 4: The crank – a key element in mechanism.
From Agostino Ramelli's *Diverse et Artificiose Machine*, 1588.

to be found in the carpenter's brace. In all these cases the mechanism was hand-operated. The first known use of the crank and connecting rod is found about 1430 when it was used in the drive of a flour mill. A useful drive mechanism was the treadle, used first for looms and, by about 1250, to drive a lathe, a cord attached to a treadle having a connection to a flexible pole above the lathe for the return stroke. Some two hundred years later, a treadle with a crank and connecting rod was used for flour milling.

Print

One of the greatest inventions of the Middle Ages, undoubtedly one that had the most widespread and long-lasting effect on the lives of every man and woman who lived after it, was the printing process devised by Johannes Gutenberg, a goldsmith of Mainz in Germany, about 1440 (see also Chapter 14). The success of this was dependent on the invention of paper, knowledge of which reached Germany about 1320. The printed book enormously stimulated the spread of knowledge, superseding the slow, costly and laborious copying of manuscripts in monastic houses on scarce and expensive parchment, made from the skins of sheep and goats, or vellum from the calf. Printing from movable type demanded a whole series of inventions in addition to those that had brought block printing into use in China, and even in Europe, for book illustrations, maps and currency. It involved the mechanical processes of cutting punches of brass or copper and later of iron, each of a single letter; the stamping of the punches into copper plate to form the moulds into which the molten type metal of tin, lead and antimony could be cast. The stems of all the letters were of the same cross-sections and the same height, so that they could be assembled in any order and were interchangeable. They were then clamped in trays to form blocks of type to make up pages, inked and then pressed against sheets of paper in a screw press. The casting of the type, the assembly into trays, the formulation of the ink and the use of the press were all steps evolved by Gutenberg over a period of years at no small cost and considerable litigation in which he lost most of his plant and process to Fust who had invested in the process and Peter Schöffer who had been Gutenberg's foreman. The popularity of the new process can be judged by its rapid spread. By 1500, only forty-six years after the first book was published by Gutenberg, there were 1050 printing presses in Europe. The first book printed in England was by William Caxton at his press in Westminster in 1474.

THE FOURTH AGE: INTIMATIONS OF AUTOMATION

Coinage – the first mass production

Coinage originated long before Gutenberg, as early as the sixth century BC. Herodotus writes that King Croesus was the first to use gold and silver coins, in Lydia, now in the southern half of Turkey but from 546 BC a province of Persia. Yet as late as the mid-thirteenth century AD Marco Polo, whose *Travels* were recorded in 1298, says of the Tibetans, 'for money they use salt' and of other eastern peoples he records the use of gold rods, white cowries, 'the Great Khan's paper money' and 'for small change' the heads of martens.

The convenience of coins over shells or the skulls of small animals, however, is not difficult to see and the practice of minting coins soon spread. At first coin blanks were cast into clay moulds to be softened by re-heating before being struck between upper and lower dies, sometimes hinged together to keep them in alignment. A collar was later placed round the blank, limiting radial expansion and, at the same time, if suitably serrated, producing a milled edge. About AD 1000 coin blanks were formed from sheets of metal, hammered to the right thickness and then cut into strips. Not until after 1500 did Bramante of Florence introduce the screw press for coining. A further sixteenth-century development was the use of small rolling mills, not only to standardize the blank thickness but with the dies, circular-faced in the rolling axis, set into pockets in the rolls. Chill cast-iron moulds were used for producing ingot blanks for rolling to the correct size. In 1797, Matthew Boulton of Soho in Birmingham started minting his own 'cartwheel' pennies on a screw press, turned by the vacuum derived from a steam engine. Several presses could be run from a single engine. Subsequently Boulton supplied plant for the Royal Mint in London and many overseas mints. Diedrich Uhlhorn's 'knuckle' press, patented in 1817, followed by Thonnelier's press of 1830, dispensed with the rather slow speed of operation inherent in the screw press and led to the era of modern coining practice.

Although not interchangeable in an engineering component sense, coins are, in fact, examples of interchangeable manufacture, as are Gutenberg's sticks of print each bearing a single letter. Moreover, both minting and printing involved a common factor: the workers used machines that belonged to their masters and were installed in premises belonging to the masters. They were the forerunners of the Factory System.

The Factory System

The mint and the printing works employed few workers, at least in the early days of both. It was not the same in the textile industry in the second half of the eighteenth century. Until this time the spinning of thread and the weaving of it into cloth had been done by outworkers in their own cottages, the raw materials being delivered and the finished products often being collected by the workmasters, who also financed the entire operation. When the new machines arrived – Kay's Flying Shuttle (1755), Arkwright's Water Frame (*c.*1790), Hargreave's Spinning Jenny (*c.*1760), Crompton's Mule (*c.*1788) and Roberts's Power Loom (1825) – they were all operated by a steam engine or, at least, a water wheel, either of which could be able to drive a number of machines: a factory (see also Chapter 17). It thus became necessary for the workers to travel daily from their homes to a central place of work.

With the steam engine as a power source, factory masters were no longer constrained to set up their enterprises on the banks of fast-flowing rivers or streams. Admittedly economies could be made by setting up close to a coalfield, for the cost of transporting boiler fuel from the pithead could be a substantial proportion of the total cost of coal. Instead of being spaced out along the river banks so as to take advantage of the available water power, factories could now huddled together cheek by jowl as close as was convenient to their owners. Passenger transport being non-existent for all but the wealthy, the workers had to give up the freedom of the countryside and move to houses within walking distance of the factories, houses often rented to them by their masters. Regular working hours were introduced and penalties strictly enforced for failure to keep to them. Thus were founded Britain's major industrial cities, Liverpool, Manchester, Glasgow, Leeds . . . Nottingham, Birmingham. A similar process, if on a lesser scale, went on around the mines, whether for coal, iron or other minerals, as well as in the other countries of Europe.

The metal-working industries followed the same pattern. Apart from the lathe, machines for boring, milling, shaping, slotting, planing, grinding and gear-cutting were among the whole family of machine tools that flourished during the late eighteenth and nineteenth centuries, and these began to be located in workshops offering general engineering facilities (see Chapter 7). Instead of the separate pole or treadle drive, a host of these would be driven by a single steam engine through line shafting, pulleys and belting. An important feature of machine tools is that, in skilled hands, they have the ability to reproduce themselves, so the machines create more machines. The work ethic already existed, for man had long become accustomed to the need for the sweat of his brow and the labour of his hands. Now he could produce much more, his hands enhanced by the machine, but in much less pleasant surroundings and circumstances which often approached slavery.

Interchangeability of components in manufacture

About 1790, Joseph Bramah in conjunction with his foreman Henry Maudslay, evolved a number of special machine tools for the production of his locks (see pp. 395–6). Individually they were of no special importance but, taken together, they are of the greatest significance. They established a completely new and revolutionary concept – that of the interchangeability of components in manufacture. So accurately were parts machined with these tools, that the barrel of one lock could be applied to the casing of another, while the sliders of one lock could similarly be inserted into the barrel of another.

The same principles were adopted in the USA. After his unprofitable invention of the cotton gin, Eli Whitney looked around for another product to manufacture. In 1798 he wrote to the Secretary of the United States Treasury proposing to supply the government with 'ten or fifteen thousand stand of arms', arms at the time being smooth-bore flintlock muskets. His offer for the lower quantity was accepted and he set up production, splitting the labour force into sections to make the different parts instead of a single gunsmith making all the components of one gun at a time, sawing, boring, filing and grinding each separately so that they would fit together. With a series of machines, jigs, clamps, stops and fixtures, each lockplate, barrel, trigger, frizzle and every other part was exactly the same as all its counterparts. Thus Whitney established the American system of mass production. It took him eight years instead of the three that he had originally projected to fulfil the contract, but the government was well pleased and placed a repeat order. Each musket was supplied complete with bayonet, powder flask and cartridge box.

A further link in the chain was forged with the construction and installation of the Portsmouth blockmaking plant of Brunel, Bentham and Maudslay in 1803 to 1805. These forty-five machines, of twenty-two different types, driven by two 22.4kW (30hp) steam engines and ranging from circular saws and mortising machines to pin turning lathes, could produce 130,000 ships' pulley blocks a year, more than enough for the entire requirements of the navy when a 74-gun ship needed as many as 922 blocks. With these machines, ten unskilled men could produce as many blocks as had previously been made by 110 skilled blockmakers. A considerable advance over Bramah's lock machinery was that Sir Marc Brunel's machines, some of appreciable size, were built almost entirely of metal, without timber beds or frames. The only operations performed by hand were the feeding of the material, the moving of part-finished components from one machine to the next and the final assembly. Only when transfer lines were introduced in the twentieth century was Brunel's concept truly surpassed.

An automatic flour mill

Oliver Evans, born in 1755 in Newport, Delaware, has been called the Watt of America, but his field of operation and inventions was much wider than that of James Watt, who concentrated on steam engines. Evans's first and possibly greatest invention was a flour mill which was entirely automatic. Bucket elevators were used for raising the corn to be ground, Archimedean screws to transfer it horizontally and a device called a hopper boy to take the moist warm meal and spread it evenly on an upper floor. He later added a 'descender', another conveyor of the belt type, and the 'drill' in which small rakes dragged the grain horizontally. The mill would run with no one in attendance so long as it was constantly fed. Evans worked on the design from about 1782 to 1790 and licensed over a hundred other millers to use his ideas.

Evans's flour mill lacked one thing. It worked at a constant speed. If the feed hopper was filled, it would grind what was put into it: if the hopper was left empty, the mill and all its functions would continue in operation without producing any meal. Speed regulation was dependent in automatic machines on the principle of negative feedback exemplified by Watt's centrifugal governor added to his rotative engines in 1788. The governor, generally regarded as the first deliberately contrived feedback device, is an example of a closed loop system. It consists of a pair of weights, generally in the form of balls, pivoted on arms so that they are free to rise by centrifugal force as they revolve. As the speed of the engine increases, the arms rise and are connected so as to operate a butterfly valve which admits and cuts off the steam supply to the engine. The more the engine tends to exceed a given speed, the less is the energy supplied to enable it to do so. The engine thus became self-regulating. Similar devices are common today in many fields of automation. In fact, Watt did not invent the centrifugal governor commonly associated with his name. It was already in use for controlling the distance between the stones in windmills, although it does date from the last quarter of the eighteenth century.

A computer too early

Charles Babbage, at one time Lucasian Professor of Mathematics at Cambridge, devoted a great deal of his time to calculating figures, astronomical, statistical, actuarial and others. At one time, he is credited with having said, 'How I wish these calculations could be executed by steam!' He devoted much of his life to the design and attempted manufacture of, first, a Difference Engine (see Figure 5), which he started in 1823, and then, from 1834, an Analytical Engine. The former was a special-purpose calculating machine, the latter a universal or multi-purpose calculator. He pursued these goals for much of his long life, but unfortunately he was ahead of his time. His machines were purely mechanical and the precision needed in their manufacture was almost beyond even such an excellent craftsman as he employed – Joseph Clement. He died a disillusioned man, but left behind him thousands of drawings that contain the basic principles upon which modern computers are built. Gears, cams and ratchets could not do what transistors or even the diode valve was capable of. The computer had to wait for the age of electronics.

THE FIFTH AGE: THE EXPANSION OF STEAM

Estimates vary, but it is generally accepted the about one-third of the population of Europe died from the Black Death which ravaged England from 1349 to 1351. The consequent shortage of labour enabled those who survived to bargain successfully for higher wages and was a great spur to investment in wind and water mills and their associated machinery. By the mid-sixteenth century any site with reasonable potential was occupied by a mill and the search for some other source of power began to occupy the minds of ingenious men. It was

Figure 5: Charles Babbage's Difference Engine, 1833. See Chapter 15.

another hundred years before the first tentative results began to appear (see also Chapter 5).

In 1606, Francesco della Porta demonstrated the suction caused by condensing steam and its power to draw up water. In 1643, Evangelista Torricelli demonstrated the vacuum in a mercury barometer. Otto von Guericke, Mayor of Magdeburg, in 1654 performed his most dramatic experiment in which two teams of eight horses were shown to be unable to pull apart two halves of a copper sphere from which the air had been exhausted by an air pump to leave a vacuum. Atmospheric pressure held them together. In 1648, Blaise Pascal showed that the weight of a column of air was less at the top of a 4000-foot (1220m) mountain than at the bottom. In 1660, the Hon. Robert Boyle formulated the Gas Laws and demonstrated the maximum height that water could be drawn by a suction pump.

Others took up the theme of producing a vacuum by the condensation of steam. In 1659 the Marquis of Worcester described experiments with boiling water in a gun barrel, the steam forcing the water out of one or more receivers connected to it. It was recorded that Sir Samuel Morland, 'Magister Mechanicorum' to King Charles II, had 'recently shown the King a new invention . . . for raising any quantity of water to any height by the help of fire alone'. Denis Papin, a Huguenot refugee from France, worked for Robert Boyle and later, in the early 1690s, constructed a small atmospheric steam engine. It worked, it is said, but was only a model, of no practical use outside the laboratory. Papin shied at the problems of building a large-scale reproduction, such as could be used for mine pumping. He devoted himself from then on to trying to harness the power of steam without the use of a cylinder and piston. His attempts led to no success.

In 1699, Captain Thomas Savery demonstrated to the Royal Society a vacuum pump with two receivers and valve gear to alternate them and later built full-sized machines. Unfortunately the maximum suction lift he could achieve was some twenty feet, insufficient for pumping in the mines. It was left to Thomas Newcomen, re-introducing the cylinder and piston but now of 21 inches diameter, to build the first practical steam pumping engine near Dudley Castle in Staffordshire (see Figure 6). It made twelve strokes a minute lifting at each stroke 10 gallons of water 51 yards. Newcomen died in 1729, but engines of this type continued to be made until the early years of the nineteenth century. At least 1047 are recorded as having been built including those in France, Hungary, Sweden, Spain, Belgium and Germany.

Many people ascribe the invention of the steam engine to James Watt, but this is far from the truth. Great though his contribution was, Watt was fundamentally an improver of the Newcomen engine which was his starting point. In 1757 he was appointed 'Mathematical Instrument Maker to the University of Glasgow', where he was allowed a small workshop. The Professor of Natural Philosophy, John Anderson, instructed Watt to put a model Newcomen engine into working order. He was not long in appreciating that the low efficiency of the engine, when he got it working, was due to the need to cool the cylinder at each stroke to condense the steam and so create the vacuum. If the steam could be condensed in a separate exhausted vessel, the cylinder could then be kept continually hot. This was his first and perhaps his greatest invention – the

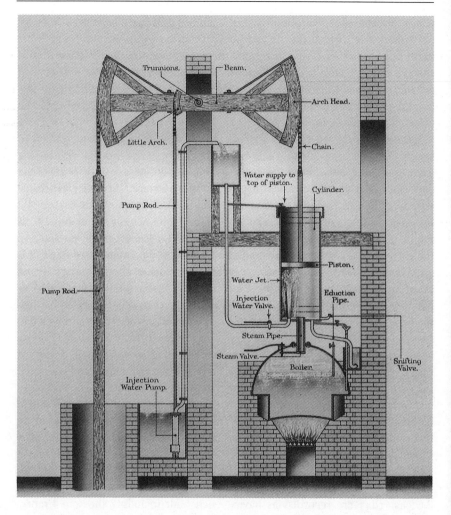

Figure 6: The first practical working steam engine of Thomas Newcomen, 1712. Erected to pump water from a mine near Dudley Castle, Staffordshire, England.

separate condenser. The first Watt engine incorporating it was erected in 1769, the same year that he was granted a patent. His other major inventions were the double-acting engine, patented in 1782, and the rotative engine, patented in 1781. The latter greatly extended the use of the steam engine from pumping to a multitude of purposes and brought additional prosperity to the partnership of Matthew Boulton and James Watt.

Throughout his life and that of his extended patent which lasted until 1800, preventing anyone else in Britain building any engine with the essential separate condenser, Watt stuck rigidly to low pressure or atmospheric engines, relying on the condensation of steam to create a vacuum against which the pressure of the atmosphere would move the piston. Many engineers were anxious to throw aside this cautious attitude, none more than Richard Trevithick, sometimes called 'the apostle of high pressure'. Once the Boulton and Watt stranglehold patent expired and the brakes were released in 1800, indeed even before this, for there were many who were willing to risk prosecution and to infringe the patent, Trevithick was ready to go ahead. In 1799 he built two engines working at 25psi (1.72 bar) for a Cornish mine. Later his pressure was to reach 100psi (6.9 bar).

So small were these engines compared with the massive Watt beam engines that Trevithick was soon using them in transport. In 1801–2 he had road carriages running in Cornwall and London. By 1804 he had built a railway locomotive to draw wagons on the plate tramway that had already been built from the Penydaren Ironworks of Samuel Homfray to the Glamorganshire Canal at Abercynon, a distance of nearly ten miles. A load of 25.5 tonnes was pulled at nearly 8kph (5mph) (see Chapter 11). In the years that followed a number of other locomotive experiments were made culminating in the Rainhill Trials of 1829. Stephenson's 'Rocket' was the winner and thus became the prototype for the traction on the Liverpool & Manchester Railway, the first public passenger railway in the world (see Figure 7). The Rainhill Trials were the highlight of the opening of the railway age over much of which George Stephenson and his son Robert presided unchallenged until in 1838 the young Isambard Kingdom Brunel began to build his broad-gauge Great Western Railway.

Steam also took to the seas over the same period (see Chapter 10). William Symington built a steamboat for Lord Dundas to run on the Forth & Clyde Canal in 1802 and this was followed by Henry Bell's *Comet* working between Glasgow and Helensburgh on the Clyde in 1812. The American Robert Fulton, however, could claim precedence with his *Clermont* on the Hudson River, running a regular service between New York and Albany from 1807. In 1815 the *Duke of Argyle* sailed from Glasgow to London carrying passengers. The *Savannah*, a steam paddleship carrying a full spread of sail, crossed the Atlantic from New York to Liverpool in $27\frac{1}{2}$ days in 1819, but only steamed for 85 hours. In 1830 the *Sirius* of the Transatlantic Steamship Company of Liverpool competed with Brunel's *Great Western* to be the first to cross the Atlantic under steam alone and won, having started four days earlier and averaging 12.4kph (6.7 knots). The *Great Western* arrived the following day, averaging 16.3kph (8.8 knots) to make the crossing in 15 days 5 hours. A new era in transatlantic steam navigation had truly begun.

The effects of the 'railway mania', which reached its height in 1845–6, only fifteen years after the Rainhill Trials, were many, various, sudden and dramatic. Providing employment for thousands of navvies, as well as surveyors, engineers and clerks, it changed the landscape of Britain and earned fortunes for contrac-

Figure 7: The 'Rocket' of George and Robert Stephenson, 1829. The advent of the high pressure, as opposed to the atmospheric, steam engine allowed it to become mobile. See Chapter 11.

tors and investors – apart from those who were drawn into many of the 'bubble' schemes that failed. The railways brought about a standardization of time throughout the country, emphasized the existing divisions between different social classes, and tended to bring about a uniformity in the materials used in buildings. They caused the decline of many towns and villages which were not served by the railway lines. They speeded up the mails and greatly accelerated the spread of news by the rapid distribution of the daily papers. They popularized seaside and other holiday resorts and improved communications by their use of the telegraph. Most of all, the railways took away business from the turnpike roads and the canals until the horse and the canal barge became almost obsolete. More and more people travelled, many of whom had never travelled outside their own villages before. Lastly they were excellent for the rapid transport of freight. Fish was added to the diet of people living inland, something they had never enjoyed before. The supremacy of the railways for carrying both passengers and freight was to last until early in the twentieth century, when the internal combustion engine began to be made in large quantities.

The reciprocating steam engine reigned supreme as a form of motive power until late in the nineteenth century when the newly invented internal combustion engine was beginning to pose a threat for the future (see Chapter 5). There was, however, another alternative. In 1884, Sir Charles Parsons patented the high-speed steam turbine which was first to make the reciprocating engine redundant in electrical power stations. In 1887 he demonstrated a small turbine-engined steam yacht, the *Turbinia*, at the Spithead Naval Review held to celebrate the Diamond Jubilee of Queen Victoria. The tiny yacht that could attain a speed of 64kph (34.5 knots) amazed all who saw her, as she darted in and out of the lines of ponderous warships. The following year the Admiralty ordered a 55.5kph (30 knot) turbine-driven destroyer from Parsons. In another three years the first turbined passenger ship was launched and, by 1907, this was followed by the 52MW (70,000hp) 39,000 tonne liner, the *Mauretania*. The steam turbine and reduction gearing was well and truly launched on the oceans.

THE SIXTH AGE: THE FREEDOM OF INTERNAL COMBUSTION

In 1884, the same year as Parsons's first patent, Gottlieb Daimler built and ran the first of his light high-speed petrol engines and in 1885, Carl Benz built his first three-wheeled car (see Figure 8). In 1892, Rudolph Diesel patented his 'universal economical engine', thereby completing the base upon which modern road transport runs. The internal combustion engine, petrol or oil fuelled, effectively ended the supremacy of the steam locomotive for long-distance transport, as well as contributing towards marine propulsion and other applications (see Chapter 5).

People had experienced a taste of freedom with the introduction of the bicycle which preceded the motor car by only a few years, the 'safety' bicycle, similar to that used today, having first appeared about 1878. They were ripe and ready for the added freedom that an engine would give them, that is those who were able to afford it. Until Henry Ford started making his first 'product for the people', the Ford Model 'A' in 1903, motor cars were luxury commodities. Ransom Olds had the same idea in 1899, but his success was nothing like that of Ford.

A motor car, or a bicycle, is of little use unless there are good roads to run it on. The pneumatic tyre, invented by Dunlop in 1887, caused much trouble as it sucked up the dust from the untarred road surfaces of the day. Bath was fortunate, for the hundred-mile road from London was watered every day from pumps situated at two-mile intervals; there was a proposal to lay a pipe up the road from Brighton to London to water it with sea-water, but this came to nothing. Many groups campaigned for improvements and, funded at first by the Cyclists' Touring Club, the Road Improvement Association was formed in 1886. Eventually the government was forced to act and the Road Board was set up in 1909. A tax of three pence a gallon on petrol provided the necessary funds

Figure 8: The Benz 'Patent Motorwagen' or Motor Tricycle of 1885. See Chapter 8.

for the local authorities which had to carry out the improvements. The era of the motor car had truly begun. In 1904 there were 17,810 motorized vehicles in Britain, including motorcycles: this had grown to 265,182 in 1914 and to 650,148 by 1920. In 1938, 444,877 vehicles were produced in Britain and 2,489,085 in the United States. There are some 22 million cars in Britain today.

The rest of the story of the motor vehicle is told elsewhere. The product of the world's most extensive industry has now reached the stage where cities are congested almost to a standstill, pollution of the atmosphere is widespread and a threat to health, and road construction programmes are said to cover the area of a whole county with concrete and tarmacadam every ten years. Pedestrian precincts have already sprung up like mushrooms in even quite small towns and by-passes proliferate to deflect the traffic from almost every town and village, but it seems that even these measures may be inadequate and sterner laws may have to be introduced to restrict the spread of the automobile and the toll of death and destruction that comes with it.

Internal combustion takes to the air

Sir George Cayley investigated the principles of flight and as early as 1809 expressed the opinion that passengers and freight could be carried at speeds of up to a hundred miles an hour (see Chapter 12). William S. Henson and John Stringfellow formed the Aerial Steam Transit Company in 1842, Henson having patented an 'Aerial Steam Carriage' in that year. The Aeronautical Society was established in 1866 and held an exhibition at the Crystal Palace in London in 1868. The French society was set up even earlier, in 1863. Men such as Lilienthal and Pilcher experimented with unpowered gliders, both being killed in the process. It was not until the internal combustion engine was available that powered flight became possible.

On 17 December 1903, Wilbur and Orville Wright flew some 165m (540ft) in twelve seconds, the culmination of four years of experiments with kites and gliders and even longer in theoretical studies. Unable to find a suitable engine on the market, they had built one to their own specification, an in-line 4-cylinder giving about 9kW (12hp) at 1200rpm with an aluminium crankcase.

The piston engine and propeller was the solitary form of air propulsion until Heinkel in Germany and Frank Whittle in Britain evolved their separate designs of jet engine, or gas turbine. The Heinkel He 178, with an engine designed by Hans von Ohain, made its first flight in August 1939. Dr Ohain's first experimental engine had been run on a test-bed in 1937 and the whole development was made without the knowledge of the German Air Ministry. In England, Whittle had taken out his first patent for a gas turbine for aircraft propulsion in 1930 while he was still a cadet at Cranwell RAF College. One of his first engines made its maiden flight in the Gloster-Whittle E28/39 at Cranwell on 15 May 1941. The performance of a plane with a Whittle turbojet engine was superior to any piston-engined machine, reaching speeds of up to 750kph (460mph). Civilian aircraft with jet engines came into service soon after the Second World War, the first service being started by BOAC with a flight by a de Havilland Comet to Johannesburg from London in May 1952. Transatlantic services started with a Comet 4 in October 1958.

The most recent chapter in man's conquest of the air is that of the supersonic airliner with the Anglo-French project Concorde. Simultaneously at 11.40 a.m. on 21 January 1976, planes took off from Paris and London to fly to Rio de Janeiro and Bahrain. The Russian factory of Tupolev made a similar aircraft which does not seem to have lasted long in service and was apparently not used for international flights. Unfortunately, the effects of the 'sonic bang', which occurs when Concorde exceeds the speed of sound, were taken by the aviation competitors of Britain and France as an excuse to prevent it landing on scheduled flights in their countries. Political rather than technical or economic considerations were foremost and the full earning potential of the plane has never been achieved, but Concorde is a magnificent technical achievement.

THE SEVENTH AGE: ELECTRONS CONTROLLED

Power on tap

The Electrical Age, which brought about power generation and mains distribution of power to every factory, office and home, was preceded by gas and hydraulic mains supply on the same basis. Experiments with gas for lighting were among the earliest reports to the Royal Society, as early as 1667. Sporadic trials were made all over Europe during the next hundred years but it was largely due to William Murdock in England and Philippe Lebon in France that the gas industry was started.

Lebon, an engineer in the Service des Ponts et Chaussées, made a study of producing gas from heating wood which he patented in 1799. He exhibited its use for both heating and lighting in a house in Paris. Commercially, his work came to nothing except, perhaps, to enthuse the German, Frederick Winzer (later Winsor) who formed the New Patriotic and Imperial Gas, Light and Coke Company and lit part of Pall Mall in London by gas in 1807. William Murdock, who was James Watt's engine erector in Cornwall, experimented with coal as the source of gas and developed it to a commercial success in the first decade of the nineteenth century. Winsor's company was chartered in 1812 as the Gas, Light and Coke Company by which name it was known until nationalization. It dispensed with its founder's services and, with Samuel Clegg as engineer, started laying mains, twenty-six miles being completed by 1816.

It was Joseph Bramah, the Yorkshire engineer, who had invented and patented the hydraulic press in 1795, who had the idea of transmitting power throughout cities through hydraulic mains. He patented this idea in 1812, envisaging high pressure ring mains fed by steam-driven pumps and weight-loaded or air-loaded accumulators. Unfortunately he died two years later, too early to put his ideas on power supply mains to practical use and municipal hydraulic mains did not come into being for more than another half-century with the work of W. G. Armstrong and E. B. Ellington in particular. A small system was started up in Hull Docks in 1877. The cities of London, Birmingham, Liverpool, Manchester and Glasgow installed municipal systems while Antwerp, Sydney, Melbourne and Buenos Aires had dockside or other installations. That in London was the biggest, having over 290km (180 miles) of mains supplying, at its peak in 1927, over 4280 machines, mostly hoists, lifts, presses and capstans.

Electricity, which was to do so much to change the world, had long been the subject of experimental investigations, at least since William Gilbert wrote his *De Magnete* in 1600 ('On the magnet and magnetic bodies, and on the great magnet, the earth'). Alessandro Volta, Professor of Natural Philosophy at Pavia some two hundred years later, took a series of discs of zinc and silver separated by moist cardboard and arranged alternately to form a pile. This Voltaic pile was the first true battery, a static source of electric power. Michael Faraday showed in 1831 that an electric current can be generated in a wire by the movement of a magnet near it and constructed a machine for producing a continuous supply of electricity, i.e. the first electric generator (see Chapter 6). Many other scientists repeated his experiments and produced similar machines. The substitution of electromagnets for permanent magnets by Wheatstone and Cooke in 1845 was the final step to bringing about the dynamo.

The development of the incandescent light bulb independently by T. A. Edison in the USA and by J. W. Swan in England brought public lighting by electricity into the realms of reality. Godalming in Surrey was lit in 1881. Edison's Pearl Street generating station in New York was commissioned the following September. Brighton's supply started in 1882 and there were many others in the same period. Ferranti's Deptford power station started operating in 1889. Thus the electricity industry was born. Its applications in the home, in industry and transport, in business and entertainment, are innumerable and contribute hugely to our comfort, convenience and well-being, One field, however, must in particular be singled out for special mention, the revolution in electronics (see Chapter 15).

The science of electronics could be said to have started with the invention of the thermionic valve by J. A. Fleming, patented in 1904. This was the diode and was followed in a short time by Lee de Forest's triode in America. Designed at first as radio wave detectors in early wireless sets, thermionic valves made use of the effect that Edison had noted with his carbon filament lamps, a bluish glow arising from a current between the two wires leading to the filament. The current flowed in the opposite direction to the main current in the filament, allowing current to flow in only one direction, from cathode to anode. The thermionic valve had many other applications. When the first computer, ASCC (or Automatic Sequence Controlled Calculator) was completed by H. H. Aitken and IBM in 1944, the necessary switching was achieved by counter wheels, electromagnetic clutches and relays. Two years later, at the Moore School of

Engineering at the University of Pennsylvania, the first electronic computer was completed. It had 18,000 valves, mostly of the double triode type, and consumed about 150kW of electrical power. To keep all these circuits in operation at the same time is said to have been extremely difficult but, if this could be done for one hour, ENIAC (Electronic Numerical Integrator and Calculator) could do more work than ASCC would do in a week.

Despite the success of ENIAC, computer engineers did not long have to depend on thermionic valves for in 1948 the point-contact transistor of Bardeen and Brattain and the junction transistor of William Shockley both emerged from the laboratories of the Bell Telephone Company in the USA. The transistor was much smaller than the thermionic valve, consumed less power and was far more reliable. The controlled flow of electrons in crystalline substances was also used in other semiconductor devices, such as the varistor, the thermistor, the phototransistor and the magnetic memory. From printed circuits, the technology soon moved on to integrated circuits in which the transistors and associated components are formed and interconnected *in situ* as thin films on a chip substratum. Up to 20,000 transistors can be integrated on a chip 6mm × 3mm and 0.5mm thick.

It is little over a hundred years since Alexander Graham Bell invented the telephone (1876) and Edison and Swan the incandescent lamp (1879). In the century that followed electronics has given us radio, television, the tape recorder, the video recorder, the pocket calculator, automation and robotics, the electron microscope, the heart pacemaker, myo-electrically controlled artificial limbs, the automatic aircraft pilot, the maser and the laser, computer-aided design and manufacture, solar cells, satellite communication and, with the rocket, man in space and unmanned space probes. Science and technology have combined to accelerate us into the Electronic Age. As J. G. Crowther wrote, 'Faraday, Henry and Maxwell would have had little influence in the world without Bell, Edison and Marconi.' Science and technology, the two arms of progress, have combined to leave us on the shore of a vast ocean of possibilities brought about by electronics.

Before and during the Industrial Revolution the ingenious powers of man were devoted to saving labour and to enhancing the capabilities of the paltry human frame. As we approach the last decade of the twentieth century, a new vista is opening before us, one in which at least the drudgery of brainwork is taken over by the electronic machine, the computer. This has already happened in fields such as accountancy and banking, in such affairs as stocktaking, engineering design and many others. But according to modern research programmes we are only at the beginning. The door is scarcely ajar, the door which opens on to life in a society where knowledge, the most prized possession, is freely available to all.

The Japanese, in a joint programme between government, industry and academics, are deliberately co-ordinating resources to develop a fifth generation of computers, computers which would be endowed with 'artificial intelligence', that is, computers which can think for themselves. This is a concept that to many people appears ridiculous and to many others a threat to the dignity of themselves as members of the human race or worse, a threat to the very existence of humanity as the only reasoning animal living on the planet. Americans who, up to now, have dominated the computer market world-wide, both technically and commercially, are expressing concern about the Japanese attack on their leadership. Perhaps the situation on a global basis is best summed up by the following quotation from Professor Edward Fredkin of Massachusetts Institute of Technology:

> Humans are okay. I'm glad to be one. I like them in general, but they're only human. It's nothing to complain about. Humans aren't the best ditch-diggers in the world, machines are. And humans can't lift as much as a crane. They can't fly without an airplane. And they can't carry as much as a truck. It doesn't make me feel bad. There were people whose thing in life was completely physical – John Henry and the steam hammer. Now we're up against the intellectual steam hammer. The intellectual doesn't like the idea of this machine doing it better than he does, but it's no different from the guy who was surpassed physically.

FURTHER READING

Armytage, W.H.G. *A social history of engineering* (Faber & Faber, London, 1961)
Boorstein, D.J. *The discoverers* (Dent, London, 1984)
Braudel, F. *Civilisation and capitalism, 15th–19th centuries: volume 1, the structures of everyday life* (Fontana, London, 1985)
Burstall, A.F. *History of mechanical engineering: technology today and tomorrow* (Faber & Faber, London, 1963)
Derry, T.K. and Williams, T.I. *A short history of technology* (Oxford University Press, Oxford and New York, 1960)
Dunsheath, P. *A history of electrical engineering* (Faber & Faber, London, 1957)
Feigenbaum, E.A. and McCorduck, P. *The fifth generation: artificial intelligence and Japan's computer challenge to the world* (Addison-Wesley, Reading Mass., 1983; Pan, London, 1984)
Gimpel, J. *Mediaeval machine: the industrial revolution of the middle ages* (Wildwood House, Aldershot, 1988)
Lilliey, S. *Men, machines and history* (Lawrence & Wishart, London, 1965)
Oakley, K.P. *Man the toolmaker* (British Museum – Natural History, London, 1972)
Pannell, J.P.M. *An illustrated history of civil engineering* (Thames & Hudson, London, 1964)
Sprague de Camp, L. *Ancient engineers* (Tandem Publishing, London, 1977)
Strandth, S. *Machines: an illustrated history* (Nordbok, Gothenburg, 1979; Mitchell Beazley, London, 1979)
Thomson, G. *The foreseeable future* (Cambridge University Press, Cambridge, 1957)
White, L. (Jr) *Medieval technology and social change* (Oxford University Press, Oxford, 1962)

PART ONE

MATERIALS

1

NON-FERROUS METALS

A. S. DARLING

NEOLITHIC ORIGINS

The birth of metallurgy is shrouded in obscurity, although weathered crystals of native copper might well have attracted the attention of ancient man because of their remarkable green colouration. Beneath this superficial patina copper in metallic form would have been discovered. Decorative and practical applications would undoubtedly have been sought for this relatively hard, heavy material, and primitive man would have been most impressed by the malleability of copper, which allowed it, unlike wood and stone, to be hammered into a variety of useful shapes.

The sharp distinction between the brightness, lustre and ductility of the interior of a crystal of native copper and the brittleness and stonelike characteristics of the green patina with which it was encrusted, would also have been noted. From this, the first users of metal might have concluded that all non-living primordial matter had originally been in this pure, bright, noble and amenable metallic state, although it had, like human beings, a natural tendency to fall from grace and, by contact with nature and the passage of time, to assume a degraded form.

Differing conclusions can, however, be drawn from the same set of evidence, and some of the earliest metallurgists, it appears, favoured a more optimistic interpretation which suggested that the natural tendency of most metals was towards rather than away from perfection. This tendency was encouraged by contact with nature so that metals buried deep in the bowels of the earth tended to mature and improve. Silver, according to this interpretation, was regarded as an unripened form of gold.

COPPER

Cold forging of native copper

The archaeological evidence indicates that although lead was also known at a very early date, the first metal to be practically utilized was copper. Small beads and pins of hammered copper found at Ali Kosh in Western Iran and Cayönü Tepesi in Anatolia date from the period between the 9th and 7th millennia BC and were made from native, unmelted copper. During the later stages of the Neolithic period more significant metallurgical developments appear to have originated in the mountainous regions north of the alluvial plains where the civilizations of the Tigris and Euphrates subsequently developed. Much use was then made of native copper which was hammered directly into small articles of jewellery or of ritual significance. At a later stage larger artefacts such as axeheads were made by melting together native crystals of copper in a crucible and casting the molten metal to the required shape. Finally, however, copper was extracted from its ores by pyrometallurgical methods.

The period during which copper was known to Neolithic man but not extensively employed is known as the Chalcolithic Age. The beginning of the Copper Age is associated with the emergence of smelting processes which allowed copper to be extracted from its ores. Native copper must undoubtedly have been worked in sites which were close to the outcrops where the metals had been found. It seems logical to assume, therefore, that copper ores were first reduced to metal, quite fortuitously, in the fires where native crystals were annealed at temperatures well below their melting point to soften them after cold forging, or in the furnaces where crystals were melted together.

Apart from gold, silver and the other noble metals, copper is the only metal which is found as native crystals in metallic form. This is because its affinity for oxygen is lower than that of most other common metals, and as a result native crystals, although not always abundant, can usually be found in weathered outcrops of copper ore.

The native copper which is still abundantly available in the vicinity of the Lake Superior deposits in North America, was worked from 3000 BC until shortly after the arrival of the Spanish invaders by the North American Indian: Giovanni Verazzano, who visited the Atlantic Coast in 1524, commented upon the vast quantities of copper owned by the Indians, and in 1609 Henry Hudson found them using copper tobacco pipes. Although unaware that copper could be melted and cast, they were working large lumps into weapons and jewellery.

The burial mounds of the pre-Columbian Indians of the Ohio valley contained many copper implements such as adzes, chisels and axes, which from their chemical composition appear to have been made from Lake Superior copper (see Figure 1.1). In more recent times the Indians on the White River in Alaska used caribou picks to dig copper nuggets out of alluvial gravels. Eskimos living in the Coppermine district on the Arctic shore of Coronation Gulf were using native copper as late as 1930, when it was reported that nuggets were occasionally found which were large enough to forge into knives with 20cm (8in) blades.

The implements and weapons produced in North America from native

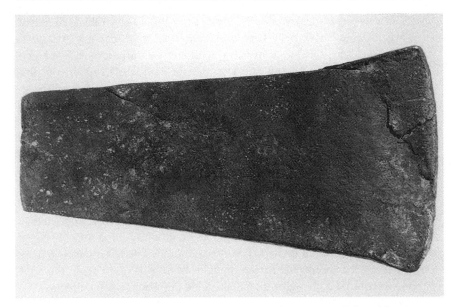

Figure 1.1: Adze blade from an unmelted nugget of native copper forged by pre-Columban Indians of the Ohio Valley Hopewell culture. Found in a burial mound at Mound City, Ohio, USA. Date 200 BC–AD 600. Note how internal flaws in the nugget have spread to the surface and failed to weld during forging.
Courtesy of the British Museum Laboratory.

copper all appear to have been produced by cold forging with frequent annealing at temperatures below 800°C. After an approximation to the final shape had been obtained in this way, the articles were finished by cold working. The cutting edges were hardened as much as possible by local hammering, and the craftsmen who produced these artefacts were obviously well aware that copper could be hardened by deformation and softened by annealing. It is difficult to understand, however, why they never succeeded in melting copper.

In Asia Minor, however, where copper was far less accessible, metallurgical development was very rapid. Cities and civilizations, which produce both wealth and demand, seem far more effective than natural resources in stimulating successful technical innovation.

Melted and cast native copper

Most of the larger copper artefacts produced in the Middle East between the seventh and fourth millennia BC have a micro-structure which is far from uniform. It consists of crystals which appear to have grown from the melt by throwing out arms into the surrounding liquid. From these arms secondary branches and spines have grown, thus producing a 'dendritic' structure characteristic of cast metal. Because of this structure, and their generally high purity, such articles could have been produced only by melting together crystals of native copper. This technique marks a great step forward, although the evidence available is very limited and it cannot be said when copper was first melted or even when the first attempts were made to utilize the effect in a practical manner. During the long period when native copper was being worked and annealed, some of the smaller crystals would, inevitably, have been accidently melted. The early coppersmiths would have attempted, instinctively, to avoid such unfortunate accidents, and would soon have developed an appreciation and recognition of those fiery conditions which would encourage copper to renounce the solid state and become mobile.

It seems most probable that when native copper was first intentionally melted it would have been heated from above by a heaped charcoal fire, and encouraged to run together and form a lens-like ingot in a clay-lined saucer-shaped depression in the ground immediately beneath the fuel bed. This arrangement might possibly have required forced draught, from bellows, to attain the temperatures required, although with suitable chimney arrangements this may not have been essential.

Crucible furnaces must soon have been employed, however, to produce items such as flat axes or mace heads which were cast directly to size. The earliest crucible furnace remains so far identified were found at Abu Matar, near the old city of Beersheba in Israel on a site used between 3300 and 3000 BC (see Figure 1.2). These furnaces appear to have had a vertical cylindrical clay shaft, supported in such a way that air could enter freely at the lower end, providing the necessary draught. The hemispherical clay crucible, about 10cm in diameter, was supported about half-way up the shaft by charcoal packed into the base of the furnace.

The slags produced by such melting processes appear, in general, to be far more enduring than the furnaces themselves. The earliest vitrified copper-bearing slags were found at Catal Huyük in Anatolia at a site dating from 7000–6000 BC, where specimens of beads and wire which appear to have been made from native copper were found. The first copper artefacts which, from their purity, appear almost certainly to have been produced by forging melted and cast native copper were found at Sialk in Iran, at a site dated around 4500 BC. These contained substantial quantities of copper oxide, and from their microstructure appeared, after casting, to have been either hot forged or cold worked and annealed. The earliest Egyptian artefacts produced from cast and wrought native copper appear to date from the period between 5000 and 4000 BC.

The archaeological evidence suggests that the technique of melting and casting native copper originated in Anatolia, and between 5000 and 4000 BC spread rapidly over much of the Middle East and Mediterranean area. Three flat axes

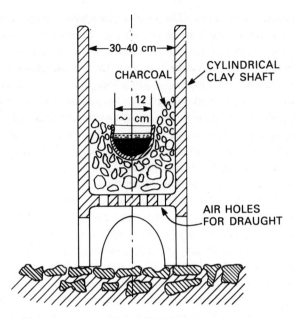

Figure 1.2: A reconstruction based on the remains of the earliest known crucible furnace, dating from 3300–3000 BC, found at the Chalcolithic site at Abu Matar, near Beersheba, excavated by J. Perrot in 1951. This appears to have been a natural draught furnace used for remelting impure copper in a hemispherical crucible which was supported on and immersed in a bed of charcoal halfway up the furnace shaft.
After J. Perrot.

produced by this approach were found between 1908 and 1961 in the Eneolithic Italian sepulchral cave Bocca Lorenza, close to Vicenza. Because of their high purity, these axes can be clearly distinguished from artefacts produced from smelted copper.

Smelting of oxide and carbonate copper ores

It seems that towards the end of the fourth millennium BC, the supplies of native copper accessible to the ancient world were incapable of satisfying a rapidly increasing demand. Most of the copper artefacts produced after 3500 BC contain substantial quantities of nickel, arsenic, iron, or other base metal impurities which indicates that they had been produced from copper which had been extracted from ore. Systematic copper mining was being undertaken well before this time, however, as early as the first half of the sixth millennium.

The Copper Age began when improved copper extraction techniques meant that primitive copper workers were no longer dependent upon supplies of relatively pure native metal. The rate of this transformation increased rapidly soon after the establishment of the Tigris and Euphrates civilizations. The wealth and specialized demand provided by these urban societies must have stimulated early copper workers to prospect in the northern mountainous regions where weathered outcrops of copper were most likely to be encountered.

The earliest copper workers appear to have extracted their metal from oxide or carbonate ores which, although not always rich or plentiful, could generally be smelted successfully in the primitive furnaces then available. The early smelters all appeared to understand instinctively that charcoal fires could be adjusted to provide atmospheric conditions which simultaneously reduced copper and oxidized iron. Methods were thus evolved which allowed relatively pure copper to be separated in the molten state from iron and other unwanted materials in the ore. These, when suitably oxidized, could be induced to dissolve in the slag. The primary ores of copper are invariably complex sulphides of copper and iron, and are generally disseminated in a porous rock such as sandstone which rarely contains more than 2 per cent by weight of copper. Such deposits were too lean to be exploited by primitive man, who sought for the richer if more limited deposits produced by the weathering and oxidation of primary ores. Thus, at Rudna Glava in Yugoslavia, a copper mine worked in the 6th millennia, did not exploit the main chalcopyrite ore body, but worked instead a thin, rich carbonate vein produced by leaching and weathering. This concentrated ore contained 32 per cent of copper and 26 per cent of iron. Quartz sand would have been added to such a smelting charge to ensure that most of the iron separated into the slag.

At Timna, in the southern Negev, copper has been mined and smelted since the dawn of history. Extensive workings, slag heaps and furnaces have remained with little disturbance since Chalcolithic, Iron Age and Roman times. These mines, traditionally associated with King Solomon, were in fact worked by the Egyptian Pharaohs during much of the Iron Age until 1156 BC.

The primary ore deposit at Timna is based on the mineral chrysocolla and is currently being exploited on a large scale. Since this ore contains only about 2 per cent of copper, it could not have been effectively smelted in ancient times. During the Chalcolithic or historical periods copper was extracted at Timna from sandstone nodules in the Middle White sandstone beds overlying the chrysocolla deposits. The nodules contain between 6 and 37 per cent of copper, which exists as the minerals malachite, azurite and cuprite. The remainder is largely silica, and the nodules contain little iron. In the fourth millennium BC copper was extracted from them in furnaces: a rough hole in stony ground, approximately 30cm (1ft) in diameter, was surrounded by a rudimentary stone wall to contain the charge, which consisted of crushed ore mixed with charcoal from the desert acacia. Controlled quantities of the crushed iron oxide haematite, was added to the charge as

NON-FERROUS METALS

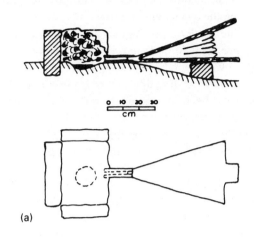

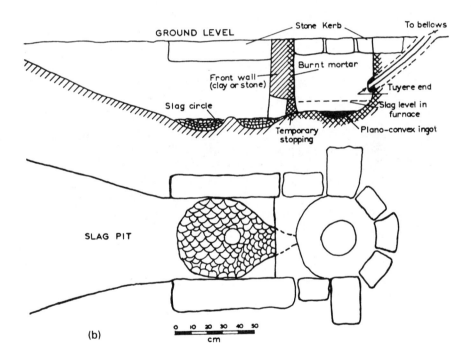

Figure 1.3: Reconstructions based on the remains of smelting furnaces used (a) in the twelfth century BC, and (b) in the eleventh century BC, at the ancient copper smelting site of Timna in the Southern Negev region of Israel.
Courtesy of the Institute of Metals.

a flux, to reduce the melting point of the silicious material and improve the separation between the copper and the slag.

The Chalcolithic smelting furnaces at Timna appeared to have had no tap hole and no copper ingots were found. High concentrations of prills and blebs of metallic copper were found in the slag, however, and it seems possible that the metal never, in fact, separated from the slag in massive form: after smelting the slag would have been broken up to remove the prills which were then remelted together in a crucible furnace.

A more highly developed smelting furnace used by the Egyptians at Timna around 1200 BC is shown in Figure 1.3(a). The slags from such furnaces contained up to 14 per cent of lime which was added to the charge as crushed calcareous shells from the Red Sea. This addition would have improved slag metal separation and allowed the reduced copper to settle to the bottom of the furnace and to solidify below the slag as plano-convex ingots.

Smelting techniques appeared to have reached their zenith at Timna around 1100 BC. After the smelting operation the slag and metal appear to have been tapped simultaneously from the furnace into a bed of sand where, as recent simulation experiments by Bamberger have shown, they would have remained liquid for about fifteen minutes, providing ample time for the molten copper to sink beneath the slag to form well-shaped ingots about 9cm (3.5in) in diameter. For Bamberger's reconstruction see Figure 1.3(b).

The rings and other small artefacts of iron found at Timna are now thought to have been by-products of the main copper refining operation. Lead isotope 'finger-printing' has shown that the source of the iron was the haematite used to flux the copper ore during refining. It would appear that when the 'as smelted' copper was remelted in a crucible furnace in preparation for the casting of axes and other artefacts, any surplus iron it contained separated at the surface of the melt to form a sponge-like mass permeated by molten copper. This layer would, in all probability, have been skimmed from the surface of the melt before it was poured. At a later stage it must have been found that the iron/copper residue could be consolidated by hot forging and worked to the shape required. The presence of copper is known to improve the consolidation of iron powder, and it would seem, therefore, that a sophisticated powder metallurgical process, utilizing liquid phase bonding, was being operated at Timna in Iron Age times.

Smelting of sulphide copper ores

From the presence of arsenic and other impurities in many of the early Copper Age artefacts it must be concluded that much of the copper used was extracted from sulphide rather than oxide or carbonate ores. In prehistory, as in modern times, the bulk of the world's supply of copper appears to have been obtained from ores based on chalcopyrite, a mixed sulphide containing equi-atomic proportions of iron and copper. Chalcopyrite must be roasted in air to convert it to a mixture of iron and copper oxides before it can be smelted. Moreover, because chalcopyrite ores contain in general less than 2 per cent of copper, and because of the presence of large quantities of unwanted earthy material, they do not respond to simple smelting processes.

The sulphide copper ores exploited at the beginning of the Copper Age appear to have been thin, localized and very rich deposits which lay some distance below the surface of a weathered and oxidized primary outcrop. Although the presence of such enrichment zones has been recognized by mining engineers and geologists for many years, their significance as ancient sources of copper has only recently been fully appreciated.

Due to the atmospheric oxidation which occurs at the surface of chalcopyrite outcrops, the sulphides are partially converted to more soluble compounds which are slowly leached away. The exposed surface of such an outcrop is therefore slowly robbed of most of its heavy metals with the exception of iron which concentrates at the surface as iron oxide, generally in the form of limonite. The iron-oxide regions above rich copper deposits are known as gossans, and the German term *eisener Hüt* to describe a gossan led to the saying, 'For a lode nothing is better than it should have a good iron hat,' and extensive gossans are noteworthy features of most of the sulphide copper ore deposits which were worked in antiquity. At Rio Tinto in south-west Spain, where copper has been extracted from the earliest times, iron is so extensively exposed at the surface that the terrain resembles that of an open-cast iron ore mine. Streams such as the Rio Tinto and Aguar Tenidas which leave this gossan owe their names to the red contaminant iron oxide. The ancient mines at Oman, which provided the Sumerians with copper, are characterized by huge ferruginous gossans, and similar terrain exists at Ergani Maden in Turkey.

From these gossans the copper content has been completely leached away, and transferred in aqueous solutions to lower horizons, where, as the dissolved oxygen becomes depleted, it is precipitated in sulphide form in the surrounding strata. In this way zones of secondary enrichment are formed which contain most of the metallic content which was originally uniformly distributed throughout considerable depths of rock. The average copper content of the thin secondary enrichment zones at Rio Tinto sometimes approaches 15 per cent and values as high as 25 per cent of copper in the enrichment zones at Ergani Maden have been reported.

The arsenical and antimonial minerals are associated with the copper in the cementation zone, and the preponderance in the Copper Age of artefacts containing substantial quantities of arsenic is therefore a further indication of the fact that much copper of this period came from these thin zones of secondary enrichment.

Arsenical copper

When an awareness that many early copper artefacts were in fact dilute alloys of arsenic in copper began to develop after 1890, it was thought that arsenic had been deliberately added to improve the mechanical properties. Bertholet, in 1906, was the first to demonstrate that the concept of alloying would have been unknown when these artefacts came into general use, and that arsenical copper, which was extensively used by the ancient Egyptians, was a natural alloy obtained by smelting arsenical copper ore.

In the cast condition arsenical copper is only marginally harder than pure copper, although its hardness increases very rapidly as a result of cold working. This must have been a factor of great importance in the Copper Age when edge tools were invariably hardened and sharpened by hammering. Arsenic deoxidizes copper very effectively and the alloys are noted for their excellent casting characteristics, a factor which early copper workers would much have appreciated. Cast billets of arsenical copper would, however, have been more difficult to work than pure copper, and it seems evident that when not cast directly to the required shape weapons and cutting tools were fashioned by a judicious mixture of hot and cold working. Such implements would certainly have retained their cutting edges for longer periods than their pure copper counterparts.

When arsenical copper ores are smelted with charcoal, the arsenic has little tendency to escape because it is held in solution by the molten copper. Similarly, when copper alloys containing up to 7 per cent of arsenic are melted in a crucible under reasonably reducing conditions, little is lost. Arsenious oxide, however, is very much more volatile than elementary arsenic, and toxic fumes must certainly have been emitted in copious quantities during the roasting of copper sulphide ores containing arsenic. Since sulphur dioxide fumes would also have been given off in vast quantities, the additional health hazards caused by arsenic would not, however, have been separately identifiable. Certainly nothing appears to have inhibited the use of these accidental arsenical copper alloys, which were extensively produced for a very long period over most regions of the ancient world. Primitive man, when he stumbled upon the thin layers of concentrated copper ores immediately below the gossan must soon have appreciated that they produced artefacts having properties vastly superior to those of the copper hitherto obtained from oxide or carbonate ores. On the basis of gradually accumulated experience he would have sought for similar or even richer ores in other localities, but in view of the absence of any concept of alloying it seems unlikely that attempts were made to control the hardness of the copper obtained by adjusting the mixture of ores fed into the smelting furnace.

The beginning of the arsenical copper era is difficult to date with any certainty. The earliest Egyptian artefacts of arsenical copper were produced around 4000 BC in predynastic times. In addition to arsenic these early weapons contained around 1.3 per cent of nickel. All the copper or bronze artefacts found by Sir Leonard Woolley during his excavations of the Royal Graves at Ur contained similar quantities of nickel which at the time these discoveries were made was considered to be a most unusual constituent of ancient copper.

It has been suggested that the Sumerians who made these artefacts came from the Caucasus and were instrumental in transferring the arts of metallurgy from the land of Elam (which now forms part of Iran) into Babylonia. The first Sumerian kingdom was destroyed by the Semitic King Sargon, and one of his inscriptions, dating from 2700 BC, indicated that the Sumerians obtained their copper from the copper mountain of Magan. Shortly after the Royal Graves excavation at Ur, the remains of extensive ancient copper workings were discovered in the vicinity of the small village Margana in Oman. The impurity spectrum of this copper ore deposit, including the nickel content, corresponded exactly with that of the artefacts of Ur.

Oman, however, is a long way from Babylonia, and since nickel bearing copper ores are also found in India and in the Sinai desert, the true origins of Sumerian copper are still uncertain. Indirect evident for direct links between Ur and Oman is provided, however, by evident similarities between the smelting techniques used to produce plano-convex copper ingots at Suza in Iran and at Tawi-Aaya in Oman, during the third millennium BC. The archaeological evidence also indicates strong cultural links between Southern Iran and Oman at this time.

Some of the slags found at Oman contained copper sulphide matte. Well-roasted sulphide ores can be effectively reduced to metal by techniques similar to those used to treat oxide or carbonate ores. Separation of the copper from the slag is facilitated by small flux additions such as bone ash, and this echoes the sea-shell additions made at Timna (see p. 54). The essential difference between the treatment of oxide and roasted sulphide ores, however, is that small quantities of copper sulphide matte appear to be produced in that furnace when the roasted sulphide ore is being treated under atmospheric conditions which reduce the copper, but retain the iron in an oxidized state. This matte, being insoluble in the slag, separates out as a thin silvery crust on the surface of the copper ingot. The presence of copper matte in ancient slag deposits is not, therefore, conclusive evidence of matte smelting.

TIN AND BRONZE

The Early Bronze Age

For well over 1500 years arsenical copper artefacts were extensively produced in the ancient world, which had come to expect from its metallic implements standards of hardness and durability which were unattainable from pure copper. After about 3000 BC, however, the arsenical content of Middle Eastern copper artefacts begins to decline, perhaps because of the progressive exhaustion of rich, accessible arsenical copper ore deposits. Artefacts containing small quantities of tin made their first appearance during the early stages of this decline although the quantities involved rarely exceeded about 2.5 per cent. Such alloys, which first appeared in Iran around 3000 BC, did not reach regions such as England and Brittany, on the outer fringes of civilization, until about 2200 BC.

These low tin bronzes must be regarded as the precursors of the later true bronzes containing 8–10 per cent of tin. They presumably emerged accidentally, either by the smelting of copper ores which contained tin minerals, or by the use of tin-bearing fluxes.

If the possibilities of alloying had been grasped at this stage it seems logical to assume that they would have been progressively exploited. No evidence for such a gradual evolutionary process has been found, and virtually no artefacts containing more than 2.5 per cent of tin have been identified which antedate the sudden emergence of the 8–10 per cent tin bronzes in Sumeria between 3000 and 2500 BC. It would appear that the copper workers of Ur suddenly discovered or acquired the concept of alloying, and rapidly developed and optimized the composition of bronze. This sudden leap forward cannot be separated from the rapid expansion of trade in the Middle East around 3000 BC, since it seems improbable that the significance of tin as an additive to copper would have first been identified in a region where its ore was not indigenous.

Sources of ancient tin

The most abundant and significant tin mineral is cassiterite, the oxide $Sn\ O_2$, which varies in colour from brown to black. Cassiterite is noteworthy in that its high specific gravity of 7.1 is comparable to that of metallic iron, and also because its hardness is comparable to that of quartz, so that it is highly resistant to abrasion and tends to concentrate in gravels and alluvial deposits. The name tin appears to be derived from the Chaldaean word for mud or slime, which implies that the tin originally used at Ur came from alluvial deposits. The Greek word *kassiteros* was taken from a Celtic term which has been literally translated as 'the almost separated islands', presumably the mythological tin islands. In classical times, *kassiteros* was loosely used to denote tin, pewter and sometimes even lead. The Greek word for Celtic, *kentikov*, was also used by Aristotle to describe metallic tin.

Four cassiterite mines have been located in the eastern Egyptian desert, one of which was accurately dated from inscriptions associated with the Pharaoh Pepi II to around 2300 BC, approximately 500 years after the appearance of the first Sumerian bronze. Copper artefacts containing significant quantities of tin were not produced in Egypt before the Fourth Dynasty, around 2600 BC, and the true Bronze Age in Egypt, when artefacts containing 8–10 per cent of tin were produced, did not begin until 2000 BC.

It seems logical to assume that at some time around 3000 BC, the concept of bronze manufacture was acquired by the Sumerians, either by trade or conquest, from a region outside Mesopotamia where tin was a readily available commodity. Many historians have claimed that by 3000 BC commerce in the Middle East had developed so extensively that supplies of tin might well have reached Sumeria by sea, up the Persian Gulf, from regions as remote as Malaysia or Nigeria. Support for this general idea is, of course, provided by the known importation by Sumeria of copper from Oman. From the Gulf, the tin would have been transported by overland caravan. Between 2600 and 2500 BC supplies of tin to Mesopotamia appear to have been interrupted, many of the copper artefacts dating from the Second Sumerian Revival being simple arsenical coppers containing little or no tin. This occurred, of course, in biblical times, when the Land of Sumer was devastated by floods. The shortage of tin, whether it was caused by natural disasters or by political upheavals, was not prolonged, however, and after about 2500 BC the use of 8–10 per cent tin bronzes expanded rapidly throughout the Middle Eastern world.

Early Bronze Age developments have also been found in southern China, Thailand and Indonesia, where alluvial tin and copper ore are sometimes found in close association. It was not until 2000 BC that bronze, or even copper, was manufactured in northern Thailand. The bronze artefacts, including spears, axeheads and bracelets, recently found at the village of Ban Chiang were obviously made locally, since stone axehead moulds were also discovered. The alloys, which contained 10 per cent of tin, must have been produced by craftsmen well-versed in bronzeworking technology; and since no evidence of earlier or more primitive metal working has been found, it would seem that this modest agricultural community suddenly acquired a fully developed facility for bronze manufacture. It probably arrived with the peaceful integration of a group of skilled foreign workers, perhaps displaced by military disturbances in China. Chinese bronze containing 8 per cent of tin was being produced in Gansu province as early as 2800 BC, apparently independently of the emergence of high tin bronzes in Sumer.

The earliest tin bronzes so far identified appear to date from the fourth millennium BC and were found in the 1930s at the site of Tepe Giyan, near Nahavand in Western Iran. This mountainous region, situated midway between the Persian Gulf and the Caspian Sea, was in the Land of Elam mentioned in the Bible, from which the Sumerians were assumed to have obtained the arts of metallurgy. Although Tepe Giyan does not appear to have a local source of tin, it is possible that there were alluvial tin deposits, now exhausted. The sudden appearance of true bronzes at Ur around 2800 BC, and in Amuq, near Antioch in Turkey, about 3000 BC would be consistent with the view that the concept of bronze manufacture originated in a community such as Tepe Giyan, in the Persian highlands, during the 4th millennium, and subsequently moved southwards to Sumeria and the Persian Gulf, and westwards to the Mediterranean seaboard, during the third millennium.

Bronze was first made in Italy between 1900 and 1800 BC, using tin from deposits at Campiglia Marittima in Tuscany, although it seems possible that cassiterite was also being obtained from mines in Saxony and Bohemia. Copper extraction started in Spain in Neolithic times and during the third millennium its copper and precious metal deposits were extensively worked (see p. 55) Exploitation of the Spanish tin deposits, however, did not begin until 1700 BC, when bronze artefacts were first produced at El Argar and other sites in the south-east. Most of the bronze artefacts of this period, which contain around 8 per cent tin, appear to have been cast roughly to shape and finished by forging. Evidence that the early Mediterranean cultures obtained significant quantities of tin from Cornwall, Brittany or Saxony-Bohemia is singularly lacking, although some of the tin used for Central European bronzes made between 1800 and 1500 BC appears to have come from the Saxony deposits.

Apart from a tin bracelet dating from 3000 BC found at Thermi in Lesbos, very few Bronze Age artefacts made of metallic tin have been discovered. This paucity is difficult to reconcile with the vast quantities of tin which must have been used for bronze manufacture over the period. It has been suggested that bronze was originally produced by co-smelting copper ore with minerals of tin, and that at a later stage, when better compositional control was required, weighed quantities of cassiterite were reduced directly with charcoal on the surface of a bath of molten copper. This implies that cassiterite, rather than metallic tin, was the standard commodity of trade. Metallic tin, however, far from behaving at all times as a highly inert material can sometimes, when buried for example in certain types of soil, disintegrate very rapidly into amorphous masses of oxide and carbonate sludge, destroying the identity of tin artefacts.

The most enduring evidence of Early Bronze Age technology is provided by a few accidentally vitrified clay tuyeres from between 2000 and 1800 BC. Many early tuyeres were 'D' shaped in cross-section and appear to have been used in primitive crucible furnaces. Here the crucible was heated from above by radiation from a glowing bed of charcoal, much of which must inevitably have fallen upon the surface of the melt.

The crucible in the furnace used at Abu Matar for remelting impure smelted copper between 3300 and 3000 BC (see p. 51) would presumably have been extracted via a hole in the side of the vertical shaft. Many crucibles used in the Mediterranean region between 3000 and 2500 BC had, to facilitate handling, a socketed boss moulded on one side. Clay-covered wooden or metal rods were

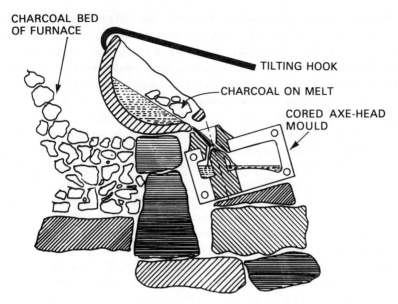

Figure 1.4: Rocking crucibles of this type were extensively used in the Greek Islands and also in Sinai between about 1600 and 1200 BC to ease the problems of lifting a heavy crucible of molten metal from a primitive furnace and subsequently pouring from it in a controllable manner. The rocking crucibles had thick hemispherical bases and were provided as shown with a pouring hole. When tilted they rolled away from their charcoal bed and discharged their content of molten metal simply and safely. One crucible of this type, found by Flinders Petrie at Serabit in Sinai in 1906 was large enough to contain nearly 8kg of bronze.

inserted into these sockets so that the crucible could be removed. During the Late Bronze Age, the crucibles used were thicker and more robust. The casting arrangement shown diagrammatically in Figure 1.4 was used between 1550 and 1200 BC in the Greek Islands and also in Sinai. Pouring was accomplished by rocking the crucible on its base, either by pulling with a hook from the front, or by pushing from the rear of the furnace.

Bronze Age casting techniques

Few plano-convex ingots of arsenical copper have been found, although the material appears to have been available at any early stage in the form of ingot torcs. These *Ösenhalsringe*, or neck-rings with recoiled ends, resemble in shape the later iron currency bars. Worked from cast ingots, they were traded as an intermediate product suitable for the manufacture of pins, jewellery and other small objects. No unworked Early Bronze Age ingots from which large artefacts such as axeheads could have been cast have yet been found.

Although many arsenical copper artefacts of the Early Bronze Age appear to have been roughly cast to shape in open stone moulds and then forged to size, this practice died out soon after the 8 per cent tin bronze alloy came into general use. Most artefacts were cast almost directly to the finished size in fairly advanced closed moulds similar to that shown diagrammatically in Figure 1.4. Many of these moulds were made from clay into which internal cavities were precisely moulded before firing.

During the early years of the third millennium foundry processes in the Land of Sumer appear to have developed rapidly. Before 3000 BC most of the copper artefacts from Ur were simply fabricated from sheet metal. By 2700 BC, however, very refined cast copper articles were being produced, probably by a version of shell moulding, using a prefired mould of thin clay. Long thin sections, such as the blades of swords and daggers, were also cast into thin shell moulds, which appear to have been heated to redness before the molten metal was introduced.

By 2500 BC the Egyptians had developed considerable expertise in the production of hollow copper and bronze statuary. Many large Egyptian statues were cast with an internal sand core, which is still present in some of the figures which have been found. Smaller components, such as the spouts of copper water vessels, were undoubtedly made by the *cire perdue* ('lost wax') technique, which had obviously been mastered by Egyptian craftsmen before 2200 BC.

Grecian bronze

Daggers with bronze blades inlaid longitudinally with niello, a black compound of sulphur with copper or other metals which formed a background for lively and naturalistic pictures in gold and silver, figured very prominently in the grave goods found by Schliemann in 1876 when he opened the Royal Bronze Age tombs of Mycenae. Such weapons, and most of the other metalwork which was found, are now believed to be of Cretan origin. The shaft graves date from the sixteenth century BC and are therefore of the Late Bronze Age, the absolute chronology of which is now well established.

By the beginning of the second millennium BC, Crete was beginning to obtain tin from Etruria and southern Spain, and their designs for spear heads and daggers were widely copied throughout the ancient world. The characteristic metal artefact of Crete was the double axe, an object of great ritual significance; the compositions of some of which range from pure copper to ductile 4 per cent tin bronzes, and finally to hard, brittle bronzes containing 18 per cent tin.

The Dorians, who invaded the Peloponnese from the north-east across the Gulf of Corinth in the thirteenth century, had primitive metal working skills of their own, and from the amalgamation of Dorian vigour and Minoan sophistication eventually emerged the metallurgical virtuosity of classical Greece. Bronze statuary soon began to demonstrate a mastery of foundry techniques which remained unsurpassed until the European Renaissance. Such castings represent what is probably the high peak of Bronze Age technology, before the later domination by Rome of the classical world.

Between prehistoric times and 500 BC, bronze technology had developed in a slow and irregular manner which does not relate in any metallurgical sense with the Early, Middle and Late Bronze Age classifications of the archaeologist. The major metallurgical discontinuity occurred around 1600 BC, in late Minoan times, when large copper ingots weighing more than 30kg (66lb) first appeared, and the usage of bronze throughout Europe and the Mediterranean region began to increase very rapidly.

The arsenical coppers and low tin bronzes hitherto employed had been gradually displaced by bronzes containing 8–10 per cent tin, similar in their general characteristics to those first utilized a thousand years previously at Ur. The emergence of the full Bronze Age, in the middle of the second millennium BC, is characterized by the almost universal employment of this type of alloy. However, after about 1600 BC the majority of the bronzes produced contained lead, which, in most cast objects of any intricacy, was present in concentrations between 5 and 10 per cent by weight, and which was added to improve casting fluidity.

Chinese bronze

The chronology of early Chinese bronze is more difficult to interpret, since there appears to have been no specific date before which no lead additions were made. The belief that China had little or no prehistorical metallurgical experience probably originated because of the massiveness and mature artistic competence of the Shang bronzes, which, having been looted from the Imperial Graves at Yin, were the first to make an impact on the artistic sensibilities of the West. It was assumed that the Chinese had waited until the middle of the second millennium BC, the culmination of bronze technology in the West, before it began, with no significant earlier experience, to produce large bronze artefacts of the highest artistic quality. However, more than 300 metallic artefacts dating from the period before 1600 BC have been discovered, and bronze was being produced in northern Thailand in 2000 BC (see p. 59).

China obviously began to produce bronze and other copper alloys at an early date. The presence in Kuangtung Province of tin deposits, both alluvial and in lode form, might well have facilitated the process. The art of bronze founding, which is now believed to have originated independently in China, appears to have first been practised in the adjacent provinces of Hansu, Shensi and Honan, where Chinese culture first began.

PART ONE: MATERIALS

Figure 1.5: Bronze ritual vessel 'Fang Ding' Shang dynasty, eleventh century BC. Decorated with a pattern of snakes and roundels.
Courtesy of the British Museum.

The Shang bronzes, produced at Anyang between 1400 and 1027 BC, are now world famous for their artistic merit and technical virtuosity (see Figure 1.5). These ritual objects, which were produced entirely by casting, were notable for the richness of their relief patterns and also for their mass and solidity, although some smaller wine vessels were also produced. One vessel weighed over 1.6 tonnes and had obviously been cast in a multipart mould made from prefired clay segments.

Most of these Shang artefacts had lead contents less than 1 per cent by weight, and some contain negligible quantities of lead. It would appear lead was an accidental impurity. The tin content of the alloys varies considerably, however, in what appears to be a random manner from about 1.8 per cent to more than 20 per cent.

The founders of Anyang could hardly have regarded tin as a rare or expensive commodity, to be used as economically as possible. However, the most impressive, highly decorated and least utilitarian items of Shang bronze tend to be those with the highest tin content, and rarely contain more than half a per cent of lead. Items of utility, however, such as weapons and accoutrements, contain smaller quantities of tin and higher concentrations of lead. It seems that the Shang foundrymen were well acquainted with the effects of composition on the characteristics of the bronzes they produced. Thus when the objective was to produce large, intricate and decorative castings which would not be subjected to mechanical ill treatment, they preferred to use a very high tin content, rather than a smaller quantity of lead to obtain the casting fluidity required. When weapons were being made, however, brittleness was a characteristic to be avoided. The tin content was then very considerably reduced and a minimal quantity of lead added to improve the castability of the alloy. The melting point of a binary 20 per cent tin bronze is approximately 880°C: a similar quantity of lead only reduces the melting point of copper to 1000°C. The high tin, lead-free alloys would therefore be easier to melt and cast in large quantities and this must have been a factor which encouraged their use in a situation where tin was plentiful and intricate castings weighing perhaps a tonne were being produced.

Bronzes containing 20 per cent tin are invariably associated with the manufacture of bells, and Theophilus in the twelfth century AD, following what was obviously an age-old tradition, recommended this composition for its purity of tone. The bell is generally regarded as being a Chinese invention, and the cast artefacts of the Shang and Chou dynasties are associated with the rituals of ancestor worship where music had a magical significance. It seems possible, therefore, that the ritual vessels were expected to be sonorous as well as artistically impressive.

The distinctive green patina of ancient Chinese bronzes has occurred by a process of slow corrosion during the long internment of the bronzes in deep loess of Central China. This yellow, porous, windblown, sandy clay contains a good deal of calcareous material, which, when taken into partial solution by circulating groundwaters, is believed to have played an essential role in the patination process.

Greek metallurgical specifications

At a very early period the Greeks attempted to standardize the composition and mechanical proportions of bronze. An inscribed stela found at Eleusis in 1893 provides what is probably the earliest known specification for bronze and cites a decree concerning the manufacture of *empolia* and *poloi*, the bronze fittings which were used for joining the drums of stone columns. The structure involved was the Philonian Stoa, a portico which was erected around 360 BC in front of the much older Telestrion at Eleusis.

It is specified that the contractor

> will use for the blocks copper from Marion, the alloy being made of twelve parts, eleven of copper to one of tin. He will deliver the blocks clean, rigid and four-

square and will round off the dowels on a lathe as in the example provided; he will fix them into the blocks straight, and perfectly rounded, so that they can be rotated without deviation. Bids for the contract are to be made at so much per mina of bronze, and the contractor will weigh out the bronze while there is constantly present one of the building commission, either the public recorder or the site supervisor.

It seems that the Greeks were aware of the weakening effect of lead on bronze and preferred the binary tin-copper alloy for structural items. The requirement in the specification that the *poloi* should be turned on a lathe is also a matter of unique importance: this Eleusinian inscription apparently provides the earliest evidence for the use of a metal-turning lathe in the Greek mainland.

The requirement of the specification that an 8.3 per cent tin bronze should be used for these castings shows that the fourth-century Greeks were well aware of the effect of tin on the properties of the alloys they employed. From the weapons which were found in the Mycaenean tombs it is evident that metal workers, even in the sixteenth century BC, knew exactly how much tin could be put into bronze before the alloy became too brittle for hot or cold forging. They improved the hardness and elastic limit of their bronze sword blades by judicious cold working, and knew that ductility could be restored, when required, by careful annealing. Pure copper rather than bronze was the preferred material where high ductility was required for the manufacture of rivets and other fixing components.

METALLURGY IN THE ROMAN WORLD

Although few metallurgical ideas or processes appear to have been truly Roman in origin, the improved standards of living associated with the rapid dissemination of the Roman way of life stimulated metallurgical demand, and encouraged the rapid diffusion of improved processes and techniques throughout the Empire. The primary metallurgical requirement of the army was iron (see Chapter 2). Bronze, in quite formidable quantities, was needed, however, both for military and non-warlike purposes, and coinage utilized vast quantities of gold, silver and copper. Roman civilization also brought with it the rapid expansion of highly organized urban life which required large quantities of lead for plumbing. The main metallurgical innovation of Roman times, however, was the general introduction of brass which was first used for coinage when tin became expensive.

Roman copper and bronze

Well before Imperial times the extraordinary mineral wealth of the remoter regions such as Spain, Germany and Britain became evident. The mines of

Spain provided gold, silver, copper, iron and zinc in great abundance, the richer deposits being situated in a region known as Tartessus, twenty-five miles north of Huelva on the Gulf of Cadiz.

At Rio Tinto (see p. 55) copper was extensively extracted in the Bronze Age, and the Phoenicians appear to have worked the deposits for silver between 900 and 700 BC. The Carthaginian empire was also dependent upon silver from Rio Tinto, and the Romans worked the Tartessian mines continuously from about 200 BC until AD 400. In Republican times the mines were worked on a limited scale, largely for silver. Under Augustus, however, the scale of operations increased very rapidly and deep tunnels were driven into the main chalcopyrite body to exploit secondary veins of rich copper ore.

The mines appear to have reached their maximum output between the second and third centuries AD. The Roman troops left Spain in AD 410, and although the Visigoths took over the Peninsula in 475 they did nothing to revive the mining industry, which was also largely ignored by the Moors. The ancient Roman workings at Rio Tinto were not reopened until the middle of the sixteenth century, during the reign of Philip II.

Rome did not begin to exploit the copper mines of Timna in the Negev (see p. 52) until the third century AD, although working continued until the middle of the seventh century when the Arabs invaded. Mining activity at Timna appears to have been controlled largely by the availability of fuel, charcoal obtained from the desert acacia.

The horses of San Marco

Most of the statuary of the Graeco-Roman world was cast in bronze. In fourth-century Greece leaded tin bronzes were generally used although zinc found its way into many of the later Roman bronzes, which by the middle of the second century AD tended to contain comparable quantities of zinc and tin. The characteristics of these alloys, which resembled those of the later gunmetals were, after all, well known, and utilized by all practical foundrymen. Thus, their low melting points and wide freezing ranges facilitated mould filling and helped to compensate for any incidental deficiencies in feeding and venting arrangements which must have been difficult to avoid when large, thin-walled castings of a complex nature were being produced in a primitive foundry. Zinc would also have deoxidized the alloys and assisted greatly in the reduction of casting porosity.

However, some artists of this period cast their statues from what we would now regard as most unsuitable material. The four horses of San Marco in Venice, when accurately analysed in 1975, were found to have been cast from virtually pure copper, containing only about 1 per cent of tin and lead. Because of its high melting point and narrow freezing range, this alloy would have been difficult to melt and cast properly.

The San Marco horses were cast by the indirect lost wax process, the technique used some 1500 years later by Benvenuto Cellini when he cast his celebrated Perseus. In this approach, the wax layer applied to the surface of the core is equal in thickness to the wall thickness of the casting to be produced. The wall thickness of the San Marco horses varies between 7.5 and 10.5mm, and originally they were gilded. From the duplex structure of the surviving areas of the gilded layer, it can be inferred that before the application of the gold leaf, the copper surface had been treated with an aqueous solution of gold mercury amalgam. When this had reacted chemically on the surface of the copper to produce the amalgamated layer required, the work would have been heated gently and the gold leaf applied.

Theophilus, writing in the twelfth century AD, warned that this technique could not be used to gild a lead-containing substrate, and the information would probably have been available in Roman times. The artist who produced the horses of San Marco might have decided to produce his castings from copper because it was easier to gild than a bronze which might well have contained substantial quantities of lead. They were popularly attributed to Lysippus of Sicyon, the official sculptor of Alexander the Great, who worked in the fourth century BC. It is possible, however, that they were cast eight hundred years later, during the reign of Constantine, although a more probable date would be some time during the second century AD.

Lead and silver

Lead is comparable to copper in its antiquity and was probably obtained in its metallic form by reduction from its ore at an earlier date. The earliest known metallic artefacts, small beads and pins of copper dating from the period between 9000 and 7000 BC, were hammered from crystals of native, unmelted copper (see p. 48). However, at 6500 BC horizons in the ancient Anatolian mound of Catal Hüyük beads of fused lead have been found in close association with hammered beads of native copper. It appears that lead must have been extracted from its ores at an earlier date than copper, especially in view of the low melting point of lead and the ease with which its oxide can be reduced by carbon.

Native lead is rarely found. Its sulphide, however, galena, upon which most lead ores are based, is fairly abundant and would have been instantly recognizable by primitive man because of its lustre, high density, and by the structural perfection of its dark grey cubic crystals. When exposed at an outcrop, galena weathers and oxidizes. Anglesite, the sulphate of lead, is a mineral formed during the intermediate stages of this reaction process which generally concludes with the formation of cerussite, a lead carbonate. Galena usually contains between 0.03 and 0.1 per cent by weight of silver, and the economics of lead extraction are generally determined by the value of the silver present in the ore.

Metallic lead has been found in Iraq, in a 6000 BC context at Yarim Tepe, and at the fifth-millennium site of Arpachiyeh. Lead artefacts have also been obtained from the fourth-millennium sites of Anan and Hissar III in Iran and Naqada in Egypt.

The earliest evidence of lead being used for aesthetic rather than practical purposes is provided by an Egyptian figurine 5cm high, dating from about 4000 BC, which is now in the British Museum. It is unusual in having been carved from a solid block rather than being cast to shape; the lead is of such high purity and contains so little silver that it could apparently have been obtained only by reducing the oxide residues from a silver cupellation process. Few leaden artefacts of comparable age have been found, although a mid-fourth millennium cemetery at Byblos in Lebanon recently yielded more than 200 silver artefacts. Silver objects of the late fourth millennium have also been unearthed in Palestine, in Ur and Warka in Mesopotamia, and also from various sites in Asia Minor. It seems probable that they emerged as a by-product of lead refining operations.

Lead artefacts of the third and second millennia BC include spindle whorls, weights for fishing, and wire which appears to have been used for repairing pottery. Lead weights, based on a unit module of 61g (2.2oz), have also been found throughout the Aegean in Middle and Late Bronze Age contexts. Shapeless lumps of lead were found in that layer of the Hissarlik ruins corresponding to the period between 3000 and 2500 BC, and many fine silver artefacts dating from about 2700 BC were found by Sir Leonard Woolley in the Royal Graves at Ur.

Numerous deposits of silver-rich lead are known in the Aegean area, several of which were described by classical authors such as Aeschylus, Herodotus and Strabo. In the fifth century BC, silver from the mines at Laurion provided the wealth needed to support the Athenian Empire. Other famous silver mines were those in the Pangaean region of Macedonia, and the Cycladic island of Siphnos. Mining at Siphnos began some time between 3150 and 2790 BC. The silver recovered from the Early Cycladic tombs appears never to have been melted down for recycling, and this has made it possible to utilize isotope abundance analysis to identify the lead mines from which the silver was obtained. Specific identifications can now be made by measuring the relative abundance of the natural isotopes in each element present in the metallic artefact. This approach allows the origin of the major alloying constituents to be identified, not from the impurities they contain, but from their own specific internal atomic characteristics. The metals extracted from every ore deposit each have an isotope distribution spectrum which is specific to, and characteristic of that deposit. Isotope abundance ratios are not influenced in any way by smelting or refining processes, or by the presence or absence of impurities introduced during alloying.

The silver artefacts in the Cycladian tombs, therefore, which appear to have been made from virgin metal and almost immediately interred, have isotope abundance ratios which make it possible to identify very precisely the lead mines from which the silver was obtained.

Of sixteen artefacts, six were found to have come from the Laurion mine near Athens and eight from the Siphnos field. A number of lead and silver artefacts from the shaft graves at Mycenae (c.1550 BC) were made from metal coming from the Laurion mine. Egyptian silver artefacts of the 10th and 11th Dynasties (2175–1991 BC) also appear to have been made from Laurion metal. Isotope ratio measurements on Cretan lead and silver artefacts of the period 1700–1450 BC indicate that about 80 per cent of the metal came from Laurion and only 10 per cent from Siphnos. Many lead and silver artefacts were preserved under the volcanic ashes which submerged Akrotiri, on the island of Thera, in 1500 BC. About 96 per cent of this metal came from Laurion and only 4 per cent from Siphnos.

Rio Tinto

At this period, Rio Tinto was being worked only for copper; silver was first exploited in this area between the twelfth and eleventh centuries BC. The chalcopyrite ore which appears to have been worked for silver around 1200 BC at Rio Tinto was taken from a thin, concentrated cementation layer, only a few centimetres thick, between the oxidized gossan and the principal mass of copper ore (see p. 55). It could have contained around 40g of silver per tonne. Up to this time the bulk of Mediterranean silver had been derived from lead by cupellation. Since Rio Tinto copper contained little lead, it is evident that lead oxide, or a lead ore must have been added to the smelting charge to take up the silver. The chalcopyrite ore might have been fused directly in a small shaft furnace to form a copper sulphide matte, from which the silver was absorbed by molten lead, and the alloy thus obtained would then have been cupelled to extract the silver.

The silver deposits at Rio Tinto were controlled successively by the Phoenicians, Carthaginians and, from about 200 BC the Romans. By this time the wealth of Tartessus, the biblical land of Tarshish, had become legendary throughout the Graeco-Roman world.

Silver production at Laurion

Silver was extracted at Laurion from a geological formation of three layers of whitish marble and limestones separated by two layers of micaceous schist.

Contact deposits of silver-bearing lead concentrated below each layer of schist. The galena-based ore from the lower contact deposit was particularly rich and is reputed to have yielded between 1200 and 1400g of silver per tonne in the time of Themistocles. This was not exploited until the beginning of the fifth century BC. Earlier mining was confined to the upper contact layer, which was closer to the surface, and yielded ore based largely upon the mineral cerussite, a carbonate of lead. The lower, richer contact layer in the mines was discovered in 484 BC and 100 talents of the silver it produced were used by Themistocles to build the fleet which destroyed the Persians at Salamis in 480 BC.

Subsequent references to the mines in plays by Aristophanes and Pherecrates suggest that they were then operated under state control. During the fourth century the mines and mining rights at Laurion still belonged to the Athenian state, who auctioned off operating franchises to private contractors and speculators. The industry declined very rapidly towards the end of the fourth century, when the silver coinage standard of the Athenian state was superseded by the gold standard introduced by the Macedonian kings. The last Athenian silver tetra-drachmas were minted in the second half of the first century BC, by which time the mines were largely exhausted, and once prosperous mining towns such as Thorikos had been virtually abandoned. Attempts made in the mid-nineteenth century AD to revive mining at Laurion were unsuccessful.

The ore extracted at Laurion in the fourth and fifth centuries BC was crushed with iron mallets and then ground to a sandy consistency in rotating stone hourglass mills or in hopper querns. The crushed ore emerged as a fine powder from which the lighter, earthy material was removed by water washing in some very ingenious helicoidal washers. A continuous current of water, during its passage around the spiral, carried the suspended particles of ore from cup to cup and classified them. The rich ore was the first to be deposited, then the poorer ore and finally the sand and silts. When the water reached the lower tank all suspended solids had been deposited and it could be returned to the stand tank, so completing the circulation process.

The argentiferous lead ore, concentrated in this way, was then taken from Laurion to smelting complexes at Thorikos, Puntazeza and Megola Pevka, where vertical shaft furnaces, using locally produced charcoal, were used to reduce the ore to metal. The silver content of the lead alloy obtained from this smelting process was subsequently recovered by cupellation, and taken to the Athenian mint, at the south-east corner of the ancient Agora, where the famous silver drachmas, the 'owls of Athene' would have been struck.

Some of the lead produced was used to a limited extent for sealing and for holding in place the iron clamps used to prevent relative movement between large marble building blocks. From the extensive deposits of litharge in the Laurion, it must be concluded that the value of lead at this time did not always justify the cost of recovery from cupellation waste.

Roman lead and silver

Roman municipal life was characterized by a prodigal use of water, for the supply of which very impressive aqueducts were constructed, and lead piping was extensively used for local distribution and plumbing.

Some of the lead piping from the ruins of Pompeii, which was submerged by volcanic ash in AD 79, has recently been examined in considerable metallurgical detail. These pipes varied in outside diameter from 30 to 40mm, and had a wall thickness of 5mm. The method used to produce the pipes was a model of simplicity. Lead sheets 6mm thick were cast on a flat stone surface, cut to size, and then wrapped around an iron mandrel in such a way that a narrow longitudinal gap was left along the pipe. Into this gap molten lead was poured at such a temperature that it was able to remelt the edges of the cast sheet before solidification. The pipes were then hammered on the mandrel over the whole of their cast surface to reduce their thickness to 5mm and to increase their diameter so that they could be withdrawn. These Pompeian lead pipes contained about 0.05 per cent copper and 0.1 per cent of zinc.

The mineral wealth of Britain provided much incentive for the Claudian invasion. Tacitus, for example, remarked that 'Britain produces gold, silver and other metals which are the reward of victory', and although the British output of gold was disappointingly low, substantial quantities of silver were soon obtained by the cupellation of lead. The Romans were working lead mines in the Mendips six years after the conquest, in AD 49. After the silver had been extracted, the lead was run into pigs weighing 77–86kg (170–190lb) which were exported to the Continent from the port of Clausentum in Southampton Water.

The lead from Derbyshire and more northern districts contained very little silver, and the lead mines of Mendip, Devonshire and Cornwall, although significantly richer, were still unable to compete with the argentiferous chalcopyrite deposits of Rio Tinto. Most of the Roman lead which has been analysed had already been desilverized by cupellation and it is difficult to estimate the silver content of the lead ores being worked in Britain. Desilverized Roman lead rarely contains less than 0.007 per cent silver. Samples taken from the Roman site of East Cliff in Folkestone and from Richborough Castle contained respectively 0.0072 and 0.0078 per cent silver, in good agreement with a silver content of 0.00785 per cent found in a lead pipe installed in Rome between AD 69 and 79. Such levels, although considerably higher than those present in modern commercial lead, probably marked the feasible limits of cupellation in Roman times.

BRASS AND ZINC

The origins of brass are almost as uncertain as those of bronze, although as a newcomer to the metallurgical world, the alloy was frequently discussed by classical authors. Zinc, in its natural deposits, is rarely associated with copper, and although some sulphide copper ores do contain small quantities of zinc, this element does not appear to have found its way, to any significant extent, into the composition of ancient artefacts. It seems most probable that in the arsenical copper and Early Bronze Age periods, any residual zinc in a sulphide ore would have been substantially removed, as a volatile oxide, either during roasting, or at a later stage during smelting, when fairly oxidizing conditions were required to prevent too much iron dissolving in the copper.

Certain ancient sources of copper, however, contained enough zinc to ensure that when smelted they produced low zinc brasses rather than copper. Thus at Cyprus, which by about 2000 BC had become one of the major sources of copper in the Middle East, many copper artefacts have been found containing zinc in random quantities up to a maximum of 9 per cent. Most of the artefacts contain between 3 and 5 per cent zinc, presumably inadvertently introduced during the smelting of a zinc-bearing copper ore. Modern Cypriot copper ores contain about 32 per cent copper and 1.5 per cent zinc. If all this zinc was retained during smelting, the brass obtained would contain around 5 per cent zinc, comparable to the higher zinc contents of copper artefacts from the Early Bronze Age site of Vounous Belapais in Cyprus.

During the Late Bronze Age, however, concentrated sulphide ores from lower horizons in the ore body were being worked, and these could have been smelted only after a roasting process which burned out most of the sulphur originally present. During this roasting zinc, as well as sulphur would have been rejected from the ore as a volatile oxide, thus accounting for the apparent anomaly that copper artefacts made in Cyprus during this period contain only traces of zinc.

Brass, containing only copper and zinc, dating from the period between 2200 and 2000 BC has been found in the Shantung province of China, where deposits of copper ore containing high concentrations of zinc occur. Zinc then disappeared from the Chinese metallurgical scene until around 220 BC, in the Han Dynasty, when some of the bronzes began to contain small quantities of zinc varying between zero and about 5 per cent. Nickel also began to establish itself as a minor constituent of bronze at this time, when, it seems reasonable to assume, the Chinese had started to utilize a complex copper-nickel-zinc ore similar to that which was used at a later date for paktong (see p. 96). Bronzes containing more than 10 per cent zinc were not made in China until about AD 1200, at which time zinc in metallic form must have become available and would have been deliberately incorporated into the alloys.

It is important, therefore, to distinguish between a group of natural alloys, usually of variable composition, where the zinc content has been inadvertently incorporated, and the true brasses which have been consciously produced. The Greeks, who appear to have been the first Western culture to appreciate this distinction, began to import brass in small quantities from Asia Minor around 700 BC. Brass was certainly being produced in Asia Minor at an early date: apart from the somewhat anomalous brass of Shantung province, the earliest artefacts so far discovered which have been produced from a binary alloy of copper and zinc are some fibulae of the eighth to the seventh centuries BC, found in 1958 at Gordion, the ancient capital of Phrygia. These alloys we should now refer to as gilding metal rather than brass, since they contain only 10 per cent zinc.

Zinc is rarely found in significant quantities in early Hellenistic bronzes, although a sixth-century BC statuette of Apollo was found to contain 6 per cent zinc. This high level could not have occurred inadvertently, and the historical evidence indicates that the sixth-century Greeks were beginning to use significant quantities of brass which at that time, because of its colour and rarity, was held in high esteem.

The first historical reference to brass as 'oreichalkos' ('mountain copper'), as opposed to 'chalkos', which was ordinary copper or bronze, occurs in the poem 'Shield of Herakles' dating from the seventh century BC: 'So he spoke, and placed about his legs the greaves of shining oreichalkos, the glorious gift of Hephaistos.' One of the later Homeric 'Hymns to Aphrodite' refers to ornaments of gold and oreichalkos being attached to the ears of the goddess. This early misconception, that oreichalkos was a metal comparable to gold in its rarity, value and desirability, persisted for a considerable period in the Greek world.

In the fourth century brass was still being imported into Greece from Asia Minor. Theopompus referred to the manufacture of oreichalkos on the shores of the Euxine, and as quoted by Strabo in his 'Geography', he mentions Andeira, a town in north-west Asia Minor, 'which had a stone which yields iron when burned'. After being treated 'in a furnace with a certain earth, it yields droplets of false silver. This, added to copper, forms the mixture which some call oreichalkos.' The mediaeval zinc reduction process produced just this type of zinc droplet by a downwards distillation/condensation process similar in general principle to the alchemical technique of *destillation per descendum*, which is obviously of great antiquity. Irrefutable evidence that metallic zinc was known and used in the ancient world is also available in the form of a sheet of zinc, 6.5cm long, 4cm wide and 0.5mm thick, dating from the third or second centuries BC, which was found in 1949 during an excavation of the Athenian agora. A Hellenistic statuette from the agora was also of 99 per cent pure zinc, although its dating was not quite so precise.

Alexander's eastern campaigns appear to have disseminated the methods of brass making used in Asia Minor throughout the Greek world. One interesting, if rather dubious, pseudo-Aristotelian work, 'On Marvellous Things Heard', compiled in the late Hellenistic period, mentions both the odour and the taste of brass, characteristics which are well known, but rarely mentioned in metallurgical works. Brass which has come into contact with sweaty hands is claimed to emit an unpleasant metallic smell, and the bitter metallic taste of brass is also mentioned. This work is also the first to refer to the tribe of the Mossynoeci, who lived on the southern shore of the Euxine, between Sinope and Trebizond. This tribe, from whose name the German word *Messing* is said to derive, made brass by smelting copper with an earth called 'calmia', presumably calamine. Pliny and also Dioscorides later referred to this substance as 'cadmia', and a material which is obviously zinc oxide is referred to as 'cadmia' by Galen in AD 166.

It has generally been assumed that the Mossynoeci made their brass by a cementation process similar to that which had been used by the Romans since the middle of the first century BC for coinage manufacture. The description is equally compatible, however, with the idea that copper and zinc ores were mined and co-smelted. This approach, although it would not have given a good yield of zinc, might well have produced brasses containing 10–12 per cent zinc, which, because of their golden colour, were greatly prized.

By the early years of the first century AD, Rome was making brass on a large scale by the cementation process even though metallic zinc was known and used in Athens three hundred years earlier. The situation is reminiscent of that which existed in eighteenth-century Bristol when, even after the introduction of metallic zinc by Champion (see p. 87), the cementation process of brass manufacture, being simpler and cheaper, continued to be used for a further century.

The Etruscans were producing statuary from brass, rather than bronze, as early as the fifth century BC. Those dating from the third or second century contain around 12 per cent of zinc. Egypt does not appear to have used brass before 30 BC, although Rome, by that time very familiar with brass, first introduced it as a coinage alloy for the manufacture of sestertii and dupondii in 45 BC. It seems to have been regarded as an attractive and economical alternative to bronze, since tin was a commodity obtainable only from Spain and Britain, whereas sources of zinc were relatively abundant within the Mediterranean area of Roman influence. The alloy, containing 27.6 per cent zinc and negligible quantities of tin and lead, was a typical product of the cementation process.

The Roman process of zinc manufacture was described by Pliny the Elder in his *Natural History*. Of the various types of copper available, he reports, the type known as 'Marian' absorbs cadmia most readily, and helps to produce the colour of 'aurichalcum' (oreichalkos) which was used for sestertii and dupondii manufacture. The fine white smoke escaping from the brass-making furnace is mentioned by Dioscorides and the use of the condensed white oxide in zinc ointment manufacture is also described by Pliny. Professor Gowland in 1912 was able to simulate the process used by the Romans and provided the following description:

The calamine was ground and mixed in suitable proportions with charcoal and with copper granules. This mixture was placed in a small crucible and carefully heated for some time to a temperature sufficient to reduce the zinc in the ore to the metallic state, but not high enough to melt the copper. The zinc being volatile, its vapour permeated the copper fragments, thus turning them to brass. The temperature was then raised, when the brass melted and was poured from the crucible into moulds.

Such a process, operated at 1000°C, will produce brass containing about 28 per cent zinc. If, instead of copper, 60/40 brass granules are added to this charge, their zinc content reduces progressively to 28 per cent. At the beginning of the Christian era, brass appears to have been made by the cementation process on a considerable scale near Aachen and Stolberg, where signs of Roman calamine working have also been found. Evidence of early brass making in Roman Britain has also been found at Colchester, Baldock and Cirencester.

The usage of brass in the first and second centuries AD increased rapidly, particularly for the manufacture of light decorative metalwork such as brooches, rings and horse trappings, although the zinc content of the coinage fell steadily from 27.6 per cent in 45 BC to 15.9 per cent in AD 79 and finally to 7.82 per cent in AD 161. After the third century AD brass coins were no longer produced. In decorative metalwork, where a golden colour was required, the usage of brass increased from 76 per cent in the first century AD to 88 per cent in the second century. These alloys, with a median zinc content around 12 per cent, contain little or no tin and were probably made by diluting brass made by the cementation process with pure copper. Producers concerned with the manufacture of cast rather than wrought alloys appeared to have remelted a great deal of the withdrawn brass coinage alloys and diluted them, for reasons of economy, with scrap bronze rather than copper. This produced the quaternary leaded tin bronzes, similar to modern gunmetals, which because of their excellent casting characteristics are still widely employed.

Zinc in mediaeval India

The original home of brass manufacture seems to have been Asia Minor where, according to Theopompus and Strabo, zinc in metallic form was being produced and used for brass making in the fourth century BC. Brass was subsequently manufactured in the Western world by the cementation process, and a feasible method of producing metallic zinc on a commercial scale was not developed in Europe until 1740. Metallic zinc was being produced by a refined distillation process at Zarwar in Rajasthan in India, certainly as early as the fourteenth century AD and probably much earlier. Extensive slag tips in this area provide evidence of silver, lead and zinc smelting activities over a long period of time. Slags produced before the fifteenth century contain zinc, cadmium, lead and only about 10 per cent of iron. This type of slag is produced when a mixed sphalerite/galena ore is smelted for lead, the remaining metallic content of the ore being of no value. Slags from close to the zinc retorts, however, contain between 20 and 35 per cent iron, and this sudden change has been interpreted as an indication that before zinc was refined the only function of the sphalerite was to reduce the melting point of the smelting slag. Where, around 1400, spharelite was required for the production of zinc, it became necessary to add iron oxide as a flux to the lead smelting charge.

The sphalerite was ground into a fine powder and roasted to convert zinc sulphide to the oxide, which was then rolled into balls with gluelike organic materials and fluxes before being inserted into clay retorts similar to those shown in Figure 1.6. The charge, being in the form of balls, did not fall out of the retort when this was inserted in the furnace, and it also facilitated the escape of zinc vapour from the mouth of the retort. The vitrified condition of the discarded clay retorts indicates that the charge must have been heated to temperatures between 1050° and 1150°C, which is considerably higher than the boiling point of zinc at atmospheric pressure (913°C). Operating conditions within the furnace chamber would have been controlled to ensure that the outlet nozzles of the retorts were cold enough to condense the zinc vapour, but hot enough to ensure that the zinc condensed above its melting point of 432°C, so that it could run out of the nozzle into a collection vessel under the influence of gravity.

This technique of zinc manufacture is described in several Indian alchemical words of the mediaeval period including the thirteenth century *Ras Ratnasmuchchaya*. The word used in this document to describe the distillation process involved is 'tirakpatnayantra', which, translated literally, means 'distillation by descending', so close to the Latin nomenclature that it is tempting to conclude that links existed at this period between European and Indian alchemical workers (see Figure 1.7(a)).

Salt must have been included in the Zarwar charge, since it has been detected in the remains of spent retorts, although not originally present in the ore. The precise function of salt in the zinc distillation process is still obscure, although its presence is known to facilitate the process, and it was occasionally used in the European horizontal zinc retort process in the nineteenth century.

The Indian production of metallic zinc does not necessarily conflict with the idea that the secrets of zinc and brass manufacture were first discovered in Asia Minor. Evidence of zinc refining on a considerable scale can still be seen near Deh Qualeh, which lies north of Kerman in eastern Iran, where Marco Polo observed the manufacture of 'tutty', or zinc oxide, in the fourteenth century, and the technical operations involved were also described by the Islamic author Hamd-allah Mustafi in 1340.

The refining process differed considerably from that used at Zarwar since it began by subjecting the ore to a high temperature distillation process which

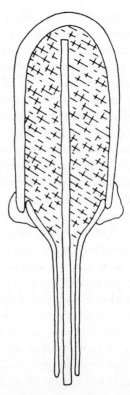

Figure 1.6: Zinc was produced in metallic form on a considerable scale at Zawar in Rajasthan from the fourteenth century by a distillation process based on the use of large numbers of clay retorts of the type illustrated. Zinc production was a major industry in India during the sixteenth century when the metal was unknown in Europe. The mines were in full production in 1760 and zinc production did not completely cease at Zawar until the opening years of the nineteenth century. Courtesy of *Mining Magazine*.

removed zinc oxide as a pure volatile vapour and condensed it as a fine powder in the cooler regions of the furnace. As a preliminary to this distillation, the ground ore was formed with some binder into cylindrical bars. The spent bars of ore from which the oxide has been removed consist largely of iron oxide, thus indicating that the ore treated was probably sphalerite. From the size of the waste heaps it is clear that vast quantities of zinc oxide must have been produced, and brass manufacture appears to be the only technology which could absorb this output, whether the zinc oxide was directly converted to metal, or used for a cementation process for which it would have been well suited.

Metallic zinc appears to have been a commodity well known to the Chinese at an early date. Zinc coins were first struck during the early years of the Ming

Dynasty (1368–1644). Some of the later Ming bronzes also contain more than 30 per cent zinc, a level which could have been obtained only by the incorporation of metallic zinc in the melt. Zinc ingots weighing 60kg (132lb) were being exported from China in 1585, and in 1785 a shipment of zinc ingots from China was lost in Gothenburg harbour. When salvaged in 1872, these ingots were found to contain less than 1 per cent of impurities.

The evidence currently available, therefore, indicates that brass certainly, and probably also metallic zinc, were first made in Asia Minor before 800 BC, and that by 700 BC golden-coloured brasses were being imported by the Hellenes from regions such as Phrygia. The Greek usage of brasses increased considerably as a result of Alexander's eastern campaigns, and the alloy may have been exported from Asia Minor to the Far East as early as 200 BC via trading cities such as Taxila in the Punjab. Brass manufacture in this region of India started in the first century AD. However, the manufacture of a natural brass alloy in Shantung province around 2000 BC, and the early usage by the Chinese of a complex copper-nickel-zinc ore for bronze manufacture, seems more consistent with the idea that the identity of zinc as a metal, and the possibilities of brass as an alloy, were indigenous discoveries owing little to the Western world.

Mediaeval Europe

European mining and metallurgical activities did not cease completely after the collapse of the Roman Empire. After a brief decline the economy began to revive, a process which achieved a momentum of its own, in Aachen, under Charlemagne, in the ninth century, Frankish miners, using Saxon slaves, first began to work the *Fahlerz* (=pale ore) copper ore deposits of Frankenburg in Saxony, and then advanced eastwards in a campaign which brought under their control the mineral wealth of Bohemia and Slovakia. Mines were initially sunk in the Hartz Mountains in the Black Forest, and then in Styria and the Tyrol. The first gold mines were in Silesia, but greater quantities were soon obtained from Hungary.

The German miners were noted for their ability to recognize the subtle differences of colour, contour and vegetation from which the presence of a rich vein of ore near the surface of the ground could be discerned. The legends however, describe accidental, almost magical discoveries. Thus, it is claimed that the fabulously rich silver deposits of Rammelsburg in the Hartz Mountains were discovered not by prospectors, but by the horse Ramelius who, when tethered to a tree by his master in AD 938, struck the ground with his hoof and uncovered a rich outcrop of silver-bearing ore. Since that time vast quantities of lead, silver, zinc and copper have been taken from the mines which still produce copper ore on a considerable scale.

The ore body at Rammelsburg was worked initially for its lead, which was rich in silver. Then, as the mine became deeper, copper ore was extracted. From the town of Goslar at the foot of the mountain came the silver which sustained Frederick II during his struggle with the papacy, and financed the Sixth Crusade. The uppermost deposit of Rammelsburg, which was grey in colour, was ignored in the early days because it contained an objectionable constituent which made it uneconomical to work for lead. This was spharelite, a zinc sulphide, which was obviously of no value to the early miners who knew nothing of the metal zinc. European brass was then made in the Aachen area, from calamine by a rather incomprehensible cementation process.

Erasmus Eberner of Nuremberg is usually regarded as the first modern European to have recognized the existence of zinc as a new metal with an identity of its own. In 1509, when working at Rammelsburg, he recovered some condensed droplets of a whitish metal from the cooler regions of a lead-smelting furnace and was able to dissolve them in molten copper to make brass. This experiment demonstrated to his satisfaction that the droplets were of the same metal which was absorbed by copper, from calamine, when brass was being made by the cementation process. It seems improbable that the German metal producers of that time were completely ignorant of the brass-making processes of their eastern contemporaries: perhaps Eberner had seen or even handled samples of this rare eastern metal from which brass could be made, and recognized it in the condensed droplets from the lead smelter. This identification had little immediate impact upon European brass making technology. Paracelsus, however, named the metal *Zinken* before 1541, and Agricola in 1556 referred to it as *liquor candidus*.

Zinc ingots of Chinese manufacture began to appear on the European metal markets towards the end of the sixteenth century, and after 1605, when the Dutch and English East India Companies became involved, zinc ingots became a standard trading commodity. Commercial zinc was generally referred to as spiautre, from which the modern term spelter derives. Other names used at the time were tutty, tutinag, and Indian tin. Although far too expensive for the manufacture of conventional brass, which was more economically produced by the cementation process, Chinese zinc soon found extensive usage for the manufacture of golden coloured brasses of the gilding metal type which were used for cheap jewellery, trappings, and accoutrements and generally referred to by exotic names such as Mannheim gold, Princes metal, pinchbeck or tomback. These required zinc contents between 12 and 15 per cent, whereas the zinc content of cementation brass generally fell between 28 and 30 per cent. The availability of zinc in metallic form also made it possible to prepare brasses of low melting point, containing 40–50 per cent zinc, which were then used as brazing alloys for joining copper and steel components.

European zinc manufacture on a small scale seems to have started at Rammelsburg shortly after Eberner's famous experiment, although the output was consumed locally and the techniques initially employed are unknown. A great deal of European brass was made in the region between Liège and Aix-la-Chapelle, following traditions established initially by the Romans. In the sixteenth century abundant calamine deposits were worked at Nouvelle Montagne, and at Vielle Montagne, situated at Moresnet. The cementation process then employed was similar to that described by Theophilus in the time of William the Conqueror, and had obviously changed little since Roman times. The brass made in the valley of the Meuse was well known, being generally referred to in Europe as 'dinanderie'. Britain, which at that time had no brass industry of its own, imported most of its requirements from this region.

Brass was referred to in Tudor Britain as latten, or sometimes latyn, from the French word *laiton*. The word 'brasse' was then used, as in the English Bible, for copper, and not until the seventeenth century was the name bronze given to the copper-tin alloy, from the Italian *brondosion*, which later degenerated into *bronzo*. It derives from the city of Brundisium, now known as Brindisi. Much confusion has stemmed, therefore, from the free and ambiguous usage of terms such as brass, bronze and latten to describe mediaeval and Tudor copper alloy artefacts.

The royal monopolies of Britain

Until 1565 brass had not been made in Britain since Roman times, and copper mining had been undertaken intermittently and on a negligible scale. Soon after the Norman conquest working rights for mines which produced gold and silver were vested in the Crown, effectively putting all metallurgical mining under government control, since the winning of silver, for all practical purposes, was inseparable from the mining of lead and copper. The Crown monopoly of 'mines royal', which included the extraction of gold, silver, lead, copper and sometimes even tin, persisted until the end of the seventeenth century.

In 1528, Henry VIII attempted to persuade the celebrated German mining expert Joachim Hochstetter to direct the English metallurgical developments. Hochstetter visited England, advised the construction of a smelting house at Combe Martin, Devon, and accepted the office of Principal Surveyor and Master of the Mines to the King. In 1529 however, he left England to develop the copper mines at Neusohl in Hungary, south of the Tatra mountains. In the reign of Edward VI, another German, Joachim Gundelfinger, was appointed to manage the silver mines at Wexford in Ireland and under Mary, Burckard Kranich was granted permission to develop the silver mines of Cornwall. Both activities were unsuccessful.

Under Elizabeth I, a large contract for refining the debased silver coinage of England was given, in 1561, by Sir Thomas Gresham to the Augsburg firm of

Haug & Co, who found the business so profitable that they began to consider the prospect of further co-operation with the English Crown. A preliminary survey of Britain's mineral resources by Daniel Hochstetter in 1563 revealed favourable prospects, and in 1564 substantial deposits of silver-bearing copper ore were found at Keswick in Cumbria.

In September 1564 the initial Patent for working copper ores in Britain was granted jointly to Daniel Hochstetter and Thomas Thurland, Master of the Savoy. The organization thus established, formally incorporated in 1568 as the Company of Mines Royal, was the first joint stock company to be set up in England for the manufacture of a commodity (copper) rather than for trading purposes only. From the legal controversies stimulated by the grant of a Royal monopoly for a 'new method of manufacture', patent law as we now know it emerged, evolved and refined itself.

A sister metallurgical enterprise, established in 1565, was the Society of Mineral and Battery Works. It was a specialist organization, concerned with the manufacture of brass and wire, and owed its existence largely to the enterprise of Sir William Humfrey, Assay Master of the Mint. Humfrey's partner in this venture, Christopher Schutz, an expert in brass manufacture, was the manager of the calamine mining company of St Annenberg, Saxony. 'Battery' was the name originally applied to all sheet metal utensils which had been formed into the shape required by beating with a hammer. By the eighteenth century the term applied specifically to beaten hollow ware of brass or copper. Before Humfrey, who was one of the original shareholders in the Mines Royal Company, applied to Lord Cecil for the privilege of introducing 'battery works' into England all the brass required was imported from Europe.

A further object of the Mineral and Battery works was to introduce improved methods of wire drawing into England, where the wool trade was then rapidly expanding and large quantities of fine brass and iron wire, needed for the manufacture of wool carding equipment, were imported from the Continent.

The Society's Patents, granted in September 1565, gave Schutz and Humfrey the sole rights to mine calamine and to make brass in England. They were also authorized to work mines and minerals in any of the English counties not already under the jurisdiction of the Mines Royal Company. By June 1566 calamine deposits were discovered at Worle Hill in Somerset, at the western end of the Mendips on land belonging to Sir Henry Wallop. Zinc ore from this deposit contained less lead and was superior in quality and yield to that which was currently imported from Aachen. A brass manufacturing site was sought within easy reach of the Mendip hills. The partners eventually erected their manufacturing plant along the Angiddy Brook at Tintern, where lead and copper working had been carried out on a small scale since mediaeval times.

Schultz estimated initially that his works would produce 40cwt (2032kg) of iron wire and 20cwt (1016kg) of brass wire per week. This output was never achieved, and not until the end of 1567 was any brass at all produced at Tintern. Although of the right colour, indicating that it contained the appropriate amount of zinc, it lacked ductility and could not be drawn into wire. An innovation introduced by Schultz had been the use of pit coal rather than charcoal as fuel for the cementation furnaces, and it has been suggested that the sulphurous fumes thus introduced embrittled the brass as they did iron. In 1569, Humfrey and Schultz gave up their manufacturing endeavours and leased the Tintern works to a man experienced in drawing iron wire.

Brass manufacture was in fact completely ignored until 1582 when a group of merchants headed by Sir Richard Martyn bought the rights to mine calamine and make brass for the sum of £50 per annum. The factory they set up at Isleworth, west of London, made brass of excellent quality for a number of years under the managership of a London goldsmith, John Brode. The manufacturing operation became very profitable, and when Brode refused to provide the Society with details of his manufacturing processes, they revoked his licence and drove him into bankruptcy. In 1605, Brode appealed against the decisions of the Society to the House of Lords, claiming that he was 'the first here in England that co-mixed copper and calamine and brought it to perfection to abide the hammer and be beaten into plates, kettles and pans by hammers driven by water'. This was probably perfectly true.

After this failure brass sheet and battery ware was imported from Europe so cheaply that few British brass-making enterprises managed to survive. Since the death of Daniel Hochstetter in 1581 the mining activities of the Mines Royal Company at Keswick had steadily declined. After the Civil War the industry declined very rapidly indeed, since Parliament refused either to reduce the duty on imported copper or to protect English manufacturers against the importation of cheap imported brassware. After 1650 hundreds of tonnes of high grade zinc ore were shipped to the Meuse valley and other European brass-making centres to be turned into brassware, subsequently to be sold in English markets at prices well below that of the domestic product. When the Society of Mineral and Battery Works became aware of the extent and profitability of this illicit commerce they did not attempt to restrict it, finding it more expedient to impose royalties, of the order of five shillings on each ton of the ore exported. By the 1680s it had become very evident that the privileges granted to the Mines Royal Company and to the Society of Mineral and Battery Works had been exercised in a repressive manner to the detriment of the non-ferrous industry. The Crown monopoly on metallurgical mining rights was finally abolished in 1689 when the Mines Royal Act was passed. Further legislation passed in 1693 enabled manufacturers to participate freely in metallurgical activities.

The British non-ferrous metal industry began to revive in the 1670s, when development work by Sir Clement Clarke at his lead works at Stockley Vale in Bristol led to the use of coal-fired reverberatory furnaces firstly for refining lead, and then in 1687 to the refining of copper. John Carter, a protégé of Clarke, subsequently went into partnership with a group of London merchants headed by William Dockwra to establish in 1692 the Copper Works of Upper Redbrook, north of Tintern on the River Wye, where deposits of copper ore had recently been discovered.

Dockwra had in 1691 become the proprietor of a brass works at Esher in Surrey which had initially been set up in 1649 by Jacob Mummer, a German immigrant. At Esher, brass made by the cementation process was cast into stone moulds to produce flat ingots weighing seventy pounds. These were then rolled to sheet, slit, drawn to wire and finally made into pins. The copper for the Esher brass works came from Upper Redbrook, then being profitably managed by John Carter. A rival copper refinery, the English Copper Company, was established soon afterwards at Lower Redbrook, only a short distance downstream. In its early days, copper refining operations at Lower Redbrook were controlled by Gabriel Wayne, who, like John Carter, had formerly been employed by Sir Clement Clarke. Wayne, however, soon perceived the limited opportunities for expansion along the River Wye, and in 1696 he set up a new copper refinery at Conham, on the banks of the River Avon two miles to the east of Bristol, where the water was still navigable. His business associate in this enterprise was the merchant Abraham Elton. The improved reverberatory furnaces built at Conham owed much to the earlier Clarke developments and to the practical lessons learned at Lower Redbrook.

The sulphide copper ores used at Conham were shipped directly up the Avon from Cornwall and also from North Molton in Devon. After the ores had been roasted at Conham they were melted with lime in a reverberatory furnace to remove silicious impurities. The iron and sulphur remaining in the purified matte thus attained was then gradually removed by reverberatory melting under oxidizing conditions until crude copper in metallic form began to separate from the melt.

At that time Bristol was a very logical centre for metallurgical activity, since it also offered a port from which metal products could be exported to all parts of the world, and locally mined coal was available at prices very much lower than elsewhere in the country. The decision to produce brass at Bristol was made in 1700 by Abraham Darby, Edward Lloyd and several other businessmen who were all Quakers of Bristol. Copper was being locally produced on a large scale by three competing refineries, and William Dockwra's works at Esher was the only brass-making establishment in England worthy of note. Darby was a manufacturer of malt mills, while Edward Lloyd was a cidermaker, and both these activities involved the extensive use of brass fittings and components.

Around 1703, it appears, Darby went to Holland, hired some Dutch or Low Country workers and set up the Brass Works at Baptist Mills. This was situated on the River Frome about 2.5km north-east of the point where this tributary emerges into the Avon. Darby's immigrant workers seem to have been skilled in casting large thin slabs of brass into flat granite moulds and in the manufacture of hollow ware and other brass utensils by standard battery procedures. In the early days at Bristol the ingots were beaten into sheet by hammering and in this respect the activities at Baptist Mills were technically behind those at Esher where brass had been rolled into sheet for a considerable time. The Bristol enterprise flourished, however, and by 1708 an additional mill had been set up on the River Avon at Keynsham.

In 1707, Darby had begun to appreciate the potentialities of cast iron. He withdrew from the brass company in 1708 and acquired an established blast furnace at Coalbrookdale, where in 1709 he demonstrated for the first time, the feasibility of making iron of high quality by using coke rather than charcoal as fuel for the blast furnace (see p. 153ff). The indications are, therefore, that Darby was the man responsible for the use of coke rather than charcoal in the brass works at Baptist Mills, to avoid sulphurous contamination of the metal from raw coal, and that he introduced the practice of turning the coal into coke by roasting it in the brass furnace itself during the preliminary stages of brass making. Between 1710 and 1712 over 400 cartloads of coal per week were used by the Bristol Brass Works.

In September 1709, as the Company for Brass and Battery Work at Bristol, they combined with their former competitors, the proprietors of the Brass Wire Works at Esher, to form the Societies of Bristol and Esher for making Brass, Battery and Brass Wire. Compared to the Bristol establishment, the works at Esher were small, and here the production of wire was concentrated. Brass sheet and battery ware were produced at Bristol. By 1712, Baptist Mills were producing 255 tonnes of brass a year, well in excess of the 210 tonnes then being imported from Europe. Between 410 tonnes and 540 tonnes per year of copper were then being produced in Bristol by two large copper refineries, both situated on the River Avon. The site at Crew's Hole, half a mile downstream from the Conham works, had been established by the Bristol Brass Company to ensure their own requirements of copper which increased greatly after their amalgamation with Esher.

In the early 1720s, Henric Kahlmeter of the Swedish Board of Mines visited England and reported that the two copper refineries at Bristol and those at Upper and Lower Redbrook were the 'most considerable' of those he had seen. By that time the four companies were working as a loose form of trade association to run a group of copper mines in Cornwall and Devon. By working together and refusing to buy copper ore until prices fell the group was able to obtain its supplies from primary producers at very low terms.

The brass works at Bristol were then being managed by Nehemiah Champion, a man of considerable technical ability. In 1723 he applied for and obtained Patent No. 454 which was concerned with the preparation of copper used for the manufacture of brass by the calamine process. Champion's leap forward was to use granulated copper rather than broken fragments. By this approach, which greatly improved the surface-to-volume ratio of the copper, the uptake of zinc during cementation increased from 28 to 33 per cent, approaching the level which is now known to be the theoretical maximum possible by cementation. Since copper was the most expensive component of the alloy, Champion's approach gave him an important commercial advantage over his competitors, and helped to ensure the prosperity and growth of his works.

The calamine brass process

The brass cementation process used at Baptist Mills did not differ significantly in principle from that used by the Romans. Fairly comprehensive details of the brass-making processes used during the sixteenth century were given in 1574 by Lazarus Ercker, a native of Annaberg in the Saxon Erzgebirge, and Chief Superintendent of Mines in the Holy Roman Empire. The cementation process he describes differs from that of Theophilus and Pliny in that alum and salt were added to the copper/charcoal/calamine mixture charged into the crucible. After smelting in eight small pots, the brass was transferred to a large crucible and cast into ingot moulds of large flat 'Britannish Stone'. It is interesting to recall that small quantities of salt were found in the zinc retorts of Zarwar (see p. 77).

The Swedish brass-making techniques described by Swedenborg in 1734 were virtually identical to those outlined 160 years previously by Ercker, with the exception that neither salt nor alum were added to the charge. The copper and calamine were still melted in eight small crucibles, the combined contents of which were then cast into sandstone ingot moulds.

In 1720, Kahlmeter reported that the thirty-six brass-making furnaces in operation at Baptist Mills were grouped into six separate brass houses, which were worked as required to produce 305 tonnes of brass per year. As in Ercker's and Swedenborg's descriptions, cementation was effected in eight small crucibles of Stourbridge clay, which were inserted into each circular furnace and emptied twice every twenty-four hours. The calamine used by Champion was carefully calcined to convert it from the carbonate to the oxide and was then ground into a fine powder before being incorporated into the cementation charge. Apart from the use of water granulated copper, the main technical innovation introduced by Champion appears to have been in the way the cementation furnaces were arranged in groups of six under large brick covers similar to those used in the glass industry. This formation, which provided improved draught and ventilation, was commonly adapted when brass making moved from Bristol to Birmingham towards the end of the eighteenth century.

William Champion and his zinc metal process

Nehemiah Champion's youngest son, William, visited most of the industrial centres of Europe before returning to Bristol in 1730 at the age of twenty. Metallic zinc was then being imported from the Far East at prices around £260 a ton, too costly for the manufacture of battery brass, although it was in great demand for manufacturing those brasses of low zinc content and attractive golden colour used for making cheap jewellery. It was also used for making brazing alloys containing 40 per cent or more of zinc which could not be obtained by cementation. William Champion's immediate objective on returning from Europe was to produce metallic zinc from English calamine at a price low enough to allow it to be used for routine brass manufacture.

His early work soon showed that zinc oxide could only be reduced by carbon at very high temperatures, so that the zinc obtained left the reaction zone in vapour form and oxidized to a blue powder as soon as it made contact with air. The essence of his reduction process, which required six years of 'great expense, study and application' before success was achieved, was to condense the vapour rapidly to metal in the complete absence of air. To do this he used a vertical retort and, as can be seen from Figure 1.7(b), the equipment he evolved is remarkably similar in its geometry and general arrangement to the vertical zinc retorts used at Zawar in the fourteenth and fifteenth centuries.

An iron tube led from the base of the reaction crucible to a cold chamber below the floor, its end being sealed by immersion in a bowl of water. This ensured that the zinc vapour did not encounter significantly oxidizing conditions before it condensed and settled as granules below the water surface. The reaction crucibles, approximately 1m high, and 90cm in diameter, were arranged in groups of six in each furnace. The distillation process took about seventy hours, during which time around 400kg (882lbs) of zinc were obtained from the six retorts.

William Champion encountered some opposition from his colleagues, and he was dismissed from his old firm in 1746. Between 1738 and 1746 he built a new factory for zinc manufacture at Baber's Tower in Bristol, where he was able to produce 205 tonnes (200 tons) of metallic zinc. This, he found, was virtually unsaleable, since the merchants who imported zinc from the Far East dropped their prices and seemed quite prepared to lose £25 per ingot ton in their efforts to drive him out of business. Moreover the city fathers of Bristol had complained about the fumes emitted by his Baber's Tower factory which he was forced to demolish. He then built a new large works at Warmley, five miles to the east of Kingswood. In complete contrast to the fragmented operations of the old Bristol Company, the Warmley plant was intended to be completely integrated, co-ordinated facilities for copper smelting, zinc distillation and brass manufacture being arranged on the new site. In 1748 the factory at Warmley

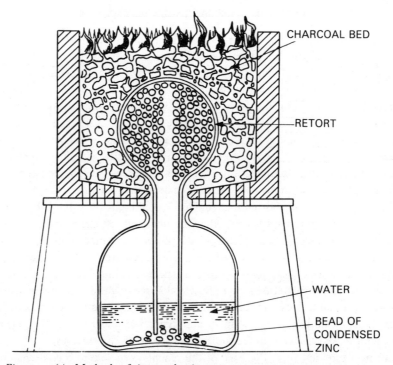

Figure 1.7 (a): Methods of zinc production.
Zinc production by the process of Tyryakpatnayantra, or 'distillation by descending' is described in the thirteenth-century Hindu alchemical work, *Ras Ratnasammuchchaya*. This recommends that the retort should be charged with ingredients such as lac, treacle, white mustard, cherry plum, resins, borax, salt and zinc ore.

was in full production. Copper ore was brought from Cornwall, refined on the site and granulated in water for the manufacture of brass by cementation according to the original patent of William's father Nehemiah. For the manufacture of ingot moulds, William seems to have departed from tradition in using granite slabs rather than sandstone.

The zinc distillation process was carried out at Warmley under conditions of great secrecy, and it was not until 1766 that the processes involved were described by Dr Watson in his *Chemical Essays*. The Warmley plant grew rapidly, and further capital was raised in 1761 for the erection of 17 new copper refining furnaces. In 1765 the Warmley Company began to manufacture brass pins on a large scale. By 1767, when the fortunes of the Warmley plant were at their peak, the pin-making operation, which was probably undertaken in the old Clock Tower building was in full operation, and the works at Warmley housed two large rotative steam engines.

Following petitions by his competitors to the Lords Committee of the Privy

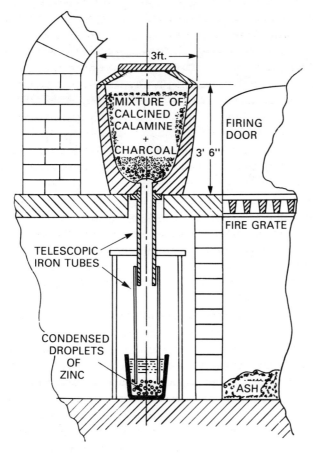

Figure 1.7 (b): Methods of zinc production.
When William Champion introduced his version of this process in Bristol in 1738 his crucibles were charged merely with a mixture of calcined calamine and charcoal.

Seal, Champion's application for a Charter of Incorporation was rejected in March 1768. This decision destroyed the Warmley Company, since it made it impossible to raise the capital required for continued operation by the issue of transferable shares. Following an unauthorized attempt to withdraw some part of his share capital from the company, Champion was dismissed by his fellow directors in April 1768. He was declared bankrupt in March 1769, when the works at Warmley were offered for sale. These were eventually acquired by the Bristol Brass Company but never extensively used.

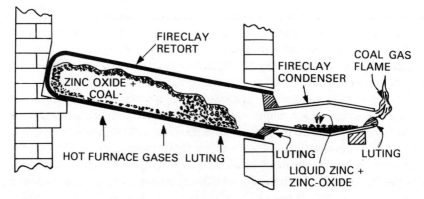

Figure 1.7 (c): Methods of zinc production.
The horizontal zinc retort process, introduced by the Abbé Dony in 1807, was far simpler and more economical to operate. Most of the zinc produced in Europe and the United States between 1820 and 1940 was made in this way.

The decline and fall of Bristol brass

The demise of the Warmley Company was soon followed by that of the 'Old Bristol Company', which had been largely responsible for its failure and destruction. The four brass companies which had prevented William Champion from obtaining his Charter of Incorporation were led by the Brass Battery, Wire and Copper Company of Bristol as it was then known. The others were John Freeman and Copper Company of Bristol, Thomas Patten and Coy of Warrington, and Charles Roe and Copper Company at Macclesfield. Brass manufacture represented only a small part of the total activities of John Freeman and Company. Together with the Bristol Brass Company, however, Thomas Patten and Charles Roe produced most of the brass used in Great Britain.

Bristol had, in fact, begun to lose its predominance as the centre of copper and brass production in 1763. In the autumn of that year, Charles Roe of the Macclesfield Copper Company first sensed the possibilities of the copper ore deposits of Parys Mountain in Anglesey, and leased part of it from Sir Nicholas Bayly. Extensive bodies of rich ore were found in 1768 and rapidly exploited. The remainder of the Parys Mountain complex was soon being worked by the Warrington Copper Company under Thomas Patten, and the Parys Mine Company which in 1780 opened its own smelting works at Ravenshead in Lancashire.

Thomas Williams, the solicitor who led the Parys Mine Company was a man of energy and vision who recognized soon after he joined the industry that the Anglesey copper ore, being abundant, close to the surface and easily worked, could be sold very cheaply at prices which would make the Cornish mines unviable. By that time the old Bristol Company had relinquished its earlier control of the copper mines in Devon and Cornwall and bought the copper ore needed by its smelting works on the open market. With Parys Mountain copper, Thomas Williams became the major producer and found himself able to control the copper market. Bristol was forced to increase the price of brass, a move which antagonized many customers and had particularly unfortunate consequences in the Birmingham area.

Birmingham began to emerge as a centre of brass fabrication at the beginning of the eighteenth century, and the vast quantity of brassware it exported to the Continent was discussed by Daniel Defoe as early as 1728. Much of the brass worked on there was initially imported from Europe, although by the middle of the century it was all obtained either from Bristol or from Thomas Patten at Cheadle in North Staffordshire. A small brass-making factory was established at Coleshill Street in Birmingham in 1740 by the Turner family. By 1780, Birmingham consumed about 1000 tonnes per annum of brass, most of which came from Bristol. By then, however, the Birmingham fabricators were beginning to resent the high price of the brass they bought from Bristol and the attitude of Brass Battery, Wire and Copper Company (derisively referred to as the 'OC', or 'Old Company') which protected its monopoly by controlling brass prices in such a way that any new brass manufacturers in the Birmingham area were soon forced out of business.

A public attack on the brass manufacturers cartel, which was headed by the Old Company, was mounted in the autumn of 1780, when, according to *Aris's Gazette*, the Birmingham fabricators 'boldly stood forth the Champions of Industry, and in Defiance of Oppression, ventured to erect Works and risque their Fortunes therein'. The result of this campaign was the formation of the Birmingham Metal Company, set up in April 1781, initially under the aegis of Matthew Boulton, although he shortly afterwards resigned from the committee. The first action of this company was to negotiate with Thomas Williams for a regular supply of copper on very advantageous terms.

Bristol, therefore, rapidly lost its hold on the Birmingham market and its fortunes dwindled. In 1784 the lease Charles Roe had taken at Parys Mountain expired and Thomas Williams then assumed control of the whole copper mining complex of Anglesey. The Bristol Brass Company had encountered serious difficulties during 1786 and the joint proprietors resolved in December of that year that the business should be dissolved and terminated. The firm at that time was owned and managed largely by the Harford family, and in February 1787 the firm was sold for £16,000 to ten Bristol merchants, six of whom were Harfords.

Relationships between the Cornish and Anglesey mining concerns had, meanwhile, stabilized. Thomas Williams had been requested by the Cornish Metal Company to market their vast stocks of unsold copper, a task which he accomplished by 1790 without any reduction in copper prices. He also assumed at this time most of the other responsibilities of the Cornish Metal Company including the purchase of ore from the Cornish mines. Arrangements for smelting this ore were made with various companies which included the Harfords and Bristol Brass and Copper Company, and also the Freeman Copper Company. To take advantage of the low price of coal and improved port facilities in South Wales, Harfords at that time moved their smelting facilities to Swansea and discontinued smelting in Bristol.

Although the Birmingham industry flourished the Bristol company continued to decline. By 1820 smelting at the Swansea site had ceased and by 1836, when the Baptist Mills site was sold, the company was no longer a manufacturing concern. Its remaining properties had been leased in 1833 to a company partner Charles Ludlow Walker. In these premises, brass continued to be made and fabricated by traditional methods until the Saltford rolling mill stopped work in 1925 and brass making at the Avon Mill premises ceased entirely in 1927.

Zinc in Belgium

Attempts to obtain zinc directly from calamine were being made in the Liège district at the same time as William Champion was experimenting at Bristol. Margraff, a somewhat legendary figure who was experimenting in 1746, appears to have passed on his results to the Liège professor Villette. He in turn confided the information to the Abbé Dony who introduced the first commercially successful zinc reduction process (see Figure 1.7(c) above).

Jean Jaques Dony was born in 1759 and, having been brought up in a brass-making locality, he was greatly interested in the metallurgy of zinc. The origins of the extraction process he evolved, however, are still very obscure. It is recorded in the *Biographie Liègeoise*, 1836, that 'before 21 March 1805, Dony had succeeded in extracting metallic zinc from calamine and had managed to melt and cast it in quantity'. In March 1806 he obtained from Napoleon an Imperial concession to exploit a deposit of calamine at Moresnet, and in 1807 he constructed a small factory in Isle. Here, it is said, zinc was first produced in quantity.

The limited room temperature ductility of zinc improves markedly if the metal is warmed slightly and Sylvester and Hobson of Sheffield first demonstrated the feasibility of producing zinc sheet by rolling at moderately elevated temperatures in 1805. Dony produced his first sheet of zinc in January 1808 and offered it, in acknowledgement of the concession he had been granted, to the Minister of the Interior. After 20 July 1809 high-quality zinc was being produced in abundance, and an Imperial decree of 19 January 1810 granted Dony a patent, giving him a monopoly for the manufacture of zinc for fifteen years.

Between 1810 and 1818 considerable technical advances were made in the factory he established in the Liège suburb of Saint Leonard. Here a rolling mill was producing zinc sheets 1.5m long by 41cm wide (59in × 16in) by 1811, and by 1813 the roof of St Paul's Cathedral at Liège had been sheathed in zinc. In the same year, in recognition of his Imperial patent, Dony sent to Napoleon his bust in Zinc, a casting weighing 74kg (163lb). Between September 1809 and February 1810 Dony had reduced the price of zinc in Liège from F8.60 to F2.60. Like all industrial innovators, however, Dony soon begun to encounter financial difficulties, and after 1818 he was forced to surrender all his commercial interests to his business partner Dominique Mosselman who, in addition to the calamine workings of Vieille Montagne, took over the entire factory at Saint Leonard. Dony died in Liège on 6 November 1819, Mosselman died in 1837. His heirs established the Vieille Montagne Company which dominated the Belgian zinc industry for many years.

Champion's vertical distillation process, which had never made zinc at a price low enough to ensure its commercial viability, was soon superseded by Dony's system. Its use of massed banks of horizontal retorts reduced considerably the quantity of fuel required. Fuel efficiency was improved even further when double-faced furnaces were introduced. The clay condensers emerged almost horizontally from the ends of the fireclay retorts and were cooled by air rather than water as in Champion's approach. Dony appears to have stumbled, probably inadvertently, on a simple and very effective geometry for his condenser. It is now known that the zinc vapour entering such a condenser from the retort is subject to a great deal of convection, and assumes a rapid revolving spiral motion round the inside of the condenser wall before condensation to the liquid state occurs.

Soon after the Belgian process emerged, Germany, which had large deposits of zinc ore in Silesia, developed a zinc furnace system which used retorts much larger and stronger than those used in Belgium. This approach, which improved both labour and fuel economy, was possible because the Silesian ore had a coarse, porous texture which facilitated the egress of gas and zinc vapour from deep beds of ore. The Rhenish retort subsequently developed was rapidly adapted throughout Europe, Britain and the United States. The retort dimensions, a compromise between those of the Belgian and Silesian systems, were the largest which could be used for treating fine high-grade ore. In cross-section the Rhenish retorts resembled an oval chain link, being about 18cm wide and 30cm high (7in × 12in) inside. Regenerative furnaces became almost universal after 1900 and by 1920 fuel consumption had been reduced to one tonne of coal per tonne of zinc distilled. Careful tests made in Swansea in 1851 on one of the last remaining Champion vertical retort plants showed that it required about 24 tonnes of coal to produce a tonne of zinc.

The bulk of the world's supply of zinc was produced in horizontal retorts until 1950. However, the process, being discontinuous, was labour and fuel intensive. The first practical continuous zinc reduction process, introduced in the 1930s by the Imperial Smelting Corporation at Bristol, utilized a large vertical retort built of carborundum bricks which was fed by briquettes of roasted ore mixed with bituminous coal. The labour requirement of such retorts was well below that of any earlier process. The output of such a unit, around 9 tons of zinc per day, should be compared with the 60lb (27kg) produced by the best horizontal retort of that period.

During the nineteenth century the possibility of devising an effective zinc blast furnace attracted the attention of many ingenious metallurgists, although little progress was made. Labour and fuel were at that time very cheap commodities, and the world as a whole seemed content with the horizontal retort process. This situation began to change rapidly after the First World War. By that time profitable metallurgical enterprise had returned to the Bristol area, and lead and zinc were being refined at Avonmouth on a large scale. The continuous blast furnace process developed by the Imperial Smelting Corporation took, in all, about seventeen years to develop and was not put into production until 1957.

Like all successful and profitable innovations, the Imperial Smelting Process depends on very simple principles. Thus, at 1000°C, the vapour pressure of lead is negligible compared to that of zinc. It is also well known that when a bath of molten lead saturated with zinc is cooled, the zinc thrown out of solution contains less than 1 per cent of impurities, and can be skimmed from the surface of the melt as a virtually pure metal. The zinc blast furnace is charged with a mixture of lead and zinc oxides which are reduced by hot coke as in a conventional iron blast furnace. The lead and slag produced by this reaction sink to the hearth of the furnace, where they are tapped and separated. The zinc leaves the reaction zone as a vapour and, mixed with the carbon monoxide formed by the reduction process, is carried into the condensing chamber where it is absorbed by streams of molten lead. Dissolved in this way it can safely be exposed to the atmosphere without serious risk of oxidation. The lead is then slowly cooled and the layer of pure solid zinc which separates out on its surface is skimmed off. The remaining molten lead is then pumped back to the condenser.

Each of these zinc blast furnaces can produce 60,000 tonnes of zinc per year, and at least ten are in operation throughout the world.

Early applications of zinc

As a new exotic metal which had only recently become an article of commerce, zinc was of great interest to the experimentalists of eighteenth-century Europe, and it began to play an increasingly important role in their physical and chemical investigations. Alessandro Volta demonstrated that the muscular contractions observed by Galvani were caused by electrochemical potentials associated with the moist contact of dissimilar metals (see Chapter 6). In one form of his voltaic pile, silver and zinc plates were used. A cheaper and more practical version, however, was his 'couronne de tasses', or crown of cups, which used zinc and copper electrodes.

Well before this time, however, it had been noted that the behaviour of zinc when immersed in water or aqueous solutions was very different to that of copper, iron or lead. As early as 1742, Dr P. J. Malouin of Rouen had shown that zinc could be substituted for tin as a corrosion-resistant coating for iron utensils. As described by Richard Watson, Bishop of Llandaff, the coating was applied to the surface of hammered iron saucepans by immersing them in an iron pot filled with molten zinc, sal ammoniac being used as a flux. For a variety of reasons galvanized iron never seriously challenged tin plate in the manufacture of cooking utensils, and little further progress occurred until the second decade of the nineteenth century.

In 1819 the French chemist Thénard began to appreciate that metallic corrosion was caused by electrochemical effects. His ideas were subsequently developed by de la Rive, who in 1830 interpreted the rapid corrosion of impure zinc in terms of the localized electrochemical cells established between the zinc matrix and segregated impurities. This led to the idea that corrosion might be inhibited by coating the metallic article with a thin layer of metal which would preferentially corrode and thus protect the substrate. Sorel attempted to apply zinc to the surface of rolled iron sheet, and in 1837 he and an associate named Ledru obtained a French patent for iron protected against corrosion by a hot dipped coating of zinc. Sorel christened his new process 'galvanizing', in acknowledgement of the electrochemical background from which his concept of sacrificial protection had emerged.

The English patent for galvanizing was granted in 1837 to Commander H. V. Craufurd RN on the basis of a communication from Sorel. It was exploited by the establishment in London in 1838 of the English, Scotch and Irish galvanised Metal Company, which started in Southwark and shortly afterwards, as demand increased, moved to Millwall. Many components such as iron sheets, chains, nails and wire were treated. In 1839 a very much larger organization, the British Galvanisation of Metals Coy, began to operate the process in Birmingham, paying royalties to Craufurd at a rate of £3 per ton of components treated. Flat sheets of galvanized iron 1 cm thick were supplied by this company on a considerable scale as roofing material for the new Houses of Parliament.

Galvanized corrugated iron is first mentioned in 1845 in a patent taken out by Edmund Morewood and George Rogers. Morewood soon established works in the Black Country, where he galvanized black corrugated iron sheet obtained from J. and E. Walker of Gospel Oak, London. When Morewood ran into financial difficulties in 1847, J. and E. Walker took over his plant and equipment in part payment of debts and transferred them to Gospel Oak. The 'G anchor O' brand of galvanized corrugated iron made at the Gospel Oak plant soon gained a world-wide reputation for its quality and reliability. Vast quantities of galvanized iron sheet and utensils were exported to the expanding colonies and the profits made from bursts of trade stimulated by the gold rushes in California and Australia were very considerable. Around 1850 the company designed and constructed a 'beautiful Gothic villa' constructed completely with galvanized corrugated iron which gained much attention.

During the nineteenth century zinc sheet became a popular roofing material in its own right, being lighter and very much cheaper than lead. Unlike corrugated iron, zinc sheet had no structural stiffness, although when properly applied to a suitably supporting surface it provided excellent protection. After about 1850 zinc sheet was extensively employed for cladding the roofs of railway stations and other buildings.

Although the hot dip galvanizing process afforded excellent protection against corrosion, it had the disadvantage of increasing significantly the dimensions of the articles being treated and could not, therefore, be easily applied to the routine coating of small components such as nuts and bolts where close dimensional control was required. Galvanizing, normally carried out between 430° and 540°C also had a tendency to soften cold-rolled and heat-treated components. Sherardizing, a process introduced by Sherard Cowper-Coles between 1900 and 1904, avoided many of the disadvantages of galvanizing. It was a cementation process, made possible by the high vapour pressure of zinc. The steel or iron components to be treated were packed in sealed retorts in contact with zinc dust, and heated for 4–5 hours at about 375°C, which is well below the melting point of zinc. At this temperature, zinc vapour soon saturated the interior surface of the retort and diffused well below the surface of the ferrous components being treated, providing a corrosion-resistant surface without serious change in external dimensions. This process, which after 1904 gained rapid commercial success on both sides of the Atlantic, is still widely employed for the treatment of small, precise components. It has the incidental advantage of producing a matte surface on the component which assists considerably the subsequent adherence of paint.

THE EMERGENCE OF NICKEL

Metallic zinc was used to produce golden brasses containing less zinc than those obtained by the calamine process. It was noted in the eighteenth century that when zinc was added to copper in higher concentrations than those obtainable by cementation, the alloys obtained, which were usually used for brazing, were very much paler in colour. Some of these were confused with another whitish alloy, paktong, which, in smaller quantities, had found its way to Europe and Britain from the Far East since the sixteenth century. It was hard, whitish in colour and, unlike the high zinc brasses, relatively ductile in the cold condition. Being resistant to corrosion and free from the unpleasant taste of brass it was much favoured for the manufacture of musical instruments. Although the metal nickel had been identified by Axel Cronstedt in 1751, the composition of paktong was unknown until 1822, when analysis by Fyffe of Edinburgh showed that it was, in essence, an alloy based on copper, nickel and zinc. It was then being sold in the British Isles for about 12 shillings per lb, and was beginning to find fairly wide application as a cutlery alloy, largely because of its white colour and resistance to corrosion.

It is uncertain when the Chinese first made paktong, although they made extensive use of a cupro-nickel coinage as early as the first century AD. Paktong was first mentioned specifically in the Chinese literature by Kung Ya in the third century and a century later it was also being used for coins. Some details of the method by which the alloy was manufactured were provided by Ho Wei in 1095. His description does not suggest that the Chinese were then aware of the existence of nickel as a metal in its own right, since the alloy was made by adding small pills prepared from the ore to a bath of molten copper held in a crucible. When a layer of slag developed over the molten metal surface, saltpetre was added to the bath, the alloy was stirred, and the ingot was then cast. It is not clear whether zinc was an ingredient of the pills prepared from the ore or whether the zinc was added at a later stage. It is mentioned, however, that the ore came from Yunan, and it seems logical to assume that it would have contained arsenic and possibly iron in addition to the nickel. During the eighteenth century cupro-nickel alloys in ingot form, which were sent from Yunan to Canton, contained approximately 30 per cent nickel and 20 per cent copper. The ingots, in the form of triangular rings about 23cm in diameter and 4cm thick (9in × 1.5in), were remelted in Canton, and zinc was added until the alloy was silvery white. During the seventeenth and eighteenth centuries, paktong was shipped from Canton to England by the East India Company, both in ingot form and also in the wrought condition.

It was difficult to manufacture paktong from first principles in Europe since a suitable source of nickel had not yet been identified. Complex cobalt-nickel-iron-arsenic ores were, however, available from the Schneeburg district of Germany, where they had been used for many years by the celebrated Blaufarbenwerke for the manufacture of cobalt blue and other pigments for the ceramic industry. By 1824 a Dr Geitner had devised a method of extracting reasonably pure nickel from the waste products of the cobalt-blue process and set up a plant for the manufacture of nickel silver which he sold throughout Europe under the name Argentan at around three thalers a pound.

In 1829, Percival Norton Johnson arranged with Geitner to operate the process in England. A works for purifying the crude nickel speiss from Schneeburg was set up at Bow Common behind the Regents Park Canal in London, and from the nickel thus obtained ingots of nickel silver containing 18 per cent Ni,

55 per cent Cu and 27 per cent Zn were subsequently cast at Hatton Garden. Between 1829 and 1832 Johnson, who was certainly the first man to refine nickel in the British Isles, produced, on average, about 16.5 tonnes of nickel silver a year. Most of this was made into cutlery, and close plated by the Birmingham manufacturer William Hutton. The cutlery was sold under the trade name Argentine. After 1830 several other firms began to make nickel silver and the price of the alloy had fallen to four shillings a pound. Johnson's most serious competitor was the Birmingham firm of H. & T. Merry, which was managed by Charles Askin and Brook Evans, the technical operations being in the hands of a very ingenious chemist, Dr E. W. Benson, who devised greatly improved methods of cobalt and nickel separation. Around 1832, H. & T. Merry changed its name to Evans & Askin and marketed a brand of nickel silver known as British Plate.

Nickel silver made in Europe was generally known as Neusilber or Argentan. The common designation in England was German Silver, although it was also known as nickel silver, Argentine, Albati, and British Plate.

The world's consumption of nickel in 1850 was approximately 100 tonnes per annum, most of this being obtained from small arsenical nickel deposits such as those at Schneeburg and Dillenberg. Usage of the metal increased slowly, until 1889 when the invention of nickel steels by James Riley of Glasgow created a demand which was difficult to satisfy. The nickel pyrrhotite deposits of Norway produced about 4600 tonnes of nickel between 1848 and 1907. This output was small, compared to that of the garnierite deposits of New Caledonia, which were discovered in 1865. These deposits, which contained initially 10–12 per cent of nickel, produced ore and mattes equivalent in total to approximately 165,000 tonnes of metallic nickel between 1875 and 1915. By 1920 the nickel deposits of the island of New Caledonia were virtually exhausted.

The famous nickel deposits of Sudbury, Ontario, are like those of Norway based on nickel pyrrhotite, although the ore is richer and more extensive. They were discovered in 1883 when cuttings were being made for the Canadian Pacific Railway. The next major nickel deposit to be uncovered was the Merensky Reef, near Rustenburg, in South Africa. Since its discovery in 1924 this has become one of the world's major sources of nickel and some of the less common metals. Very large cupro-nickel deposits in the north-western province of Gansu in China were discovered in 1958, and this area, where mining has developed rapidly since 1965, might well become a major source of nickel in future years.

The world's usage of nickel increased slowly in the second half of the nineteenth century. Several European governments were attracted to the use of nickel silver as a coinage alloy, but were deterred initially by the difficulty of reconciling the nominal and actual values of the metals from which the coins were composed. Switzerland, which introduced a nickel silver 'billon' coinage in 1850, solved this problem by including specified quantities of silver in the alloys used for the various denominations. In 1879 the silver content was omitted and Switzerland adopted the 75:25 copper–nickel alloy which was already being used by Belgium, the United States and Germany.

In the years which followed the introduction of pure unalloyed nickel as a marketable commodity, it came to be regarded as a metal which was difficult to melt, cast and fabricate. Although no serious problems were encountered in melting the metal in the coke-fired pot furnaces usually employed for melting carbon steel, the ingots obtained were often too brittle to be worked either hot or cold. Sometimes it was found that an ingot would crumble into fragments when hot working was attempted, whereas other ingots from the same batch could be cold rolled.

This problem, which seemed to be analogous to the hot shortness of Bessemer steel (see p. 167) was solved by adding spiegeleisen, an alloy of iron, manganese and carbon, to the melt, although this addition had a deleterious effect upon the corrosion resistance of the nickel sheet produced and it became necessary to use pure manganese. In the Birmingham works of Henry Wiggin & Co, nickel wire being produced in 1880 required between 1.5 and 3.0 per cent manganese to ensure ductility. The standing melting procedure at that time was to reduce the carbon content of the melt to about 0.10 per cent by additions of nickel oxide, and then to complete the scavenging process by adding manganese in concentrations up to 3 per cent.

By that time it had been found by Thomas Fleitman and other workers in the United States that magnesium was far more effective than manganese in improving the ductility of nickel and that lower concentrations were required. The use of magnesium as a scavenger was confined initially to the United States until 1924 when the mechanism responsible for the variable ductility of nickel and its alloys was elucidated by Paul Merica and his co-workers at the National Bureau of Standards in Washington. They were able to show that nickel was embrittled not by oxygen but by sulphur contamination of anything above 0.005 per cent. The secret of success was to treat the molten nickel with a metal such as magnesium which has a higher affinity than nickel for sulphur. Any excess magnesium not required for desulphurization removed itself by evaporation from the melt.

Claims to most of the significant ore bodies in the Sudbury district had been established by numerous prospectors, mining engineers and small companies soon after the initial discovery of ore in 1883. By 1886 most of the metallurgical assets had been acquired by the Canadian Copper Company under the leadership of S. J. Ritchie. This company shipped most of its roasted ore to the Orford Copper Company of Bayonne, New Jersey, for refining. Difficulties were encountered, however, in extracting the nickel since no satisfactory method existed at that time for separating the constituents of a copper iron nickel matte. By 1891, however, John L. Thomson, the Superintendent of the Orford Copper Company, had devised an elegant method of extracting nickel from the Sudbury ore. He utilized the ability of nitre cake, or acid sodium sulphate, to split up the molten copper nickel matte into two distinct layers. The copper-sodium sulphide, being lightest, separated into the upper layer, while the bottom layer contained nearly all the nickel in the matte. When the molten matte, thus treated, was allowed to cool very slowly in large pots, the two layers separated very sharply. Thus was born the celebrated 'tops and bottoms' process. The tops were transferred to convertors and blown to blister copper. The bottoms were roasted to remove all sulphur. The black nickel oxide remaining was then reduced to metallic form in an open hearth furnace, from which ingots suitable for electrolytic refining were cast.

The patent covering the Orford process was filed in May 1891, Thomson's assistant C. C. Bartlett being named as the inventor. Two years earlier, however, a very much simpler method of extracting nickel from cupro-nickel ores had been discovered in London by Ludwig Mond. After settling in London at the age of twenty-three, Mond had become deeply involved in the world of the chemical industry and in his forties had established the industrial empire of Brunner Mond & Co. In the early 1880s he set up a private research laboratory at his London home, and here, in October 1889, his young Austrian research colleague Carl Langer, who had been heating finely divided nickel in a tube furnace under a current of carbon monoxide, became aware of a remarkable effect. The colour of the carbon monoxide flame, burned at a jet as it left the furnace tube, changed from blue to a sickly green colour when the temperature of the nickel in the furnace fell somewhat below that of the boiling point of water.

Mond was immediately called to see this strange effect, and found that a nickel mirror was deposited on the surface of a white porcelain tile thrust into the flame. It was soon demonstrated that the gas leaving the furnace contained a hitherto unknown volatile compound of nickel which was named nickel carbonyl. It was found to form when carbon monoxide contacted pure nickel below about 110°C and to decompose above 180°C, when pure nickel, free from carbon was deposited, Mond appreciated very quickly that carbon monoxide gas could very well be the ideal leachant for selectively removing nickel from the Sudbury ores. By 1892 he had demonstrated the effectiveness of this approach in removing nickel from most types of ore in his London laboratory, and by the end of that year he had established a large-scale pilot plant at Smethwick, near Birmingham, on land belonging to Henry Wiggin. This plant was soon extracting nickel from the Canadian ore at a rate of about 1400kg (3080lb) a week.

By the end of the century the Mond Nickel Company had been established, and had acquired the Garson and Levack Mines in the Sudbury nickel complex. By that time the Canadian Copper Company, united with its former allies and associates such as the Orford Copper Company, had transformed itself into the International Nickel Company and had transferred its refining operations from New Jersey to Port Colborne, Ontario, where nickel was extracted from the matte by the Orford process. By 1921 the International Nickel and the Mond Nickel Companies were the only two organizations treating the Sudbury ores. Mond Nickel had a smelter at Coniston from which the matte was shipped to Clydach in South Wales for refining by the carbonyl process. Carbonyl nickel as originally produced at this plant contained less than 0.05 per cent metallic impurities, and since 1900 has been the purest grade of nickel commercially available.

In 1905, fourteen years after the Orford process had been introduced, it was suggested by Robert Stanley of International Nickel that if the copper nickel iron converter matte then being received by the refinery was converted directly to metallic form instead of being laboriously separated into its constituents, it would yield a ductile, corrosion-resistant cupro-nickel alloy which could be of considerable value to the chemical industry. Monel metal was named after Ambrose Monell, at that time president of the International Nickel Company. It contained approximately 67 per cent nickel, 28 per cent copper and 5 per cent iron plus manganese. In this natural alloy the copper and nickel retained the same relative proportions as in the converter matte. The iron also came directly from the matte, the only intentional addition being the manganese which was incorporated to remove the last traces of sulphur from the alloy.

Monel metal is still widely used where a ductile corrosion-resistant alloy is required. In its colour, mechanical properties and ductility it resembles nickel, and it is resistant to anhydrous ammonia, sea water and a wide range of organic chemicals. It is used for marine propellors and much marine hardware and at one time was used for roofing train sheds and railway stations in the United States. Between the two world wars it was also used extensively for the manufacture of kitchen sinks and for the heads of golf clubs. From such applications it has been largely displaced by austenitic stainless steels which are significantly cheaper.

The Orford process was not superseded until 1950 when it was found that the converter matte, when allowed to solidify in its equilibrium condition, crystallized as a mechanical mixture of the copper and nickel phases. These were separated by crushing the matte to a fine powder from which the nickel-rich phase was removed by magnetic and flotation processes.

The International and Mond Nickel Companies have now combined their resources, and the nickel oxide produced by the new Inco refining process is turned into metallic form either by electrolytic refining or by the carbonyl process. Monel metal, for reasons of control and economy, is no longer a natural alloy, but is produced by alloying copper, nickel and iron together in the desired proportions.

THE LIGHT METALS, ALUMINIUM AND MAGNESIUM

In the middle years of the nineteenth century, most of the metals in the periodic table of the elements had been chemically identified, although few had been produced in the pure metallic state and engineers were still dependent upon iron and a few copper-based alloys for the construction of machinery. The metals which were used had oxides which could be reduced by carbon at atmospheric pressure. The other metals which were known could not be obtained in metallic form because of their high chemical activity.

A particularly stable metallic oxide which could not be reduced was extracted from alum in 1760 and named alumina by the French chemist L. G. Morveau. By 1807, Sir Humphry Davy had concluded that even the most stable chemical compounds should be electrolytically reducible with the aid of the newly available voltaic cell, and had succeeded by this approach in obtaining sodium, potassium, barium, strontium and calcium in metallic form. For this remarkable demonstration of the power of electrochemistry, Davy was awarded a prize of 50,000 francs by Napoleon. Although he failed in his endeavours to obtain the element he first named 'aluminum' and then 'aluminium' in metallic form, it seemed evident that the other reactive metals he had obtained might well, under appropriate conditions, prove to be more powerful reductants than either carbon or hydrogen. In 1808 he succeeded in obtaining pure elementary boron for the first time by reducing boric oxide with electrolytically obtained potassium.

The search for metallic aluminium was continued by the Danish chemist Hans Christian Oersted, who in 1825 described to the Imperial Danish Society for Natural Philosophy a method of reducing aluminium chloride to metallic form with a mercury amalgam of potassium. The mercury from the amalgam was subsequently removed by distillation, leaving behind a grey powder which was described as aluminium, although it must have contained a good deal of oxide.

In 1827, Wöhler, who was then a teacher of chemistry at the Municipal Technical School in Berlin, improved on Oersted's reduction method by using a vapour phase process in which volatilized aluminium trichloride was reacted with potassium in metallic form. Potassium was a rare and costly reactive metal, and aluminium trichloride, because of its hygroscopic characteristics, was also a very difficult material to work with. Wöhler's initial experiments, therefore, although they produced small quantities of aluminium powder, did not provide a basis for a viable aluminium production process. His early work on aluminium was abandoned until 1854, when he was able to modify his process so that it produced a quantity of small shiny globules which were sufficiently pure to allow the low density of aluminium to be confirmed, and the ductility and chemical characteristics of the metal to be established.

In 1854, however, Henri Lucien Sainte-Claire Deville had already delivered an address to the Académie des Sciences of Paris on the subject of aluminium, and had been awarded a grant of 2000 francs to continue his research. Bunsen had also published a paper in Poggendorf's *Annalen* on improved methods of obtaining aluminium by the electrolytic reduction of fused salts. This approach, although of great theoretical interest could not then be used on an industrial scale because no satisfactory sources of heavy electric currents were then available.

Deville had initially studied chemistry at the Sorbonne and after several years of private research was appointed Professor of Chemistry at the Ecole Normale, where he began to work on aluminium. His first step was to substitute sodium for the reactive and expensive potassium which Wöhler had used. Sodium had the additional advantage that the sodium chloride formed by the reaction fluxed the surface of the globules of aluminium obtained, and assisted them to fuse together in an adherent lump. By 1855, Deville had a small chemical works at Javel, and bars of aluminium he produced there were exhibited at the Paris Exhibition of 1855.

It was soon found that the aluminium trichloride used by Wöhler was far too deliquescent and temperamental for use in a routine production process, and sodium aluminium trichloride ($AlCl_3 NaCl$) was generally used by Deville. He also devoted considerable attention to the production of cheaper sodium, since the cost of this reagent largely determined the selling price of any aluminium produced. Having seen the new light metal at the 1855 Exhibition, Napoleon III appreciated its military possibilities. He asked Deville to make him a breastplate of aluminium, a service of spoons and forks for state banquets and other items. As an artilleryman, he was also interested in the promotion of aluminium as a material for gun-carriage wheels.

In 1854, Deville had established a small company, the Société d'Aluminium de Nanterre, to exploit his reduction process. Further developments took place at Salindres, near Arles, at a factory belonging to Henri Merle & Co. In 1860, Deville sold his aluminium interests to Henri Merle who died soon afterwards. The Salindres plant was subsequently acquired by Alfred Rangod Pechiney, who eventually came to dominate the French aluminium industry via his Compagnie de Produits Chimiques et Electro-metallurgique Alais Froges et Camargue. In the 1860s, however, Pechiney appears to have been somewhat unenthusiastic about the future possibilities of aluminium which, he felt, was redeemed only by its lightness.

The manufacturing process at Salindres began with the production of pure alumina. This had originally been accomplished by the calcination of ammonium alum. In 1858, however, Deville had been introduced by the mining engineer Meissonier to the mineral bauxite, found at that time as a band of red earth in the limestone formations of Les Baux in Provence. Since this material contained a good deal of iron, it required extensive purification after which it was mixed with sodium chloride and carbon and reacted with chlorine at a good red heat to sodium aluminium trichloride. This, being a vapour, distilled away from the reaction zone, and condensed as a crystalline deposit at temperatures below 200°C. The double chloride was then mixed with cryolite and reacted with metallic sodium in a reverberatory furnace. The furnace charge consisted of 100kg (220lb) of the double chloride, 45kg (99lb) of cryolite and 35kg (77lb) of sodium. The function of the cryolite was to act as a flux and dissolve the alumina on the surface of the aluminium globules produced, so that they were able to coalesce. It also produced a slag which was fluid enough and light enough to let the reduced globules of aluminium sink to the base of the reaction bed and unite.

The reverberatory furnace reaction was accomplished at a good red heat and once started was accompanied by a series of concussions which persisted for about fifteen minutes: it was necessary to brace the brickwork and roof of the furnace with iron rods. After three hours at red heat the reaction had been completed and the products had settled down into two layers at the bottom of the furnace. The upper layer was a white fluid slag free from aluminium which was easily tapped off. The molten aluminium from the lower layer was then run into a red hot cast-iron ladle from which it was cast into ingots.

In 1872, 3600kg (7935lb) of aluminium were made at Salindres at a cost of 80 francs per kilo. Since the selling price of aluminium at that time was only 100 francs per kilo, profit on this activity was not large. Deville had visited London in 1856, when he demonstrated his aluminium reduction process to the Prince Consort and to Michael Faraday. His first contact with the London firm of Johnson and Matthey was in 1857, and from 1859 they acted as the British agents for the sale of his metal. At that time sodium reduced aluminium was around 98 per cent pure, and was being sold in Paris at 300 francs per kilo. By 1880 the price of aluminium in most parts of Europe had settled down to about 40 francs per kilo. The demand for the metal was not large, however, and in 1872, for example, the total quantity of sodium reduced aluminium sold by Johnson and Matthey was only 539 ounces (15kg).

The possibility of using cryolite not merely as a flux for the reaction process, but as the primary raw material for aluminium production, was first investigated by H. Rose in Berlin around 1856. Shortly afterwards William Gerhard established a plant in Battersea, London, for producing aluminium in this manner. Unexpected technical and economic difficulties were encountered, however, and aluminium production at Battersea was discontinued at the end of 1859.

At the beginning of 1860 the firm of Bell Brothers started to produce aluminium by a variant of the Deville process at Washington, County Durham. The driving force behind this enterprise was the celebrated ironmaster Sir Lowthian Bell who, although primarily concerned with the rapidly developing iron and steel industry of Teesside, was also interested in the Washington Chemical Company. In preparation for this venture into aluminium production, he sent his son Hugh to Paris in 1859 to study under Deville. At a later stage, Hugh Bell worked with Wöhler who, by that time, had become Professor of Chemistry at Göttingen.

The works at Washington, using a process very similar to that developed at Salindres, produced aluminium from 1860 to 1874. Sodium reduced aluminium was also being produced by James Fern Webster, who began to experiment with the metal in 1867 at his private house in Hollywood near Birmingham. Webster's aluminium was very much purer than Deville's, generally containing only about 0.8 per cent impurities. In 1882 he established the Aluminium Crown Metal Company at Hollywood. Hamilton Y. Castner brought his sodium reduction process to England in 1887, and shortly afterwards he established a manufacturing plant at Oldbury near Birmingham. Webster seized this opportunity to obtain cheap sodium and chlorine. He acquired a site adjacent to that of Castner, raised a working capital of £400,000 and established a factory intended to produce 50 tons per year of aluminium. Chlorine was bought by pipeline into Webster's plant from Castner's.

The aluminium produced by Webster was rolled into sheet, and aluminium foil of high quality was produced by beating. It had, unfortunately, been established too late to succeed. The manufacture of aluminium by the sodium reduction process in Britain became obsolete overnight in July 1890 when the Aluminium Syndicate at the Johnson Matthey site at Patricroft began to produce aluminium by Hall's electrolytic process on a considerable scale (see p. 108).

The other British firm producing sodium reduced aluminium at this period was the Alliance Aluminium Company at Wallsend on the River Tyne, which used the Netto process devised by Professor Netto of Dresden. Sodium was produced by spraying fused caustic soda on to an incandescent bed of coke held in a vertical retort. Reduction occurred very rapidly, and sodium vapour escaped from the reaction chamber before entering a water-cooled condenser. The sodium thus obtained was used to reduce cryolite in a modified version of the Deville process. The reaction bath was in fact similar to that used by Hall and Héroult, since it consisted of cryolite in which alumina was dissolved. Lumps of sodium weighing 2.25kg (5lb) were immersed in this fused salt solution.

Sodium reduced aluminium was also produced at this time in the United States by Colonel Frishmuth of Philadelphia, whose firm cast the famous aluminium pyramid used to cap the Washington Monument which has been in service since December 1884. When last examined in 1934, this cap had been fused near the top by a lightning flash but was remarkably free from corrosion. Like most aluminium produced by Deville's process, it was only 98 per cent pure, and contained about 1 per cent iron and 0.75 per cent silicon. At the Alumimium- und Magnesiumfabrik at Hemelingen in north Germany, which was established in 1886, magnesium was used to reduce cryolite, the final result being a silicon aluminium alloy containing 1–2 per cent iron.

Electrolytically produced aluminium

Generators capable of producing heavy electrical currents did not become generally available until the mid-1870s (see Chapter 6). By this time, interest in electro-metallurgical possibilities had begun to revive, at a time when the limitations of existing metallurgical production techniques were becoming very obvious. One interesting application of electrical power to metallurgical production was patented in 1885 by the brothers Eugene and Alfred Cowles of Cleveland, Ohio, who utilized the newly developed electric arc furnace to produce the high temperatures required for the direct reduction of alumina with carbon.

In general, the Cowles process was used for the manufacture of aluminium bronze, for which at that time a greater demand existed than for aluminium in its pure condition. A suitably proportioned mixture of alumina, carbon and copper was smelted with the arc on the furnace hearth. The function of the copper, which did not participate directly in the decomposition of the alumina, was to absorb the aluminium vapour immediately it was liberated and to remove it from the reaction zone before it could reoxidize. Very clear analogies can be discerned between this process and the cementation method of brass production, where copper was used to dissolve the zinc liberated by the reduction of calamine with carbon. In both instances, the product of the reduction process, either aluminium or zinc, was held in a state of low activity, thus allowing the reduction process to be driven to completion.

The Electric Smelting and Aluminium Company, set up in 1885 by the Cowles brothers, established production units at Lockport, NY, and at Milton, near Stoke-on-Trent in Staffordshire, which produced copper alloys containing between 15 and 40 per cent of aluminium. These were subsequently diluted with copper for the manufacture of aluminium bronze.

Between 1885 and 1890 the process enjoyed considerable success, since the alloys it produced cost far less per pound of contained aluminium than pure aluminium produced by other methods. This advantage declined when cheap electrolytically produced aluminium became generally available.

Electrolytic aluminium

Since the early work of Davy, chemists and metallurgists had come to equate the voltage required to decompose compounds in electrochemical experiments with the strength of the bond holding the atoms of the compound together. From observations of the differences in electrical potential which developed when dissimilar metals were brought into moist contact, the metals themselves were also sorted out into a well defined voltage series. As early as 1854, when Bunsen and Deville himself had employed the electrochemical route, it was evident that aluminium was an extremely electronegative element. Its affinity for oxygen was higher than that of any of the metals known to antiquity, and the electrochemical route to its production was the only feasible alternative to the hopelessly expensive sodium reduction process.

The first serious attempts in the United States to obtain aluminium by the electrolysis of fused salts appear to have been made by Charles S. Bradley of Yonkers, NY. His ideas and conceptions were similar to, and anticipated in several ways those of Hall and Héroult. Bradley was very unfortunate, however, because for reasons which are difficult to understand, his Application encountered much opposition from the US Examiners, and his Patent was not granted until 1892. By that time it was difficult for him to contest the validity of the Patents on which the Hall/Héroult process was based since that process was already in commercial operation in several countries.

The Hall process

Charles Martin Hall was instructed in chemistry at Oberlin College by Professor F. F. Jewitt, who in his youth had studied in Germany where he had met and had been strongly influenced by Wöhler. One of the undergraduate projects he gave Hall was concerned with the chemistry of aluminium, and he encouraged Hall to believe that the world was waiting for some ingenious chemist to invent a process for producing aluminium cheaply and reliably on a large scale.

Immediately after graduating in 1885, Hall began to investigate various electrolytic approaches in his private laboratory and soon concluded that fused salt baths would be essential. On 10 February 1886, he found that alumina could be dissolved in fused cryolite 'like sugar in water' and that the alumina/cryolite solution thus obtained was a good electrical conductor.

It was well known that cryolite would dissolve alumina. Deville, for example, had added cryolite to his reaction mixtures at Salindres to reduce the melting point and viscosity of the slag, and to dissolve the thin layers of alumina which formed on the surface of the reduced globules of aluminium, thus enabling them to coalesce. Cryolite also figured prominently in the sodium reduction processes devised in the late 1850s by Rose in Berlin and Netto in Dresden.

By dissolving 15–20 per cent of alumina in cryolite Hall obtained a bath whose melting point was between 900 and 1000°C, at which temperature its electrical conductivity was high enough to permit electrolysis (see Figure 1.8(a) below). The only difficulty was that the bath rapidly dissolved silica from the refractory materials used to contain it. By 16 February 1886, Hall had solved this problem by containing his melt in a graphite crucible and had obtained a number of small globules of aluminium, which formed close to the crucible which acted as his cathode.

Hall experienced difficulties both in establishing his patent rights and in finding backers for the production of aluminium by his process. He finally gained the financial support of Captain A. E. Hunt, who owned the Pittsburg Testing Institute. The Pittsburg Aluminium Company was established in 1889 and set up works at Smallman Street in Pittsburg which by September 1889 were producing about 385 pounds (173kg) of aluminium a day at a cost of only 65 cents per pound. This figure contrasts strongly with the $15 per pound which Colonel Frishmuth had been charging for his sodium reduced aluminium. By 1890, however, the Pittsburg Aluminium Company was still not paying a dividend. Hall was on a salary of $125 per week. At this time the firm, being dangerously short of capital, sold 60 of its shares to Andrew Mellon, thus bringing the Mellon family into the aluminium business.

Also in 1890, Hall contacted Johnson Matthey, who were at that time the main British dealers in aluminium. The Magnesium Metal Company, owned by Johnson Matthey, produced magnesium in a factory at Patricroft, close to Manchester on the banks of the River Irwell. Here the Aluminium Syndicate, owned by the Pittsburg Aluminium Company rented land and erected a factory, in which two large Brush engines and dynamos were installed. By July of 1890, this plant was producing about 300lb of aluminium per day by Hall's process. This metal was sold by Johnson Matthey until 1894, when aluminium production at Patricroft was discontinued.

The statue of Eros in London's Piccadilly Circus, cast in the foundry of Broad, Salmon and Co. of Gray's Inn Road and erected in 1893, appears to have been made from electrolytic aluminium supplied by the Aluminium Syndicate of Patricroft. The composition of the metal (99.1% Al, 0.027% Fe, 0.6% Si and 0.01% Cu) is incompatible with the general assumption that sodium reduced aluminium, which generally contained about 2 per cent impurities, had been employed. The foundry where the statue was cast was only a few hundred yards from the Hatton Garden establishment which was at this time handling a considerable quantity of the aluminium output of Patricroft. Moreover, it is well known that Sir Alfred Gilbert chose aluminium for his statue because it was cheaper than copper or bronze, which would not have been the case if the only aluminium he had been able to obtain had been the sodium reduced variety.

In 1895 the Pittsburg Aluminium Company, now known as the Pittsburg Reduction Company, moved to Niagara Falls to take advantage of the large supplies of cheap electric power which were available there. By 1907 the Company was producing around 15 million lb of aluminium per year, compared to the 10,000 lb produced in 1889 when operations were first started. In 1907 the company changed its name to the Aluminium Company of America.

The Héroult process

Paul Louis Toussaint Héroult was born at Thury-Harcourt near Caen in 1863. At the age of 15 he read Deville's book on aluminium and became obsessed with the idea of developing a cheap way of producing the metal. After studying at L'Ecole Ste Barbe in Paris he returned to Caen and began to experiment privately in a laboratory in his father's tannery. His first French patent, applied for on 23 April 1886, described an invention which was virtually identical to that of Hall: 'a method for the production of aluminium which consists in the electrolysis of alumina dissolved in molten cryolite, into which the current is introduced through suitable electrodes. The cryolite is not consumed, and to maintain a continuous deposition of metal it is only necessary to replace the alumina consumed in the electrolysis.' See Figure 1.8(b).

A further patent, filed in 1887, describes the production of an alloy, such as aluminium bronze by collecting the aluminium liberated by the electrolytic process in a molten copper cathode. This process was developed by Héroult with the encouragement of Alfred Rangod Pechiney who had taken over the works at Salindres where Deville's process for making sodium reduced aluminium was still being operated (see p. 103). It was operated for a short while at Neuhausen in Switzerland, but was abandoned in 1891.

Although an industrial expert from Rothschild's Bank reported unfavourably on Héroult's process in 1887, he was able to gain support from the Swiss firm of J. G. Neher Söhne, who had a factory at the Rhine Falls with plentiful supplies of water power. The ironworks at this site, established in 1810, had become unprofitable and new metallurgical applications for water power were being sought. In 1887 the Société Metallurgique Suisse was established, Héroult being the technical director. In the following year the Société joined the German AEG combine to establish the Aluminium Industrie Aktiengesellschaft with a working capital of 10 million francs. This consortium subsequently established large aluminium production units, at Rheinfelden (1897), Lend in Austria (1898) and Chippis, near Neuhausen (1905). The combine soon amalgamated with the French producers to form L'Aluminium Français.

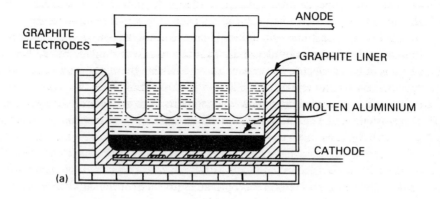

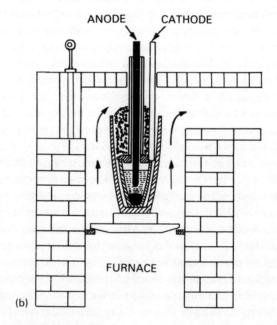

Figure 1.8: Electrolytic aluminium.
(a) The electrolytic process of aluminium production introduced by Charles Martin Hall in 1886 (U.S. Pats 400664, 400665, 400666, 400667, and 400766, April 1889) used a bath of fused cryolite in which pure alumina was dissolved. This bath was maintained at temperature in the molten condition by passage of the electrolysing current.
(b) Paul-Louis-Toussaint Héroult also filed patents in 1886 for a similar process. Figure 1.8(b) is based on the drawing in Héroult's French Patent 175,711 of 1886 when he did not appreciate that external heating was unnecessary. In a subsequent patent of addition he claimed the same electrolytic process without external heating.

Aluminium developed very rapidly in Europe where it was well appreciated as a light and corrosion-resistant metal which had also a very high thermal conductivity. This encouraged its use for cooking utensils. The first authenticated use of aluminium as a roofing material is provided by the dome of the Church of San Gioacchino in Rome which was roofed with Neuhausen aluminium in 1897. When examined by Professor Panseri in 1937 this sheeting was still in excellent condition. Its composition agrees closely with that of Eros in Piccadilly Circus (see p. 108); both were typical of electrolytic aluminium produced between 1890 and 1897.

Much discussion has centred on the remarkable coincidence that two young inventors, several thousand miles apart, should independently, at the same age, have defined identical technical objectives: after considerable experimentation, each in his own private laboratory, they arrived at the same technical solution to their problem and applied for patent cover within a month's interval. Even more remarkably, Hall and Héroult were both born in 1863 and both died in 1914.

The Bayer process

The extraction metallurgy of iron, copper and the other metals was characterized by the use of smelting and refining processes which accepted, as their raw materials, grossly impure ores, and produced, as an intermediate product, impure metal or pig which was subsequently refined to the required level. The impurities present in aluminium, however, having far less affinity for oxygen than the parent metal itself, could not be selectively removed after the initial extraction process had been accomplished, and it was soon appreciated that the only feasible philosophy of aluminium manufacture was to produce directly, in one stage, molten aluminium having the highest state of purity in which it was likely to be required.

Since the purity of the aluminium obtained, either by sodium reduction or by electrolytic dissociation, is largely determined by the purity of the alumina from which it is derived, the industry has, since its earliest days, been dependent upon its ability to purify minerals such as bauxite cheaply, and on a very large scale. One tonne of aluminium requires two tonnes of alumina and this requires four tonnes of bauxite. Approximately 20,000kWh of electrical power are needed for the production of one tonne of aluminium by the Hall process, so the electrolytic process can be economically undertaken only where cheap electricity, generated by water power or by nuclear reactors, is available. Bauxite, however, can only be purified economically in countries where fuel such as coal, gas or oil is cheap and plentiful. Many of the most modern electrolytic refineries are dependent, therefore, upon the importation not of bauxite but of foreign alumina.

Deville, at Salindres, found that only after very carefully purifying his alumina was he able to produce aluminium containing less than 2 per cent impurities. The technique he employed, the Deville-Pechiney process, was developed by his associate Paul Morin; it is still occasionally used for the treatment of bauxites which contain a lot of iron although it was not too effective in removing large quantities of silica.

The Deville process was, however, superseded in 1887 by a cheaper and simpler approach devised by the Austrian chemist Karl Joseph Bayer which is now almost universally employed. Unlike the expensive Deville technique, it depended entirely upon wet chemistry and involves no fusion process. Bauxite was digested under pressure by a caustic soda solution in an autoclave at temperatures between 150° and 160°C. This reaction produced a soluble solution of sodium aluminate, the major impurities, such as iron oxide, titania and most of the silica, being left behind as a red mud. Alumina was precipitated from the caustic soda solution when it was cooled and diluted with water. After calcination it was then suitable as a feedstock for the electrolytic cells.

The British Aluminium Company (BAC) Limited was floated on 18 December 1894 to acquire the British rights to the Bayer and Héroult processes and others including the patents and factory site of the Cowles Syndicate at Milton in Staffordshire where a rolling mill was installed. Lord Kelvin was appointed as Scientific Adviser to the company and in 1898 he joined the board of directors. The progress of this company, however, is inevitably associated with William Murray Morrison, who was appointed as chief engineer at the beginning of 1895 and served BAC for half a century. Over this time the world output of aluminium increased from 200 tonnes per annum in 1894, to 5000 tonnes in 1900, and in 1945 to well over two million tonnes. The world output of primary aluminium in 1980 reached a peak in the vicinity of 16 million tonnes per annum.

The first BAC plant was established at Foyers, close to the Falls of Foyers on the southern side of Loch Ness, in 1895. It produced about 3.7MW (5000hp) of hydroelectric power and by June 1896 it was extracting about 200 tonnes per year of aluminium, most of which could not be sold. In 1896 about half the power generated at Foyers was sold to the Acetylene Illuminating Company which made calcium carbide by fusing lime and carbon in an electric furnace, a process which had been invented by Moissan in 1892. When after the turn of the century the world demand for aluminium began to increase significantly, BAC started to build an additional hydroelectric plant at Kinlochleven. This took about five years to build.

The Lochaber scheme, which commenced in the mid-1920s, was far more ambitious, since it took water from a catchment area covering more than 300 square miles. Waters draining from the Ben Nevis mountain range and the waters of Lochs Treig and Laggan were collected and fed to a powerhouse situated only a mile from Fort William. Further hydroelectric plants were established in other regions of the Western Highlands as the demand for aluminium increased after the 1930 period.

Magnesium

Magnesium was first isolated in 1808 by Humphry Davy, who electrolysed a mixture of magnesia and cinnabar in naphtha. The magnesium liberated was absorbed into a mercury cathode to form an amalgam. The first chemical reduction was accomplished in 1828 by Bussy, who used electrochemically produced potassium to reduce anhydrous magnesium chloride. Magnesium, however, remained a chemical curiosity until 1852 when Bunsen, at Heidelberg, devised a method of producing it continuously by the electrolysis of fused anhydrous magnesium chloride. In 1857 the metal was produced for the first time in quantities large enough to allow its properties to be evaluated by Deville and his colleague Caron. Using the technique he had already perfected for aluminium production (see p. 103), Deville reacted sodium with fused magnesium chloride to obtain magnesium, which was found to be a very light reactive metal. It was also volatile, and could readily be separated, by distillation in hydrogen, from the mixture of fused sodium and magnesium chloride left behind in the reaction vessel.

By this time it was also known that magnesium was a metal which burned readily to produce a very intense white light. Bunsen, who studied this effect, found that the light emitted had powerful actinic qualities which rivalled those of sunlight, so that the metal could therefore be of value to the new science of photography. In 1854, Bunsen and a former pupil of his, Dr Henry Roscoe, published a paper on the actinic properties of magnesium light in the *Proceedings* of the Royal Society and this stimulated a great deal of interest in the metal. In 1859, Roscoe became Professor of Chemistry at Owens College, Manchester. The paper stimulated the inventive genius of Edward Sonstadt, a young English chemist of Swedish descent who is known today largely because he was the first analyst to determine accurately the concentration of gold in sea water. Between November 1862 and May 1863, Sonstadt applied for patents which covered an improved process for producing magnesium, and for purifying it by distillation. By the summer of 1863 Sonstadt was able to claim that his 'labourer and boy' had produced several pounds of magnesium metal.

During 1863, Sonstadt met Samuel Mellor, from Manchester, who made his living in the cotton industry. Mellor was an enthusiastic amateur chemist, who had, as a part-time student at Owens College, worked under Roscoe and had established close personal contact with him. Sonstadt and Mellor became partners, and Sonstadt moved his laboratory from Loughborough to Salford. In Manchester, Mellor introduced Sonstadt to the pharmaceutical chemist

William White, who had developed an improved method of making sodium metal, needed by Sonstadt for his magnesium reduction process.

On 31 August 1864 the Magnesium Metal Company was incorporated with a working capital of £20,000. Land for a production plant was acquired at Springfields Lane at Patricroft, with a frontage on the River Irwell. One of the directors appointed at this time was William Matther who also acted as chief engineer of the company. At a later stage, as Sir William Matther, he headed the well-known firm of Matther and Platt. At Salford he developed improved methods of drawing magnesium wire which was then rolled into ribbon. Magnesium ribbon was a major innovation, since it was simpler and cheaper to produce than fine round wire. A very important advantage from the photographic viewpoint was that it burned far more consistently than wire of circular cross-section.

Magnesium production at Salford commenced in 1864 and in 1865 the plant produced 6895 ounces (195.5kg) of the metal. A peak output of 7582 ounces (215kg) of metal was reached in 1887. In 1890 magnesium was being imported into Britain from Germany at 1s 6d per lb: Salford could no longer compete and the works were closed.

Just before the First World War, Britain had no indigenous source for the magnesium it required, and in 1914 Johnson Matthey, in association with Vickers, set up a magnesium plant at Greenwich. Using the sodium reduction process originally developed by Sonstadt, this plant produced most of the magnesium needed by the Allies for pyrotechnical purposes. After it was closed down in 1919, no more magnesium was produced in the United Kingdom until 1936.

German magnesium

Bunsen had appreciated as early as 1852 that the cheapest and most favourable production route would involve the electrolysis of a fused anhydrous magnesium salt. Large deposits of carnallite were found in the salt beds of Stassfurt, and most of the earliest German magnesium production operations attempted to utilize this readily available compound.

Carnallite, which is found in many evaporite deposits is extremely hygroscopic. Magnesium metal was first obtained on a production scale by the electrolysis of fused carnallite in 1882 by the German scientists Groetzel and Fischer. In 1886 the Aluminium- und Magnesiumfabrik Hemelingen established a plant for the dehydration and electrolysis of molten carnallite using cell designs based almost completely on Bunsen's original conceptions. Much of the magnesium originally produced at the Hemelingen plant was used for the manufacture of aluminium by a variant of Deville's process (see p. 106).

Shortly after the turn of the century, the Hemelingen plant was acquired by Chemische Fabrik, Griesheim Elektron, who transferred it to Bitterfeld in Saxony which soon became the centre of the magnesium world. Before this time magnesium had been regarded merely as an inflammable metal which had applications in photography and pyrotechnics. Griesheim Elektron, however, was led by the energetic and far sighted Gustav Pistor who was one of the first to appreciate the engineering possibilities of a metal with a density of only 1.74g/cm^3 (0.063lb/in^3). Aluminium, the only comparable alternative, had a density of 2.70g/cm^3 (0.098lb/in^3). Magnesium could be of great value, it was felt, particularly in the aeronautical field. The first hurdle to be overcome was to develop a magnesium free from the corrosion difficulties which had hitherto inhibited its industrial application.

Pistor, supported by the very able chemist Wilhelm Moschel, concentrated initially upon the existing carnallite process and evolved an electrolytic bath based on the chlorides of sodium, calcium and magnesium, mixed in such proportions that the eutectic mixture formed melted at approximately 700°C. This type of bath, although melting at a lower temperature than the cryolite/bauxite mixture used for aluminium production, was and still remains far more difficult and expensive to operate. The main difficulty is that the magnesium metal obtained, being lighter than the electrolyte, floats to the surface of the bath where it must be collected without shorting the electrodes. Furthermore, the gas liberated at the anode is chlorine rather than the oxygen given off in the aluminium cell. The chlorine liberated in magnesium production is collected in bells surrounding the anode and utilized in the production of fresh magnesium chloride.

These disadvantages were soon appreciated and after 1905 determined attempts were made to evolve a magnesium bath using an analogy of the Hall/Héroult method in which magnesia was dissolved in a fused salt which, like cryolite, did not participate in the electrolytic process. A process developed in 1908 by Seward and Kügelgen utilized magnesia which was dissolved in a mixture of magnesium and lithium fluorides. A cathode of molten aluminium was used and the cell produced a magnesium-aluminium alloy. A variation of this approach was introduced by Seward in 1920 (see Figure 1.9) and was used in the 1920s by the American Magnesium Corporation: magnesia was dissolved in a bath of fused sodium, barium and magnesium fluorides and electrolysed to obtain magnesium of 99.99 per cent purity. The fundamental obstacle in such processes is the low solubility of magnesia in fluoride baths, so that it is difficult to ensure its constant replenishment.

A problem encountered at Bitterfeld was caused by the reluctance of the magnesium globules produced to coalesce. To improve this, small quantities of calcium fluoride were added to the bath. This addition had no beneficial effect, however, upon the hygroscopic tendencies of the carnallite-containing baths,

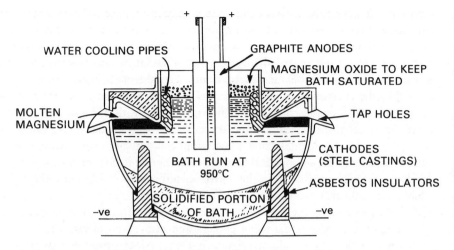

Figure 1.9: Electrolytic magnesium. A sectional view of the cell subsequently introduced by Seward in 1920 (US Pat. 1408141, 1920). In this process, operated by the American Magnesium Corporation, pure magnesia was dissolved in an electrolyte consisting of the fluorides of magnesium, barium and sodium. The cell shown used currents of 9000–13,000 amperes at emfs of 9–16 volts. The Dow process by then was producing purer and cheaper magnesium, and the American Magnesium Corporation ceased production in 1927.

and special techniques were required to dehydrate the fused salts so that magnesium oxychloride was not produced.

It was against this difficult electrochemical background that Pistor and Moschel developed the first practical magnesium alloys, which contained aluminium and zinc for strengthening purposes and were introduced by Griesheim-Elektron to the general public at the International Aircraft Exhibition at Frankfurt in 1909. These alloys were used in considerable quantities by Germany during the First World War, by which time Griesheim Elektron had merged with IG Farben Industrie.

After struggling with the carnallite process until 1924, Pistor and Moschel decided to synthesize a truly anhydrous magnesium chloride using the chlorine which was readily available from IG Farben. Chlorination was accomplished in large retorts where magnesite was reacted with chlorine in the presence of carbon at temperatures above 708°C, the melting point of the chloride. This accumulated as a liquid in the base of the reactor, and was transferred, still in the molten state, to the electrolytic cells. Since 1930 most of the world's magnesium has been produced by the electrolytic route of Pistor and Moschel, now known as the IG electrolytic process. Although some technical improvements have been introduced, the general philosophy of this approach has never been seriously challenged in situations where chlorine and cheap electric power are readily available and continuous magnesium production over a long timescale is envisaged.

Revival of magnesium production in Britain

After the First World War, British interest in the future of magnesium was kept alive by two firms, F. A. Hughes and Co. and British Maxium, later Magnesium Castings and Products.

F. A. Hughes was led by Major Charles Ball, who originally established contact with IG Farben and Griesheim Elektron while he was still in the post-war Control Commission. F. A. Hughes and Co. gradually developed the market for magnesium, although the demand was not great until the mid-1930s when industry began to revive. By that time it was obvious that Britain would soon require an indigenous source of magnesium, and in 1935 F. A. Hughes, supported by the British government, acquired the British and Commonwealth rights to the IG patents covering the Pistor/Moschel process of magnesium production, and then, in partnership with IG and ICI, set up Magnesium Elektron Ltd (MEL) as an operating base.

The site of the works at Clifton Junction, near Manchester, was determined by its proximity to the ICI works from which the large quantities of chlorine needed for the production of anhydrous magnesium chloride could be obtained. The Clifton Junction plant began production in December 1936. It was initially intended to produce 1500 tons of magnesium a year, although by government intervention in 1938 this capacity was increased to 4000 tons per annum. In 1940 another unit of 5000 tons per annum was added, and a further plant, capable of producing 10,000 tons per annum of magnesium commenced production at Burnley in 1942.

The MEL production process soon began to use sea-water-derived magnesia as a raw material. Before 1938 it had been dependent upon imported magnesite. The sea-water magnesia was produced by the British Periclase Company at Hartlepool, by treating sea water with calcined dolomite so that both constituents precipitated magnesia. During the period 1939–45, 40,000 tonnes per annum of magnesia were produced in this way for the UK.

Magnesium in North America

Magnesium appears to have first been manufactured in the United States by an offshoot of Edward Sonstadt's Magnesium Metal Company in Salford. The American Magnesium Company (AMC) of Boston, Massachusetts, was incorporated by Sonstadt on 28 April 1865, shortly after the grant of his US Patent, to which the company was given an exclusive licence. Sonstadt himself appears to have been the head of this company, which continued to produce magnesium by the sodium reduction process until it ceased operations in 1892, two years after the demise of the parent company in Lancashire.

Magnesium production in the United States then ceased until 1915 when a number of prominent American companies, such as General Electric, Norton Laboratories and the Electric Reduction Company, were persuaded to produce some part of the magnesium requirements of the Allies. American production in 1915 totalled 39 tons. Three producers were involved in the production of magnesium stick and ingots which sold for $5 per lb.

The Dow Chemical Company started to produce magnesium in 1916. By 1920 only Dow and the American Magnesium Corporation remained in operation. The AMC, then a subsidiary of the Aluminium Company of America, operated the oxide/fluoride process developed by Seward and Kügelgen. AMC went out of business in 1927, leaving Dow as the sole primary producer of magnesium in the United States.

The Dow magnesium process

Dr Herbert Dow initially established his company to extract the alkali metals sodium and calcium, together with the gases chlorine and bromine, from the brine wells of Michigan. No outlet was found for the magnesium chloride solutions which were originally run to waste. As the demand for magnesium began to increase, however, Dow devised a very elegant method of making anhydrous magnesium chloride in which the hydrated chloride was dried partially in air and then in hydrochloric acid gas, which inhibited the formation of the oxychloride. The anhydrous chloride thus obtained was then electrolysed in a fused salt bath. This magnesium extraction process, which ran as an integral part of the existing brine treatment operation, produced magnesium so cheaply that Dow soon emerged as the major US manufacturer.

In 1940, Dow moved his magnesium plant from Michigan to Freeport in Texas where anhydrous magnesium chloride was obtained from sea water by processes similar to those used by MEL in Britain. The Texas site had the advantage of unlimited supplies of natural gas, which allowed it to produce magnesium very cheaply indeed.

The strategic significance of magnesium as a war material had become very evident by 1938, and between 1939 and 1943 the United States Government financed the construction of seven electrolytic and five ferrosilicon plants so that any foreseeable demands for the metal could be satisfied. The biggest plant, Basic Magnesium Inc., was jointly built and managed by Magnesium Elektron from Clifton Junction and by the US Company Basic Refractories. This plant, in the Nevada Desert, used power from the Boulder Dam, and had a rated capacity of 50,000 tonnes per annum of magnesium. All these government-financed plants ceased production in 1945, leaving the Dow Chemical Company unit at Freeport as the sole US manufacturer, although some were reactivated for short periods during the Korean war.

Thermochemical methods of magnesium production

Magnesium, unlike aluminium, is a fairly volatile metal which at high temperatures behaves more like a gas than a liquid. It can, therefore, be obtained from some of its compounds by thermochemical reduction processes to which alumina, for example, would not respond. Because of its volatility, magnesium is able to vacate the vicinity of a high temperature reaction zone as soon as it is liberated, and this provides the reduction mechanism with a driving force which is additional to that provided by the reduction capabilities of carbon or of a reactive metal. Some of these processes are capable of producing high quality magnesium directly from magnesite without consuming vast quantities of electric power, although the total energy requirements of such processes are usually greater than the electrolytic technique, and they tend also to be highly labour intensive. During the war years, when magnesium was urgently required, the economics of thermal reduction processes were of less importance than the simplicity of the plant required and its ability to produce metal quickly, using indigenous resources without the benefit of hydroelectric power. Several processes were pressed into service and some proved surprisingly effective, although few survived into the post war era.

The ferro-silicon reduction process was developed at Bitterfeld by Pistor and his co-workers in parallel with electrolytic routes to magnesium production. As with the electrolytic process, British and Commonwealth rights to the IG Ferrosilicon process were acquired by MEL at Clifton Junction in 1935.

The process depended upon the reaction which occurs between dolomite and silicon, in which, as indicated below:

$$2\,MgO\,CaO + Si \rightarrow 2\,Mg + (CaO)_2\,SiO_2$$

In this reaction, the tendency of silicon to reduce magnesia is assisted by the high affinity of lime for silica. Because the product of the reaction, calcium silicate, is solid, it has little tendency to reoxidize the liberated magnesium, which escapes from the reaction zone as a superheated vapour and is then rapidly condensed directly to the solid state.

Rocking resistor furnaces were used at Bitterfeld to obtain magnesium from dolomite in the 1930s. The coaxial resistance element provided temperatures in the reaction zone of 1400°C and the furnace was run in a vacuum. Ferro-silicon was found to be the cheapest and most effective reduction agent, and the magnesium liberated from the reduction zone condensed directly to the solid state in a cooled receiver at the end of the furnace. Furnaces of this type, which produced about a tonne of magnesium a day, were built by IG Farben for the

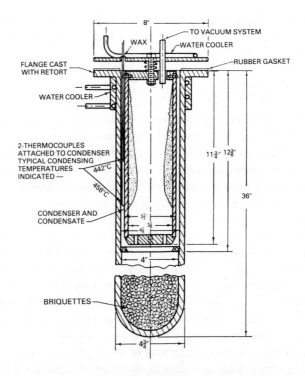

Figure 1.10: One of the earlier experimental retorts used by M. Pidgeon in Canada in 1941 for his thermo-chemical magnesium production process. In this approach dolomite was reduced by ferrosilicon, in nickel-chromium retorts, in vacuo, at temperatures around 1100°C. The magnesium vapour which distilled from the reaction zone condensed, as indicated, on the inside of a tube maintained at 450°C. Gas pressure inside the retorts were kept below 0.2 mm of Hg.
From L. M. Pidgeon and W. A. Alexander, *Magnesium*, American Society for Metals, 1946, p. 318. With permission.

Italian government and used at Aosta until 1945. Magnesium is still produced from dolomite in Italy by variants of the ferrosilicon process.

The ferrosilicon process was also used, very successfully, by the Allies during the Second World War, more particularly in North America. The process introduced by M. Pidgeon in Canada in 1941 produced high purity magnesium from dolomite by reacting it with ferrosilicon in horizontal tubular metal retorts which in the first pilot plant were only about 10cm in diameter. These retorts were disposed horizontally, as shown in Figure 1.10, in a manner which resembled closely that of the Belgian zinc retort process (see p. 93). A plant built by the Dominion Magnesium Company at Haley, Ontario, began to produce magnesium at a rate of about 5000 tonnes per annum in August 1942, and five similar plants were subsequently established in the United States.

Magnesium can be obtained by the direct reduction of magnesite with carbon, although at atmospheric pressures the reduction does not commence at temperatures below about 2000°C. The magnesium escapes from the reaction zone as a superheated vapour, and to prevent reoxidation it must then be rapidly cooled, when a fine pyrophoric powder is obtained.

The reaction involved in the carbo-thermic process is as indicated by the equation below completely reversible.

$$MgO + C \rightleftharpoons Mg + CO$$

If, therefore, the magnesium vapour released from the reaction zone is allowed to establish contact with carbon monoxide at a lower temperature, it will reoxidize. In this respect, therefore, the vapour of magnesium resembles that of zinc. After carbo-thermic reduction the magnesium vapour must be rapidly cooled and a variety of approaches to this difficult problem have been explored.

Most of the plants which attempted to operate the carbo-thermic process used variants of the approach patented by Hansgirg in Austria in the 1930s. This involved the reaction of magnesite with carbon in a carbon arc furnace at temperatures which exceeded 2000°C. The magnesium vapour leaving the furnace was rapidly cooled by a recirculating curtain of gas or by an appropriate organic liquid, and was obtained in metallic form as a fine pyrophoric powder.

A good deal of magnesium was made this way in Austria by the Osterr-Americk Magnesit AG before and during the Second World War. In 1938 the Magnesium Metal Corporation was jointly established by Imperial Smelting and the British Aluminium Company to produce magnesium by this approach. The factory, at Swansea in Wales, produced magnesium at a high cost and was shut down in 1945. The magnesium powder produced was handled under oil to avoid pyrophoric combustion, and required pelleting before it could be melted down. A large carbo-thermic plant built privately by Henry Kaiser during the Second World War at Permanente, California, had a rated capacity of nearly 11,000 tonnes of magnesium per annum. The magnesium vapour leaving the reaction furnaces at this plant was condensed in a curtain of oil. The lethal potentialities of this type of mixture were soon appreciated. Under the trade name of Napalm it was soon in great demand.

Although the bulk of the world's requirement of magnesium is still produced electrochemically, thermochemical reduction processes continue to attract a good deal of research effort and it is possible that a cheaper and more viable alternative to electrolysis will eventually emerge.

AGE HARDENING ALLOYS

At the beginning of the twentieth century, steels were the only alloys which were intentionally strengthened by heat treatment. In 1909, however, Dr Alfred

Wilm noted that an aluminium alloy he was working on had the remarkable ability to harden slowly at room temperature after having previously been quenched from a temperature just below its melting point. The effect was observed, quite fortuitously, while Wilm, working on a contract from the Prussian government, was attempting to develop alloys for cartridge cases which were lighter than the 70/30 brass which was normally employed.

Between 1909 and 1911 Wilm filed several Patent Applications, and assigned his rights in this 'age-hardening' invention to the Durener-Metallwerke in Duren, who subsequently marketed the alloys under the trade name Duralumin. During the First World War large quantities of age-hardened aluminium alloys were used by the combatants, first for Zeppelins and then for other types of aircraft. It was found that the strengthening reaction could be slowed down, very significantly, by refrigeration. This made it possible to quench aluminium alloy rivets, and store them in the soft condition in a refrigerator. The age hardening process began only after the rivet head had been closed after insertion into the aircraft structure.

A great deal of the metallurgical research stimulated by the introduction of Duralumin was co-ordinated in Great Britain from 1911 by the Alloys Research Committee of the Institution of Mechanical Engineers. The 11th report of this committee, published in 1921, summarized eight years of work by the National Physical Laboratory (NPL). The outstanding result of this work was the development of 'Y' alloy (4 per cent Cu, 2 per cent Ni, and 1.5 per cent Mg), which retained its strength to moderately high temperatures and was extensively used for pistons and cylinder heads. Y alloy was originally used in the cast form. High Duty Alloys was established in 1928 by Colonel W. C. Devereux to produce variants of Y alloy in the forged condition for aircraft engine use. From these experiments stemmed the RR series of light alloys, jointly developed by Rolls Royce and High Duty Alloys.

The first rational explanation of the age hardening process in light alloys such as Duralumin and Y alloy was provided in 1919 by Paul Merica and his colleagues Waltenberg and Scott at the National Bureau of Standards in Washington. They found that when an alloy such as Duralumin was heated to temperatures close to its melting point most of the alloying constituents were taken into solution by the matrix. Quenching retained the dissolved metals in this supersaturated solid solution which was, however, somewhat unstable at room temperature: minute crystals of various intermetallic compounds were slowly precipitated. Provided these crystals were below a certain critical size, invisible to the optical microscope, they strained and distorted the aluminium lattice and acted as mechanical keys, which inhibited plastic flow in the alloy and increased its strength. The aluminium alloys which behaved in this way were unique only in the sense that the precipitation effects manifested themselves, quite fortuitously, at room temperature. A wide range of other alloys were soon identified in which precipitation could be induced by ageing at elevated temperatures and new precipitation hardening alloys of all types were rapidly developed. In 1934 Dr Maurice Cook of ICI was able to review the precipitation hardening characteristics of 37 copper alloy systems.

Beryllium and beryllium alloys

Beryllium, first identified as an element by Wöhler in 1828, was at first regarded merely as a chemical curiosity. Soon after the introduction of the incandescent gas mantle by Auer von Welsbach in 1885 it was found that the strength and durability of the ash skeleton could be greatly improved by adding small quantities of beryllium nitrate to the mixtures of thorium and cerium nitrates used to impregnate the fabric of the mantle. This treatment was probably the major outlet for beryllium until the end of the First World War.

Attempts to produce beryllium by the electrolysis of a fused bath of beryllium chloride were first made in 1895 by Wilhelm Borchers. Goldschmidt in 1915 found that fused fluoride baths offered better production prospects. This approach, refined by Stock, Praetorius and Priess, yielded relatively pure beryllium for the first time in 1921. The properties of the metal obtained confirmed theoretical predictions that beryllium would be a light metal with a density around 1.8 grams per cm^3 and that its direct modulus of elasticity would be significantly higher than that of steel. If its ductility could be improved, it was likely to be the light, stiff metal which had long been sought by the aircraft industry.

The alloying behaviour of beryllium was studied by Regelsberger in 1924. Additions of up to 10 per cent by weight of beryllium improved the hardness and tensile strength of magnesium without making it brittle. Up to 10 per cent of beryllium could also be dissolved in copper to produce a pale yellow alloy. The colour of the alloy improved as the beryllium content decreased, and Regelsberger was the first to mention the beautiful golden colour of copper containing around 2 per cent of beryllium.

Research on beryllium in the United Kingdom was initiated by the Imperial Mineral Resources Bureau at the Metallurgy Department of the NPL in 1923. By 1926 Dr A. C. Vivian had produced solid deposits of metal by the electrolysis of fused baths of beryllium and sodium fluorides similar to those used by Stock, Praetorius *et al.*, whose detailed results had been published the previous year. The NPL beryllium was 99.5 per cent pure and, like the German product, was completely brittle when cold.

By 1926 popular interest in beryllium had started to develop. With a density equivalent to that of magnesium, a stiffness higher than that of steel, and a melting point approaching 1300°C, this seemed destined to become the light metal of the future. Towards the end of 1926, when facilities for the production of relatively pure beryllium existed in Germany, the USA and the United Kingdom, it was estimated that the metal could, if required, be produced for approximately 20s per lb. The electrolytic beryllium available in 1927, although reasonably pure, had a Brinell hardness of 140 and was still brittle when cold. In 1928 when the NPL improved the purity to 99.8 per cent, beryllium was still brittle and no improvement was noted when the purity level was gradually increased to the 99.95 per cent level. By 1932 attempts to produce ductile beryllium at the NPL were abandoned, and it was generally concluded, throughout Europe and the United States, that beryllium would in future find its major use as an alloying ingredient where the problems of purity and ductility were not so critically involved.

Beryllium copper and beryllium nickel alloys were first produced commercially for Siemens and Halske by Dr W. Rohn at Heraeus Vacuumschmelze at Hanau, which began to operate just after 1923 at the height of the inflationary period in Germany. Also in the early 1920s, Michael Corson in the United States was pursuing similar lines of development. In 1926 he patented the composition of an age hardening beryllium-nickel-copper alloy, while Masing and Dahl of Siemens also protected the compositions of beryllium copper and beryllium cobalt copper alloys which had higher electrical conductivities than those of the American alloy and were therefore of greater industrial potential.

Beryllium copper displayed a degree of age hardening far higher than that of any other copper-based alloy, and in spite of its cost beryllium copper soon found extensive industrial applications. It was particularly valuable as a spring material: the alloys were soft and ductile in the solution treated conditions and could then be fabricated into complex shapes. After heat treatment at temperatures between 300 and 350°C the best alloys developed tensile strengths approaching 100 tons per square inch, and as a result of the precipitation which had occurred the electrical conductivities improved to about 40 per cent that of pure copper. Even higher mechanical properties were obtained with beryllium nickel alloys, although these were not suited to electrical work because of their higher resistivity.

DEVELOPMENT OF HIGH TEMPERATURE ALLOYS

One effect of the rapid introduction of mains electricity into the domestic environment was an increasingly urgent requirement for improved alloys for electrical heating applications. The alloy required was one which combined a high electrical resistivity with extreme resistance to oxidation at high temperatures and mechanical properties high enough to ensure that it did not fail by creep after prolonged use at a good red heat. Prior to 1900 the only high resistivity alloys available were cupronickels such as Ferry and Constantan and iron alloys containing up to 20 per cent nickel.

Nickel-chromium alloys

The first satisfactory electrical heating alloys, introduced by A. L. Marsh in 1906, were based on the nickel-chromium and nickel-chromium-iron systems. Wires of these alloys had an electrical resistivity around 110 microhms per cm^3, more than twice that of the best cupro-nickel alloys. They were, moreover, far more resistant to oxidation and stronger at high temperatures. Previous investigators had found that chromium additions tended to increase the oxidation rate of nickel. Marsh found that chromium additions in excess of 10 per cent rapidly decreased this rate and that alloys containing around 80 per cent nickel and 20 per cent chromium showed a good balance between oxidation resistance and resistivity. They were also ductile enough to be drawn into wire. It is now known that the oxidation resistance of alloys of this composition depends on the formation of a protective oxide layer.

These alloys were initially induction melted in air and deoxidized with manganese. Air melted material was cheap to manufacture, although it was not always easy to draw into fine wire. At the end of the First World War, Siemens and Halske found that vacuum melted nickel-chromium alloys were easier to draw into wire and also had a longer high-temperature working life, compensating in some degree for the added expense of vacuum melting. Heraeus constructed at Hanau in 1921 a 300kg (660lb) capacity low frequency vacuum melting furnace of the Kjellin type for melting nickel-chromium based resistance alloys. By 1928 two 4000kg (8800lb) furnaces capable of casting two 2000kg (4400lb) nickel-chromium ingots were in regular operation. Siemens and Halske acquired Heraeus Vacuumschmelze in 1936, after which considerable quantities of nickel-chromium alloys, beryllium copper, and other specialized materials for the electrical industry were produced.

The rare earth effect

In the early 1930s it was found that some of the vacuum-melted alloys which had been processed from melts deoxidized with *mischmetall* (a rare-earth mixture) were remarkably resistant to high temperature oxidation. Rohn was able to show that the metal responsible for this effect was cerium, and in 1934 Heraeus applied for patents covering the manufacture and use of heating elements to which small quantities of cerium and other metals of the rare earth group had been added. This 'rare earth effect' was followed up in the Mond Nickel Laboratories at Birmingham, and led to the introduction of the well-known 'Brightray C' series of alloys. The mechanism which permits small quantities of the reactive metals, such as cerium, yttrium, zirconium, lanthanum and hafnium to improve the protective nature of the oxides which form on the surface of nickel chromium and other high temperature alloys is still imperfectly understood, although the effect is widely employed.

Aluminium containing nickel-chromium alloys

It was found in 1929, by Professor P. Chevenard of the Imphy Steelworks, that small quantities of aluminium improved the oxidation resistance and high temperature strength of nickel-chromium alloys and made the alloys responsive to age-hardening. In 1935 he showed that the strengthening effect of aluminium could also be augmented by small quantities of titanium.

When, around 1937, Britain and Germany both began to develop a workable gas turbine for aircraft propulsion, the main technical problem they encountered was that of producing turbine rotor blades which were strong enough at high temperatures to withstand the high, centrifugally imposed tensile stresses. The alloy selected was the 80/20 nickel-chromium solid solution alloy, strengthened in accordance with Chevenard's findings by small quantities of titanium and aluminium. Work on this material started in 1939, and the first production batch was issued in 1941. Nimonic 80, as the alloy was called, was the first of the long series of nickel-based 'superalloys' produced by Mond. All of these depend for their high temperature strength upon substantial quantities of the precipitated γ (gamma prime) phase which, although based on the compound Ni_3Al, also contains a good deal of titanium.

The first Nimonic alloys were melted and cast in air and hot forged. As the alloys were strengthened and improved, however, very reactive alloying constituents were being added, and vacuum melting became mandatory. By 1960, most of the stronger alloys were being worked by extrusion rather than by forging. The use of molten glass lubrication, introduced by Sejournet in the 1950s, made it possible to extrude even the strongest Nimonic alloys down to rod as small as 20mm in diameter in one hot working operation. The alloys were progressively strengthened, initially by increasing the aluminium content and subsequently by additions of the more refractory metals such as tungsten and molybdenum. Nimonic 115, the last of the alloys to be introduced, marked the practical limit of workability. Stronger and more highly alloyed materials could not be worked and by 1963 a new generation of nickel-based superalloys was being developed. There were not worked, but were cast directly to the shapes required.

Since the early years of the century it had been known that the grain boundaries of metals, although initially stronger than the body of the grains, tended to behave in a viscous manner as the temperature increased and lost their strength far more rapidly than single crystals. The concept of an equi-cohesive temperature, applicable to any alloy, above which the grains themselves were stronger than the boundaries and below which the boundaries were stronger than the grains, was first advanced in 1919 by Zay Jeffries. This suggested that the strongest high temperature alloys would have large rather than small grains.

Difficulties were, however, caused on odd occasions by very large randomly orientated grains. After 1958 it became customary to apply a thin wash of cobalt-aluminate to the interior of the moulds in which the turbine blades were cast. This coating helped it to nucleate a uniform grain size in the castings. It also seemed logical to produce blades having as few grain boundaries as possible aligned at right angles to the axis of the tensile stress to which the blade would eventually be subjected. Methods of directionally solidifying superalloys in such a way that the structures obtained formed a bundle of longitudinally aligned columnar crystals were first described in 1966 by B. J. Piercey and F. L. Versnyder of the United Aircraft Corporation. This process, now known as directional solidification, resulted in an immediate improvement in high temperature blade performance even without change in alloy composition. Single crystal blades, which contained no boundaries, were soon produced as a logical development of the directional solidification concept. Because grain boundary strengthening additions such as hafnium were no longer required in single crystal blades, and tended to interfere with the perfection of their growth, single crystal turbine blades are now manufactured from alloys which have very much simpler compositions than the conventional casting alloys.

Cobalt-base high temperature alloys

While the British were using the 80/20 nickel-chromium resistance wire alloy as a base for their first turbine blade material American manufacturers were adopting a completely different approach to the problem of high temperature alloy strength. The blades and other components of gas turbines used for driving the superchargers of large piston engines had, for a considerable time, been very effectively and economically manufactured by casting them from cobalt chromium alloys.

Alloys based on the cobalt chromium system had originally been developed by Edward Haynes. His original 'Stellite' alloy, which would 'resist the oxidizing influence of the atmosphere, and take a good cutting edge' was patented in 1909. Further improvements in hardness were obtained by tungsten and/or molybdenum additions. Although originally intended for dental and surgical instruments, the alloys soon found a considerable industrial market when it was found that they could be used for heavy turning operations. During the First World War they were used extensively by the Allies for shell turning, and in 1918 approximately four tonnes a day of the alloy were being cast. The Haynes Stellite Company was acquired by Union Carbide in 1920, and the alloy Vitallium was developed in the late 1920s. This was successfully cast into turbine blades used to supercharge the engines of the Boeing Flying Fortress.

The American engine manufacturers understood that no long-term future existed for workable turbine blade alloys: the high alloying content needed to achieve high temperature strength inevitably lowered the melting point of such alloys and simultaneously reduced their workability, and it was illogical that an alloy specifically designed to withstand creep failure at high temperature should also be expected to undergo severe plastic deformation during manufacture. From the very beginning of superalloy development, therefore, turbine blades were produced in the United States by vacuum melting and investment casting. The alloys employed were based on the cobalt chromium system, being strengthened, not by the 'gamma prime' phase on which the British nickel base alloys depended, but by the presence of substantial quantities of stable carbides.

The way in which the high temperature capabilities of superalloys has increased with time is illustrated diagrammatically in Figure 1.11. It is significant that the safe operating temperature for the best single crystal alloys is only marginally higher than that of directionally solidified material. In view of the limitations imposed by the known melting points and oxidation resistance of existing superalloys it seems unlikely that new alloys having high temperature capabilities greatly superior to those currently available will be developed. Improved gas turbine performance will, most probably, result from the exercise of engineering rather than metallurgical ingenuity.

POWDER METALLURGY

Prehistoric iron must have been the first metal to have been consolidated from a spongelike mass (see Chapter 2). The legend of Wayland the Smith, which appears to date from the fifth century AD, obviously embodies a good deal of the folklore on which the whole fabric of powder metallurgy has since been erected. Various accounts describe him as making a steel sword by conventional blacksmithing techniques, then reducing it completely to filings, which were mixed with ground wheat, and fed to domestic geese or hens. The droppings of these birds were collected, reheated in his forge, consolidated, and then hammered into a sword. This again was reduced to filings and the process repeated, the end product being a weapon of unsurpassed strength, cutting power and ductility.

Apparently even the early smiths were aware that metallurgical quality could be improved by subdivision, and that by repeating this process of subdivision a product would be eventually obtained which would combine hardness and ductility. The improvements which ferrous metallurgists have attributed to the passage of steel filings through the alimentary canal of a domestic fowl are more difficult to account for, however, as it is difficult to believe that this would reduce the phosphorous content of the metal to any significant effect. It is possible, however, that the high ammonia content of bird droppings might help to nitride the metal particles during the early stages of consolidation, although in the absence of reliable confirmation of this effect it seems logical to attribute the effectiveness of Wayland's working approach to the way in which it was able to progressively refine an initially coarse metallurgical structure.

The use of powder metallurgy by Wayland the Smith was obviously one approach to the problem which all swordsmiths have encountered: swords which were strong and hard had a tendency to break in service, while those which were free from this defect bent too easily and did not retain their cutting edge. Wayland's technique must have produced a refined crystal structure which, like that which existed in Samurai blades, combined strength and ductility with the ability to maintain a fine cutting edge. It provides an effective demonstration of the ability of powder metallurgy to improve the quality of an existing product even when the metals involved could, if required, be melted together.

Powder metallurgical techniques were used in pre-Columbian times by the Indians of Ecuador to consolidate the fine grains of alluvial platinum they were unable to melt. These platinum grains were mixed with a little gold dust, and heated with a blow pipe on a charcoal block. The platinum grains were, therefore, soldered together by a thin film of gold and it was possible by this method to build up solid blocks of metal which were malleable enough to withstand hot forging, and were fabricated into items such as nose rings and other articles of personal adornment.

Although platinum was a metal unknown to the ancients, fragments of this metal and its congeners (elements belonging to the same group in the periodic table) were unwittingly incorporated into many of the earliest gold artefacts from the Old and New Worlds. The first important source of the metal was Colombia on the north-western corner of South America. Here, the grains of native platinum were regarded initially as the undesirable component of South American gold mining operations. The identity of platinum as a new element was recognized in 1750, by William Brownrigg. In 1754, William Lewis of Kingston upon Thames found that the grains of platinum could generally be dissolved in aqua regia, and that the 'beautiful brilliant red powder' now known to be amino-platino-chloride was obtained when sal ammoniac was added to the aqua regia solution. This precipitate, when calcined, decomposed to provide a dark metallic powder which appeared to be pure platinum. No satisfactory method of melting this platinum powder or sponge was found, however. The first truly ductile platinum was produced in 1773 by Romé Delisle, who found that if the platinum sponge, after calcination, was carefully washed and then reheated in a refractory crucible it sintered to form a dull grey mass which could then be consolidated by careful forging at a good red heat. The density of the mass changed, as a result of this treatment, from 10.05 to 20.12g/cm^3. This was the first time that the high theoretical density of platinum had been demonstrated.

This sintering process was described by Macquer in 1778 in his *Dictionnaire de Chimie*, and in doing so he provided what is probably the first interpretation

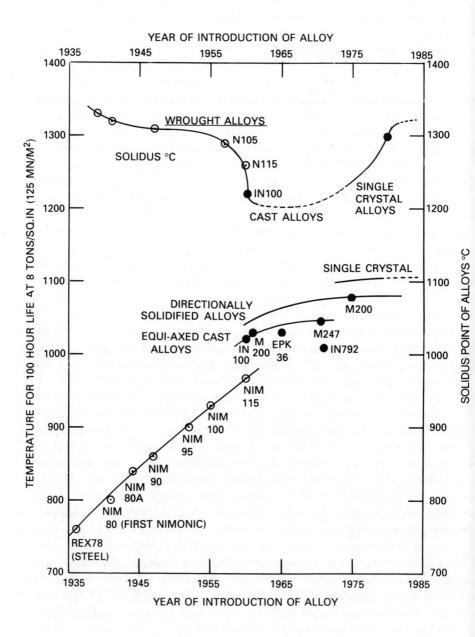

Figure 1.11

of the scientific principles upon which powder metallurgy depends: 'a metallic mass quite compact and dense, but it completely lacks malleability when it has been exposed to a moderate heat, and only assumes it when subsequently subjected to a much greater degree of heat'. He goes on to report that:

> Particles of platina being infinitely divided in the precipitation it is not surprising that the heat penetrates such very small molecules more effectively than the ordinary grains of platina which, in comparison, are of enormous size; and their softening occurring in proportion, they should show the extraordinary effect on their agglutination in the proportion of their points of contact; moreover, these points being infinitely more numerous than can be those of much greater molecules, solid masses result which have all the appearance of quite dense metal, melted and solidified on cooling, but they are nothing but the result of a simple agglutination among an infinite number of infinitely small particles, and not that of a perfect fusion as with other metals.

This powder metallurgy route was subsequently refined and applied by William Hyde Wollaston who in 1800 at the age of thirty-four retired from medical practice and decided to make his fortune by producing and selling ductile platinum on a commercial scale. Wollaston was a friend of Smithson Tennant, who between 1803 and 1804 had shown that the black residue which was left behind after native platinum had been dissolved in aqua regia contained the two hitherto unknown noble metals, osmium and iridium. The presence of the noble metal palladium in the soluble fraction was demonstrated by Wollaston in 1803, and he announced the discovery of another noble metal, rhodium, in the following year. Ruthenium, the sixth and last member of the platinum group was identified by Claus in 1844.

Wollaston's process of platinum production was probably the first to utilize a chemical extraction and refining process so controlled that it produced pure

Figure 1.11: Improvement in the high temperature capabilities of nickel based superalloys improved since their introduction in 1940. The strongest wrought alloy to be developed was Nimonic 115 which was introduced in 1960. Stronger alloys could not be worked, and cast alloys were then introduced. These were subsequently improved, firstly by directional solidification and finally by turning them into single crystals.

The upper curve on the diagram shows how the solidus of the workable alloys began to decrease rapidly as the alloys were progressively strengthened by alloying. By 1960 the gap between the melting point of these alloys and their working temperature capability had decreased to about 100° C. It was found, however that single crystal alloys could be based on very simple compositions, since no alloying additions were required to impart grain boundary strength. As these alloys developed, therefore, their melting points began to increase, thus providing a much needed margin of high temperature safety.

fine metallic powder which was specifically intended for consolidation by the powder metallurgical method. He soon became well known for the high quality and ductility of the platinum he produced in this way. In the early years of the nineteenth century, large quantities of platinum were needed for the construction of items of chemical plant such as sulphuric acid boilers. Between 1803 and 1820, Wollaston produced about 7000 ounces (197kg) of ductile platinum for sulphuric acid boilers alone.

Platinum deposits were discovered in the Ural mountains in Russia in 1819 and rich alluvial platinum deposits were found in 1824. In 1825, deposits at Nizhny Tagil north of Ekaterinburg (Sverdlovsk) were found to yield 100 ounces of gold and platinum per ton (3.6kg per tonne) of gravel, and Russian platinum continued to satisfy the world's demands until 1917, when political changes encouraged the extraction of this metal from the nickel deposits at Sudbury, Ontario (see p. 98). In 1924 platinum was discovered by Dr Hans Merensky in a South African reef near Rustenburg in the Transvaal, now known to be one of the world's richest platinum deposits. Vast quantities of copper, nickel and cobalt, and also gold and silver, are also associated with the rich metalliferous vein.

The very large cupro-nickel ore deposits discovered at Jinchuan in Ganzhou Province in China in 1958 also contain considerable quantities of both gold and platinum. Rich palladium-platinum deposits have also been found in the Mitu area of Yunan Province which is only about 80km (50 miles) away from the nearest railway line and is therefore capable of being rapidly developed.

Powder metallurgy in electrical engineering

The first critical metallurgical requirement of the electrical engineers was a lamp filament which was more robust and could run at higher temperatures than the carbon filament of Edison (see Chapter 6). By the end of the nineteenth century it was appreciated that the light emitted by an incandescent filament varied as the twelfth power of its temperature. The most efficient lamp, therefore, was that which could operate at the highest possible temperature, and attention was concentrated upon the melting points of the most refractory metals then known. These melting points reached a peak at tungsten. Rhenium and technetium were unknown at that time and hafnium and zirconium had never been made in pure metallic form.

Osmium, however, was known to have a high melting point and could readily be produced as an exceedingly pure finely divided powder. The use of osmium as a light filament material was first proposed in 1898 by Welsbach, ironically the inventor of the incandescent gas mantle.

The 'Osmi' lamp, first produced in Berlin, although more economical in operation than the carbon lamp was more expensive and rather delicate. It was soon overtaken by the tantalum lamp which was introduced by Siemens and Halske of Charlottenburg in 1903. Although the melting point of osmium, 3050°C, was marginally higher than that of tantalum at 2996°C, the latter metal had the tremendous practical advantage of being ductile so that it could readily be drawn into fine wire. Ductile tantalum was first produced in 1903 by W. von Bolton. The gaseous contaminants in the nominally pure tantalum sponge were removed by melting buttons of the metal under high vacuum conditions. The prototype vacuum furnace used by von Bolton for this work is shown in Figure 1.12. These arc furnaces had water-cooled metal hearths, and were so successful that scaled-up versions were designed and used by Kroll during the Second World War for melting titanium (see p. 143). The tantalum lamp met with great and virtually instantaneous success. Between 1905 and 1911 over 103 million tantalum lamps were sold.

Tungsten, which melted at 3422°C, was of course the ultimate lamp filament material, although tremendous practical difficulties were initially encountered in producing ductile tungsten wire. The first tungsten filaments, produced in 1904 by a squirting process similar to that used for the osmium filaments, produced a great deal more light than carbon filament lamps, but they were brittle and expensive. The process was developed by Just and Hanamann in 1904 at the Royal School of Technology in Vienna and lamps were produced in Britain from 1908 in the Hammersmith Osram-GEC Lamp Works, although the operation was never a commercial success.

Ductile tungsten filaments were first produced by W. D. Coolidge of the US General Electric Company at Schenectady in 1909. The process he patented in 1909 is still used. Fine tungsten powder was pressed into bars about one square inch in cross-section. These bars were then sintered by heating them electrically in a pure hydrogen atmosphere to temperatures approaching their melting point. They were then reduced into wire by a process of hot fabrication within a well defined range of temperatures so that a fibrous structure was gradually developed within the rod or wire. See Figure 1.13. The Coolidge tungsten wire process was the first to make extensive use of the newly developed rotary swageing machine. Swageing was continued until the bar had been reduced to rod about 0.75mm in diameter. This was then hot drawn to wire through diamond dies. By 1914, over 100 million lamps had been sold in the United States alone, and manufacturing licences had been granted by the US General Electric Company to most of the developed world. The trade name Osram, first used by the Osram Lamp Works in Berlin, derives from the words osmium and wolfram. After 1909 it became associated with those lamps using drawn tungsten filaments made in accordance with GE patents.

After about 1913 ductile tungsten, and then ductile molybdenum, became the dominant materials of the electrical industry. When tungsten contacts were introduced the reliability of motor-car ignition systems improved considerably.

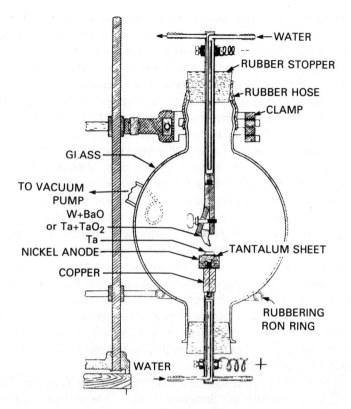

Figure 1.12: The vacuum arc furnace first used in 1902 for the production of ductile tantalum. The pressed powder pellet, weighing 80–100g was melted on a water cooled nickel hearth by a direct current arc of 50–300 amperes. During melting, pressure in the furnace chamber increased from 5×10^{-3} to 5×10^{-2} mm of Hg. This process was devised by Dr W. von Bolton of the Charlottenburg lamp factory of Siemens and Halske. He was assisted in this development by the engineer Otto Archibald Simpson and also by Dr M. Pirani who later invented the thermal conductivity vacuum gauge.

Between 1903 and 1912 more than 60 million tantalum lamps were produced at Charlottenburg. For this about one ton of vacuum melted tantalum was processed. The furnace diagram shown here was first published by Dr Pirani in 1952.

Courtesy *Vacuum*.

The Coolidge X-ray tube, introduced in 1913, had a tungsten target and was the first to permit the long exposures required by the new technique of X-ray crystallography.

Between the wars, most of the serious metallurgical research in Great Britain was undertaken by firms such as Metropolitan Vickers, GEC, Telcon, Standard

Telephones and the major cable companies. Bell Telephones, the US General Electric Company and Westinghouse exercised a similar influence in the United States. The influence of Siemens and Heraeus in Germany, and the Philips Laboratories at Eindhoven was equally profound. The healthiest and most vigorous offspring of this marriage between metallurgy and electrical technology was the sintered carbide industry which must, by any standards, be regarded as one of the major metallurgical innovations of the twentieth century.

SINTERED CARBIDE CUTTING TOOLS

Tungsten wire had to be hot drawn, and the deficiencies of the steel dies which were first used to reduce the diameter of the hot swaged tungsten rod to the dimensions of wire which could be handled by diamond dies soon became apparent. Tungsten carbide seemed to fulfil most of the characteristics of the material required. The extremely hard carbide W_2C had first been prepared by Moissan in 1893 when he fused tungsten with carbon in his electric furnace. Sintered carbides were first produced in 1914 by the German firm of Voigtländer and Lohmann. These compacts, which were rather brittle, were produced by sintering mixtures of WC and W_2C at temperatures close to their melting point.

As Moissan had shown, tungsten carbide could be melted and cast from an electric arc furnace. Cast tungsten carbides dies were produced by this method before 1914 by the Deutsche Gasglühlicht Gesellschaft, although the cast product had a very coarse grain size and the dies were again very brittle. By 1922 Schröter of the Osram Lamp Works in Berlin had shown that tungsten carbide, sintered in the presence of a liquid cement could be very tough. The three metals, iron, cobalt and nickel all provided a satisfactory molten cement, although cobalt provided the best combination of hardness and toughness.

The new sintered alloy was rapidly adopted in Germany for the manufacture of wire drawing dies and between 1923 and 1924 was used and sold by the Osram Group under the trade name Osram Hartmetall for wire drawing and also for cutting tools. Friedrich Krupp AG of Essen were granted a manufacturing licence for this exciting new sintered product in 1925 and introduced the first successful sintered cutting tools to the world in 1927. Widia (*wie Diamant* = like diamond) consisted of tungsten carbide powder sintered together with a cement consisting of about 6 per cent cobalt. It was exhibited in 1927 at the Leipzig Fair, where its ability to machine cast iron at unheard-of speeds was demonstrated. A. C. Wickman of Coventry acquired the sole rights to import Widia and sell it in the United Kingdom. In association with Wickman, Krupps started to manufacture tungsten carbide in Britain in 1931. Krupp, originally the major shareholder in the Tool Manufacturing Company, eventually became the sole owner. When the Second World War started in 1939, A. C. Wickman

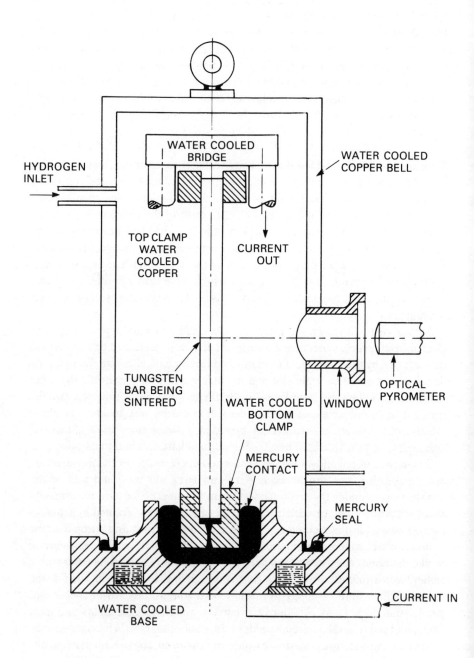

Figure 1.13

took over control of the factory and changed its name to Hard Metal Tools Ltd. Manufacturing licences to produce sintered carbides had by the mid-1930s been granted by Krupp to British firms such as BTH Ltd, Metro Cutanit, Firth Brown Tools and Murex Ltd.

The American General Electric Company acquired the sole American rights to the Widia process in 1928 and issued many manufacturing licences. From this operation emerged nearly all the well-known proprietary grades of carbide. In order to circumvent the German patents, firms such as Fansteel attempted in 1932 to introduce tantalum carbide tools which were sintered with nickel rather than cobalt. Such materials were inferior in toughness to the cobalt bonded composites and were never very successful.

Other powder metallurgical innovations

Other products of the powder metallurgy industry, which began to develop very rapidly after 1910, are too numerous to describe exhaustively. Sintered porous self-lubricated bronze bearings were introduced by the American General Electric Company in 1913, and this was followed by sintered metallic filters, first produced by the Bound Brook Oilless Bearing Company in 1923. Powder metallurgy was also used to produce a variety of magnetic alloys, since it was found that fine powders of metals such as iron, cobalt and nickel could be cheaply produced in a high state of purity by chemical methods. Such powders could be fully or partly consolidated into any desired shape by powder metallurgical methods without the contamination inevitably associated with conventional melting and alloying procedures. Permalloy, a very soft iron-nickel magnetic powder, was developed in 1928 by the Bell Telephone Laboratories. In the 1940s when the domain theory of magnetization suggested that very fine powders could be used to produce permanent magnets of very high coercive force some very elegant powder metallurgical techniques were developed.

> *Figure 1.13*: The ductile tungsten wire, first produced by Coolidge of the General Electric Company of Schenectady in 1909 was obtained from bars of tungsten sintered in apparatus similar to that illustrated. After a preliminary low temperature sintering process, pressed bars of tungsten powder about in square cross-section were gripped between water-cooled copper electrodes. The lower electrode floated on a bath of mercury to allow for the considerable shrinkage which occurred as the bar sintered. The apparatus was then capped by the water cooled copper bell shown, and all air purged from the vessel by a flowing current of hydrogen. Sintering was accomplished by passing a heavy alternating current through the bar so that its temperature was raised to about 3300°C. Bars thus sintered were then hot worked by rotary swaging and finally drawn to wire in diamond dies.

Powder metallurgy also made it possible to produce, merely by mixing the appropriate powders, a whole range of composite materials which could have been manufactured in no other way. In this way a whole generation of new electrical contact materials was developed between the wars. These included, for example, composite contacts which incorporated insoluble mixtures of silver and nickel, silver and graphite, silver-tungsten, copper tungsten and, probably the most important, silver-cadmium oxide. The last formulation is still widely employed because it combines the low contact resistance of silver with the ability of cadmium oxide to quench any arcs formed when the contacts open under load.

The compaction of powdered metals

As in the time of Wollaston, metal powders are still compacted in steel dies since this is generally the cheapest and most convenient method of producing large quantities of components on a routine basis. The technical limitations imposed by die wall friction are generally accepted and various expedients have been devised to obtain more uniformly pressed components and to reduce the incidence of interior cracking caused by concentrated internal stresses in the compact.

Considerable difficulties were encountered, however, in the early years of the powder metallurgy industry when the production of larger and more complex components was attempted. It was then appreciated that such products would most effectively be compacted under a pure isostatic pressure but it was not until 1930 that F. Skaupy devised a method of hydrostatic pressing which was simple, cheap and industrially acceptable.

Figure 1.14 illustrates a typical arrangement which was used for pressing relatively thin-walled tungsten carbide tubes, an operation which would not have been feasible in a conventional steel die. The carbide powders were contained in a sealed rubber bag, which was then subjected to hydrostatic pressure from a fluid pumped into the pressure vessel shown. The shape and configuration of the powder compact was ensured by the use of polished steel tubes which were so arranged that they allowed the working fluid to act on the compact from all directions. Compacts with high pressed densities and low internal stresses were thereby obtained. With arrangements of this type, compaction pressures of the order of 9250 bar (60 tons per square inch) could be safely utilized. Pressures in a steel die were usually limited to about 2300 bar (15 tons per square inch) to avoid the formation of internal cracks in the compact.

In the years immediately after the Second World War it was established by organizations such as the Nobel Division of ICI that large metallic compacts could be effectively compacted by surrounding them with a uniform coating of high explosive which was then detonated within a large bath of water. This technique was among those investigated for the consolidation of large ingots

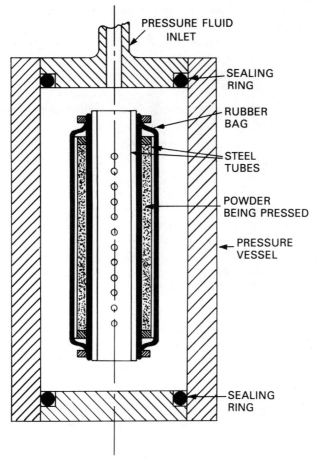

Figure 1.14: The first satisfactory method of consolidating metal powders under a hydrostatic pressure devised by F. Skaupy in 1930.

of titanium from the sponge then being produced by the Metals Division of ICI. The explosive approach, though versatile, is expensive and not well suited for routine manufacture, although its value in the research and development area has been amply demonstrated. Isostatic pressing is now widely employed for the manufacture of smaller components: vast quantities of pressings such as the alumina insulators of sparking plugs are economically produced in this way.

In 1958 the Battelle Memorial Institute of Columbus, Ohio, began to modify and improve the isostatic pressing process so that it could be used to consolidate metal powders at high temperatures. See Figure 1.15. Laboratory gas pressing equipment was developed which allowed metals, alloys and ceramics to be consolidated in metal capsules at temperatures up to 2225°C and pressures

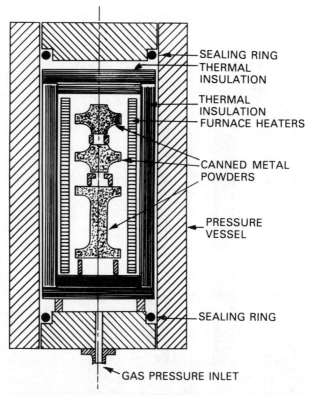

Figure 1.15: The technique of hot isostatic pressing, introduced in 1960 by the Battelle Memorial Institute is shown in the second diagram. In this approach the metal or alloy powders are poured into shaped sheet metal cans which are subsequently evacuated and sealed by welding. The can and its contents are then subjected to the combined effects of heat and an external pressure of an inert gas in a cold-walled autoclave. The hot isostatic pressing 'HIP' process is now widely employed, particularly for consolidation of complex shapes from superalloy powders.

up to 3900 bar (25 tons per square inch). The commercial units available in 1964 operated at similar temperatures, although routine operating pressures were limited to about 1100 bar (7 tons per square inch). Pressing was accomplished in a cold wall autoclave, with the furnace used for heating the compact enclosed within this chamber and thermally insulated from it so that the chamber walls remained cold. The unit was pressurized with an inert gas such as argon. HIP (hot isostatic pressing) is now extensively used to compact either one very large component or a multiplicity of relatively small specimens. Units with an ability to consolidate components up to 22 inches in diameter and 108 inches long (56cm × 274cm) were being used in the USA in 1964 and much larger facilities are now providing routine service throughout the world. 'Hipping', as a metallurgical technique, is no longer confined to powder metallurgy. Many critical and expensive castings such as those used for gas turbine blading are routinely processed in this way to seal up and eliminate any blowholes or other microscopic defects which might possibly be present. Higher operating pressures are now available; the economics of the process are largely dictated by the fatigue life of the pressure vessel.

TITANIUM AND THE NEWER METALS

Titanium was first produced in metallic form in 1896 by Henri Moissan, who obtained it by electric furnace melting. His product, being heavily contaminated by oxygen, nitrogen and carbon, was very brittle. In the early years of the twentieth century ferro-titanium alloys were widely used for deoxidizing and scavenging steel. Metal containing less than 1 per cent of impurities had been made, however, and such material had a specific gravity of only $4.8g/cm^3$. The metal had a silvery-white colour and was hard and brittle when cold, although some material was forgeable at red heat. Widely varying melting points were initially reported, although work by C. W. Waidner and G. K. Burgess at the US Bureau of Standards indicated that the true melting point was probably between 1795° and 1810°C. The extraordinarily high affinity of titanium for nitrogen was noted in 1908 and this led to a number of proposals that the metal could be used for the atmospheric fixation of nitrogen. It was claimed that the titanium nitrides reacted with steam or acids to produce ammonia.

Pure ductile titanium was first produced at the Schenectady Laboratories of the General Electric Company in 1910 by M. A. Hunter, who reduced titanium tetrachloride with sodium in an evacuated steel bomb. This approach was used for a considerable time for producing small quantities of relatively pure titanium for experimental purposes.

Titanium of very high purity which was completely ductile even at room temperature was first made in 1925 by Van Arkell and De Boer who prepared pure titanium iodide from sodium reduced titanium and subsequently decomposed this volatile halide on a heated tungsten filament. It had a density of $4.5g/cm^3$, and an elastic modulus which was comparable to that of steel. For the first time titanium was seen as a promising new airframe material which was free from many of the disadvantages inherent to beryllium.

W. J. Kroll began to work on titanium in his private research laboratory in Luxembourg in the decade before the Second World War, using an approach which was identical in principle with that employed by Wöhler in his work on aluminium in 1827 (see p. 102). He found that the metal obtained by reducing titanium tetrachloride with calcium was softer and more ductile than that obtained from a sodium reduction, and applied for a German patent for this

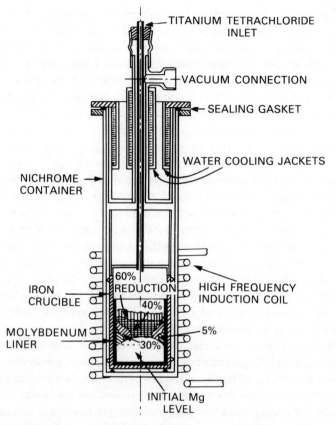

Figure 1.16: This cell was used by W. J. Kroll in 1940 to obtain ductile titanium by reacting titanium tetra-chloride with molten magnesium. This approach, which was developed during the war years by the US Bureau of Mines, was soon generally adopted as the most feasible method of producing titanium on an industrial scale. Much of the titanium now being made is reduced to metallic form by sodium rather than by magnesium.

process in 1939. High purity calcium, however, which was required with a low nitrogen content, was an expensive commodity. Kroll found that titanium tetrachloride could be very effectively reduced by pure magnesium, which was cheap and readily available. Details of Kroll's magnesium reduction process were first published in 1940 in the *Transactions* of the Electrochemical Society of America. By this time Kroll had left Europe to join the United States Bureau of Mines.

A simplified version of the reaction vessel in which magnesium reduced titanium was first obtained is shown in Figure 1.16. Liquid titanium tetrachloride was dropped on to a bath of molten magnesium held between 850° and 950°C in the molybdenum container at the bottom of the cell. Once the reaction started no further heating was required, and the temperature was controlled simply by adjusting the rate at which titanium chloride was fed into the reaction vessel. The product of the reaction was titanium sponge, which built up in the reaction vessel. Apart from the small quantities of iodide titanium produced by the Van Arkell process, Kroll's magnesium reduced titanium was the first which had shown a high degree of room temperature ductility.

Titanium is the fourth most abundant metal in the earth's crust, after aluminium, iron and magnesium. The most valuable deposits are those based on the minerals rutile and ilmenite, first found in the Ilmen Mountains of the USSR. The US Bureau of Mines research programme was initially concerned with the exploitation of the large ilmenite deposits in North Carolina. The Kroll process as first developed was based on powder metallurgy. The sponge, after careful washing and purification was crushed, sieved and consolidated in steel dies. The pressed ingots so obtained were vacuum sintered, after which they were worked, either by hot or cold rolling. It was found, however, that the titanium powder did not consolidate very well. Pressures of the order of 4650–7725 bar (30–50 tons per square inch) were required to produce components having a density high enough for effective sintering, and large components were not obtainable from vacuum sintered bars.

By the early 1940s it was known that the true melting point of titanium, approximately 1670°C, was well below the level of earlier estimates. Even so, titanium could not be melted by conventional methods, since it reacted strongly with all known refractories. This was a problem which von Bolton had encountered and solved with tantalum in 1903. The trick was to melt the metal on a water-cooled crucible with an electric arc. When, in 1940, Kroll adopted this approach, he found that titanium could be readily melted on a water-cooled copper hearth, providing the non-consumable tungsten electrode he used was made the cathode. Such furnaces have now been superseded by large consumable electrode furnaces capable of producing titanium ingots weighing many tonnes.

Titanium has been commercially available, in pure and alloyed form, for over forty years, although a mass market has not yet developed. Titanium alloys are beginning to find a place in the construction of new supersonic aircraft, but as a constructional material, titanium suffers from one insuperable defect: an allotropic change in structure occurs at 882°C and even the best and strongest alloys begin to weaken catastrophically at temperatures above 800°C. Titanium is therefore unlikely to become a high temperature material. However, the corrosion resistance of pure titanium is comparable to that of stainless steel, and this should eventually result in its wider usage in the chemical industry. It is extremely resistant to prolonged exposure in sea water and, being resistant to both cavitation – erosion and corrosion – fatigue, has found many applications in shipbuilding and marine technology. The possible use of titanium for the construction of deep-water submarines has recently attracted much attention. Submarine hulls, like all thin shells, fail by compressive buckling at extreme depths. By constructing the hull from titanium rather than steel, the outer shell can be doubled in thickness without serious increase in weight, thus permitting safe descent to considerably greater depths without danger of collapse.

Thin adherent films of oxide develop on the surface of titanium when exposed to air even at ambient temperatures, and this accounts for the metal's resistance to corrosion which can be greatly improved by anodization. Many proprietary anodizing processes have been developed for titanium and its alloys, and these are widely used for the application of coatings resistant to both corrosion and abrasion. The films developed on titanium are roughly proportional in thickness to the anodization voltage applied. Since they are transparent, interference colours are formed as the film thickness increases, and this allows for the selective and intentional production of very attractive colour effects, which are widely used, particularly on titanium costume jewellery. Anodized pictures can also be painted on titanium sheet by a cathodic brush supplied with an appropriate electrolyte. A potential controlled power source is connected to the brush so that any desired colour can be readily selected.

Because of its lightness, stiffness and abrasion resistance, thin anodized titanium sheet is now the preferred material for low inertia camera shutters. Titanium electrodes are employed in those cathodic protection systems used to inhibit the corrosion of ships and other marine structures immersed in sea and brackish waters. Here the anodes consist of a core of titanium supporting externally a thin layer of platinum. This combination permits the safe discharge of very heavy currents to the sea-water at voltages which are well below those likely to break down the anodized layer on the titanium.

NIOBIUM

Niobium was one of the last new metals to emerge into the industrial arena. It resembles in many ways its sister metal tantalum, although it has several unique characteristics which it was felt, in 1960 when the metal first became commercially available, would allow it to assume a far more important role in the technologies which were then emerging. Niobium, it was then believed, being a perfectly ductile refractory metal with a density only half that of tantalum, would provide a basis for the development of a new group of high temperature alloys, capable of operating effectively at temperatures far in excess of those which nickel-base alloys could resist. Niobium also had unique superconducting characteristics. The hard superconducting alloys such as niobium-tin and niobium-titanium had transition temperatures very much higher than those of the alloys hitherto available and it was felt in the 1960s that such materials would soon be needed in large quantities for constructing the magnets needed for the magneto-hydrodynamic generation of electrical power. The third unique characteristic of niobium was its low neutron cross-section, only about 1.1 barn per atom. In 1960, therefore, it was also considered that niobium would be required in large quantities as a fuel canning material in the new generation of atomic reactors.

Unfortunately, however, no large-scale industrial applications for niobium emerged, in spite of a prodigal expenditure of money, enthusiasm and research ingenuity. Superconducting alloys were never required in significant quantities after the magneto-hydrodynamic approach to power generation was abandoned, and the role for niobium as a canning material for fuel elements disappeared when it was found that stainless steel was perfectly adequate. Interest in the prospects of niobium and its alloys as high temperature materials began to fade rapidly after 1965 when it became very clear that they had no inherent resistance to oxidation, and could not be relied upon to function in the hotter regions of the gas turbine even if protective coatings could be developed.

FURTHER READING

Agricola, Georgius (= Georg Bauer) *De re metallica* (Basle, 1556); translated *On the business of metals* by H. Hoover (Dover Press, New York, 1950)
Beck, A. *Technology of magnesium and its alloys* (F.A. Hughes & Co., London, 1940)
Biringuccio, Vanoccio *De la pirotechnica* (Venice, 1540); translated by C.S. Smith and M.T. Gnudi (New York, 1942)
Ste Claire Deville, H. *De l'aluminium* (Mallet-Bachelier, Paris, 1859)
Day, J. *Bristol brass* (David & Charles, Newton Abbot, 1973)
Erker, Lazarus *Beschreibung Allerfürnemisten Ertzt* (Prague, 1574); translated by A.G. Sisco and C.S. Smith (Chicago, 1951)
Hamilton, H. *The English brass and copper industries to 1800* (1926; reprinted Cassell, London, 1967)
McDonald, D. *A history of platinum* (Johnson Matthey and Co. Ltd, London, 1960)
Raymond, R. *Out of the fiery furnace* (Macmillan, London, 1984)
Rickard, T.P. 'The early use of metals'. *J. inst. metals: XLIII (1)*, 1930, pp. 297–339
Tylecote, R.F. *A history of metallurgy* (The Metals Society, London, 1976)

2

FERROUS METALS

W. K. V. GALE

INTRODUCTION

Iron and steel are an essential feature of the industrial civilization in which we live. They were largely responsible for it, and they remain an indispensable part of it.

The metal iron, which is derived from one or other of several naturally-occurring ores, can be made to take on different characteristics according to the way it is processed. It has the useful property of being able to combine with other elements to produce an alloy, and a small quantity of some elements will have remarkable effects on its properties. Steel, which is itself a form of iron, can exist in even more forms, all having different chemical or physical properties, or both, and some properties can be varied considerably without changing the chemistry. Thus carbon steel (an alloy of iron with a small amount of carbon) can be soft enough to be cut by a file, hard and brittle, or hard and tough. Which of these states is obtained depends on how it is processed. A simple tool like a metal-cutting chisel, for example, must be hard at the cutting end, but not so hard that it breaks. At the other end, where it is struck by the hammer, it must be soft, so that there is no risk of pieces breaking off and perhaps injuring the user. By heat treatment (very careful heating to a chosen temperature and then cooling) the chisel is made hard at one end and gradually getting softer towards the other.

Steels, like other metals, have some strangely human characteristics, too. They can be toughened up and have their strength increased by hard work (that is by subjecting them to external forces such as squeezing them between rolls, hammering them, or stretching them by machinery). But if they get too much work they suffer from fatigue. In the end they will break, unless they are given a rest and subjected to processes which remove the fatigue. Other steels are extremely strong and will put up with a tremendous amount of hard work. Three fundamental types of iron are used in commerce: wrought iron (the oldest, historically, but now virtually extinct, although some decorative metalwork is incorrectly described as wrought iron); cast iron (the next in age and still in use); and steel (the youngest historically).

It was man's ability to make and use tools that first distinguished him from other animals, and iron was crucial in this respect. Other metals were used before iron, the most important being bronze (see pp. 57–67) but when iron came on the scene it gradually took over, since it is better, stronger and more abundant. It was a very good material for weapons as well. Given weapons for the hunt man was assured of food, and the same weapons gave him some protection against his natural enemies. With tools he could more readily cultivate crops and prepare his food, clothes and shelter. So he gained a security which could never have been his had he relied on his hands alone. With this security the human race was able to settle and develop; as it did it found a greater need for tools, and it discovered, too, a multiplicity of new uses for iron.

As the art of ironworking progressed it became possible to harness natural forces more effectively. A windmill or waterwheel could be made of stone or brick and timber (although metal tools were needed to build it), but when ways were found to use the power of steam only metal was strong enough for the machinery involved. And if iron made steam power practicable – and with it the industrial revolution – steam made possible the production of iron on an industrial scale and turned a domestic craft into an important industry. The demand for iron increased as well, for the availability of mechanical power brought a boom in the demand for machinery. Iron, steam power and machinery all helped each other; more of one meant more of the other two.

WROUGHT IRON: THE PREHISTORIC ERA TO AD 1500

Iron has been made for at least 4000 years. The discovery may well have been accidental and have been made in several different places over a long period.

Throughout history iron has been produced from naturally-occurring iron ores. Very small amounts of iron – more accurately, a natural alloy of iron and nickel – have been found as meteorites, and they were hammered out into useful shapes, but the quantities were so small that meteoric iron has never been more than a curiosity. Specimens can be seen in some museums. Most iron ores – there are several varieties – are a dusty reddish-brown rock, though some are darker in hue, almost black or purple. Iron is the fourth most abundant element in the world, and reddish-coloured earth gives a clue to its presence. The red soil of Devon, for example, shows that iron is present, though in fact the few ores there are not rich enough to be worth working.

Iron ores are all, basically, a chemical mixture of iron and oxygen (iron oxide), with small quantities of other elements, and as found in the earth they also have varying amounts of contaminants such as clay, stone, lime and sand mixed with them. Some of the impurities are easily removed; others are more difficult, and many of the important inventions in the history of iron and steel have been connected with the removal of impurities.

Iron ore is an oxide because iron has a strong affinity for oxygen and there is always a supply of oxygen available in the air. If metallic iron is left exposed to the air it will slowly become an oxide again: it will rust. Fortunately for the ironmaker, carbon has an even greater affinity for oxygen than iron. If iron ore is heated strongly in contact with carbon, the oxygen and carbon will unite to form a gas, which burns away, leaving the iron behind. That is the basis of iron ore conversion into iron – reduction or smelting.

One of man's earliest technical achievements was to make fire, and it could be that when a fire was started – for protection, warmth and cooking – somebody noticed a change in the nature of the stones used to surround and contain the fire. If two of the stones were banged together, they gave off a dull sound and did not crack or splinter: the charcoal (which is a very good and pure form of carbon) of the wood fire, urged perhaps by a strong wind, had reduced to iron some of the stones, which were actually iron ore. It would not be long before somebody had the curiosity to try other likely-looking stones round the fire, then it would only be a matter of time before somebody tried hammering one of the changed stones while it was red hot. He would find that he could beat it out into useful shapes which, when cold, were strong and did not break or bend easily. By hammering the material into, say, a knife or a spearhead, and rubbing the point on a rough stone to sharpen it, our early experimenter could make a much better tool or weapon than he had ever had before. Such speculation is justified in the absence of known facts. At all events, ironmaking had spread to Europe by about 1000 BC from the Middle East, where it apparently began much earlier.

At first, and for many centuries, the equipment used was very simple and the production of iron extremely small. A group of men working for several hours could only make a piece of iron perhaps not much bigger than a man's fist, and weighing no more than one or two kilograms. But the trade of ironmaking had started, and villages began to get their ironmakers – just as they had their millers, potters and weavers – wherever iron ore could be found. In those parts of the world where there was no iron ore, traders began to take iron goods to exchange for other products and international trade in iron began to spread. Iron was still scarce, however, and used only for such things as tools and weapons.

The product made by the early workers in iron was wrought iron. Pure iron as such is only a laboratory curiosity and has no commercial use, but wrought iron is quite close to it. It has a fibrous structure: if a piece of wrought iron is nicked with a chisel on one side and then hammered back on itself it will tear and open out to show a structure that looks very much like that of a piece of wood. Wrought iron can be shaped by hammering it while it is hot (or in later years by passing it between rotating rolls) and if two pieces at the right temperature are hammered together they weld into one piece. It is possible to melt wrought iron but of no practical value, so it was never melted in practice: the iron was converted, or reduced, directly from the ore in what is therefore termed the direct reduction process.

The early equipment used to make wrought iron was as simple as the metal itself, consisting of a small furnace, heated by charcoal and called a bloomery, hand- or foot-operated bellows to blow the charcoal fire, and some tongs to hold the hot metal while it was forged into shape. Bloomeries varied in shape and size, though they all functioned in the same way. They were made of clay, which would resist the heat of the fire. Charcoal was lighted inside the bloomery and then, while a continuous blast of air was kept up by the hand or foot bellows (the operators taking turns), more charcoal and some iron ore were fed in by hand through a small aperture in the top. As the oxygen in the ore united with the carbon of the charcoal it became a gas, which burned off at the top of the bloomery as a light blue flame. After a few hours all the oxygen had gone from the iron ore, and a small, spongy ball of iron, the bloom from which the bloomery took its name, remained. Then the front of the bloomery was broken open and the bloom was raked out and taken to an anvil for hammering to whatever shape was required. In common with workers in other trades, ironworkers relied on their practical skills, not on theoretical knowledge. Apprentices learned from their masters, or their own experience, how to judge when the bloom was ready inside the enclosed furnace, or how to choose the best ores from their appearance. Such craftsmanship was the basis of their operations until comparatively recent years, when scientific methods took over.

The bloomery could never have been operated on a large scale, even if mechanical power had been available. Some modifications were made to the process in some parts of the world and sometimes a waterwheel was used instead of manpower to work the bellows. Individual bloomery outputs grew a little, too, but no essential change in technology occurred in the three thousand years up to the fifteenth century AD.

CAST IRON: 1500–1700

The blast furnace, introduced near Liège in what is now Belgium some time towards the end of the fifteenth century, reached Britain by about 1500 and spread slowly throughout Europe. Eventually it came to be used all over the world, as it still is.

Externally the blast furnace looked like a short square chimney, standing straight on the ground. It was built of brick or stone – whichever happened to be available on the particular site – and internally it had a lining of bricks or stones chosen for their ability to resist fire. The furnace, at 3–5m (10–16ft) tall, was much bigger than anything used previously for ironmaking, though still tiny by today's standards.

The blast furnace brought several changes, technical, economic and social. Technically it introduced a new product, cast iron, an alloy of iron and carbon which, unlike wrought iron, is quite easily melted. When molten it will flow into a cavity where it solidifies to produce a faithful copy of the mould. It can, in short, be cast – hence the name – and moulds can be made in sand or certain other materials to produce simple or complicated castings as required. Cast iron is very different from wrought iron. It is strong in compression – that is, it will support heavy loads resting on it – but it is comparatively weak in tension – it will not carry heavy loads suspended from it. In addition it is relatively brittle and cannot be forged or shaped by hammering, so its uses were limited in comparison with wrought iron. But cast iron could be made in much larger quantities in the blast furnace, and it can be converted into wrought iron by a second process, so the needs of the market were met.

A blast furnace, like a bloomery, needs a continuous blast of air to keep the fire burning, but its greater size demanded mechanical power to work the bellows. This meant, at the time, a waterwheel, and blast furnaces were built alongside streams where water was available. Nature does not always put streams and rivers close to deposits of iron ore – and blast furnaces needed both, plus forests to provide timber for making charcoal. One district in Britain which had all these requirements was that part of Surrey and Sussex known as the Weald, and it was also close to an important market, London. The iron trade became important there, and many signs of its former importance survive in place names like Furnace Mill or Farm, Forge Wood, Minepit Field and Hammer Brook. Several of the old waterwheel ponds remain, some of them now used for commercial watercress growing.

Economically the introduction of the blast furnace meant that ironmaking took the first real steps towards becoming an industry as distinct from a craft. It also brought about a change in the organization of the trade. A bloomery worker needed little more than his skilled knowledge: everything else he required to work his furnace he made himself. Stonemasons and bricklayers were needed to build and maintain a blast furnace; millwrights were necessary to make the waterwheels and keep them in repair, and numbers of other specialized workers were also required. All this called for a new type of organization, and investment. Some landowners were able to finance the building of blast furnaces themselves; otherwise groups of men formed partnerships, sharing the funding and the profits. Partnerships in business were not new, but they were novel in the iron trade.

The blast furnace also brought about social change. It had to work continuously, twenty-four hours a day, seven days a week, and the workers had to be organized accordingly. Two teams of men, each working twelve hours, were needed to operate the furnace, so shift working, now general in many industries became common. These men would have to adjust their home lives to a programme which meant that sometimes they worked all night and at other times all day.

The blast furnace was in some respects like a bloomery (though much bigger) and it still used charcoal as its fuel. The major difference was that, because the furnace operated at a higher temperature and the ratio of charcoal to ore was greater, the iron absorbed a greater amount of carbon; therefore it produced, instead of a spongy piece of wrought iron ready for forging, molten cast iron. This was allowed to accumulate in the bottom (or hearth) of the furnace and taken out, or tapped every twelve hours or so. The molten iron was allowed to run into channels moulded in a bed of sand, where it solidified. To produce pieces a man could lift, a main channel was made, with others branching off it at right angles and from these a number of short, dead-ended channels branched off again, looking from above not unlike a number of combs. The side channels also looked, some people thought, like a litter of pigs lying alongside a sow: pig iron is now made by machine, but the individual pieces are still called pigs.

As the charcoal and iron ore were used up in the furnace, more were tipped in at the top. The earthy materials and other rubbish mixed with the iron ore also melted and, being lighter than the molten iron, floated on top of it; they were also run off at intervals. Some limestone was also charged into the furnace, along with the iron ore and charcoal, to act as a flux, that is, to combine with the waste materials and help to form a molten waste called slag. At first, and for very many years, slag had no real use – except perhaps to fill up holes in the ground – so it was tipped in heaps and forgotten. Old furnace sites could often be traced by slag heaps or the remains of them, but this is becoming more difficult as the heaps are bulldozed away to tidy up the area, or for use as hard core in road or motorway construction. Slag was also left by bloomeries and some slag heaps are known to result from Roman or even earlier ironworking. The presence of the right kind of slag will always indicate that there has been ironworking of some kind nearby, but interpretation of slag heaps calls for expertise. Other metals besides iron produced slag, and some so-called 'slag' heaps are not of slag at all – colliery waste heaps are an example.

The blast furnace spread gradually; there was no dramatic change, and a number of bloomeries still remained in use; some survived into living memory in remote areas of Africa and Asia.

A few uses were found for cast iron as it came from the blast furnace – cast iron cannons were being made in Sussex by 1543, and decorative firebacks for domestic fireplaces are among the oldest existing forms of iron castings – but most of the iron made in blast furnaces went through a second process for conversion into wrought iron. This apparent duplication of effort was in the interests of increased production: even in the early days of the blast furnace it would make as much iron in a day as a bloomery could make in weeks. Conversion of the cast iron from the blast furnace into wrought iron was done in a furnace called a finery, which also used charcoal for fuel and was blown by waterwheel-driven bellows. Here the solid pieces of cast iron were remelted in the fire and the carbon, which had combined with the iron in the blast furnace, was driven off, leaving wrought iron. Because this iron was made in two stages instead of one as in the bloomery, the name indirect reduction is given to the process.

With the blast furnace and the finery, both worked by water power, it was not only possible to make more iron, it was possible to make bigger pieces and the waterwheel-driven hammer came into use for hammering the wrought iron pieces (still called blooms). Records exist of power hammers before the blast furnace and finery, but before 1500 they were rare.

The making of iron in bigger pieces, although it was more economical, brought its problems, for many iron users still needed long thin bars, blacksmiths for horseshoes, for example, and nailmakers. The power hammer could not forge a bar smaller than about 20mm square, simply because when hot the iron became too long and flexible to be handled. Furthermore, the long thin lengths cooled down too quickly, and no furnace available at that time could reheat them.

A very effective answer to the problem was provided by the slitting mill. This machine, driven by water power, cut up long thin strips of iron into a number of small rods as the strip passed between rotating discs or cutters, and it could be adjusted to slit various sizes. To prepare the long thin strip for slitting the machine incorporated another device, a pair of smooth rolls which were, in the long run, even more important than the slitting mill itself. They were the forerunner of all the rolling mills which are vital in steel processing today. A piece of iron hammered out as long and thin as possible was passed between the rolls while it was still red hot and they squeezed and elongated it to make the required strip.

Some slitting mills were in use in Britain by 1588, having been introduced from Liège, where the blast furnace originated, and the rolling mill came with them. The idea of rolling metals was not new; Leonardo da Vinci drew a sketch of a mill in about 1486 and simple mills had been used on soft metals such as gold long before the slitting mill was devised.

Ironmaking was now properly on the way to being an industry and ready to take part in the general industrialization of Britain and some of the continental countries. By 1600 there were about 85 blast furnaces in various parts of Britain, notably in Surrey and Sussex. For every blast furnace there would be three or four wrought iron works or forges and quite a few bloomeries still remained in use. But expansion of output demanded increased supplies of raw materials, and charcoal was becoming scarce. Although there were large forests in various parts of Britain, there were many uses for the timber besides charcoal making, notably for house building, shipbuilding, and as household fuel. When it seemed that timber supplies for building naval ships might be affected, laws were passed in Britain to control the use of timber but they were not very effective. In spite of increased tree planting, forests were being felled much faster than new ones were growing. By the early part of the seventeenth century the charcoal shortage was serious. Blast furnaces often had to be stopped for a time, or blown out, and wait while more trees were felled and a stock of charcoal was built up. Just at the time when the ironmakers had the technical processes suitable for expansion it was hard to see how they could carry on at all, let alone expand.

MINERAL FUELS

It was known that coal would burn and make a fierce fire if a blast of air were blown at it. Coal had been used long before the seventeenth century for a few purposes, and there was plenty of it. However, for various reasons, raw coal, just as it was mined, could not be used at that time in the blast furnace. The chief difficulty lay in the fact that coal, as found in the earth, contains impurities. It has been pointed out that iron unites easily with some elements. Sulphur is one of them and there is sulphur in coal. If sulphur gets into iron – even a very small amount – it makes the metal brittle (or hot short). Worse, sulphur makes iron brittle when it is hot, so if an ironworker hammered it, it would crumble and he would find it impossible to shape the metal at all.

Several people tried using coal for making iron and a few actually took out patents, but none was successful. Dud Dudley, a Midlander, not only took out a patent but also wrote a book about using coal in the blast furnace, and because he claimed to have been successful his name has gone on record as the first man to use coal for making iron. However, if he made iron at all (which he could have done) it could not have been useable.

The year 1709 marks the second great step forward in the history of iron after the introduction of the blast furnace in about 1500. It is not an exaggeration to say that the industrial revolution really became possible after that date, for in that year Abraham Darby succeeded in making iron in the blast furnace with mineral fuel – not, that is, with raw coal, but with coke. Darby did not invent coke – it was known and used before his time for a few purposes such as making malt for brewing – but he did invent the idea of using it in a blast furnace. Coke was made at the time by burning coal in large heaps until all the unwanted impurities had gone off in smoke, and then cooling it quickly with large quantities of water. It was a similar method to that used for making charcoal from wood: in both cases what was left behind was really carbon, an essential constituent of ironmaking. By a stroke of good luck the coal Darby used had a low sulphur content, most of which was burned off during coke making.

Darby was a maker of cast iron cooking pots who had learnt his trade in Birmingham and Bristol. In 1708 he took over a small charcoal blast furnace at Coalbrookdale in Shropshire and in 1709 he made his successful experiments there. Today the site is a museum of ironmaking (part of the Ironbridge Gorge Museum). Darby's original furnace has not survived, but a later one, rebuilt in 1777, is more or less complete and it is possible to see the sort of equipment Darby used. There are also many examples of objects made from cast iron in the Darby works which, after nearly 280 years, is still in operation as a foundry. It no longer has any blast furnaces.

It took some time for Darby's coke-smelting process to spread to other parts of the country, and there were very good reasons why development was slow. Coalbrookdale was a very remote place then and news of developments there leaked out slowly. Secondly, Darby was only interested in the process for his own use: he did not patent it, but neither did he publicize it. However, by 1788 there were 53 coke-fired blast furnaces in England and Wales and only 24 using charcoal. By early in the nineteenth century the last of the charcoal furnaces had stopped.

STEAM POWER AND EARLY INDUSTRIALIZATION

Coke smelting made it possible for the blast furnace to develop and the iron industry took advantage of its new freedom. It was no longer necessary to choose sites that were near to woodlands and sources of iron ore: blast furnaces could be built near to the new fuel, coal, and where coal was to be found, iron ore was usually available as well. The iron trade began to expand in different parts of the country. But the blast furnace needed power. Windmills, used for centuries for grinding corn and a few other industrial processes (see Chapter 4), were of no use for ironmaking. The wind varies in strength and sometimes it does not blow at all, and the blast furnace needs a constant and continuous amount of power. Streams and rivers suitable for driving waterwheels were scarce and often in the wrong place. Moreover, water has an unfortunate habit of drying up in the summer and freezing in the winter. Some protection against the failure of water supplies in a drought could be provided by building dams to store water and this was usually done. Against frost there was no defence at all. Forced stops when the water supply failed for any reason meant, naturally, that the furnace produced no iron. It takes several days to raise the temperature of a furnace to the working level, and also to blow it out, so there were periods beyond the actual stop when the furnace was unproductive.

It was not practicable to locate a furnace in a place where there was a good stream but no coal or iron ore and then transport the raw materials to it. There were no railways and the roads were very bad; some of them were no more than tracks. The cost of transporting tonnes of ore and coal over even a few kilometres would have been much too high.

A further difficulty was that, although the blast furnace could now use coke fuel, charcoal was still needed at the finery to convert the blast furnace cast or pig iron into wrought iron. There was no point in making great quantities of cast iron if it could not be converted into the wrought product, which was still the one in greatest demand.

The iron trade needed a new source of power and a new way of making wrought iron from pig iron. Both came at about the same time. They were James Watt's improved steam engine and Henry Cort's wrought ironmaking process, which became known as puddling.

Both were of great importance but of the two the steam engine was the greater. It had a tremendous effect on the iron industry, and it was also responsible for many changes in the life and standard of living of Britain and indeed of the whole world. It laid the foundations of the industrial towns – not just the ironmaking centres but the others as well. For steam power could operate any kind of machinery and there was a great surge of inventions which could never have succeeded without mechanical power.

The Darby works at Coalbrookdale supplied a number of cast-iron cylinders for the engine devised by Thomas Newcomen in 1712 (see p. 275). Cast iron was ideal for the purpose. It would withstand the heat of steam and it could be cast into the cylindrical form required. The Newcomen engine was used at a number of coal mines and it also found a limited use at a few ironworks, where it pumped water from the waterwheel back into the reservoir, from which it could be reused.

Watt went into partnership with a Birmingham manufacturer, Matthew Boulton, in 1775, to market his improved steam engine and the first one was supplied for draining a Midland coal mine. The second went to a famous ironmaster, John Wilkinson, to blow one of his blast furnaces at Broseley, in Shropshire. Wilkinson, in fact, built the engine, by arrangement, to Watt's design.

The association between Wilkinson, Boulton and Watt was not only of great importance to the iron trade; it was an interesting example of three remarkable men, each of whom could contribute something vital to success. Watt was purely an inventor; Boulton was a businessman; Wilkinson was both. He was a fanatic about iron. He used it for everything he possibly could, even making himself a cast iron coffin and joking about it to his friends. (He was not, in fact, buried in it, for when the time came he was too fat.) Naturally, people scoffed at him – he was called 'Iron-mad Wilkinson' – but apart from his personal publicity stunts he did some valuable work. He invented a machine for boring cast iron cannons, but which would also bore engine cylinders to a much greater degree of accuracy than the earlier ones. For a time, Wilkinson was really the only person who could cast and bore a cylinder suitable for the Watt engine.

He went on to pioneer other iron developments including, in 1787, a wrought iron boat. Wilkinson was also associated with the building of the famous cast-iron bridge which still stands at Ironbridge in Shropshire, though he was not the actual builder of it. The credit for this, the first iron bridge in the world, goes to Abraham Darby III, the grandson of the Abraham Darby who invented coke smelting. The bridge was cast at the Coalbrookdale works in 1779.

When Watt adapted his engine to produce rotative power in 1781, and improved it to become a better mechanical job in 1784, the forges and rolling mills could use it as well as the blast furnaces, and the problems of water power were over. Many other industries also took advantage of this new type of engine – textiles particularly – and all this development was good for the iron trade. The ironmakers not only had the means for producing more iron; they now had an increasing number of customers for it.

Because the trade was no longer dependent on charcoal and water power, but did need iron ore and coal, it began, naturally, to move to sites where these two minerals could be found. In Britain these included South Wales and parts of Staffordshire and the West Midlands. The latter was to become, in the first half of the nineteenth century, the biggest ironmaking area of the world and provides an example of how the iron trade changed an area completely, once it had the technical facilities to use what natural resources the area offered. It was known long before the eighteenth century that there was plenty of coal and iron ore in the area, as well as limestone (for flux) and fireclay (for building furnaces), and they were all either at or near to the surface of the ground. The one thing it lacked was water power, but when the steam engine was perfected the district had advantages no other area could equal for a time. Blast furnaces and ironworks were built in large numbers and other industries using iron and coal came into the area for the sake of these raw materials. From the vast numbers of smoking chimneys the area soon acquired the name of the Black Country, which it still has, though its smoke has almost gone and its trade has altered. Much the same happened in parts of Wales, where the steep valleys, which had never before housed more than a few farms and villages, found themselves, almost overnight, industrial areas.

While new ironmaking communities sprang up, changing the face of the countryside, the older areas such as the Weald were in decline. The last blast furnace there, at Ashburnham, Sussex, using charcoal to the end, was blown out in about 1810. A forge making wrought iron carried on there for a few more years and then the Weald iron industry came to an end. Here and there a few charcoal ironworks carried on but they were not important; the charcoal iron industry was effectively dead by about 1800.

The change meant prosperity for many people, and great unheavals for some. Migration of population into the industrial areas was to increase greatly during the nineteenth century, not because of iron and coal alone, but these materials certainly played their part in this great social transformation.

Concurrently with these developments there was one more, which did not arouse much interest at first but came to be of vital importance. This was the invention, in 1784 by Henry Cort, of a process for making wrought iron from cast iron with mineral fuel, which was to free the ironmakers from their last dependence on charcoal. Others had tried to use coal for converting cast into wrought iron, but Cort was the first to succeed.

It has been pointed out (see p. 153) that if sulphur in the coal combined with the iron, it spoilt it. Cort solved the problem by burning the coal in a separate part of his furnace, allowing only the flames to come into contact with the iron. He used a kind of furnace (known before his time) in which the flames were reflected or reverberated down on to the iron to be treated by a specially shaped, sloping roof. From this fact the furnace was known as a reverberatory furnace.

A quantity of cast iron in the form of pigs, totalling about 250kg (550lb) was put into the furnace, the coal fire being already burning brightly and, in about two hours, the charge was decarburised and converted to wrought iron. This, by the nature of the process, was not molten, but in the form of a spongy red-hot mass. It was then puddled, that is, worked up by the puddler, a man using a long iron bar, into four or five rough balls and the balls were pulled out of the furnace with a large pair of tongs. Each ball was taken to a heavy power hammer and shaped into a rectangular lump called – keeping up the old name – a bloom. This was wrought iron and it could be rolled or forged into whatever shape was needed.

Cort's puddling process was wasteful and inefficient (see p. 162) but it was very much better than any of the earlier processes, and it used coal instead of charcoal; it was taken up by many ironmakers and eventually the old finery died out.

No output figures are available for the period around 1700, but by 1788, when the coke-fired blast furnace was beginning to spread, British production was about 68,000 tonnes of iron a year. By 1806 the annual production had risen to 258,000 tonnes. These figures may be unimpressive by today's standards (British steel production in 1988 was 19 million tonnes and world output is more than 700 million tonnes a year) but they were remarkable at the time.

A typical blast furnace of the period 1790–1820 was similar in shape to what it had been more than 100 years earlier, but it was much bigger, being perhaps 10m (33ft) or so high. It had to be fed at regular intervals with iron ore and coke, which were wheeled to the top in barrows and tipped in. Wherever possible the furnace was built close to a hillside, so that the raw materials could be collected on the hill, loaded into the barrows and wheeled over a short wooden, stone or brick bridge to the furnace top. The furnace at Coalbrookdale, now part of a museum, shows this arrangement very well, though the actual bridge has not survived. So do the furnaces at Blaenavon, in Gwent, and Neath Abbey in West Glamorgan.

When a furnace was built on flat land, a ramp had to be provided for the men to wheel the barrows to the furnace top. Before the nineteenth century was very old, steam power was used to raise the barrows to the furnace top, either by pulling them on a wheeled carriage up the ramp, or by a vertical lift. By the end of the eighteenth century an average furnace made about 1000 tonnes of iron a year, and it would need 3000–4000 tonnes of iron ore (depending on how pure the ore was). It would also consume at least 2000 tonnes of coke – possibly much more. Thus some 7000 tonnes of raw and finished materials had to be moved in the course of a year, all by hand, with shovels and wheelbarrows.

A little-known fact is that of all the materials needed to operate a blast furnace the air for the blast is the largest in quantity. For every tonne of iron made, the furnace would use about 5 or 6 tonnes of air: this is where the steam engine was essential. Steam power also made possible what is now called the integrated ironworks; that is, a works which produced everything from the ore to the finished product, all on the same site. In the waterwheel-powered ironworks it was often necessary to build the blast furnace and the wrought-iron works some distance apart and cart the pig iron to the finishing works. This was unavoidable when the water power was limited, as there would be insufficient power at any one point to work a blast furnace bellows, a finery, a hammer and possibly a rolling mill. The stream would be used for a blast furnace and the water allowed to run away to be caught and used again a few kilometres downstream for the forge. Steam power eliminated such waste of effort, time and money. An ironmaker could set up his works close to the iron ore and coal pits and build as many steam engines as he needed to carry out all the processes on the same site. Not every ironworks was integrated, but the arrangement became quite common.

The famous Coalbrookdale works developed in this way, and so did many in the Black Country. In some places, such as Ebbw Vale, Gwent, the effects can still be seen today. When the first small ironworks was started in the valley above the present town of Ebbw Vale in 1779, the population of the area was only about 140. As the industries developed, so a complete town, with houses, shops, churches and chapels and, later, schools, grew up to house several thousand people, where formerly there had been only a few farms. By 1900 the population was over 20,000. Ebbw Vale owes its entire existence to coal and iron and it is by no means unique: there are several other places where the same process of development can easily be traced.

By the beginning of the nineteenth century the iron trade was prepared to meet the increased demands that the fast-developing industrial revolution was to make. The next 50 years or so saw iron as the supreme metal for making countless types of machinery and products of use to man.

STEEL

If the right amount of carbon is added to iron, the metal is capable of being hardened to any required degree. This fact was known some thousands of years ago. It was discovered that if a piece of iron were heated in contact with charcoal, the surface became quite hard, a process now called casehardening. The trouble was that only a very thin skin – as little as a few hundredths of a millimetre – was hard. When this wore off, as it would in time, the softer part of the iron was exposed.

By the early years of the eighteenth century it was well known that the longer iron was heated while packed in carbon, the deeper the carbon penetrated into the iron, giving a thicker hard skin. The process was known as cementation. Some carbon steel was made in this way but it took a long time to get the composition right. Iron could be in the furnace for as much as three weeks to produce what was called shear steel (because it was widely used for making shears for the textile industry). It was not of very good quality, but it was the best that could be done at the time.

Some further processing would improve the quality and this was practised to make the so-called double shear steel. Because it was so slow and costly to make, shear steel was only used when it had to be, and even then it was employed very economically. Such things as scythe blades and other tools were made by heating a thin strip of carbon steel and a thicker one of wrought iron together and welding them into one piece quickly under a power hammer. In this way the expensive steel was only used at the cutting edge, where it was essential. The process was satisfactory for some things but it was of little use for a type of tool which was going to be in increasing demand as the industrial revolution developed, the engineer's cutting tool.

Cast and wrought iron were excellent materials for making the many types of machinery needed, but they could not be cast, forged or rolled to the precise shape and dimensions required. Parts of iron components, at least, had to be machined to shape. A steam engine cylinder, for example, could be as it was cast on the outside, but the inside had to be machined to fit the piston, and the piston itself had to be machined also.

The man who found the first way of making better carbon steel was neither an engineer nor an ironmaker. Benjamin Huntsman, a Doncaster clockmaker, was dissatisfied with the steel available for making clock springs. In 1740 he took some shear steel and melted it in a clay pot or crucible. He poured out the molten metal, let it solidify and tried working it: it was better than anything he had ever seen before.

Although the process was not understood at the time, the carbon had spread itself throughout the molten metal, and the resulting steel was much more uniform. Huntsman tried in vain to keep his discovery secret, but it was taken up widely in Sheffield, where Huntsman started a small works to make what soon became called crucible steel.

In time it was found that the quality of the steel could be not only controlled but varied according to need. Steel for making, say, a wood chisel, or a chisel for cutting metal, or a razor, needed different grades. Crucible steel could be made to suit the application, and the steelmakers of Sheffield in particular became very expert at supplying exactly what their customers needed.

Crucible steelmaking lasted all through the nineteenth century and into living memory, but it is now extinct. A crucible steel-melting shop has been preserved at Abbeydale industrial museum, Sheffield (where some very good examples of waterwheel-driven hammers can also be seen, in proper working order).

THE INDUSTRIAL IRON AGE: 1800–1850

The many inventions and developments in the period we have been considering – especially during the eighteenth century – had not only changed the whole structure of the iron industry: they had brought into being a new breed of men. The small teams of men needed to work the bloomeries were largely interchangeable and could do most, if not all, of the jobs. As the blast furnace processes became more mechanized, so specialist workmen became essential. Of course there was still plenty of work for an unskilled labourer: the man who loaded barrows with iron ore and coke and wheeled them to the furnace could just as easily wheel barrowloads of pig iron away from it. But a large blast furnace plant of the early nineteenth century needed steam engine drivers; blacksmiths, to make and repair tools; mechanics and engineers to keep the machinery in order; and many more trades as well. In an integrated works the number of specialized craftsmen were even greater. All these were needed in addition to the men who worked at charging and tapping the blast furnaces, at the puddling furnaces and at the rolling mills on the actual manufacture and shaping of iron. As the ironmaking centres developed into complex organizations of craftsmen in many trades, specialization among the workers arrived.

Expansion in ironmaking, and in all the other industries in which iron played its part, soon began to have outside effects, among which the need for better communications was particularly strong, Improvement of roads was the first development, from the middle of the eighteenth century onwards (see Chapter 8) Ironmakers took no significant part in this, but they did to some extent make use of the better facilities. Areas such as the Black Country and South Wales were a long way from the established commercial centres like London and Bristol, so any improved way of transporting the iron was welcomed. Water transport was used as much as possible, but it was still necessary to get the iron to a seaport or to a navigable river. Most of the Shropshire ironworks were fairly close to the River Severn, and the ironworks of South Wales were relatively near the sea, but the rapidly-developing Black Country was a long way from both rivers and sea.

When an ironworks was a short distance from water transport it was often connected by means of a tramway with wooden rails and wagons hauled by horses. The iron industry was not the inventor of tramways, nor the only user of them, but it was responsible for the iron rail. Richard Reynolds, a Shropshire ironmaster, experimented with cast-iron rails in the Coalbrookdale area in 1767 and by 1785 he claimed to have over 32km (20 miles) of iron tramways. In this development we can see the beginnings of the railway age: see Chapter 11.

The other great transport system which grew up in the second half of the eighteenth century was the canal (see Chapter 9). The first Midland canal, the Staffordshire and Worcestershire, built under an Act of Parliament of 1766, passed outside the coal and iron areas of the Black Country but it was not long before a connection was made, and canals were constructed in the Black Country itself from 1769 onwards. This form of transport was vital to the rapidly expanding iron industry.

Ironfounders were acquiring new skills. Developing industries wanted bigger castings and more complex shapes, steam engine cylinders and valve gears, for example. Many other kinds of machinery called for castings which had to be accurate as well as of complicated shape. Mining, too, called for quantities of iron castings, especially in the form of pipes for pumping water to the surface. So did a new industry, gas manufacturing. William Murdock had shown that an inflammable gas could be made by heating coal in a closed retort. An ideal material for making the retorts and for the main gas pipes was cast iron. Steam engines generally were originally made largely of wood with the stonework or brickwork of the building playing an important structural part, and with iron only where it was essential. By the early 1800s iron had replaced wood. So new skills were coming into the iron trade to match the developing skills of the engineers whose main material of construction was now iron.

Inventors and improvers were active in the iron trade as elsewhere. The puddling process (see p. 157), for instance, was made more efficient in about 1816, when Joseph Hall, of Tipton, Staffordshire, introduced a method which was soon to supersede Cort's, not only in Britain but in every other industrial country. The older method, dry puddling, lingered on in places, like the old charcoal blast furnace – indeed, it could be found, rarely, in living memory– but for all practical purposes Hall's process quite quickly became the standard one as long as wrought iron was made.

In Cort's process he used atmospheric air as a decarburizing agent and ignored the other chemical reactions in his furnace; chemistry, like other sciences, was far from advanced at that time. The amount of oxygen available for decarburization was limited, so the process had to be slow and the longer it took the more fuel would be needed. Much worse, however, was the inefficiency of the process: for every tonne of wrought iron made, about two tonnes of pig iron were needed. When the pig iron was brought up to melting point, some of it formed magnetic oxide ($Fe_3 O_4$) which made slag with the silica in the sand lining of the furnace. This was the source of the waste; about half of all the pig iron went away with the slag.

Hall's wet puddling used the same reverberatory furnace as Cort, but a flux was added with the iron which helped the chemical reaction by providing a better source of oxygen than could be provided by air alone, and made the whole process much more efficient. The reaction was a violent one, with the iron at one stage literally boiling; at his first trial Hall was alarmed and thought the whole furnace would be destroyed. But after a short time the boiling stopped and the iron, when taken out, was found to be excellent. Hall started his own works to use the new idea and his iron became famous for its high quality. Others followed and before long the old Cort process was dropped by most ironmakers in favour of Hall's process, known from what happened at one stage as 'pig boiling'.

The man responsible for improving the blast furnace was not an ironmaker but the manager of Glasgow gasworks, J. B. Neilson, who put his idea to work in 1828. Before that date the air for the blast furnace was blown in at its natural temperature: Neilson tried heating the air before it was blown into the furnace. In fact the current general theory was that the colder the blast could be, the better. This was later shown to be quite wrong, and eventually Neilson's hot blast was generally adopted. Today every blast furnace uses hot blast, which can now mean 1000°C or more. Neilson's blast temperatures were very much lower, but he succeeded in showing that they resulted in fuel economy as well as some technical advantages. When Neilson started his experiments the amount of coal, converted into coke, needed to produce a tonne of iron was about 8 tonnes. Neilson brought this down to 5 tonnes very quickly and when he had improved his apparatus, it came down again, to 2.25 tonnes. The use of hot blast spread, though slowly at first, especially in those areas where coal was cheap. The Black Country, with its famous ten yard (9m) thick coal seam was one such: coal was certainly cheap there and it was thought to be inexhaustible. Nevertheless, hot blast was attractive not only because it saved money. It enabled outputs to be increased as well, and it was adopted by progressive ironmasters.

Among other developments in the first half of the nineteenth century was alteration in the internal shape of the blast furnace. A Midland ironmaster, John

Gibbons, had several blast furnaces under his care, all made in the traditional shape, square inside at the bottom, or hearth. He noticed that every furnace which was blown out for repair had altered its shape. It was no longer square, but roughly round. In 1832, Gibbons built a furnace which had a round hearth from the start. The new furnace was run alongside a traditional one, using the same raw materials and with all other conditions unaltered, and made 100 tonnes of iron a week, then a record. The old furnace, which was for the time a big one, could only make 75 tonnes.

The most important development in the blast furnace was the use of waste gas. A blast furnace makes large quantities of inflammable gas as a result of the iron-making reactions. This gas came out at the top and burned away. Any area which had a number of blast furnaces was very impressive at night, with every furnace giving out great flames three or four metres (10–13ft) long. If the gas could be trapped and taken down to ground level, it could be burned there under steam boilers or to heat the air blast. However, the top of the blast furnace, where the gas vented, was also the place where the raw materials went in. It had to be open for this purpose, and charging of raw materials carried on day and night. As early as 1834 an idea was tried for collecting the gas on a furnace at Wednesbury, Staffordshire, but it was not a success and nor were several other attempts.

In 1850, G. Parry, of Ebbw Vale, produced a device known as the bell and hopper (or cup and cone) which is found in modified form on most blast furnaces today. Parry fixed at the top of the furnace a large cast-iron hopper, roughly funnel-shaped, with the narrow neck facing down into the furnace. Raw materials tipped into this hopper fell down into the furnace in the normal way. To close the hole in the hopper when raw materials were not being tipped, a cast iron 'bell' was fitted inside it and connected to a lever by which it could be raised or lowered. When it was raised, the bell sealed off the hopper completely and no gas could escape. When it was lowered, the gap between it and the hopper allowed the raw materials to enter the furnace (see Figure 2.1).

To collect the gas a large pipe was fixed in the side of the furnace, just below the top. It led down to ground level where it branched off into smaller pipes to feed the gas to steam boilers and to stoves to heat the blast. Although some of the gas was lost when the bell was lowered, the greater part was made available as fuel, bringing marked improvements in economical working.

Parry's idea is, in essence, still in use today although a modern furnace has two bells and hoppers, one above the other to prevent loss of gas. The top bell is opened and the raw materials fall on the lower one, which is closed and sealing the furnace. Then the top bell is closed and the bottom one opened, while the top bell does the sealing. Another design, the bell-less top, using two hoppers with valves, is now beginning to supersede the double bell top.

By the middle of the nineteenth century iron was at its peak. Cast iron has many uses, but wrought iron was still the most important form of the metal and

PART ONE: MATERIALS

Figure 2.1: From *c.*1900 the mechanically-charged blast furnace began to be adopted widely. These two were at Corby, Northamptonshire. The photograph was taken in 1918.
Author's collection.

by far the most important metal of commerce. Not only was there a better method of making it – Hall's process – there were several improved ways of processing it by rolling and forging. And the market was still expanding.

In 1822, Aaron Manby built an iron steamboat at Tipton, Staffordshire: it was sent in sections by canal to London and assembled and launched there. The *Aaron Manby* was not the first iron boat, nor was it the first to have a steam engine, but it was strikingly successful, trading to France for many years. Steamboats of iron, and engines for steamers, opened up a whole new market for the iron trade.

But the biggest new market of all was that provided by the railways. The efforts of Richard Trevithick, George Stephenson and others, helped by the ironmakers, who produced wrought iron rails to stand the heavy loads, made railways a practicable proposition. The first half of the nineteenth century saw tremendous railway expansion.

Up to about 1850 wrought iron was still the king of metals and by far the greatest amount was made by Hall's puddling process. Each puddling furnace

Figure 2.2: The product of the puddling furnace was a white-hot spongy mass of wrought iron. It was taken out as four or five balls and hammered (shingled) to blooms.
Author's collection.

was operated by two men – a puddler in charge of an underhand or assistant who was often learning the job (see Figure 2.2). A furnace charge of about 250kg (550lb) of pig iron took about two hours to work up to wrought iron and then the process started all over again. As the men worked for 12-hour shifts, each furnace produced about 1500kg (3300lb) of wrought iron per shift. It was not a very big output compared with what the mild steel makers were to achieve later (see p. 169), but the needed quantities were obtained by having large numbers of puddling furnaces. There are no official records for wrought iron production in the middle of the nineteenth century but it has been estimated that in 1852 the British output was about 1,270,000 tonnes. Sizes of wrought iron works varied considerably. Some had only a few puddling furnaces, others had as many as a hundred – but size was the only real difference between the various works: they were all very much the same otherwise, and up to a certain point in the manufacture the process was also the same. However, there were now several different grades, or qualities, of wrought iron in production. The wrought iron balls as they came from the furnace were always hammered out into a bloom, but the next stage depended on the grade of iron being made.

PART ONE: MATERIALS

For the first grade – merchant, or common iron – the balls were rolled out to bars. These were cut up into short pieces, put together in a cube-shaped block, reheated and rolled out to the finished shape and size required. There had been changes in this part of the works. The old water-powered hammer had, of course, given way to the steam-driven hammer – known as a helve. This, in turn, had been superseded in most works by the direct-operating steam hammer invented by James Nasmyth in 1839. Merchant iron was good enough for many purposes. A large part of the output went to iron merchants all over the country – which is the reason for the name – and for export. For some applications a better quality was needed, and merchant iron was processed to provide it.

Up to a point, wrought iron improves as it is reprocessed. It becomes tougher and stronger and more able to resist shocks and stresses. Merchant iron was reheated and rerolled to produce 'Best' iron and 'Best' iron was treated in the same way to produce 'Best Best' (or BB) iron. A few firms went even further and made 'Best Best Best' (or BBB) iron. Iron used to make, say, a fence, need not be of such high quality as that used to make the piston rod of a steam engine. Iron for such things as steam boilers, for the axles of railway engines, for chains used in collieries and for anchoring ships, had to be of good quality. There had been improvements, too, in the rolling of iron. More sizes and shapes were now available, and their accuracy and consistency were better.

One important fact must be noted, however. Although power had been applied very widely in all branches of the iron industry, it was still used only for tasks that were beyond human muscle power. At the blast furnace, for example, men shovelled ore and coke into the charging barrows. Steam power then hoisted the barrows to the furnace top, where the men took over again, wheeled the barrows to the mouth of the furnace and tipped the contents in. At the pig beds the cast iron pigs, when solid but not cold, were broken off the sows by a man using a heavy sledgehammer or a long iron bar, walking on the hot iron. When the pigs, each of up to 50kg (110lb) were cold, they were lifted by hand from the pig bed and wheeled away. The only steam power at the blast furnaces was for providing the blast and hoisting the barrows; and in the wrought iron works the same applied. Steam drove the hammer and the rolling mills and operated cutting machines (or shears) for cutting the rolled iron to length. Everything else was done by hand, including feeding the red-hot iron to the rolling mills. An ironworker had to be strong and agile, and prepared to work long hours in hazardous conditions.

The iron industry was not behind other industries in the use of mechanical power: it was neither better nor worse than any other. Certainly it used power only where it had to, but so did other industries. Ironmaking, again in common with several other industries, also offered plenty of scope for a man to develop skill and craftsmanship if he was not restricted to a purely manual job. Coke and iron were not tipped casually into a furnace, for example. They had to be charged according to a plan, otherwise the working of the furnace could be upset. Various factors could cause this plan to need changing and the man in charge had to use his own judgement to decide when and how to change it. The puddler was another man who was very much on his own in the matter of deciding how and when to act. His own skill and judgement were all he had to guide him.

By about 1850 the three main iron-producing areas in Britain were the Black Country, South Wales and Scotland. All were based on local supplies of iron ore and coal. All produced both pig and wrought iron, though the Black Country was the biggest maker of the wrought product. It had a high reputation for quality and a few of its firms had become world-famous for their iron. At that time Britain was the world's leading producer of iron and in many respects was almost alone. But abroad there were also signs of development. In Belgium and France there were some important ironworks, with the English ones as a model. Germany had some works of note and, to back them up, developments in railways and in industry generally. In the USA ironmaking was growing as well. Sweden had an iron industry which, although it was not large by the standards of the time, was in a special position because that country's very pure iron ore made it possible to produce iron of very high quality. It was in demand for special purposes such as carbon steel manufacture (see p. 159) and in fact Sheffield used it almost entirely; smaller quantities of Russian iron were also imported for the same purpose.

THE STEEL AGE

The Bessemer process

Sir Henry Bessemer was well known for his inventions before he started to take an interest in iron. Before he was twenty he had invented a stamp for government documents which could not be forged. Then he improved lead pencils and printers' type, devised a better means of embossing velvet, and found a new way of making bronze powder, which brought him a useful sum of money. Machinery for crushing sugar cane came next and then a method of making plate glass.

When the Crimean War started, in 1854, Bessemer invented a new type of gun, which he offered to the War Office, but got no response. His gun, however, showed the need for a better type of iron to withstand the stresses set up. Although he only began to consider the matter just before Christmas 1854, by 10 January 1855 he had taken out his first patent for 'Improvements in the Manufacture of Iron and Steel'. Bessemer says in his autobiography that his knowledge of iron and steel at that time was very limited, but that this was in some ways an advantage for he had 'nothing to unlearn'. His earlier inventions had brought him enough money to be able to experiment on the new idea, working at his bronze-powder factory in London – near to where St Pancras railway station now is – and in a fairly short time he had made what we now know was a new metal, by a process so novel that many people thought it impossible.

More patents followed and then, when Bessemer felt that his ideas were properly protected, he read a paper called 'The Manufacture of Malleable Iron and Steel without Fuel' to the British Association meeting at Cheltenham in August 1856.

This historic occasion marked the beginnings of what is now called the Steel Age. In fact, after a flying start, the process ran into serious technical trouble and it was a long time before it became widely adopted. Nevertheless the process caused a great stir in ironmaking circles from the start. There were ironmakers in the audience when Bessemer read his paper and straight away they became divided into two camps. Some tried to dismiss the whole idea; others saw it as having great potential and were anxious to try it out in their own works. Bessemer, who had watertight patents, prepared to grant licences to use it, on payment of a royalty.

It is easy to understand why many people scoffed at Bessemer's idea. He took molten cast iron and blew a blast of cold air through it. Surely, people reasoned, all this would do would be to cool the molten iron down. In fact the iron actually got hotter. Even Bessemer, who was not easily surprised, was alarmed at what happened when he set his experimental apparatus – or converter – to work: he said later that it was 'like a veritable volcano', with flames, slag and bits of molten metal shooting up into the air: see Figure 2.3. (This was no exaggeration. The Bessemer process remained to the end – it is virtually extinct now – the most spectacular sight in the iron and steel industry.) He let the process go on, indeed nobody could approach the converter to turn it off, and after some minutes the fireworks stopped and there was nothing but a clear flame from the mouth of the converter. Bessemer tapped the metal he had made and found it behaved like good wrought iron.

It will be remembered that to convert cast iron into wrought iron the carbon has to be removed. In the puddling furnace this was done by heating the iron in contact with fluxes containing oxygen. Bessemer used the cheapest form of oxygen there is, ordinary air, which contains about 21 per cent of oxygen.

Because there was so much oxygen, the carbon in the cast iron reacted very strongly with it. The reaction was exothermic, that is, it actually generated heat; so the iron became hotter instead of colder, and produced the pyrotechnics.

Bessemer had achieved his object of making what he called 'malleable iron' from molten cast iron without fuel, but what kind of metal, in fact, had he made? His term 'malleable iron' is confusing, for strictly it describes a kind of cast iron so treated that it became to some extent capable of being bent or formed with-

Figure 2.3: The Bessemer process is now obsolete. This 25 tonne converter, shown at the start of the blowing cycle, was formerly at Workington, Cumbria. British Steel Corporation.

out breaking. In Bessemer's time the words were often incorrectly applied to wrought iron. Bessemer was really trying to make a better form of wrought iron. In many ways he succeeded, for his new metal – which we now call mild steel – would do almost everything that wrought iron would. And since he could make it much quicker than in the puddling furnace, and in bigger quantities at a time, it could be cheaper once the plant was set up.

His experimental converter was only big enough to deal with about 350kg (770lb) of iron at a time, but it did so in about thirty minutes, compared with a production of about 250kg (550lb) in two hours in a puddling furnace. It was easy to make a bigger converter and, as the time taken by the process was no longer whatever size the converter was, larger outputs were soon possible. It was realized that Bessemer metal was not the same as wrought iron – technically it is different in several ways but it could be used for most purposes where the older metal had been supreme for so long, it was cheaper to make, and the demand for all kinds of iron and steel was still growing. Licences were granted, the necessary apparatus was set up and it seemed that Bessemer would confound his critics. But when the licensees started production there was trouble: all the steel they made was useless. These licensees included the Dowlais and

Ebbw Vale works in South Wales and, with a number of others, they had paid thousands of pounds for the privilege of using a process that would not make good steel. Yet Bessemer had done so, and had demonstrated the fact at his St Pancras works. It took more than twenty years for the cause of the trouble to be found and a remedy devised: the Bessemer process was first announced in 1856 but it was not until 1879 that it was modified so that it could be used on a really large scale. Then it spread all over the industrial world.

The Thomas process

The most important problem affecting the Bessemer process derived from the presence of phosphorus, which occurs naturally in small quantities in most iron ores found in Britain and on the Continent. As with sulphur (see p. 153), if even a minimal amount of phosphorus combined with wrought iron or steel, the product became weak and brittle. Processing in the puddling furnace removed the phosphorus; the Bessemer process, as at first used, did not. By pure chance the early experiments at St Pancras were carried out with a cast iron made from an ore mined in Blaenavon, Gwent, one of the very few British irons with negligible phosphorus. When the licensees tried to apply the process using iron which contained phosphorus, the result was failure; and most of the iron ores then available were phosphoric. So, once the problem was identified, the Bessemer process developed slowly, and only in places where suitable iron was available. One of these places was Sweden, where the ores, and therefore the iron, were very pure.

Bessemer, a good businessman as well as an inventor, set up his own works at Sheffield and made himself another fortune (see Figure 2.4). His steel was gradually accepted and in 1860 steam boilers were made of Bessemer steel. Railway rails followed in 1863 and two years later the London and North Western Railway started making Bessemer steel and rolling rails for its own use, but the wrought iron trade was still far from threatened by the new material.

In the 1870s a completely unknown man, P. G. Thomas, started to look into the Bessemer process. He was a police-court clerk in London, who had studied chemistry at night school and heard it said that whoever could solve the phosphorus problem in the Bessemer process would make a fortune. By 1879 he had modified the process so that it could use phosphoric iron and the way was wide open for a great expansion of steelmaking.

In his converter Thomas used a special form of lining such as dolomite, which, being chemically basic, united with the phosphorus and left the metal free of this troublesome element. The phosphorus went away with the slag, and he was able to sell this as an agricultural fertilizer. Phosphorus is necessary for plant life, the Thomas process slag was a cheap way of providing it, and there was a ready market for all the slag that was made.

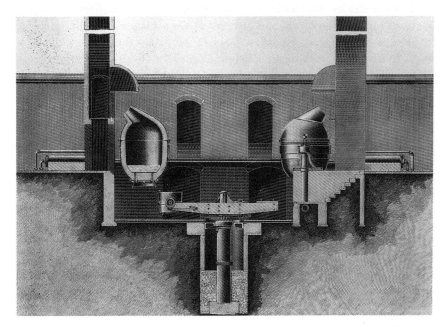

Figure 2.4: Bessemer plant at Sheffield. Converters, ladle and casting pit. From Sir Henry Bessemer's *Autobiography*, 1905.

There were now two bulk steelmaking processes: Bessemer's original, which suited non-phosphoric iron, and Thomas's, which would deal with the phosphoric types. As the latter were the most common the Thomas process spread rapidly, in Britain and on the Continent: in France and Belgium it became so common that the steel was generally called Acier Thomas, or Thomas steel. In the English-speaking world the two processes were distinguished by the names acid Bessemer (the original) and basic Bessemer (Thomas's) because of the chemistry involved.

The Siemens–Martin process

Meanwhile the open-hearth steelmaking process had been developed. It was the work of C. W. Siemens, a German, who came to Britain when he was 20 and eventually became naturalized. Siemens had a scientific education and he applied his knowledge systematically. He was initially concerned with improving furnaces – any furnaces – and his first successful one was used for making glass. The novelty lay in the fact that the waste gases, which normally went up the chimney stack, were used to heat the air used to burn the fuel. By 1857, Siemens was able to claim that he could save 70–80 per cent of the fuel previously used in glassmaking. The Siemens furnace was first applied to making steel in France, by Emil and Pierre Martin in 1863. Siemens set up a small works in Birmingham in 1866 to demonstrate how steel could be made in his furnace, and by 1869 a company at Swansea was producing about 75 tonnes a week. By 1870 the Siemens process (often called the Siemens–Martin process for the obvious reason) was fully established.

Wrought iron now had some really serious competition. The Siemens and Bessemer processes were complementary, but both could use phosphoric or non-phosphoric iron. Bessemer was cheaper since it used no fuel, but it needed to be charged with molten iron. This was easy when the Bessemer plant was adjacent to blast furnaces. Molten iron could be used in the Siemens open-hearth furnace, and it often was, but an advantage of this process was that it could melt scrap iron. With the spread of industry, scrap had become a useful raw material as machinery of all kinds wore out or was replaced by new and better types and it was cheaper than pig iron. See Figure 2.5.

The Siemens process was slower than the Bessemer. A Bessemer charge of iron took about thirty minutes to convert to steel; in the Siemens furnace it took eight to twelve hours. This could be an advantage, for it enabled the furnace operator to make frequent checks on the steel and to adjust its composition as required. Various grades were now demanded, and the new steelmaking processes were able to provide them.

In the second half of the nineteenth century steel began, slowly but surely, to push the wrought iron trade out of existence; although it survived in a small way until recently, from about 1870 onwards steel was what really mattered. Cast iron continued to be made, partly for castings but increasingly as the raw material for steel and this is still the position today.

Alloy steels

Before 1856 there were only two important iron products, cast and wrought iron. A third, carbon steel, was essential for some purposes but its output was small. Bessemer added a fourth, mild steel. In 1868 appeared the first of a fifth group, now called special, or alloy, steel, made by R. F. Mushet in his small works in the Forest of Dean in Gloucestershire.

More engineering products meant that more machining had to be done and this needed cutting tools. Before Mushet's time the only means of cutting iron and steel was a tool made of carbon steel. This can be hardened by making it red hot and cooling it quickly; the obvious way was to plunge it into cold water. Such treatment makes the steel very hard and brittle; it has to be heated again to a lower temperature, and cooled once more to make it both hard and tough. Carbon steel which has been hardened can be softened again by making it red hot and letting it cool naturally. When a carbon steel tool wore or was damaged, it was useful to be able to soften it, file it or machine it back to its original shape

Figure 2.5: Like the Bessemer process, open hearth steelmaking is now obsolete. Open hearth furnaces could range in size from 5 or 6 tonnes in the early days to as much as 500 tonnes capacity at the end.
British Steel Corporation.

and reharden it. However, softening was a distinct disadvantage when it occurred accidentally. Heat is always generated during cutting and when iron or steel are cut the temperature can easily reach a point where the tool is softened. Then it is useless until it has been rehardened. Thus there is a limit to the speed or depth of cut which can be made by a carbon-steel tool.

Many improved machine tools had appeared in the first half of the nineteenth century. Driven by steam, they were strong and powerful, and capable of heavier work than their cutting tools could achieve. Mushet was asked to make somebody else's patent tools and, finding them a failure, invented one of his own. He took advantage of the fact that iron and steel will unite with many other elements and experimented with tungsten. Mushet's tungsten steel could be forged to shape and left to cool naturally in the air, when it became very hard and tough. It only needed grinding to a sharp cutting edge to be ready for use, and when it became blunt from use it was reground. The most useful feature, however, was that the new steel did not soften even at a dull red heat. Engineers and machinists welcomed it, although it was much more expensive than carbon steel. Tungsten steel is still used today, although its detailed composition is often a little different. The best engineers' drills are usually made of a form of tungsten steel called High Speed Steel introduced in 1900.

Others followed Mushet with different alloys for special purposes. R. A. Hadfield, for example, used manganese, in 1887, to make a steel which was particularly tough and wear-resisting. It was used in such things as railway points and crossings, and in rock-crushing machinery. Over the years very many more alloy and special steels have been added to the list, all of them with some special application or group of applications. Stainless steel, in particular, has affected all areas of everyday life.

Stainless steel was invented by Harry Brearley, in Sheffield, in 1913. While experimenting with a better steel for rifle barrels he noticed that one of the steels he made was unaffected when he tried to treat it with acid so that he could examine it under the microscope. Attack by acid, or etching, is a form of corrosion. If the steel would not etch it would not corrode either, at least under many conditions where ordinary steel would. Some possible uses for the new steel suggested themselves at once, including cutlery.

Many foods contain weak acids which do not harm people but will stain and corrode ordinary steel. Brearley had some table knives made and found that they stood up to use very well, without losing their bright surface. His original experimental steel contained nearly 13 per cent of chromium and Brearley tried out other proportions, and also added different elements such as nickel. After his experiments had been suspended during the First World War, he and others went on to develop several kinds of stainless steel. One problem was that of hardening the steel: Brearley's first knives were thick and blunt and he once said that he got a name for making knives that would not cut. Over time, several different kinds of stainless steel were developed for particular purposes. One of the commonest for table ware such as dishes and coffee pots contains about 18 per cent chromium and 8 per cent nickel, hence the figures 18/8 which are often found stamped on such goods. Other types, of different composition but all correctly called stainless steel, are used for such things as knives and razor blades, and surgeons' scalpels. Still further varieties are used in industry to resist the heat of furnaces, or in other severe conditions.

Steel manufacturing after 1860

While new types of steel were being invented and improved, developments were also taking place in the manufacture and handling of the product. Rolling mills, in particular, changed drastically. A rolling mill in its simplest form consists of a pair of rolls (never called rollers in the iron and steel industry) mounted one above the other in a strong framework. Their surfaces may be flat, for rolling strip, plates or sheets; or they may have grooves cut in them to shape the iron or steel into, say, a round bar, railway rail or girder. Only a limited amount of shaping can be done in a single passage through the rolls; the metal must be passed through several times. This was done by men with tongs holding the red-hot metal, hard but skilled work. With the Bessemer and Siemens processes steel could be made not only quicker but in bigger individual pieces. Molten steel was cast into rectangular blocks, or ingots, which were rolled to the required shape. As ingots became bigger, men could not handle them and machinery had to take over.

Sir John Alleyne, of Butterley in Derbyshire, patented a mechanical device for moving ingots in 1861. He also devised a rolling mill which could rotate the rolls in either direction, enabling the metal to be passed back and forth, some rolling being carried out at each movement, until it was of the required size and shape. These two devices saved time and took over work which was too heavy for men to do. Alleyne's mill was improved upon by John Ramsbottom, of Crewe railway works, who made the first really successful reversing mill in 1866 by coupling the engine of a railway locomotive to a rolling mill. A railway engine, of course, has to be capable of running in either direction, so it could just as easily rotate the rolls either way; all the man in charge had to do was to pull a lever.

These inventions were the basis on which many modern rolling mills operate today. There were several other changes in steel rolling, some of them matters of detail, but all adding up to a general improvement. One development, however, was a completely new idea. This was the continuous rolling mill invented by George Bedson, of Manchester, in 1862. Bedson built a mill, for rolling small iron rods for making into wire, with sets of rolls in a long straight line, so that the metal being rolled came out of one pair of rolls, passed immediately to the next pair and so on down the line until the finished size and shape were achieved. It saved time, needed fewer men and made a better product. The first Bedson mill had sixteen pairs of rolls so the metal was rolled sixteen times before it came to the end of the line. Later mills had even more pairs of rolls and could produce rods at very high speed. The principle was adapted to roll other steel products besides rods, and it is the basis of many modern mills.

By the end of the nineteenth century iron- and steelmaking were very much more mechanized, but the source of power was still steam. Electricity had appeared in the iron- and steelworks, but at first only for lighting: some works had electric light in the 1880s. Today it is the source of power in every works and the steam engines survive only in museums.

Electricity is also used today in steelworks as a direct source of heat, and its beginnings for this purpose go back a long way. If an electric arc is struck very intense heat is generated; Siemens suggested that it could be used for steelmaking in 1878, but nothing came of it at the time. Nor was there much interest in the arc furnace designed by the Frenchman Paul Héroult in 1886, though he used it himself for making aluminium (see p. 109), and in 1900 actually made steel in it. The electric furnace was ahead of its time, and it was a national emergency which really started it going. In the First World War, munition manufacture produced quite large quantities of small shreds and scraps of metal called swarf, which was very valuable raw material for steelmaking if it could be melted. Neither the Bessemer converter nor the Siemens open-hearth furnace would do this adequately, but the electric furnace would.

MODERN STEELMAKING

During much of the first half of the present century two world wars and several years of international trade depression hindered development in the steel industry throughout the world. But changes were still taking place and many technological improvements were on the way.

In the last twenty years or so the industry has altered at an enormous rate, technologically and economically. Iron- and steelmaking are now, technologically, international. Any large modern steelworks, no matter where it is, will use plant and processes from many countries.

The Anchor development of the British Steel Corporation (BSC) at Scunthorpe provides a typical example. Anchor – a code name – was completed at a cost of some £236 million. A multi-product iron and steel complex which will make about 5.2 million tonnes of steel a year, it is not the biggest project in Britain and far from being the biggest in the world. The BSC works at Teesside was designed for an annual output capacity of 12 million tonnes, and the planned capacity at the Fukuyama works in Japan was 16 million tonnes a year.

But Anchor is as modern as any and serves as a good example of how international a steelworks is today.

The route to steel at Anchor is the same as at other large integrated works and this pattern is likely to continue for some time. Briefly, iron ore is reduced to cast iron in large blast furnaces, converted to steel by a new method, the Basic Oxygen Steel process (see below), cast into slabs or ingots and rolled into blooms, plates, billets or sections. Some idea of the size of the works can be gained from the fact that the bloom and billet rolling mill building alone is more than 1.6km (1 mile) long.

The blast furnaces are fed with a mixture of British and foreign iron ores, all in the form of sinter, an agglomerated mixture of fine ore and other iron-bearing substances which produces a very uniform raw material. Sinter originated in the USA; the use of 100 per cent to form the charge is a British development which took place on an earlier plant at Scunthorpe.

Steel is made by the BOS (Basic Oxygen Steel) process (see p. 178) perfected in Austria, with some background work in Germany and Switzerland. The original idea was Bessemer's, though he never tried it out. The process needs very large supplies of pure oxygen, which were not available in Bessemer's time.

Much of the steel is made into large slabs by a process called continuous casting. This again was first suggested by Bessemer, but his idea was not successful. As used today the machines are based on the work of S. Junghans in Germany, with important contributions from Britain and the Soviet Union. The first commercial machine was built at Bradford in 1946. The Anchor continuous casting plant was built by a German firm, rolling of the steel is carried out in mills of British and German design, and some of the electrical equipment is Swedish.

Economics, always important in iron- and steelmaking, are even more so today and this is the reason for the changes in blast furnace ore practice and steelmaking. The British, German, French, Belgian and other Western European iron industries developed because there were large local supplies of iron ore and coal, as did those of the USA and the Soviet Union. In most cases there are still large – sometimes very large – reserves of iron ore left, but they are often of low grade, and in Europe and some other parts of the world it is now more economic for ironworks to import high-grade foreign ores. At the Anchor site, for example, the local ores contain about 20 per cent of iron. Vast new deposits have been opened up in Africa, Australia, Canada and South America, and these ores can contain as much as 60 per cent of iron. With supercarriers – ore-carrying versions of the giant oil tanker – these ores can be brought halfway round the world at low cost. Some of the local ores, however, although they are of low grade, are very close to the surface, and can be quarried cheaply by modern earthmoving machines. In some places, therefore, a certain amount of local ores will be used, mixed with larger quantities of imported ores.

Consequently, British bulk iron- and steelmaking are being concentrated at five sites, all near to deep-water ports capable of taking supercarriers of 100,000 tonnes now and expected to become twice as big or even bigger in the near future. These sites are: Scunthorpe, near to the Immingham ore terminal; Port Talbot and Llanwern, both served by the Port Talbot terminal; Lackenby, close to Redcar terminal; and Ravenscraig, Scotland, near Hunterston terminal. Some steelmaking will continue away from the ports and so will a lot of processing, but the bulk tonnages will come from these places. A similar pattern obtains in Europe and Japan, though the latter country, having no local ore reserves of any value, is even more dependent on imports. The Soviet Union alone, among the big steelmaking nations is likely to remain independent of imports.

The blast furnace remains the same in principle, although very much bigger. A blast furnace of the nineteenth century had a hearth – or lower working part – about 1.8–3m (6–10ft) diameter and made 100 tonnes or so of iron a week. Number three new blast furnace at BSC's Llanwern Steelworks, Gwent, has a hearth diameter of 11.2m and will produce at least 5000 tonnes of iron a day, and BSC's Redcar furnace, Cleveland, can produce 10000 tonnes a day. All modern blast furnaces are of course mechanically charged, the machinery being under push-button control or even automatic, according to a present programme. Mechanical charging originated in the USA towards the end of the nineteenth century and spread in time to all ironmaking countries.

The BOS process is the biggest single development in steelmaking in the present century. It looks rather like the Bessemer process, but it uses pure oxygen instead of air and the oxygen is blown on to the surface of the molten iron at a very high speed, instead of through it as the air was blown in the Bessemer converter. Some modern BOS vessels are very large; there are three of 300 tonnes capacity each both at Port Talbot steelworks, for example, and at Anchor.

BOS steelmaking is very fast; a charge of 300 tonnes of iron can be converted into steel in about thirty minutes. Enormous quantities of oxygen are used and the BOS converters have to have their own oxygen plant in a nearby building. The process gives off a vast amount of gas and, under modern anti-pollution regulations, this has to be cleaned and disposed of harmlessly instead of being discharged to the atmosphere, a costly but essential procedure. In fact everything to do with the BOS process is expensive. The BOS plant at Port Talbot, for instance, cost more than £28 million, but the process is so fast, and uses comparatively so little labour, that it is more economic than any other process for bulk steelmaking today.

Oxygen steelmaking is another example of the international nature of the industry. It was first put to practical use in Austria in 1953, because it happened to suit local conditions there at that time. From the fact that it was used in the

Austrian towns of Linz and Donawitz the process was often called by the initials L–D. However, Basic Oxygen Steel (BOS) is now the accepted term in English-speaking countries.

The oxygen process has caused a complete revolution in bulk – or to use another common name – tonnage steelmaking. The Bessemer process had been dying out for some time. The last British basic Bessemer converters – at Corby, Northamptonshire – closed down in 1966 and only two converters remained in Britain, at Workington, Cumbria, both of the acid type. They finally closed in 1974. The Bessemer process is now extinct in Britain and practically so throughout the world, and now the BOS process has sealed the fate of the open-hearth furnace, which is being phased out fairly quickly.

Speed of production, attractive though it is, brings its problems, one of which is the control of the process. It calls for a very high order of instrumentation. In the days of the open-hearth furnace the more leisurely pace allowed the operator to take samples of steel, have them analysed in the works laboratory, and make adjustments to the furnace charge as required. Now the analysis is done automatically: a small sample of steel is cooled down and polished, then analysed by an automatic spectrograph.

Disposing of the steel made in the BOS converter calls for equipment which will work at the same rate. One of the ways of doing so is continuous casting, as revolutionary in its way as the BOS process. Ever since steel was first made in bulk it has been poured molten into moulds to produce rectangular pieces (ingots) or heavy flat ones (slabs). These could then be processed into finished products by rolling. Ingots can now weigh anything from about 45kg (99lb) or so in the case of special steels to as much as 20 tonnes or more in tonnage steels. For many purposes continuous casting, simple in principle but needing very accurate control, has replaced ingots. Liquid steel is poured into an open copper mould of the required size for a billet, bloom or slab. The mould is water-cooled and the outer skin of the metal solidifies rapidly. At the bottom of the mould the metal is solid enough for it to be drawn out at the same rate as the molten metal enters at the top. Water sprays below the mould complete the solidification. As long as molten steel is poured in, a solid product comes out in a continuous length. This is cut up automatically by a flame-cutting head into the lengths required for further processing. Continuous casting is not literally continuous in that it never stops, and produces a billet or slab which would be of infinite length if it were not cut up: in fact steelmaking itself is not continuous, so the casting machine has to deal with batches of molten steel. But it is sometimes possible to get several batches in succession so that the casting machine can have a satisfactorily long run.

If steel is cast into ingots there is always a certain amount of unsound metal at the end, which has to be cut off. In continuous casting there is also a small piece at the start and finish of the cast which has to be discarded, but this means that there are only two pieces of scrap for every batch. If a batch of, say, 100 tonnes were cast into twenty 5-tonne ingots there would be twenty pieces to discard. The saving is obvious, so it is not surprising that continuous casting has spread all over the industrial world.

Compared with steelmaking and casting, rolling mills have evolved in a rather less spectacular way, but development has been impressive nevertheless. Continuous wide strip rolling now produces all the steel sheet used for motor-car bodies, cookers, refrigerators, washing machines and a host of other industrial and domestic purposes.

Bedson's continuous mill is still used in modified form for rolling rod and the idea was adapted for wide strip. It seems logical that this development should have taken place in the USA, for it solved the problem of the time. In the 1920s the growth of the mass-produced car trade, pioneered there, coupled with increased demand for kitchen equipment and canned goods, created a vast market for steel sheets and the old hand-rolling methods of sheet production could not cope with the demand. The answer was mechanization and the Americans provided it in the form of the wide strip mill.

A steel slab is heated to rolling temperature and passed through a series of rolls in line until it is reduced to a very long thin plate. This is automatically wound into a coil weighing several tonnes. Then the coils are pickled in acid to clean the surface and passed through a further series of rolls, this time cold, to become finished coils of thin, wide strip.

Coils of strip can be used in several ways. They can go to cut-up lines, where they are unwound and the continuous length is cut automatically, without stopping the line, into sheets of a required length. Alternatively the coils can go direct to the user, where they become the raw material for automatic production lines. Or the coils can be passed through an automatic tinning line, where they are given a microscopically thin coating of tin to become tinplate, which is used for making cans (or, as they are often, and wrongly, called, tins).

Some coils go through a different automatic production line to be coated with zinc (galvanized). The product, like plain wide strip or tinplate, can be sold in coil form or cut automatically into sheets. If galvanized strip is cut into sheets, many of them will be corrugated to form the product still familiarly and incorrectly known as corrugated iron. It has been made of steel for many years now, but the old name has stuck, as it has with 'L'-shaped or angle sections which are now of steel, although they are popularly called angle iron.

Two other means of finishing strip steel – wide and narrow – are of growing importance: plastic coating and prepainting. Sheets and coils can now be coated on either or both sides with a thin layer of plastic film. It is bonded on to the steel and adheres so tightly that it cannot be pulled off without destroying it. Several different colours and surface finishes are available and plastic-coated sheets have many uses. Office furniture and partitioning are common outlets, as are some domestic appliances such as electric fan heaters. Another use is for the sheeting on the outside and roofs of buildings. In prepainted sheets and coils there is also a wide range of colours and types of paint available. All the paints are chemically based and adhere very firmly. Typical uses for prepainted sheets are for car instruments and clock faces: all the customer has to do is to cut out the required shape and print on the figures. Both types of sheet can be shaped by metalworking machinery without damaging the finish, and they save the user having to install expensive finishing equipment.

The continuous wide strip mill did not attract much attention outside the USA at first, but as other countries stepped up their demands for more sheet it began to spread. The first outside the USA was put down at Ebbw Vale, Gwent, in 1938. It produced about 5000 tonnes a week which, though not particularly impressive by current standards, was a great advance over anything possible in Britain before. Strip today can be either 'narrow' (up to and including 600mm wide) or 'wide' if it exceeds that figure. Thickness is now usually stated in millimetres and decimals of a millimetre: it is technically sheet if it is up to and including 3mm; above that it is plate. So tinplate, which is much less than 3mm thick, should really be called sheet – it is neither tin nor plate – but the name is very old and it shows no sign of disappearing.

Like any other process operating at high speed the wide strip mill demands very accurate control. Strip can issue from the finishing end of the mill at about 95km/h and the mill could make scrap very fast if anything went wrong. One of the problems found in any rolling mill is that the component parts can stretch and distort when the metal is rolled. The stretch may be very small. It may only cause inaccuracies in the finished strip thickness of a fraction of a millimetre, but such inaccuracies are unacceptable to some buyers of strip and sheet, and allowance has to be made for this problem in the design. A British invention, now used all over the world, which deals with this problem is automatic gauge control. With this device variations in strip thickness are measured as they are actually occurring, and corrective action is taken automatically before there is time for a serious fault to develop. Very remarkable strip accuracy is possible: consistency of strip thickness to within 0.005mm is achieved regularly in commercial production.

Such control is automation in the correct sense of the word. Automatic operation of many types of machine – not only in the steel industry – is very common and quite old. It is not, however, automation, for automatic operation merely goes through a preset programme with workers to check the products and adjust the machinery. With true automation the machinery checks its own products and makes its own corrections. This is exactly what automatic gauge control does. As production rates increase, automation is becoming more essential, particularly as customers are now demanding higher standards in their steels than ever before. For some purposes the computer is providing the answer; the steel industry shares with the oil refineries the credit for much pioneering work in this field.

Apart from the obvious office use of computers for anything from preparing wage statements to scheduling orders so that they go to the works in the most economic way, computers are now being used 'on-line', to control actual production processes. A simple example comes from the cutting of billets to suit customers' orders. Billets of special and very expensive alloy steel have to be cut into many different lengths to suit individual customers. The billets do not come from the mill in exact lengths, and the practice was always for the operator who controlled the billet saw to be given the measured length of each billet as it came to him. He then had to decide the best way to cut it up, according to all the orders he had to satisfy, and set the saw accordingly, so that as little as possible of the expensive material was wasted. However skilled the operator, human error was unavoidable.

With computer control the billets are measured automatically as they are rolled and this figure is fed into the computer with details of the customers' orders. It equates these two sets of information and produces figures for each billet which will use it most economically. There are numerous other examples of computer control in steelmaking and processing, and the trend is increasing.

In special and alloy steelmaking there is also a great deal of mechanization and process control is highly instrumented, but the scale of operations is smaller and the range of products wider. Today all alloy and special steels are made in electric furnaces. The old crucible process is extinct, though there is a crucible melting shop preserved in working order at Abbeydale Museum, Sheffield. Electric furnaces can be of two types, arc or induction. In the arc furnace heat is generated by means of an electric arc and metal is melted by this heat in a refractory-lined vessel of drum shape, which can be tilted mechanically to tap the finished steel (see Figure 2.6). Arc furnaces are now of many sizes, holding from a few tonnes to 150 tonnes or more. The induction furnace was invented in Italy in 1877 but the time was not ripe for it to develop and the first one in Britain was installed in 1927. In this furnace there is no arc; an electric current induces a secondary one inside the furnace itself and generates sufficient heat to melt steel. Induction furnaces are generally used for the very special and expensive alloy steels and may range in capacity from a few kilogrammes to 5 tonnes or more.

Neither type of electric furnace uses liquid, solid or gas fuel, so the steels made in them cannot be contaminated from these sources, an important consideration in the higher-grade steels. Any contaminant – even in a very small percentage – might be disastrous in a gas-turbine disc, for example, so a very 'clean' furnace is highly desirable.

Unfortunately, some gases also act as contaminants in steel, and since these

Figure 2.6: Model of an electric-arc furnace.

gases are present in air it is impossible to avoid their getting into the steel. They must then be removed and there are several ways of doing so.

One is vacuum melting. Steel is made as carefully as possible in an ordinary electric furnace and then remelted in an induction furnace which is contained in a sealed chamber under a vacuum. As the steel melts, the gases are given off and drawn away by the vacuum pumps. Vacuum furnaces are complicated pieces of machinery, with equipment for taking samples of molten steel, making corrective additions to the melt, and casting the steel into an ingot, all without disturbing the vacuum. They are naturally expensive to buy and they have small outputs, but they are justified when very high quality is essential.

Among other methods of cleaning and purifying high-grade steels, is electro-slag refining, which is of growing importance. In this process a long bar of steel is made and the end of it is gradually melted off under a pool of molten slag. The slag is specially prepared so that the impurities pass into it; in effect they are 'washed' out of the steel.

Iron and steelmaking today is a pattern of highly sophisticated and very expensive plant and, for the bulk producers, of very large outputs coming from a few works in coastal areas. The pattern is world-wide; almost every major industrial country is concentrating its bulk steelworks in this way.

There are, however, two exceptions to this trend towards ever-increasing size. In the first place the alloy and special steelworks have not followed the pattern and are not likely to. A new giant steelworks could cost £1000 million or more to build today, a cost which could only be justified by the very large outputs achieved for tonnage steels. Alloy steels are not wanted in such great quantities, so the alloy steelworks, though bigger and costing considerably more than they used to, will never reach giant size.

In the second place, there is a new type of steelworks, not at all large by modern standards, which is proving very successful. This is the so-called mini-mill, made economically possible by the electric furnace and continuous casting. There are several in Europe and quite a number in the United States and Japan, where the giant works is often thought to be supreme. One has been built in Britain and others are under construction or planned.

A mini-mill is a steelworks based on scrap steel, collected over a fairly small area locally, melted in an electric-arc furnace, cast in a continuous casting machine and then rolled in a mechanized mill. Its products will be few in number; sometimes there will be only one finished product. Concrete reinforcing bars are a typical mini-mill product. The output can range from as little as 50,000 up to about 400,000 or more tonnes a year.

The size of a mini-mill is determined by that of the local market and that of the area from which the raw material, scrap, is collected. In the USA, with its great land mass, a market can be a long way – and in terms of transport an expensive way – from the traditional steel-producing areas, so a locally-based works can do well. But even in Japan and Britain, mini-mills can prosper if they get their economics right. The first British mini-mill, at Sheerness in Kent, was designed for expansion from an annual capacity of 180,000 tonnes to 400,000 tonnes.

Discounting the possibility that steel may in the foreseeable future be superseded by some other material, it is difficult to imagine any major changes in the metal itself. Specifications will change, new alloys will be developed for new, unforeseeable applications, but steels will still be recognizable as such.

It is in the methods of manufacture that the greatest changes are likely. The blast furnace could be the first to face this effect. At present it is the most economic means for converting iron ore to metal. But coke is getting more expensive and scarcer: not every type of coal is suitable for making coke and the world's reserves of those that are are dwindling. There are alternative methods of reducing iron ore. Some modern plants using oil or natural gas as fuel can produce relatively pure iron pellets which are suitable for melting down to make steel. It is also possible to smelt electrically. In a few parts of the world, where coke is too expensive, they are already in use: in Mexico, for example, where there is iron ore, but coke would have to come from the USA and the transport charges alone would make it very expensive.

Perhaps the ultimate dream of the steelmaker is the fully-continuous production of steel. Raw materials would come in at one end of the works, flow through the various processes in a continuous line and come out at the other end as finished products. Parts of the production line are already continuous, but there are major technical problems to be solved before truly continuous steelmaking is practicable.

But there are quite a lot of people in many countries, trying to fill the gaps between theory and practice and nobody can predict what might happen, or when. The one certainty is that we have not yet heard the last of steel or of its basis, the element iron.

3

THE CHEMICAL AND ALLIED INDUSTRIES

LANCE DAY

INTRODUCTION

The chemical industry is that by which various kinds of matter are transformed into other kinds that are needed in manufacturing or in everyday life. Its history falls into two periods. The first, the pre-scientific stage, stretches back into the distant past, to man's earliest attempts to deal with materials. The second, scientific, period is of quite recent origin, in the late eighteenth century, when chemical science began to be usefully applied to chemical technology.

In the earliest period up to the end of the neolithic age, that is around 3000 BC, practice of the chemical arts was restricted almost entirely to the making of fire, alcoholic fermentation and the baking of pottery. This was succeeded in various parts of the world by what are termed the ancient civilizations, such as those of Egypt, Mesopotamia, the Indus Valley and, somewhat later, China. Here city life began, communication, above all writing, developed to make it possible to keep records and disseminate knowledge, and new techniques were developed. Some of these, such as medicine, surveying and astronomical observation, were carried out systematically. At the same time the range of materials available and the processes by which they were treated widened, with the effect of improving the way of life for the citizens. Most important of these materials were the metals, gold and silver, copper and tin, alloyed in bronze, and later iron (see Chapters 1 and 2).

Throughout this period, and in the succeeding ages of Greece and Rome, the rise of Islam and on into Western Europe, the extraction and preparation of useful substances was essentially a craft, carried on, like any other, by skilled artisans who learned their trade through apprenticeship and experience and not from a corpus of literature; there was none, and if there had been, very likely they were unable to read. Such accounts as there are of the ancient chemical arts were drawn up by those not engaged in the craft. For example, the encyclopaedic *Historia naturalis* of the Roman official Pliny has embedded in it many descriptions of chemical processes, some accurate, some less so, for they are based on secondhand reports rather than original observation. Many recipe books survive to give us some idea of what went on, like the chemical tablets of seventh-century BC Assyria, although these are no more than lists of ingredients.

With the coming of printing and a mercantile and practical class of reader, a demand developed for clear accounts of the making of various substances. These begin to appear early in the sixteenth century, some being fine examples of the art of book-making. Hieronymus Braunschweig's books on distilling were printed early in the 1500s and were the first to include illustrations of chemical apparatus. Neri's sober account of glass-making followed and there were the metallurgical treatises of Vannoccio Biringuccio (1540), Agricola (Georg Bauer) (1556) and Lazarus Ercker (1574). This literature is severely practical and shows little trace either of magical or superstitious elements on the one hand or, on the other, of the current philosophical ideas about the nature of matter. It was the Greek philosophers of the sixth century BC onwards who began to seek an underlying unity in the variety of materials in nature and a few fundamental principles or elements from which this variety could be derived. The explanation that gained widest acceptance was the four-element theory propounded by Aristotle and his followers from the fourth century BC, which held that all materials consisted of varying proportions of the elements fire, air, water and earth. This theory was not seriously criticized until the seventeenth century and fell into disuse during the following century, surviving today only in such phrases as 'the fury of the elements'. This and certain other ideas about the nature of matter and the ways it could undergo change were applied by the alchemists in the course of their work attempting to make gold.

The artisan did not think in philosophical terms because he had not been educated in the schools, and if he had been, it would not have been the slightest help in his craft of making useful materials. The lack of a genuine theoretical understanding was a great handicap particularly in identifying and valuating materials. The glass-maker, for example, did not know that silica, sodium carbonate and lime as such were needed to make glass; the first glass-maker discovered by accident, and his followers knew from experience, that sand melted with the ashes of certain plants would yield glass. They knew that the whiter the sand, the more colourless the glass. Likewise, it had been found from experience that the ash from some maritime plants produced the best glass, but not because they understood that these contained soda, potash and lime and that these were necessary for glass making. Materials were recognized by their look and feel, learned from those who already knew. Knowledge of materials and processes tended to be kept within rather closed communities and not widely disseminated. Communications were difficult enough without the deliberate secrecy that was sometimes practised, as when the earliest Venetian glass-makers sought, albeit vainly, to prevent a knowledge of their art from spreading.

Lacking a means of identifying substances correctly, the early chemists could be so confused about them as sometimes to use the same name for different substances, such as 'nitrum', which could mean both sodium carbonate and potassium nitrate. On the other hand, different names could unwittingly be used for the same substance: 'vitriolated tartar' and 'vitriolated nitre' were both used at times for more or less impure potassium sulphate, apparently with no awareness that the substances so designated were essentially the same.

Inability to identify materials made it impossible to evaluate them, that is, to determine how much of them was present. Finding out what and how much is the object of chemical analysis and not surprisingly it arose in connection with those materials that could be recognized, namely the metals, particularly gold and silver. The printed books on the assay of metals which began to appear early in the sixteenth century are evidence of a practical tradition in the quantitative evaluation of gold and silver. But for other materials it was hit or miss. The ironworker, for example, had no way of knowing whether he had extracted all the iron from a charge of ore. Very likely more than half would have been left in the refuse or slag, so that later workers often found it worth while to rework them. As for determining the quality of the product, if the customer was satisfied, that was enough; there was no other criterion.

Not understanding what was going on, the artisan found it difficult to regulate his processes and distinguish significant from irrelevant factors. Adding a new ingredient one day, or giving the mixture a good stir, might appear to have improved the result and the new procedure would have passed into time-honoured practice until somebody accidentally omitted it without adverse effect. Nobody would have known or even asked why the new procedure seemed to work. There was thus no understanding of the effects of temperature, pressure and all the other conditions that are now known to influence chemical changes. Temperature was in any case difficult to control. The mainly charcoal-fired furnaces were awkward to regulate and things could easily get out of hand, as illustrated by the explosive mishaps that befell the alchemists of Chaucer and Ben Jonson. The poor quality of many of the reaction vessels was also a hindrance and led to much waste. Considering all the handicaps, it is indeed remarkable that such a range of useful materials was produced with an acceptably high quality. It was all achieved by craftsmen relying on skill of eye and hand gained through years of practice and inherited from generations of work in the industries concerned.

All this was to change dramatically within a relatively short space of time. The idea of increasing natural knowledge gained from observation and applying it to industry or the useful arts developed during the seventeenth century. The Fellows of the Royal Society, from its foundation in 1662, took a considerable interest in industry and made some useful suggestions for improvements in chemical processes. One of the founder members, the Hon. Robert Boyle, sharply criticized the prevailing ideas in chemistry and urged that it shed its disreputable alchemical connection and apply the concepts of the new mechanical philosophy. This criticism lacked precision, however, and a further century was to elapse before a chemical theory was established which actually corresponded with reality, at the hands of Antoine-Laurent Lavoisier from around 1780. The processes involving oxygen, such as combustion, were correctly explained, the nature of acids, bases and salts was put on a sounder footing, and in particular a clear definition of a chemical element was not only stated but usefully applied to draw up the first list of elements in the modern sense. A beginning was made in chemical analysis and after 1800 great improvements were made in quantitative analysis. Soon after 1800 rules for the way in which elements combined to form compounds were first enunciated, and with the atomic theory of John Dalton, chemists could visualize and explain chemical reactions in terms of the ultimate particles forming the basis of all matter.

Early beneficiaries of the chemical revolution were the manufacturers of cheap sulphuric acid, caustic soda, and chlorine for the textile industry. Developments in the industry gathered pace, informed by discoveries on the theoretical side. The nineteenth century was the era of pure and applied chemistry. The pure chemist was concerned to advance chemical knowledge for its own sake, irrespective of its possible practical use. The applied chemist, meanwhile, was employed to improve the processes for producing commercially useful substances, seeking new exploitable materials and, above all, in chemical analysis to monitor processes and the quality of products. Too often the two kinds of chemist worked in isolation from each other, the former being blissfully unaware of the needs of industry and the latter prevented from carrying out research that did not show an obvious profit. This division of role has, however, become increasingly blurred with the growth from the beginning of this century of the great chemical firms: indeed, the terms 'pure' and 'applied' chemistry can be said to belong to a bygone age. Improved contacts between the universities and industry make the former's research departments more aware of problems in industry, while much research in industry is in areas wider than those for which there is an immediate cash return. In addition, in most industrialized countries the state sponsors research and itself carries it out, in government laboratories, and without rigidly restricting its attention to problems of public concern. It is a melancholy fact that, in Britain, state, industry and the universities combined to deal with common needs never so effectively as in the two world wars. The production of the first atomic bomb is the prime example of such co-operation on an international scale.

By and large the chemical industry in the developed countries has been in the hands of private commercial firms and, however altruistic some of their activities may be at times, the ultimate reason for a process or product to be developed is that it will make a profit. It is worth noting that this profit is the source of funds for research by the state and the universities whether by direct sponsorship or indirectly through taxes. Because of the successful and systematic application of theoretical chemistry, first in inorganic then in organic chemistry and physical chemistry, especially the mechanism of reactions, the range of substances which the chemical industry has produced for man's use, with ever-improving quality, has been truly remarkable. The comparison with several millennia of near stagnation makes the progress of the last two centuries all the more striking. In 1800 the chemical industry was important, but on a small scale, its products limited to metals, acids, alkalis, pigments, tan-stuffs, medicines and a few other chemicals, some made on a scale not much greater than in the laboratory. Now the scale is vast, yet the industrial chemist exercises a precise control over the processes to yield an exactly predictable result. The source of this progress has been research. Sometimes progress has come by directing research to solving a particular problem, such as making a substance with certain required properties. But the more fruitful source has been to apply discoveries not made with a particular practical end in view. An example of the first is presented by Alfred Nobel and his intention to make nitro-glycerine a safe explosive. In the course of this he invented dynamite and blasting gelatine (see p. 223). But the more remarkable discoveries have been those that were not intended. Thus Perkin, while trying to synthesize quinine lighted on something quite unexpected, the first aniline dye, mauve – which led to a whole new industry (see p. 201). An example of the deliberate application of the results of pure research can be seen in the hydrogenation of oils to make fats like margarine, stemming from the study of the catalytic hydrogenation of unsaturated compounds in the presence of a metallic catalyst by Sabatier and his colleagues around 1900. Until then, the production of margarine, invented in 1869 by the French chemist Hippolyte Mège Mouriès, had been limited by the availability of raw materials, but the hydrogenation process enabled almost unlimited quantities of oils such as cottonseed oil to be converted into solid fats.

This chapter follows the history of the making of the more important substances or groups of substances that are a help to man, in one way or another, in his everyday life.

POTTERY, CERAMICS, GLASS

Pottery and ceramics

The hand-forming of plastic clay and changing it by heating into a hard body impermeable to water is a technique that goes back to the dawn of civilization, that is, to before 6000 BC. Indeed, primitive man, whether in prehistoric times or the present, fashions clay by hollowing out a ball and leaving it to dry in the sun or heating it on an open fire. Such simple means can hardly be classed as even primitive industrial chemistry, but with the rise of the ancient civilizations and the settled urban life that made tolerable the fragile nature of pottery, a number of materials began to be used for a variety of decorative effects. Potters learned, too, to control the temperature of their kilns to produce different colours. In modern parlance they employed reducing and oxidizing conditions to achieve various effects, without of course understanding the reason for this.

Clays are, chemically speaking, hydrated aluminium silicates with other substances such as alkalis, alkaline earths and iron oxide. It is this last that gives the commonest clay its characteristic red colour. The clays commonly found in nature are plastic when mixed with water and can be formed into a variety of shapes. When left to dry until the water content is 8–15 per cent the clay can still be worked, by scraping or turning, but lacks mechanical strength. After further drying and firing at 450–750°C the chemically-combined water is driven off, the clay can no longer combine with water, and it becomes like moderately hard stone. Firing at a higher temperature eventually causes the clay to vitrify and fuse, but that stage was rarely reached in the ancient world. It is impossible to date the technical advances made during the early civilizations of Egypt, Mesopotamia and the Indus, but the art of throwing pots on the potter's wheel evolved at this time, as also the firing of the ware in kilns, fuelled with wood or charcoal, in place of the open fire. Temperatures of just over 1000°C could occasionally be reached and much greater control of the draught and therefore the heating conditions was achieved. Pottery could be rendered sufficiently non-porous by burnishing, that is, smoothing the unbaked surface by rubbing, but a better surface could be obtained by dipping the ware in a 'slip' or a slurry of fine clay and firing, or by glazing, that is, painting on to the surface a substance which on firing would turn into a thin layer of glass. The Egyptian blue glaze was a notable example. It was made from white sand, natron, limestone and a copper compound, perhaps malachite, which imparted a blue colour to the mixture. This was heated for two days at around 900°C, powdered and applied as a glaze to a siliceous body.

The Assyrians, about 700 BC, introduced lead oxide-based glazes, an important development as this was the first glaze that would adhere to a clay base. They were able to obtain a yellow colour by roasting antimony sulphide with lead oxide, and blue and red from copper compounds. The Greeks and Romans made progress in fine workmanship and artistic design rather than in technology. The Greeks, from about 600 BC, did however develop the technique of black and red ware achieved by using reducing and oxidizing conditions to produce two different states of the iron oxide in the red clay.

PART ONE: MATERIALS

The most interesting development over the next millennium was that of lustre ware. A paste formed from powdered sulphides of copper and silver was applied to a body that had already received a tin-lead glaze, then heated to leave a thin, lustrous layer of copper and silver. The technique arose in the Middle East in the ninth century AD and spread through Islam, reaching Moorish Spain by the fourteenth century. From here it was exported throughout Europe and highly prized by those who could afford more sophisticated tastes, while of course simple peasant ware continued to be made. In Italy it was doubtless the spread of lustre ware that stimulated the tin-glazed ware known as majolica which flourished particularly over the period 1475–1530. Applying coloured tin–lead glazes to sculpture, Luca della Robbia achieved delightful results. A manuscript by one Picolpasso gives details of the glazing and colouring materials and processes that were applied to the white clay base.

Now, however, an entirely new product was about to make an impact on European taste and fashion, Chinese porcelain. It was known to the Muslims, but examples did not percolate into Europe until the sixteenth century. The trickle became a flood after the eastern trading companies were set up, in the wake of the voyages of exploration, in particular the Dutch East India Company founded in 1609. Chinese pottery is of great antiquity, going back to the third millennium BC. Glazed pottery appears in the third century BC and lead glaze soon afterwards in the Han dynasty, a little earlier than Roman practice in the West. But the great Chinese discovery was that of porcelain, of which the main constituents are kaolin or china clay, which is infusible, and a fusible mixture of feldspar, clay and quartz. This had to be fired at a higher temperature – around 1400°C. Various colours were applied, but above all blue from cobalt minerals. A mineral with just the right amount of impurities, imported probably from Persia in the fourteenth and fifteenth centuries, produced a particularly lovely blue. Thereafter a local mineral had to be used, giving a rather inferior colour. The earliest porcelain is of the eighth or ninth century, it came to maturity during the Sung dynasty (960–1127) and reached its glorious perfection in the Ming dynasty (1368–1644).

The energies of European potters were now to be directed to discovering the secret of Chinese porcelain and to making something that looked like it. Dutch potters centred on Delft produced the first successful imitation, using carefully prepared clay and tin-enamel glaze. By the end of the seventeenth century delftware had spread to England and was being manufactured at Lambeth, Bristol and Liverpool.

Porcelain is fired at a temperature that produces vitrification, that is, the formation of a glassy substance, hence its translucent appearance. It seemed to the early experimenters to be half-way between pottery and glass, so they tried using glassmaking materials and fair imitations were the outcome. But the first true porcelain, using china clay, was achieved by the German Johann Friedrich

Böttger, who began working on the problem in 1701 in the royal laboratory of Friedrich August II, Elector of Saxony, in the town of Meissen. After ten years he came across kaolin, being sold as a wig powder, and feldspar, which often occurs with it. With these he obtained a true porcelain. By 1716 he had perfected the technique to such an extent that the products could be marketed. The Elector was anxious to keep the secret to himself and kept Böttger a virtual prisoner, but to no avail. The knowledge spread with the wares, and Meissen was soon to be overtaken by Sèvres. Originally at Vincennes, the French state factory moved to Sèvres in 1756. It always enjoyed royal patronage and by 1759, Louis XV had become proprietor. At first, soft paste substitutes were produced, but eventually, from 1768, under the superintendence of the chief chemist P. J. Macquer, a true hard porcelain was produced.

In England the pattern was repeated. A soft paste porcelain, using powdered glass with a white clay, was produced at factories established at Stratford-le-Bow in east London, in the late 1740s, followed by Chelsea, Derby, Lowestoft, Longton Hall in Staffordshire and Worcester. At the same time, William Cookworthy, the Plymouth chemist, had been experimenting with clays in Devon and Cornwall and in 1768 he felt sufficiently confident to take out a patent for the production of a true porcelain. His technical prowess was not, however, matched by business acumen and he disposed of the patent to Richard Champion. The latter found difficulty in renewing the patent in 1775 when it was successfully challenged by a group of Staffordshire potters including Josiah Wedgwood, who began to make hard paste porcelain from 1782 at New Hall, near Shelton.

Most of the factories mentioned above were not particularly well sited in relation to sources of raw materials and fuel. Those in north Staffordshire were much better in this respect and so it was here that the great industrial expansion of porcelain manufacture took place. The momentous changes that came about were largely the work of one of the greatest potters of all time, Josiah Wedgwood. One major change was the substitution of a white-burning clay and calcined crushed flint (silica) to give a ware that was white through the whole body in place of the common and buff clays. The preparation of the raw materials, including the crushing of the flint, required mechanical power, first supplied by water power, then by steam. Wedgwood had seen a Newcomen engine at work when visiting clay sites in Cornwall and he was the first potter to order an engine for his works from Boulton and Watt (see p. 276). Wedgwood evolved a ware consisting of four parts ground flint and 20 to 24 of finest white clay, glazed with virtually flint glass. Having secured royal patronage, it became known as Queen's Ware and was widely used for all kinds of table ware.

Wedgwood pioneered the application of steam power in the pottery industry in 1782. As elsewhere, this changed the pattern and scale of operations and led to the factory system with a central source of power for a variety of mechanical operations. Wedgwood's Etruria works were the first on these lines. But he was also concerned to apply the scientific knowledge of the time to materials and processes, as with his clay pyrometric cones which contracted on heating, enabling the temperature in the kilns to be more accurately measured and therefore more effectively controlled.

By 1787 there were some 200 master potters, employing 20,000 in north Staffordshire, making it the foremost pottery manufacturing area in the world. A new product was introduced by Josiah Spode: bone ash and feldspar with white clay to produce that characteristically English ware, bone china. As the Industrial Revolution gathered pace, there was an increasing demand for porcelain or similar ware, otherwise called 'china ware', and two inventions, both originating in the mid-eighteenth century, helped to meet it. One was mould forming, in place of the traditional potter's wheel, while the other was transfer printing instead of decorating each piece freehand. A compromise here was to transfer only a faint outline of the design on to the piece, leaving the craftsman to paint in the detail. This method was practised from the 1830s especially at the Coalport works in Shropshire and the Rockingham works at Swinton in Yorkshire. Mass production methods and improved transport brought cheap china to most tables in the industrialized countries, although, as often but not necessarily happens, there was a decline in the quality of design. On the other hand, after the chemical revolution, a better understanding of the nature of the potter's materials produced better bodies and glazes. This among other things ended the dependence on lead glazes, to the great benefit of the health of those who had to use this harmful material. New effects were produced, such as the celebrated Persian turquoise blue of J. T. Deck of Paris in 1861 (*bleu de Deck*). New colouring agents arrived, like uranium (1853), and new effects such as flame-mottling by controlling conditions in the kiln.

Meanwhile developments in other industries found new uses for ceramic materials, above all the electrical and chemical industries. The word 'ceramics' also came into common use during the last century, denoting articles made by forming and firing clay, from the Greek *kerameikos*, the potters' quarter of Athens. The mechanical, weathering and electrical properties of porcelain made it an ideal material for insulators and resistors, still largely made from ceramics. In the 1850s bell-shaped insulators for telegraph poles came into use throughout the world. In the chemical industry, ceramic-lined vessels became a necessity for certain processes and to contain such materials as acids: chemical stoneware is resistant to cold acids, except hydrofluoric, and most hot acids.

Progress in the metallurgical industries could not go far without improving on the crude clay-lined furnaces of earlier times. From 1860, Austrian magnesite bricks came into wide use for iron and the new steelmaking furnaces. An understanding of the acid or alkaline nature of refractory furnace linings was crucial to the success of the Bessemer steelmaking converter. When the outbreak of the First World War interrupted the supply of magnesite, the drawbacks in the use of the cheaper dolomite were overcome.

The effect of mechanization in raising output has already been mentioned. A further boost was provided by improvements in kiln design. Efforts were first made to economize in fuel consumption and to cut down the smoke that poured from the kilns, making the pottery districts most insalubrious. But the major development was, as in other industries, to replace batch, or non-continuous, heating by continuous firing, as in the tunnel kiln. The first was built in Denmark in 1839 and, although not really satisfactory, its importance was recognized. Improvements followed and a kiln fired by producer gas was erected in 1873 and patented four years later. In 1878 a tunnel kiln was installed in London and the first in the USA was at Chicago in 1889.

During the last half of the nineteenth century, the ceramics industry changed further into a science-based technology, as the materials used in the industry and the processes they underwent were subjected to systematic scientific examination. The credit for much of the pioneer work on the clays belongs to the chemist H. Seger.

In the present century the range of ceramic materials and their application throughout industry has greatly widened. In fact over the last fifty years the traditional definition of ceramics as clay-based products has had to be abandoned. The term is now broadened to cover any inorganic substance which, when baked, attains the familiar rock-like hardness with other special characteristics. Silicon carbide is such a material, with important applications in the abrasive industry.

Glass

Glass is one of the most familiar of materials, with a wide range of applications in the modern world, yet with a history stretching back into antiquity. It is formed by melting mixtures of various inorganic substances and cooling them in a way that prevents crystallization – the molecules do not, as with most solids such as metals, arrange themselves in regular crystalline patterns. It is in fact more accurate to speak of glass as a rigid liquid than a solid. The basic ingredients of common or soda-lime glass are sand (15 parts), soda ash (5 parts) and lime (4 parts). Instead of sand, the silica could be in the form of quartz or crushed flint. In pre-industrial eras the alkali was provided by the ash of certain plants, fern being particularly preferred.

Primitive man sometimes fashioned naturally occurring glasses such as obsidian, a glassy volcanic rock, into useful objects, like arrowheads. The earliest artificial glass dates from around 4000 BC in Egypt, in the form of a coloured, opaque glaze on beads. During the second millennium BC small hollow vessels could be produced by core moulding. A clay core was covered in successive layers of molten glass and the core scraped or washed out.

During the first century BC came one of the technological breakthroughs of the ancient world, the invention of the blowing iron. This made possible the art of glassblowing, either free or into a mould. The art spread rapidly through the Roman Empire and, with tools for decorating the surface of the glass and materials for colouring it, a wide variety of useful and often beautiful ware was produced. In essentials the techniques employed have survived throughout the period of hand-made glass even to this day.

After the collapse of the Roman Empire the tradition of glass-making in its most sophisticated form survived in the Near East and later in Islam. In Europe during the so-called Dark Ages, the tradition remained alive in a simpler form. The compilation of *c.* 1100 by the German monk Theophilus of Essen, *Schedula diversarum artium* (Account of various arts), gives details of the glass-making methods in use at the time, including window glass. This was formed by blowing a vessel like a 'long bladder', opening it out and flattening it. It was then cut into the required sizes and shapes. Very early ecclesiastical stained glass can be seen in the churches in Ravenna, but it reached its perfection during the Middle Ages. The glory of mediaeval glass is the richness of its colouring, produced by chance combinations of impurities in the colouring materials used. These combinations have long since been lost, so the colours of mediaeval stained glass have hardly been matched.

The Roman glass-making techniques were brought to Europe, possibly as a result of the Crusades, especially to Venice, where the art began to flourish during the thirteenth century. The Venetian craftsmen established themselves on the island of Murano, at first in conditions of strict secrecy; but as their ware became renowned throughout Europe, so knowledge of their materials and methods spread too. Apart from the quality and intricacy of the glass, prized above all was the cristallo, a colourless glass produced, like the Romans before them, by adding to the melt manganese dioxide, which oxidized the iron in the sand to the colourless ferric state.

The furnaces and tools in use during the sixteenth century are described and illustrated in the celebrated *De re metallica* of Agricola, printed in 1556. The furnaces were in two portions; in the lower the materials were melted in pots from which the glassblower gathered a 'gob' of molten glass on his blowing iron. The upper part was the annealing chamber, where the finished ware was allowed to cool slowly, to ease out the strains which would be caused by rapid cooling and result in breakages.

Glass-making was scattered throughout Europe, where the raw materials were to hand, but Venetian clear or 'cristallo' glass remained highly prized. The Worshipful Company of Glass Sellers of London commissioned one George Ravenscroft to find a recipe for a comparable glass using local materials. At first he took crushed flint as his source of silica, but this produced numerous fine cracks in the glass, known as 'crizzling'. To overcome this defect he used increasing amounts of lead oxide and obtained a relatively soft, heavy glass with high refractive index and dispersive power. This made it amenable to deep cutting and this, with its optical properties, produced the brilliant prismatic effects of cut glass. Ravenscroft's lead glass was patented in 1673 (it has also been called 'flint glass' from his original source of silica) and from it was formed the sturdy baluster ware, the later engraved Jacobite ware and the familiar cut glass.

Glass has been made for optical purposes since the Chinese began to make magnifying glasses in the tenth century. Spectacles to correct long sight appeared in thirteenth-century Italy. During the seventeenth century, the period of the scientific revolution, the invention of the telescope and the microscope made much greater demands on optical glass, but glass of satisfactory quality for lenses was not consistently made until Guinand's invention in 1805 of a porous fireclay stirrer to bring about a proper mixing of the glass melt and eliminate gas bubbles.

Another use of glass with a long history is for windows. In Roman times, only small pieces of flat glass could be produced, by casting in a mould. From the Middle Ages until the present century window glass was formed by blowing, following one of two processes. The crown glass method involved the blowing of a cylinder which was opened at the bottom; after heating the open end at the furnace mouth, or 'glory hole', the blowing iron was rotated rapidly until by centrifugal force the bottle suddenly flared out to form a flat disc. This was then cut into rectangular pieces measuring up to about 50 cm (20 in). The glass at the centre or crown of the disc (hence the name crown glass), where the iron was attached, was too thick to be used in windows except in lights above doors where light was to be admitted but transparency not required. The other process also entailed blowing a cylinder, but this was then slit down the side and the glass gently flattened while still in a plastic state. In the nineteenth century very large cylinders could be blown and these were the source of the glass for such structures as the Crystal Palace and the large railway station roofs that were such a feature of Victorian structural engineering. Later, in the 1920s the drawn cylinder process was developed whereby a circular plate was dipped into molten glass, then slowly drawn up. The flat glass produced by these methods retained a fire-polished finish but was never perfectly flat. To achieve that, the cast plate process was invented in seventeenth-century France, particularly for the large windows and mirrors for the Palace of Versailles. In the 1780s the process was established in England at the Ravenhead works near St Helens in Lancashire. Some forty years later the firm was rescued from the low ebb into which it had sunk by a Dr Pilkington, one of the most illustrious names in glass-making history.

Cast plate glass was certainly flat, but removing it from the casting tray destroyed the fire finish and this had to be restored by grinding and polishing. The age-old dilemma, between nearly-flat glass with a fire finish and flat glass without it, was eventually resolved in the 1950s by perhaps the most notable advance in glass technology this century, the invention of the float glass process. Patents for float glass date from the early years of the century but came to nothing. It was the invention by Sir Alastair Pilkington FRS that succeeded, working at Pilkington Bros (he is a namesake, not a relative of the family). In 1952 he conceived the idea of floating a layer of molten glass on a bath of molten tin in a closed container, in an inert atmosphere, to prevent oxidation of the tin. The product is flat glass that retains a polished fire finish. After seven years of development work, the new product was announced and became a commercial success.

During the nineteenth century the increased wealth generated by the Industrial Revolution led to an increased demand for glassware of all kinds and in 1845 the repeal of the excise duty that had been hampering the British industry since 1745 stimulated growth still further. The old furnaces with their small pots for making glass were outpaced and outmoded. They were replaced by the large-scale, continuous operation tank furnaces, developed by Siemens and others. The pot furnace survived only for small-scale handmade glassworking.

The glass bottle had begun to replace stoneware to contain wine and beer around the middle of the seventeenth century. The earliest wine bottles were curiously bulbous in shape but as the practice grew of 'laying down' wine, the bottles had to take on their familiar parallel-sided form, by about 1750. The use of glass as a container for food and drink grew considerably from the middle of the nineteenth century and improvements were made in form and process. Codd in 1871 invented an ingenious device for closing bottles of mineral water, by means of a marble stopper in a constricted neck. So far bottles were hand blown into moulds but in 1886 Ashley brought out a machine that partially mechanized the process. The first fully automatic bottle-making machine appeared in the USA in 1903, invented by Michael Owens of Toledo, Ohio. Further development came with the IS (Individual Section) machine from 1925, in which a measured amount or 'gob' of molten glass was channelled to the bottle moulds.

The trend throughout this century has been to greater mechanization; stemware, for example, could be produced automatically from 1949. The other trend, from the last two decades of the nineteenth century, was to develop glasses with special properties with different compositions. One of the best-known examples is borosilicate glass, formed essentially from silica, boron trioxide and alumina, magnesia or lime. It has a high resistance to chemical attack and low thermal coefficient of expansion, making it very suitable for laboratory and domestic ovenware.

New forms of glass have appeared, such as glass fibre, with the interesting development of fibre optics. Experiments were made in the 1920s on the transference of images by repeated internal reflection in glass rods and these led in 1955 to the fibrescope. Bundles of fibres cemented together at the ends can be used to transmit images from objects otherwise inaccessible to normal examination.

TEXTILE CHEMICALS

Dyestuffs

Man's attempts to brighten himself and his surroundings with the use of colour go back to prehistoric times and many kinds of plants were used to stain skins and textiles with variable success. To dye cloth successfully so as to withstand the action of air, light and wear, dyers settled down to use scarcely more than a dozen or so dyestuffs and this limited range lasted up to the middle of the nineteenth century. It was only then that progress in organic chemistry enabled the secrets of the structure of the chemicals concerned to be unravelled and this paved the way for the synthetic dyestuffs industry, adding enormously thereby to the stock of colouring materials at man's disposal.

It is useful here to distinguish between vat and mordant dyes. In vat dyeing, the dye substance, insoluble in water, is converted by chemical treatment into a substance that is soluble; the cloth is then steeped in a solution of the latter and left to dry. The action of the air forms, by oxidation, the colour of the original substance on the fibres of the cloth. Indigo and woad were used in this way. With mordant dyeing the cloth is first boiled with a solution of the mordant, usually a metallic salt, and then again in a solution of the dyestuff. Different mordants can form different colours with the same dyestuff. Since antiquity the mordant used above all was alum, a term now commonly taken to mean a double sulphate of aluminium and potassium crystallized with 24 molecules of water. Before its composition was known, towards the end of the eighteenth century, the word was more loosely used to mean a white astringent salt to cover several different substances. The use of alum as a mordant for dyeing cloth red with madder can be traced back to around 2000 BC in Egypt.

The mediaeval dyer used alum in large quantities, sometimes from native alum-rock from Melos or other Greek islands, a source since Roman times. Various kinds of alum were imported from Middle Eastern regions, or the mineral alunite was converted to aluminium sulphate by roasting. Alternative sources were eagerly sought, particularly after eastern supplies were cut off by the advance of the Turkish Empire in the fifteenth century. Fortunately a large deposit of the mineral trachite, which yielded alum on being treated with sulphurous volcanic fumes, was found at Tolfa in the Papal States. This led to a highly profitable papal monopoly in the alum trade. After the Reformation,

Protestant countries sought other, local, sources. In England, for example the shale found in Yorkshire was used; roasting oxidized the iron pyrites it contained to ferrous sulphate, which at a higher temperature decomposed and converted the aluminium silicate also present to alum. Similar processes were carried on all over Europe. They remained inefficient until they were better understood; when it was realized that the aluminium ion was the active agent in mordanting, aluminium sulphate gradually replaced alum.

Of the dyestuffs themselves, the only successful blue dye known before the last century was indigotin, derived from the indigo plant in tropical areas and from woad, with a lower content of indigotin, which grew widely throughout Europe. Indigotin is insoluble and so could not be used in a dye bath. The plants were therefore left to ferment to produce the soluble indigo-white. The cloth was steeped in the liquor and as it was hung out to dry, oxidation to indigo-blue took place. The Egyptians of 1500 BC were dyeing with indigo and some of their fabrics have retained their blue colour to this day. In Graeco-Roman and mediaeval times, woad was chiefly used to dye blues.

For yellow, the earliest dye appears to have been safflower. It had a long history for it has been detected in mummy wrappings of 2000 BC and was still in use until quite modern times. Saffron, the stigmas of the saffron crocus, was also used, hence for example the name of the town Saffron Walden in Essex, a noted centre for the flower in the Middle Ages. It has long been used in India for dyeing the robes of Buddhist monks. The most widely used yellow dye in mediaeval times was weld or dyers' weed, which gave a good yellow on cloth mordanted with alum.

The madderplant, derived from a species of *Rubia* which grows wild in the Mediterranean and Near East regions, also has a long and distinguished history. Again, the Egyptians were using it around 1500 BC, to dye cloth red. From Graeco-Roman times a red dye was also obtained from various species of *Coccus*, insects parasitic on certain plants, such as cochineal. The Arab word for coccus was *kermes*, from which our word crimson is derived. When cardinals began to sport their red-coloured robes in 1464, the hue was produced from kermes with alum as mordant – a crimson rather more subdued than the brilliant scarlet we know today. The latter became possible in the seventeenth century using cochineal mordanted with a tin salt. A highly prized colour in the ancient world was a dark brownish-violet produced from several species of shell fish including *Purpura*, hence the name purple. The purple-dyeing industry tended to be located in Mediterranean coastal regions and the Phoenician towns of Tyre and Sidon were notable centres of the trade.

Of the dyer himself and his methods rather less is known. His was a messy, smelly occupation and he tended to keep to himself the secrets of his craft. It was a skill learned from others and by practice which probably changed little until the first printed accounts began to appear. The craft was virtually static until the late eighteenth century when rapid changes in the textile industry demanded improvements in dyeing techniques. But the range of dyestuffs available remained the same until the mid-nineteenth century, when the ancient craft began to be transformed into a science-based technology.

Although chemical theory had been put on a sound footing around 1800, the structure of colouring matters was too complicated to be quickly resolved. Starting with Lavoisier in the 1780s, then Jöns Jakob Berzelius, and Justus von Liebig from 1830, substances found in the plant and animal kingdoms, hence known as organic compounds, were analysed. Lavoisier had established that carbon was always present and usually hydrogen and oxygen. Later the presence of other elements such as nitrogen or sulphur was recognized. By the middle of the century the formulae of many organic substances had been ascertained, that is, the numbers of the different kinds of atoms contained in the molecules. It was found that compounds could have the same 'molecular formula' but possess different properties because their atoms were combined in a different way. This was expressed in structural formulae or diagrams showing how the atoms were imagined to be combined. Certain groups of atoms, like a carbon atom linked to three hydrogen atoms (CH_3), were found to be present in many different compounds, producing a particular effect on its properties.

At the same time as progress was being made on the theoretical side, many of the constituent compounds were being extracted from natural substances. One of the most important of these was benzene, found to be present in coal tar in 1842, which became a subject of research by the brilliant group of chemists which August Wilhelm von Hofmann gathered round him at the Royal College of Chemistry, founded in 1845. Prince Albert had been instrumental in securing Hofmann's appointment as the first professor there. Benzene was the starting point for many compounds, including an oil called aniline (first prepared from the indigo plant for which the Portuguese name is *anil*). One of Hofmann's keenest and brightest students was the eighteen year-old William Henry (later Sir William) Perkin. In 1856, in a laboratory fitted up at his home, he was trying to prepare quinine from aniline and its derivatives, as they appeared to be related structurally. The result, not unfamiliar to chemists, was an unpromising black sludge. On boiling it in water he obtained a purple solution from which purple crystals were formed. He tried dyeing a piece of silk with this substance and found it produced a brilliant mauve colour, resistant to washing and fast to light. It was the first synthetic, aniline dye. Perkin sent a specimen to the dyers Pullars of Perth, who reported favourably. He then set about exploiting the discovery, first on a back-garden scale, and then in a factory opened the following year at Greenford Green near Harrow, from family capital. The new 'aniline purple' swept the board in England and abroad – the French seized on it, naming it 'mauve'. Queen Victoria wore a mauve dress at the opening of the International Exhibition of 1862; penny postage stamps were dyed mauve. Perkin's commercial success was such that he was able to retire from business at the ripe old age of 35 and devote his time to chemical research. Other chemists followed in his wake with other dyestuffs derived from aniline. It was found that aniline could be subjected to the diazo reaction, lately discovered by Peter Griess and so named because two nitrogen atoms were involved. When the product of this reaction is treated with phenol, highly coloured substances are formed, many yielding satisfactory dyes. The first azo dye was Bismarck brown, prepared by Carl Alexander Martius in 1863.

The next nut to crack was the synthesis of the red colouring matter in madder, alizarin. The elucidation of its structure had to await further progress on the theoretical side and this was forthcoming when August Kekulé realized that benzene had a cyclic structure, that is, the six carbon atoms in the benzene molecule were joined up in a ring, visualized as the famous hexagonal benzene ring. Following on this, Graebe and Liebermann were able to work out the structure of alizarin and then to devise a way of synthesizing it on a laboratory scale. It was however Heinrich Caro, a chemist responsible for many advances in this field, who worked out a manufacturing process for synthetic alizarin, involving sulphonation of anthraquinone with concentrated sulphuric acid, while working for the firm Badische Anilin- und Soda-Fabrik. Perkin was working along the same lines and was granted a patent for his process on 26 June 1869 – one day after Caro received his. A friendly settlement was reached, allowing Perkin to manufacture alizarin in Britain under licence from BASF. The synthetic dye was much cheaper than the natural version, so the madder-growing industry fell into rapid decline and expired.

Success with alizarin stimulated chemists to turn their attention to indigotin. After many years of research, in which Adolf van Baeyer figured prominently, the molecular structure was found and in 1880 a method of synthesizing indigotin described. Again, the transition to manufacturing scale proved difficult; it was only in 1897 that success was achieved, after long and expensive research supported by BASF.

By 1900 and beyond, the industry had not only achieved cheaper and more consistent production of the dyes previously found in nature, but at an ever-increasing rate added enormously to the range of dyestuffs and colours available. In this great advance Britain had been given a head start by Perkin's discovery, but the initiative was let slip and passed to Germany. The British industry lagged behind to such an extent that by the outbreak of the First World War, Britain had to import all but 20 per cent of her dyestuffs, mainly from Germany. The sudden removal of German competition had a tonic effect on the home industry, which rose to the occasion to meet the need. British, and also American, industry soon matched the Germans in this field. Two main factors contributed to the German pre-war pre-eminence in this and other areas: one was the sound education offered in school, university and polytechnic, heavily subsidized by the state, to ensure a good supply of well-trained chemists, and the other was the willingness of industrial concerns to employ chemists and fund research on a quite lavish scale. The synthetic aniline dye industry was the first really science-based industry and demonstrated the spectacular progress that could be achieved by the direct and deliberate application of scientific knowledge to industry.

The twentieth century has become the most riotously colourful period in history thanks to the researches of the dyestuffs chemists. Apart from improving the properties of the existing dyestuffs, research has been directed to finding new dyes and ways of applying them to the new man-made textile fibres (see Chapter 17) and to a wide range of other products – plastics, rubbers, foods, printed materials and so on. Of particular lines of progress, the evolution of rigorous, internationally standardized colour-fastness tests is notable, also the widening of the range of vat dyes, including the first really fast green dye, Caledon Jade Green, announced by James Morton of Scottish Dyers Ltd in 1920. The introduction of artificial fibres was held up for a while when it was found that the water-soluble dyes that had so far been used could not be applied to them. The British Dyestuffs Corporation was particularly effective in the 1920s in solving the problem. Certain insoluble amino-azo compounds could dye acetate fibres when used in a finely divided state; later anthraquinone dyes used with a dispersant gave a good range of fast colours for these fibres. Another development was that of the metal phthallocyanine dyes, giving brilliant, fast colours. Manufacture was begun on a small scale by ICI in 1935–7 and, after the Second World War had held up development, these became commercially significant during the 1950s.

Soaps and detergents

Man's efforts to keep himself and his clothes clean go back to ancient times. The Egyptians used natron, an impure form of sodium carbonate from lake deposits, as a cleansing and mummifying agent, and some other alkaline materials obtained by extracting with water the ashes of burnt plants, yielding potash (impure potassium carbonate). Oils were also used and if these were boiled with lye, or alkali solution, a soap would have been formed, as is mentioned by the Ebers papyrus of 1550 BC. There is, however, no clear reference to soap and its use by the ancients must remain conjectural. In Graeco-Roman times, oil was much used and also abrasive detergents such as ashes or pumice stone. Soap, it would seem, still was not used, although the roots of certain plants contained saponin and would therefore form a lather.

The earliest clear reference to soap occurs in writings of the first century AD. The word itself is probably of Teutonic or possibly Tartar origin; the latter may, indeed, have invented it. Soap was certainly widely used in mediaeval times, for washing clothes rather than people. The lye was 'sharpened' by adding lime (this would convert it to the caustic form) and the clear solution boiled with oil or fat. Lye prepared from wood ashes (potash) yielded soft soap, while that from natron, barilla or rocchetta, forming soda, produced hard soap. A French manuscript of *c.* 800 has the first European mention of soap. In northern countries it was soft soap that was produced, by boiling wood ash lye with animal fats or fish oils. Its smell would not have been pleasant, which explains why it was used for washing clothes and not people. But in Mediterranean regions, the ash of the soda-plant was used, with olive oil, to give a hard, white, odourless soap. Its manufacture flourished in Spain from the twelfth century – Castile soap has an eight-hundred-year history. Marseilles became a leading centre in the fourteenth century and later Venice. From these areas, hard soap was exported all over Europe as a luxury item.

In England, by the end of the twelfth century, Bristol had become the main centre for the making of soft soap. Three hundred years later, this had become a major industry, for which ash had to be imported to supplement local sources, to meet the needs of the flourishing woollen cloth industry. The Royal Society, founded in 1662, took an interest in soap manufacture, among many other technical processes, and soon afterwards occurs the first mention of 'salting out', that is, the addition of salt to the hot soap liquor to throw the soap out of solution, on which it floats and solidifies on cooling. It can then easily be removed. This had the effect of speeding up the whole process. The imposition of the salt tax hindered the spread of this improvement but its removal early in the nineteenth century was a fillip to this and other branches of the chemical industry.

Soap reigned supreme in the nineteenth century, but it had its drawbacks. It broke down, and so was ineffective in acid, but the main problem was the scum formed on textile fabrics in hard water. The textile industry had a definite need for non-soapy substances with soap-like properties. Again, it was the academic chemists, in Germany, Belgium and Britain over the period 1886 to 1914, who showed the way, by finding that compounds with a long hydrocarbon chain ending in a sulphonate group could act as detergents without the undesirable properties of soap. Industry then had the task of converting these findings into commercially viable production. The first synthetic detergent appeared in Germany in 1917, stimulated by the extreme shortage of animal fats during the war, and was marketed under the name Nekal, but it was not altogether satisfactory. Further work during the 1920s on sulphated fatty alcohols led at last to a product that was commercially successful for wool, but less so for cotton. During the 1930s American, German and other European chemists developed new detergents incorporating complex phosphates with the previous constituents. Another major development of the 1930s was the production of detergents from petroleum-based materials, in which the oil companies figured.

Another innovation, generally adopted after 1945, was the use in minute quantities of fluorescent whitening agents – 'blue whiteners' as they are known. Krais, a German chemist, noted in 1929 that certain substances converted the ultra-violet rays in sunlight to visible blue rays. The first patents were taken out by IG Farbenindustrie in 1941 and led to the commercial use of this means of offsetting the yellowing of whites in the wash.

Shortage of the raw materials for soap-making during the Second World War hastened the steady replacement of soap by detergents; now soap is almost entirely restricted to personal washing.

Bleaching

The ability to remove unwanted colour is as important as the means to introduce wanted colour. The revolution in the textile industry would have been much impeded had it not been for the improvements in the bleaching process, speeding up what had been until then a very slow process. Linen was boiled in lye, washed and laid out in the sun for a few weeks. This was repeated several times, until a final treatment with sour buttermilk. The whole process took six months, and a good deal of space, for which the rent was an appreciable cost. Cotton took up to three months to bleach. Around the middle of the eighteenth century, sulphuric acid was substituted at the 'souring' stage, which was thereby reduced to a matter of hours. Then, in 1774, the Swedish chemist Carl Wilhelm Scheele discovered chlorine, among many other substances of fundamental importance, and noted its bleaching properties, but it was Claude Louis Berthollet who in 1785–6 experimented in the practical use of chlorine for bleaching. The method was quickly taken up in England, Scotland and continental Europe. The gas was first prepared by the action of sulphuric acid on manganese dioxide. Water was then saturated with the gas and cloth was bleached by soaking it in the heated chlorine water. This dangerous process was still in use in 1830, but better methods soon came in. The chlorine was dissolved in alkaline solutions, to give hypochlorites, which were safer to handle. More widely used was Charles Tennant's bleaching liquor from 1798, made from chlorine and lime-water. A year later this was converted into solid bleaching powder, which is used to this day.

Between 1866 and 1870 the Weldon and Deacon processes for producing chlorine began to supersede the earlier methods, and in turn gave place to the electrolysis of fused sodium chloride (salt). For materials such as wool and silk that are damaged by chlorine, sulphur dioxide was the normal agent.

The only additional agent for bleaching has been hydrogen peroxide, discovered by Louis Jacques Thenard in 1818. It was usually prepared by the action of dilute sulphuric acid on barium peroxide.

PART ONE: MATERIALS

FUELS

Wood and charcoal

The origins of fire-making are lost in prehistoric times, even antedating our own species, for it can be traced back to Peking man. The most widely used fuel was wood, although in the regions of two of the earliest civilizations, Egypt and Mesopotamia, where wood was in relatively short supply, other materials were used, such as dried ass's and cow's dung and the roots of certain plants, including the papyrus plant. But for the more demanding processes, such as the firing of pottery or the smelting of metals, wood and charcoal had to be imported. Charcoal is a porous form of amorphous carbon, made by burning wood with a supply of air insufficient to secure complete combustion. It is the almost perfect solid fuel, for it burns to give a high temperature with little ash and no smoke. It was used widely for domestic heating, being burned on shallow pans, but, much more important, it was the fuel *par excellence* for industrial furnaces from ancient times until the seventeenth century, when it began gradually to be replaced by coal. The environmental, social and economic effects of the charcoal burner's trade were profound, for vast tracts of forest land were laid waste to satisfy the voracious appetite of industry for fuel.

The method of making charcoal hardly varied over the centuries. Logs, cut into 90cm (3ft) lengths, were carefully stacked around a central pole into a hemispherical heap, up to 9m (30ft) in diameter at its base. The heap was then covered with earth or, better, turf, the central pole was removed and burning charcoal introduced down the centre to set light to the mass. Combustion was controlled by closing or opening air-holes in the outer covering.

The charcoal for iron smelting was often made from oak or ash, while alderwood stripped of its bark was used for charcoal that was to be ground fine for gunpowder. From the early nineteenth century the firing was carried out on a hearth that sloped towards the centre so that liquid products, particularly pitch, could be drained off. Pine and fir were preferred where these products were important, as they were for making timber preservatives, above all for the shipping trade, which required pitch for timbers and ropes.

Another important product of the combustion of wood was the inorganic, alkaline constituent of the wood, required in large quantities by glass- and soap-makers (see pp. 195, 203). All in all, the demands on the forest resources were great and ever-increasing; a substitute would sooner or later have been essential, but a rival claimant for timber made matters worse. Throughout this period timber was the main construction material and the demands of shipping, particularly the strategic requirement of navies, hastened the arrival of coal on the scene.

THE CHEMICAL AND ALLIED INDUSTRIES

Coal

Coal is formed from tree and plant remains under the influence of high temperature and pressure over a period of thousands of years. The formation of peat and brown coal or lignite are stages in the process. It was first used as a fuel in either India or China about 2000 years ago and it was certainly known to the Greeks and Romans. In mediaeval Europe it was in use for industrial purposes, especially in dyeing and brewing, but not for domestic heating. In the absence of chimneys, the fumes from burning fuel had to find their way out through windows or any other available opening; the products of combustion of wood and charcoal were apparently tolerable, but those of coal were too offensive.

From the middle of the sixteenth century coal production in Britain increased considerably, in mining areas in South Wales, Scotland and, above all, Northumberland and Durham, all coastal regions whence the coal could be transported to London and other centres by sea (hence the term 'sea-coal'). In Elizabethan times, town skylines began to be punctuated by chimneys, to enable the domestic user to burn coal, with effects on the environment that provoked a sharp reaction from the authorities. From the early seventeenth century, anxieties about timber supplies led to prohibitions of its use in certain industries. In 1618 glass-makers had to turn to coal, covering their pots to prevent harmful fumes affecting the molten glass. The iron masters were considerable potential customers, but periodic efforts to smelt iron ore with coal during the century ended in failure. The sulphur often present in coal transferred itself to the iron, rendering it brittle and useless. The problem was overcome by using coke instead of coal (see p. 153). Brewers had found in the seventeenth century that coal fumes affected the brew unpleasantly, but substituting coke left the flavour unimpaired. The real technological breakthrough came in 1709 when Abraham Darby, at Coalbrookdale in Shropshire, succeeded in smelting iron ore by first converting the coal into coke. For economic reasons as much as innate conservatism the new process made slow headway but, with increasing demand and technical improvements in iron production, surged ahead in the 1750s, and a decade later coke-smelted iron overtook charcoal iron, which was virtually extinct by the end of the century. In other countries, where the balance of timber and coal supplies was different from Britain, charcoal remained in use much longer.

British domestic and industrial users stimulated coal production to such an extent that by 1800, Britain was producing 80 per cent of the world's coal. This had far outstripped the resources of open-cast and drift mines and by the end of the seventeenth century, miners had penetrated to the limit of what could be drained by animal or water power. There would have been a real impasse had it not been for the invention of the steam engine by Thomas Newcomen in 1712, to enable deeper mines to be pumped dry. But it was the nineteenth century that was the age of coal, as a fuel and as a source of other useful materials.

The possibility of extracting tar from coal had been envisaged as early as 1681, and cropped up several times during the next hundred years, but coal tar really made its presence felt when vast quantities of it were obtained as a by-product from the early coal gas plants (see below). In desperation, the Gas, Light and Coke Company obtained permission to dump tar in the Thames. Others made attempts to use it as fuel or, as in 1838 by Bethell, as a preservative for railway sleepers. But its use as a source of useful chemicals had to wait until the 1860s, that is, until the theoretical and practical equipment of organic chemists was equal to the task of separating and identifying the pure constituents of tar. Hofmann and his group of chemists at the Royal College of Chemistry turned their attention to coal tar and it was Mansfield who published a classic paper on the extraction of benzene from tar, by distillation and fractional crystallization. He recommended benzene as a solvent of grease and so paved the way for dry-cleaning. Sadly, benzene's high inflammability led to Mansfield's untimely death when one of his benzene stills caught fire.

Besides benzene, coal tar contains many other organic compounds, mainly hydrocarbons, including toluene, naphthalene and anthracene. These, with other constituents such as phenol, were not only useful in themselves but were the starting-points for the synthesis of a whole range of useful substances. Synthetic dyestuffs were the first area opened up by coal tar derivatives (see p. 201). However, with the rise of the petroleum industry, petroleum has supplanted coal tar as a source of organic chemicals.

British coal production rose steadily, showing a marked increase just before mid-century, reaching its zenith in 1913. The coal industry in other countries, especially Germany and the USA, took off at this time, so that Britain's proportion of world output had fallen to 35 per cent by 1900.

Coal gas

By the end of the eighteenth century, the importance of the liquid and gaseous products of coal was beginning to be appreciated. Several investigators had made experiments with producing an inflammable gas by heating coal, but it was William Murdock (born Murdoch) who made the first successful large-scale attempt. In 1792 he succeeded in lighting part of his house at Redruth in Cornwall by gas produced by distillation of coal in an iron retort and washing the gas produced by passing it through water to remove some impurities. In bare essentials, the process has remained the same ever since. A few years later he joined the celebrated firm of Boulton and Watt and, although failing to enlist their help in securing a patent, installed a plant to produce gas for lighting their factory in 1802. Other factories received the same treatment and in 1808 Murdock was awarded the Rumford Gold Medal. He had been helped by the chemist Samuel Clegg, who, on leaving the firm, carried out installations in other factories and in 1811 the first non-industrial establishment to receive gas lighting, Stonyhurst College, the Jesuit seminary and school in Lancashire. Meanwhile in France, Philippe Lebon had made a striking demonstration of gas lighting, using gas from wood distillation, although the patent granted him in 1799 envisaged coal gas. His efforts were cut short by his untimely murder in 1804.

So far, there was no plan for large central plants serving a whole neighbourhood; indeed Watt, Murdock and later Sir Humphry Davy, no less, opposed the idea. This larger concept was first developed by the flamboyant Friedrich Albrecht Winzer, anglicized to Winsor, who, failing to stimulate interest in Germany, migrated to England in 1803 and was soon giving striking demonstrations of gas lighting. In 1807 he lit one side of Pall Mall in London from an installation in his house there. Winsor knew little chemistry but was energetic in urging his ideas of large central installations serving a wide area. Parliament eventually, in 1812, approved a more modest scheme than the one he originally had in mind and the Gas, Light and Coke Company came into being. After an uncertain start, Clegg joined it as Chief Engineer in 1815. He made a number of important contributions to the development of the industry; for example, he invented the gasholder, and introduced lime washing to remove hydrogen sulphide and sulphur from the gas. The demand for gas lighting was immediate and widespread; by 1820 fifteen of the principal cities of England and Scotland were equipped with it and at mid-century hardly a town or village of any consequence lacked a gas supply. Other countries soon followed suit.

The method of production remained in principle that devised by Murdock, but improvements in detail raised quantity and quality. The horizontal oven or retort remained supreme until after 1890 and in places survived until the end of coal gas in Britain in the 1960s. Iron retorts had some advantages but a short life and after 1853 gave way to clay retorts. In the 1880s, the inclined retort was introduced and soon after 1900 the vertical, together with mechanical handling of solid charge and product and improved treatment of the gas. By the end of the century a rival had appeared, electric light (see Chapter 6), but two innovations enabled gas lighting to keep its place for a while longer – the incandescent mantle produced by Welsbach in 1887, using the property of rare earth oxides to glow brightly in a gas flame, and the penny-in-the-slot meter introduced by Thorp and Marsh in 1889; this increased the number of potential customers by bringing gas lighting within reach of the working classes.

After centuries of rush light, candle and torch, gas light burst upon the scene with staggering effect, and with profound social and economic consequences. In commerce and industry the working day, especially in winter, was lengthened. Streets became much safer to frequent after dark; it was indeed the police in Manchester who promoted their gas lighting system. It made it easier for dinner, until now taken at around three in the afternoon, to slip into the evening. Not least important, evening classes became a possibility, enabling working people to gain an education after the day's work was done.

But water and space heating with gas came slowly. After many experiments with ceramic radiant materials, the first successful gas fire was evolved by Leoni in 1882 using tufts of asbestos fibre embedded in a firebrick back. The 'cookability' of gas was established rather earlier. Alexis Seyer, the great chef at the Reform Club in London, was cooking with gas in 1841. Bunsen's famous burner of 1855 greatly improved the design of burners in all appliances, but it was not until the 1870s that gas cookers became widespread. In fact, it has been these applications that have made great strides, while gas light was gradually supplanted during the first quarter of this century.

Natural gas

Other materials were tried for gasification, but through the period of cheap and accessible coal, some 100 years, it was the major fuel. It had the advantage of yielding coke as a by-product, not of a quality suitable for metal smelting but still useful as a fuel in its own right. But the rise of the petroleum industry (see below) brought a new feed-stock for gas production and in the 1950s the change-over from coal to oil was virtually complete. A wider range of gases could be produced, without the toxic carbon monoxide present in coal gas. The plant to produce town gas was more compact and less capital-intensive. This upheaval had hardly been settled when another newcomer appeared – natural gas. Large reservoirs of natural methane gas, usually associated with oil, occur in various parts of the world, including North America, the Soviet Union, Mexico and Africa. Britain was importing liquefied natural gas for some years but in 1959 came the first major find in Western Europe – the Netherlands Slochteren field. In 1965, Britain began to develop the North Sea field and it became possible to pipe in gas, with minor treatment, direct to the consumer. Methane has different burning characteristics from the hydrogen that is the main constituent of coal gas. The former has roughly double the calorific value of town gas, yet needs a little greater air supply for combustion. It has therefore to be supplied at a greater pressure and a different design of burner is required. The change to natural gas in Britain thus entailed a mass conversion programme of some forty million appliances, a scale unparalleled anywhere.

Petroleum

Gaseous and liquid seepages of petroleum were known in the ancient world, the greatest concentrations being in Mesopotamia. The Babylonians gave to an inflammable oil, of which no use was made, the name 'naphtha', or 'the thing that blazes'. A use was found for rock asphalt and the thicker seepages, to form bitumen, preparations of which were used for many centuries for waterproofing. Petroleum also had medicinal applications, in which a trade developed in Europe in the fifteenth and sixteenth centuries. There were scattered instances of the treatment of naturally occurring oils to obtain useful materials, such as the sandstone shales at Pitchford-on-Severn in Shropshire, patented in the late seventeenth century. By distillation and boiling the residue with water, a turpentine was extracted and sold as medicine and as pitch, useful for caulking ships. Elsewhere, there were rock asphalt workings which after 1800 became important for pavements and roads.

The more serious search for mineral oils was stimulated by the need for improved lighting during the Industrial Revolution. Although gas lighting could satisfy this need, it was not always a practical or economic proposition in rural areas. Better lamps improved the lighting quality, like the circular burner with cylindrical wick and glass chimney invented by Argand in 1784 and developed over the years. Even so, the vegetable and animal oils available produced a rather poor illumination. Salvation was looked for from the mineral kingdom. In the 1850s, James Young was manufacturing a paraffin oil by distilling a brown shale in Lothian in Scotland and after 1862 went on to develop the Scottish shale oil industry. A better product, however, was developed in the USA by Abraham Gesner, a London doctor with geological leanings, by treating and distilling asphalt rock, to obtain kerosene (in English, paraffin). This was sold with a cheap lamp and by 1856 seemed a promising answer to the lighting problem. But a complete transformation of the scene was soon to take place, brought about by the drilling of the first American oil well. This historic event was the outcome not only of the stimulus of demand, but of various technological factors. In 1830 the derrick was introduced to make it easier to manipulate the drilling equipment, the steam engine by 1850 was providing adequate power and sufficiently hard drills were available. A few accidental 'strikes' were made in the 1840s and 1850s, but in 1859, the industrialist G. H. Bissell began a deliberate search for oil. He had samples of oil seepages in Pennsylvania examined by Benjamin Silliman Jr, Professor of Chemistry at Yale, who showed that illuminating gas, lubricating oil and, most interesting then, an excellent lamp oil could be obtained. Bissell's contractor, Edwin L. Drake, drilled $69\frac{1}{2}$ft (21.18m) through bedrock and struck oil on 27 August 1859. This event not only opened up the Pennsylvania oilfield; it began a new chapter in world history.

Progress was rapid in the USA; within fifteen years, production in the Pennsylvania field had reached 10 million 360lb (163.3kg) barrels a year. Oilfields were developed in Europe too, particularly in Russia, where the first well was drilled at Baku in 1873. Flush drilling with hollow drill pipes was first used in France in 1846, enabling water to be pumped down to clear debris and thus speed drilling. It also enabled rock samples to be removed to reveal the structures of underground formations, and so assist prospecting. Drilling bits were first of wrought iron with steel cutting edges, but these were replaced by cast steel bits, and then diamond drills, introduced by the French in the 1860s. The crude oil was heated to yield useful products by means of distillation, a process long familiar to chemists. Further refining was achieved by treatment with various chemicals, such as sulphuric acid or caustic soda. The most volatile fraction of the oil to distil over first was petroleum or gasoline, for which at first there was no use; indeed, it was a nuisance because it was highly inflammable. The second fraction, boiling at 160–250°C was paraffin (kerosene) and for the rest of the nineteenth century this was the useful product, as an illuminant. The final or heavy fraction became valuable as a lubricant, replacing the animal and vegetable oils that had previously satisfied the ever-increasing lubrication demands of machinery of all kinds.

The invention of the motor car changed the balance: now, it was the light, petrol, fraction that was in demand, leading to rapid expansion of the petroleum industry. To boost petrol production, cracking was introduced and became widespread during the late 1920s. Here, the heavier are converted into the lighter fractions by subjecting them to high temperature and pressure to break down the chains of carbon atoms into shorter ones. On the other hand, lighter, gaseous products can be formed in the presence of catalysts into motor fuel, as in platforming or re-forming with a platinum catalyst. A small but extremely important proportion of the output, about 1 per cent, is a source of organic chemicals; by 1900 it had accounted for a third of the organic chemical industry. Before the First World War, the petrochemical industry produced mainly simple olefins, such as ethylene and its derivatives including ethylene glycol, the first antifreeze for motor cars, available from 1927. The range of chemicals widened rapidly after 1940, stimulated by the demands of the synthetic rubber, artificial fibre and plastics industries (see p. 217).

Nuclear energy

The possibility of using the energy locked up in the atomic nucleus is the direct result of research into the nature of matter, stemming ultimately from the speculations of the materialist philosophers of ancient Greece. Leukippos and Demokritos of the fifth century BC conceived of matter as consisting of myriads of minute, indivisible particles called atoms. The variety and behaviour of matter was explained in terms of the arrangement and motions of these atoms. The concept was elaborated in the poem *De Rerum Natura* (On the nature of things) by the first-century BC Roman poet Lucretius, but fell into obscurity, from which it was not rescued until the revival of atomism in the seventeenth century. From 1803, John Dalton developed his atomic theory, that is, that the atoms of each chemical element were identical but different from those of other elements in one respect: they had different weights (now called relative atomic masses). He also indicated the ways in which compound atoms, or molecules, could be formed by combinations of atoms and embarked on the determination of the weights of the atoms of various elements, that is, the number of times heavier they were than the lightest known element, hydrogen. Progress was promising at first, but some anomalous results cast doubt on the possibility of working out atomic weights and led to half a century of confusion, when it was said in despair that each chemist had his own system of atomic weights. At the Karlsruhe Conference of 1860, sound rules were finally established and from then on, reliable and accurate values for atomic weights became available to chemists. In 1869, Dimitri Mendeleev in Russia and Lothar Meyer in Germany arranged the elements in ascending order of atomic weight, noting that elements with similar properties formed themselves into classes or group. This was a step towards understanding the relationship between the elements, which hitherto seemed to be unrelated, isolated individuals. But the atoms were still regarded, until the end of the century, as solid, indivisible, indestructible particles of matter, obedient to the traditional Newtonian laws of motion.

A series of discoveries around the turn of the century shattered this image. By 1897, J. J. Thomson had established the existence of the electron, a particle with a negative electric charge only a minute part of the weight of an atom. From 1895, Henri Becquerel and Marie and Pierre Curie explored the radioactivity of the heavy elements such as uranium and radium, which were disintegrating spontaneously with the release of energy and minute particles of matter. Albert Einstein showed that matter and energy are interchangeable, a very little mass being equivalent to a very large amount of energy, related to each other by the now celebrated equation $E = mc^2$, where E is the energy, m the mass and c the velocity of light. After the work of Frederick Soddy and Ernest Rutherford during the first decade of this century, Niels Bohr was able to propose in 1913 a model for the atom entirely different from the traditional view. The mass of the atom was concentrated in a positively charged nucleus at the centre and sufficient negatively charged electrons circling round it to leave the atom electrically neutral. During the 1920s a new system of mechanics, quantum mechanics, was developed as, at the atomic level, traditional mechanics no longer applied. The neutron, a particle identical in weight to the proton but having no electrical charge, was discovered by Chadwick in 1932. From then on, the atom was visualized as a nucleus made up of protons and neutrons with, circling round it in orbits, a number of electrons equal to the number of protons. The latter is the atomic number and determines the chemical identity of the element. The number of protons and neutrons is the mass number.

Atoms of the same element with the same atomic number can have a slightly different number of neutrons: these are known as isotopes. Thus, most uranium (atomic number 92) has in its nucleus 146 neutrons, giving a mass number of 238. But there is another isotope of uranium with only 143 neutrons, known as uranium (U-235). Around the mid-thirties, scientists were trying to obtain artificial elements, heavier than uranium, with nuclei so unstable that they could not occur in nature. The most promising method was to bombard uranium atoms with slow neutrons in a particle accelerator. Hahn and Strassman were adopting this method in 1938 and expected to detect the element with atomic number 93 but instead found indications of an element similar to barium, with an atomic number about half that of uranium. This was puzzling and it was the Austrian physicists Lise Meitner and Otto Frisch who published the correct explanation in two famous letters in *Nature* in February 1939: the neutrons had split each uranium atom into two medium-sized ones with the liberation of further neutrons and a considerable amount of energy. Later that year, Bohr and Wheeler suggested that the liberated neutrons could split more atoms, with production of yet more neutrons, and so on – a chain reaction. Two days after this publication appeared, the Second World War broke out.

The sudden and immense release of energy upon the splitting of fissile (easy to split) atoms had military possibilities that were not at first realized, but found recognition in the setting up of the Manhattan Project in the USA. Leading scientists in this field were assembled from the Allies and from the distinguished refugees from Nazi and Fascist oppression. International co-operation in science on this scale was unparalleled and has not since been matched. The objective was to develop knowledge and processes relating to nuclear energy and to make an atomic bomb. Two kinds of fissile material were chosen, uranium-235 and plutonium-239, an artificially made element with atomic number 94. Small quantities of the latter had been produced on a laboratory scale, but this was quite inadequate. Enrico Fermi, who had come over from Italy, designed and built at Chicago University the first nuclear reactor in which plutonium could be produced by a controlled reaction. It was a historic moment on 2 December 1942 when the reactor first went critical, that is, the chain reaction continued spontaneously. The Project reached its objective. Two atomic bombs were dropped on Japan in August 1945, bringing the war to a swift conclusion and raising two mushroom-shaped clouds that have haunted mankind ever since.

The spirit of international co-operation and exchange of knowledge did not long survive the war and soon each country with the will and the means pursued its own programme of research and development. In Britain, two bodies under the Ministry of Supply were set up in 1946, to carry out research, which was the prime purpose of the establishment at Harwell, and to manufacture atomic bombs. It was decided that the plutonium route to the bombs would be more practicable and economic than the U-235 route; two large graphite-moderated, air-cooled reactors were constructed at the Windscale site in Cumbria and early in 1952 began to produce plutonium which was then passed to the weapons establishment at Aldermaston. In October of that year Britian's first nuclear weapon was tested. More plutonium was needed and design started on a second station alongside Windscale, known as Calder Hall. Here, it was decided to make use of the heat generated by the process and compressed carbon dioxide was circulated by powerful fans through the reaction vessel and through boilers to produce steam, which could drive a conventional electric generator. To do this, it had to work at a higher temperature than Windscale; among various design changes, the cladding of the uranium-metal fuel rods had to be altered from aluminium to a specially developed alloy of magnesium, known as Magnox, a word that came to be applied to the power stations based on the Calder Hall design. In October 1956, the Queen opened Calder Hall, the world's first commercial atomic power station. The British government announced a programme, the world's first, for building atomic power stations, and by 1971, eleven had been constructed on the Magnox pattern with a combined output of electric power of some 4000MW. In 1964 a further programme was begun, to build power stations with an improved reactor developed from the Magnox – the Advanced Gas-cooled Reactor (AGR). Working at a much higher gas temperature and pressure, these gave steam conditions and performance matching the most efficient oil and coal-fired stations. Changes in the materials used had to be made to enable them to withstand the more demanding conditions; instead of cladded uranium metal fuel, uranium oxide pellets were used.

Meanwhile the USA was pursuing a different line to the same end. The requirement to create small reactors for powering submarines determined the development of the Pressurized Water Reactor (PWR), in which water under high pressure acted as both moderator and coolant. Even so, the temperature of the water had to be kept below 280°C to ensure it did not boil. In the less demanding conditions of on-shore power stations, the water was allowed to boil and the steam passed direct to the turbo-generator – the Boiling Water Reactor (BWR); one drawback is that the water becomes somewhat radioactive and so special precautions are needed in the turbine as well as the reactor area. Through vigorous promotion, the American system has been more widely taken up by other countries than the British. France at first followed the British in using Magnox reactors but changed over to PWRs from the mid-1960s. The Soviet Union early entered this field, achieving a nuclear explosion in 1949. Again, power generation was first based on Magnox but in view of the Soviet nuclear-powered submarine programme, moved over to PWR. A unique application here is the nuclear-powered ice-breaker.

Many countries embarked on nuclear energy programmes in the 1960s and 1970s, and by the end of 1986 there were 394 nuclear reactors producing electricity in 26 countries, with more under construction. Nuclear power accounts for over 15 per cent of world electricity production, although there is considerable variation between one country and another. In a number of countries, the proportion is very small. In the USA it is just over 16 per cent, Britain rather more at around 20 per cent, while France and Belgium attain nearly 70 per cent.

POLYMERS: RUBBERS, PLASTICS AND ADHESIVES

Everyday and industrial life has been transformed by the introduction of a large group of substances quite different from the metals and non-metals in use over the centuries. They have in common a certain type of complex chemical structure, in which large molecules are formed by linking up small groups of atoms into long chains, known as polymers. Some occur in nature, like cellulose, and the first materials of this kind to be made were derived from natural materials. But from the 1920s, when the chemistry of their structure and formation became clearer, an ever-increasing range of materials was produced, from organic chemicals derived first from the coal-tar industry, then from petrochemicals.

Rubber

In Central and South America, rubber trees were tapped for latex, a milky emulsion of rubber and water, from which the rubber can be coagulated and separated by heating. In the thirteenth century the Mayas and Aztecs used articles made from rubber, such as balls for games, but the material remained unknown to Europeans until the Spanish conquerors descended on the Americas. Even so, they made little use of it and it was left to the French Academy of Sciences to make a systematic examination of *caoutchouc*, as it was called, published in 1751. Joseph Priestley in 1770 noted its use in rubbing out pencil marks, hence the word by which the material is known in English. The uses of rubber remained limited until Thomas Hancock introduced improved methods of making sheets, using spiked rollers turning in hollow cylinders. His 'masticator' dates from 1820. Soon afterwards, Charles Macintosh found that rubber dissolved in naphtha, a product of the new gas industry, could be brushed on to clothing to make it waterproof (see p. 849). During the 1830s rubber imports to Britain rose sharply and it came into wide use for garments and shoes, for miscellaneous engineering uses such as springs, and for various surgical purposes. Its use spread to France and to the USA. The untreated rubber so far used was found to be unsatisfactory in the wide extremes of temperature met with in the USA. Charles Goodyear, a hardware merchant of Philadelphia, found that heating rubber with sulphur greatly improved its properties, a process that came to be known as vulcanization. Finding little interest in the process in the USA, Goodyear passed samples of vulcanized rubber to Hancock, who developed a process for producing it, patented in 1843. The growth in the use of rubber during the rest of the century is indicated by the rise in output by Brazil, the main supplier, from 31.5 tonnes in 1827 to nearly 28,100 tonnes in 1900. This, however, was not enough to meet the demand and plantations of rubber trees were established in the Far East, including Malaya, and in the West Indies, Honduras and British Guiana, becoming effective after 1895, so much so that they led to the collapse of the Brazilian trade. But the application that was to swamp all others, transport, made a slow start. Hancock was making solid tyres for road vehicles in 1846 and there was an abortive use of pneumatic tyres around the same time. In the 1870s, bicycles were equipped with solid tyres and in 1888, Dunlop introduced, then improved his pneumatic tyres. This development was timely for it coincided with the invention of the motor car. Michelin's first motor tyre appeared in 1895, Dunlop's in 1900, and production rose rapidly as motoring increased in popularity (see p. 449).

Plastics

The word denotes an organic substance that on heating can be shaped by moulding and retains its shape on cooling. Some plastics, after being softened by reheating, become hard again on cooling; these are thermoplastics. Others undergo some chemical modification on heating and can not be softened by reheating; these are thermosetting plastics.

Some natural substances could be formed in this way; gutta percha, a latex derivative imported from Malaya after 1843, was moulded into small ornamental objects. Next, chemists experimented with organic substances of natural origin to produce plastic materials. Christian Friedrich Schönbein, of the University of Basle, produced cellulose nitrate by the action of nitric and sulphuric acids on paper (cellulose) and this could be shaped into attractive vessels. This led the metallurgist and inventor Alexander Parkes to develop the first commercial plastic, cellulose nitrate with camphor as a plasticizer. He exhibited this Parkesine at the International Exhibition of 1862, but the company he set up to manufacture it failed in 1868. More successful was the American printer John Wesley Hyatt, whose attention was turned to cellulose nitrate by the offer of a prize of $10,000 by Phelan & Collander, makers of billiard balls who had run short of ivory, thanks to the efforts of the elephant hunters, and were desperate for a substitute. After experimenting Hyatt filed a patent covering the use of a solution of camphor in ethanol as a plasticizer for the cellulose nitrate, or Celluloid, as it came to be called. It could be shaped and moulded while hot and on cooling and evaporation of the camphor, became 'hard as horn or bone'. Hyatt prospered and set up plants to make celluloid in Germany, France and Britain, where it became a popular material, particularly for detachable collars and cuffs. It was found that cellulose acetate could also be used, with the advantage that it was non-inflammable. Towards the end of the century it was available in thin film, and could be used as a base for photographic emulsions. Photography and the new art of cinematography made increasing demands on cellulose nitrate, which reached a production peak in the 1920s, when it began to be replaced by other less flammable plastics derived from cellulose. A notable use of cellulose acetate was as a covering for aircraft wings, rapidly developed during the First World War.

The second semi-synthetic plastic was formed from the reaction between casein, the main protein in milk, and formaldelyde, announced in 1897 by Spitteler & Krische in Germany. The manufacture of the first casein plastics, giving a hard, horn-like material, began three years later and has continued ever since, being especially suitable for buttons. More important was the announcement in 1909 of the first thermosetting plastic by a Belgian who had settled in the USA, Leo Hendrik Baekeland. The German chemist Baeyer had observed in 1872 that phenol and formaldehyde formed a hard, resinous substance, but it was Baekeland who exploited the reaction to produce commercially Bakelite, a versatile material resistant to water and solvents, a good insulator, like other plastics, and one which could be easily cut and machined.

Chemists were now investigating the structure of such substances as cellulose, produced in plants, with long-chain molecules. This led to the notion that such molecules might be produced in the laboratory. Also there was a growing understanding of the relationship between physical properties and molecular structure, so that it might be possible to design large molecules to give materials of certain desired characteristics. More than any other, it was Hermann P. Staudinger in Germany who achieved an understanding of the processes of polymerization, or forming large molecules from repeated additions of small, basic molecules, upon which is largely based the staggering progress of the plastics industry since the 1930s. For this work Staudinger was awarded the Nobel Prize in Chemistry in 1953. The other great name in fundamental research in this field is Wallace H. Carothers, who was engaged by the Du Pont Company in the USA in 1928 to find a substitute for silk, imports of which from Japan were being interrupted by the political situation. Carothers developed a series of polymers known as polyamides; one of these mentioned in his patent of 1935 was formed from hexamethylenediamine and adipic acid. Production of this polyamide, known as Nylon, began in 1938 and the first nylon stockings appeared the following year; during the first year, 64 million pairs were sold.

Another extremely important result of Carothers's researches was synthetic rubber. He found that when polymerizing acetylene with chlorine, the product, polychloroprene, was a superior synthetic rubber, better known as Neoprene. Commercial production began in 1932. Meanwhile, in Germany a general purpose synthetic rubber was developed as a copolymer of butadiene and styrene.

These products assumed desperate importance after the Japanese overran the Asian rubber plantations in 1941, cutting off the Allies' source of this essential material.

For the rest, it is not possible to do more than indicate a few of the better-known plastics. Vinyl acetate had been synthesized in 1912 in Germany and by 1920 polyvinyl acetate (PVA) was being manufactured, used mainly in adhesives, emulsion paints and lacquers. An important discovery by W. L. Semon of the B. F. Goodrich Company in the USA in 1930 showed that placticized polyvinyl chloride (PVC) produced a rubber-like mass, which found many applications, as cable insulation, substitute leather cloth, and in chemical plant and packaging.

Various chemists had described glass-like polymers of esters of acrylic acid (produced by the action of an alcohol on the acid). In 1931, Rowland Hill of ICI began to study the esters of methacrylic acid. (Imperial Chemical Industries Ltd had been formed five years earlier by the merger of several chemical companies, forming one of the world's great chemical concerns from which a succession of notable discoveries have flowed.) It was found that the polymer of the methyl ester was a clear, solid glass-like material which could be cast in a mould, and was light, unbreakable and weatherproof. Another chemist at ICI, J. W. C. Crawford, worked out a viable method of preparing the monomer and commercial production of polymethyl methacrylate, or Perspex began in 1934. This was another product which had valuable wartime uses; nearly all the output of Perspex sheet up to 1945 was required for RAF aircraft windows. Since then, many other uses have been found.

Another important ICI discovery was polyethylene (polythene). It stemmed from investigations into the effect of high pressure on chemical reactions, begun at the Alkali Division in 1932. Fawcett and Gibson heated ethylene with benzaldehyde, aniline and benzene at 170°C at 1000-2000 atmospheres pressure and noticed the formation of a white, waxy coating on the walls of the reaction vessel. It was found to be a polymer of ethylene. In 1935 polymerization of ethylene alone was achieved, further development followed and commercial production began days before the outbreak of war. Polythene was found to be a very good electrical insulator, was chemically resistant and could be moulded or made into film or thread. Yet again, it proved to be a vital material in war, mainly on account of its electrical properties. As a dielectric material, it made manageable tasks in the field of radar that would otherwise have been impossible; that testimonial comes from Sir Robert Watson Watt, the inventor of radar. After the war, polythene became one of the major commercial plastics, with a wide range of domestic and industrial uses.

One final ICI contribution may be mentioned. Carothers and Hill in 1931 described some aliphatic polyesters which could be extruded and cold-formed into strong fibres. During the war, Whinfield and Dickson of Calico Printers

Association followed this up, with a strong fibre and lustrous film, identified as polyethylene terephthalate. ICI evaluated this in 1943 and the possibilities of Terylene, as it was named, were realized; commercial production at ICI's Wilton plant started in 1955, with the result that most people wear some garment made from Terylene fibre. A year earlier, the Du Pont Company in the USA were producing the material under the name of Dacron (see Chapter 17).

Adhesives

Natural products have been used to bond surfaces together from earliest times. A number of animal, plant and mineral products were used in the ancient civilizations for particular applications. For example, the Egyptians were employing animal glues made by boiling bones, hooves and hides, to join veneers to wood surfaces around 3000 BC, while they used flour paste to bind layers of reeds to form papyrus as a writing material. Bitumen and pitch were also in use in the ancient world. Such materials, and the methods of making them, survived virtually unchanged until the present century. Then, within a few decades, chemists produced not only a spate of new synthetic adhesives, but also a theoretical background against which more appropriate and effective adhesives can be produced. The science of adhesives began to take off during the 1940s, notably with the work of Zisman. The first purely synthetic adhesive was due to Baekeland and his development of phenol formaldehyde resins in the early years of this century, although their large-scale application to the plywood industry did not take place until the 1930s. Then IG Farben introduced urea formaldehyde resins, also as a wood glue. After the Second World War, polyvinyl acetate emulsions began to supplant animal glues for woodworking purposes.

The growth of the synthetic polymer industry has enormously increased the range of adhesive available, notably the epoxy resins from the 1950s.

HEAVY INORGANIC CHEMICALS

Acids and alkalis

During the early period of industrial chemistry, the alkalis were the most important inorganic substances, that is, those that do not form part of living matter, whether plant or animal. Their preparation and uses in the textile, glass and soap industries have already been mentioned. Another important group of substances were the mineral acids – nitric, sulphuric and hydrochloric. Nitric acid and *aqua regia* ('royal water' – a mixture of nitric and hydrochloric acids) were probably discovered during the second half of the thirteenth century in Europe. Their preparation is first mentioned in a chemical work compiled soon after 1300, the *Summa Perfectionis* attributed to Geber, the latinized name of the eighth-century Arab alchemist Jabir; the *Summa* appears to be a European compilation derived from earlier Arabic sources. Nitric acid was prepared by heating green vitriol (ferrous sulphate) and saltpetre (potassium nitrate). If ammonium chloride is added, chlorine is formed and remains dissolved in the acid, giving it the ability to dissolve gold, hence the royal association. The main use of these acids was in metallurgy, in the purification of gold. Saltpetre soon acquired considerable importance, for it was the major ingredient in gunpowder; invented in Europe some time in the thirteenth century but known in China centuries earlier, its preparation became a major industry. It was commonly to be found in earth saturated with animal refuse and excrements; its soluble salts were extracted with boiling water and fairly pure saltpetre separated by successive crystallizations.

Sulphuric acid, or vitriol, was little used until it became important for the manufacture of soda in the eighteenth century. Recipes for preparing it begin to occur in the sixteenth century, by strongly heating sulphates of copper and iron or burning sulphur, and absorbing some of the gaseous product in water. In the following century, hydrochloric acid was recognized as a distinct substance and Glauber set out in 1658 the standard method of preparation, from common salt and oil of vitriol. This acid too found little use in the chemical industry until the nineteenth century.

But it was the increasing demand for alkalis, particularly from the rapidly developing textile industry, that stimulated the major advances in the chemical industry. Overseas natural sources were exploited to the full, such as wood ash in bulk from Canada, as a source of potash. In France, international conflicts had led to an acute shortage of alkali and in 1775 the Academy of Sciences offered a prize of 2400 livres for a successful method of making soda from salt. Nicolas Leblanc, physician to the Duc d'Orléans met the challenge, patenting in 1791 a process that was to be of fundamental importance to the industry for over a hundred years. The process consisted of treating common salt with sulphuric acid, to form a salt-cake (sodium sulphate) which was then roasted with coal and limestone. The resulting 'black ash' was leached with water to extract the soda, finally obtained by evaporating the solution in pans.

In Britain the salt tax inhibited the spread of the Leblanc process until its removal in 1823. James Muspratt set up a plant near Liverpool, where the raw materials lay close at hand. In 1828, in partnership with Josias Gamble, he established new works near St Helens, ever since an important centre of the chemical industry in Britain. Charles Tennant had meanwhile started making soda at his St Rollox works outside Glasgow, which became for a time the largest chemical works in Europe.

One of the main ingredients of the Leblanc process was sulphuric acid. Improvements in its manufacture had been made in the eighteenth century. In 1749, Joshua Ward (otherwise known as the quack in Hogarth's *Harlot's*

Progress) patented the making of sulphuric acid by burning sulphur and saltpetre in the necks of large glass globes containing a little water. That was converted to sulphuric acid, concentrated by distilling it. The price fell from £2 to 2s a pound, but the fragile nature of glass limited the scale of the process. Large-scale manufacture only became possible when John Roebuck substituted 'lead chambers', consisting of sheets of lead mounted on wooden frames, lead being both cheap and resistant to the acid. Roebuck, a student of chemistry at Leyden and Edinburgh set up his lead-chamber process outside the latter in secrecy, a condition that lasted as long as it usually does. Knowledge of the process spread rapidly; France had its first lead-chamber factory at Rouen around 1766. The process was improved and enlarged in scale; Roebuck's chambers had a capacity of 200ft^3 (5.66m^3) but by 1860 Muspratt had achieved one of 56,000ft^3 (1584.8m^3). The rise in scale of course lowered the price; by 1830 it was $2\frac{1}{2}$d a pound.

The first stage of the Leblanc process produced large quantities of hydrochloric acid gas, both poisonous and destructive. Not for the first or last time, the chemical industry made itself unpopular by its unfortunate environmental effects. In this case, from 1836 the gas began to be absorbed by a descending stream of water. The Alkali Act of 1863 required manufacturers to absorb at least 95 per cent of the acid. Important as the Leblanc process was, it had other drawbacks, principally the problem of disposing of the unpleasant 'galligu' or residue after the soda had been extracted. This led to a long search for an alternative and the gradual emergence of the ammonia-soda process, which eventually achieved success at the hands of the Belgian brothers Ernest and Alfred Solvay, with a patent in 1861 and satisfactory working four years later. It was introduced into Britain in 1872 by Ludwig Mond, who set up a works at Winnington in Cheshire in partnership with John Brunner which was to become part of Imperial Chemical Industries.

Improvements were also made in the utilization of the hydrochloric acid formed in the Leblanc process. Henry Deacon oxidized it to chlorine using a catalyst (a substance that facilitates a reaction but can be removed unchanged at the end of the reaction). Also, Walter Weldon introduced the successful oxidation to chlorine using manganese dioxide, which enabled the output of bleaching powder to be quadrupled, with a considerable reduction in price.

A new method of producing sulphuric acid had been suggested in 1831, by oxidizing sulphur dioxide to the trioxide using a platinum catalyst, but the platinum was found to be affected or 'poisoned' by the reaction and progress came to a halt. There was then little incentive to solve the problem, until the new organic chemical industry in the 1860s began to make demands for not only more but stronger sulphuric acid. So far, the only source of oleum, a form of strong sulphuric acid with an excess of sulphur trioxide, had been Nordhausen in Saxony, where it had been produced in limited quantities since the late seventeenth century. The contact process, as it came to be called, was revived and at the hands of the German Rudolf Messel became, from around 1870, a practical proposition. The Badische Anilin- und Soda-Fabrik (BASF) was very active in pursuing research into this process. Various catalysts were tried, but from 1915 BASF used vanadium pentoxide and potash, and this became the most widely used material.

Explosives

Another important branch of the chemical industry was the manufacture of explosives (see Chapter 21). Until the middle of the nineteenth century, the only important explosive was gunpowder, but in the 1840s two other substances were noted. Schönbein found that the action of nitric acid on cellulose produced an inflammable and explosive substance and with John Hall at Faversham began to make gun-cotton, so called because cotton was the source of cellulose. In the same year nitroglycerine was discovered, formed by the action of nitric acid on glycerine, a product of soap manufacture. But these two substances were found to be too explosive to make or use. Alfred Nobel, however, showed that nitroglycerine could be made by absorbing it on kieselguhr, a kind of clay; in this form it was called dynamite. It was found that it would explode violently when touched off by a mercury fulminate detonator. In 1875, Nobel invented blasting gelatine, consisting of nitroglycerine with a small quantity of collodion cotton. The first application of these materials was in blasting in mines and quarries; their use in munitions became important in the 1880s.

Fertilizers

A less destructive application of the chemical industry was the manufacture of fertilizers. Apart from carbon, hydrogen and oxygen, there are three elements needed in relatively large quantities for plant nutrition: nitrogen, phosphorus and potassium. Until the end of the nineteenth century, the most important source of nitrogen was natural organic materials, but mineral sources were also important. Of these, by far the most significant were the sodium nitrate deposits in Chile, providing some 70 per cent of the world supply. By the 1960s this figure had shrunk to 1-2 per cent. The reason for the decline was the successful tapping of the richest source of nitrogen of all: the air we breathe. The 'fixation', or chemical combination, of nitrogen was known to be chemically feasible by the end of the eighteenth century and from around 1900 several processes had been developed on an industrial scale. But by far the most important of these was that worked out by Fritz Haber. It consisted of the synthesis of ammonia from its two constituent elements, nitrogen and hydrogen. The reaction had been studied spasmodically for many years but it was Haber who transformed it into an industrial proposition. High pressures were at first avoided, but Haber found during his researches from 1907 to 1910 that a pressure of 200 atmospheres produced the highest yield. At that stage BASF once again entered the scene and engaged in research to find the most suitable catalyst. In 1913 the first full-scale plant for the synthesis of ammonia by the Haber process was built at Oppau, with a second at Leuna, near Leipzig three years later. By 1918, the process contributed half of Germany's output of nitrogen compounds. Hostilities with Germany hindered the spread of knowledge of the process and it was only in the 1920s that manufacturing plants were set up in other leading industrial countries. Progress was then rapid and by 1950 four-fifths of nitrogen fixation was by this process. The catalyst most widely used from 1930 was finely divided iron mixed with various oxides. Nitrogen fixation became important not only for the production of ammonium sulphate and nitrate fertilizers but for the manufacture of nitric acid, much in demand before and during the wars for making explosives.

As to phosphorus, the main source until around 1900 was ground bones or bone meal, but in the last decades of the nineteenth century large deposits of calcium phosphate were discovered in northern Africa and, later, other major producers were the USA, USSR and the Pacific island of Nauru. The calcium phospate is converted to 'superphosphate' by treatment with sulphuric acid, first achieved on a large scale from 1834 by John Bennet Lawes at Rothamsted in Hertfordshire. 'Triple superphosphate', or monocalcium phosphate produced by treating the mineral form with phosphoric acid, attained equal importance with the 'super' variety in the USA in the 1960s.

Potassium fertilizers (potash), mainly potassium chloride, have been applied as they were mined, the Stassfurt region in Saxony being the leading source for some 130 years.

Electrolysis

An entirely new way of producing chemicals arose towards the end of the last century. Electricity had been used to decompose substances, for example by Sir Humphry Davy in 1807 in obtaining sodium metal for the first time, but it was not until cheap supplies of electricity were available that electrolytic methods of preparing chemicals became commercially viable. In 1890 an American working in Britain, Hamilton Castner, developed a method of producing sodium by electrolysis of molten caustic soda, for use in the making of aluminium. At the moment of success, an electrolytic method of preparing aluminium was achieved by Hall and Héroult (see pp. 107–9), rendering the sodium superfluous. But relief was at hand. The 'gold rushes' of the 1890s dramatically increased the demand for sodium, in the form of its cyanide, used in the purification of gold and silver.

Castner then worked out a cell for making high-purity caustic soda, by electrolysis of brine over a mercury cathode and carbon anodes. The sodium released formed an amalgam with the mercury which, by rocking, came into contact with water in a central compartment; there, it reacted with the water to form caustic soda. An Austrian chemist, Carl Kellner, was working along similar lines and to avoid unpleasant patent litigation, the two came to an arrangement, with the result that the cell is known as the Castner–Kellner cell. The Castner cell was later modified, particularly by J. C. Downs's patent of 1924, defining the electrolysis of molten sodium chloride with graphite anode and surrounding iron gauze cathode, and using calcium chloride to lower the melting point of the electrolyte. This was a more efficient process electrically, although until 1959 both cells were in use, to provide cheap sodium. Other heavy chemicals were made by electrolytic methods from around 1900, such as sodium chlorate, much used as a herbicide.

FURTHER READING

Chandler, D. and Lacey, A. D. *The rise of the gas industry in Britain* (Gas Council, London, 1949)

Clark, J. A. *The chronological history of the petroleum and natural gas industries* (Clark Book Co, Houston, Texas, 1963)

Douglas, R. W. and Frank, S. *A history of glassmaking* (G. T. Foulis, Henley-on-Thames, 1972)

Haber, L. F. *The chemical industry during the nineteenth century* (Clarendon Press, Oxford, 1958)

—— *The chemical industry 1900–1930* (Clarendon Press, Oxford, 1971)

Hardie, D. W. F. and Pratt, J. D. *A history of the modern British chemical industry* (Pergamon Press, Oxford, 1966)

Kaufman, M. *The first century of plastics* (Plastics Institute, London, 1963)

Longstaff, M. *Unlocking the atom: a hundred years of nuclear energy* (Frederick Muller, London, 1980)

Nef, J. U. *The rise of the British coal industry* (Routledge, London, 1932)

Russell, C. A. *Coal: the basis of nineteenth century technology* (Open University Press, Bletchley, 1973)

—— 'Industrial chemistry' in *Recent developments in the history of chemistry* (Royal Society of Chemistry, London, 1985)

Singer, C. et al. (eds.) *A history of technology*, 7 vols., each with chapters on industrial chemistry (Clarendon Press, Oxford, 1954–78)

Taylor, F. S. *A history of industrial chemistry* (Heinemann, London, 1957)

Warren, K. *Chemical foundations: the alkali industry in Britain to 1920* (Clarendon Press, Oxford, 1980)

Williams, T. I. *The chemical industry past and present* (Penguin Books, Harmondsworth, 1953)

—— *A history of the British gas industry* (Oxford University Press, Oxford, 1981)

PART TWO

POWER AND ENGINEERING

4

WATER, WIND AND ANIMAL POWER

J. KENNETH MAJOR

The three main forms of natural power have a long history of development, and since classical times the development of water and wind power has been interrelated. The use of all three forms has not ceased, for water and wind power are gradually coming back into use as alternatives to the fossil fuels – oil and coal – and to nuclear fission. The need to develop rural communities in the Third World has brought about a rediscovery of the primitive uses of water, wind and animal power which can remain within the competence of the rural craftsmen.

WATER POWER

The ancient world

The first confirmed attempts to harness water to provide power occurred in the Fertile Crescent and the countries that border the eastern Mediterranean, in the centuries before the birth of Christ. The harnessing of these natural forms of power grew out of the difficulties of grinding grain by hand or raising water for irrigation laboriously by the bucketful. Slaves were not cheap, and the milling of enough flour by hand became increasingly expensive. At first the grain was ground by being rubbed between two stones known as querns. The grain would rest on a stone with a concave upper face and would be rubbed with a large smooth pebble. The next stage was to shape a bottom quern and to have a top stone which matched it and which could be pushed from side to side over the grain. By making both upper and lower stones circular, and by fixing a handle in the upper stone, a rotary motion was imparted to the hand quern. From that it is a short step to mounting the quern on a frame and having a long handle to rotate the upper millstone.

The first attempt to power millstones with water resulted in a form of watermill which we now call the Greek or Norse mill (see Glossary). In this the two millstones – derived from rotary hand-processing – were mounted over a stream with a high rate of fall. The lower millstone was pierced and mounted firmly in the mill, and the upper (runner) millstone was carried on a rotating spindle which passed through the lower millstone. This spindle was driven by a horizontal waterwheel which was turned by the thrust of water on its blades, paddles or spoon-shaped buckets. The stream was arranged, possibly by damming, to give a head of water, and this head produced a jet of water which hit and turned the paddles of the horizontal waterwheel. The horizontal watermill was a machine, albeit primitive, which ground grain faster than it could be ground by hand, and soon improvements began which further increased the speed of grinding.

The Romans adopted the Greek mill and made the hand mill more profitable by turning it into a horse-driven mill. Roman Europe became a civilization in which the countryside supported a growing number of larger towns. The use of slaves or servants to grind the grain for the family became too expensive, and there was a real need to increase the production of meal from the millstones. The often-quoted example of the range of ass-driven hourglass mills in Pompeii (see p. 262) shows how a town bakery was able to produce large quantities of meal for sale or for a baker's shop. The Romans are thought to have been the inventors of the vertical waterwheel. In this system a waterwheel is turned in a vertical plane about a horizontal shaft and the millstones are driven from this by means of gear wheels. About 25 BC Vitruvius wrote *De Architectura*, and in the tenth book he describes the vertical waterwheel and its gear drive to the millstones. The earliest form of mill in which a single pair of millstones is driven by a waterwheel is therefore known as the Vitruvian mill.

Many examples of the Vitruvian mill have come to light in archaeological excavations. The most famous of all the Roman Vitruvian mills is the group at Barbegal near Arles, in the Bouches-du-Rhône department of France. Here an aqueduct and channel delivered water to a chamber at the top of a steep slope. The water descended in two parallel streams, and between the streams there were the mill buildings. There were sixteen overshot waterwheels in two rows of eight, the water from one wheel driving the next one below it. This example from *c.* AD 300 shows a sophistication in millwrighting design which typifies the Roman approach to many of their problems of engineering and architecture. There are representations of vertical waterwheels in Byzantine art and in the Roman catacombs. By Hadrian's Wall in 1903, Dr Gerald Simpson excavated a Roman watermill which dates from *c.* AD 250. Lord Wilson of High Wray prepared drawings of this which were published in *Watermills and Military Works on Hadrian's Wall* in 1976. Similar and more important excavations, carried out in Rome in the 1970s by Schiøler and Wikander, show how universal the spread of the Vitruvian mill was in the Roman period. Further archaeological evidence confirms the presence of Roman watermills in Saalburg, near Bad Homburg in Germany, and Silchester, Hampshire.

The Romans and their colonials were among the first people to dig mines for metals. In some of these deep mines there was a need to install mechanical water-raising devices and in the mines at Rio Tinto in south-west Spain a substantial water-lifting wheel has been found. While this is not strictly a water-driven wheel it is analogous. The large rim carries a series of wooden boxes which have side openings at the upper ends. The wheel is turned by men pulling on the spokes and the boxes lift the water so that it empties out at high level into a trough which leads it away from the wheel. This wheel is one form of water-raising device and its derivatives still exist in parts of Europe and the Near East. The best-known examples are those at Hama in Syria, where water-driven wheels carry containers on the rim which raise the water to the top of the wheel where it empties out into irrigation channels. The biggest of these is 19.8m (65ft 6in) in diameter.

Mediaeval and Renaissance Europe

As a result of archaeological excavation, the Dark Ages are now producing examples of water-powered devices, and more attention is being paid to these examples. At Tamworth in Staffordshire the excavation of a watermill shows evidence of a well-designed mill of the Saxon period. These excavations revealed that the mill had been powered by two horizontal waterwheels housed in wooden structures.

The mediaeval period gives us our first real insight into the growth of water power in Europe. In England, the Domesday survey of 1086 gives a record of the number of mills in England – some 5000. While all the counties in the country were not surveyed, those that were show a surprising number of mills in relation to the manors, villages and estates. It must not be taken for granted that all the mills were water driven: hand mills may be indicated by the low level of their rents. Similarly, it must not be assumed that the mills were separate buildings, just that there was more than one waterwheel. The documents, cartularies, leases and grants of land give a greater insight into the way in which the use of water power was developing. All over Europe the abbeys, manors and towns were building watermills, and while most of these were corn mills, there is evidence of the construction of fulling mills, iron mills and saw mills. For example, the most famous drawings of a saw mill were made by Villard de Honnecourt in 1235. A mill built to his drawings was erected as a monument to him at Honnecourt sur Escaut, France. In this mill the waterwheel rotated as an undershot or stream wheel, and by means of cams the reciprocal motion was given to the saw blades. Although it cannot be assumed that this was the first saw mill, it is the first positive document showing the mechanism to have survived.

Although drawings may always have been few in number, it must not be assumed that the knowledge passed slowly from one centre to another. It is well established that there was a great deal of movement of master masons about their own countries, and also about Europe, and it is likely that the knowledge of new methods of millwrighting was passed around in the same way. It is also true that the industrial development of monastic orders, and in particular that of the Cistercians, enabled processes to take place on several of their lands, as industrially-minded monks would be moved about to take their technology to other sites. The working of iron and lead, in the Furness district of Cumbria and the Yorkshire Dales respectively, is an example of the great industrial development pursued by the Cistercians. These religious orders also crossed national boundaries quite easily, and so the development would take place in related sites in other countries. In these countries the local landowners would also take pains to copy the latest monastic developments in machinery.

In terms of the movement of technologists, in Britain there is the example of the deliberate invitation of Queen Elizabeth I to the German miners of the Harz, such as Daniel Hochstetter, to start up the Cumbrian lead, silver and copper mining industry in the Vale of Newlands, with water-driven smelt mills at Brigham near Keswick. From that settlement further members of the German community moved to start smelt works in the Vale of Neath and Swansea in South Wales. The site at Aberdulais (National Trust) is one started by German mining engineers from Keswick in about 1570. The production of iron in England required furnaces which had water-powered bellows and hammers for the refining of the iron blooms produced by the furnaces (see Chapter 2). The large number of hammer ponds in the Weald of Kent and Sussex give an indication of the scale of water power required in mediaeval England to produce wrought iron and the cast-iron guns and shot. The hammer ponds were created to supply the water power for the furnace bellows and for the tilt and helve hammers.

In 1556, the German author Georg Bauer, writing under the pseudonym 'Georgius Agricola', wrote *De Re Metallica* which is effectively a text-book of metal mining and metallurgy (see p. 145). In this large book, well illustrated by wood-block pictures, he sets out the whole process of mining and metal refining on a step-by-step basis. His illustrations show the various stages through which the mining engineer finds his mineral veins, how he digs his shafts and tunnels, and how he uses waterwheels, animal-powered engines and windmills to drain the mines, raise the ore and ventilate the workings. It is quite clear that Bauer was not the inventor of these systems, just that he recorded them from his own studies of central European practice, particularly in the German lead and silver mines. In these areas there are fifteenth- and sixteenth-century religious pictures which are as detailed as the illustrations in *De Re Metallica*. The painting by Jan Brueghel of *Venus at the Forge* of about 1600, shows several forms of water-driven forge and boring mills. Obviously, these painters could take only existing installations as their models.

In the English Lake District there are some sites of mineral-dressing works which date from the late sixteenth century. While some have been overlain by later developments, it could be possible to identify waterwheel sites, water-driven buddles (ore-washing vats) and the like, by archaeological excavation. The dressing works at Red Dell Head, on the flanks of Coniston Old Man and Wetherlam, were abandoned quite early in the 1800s. As the mines grew the mill streams were diverted to other sites where the workings have not been obscured by later developments.

The construction of waterwheels is quite clear in *De Re Metallica*. Obviously the wheels were made of wood with only the very minimum of iron being used for bearings. Joints would be pegged with dowels rather than fixed with nails. The construction of the millwork, according to these German precedents, would be seen and copied by the local millwrights, when they were concerned with corn mills. This sixteenth-century pattern continued with little improvement until the beginning of the eighteenth century.

The eighteenth century

The corn mill of the late mediaeval period followed the Vitruvian pattern in which each pair of millstones was served by a separate waterwheel. At Dowrich Mill, near Crediton in Devon, this mediaeval arrangement can be seen. There are two holes for the shafts of two waterwheels, each of which served a pair of millstones; these have been lost, and have been replaced by a conventional arrangement of two pairs of millstones driven by stone nuts off a single great spur wheel and a single waterwheel. The water-driven corn mill at Barr Pool in Warwickshire, shown in an illustration in the *Universal Magazine* published in 1729 (Figure 4.1), shows how the Vitruvian arrangement worked in the case of the pair of millstones over the shaft. The same illustration shows a variant on the Vitruvian mill in which a second pair of millstones was driven off a lay shaft, and not by a great spur wheel. In the Barr Pool example it is clear that at the beginning of the eighteenth century the millwrights were still working entirely in wood, the only metal parts being the bearings and gudgeons.

In France, Germany and the Netherlands, the beginning of the eighteenth century saw an upsurge in the study of millwrighting and mechanical engineering. The professional millwright was becoming an engineer and he was approaching millwork design scientifically rather than empirically. In France, in 1737, Bernard Forest de Belidor produced his classic volume *Architecture Hydraulique* in which he showed designs for improved waterwheels. Buckets in overshot waterwheels, though still made of wood with wooden soles to the back of the bucket, were angled so that the water would flow in more smoothly, and so that the water was held in the bucket for longer, therefore giving an increased efficiency to the waterwheel. He worked out designs for all forms of

Figure 4.1: The water-driven corn mill at Barr Pool in Warwickshire. This is the illustration from the *Universal Magazine* of 1729.

floats and buckets for the waterwheels, and he improved the way in which the water was led from the mill race through hatches, or launders, on to the waterwheels. It is thought, too, that Belidor first formulated the idea that the wheel would be better if the buckets were built between the rims so that the water did not spill out at the side. He was working towards a greater efficiency in the use of water power by also improving the design of dams and water controls.

One particularly important use of water power which grew in scale in the seventeenth and eighteenth centuries was the supply of water for drinking purposes in towns. In Paris, waterworks had been erected on the Pont Neuf about 1600 and these were rebuilt by Belidor in 1714. In London, a similar series of waterwheels was built under the northern arches of London Bridge by George Sorocold about 1700, to replace an earlier set inserted by Peter Morice, a

Dutch engineer, in 1582. Sorocold had been responsible for the installation of several other water-driven water supply systems in English towns, including Derby, Doncaster and Leeds. The system at Pont Neuf was known as a 'moulin pendant'. The Seine rises and falls quite severely and so the waterwheel has to rise and fall with it. Since the moulin pendant is a stream wheel, which is turned only by the flow of the water, it is important that the floats retain the same relationship to the flow of the water at all levels of the river. At Pont Neuf the whole body of the pumps and waterwheel was raised on four large screws as the water rose, so that the pumps could continue to work. The waterwheels at London Bridge had a slightly different set of conditions to deal with. The bridge spanned the tidal Thames and the starlings (foundations) of the bridge piers reduced the water passage to 50 per cent of the river's width. At high water the difference in level across the width of the bridge was 25cm (1ft) and at low water 1.38m (4ft 6in). To meet these differences in level the shafts of the waterwheels moved up and down on hinged levers and the gears continued to be engaged with the pumps since they moved about the centre of the shaft on the hinged beams. Later waterwheel-driven pumps were installed at Windsor (The King's Engine), Eton and Reading and these continued in use, in some cases, until the beginning of the twentieth century.

In Germany there were similar pumps for pumping the town water at Lüneburg, but more important examples existed to pump water for the fountains in the Nymphenburg gardens, near Munich. The idea of the water supply of formal gardens being raised from nearby rivers was developed to its fullest extent in the Machine de Marly, built about 1680 to supply water to the gardens and fountains of Versailles. Fourteen waterwheels were built below a dam on the River Seine and the water was brought on to these wheels through separate mill races. These undershot waterwheels were 11m (36ft) in diameter and 1.4m (4ft 6in) wide. In addition to 64 pumps adjacent to the wheels, connecting rods took a further drive a distance of 195m (600ft) to a second group of 49 pumps which lifted the water a further 57m (175ft) above the first reservoir. In all, the fourteen waterwheels operated 221 pumps and lifted the water 163m (502ft) above the level of the river.

While the waterwheels of the period, such as those at Marly, appear to have behaved well, they were clearly cumbersome and inefficient. The millwrights were able to build large wheels with high falls, such as those in the mountainous metal-mining areas, but these were still empirical in design. The designs shown in the books of this period, such as Jacob Leupold's *Schauplatz der Meuhlen-Bau-Kunst* published in Leipzig in 1735, bridge the gap between the apprentice system of training millwrights and the scholarly, scientific approach which came later in the eighteenth century. Leupold's book shows, by means of copperplate engravings, exactly how water-driven mills could be built. The plans are very accurately set out, with useable scales, so that the millwright could build his mill. The associated explanations, to be read with key letters on the plans, explain every step which has to be taken. The illustrations show grain mills, 'panster' mills with rising and falling wheels, mills with horizontal waterwheels, boat mills, paper mills, oil mills, fulling mills and saw mills. They are a design guide to every conceivable form of mill which the millwright could be asked to construct. There are tables showing how the lantern gears and pinions should be set out, so that the millwright could almost work with the book propped up in front of him. There are other German text-books of a similar character which produce even greater detail for the millwright. A good example is the text-book on water-driven saw mills *Von der Wasser-Muehlen und von dem inwendigen Werke der Schneide-Muehlen* by Andreas Kaovenhofer, which was published in Riga in 1770. This book details all the joints and fastenings required in a waterwheel, for example, and as in Leupold's book details of dam and watercourse construction are also included.

In France, the great encyclopaedia of Diderot, with its associated eleven volumes of plates, was published between 1751 and 1772. These plates, like Leupold's, showed the methods of construction and manufacture of every trade. Thus, in the chapter on the making of black powder or gunpowder, the nineteen plates show not only the various machines required and the stages to be undertaken, but also the way in which that machinery was driven. While the purchase of a set of Diderot volumes would have been beyond the purse of a master craftsman, enough copies would exist in manor houses and stately homes for these plates to have been seen, and used, by the millwrights of the locality. Indeed, it may well have been that the landowner as client would show such books to his millwright. Modern understanding, based on mass communication and transportation, finds it hard to realize how much craftsmen moved about, and equally how the intelligent gentry absorbed everything they could see and find on their travels abroad or on their 'Grand Tours' with their tutors. If they had a mechanical bent they would follow this up in the workshops and libraries of the countries they visited.

In the mid-eighteenth century there was an upsurge of understanding in mathematics and science. In terms of millwork, one breakthrough concerning the efficiency of the water-powered or wind-powered mill was the move, in Britain, away from cog and rung gears to cycloidal gears (see Figure 4.2). This was partly the result of the application of scientific and mathematical thought by scientists like Leonhard Euler. In Britain, the application of science to the profession of millwright was to be seen in the work of John Smeaton – a civil engineer in the modern sense of the word. He was the designer of many types of civil engineering works but he also designed forty-four watermills between 1753 and 1790, ranging from corn mills to iron-rolling and slitting mills. He also carried out research into windmills and watermills which was published in his paper 'An Experimental Enquiry Concerning the Natural Powers of Water

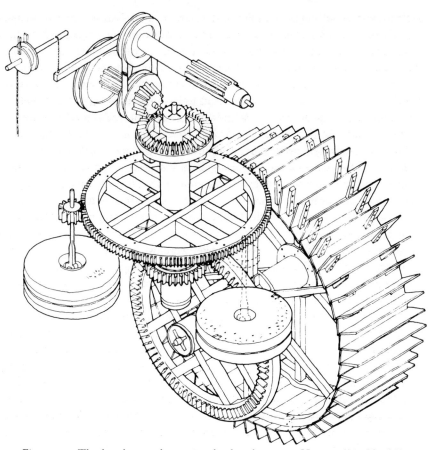

Figure 4.2: The low-breast shot waterwheel and gears at Hunworth in Norfolk. This watermill still has all its wooden gears which date from about 1775. Drawing by J. Kenneth Major.

and Wind to turn Mills' in 1759. His work was parallel to that of Christopher Polhem in Sweden, and they could well have been aware of each other's work. Smeaton's experiments set out to analyse the relationship between the various waterwheel types, the head and flow of water, and the work these could do. This had a great influence on the design of waterwheels for given situations of fall, flow and power required. No longer was an empirical solution the only one which answered a given problem, and design in the fullest sense of the word came into the process of creating a water-powered answer to the requirements of a factory.

Smeaton's work was closely studied abroad, and the newly-created United States of America in particular accorded new reverence to the scientific solution of problems. The corn millers who had arrived in the eastern states brought with them the old empirical solutions. Many of them had escaped from the repressive laws controlling milling in Europe, and had brought their old technology with them. However, they moved from small village mills, the result of the ancient institution of milling soke, to create much larger mills. Some of these were trading mills; others still worked on a toll-milling system, but without the imposition of a landlord who had to have his 'rake-off'. These millers frequently settled in an area where millwrights were not readily available, and so books such as *The Young Mill-Wright and Miller's Guide* by Oliver Evans, first published in Philadelphia in 1795, were invaluable as guides to the 'do-it-yourself' miller. He explains the science behind his designs, but because of the problems in the emergent states, his machinery is still made of wood. Evans, too, was the innovator of many changes in corn-milling practice. The most important plant in the corn mills following the publication of his book, was his grain elevator and horizontal feed screw, both of which cut down on the amount of labour needed to run the mill. No longer were two men required to hoist the grain sacks up the mill, nor to take the meal sacks back up so that the meal could be dressed. As the meal fell from the millstones, it was deposited at the foot of the bucket elevator to be taken up the mill to the bins, from which it would pass through first one dressing machine and then another, until it was properly graded. The screw crane for lifting and reversing millstones when they had to be redressed is also an example of an arrangement needing only the operation of one man. This was important designing in a country which was short of labour.

While Smeaton and Evans worked mainly with wooden machinery, by the end of the eighteenth century cast iron had become cheap and was used for millwork in Britain (see Figure 4.3). In Scotland, Andrew Gray published *The Experienced Millwright* in 1803, and this was filled with many details of millwork in which cast iron was the predominant material, particularly for gears, wheels and shafts. At this time the machining of gears was not easy, so many arrangements of gearing were built up using a large mortice wheel, in which wooden teeth were mounted in sockets in an iron wheel, and a small all-iron gear wheel. In the text-books of a parallel date in Germany, the millwork was still made of wood. In fact, in Holland and North Germany iron millwork was never used to any great extent before the water-driven mills ceased to work. The use of cast iron enabled the mills to be better set out, as the iron gears occupied less space. The change to cast iron also meant that the millwrights either owned foundries, such as Bodley Brothers in Exeter, or had to send designs or wooden patterns to the foundries.

Although the steam engine began to be used in factories in the 1750s, a large growth in water-driven factories took place throughout the eighteenth century to reach a climax in about the 1830s, at which time steam-powered factories became universal (see Chapter 7). Water-driven factories for the production

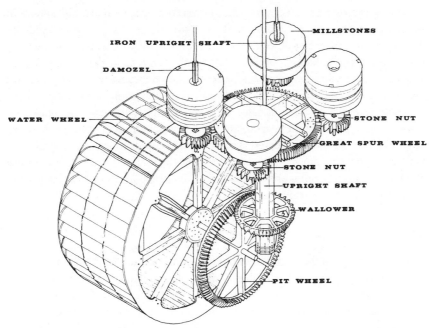

Figure 4.3: The conventional arrangement of a waterwheel and gears at Heron Corn Mill, Beetham, Cumbria. This dates from the early 19th century and is a combination of wood and iron gears with an iron waterwheel.
Drawing by J. Kenneth Major.

of woollen cloth sprang up in the Yorkshire valleys and in the steep valleys of the west face of the Cotswolds, while cotton factories were built on the western flank of the Pennine chain and in the Derwent valley in Derbyshire (see Chapter 17). Here the power requirements were larger for each factory than for the humble corn mill. Where the corn mill had managed with, perhaps, 9–11kW (12–15hp), the cotton mill would need five times that amount. The Arkwright cotton mills in Cromford, Derbyshire, and the Strutt cotton mill at Belper in the same county, had huge waterwheels. That at Belper, designed and built by T. C. Hewes of Manchester, was 5.5m (18ft) in diameter and 7m (23ft) wide. While the iron wheels at Belper appear at first sight to be iron versions of wooden patterns, there were many innovative features about them. In the first place, they were suspension wheels, in which the rim was held equidistant from the hub by tension rods rather than by stiff spokes. The buckets ceased to be angular but had outer sides made of sheet iron in smooth parabolic curves which were joined to the iron sole plates. The brick or stone casings to the wheel pit could fit more perfectly because the iron wheel could be held in a circular shape more easily than a wooden one, and in the head races new iron hatches gave a much more sophisticated control of the water flow.

During the eighteenth century water power also became more widespread in the mining fields and in the iron- and metal-working areas. The preserved site at Abbeydale, on the southern side of Sheffield, is an example of a water-powered edge-tool factory, and it was typical of dozens in the Sheffield river valleys. A large mill pond was created across the valley, and on the downstream side of the dam the various low buildings of the edge-tool factory hide. There are four waterwheels: for furnace blowing, for tilt hammers, for edge-tool grinding and to power the workshop. Apart from the forges with their blowing engines, the edge-tool industry of Sheffield required grinding and polishing workshops for the finish to be added to the tools. The preserved Shepherd Wheel is an example of a grinding and polishing shop in which the waterwheel drove twelve or more grindstones.

The nineteenth century

In the mining fields the introduction of cast-iron millwork enabled better use to be made of the potential of water power. The use of waterwheels for mine drainage and mine haulage had become well established in mining areas throughout the world by the nineteenth century. In Britain there were some very large waterwheels for mining purposes. The Laxey waterwheel on the Isle of Man, built in 1854 by Robert Casement, is the largest surviving waterwheel. This is a pitch-back waterwheel where the water is delivered on to the top of the waterwheel in the opposite direction to its flow in the launder, and it is 22.1m (72ft 6in) in diameter and 1.8m (6ft) wide. In the Coniston copper mines in Cumbria there were several large haulage wheels. The biggest was 13.4m (44ft) in diameter and 2.75m (9ft) wide, and there were others of 9.1m (30ft) and 12.8m (42ft) in diameter. Down in the Paddy End copper ore dressing works there were several waterwheels of which the biggest, 9.75m (32ft) in diameter and 1.5m (5ft) wide, was replaced by a turbine in 1892. So much water was used that the main streams were interlinked by four principal mill races, and Levers Water was turned into a large holding reservoir by means of a 9m (30ft) high dam. In a similar way the slate industry had water power for its machinery, and at Llanberis in North Wales the factory and maintenance works of the huge slate mines and quarries were powered by a waterwheel which was 15.4m (50ft 5in) in diameter and 1.6m (5ft 3in) wide, built in 1870, later to be replaced by a Pelton wheel (see p. 244).

On the continent of Europe and in the United States of America factories began to spring up along the larger rivers. The New England states began the concept of the factory town in which the factories were water powered. In these towns complicated water power canals were arranged so that many factories could be supplied in sequence as the water flowed through the town. Lowell, Massachusetts, had an extensive power system of which the first part dates from

1820. Here the Pawtucket Falls on the Merrimack River were of sufficient height to give an overall head across the town of 11m (35ft). An existing barge canal was taken over, a new dam was constructed, and the canal became the head race for the mills. Lateral canals supplied individual cotton mills and the tail races were collected to form the head races of further mills. At first waterwheels were used, but these were very soon replaced by turbines. The Lowell system was followed in Lawrence, Massachusetts and Manchester, New Hampshire. In parallel with the creation of Lowell's water-power system, one was brought into being at Greenock in Scotland where, in the early 1820s, Robert Thom designed a system of dams and feeder canals. This system was completed and at work in 1827 supplying 33 mills over a fall of 156m (512ft). Similar schemes were put in hand on the river Kent, above Kendal in northern England, for the corn mills, bobbin mills, woollen and powder mills of that valley, and in the area around Allenheads in Northumberland, a further scheme was created to serve the needs of lead mines and mineral dressing works. John F. Bateman was the civil engineer responsible for the river Kent scheme, and as an engineer he achieved a name for many water supply schemes in Britain.

The need for water power went on increasing as the factory units grew in size and a number of large millwrighting and engineering firms grew up in Britain to meet the needs of the textile industries. Hewes has been mentioned above for his work at Belper and, in combination as Hewes and Wren, supplied the huge waterwheel for the Arkwright mill at Bakewell. Hewes had a draughtsman named William Fairbairn in 1817, who left him in that year to join in partnership with James Lillie. Lillie and Fairbairn were responsible for a large number of big waterwheels in textile factories in Britain, and their partnership was to run for fifteen years. One of their big wheels was at Compstall, near Stockport in Cheshire. This breast-shot waterwheel, situated between two halves of the mill, was 15.25m (50ft) in diameter and 5.2m (17ft) wide. Other wheel diameters available as stock patterns in the Fairbairn works were 18.4m (60ft 4in), 14m (46ft), 12.1m (39ft 9in), 11m (36ft) and 9.15m (30ft), and of course other sizes were also made. In using waterwheels of this size a change had been made in the way in which the power was delivered to the machinery. No longer was the shaft of the waterwheel extended into the mill so that a pit wheel could provide the power take-off; instead, a rim gear, often of the same diameter as the rim or shrouds, would engage with a pinion on the shafts going into the mill. William Fairbairn was a famous civil engineer and millwork formed only a small part of his business, but he did make several changes in the design of waterwheels and the arrangements of the water controls for those wheels. One innovation was the ventilated bucket on the waterwheel. As the water went into an unventilated bucket, a cushion of air was built up beyond the water which prevented the water from entering the bucket smoothly. By providing a ventilation slot at the back of the bucket on the sole, the air was forced out of the bucket and the water filled it properly. Fairbairn was also responsible for the introduction of a workable system of governors to control the flow of water through the hatches and on to the waterwheel, and by using a shaped series of slots in the hatch a smoother flow of water was delivered to the buckets.

William Fairbairn was knighted for his engineering work, and was recognized for his scientific approach to structures by being elected a Fellow of the Royal Society. His book *Treatise on Mills and Millwork*, first published in 1863, became the classic text-book in Britain on the construction of these 'modern' waterwheels. On the continent of Europe, Armengaud the elder published his *Moteurs Hydrauliques* in 1869, Heinrich Henne his *Die Wasserräder und Turbinen* in 1899, and Willhelm Müller his *Die eisernen Wasserräder* in 1899, while in the United States, *Practical Hints on Mill Building* by James Abernathy, published in 1880, was of great importance. By 1900 the emphasis on water power was switching from large, efficient waterwheels to the smaller and even more efficient water turbine.

WATER TURBINES

In France, the design of waterwheels had been given considerable attention at the beginning of the nineteenth century, but there was always a search for greater efficiency. J. V. Poncelet had taken the old form of vertical undershot waterwheel which had straight blades or floats made of wood and set radially, and by curving the blades and constructing them of metal, had produced much greater efficiency. By using tight-fitting masonry walls and floors in the wheel pits, he ensured that all the water would be swept into the space between the blades. He used formulae to determine the size of the floats in relation to the wheel and the water flow. A further vital point, particularly with an undershot wheel, was that the water flowing out of the floats fell clear of the wheel so that it did not run in tail water. This interest in waterwheel performance led to the first viable turbine designs being produced in France: it is significant that the south of France, Spain and Portugal had large numbers of water-driven corn mills in which the waterwheel ran horizontally, often forced round by a jet of water. Benoît Fourneyron produced a successful water turbine in 1827, and the design of other water turbines was proceeding in Germany and in the United States, but not in Britain where the waterwheel designs were reaching their peak. The Fourneyron turbine consisted of an inner fixed ring of curved gates set in one direction, and an outer ring, which had curved blades set in the opposite direction, mounted on the drive shaft of the mill or factory. The water flowing into the fixed ring was controlled by a circular iron hatch which moved up and down in the water to control the flow. The water flowed over the gate and through the fixed gates to impinge on the rotating outer ring of blades, and thereby revolve it at a considerable speed; thus a small turbine could produce more power at greater speed, using less space than the equivalent waterwheel. The Macadam brothers of Belfast, Northern Ireland, produced a very efficient version of the Fourneyron turbine, of which the preserved example from Catteshall paper mill in Godalming, Surrey, is typical. The stator (fixed inner ring) and rotor (rotating outer ring) each have forty-eight vanes, and the inside of the stator is 2.5m (8ft 3in) in diameter. The Catteshall turbine developed 37kW (49.6hp) at approximately 25rpm, and in 1870, when it was built, it was among the biggest then in use. The principle of the Fourneyron is that of an outward-flow reaction turbine.

In England, between 1734 and 1744, an invention called the Barker's Mill was introduced. In this water is led into a vertical tube, which rotates, and at the bottom of this tube two arms project which have nozzles at their tips. As the water flows, so the jets spout out at the ends of the arms, and the whole is pushed round by the reaction of the jets against the pressure of the air. While one or two examples of this are known, such as that at the Hacienda Buena Vista at Ponce, in Puerto Rico, it must have been hopelessly inefficient. James Whitelaw took the principle of the Barker's Mill, improved the shape of the arms and introduced the water from below, to give a much more efficient machine known as the Scotch turbine. The arms were in the form of an elongated 'S' with the inlet pipe in the centre. The new shape induced a better flow out of the nozzles at the ends of the arms. The Whitelaw turbine had governors to control the speed of rotation. After 1833, when Whitelaw built his prototype, several of these were installed in factories in Scotland. One example is quoted as having 1491.4kW (200hp) with a fall of 6.7m (22ft) and a speed of 48rpm. Escher Wyss & Co. of Zurich installed Whitelaw turbines in 1844, and these were up to 45kW (60hp).

James Thomson trained as an engineer in the works of Sir William Fairbairn and became a professor in Belfast. In 1846 he was at work on the design of a new form of turbine which, when tested in 1847, worked at one-tenth of a horse power at an efficiency of some 70 per cent. He went on to patent this turbine in December 1850. Described by its inventor as a Vortex turbine, it was an inward-flow turbine in which the water came into the casing and was taken through a spiral path to be discharged through gates on to the rotor. At the same time J. B. Francis was working on a similar arrangement in Lowell, Massachusetts, which he called a centre-vent turbine. His work was published in *Lowell Hydraulic Experiments*. He was associated with Uriah A. Boyden in producing the Boyden turbine which was an adapted Fourneyron, designed for the particular needs of the Lowell cotton mills, and the horse-power results were extremely good. In 1854, the Merrimack Company's mills had Boyden turbines of 2.75m (9ft) diameter which generated 522kW (700hp) under a 10m (33ft) head.

Meanwhile, the development of other turbines was proceeding in France, owing to a lack of coal to power steam engines, and in Switzerland and

Germany. The ability to use a high head of water, as one would get in Switzerland or Germany, led to the design of other forms of turbine. The most famous of these high-head turbines, requiring only a small flow, is the Pelton wheel. This is an impulse wheel, patented in 1880, in which a jet of water is focused on to a bucket on the diameter of a small wheel. The bucket is cast in the form of two cups which receive water equally when it comes out of a jet at a tangent to the wheel. The water is turned back on itself as the bucket moves forward. The Pelton wheel can achieve high speeds and is easily controlled by the amount of water allowed out of the nozzle which opens or closes the size of the jet. On the Coniston Fells in Cumbria there was a Pelton wheel made by Schram and Marker of London, 0.6m (2ft) in diameter with approximately a 150m (500ft) head, which drove the air compressor in the slate quarries high up on the fell.

Hydro-electric power

Professor Thomson's first English turbine, made for him by Williamson Brothers of Kendal in 1856, was a 3.73kW (5hp) Vortex made for a farm. (This turbine, Williamson No. 1, can be seen in the Museum of Lakeland Life and Industry in Kendal.) In 1880 this firm supplied their 428th turbine to Sir William Armstrong of Cragside, Northumberland. Here Armstrong, himself a prominent engineer, installed the turbine on an 8.8m (29ft) head of water to evelop 9hp, transmitted by belt to a 90V dynamo from which Cragside was lit by Swan's electric lighting (see Chapter 6). This was the first hydro-electric plant in Britain. In 1884, W. Günther of Oldham built his turbine number twelve, a 'Girard' impulse turbine of 30kW (40hp), which was used for the electric lighting of Greenock, Scotland.

Although factories continued to require turbines to power their machinery, the main demand for water turbines lay in the production of hydro-electric power. Turbines grew bigger, more efficient, and increased in horse power to meet the growing demand for electricity in hilly countries where coal was not available. The upper Rhine, between Basel and Strasbourg, has eight hydro-electric stations, the dams of which make the river fully navigable to Basel all the year round. Kems, built in 1932, is the oldest of this series, with six turbines each developing 18.6MW (25,000hp) on a 16.5m (54ft) head. The whole suite of hydro-electric stations between Basel and Strasbourg can produce 895MW (1,200,000hp). Similarly, the Tennessee Valley Authority (TVA) water control requirements enabled the Federal Government of the United States to build a series of dams with hydro-electric plant in the years of the Roosevelt 'New Deal', and in the immediate post-war years. The basic turbine type for most of the high-powered hydro-electric plant has been the Kaplan. This is a reaction turbine which uses a large flow on a low head, and which is made like a ship's propeller with variable pitch vanes running in a close-fitting casing. Victor Kaplan patented his turbine before the First World War in Brno, Czechoslovakia, but it was neglected until the 1920s. These turbines have been capable of some 85 per cent efficiency, and modern developments in their design have made them even more efficient.

Perhaps the most impressive use of the Kaplan derivatives is in the great tidal-power system of the Rance estuary by St Malo in France. This scheme was completed in 1966 and presents many additional factors. As turbines they must work on a variable head yet give a constant speed. Thus the Rance design must represent the most flexible of all turbine solutions. It is interesting to note that, in the new proposals to use the flow of the tides in areas such as the Bay of Fundy in the United States, and the Severn estuary and Morecambe Bay in England, there is a harnessing of a water-power source first used in the late twelfth century in England.

Water power has been harnessed to great effect for 2000 years to provide power for the many requirements of life. It started in order to ease the grinding of grain for food, and grew to power all man's industrial needs before steam came into use as the power behind the burst of development which we know as the Industrial Revolution. It is developing again to provide electricity on a huge scale, as well as being used as a means of giving minor industrial development to the Third World.

WIND POWER

The ancient world

While there is some certainty about the timescale and geographical distribution of the water-powered corn mill in antiquity, there is far less knowledge about the geographical spread of windmills. Given that the grinding of grain took place between a pair of quern stones which rotated in a horizontal plane, then it would seem to be easy to accept that the upper (runner) millstone could be fixed at the bottom of a vertical shaft which could be turned mechanically. If that shaft were to be of a fair length, and if it had sails attached parallel to it, then these could be blown round by the wind. The problem is, of course, that the sails present the same area of face to the wind as they are blown by it and as they come towards it. This would seem to create equilibrium and prevent the mill from turning. We are fortunate to have survivors of such very primitive horizontal windmills at Neh, in the area of eastern Iran which is near the Afghan border. Here there are rows of these mills built together to economize on the construction of the supporting walls. The arrangement of a typical windmill in this area is that two walls are built, some 6m (20ft) high, on top of a roofed mill room containing the single pair of millstones. The shaft from the runner millstone rises through the roof of the mill room to a bearing in a beam between the tops of the two walls. On the side facing the prevailing wind a wall is built up to the top of the two walls but only across half the space between them. The wind, its force concentrated by being funnelled through the row of narrow openings, turns the shaft by means of the six or eight sails mounted on it. These sails are made of wooden slats or reed mats fixed to the five or six sets of spokes on the shaft. As the sails turn, the lee surfaces are never opposed by the wind because of the protective wall. These windmills were first recorded in use in AD 644 in a Persian manuscript of the early tenth century, but it is thought that they may have existed in Egypt at the time of Moses.

The Persian horizontal windmill appears to have remained a static concept, unchanged to the present day. What is missing is the link to the Mediterranean windmill which is thought to be the precursor of all our vertical windmills. The Mediterranean windmill is a two-storeyed stone tower mill with one pair of millstones driven from the windshaft. The wooden windshaft projects beyond the face of the tower and carries six, eight or twelve triangular cloth sails which are set like jib sails from the sail stock to the opposite sail stock. Indeed, the vertical windmill used in Europe may be a quite independent innovation which derives from the Vitruvian arrangement of the water-powered corn mill and not from the Persian windmill. The arrangement of the drive in a vertical windmill is that the nearly-horizontal shaft through the sails, which turn in a vertical plane, has a gear wheel mounted on it which engages with a second gear wheel on the drive shaft of the runner millstone. It we substitute the windmill sails of this concept by a waterwheel, we can see the validity of the argument to support the lack of a technological link between the Persian horizontal windmill and the European vertical windmill.

Mediaeval and Renaissance Europe

There were windmills in England in the years just before AD 1200. The charters of St Mary's Abbey at Swineshead in Lincolnshire give ownership of a windmill there in 1179, and at Weedley in Yorkshire there was also evidence of a windmill. Pope Celestine III ruled that windmills should pay tithes, so the use of the windmill would seem to have been well established if they were worthy of concern over their tax and tithe dues. The windmill was therefore well established in northern Europe by the end of the thirteenth century, and had an advantage in that it did not freeze up in winter as the watermill was prone to do.

One great problem with the windmill in Europe is that the wind has no prevailing direction as it clearly has in Persia. This means that the windmill must always face the wind and that the wind must never be presented to the back of the mill. The post mill would appear to be the first type of windmill to do this, and is the type of which we have the largest number of windmill illustrations from the Middle Ages.

The post mill in the mediaeval period was probably quite small and simply arranged. Stained glass and documentary representations show considerable variation in form and construction, much of which must be discounted as fanciful, but enough remains for the mill student to be able to learn the form of the early windmill. All post mills consist of a wooden body, known in England as the buck, which carries the sails, the windshaft, the gears and the millstones. This can be turned into the wind, because it is pivoted about the top of the king post. The mill body is heavy, and as it should not be backwinded, it is fitted with a tail pole coming out of the back of the bottom storey of the body. The miller puts his back to the tail pole and pushes it round when there is a variation in the wind direction. The king post could not be a single post as it would blow over, but would have to be propped by other timbers to give a broad base to prevent the whole mill being overturned. Archaeological excavation, such as that at Great Linford in Buckinghamshire, shows the horizontal crosstrees which had the king post at their centre and the diagonal quarter bars at their ends which provide the props to the king post. This frame of the crosstrees, quarter bars and king post is so made that it is a rigid entity which sits either on a cross of masonry on the ground, or on four pillars at the ends of the crosstrees.

Post mills are shown in several mediaeval documents and in the illustrations in the margins of manuscripts. From these we can see that the mill was small: probably only big enough to house one pair of millstones with space over them for the windshaft, brake wheel and gears on the runner millstone spindle. The body would be suspended one floor up on the king post propped by the quarter bars. In the picture in the fourteenth-century *Decretals of Pope Gregory IX* the mediaeval post mill is properly understood, for here the post is seen to go up to the underside of the second level within the body. Clearly, the body cannot just sit on top of the post, for it must have some lateral hold on the post where it passes through the framing of its bottom floor.

The sails of these primitive post mills were made of a frame of laths over which cloth was stretched and tied in place. The sails were short, stretching to the ground so that the miller could set the sails on the sail frame, and climb up the frame if he wanted to remove the sail cloth completely. To adjust to variable wind speeds he would set less sail, and he was possibly aware that by adjusting the space between the millstones he would also meet the problems associated with variable wind speeds. The windshaft (holding the sails) may or may not have been inclined so that the plane of the sails was tipped slightly backwards. A refinement such as that, which made for better balance in the sails and windshaft and stopped a tendency to tip forward, is clearly a matter of empirical knowledge. The gears would be heavy cog and rung gears, and the method of braking is unknown; possibly the miller turned the mill out of the wind in order to stop it. No storage was available in this early post mill and probably none was needed, because the mill would certainly have been held by a landlord, such as a manor or an abbey, and the miller would be an employee or a tenant operating under the milling soke and receiving his payment, or multure, by removing a proportion of the meal.

Tower mills followed the post mill and appear to be shown in pictures and in stained glass from the fifteenth century. In the tower mill only the cap and sails rotate to face the wind while the mill body remains stationary. The principal advantages of the tower mill over the post mill are in its stability and in the fact that the portion to be turned to the wind is so much lighter. The cap is built up on a frame which has a beam under the neck bearing supporting the front end of the windshaft. The rear of the frame is usually extended to support a tail pole which goes down to ground level. When the sails are turned to face the wind the miller pushes the tail pole, and the cap, windshaft, sails and brake wheel turn. The turning is usually achieved by the frame being supported on rollers or sliders on top of a rigid circular track on the top of the tower.

The post mills of the Renaissance period are quite accurately depicted by artists such as Brueghel the Elder. The picture known as *The Misanthrope*, painted in 1568, shows a post mill very like those to be seen today in the Low Countries. This has a tallish body on a post and quarter bars, with a ladder and tail pole at the rear and four sails. An even more acceptable representation is the picture in the National Gallery, London, *A Windmill by a River*, which was painted by Jan van Goyen in 1642. Here the post mill stands on a tall frame of multiple quarter bars on tall brick posts, and has the steep tail ladder which characterizes many Dutch post mills. The tower mill depicted in *The Mill at Wijk bij Durstede* by Jacob van Ruisdael, shows a vertically-sided tower mill with a reefing stage, from which the cloth sails were set, and to which the strutted tail pole extended. By the time this was painted, about 1670, the windmill for grinding corn had become a much larger building. In the case of the post mill, it could well be built so that there were two pairs of millstones, each driven off the windshaft by its own set of gears. The tower mill would be made bigger in diameter and taller, with a storage floor at the top of the mill, and then, with the use of a great spur wheel and stone nuts, there could be more pairs of millstones on the stone floor.

Dutch mills and millwrights

The Netherlands were situated well below sea level throughout the early mediaeval period. It was a country beset by continual flooding, of which the floods of 18 and 19 November 1421 were perhaps the worst, when seventy-two villages were destroyed. Sea defences were constructed to keep out the sea flood water; solid land was formed by draining the spaces between these sea dykes; residual pools and lakes remained to be emptied; and windmills were brought in for this purpose. When the polders were dry it became necessary to

Figure 4.4: A wip mill for pumping water. This is not the Dutch type but one in which the water is pumped by cylinder pumps.

retain the windmills to drain the rainwater off the land and keep a level water table.

The first drainage mills were a variant of the post mill. The drainage was carried out by scoop wheels, i.e. waterwheels in reverse, at ground level. The problem was to get the drive from the sails down to the scoop at ground level when they were, in fact, separated by the buck and the post. The hollow post mill was created to solve that problem. Here the drive is taken down through the centre of the post, which is made up of four pieces, to gears at ground level which turn the drive at right angles to power the scoop wheel. This hollow post mill, *Wipmolen* in Dutch, was small but efficient and can still be found on many of the polders. A version of the drainage wip mill, using cylinder pumps instead of scoops, is illustrated in Figure 4.4. The tower drainage mill was introduced about 1525. This was, in fact, an octagonal smock mill made of timber and thatched, and it had a conventional rotating cap. Smock mills are not really a different type of mill; they are effectively tower mills in which the stone or brick is replaced by wood and weather-boarding or thatch.

One of the great names in the history of millwrighting is Jan Leeghwater. In the first half of the seventeenth century he was renowned for his dyke building and hydraulic engineering, became a consultant, and travelled to Holstein, Flanders, France and England to advise on drainage schemes. In Holland, the most famous of his proposals – the drainage of the Haarlemmermeer – would have required 160 windmills to do the work. The hydraulic designs demonstrated in the Haarlemmermeer study were much in demand for use elsewhere in Holland. Leeghwater followed Cornelius Vermuyden, who came to England in the reign of James I to carry out drainage schemes in the Fens of East Anglia and in the Yorkshire Carrs. He also brought with him the principles of the drainage smock and tower mills.

Although we have little or no knowledge of the work of the British millwrights in the seventeenth century, we have a few remains of mills which were built in that period. While the envelope of the mills must remain essentially as it was constructed, it would be extremely difficult to say that these mills represented seventeenth-century practice. No cycloidal gears were in use, so the windmill gearing would have been cog and rung. The post mills, which are usually the only dated seventeenth-century examples, may have had only one pair of millstones at that time, whereas the survivors frequently have two pairs, and they are spur wheel driven. The efficiency of the new millwork of the late eighteenth and early nineteenth centuries would appeal to the millers who would insert it in the older envelopes of their mills.

The great period of the windmill is the early eighteenth century in the Netherlands. The country had shaken off the oppressive yoke of the Spanish empire, and a new prosperity was spreading throughout the land with its many wealthy towns. The towns grew and had to be supplied with meal. Often the Italian-style fortifications around these towns had a windmill on each of the star-shaped bastions. De Valk in Leiden is a working survival from this period standing on a bastion to the north of the town and typifies the great wind-driven mills of the Netherlands. It is a brick tower mill, 30m (98ft) to the cap, reputedly containing 3,000,000 bricks, and it was completed in 1743. The common (i.e. cloth covered) sails are 27m (88ft 6in) in diameter and are reefed from a stage at fourth-floor level. The two lowest floors are the miller's house, and then there are two floors of storage space before the floor at which the reefing stage is mounted and which was used as the meal floor where the meal was stored after it had come from the millstones. The four pairs of millstones on the stone floor (the fifth floor) were driven from the great spur wheel mounted at the bottom of the upright shaft which comes down to this level. The sixth floor contains the sack hoist and was where the sacks of grain were stored before milling. The top floor, which is open to the cap, contains the gearing which takes the drive from the windshaft to the upright shaft: the brake wheel and wallower. The sails are mounted in a cast-iron canister on the windshaft which must have been inserted at some time after 1743. The cap is turned to the wind by means of the tail pole which drops from the cap to the reefing stage. On the tail pole there is a spoked wheel which turns a windlass. The windlass winds up a chain, which is anchored to bollards on the stage, to pull the cap round.

The detailed design of De Valk is repeated in the books which gave the Dutch millwrights text-books of windmill construction at this time. These large folio volumes are invaluable documents for they are the source of our knowledge of the windmill at the beginning of its great period. The most important of these are the *Groot Volkomen Moolenboek* by Natrus and Polly, published in 1734, and the *Groot Algemeen Moolen-Boek* of van Zyl, published in 1761. The *Groot Volkomen Moolenboek* contains precise instructions, illustrated with scale drawings and projections, of the way in which gears are made to relate to each other properly, and the way in which the sail stocks are socketed at angles to produce the correct curving sweep (known as the weather) from shaft to tip. The notes explain the stages whereby the detailed work is carried out. A nice touch is the presence of the men on these drawings which gives an added reference to the scale of the mills; the various tools and pulley blocks required for the erection of the frames are also shown.

What is more important in the *Groot Volkomen Moolenboek* is the pattern of windmill usage which was available in Holland in 1734. The list contains saw mills, paper mills, oil mills, corn mills, glass-polishing mills and the various types of drainage mills. In the Netherlands industrial windmills form a very large element of windmill history, although there are very few among the several thousand preserved windmills, because drainage and corn mills remained financially viable after the industrial processes had become factory based.

It is useful to examine some of these processes in the industrial windmill in order to realize the heights which the Dutch windmill had achieved in the first half of the eighteenth century. The windmills used for wood sawing come in two forms: the tower mill and the 'paltrok'. The paltrok is a windmill in which the whole body is turned to face the wind on a series of rollers mounted at ground level. The windmill sails are mounted in the tower or paltrok cap. The windshaft carries the large brake wheel which is braked by means of curved wooden blocks around the outside of the wheel which are tightened by means of the brake strap. The peg teeth of the brake wheel engage with a lantern cog mounted on a horizontal shaft. This horizontal shaft has a crank on either side of the lantern cog, and each of the cranks moves a saw frame up and down. A supplementary crank on this upper shaft works levers which drive the timber carriage forward when each cutting stroke of the saw is finished. The timber carriage is mounted on rollers right across the body of the mill. The timber is

Figure 4.5: The paltrok 'De Poelenburg' at Zaandam. This saw mill was built in 1869. The brick base on which the whole mill rotates is visible behind the ladder and winding windlass. The crane for lifting the logs onto the stage is on the left. Anders Jespersen.

hauled up on to the carriage by means of a crane at the entrance end of the carriage. The chain on the crane is wound up by a windlass driven by the sails through the drive shafts of the carriage. The saw frames consist of double top and bottom bars separated by about 25mm (1in). In this gap the saw ends are slotted and are separated by amounts equal to the thickness of the timbers to be cut. The saws are tensioned by wedges driven into sockets in the ends. These saws are extremely powerful, and at d'Heesterboom, the smock saw mill in Leiden, built in 1804 and still used, the wind-driven saws are used for logs of 600mm (24in) or more in diameter. Figure 4.5 shows the paltrok 'De Poelenburg' of 1869.

The paper-making windmills are exemplified by De Schoolmeester at West-zaan in North Holland, a thatched smock mill with a reefing stage, a tail pole and windlass, where fine white paper is made from rags. The drive comes from the cap right down to ground level where the machinery is situated. The rags are sorted and cut by hand against fixed knives in order to reduce the pieces to a manageable size. The rags are then chopped to finer particles in a tub with a strong wooden bottom in which four knives on poles are lifted and dropped by means of cams, and they are then soaked in a caustic solution and placed in the 'hollander'. The hollander is a long tub with semicircular ends and a spine down the centre. On one side of the spine the floor of the tub rises to meet the underside of a drum which is covered with blades along its length and around the rim. The fluid of rags, chemicals and water is pushed around the tub by the motion of the drum. As the fluid passes under the drum, the blades reduce the rags to their constituent fibres. The resultant fibres are then taken to the vats, where the vatman makes the individual sheets of paper by hand. He takes a wire frame of gridded brass wire, possibly with a pattern on it which becomes the watermark, shakes an even layer of fibres on to his wire frame, lifts it out of the water and drains it. He then flops the fibre out of the frame on to an adjacent sheet of felt, and this is the basic material for one sheet of paper. The paper with its sheet of felt is lifted on to a pile of felts which are gradually draining. The felts are then put into a screw press to be reduced to the normal thickness of that quality of paper. The screw is turned by hand until it needs more force to make it go down further. At that point a horn is blown and all the staff of the paper mill drop whatever they are doing and come to pull on a long lever arm on the press. They may even be on a rope attached to the end of the arm. They pull until the press is right down, the water has drained out, and they have the sheets of paper between the felts. The pile is opened out and the sheets are hung, one at a time, over cow-hair ropes (which do not stain) to dry. This explains the presence of the long shed which extends to the base of the windmill across the prevailing wind.

A third type of industrial windmill, the oil mill, is almost the only industrial windmill known in England. The oil mill is usually a smock mill, and Het Pink at Koog an de Zaan is a preserved example. See Figure 4.6. The drive comes down from the cap to drive the machinery, which is so heavy that it must stand at ground level. There are three wind-driven elements in the oil mill: the kollergang (or roller crusher), the stamps and the heating plate. The kollergang consists of two huge stone edge-runners which roll, on different paths, around a circular stone base. The oil seed is put on this base and the drive to the kollergang is engaged. The rollers turn ponderously round and round on the base. The oil seeds are crushed while wooden rakes, fixed ahead of the rollers, keep the seed in the tracks. The seed is then taken to the heating plate where it is heated to release the oil. To circulate the heat the seeds are turned over and over by an 'S'-shaped blade rotating over the plate. The heated seeds, releasing

Figure 4.6: The cross section of a smock windmill for crushing oil seeds. The stamps, kollergang and heating bowl are shown in this picture.

oil, are put in horse-hair pockets which are held in a great wooden beam by blocks and wedges. Each block has a wooden stamp above it which is raised and lowered by a cam which is driven from the upright shaft. After the stamps have dropped a set number of times, a further stamp is released which hits a reverse wedge, and this loosens the hair pockets and the seed is lifted out. As the stamps and blocks squeeze the seed in the hair pockets, so the oil runs out to be collected. The first pressing of oil is set to one side before the crushed seed is subjected to the same process again.

John Smeaton designed a wind-driven oil mill at Wakefield. This is like the Dutch oil mill, of which Smeaton was well aware, but has several refinements which indicate the state of the windmill millwrights' craft in the mid-eighteenth century. Wakefield oil mill was completed in 1735 and was a smock mill on a square brick base. The base consisted of two tall storeys containing all the machinery, while the smock was a means of giving height to the sails and had no floors or machinery in its height. The machinery consisted of a kollergang, which was right in the centre of the mill so that the two rollers were turned by the upright shaft without intermediate gearing, and to one side of this there was the bank of stamps. In Smeaton's design in the Royal Society Library there is no hot plate shown, but this was certainly a requirement of this type of mill, as it is needed in the cracking process. Above the first floor there are sack hoists and hoppers. The other variation from conventional windmill design lies in the sails, of which there were five, mounted on a cast-iron windshaft by means of a cross, a series of flat channels into which the sail stocks are bolted. The other way of mounting the sails is in a canister, or poll end, through which the sail stocks are wedged. The shafts and gear wheels were all of wood and only the windshaft was made of cast iron.

Eighteenth- and nineteenth-century innovations

Smeaton carried out experiments in watermill design (see p. 236) and his curiosity also extended to improvements in windmills. There is a model of his experimental design for windmills in the Science Museum in London. It is interesting to note that the 'weather' detailed for the sails in his experiment is the same as that shown in the diagram in the *Groot Volkomen Moolenboek*. Of course other millwrights were working at the same time as Smeaton to improve the efficiency of the windmill. Perhaps it is in the development of the sails that the largest number of types and patents is found.

A common sail is one in which a sheet of canvas is stretched over a sail frame. A common sail frame which is not clothed consists of square panels formed by the laths and hemlaths over which is draped the rolled-up sail. At the inner end of the sail there is a 'curtain rod' with rings on it to which the sail cloth end is fastened. All down the stocks there are cleats to which ropes in the hems of the sail cloths are tied. The other edges of the sail cloths are tied around the laths. At low windspeeds the sail is stretched out fully, but as the windspeed increases, the area of the sail is reduced by making the sail cloth dagger-shaped, with the point towards the tip. One great disadvantage of the common sail is that the miller cannot set four sails at once; each sail has to be brought to the ground, its cloth set, the brake released, and the next sail brought into the vertical position. This cannot be hurried, and is a real problem, or even a danger if a squall should develop suddenly: we can imagine the miller's horror if improperly braked sails started to turn while he was on the bottom one.

Andrew Meikle, a famous Scottish millwright, produced the spring sail in 1772. In the spring sail, a number of hinged shutters made of wood and canvas take the place of sail cloths. The shutters are all connected to each other by means of a shutter bar running the length of the sail. The movement of the bar, therefore, moves all the shutters open or closed together. The movement is controlled by a spring at the tip of the sail which can be pre-set by means of a slotted strap. As the wind blows, the tension in the spring causes the shutters to open or close according to the wind pressure, letting the wind spill through the openings and slowing the sail to the required speed. The disadvantage of this arrangement is that the tensioning has to be done to a sail at a time when it is stopped in the vertical position.

In 1789, Captain Stephen Hooper patented an automatic sail system which is known today as the roller-reefing sail. In this type of sail a number of small roller blinds replace the shutters of the spring sail. Each roller blind is attached to the one above it and below it by means of webbing straps. The blinds are connected to two wooden bars which run along the sail. The pressure of the wind adjusts the extent to which the rollers open or close the sail. It was not a great success, as too many parts could fail or decay. Sir William Cubitt, a millwright and engineer, introduced his 'patent sail' in 1807, and this is the shuttered sail most commonly met in England. The shutters, made of wood and canvas, are interconnected as in the spring and roller-reefing sails, and all their inner ends are connected by means of bell cranks to a 'spider' mounted in front of the cross or canister. The windshaft is hollow and contains a rod, called a striking rod, which projects from the back of the windshaft and is joined to the spider at the front of the windshaft. At the rear, an endless chain and a weight hang from a chain wheel on a pinion. The pinion engages with a rack on the striking rod, and as the sails respond to the wind the weights control the extent of opening to meet its pressure. This form of shutter is most useful because it can be adjusted while the mill is running and without stopping the sails.

On the continent of Europe, the shuttered sail in the English form is not met with except in north Germany and Denmark, where it derives from English practice and millwrights. In the Netherlands there are some shuttered sails in the north-east, near the German border, but most of the mills still have cloth sails. There are several variations on sail forms, in particular in the design of leading boards and stocks, but these are incidental to all the processes of sail design. In France, the sail form most usually met with is the Berton sail. In this the four sails are made of wooden slats set parallel to the sail bars. The slats can open out or close according to the wind speed and pressure. The controls run down inside the windshaft and can be adjusted by the miller inside the cap according to the requirements of grinding and windspeed.

Another contribution to ease the miller's duties was the introduction of an automatic means of turning the mill to face the prevailing wind (known as winding and pronounced 'win-ding'). In 1745, Edmund Lee patented a 'Self-Regulating Wind Machine'. The patent drawing shows a fantail geared down to a 'travelling wheel' at the rear of the mill on the tail pole which moved around the base of the tower and turned the cap when the fantail turned. By modern standards this would have been ineffective, as the fantail was shielded to a considerable extent by the body of the windmill. In 1782 John Smeaton observed that . . . 'in this part of the country' (Yorkshire, near his home) 'it is a common thing to put Sail Vanes that keep the mill constantly in the wind without attention or trouble to the millman'. In 1782 he used a fantail in the construction of a five-sailed windmill, Chimney Mills, Newcastle upon Tyne, and this is thought to be his first use of this detail. The post mill could also be winded in the same way, using a fantail mounted on the end of the tail pole or ladder strings, and driving the wheels on a cast-iron track on the ground. This system grew until the large Suffolk-pattern post mills, of which the preserved mill at Saxtead Green is an example, reached the peak of efficiency and performance. This type of post mill has a body mounted above a two-storey roundhouse with the quarter bars at high level and a long ladder with a fantail coming down to ground level.

The grinding of the hard wheat of the European prairies in Hungary and elsewhere was a problem which was solved in Hungary in 1829 by the invention of the roller mill. In this the grain was ground between successive pairs of steel rollers in a continuous reduction process, being carried up the mill by elevators between each grinding. The mill could be as big or as small as capital would allow, since most of these mills were, from the start, steam powered. The amount of grain ground even in the smallest plant would be six times the amount ground between a pair of millstones. The miller with a windmill could not compete with the roller mills because of the uncertainties of his trade: low windspeeds on many days in the year, and storm conditions when the mill would have to shut down completely. The end came gradually, but by 1939 the demise of the windmill in Britain was virtually complete.

PART TWO: POWER AND ENGINEERING

Wind engines and electric generators

In Britain a change in the form of the windmill took place and reached one or two windmills: those at Haverhill and Boxford in Suffolk for example. This was the annular sail. Here a ring of shutters is mounted at the tip of the sail stocks and these provide the driving force to the windshaft instead of the conventional sails. In the case of Boxford, there were 120 shutters in eight units between eight stocks. By having all the shutters in the approximate position of the tips of the conventional sails the greatest use can be made of the wind, for it is at the uppermost tip that the work of the wind is most effective. Structural difficulties prevented the conventional windmill from being adapted to take annular sails: the pressure on the sail stocks would be greater, with more risk of failure, for example. The originator of the Suffolk annular sail was Henry Chopping. He built one at Richard Ruffle's mill at Haverhill where the sails were 14.6m (48ft) in diameter on a tower 20.1m (66ft) high.

Chopping had a provisional patent which he assigned to John Warner & Sons of Cricklewood who were pump and machinery manufacturers – for example they made a low-level horse-driven pump. They produced a pumping wind engine which consisted of a skeletal tower carrying an adjustable-shuttered annular sail in which the shutters radiated from the centre and were not in a ring at the outside of the diameter. Haverhill was equipped with its annular sail in 1860–1, by which time annular-sailed windmills were beginning to appear in the United States. (In American usage, the word 'windmill' has come to mean an annular-sailed windmill on a skeletal tower, but in Britain the windmill is a post or tower mill with four sails. It is more appropriate, therefore, to use the term 'wind engine' when referring to the American version.)

The first commercially successful wind engines to appear in the United States were invented in 1854 by Daniel Halladay of Marlbro, Vermont. He had been in London in 1851 to visit the Great Exhibition in Hyde Park which had such a great effect on the development of invention in both Europe and the USA. Halladay worked on a wind engine which had self-reefing sails and which was turned into the wind by a wind vane at the back so that it turned to the wind like a weathercock. He quickly moved from paddle-shaped blades to thin blades slotted into wooden rims. As the wind pressure varied, so the angle of the vanes to the plane of their mounting varied. At low speeds the vanes would be fully extended and flat with the face of the mounting, but at high speeds the vanes would be at right angles to this face. There are many variants in a whole series of adjustable-vaned wind engines, but the common American wind engine which was quickly brought into being was one with fixed vanes which met the variations in wind pressure by the way in which the tail vane, held by a tensioned spring, turned the vanes away from the wind. This type of wind engine was erected in tens of thousands all over the United States and in particular in the prairie states, and is still to be found in use. In the ranch country, one duty of the cowboy was wind engine maintenance, a role which is never part of his film image. One peculiar variant of the wind engine in the United States was the creation of horizontal wind engines, such as the Gladden Mill preserved in Randolph, New York.

In Europe, the wind engine followed the fixed-vane pattern common in the United States, and perhaps the best-known British example is the Climax, built in its thousands by Thomas of Worcester. This is a fixed-blade wind engine with the gears encased in a box filled with lubricating oil. This oil-bath wind engine was adapted by many manufacturers and was popular because lubricating was reduced to a once-a-year task. In addition to the fixed-blade wind engines several forms of adjustable-vaned or shuttered wind engines were produced in England and in Europe. One particularly notable group of these, produced by John Wallis Titt of Warminster, achieved a very large diameter – up to 12.2m (40ft) – and were exported all over the world for water pumping. In Germany and Denmark, several other types of shuttered wind engines were produced, such as those of G. R. Herzog & Karl Reinsch of Dresden and Reuter & Schumann of Kiel. Other experiments with wind engines saw a return to the principles of the water turbine, in which the wind was focused by the fixed non-rotating vanes of a stator on to the rotating vanes which represent the rotor. One group of these wind engines was the invention of the Bollée family of Le Mans, France, who also made pumps, steam cars and, later, automobiles.

While the majority of wind engines were used for pumping, they were also used for conventional farm duties and to generate electricity. The wind engine is now being examined seriously for the development of electric power and water pumping. Perhaps the most famous of all the wind-driven generators was that at Grandpa's Knob in Vermont, which began to supply electricity on 19 October 1941. In a 40kph (25mph) wind it had settled down and was producing 700kW (940hp). However, by 26 March 1945 it had failed structurally. This was a two-blade windmill with adjustable aerofoil section blades mounted in front of a generator housing on top of a heavy structure. The two blades were 53.3m (175ft) from tip to tip on a 32.6m (107ft) high tower. There were experiments going on at the same time in Britain, the Soviet Union, Denmark and Germany with very big wind engines. A formidable variant of this is the Enfield–Andrew 100kW (134hp) generator which was erected in St Albans in Hertfordshire, following development of a prototype in France. In this the 24m (79ft) diameter blades drew air through themselves at considerable velocity and threw this out at their tips by centrifugal force. The air was drawn up the tower through the turbine and generator from the bottom. The advantage of this was that the turbine and generator were in the fixed base of the machine, which eased the problem of getting the electricity out of the generator. On other types of wind engine the generator rotates with the crown and blades, making the supply of electricity very difficult.

Since the mid-1970s the supply of electric-generating wind engines has grown in Europe. In the southern Netherlands there are modern wind engines around the towns, but because these are very new no one form has become established as the most appropriate type. The conventional Dutch smock mill has been used to generate electricity on the island of Texel, and several other forms of wind generator derive from scientific studies which have been carried out in Denmark and in the United States. One important example is at Gedsermolen, on a cape in the south-east of Denmark, with a 24m (69ft) diameter for three blades, and this and many other small examples are used to power greenhouse sites, private houses and objects which can have no attachments to power stations such as the lights on sea marks and buoys.

Research has also been under way in Europe to develop small wind engines for use in the Third World. This work has been undertaken in many universities and development centres and is successful in producing small low-powered pumps for irrigation and town water supplies. This is in some way a logical development because the small unit is within the means of the Third World, but the capital costs of large wind engines inhibit their development as an alternative power source to oil or coal in the western world.

Wind power has a long history and it can go on contributing to the well-being and needs of the world in many ways for the foreseeable future.

ANIMAL POWER

Classification

While animal power has made a significant contribution, it stands to one side of the main stream of development. It may be true to say that in terms of the development of power the use of animals as a power source came first.

Animal-powered machines fall into two categories: those in which the animal is working in a vertical plane around a horizontal shaft, and those in which the animal moves on a horizontal path around a vertical shaft. Within these two classifications there are several types of animal-powered engine. The vertical engine is the simplest classification, for in this there are only two types. In the more important, the treadwheel, farm animals, or men, trod the inside of the boarded rim of a wheel. The treadmill is a minor form in which men, or more rarely animals, trod the outside of the rim of a wheel.

Horizontal animal-powered engines are far more varied and fall into several types. The first division contains those machines which have a direct action, i.e.

Figure 4.7: A two-horse winding engine for a colliery. This illustration from Pyne's *Microcosm* of 1803 is of a one-up one-down arrangement.

there is no gearing between the animal and the work it does. One well-known form of direct-action machine is the cider mill, in which a large edge-runner stone rolls over the cider apples in a circular trough. The crushing of ore or brick clay by a similar edge-runner mill is another form of the direct-action animal machine. The 'direct action' designation can also be applied to the horse-driven winding drum used in coal or ore mines. In this engine a large-diameter drum is mounted at the head of a vertical shaft, and underneath this a horse arm is mounted to which one or two horses can be harnessed. As the horses walk in a circle so the cage, or kibble, is drawn up out of the mine shaft. This type of machine often has two cages so that by working on a one-up one-down principle, the weight of the cage and the rope is counterbalanced (see Figure 4.7).

In the second division there is a train of gears between the motion of the animal and the work it does. The geared machine was used for an even greater number of agricultural and industrial purposes, and there are two types of these. The older type is one in which a large-diameter gear wheel is mounted on a vertical shaft at high level above the height of a horse. The horse is harnessed inside the circumference of the wheel or outside it, according to the diameter of the gear wheel. The gear wheel engages with a second gear on a lay shaft which takes the power from the 'horse wheel' to drive other machinery by means of further gears or pulley belts. The later type, which was really only made possible by the universal use of cast iron for gearing and machinery design, is the low-level machine, in which a small gear wheel is encased in a frame from which the vertical shaft rises to be harnessed to the horse. The animal has to step over the gear shaft which runs at ground level from the engine in the centre of the horse path to the equipment which is to be powered outside the horse path. This form of horizontal machine is often to be found as a portable machine which can be taken round the farm to do various jobs.

To the vertical and horizontal machines should be added two forms of animal-powered machines which are outside the above classification. In the oblique treadmill the animal, usually a horse, is harnessed firmly between two frames and stands on an obliquely-mounted belt which moves off from under his feet. The upper end of the belt is on the same shaft as a pulley wheel so that there is a considerable speeding up between the engine – often called a paddle engine – and the machinery, which could be a circular saw or a threshing drum. The other 'oddity' is the oblique treadwheel, which consists of a fairly small-diameter circular tread plate which has treads fixed radially on its upper face. The underside of the oblique treadwheel carries a gear wheel from which the drive is taken, and which is used for small tasks like churning butter.

The ancient world

We have considerable knowledge of the use of animal-powered engines in prehistoric and classical times, but no certainty of their form until the Roman period. The use of the vertical treadwheel for hoisting purposes is shown in a relief in the Lateran Museum. This shows a treadwheel with at least four operatives inside. The windlass on the shaft of this treadwheel is lifting a stone by means of two two-sheave pulleys in order to give a further 4:1 lift in addition to the advantage of the treadwheel. Ropes pass from the treadwheel at the bottom of a single boom to lift a stone on to the roof of a temple. While this is not the only example, the quality of this relief shows how skilled the Roman engineers were in being able to set up the treadwheel in a temporary setting such as a building site.

The Romans were probably the originators of the hourglass animal-powered corn-grinding mill. There are examples of this form of direct-drive animal-powered mill in London, Pompeii, Capernaum in Israel, and Mayen in the Eifel region of Germany, the best being the row of four set up in a bakery in Pompeii; the example in the Museum of London, from a London site, is similar. The fixed stone of the mill is a single stone, circular in plan and finished in a long cone on a short cylindrical base. The runner stone, to use the analogy of the 'normal' pair of millstones, consists of a large stone block which is carved to form two connected shells in the form of an upright and an inverted cone. This sits over the base and is cut to be a close fit. The middle of the runner stone has sockets on either side of the waist to take the fixings of the rigid wooden frame to which the animal, usually an ass, was harnessed. The grain, fed into the inverted cone – in effect a hopper – works its way down and round the lower cone, and is ground by the motion of the runner stone on it. The meal is collected from the base of the mill after it is ground, but there appears to be no means of collecting it mechanically; it just remained to be swept up.

Another direct-drive animal-powered engine used in Roman times is the trapetum. This is an olive-crushing mill, and it has a successor of more recent times: the cider apple crusher (see p. 260). In the trapetum two crushing rolls, in the form of frustrums or spheres, rotate and move around a circular trough. The spheres are suspended just clear of the trough so that the olive pips are not crushed, making the oil bitter. The animal is harnessed to arms which are inserted in sleeves in the roller crushers so that the rollers rotate about the sleeves while describing a circular path in the trough. The central part of the trough is raised to form a pillar on which the arms and rollers are pivoted.

In the fertile areas of the eastern Mediterranean there were several forms of water-raising device which were used principally for irrigation. Some of these are still in use today in the same manner as they were some 2000 years ago, except that they are more sophisticated in their construction. The chain of buckets, raised at first by the action of men on levers, was fitted with a crude arrangement of gears at quite an early date. An ox or camel could be harnessed to the arms on top of the gear wheel, and by the movement of the animal walking round the central point the gear wheel, engaging with the gears on the same shaft as the chain of buckets, could raise these to the surface, where they discharged automatically into irrigation ditches. Their successors, which can be found in Portugal, Spain and the Greek islands, are now made of metal, but are rarely to be found in use as they have been superseded by electric or diesel pumps. In a similar way a bucket was wound out of a well on a drum windlass mounted on an upright shaft which was turned by an animal walking round the shaft in a circular path. The animal would be harnessed to one end of an arm socketed into or housed around the shaft. This form of water-raising device can still be found, regrettably not in use, on the North Downs, between Canterbury and Maidstone, in Kent.

Mediaeval and Renaissance Europe

As with waterwheels and windmills, the sources of information on the use of animal-powered engines in the mediaeval period are only archival or iconographic. There are one or two examples of treadwheels in place in the cathedrals of Europe where they were installed as lifting devices for the maintenance of the stone walls, roofs and timbers of the tower, or of spires and high roofs. These date back to the mediaeval period, and we know that they also played a large part in the construction of the buildings themselves. As the great nave vaults were gradually being built the scaffolding followed the building from bay to bay. A treadwheel crane would be mounted on the scaffolding, or on the top of a completed vault, and moved along as the work proceeded. Those which still remain are, of course, well constructed and properly built into permanent housings, but if they were used as part of the moveable scaffolding they could have been more crudely constructed. Some of the English survivors are interesting and worthy of note. At Tewkesbury there is a single rim supported off the shaft and windlass by a single set of spokes. A man worked the wheel by treading on the outside of rungs which project on either side of the rim. This example may date from about 1160 when the tower was completed. At Salisbury, a similar wheel in which the simple single rim is replaced by three rims, but which still has only one set of spokes, dates from about 1220. The conventional treadwheel, in which a man walks on the inside of a drum, can still be seen at Beverley Minster and at Canterbury. On the Continent examples of the treadwheel may be seen in Haarlem in the Netherlands and at Stralsund and Greifswald in the German Democratic Republic.

While slaves and workmen had used the treadwheel to lift materials and water in the Roman and mediaeval periods, the deliberate use of machines as a punishment did not become general practice in Britain until the nineteenth century. Sir William Cubitt is thought to have invented the prison treadmill in about 1818, although he did not patent it. In this machine the prisoners trod the steps on the outside of a wheel which turned away underneath them. The power generated was used to grind grain into meal, to pump water or merely to work against a brake. The treadmills, of which the only survivors are at Beaumaris gaol in Anglesey and the one from York now at Madame Tussauds in London, appear to have been made from standard castings so that the length could be determined by the number of prisoners required to tread them. This form of punishment existed in nearly all British prisons until it was abandoned in about 1900.

As with the watermill, the documentation of the form of the animal-powered engine can be seen in Agricola's *De Re Metallica* of 1556 (see p. 232). The use of animal-powered machines in mining in the 1550s is also shown in the Kutna Hora *Gradual* which dates from the last years of the fifteenth century. This shows particularly accurately the use of a large-size windlass for winding ore out of mine shafts. The illustration shows four horses harnessed in this engine. What is more interesting in the history of technology is that this machine shows the horses to be harnessed outside a large-diameter gear wheel which engages with the windlass. The vertical horsewheel house containing the engine has a parallel in the preserved example at Kiesslich-Schieferbruch bei Lehesten in the DDR. While these iconographic examples are artistic, the details provided in *De Re Metallica* are factual, and appear quite workable. The horsewheel house (Book VI) with a conical roof has four arms to which eight horses can be harnessed. The upright shaft passes below ground where a crown wheel with peg teeth engages with a pinion in which the teeth are carved out of a cylinder, rather like a lantern gear. The pinion then drives the lifting drum of a chain pump which is mounted on the same shaft. Again in Book VI there is a conventional treadwheel driving a chain pump through a train of a spur gear and a lantern gear. The evidence is all there to demonstrate a substantial use of animal-powered machines in the mediaeval period, and it is unfortunate that only very few examples survive to show how efficient they were.

On ships, and sometimes on land, heavy loads such as the anchor and its cable or packages for transportation, were lifted by means of the capstan. This is usually a long wooden drum mounted on one deck with its head projecting above the main deck. This gives it stability when the load is taken up. Capstan bars are inserted in the head so that a large number of sailors can push on the bars and turn the cylinder. The load is drawn up by the chain or cable being wound on the lower portion of the drum. A ratchet stops the load pulling the chain off the drum again as it tries to sink back under its weight.

As the industrialization of Europe grew in pace in the period following the 1500s, so the use of animal-powered machinery grew as well, although the number of these machines could never have been as great as the number of waterwheels, for they are more cumbersome and not necessarily as powerful. With the growth of printing, and with it the proliferation of wood-block illustrations (see Chapter 14), the designs of the various machines were published throughout Europe. Similarly, people travelling in Europe, either on the Grand Tour or as master craftsmen moving from job to job, were able to take the information on the many machines for use in other countries. One important book was Ramelli's *La Diverse et Artificiose Machine*, published in 1588 in Paris. In this book there are many examples of animal-powered engines; some may be regarded as purely fanciful, but others, such as the horse-driven corn mill in figure CXXII, are examples of workable machines. This was followed by Zonca's *Novo Teatro di Machine et Edifici*, published in 1607 in Padua. This showed similar examples to those above, and also the first example of the oblique treadwheel, driven by an ox, for grinding grain.

The eighteenth century

In the early 1700s the first really good millwrights' books were published, true text-books with scale drawings and fairly complete constructional details. The great millwrighting books of Holland (see p. 251) were in circulation in Europe: it is known that John Smeaton had his own copies and used them. They were clearly intended for the new professionals, for they were written by millwrights and engineers. While these millwrights' books were comprehensive in that they dealt with windmill and watermill construction, they also gave details of the construction of horse-driven corn mills and other horse engines. The use of horses to drive corn mills in Dutch towns is unexpected, as there were windmills all round the towns and on the town walls in many cases. These horse-driven corn mills, called *grutterij*, were established at the back of bakers' shops to grind buckwheat for the production of *poffertjes* and pancakes.

In the eighteenth century the use of the high-level horse-driven engine became well established for industrial purposes, although it is now impossible to determine the scale of this introduction. In Europe the established use of the horse-driven corn mill is well documented, particularly by the presence of preserved examples in the open-air museums of the Netherlands, Germany and Hungary. In Hungary, for example, there are several horse-driven corn mills and a preserved example of the oblique treadwheel, as shown in Zonca, drove the corn mill from Mosonszentmiklos in the open-air museum at Szentendre. These appear to date from the eighteenth century, for although they do not have sophisticated cycloidal gearing, the cog and rung gearing has reached a high standard. In England there is a fine example of a horse-driven corn mill at Woolley Park in Berkshire, but there are not many records to show that other examples existed in substantial numbers. In industrial terms, too, there are not many records to identify particular sites where these machines were used. It is known, for example, that Strutt and Arkwright had a horse-driven cotton mill in Derbyshire before the foundation, by Arkwright, of the large water-driven mills in the Derwent valley. In Holland, the use of horse-driven machinery to drive oil-seed crushing mills was well known. In these mills, which exist in museums, the horse pulls round the great edge-runner stones, and by the same motion drives the oil stamps and the rotating 'cracking' plates, by means of trains of gears on the head of the shaft. Similar machines existed for the grinding of black powder in gunpowder works and for the grinding of pigments in colour mills. In the national open-air museum at Arnhem there is an example of a horse-driven laundry which was rescued from a site between Haarlem and Amsterdam. In going round the vertical shaft, the horse turned a shaft on which cams were mounted which effectively rotated and squeezed the clothes in three wooden tubs. These laundries, of which there were several near Amsterdam, provided a service for the twice-yearly spring and winter washes. The normal weekly wash of a town house was not done here, for after washing in the mechanical laundry, the linen would be bleached in the fields.

High-level direct-drive mine windlasses must have been fairly numerous, as the illustrations from the English mining districts show many examples. The drawings of T. H. Hair in his book *Sketches of the Coal Mines in Northumberland and Durham* show one or two examples of these high-level winding drums with a good example as a vignette on the title page. This was published in 1839, long after the introduction of the steam-driven winding engine to the mines. We know from the insurance papers of 1777 from Wylam colliery, Northumberland, that there were five horse-driven winding engines (locally called 'whin gins') there. Horse engines, erected temporarily on scaffolding, were needed when mine shafts had to be dug. There was one at East Herrington, near Sunderland, which was left in place to service the pumps in the pump shaft of this mine, and there is the preserved example at Wollaton Hall Industrial Museum, Nottingham, known to have been built in 1844 as a colliery gin, and which ended its days being used for shaft inspection and repair. The same type of horse-driven winding drum continued in use until much more recently in the Kimberley diamond mining field in South Africa and in the Australian goldfields. At the turn of the century, the 'Big Hole' in Kimberley was ringed with these winding drums.

While the industrial use of horse-driven machinery remained low, the real growth in the use of the horse-driven engine was on farms in the last quarter of the eighteenth century. Andrew Meikle patented a horse-driven threshing drum in 1788 in Scotland, and this helped to overcome the shortage of labour which existed on farms in Scotland and the north of England, from which the country labourer had been driven to more remunerative work in the coalfields and the expanding industrial towns. These machines did not grow in number in the south of England where labour was still available to work on the land and where the 'Captain Swing' riots of the 1830s took place to protest against mechanization. Meikle's horse engine was a high-level machine with a large gear wheel mounted so that the horses could be harnessed underneath it. Clearly, while it was set up in permanent structures attached to the barns, it could be used for other purposes such as chaff or turnip cutting, wood sawing and water pumping. The number of these high-level wooden horse engines in existence is extremely small, but the buildings which housed them can still be found in quite large numbers in north-east England. More than 1300 horse-wheel houses have been identified in Kenneth Hutton's paper 'The Distribution of Wheelhouses in the British Isles', in the *Agricultural History Review*, vol. 24, 1976.

The nineteenth century

The large number of horsewheel houses indicates the number of high-level horse-driven gear wheels which must have existed at one time. They were manufactured throughout the eighteenth and nineteenth centuries. Early in the nineteenth century the universal use of cast iron made it possible to introduce the low-level horse engine. In this the horse or horses went round the central shaft harnessed to the end of the horse arm. The central shaft was usually short and carried a small-diameter crown wheel of cast iron. The first gear wheel would engage with a smaller bevel wheel which could be connected with the farm machinery to be driven by means of a universal joint. Sometimes the frame carrying the horse wheel would carry a train of gears so that the shaft, which is the final element in the engine, could rotate extremely fast in relation to the 4kph (2.5mph) which the horse imparted to the ends of the horse arms. If the machinery required to be driven with considerable rotational speed, then the horse engine could be connected with it by means of intermediate pulleys and belts. Examples of the low-level horse engine exist in which the engine is mounted on a frame with wheels so that it can be pulled around as necessary. One or two designs for threshing engines with integral horse engines are known and these were dismantled and placed on a frame so that the whole unit could be taken from farm to farm by contract threshers. These, of course, were displaced by the portable steam engine.

The production of the low-level horse engine must have reached enormous numbers during the nineteenth century; so much so that competitions were held at the shows of the Royal Agricultural Society to evaluate the work done by these machines. The low-level gear came in several forms, but the type usually met with is the one in which there is no casing to protect the user from being trapped in the gear. Later, safety models were introduced in which all the gears were enclosed in a casing, so that all that was to be seen was the horse arm and upright shaft at the top, and the drive shaft and universal joint coming out at the bottom for connection to the machinery. The example of the safety gear produced by the Reading Iron Works in England, was a cylinder of iron some 90cm (3ft) high and 60cm (2ft) in diameter. Examples of this Reading safety engine have appeared in Western Australia, together with the 'standard' Reading Iron Works corn mill. This corn mill had one pair of stones, mounted on an iron hurst frame, driven by gears from the associated horse engine. The Reading Iron Works, which had previously been the firm of Barratt, Exall & Andrewes, maintained offices in Berlin and Budapest in the nineteenth century, and even printed their catalogues in Russian as well as other European languages. England exported a great many low-level horse engines. Hunt of Earls Colne and Bentall of Malden, both in Essex, and Wilder of Reading are names of manufacturers whose products have been found overseas. In the USA and Canada horse engines were required in large numbers and these countries produced their own models. The manufacturers were usually based in the prairie states, such as the Case Co. of Racine, Wisconsin. In the USA horse engines were considerably larger than in Europe, for on the vast prairies reaping machines had large blades and were hauled by teams of up to twenty-four horses. When the reaping had been completed, the horses were harnessed to low-level horse engines to drive the threshing machines. As a result of this, the horse arms grew in size and number so that they could take twelve horses rather than the three or four catered for by European designs. The horses were harnessed to the horse engine and the teamster then stood on the machine to ensure that the horses pulled their weight. In the USA, too, the paddle engine, or oblique treadmill (see p. 262), was frequently to be found in farmyards where it was attached to saw benches or to farm machinery. These were worked by one or more horses and were usually made to be portable. The oblique treadmill for churning butter, and worked by dogs, was also known.

On the dry chalk uplands of England low-level horse engines are occasionally to be found operating pumps to deliver water from deep wells to isolated houses. In other places they are to be found in use for haulage from small mine shafts such as the copper mines on the Tilberthwaite Fells in the English Lake District.

Nowadays, animal-powered machines can still be found at work in countries where irrigation is necessary, such as Egypt, Syria and Iran, but these survivals are rare. Occasionally they are to be found in use on farms in Europe. It is to be regretted that so little remains of this element of natural-power machinery for it played a significant part in the history of agriculture, and a minor part in the development of industry.

GLOSSARY

Bell crank a means of turning a pulling motion at right angles. Two arms at right angles are mounted to pivot about a point between their tips.

Cog and rung gears a gear system in which plain, unshaped teeth engage with a set of staves held between two flanged rings.

Cycloidal gears gears formed to a precise profile so that when they engage they roll along the face of the teeth to give a smooth motion and not the striking effect of the cog and rung gear.

Fantail a means of turning a windmill into the wind automatically. This consists of a set of blades mounted at the back of the mill at right angles to the sails, which rotate the cap by means of a gear train.

Grain elevator a series of small rectangular buckets mounted on a continuous belt inside a double wooden shaft. Grain is poured in at the bottom of the rising shaft so that it falls into the buckets which empty themselves into a hopper at the top when the belt goes over the top pulley.

Great spur wheel mounted on the upright shaft in a mill, drives the millstones by means of the stone nuts and stone spindles.

Greek mill a mill in which a horizontal waterwheel is mounted on the same shaft as the runner millstone. As the horizontal waterwheel turns so the runner millstone is turned at the same speed. This type of watermill is in common use from Portugal across the Mediterranean area.

Hemlath the longitudinal member at the outer edge of a sail frame.

Horizontal feed screw a means of carrying meal horizontally in a watermill. A rotating shaft in a long square box or sheet metal tube carries a continuous screw of sheet metal or a series of small paddles set in a screw form. The meal is pushed along to the appropriate opening by the motion of the screw.

Horizontal waterwheel a waterwheel mounted to rotate in a horizontal plane so that its rotation is transmitted into the mill by means of a vertical shaft.

Hurst frame the frame of stout timbers which carry the millstones. Often this frame is independent of the structural timbers of the mill.

Impulse wheel waterwheels, or more particularly turbines, which are driven by the pressure of the water being forced on them through a nozzle, rather than by the weight of water flowing on to them directly.

Lantern gears and pinions gears (resembling a lantern) having staves between two flanges, turned by pegs on the rim of another wheel.

Mortice wheel a cast-iron wheel in which sockets in the rim are set to receive wooden gear teeth.

Moulin pendant a form of watermill still to be found in France in which an undershot waterwheel is suspended in a frame so that the whole can be raised or lowered to meet variations in the water level caused by flood water.

Norse mill another name given to the drive system found in the Greek mill (q.v.). This form of mill still exists in Scandinavia and was common in the north of Scotland and the Faroes.

Overshot waterwheel a waterwheel in which the water is delivered to the top of the wheel so that it turns in the direction of flow.

Panster Muhle the German equivalent of the French moulin pendant (q.v.) which can still be found in the German Democratic Republic. In some instances the frame carrying the waterwheel is hinged and raised by chains and large man-operated lifting wheels.

Pit wheel name given to the first gear wheel in a watermill mounted on the waterwheel shaft. Because of its large size it usually runs in a pit on the inside of the wall which separates the waterwheel from the mill machinery.

Querns name given to primitive hand-operated millstones. These can be stones between which the grain is ground by rubbing or in which an upper flat-faced circular stone is rotated over a fixed flat-faced stone.

Stream waterwheel waterwheel in which there is no head of water but in which the floats are driven round by the flow of water striking them.

Tail water the water emerging from the bottom of a waterwheel while it is turning.

Undershot waterwheel a waterwheel in which there is a small head of water driving the wheel around. The water hits the wheel at about 60° below the horizontal line through the centre of the wheel.

Vertical waterwheel waterwheel mounted on a horizontal axis and rotating in a vertical plane.

Vitruvian mill the simplest form of watermill with a vertical waterwheel. The waterwheel is coupled by a pit wheel to a single runner millstone by means of a gear on the stone spindle. It is so called because it is described by Vitruvius in the tenth book of *De Architectura* which was published in about 20 BC.

Wallower a gear wheel which transmits the drive from the pit wheel to the upright shaft in a watermill or from the brake wheel on the windshaft to the upright shaft in a windmill.

FURTHER READING

Baker, T. L. *A field guide to American windmills* (University of Oklahoma Press, Norman, 1985)

Hunter, L. C. *Waterpower, a history of industrial power in the United States, 1780–1930* (The University Press of Virginia, Charlottesville, 1979)

Major, J. K. *Animal-powered engines* (B. T. Batsford, London, 1978)
Major, J. K. and Watts, M. *Victorian and Edwardian windmills and watermills from old photographs* (B. T. Batsford, London, 1977)
Reynolds, J. *Windmills and watermills* (Hugh Evelyn, 1970)
Reynolds, T. S. *Stronger than a hundred men, a history of the vertical water wheel* (Johns Hopkins University Press, Baltimore, 1983)
Syson, L. *The watermills of Britain* (David and Charles, Newton Abbott, 1980)
Wailes, R. *Windmills in England,* (The Architectural Press, London, 1948)
—— *The English windmill* (Routledge & Kegan Paul, London, 1967)

5

STEAM AND INTERNAL COMBUSTION ENGINES

E. F. C. SOMERSCALES

INTRODUCTION

This chapter outlines, in a more or less chronological sequence, the history of power production by steam and by internal combustion. Steam power is represented by the reciprocating steam engine and the steam turbine. The internal combustion engine and the gas turbine are examples of the other class of thermal prime movers.

The devices described in this section are practical realizations of the theoretical concept called a heat engine. This is the ideal against which all practical engines are evaluated, and with which the inventor formulates his original concept and the performance of which he strives to attain. Because of the central role of the heat engine and thermodynamics (the relation between heat and work), a brief review is given in the Appendix, p. 342.

The historical progress of thermal prime movers is concisely summarized by a chronological record of the important performance parameters. Appropriate graphs are given for each class of prime mover. The parameters displayed are maximum temperatures and pressures, the power output, and the thermal efficiency or specific fuel consumption. For the reciprocating internal combustion engine the mean effective pressure (P) and the piston speed (= 2 × stroke × rpm) are also presented. These are, in turn, related to the power output (hp) of the single-cylinder engine through hp = PLAN. (P = mean effective pressure (N/m^2 or lbs/in^2), L = stroke (m or ft), A = piston area (m^2 or in^2), N = number of effective strokes per minute = rpm/2 (four-stroke internal combustion engine) or rpm (two-stroke internal combustion engine or the single-acting steam engine). For the opposed piston or double-acting engines the power given by the formula must be doubled.)

STEAM ENGINES

The role of the steam engine in bringing about the industrialization of a rural world is well known. It also had an important influence on the development of materials and of design methods. Another remarkable feature is its longevity. The first steam engine was built in 1712, and steam engines continued to be built, admittedly in decreasing numbers, into the Second World War and after. See Appendix for a discussion of the operating principles of the steam engine.

Developments before 1712

The discoveries that, when brought together by Thomas Newcomen in 1712, resulted in the steam engine were (a) the concept of a vacuum (i.e. a reduction in pressure below the ambient); (b) the concept of pressure; (c) techniques for generating a vacuum; (d) means for generating steam; (e) the piston-and-cylinder.

The concept of the vacuum is very old and appears to have been most clearly enunciated by Strato, an Athenian scientist of the second century BC, but it has come down to us through Hero, an Alexandrian scientist who probably lived in the first century AD. It was a theoretical idea that played an important part in Greek science, but its application in the conversion of heat into work had to wait until the Neapolitan scientist Giambattista della Porta described in 1601 the way in which a vacuum could be produced by the condensation of steam.

The ability to produce a force by means of a piston and cylinder, with the piston exposed on one side to the pressure of the atmosphere (a concept due to Evangelista Torricelli) and on the other to a vacuum, was demonstrated in 1672 by Otto von Guericke. However, his device could not operate very conveniently or very rapidly because the vacuum was produced by filling a container with water and then allowing it to empty. The use of steam to accomplish this had to await the invention by Denis Papin, sometime between 1690 and 1695, of the piston and cylinder. In the meantime, devices in which the pressure of the steam acted directly on the surface of water that was to be moved were invented by Edward Somerset, second Marquis of Worcester, and by Thomas Savery, in 1663 and 1698 respectively. These apparatus were used to raise water by a vacuum, produced by the condensation of steam, and then to raise it still further by the action of the steam on the free surface of the water. Somerset was unable to obtain financial support for his device, but Savery, the shrewder individual, was successful, and his apparatus was used in mine-pumping.

Finally, in 1690, Papin demonstrated a small piston-and-cylinder device (Figure 5.1) in which the piston is raised by the evaporation of water due to heat applied to the outside of the cylinder. With the demonstration of Papin's engine, all the elements were available for manufacturing a reciprocating steam engine.

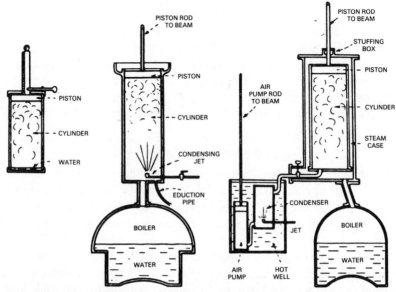

Figure 5.1: Schematic arrangements of the steam engines of Papin, Newcomen, and Watt.

Papin's engine. A small quantity of water in the cylinder is heated by an external fire. The steam formed raises the piston to the top of the cylinder where a latch engages a notch in the piston rod. The fire is removed and condensation (perhaps assisted by an external water spray on the cylinder walls) of the steam produces a vacuum below the piston, on removing the latch the piston is driven downward by the atmospheric pressure. This raises a weight attached by a rope, passing over a pulley, to the piston rod. The cycle may be repeated.

Newcomen's engine. Steam passes from the boiler through a valve into the cylinder, this balances the atmospheric pressure on the upper side of the piston and allows the weight of the pump rod, connected to the piston rod by a beam balanced on a fulcrum, to raise the piston. The valve between the boiler and the cylinder is closed and a jet of water is sprayed into the cylinder. This produces a vacuum by condensing the steam, so the atmospheric pressure on the upper side of this piston forces it downward, lifting the pump-rod by means of the beam. The cycle may be repeated. See also Figure 6, p. 34.

Watt's engine. Steam is admitted, through a valve, into the steam case surrounding the cylinder. The separate condenser is then connected to the cylinder, the condensation of the steam produces a vacuum and the pressure of the steam in the steam case acts on the upper side of the piston and forces it downward. When the piston reaches the base of the cylinder, a valve is opened that equilibriates the pressure on either side of the piston so the weight of the pump rod, acting through the beam and piston rod, raises the piston. The cycle is then repeated.

Reproduced with permission from H. W. Dickinson *A Short History of the Steam Engine*, 2nd edn (Cass, London, 1963).

Newcomen's engine

The practical steam engine (Figure 5.1) was introduced in 1712 by Thomas Newcomen. A crucial element of this machine was the overhead beam, which allowed mechanical effort to be obtained from the engine. Other important features were: (a) a jet of water to condense the steam inside the cylinder, which would increase engine speed by decreasing the time required for condensation; (b) valve gear that makes the engine self-acting (the second example, after the clock escapement, of a self-acting device).

The first Newcomen engine was erected in 1712 to operate a mine drainage pump near Dudley Castle in Staffordshire (see p. 34). The subsequent chronology of Newcomen engine erection is impossible to determine with certainty, but engines were erected in Britain and on the European continent chiefly for water pumping (mine drainage, recovery of flooded land, emptying dry docks, water supply etc.). They were built by various engineers, many English, operating under licence from a committee of proprietors who owned the master patent (originally belonging to Savery).

Watt's engine

In 1764, James Watt, the most significant figure in the history of the steam engine, was employed as an instrument maker and was retained to repair a model Newcomen engine owned by the University of Glasgow. He noted the high internal energy of steam compared to liquid water (a ratio of about 6:1 for the conditions in Watt's model engine), which demonstrated to him its great economic value and the need to conserve its internal energy. Accordingly, Watt insulated the boiler and steam pipes, and applied a steam jacket to the engine cylinder. However, he quickly appreciated that even more significant losses occurred, namely: (a) energy lost in cooling the piston and cylinder when water was sprayed into the cylinder to produce condensation and a vacuum; (b) loss of power due to the pressure of vapour below the piston as a consequence of incomplete condensation.

He realized that these losses could be avoided if condensation was carried out in a separate chamber that was connected to the engine cylinder (see Figure 5.1). Immediately, in May 1765, he built an improvised engine, using a large (44mm (1.75 inch) diameter by 254mm (10 inch) long) brass surgeon's syringe, to demonstrate the correctness of his ideas. Two other models (1765, 1768) were built, and in 1768 Watt applied for a patent. A fourth, much larger (450mm (18 inch) stroke by 1500mm (5 feet) bore), engine was built with the financial assistance of Dr John Roebuck at Kinneil, Scotland, probably in 1769.

In 1768, following a visit to London to obtain his patent, Watt met Matthew Boulton, a Birmingham manufacturer, owner of the Soho Foundry. Boulton was immediately interested in Watt's invention and in 1769 the two became partners (although Watt did not move to Birmingham until 1774). The combination of Watt's scientific and engineering talents, the business acumen of Boulton, and the compatability of their personalities ensured the success of the partnership and of Watt's engine.

Interest in Watt's engine among engineers concerned with mine drainage and water supply was immediate, both because of Boulton's reputation for sagacity and because of its manifestly superior efficiency and higher speed compared to the Newcomen engine. However, engines of the latter type continued to be built for many years after Boulton and Watt commenced manufacture in 1774.

The initial application of the Watt engine was to mine pumping, but Boulton recognized that there was a large market for engines that could drive mill machinery. This required a rotary output from the engine.

On 25 October 1781, Watt obtained patent on his sun-and-planet gears that provided a rotary output from the engine without the use of a crank. Watt wanted to avoid the use of a crank which is the obvious, and currently conventional, method of achieving his objective, not, as tradition has it, because others had patented this device (in fact, Watt considered it unpatentable because of its prior application in the foot lathe by an unknown person), but because he did not wish to become involved in lawsuits that might overthrow his 1781 patent or others he owned. This could have happened if patents owned by Matthew Wasbrough and James Pickard, which had employed the crank, without patenting it, as part of various mechanisms to convert reciprocating to rotary motion, had been declared invalid.

While the sun-and-planet gear will in principle convert reciprocating motion to rotary motion it produces a very irregular speed at the engine output when combined with the single-acting single cylinder engine. The double-acting engine, in which steam is supplied to both sides of the piston, is essential for uniform speed. This was covered by a patent issued to Watt on 17 July 1782 (the patent was also concerned with the expansive use of steam and a rotary engine). The first double-acting engine was built in 1783 for John Wilkinson, and another for the Albion Flour Mill, London, in 1784.

The double-acting engine requires a rigid connection between the piston rod and the oscillating beam that transmits the power from the cylinder to the engine output, in order to transmit the piston thrust on its upward stroke. In June 1784, Watt devised a suitable mechanism called by Hartenberg the 'perpendicular motion'. This particular linkage had the disadvantage that it lengthened the engine by about half a beam-length, so only two engines were fitted with this arrangement. Watt then devised the so-called 'parallel motion', which connects the end of the beam to the piston rod by a pantograph. So, nearly three-quarters of a century after Newcomen built his first engine, the steam engine was able to provide a rotary output, but, more than that, because it was now of necessity double-acting, its power output had doubled.

1800–1850: new types of engines

In 1800, Watt's patents expired, opening the way for the development of new types of engines. These can be broadly classified into three groups. First, the high pressure engines, in which the boiler pressure was increased above the 0.34bar (5psig) used in the Watt engines, and exhaust was to the atmosphere instead of into a condenser. The second type was the Cornish engine, which was essentially a non-condensing Watt pumping engine. The third group (not described because of space limitations) comprises those engines that cannot be classed as high pressure or as Cornish engines.

The high pressure engine

High pressure non-condensing engines appear to have been first proposed in 1725 by Jacob Leupold of Leipzig. Watt, as well as his assistant William Murdock, also considered the concept. The advantages of this type of engine are: (a) it can operate at a higher speed because there is no need to allow time for the condensation process to occur; (b) the valve gear is simpler than that of the atmospheric engine; (c) the dimensions for a given piston force are independent of the atmospheric pressure, so the higher the boiler pressure the smaller the piston diameter; (d) the engine is easy to work because of its simple valve gear, and because there is no longer any need for constant vigilance over the condenser; (e) because of its light weight and small size the engine is low in manufacturing and installation costs. Disadvantages of the high pressure engine are: (a) it is of lower efficiency than its condensing counterpart, but this could well be of secondary importance where fuel is cheap (e.g. at a coal mine); (b) at the beginning of the nineteenth century boilers suitable for operating 3.4bar (50psig) did not exist.

The first engineers to build and operate successful high pressure engines were Richard Trevithick, a Cornish mining engineer, in 1798, and Oliver Evans in the United States in 1804.

The Cornish engine

The high pressure engine entered into engineering practice by way of the low-speed Cornish engine rather than the high-speed engine of Trevithick and Evans, although Trevithick was intimately connected with the development of the Cornish engine. This engine was essentially a non-condensing Watt engine, operated expansively, and supplied with steam at a higher pressure than the 0.34bar (5psig) favoured by Watt. It had a profound effect on the history of the steam engine because it demonstrated the possibilities of the high pressure, non-condensing engine as an efficient prime mover. Its immediate impact was in the mines of Cornwall where, following the installation of the first Cornish engine by Trevithick in 1812 at the Wheal Prosper mine, the reported average thermal efficiency of pumping engines increased more than three times between 1814 and 1842.

This ever-improving performance attracted the attention of engineers elsewhere, and in 1840 a Cornish pumping engine was installed at the Old Ford Water Works of the East London Waterworks Company. A slightly smaller Boulton and Watt engine was also placed in service at the same site with the intention of comparing the performance of the two engines. Tests showed that the Cornish engine was more than twice as efficient as the Boulton and Watt engine. Consequently, the use of Cornish engines spread to other water supply companies and mines.

The power output of any non-condensing steam engine (or steam turbine) can only be increased by raising the boiler pressure. Because of the direct relationship between steam density and pressure, the mass of steam admitted to an engine cylinder of a given size will increase as the boiler pressure increases. Consequently, with ever-increasing boiler pressure larger and larger volumes of steam have to be accommodated in the cylinder at the end of the piston stroke. If the cylinder is of fixed size it may not be possible to expand the steam to atmospheric pressure, with a resulting loss of energy. To avoid this waste, the engine cylinder can be increased in size, but there is a practical limit to this, fixed by manufacturing facilities and means of transporting the massive cylinder, and this had been reached by 1850 when Cornish engines with cylinder bores in excess of 2540mm (100 inches) and strokes longer than 3000mm (10 feet) were being constructed. The only way to avoid this situation was to carry out the expansion in two or more cylinders: that is, the compound had to be introduced.

The compound engine

In the compound engine the steam supplied by the boiler is expanded to the exhaust pressure by passing successively through a number of cylinders, where each cylinder operates between decreasing inlet and outlet pressures (see Figure 5.2).

The compound engine was patented by Jonathan Hornblower in 1781 in an attempt to avoid the Watt patent. He erected the first one at Radstock Colliery, near Bristol, in 1782 and a second in 1790 at the Tincroft Mine in Cornwall. Neither engine was superior to the contemporary simple (non-compound), atmospheric engines, because the boiler pressure (0.34bar, 5psig) then in use was too low to justify the application of the compound principle.

The first engine to realize some of the theoretical advantages of compounding was an existing simple engine that was modified in about 1803 by the addition of a cylinder by Arthur Woolf. Probably Woolf was aware of Hornblower's work with compounding because he was employed by the latter in erecting steam engines. He appreciated that high pressures were essential to the successful operation of a compound engine, and in its final form his engine incorporated a cast-iron high pressure boiler that produced steam at about 3.1bar (45psig). Unfortunately, Woolf employed completely incorrect cylinder proportions in designing his engine, but apparently it worked well enough to convince him and his employers, a London brewery, that a larger engine should be built. This was completed in 1805 (HP: 203mm (8 inch) bore × 914mm (36 inch) stroke; LP: 762mm (30 inch) bore × 1524mm (5 feet) stroke; boiler pressure 2.8bar/40psig), but it was unable to produce its design power output of 27kW (36hp). Woolf, in consequence, left the brewery, and went into partnership with Humphrey Edwards, and by 1811 he had apparently obtained a satisfactory design for his compound engine.

The partnership between Edwards and Woolf was dissolved in 1811 and Edwards emigrated to France. There he built Woolf compound engines and sold them throughout Europe. The Woolf compound engine was characterized by having the cranks on the high and low pressure cylinders set at 180°.

In spite of the success of the Woolf engine on the Continent, the compound engine did not reappear in Britain until 1845, when William McNaught patented an arrangement where a high pressure cylinder was added to an existing beam engine in order to increase its power output. This system was widely employed to increase the power output of stationary engines and was known as McNaughting.

1850–1900: the steam engine in its final form

By 1850 the steam engine was a well-established machine, in which the particular forms most suitable for each application had been identified, and in which the manufacturing techniques and tools to ensure reliable operation had been established. In the second half of the nineteenth century the emphasis in steam engine development and design was on techniques for improving the efficiency of the engine, on increasing the crankshaft speed, and on ensuring the maintenance of a steady speed. High efficiency was particularly important in marine engines (see Chapter 10). A high and uniform speed was important in driving textile mill machinery (see Chapter 17), and in electric power generation, which spread rapidly after Edison began operating the Pearl Street generating station in New York in 1881 (see Chapter 6).

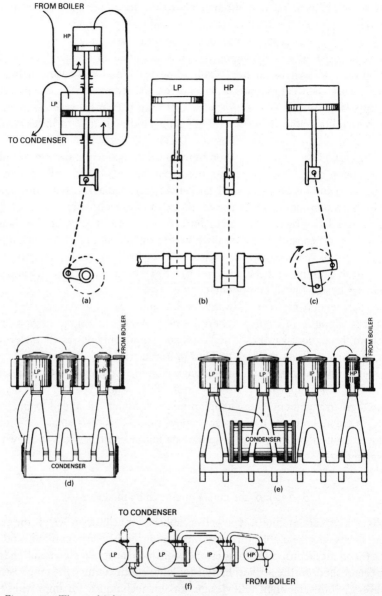

Figure 5.2: The multiple expansion steam engine.
(a) Tandem compound engine. This works on the Woolf principle with steam passing directly from the high pressure (HP) cylinder to the low pressure (LP) cylinder during the whole stroke. A receiver is not required between the HP and LP cylinders because the cylinder events are appropriately synchronized.
(b) Compound engine with side-by-side cylinders. The arrangement shown has cranks at 90°. This makes the engine easier to start and to reverse than one working

on the Woolf principle with cranks at 180°. However, a steam receiver must be placed between the HP and LP cylinders because of the relative order of events in the two cylinders.
(c) Side elevation of (b).
(d) Three crank triple expansion engine. Note the use of a piston valve on the high pressure (HP) cylinder, and slide valves on the intermediate pressure (IP) and low pressure (LP) cylinders. Cranks would usually be set at 120° and receivers would be placed between each stage of expansion.
(e) Four-crank triple expansion engine. The low pressure (LP) expansion is divided between two cylinders in order to avoid a low pressure cylinder of impracticably large diameter.
(f) Plan view of (e). Note the pipes connecting the various stages of expansion are sufficiently large in this case to act as receivers.
Reproduced with permission from: W. J. Goudie, *Ripper's Steam Engine Theory and Practice*, 8th edn (London, Longmans, Green, 1932).

The beam engine, except for a few special cases, disappeared between 1850 and 1870 with the increasing adoption of the crosshead (possibly introduced by Trevithick in his high pressure engine, above) for converting reciprocating motion into rotary motion. The most notable application of the beam engine during this period was in American river and coastal steamships, where it was known as a walking beam engine. Engines of this type continued to be built for this purpose until the late 1880s. As an example of the sizes attained, the Long Island Sound steamship *Puritan*, launched in 1889, used a compound beam engine of 5600kW (7500hp) output with cylinders of 1900mm (75 inch) bore × 2700mm (106 inch) stroke, and 2800mm (110 inch) bore × 4300mm (14 feet) stroke. Beam type pumping engines were built for water supply purposes in Europe until at least 1878.

Automatic variable cut-off

Expansive working, in which steam is admitted to the cylinders only during the first part of the piston stroke, is essential for maximum efficiency. This type of operation is most effective if the point of steam supply cut-off is related to the engine load. However, from the time of Watt variations in engine load had been accommodated by adjusting the opening of the throttle valve between the boiler and the engine which controlled the steam pressure at inlet to the engine. In most cases (except steam locomotives) this was done automatically by a governor, and the engine was then said to be throttle governed. However, the energy lost (wire drawing) by the steam to overcome the pressure drop in the throttle valve was no longer available to do work in the cylinder.

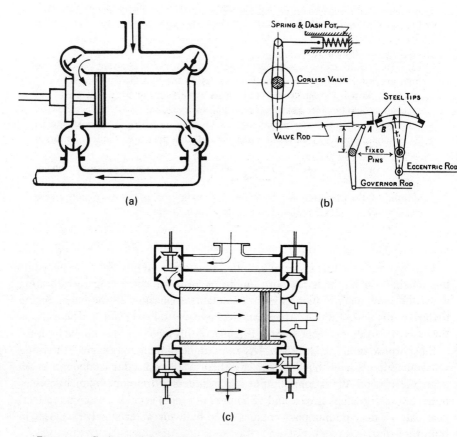

Figure 5.3: Corliss and drop valves.
(a) Schematic arrangement of Corliss valves. The valve is a machined cylinder oscillating about an axis lying at right angles to the piston stroke. The upper valves are the steam and the lower valves the exhaust valves. The valves are loose on the valve spindle so they are free to find their seats under the action of the steam pressure, and free to lift off their seats to allow trapped water to escape.
Reproduced with permission from R.A. Buchanan, and G. Watkins *The Industrial Archaeology of the Stationary Steam Engine* (Allen Lane, London, 1976).
(b) Trip gear that controls the cut-off of the Corliss valve. The valve is opened by means of the oscillating link driven by the eccentric. When the connection at the points A and B is broken the valve closes under the action of the powerful spring. The moment of closure depends on the height h of the lever. As the engine speed increases, h increases and the valve closes earlier. Closure of the valve is assisted by atmospheric pressure acting on the exposed side of the dash-pot piston and the vacuum formed on the other side when the valve is opened. The exhaust valves are not provided with trip gear and the angular motion in one direction is an exact repetition of the motion in the other direction.

Reproduced with permission from D. A. Wrangham *The Theory and Practice of Heat Engines* (Cambridge University Press, Cambridge, 1951).
(c) Schematic arrangement of equilibrium drop valves. The valves are mushroom shaped and seat as shown. Motion is in the vertical direction. Note that identical valves are mounted on the same stem so the pressures on the valve faces oppose one another and a balanced arrangement is obtained. Consequently, the valve operating force is only required to overcome friction and inertia. Because there are no sliding parts (cf. Corliss valve) the valve is well adapted to use with superheated steam.
Reproduced with permission from R.A. Buchanan and G. Watkins *The Industrial Archaeology of the Stationary Steam Engine* (Allen Lane, London, 1976).

Admission and exhaust of the steam was controlled from the beginning of the nineteenth century in most cases by the slide valve or in some cases, by the poppet valve. Both types of valve were subject to wire drawing because they did not close sufficiently quickly at the points of cut-off.

In 1842, F. E. Sickels took out a patent on a quick-closing valve gear using poppet valves, and a trip gear to control the cut-off (see Figure 5.3), with gravity assisted closure (called 'drop-valves' by Oliver Evans). To save wear and tear on the valve seat and value face, Sickels used a water-filled dashpot to decelerate the valve smoothly as it approached the end of its travel.

Sickels was not the only engineer who understood the advantages of rapid valve operation, and in about 1847, G. H. Corliss invented a quick-closing valve gear (patented in 1849) consisting of four flat slide valves, one inlet and one outlet at each end of the cylinder, but he did not persist very long with this valve gear. In order to simplify manufacturing and to reduce valve friction, engines built by his company after about 1850 used the oscillating rotary valve (see Figure 5.3) that is normally associated with his name. Corliss made an even-more fundamentally important contribution to steam engine technology by replacing the inefficient method of throttle governing with the better method of adjusting to load variations by using the governor to control the cut-off, so-called cut-off governing.

Engines using Corliss valve gear (commonly called Corliss engines) were built by Corliss, and his licencees from 1848, and by many others after the Corliss patent expired in 1870. This type of engine was extensively used for driving textile mill machinery where the close regulation of engine speed, that was ensured by the governor control of cut-off, was essential.

The medium speed engine

Automatic trip gear mechanisms do no operate satisfactorily at rotational speeds in excess of about 150rpm. Higher speeds became possible in 1858 when C. T.

Porter combined his very sensitive governor with a positive action, variable cut-off valve gear that had been invented in 1852 by J. F. Allen. In 1861, Porter and Allen formed a partnership to build engines using their two inventions.

The outstanding feature of the Porter–Allen engine was its quiet, vibration-free operation at all speeds, which resulted from Porter's careful study of the dynamics of the reciprocating engine. Because of their high and closely controlled speed, engines of this type were extensively used from about 1880 onward for driving electric generators, e.g. Edison's Pearl Street Station.

The high speed engine

Steam engines operating with piston speeds in excess of 3m/s (600ft/min) could be made sufficiently small, for a given power output, so that a significant reduction in first cost could be realized compared to slower-running engines, by the invention in the early 1870s of the shaft governor, which is usually mounted on the flywheel shaft. This operates by balancing the force of the governor spring against the centrifugal force. These had a tendency to hunt (an inability of the governor to locate a steady operating speed) because the rate at which they acted was independent of the rate of change of the load. This fault was overcome by the invention in 1895 by F. M. Rites of the inertia governor in which inertia forces augmented the centrifugal forces in the governor. With this modification the high speed engine could be applied where very close control of the engine speed was required, e.g. electric power generations.

High speed engines were characterized by stroke:bore ratios less than unity, and power outputs that did not usually exceed 370kW (500hp). Typically, piston valves and cam-driven poppet valves were used for steam distribution. Lubrication was particularly important and in 1890 the Belliss & Morcom Co. in England introduced forced lubrication, which is now a standard feature of any high speed machinery.

Many of the characteristic features of the high speed engine were carried over into the early internal combustion engines, so that contemporary engines of both types bear a considerable resemblance, both in superficial features and in certain details.

Compound engines

The zenith of steam engine design was in the multiple expansion engines that were developed in the second half of the nineteenth century. Such engines were used in the largest numbers for water supply system pumping and as marine engines, because efficiency was an important consideration in these applications. The inverted engine, in which the cylinders are arranged above the crankshaft with the piston rod acting downwards, was the universal type in large multiple expansion engines, because of the need to economize on floor space.

Stationary engines

Stationary compound engines were employed for operating the pumps of public water supply systems and for turning electrical generators, as well as in textile mills, as rolling mill drives, mine hoisting engines and blast furnace blowing engines.

In 1866 a pressurized water supply system using a pump, rather than an elevated reservoir or standpipe, started operating at Lockport, New York. This development, which was to have a significant effect on the history of steam pumping engines, was due to Birdsill Holly. The Lockport pumps were driven by a water-wheel, but a second installation in Dunkirk, New York, used steam-driven pumps.

The compound engine was first applied to water pumping in 1848 by the Lambeth Water Works, London, but the employment of this type of engine was not extensive until E. D. Leavitt installed such a machine at Lynn, Massachusetts, in 1873. These engines were landmarks in both capacity and efficiency, and their performance was not surpassed until H. F. Gaskill introduced (1882) a pump driven by a Woolf compound steam engine at Saratoga Springs, New York. This was a compact, high capacity, efficient steam pump, which was very popular in the United States.

The first triple expansion pumping engine (see Figure 5.2(d) above) was built in 1886 for the City of Milwaukee. It was noteworthy in being designed by E. T. Reynolds of the E. P. Allis Co. (later Allis–Chalmers). He joined this company in 1877 from the Corliss Steam Engine Co., and was one of the chief proponents of engines using the Corliss valve gear after the Corliss patent expired in 1870.

Two 7500kW (10,000hp) double tandem compound engines were designed in 1888 by S. Z. de Ferranti for use in the Deptford Power Station of the London Electricity Supply Company. This was a pioneering AC distribution system, but because substantial difficulties were encountered in placing it in service, Ferranti's connection with the company was terminated, and the engines were never completed.

Somewhat later, in 1898, the Manhattan Railway Co., which operated the overhead railway system in New York, purchased from the Allis–Chalmers Co. compound engines with horizontal high pressure and vertical low pressure cylinders. These unique engines, which were undoubtedly among the largest stationary engines ever built, and probably the most powerful (6000kW; 8000hp), are usually known as the Manhattan engines.

Marine engines

It was the application of the compound engine that allowed the steamship to take over from the sailing vessel on the longest voyages. The consequent reduction in coal consumption made more space available on the ship for passengers and cargo, and fewer stops were needed to replenish the bunkers.

Compound engines had been tried in small vessels between 1830 and 1840, but they were not fitted in ocean-going ships until 1854, when the *Brandon* was launched by Randolph, Elder & Co., of Govan on the River Clyde in Scotland. This had a Woolf compound engine with inclined cylinders and an overhead crankshaft, designed by John Elder. Saturated steam was supplied at 2.8bar (40psig), and in service the coal consumption was 2.13kg/kWhr (3.5lb/ihphr) which was about a 30 per cent improvement over the performance of vessels fitted with simple engines.

The compound engine was quickly adopted by British shipping companies operating to ports in the East and in Africa. However, it was not until the 1870s that vessels operating on Atlantic routes were fitted with engines of this type. By about 1880 the compound engine was used almost universally in marine service, but between 1875 and 1900 triple expansion, and then quadruple expansion engines, were adopted for the largest vessels.

The triple expansion engine was originally proposed in 1827 by Jacob Perkins, but no engine of this type was built until 1861 by D. Adamson; this was a stationary engine. The first application to a sea-going ship was by John Elder & Co., who in 1874 fitted an engine using steam at 10.3bar (150psig) with cylinders 580mm (23 inch) × 1040mm (41 inch) × 1550mm (61 inch) bore by 1070mm (42 inch) stroke in the *Propontis*.

However, it was the *Aberdeen*, launched in 1880 by Robert Napier & Sons, that had the greatest influence on the history of the marine engine. The engine had cylinders 760mm (30 inch) × 1140mm (45 inch) × 1780mm (70 inch) bore by 1370mm (58 inch) stroke. It used steam at 8.6bar (125psig) and had an output of 1340kW (1800ihp). On its first voyage the coal consumption was 1.0kg/kWhr (1.7lbs/ihphr). This engine was the prototype for thousands of marine engines that were built until the middle of the twentieth century: Liberty ships were fitted with triple expansion engines.

The first quadruple expansion marine engine was fitted in the *County of York* built at Barrow, Lancashire, in 1884, but this type of engine was not tried again until 1894 when the *Inchmona* was launched. this was a 707kW (948ihp) engine supplied with steam at 17.6bar (255psig) 33°C (60°F) superheat.

The reciprocating steam engine undoubtedly reached its highest level in some of the quadruple expansion engines built for the very large ocean-going liners at the close of the nineteenth century. One of the most outstanding was the 29,830kW (40,000hp) engine produced for the twin-screw *Kaiser Wilhelm II*.

Uniflow engine

The Uniflow engine, which represents the final stage of the development of the steam engine, was motivated by the problem of cylinder condensation and re-evaporation, which was a serious cause of energy loss in the engine. Steam enters the engine cylinder and is immediately exposed to cylinder walls that have been cooled by the previous charge of steam, which had itself been cooled in consequence of its expansion during the working stroke of the piston. If the cylinder wall temperature is low enough, the incoming steam will condense on the cylinder walls, and energy is given up to the walls. As the expansion proceeds in the cylinder the steam temperature can fall below the cylinder wall temperature, resulting in re-evaporation of the condensed steam. However, because this occurs near the end of the stroke, very little of the energy thus returned to the steam is available to do work and it is carried away as the steam leaves the cylinder.

The question of cylinder condensation and re-evaporation became the central concern of steam engine engineering from 1855 to 1885. Several developments alleviated this problem, either deliberately or incidentally: compounding; steam jacketing of the cylinders; superheating; and increasing inlet steam pressure. While these techniques were used on large marine and stationary engines, they were prohibitively expensive for the lower power single-cylinder engines that were widely used in industrial applications: an economical solution for engines of this type required a radically new design. The requisite development was the introduction of one-way steam flow (hence 'Uniflow' or 'Unaflow') in the cylinder, so that steam was admitted at each end of the cylinder and exhausted in the centre through circumferential ports in the cylinder wall uncovered by the piston (Figure 5.4).

The Uniflow principle appears to have been proposed quite early in steam engine history (Montgolfier, 1825; Perkins, 1827), but serious consideration of the idea did not occur until T. J. Todd took out a British patent in 1885 on an engine of this type (he called it the 'Terminal-exhaust' cylinder). It is not clear if such an engine was ever built, so the practical realization of the Uniflow engine is usually credited to J. Stumpf.

Uniflow engines were built in Europe, Britain and the United States. One of the most successful builders was the Skinner Engine Company of Erie, Pennsylvania. This company built Uniflow engines until the mid-1950s, and during the Second World War supplied them for naval craft (where they were popular because they generated less vibration than other types of reciprocating engines).

The progress in the steam inlet pressure and the heat rate between the time of Newcomen (1712) and the end of the nineteenth century is indicated in

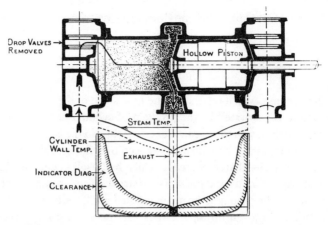

Figure 5.4: Schematic section of a Uniflow engine. The diagram shows the Uniflow steam path and the small difference in temperature between the steam and the cylinder wall throughout the piston stroke. Since the exhaust ports must not be uncovered until the piston reaches the end of the expansion stroke the piston length must equal its stroke. Hollow construction is used to reduce the piston weight. Because the return stroke of the piston is a compression stroke and because the compression ratio (expansion ratio for the expanding steam) is high, typically 46, there is a danger that the pressure of the residual steam in the cylinder could become high enough to dislodge the cylinder head cover. This is avoided by providing additional clearance space, in the form of a cavity in the cylinder head connected to the cylinder by a spring-loaded valve, with manual override for starting the engine, or by using various types of automatic and manual valves that divert the residual steam into the exhaust.
Reproduced with permission from D. A. Wrangham, *The Theory and Practice of Heat Engines* (Cambridge University Press, Cambridge, 1951).

Figure 5.5. The data for the earliest years do not have the precision and accuracy of the later period, nevertheless they are indicative of the general trend. In the final decade of the nineteenth century, the introduction of regenerative feed-water heating, in which steam used in the cylinder jackets was returned to the boiler, had a marked effect in lowering the best values for the heat rate. In fact, the best of the final generation of reciprocating steam engines with cylinder steam-jacketing had heat rates that were comparable with those of contemporary non-regenerative steam turbines (typically 23.0×10^3 btu/kWhr; see Figure 5.10).

STEAM TURBINES

The steam turbine is a device that directly converts the internal energy of steam into rotary motion (see Appendix). It is also characterized by uniformity of

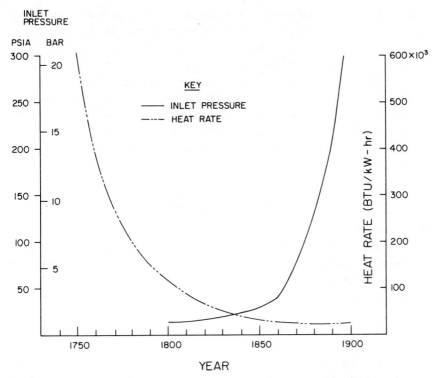

Figure 5.5: The historical trend of reciprocating steam engine inlet pressure (boiler pressure) and power plant heat rate between 1800 and 1900. The data shown are based on marine practice, but are also representative of stationary engines. The upward trend beginning between 1850 and 1860 marks the introduction of the compound engine. To convert heat rate to efficiency multiply by 2.93×10^{-4} and invert.
Adapted with permission from R. H. Thurston, *A History of the Growth of the Steam Engine*, Centennial edition (Cornell University Press, Ithaca, N.Y., 1939).

turning moment and by the possibility of balancing it perfectly, which is important where high powers are involved, when a reciprocating engine would have correspondingly massive dimensions that could produce vibrations of an unacceptable magnitude.

Because the steam flow through the turbine is continuous rather than cyclic, it is able to expand steam from a high pressure to a very low pressure, thereby maximizing the efficiency. This, because of the low density of the low pressure steam, would require the cylinder of a reciprocating engine to have impracticably large dimensions.

The continuous flow characteristics of the steam turbine allow it to avoid the complex valve gear necessary in the reciprocating engine, which should have made it attractive to the ancients, and indeed there is some historical evidence that a turbine was built in the first century AD by Hero in Alexandria. Later, in the seventeenth century, an Italian, de Branca, proposed another type of steam turbine. Unfortunately, these primitive machines suffered from the fatal defect that, for their times, their speed was too high to be either needed or useable. Consequently, the practical steam turbine had to await the development of the ability to design and construct high speed gears, or high speed electrical generators, or the formulation of such principles of fluid mechanics as would allow the energy of the steam to be utilized without high rotational speeds. None of this occurred until the end of the nineteenth century.

1884–1900: early history

The simplest form of turbine (Figure 5.6(a)), which was first demonstrated by the Swedish engineer Gustav de Laval in $c.1883$, is constructed by arranging the blades on a single revolving wheel in such a way that they turn the steam through an angle, thereby imparting a portion of the kinetic energy of the steam jets to the rotating blades with no change in steam pressure as it passes through them. This last is an important distinguishing characteristic of this type of turbine, which is sometimes known as an impulse turbine. In about 1888, de Laval made the crucial discovery that in order to extract the maximum amount of energy from the steam, the nozzle had to be of a converging–diverging form (see Appendix). This results in a very high steam velocity, leading to: (a) high rotational speeds (10,000–30,000rpm), which usually requires a speed reducing gear for its practical utilization; (b) wasted kinetic energy in the leaving steam; (c) large friction losses between the steam, the blades, and the turbine rotor. The losses, items (b) and (c), make the efficiency of the de Laval turbine inferior to that of a reciprocating engine operating with identical inlet and exhaust conditions.

The principles of the impulse turbine, and its problems, were well known to nineteenth-century engineers, including James Watt. Consequently, until the close of the century it was never considered a serious competitor to the reciprocating engine. The steam turbine was first placed on a practical basis by C. A. Parsons. He took out a patent in 1884 on a turbine which avoided the limitations of the impulse turbine by using the steam pressure drop inlet to exhaust in small steps rather than one large step (see Figure 5.6(b)), resulting in a much lower steam speed. However, there is an important difference between the Parsons and impulse turbines in the way in which the pressure drop is arranged in a stage (a pair of moving blades and stationary nozzles): in the former the pressure drop occurs in both the rotating and stationary elements, while in the impulse turbine the pressure drop only occurs in the stationary element. The Parsons turbine is known as a reaction turbine, and since Parsons arranged for equal pressure drops to occur in both the stationary and rotating elements of each stage, it is called a 50 per cent reaction turbine.

Parsons commenced experimental work on his turbine shortly after becoming a partner in 1883 in the firm of Clarke, Chapman & Co. In 1885 he constructed his first working turbine Figure 5.7 which drove an electrical generator at 18,000rpm having an output that appears to have been between about 4kW (5.4hp) and 7.5kW (10hp). The surprisingly high speed, considering the Parsons turbine was intended to avoid this feature of the de Laval turbine, is a consequence of the small dimensions of the first Parsons turbine. Thus, an increase in the mean blade radius from 4.4cm (1.75 inches) (estimated for Parsons's first turbine) to 26.7cm (10.6 inches) would reduce the rotational speed to 3000rpm. The essential point is the lower steam speed in the Parsons turbine compared to the de Laval turbine. Thus, for comparable steam conditions, the steam speed in the de Laval turbine is about 760m/s, (2500ft/sec), whereas in the Parsons turbine it is about 116m/s (380ft/sec).

From 1885 until 1889 development of the Parsons turbine was very rapid with machines of maximum output 75kW being ultimately produced. However, Parsons's partners did not have his faith in the steam turbine, so in 1889 the partnership was dissolved. Parsons founded C. A. Parsons & Co. and proceeded to develop a radial reaction turbine that circumvented the patents controlled by his former partners (a truly remarkable achievement). In 1894, Parsons came to an agreement with Clarke, Chapman that allowed him to use his 1884 patents for the axial flow steam turbine, and, as a result, construction of the radial flow steam turbine stopped.

Although the Parsons turbine was in many ways significantly superior to the de Laval turbine, it did have some disadvantages. In particular, it was difficult to build with outputs greater than 2000kW (2682hp). This was a consequence of the great length of the machine, which was necessitated by the numerous rows of moving and stationary blades that were required to keep the steam velocity low. As a result the rotor was particularly sensitive to slight mechanical and thermal imbalances that could lead to distortion and, hence, damage to the moving blades if they touched the turbine casing. The techniques that were eventually devised to overcome this problem are discussed below.

In addition to the limitations in size that were encountered in the early Parsons turbine, there is a substantial loss in power output due to leakage through the clearance space between the rotating blades and the turbine casing, which is particularly serious in the high pressure sections of the turbine where the blade heights are small. In order to avoid this leakage problem and yet retain the advantages of the Parsons multi-stage concept, Auguste Rateau, a French engineer, designed sometime between 1896 and 1898 a multi-stage impulse turbine that is now commonly known by his name. The principle of this turbine is illustrated in Figure 5.6(c). As in the Parsons turbine, the pressure drop is

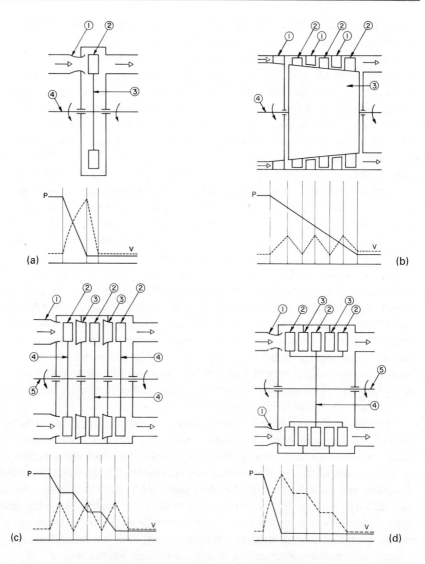

Figure 5.6: Schematic arrangement of the basic steam turbine types.
(a) Simple impulse turbine: de Laval (*c*.1883). *Key*: 1: stationary nozzles; 2: rotating blades; 3: wheel; 4: rotating shaft. The lower diagram shows the change in pressure (P) and velocity (V) of the steam as it passes through the turbine.
(b) Reaction Turbine: Parsons (*c*.1884). *Key*: 1: Nozzles and guide vanes; 2: rotating blades; 3: rotating drum; 4: rotating shaft. The conical drum counteracts the decrease in density of the steam as it passes through the turbine. This arrangement was not used by Parsons who employed a limited number of step changes in drum diameter. The lower diagram shows the change in pressure (P) and velocity (V) of the steam as it passes through the turbine.

(c) Pressure-compounded impulse turbine: Rateau (c.1900). Key: 1: nozzles; 2: rotating blades; 3: nozzles in diaphragm; 4: rotating wheels; 5: rotating shaft. The lower diagram shows the change in pressure (P) and velocity (V) of the steam as it passes through the turbine.
(d) Velocity-compounded impulse turbine: Curtis (1896). Key: 1: nozzles; 2: rotating blades; 3: stationary guides; 4: rotating wheels; 5: rotating shaft. The lower diagram shows the change in pressure (P) and velocity (V) of the steam as it passes through the turbine.
Reproduced from E. F. C. Somerscales, 'Two Historic Prime Movers', ASME Paper number 84-WA/HH-2 (1984).

divided between a number of stages, thereby limiting the steam speed. This type of turbine avoids the leakage problem of the Parsons turbine by only using each stage pressure drop in the stationary nozzles. The Rateau turbine is classed as a pressure-compounded impulse turbine.

The Rateau turbine differed from the Parsons turbine in another important respect. In the latter the moving blades are attached to the periphery of a rotating drum, with the seals between the stages at its outer surface and at the outer ends of the rotating blades. In the Rateau turbine the blades are mounted on the periphery of a disc carried on the turbine shaft, so there is only one interstage seal and it is where the rotating shaft passes through the diaphragm separating the stages (see also Figure 5.6(c)). This type of construction has two advantages over the system used in the Parsons turbine: the leakage area, and, hence the leakage flow, is smaller because the seal is at the shaft, which has a smaller diameter than the Parsons drum; and a shaft seal involves parts with heavier dimensions than a ring of stationary blades, so it can be made more effective.

The disc construction of the Rateau turbine would, because of the pressure drop across the moving blades, result in an unacceptably large axial thrust on the shaft bearings if applied to the Parsons turbine. The axial thrust that does exist in the latter is smaller because it only acts over the blade area, and is accommodated by thrust balancing cylinders, connected externally to points of higher steam pressure, at the high pressure end of the rotor.

Because the solid construction of the drum is better suited to the high temperatures encountered in the high pressure stages of modern turbines, the disc-on-shaft construction has been effectively abandoned in these sections and the blades are mounted on vestigial discs machined out of a forged cylindrical rotor.

Although the Rateau turbine mitigated the leakage problem of the Parsons turbine, it still required a long shaft to accommodate the large number of pressure compounded stages. In 1896 the American C. G. Curtis patented two concepts that resulted in a substantial reduction in the overall length of the turbine.

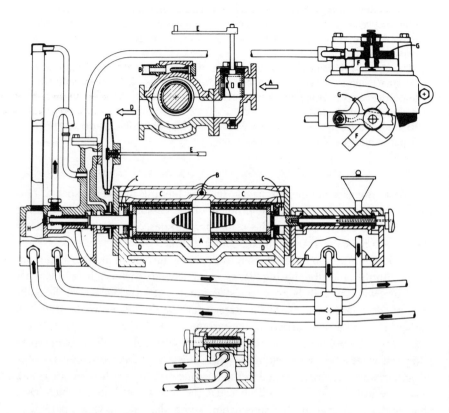

Figure 5.7: Cross section of the first Parsons steam turbine. This was supplied with saturated steam at 5.5bar (80psig), ran at 18,000rpm and produced between 4 kW and 7.5 kW at the generator terminals.

Steam was admitted through the inlet (A) and flowed axially to right and left through alternating rows of moving and fixed blades (shown schematically). The steam exhausted into the passages C and D which were cast in the turbine casing. The exhaust pressure (about atmospheric) was maintained by a steam ejector (B). The diagram also shows the governor (designed because available rotary governors could not withstand the high shaft speed), which consisted of a small centrifugal fan, located just to the left of the item lettered C in the diagram, rotating at the turbine shaft speed, and drawing air from one side of a diaphragm (located just above the item lettered C). The motion of the diaphragm, as the vacuum varied with turbine speed, was conveyed by the link (E) to the governor valve controlling the steam admission. The sensitivity of this governor was increased by sensing the generator voltage (actually the magnetic field strength, which is proportional to the voltage) and bleeding more or less air into the vacuum system. In the figure, F is the soft-iron field sensor mounted on the dynamo pole pieces, and coupled to this is the spring loaded brass arm (G) that controls the flow of the bleed air. Forced lubrication using a screw pump (H) was used. The arrows show the direction of oil flow. The small diagram at the foot of the figure shows one of the bearings that were designed to allow some transverse movement of the shaft, due to small residual imbalances. The shaft ran in a long thin sleeve, surrounded by a large number of washers alternately fitting the sleeve and the casing. The washers were held in contact with each other by the pressure of a short helical spring surrounding the sleeve near its end and tightened up by a nut threaded on the sleeve. The bearing oil was supplied by the screw pump.

Reproduced with permission from W. G. S. Scaife, in *First Parsons International Turbine Conference* (Parsons Press, Dublin and London, 1984).

The first of these was the recovery of the velocity energy in the steam jet leaving a converging–diverging nozzle in several rows of moving blades with stationary turning vanes between each row (see Figure 5.6(d)). This multi-row de Laval turbine is known as a velocity-compounded impulse turbine. The second arrangement patented by Curtis, which was his special contribution to turbine engineering, was to combine velocity- and pressure-compounding (Figure 9).

To obtain the necessary funds to support the development of his turbine, Curtis sold most of his patent rights to the General Electric Company of Schenectady, New York, in 1897. However, a satisfactory working machine was not produced until 1902.

The Curtis turbine, like the Rateau turbine, was a disc turbine. Eventually, as discussed below, the Curtis and Rateau turbines were combined, so that later forms of impulse turbines that have derived from these two types generally tend to be of the disc construction, or at least have vestigial discs machined from a drum.

By 1900 the four basic types of steam turbines had been developed into practical machines. Although there was no general and immediate tendency to adopt this prime mover, in spite of its obvious advantages, by the end of the following decade the steam turbine had established itself in the electric power generation industry, and had also provided a number of convincing demonstrations of its usefulness as a marine power plant.

1900–1910: turbines take over

During the first ten years of this century the application of steam turbines to both electric power generation and marine propulsion spread very rapidly in consequence of a number of influential demonstrations of its capabilities. The first of these was in 1891 when a measured steam consumption of 12.2kg/kWhr (27lb/kWhr) was reported for one of the 100kW condensing (the first such) machines installed at Cambridge. This was a record for steam turbines and equalled the performance of the best simple expansion reciprocating steam engines.

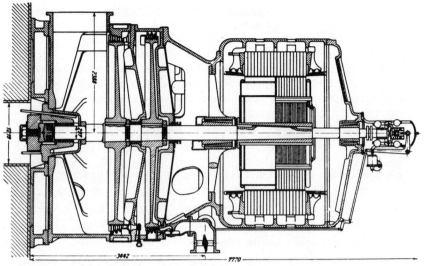

Figure 5.8: Vertical section through the 5000kW vertical Curtis steam turbine supplied by the General Electric Co. to the Commonwealth Electric Co. for installation in the Fisk Street Station, Chicago in 1903. Steam inlet conditions: 11.9bar (175psig), 271°C (520°F); exhaust: 0.068bar (28in Hg vacuum); minimum steam consumption 6.81kg/(kW/hr) (15.0lb mass/(kW/hr)). The turbine had two pressure compounded stages consisting of a nozzle, three rows of stationary guide and four rows of moving blades attached to one wheel. The vertical arrangement was discontinued by the General Electric Co. built after 1913.
Reproduced with permission from C. Feldman 'Amerikanische Dampfturbinen', *Zeitschrift des vereines deutscher Ingenieure*, vol. 48 (1904), p. 1484.

A second significant installation was the two 1000kW turbines, the largest built to date, that were delivered in 1900 by C. A. Parsons & Co. to the city of Elberfeld in Germany. These had a steam consumption of 9.12kg/kWhr (20.1lb/kWhr). These turbines had many of the features of the modern unit: they were two cylinder tandem compound (see below), and the condenser was placed below the level of the operating floor, immediately under the exhaust from the low pressure cylinder.

Another very influential steam turbine, which was placed in service in 1903 at Chicago's Fisk Street station (see Figure 5.8), was the third Curtis steam turbine sold by the General Electric (GE) Company. It had a power output of 5MW and was the most powerful steam turbine built up to that time. It was unusual in being arranged with its axis vertical (a type of construction abandoned by GE in 1913) and with the alternator above the turbine. Pressure-and-velocity-compounding were used, and each of the two pressure-compounded stages consisted of a row of nozzles, three rows of stationary turning vanes, and four rows of moving blades attached to one wheel.

1910–1920: blending of types

As the patent protection on various types of turbines expired the manufacturers devised hybrid machines. One of the most significant of these combinations was the addition to the Parsons turbine at the inlet of a single velocity-compounded stage, sometimes called a control stage, in place of a number of the high pressure reaction stages. The rotor of this combined Curtis–Parsons turbine is much shorter than the rotor of the pure Parsons type of the same power output. This arrangement confines high temperatures and pressures to a shorter portion of the turbine, minimizing expansion effects due to temperature gradients, which can lead to eventual failure of the machine. Although the Curtis stage is not as efficient as the reaction stage it replaces, the minimization of leakage and improvement in reliability result in a net gain to the user of the turbine.

Another technique for minimizing the turbine shaft length, which was probably first introduced into regular steam turbine design practice with a 25MW turbine built in 1913 by C. A. Parsons & Co. for the Fisk Street station in Chicago of the Commonwealth Edison Company, was the tandem compound. In this machine two rotors are arranged in two separate cylinders, with bearings in each cylinder. The steam passes in succession through the two cylinders. The rotors are coupled between the cylinders and the load is connected to one end of the shaft, usually at the low pressure cylinder.

1920–1930: increasing size

The period between 1920 and 1930 was characterized by a very rapid growth in steam turbine power output; preliminary attempts to use very high steam pressures; and a large number of serious mechanical failures.

Increasing power output implies increasing mass rate of flow of steam, and increasing dimensions. The largest dimensions are encountered at the exhaust from the turbine, where the steam has the largest volume, and the critical dimension in this region is the length of the last row blades, which are the longest in the turbine and, therefore, are subjected to the highest centrifugal stress. Consequently, very careful consideration must be given to the mechanical design of the rotor and the last row blades, their manufacture and the materials used, and the means of attaching the blades to the rotor. Since sound forgings for the rotor can be assured with greater certainty the smaller their size, there was a tendency during the period from 1920 to 1960 to use the disc construction in the low pressure sections of the turbines.

Where the length of the last row blades could not be increased a number of alternative techniques were developed in the 1920s including Baumann multiple exhaust, due to K. Baumann of the Metropolitan-Vickers Co., multiple exhaust flows, and multiple low pressure turbines (not used until the 1930s). Extreme forms of this arrangement, involving three or four low pressure turbines, each with a double exhaust flow, have been used on the very high output machines constructed from 1960 onward.

In order to accommodate the largest outputs a combination of decreasing the turbine speed and multiple exhaust flows was used in the 1920s. This was in the form of a cross-compound turbine in which the high pressure cylinder operated at 3000 or 3600rpm and the low pressure cylinder ran at 1500 or 1800rpm, the two cylinders being coupled to separate alternators. This arrangement appears to have been first used for three 30MW turbines installed by Westinghouse in 1914 at the 74th Street station in Manhattan of the Interborough Rapid Transit Co.

The cross-compound tends to be an expensive solution to the problem of building turbines with large outputs, and European machines, with their lower speeds (1500 and 3000rpm), did not employ it as extensively as turbines built in the United States. The two-speed cross-compound has not been used since the 1970s (except to accommodate very large turbines with outputs of 1000MW or more), because improved materials of construction removed the incentive for its use.

The steady growth in turbine power output in the 1920s culminated in a 208MW turbine that was constructed by GE for installation in the State Line station near Chicago in 1929. This was a three-cylinder cross-compound turbine, and it remained the turbine with the world's largest output until 1955.

The application of very high inlet pressures to steam turbines was first attempted in 1925, in order to increase cycle efficiency. Three such turbines were installed at the Edgar Station of the Edison Electric Illuminating Company of Boston (now the Boston Edison Company). The first (3.15MW) had an inlet pressure of 83bar (1200psig) and the other two (10MW) operated at 97bar (1400psig). They exhausted at 24bar (350psig) into the steam line connecting the low pressure boilers to two 32MW units.

In many instances in the 1920s and 1930s new high pressure turbines were employed in connection with existing low pressure systems; a method known as superposition. (In a few cases the high pressure turbine was actually mounted on the low pressure turbine casing.) It allows existing, serviceable plant to be increased in efficiency and power output for minimum cost, and for this reason it was popular in the post-depression years of the 1930s.

The rapid advances in steam turbine technology in the 1920s were not achieved without some cost. Early in this period the manufacturers of disc turbines experienced an exceptional number of failures. It was discovered that the discs were subject in service to vibrational oscillations leading to fatigue failure. This was overcome by employing better design theories and also by using heavier discs, including discs forged integral with the shaft. Important contributions to the solution of this problem were made both in the United States and in Europe; particularly noteworthy was the work of Wilfred Campbell of the

General Electric Company, after whom one of the important design tools developed at that time, the Campbell Diagram, is named.

1930–1940: refinement

Between 1930 and 1940 no really striking advances of the type seen in the 1920s occurred. The depression following the stock market crash in 1929 did not provide a business climate in which large orders for steam turbines could be expected. Consequently, turbine manufacturers turned to improving the detailed design of steam turbines and a number of features that are now common practice in steam turbine construction were introduced.

As a result of the desire to increase the inlet pressure the double shell construction shown in Figure 5.9 was introduced. This decreases the load on the turbine casing and fastenings by dividing the pressure difference from the turbine inlet pressure to the ambient pressure between two casings, one inside the other.

In the 1930s creep, the slow 'plastic movement' of steel subjected to high temperatures and pressures, which was first identified in the 1920s, began to be considered in steam turbine design. This phenomenon can significantly alter stress distributions during the time of exposure of the turbine parts to operating conditions. Its effects can be minimized by adding suitable alloying materials to the steel that stabilize the material, and by developing extensive empirical data on the material's properties for use in design.

Warped, or twisted, low pressure blades, were introduced in the 1930s. These compensate for the effect of variations in the blade tangential velocity with radius so as to ensure that the steam impinging on the blade enters the blade smoothly and with the minimum flow disturbance.

1940–1950: increasing speed

Steam turbine progress in the 1940s was constrained by the outbreak of the Second World War. After the war ended turbine design showed a definite trend away from the standard speed of 1500/1800rpm and the establishment of the high speed turbine operating at 3000/3600rpm. Higher speed turbines are smaller and lighter than comparable low speed machines. Smaller turbines expand less when heated, so distortion is decreased, which improves the turbine's long-term reliability. Consequently, the high speed machine is better adapted to increasing inlet pressures and temperatures and to the application of reheat, which were features of steam turbine design from the late 1940s onwards.

In about 1948 there was a revival of interest in the reheat cycle (first introduced in Britain at North Tees in 1920, and in the United States at Philo,

Ohio, in 1924) which was an added incentive for the introduction of the high speed turbine. In this cycle the temperature of the partially expanded steam is raised to about the original inlet temperature by passing it through a special heat exchanger, the reheater, which is heated either by the boiler combustion gases or by high temperature steam (the latter arrangement was only used in the 1920s), and then returned to the turbine for further expansion to the condenser pressure. The motivation for the earlier application in the 1920s had been to decrease the moisture, which causes blade erosion, in the low pressure sections of the turbine. Because the required additional valves and piping could not be economically justified at that time, the construction of new reheat turbines ceased in the early 1930s. The revival of interest in the late 1940s was stimulated by a need to increase plant efficiency in order to counteract rising fuel costs.

Reheat produces a 4–5 per cent improvement in cycle efficiency, and has a number of other advantages compared to non-reheat operation. Thus, there is a reduction in the mass rate of flow of the steam, which, in turn, leads to a decrease in the size of the boiler feed pump, the boiler, the condenser, and of the feed water heating equipment. This, together with the ability to reduce the wetness of the steam at the exhaust makes reheat an attractive feature, and is widely used in modern steam cycles.

1950–1960: very high pressures and temperatures

The first turbines handling steam at supercritical pressures (pressures in excess of 221bar/3200psig) were built in the 1950s with the aim of improving cycle efficiency, while avoiding the problems associated with increasing the turbine inlet temperature, which requires the development of new materials, an expensive and time-consuming process.

The first supercritical turbine was installed in 1957 at the Philo station of the Ohio Power Company. It had an output of 125MW and used steam at 310bar (4500psig) and 621°C (1150°F). Double shell construction was used, and, because of the exceptionally high pressure, the outer casing is almost spherical in form.

A significant feature of these very high pressure turbines is the small blade lengths in the high pressure stages (0.95cm in the Philo turbine), which results from the high density of the steam. The leakage space is, in consequence, a large fraction of the blade length, so turbines operating at very high pressures should, to offset the leakage loss, be designed for large outputs, e.g Philo (1957), 125MW; Philip Sporn (1959), 450MW; Bull Run (1965) 900MW.

In about 1955 the 'average' steam temperature reached its present plateau of 566°C (1050°F). The attainment of this temperature was made possible by the introduction of the ferritic stainless steels (11–13 per cent chromium), and required the adoption of special design features. In the high pressure and intermediate pressure sections all parts were designed to allow free expansion and contraction. For example, the steam chests and valves were mounted separately from the turbine casing and connected to the inlet nozzles by flexible piping. To reduce temperature gradients in the turbine, the partially expanded steam was arranged to flow through the outer space of the double shell construction.

1960–1980: increasing size

During the 1960s a number of extremely large output machines were placed in service. The mass rate of flow steam is so great (for a typical 660MW turbine about 2.1×10^6kg/hr or 4.7×10^6lb mass/hr) that multiple low pressure sections have to be provided.

The 1960s was a period when many nuclear power stations commenced operation, and often these used either the boiling water or pressurized water cycles in which steam is supplied at pressures ranging from 31bar (450psig) to 69bar (1000psig), with the steam dry and saturated at about 260°C (500°F). To compensate for the relatively low energy content of steam, the turbines have large outputs (1300MW), with correspondingly large steam mass flow rates (e.g., a 1300MW 'nuclear turbine' handles about 7.3×10^6kg/hr (16.0×10^6lb mass/hr)). The resulting long (1.143m, 3.75ft) last row blades in the low pressure sections have forced American practice to adopt the 1800rpm tandem-compound design. Because of the lower speed, European designs of 'nuclear turbines' have sometimes been able to employ 3000rpm machines, but most examples of this type of turbine have been tandem-compounds operating at 1500rpm.

The saturated inlet conditions result in a high moisture content in the turbine low pressure sections, leading to blade erosion unless reheating is used. Because the water cooled reactor provides only a low temperature heat source, reheating is not as effective as it is in fossil fuel-fired plants, so mechanical moisture separation must be used in addition to reheating.

In the 1970s some even larger steam turbines, with power outputs in excess of 1000MW, came into service. To handle the large quantities of steam required by these machines, multiple low pressure stages were arranged in parallel. Figure 5.9 shows the section of one of the two 1300MW turbines completed by Brown Boveri in 1974 for the Gavin station of the American Electric Power Company. Machines of this type are the largest ever to be used on a fossil fuel-fired cycle.

The historical development of the steam turbine can be summarized in a number of ways. Figure 5.10 shows the progress in power output, inlet temperature and inlet pressure from 1884 to 1984.

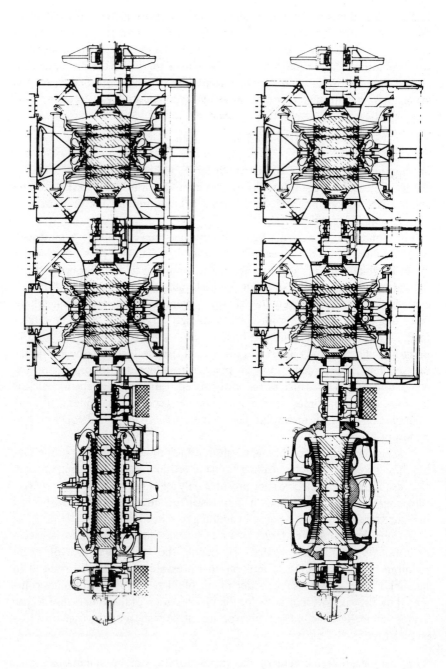

Figure 5.9

Figure 5.9: Vertical section of the 1300MW, 3600rpm cross-compound turbine built by Brown, Boveri & Co. and installed at the Gavin station of the American Electric Power Co. in 1974. Steam inlet conditions: 241bar (3500psig), 538°C (1000°F) with single reheat to 538°C (1000°F). This is the largest steam turbine in the world supplied by a fossil-fuel fired steam generator.

Although this turbine is described as a cross-compound (introduced *c.*1914) because sections of the machine are divided to drive two separate loads, it also incorporates tandem compound features (introduced *c.*1913) because the high pressure (upper left) and intermediate pressure (lower left) sections are coupled to the low pressure sections (on the right). All sections are double flow (introduced *c.*1906) with steam supply at the centre point of the section. The steam flows in opposite axial directions balancing any thrust on the rotors from the stage pressure drops. The high pressure section is supplied with steam from the steam generator. This section then exhausts, through the reheater, to the intermediate pressure section, which in turn exhausts to the four low pressure sections. Double shell construction (introduced 1937) is used in all sections.

The increasing inlet temperatures and pressure shown in Figure 5.10 are a consequence of the desire to increase the efficiency of the thermodynamic cycle of which the steam turbine is a part. The result has been a considerable improvement in cycle efficiency over the years as indicated by the decreasing plant heat rate. Reliability is as important as efficiency, and the continual advances in steam conditions and power output have required a corresponding effort in metallurgy, material behaviour and inspection, blade and disc vibration, and fluid mechanics. The fact that reliability has been maintained with the continuing need to increase the turbine operating parameters is a monument to the engineers who have made it possible.

INTERNAL COMBUSTION ENGINES

The reciprocating internal combustion engine was the second type of thermal prime mover, after the reciprocating steam engine, to be developed, with the first practical example, the Lenoir engine, being built in 1860. There were three incentives to replacing the steam engine: (a) elimination of the boiler and condenser, and the need for good water; (b) the increasing availability of suitable fuels: coal-gas and petroleum derivatives; (c) the potentially higher thermal efficiency of the internal combustion engine, as indicated by thermodynamics, resulting from the higher allowable maximum temperatures compared to other prime movers.

Three fuels, gas, petrol (gasoline) and oil are in normal use in the internal combustion engine. In addition, there have been, and there are, efforts to use coal as a fuel (although contrary to myth, Rudolf Diesel did not originally intend this for his engine). Engines can, therefore, be classified by the fuel they employ. Alternatively, engines may be categorized by the method used to ignite

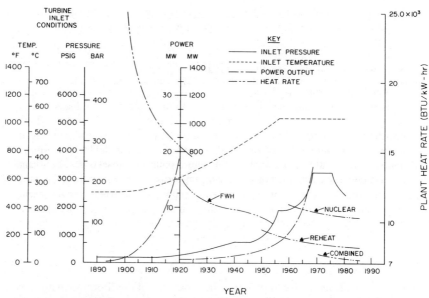

Figure 5.10: Historical trend of the performance parameters of steam turbines in electric power generation service 1890–1985.

The plotted curves represent the 'average' or 'typical' machine. The progressive increase in inlet steam temperature is particularly dependent on the materials used in the rotors and casings. Periods of application of different materials are roughly as follows: cast iron: 1883–1913; cast steel (ferritic): 1913–37; alloy steel (chrome ferritic): 1930 to date; alloy steel (austenetic): 1947–76. To convert the heat rate values to efficiency multiply by 2.93×10^{-4} and invert. The curves are based on data in K. Baumann, *Journal of the Institution of Electrical Engineers*, vol. 48 (1912), pp. 768–877; vol. 59 (1921) pp. 565–663, and *The Metropolitan-Vickers Gazette*, vol. 12 (1930), pp. 212–20, and the annual power plant surveys appearing under various titles in *Power* between 1931 and 1985.

the fuel; a hot surface, a spark or compression of the working substance. The modern tendency appears to favour the last method. Accordingly, the following nomenclature has been adopted here: the spark ignition engine describes essentially the modern petrol engine and its forebears; the compression ignition engine refers to what is now commonly known as the Diesel engine (which is not the machine originally invented by Diesel). Two additional classes, described further below, are the gas engine and the 'hot bulb' engine.

Proposals and developments before 1860

The development of the internal combustion engine began in the seventeenth century with attempts by Christian Huygens to harness the energy released in a cannon on firing. This was done by exploding a charge of gunpowder at the base of a vertical cylinder, closed at its lower end; the expanding products of combustion raised a free-piston fitted in the cylinder. If some of the products of combustion were released when the piston reached its point of maximum travel, then the residual gases would, on cooling and contracting, produce a pressure less than atmospheric (a vacuum) in the cylinder below the piston. The application of this vacuum to power production was the same as that in the steam engine (see p. 273).

The free-piston engine is not convenient for producing mechanical power because on the upward (expansion) stroke the piston has to be disconnected from the output shaft. An atmospheric engine with a conventional permanent connection to its load was first demonstrated by William Cecil in 1820. This was a hydrogen fuelled, single acting, atmospheric engine that drove its load through a rocking beam (like Newcomen's steam engine: see p. 275) and crank. It was spark ignited, and Cecil designed an automatic device for metering and mixing the air–fuel mixture.

Samuel Brown, a Londoner, constructed between 1823 and 1833 the first commercially successful internal combustion engine. A vacuum was produced in a constant volume container by burning coal-gas and cooling the products of combustion with a water spray (as in Watt's steam engine: see p. 275). The vacuum was then used in a separate power cylinder, fitted with a piston.

A free-piston engine was designed and built, between 1854 and 1864, by the Italians Eugenio Barsanti and Felice Mattucci. This was a return to the original concept of Huygens, but the fuel was now coal-gas with spark ignition. For various reasons it was not a commercial success.

1860–1880: the early gas engine

The first internal combustion engine that could be said to provide a reliable and continuous source of power was the gas engine (using coal-gas) introduced in France in 1860 by Etienne Lenoir. The air standard Lenoir cycle is shown in Figure 5.11(a). It was a double-acting, two-stroke cycle engine (see below) that used slide valves to control the admission and exhaust processes. It was very popular, being made in sizes between one half and six brake horsepower, but the thermal efficiency was low (5 per cent).

The most important contribution to the identification of the principles that should be followed in the design of internal combustion engines was made by Alphonse Beau de Rochas and given in a French patent filed in 1862. In addition, Beau de Rochas advocated the use of a four-stroke cycle (see Figure 5.12(a)) for maximum efficiency, rather than the two-stroke cycle (see Figure 5.12(b)) that was more popular at that time.

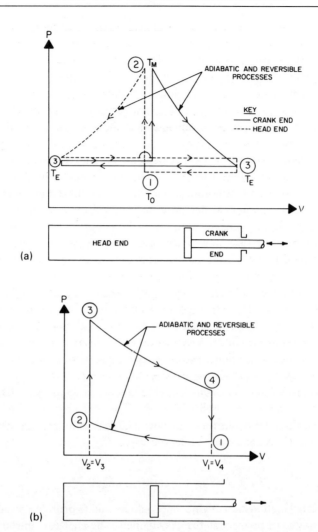

Figure 5.11

(a) The air-standard Lenoir cycle plotted on coordinates of pressure (P) and volume (V) for a double acting engine showing suction and exhaust strokes. In the actual engine an air and coal-gas mixture is drawn into the cylinder (1)–(3). When the piston reaches mid-stroke the mixture is ignited, followed by a steep rise in pressure (1)–(2). The pressure decreases (2)–(3) as the piston continues its stroke. The exhaust stroke is (3)–(3). This cycle is sometimes known as the non-compression cycle because the maximum pressure at point (2) is obtained from heat addition (by combustion in the practical engine) not from motion of the piston. T_M = maximum temperature at the conclusion of the heat addition (combustion) process, T_o = temperature at the conclusion of the suction stroke (ideally equal to the exhaust/admission temperature T_E).

(b) The air-standard Otto cycle plotted on coordinates of pressure (P) and volume (V). Exhaust and suction strokes are not shown. The working substance (air–fuel mixture in the actual engine) is compressed adiabatically (no heat transfer) and reversibly (with no friction or other dissipative effects) from (1)–(2). Heat is added at constant volume (2)–(3) (by combustion of the air–fuel mixture in the actual engine). Reversible and adiabatic expansion (3)–(4) follows. From (4)–(1) the working substance is cooled at constant volume (in the actual engine the products of combustion are discharged to the ambient at P_1).

The free-piston internal combustion engine reappeared in 1867 when N. A. Otto and Eugen Langen demonstrated one at the Paris Exhibition. Its fuel consumption was less than half that of the Lenoir engine. Although this engine sold widely, it was heavy and noisy, and in 1876, Otto produced an engine working on the four-stroke cycle of Beau de Rochas. Otto based his engine on an air standard cycle (Figure 5.11(b)) that is nowadays identified by his name, which is applicable to both two- and four-stroke cycle engines.

An important feature of Otto's four-stroke engine was its incorporation of the concept of the stratified charge, which is applied in some modern engines to minimize the production of undesirable pollutants in the exhaust gases (see p. 319). His objective was to provide smooth operation by eliminating combustion 'knock', or detonation (see below). In the stratified charge engine the air and coal-gas mixture was introduced into the cylinder in such a way that it was lean near the piston with increasing richness toward the ignition source (a gas pilot-flame). This was accomplished by a special valve gear that first admitted air and then admitted the gas when the piston was about half-way through the induction stroke.

The Otto engine, called by its manufacturers the Otto Silent Engine (compared to the very noisy Otto and Langen engine), was a landmark in the history of internal combustion engines, because it incorporated all the essential features of the modern internal combustion engine.

Four-stroke and two-stroke cycle engines

The cycles forming the basis of the reciprocating internal combustion engine may be classed as either two-stroke or four-stroke (Figure 5.12), with the former having a power stroke every revolution and the latter on every second revolution. The four-stroke cycle is in practice (although not in theory) more efficient than the two-stroke cycle, but nineteenth-century engineers continued to be interested in the latter because of its theoretical advantages, and, more importantly, because it circumvented Otto's patents on the four-stroke cycle. The first engine operating on the principle was built by Dugald Clerk in 1878.

Experience showed that the power output of the two-stroke cycle engine was only about 30 per cent greater than that of the corresponding four-stroke cycle

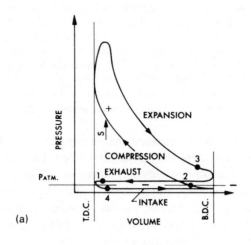

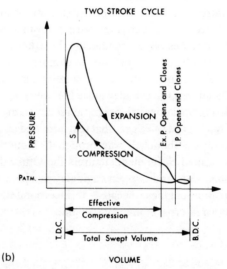

Figure 5.12

(a) Indicator diagram (actual pressure-volume diagram for one complete cycle) of a four-stroke cycle internal combustion engine.
This is the cycle advocated by Beau de Rochas in his 1862 patent. The cycle events are as follows: (4)–(2) intake during an entire outward stroke of the piston; (2)–(S) compression during the following return stroke of the piston with ignition at point S; (S)–(3) increase in pressure due to combustion of the air–fuel mixture followed by expansion to point (3) on the outward piston stroke; (3)–(1) exhaust on the fourth, and last, stroke of the cycle.

(b) Indicator diagram (actual pressure-volume diagram for one complete cycle) of a two-stroke internal combustion engine. The cycle of events is similar to that in the four-stroke cycle except for the exhaust processes. Close to the end of the expansion stroke the exhaust ports (Ex.P.) open and the products of combustion are discharged from the cylinder into the exhaust manifold. Before the outward stroke of the piston is complete and before the exhaust ports close the inlet ports (I.P.) open. On the return stroke of the piston, after it passes bottom dead-centre (BDC), the inlet ports close followed by the exhaust ports. This shows the intake process in the two-stroke cycle differs from that in the four-stroke cycle in the following respects: (i) as the piston returns from BDC it is opposing the intake process; (ii) intake only occurs during part of the piston stroke which can inhibit the induction of the charge. For these reasons engines working on the two-stroke cycle are not self-aspirating and a compressor (called the scavenge pump) is required to force a fresh charge into the cylinder and to push the products of combustion out. Because there is a tendency for the incoming charge to flow straight out of the exhaust ports, two-stroke spark ignition engines, where the air is carburetted outside the cylinder, have a high fuel consumption compared to the corresponding four-stroke cycle engines.

Reproduced with permission from D.J. Patterson and N.A. Heinen, *Emissions from Combustion Engines and their Control* (Ann Arbor, Ann Arbor Science Publishers, 1973).

engine, instead of being 100 per cent greater, as predicted by theory. This is because the two-stroke cycle engine is not self-exhausting (technically, self-aspirating, see Figure 5.12(b)). The fresh incoming charge (air or air-fuel mixture) must drive out the residual gases, a process called scavenging. Incomplete scavenging of the burnt gases decreases the amount of fresh charge that can be introduced, so the power output of the engine is less than predicted. The scavenging process is assisted by ensuring that the pressure of the incoming fresh charge is slightly higher than the pressure of the burnt gases in the cylinder at the end of the expansion stroke (see Figure 5.12(b)). A scavenge pump provides the necessary pressure increase in the engine inlet manifold. Various types of reciprocating and rotary pumps are used, and in some cases the under side of the piston in a single-acting engine (see Figure 17(b)). A supercharger can, in principle, also act as a scavenge pump, but because the efficiency of a two-stroke cycle engine is very sensitive to the design of its exhaust system, it is usual, particularly with an exhaust gas turbine driven supercharger (turbocharger), to provide a separate scavenge pump (see Figure 17(b)).

The two-stroke cycle is used only by the lowest power (15kW or 20hp) spark ignition engines or by the highest power (7500kW or 10,000hp or higher) compression ignition engines. In both situations the two-stroke engine is used to provide the maximum power from an engine of minimum volume. This is accomplished in the spark-ignition engine with the very simplest techniques (crankcase compression), and in the large compression-ignition engine the most sophisticated methods are applied (see below).

1880–1900: different types of fuel

By about 1880 the principles of the practical internal combustion engine were established. These early engines used coal-gas as their fuel, but this was inconvenient where the engine was to drive a vehicle, and, for stationary applications, because ready access to the gas mains was not available at all locations. Liquid fuels provided a solution to this problem, but satisfactory combustion required them to be vaporized before they were ignited.

This was (and is) accomplished in one of three ways: (a) carburetion: engine induction air passed over or through the fuel in a carburettor, which was independently invented in 1885 by Wilhelm Maybach and Karl Benz; (b) hot bulb engine: spraying the fuel on to a hot surface and passing the engine induction air over it (see below); (c) compression ignition: spray the fuel into the cylinder, relying for evaporation on the hot gases produced by compression of the air in the cylinder (the Diesel engine, see p. 311).

The application of the first method was limited between 1880 and 1900 because the necessary low volatility fuels (flash point $-12°C$ to $-10°C$; $10°F$ to $14°F$) were hazardous, resulting in legislation that restricted their use. This made hot bulb engines with fuels (e.g. kerosene) of high flash point (above $23°C$ or $75°F$) the most practical form of liquid-fuelled engine until the relevant legislation was changed.

The 'hot bulb' engine

The first liquid fuel engine was constructed by G. B. Brayton in 1872 in Boston, Massachusetts. It used a carburettor, and the fuel–air mixture, which was compressed before admission to the engine cylinder, was ignited by a flame. The next liquid fuel engine was built by W. D. Priestman of Hull, Yorkshire, in about 1885. This operated on the four-stroke Otto cycle and employed an external, exhaust gas heated vaporizer (flame heated for starting) into which the fuel was sprayed. The induction air passed through the vaporiser and the resulting mixture was ignited in the cylinder by an electric spark. Thermal efficiency was about 13 per cent (specific fuel consumption about 1lb mass/hphr; 0.61kg/kWhr).

Herbert Akroyd Stuart was the first (1890) to invent an engine, operating on the four-stroke Otto cycle, that made no use of an ignition source (spark or flame) and is, therefore, clearly related to the modern compression ignition engine. The vaporizer, or 'hot bulb', into which the fuel was sprayed was mounted on the cylinder head and connected to the cylinder by a narrow passage. It was heated either by hot cylinder cooling water, or by the exhaust gases (an external flame was used for starting). The induction air was drawn into the cylinder, and compressed, through the narrow connecting passage, into the vaporizer, where ignition occurred when a combustible fuel–air mixture was attained. This was self-ignition resulting from contact between the mixture and the hot walls of the vaporizer, and should not be confused with ignition due to the high air temperatures encountered in the compression ignition engine. The fuel consumption of Akroyd Stuart's engine was comparable to that of Priestman's, but it avoided the spark ignition (unreliable in those days) of the latter.

The hot bulb engine lasted in various forms until the late 1920s (often being called a semi-diesel, no doubt for advertising purposes) even though it was not as efficient as the compression ignition engine. It had the advantage of simplicity because it did not require the air compressor used in the early compression ignition engines, since the fuel was injected mechanically (so-called solid injection) near the beginning of the compression stroke, at a much lower pressure than the injection pressure of the compression ignition engine.

The petrol engine

The application of the internal combustion engine to transport requires, besides a liquid fuel, a high power–weight ratio, which in turn requires an engine operating at a high speed. Increasing the speed from 170rpm to 800rpm should reduce the engine weight by about 80 per cent. This was the objective of Gottlieb Daimler and Wilhelm Maybach.

Almost at the same time in 1886, Karl Benz and Daimler and Maybach produced single cylinder Otto cycle engines, using petrol and a carburattor, operating on a four-stroke cycle, that was light enough for use in the automobile. One important difference between the engines was the method of ignition. Daimler and Maybach employed the hot-tube igniter; Benz used spark ignition provided by a battery and an induction coil (the coil contacts were opened and closed independently of the engine speed, producing sparks at a steady rate).

Ignition and carburation underwent significant advances before the close of the century. The float-fed spray carburettor was introduced by Maybach in 1893 on the Daimler engine. In 1899, Daimler engines were converted from the hazardous flame-heated hot tube to Priestman's electric spark ignition with the spark generated by the low tension magneto invented by Robert Bosch.

The Diesel engine

A most important advance in internal combustion engine design was made by Rudolf Diesel with his invention of the compression ignition engine, so called because the fuel is introduced directly into the cylinder, in the form of a finely

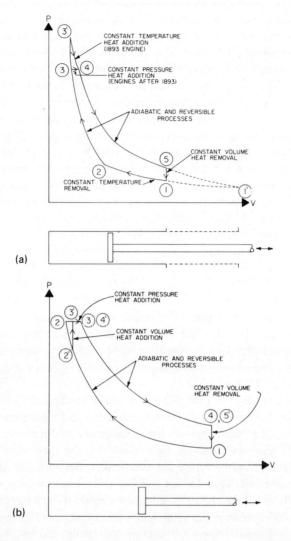

Figure 5.13

(a) The air-standard Diesel (modified Carnot cycle) plotted on coordinates of pressure (P) and volume (V). The true Carnot cycle (1'), (2), (3'), (1') was modified by Diesel as follows: (i) expansion was terminated at point (5) in order to provide a cylinder of reasonable dimensions (shown solid in the lower diagram, with the dashed lines showing the cylinder dimensions required by the true Carnot cycle); (ii) the constant temperature heat addition process (3')–(4) of the true Carnot cycle was replaced by the constant pressure heat addition process (3)–(4) in engines built after 1893 when it was found in the prototype to be impossible to produce a constant temperature combustion process.

(b) The air-standard compression ignition cycle plotted on coordinates of pressure (P) and volume (V). The heat addition (combustion in the real engine) processes tend to be different depending on the method of fuel injection. Research by Ricardo showed that fuel injected by air is much more finely divided than that resulting from solid injection, so burning is very fast and the rate of combustion can be controlled by the rate of fuel admission. Consequently, the pressure–volume diagram follows the path (2)–(3). With solid injection there is a delay before combustion starts, this is followed by a rapid rise in pressure (2')–(3'). At point (3') the pressure and temperature of the cylinder contents are so high that the remaining fuel ignites on entry, and, as with air injection, the rate of burning depends on the rate of fuel supply (3')–(4').

divided spray, at the end of the compression process, and is ignited by contact with the high temperature compressed air in the cylinder.

Diesel's original proposal, made in about 1890, was for an engine working on the Carnot cycle (see Appendix, p. 342), with a compression ratio of 50:1, a maximum pressure of 253bar (3675psi) and temperature of 800°C (1470°F), giving an air standard efficiency of 0.73 (Figure 5.13(a)). However, the mean effective pressure of the proposed cycle was very small, which meant that a slight deviation from the design conditions would prevent the engine from operating because it was unable to overcome the internal friction.

In 1892 the Maschinenfabrik Augsburg (later the Maschinenfabrik Augsburg-Nürnberg or MAN) contracted with Diesel to construct an experimental engine with a 15:1 design compression ratio in exchange for manufacturing rights. This engine was tested between July and September 1893. It was not able to run under its own power because the friction power exceeded the engine power, but it did demonstrate firstly that a charge of fuel could be ignited by compression alone; and, secondly, that when the fuel is injected it must be broken up into minute droplets, and that this was best achieved (at that time) if the fuel was sprayed into the cylinder by compressed air.

In spite of the problems, the engine was sufficiently promising to attract a number of licensing agreements in 1893. In February 1897 a fourth experimental engine (see Figure 5.13(a)) with a compression ratio of 11:1 was built, which, when tested, had a brake thermal efficiency of 30.7 per cent (fuel consumption 0.44lb mass/hphr; 0.27kg/kWhr), which was substantially higher than any contemporary heat engine. Further licences were taken up, but the move was premature because no licensee was able to build a reliable engine. Consequently, between 1897 and 1902 a major development programme was carried out by MAN to produce a satisfactory engine.

During the next ten years or so, the Diesel engine gradually found its place in the stationary engine field with engines of 15–75kW (20–100hp) that were more convenient and more economical than gas engines. However, mobile application did not appear practical because of the bulky air compressor required to produce the fuel injection 'air blast' and because of the low engine speed (piston speed: 3m/s or 590ft/min).

The modern descendant of the Diesel engine is better called a compression ignition engine because it does not operate on Diesel's theoretical cycle. Furthermore, beginning in the 1920s, when solid fuel injection was introduced (see p. 322) the mode of operation differed even more from the original Diesel concept (see Figure 5.13(b)).

The gas engine: 1880–1986

The history of the gas engine from 1880 onwards has two phases, the first lasting until about 1914, and the second from about 1925 to the present.

Until 1914 the history of the gas engine is very closely tied to the available fuel. In its earliest days (1860 to 1880) coal-gas was used (LHV: $19,371kJ/m^3$ or $520BTU/ft^3$), but it was expensive and restricted the engines to power outputs of no more than 55kW (75hp). In 1878, J. E. Dowson invented the suction gas producer which generated a gas mixture (LHV: $5029kJ/m^3$ or $135BTU/ft^3$) by the partial combustion of coal with air and steam, and costing about one-tenth the price of coal-gas. Because its lower cost more than compensated for its smaller heating value compared to coal-gas, the use of the gas engine increased, along with its size (up to 75kW/100hp).

In 1895, B. H. Thwaite demonstrated at the Glasgow Iron Works that blast furnace gas (approximate composition CO 35%, N 65%; and LHV $3725kJ/m^3$ or $100BTU/ft^3$) was a suitable engine fuel, and since it was otherwise a waste product of the iron smelting process, its small heating value was no disadvantage. The availability of large quantities of cheap gas and the need to drive big blast furnace blowers, resulted in the construction of extremely large gas engines for use in iron and steel works. One of the largest was a four-stroke cycle engine of 7500kW (10,000hp) that was installed at the South Chicago Works of the Illinois Steel Company in 1931. Nevertheless, by about 1914 it was clear that the gas engine could not compete with the compression ignition engine or the steam turbine.

It was the commercial exploitation of natural gas (mostly methane, with smaller quantities of other hydrocarbons; LHV: $39,113kJ/m^3$ or $1050BTU/ft^3$) in the United States, starting in the mid-1920s, that revived the fortunes of the gas engine. This gas has to be conveyed by pipeline from its source to its point of use, requiring compressors located at intervals. These can be driven by reciprocating engines (sometimes gas turbines) burning as fuel the gas being pumped. The availability of well-developed compression ignition engines, together with a fuel in the pipeline at a high pressure, suggested that these engines could be adapted to burn natural gas by replacing the compression ignition cylinder head by one incorporating a fuel injector, and either a spark plug or an oil fuel injector to provide a pilot flame, as an alternative to the electric spark. Modern forms of this engine operate with compression ratios as high as 13:1.

Supercharging: 1909–1930

Supercharging describes any technique intended to increase the density of the air (oxygen) supplied to an internal combustion engine. It can be accomplished by either raising its pressure by a compressor ('blower') or decreasing its temperature by the addition of volatile liquids; however the name is usually associated with the first method. So-called charge-air cooling has also been considered for compression ignition engines.

Supercharging was apparently first proposed by Daimler, but was not tried until 1878 when Dugald Clerk employed it on his two-stroke gas engine (see p. 307). The first significant use of supercharging was in aircraft engines, which operate at a low ambient pressure. This was first proposed by Auguste Rateau in 1914, and the idea was taken up about 1918 by, among others, S. A. Moss in the United States and A. Büchi in Switzerland. The supercharger has also been used extensively in racing cars, since its first application by Mercedes in 1922, to increase the power output by an engine of a given size. High power compression ignition engines have adopted supercharging so extensively, in order to increase the power-to-weight ratio, that supercharged engines of this type are the rule rather than the exception.

Supercharging can be provided by piston pumps, vane pumps, Roots compressors and the centrifugal compressor. They can be driven mechanically off the engine crankshaft, or, where space is not limited, electric motors and even steam turbines have been used. However, it is the gas turbine, as originally proposed by Rateau, using the exhaust gases from the engine, and driving a centrifugal compressor that has been the most significant development in supercharging (nowadays commonly called turbocharging).

The application of the gas turbine driven supercharger is not simple. Firstly, if the engine is to produce power, the gas turbine–compressor combination must have a high efficiency (more than 55 per cent) and the temperature of the exhaust gas entering the gas turbine must be greater than about 400°C (750°F). Secondly, difficulties can arise because of the overlap of the opening of the intake and exhaust valves, which is required to minimize the effects of valve and gas inertia and is, furthermore, required in the two-stroke engine to provide adequate scavenging. If, when this situation occurs the pressure in the exhaust manifold is higher than the intake pressure, then back flow can occur. This situation can be overcome in two ways. In constant pressure supercharging the volume of the exhaust system can be made large enough to keep its pressure essentially constant and never exceeding that in the inlet manifold. This method, invented by Rateau in 1914, is used in aircraft engines. In pulse supercharging the exhaust manifold is divided so that cylinders exhausting into a particular manifold do not produce interfering pressure pulses. Büchi devised this method in about 1925 as a result of tests that he started in 1909.

The reasons for choosing a particular supercharging technique are complex, but in general pulse supercharging is preferred in two-stroke cycle engines, because it makes better use of the energy in the exhaust gas, thereby compensating for the decrease in the gap temperature caused by the excess air necessary to ensure efficient cylinder scavenging. Pulse supercharging involves complicated exhaust manifold arrangements, so the constant pressure method is desirable where first costs must be minimized, as in automobile applications.

The spark ignition engine

The spark ignition engine is in almost universal use for automobiles, and it is also used as an aircraft engine when the power and speed capabilities of the gas-turbine are not required. The modern spark ignition engine has been subjected in its lifetime to some of the most intensive scientific study of any thermal prime mover because of its sensitivity to the fuel used, and because it is a potent and very widespread source of atmospheric pollutants.

1900–1920: The high speed spark ignition engine

The need to produce a simple and reliable automobile engine led to changes in combustion chamber shape, and increase in the number of cylinders (from 1900 to 1915), improved ignition systems, decreasing gasoline volatility, better fuel supply systems, and forced lubrication (1908).

Combustion chambers were continually changing in order to provide an engine that was simple to manufacture and maintain, had satisfactory power output and fuel economy, and avoided detonation (see below). The most widely used combustion chamber was the side-valve or L-head, which, however, had a strong tendency to produce detonation. Research conducted by H. R. Ricardo between 1912 and 1918 showed that the compression ratio at which detonation first occurred could be raised from 4.0 to 4.8 if the turbulence of the air–fuel charge was increased and the spark plug was located nearer the centre of the cylinder.

1920–1945: detonation

This particular phenomenon produces an unpleasant noise, but more importantly, it indirectly has a deleterious effect on the fuel consumption, because it limits the maximum useable engine compression ratio.

Detonation ('knock') was first identified and differentiated from pre-ignition (due to hot surfaces) in 1906 by Dugald Clerk and H. R. Ricardo. It is defined as self-ignition of the end gas (the unburnt portion of the charge) as a consequence of its temperature and pressure having been increased by compression resulting from the expansion of the burning portion of the charge. This produces high frequency oscillations in pressure, and hence the characteristic engine noise.

Detonation first became a significant consideration in 1905 as more volatile petrol came into use, but it was not until the First World War that it became critical as a result of its effect on aircraft engine performance. Consequently, a group in the United States, under the direction of C. F. Kettering of the Dayton Metal Products Company, developed a special aviation fuel consisting of a mixture of cyclohexane and benzene which allowed aircraft engines to operate without detonation at compression ratios up to 8:1.

Following the war, work on solving the detonation problem adopted two approaches: firstly, the search for a chemical additive (dope) which would suppress detonation, was undertaken by Kettering, T. A. Boyd, and Thomas Midgley, all of General Motors. The second approach, involving the best combination of engine design parameters and fuel to minimize detonation, was initiated by Ricardo. Boyd, in 1919, starting from his discovery in 1916 that iodine could suppress detonation, found that aniline was also suitable. However, neither compound was entirely practical (iodine was corrosive and aniline had a bad smell). Eventually in 1921, as a result of a lengthy search, lead tetraethyl was identified as a suitable detonation suppressant. This historic discovery made the modern high compression spark ignition engine possible (see Figure 5.14).

In England, Ricardo formed his own company in 1917 with the objective of finding the combination of fuel and combustion chamber form that would eliminate detonation. As a result he decided that the detonation tendencies of engines and of fuels must be defined in terms of standard fuels and standard engines. In 1924 he introduced a mixture of toluene (resists detonation) and pure heptane (readily detonates). The detonation characteristics of a candidate fuel could then be expressed in terms of the percentage of toluene in the standard fuel that matches the detonation characteristics of the given fuel. Toluene was later (1927) replaced by octane, hence the now-familiar octane number.

Since detonation depends on the engine compression ratio, as well as the fuel, the candidate fuel must be tested in a standard engine in which the compression ratio may be varied while the engine is running. Ricardo was first to build such an engine (in 1919). Eventually in 1931 a committee (the Co-operative Fuel Research Committee) of interested parties (engine manufacturers, oil refiners) in the United States, under the leadership of H. L. Horning of the Waukesha Engine Company, produced a standard engine design, known

PART TWO: POWER AND ENGINEERING

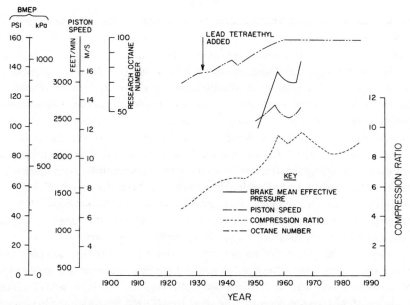

Figure 5.14: Historical trend of the performance parameters of four-stroke cycle spark ignition automotive engines 1920–85. Unfortunately, the data available to the author were limited as shown.
The curves are based on data in C. F. Taylor *The Internal Combustion Engine in Theory and Practice*, vol. 2, revised edition (The MIT Press, Cambridge, Mass., 1985) and *Automotive Industries*, vol. 165, no. 6 (1985), p. 521.

today as the CFR Fuel Research Engine, which is now a world-wide standard for detonation research and testing.

1945–1986: power, efficiency and cleanliness

Initially, after the war ended in 1945, engines of ever increasing power, and mean effective pressure were produced (Figure 5.14). There was also a progressive increase in the compression ratio (Figure 5.14) and the octane number of available petrols. This trend was arrested by the combined effects of environmental concerns (since 1963) and the 'energy crisis' (1973).

Scientific investigations, particularly in southern California, in the late 1940s and early 1950s had demonstrated that hydrocarbons, carbon monoxide, and various oxides of nitrogen, produced by automobile spark ignition engines, were important air pollutants. It was clear that with the anticipated increase in the number of motor vehicles, particularly in large urban areas, control of emissions from their engines was essential. The government of the State of California, and then the United States government, enacted legislation on automobile emissions, and comparable regulations were imposed in many countries outside the United States.

The environmental regulations imposed by the US government on the automobile engine were embodied in the so-called 1963 Clean Air Act, amended in 1968, 1970 and 1977. The practical realization of the various techniques devised to meet the legal requirements has resulted in the need to solve some of the most challenging technical problems presented by the internal combustion engine.

There are five main emission control methods. (a) Air injection into the exhaust manifold to burn hydrocarbons and carbon dioxide. (b) Exhaust gas recirculation (EGR), which lowers nitrogen oxides emitted by the engine by diluting the intake fuel–air mixture, thus decreasing the maximum cylinder gas temperature. (c) Oxidizing catalytic converter, which is located in the exhaust line and assists the oxidation of hydrocarbons and carbon dioxide. In 1981 a so-called three-way catalytic converter, which additionally converts oxides of nitrogen to nitrogen, was introduced. This has allowed the engine to operate at conditions suitable for minimum fuel consumption. (d) Electro-mechanical carburettor which controls the engine fuel supply by sensing the oxygen level in the exhaust gas before it enters the three-way catalytic converter. The sensor and the carburettor are linked through a microcomputer to provide a control system that maintains the air–fuel ratio at 14.6 ± 0.2. (e) Fuel injection into the inlet manifold or at the cylinder intake ports, which allows even more precise control of the air–fuel ratio than the electro-mechanical carburettor.

The effort to meet emission standards in mass-produced engines has resulted in a complicated engine that has a higher first cost and is expensive to maintain. It is possible that this situation could be avoided if the combustion processes in the engine cylinder could be fundamentally changed. It has been known for some time that a marked reduction in carbon monoxide and oxides of nitrogen emissions (but not hydrocarbons) can be obtained if the fuel–air mixture was weaker than that required by purely chemical considerations (the so-called stoichiometric mixture): however, the engine has a low efficiency and runs very roughly. This can be avoided by using a stratified charge (see p. 306–7), and a substantial effort has been made by automobile manufacturers to develop an engine of this type. Because its operation is sensitive to variations in design and conditions of use, it has been found difficult to employ this type of engine in a mass-produced vehicle: the only commercially available engine in the late 1980s was the Honda Compound Vortex Controlled Combustion (CVCC) engine.

Aircraft engines

The contribution of the aircraft engine to the development of the internal combustion engine has been particularly important in producing engines of low specific weight (kg/kW). Many of the features that have contributed to this development have reappeared in the 1980s in low specific weight automobile engines with minimum fuel consumption and pollution.

The automobile engines available to the earliest aircraft builders (c.1900) were too heavy and of too low power, and the first practical engines produced by the Wright brothers (see p. 622) and by C. M. Manly for Langley's 'Aerodrome' of 1903, were designed specifically for use in aeroplanes. Manly's engine was remarkable in having a specific weight of 1.45kg/kW (2.38lb mass/hp), which was not improved on until the American 'Liberty' engine of 1918. The Manly engine was unique in its anticipation of many features that became standard on later aircraft engines: a radial arrangement of the cylinders with a master connecting rod, the cam and valve gear arrangement, and crankcase, cylinders and other parts that were machined all over to carefully controlled dimensions.

Aircraft engines following the initial period (c.1903–14) were of two main types. The air-cooled radial engine was based on the Gnome rotary (cylinders rotated around the crank shaft) radial engine, which was the most popular aircraft engine up to the First World War and was used by both sides in the conflict. It was mostly, but not exclusively, used by civilian aircraft operators, because of its simpler cooling arrangements. The first large radial air-cooled engine of modern design was the Pratt and Whitney 'Wasp' (1927). It employed a mechanically driven centrifugal supercharger, as well as a forged and machined aluminium crankcase and cylinder head.

The liquid-cooled vee engine had its origins (c.1915) in the Hispano-Suiza engine (1.9kg/kW or 3.1lb mass/hp). The basic structure consisted of a cast aluminium crankcase with en bloc water jackets. Initially, aircraft engines used water for cooling, but this was changed c.1932 by the Curtiss Company in the US when they introduced a water–ethylene glycol mixture. This boiled above the boiling point of water, so that the higher temperature difference between the coolant and the ambient air allowed the use of a smaller, lower drag radiator.

The Curtiss Company, starting in 1920, developed a succession of in-line engines, based on the Hispano-Suiza engine, which had considerable success in racing, particularly in the Schneider Trophy sea-plane races (see p. 630). This led the Rolls-Royce Company in Britain to design the 'Kestrel' engine in 1927 (1.1kg/kW or 1.77lb mass/hp), followed by the 'R' (for racing) type, and then the 'Merlin' (0.78kg/kW or 1.28lb mass/hp). This last engine, together with the Wright 'Cyclone' and Wright 'Wasp' radial engines, represent the pinnacle of the reciprocating spark ignition aircraft engine development.

Figure 5.15 summarizes the development of the aircraft piston engine between 1900 and 1960.

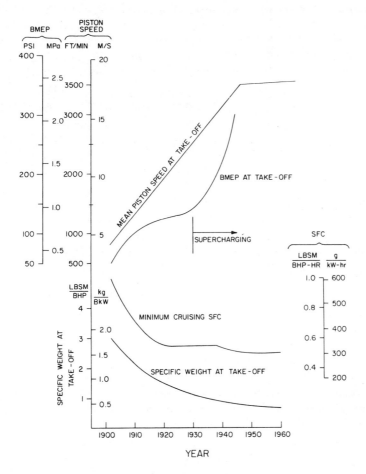

Figure 5.15: Historical trend of the performance parameters of aircraft spark-ignition engines 1900–60.
Adapted with permission from: C. F. Taylor 'Aircraft Propulsion: A Review of the Evolution of Aircraft Power Plants', *Smithsonian Report for 1962* (Smithsonian Institution, Washington, D.C., 1962).

The compression ignition engine

The modern compression ignition engine is the most efficient thermal prime mover available today, and the pollutants in the exhaust are less than those produced by spark ignition engines. It is the engine of choice in non-aircraft transportation applications where low operating costs are more important than the first cost. However, the compression ignition engine was not readily adaptable to transport applications until about 1930 when engine speeds increased and solid injection replaced air-blast fuel injection.

PART TWO: POWER AND ENGINEERING

1920–1945: The high speed compression ignition engine

Compression ignition engines for application to transport require a compact, lightweight, high speed (piston speeds exceeding 7.6m/s; 1500ft/min), high efficiency internal combustion engine, with solid injection of the fuel. Such an engine was developed between 1920 and 1930. At the time it was a widely held belief that high speed in the compression ignition engine was impossible, because the combustion processes were too slow. In addition, Diesel had found, and experience in the 1920s confirmed, that solid injection methods required careful study, otherwise the engine ran roughly, with poor fuel consumption.

The requirements for a successful high speed, solid injection engine were elucidated as a result of careful studies of the details of the combustion process. One of the most important contributions to this work was H. R. Ricardo. He found among other things that careful control of the air motion in the cylinder was important, consequently the introduction of solid injection coincided with development of many different types of combustion chamber.

1945–1986: Power, efficiency and cleanliness

In the years following the Second World War the history of the compression ignition engine has been dominated by the development of engines of ever-increasing specific power output (kW/kg or hp/lb mass), which has been due to the rising mean effective pressure as a result of the universal adoption of turbocharging (Figure 5.16). Consequently, attention has to be paid in design to handling high mechanical and thermal loads.

The design of the compression ignition engine has settled into a pattern where most low speed (less than 250rpm), large stationary or marine engines operate on the two-stroke cycle. High speed engines, with speeds in excess of 1200rpm are four-stroke, and medium speed engines (250–1200rpm) usually operate on the four-stroke cycle, but occasional examples of engines using the two-stroke cycle can be found (e.g. General Motors locomotive engines).

Because the fuel consumption of the compression ignition engine is not as sensitive to pollution control measures as the spark ignition engine, a number of passenger automobiles were offered in the mid-1970s with the former. This interest seems likely to revive as oil supplies dwindle toward the close of the twentieth century.

The need to develop a compression ignition engine with a high thermal efficiency has resulted, among other things, in a re-appraisal of the combustion chamber form. In particular, the direct injection combustion chambers, with their lower heat losses compared to the indirect injection chambers (e.g. Ricardo Comet), have been receiving considerable attention. However, prob-

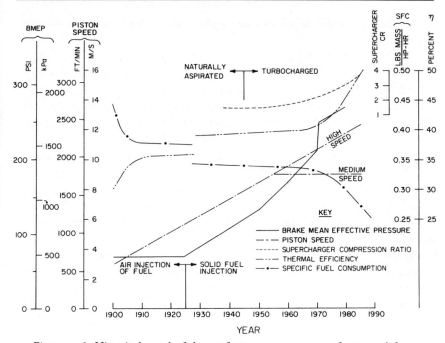

Figure 5.16: Historical trend of the performance parameters of two- and four-stroke cycle compression ignition engines 1900–85.
The curves are based on data in J. W. Anderson, 'Diesel – Fifty Years of Progress', *Diesel Progress*, vol. 14, May (1948), pp. 5–331, and Anonymous, 'Fifty Years of Diesel Progress, 1935–1985', *Diesel Progress*, vol. 37, 50th Anniversary Supplement, July (1985), pp. 12–182.

ably of greater importance is the clear indication, as has already occurred in the spark ignition engine, of a trend towards microprocessor control of the engine.

Marine compression ignition engine

Approximately half the world's tonnage of shipping is now propelled by compression ignition engines, and the percentage is increasing. This form of power was first applied in 1912 to ocean-going shipping in the Danish *Selandia*. This was a 7500 tonne freighter fitted with two Burmeister & Wain four-stroke, single acting, eight-cylinder engines of 1340SkW (1800shp) each.

There are a number of advantages in marine applications of the compression ignition engine compared to the reciprocating steam engine. (a) The oil fuel is easier to handle and store than coal: it can be carried in the bilges, whereas coal bunkers use cargo space. (b) The oil fuel has a larger heating value than coal (about 25 per cent), and weighs less. (c) The compression ignition engine is about half the weight of a boiler and reciprocating steam engine of the same power.

The 1920s was the heyday of the four-stroke single-acting crosshead engine with air injection, and probably the most important manufacturers of this type of engine at that time were Burmeister & Wain of Copenhagen and their licensees Harland & Wolff, the Belfast shipbuilding and engineering company. However, by the mid-1920s these two companies had turned their attention to the four-stroke cycle double-acting engine, and they jointly developed an experimental single cylinder engine that produced about 745SkW (1000shp), about 150kW (200hp) more than the then probable maximum value for a single-acting four-stroke engine. However, in spite of considerable engineering effort, this type of engine was never successful. The main difficulty was to design the complicated lower cylinder head to withstand the thermal stresses induced by its form.

By the mid-1920s the two-stroke single-acting engine started to compete seriously with the four-stroke engine. The first engine of this type in an ocean-going vessel was fitted in the *Monte Penedo*, which was built for Brazilian owners, and had two 630SkW (850shp) Sulzer engines. By 1935 the two-stroke engine had essentially eliminated the four-stroke engine from consideration for marine use when engines of more than 1500kW (2000hp) were required. (By 1930 Burmeister & Wain and Harland & Wolff, established producers of four-stroke engines, were offering the two-stroke type.)

Inevitably, demands for even greater power output directed the attention of owners and builders to the double-acting two-stroke engine. An engine of this type was first built by Krupp in 1909; this firm and Maschinenfabrik Augsburg-Nürnberg were each commissioned by the German navy in 1910 to produce a double-acting lightweight engine for use in warships. The MAN engine was a 8950SkW (12,000shp) machine with six cylinders. Work ceased on the project in 1914 and it was broken up in 1918 by order of the Allied Control Commission. However, this type of engine was revived in the 'Deutschland' class of pocket battleship (the *Graf Spee* was a member of the class), which were fitted with eight double-acting two-stroke solid injection nine-cylinder engines each rated for 4550BkW (6100bhp) at 450rpm.

The double-acting two-stroke engine suffered somewhat from the faults of the four-stroke double-acting engine, except that the use of port scavenging reduced the number of valve openings required in the lower cylinder head. Although this type of engine was installed in ships up to the mid-1950s, it has now, because of the lower head problems, been displaced by single-acting and opposed piston types of engine.

The opposed piston engine was introduced in 1898 as a two-stroke gas engine by W. von Oechelhäuser and was adapted to burn oil by Hugo Junkers. It had been applied to ship propulsion in Germany without success, but in spite of this the British firm of William Doxford & Sons, marine steam engine builders, decided in about 1912 to investigate the possibility of using it for marine service. The first production engine (four cylinders, 2240IkW/3000ihp, 75rpm) went to sea in 1921 in the 9140 tonne *Yngaren* (Swedish Transatlantic Lines). The Doxford engine (Figure 5.17(a)) has probably been one of the most successful marine internal combustion engines ever produced, but in spite of this, production ceased in 1980 (undoubtedly as a consequence of the shift in the centre of shipbuilding activity to the Far East). Opposed piston engines were also built by Burmeister & Wain–Harland & Wolff, and Cammell-Laird, among others.

The development of the marine compression ignition engine since 1945 has concentrated on increasing the power output per cylinder, the thermal efficiency, and the ability to burn low-quality fuels. Two-stroke single-acting engines of this type (Figure 5.17(b)) produce up to 3505bkW per cylinder (4700 bhp per cylinder) with a specific fuel consumption of about 157 grams/BkWhr (0.258lb mass/bhphr) which for a fuel with a heating value of 44,184J/gram (19,000btu/lb mass) corresponds to a thermal efficiency of 52 per cent. This performance makes engines of this type the most efficient of all thermal prime movers.

Wankel engine

Rotary internal combustion engines in which all the processes of the cycle occur in one structural containment (or body) are attractive because they have the inherent dynamic balance characteristics of the turbine, but avoid the need to provide separate components for each of the cycle processes. The first attempt to produce an engine of this sort is probably due to an English engineer Umbleby who proposed the adaptation of a steam rotary engine that had been invented by J. F. Cooley of Alston, Massachusetts, in 1903. This particular engine was essentially the last in a long line of steam rotary engines that appears to have its origins in a proposal due to Watt in 1782, which was the last of various types of rotary engines considered by Watt and his partner Boulton between 1766 and 1782 in a frustrating, and eventually unsuccessful, attempt to obtain rotary motion without using the piston and cylinder. Rotary steam engines never successfully competed with the reciprocating steam engine or, for that matter, the steam turbine. In general the necessary technological sophistication, materials and financial resources were not available. This is confirmed by the development history of the most successful of the rotary internal combustion engines produced so far, the Wankel engine. Substantial resources have been devoted to the development of this engine since its original conception (1954), but it still (1988) cannot be said to be the equal of or superior to the reciprocat-

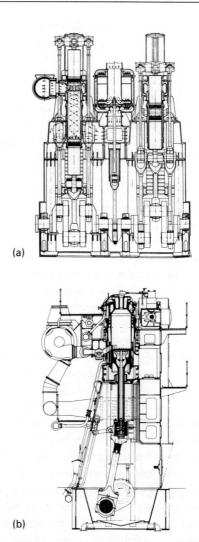

(a)

(b)

Figure 5.17
(a) Doxford opposed piston two-stroke compression ignition engine, *c*.1950. The figure shows two cylinders with a piston-type scavenge pump between the power cylinders. The arrows trace the path of the scavenge air through the pump, to the inlet manifold, through the intake ports, traversing the space between the pistons in a spiral uniflow pattern, and leaving through the exhaust ports into the exhaust manifold. In the right hand cylinder the pistons are at inner dead centre and their faces are so shaped that an almost spherical combustion space is formed.
Reproduced with permission from W. H. Purdie, 'Thirty Years' Development of Opposed-piston Propelling Machinery', *Proceedings of the Institution of Mechanical Engineers*, vol. 162 (1950), pp. 446–64).

(b) End section of a two-stroke, single acting, crosshead type compression ignition engine, 760 mm (30 in.) bore by 1550 mm (61 in.) stroke, with power output per cylinder of 1340kW (1800hp). This is actually a single cylinder experimental engine (type 1RSA76) built by Sulzer Brothers in the late 1950s. The engine is fitted with a turbo-charger (top left-hand) supplying air (through an after-cooler, shown dashed) to the inlet plenum located on the right-hand side of the engine. The lower side of the piston acts as a scavenge pump. Loop scavenging is used with incoming air introduced through the right-hand ports and the exhaust leaving through the left-hand ports. Super-charging is by the pulse (Büchi) method. Direct fuel injection is by the injector mounted in the cylinder head. The piston is water cooled with coolant supply and removal through the hollow piston rod and a swinging link connected to the crosshead.

Reproduced with permission from A. Steiger, 'The Significance of Thermal Loading on Turbocharged Two-Stroke Diesel Engines', *Sulzer Technical Review*, vol. 51, no. 3 (1969), pp. 141–159.

ing engine, except in regard to the previously noted characteristics of good balance and high power-to-weight ratio.

The rotary engine is an elegant concept that undoubtedly appeals to the aesthetic sense of the engineer. It has attracted a number of inventors, but the difficulties of translating the idea into a working engine have been such that only the engine conceived by Felix Wankel in 1954 has been developed to the point where it has been extensively used in automobiles (most notably by the Mazda Motor Corporation, formerly Toyo Kogyo Co. Ltd).

In the rotary engine the piston and cylinder of the reciprocating engine are replaced by working chambers and rotors enclosed by a stationary housing. The rotor and the working chamber are specially shaped to allow the working substance that is trapped between these two components to undergo the processes of some internal combustion cycle. The exact shapes of the rotor and the working chamber depend on the kinematics of the linkage that connects the rotor to the drive shaft. In the case of the Wankel engine the working chamber is shaped like a figure of eight with a fat waist (actually an epitrochoid), and this contains a three-cornered rotor that orbits the drive-shaft so that all three corners are permanently in contact with the walls of the working chamber. Ports allow for the admission of the mixture (carburetted engine), or air (engine with in-cylinder injection), and the exhaust of the products of combustion. For a spark ignition engine the necessary spark plug is located at an appropriate point in the wall of the working chamber.

When working on a four-stroke Otto cycle with spark ignition, the space between the rotor and the inner surface (the working surface) of the working chamber continually changes in size, shape and position. These volumetric changes provide for charge induction, compression, expansion and exhaust. The cyclic sequence of these processes is followed by all three faces of the rotor, so a single rotor engine is acting like a three-cylinder reciprocating engine. As in the two-stroke reciprocating engine, no valves are required on the inlet and exhaust ports, since the gas flow is controlled by the rotor motion (cf. piston motion in the two-stroke reciprocating engine). This feature of the engine, among others, makes it mechanically quite simple.

The Wankel engine had its origins in a rotary compressor that was invented by Felix Wankel between 1951 and 1954. He operated a research and development company that carried out contract work for various German engine manufacturing companies. He had little formal engineering training, but had served German industry since 1922. One of his most successful inventions prior to the Wankel engine was a rotating disc valve, invented in 1919, that replaced the conventional poppet valves of the reciprocating engine and required the solution of particularly difficult sealing problems; this background was probably of great assistance in the development of the Wankel engine, where the sealing at the rotor apex, its point of contact with the working surface, is especially critical.

The German NSU company obtained the rights to the Wankel disc valve for use in racing motor-cycles, and at about the same time (1954) they also considered the use of a Wankel compressor as a supercharger. It was only a short and obvious step to investigate the possibility of adapting the epitrochoidal chamber and triangular rotor of the compressor for use as an internal combustion engine. NSU and Wankel then spent the period between 1954 and 1958 in an intensive development effort on the Wankel engine. Ultimately a car, the NSU Wankel Spider (1964–7), was produced.

The Curtiss-Wright company in the United States acquired the American rights to this engine, as have a number of other companies in various parts of the world. The most successful of these licensees has been the Mazda Motor Corporation in Japan. This company has manufactured since 1967 a line of very successful cars using their version of the Wankel engine.

Experience with the Wankel engine has shown that its fuel economy and emissions performance are not as good as those of the reciprocating engine. The former is strongly affected by heat losses (the rotary engine has a larger combustion chamber surface-to-volume ratio than the reciprocating engine), by friction at the rotor seals, and by difficulties in optimizing the inlet port configuration for all engine speeds. The Wankel engine has a lower oxides of nitrogen content in the exhaust gases, but more hydrocarbons than the reciprocating engine. Initially the high hydrocarbon level was reduced by oxidizing this component in a thermal reactor ('afterburner') located at the exhaust port, but more recently a two-bed catalyst in the exhaust system has been found to reduce the hydrocarbon content of the engine exhaust gases.

The Mazda company appears to be still interested in the Wankel engine, so it appears likely that this type of rotary engine will continue to be built for the foreseeable future.

Figures 5.14 and 5.16 summarize the progress in internal combustion engine performance from 1900 to 1986. The graphs refer to spark ignition and compression engines respectively, with performance data for both two-stroke and four-stroke cycle engines included. The important effect of the fuel injection method and the use of supercharging on the performance of the compression ignition engine should be noted.

GAS TURBINES

The gas turbine appears to be the ideal prime mover, both mechanically and thermally, since it involves only rotary motion, with its consequent advantages, together with internal combustion, which avoids the drawbacks associated with steam boiler plant. Because of this superiority to other forms of prime mover, the gas turbine has been of interest to engineers from an early date, with the first patent being issued to one John Barber in England in 1791, although it is very unlikely that a practical device based on his proposed machine was ever built.

The gas turbine can, in principle, be based on either of two air standard cycles, see Figure 5.18: the constant volume heat addition (Lenoir) cycle and the constant pressure heat addition (Brayton or Joule) cycle. The latter is used exclusively as the basis for modern gas turbines, although considerable effort was expended in the early years of this century on developing machines based on the Lenoir cycle.

Both of the air standard gas turbine cycles incorporate a heater, an expansion turbine and a cooler (see Figure 5.18). The constant pressure cycle (Figure 5.18(a)), requires, in addition, a compressor. The lack of this latter component in the constant volume cycle (Figure 5.18(b)) gives it two advantages over the constant pressure cycle: the air standard cycle efficiency is higher for comparable conditions; and it avoids the need for a high efficiency rotary compressor (very difficult to build with satisfactory efficiency), which are used in gas turbines working on the constant pressure cycle. The disadvantage of the constant volume cycle is the unsteady nature of the heating process, which leads to unattractive mechanical complications.

Working gas turbines have been built in which heat exchangers are employed as the heaters and coolers, respectively, and the working fluid (usually air or helium) is continually circulated in a closed system, i.e., a closed cycle. A commercial closed cycle machine (12.5MW), using air as the working fluid, which is no longer operational, was placed in service at St Denis, France, in 1952. More recently a 50MW helium machine, using coal combustion as a heat source, was built at Oberhausen, West Germany. The motivation for using helium is the possibility of using the gas turbine to extract energy from a power-producing

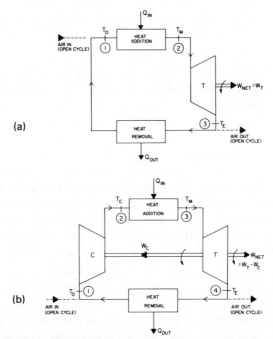

Figure 5.18: Schematic arrangement of the components of the air standard gas turbine cycles. In the open cycle the heat removal process is omitted. An actual gas turbine working on the open cycle would replace the heat addition process by a combustion chamber. (a) Constant volume heat addition (Lenoir) cycle. (b) Constant pressure heat addition (Brayton or Joule) cycle.

nuclear reactor, because, unlike air, helium is not activated in its passage through the reactor core.

The majority of gas turbines are based on the so-called open cycle (see Figure 5.18), where air is drawn into the machine from, and discharged to, the ambient. In the practical open cycle gas turbine, the heater is replaced by a combustion chamber (or combustor) in which a fuel (liquid, gaseous or solid) is burned in the air supplied by the compressor. This means that the expansion turbine handles the products of combustion, which, because of their temperature and tendency to react chemically with the materials of this component, has resulted in a major effort since 1940 to develop suitable turbine blade and disc materials.

For non-aircraft gas turbine applications the products of combustion are discharged at low velocity through an exhaust stack to the surroundings. In the aircraft turbojet the gases leaving the expansion turbine are accelerated by discharging them to the ambient through a nozzle, and this provides the desired thrust.

Early attempts (1900–10) to build gas turbines based on the constant pressure heating cycle were unsuccessful because it was impossible, at that time, to build a compressor of sufficiently high efficiency. The strong influence of the compression process on the cycle efficiency is a result of the large volume of air that must be handled by the compressor. This, in turn, results from the need, because of material limitations at that time, to limit the cycle maximum temperature (T_M) to 538°C (1000°F). As an alternative to employing large amounts of air, the combustor and expansion turbine can be cooled. This was the approach used by Holzwarth, who constructed a number of gas turbines based on the constant volume heating cycle between 1908 and 1933 (see below). He was motivated to use this method because of the inadequacies, already referred to, of contemporary compressors.

The necessary compressor efficiency requirements that had to be attained by the early gas turbines can be demonstrated as follows: suppose an air standard cycle thermal efficiency of 20 per cent is desired, which is comparable to that achieved in actual steam power plants in 1910 (see Figure 5.10). Further suppose the maximum cycle temperature is to be 530°C (1000°F), which is attainable if the materials used in the expansion turbine are chosen with care, and if the turbine blades are cooled (although it is doubtful that appropriate methods were available at the time). Then the efficiency of the compressor (assumed, for simplicity, equal to that of the expansion turbine) would have to be 88 per cent. This is high for a centrifugal compressor now, and was certainly not attainable in the early years of this century when typical efficiencies were near 60 per cent.

The most successful of the early gas turbines were based on the constant volume heating cycle in which high pressures at the expansion turbine inlet are obtained by constant volume heating instead of using a compressor. In theory, no compression work is required, which avoids the limitations, mentioned above, of the constant pressure cycle, but in practice it is desirable to employ a compressor, albeit of very modest performance, in order to introduce a fresh charge into the combustor and to scavenge the 'end gases'.

The apparent advantages, in the cycle thermal efficiency and in the compression process, that made the constant volume cycle attractive to the early gas turbine engineers, unfortunately brought with it some very undesirable complications. Constant volume combustion must take place in a closed chamber with valves for admitting the air and discharging the products of combustion to the expansion turbine. Furthermore, the unsteady combustion process is not well matched to the steady flow expansion process. In spite of this a number of moderately successful turbines based on the constant volume cycle were built.

Early development

At the time (about 1900 to 1910) that engineers started seriously to consider the gas turbine as a candidate prime mover, neither the necessary understanding of turbo-machinery fluid mechanics nor satisfactory materials were available. Nevertheless, the advantages of the gas turbine were such that a number of machines were constructed and operated, notably by Armengaud and Lemale (1903–6), Holzwarth (1908–33), Karavodine (1908), and Stolze (1900–4). The Armengaud and Lemale gas turbine, and that due to Stolze, used the constant pressure heat addition cycle, while the Holzwarth and Karavodine turbines used the constant volume heat addition cycle. A significant feature of these machines was the use of water cooling of the expansion turbine in order to decrease the amount of air that would otherwise have been required in order to control its inlet temperature.

Holzwarth's was the most successful of the early gas turbines and a number of versions were built between 1908 and 1933 under his supervision. The final model, a 5000kW machine, was built by the Brown Boveri Company and placed in service in a steelworks at Hamborn, Germany, in 1933. This experience apparently demonstrated to Brown Boveri the superiority of turbines based on the constant pressure combustion cycle and led to their involvement in the development of this type; the company was to become one of the major contributors to gas turbine technology.

1933–1946: beginnings of the modern gas turbine

There was a substantial hiatus in gas turbine activity following the early attempts to construct the machines described above, while efforts were made to develop high efficiency compressors with an adequate compression ratio (at least 4:1). In addition, during this period, as a consequence of their work with Holzwarth's Hamborn gas turbine, the Brown Boveri Company took an important step toward the development of a practical gas turbine based on the constant pressure heat addition cycle. It was noted that in the Hamborn machine, because of the high velocity and pressure of the combustion gases, the heat transfer rates in the combustion chambers and turbine inlet nozzles were extremely high. This suggested that it might be possible to construct a boiler with a large output per unit of heat transfer surface. The gas (expansion) turbine became an essential element of the device by employing the products of combustion leaving the boiler to drive the compressor, which, in turn, supplied the high pressure, high velocity air to the boiler. Brown Boveri called this supercharged boiler the Velox boiler: it sold in substantial numbers between 1933 and 1965. The supercharging principle, and hence the gas turbine, found a further application in 1936 when it was used in the Houdry catalytic converters being installed at the Marcus Hook, Pennsylvania, refinery of the Sun Oil Company.

While these applications of the gas turbine did not involve the production of useful amounts of work, the knowledge gained by Brown Boveri in the development of the Velox boiler, and their association with the Houdry process, placed it in a strong position to construct a power-producing gas turbine. This came about in 1939 when the first commercial gas turbine-driven electric generator was placed in service in Neuchâtel, Switzerland. The set was for emergency standby purposes and it was only expected to run on an intermittent basis, an ideal application for a new and untried machine.

During the time that Brown Boveri were developing their land-based gas turbine, Frank Whittle, a British air force officer, was working on aircraft gas turbines. (A comparable or even larger effort was under way in Germany at this time, led by Hans von Ohain, to develop an aircraft gas turbine. However, its influence on gas turbine history appears to have been slight, probably because of Germany's defeat in the Second World War and the consequent dominance of Anglo-American technology.) Whittle's work, which was to revolutionize aircraft propulsion and gas turbine technology in general, was a consequence of his recognition that the thermal efficiency of the gas turbine increases with altitude, which is the opposite of the behaviour of the reciprocating internal combustion engine.

In 1930, Whittle obtained a patent on an aircraft gas turbine jet propulsion engine. He was unable at that time to interest the British government in supporting the effort to develop a working engine, but eventually, in 1936, he raised some private capital. At this point the British air force recognized the potential in Whittle's ideas and released him from his regular duties so that he could devote all his time to the development of the gas turbine engine.

Whittle commenced work, with the help of the British Thomson-Houston Company, on his first engine (see Figure 5.19(a)) in 1936. It employed a double-sided centrifugal compressor, a combustion system, and a single stage impulse expansion turbine with an output of about 3000hp. The compression ratio was 4:1, much higher that the 2:1 available in contemporary centrifugal compressors. Even more remarkable was the combustion system, which involved volumetric heat release rates far greater than previously experienced in a continuous flow combustor. The engine first ran on 12 April 1937.

In the summer of 1939 the British government placed an order for an engine, called W1, that would be tested in an aircraft built specially for the purpose. It first flew on 14 May 1941. (It was not the first jet propelled aircraft to fly: a German Heinkel HeS3 engine first flew in August 1939.) This engine, which is shown in Figure 5.19(b), was, together with its successor the W2, a remarkable technical achievement. A modified version of the latter engine, the W2B, became the basis of the American version of the Whittle engine, the IA, which was built by the General Electric Company at Lynn, Massachusetts.

Meanwhile, the success of Brown Boveri with the gas turbine for its Velox boiler stimulated the interest of other engineers, notably in the Large Steam Turbine group of the General Electric Company in Schenectady, New York.

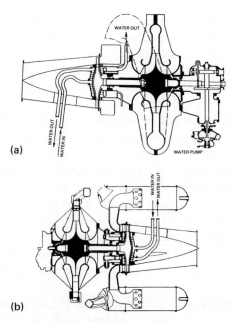

Figure 5.19: Early examples of Whittle gas turbine engines.
(a) First model of the Whittle engine which was operated between April and August 1937. A single combustion chamber was located in the air duct (chain dotted) carrying the air from the compressor to the expansion turbine. Note the water-cooled turbine disc.

(b) W1 engine, the first flight engine, which was operated between January 1940 and May 1941. Ten separate combustion chambers were used. It is not clear if water cooling was used during flight.
Reproduced with permission from F. Whittle 'The Early History of the Whittle Jet Propulsion Gas Turbine', *Proceedings of the Institution of Mechanical Engineers*, vol. 152 (1945), pp. 419–35.

This interest by GE was natural, both because of the company's extensive experience with steam turbines and because it had a long-standing interest, dating from 1903, in the gas turbine and, since 1918, in the gas turbine-driven supercharger for reciprocating internal combustion engines. In the late 1930s design studies for a locomotive gas turbine 3355kW (4500hp) were initiated by GE. Before work on this had proceeded very far the engineers concerned became involved in the American wartime effort to develop an aircraft gas turbine, stimulated by information conveyed secretly in 1941 by the British to the Americans. Consequently, a major engineering programme involving the Westinghouse Electric Company, General Electric and various aircraft manufacturers was initiated. This led to the production by Westinghouse (1941–3) of the Type 19 gas turbine, and by GE of two engines, one a jet engine, the

TG180 (1943–6) and the other a turboprop engine, the TG100 (1941–5). The two GE engines are of considerable historical significance because they were, apart from the German engines, probably the first engines using axial compressors to fly, as well as being the direct forerunners of a long line of GE aircraft and land-based gas turbines.

Post-war development of individual types

The successful demonstration of the aircraft gas turbines by Whittle in the mid-1940s turned the thoughts of engineers to other applications of this prime mover. Its coming had been long awaited and, as hindsight shows, unreasonable expectations were raised. Numerous applications were proposed which, when tried, were found to be inappropriate for the gas turbine. Furthermore, many engineers appeared to forget that the development time for such a radically new prime mover could be expected to be lengthy.

Probably the most important factor constraining the application of the gas turbine has been fuel; it is just not possible for the gas turbine, as was first thought, to use any fuel. These early opinions about fuel were undoubtedly influenced by the experience in the 1920s and 1930s with fuels for gasoline engines, where the problem of engine knock was a serious constraint on both automotive and aviation applications of the spark ignition engine (see p. 316), and the struggle to overcome it made a strong impression on engineers with an interest in prime movers. Since the gas turbine introduces the fuel into the cycle after compression, knock is absent. However, this difficulty has been replaced by the effects of corrosion and ash on the performance of the expansion turbine blades (buckets).

It would appear that gaseous fuels are the most desirable, particularly natural gas. Liquid fuels have to be refined and corrosive constituents removed, or their activity suppressed by additives. Solid fuels have not so far been applied successfully because of their tendency to foul the expansion turbine blades with ash.

Historical trends show that the gas turbine is most useful in aircraft propulsion and electric power generation, as well as in gas pipeline compression.

Aircraft

Aircraft propulsion is by far the largest use of the gas turbine. This has come about because it has the following advantages compared to the piston engine: (a) the efficiency increases with height and speed; (b) the engine weight and volume are smaller for a given power output; (c) lack of vibration; (d) greater reliability.

Application of the gas turbine to aircraft propulsion was initially motivated by the desire to fly at heights (11,000m or 36,000ft and above) at which the piston engine, even with supercharging, could not produce adequate power because of the low air density. Whittle, as noted earlier, appears to have been one of the first to appreciate that the gas turbine could provide a suitable engine for high altitude flight.

The aircraft gas turbine is applied in one of three forms, the turbojet, the turboprop, and the turbofan or ducted fan. This classification is based on the way in which the engine output is used, as illustrated in Figures 5.20 and 5.21. Although these different modes of using the gas turbine had been known for some time (at least since the mid-1930s, when Whittle described the turbofan engine, and work started in Britain on the turboprop), attention was concentrated initially on the turbojet engine, together with a somewhat lower level of interest in the turboprop engine. Following the Second World War there was a marked increase in the development effort applied to the turboprop engine, and particularly noteworthy aircraft that flew using this type of engine were the Vickers Viscount (1953), using four Rolls-Royce Dart engines, and the Lockheed Electra (1959) with four Allison T56 engines. However, since about 1960 the turboprop has been mainly applied in helicopter engines and in the smaller pressurized aircraft used by feeder airlines. The most important type of aircraft gas turbine, because of its high propulsion efficiency, is now the turbofan engine.

A high propulsion efficiency is ensured if the jet speed and aircraft forward speed are about equal. However, under these conditions the engine has a very low thrust (which is proportional to the difference between the two speeds). But, because the thrust is also proportional to the mass rate of flow of air through the engine, it is possible to have both a high thrust and high propulsive efficiency by designing the engine to operate with a high air mass flow rate.

In principle, the turboprop could provide the required characteristics because the effective mass flow ($\dot{m}$) handled by the engine can be controlled by suitably sizing the propeller. Unfortunately, at the time this approach to improving aircraft engine performance was under consideration (the early 1940s), compressibility effects at the tips of the propeller at speeds exceeding 560km/hr (350mph) greatly degraded its performance. In view of this limitation, it occurred to Whittle that very short propeller blades, fan blades, arranged in an enclosure, so that the incoming air was decelerated to avoid compressibility effects, would be satisfactory. The fan would be driven by the expansion turbine and the engine would be called a ducted fan, or by-pass engine.

Preliminary experiments with ducted fan engines were carried out by Whittle and others, but these were premature because many other details of the turbojet engine had to be perfected before serious attention could be given to the by-

STEAM AND INTERNAL COMBUSTION ENGINES

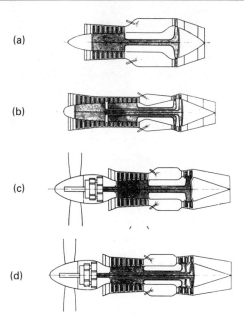

Figure 5.20: Turbojet and turboprop aircraft engines. (a) Single shaft turbojet. (b) Twin spool turbojet. (c) Direct connected turboprop. (d) Free-turbine turboprop.
Reproduced with permission from R. H. Schlotel, 'The Aircraft Applications' in H. R. Cox (ed.), *Gas Turbine Principles and Practice* (Van Nostrand, New York, 1955), pp. 22–1 to 22–120.

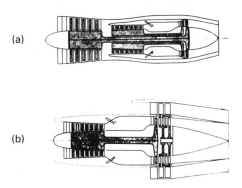

Figure 5.21: By-pass and fan turbojet aircraft engines. (a) By-pass engine (twin spool). (b) Rear fan engine (twin spool).
Reproduced with permission from R. H. Schlotel, 'The Aircraft Applications' in H. R. Cox (ed.), *Gas Turbine Principles and Practice* (Van Nostrand, New York, (1955), pp. 22–1 to 22–120.

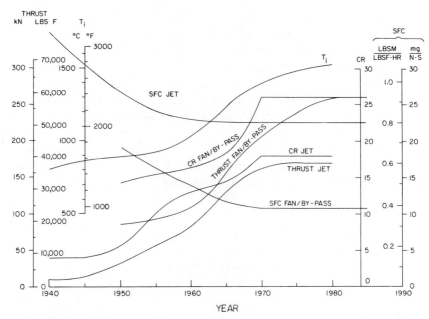

Figure 5.22: The historical trend of aircraft gas turbine engine performance parameters 1940–85.
The plots are based on data in O. E. Lancaster, 'Aviation Gas Turbines', in R. T. Sawyer (ed.), *Gas Turbine Engineering Handbook*, vol. II, (Gas Turbine Publications, Stamford, Conn., 1976), pp. 211–32 and *Jane's All the World's Aircraft* (Jane's Publishing Co., London, 1969–85).

pass concept. Consequently, this type of engine did not begin to supersede the pure jet until the late 1950s.

As a consequence of difficulties experienced with the rear fan engine, interest shifted in the early 1960s to the forward fan engine (see Figure 5.21). This type of engine is a standard item of equipment on the wide-body ('jumbo') jet aircraft that were introduced into airline service in the mid-1970s. In fact, its high thrust, particularly at take-off, and good fuel consumption (propulsive efficiency) can be said to be essential to the operation of this type of aircraft.

The history of the progress in the performance of aircraft gas turbine engines between 1939 and 1985 is shown in Figure 5.22.

Electric power generation

The application of the gas turbine to the generation of electric power represents the largest use of this type of prime mover after aircraft propulsion. This historic trend of the application of the gas turbine to electric power generation is illustrated by the growth in the cumulative number installed in the United States from one in 1949 to about 1600 units by 1985.

As noted above, the gas turbine was first applied to electric power generation in 1939 when a 4000kW stand-by plant was installed at Neuchâtel. After the war a number of gas turbine-driven electric generating plants were built. Because the gas turbine at that time had a significantly lower thermal efficiency (17 per cent) than the steam power plant (30 per cent), complicated cycles involving various stages of compression and expansion with heat recovery from the exhaust were devised. This approach was eventually abandoned because the plant reliability and cost were adversely affected by its complexity, and because the performance of machines based on the simple cycle improved as a result of the development of materials and cooling methods that allowed the inlet temperature to the expansion turbine to be increased. In addition there was a steady decline in fuel costs in the 1950s and 1960s. However the increase in fuel costs from 1973 reawakened interest in more complex plants.

The gas turbine is currently most widely used in electricity generation to meet peak loads, because: (a) it starts quickly (full load can be attained within a few minutes and in no more than about 30 minutes); (b) it can be remotely started with automatic equipment; (c) first, installation, and maintenance costs are low; (d) the short duty-cycle associated with peak load operation means that the relatively low efficiency and relatively high fuel costs of the gas turbine are inconsequential to the overall economics.

Until about 1965 the interest in gas turbines by the power companies was not very great. However, a number of events, particularly the power failure in the north-eastern United States that occurred in 1965, which indicated the need for generating equipment that could be started without outside power, stimulated interest in this type of prime mover. Other factors were delays in the completion of nuclear power plants, unexpected breakdowns of generating equipment and, in the United States, a sharp increase in the air conditioning load.

Applications of the gas turbine to base load operation require that it be incorporated as part of a steam power plant. Such combined cycles, which can have overall thermal efficiencies as high as 40 per cent, are essential in this case because of the comparatively low efficiency and high fuel cost of the gas turbine. In these cycles heat is either extracted from the gas turbine exhaust in an economizer (last stage feedwater heater), or the gas turbine exhaust is used in a steam generator (this would include an economizer).

Figure 5.23 shows the change in performance of gas turbines used in non-aviation applications, including electric power generation, from 1949 to 1985. The improvement in specific fuel consumption is particularly noteworthy and is indicative of the steady improvements in the performance of the individual components of the gas turbine.

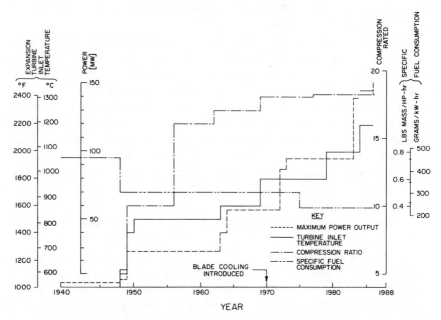

Figure 5.23: The historical trend of industrial gas turbines 1940–85. The plots are based on data in the annual power plant surveys appearing under various titles in *Power* between 1931 and 1985.

Other applications

In spite of optimistic statements made in the mid-1940s, the gas turbine has only slowly been adopted for industrial purposes. This appears to be partly a consequence of its apparent advantages not outweighing the proved reliability of other prime movers, and partly because the range of fuels that can be used in the gas turbine is much more restricted than was at first realized. The most important industrial applications have been to gas pipeline compression, and as power sources on off-shore oil and gas drilling rigs.

The gas turbine was applied experimentally in a number of merchant ships, but it has shown itself to be much more useful as a propulsion unit for naval vessels: since 1969 the British navy has adopted a policy of using gas turbines as the only form of propulsion in all classes of major warships.

The gas turbine was first applied to a railroad locomotive when Brown Boveri placed a 1640kW (2200hp) locomotive in service on the Swiss railways in 1941. The most extensive application was between 1948 and 1969 in the United States. However, it was found that fuel costs were higher than expected and that the increasing power output of contemporary diesel locomotives tended to reduce the advantage of the gas turbine in this application. The most important current use appears to be in special purpose, lightweight, permanently coupled, multiple unit trains.

Immediately following the Second World War the gas turbine was enthusiastically promoted as a new automobile power plant. A demonstration model, Jet 1, was introduced by the Rover Company in Britain in 1950. Unfortunately, the apparent promise was never realized because of high first and operational costs. Particularly significant was the inability to devise a satisfactory heat exchanger, which was required for good fuel consumption, by exchanging heat between the leaving exhaust gases and the entering air.

The development of the gas turbine as summarized in Figures 5.22 and 5.23 has been very rapid and has been a consequence of two factors, the experience provided by the development of the steam turbine between 1884 and 1935, and the demands of the aircraft industry for engines of increasing power and efficiency. The fluid mechanical and materials problems of the gas turbine could not have been solved without the foundation of knowledge generated by the development of the steam turbine; and the difficulty of the engineering problems presented by the gas turbine are so great that it is unlikely that it would have reached its current state of development without the impetus provided by the demands of the aircraft industry for engines of high power and good fuel economy.

EXTERNAL COMBUSTION ENGINES

The use of a substance that does not undergo the phase change associated with water when used in the steam engine appears to have originated with Henry Wood in 1759. He obtained a patent to use hot furnace gases in the engine cylinder, instead of steam, thereby avoiding the steam generation process with its associated inefficiencies. Such a device is commonly called a hot air engine, but might better be called an external combustion engine. It is appreciated that the question of nomenclature is difficult in this case. First, the hot air is not restricted to air as a working substance, but can use any gas that does not undergo a phase change in the course of the working cycle of the engine. Second, the nomenclature is applicable to the steam engine, because the combustion in the furnace of the steam engine is external to the engine. Third, the first example, the engine proposed by Wood, is not an external combustion engine, because the working substance is the combustion gases of the heat source. However, it is an external combustion engine in the sense that the process takes place outside the piston and cylinder. In spite of the unsatisfactory nomenclature, the term 'external combustion engine' serves to differentiate this engine from the steam engine and the internal combustion engine.

Sir George Cayley was the first to build an engine of Henry Wood's type in 1807. Two other cycles that have been employed in external combustion engines are one due to John Ericsson in 1826 and another named after Robert Stirling, who patented an engine based on the cycle in 1816. Engines based on the Wood, Ericsson and Stirling cycles were used quite extensively in the nineteenth century because they did not need a steam boiler, which required space and involved the danger of explosion. Although external combustion engines were extinct by the early years of this century, engines based on the Stirling cycle have experienced a substantial revival of interest beginning in the late 1960s.

Stirling engine

The ideal Stirling cycle involves constant temperature heat addition and removal, and constant volume heat addition and removal. It has an efficiency equal to that of the Carnot cycle operating between the hot and cold reservoirs of the Stirling cycle, but without the need to accommodate the very large changes in volume of the working substance that are characteristic of the Carnot cycle.

A particular feature of the Stirling engine is the use of two cylinders and two pistons. One of these, the displacer piston, serves to transfer the working substance between the hot and cold reservoirs, while the other, the power piston, is connected to the surroundings by some appropriate means (mechanical or electrical).

The practical Stirling engine differs from the thermodynamic ideal by replacing the hot and cold reservoirs attached to the displacer cylinder by heat exchangers located in the transfer connection between the two ends of the displacer cylinder. A regenerator in the transfer connection is also added to improve the thermodynamic efficiency.

APPENDIX

Heat engines and thermodynamics

A heat engine is a fixed mass of material, e.g. air, water or steam, called the working substance, that undergoes a series of processes in such a way that it converts heat into work. The processes are arranged so that the working substance returns to its original state, i.e., it has undergone a cycle. The cycle is also characteristic of a practical heat engine and in that case it is repeated as often and as rapidly as the operator desires or is practical.

The history of the heat engine is closely associated with the first and second laws of thermodynamics. The first law states that energy can be neither created nor destroyed so, in theory, all the chemical energy in the coal burnt to produce steam could produce work in a steam engine. The engine would then be said to be 100 per cent efficient. However, this is not observed in practice and Sadi Carnot showed that it was impossible. Carnot also showed that if the heat was supplied to the heat engine at a constant temperature T_h (K or °R) and rejected at a constant temperature T_c (lower than T_h), then the efficiency (η) of the heat engine would be $\eta = 1 - (T_c/T_h)$. The corresponding heat engine cycle is called the Carnot cycle. This result is important in the theory of thermal prime movers because it indicates that the efficiency of a heat engine using the Carnot cycle (and, by implication, any other cycle) is increased if T_c is decreased, T_h is increased, or both. In practice the minimum available value for T_c is 22°C (72°F), so the goal of all heat engine designers is to increase T_h and this is the common thread that runs through the history of thermal power production. However, this ambition has usually been frustrated by the inability to obtain materials that can stand the elevated temperatures and pressures of the working substance.

The Carnot cycle is said to be an ideal cycle because all its processes are reversible, that is, they occur infinitely slowly without friction, without fluid turbulence, and heat exchange employs minute temperature differences. Practical heat engines that are intended to work on a close approximation to the Carnot cycle have impractical dimensions, so ideal cycles have been defined that conform more nearly to the characteristics of the practical prime mover. A particularly important group of such cycles are the air standard cycles, which use air as the working substance. These are the ideal cycles associated with the internal combustion engine and the gas turbine.

Reciprocating steam engines

The reciprocating steam engine comprises two essential parts: the cylinder and the valve or steam chest. The cylinder has two ports at each end. These are opened and closed while the piston moves from one end of the cylinder to the other end, and back again, with the cycle of events repeated at each revolution of the crankshaft.

The piston, which fits in the cylinder, is a circular disc with grooves around its circumference, which hold spring rings in position (Figure 5.24). The latter are free to expand outward and, thereby, fit the cylinder so tightly that steam cannot leak past the piston. The piston is secured to the piston rod. On both end faces of the cylinder there are a number of studs that pass through corresponding holes in the cylinder covers so that the latter can be secured by nuts screwed on to the studs. The rear cover has an opening for the piston rod that is sealed by a stuffing box filled with packing held in place by the gland.

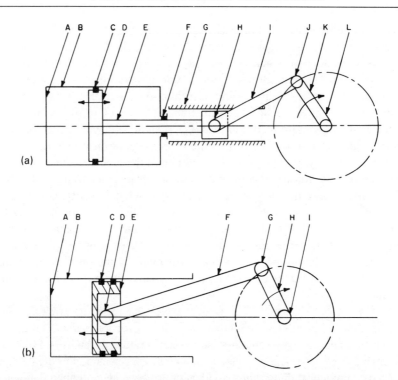

Figure 5.24: The reciprocating engine (steam or internal combustion).
(a) Crosshead engine (double acting). Key: A cylinder head; B cylinder wall; C piston ring; D piston; E piston rod; F gland; G crosshead guides; H crosshead; I connecting rod; J crank-pin; K crank; L crank shaft.

(b) Trunk piston engine (single acting). Key: A cylinder head; B cylinder wall; C piston rings; D gudgeon pin; E piston; F connecting rod; G crankpin; H crank; I crankshaft.

A space is always left between the piston and the cylinder covers when the piston is at either end of the cylinder to ensure that the piston does not strike the cylinder cover.

The piston moves to and fro with a reciprocating motion and this linear motion is converted into a circular one by a connecting rod, which is joined to the piston rod at the crosshead. The latter slides on the guides or slidebars.

The connecting rod big-end is located at a distance, equal to half the piston stroke, from the centre of the crank shaft.

The valve or steam chest has a plain flat surface machined parallel with the axis of the cylinder in which there are three ports, the outer ones connected to the corresponding ports in the cylinder, and the middle one the exhaust. Steam is supplied from the boiler through the pipe. The slide valve moves back and

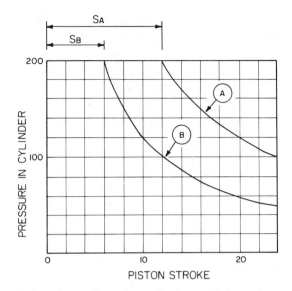

Figure 5.25: Effect of cut-off on the performance of the reciprocating steam engine. In the diagram each square equals one unit.

Curve	Cut-off	Steam admitted to cylinder	Units of work obtained	Work obtained per unit of steam
A	50%	$S_A = 60$	101	1.66
B	25%	$S_B = 30$	74	2.5

Adapted with permission from C. S. Lake and A. Reidinger, 'Locomotive Valves and Valve Gears' (Percival Marshall, London, n.d.).

forth on the machined surface and opens and closes the ports to admit and release steam in accordance with the cycle of piston movements. This is ensured by driving the valve by a crank, with the two connected through a more or less complicated system of linkages known as the valve gear. (Modern steam engines, particularly when working with high temperature steam, used piston valves. Other important types are the Corliss valve, the poppet valve and the drop valve.)

It is important to note that the valve does not admit steam throughout the piston stroke, but only for a short period at the beginning. The pressure and energy of the steam decreases as the piston completes its stroke and, in an ideal engine with no friction, or other losses, the energy given up by the steam appears as a moving force (work) at the piston rod. The steam is then said to be used expansively. The same effect could be produced by admitting steam to the cylinder throughout its stroke, but such non-expansive working would not be as efficient as expansive working. This is clarified in Figure 5.25.

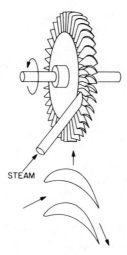

Figure 5.26: Principle of the steam turbine. Steam passes through the stationary nozzle and is directed as a high velocity jet onto the blades attached to the periphery of the rotating wheel. The steam experiences a drop in pressure as it flows through the moving blades. The blades are so shaped that the steam, which is flowing axially in this portion of the turbine, is turned (see the enlargement). The change in steam direction and pressure (if employed) as the steam passes through the moving blades imparts a force in the tangential direction to the wheel that causes it to turn.
Reproduced with permission from W. G. Scaife, 'The Parsons Steam Turbine', in *Scientific American*, vol. 152, no. 4 (1985), pp. 132–9.

Steam turbines

The steam turbine converts the internal energy of the steam into rotary motion by accelerating it to a high velocity in a specially shaped stationary passage called a nozzle. The steam leaving the nozzle is then directed on to a row of blades or buckets attached to a rotating wheel (see Figure 5.26). The flow cross-section of the blade passages is specially designed to change the direction of motion of the steam, and, in the reaction steam turbine, the pressure of the steam. Because the linear speed of the blade increases with the radius, warped twisted blades were introduced in the 1930s. This ensures that the steam is incident on the blade at all radii with the minimum of losses due to turbulence and friction.

The form of the stationary nozzle that accelerates the steam and directs it on to the rotating blades depends on the desired speed of the steam at the nozzle exit. De Laval discovered in 1888 that for very high steam speeds the nozzle must have a converging–diverging form. The form of this nozzle is contrary to expectation, in that it might be assumed that the continually decreasing cross-section of a purely converging nozzle would be required to accelerate the steam. However, the paradox arises from our everyday experience with incompressible fluids, particularly water. Steam is a compressible fluid, that is, its density depends on its pressure, whereas the density of water is, for practical purposes, independent of pressure. A fluid, as it passes through a nozzle, experiences a decrease in pressure, so, if it is steam, its density decreases. At first the density change is small and the velocity increase, induced by the drop in pressure, is large. Since, for a constant mass rate of flow at all points along the nozzle, the cross-sectional area is inversely proportional to these two quantities, a decrease in nozzle cross-section is required. However, at a point where the so-called critical pressure is reached in the nozzle, the density starts to decrease much more rapidly with the decrease in pressure along the nozzle, and, since the steam velocity is still increasing, the cross-sectional area has to increase in order to accommodate the same mass rate of flow. Hence, if the nozzle operates with an exit pressure less than the critical pressure, it has the initial converging form followed by the diverging sections.

ACKNOWLEDGEMENTS

The author would like to thank Ian McNeil for the invitation to prepare this chapter, and for his continued encouragement and patience during its preparation. At the Rensselaer Polytechnic Institute: Mary Ellen Frank and Susan Harris for assistance with the typing, the staff of the Folsom Library for obtaining numerous papers and for facilitating the preparation of illustrations, Professors F. F. Ling and M. Lai, chairman and acting chairman of the Department of Mechanical Engineering, Aeronautical Engineering and Mechanics.

FURTHER READING

Steam engines

Buchanan, R. A. and Watkins, G. *The industrial archaeology of the stationary steam engine* (Allen Lane, London, 1976)

Dickinson, H. W. *A short history of the steam engine*, 2nd edn (Cass, London, 1963)

Hunter, L. C. *A history of industrial power in the United States 1780–1930*. Volume two: steam power (University of Virginia Press for the Hagley Museum, Charlottesville, VA, 1985)

Matschoss, C. *Die Entwicklung der Dampf-Maschine*, two vols. (Julius Springer, Berlin, 1908)

Rolt, L. T. C. and Allen, J. S. *The steam engine of Thomas Newcomen*, (Moorland, Hartington and Science History Publications, New York, NY, 1977)

Thurston, R. H. *A history of the growth of the steam engine*, Centennial Edition (Cornell University Press, Ithaca, NY, 1939)

Steam turbines

Harris, F. R. 'The Parsons centenary – A hundred years of steam turbines', *Proceedings of the Institution of Mechanical Engineers*, vol. 198, no. 9 (1984), pp. 183–224

Internal combustion engines

Büchi, A. J. *Exhaust turbocharging of internal combustion engines*, Monograph no. 1 (published under the auspices of the *Journal of the Franklin Institute*, Lancaster, PA, 1953)
Cummins, C. L. Jr. *Internal fire* (Carnot Press, Lake Oswego, Oregon, 1976)
Delesalle, J. 'Les Facteurs du progrés des diesel', *Entropie*, vol. 21, no. 122 (1985), pp. 33–9
Hardy, A. C. *History of motorshipping* (Whitehall Technical Press, London, 1955)
Jones, J. 'The position and development of the gas engine', *Proceedings of the Institution of Mechanical Engineers*, vol. 151 (1944), pp. 32–53
Mondt, J. R. 'An historical overview of emission-control techniques for spark-ignition engines', *Report No. GMR–4228* (General Motors Research Laboratories, Warren, MI, 1982)
Norbye, J. P. *The Wankel engine: design, development, applications* (Chilton Book Co., Radnor, Pa., 1971)
Pattenden, R. F. S. 'Diesel engine research and development', *Chartered Mechanical Engineer*, vol. 9, January (1962), pp. 4–12
Ricardo, H. R. 'Diesel engines', *Journal of the Royal Society of Arts*, vol. 80 (1932), pp. 250–62, 267–80
—— *The high-speed internal combustion engine*, 4th edn (Blackie, London, 1953)
Taylor, C. F. 'Aircraft propulsion: a review of the evolution of aircraft power plants', *Smithsonian report for 1962* (Smithsonian Institution, Washington, DC, 1962), pp. 245–98

Gas turbines

Baxter, A. D. 'Air flow jet engines', in O. E. Lancaster, (ed.), *Jet propulsion engines*, Vol. XII, High Speed Aerodynamics and Jet Propulsion (Princeton University Press, Princeton, NJ, 1959), pp. 29–53
Cox, H. R. 'British aircraft gas turbines', *Journal of the Aeronautical Sciences*, vol. 13, (1946), pp. 53–87
Denning, R. M. and Jordan, T. 'The aircraft gas turbine – status and prospects', in *Gas turbines – Status and prospects*, (ME Publications, New York, 1976), pp. 17–26
Keller, C. and Frutschi, H. 'Closed cycle plants – Conventional and nuclear-design, application operations', in *Gas turbine engineering handbook*, Vol. II, (Gas Turbine Publishers, Stamford, CT, 1976), pp. 265–83
Meyer, A. 'The combustion gas turbine: its history, development and prospects', *Proceedings of the Institution of Mechanical Engineers*, vol. 141, (1939), pp. 197–222
Moss, S. A. 'Gas turbines and turbosuperchargers', *Transactions of the American Society of Mechanical Engineers*, vol. 66, (1944), pp. 351–71

Whittle, F. 'The early history of the Whittle jet propulsion gas turbine', *Proceedings of the Institution of Mechanical Engineers*, vol. 152, (1945), pp. 419–35

External combustion engines

West, C. D. *Principles and application of Stirling engines* (Van Nostrand Reinhold, New York, 1986)

6

ELECTRICITY

BRIAN BOWERS

STATIC ELECTRICITY

The attractive power of lodestone, a mineral containing magnetic iron oxide, was known to Lucretius and Pliny the Elder. The use of the magnetic compass for navigation began in medieval times. A letter written in 1269 by Peter Peregrinus gives instructions for making a compass, and he knew that it did not point to the true North Pole. Knowledge of static electricity is even older, dating back to the sixth century BC: Thales of Miletus is said to have been the first to observe that amber, when rubbed, can attract light bodies.

The scientific study of electricity and magnetism began with William Gilbert. Born in Colchester and educated at Cambridge, Gilbert was a successful medical practitioner who became physician to Queen Elizabeth I in 1600. In that same year he also published his book *De Magnete*, which recorded his conclusions from many years' spare-time work on electrostatics and magnetism, and, for the first time, drew a clear distinction between the two phenomena.

In a very dangerous experiment the American statesman Benjamin Franklin showed that a kite flown in a thunderstorm became electrically charged. His German contemporary Georg Wilhelm Richmann was less fortunate: he was killed trying the same experiment at St Petersburg in 1753.

Franklin also studied the discharge of electricity from objects of different shapes. He suggested protection of buildings by lightning conductors and, in the light of his discharge experiments, said that they should be pointed. The value of lightning conductors was not fully accepted at first. Some argued that they would attract lightning which would not have struck if the conductor had not been there. Some people preferred ball-ended lightning conductors to pointed ones. Among these was King George III, whose reasoning seems to have been that since Franklin was a republican his science must be suspect too.

The distinction between conductors and insulators; the fact that there were two forms of static electricity, later called 'positive' and 'negative'; and the principle of storing electric charges in a Leyden jar – a capacitor formed by a glass jar coated with metal foil inside and out – were all worked out during the eighteenth century. Most electrical experiments at that time used electricity produced by frictional electric machines. There were many designs of such machines, but typically a globe or plate of glass or other non-conducting material was rotated while a cloth rubbed its surface.

The discovery of the electric current, about 1800, did not end the story of static electricity. Two important machines of the nineteenth century were Armstrong's hydro-electric machine and the Wimshurst machine. William Armstrong was a solicitor and amateur scientist who founded an engineering business in Newcastle upon Tyne. His attention was drawn to a strange effect noticed by an engine driver on a colliery railway. The driver experienced 'a curious pricking sensation' when he touched the steam valve on a leaking boiler. Armstrong found that the steam, issuing from a small hole, became electrically charged. He then built a machine with an iron boiler on glass legs and a hardwood nozzle through which steam could escape. He found the steam was positively charged, and he then made a larger machine which was demonstrated in London producing sparks more than half a metre long. This 'hydro-electric machine', as he called it, established Armstrong's scientific reputation. A War Office committee on mines suggested in 1857 that Armstrong's machine, with its very high voltage output, could be used for detonating mines. In practice magneto-electric machines were soon available, and Armstrong's machine never saw practical use.

During the nineteenth century numerous machines were made which multiplied static electric charges by induction and collected them in Leyden jars or other capacitors. Best known was the Wimshurst machine, made in 1883 by James Wimshurst, a consulting engineer to the Board of Trade.

CURRENT ELECTRICITY

An electric current, as opposed to static charges, was first made readily available in 1800 as a result of the work of the Italian Alessandro Volta who later became Professor of Natural Philosophy at the University of Pavia. He was following up work done by his fellow-countryman Luigi Galvani, Professor of Anatomy at the University of Bologna. Galvani had been studying the effects of electric discharges from frictional machines on the muscles of dead frogs. In the course of this work he noticed that a muscle could be made to twitch with the aid of nothing more than two pieces of different metals. Galvani thought the source of the phenomenon he had discovered was in the muscle; Volta showed that it was the result of contact between the two different metals.

Volta announced his discovery in a letter to the President of the Royal Society in London, and the letter was published. Accompanying the letter was a drawing of pairs of metal discs interleaved with pieces of leather or card soaked in salt water. He referred to the discs as making a 'pile' (*colonne* in the original French of the letter), and the term Volta's pile has remained.

The first mass-produced battery was designed by William Cruickshank, a lecturer at the Royal Military Academy, Woolwich, soon after 1800. He soldered together pairs of copper and zinc plates and set them in wax in grooves across a wooden trough. The trough was then filled with acid. Michael Faraday used Cruickshank batteries in his early electrical researches, and in his book *Chemical Manipulation*, published in 1828, he went into detail about the right acid to use. (He recommended one volume of nitric acid, three volumes of sulphuric acid, mixed with water to make one hundred volumes.)

The availability of a steady electric current opened up new possibilities for research. Many people studied the properties of an electric current, especially its chemical properties. Among the first was Humphry Davy. Born in Penzance, Davy was first apprenticed to a surgeon-apothecary, but he was released from his apprenticeship to take a post as assistant at the Pneumatic Institution in Bristol, a body devoted to the study of the physiological properties of gases. Davy's main work there was a study of the effect of breathing nitrous oxide ('laughing gas'), which Davy said gave him a sensation 'similar to that produced by a small dose of wine'.

Davy's future lay not in Bristol but at the new Royal Institution, founded in 1799 in London, where in 1801 he was offered a post as lecturer. The purpose of the Royal Institution was 'to encourage the diffusion of scientific knowledge and the general introduction of useful inventions'. It was equipped with a lecture theatre and laboratories. Davy gave lectures which attracted large audiences from all levels of society, but as well as teaching chemistry he advanced it. He established the science of electrochemistry, and within a few years had isolated the metals potassium, sodium, strontium, barium, calcium and magnesium by electrolysis of their compounds.

In the course of his work Davy discovered the brilliant light produced by an electric arc between two pieces of carbon connected to a suitable source of electricity. He used the arc as a source of heat, but it seems unlikely that he really envisaged it becoming a practical source of illumination – the limited batteries at his disposal would have made arc lighting prohibitively expensive.

Despite their expense, chemical cells were the main source of electricity until the development of practical generators in the 1860s. Batteries such as Cruickshank's had two serious defects. One was the phenomenon known as local action, in which impurities in the zinc plate reacted electrolytically with the zinc itself, eventually destroying the plate. The initial solution was to remove the plates from the battery, or drain off the acid, whenever it was not in use. However, it was found that if the zinc plates were 'amalgamated' by rubbing them with mercury, then local action is prevented. The probable explanation is that the surface of the plate becomes coated with a mercury–zinc compound which acts as a perfectly acceptable electrode for the cell and does not participate in local action but shields the particles of impurities from the acid. The more difficult problem was polarization. With a copper–zinc cell, bubbles of hydrogen gas are released at the copper anode, and the result is a layer of bubbles which increases the resistance of the cell, thus reducing the effective output voltage. The solution to this problem was to interpose between the anode and the electrolyte a substance that removed the hydrogen but did not impede the current. The first practical cell which did not polarize was the Daniell cell, developed in 1836. John Frederic Daniell was Professor of Chemistry at King's College London. In his cell a porous pot containing copper sulphate surrounded the copper anode, and the acid electrolyte was between the outside of the porous pot and the zinc cathode. The hydrogen generated then reacted with the copper sulphate solution, and that in turn deposited copper on the anode. Other 'two-fluid' cells were made, using different chemical combinations. The Grove cell devised by William Grove, the chemist who became a High Court Judge, used a platinum electrode in strong nitric acid and a zinc electrode in dilute sulphuric acid. The Bunsen cell was similar, except that Bunsen replaced the expensive platinum with carbon, which was cheap and equally effective.

By far the best known of all the primary cells was the one developed by Georges Leclanché in 1868. It is a carbon–zinc cell with ammonium sulphate as the electrolyte, and the depolarizing agent is manganese dioxide which is packed around the carbon. Most modern 'dry cells' are Leclanché cells with the electrolyte made into a paste.

Secondary batteries, or 'accumulators', which can be re-charged from an electricity supply, may conveniently be mentioned here. Best known is the lead–acid battery, familiar as the car starter battery. In its earliest form it is due to the Frenchman Raimond Louis Gaston Planté, whose first accumulator was simply two lead plates in a vessel of acid. When a current was passed through his cell to charge it, hydrogen was released at one plate and oxygen at the other. The hydrogen escaped, but the oxygen reacted to form lead peroxide. When the cell was discharged, both plates became coated with lead sulphate. Planté found that his cell became better after it had been charged and discharged several times, a process that became known as 'forming'. See Figure 6.1.

Camille Faure, another Frenchman, found a better way of forming the plates than the slow and expensive method of repeatedly charging and discharging the cell. He coated the plates with a paste of lead oxides: Pb_3O_4 for the positive plate and PbO for the negative. When the cell was charged the PbO on the negative plate was converted into a very spongy mass of lead which presented a very large effective surface area to the electrolyte. Lead paste tended to fall off the plates, and Joseph Swan (see p. 366) made plates in grid form, with the paste

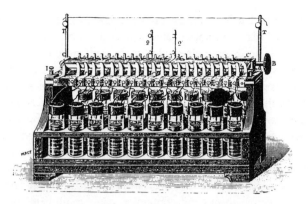

Figure 6.1: Battery of Planté cells arranged for high-voltage experiments. From Guston Planté *Recherches sur l'électricité*, Paris 1883, p. 97.

pressed into the holes in the grid. This kind of construction continues to be used, although many detailed improvements have been made. Lead–acid batteries are heavy, which is their main disadvantage. The principal alternative is cells based on nickel and iron or nickel and cadmium, with sodium hydroxide as the electrolyte. Such cells are more robust than the lead–acid ones and not so heavy, but neither are they so efficient. They are often used in electric vehicles such as milk floats, and also as stand-by batteries for emergency lighting systems. In recent years they have also been made in very small sizes for use in battery-powered torches, radios and other small appliances.

MICHAEL FARADAY

Michael Faraday, who has been called 'the father of electricity', was born in the Surrey village of Newington, now part of Greater London, the third child of a blacksmith who had recently moved from Westmorland. His formal education was minimal – in his own words 'little more than the rudiments of reading, writing and arithmetic at a common day school'.

Faraday's real education began when, at the age of fourteen, he was apprenticed to a bookseller and bookbinder. He became a competent bookbinder, which was valuable practical training for later years when manual skills were vital in the laboratory. Even more important was that he read many of the books that he bound, including some dealing with electricity. The young man impressed one of the customers in the shop with his interest in science, and he was given tickets to hear some of Davy's lectures at the Royal Institution. He took detailed notes, bound them, and sent them to Davy asking for work in any scientific capacity. There was no work available at the time, but later, when Davy was looking for a laboratory assistant, he remembered Faraday and gave him the post. From 1813 to 1815, Davy was travelling in Europe, and Faraday accompanied him as his assistant, meeting many of the leading scientists of his day and becoming familiar with their work.

Faraday's early scientific work was in chemistry. He conducted research into the properties of steels of different compositions, and also into optical glass. This work was paid for by sponsors (the finances of the Royal Institution were such that sponsored research was essential to keep the Institution solvent). His chief interest, however, was electricity and he was intrigued by Oersted's discovery that a compass needle could be deflected by an electric current. H. C. Oersted was Professor of Physics at the University of Copenhagen. The significance of his discovery was that it demonstrated a link, which had long been suspected, between electricity and magnetism.

In 1821, Faraday was invited to write a historical account of electromagnetism for the *Annals of Philosophy*. While preparing the article he repeated all the important experiments and he became convinced that it ought to be possible to produce continuous circular motion by using the circular magnetic force around a current-carrying wire. Success came in September 1821 when he made two devices which, with a little imagination, may be called the first electric motors. Both devices had a basin filled with mercury. In one a bar magnet was fixed vertically and a loosely suspended wire was hung from a point above the bar magnet so that it dipped into the mercury. When a current was passed through the wire (and through the mercury), the wire moved in a circular path around the magnet. In the second device the wire was fixed centrally and the magnet (which floats in mercury) had one end tied to the bottom of the basin. When a current flowed in the wire then the magnet moved in a circular path around it.

Oersted's experiment had shown that an electric current produced magnetism. The question in Faraday's mind was, could magnetism be made to produce electricity? It was not until the autumn of 1831 that he had time to pursue the matter properly, but he then succeeded in establishing the principles that relate electricity and magnetism in a series of three crucial experiments. The first was carried out on 29 August 1831. He had made a soft iron ring about 2cm (1in) thick and 15cm (6in) in diameter and wound two coils on the ring. Since the only wire available to him was bare metal, he insulated the turns by winding a layer of calico under each layer of wire and a piece of string between adjacent turns. The two coils were in several parts, so that he could change the effective number of turns. He drew the arrangement in his notebook, calling the coils A and B, and noted:

> Charged a battery of 10 pr. plates 4 inches [10cm] square. Made the coil on B side one coil and connected its extremities by a copper wire passing to a distance and just over a magnetic needle (3 feet [91cm] from iron ring). Then connected the ends of the pieces on A side with battery. Immediately a sensible effect on needle.

It oscillated & settled at last in original position. On *breaking* connection of A side with Battery again a disturbance of the needle.

Faraday then showed that the iron ring was not essential – he could produce the effect with two coils wound on a tube of cardboard. He also studied the effect of changing the number of turns in the coils, and showed that the deflection of the needle varied.

The experiment without the iron core was virtually the same as one he had tried some years earlier, without success. Why had he failed then? The important difference was that his understanding of what he was looking for had changed. He had originally expected the mere presence of current in one wire to produce an effect in the other, but by 1831 he was expecting another factor to be involved. That additional factor was motion, or change. He expected an effect at the moment he completed, or broke, the battery circuit, and because he was looking for a transient effect he found it. His experiments continued with a variety of coils, magnets and pieces of iron. On 24 September 1831 he described an arrangement with two bar magnets and a piece of soft iron arranged in a triangle. A coil connected to a galvanometer was wound on the soft iron, and he found that when the magnets were pulled away the galvanometer recorded a brief current in the coil. Faraday noted: 'here distinct conversion of magnetism into electricity'. He then arranged to use the most powerful magnet in London, which belonged to the Royal Society. On 29 October 1831 he rotated a copper disc between the poles of this magnet, and showed with his galvanometer that a current was produced between the axis and the edge of the disc.

That arrangement of disc and magnet was the first generator, in the sense of a machine which rotates conductors and magnets relative to one another and produces electricity. Faraday himself seems not to have developed the idea further. His next research interests were to show that the 'magneto-electricity' produced in his experiments was indeed the same electricity as that produced by chemical cells, by frictional machines and also by electric fish. Having satisfied himself on that he went on to study electro-chemistry.

GENERATORS

Faraday's demonstration that electricity could be produced mechanically was followed up by the Parisian instrument maker Hippolyte Pixii, who was closely associated with the Academy of Sciences in Paris. Pixii realized that the output of Faraday's machine was limited because only one conductor – the radius of the disc – was passing through the magnetic field at any one time. He made an arrangement with two horseshoes end to end with their poles nearly touching. One was a permanent magnet and the other a piece of soft iron with a coil wound on it. The soft iron was fixed and the permanent magnet rotated about its axis by a handle and gearing. As it turned it magnetized the soft iron, first in

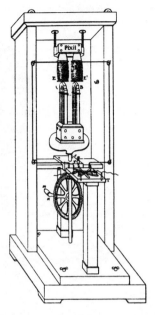

Figure 6.2: Hand turned magneto-electric generator made by Pixii in 1832 or shortly after.

one direction then in the other, and at each change of the direction of magnetization a current was induced in the coil. The resulting current alternated to and fro, but at that time no one could conceive of any use for alternating current. At the suggestion of Ampère, Pixii fitted a rocking switch, operated by a cam on the axis, which reversed the connections to the coil at each half turn of the magnet (see Figure 6.2). The output current was then uni-directional, and Pixii showed that electricity from his machine could do the things that physicists were then doing with electricity from other sources. Pixii made a number of similar machines, which were sold with a printed leaflet describing experiments that could be done; though only two of his machines survive.

Other scientific instrument makers were soon making similar machines, and many designs appeared during the 1830s. See for example Figure 6.3. William Sturgeon invented the metal commutator, used ever since on most rotating electrical machines in place of Pixii's rocking switch.

The first attempt to use these 'magneto-electric machines' (soon shortened to 'magnetos') for practical purposes was in the telegraph. The first practical electric telegraph was installed by Cooke and Wheatstone in 1838, and in 1840 Wheatstone was using a magneto for a new telegraph he was developing (see p. 714).

In the 1840s a Birmingham chemist, J. S. Woolrich, and the firm of

Figure 6.3: Magneto-electric generator by Saxton, 1833. Note that no wires come from this machine. It was used simply to demonstrate that rotating a coil in front of a magnet could generate electricity, which appeared as a spark when the circuit was broken as the contact strip at the end of the shaft came out of mercury in the small cup mounted on a pillar.

Elkingtons used a large magneto for electroplating (see Figure 6.4), and in 1852 the first electric lighting company was formed. This was the Anglo-French Société de l'Alliance which intended to use magnetos to electrolyse water, yielding hydrogen which would then be used in a limelight. That was not successful, but in 1857 an electric arc lamp (see p. 362) supplied from a magneto weighing two tonnes was demonstrated in a lighthouse. At the request of Trinity House, Faraday supervised the demonstrations, which took place at Blackwall, and was very pleased with the results. The machine was made by F. H. Holmes, who then received an order for two machines for the South Foreland lighthouse. These were working in December 1858, and several more machines were installed in other lighthouses in subsequent years.

The output of a magneto is limited by the strength of the permanent magnets, and until that limitation could be overcome there could be no large-scale

Figure 6.4: A replica of Woolrich's generator of 1844, used for supplying current for electroplating.

generation of electricity. In 1845, Wheatstone and Cooke patented the idea of using an electromagnet, supplied from a battery, in place of the permanent magnet in a magneto for a telegraph. That was a step in the right direction, and other people later made machines in which a magneto supplied electricity to energize an electromagnet on another machine which then gave a considerably higher current output. The real answer to the problem, however, was the self-excited generator in which the current for the electromagnet is supplied from the output of the machine itself. Several people made such machines in 1866. One of them, C. W. Siemens, expressed their advantage succinctly: 'it is thus possible to produce mechanically the most powerful electrical effects without the aid of steel magnets.'

With its prospect of virtually unlimited electricity, the invention of the self-excited generator stimulated electrical developments generally. The idea of the electric arc light was well known, though until practical generators were

Figure 6.5: An 'A' pattern Gramme generator, as used for small lighting installations in the 1870s.

available there had been little encouragement to develop it. The first large-scale manufacturer of practical generators was the Belgian engineer Z. T. Gramme who worked in Paris. His machines used the ring armature, known ever since as the Gramme ring, although similar armatures had been used earlier. This had a toroid of iron wrapped round with a number of coils all connected in series (see Figure 6.5). Many of these machines were sold from the 1870s, mainly for arc lighting.

The first British generator manufacturer was R. E. B. Crompton. After serving in the Indian army for some years he had returned to England and bought a partnership in an agricultural and general engineering firm at Chelmsford, in Essex. He intended to pursue a longstanding interest in steam transport, but found himself involved in electric lighting. Crompton designed a new foundry for relatives who owned an ironworks and sought lighting equipment so that it could be worked day and night. He visited Paris and bought some of Gramme's equipment for the ironworks, then realized there was a market for electric lighting apparatus in Britain. Initially he imported equipment from Gramme and others in Paris, but soon decided that he could improve upon it, and he began manufacturing his own.

The first generators Crompton made were based on the designs of the Swiss engineer Emile Bürgin (see Figure 6.6). Burgin had improved on Gramme's machines by using a series of iron rings for the armature, arranged in parallel

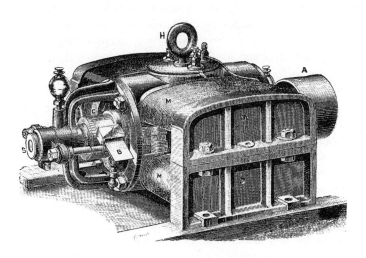

Figure 6.6: Crompton-Burgin generator of about 1880.

along the axis, where Gramme had used a single ring. This led to a much stronger construction, and one which was easier to cool than Gramme's because there were plenty of air passages through the armature coils. Crompton made many detailed improvements in the mechanical design of generators, and jointly with Gisbert Kapp he patented the compound field winding in 1882.

The output voltage of early generators varied considerably with the load. As the load increased the voltage would fall, and, if the load were shed suddenly (possibly as a result of a fault) the voltage would surge. This did not matter when the load was arc lighting, but with filament lamps (see p. 365) it was critical. Compound winding sought to overcome this problem: the field magnet had two sets of windings, one carrying the main magnetizing current and the other carrying the load current. The second, compounding, winding would boost the magnetic field as the load current increased, thus increasing the output voltage. Another benefit of compound winding was that the magnetic field of the extra coils compensated for the distortion of the magnetic field caused by the current in the armature. In the simple generator, without compounding, the brush position had to be varied as the load changed, to avoid excessive sparking and consequent wear of the commutator and brushes. Compounding resulted in a machine that did not need such constant attention.

During the 1880s generator design gradually became a science rather than an art. Gisbert Kapp, who had worked for Crompton but later became the first Professor of Electrical Engineering at the University of Birmingham, sought to design machines mathematically, while Crompton said that he always designed 'by eye'. The detailed design of the field magnets and coils was studied by John

Hopkinson, Professor of Electrical Engineering at King's College London, and a consultant to the American electrical inventor Thomas Alva Edison. Seeking to introduce a complete electric lighting system, with lamps, generators and other equipment all of his own design, Edison had been making generators with very long field poles and coils. Hopkinson made a number of small models of field systems of different shapes, and measured the magnetic field produced at the armature. As a result he concluded that Edison's poles were much too long, and he designed a fairly squat machine whose general proportions were followed by many manufacturers.

The machines described above were for direct current. A different pattern was adopted for the early alternating current generators. These normally had the armature coils arranged around the edge of a fairly thin disc and moving between the poles of a multi-polar field system. This design gave a machine whose reactance was low – important on an alternating current system – and allowed a sufficient number of poles to be used. It was essential to use multi-polar machines if the generator were to be coupled directly to a steam engine. Even the fastest reciprocating engines ran at only about 500rpm. A twelve-pole generator running at that speed would give a 50Hz output. In practice supply frequencies varied from 16.6 to 100Hz. The disadvantage of the disc generator was that it was impossible to make such a machine for three-phase operation. However, before three-phase supplies came into general use the turbine had replaced the reciprocating steam engine, and generators were being designed for the higher running speed of the turbine.

ARC LIGHTING

Although the possibility of electric arc lighting had been demonstrated very early in the nineteenth century, it could not be a practical proposition until a supply of electricity was readily available. The development of satisfactory generators in the 1870s stimulated fresh interest in the possibility of electric lighting.

In an electric arc lamp two carbon rods are connected to the opposite poles of the supply. The rods are briefly touched together and then drawn a few millimetres apart. This draws a spark, or 'arc', which continues as long as the electricity supply is maintained. The current in the arc produces considerable heat, and the contact points on the carbons quickly become white hot. These white hot places on the carbons are the source of the light. White hot carbon burns in air, and so some arrangement is necessary to feed the carbons closer together so that the gap is kept constant. Without any adjustment the gap widens and within a minute or two the electricity supply will be unable to maintain the arc across the wider gap and the lamp will be extinguished.

The first arc lamps were manually adjusted, and for applications such as theatre spotlights there was no problem about having a man constantly in attendance. For general lighting, however, the arc would only be acceptable if some reliable method could be found for adjusting the carbons automatically. Much inventive ingenuity went into devising suitable mechanisms for the purpose.

The first arc lamps to be used in quantity for general lighting were known as Jablochkoff candles. Paul Jablochkoff was a Russian telegraph engineer who set out from Russia in 1875 intending to visit the United States centennial exhibition at Philadelphia in 1876. He only got as far as Paris, where he became interested in electric lighting, and it was in Paris that he invented his 'candle'. Jablochkoff's candle consisted of two parallel carbon rods placed side by side but separated by a thin layer of plaster of Paris. At one end the rods terminated in brass tubes which secured the candle in a holder and made the electrical connections, at the other end they were joined by a thin piece of graphite. When the candle was in its holder and the current was switched on, the graphite fused, starting an arc between the ends of the carbons. As the carbons burned the plaster crumbled away in the heat, exposing fresh carbon. Provided the candle had been well made the carbon and the plaster were consumed at the same rate and the result was a steady light. However, once the light was extinguished, for whatever reason, it could not be restarted. For street lighting this did not necessarily matter: the candle would last for an evening, and during the next day a man could go round putting in fresh candles. Automatic mechanisms were made which brought a new candle into the circuit when the first was extinguished, but the candle itself soon became obsolete as regulating mechanisms were devised which could be mass produced.

The Jablochkoff candle was only used for a few years, but it was used prominently. It was first installed in Paris, attracting much attention. In July 1878 the London technical journal *The Electrician* complained that 'The application of the electric light is in Paris daily extending, yet in London there is not one such light to be seen.' In October the same year the Metropolitan Board of Works arranged a trial of electric lighting, using Jablochkoff candles, on the Victoria Embankment. About the same time the City of London authorities arranged some trials in front of the Mansion House, on Holborn Viaduct and in Billingsgate fish market. All these installations were working by Christmas 1878.

Arc lamp regulators have to perform two distinct functions. First the carbon rods must be brought together and then drawn apart when the current is turned on, and secondly the spacing of the rods must be maintained. The first function was quite easy to achieve: the upper carbon was allowed to fall under gravity and make contact with the lower. An electromagnet connected in series with the lamp then pulled the lower carbon down a few millimetres to start the arc. The rate of fall of the upper carbon was controlled by a brake, and the second function of the mechanism was to control the brake. The smooth operation of the

Figure 6.7: Compton arc lamp mechanism of about 1878.

lamp depended on the brake, and any sudden movement of the upper carbon would cause the light to flicker. Most of the earlier lamps used the series electromagnet to control the brake. This was easily arranged: as the gap widened and the arc lengthened the current would fall, and the electromagnet, which had been holding the brake on, would weaken, releasing the brake until the gap was restored (see Figure 6.7).

The disadvantage of using the series electromagnet to control the arc lamp was that only one lamp could be connected to the supply. If two lamps were connected in series to one generator, then a fall in current due to one gap widening would affect both regulators, and the brake on one lamp would be released too soon. (Arc lamps will not operate in parallel, either, because the arc is a negative resistance and one lamp would take all the current while the other went out.)

More satisfactory arc lamps were made by introducing another electromagnet, connected in parallel with the arc. As the arc widens the current falls, but the voltage across it increases. The parallel electromagnet therefore became stronger as the gap widened, and it was used to release the brake which was normally held on by a spring. Most arc lamps from the mid-1880s onwards worked in this way. Many designs were made, with the object of producing a cheap, reliable, yet sensitive control mechanism.

Arc lighting was widely adopted in places where it was suitable – in large buildings like markets and railway stations and for street lighting. King's Cross Station in London, for example, was lit by twelve lamps in 1882. The lamps, rated at 4000 candle power, were hung ten metres above the platforms and supplied from four Crompton-Burgin generators all driven by a single steam engine.

Two improvements in arc lighting were made during the 1890s. One was the enclosed arc, which had the arc contained within a small glass tube that restricted the air flow. The effect of this was to reduce the rate of burning of the carbons. The second improvement was the addition of cores of flame-producing salts, mainly fluorides of magnesium, calcium, barium and strontium, to the carbon rods. They increased the light output, and also gave some control over the colour of the light.

In 1890 there were reported to be 700 arc lamps in use for street lighting in Britain, and probably a similar number in private use. About 20,000 were installed in the following twenty years, but by then the filament lamp had been developed to an efficiency at least equal to that of the arc. Although few, if any, further arc lamps were installed, those that were already in place continued in use. London retained some arc street lighting into the 1950s.

THE FILAMENT LAMP

There was no call for a public electricity supply until the invention of a satisfactory filament lamp. Electric arc lighting was proving its worth for streets and public buildings, but it was quite unsuited to domestic use. The individual lamps were far too bright, too complex, and too large physically for use in the home, and they would probably also have been a fire hazard.

The idea of the filament lamp was almost as old as the arc lamp. Many people had tried to produce light by heating a fine wire electrically so that it glowed, but they all faced a series of seemingly insuperable problems. They had to find a material that would stand being heated repeatedly to white heat and then cooled, then it had to be sealed into a glass vessel in such a way that the glass did not crack when the wire was hot, and finally the air had to be pumped out so that the filament did not oxidize.

The early attempts at making a practical incandescent filament lamp all failed, mainly because of the difficulty of obtaining an adequate vacuum. After the invention of the Sprengel mercury pump in the mid-1870s several inventors succeeded in making viable lamps, the best known being Edison and Swan. At the first International Electrical Exhibition, held in Paris in 1881, four manufacturers had filament lamps on display: Swan and Lane-Fox from Britain and Edison and Maxim from the USA. There was little to choose between the four lamps at the exhibition, though Swan and Edison soon captured the market while the others disappeared from the scene.

Swan and Edison were very different men, with different approaches to their common objective of developing an electric light suitable for domestic use. As

Sir James Swinburne later remarked, 'Edison and Swan were hardly racing, as they were on different roads.'

Joseph Wilson Swan, later Sir Joseph, was born in Sunderland and apprenticed to a firm of druggists there. Subsequently he set up in business with his friend John Mawson as chemists and druggists in Newcastle upon Tyne. A man of wide scientific interests, he was intrigued by some of the early experiments towards an incandescent filament lamp, and made several experimental lamps himself in the 1840s and 1850s. He made carbon filaments from strips of paper treated with sulphuric acid to give a very smooth material called 'parchmentized paper' because of its resemblance to parchment. He carbonized the paper and mounted it in a glass vessel closed with a rubber stopper and then evacuated. However, he could not obtain a sufficient vacuum, and left the experiments for twenty years during which time he worked on other matters, especially photography: the most important of his inventions in that field was the silver bromide photographic paper still used for black-and-white prints. In 1867, John Mawson, who was by then his brother-in-law as well as close friend and colleague, was killed in an accident. Swan found himself responsible for the families and for a large chemical business.

About 1877, Swan was able to resume his interest in electric lighting, and the new Sprengel airpump gave him fresh impetus. He first spoke publicly about it on 19 December 1878, at an informal meeting of the Newcastle Chemical Society when several members gave brief talks. It seems probable that he did not have a working lamp to demonstrate then, but he certainly did so at several public meetings in the area in January to March 1879. During 1879 he worked to improve his lamp. The main problem, which was discussed at another Newcastle Chemical Society meeting on 18 December 1879, was that residual gas was occluded in the carbon and came out when the filament became hot. This gas carried particles of carbon which were deposited on the cooler glass, causing blackening. The solution was to pump the lamps while the filament was hot, and Swan applied for a patent for that process on 2 January 1880. He never tried to patent the basic idea of a carbon filament lamp, since he considered that there was nothing novel in that. During 1880 he worked on the filament material. While studying the problems of evacuating and sealing the lamp he had used mainly thin arc-lamp carbons for his filament – they were available only about one millimetre in diameter, and were relatively strong. The carbonized paper and carbonized parchmentized paper were not entirely satisfactory. He tried other substances and found a suitable material in parchmentized cotton, which was ordinary cotton thread treated with sulphuric acid, as the parchmentized paper had been. This gave a compact and uniform material which could be carbonized in a furnace to give satisfactory filaments. Swan applied for a patent for this filament on 27 November 1880, and went into commercial production in 1881 (see Figure 6.8). He never sought to manufacture

ELECTRICITY

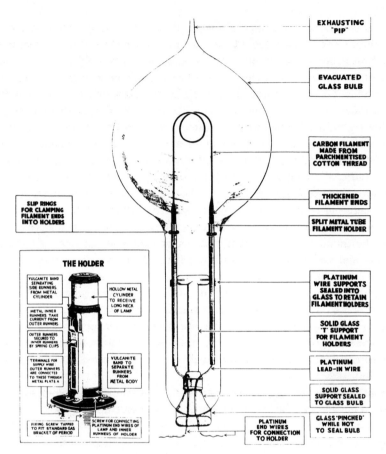

Figure 6.8: Advertising drawing published by the Ediswan company in the late 1930s, showing Swan's first successful filament lamp of 1881.

other components for an electric lighting system, although he worked closely with Crompton who was making generating equipment and arranging electrical installations.

Edison's approach was quite different. He had already made a name for himself as an electrical inventor, and had built up a research organization. He became interested in electric lighting late in 1877, after visiting William Wallace's electrical factory in Connecticut. Wallace made arc lighting equipment. Edison thought that a viable electric lighting system should have lamps of about the same power as the gas jets then in use and that electricity should be distributed in a similar way to gas, with each light being independently controlled. He wanted to produce both the lamps and the electricity supply system to feed them, and all the resources of his laboratory and staff were turned to the subject.

Edison's friend, the lawyer Grosvenor P. Lowrey, put up $300,000 to establish the Edison Electric Light Company in October 1878. It was in that month that Edison announced publicly: 'I have just solved the problem of the subdivision of the electric light.' The search for a viable filament lamp was often called the problem of 'subdividing' the electric light because the perceived need was for a much smaller lighting unit than the arc light. Edison's announcement, which received wide publicity and caused an immediate slump in gas shares, was based on a lamp with a platinum filament. This lamp contained a thermostat which momentarily cut off the current when the filament was in danger of overheating and melting. It was not until late 1879 that Edison turned again to carbon as his filament material. When he began commercial production of filament lamps in 1880 he used filaments made from Bristol-board – a thick paper with a very uniform texture. Continuing the search for better materials, however, he found that fibres from a particular variety of bamboo gave the best results, and used that from mid-1881 to about 1894.

Several other people made workable filament lamps, and at least two of them went into commercial production. Hiram S. Maxim, who is better known now for his work on guns and on aerial navigation, was an American by birth, but later became a naturalized Briton and was knighted. The other was an Englishman, St George Lane-Fox, who also designed a complete distribution system. His patents were acquired by the Anglo-American Brush Electric Light Corporation.

The manufacture of Swan's first commercial lamps was a complex enterprise. The ladies of the Swan household in Newcastle upon Tyne prepared the filaments and Swan himself carbonized them. The bulbs were blown by Fred Topham and all the components were conveyed to Birkenhead where C. H. Stearn mounted the filament assemblies in the bulbs and evacuated them.

A catalogue published by the Swan United Electric Light Company in 1883 lists more than a hundred houses and other buildings and twenty-five ships lit with Swan's lamps. Probably the most prestigious contract was for lighting the new Law Courts in London, which opened in December 1882. Crompton supplied the generators and six arc lamps for the large hall, and Swan supplied filament lamps for the courts and other rooms.

A number of large private houses were lit electrically, using current supplied from their own generating plant. Sir William Thomson lit his house in Glasgow. In a letter to Sir William Preece, who was lighting his house in Wimbledon, Thomson noted the need for lampshades. 'The high incandescence required for good economy is too dazzling and I believe would be injurious to the eyes if unmitigated. I have found that very fine silk paper round the globe spreads out the light quite sufficiently to make it perfectly comfortable to the eye while consuming but a small percentage of the light and Lady Thomson has accordingly made little silk-paper globes for nearly all our lights.' He goes on to say that there were 112 lights in the house.

CENTRAL POWER STATIONS

The first central electricity generating station offering a supply of electricity to the general public was probably the one that began operating at Godalming, Surrey, in the autumn of 1881.

Until that year the streets of Godalming were lit by gas, under a contract that expired at the end of September. In 1881 the town council and the gas company were unable to agree on the price to be charged for the coming winter's street lighting. An influential figure in the town was John Pullman, of R. & J. Pullman, leather dressers, who had a business based at Westbrook Mill, on the River Wey. It was probably Pullman who suggested that the town should have the new electric light rather than gas; he offered the use of his waterwheel to drive a generator in exchange for free light at the mill. The apparatus was soon installed. A Siemens generator at the mill supplied seven arc lamps and about forty Swan filament lamps. A few of each type of lamp were at the mill, the remaining arc lights were in the main streets of the town and the filament lamps in the side streets. It was announced that people who wanted electric lighting could have the wires taken into their homes. Although the lighting created great local interest and was reported in the local and national press, very few people took up the opportunity. In May 1882, Sir William Siemens said there were only 'eight or ten' private customers with a total of 57 lamps between them. The installation was never a commercial success, though the Siemens company felt that they learned useful practical lessons from it before, in 1884, the town reverted to gas lighting.

The River Wey proved to be an insufficiently reliable source of power for electric lighting, and within a few months the generator was removed from Pullman's mill and set up in the town centre, where it was driven by a steam engine. That move also helped with the solution to another problem, volts-drop in the wires. It was soon found that the voltage at the end of the supply cables was less than the voltage at the generator, and that therefore if the voltage was right close to the generator, then lamps at the other end of the circuit only glowed dimly. This problem was reduced when the generator was moved to the town centre, but it lead Sir William Siemens, in evidence to a House of Commons Select Committee, to express the opinion that no electricity supply station could be more than about half a mile from its most distant customer.

The Select Committee was considering the Bill which became, in August 1882, the world's first Electric Lighting Act. This Act laid down a general legislative framework for electricity supply, and was drafted on the assumption that an electricity supply undertaking would be a fairly local matter, and that the local authority should, if it so wished, have a large measure of overall control. At least three other public supply systems were in operation before the Act came into force. Edison had an experimental steam-powered station in London, at Holborn Viaduct, which began supply in January 1882 and was really a trial for his first New York station, which opened in September of the same year. Crompton extended a street lighting and market lighting installation in Norwich to supply private houses from about March 1882. Perhaps the most important of these early schemes, however, was the one at Brighton, Sussex. The electrical pioneer Robert Hammond had visited Brighton in December 1881 to give an exhibition of Brush arc lighting. The demonstration was so successful that Hammond was asked to extend it, and on 27 February 1882 he opened a permanent supply undertaking. Brighton has had an electricity supply for longer than anywhere else, for all the other very early undertakings closed within a short period.

A supply undertaking that did not use overhead wires and did not need to break up the streets could operate outside the provisions of the Electric Lighting Act, 1882. The largest company to use that loophole was the Kensington Court Electric Light Company. Kensington Court was a new housing development just south of Kensington High Street, in west London. About a hundred houses on the estate were linked by a system of subways, in which the company laid its mains.

Crompton was the leading figure in the Kensington Court Company, which was registered in June 1886 and commenced supply in January 1887. Initially they had only three customers, and by the end of the year there were still only nine. Requests for a supply soon came from people outside the estate, and the company obtained a licence from the Board of Trade to increase their area of supply. The initial generating plant was rated at 35kW (47hp), but additional plant was soon added, and by 1890 the generating capacity was 550kW (738hp).

TRANSMISSION: AC v DC

Except for very small local systems, all supply undertakings had to solve the problems of transmitting electricity at high voltage and then reducing and stabilizing the voltage at a point near the customer. A great rivalry, the 'battle of the systems', ensued between the proponents of alternating current and those of direct current. The advantages of AC were that it was easy to change the voltage up and down by means of transformers, and the voltage could be adjusted by tap-changing on the transformers. If, however, it was desired to maintain the supply day and night, then at least one generator had to be kept running all the time. With a DC system batteries could be used to maintain the supply at times of low demand. Furthermore, DC generators could easily be operated in parallel: parallel operation of AC machines was always difficult. Another disadvantage with AC, initially, was the lack of a practical electric motor, but after induction motors were developed around 1890, AC supplies became attractive to potential industrial customers.

The leading pioneer of AC electricity supply was Sebastian Ziani de Ferranti, a Liverpudlian of Italian extraction. At the age of 17 he was working for Siemens Brothers at Charlton, but he soon branched out on his own. His first company, Ferranti, Thomson and Ince Ltd, was formed in September 1882, with Alfred Thomson, another engineer, and Francis Ince, a lawyer, to manufacture generators; the following year the firm was dissolved and Ferranti, still only 19 years old, set up in business on his own, manufacturing generators, meters and other equipment in London.

Within a few years he was chief engineer of the Grosvenor Gallery Company, which had sought his help with technical problems. This company had been formed by Sir Coutts Lindsay to light his art gallery in New Bond Street. Neighbours had been impressed and sought a supply, with the result that the company was soon supplying electricity over a substantial area of Westminster and surrounding districts. They employed the Gaulard and Gibbs distribution system in which electricity was distributed at high voltage to transformers (known as 'secondary generators') at or near to each customer. The primary windings of all the transformers were in series and the current in the system was maintained constant at 10 amps. Individual loads were fed from secondary windings on the transformers, and the transformer design was such that the secondary voltages were roughly constant whatever the load.

As the Grosvenor Gallery Company's network expanded, every additional secondary generator required an increase in the circuit voltage, and the practical limit was soon reached. When Ferranti took charge he rearranged the system for parallel working, using distribution at 2400 volts and a transformer on each consumer's premises to reduce the voltage to 100. He also replaced the Siemens generators originally used by machines of his own design, each able to supply 10,000 lamps of 10 candle power, which required about 350kW. The supply was not metered: each customer paid £1 per 10 candle power lamp per year.

As business expanded still further a new company, the London Electric Supply Corporation Ltd, was formed in 1887 with a capital of one million pounds. Ferranti planned a massive power station on the banks of the River Thames at Deptford. Land there was relatively cheap; there was easy access for coal barges, and there was ample cooling water. He designed generators of 10,000hp (7460kW) and planned to transmit electricity into London at 10,000 volts. Such a pressure was quite unprecedented, and he had to design everything himself, including the cables. When the Board of Trade questioned the safety of the system he arranged a demonstration in which a concentric cable of his own design was made live at 10,000 volts and a workman drove a chisel through it. The workman was unhurt and the Board of Trade were satisfied. The main conductors were concentric copper tubes separated by paper impregnated with ozokerite wax. They were entirely successful, some remaining in service until 1933.

The generators, however, were less successful and, to make matters worse for Ferranti, the Board of Trade would only allow the company to supply a much smaller area of London than he had hoped. The directors decided to build only a small part of Ferranti's planned generating plant, and in 1891 he resigned from the Corporation.

The Deptford scheme was very ambitious, and although the subsequent history of electricity supply shows that Ferranti was working on the right lines, he may well have been more immediately successful if his plans had been rather more modest.

All public electricity supply systems are now entirely AC, although the last DC supply in Britain – to a Fleet Street newspaper – remained in use until 1985. For transmission over long distances DC systems are sometimes preferred. Very high power mercury arc valves, developed in Sweden in the 1930s, permit very heavy currents to be rectified, transmitted over a DC line, and then inverted back to AC. DC transmission is also used when it is desired to link two AC systems which operate at different frequencies or which operate at the same frequency but cannot be kept in synchronism. An example of the former is in Japan, where part of the country operates on 50Hz and part on 60Hz, and the two networks are interconnected through a DC link. The cables under the English Channel linking the British and French grids also operate on DC. The first Anglo-French link, opened in 1962, had a capacity of 160MW. New cables, laid in 1986, with a capacity of 2000MW, are buried in a trench cut in the sea floor to avoid damage by ships' anchors or fishermen's trawls.

ECONOMICS: OFF-PEAK LOADS

At first the public electricity supply was used almost exclusively for lighting, but this meant that the system was not used to its maximum efficiency because the demand was concentrated in to a few hours each day. In 1888, Crompton produced a graph of the demand for electricity over an average day in which he showed that the load exceeded 50 per cent of the peak load for only five hours out of the twenty-four, and exceeded 75 per cent of peak for only about two hours. The peak load determined the plant costs, and the plant had to be manned all day and all night. Any load that could be attracted at times other than the peak hours could therefore be supplied economically at a price barely greater than the fuel cost.

The potential sources for such off-peak loads were electric heating and cooking, and industrial uses. Electric cooking was vigorously promoted, by advertising and by articles in the press. Most supply undertakings offered electricity at reduced price – often half-price – for cooking and hired out appliances at low rentals. A school of electric cookery was opened in Kensington, London, in 1894 to promote the idea. In 1912, A. F. Berry opened his 'Tricity House' which combined a restaurant serving 700 meals a day with a showroom for cooking equipment. Berry also gave lectures on the advantages of electric cooking, in which he argued that one ton of coal in a power station could do as much cooking as ten tons of coal delivered to the house (see Chapter 19).

The reduced prices for electricity used other than for lighting continued into the 1930s. Today off-peak electricity is still available in the UK at about half-price, but lighting is now only a fraction of the total demand and the off-peak hours are in the night.

MEASUREMENT

The rapid progress of electric lighting in the 1880s, created a demand for practical and reliable measuring instruments. Most important were devices for measuring voltage and consumption. Accurate voltage measurements were essential because the life of filament lamps was critically dependent on the supply voltage. Devices to measure electricity consumption were required so that customers could be charged according to the electricity they had used.

The first scientific attempts at electrical measurements were in the eighteenth century. Electrostatic forces between charged bodies were measured as early as 1786 when the Revd Abraham Bennet wrote a letter to the Royal Society describing the gold leaf electroscope. In this two pieces of gold leaf hanging side by side are in electrical contact with a conductor whose potential is to be observed. Since the gold leaves become similarly charged they repel each other, and hang apart at an angle which is a measure of the potential. The most important of the early workers on electrostatic measurements was Charles Augustin de Coulomb, who invented the torsion balance electrometer and used it to establish the inverse square law relating the force of attraction between charged bodies and the distance between them. Coulomb's instrument had a long silk thread supporting a horizontal straw inside a large glass cylinder. The straw was covered with sealing wax and the silk thread was fixed to a cap at the top of the cylinder which could be turned. Also inside the cylinder was a metal ball connected to an external knob which could be charged. The device was so sensitive that Coulomb could measure a force of a few milligrammes.

More immediately relevant to the practical business of electricity supply, in the late nineteenth century Sir William Thompson designed electrostatic voltmeters which relied on measuring the force of attraction between charged elements of the instrument. Two patterns were in widespread use. The quadrant electrometer, which was sold commercially from 1887, had two vertical sheets of metal with a narrow space between them into which a pivoted vane was drawn by electrostatic attraction. This instrument could measure up to 10,000 volts. His multi-cellular meter, made from 1890, had what were essentially a number of 'quadrant' movements side by side on the same axis, and these meters were used for voltages of a few hundred. A modified form of the quadrant electrometer, with additional vanes, was used as a wattmeter.

Until the relatively recent advent of electronic measuring instruments, most current measurements have depended on measuring the magnetic force created by a current in a wire. This magnetic force has been balanced against a controlling force, which may be produced by a spring, by gravity, or by another magnet. Simple galvanometers, in which a large diameter coil surrounded a pivoted magnetic needle, with the earth's magnetic field providing the controlling force, were made in 1837 by Claude-Servain Pouillet. Such instruments were accurate and straightforward to use, but they were bulky and the needle took a long time to come to rest, so measurements were slow. The electricity supply industry wanted instruments that were portable, quick to read, and stable in the readings they gave. In the 1880s there were no really stable magnetic materials, and simple instruments making use of a magnetized needle had to be recalibrated from time to time.

A current measuring instrument which did not depend on the vagaries of magnetic materials was the electrodynamometer. This measured the forces between current-carrying conductors, so it was really only suitable for high current work. Being very reliable, however, it was frequently used in laboratories and for calibrating other instruments. The principle of the electrodynamometer is due to Wilhelm Weber, who was following up Ampère's mathematical study of the forces between currents. A moveable coil, usually supported by a torsion suspension, hangs within a fixed coil. When current flows in both coils magnetic forces tend to twist the moveable coil. Usually the moveable coil is returned to its original position by twisting the top of the suspension through an angle which is a measure of the current. An instrument of this kind was used to monitor the operation of the generator in the first public electricity supply system, at Godalming.

A laboratory version of the electrodynamometer was the Kelvin Ampere Balance, in which the attractive force between fixed and moving coils was balanced by a sliding weight on a balance arm. The arrangement was similar to a balance for weighing. In 1894 two such instruments were held by the Board of Trade as legal standards for the measurement of current. The accuracy was stated to be within 0.2 per cent – a matter for great satisfaction at the time!

The need for calibration, and all the problems attendant on changing calibration, can be avoided if the instrument is used simply to determine the presence or absence of a current. This is the principle of the potentiometer and bridge measuring circuits.

In the potentiometer, which was first described in 1841, a constant current was passed through a straight uniform wire placed alongside a linear scale. The potential between two points on the wire was therefore proportional to the distance between them, and this could be read from the scale. Once the scale had been calibrated by means of a voltage standard, other voltages could be measured accurately. The unknown voltage was connected through a galvanometer to a length of the potentiometer wire, and when the galvanometer showed that no current flowed then the unknown voltage was equal to the voltage across that length of wire.

The potentiometer was used in the bridge circuit described by Wheatstone in a long Royal Society review paper on electrical measurements in 1843. (Although it has been known ever since as the Wheatstone bridge, he clearly attributed it to S. H. Christie.) The bridge circuit was intended for comparing resistances. An unknown resistance and a standard were connected in series, carrying the same current, and in parallel with a potentiometer. A galvanometer was used to determine the point on the potentiometer wire at which the voltage was the same as at the junction of the two resistances, and the ratio of the resistances was then the same as the ratio of the two sections of the potentiometer.

The rapid progress in electric lighting around 1880 created a demand for direct reading instruments. The new filament lamps were far more critically dependent on the supply voltage than the older arc lamps, which could tolerate wide fluctuations without damage. The first practical moving iron ammeter was devised by Professors W. E. Ayrton and John Perry in 1879. It had an iron core which was drawn into a circular coil carrying the current to be measured. In the 'magnifying' version the core was restrained by a spring wound helically from flat metal strip. As the spring was stretched the free end rotated, turning the pointer. Other moving iron instruments of the period included Schuckert's, in which an asymmetrically mounted piece of soft iron was made to twist by the magnetic field within a current-carrying coil.

The moving coil meter, with a light coil hanging in the annular air-gap between the poles of a permanent magnet and an iron cylinder, was devised by J. A. d'Arsonval. Such instruments were first made on a large scale by the Weston Electrical Instrument Company of New Jersey. Weston introduced the idea of 'damping' the movement by mounting the coil on a copper former, so that its motion was damped by eddy currents and it came to rest quickly when a reading was taken. He also used jewelled bearings to reduce friction. The unipivot instrument, in which the moving parts are carried on a single bearing to give further reduced friction and greater sensitivity, was introduced by the London instrument-maker R. J. Paul in 1903.

Hot-wire instruments make use of the fact that a wire expands when heated. Philip Cardew designed the first commercial hot-wire instrument in 1883. It was reliable and accurate, but rather unwieldy, having three metres of platinum–silver wire stretched over pulleys in a tube almost a metre long. One end of the wire was fixed, the other held taut by a spring and connected by gearing to a pointer. The hot-wire principle was adapted by Hartmann and Braun in the 1890s to produce a more manageable instrument. They utilized the sag of a taut wire rather than its change in length. The pointer of their instrument was driven by a wire attached to the middle of the current-carrying wire and pulling it sideways.

Arthur Wright, the engineer of the Brighton electricity undertaking (see p. 370), made the first recording meter about 1886 to monitor the load on his system. He used a strip of paper coated with lampblack which was pulled along by clockwork and marked by a pointer on a simple moving iron ammeter.

Commercial electricity supply required meters to measure the energy consumed, so that customers could be charged accordingly, although a few early systems simply charged according to the number of lamps connected. Edison made electrolytic meters, in which the current to be measured passed through a cell containing two zinc plates in zinc sulphate solution. The plates were weighed periodically to measure the transfer of zinc and hence the current that had flowed.

The first direct reading supply meter was devised by Ayrton and Perry, though usually known as the Aron meter after the man who improved and then manufactured it. These instruments made use of the fact that the speed of a pendulum clock depends on the force of gravity. Two pendulum clocks were coupled together through a differential gearing connected to dials which showed the difference in 'time' measured by each clock. Coils were arranged adjacent to the iron pendulum bobs so that the apparent force of gravity varied in accordance with the supply being measured.

Most energy meters have what is in effect an electric motor driving a disc which is restrained by an eddy current brake. The motor torque is arranged to be proportional to the electrical energy passing; the braking torque is proportional to speed. The total number of revolutions of the disc is then a measure of the energy consumed. Elihu Thomson, born in England but brought up in the USA, devised the first motor meter in 1882. He founded the Thomson-Houston company in 1879, jointly with his former teacher E. J. Houston, to make arc lighting equipment, but they soon expanded into the whole range of electrical manufactures. Ferranti's mercury motor meter had a copper disc rotating in a bath of mercury, with current flowing radially in the disc.

The Edison, Thomson and Ferranti meters described were essentially DC instruments (although Thomson's would work on AC). After the invention of the induction motor (see p. 384) induction type instruments were adopted for all AC systems and are now used in virtually all electricity supplies.

With alternating current systems there was interest in the nature of the waveforms involved. A simple contact operated synchronously with the supply by a cam made it possible to monitor the supply voltage (or current) at any specific point in the cycle, and by taking a series of such measurements it was a simple though tedious process to build up a picture of the waveform. Wheatstone had developed the method in the course of his telegraph researches, and the idea was reinvented by Joubert in 1880 when studying the behaviour of arc lighting circuits. A mechanical oscillograph which would give a visual display of a complete waveform was suggested in 1892 but only achieved in 1897 when William du Bois Duddell succeeded in making a galvanometer whose movement was light enough to follow the variations of an alternating current waveform. In another field, medicine, such instruments were developed to study the electrical action of the heart.

The discovery of the electron in 1897 led to the cathode ray tube. Tubes were made which produced a beam of electrons on a screen and could deflect the beam in two axes at right angles. Circuits were developed to give a deflection varying with time on one axis. The voltage being studied was then applied to deflect the beam in the other axis, and a complete waveform drawn out on the screen. A review of such instruments published by the Institution of Electrical Engineers in 1928 gave more space to mechanical than to cathode ray oscilloscopes, but during the 1930s the cathode ray instrument completely replaced the mechanical.

The first internationally agreed electrical standards were drawn up at the International Congress of Electricians that met in Paris in 1881. Before that many workers used their own standards, though the British Association had been considering national standards for some years. The Congress defined units for resistance, current and voltage, and since that time electrical science and engineering has benefited from universally accepted standards of measurement. This work is now the responsibility of the International Electrotechnical Commission, in Geneva.

ELECTROMAGNETIC ENGINES

The story of the electric motor really begins with Oersted's discovery in 1819 that a compass needle could be deflected by an electric current. In 1821, Michael Faraday showed that it was possible to produce continuous rotary motion by electromagnetism (see p. 355). During the nineteenth century many machines were designed that produced mechanical motion from electromagnetic effects, collectively known as 'electromagnetic engines'. One early machine was made by William Sturgeon in 1832, and it was probably the first electromagnetic engine to be put to practical use – to turn a roasting spit. Sturgeon's machine had a vertical shaft carrying two compound permanent magnets, one at the top, one at the bottom, with their north poles pointing in opposite directions. As the shaft turned, the ends of the permanent magnets passed close to the poles of the four vertical electromagnets fixed on the base board. The commutator was an elaborate arrangement with two concentric mercury cups carried round by the shaft, into which wires dipped, and a horizontal disc cut into four quadrants, with wiper arms pressing on them. Another early but little known maker of electromagnetic engines was Sibrandus Stratingh, a medical doctor and Professor of Chemistry at Groningen in the Netherlands, who wanted to make an electric road vehicle. In 1835 he constructed a table-sized model, but he never achieved a full-size version. Like a number of other early inventors, however, he did make an electric boat and he managed to take his family in a boat, electrically powered, in 1840.

The first person to obtain patents for electromagnetic engines was Thomas Davenport. He patented a machine in the USA in 1837, and later the same year obtained an English patent also. A model of his machine, now in the Smithsonian Institution in Washington, has a rotor consisting of four coils on a cruciform frame fixed to a vertical shaft. Opposite pairs of coils are connected in series and the ends of the wires go to simple brushes which press on a two-part commutator consisting of two semi-circular pieces of copper. The battery is connected to the copper pieces. The stator is two semi-circular permanent magnets with their like poles adjacent.

Another American inventor was W. H. Taylor, who exhibited a motor in London in 1840. Taylor's machine was written up enthusiastically in the *Mechanics Magazine* (see Figure 6.9). The construction was quite simple. An arrangement of four electromagnets on a frame surrounded a wooden wheel with seven soft iron armatures around its edge. A simple commutator on the axis switched on each of the four electromagnets in turn. Taylor claimed that earlier ideas for electromagnetic engines had depended on reversing the polarity of electromagnets. He said that his invention was the idea of switching the magnets so that they were simply magnetized and demagnetized, but not reversed in polarity. It seems that he realized that it took a significant time to reverse the polarity of an iron-cored electromagnet.

The Scotsman Robert Davidson made motors which also operated by switching the electromagnets on and off, not reversing them. In the winter of 1841–42 the Royal Scottish Society of Arts gave him financial help with his experiments, and in September 1842 he made an electrically driven carriage which ran on the Edinburgh and Glasgow railway. The four-wheeled carriage was nearly 5m long and weighed about 5 tonnes. The motors were wooden cylinders on each axle with iron strips fixed on the surface. There were horseshoe magnets on either side of the cylinder which were energized alternately through a simple commutator arrangement on the axis. The batteries, a total of 40 cells each with a zinc plate between two iron plates, were ranged at each end

ELECTRICITY

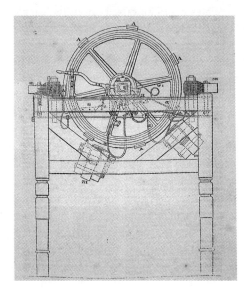

Figure 6.9: Engraving of Taylor's electromagnetic engine. Reproduced from *Mechanic's Magazine*, 9 May 1840.

Figure 6.10: Electromagnetic engine by Wheatstone at the time Taylor's machine was exhibited.

of the carriage. The plates could be raised out of the wooden troughs containing the electrolyte by a simple windlass. The carriage could run at about 6.5kph (4mph) on level track. There is no contemporary record of the distance actually travelled by the vehicle: presumably it did not actually travel all the way from Edinburgh to Glasgow. In 1839 the Tsar gave a grant to Professor M. H. Jacobi of St Petersburg for work on an electric motor, probably the first government grant ever given for electrical engineering research. In a letter to Michael Faraday, Jacobi described how he had arranged for an electromagnetic engine to drive the paddlewheels on a boat and travelled for days with ten or twelve people aboard on the River Neva. When supplied from a battery of 128 Grove cells, the vessel travelled at about 4kph.

In some cases enough data is given in contemporary records for the efficiency of these machines to be calculated, and figures of 10 to 20 per cent are obtained. The efficiency of electromagnetic engines was possibly a matter of some interest in the 1840s. In 1843, Charles Wheatstone was describing a variety of electrical devices in a paper to the Royal Society. One was the rheostat, which developed initially as a measuring device, but Wheatstone said that it could be used for controlling the speed of a motor; he also said, wrongly, that a rheostat in series with a motor could control its speed without any loss of efficiency.

In 1849 the United States Commissioner of Patents, Thomas Ewbank, included some thoughts on the subject of electric motors in his annual report to Congress. He said: 'The belief is a growing one that electricity . . . is ordained to effect the mightiest of revolutions in human affairs.' He referred to various experiments with electric motors and then continued, somewhat pessimistically, 'but these experiments, interesting as they certainly were, have brought no marked results, nor afforded any high degree of encouragement to proceed. It might be imprudent to assert that electromagnetism can never supersede steam, still, in the present state of electrical science the desideratum is rather to be hoped for than expected.' Ewbank's pessimism was not shared by Congress. In 1850 the United States Congress gave $20,000 to Professor Charles Page of Massachusetts, to develop electromagnetic engines, apparently with the navy mainly in mind. In a report to the Secretary of the Navy, Page said that he had made machines of one and four horsepower and asked for a grant to build a machine of 100hp.

The most ingenious of all the early electromagnetic engines must surely be Allan's machine made in 1852. This is basically a reciprocating engine with four cranks and four 'piston rods'. Each piston rod carries four armatures which press on collars on the rod but are not otherwise fixed to it. There are 16 sets of coils, one for each armature, and the coils are energized one at a time by a commutator. Each electromagnet is therefore energized for $\frac{1}{16}$ of a revolution. As each armature reaches its electromagnet it is stopped by it, while the piston rod continues its travel.

All these machines were, in any commercial sense, failures. They were entirely dependent on expensive chemical batteries, but if that were the only reason then the electromagnetic engine would have flourished when generators made electricity readily available in the 1880s. In modern terminology these engines were all 'magnetic' machines which depend on direct magnetic attraction between stator and rotor. Most modern motors, such as the induction motor (see p. 384), are electromagnetic machines in which the fields on one side of the air gap generate currents on the other. Conventional machine theory tells us that magnetic machines get better as they get smaller, while electromagnetic machines get better as they get bigger. Therefore, the future of electric power lay with electromagnetic machines. This was beginning to be appreciated by about 1880, although no theoretical reasoning had been set out.

PRACTICAL ELECTRIC MOTORS

The fact that the machines used as generators could be used also as motors was recognized quite early. The Siemens brothers, for example, noted in 1872 that the 'small rotating machine runs just as well as a motor as it does as a generator'. The firm of Siemens & Halske set out to find customers requiring electric power transmission. The first public exhibition of the electrical transmission of power by means of a generator and motor was probably the demonstration at the Vienna Exhibition in 1873. There the Gramme Company showed two identical machines linked by wires but 500m apart; one was used as a generator, the other was used as a motor driving a pump. Similar demonstrations were given at other exhibitions in the USA and in Britain in the following few years. By 1874, Gramme had electrically-driven machinery in his Paris factory – though he used a single large motor driving a line shaft, as in a steam powered factory, not an individual motor for each machine.

Several exhibitors at the first International Electrical Exhibition, held at Paris in 1881, demonstrated motors. Marcel Deprez, for example, had a motor-driven sewing machine, lathes, a drill and a printing press. Siemens exhibited a lift within the building and a tram running towards it along the Champs-Elysées.

One of the first Gramme machines to be sold as a motor was used to drive a conveyor for sugar beet in a factory at Sermaize, France. It was so successful that in May 1879 the factory owners, Messrs Chrétien & Félix, decided to try electric ploughing. The plough was hauled across a rectangular field by ropes between two wagons, each carrying a motor and a winding drum.

The first company to exploit electric motors on a large scale was Siemens, and their first major customer was the Prussian state mines. Electricity in mining, however, developed fairly slowly. Although some mine winders were electric before the First World War it was only in the 1920s that mine electrification became widespread. A major industrial use of electric power was the iron and steel industry. Several ironworks adopted electric lighting quickly because it facilitated all-night working, so perhaps ironmasters were easily alerted to the possibilities of the new power. Electric motors proved to be very good for driving rolling mills, where the combination of power and precise control was valuable. Thereafter electric motors were gradually introduced for driving machine tools, and the advantages of individual drives over line-shaft systems were readily appreciated.

The first permanent public electric railway was opened at Lichterfelde in Germany in 1881. Built by Siemens, it ran for about three kilometres. Each carriage had a motor under the floor connected to the wheels through a belt drive. The first electric railway in the United Kingdom ran between Portrush and Bushmills, in Ireland. Electric traction was chosen there because abundant water power was available and a hydro-electric generating station was built on the River Bush. England's first electric railway was the Volk's Railway which still runs on the sea front at Brighton.

The railways just mentioned were all small systems. The first really practical electric tramway system was built by Sprague in Richmond, Virginia, in 1888. Forty cars powered from overhead conductors ran over 20km (12.5 miles) of streets. Frank J. Sprague trained as an engineer with the US navy, then set up his own electrical engineering company in 1884. His most important contribution was the multiple unit control system, which made it possible to have motors distributed along the length of a train and supplied from current collectors on each coach but all controlled from the driver's cab. He also introduced the motorized bogie construction in which one end of the motor is pivoted on an axle and the other supported on springs.

Electric trams were introduced only slowly in Britain, partly because the established horse-drawn tramways were approaching the date when, under the Tramways Act, they could be purchased compulsorily by the local authorities. New tramways after 1890 were virtually all electric.

Deep tube railways in London and other cities only became practicable with electric traction. The first was the City and South London Railway which ran initially from Stockwell to the Bank. The original rolling stock was fourteen locomotives each of which could haul three carriages with thirty-four passengers in each. Five million passengers were carried in the first year. Electricity was generated in a specially built power station at Stockwell, and supplied through a third rail. The service interval at periods was just under four minutes.

Most early electric traction systems used direct current because the DC series wound motor has good operating characteristics for the purpose. In 1920 the Ministry of Transport adopted 1500 volts DC as the standard for main line railways. Technical advance soon forced a change in the standard, however. The introduction of the mercury arc rectifier in 1928 made it possible to transmit AC and convert to DC on the train; most of British Rail now uses this system, with 25kV overhead lines. Since about 1960 semiconductor rectifiers have been available for high powers, and are replacing the mercury arc rectifiers. The Southern Region, where the traffic density is higher than in most of the country, uses a third rail for the conductor and operates at about 700 volts DC. Other countries have used a variety of systems and frequencies. Switzerland and Italy have some three-phase railways, with one phase earthed to the running rails and the other two phases connected through two overhead wire systems.

The advantages of alternating current transmission encouraged engineers to develop AC motors. Most DC motors will in fact operate on AC supplies, provided that the iron cores in their fields are laminated, but they are not so efficient. Such machines are called universal motors, and are often found in small domestic appliances such as vacuum cleaners and food mixers.

The first practical AC motors were developed by Nicola Tesla in 1888. He was born in Austria-Hungary but in 1884 emigrated to the USA where he spent most of his life. He worked for a time for Edison, the leading exponent of DC systems in the USA, then joined Westinghouse, the leading AC man.

In his machines Tesla made use of the fact, discovered by Arago in 1824, that a piece of magnetic material free to turn will follow a rotating magnetic field. He created a rotating magnetic field by using two coils energized from supplies that were in synchronism but not in phase. His first machine had two coils with axes at $90°$ to each other supplied with alternating currents also $90°$ out of phase. He showed that the resultant of the two oscillating magnetic fields was a rotating field, and he performed the same analysis for a three-phase system at $120°$. Tesla's first motor was a synchronous one – that it, the rotating member either was or became a permanent magnet, and its poles followed the rotating field round, keeping in synchronism. He also made induction motors, in which the rotating member is not a permanent magnet and turns at a speed slightly lower than the speed of the rotating field. Currents are then induced in the rotor and these interact with the rotating field to provide the driving force.

Other people were also working on the idea of a motor driven by a rotating magnetic field, and Tesla's patent claims were challenged in the US courts, but his claims were upheld. He also showed that an induction or synchronous motor can be run from a single phase supply if part of the field winding is connected through a capacitor or inductor to give a second phase. Once started, such a motor will run satisfactorily on a single phase supply. The Westinghouse company bought Tesla's patents, and from 1892 they were manufacturing AC motors and promoting AC supply systems.

In the twentieth century most of the world's electric motor power comes from induction motors. Their disadvantage is that they are essentially constant speed machines, but for many applications that is perfectly satisfactory and the inherent simplicity and robustness of induction motors, which usually have no brush gear, make them the first choice for many applications. Most domestic washing machines are driven by an induction motor with capacitor start.

A motor that is sometimes confused with the induction motor is the repulsion motor, developed by Elihu Thomson and Professor J. A. Fleming. Fleming, better known for his work on radio, studied the forces between conductors carrying alternating currents. In 1884 he showed that a coil carrying alternating current tries to position itself edge on to a magnetic field, and the force produced in this way provides the basis for the repulsion motor. A typical repulsion motor has a single field coil connected to the supply and a wound multi-coil armature with commutator. Two brushes on opposite sides of the commutator are connected together (but not to the supply) and short circuit the armature coils which at that instant are across the magnetic field. There is then a turning force which moves the armature. Repulsion motors have been used for electric traction. They have a good starting torque and some speed control is possible by moving the brushes.

MODERN ELECTRIC MOTORS

By the mid-twentieth century it seemed reasonable to say that electric motor development was complete, but in 1957 Professor G. H. Rawcliffe developed the pole amplitude modulated, or PAM, motor. This is a synchronous or induction motor whose field coils are so arranged that by interchanging a few connections the number of poles can be changed. Since the speed is determined by the number of poles (as well as by the supply frequency) this gave a motor whose speed could be switched between two distinct values. A PAM induction motor therefore retains the reliability and robustness of the conventional induction motor but can work at two different speeds.

The other approach to variable speed control is to change the supply frequency. With power semiconductors that is becoming possible. By 1960 semiconductor devices were available capable of controlling a few tens of amperes, but progress in the following decade was so rapid that by the end of it semiconductor frequency convertors were available capable of supplying the largest motors – and controlling their speed.

Another area of motor research that remains active is linear motors. Often described as conventional motors that have been slit open and unrolled, linear motors have been a subject of research at least since 1841, when Wheatstone made one. The idea was revived in 1901 when the Norwegian Kristian Birkeland tried to use a linear motor as a silent gun. Their best-known modern exponent is Professor Eric Laithwaite, who was first interested in them as a means of driving a weaving shuttle. Linear motor research continues, and the widespread application of these machines will probably use semiconductor controls also.

THE STEAM TURBINE

In nearly all the early power stations the prime movers were reciprocating steam engines. The technology was well established, and the electrical designers made generators to be driven by the available steam engines even though their rotational speed was less than ideal. Some stations used a belt drive to increase the speed, though Willans and Belliss & Morcom high speed engines were often directly coupled to the generator. The higher rotational speed of the turbine made it the ideal prime mover for power stations.

Generators for use with turbines have usually only a single pair of field poles, rather than the multi-polar machines used with reciprocating drives. Initially the armature windings, in which the current was generated, were the rotating member and the field poles were static. As machines became larger, it became difficult to make brushes and slip rings or commutators adequate to take the current. The solution was to 'invert' the machine, having the armature static and the field rotating. The brushes then had to carry only the magnetizing current for the field. The last large rotating armature machine was a 1500kW set for Neptune Bank power station on Tyneside in 1901.

The first rotating field generators had salient poles built up on the rotor shaft. The Anglo-Swiss engineer Charles Brown, of the Brown-Boveri partnership, proposed that the rotor should be a single forging and that the windings should be carried in slots milled in the surface. This basic design has been used for large generators ever since.

The largest turbines and generators now used by the Central Electricity Generating Board, serving England and Wales, are rated at 660MW. Today the entire electricity demand of the United Kingdom could be supplied from only one hundred generators.

ELECTRICITY TODAY

Modern life depends on electricity. Virtually every home in Britain is connected to the public electricity supply, though that has been achieved only since the Second World War. In 1920 the supply industry had under a million customers in England and Wales. The figure reached 10 million by 1945 and 15 million by 1960. Now there are 21 million, and they use about two hundred thousand million units of electricity per year. Of this consumption 36 per cent is used at home, 38 per cent in industry, 22 per cent in commerce, and the remaining 4 per cent in such diverse applications as farming, transport and street lighting. Of the domestic electricity, 21 per cent is used for space heating, 18 per cent for water heating, 11 per cent for cooking and 17 per cent for freezing and refrigeration. Everything else, including lighting, comes out of the remaining 33 per cent.

At first electricity was only for the well-to-do. The major expansion came during the 1920s and 1930s, and during that period the average consumption per household fell, reflecting the fact that new consumers used electricity mainly for lighting, and not much for other purposes. The range of domestic electrical appliances with which we are familiar today have in fact been available almost from the beginning. Catalogues of the 1890s include electric cookers, kettles, saucepans, irons and fires. Early electric fires used carbon filament lamps as the heating member because there was no metal (except platinum) which could be heated to red heat in air without oxidizing. A great advance came in 1906 with the alloy nichrome, a mixture of nickel and chromium. This does not oxidize when red hot, and most electric fires since that date have used nichrome wire elements on fireclay supports.

Storage heaters for room heating were introduced on a small scale in the 1930s. In the 1960s the Electricity Council conducted research to improve their design, seeking longer heat retention, and modern storage heaters are much smaller than their earlier counterparts.

Motorized appliances generally came later than lighting and heating, though an electric table fan was on sale by 1891. The first electric vacuum cleaner was made in 1904. Early electric washing machines had a motor fixed underneath the tub. Usually there was a mangle fitted on top (spin driers came later) and a gearbox that permitted the user to couple the motor either to the agitator in the tub or to the mangle. Food mixers and refrigerators came after the First World War, though they were rare until the 1950s (see Chapter 19).

Electric space heating and refrigerators have changed house design. Before the mid-1930s it was normal to have a fireplace in every bedroom, and into the 1950s every house was built with a larder. Many modern houses have no fireplace, except possibly one in the living-room for effect. Larders have become obsolete since it is assumed that food which might go bad will be kept in the refrigerator.

Lighting has also progressed. The carbon filament lamps that were such a wonder in the 1880s and 1890s encouraged the gas industry to develop the mantle, and for a time gas lighting undoubtedly had the edge over electricity. The electric lighting industry sought a better filament material. Three metals seemed promising: osmium, tantalum and tungsten. Osmium filament lamps were on sale from 1899, but since 1909 all metal filament lamps have used tungsten. Carbon lamps continued to be made for some years since they were cheaper in first cost, but the metal filament lamps were more efficient, giving cheaper light when the cost of electricity was taken into account. The latest development in high-power filament lamps is the inclusion of a halogen gas (usually bromine or iodine). This reacts chemically with tungsten that evaporates from the filament and is deposited on the glass. The resulting tungsten halide is a gas which decomposes close to the hot filament, depositing the tungsten back on to the filament. Such lamps can be run at a higher temperature and are therefore more efficient.

Various gas discharge lamps were made in the 1890s, and neon lamps were introduced about 1910. The widespread use of both mercury and sodium discharge lamps dates from the 1930s. The low pressure sodium lamp, with its extremely monochromatic yellow light, has been popular for street lighting because it is the most efficient of all. Since the early 1970s, however, the high pressure sodium lamp has been taking over. It is almost as efficient, and although its light has a yellow-pink tinge its colour rendering ability is fairly good.

Fluorescent lamps, developed in Britain just before the Second World War, have an efficiency in between that of filament lamps and discharge lamps. A low pressure mercury discharge within them produces ultra-violet light which acts on the fluorescent coating of the tube to give visible light. The choice of phosphor determines the colour and the efficiency of the lamp, and they are widely used in commercial applications.

One great advantage of electricity is its easy controllability, and with time-switches, thermostats and semi-conductor dimmers that is even more true than before. Other technologies have done much for mankind: electricity has put virtually unlimited power at the disposal of all.

7

ENGINEERING, METHODS OF MANUFACTURE AND PRODUCTION

A.K. CORRY

INTRODUCTION

'Man is a tool using animal' said Thomas Carlyle but, additionally and more significantly, man has progressively designed and made tools for use in meeting his developing needs and adapted his techniques and social organization to make the best use of his skills of mind and body in manipulating materials to his advantage. Side by side with the development of hand tools the principles of work organization were being realized in tribal cultures by making the best use of the special skills of individuals in making and maintaining tools for the experts using them: the relationship between hunter and spearmaker is an example of this division of labour. Progressively it was also recognized that there exist three basic elements in tool design. The first of these, the need for a cutting edge harder than the material to be worked, is the most fundamental and delayed wider development until the discovery of metals. The other two factors (which are also dependent to an extent on cutting tool materials) are the need to minimize demands on human energy and the search for substitutes for manual skill; all three are the subject of research and development to this day.

BRONZE AND IRON AGE TOOLS

It is probable that bronze was the first metal used as a tool material although iron was known about at an early date. There are many references to iron in the Bible and its superiority to bronze. Goliath is described as wearing bronze armour and having a spear pointed with iron, and there are also mentions of steel. Aeschylus, c. 480 BC, writes about the superiority of iron for weapons.

The greater difficulty of extracting and forging iron, and the effect of corrosion, extended the use of bronze in weapons, armour and tools until the techniques of working iron were developed to demonstrate its superior qualities in providing and keeping a keen cutting edge. By the end of the Roman Empire virtually all the forms of modern hand tools had been devised; the second major step in the development of manufacturing was already under way by the introduction of mechanical means of enhancing the force of cutting and forming to take advantage of the high cutting speed possible with iron tools.

The first mechanically assisted cutting operation was drilling and the earliest example is the rocking drill, the most important method being the cord drive whereby an assistant manipulates a cord wrapped round the vertical drill spindle to give it an alternating rotary movement. This system, applied horizontally, was probably used to drive the lathe spindle which produced the first extant example of turning: an Etruscan wooden bowl found in the Tomb of the Warrior at Corneto, c. 700 BC. The earliest illustration of this type of lathe, on a wall of the Egyptian tomb of Petosiris (third century BC), shows an assistant holding each end of the cord to give the rotational movement to the spindle. The Egyptian figurative convention confusingly shows the spindle vertical, but it illustrates the provision of bearings and tool rest for accurate positioning of the cut being made and to take the load imposed on the workpiece by the cutting action. These requirements, for load bearing and tool rigidity in relation to the workpiece, have continued to be principal elements in machine tool design, together with work holding and spindle drives.

The Kimmeridge 'pennies' discovered at the Glastonbury Lake Village, Somerset, were turned from soft stone c. 100 BC and show interesting methods of attaching and driving the workpiece from the spindle. One has a roughly drilled hole used to mount the work on a shaft or mandrel which in turn is held between centres similar to the Egyptian lathe. Others show the use of a squared hole to permit driving from a similarly squared spindle nose and the small centre holes necessary for head and tail centring. Spindle driving methods advanced slowly. Although the ability to turn in stone led to the spindle-mounted grindstone with a turned true outer surface, for maintaining cutting edges on tools and weapons, the cranked arm used for driving the wheel in the earliest illustration of it on the Utrecht Psalter, AD 850, was not used for a lathe spindle until the second half of the fifteenth century. Similarly the bow replacement for the cord drive operated by an assistant was not used for lathe drive until the Roman Empire. This method is still used in the Middle East by wood turners and watchmakers in the Western world were using bow drills in the early part of this century. The difficulty of manipulating the bow while guiding the tool with intermittent cutting, calls for a very high degree of manual skill and dexterity and it is only possible to make light cuts.

Figure 7.1: Miniature painting of a pole lathe, *c.* 1250, from the Bible Moralisée.

EARLY MACHINES

As the size and complexity of work for the lathe increased so did the need for increased rigidity and more power, which was met by heavier construction of the wooden frames of lathes and the development of the pole lathe. This still gave intermittent cutting, but freed the turner's hand to concentrate on guiding the tool by the use of a spring pole to which one end of the cord was attached and the other end to a foot operated treadle after passing round the work to be turned. When the treadle was pressed the work revolved towards the turner for cutting, and on completion of the treadle throw the spring pole returns the work. This type of drive probably existed in the twelfth century; the best early illustration of the pole lathe occurs in the Bible Moralisée, *c.* 1250 (see Figure 7.1). The use of this type of lathe has continued in similar form until well into the twentieth century with the 'chair bodgers' at work in the woods around High Wycombe in Buckinghamshire.

The pole lathe, with its intermittent cut, was not adequate for turning metal and, as the need for machined metal products increased, the continuous method of driving was developed, first of all through the use of a large wheel in separate bearings carrying round its periphery a cord which also passed round the work spindle. The large wheel was turned by an assistant using a cranked arm, first illustrated *c.* 1475 and also in Jost Amman's *Panoplia* of *c.* 1568. The next method of continuous driving using a treadle and crankshaft, shown by Leonardo da Vinci, *c.* 1500, in the *Codice Atlantico* and developed by Spaichel, *c.* 1561, still gives a satisfactory system for the ornamental turners, sewing machines and other machines where only human power is available; but it was the large wheel and continuous band method which was to enable the use of other power sources: horse gins, water wheel, steam engine and electric motor.

The development of continuous drive also made possible the control of the cutting tool relative to the work through systems of gears, screws and guides progressively to eliminate the skill required in holding and guiding the cutting tool and make use of the ideas first expressed according to drawings in the *Mittelalterliche Hausbuch, c.* 1480, of tool holder and cross slide with screw cutting lathe.

MEASUREMENT

From the earliest days of man's use of tools, measurement of the size and shape of things produced has been of prime importance to satisfy the performance required. The earliest standards were those designed to meet individual needs, but these were gradually developed to use units of measurement which could be employed to reproduce articles in a range of sizes. The first 'standard' was the Egyptian Royal Cubit, equivalent to the Pharaoh's forearm length plus palm and made of black granite. This master standard was subdivided into finger widths, palm, hand, large and small spans, one remen (20 finger widths) and one small cubit, which was equivalent to six palms. The small cubit was used for general purposes and made in granite or wood for working standards. These were regularly checked against the master, and many Egyptian temples and other buildings had reference measures cut into walls to check the wooden cubit which, being much easier to handle than stone although more prone to variation, came into more general use. These principles and the use of cubits and their subdivisions became the basis of Roman, Greek and Middle Eastern measures and later European measures, although the actual 'reference factor', the forearm, produced some alternative cubits in different parts of the world. All these standards were 'line standards' involving measuring between engraved lines and this remained the basis of national and international standards until 1960, when the concept of keeping a physical standard was abandoned in favour of the wavelength of krypton 86 which is a readily reproducible and constant reference factor.

Once standards of length were established, these were used to check parts for accuracy and measuring tools, for use in transferring sizes for comparison with the standard, were developed. Calipers, dividers and proportional dividers were evolved by Greeks and Romans some 3000 years ago. These instruments, with the cubit measuring stick, made possible accurate calculations, by measuring the shadow cast by the sun, to determine heights of buildings, agree the time of day, establish the calendar and navigate by reference to the stars. From this point the development of metrology has been in the direction of increasing the accuracy of measurement and in this the screw is of major importance. The ability to manufacture accurate screw threads has been critical in the design of instruments for astronomy, navigation, time measurement, the production of screw controlled machine tools and engineering inspection devices.

The measurement of time, based on observation of the sun's passage, has always exercised the mind of man and many methods have been explored. By the ninth century AD the Chinese had combined hydraulic and mechanical methods to make a large water clock which showed the time by ringing bells and the appearance of various figures. The first all-mechanical clocks were made in Europe, the most outstanding example being that produced by Giovanni de Dondi, c. 1364, which incorporated elliptical gear wheels and sun and planet gears, all cut by hand. A replica has been exhibited in the Science Museum, London. The oldest clock in England is that at Salisbury Cathedral, constructed c. 1386 in a similar way with forged iron gears and lantern pinions, all held in an iron framework joined by rivets and wedges. These construction methods, while adequate for large public clockwork, were not suitable for the smaller timekeepers which the people sought, and the size of this demand and the development of the designs led to improvements in machining, including accurately cut gears, turned spindles and screws which were in advance of general manufacturing by 300 years. During this period the spring-driven pocket watch made its appearance and with it the need for a device to control the pull of the spring to exert a constant force throughout its unwinding. The fusee, invented c. 1450, consists of a conical drum with a spiral groove carrying a chain or cord which, attached to the end of the spring, controls its force by unwinding from a different part of the drum as the spring runs down. Manufacturing fusees called for an extension to the machining capacity of the small all-metal lathe already used by the clockmaker to make accurate spindles and called a turn. This rigid bow-driven device gave the precision necessary to produce small parts with the repeatability required for production purposes and exemplified one of the key features of modern machine tools. Simple fusee engines, employing a screw-controlled linear motion but hand-controlled in-feed to the profile, were designed from 1740, but by 1763 Ferdinand Berthoud had made an engine which was automatic in fixing the relationship between cutting tool and work, thereby illustrating our second principle of machine tool design by eliminating hand skill. Figure 7.2 shows an example of this type of machine.

GENERAL MACHINE TOOLS

Clockmakers also required accurate screw threads and these had been cut by machine since c. 1480 by the master screw method. The earliest representation, in the *Mittelalterliche Hausbuch*, shows this design, which produces a thread if a cutting tool with a suitable 'V' profile is held against the workpiece and the

Figure 7.2: A watch-maker's fusee and screw-cutting lathe of the eighteenth century.

crank handle turned to advance the work according to the pitch of the master thread. This method will only cut a thread of the same pitch as the master screw; the artist did not understand its function, as the threads shown are opposite handed. Many similar machines were constructed on the master screw principle, such as that of Emanuel Wetschgi, *c.* 1700. However, greater flexibility in screw making was achieved by the use of a sliding spindle controlled by a set of master threads brought into mesh as required. Plumier's lathe of 1701 used this system, which was continued in Holtzapffel's lathe of 1785. The use of gearing and leadscrew to obtain alternative screw pitches is attributed to Leonardo da Vinci in his machine design of *c.* 1500, which shows the principles of an industrial machine capable of producing machine leadscrews. Developments from this are first shown in the design of Jacques Besson, the successor to Leonardo as engineer to the French court, in his machine of 1578. This machine is the first example of the leadscrew and nut guidance and drive combination of later machines; however, despite its massive wooden frame and resultant rigidity, its driving system would not provide a high degree of precision and this line of development represents the beginning of the ideas of ornamental turning. Another machine of 1578 by Besson has cams and templates, enabling copies of many shapes to be produced, and the use of similar techniques and machines became a hobby of high society in Europe. Beautiful and intricate decorative objects were produced on machines of increasing complexity, involving: sliding spindles, rosette cams, overhead drive to tools held in the cross slide, and gear and cam controlled geometric work-holding devices and cutting frames. Ornamental lathes, still in mainly wooden frames and with treadle drives, were being produced in the twentieth century by firms such as Holtzapffel, see Figure 7.3.

Figure 7.3: Holtzapffel ornamental turning lathe of 1815.

Instrument makers had special problems involving the need for accurate screws to be used in obtaining precise linear and circular divisions. Jesse Ramsden's screwcutting lathes of 1777 and 1778 were designed for this purpose, the latter to make a very accurate screw for his dividing engine to measure off accurate intervals in straight lines. The tangent screw, cut on his lathe of 1777 using gear wheels with large numbers of teeth to obtain the fine correction of pitch required, was used in turn to drive a very large gear wheel with a central boss carrying one end of a steel strip which, when it unwinds, controls the travel of the cutting tool. Other gears control the relative rotation of the tangent screw shaft and the workpiece to generate screws of high precision and any length within the capacity of the machine.

While these developments of lathes to meet the special needs of clockmaker, instrument maker and ornamental turner were taking place, the general industrial lathe for work on wood, ivory or soft metals continued in a simple wooden frame with band drive until the end of the seventeenth century. In his book *L'Art du Tourneur*, published in 1701, Plumier describes special features of machine design for an iron turning lathe, specifying spindle materials, bearings and the shape and sharpening of the cutting tool. Mounting of the work between centres, and the strength and rigidity to maintain the precision required under the heavy cutting loads in machining metal, were also to form the basis of the design of metal cutting machine tools thereafter. Christopher Polhem of Sweden in about 1716 produced an iron cutting lathe which incorporated a screw drive to the cutting tool driven by the same water wheel that turned the work spindle and was of the heavy construction required to give precision in the turning of rollers for his mills. In 1760, Vaucanson produced an industrial lathe within a heavy framework of iron bars solidly bolted together and carrying substantial centres for work holding, although the size was limited by the framework. The carriage for the toolholder is mounted on square bars utilizing the 45° faces for the most precise support and driven by a leadscrew manually operated. Senot's lathe of 1795 includes the change wheels envisaged by Leonardo and used later by clockmakers to give a screw pitch choice from the leadscrew, and incorporates lubricated bearings to take the heavy cuts of screw cutting in hard metal; it was much more convenient to use than that of Vaucanson.

In 1797 the true industrial lathe came into being through the genius of Henry Maudslay, who created a machine which was the synthesis of all previously recognized desirable features: rigid construction, prismatic guide bars for the carriage, change gear drive from spindle to leadscrew, graduated in-feed for the tool holder and substantial centres for work holding, readily adjustable for length by moving the tailstock. With this lathe Maudslay was able to cut precision screws of a range of pitches to a high degree of repeatability and he is regarded as the 'father of the industrial lathe'. Figure 7.4 illustrates Maudslay's lathe. At about the same time David Wilkinson in the USA produced his screw-cutting industrial lathe which bears a strange resemblance to that of Leonardo da Vinci nearly 300 years earlier. Although of heavy construction throughout, with guideways 5.5–6m (18–20ft) long, it could only cut screws of the same pitch and length as its leadscrew as no change gears were incorporated and little adjustment to the centre position. In 1806, Wilkinson made a large general purpose lathe with screw drive to the carriage which was used extensively and earned him regard as the founder of the American machine tool industry.

While this general development of the lathe was in progress, special needs of manufacturing in different fields was being met in a variety of ways. The need for accurate gear wheels in clock and instrument making brought into being the dividing engine: the example from *c.* 1672 in the Science Museum, London, is the earliest machine tool in the collection. A formed rotary file cuts the teeth of the gear in turn indexed using a plate containing a number of hole circles. Very little difference is apparent between this machine and those of 100 years later, although in 1729 Christopher Polhem had developed a hand-operated gear cutting machine of a production type using reciprocating broaches to cut the teeth of a number of wheels on separate vertical spindles. The broaches were

PART TWO: POWER AND ENGINEERING

Figure 7.4: Henry Maudslay's original screw cutting lathe of 1797.

later changed for rotary cutters and formed part of a series of machines for large-scale production of clocks. The shape of teeth for gearing had exercised the minds of mathematicians and clock and instrument makers since the fifteenth century and the work of Phillipe de la Hire in 1694 contains a detailed mathematical analysis of different forms, concluding that the involute was the best profile. This form was not adopted in practice for 150 years and the clockmakers' products were only adequate for their time until Rehe's engine of 1783 which used formed cutters instead of rotary files and was regarded as a real gear cutting machine.

Engineering grew out of military campaigning and machine tool developments owe a great deal to the ingenuity exercised in producing the engines of war. The Pirotecnica of 1540 by Biringuccio shows a water-powered cannon boring mill, with the cannon mounted on a movable carriage and the tool entering and finishing the cored hole in the casting. This method, although inaccurate, continued until the eighteenth century. About 1713, Maritz invented a vertical boring mill which was capable of boring cannon from solid, but, dissatisfied with the operation of this type of machine in the Netherlands State Gun Foundry at The Hague, Jan Verbruggen, the Master Founder, produced in 1795 a horizontal boring mill which employed a stationary boring bar and

Figure 7.5: A model of John Wilkinson's cylinder boring mill of 1776.

rotated the cannon in bearings. This type of machine was introduced in the Royal Arsenal at Woolwich when Verbruggen was appointed Master Founder there in 1770.

With the development of the steam engine (see Chapter 5) the need for machines to bore much larger diameters were met by designs based on the cannon borers with devices to reduce the inaccuracy inherent in the use of a long, unsupported boring bar. John Smeaton's mill at the Coalbrook Company's Carron Ironworks employed a travelling carriage running inside the casting to support the end of the boring bar. These methods were all unsatisfactory, as the cutter tended to follow the cast profile and it was not until John Wilkinson produced his cylinder boring machine in 1776 (Figure 7.5) that it became possible to correct run-out in the bore to produce a cylinder sufficiently accurate to satisfy the design of James Watt's steam engine. Wilkinson's Bersham Ironworks and his other foundries were in turn the first to use Watt's engines for industrial power: he was the greatest ironmaster of his generation and the father of the Industrial Revolution, with his cylinder boring mill the first truly industrial machine tool. The machine consisted of a carriage on which the cylinder to be bored was securely mounted with the boring bar, running through its centre, fixed in bearings at each end, one of which was driven through gearing from a water wheel. The boring bar was hollow, with a rod running through its centre which advanced the cutter head along the revolving bar by means of a longitudinal slot and a slider attached to the cutter head and traversed by a rack and gear, with a weighted lever to provide the moving force. The use of rack and slotted keyway for tool traverse are developments widely used in machine tools, notably the lathe with its feedshaft and rack for moving the saddle.

As the output of the ironmaster's works, such as that of Abraham Darby at Coalbrookdale, increased tremendously by the use of coke in smelting (see

p. 153–4), the metal working machine tools had to keep pace and better cutting tools devised. The invention by Benjamin Huntsman of his process for making crucible steel in 1746 was of paramount importance, affecting the design of the feeds and speeds of machine tools to take advantage of the improved cutting capacity of this carbon steel and enable the machine tool makers to meet the demand for machined products.

The workshops of Henry Maudslay were outstanding in this respect, producing general machine tools, including lathes based on his original design, and special purpose machines, such as the Portsmouth block-making machinery designed by Mark Isambard Brunel in 1800 and completed in 1810 to provide for the mass production of rigging blocks for the Royal Navy (see p. 30). Maudslay was also interested in many other types of machine. In 1805 and 1808 he patented calico printing machines, in 1806 a differential gear hoist with Bryan Donkin and a water aerator with Robert Dickinson, in 1824 a marine boiler water changer with Field, and in 1807 his famous Table Engine. Very many engines of this last type were constructed and were very successful in providing power in workshops. Marine engines were also constructed and the company became renowned for this, beginning in 1815 with the 12.7kW (17hp) engine for the *Richmond*, the first steam passenger vessel to ply on the Thames. Altogether the company produced engines for 45 ships in Maudslay's lifetime. Mill machinery for Woolwich Arsenal, gun boring machines for Brazil, hydraulic presses, coin minting machines, pumping engines, foundry equipment and the shields and pumping equipment for Brunel's Thames Tunnel in 1825 were all among the products of this company which became one of the greatest engineering businesses of the nineteenth century. In addition, and perhaps more importantly, it was the training ground for many of the important engineers of the day, including Maudslay's partner Joshua Field, Richard Roberts, Joseph Clement, James Seaward, William Muir, Joseph Whitworth and James Nasmyth.

Much of Maudslay's success was due to his ideas of precision working and workshop management. He went to great lengths to produce accurate screw threads as the basis for precise division and measurement and developed the use of surface plates to obtain the true plane surfaces he regarded as essential in building his machines: these plates were placed on the fitters' benches to test their work. Although Maudslay did not invent the micrometer (Gascoigne in 1638 and James Watt in 1772 (Figure 7.6) had made their devices), he constructed in about 1805 a bench micrometer called the Lord Chancellor which was the ultimate standard in his workshops. This instrument consists of a gunmetal bed carrying two anvils with end measuring faces. The movable anvil connects to a screw, cut to 100 threads per inch, through a slot to a split nut. Both anvils have bevel-edged slots through which the scale divisions on the bed can be read, and the milled head to turn the screw has 100 graduations, giving a

Figure 7.6: James Watt's micrometer, 1772.

movement of the measuring face of 0.0001in (0.00254mm) per division. There is also an adjustment to eliminate end play. Recent tests of the Lord Chancellor's accuracy by the National Physical Laboratory justify the opinion that Maudslay was the first engineering metrologist of significance.

Joseph Clement, who had followed Maudslay from the workshop of Joseph Bramah, was employed by him as chief draughtsman and later went on to improve the design of taps and dies in his own works where he specialized in fine mechanisms such as that in the calculating engines of Charles Babbage which he constructed from 1823. In 1827, Clement designed and made a facing lathe of such excellence that it won him the Gold Medal of the Royal Society of Arts. It had many features in advance of other lathe design and in particular it incorporated a mechanism to ensure a constant cutting speed across the face of the workpiece. He won many other awards for ingenious fitments for the lathe and in 1820 produced a planing machine in which the work was clamped on a reciprocating table passing beneath a cutting tool which could be moved horizontally and vertically. Matthew Murray, James Fox and Richard Roberts also produced planing machines at about this time, as this sort of machine was the key to satisfying the need for accurate plane surfaces in the building of lathes, special purpose machines and engines, the production of accurately guided crossheads being specially important for the development of steam engines in locomotive and marine applications. Clement's most ingenious and lucrative machine design, however, was the immense planer he built in 1825, which employed a table moving in rollers on a masonry foundation to support the planing of workpieces 6 feet (1.82m) square. This was the only machine capable of handling work of this size for 10 years and at eighteen shillings per square foot of machining gave him a comfortable income.

Figure 7.7: Richard Roberts' planing machine, 1817.

After two years in Maudslay's works, Richard Roberts left in 1816 and set up his own works in Manchester where he produced in 1817 his planing machine (Figure 7.7) and his large industrial lathe which incorporated a back gear device to allow changes in spindle speed which is still in use today on simple lathes. The lathe also provided for a choice of relative speed between spindle and leadscrew, through a crown wheel arrangement, forward and reverse of the leadscrew and automatic stop. He also made many other machine tools and was the first to introduce plug and ring gauges to obtain uniformity in manufacture of textile machinery (see Chapter 17). His most important inventions were the automatic spinning mule and differential gear, in 1825, and his power loom of 1855 was the first effective weaving machine. Roberts, the most prolific and inventive engineer of his time, went on to develop steam locomotives; a steam road carriage including his differential gear in 1833; gas meters; clocks; and in 1847 the multiple punching machine, using the 'Jacquard' principle, designed to punch the rivet holes in plates used in the construction of Robert Stephenson's bridges at Conway and Menai Straits.

James Nasmyth was appointed by Maudslay as his personal assistant in 1829 and left to start his own workshop when Maudslay died in 1831. He was enthralled by the 'master', wrote about his workshop and methods and followed these perfectionist practices in his own designs and workshops in Manchester. Here the firm of Nasmyth & Gaskell, later Nasmyth, Wilson & Co., manufactured locomotives and machine tools, many of the latter being exported to St Petersburg to equip the Russian locomotive works. Nasmyth introduced many ingenious devices to improve the operation of various machines but is principally famous for his invention of the steam hammer, *c.* 1839, which was produced to forge the 30in (75cm) diameter paddle wheel shafts for the *Great Britain*, the second steamship designed and built by Isambard Kingdom Brunel. Finally screw propulsion was chosen, and the massive shaft not required, but the hammer revolutionized the production of heavy forgings elsewhere. The original invention was modified soon after to use steam on the down stroke of the hammer, making it double-acting; the automatic valve gear for regulating the stroke was invented in 1843 by Robert Wilson, one of Nasmyth's partners. A working model of the hammer is shown in Figure 7.8. Nasmyth's greatest commercial success was in the small shaping machine invented in 1836 which has remained in similar form in engineering workshops to this day.

Of all the engineers who worked with Maudslay and emerged from that famous workshop to found their own, none adopted the ideas on standardization, accuracy and interchangeability, or developed them more effectively, than Joseph Whitworth. He worked for a time with Holtzapffel and then Clement before returning to Manchester to set up his own business in 1833, improving the design of machine tools, constructing them with greater precision using standard gauges based on accurate measuring machines and becoming the dominant figure in the world of machine tools of the nineteenth century. To provide for the accurate measurement required for the production of standard gauges used in manufacturing, Whitworth constructed two measuring machines: the first, the Workshop Measuring Machine, was designed to read to 0.0001in (0.00254mm) and the second, the Millionth Machine, to read to 0.000001 in (0.0000254mm). Both were comparators, set first to a standard and then on the gauge to be measured. A gravity 'dropping piece' was used to indicate the critical contact point. The Millionth Machine was exhibited at the Great Exhibition of 1851 along with a wide variety of his 'self acting' machine tools, including lathes; planing, drilling, boring and shaping machines; a wheel cutting and dividing engine; punching and shearing machines; and screw cutting equipment. Whitworth was a principal exponent of decimalization and standardization: the most famous example of his work, and that by which he is most remembered, is the standardization of screw threads. His paper 'On a

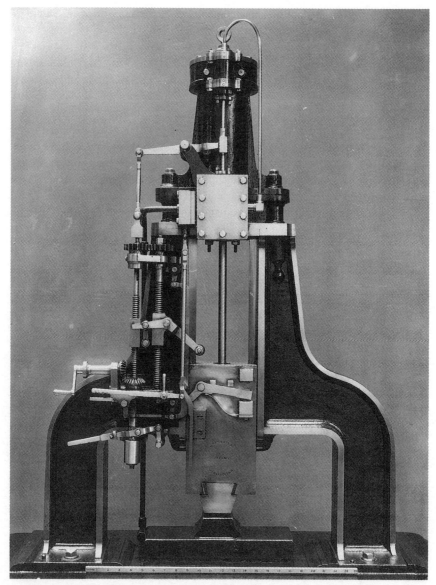

Figure 7.8: A model of James Nasmyth's original steam hammer of 1839.

uniform system of Screw Threads', read to the Institution of Civil Engineers in 1841, was the beginning of rationalization in the manufacture of screwed fastenings. A compromise system was eventually worked out based on the average pitch and depth of thread in use by leading engineers, and a table was produced giving the pitches of screws of different diameters and a constant proportion between depth and pitch by adopting an angle of 55° for the 'V' profile. By 1858, Whitworth could claim that his standardization of screw threads had been implemented, although his advocated decimal scale was not accepted except where it coincided with fractional sizes and in Europe, where it competed with the metric thread. (A similar fate greeted his decimal Standard Wire Gauge which was never adopted as a national standard.)

In 1853, Whitworth joined a Royal Commission visiting the New York Exhibition and reported that American machine tools were generally inferior to English, although their eagerness to use machinery whenever possible to replace manual labour appealed to him. A request by the Board of Ordnance to make machinery for manufacturing the Enfield rifle in 1854 turned his interest towards the manufacture of firearms. He produced his own rifle and later cannon which were superior to their competitors in performance but they were rejected by the official committees, although he obtained large orders from abroad. His visit to America also confirmed his belief in the value of technical education, first shown in his support for the Mechanics Institutes and Manchester School of Design in the late 1830s, and led to the launch of Whitworth Scholarships in 1868–9. Unfortunately he was a supreme egotist, which led to conflict with authority, and his rigorous, authoritarian control of the details of his manufactures stultified later development.

The greatest users of machine tools of the day were the engine builders. Concurrently with Henry Maudslay's, the works of Boulton and Watt at Soho, Birmingham, and Matthew Murray's Round Foundry at Leeds expanded rapidly at the end of the eighteenth century. Such growth was only possible by the injection of capital and it was at this stage that the business men, Matthew Boulton at Soho and James Fenton at Leeds, began to make their impact on machine-tool building by financing the inventiveness of the engineers to take advantage of the great demand for engines and selling the tools developed for this purpose to other manufacturers. The Soho Foundry was completed in 1796 and William Murdock was put in charge in 1798. He constructed a massive horizontal boring mill of Wilkinson pattern, but with his own worm and wheel drive and an attachment to machine the end face of cylinders. A 64in (162.5cm) diameter cylinder was completely machined on this mill in $27\frac{1}{2}$ working days in 1800. A vertical boring mill was also built at Soho in 1854 to bore four cylinders for Brunel's *Great Eastern* steamship to 7ft (2.13m) in diameter. This engine produced 2000hp and was the most powerful in the world at that time. Many tools were purchased from smaller makers for general purposes and William Buckle, who became manager at Soho in 1825, introduced the first large screw cutting lathe to the works: previously large screws had been cut by hand methods. Matthew Murray at Leeds was more inventive and a better production engineer, making work of higher quality than that of Soho, and was one of the first manufacturers of high quality machine tools for sale in addition to engines. He devised the 'D' slide valve for steam engines and designed and built a planing machine to make it which was so ahead of its time that it was kept secret. Many designs of boring mill were produced by Murray and sold in Britain and abroad in which screw drive to the cutter head was incorporated by 1802. Murray's steam locomotives were the first commercially successful in the world (see p. 559).

Other machine tool designers had different primary objectives. One of the most outstanding was James Fox who gave up his post as butler to a country parson to make improved textile machinery. In order to do this he had first to design and construct his own machine tools and he was highly successful in this. His detailed improvements in carriage traverse became standard practice on the lathe. His first planing machine was built in 1814 and by 1817 it included power drive to the table with automatic reverse and automatic feed to horizontal and vertical tool travel. The machine tools built by Fox were ahead of other manufacturers in the detailed design of traverse mechanisms and precision guideways and he was a successful exporter to France, Germany, Russia and Poland. Several of his machines can be seen in the museum at Sielpia Wielka in Southern Poland.

MASS PRODUCTION

Special purpose machine tools are designed to perform a particular operation repetitively in the manufacture of numbers of specific products. Clockmaking was the first important application of these techniques and some of the machines developed had features used later in general purpose machines. Joseph Bramah's lock, which he invented in 1784, required many small parts in the construction which would have been difficult and uneconomic to produce accurately by hand methods, so he engaged Maudslay, following his training at Woolwich Arsenal, to help in devising and constructing suitable machines. A sawing machine, (*c.* 1788), quick grip vice, milling cutters, drilling templates and a spring winding machine (1790), were made and still survive in the Science Museum, London. The last of these exhibits, a screw method of traversing the spring wire winding head, appeared later in the screw cutting lathe. Bramah's other inventions included: hydraulic press, fire engine, beer pump, extruded lead pipe, water closet, fountain pen, banknote numbering machine, and a wood planing machine with hydraulic bearings and feed system (*c.* 1809).

Maudslay's experience in special machine construction was useful, and laid the foundation of his fortune, when he was selected to manufacture the Portsmouth block-making machines designed by Marc Isambard Brunel. Working models of these machines constructed by Maudslay (now in the Maritime Museum, Greenwich) were useful in persuading the Admiralty to set up its own block-making factory. Maudslay constructed all the full-size machinery between 1802 and 1809. The plant consisted of 45 machines of 22 different kinds. When it came into full operation making three sizes of block in 1810, it was producing 100,000 blocks per annum and was the first large-scale plant employing machine tools for mass production. With these machines ten unskilled men could do the work of 110 skilled blockmakers. Apart from two large sawing machines, all the others were of metal and precise in operation to allow the assembly of component parts. They were used up to the mid-twentieth century and several are now exhibited in the Science Museum in London: Mortising, 1803; Block Shaping, 1804; Scoring, 1804; Coaking, 1804; Circular Saw, 1803.

The manufacture of firearms, particularly small arms, was another area requiring a high rate of production with the special problem of producing many small precision parts which could be made to fit together easily. Traditional methods of hand production were slow; and weapons were made of individual, non-standard parts, so that if a component was lost or broken a replacement had to be specially made by a skilled gunsmith. Large numbers of infantry could be made ineffective in this way and governments became concerned about this weakness should long campaigns have to be waged. In 1811 the British army had 200,000 useless muskets awaiting lock repairs.

A French gunsmith, Le Blanc, was the first to propose in 1785, a system of manufacturing in which the components of firearms should be made so accurately that all parts would be interchangeable and thus allow simple replacement when parts failed in action. The French government appointed several committees to examine this proposal and approved its development, but the political difficulties and revolution in France prevented its realization. Thomas Jefferson, the American Minister to France in 1785, saw the products of Le Blanc's workshop and tested the parts of the musket lock for interchangeability, reporting its value to his government, but it was not until 1798 that a contract to manufacture arms for the Federal Government on these principles was given. Eli Whitney, the inventor of the cotton gin (see Chapter 17), was the first to be awarded a contract from the $800,000 voted by Congress for the purchase of cannon and small arms. He was to provide 1000 stands of arms for which a payment of $15,000 would be made, with continuing payment up to 10,000 stands of arms. Whitney was unable to achieve the targets set in the contract and it is doubtful if he employed new methods of machining the parts rather relying on the use of filing jigs and the 'division of labour' to produce the 4500 muskets delivered in September 1801.

The US government armoury at Springfield was also producing firearms and employing new machinery and methods of assembly and work organization which enabled production to increase from 80 to 442 muskets per month in 1799. John Hall designed his rifle in 1811 to be made by interchangeable parts at Harpers Ferry Armoury by 1817 and he introduced many new machines, a system of dimensioning from a single datum, and gauges at each stage of manufacture to ensure accuracy. He also used secondary and tertiary gauges to check the bench gauges. Other American manufacturers began using these methods, including pistols by Simeon North and Elisha K. Root's arrangements for making Colt's revolver. By 1815 all firearms made to United States government contract were required to be interchangeable with weapons made at the National Armouries and the system was proved in 1824 when 100 rifles from different armouries were brought together, disassembled and reassembled at random successfully.

The ideas of Hall were developed by Root at the Colt Armoury at Hartford which contained 1400 machines and proved a tremendous success, its influence being spread by the outstanding engineers who worked there in similar fashion to those from Maudslay's workshop in England. Two of these men, Francis Pratt and Amos Whitney, were to form the great engine company and became great exporters of machine tools for gunmaking as well as principal manufacturers of engine lathes. The company also supported the research and development by William Roper and George Bond of a line standard comparator employing a microscope with micrometer adjustment to calibrate their gauges. Another armoury of importance was the Robbins & Lawrence shop in Windsor, Vermont, run by Richard S. Lawrence, Frederick W. Howe and Henry D. Stone. In 1851 they sent a set of rifles to the Great Exhibition in London and demonstrated the ease with which parts could be interchanged, creating the official interest which led to the Commission to America which was chaired by Nasmyth and had Whitworth as one of its members (see p. 403). Their successful visit and recommendations led to the order for Robbins & Lawrence and the Ames Manufacturing Company to supply similar machinery for the Enfield factory to manufacture the 1853 pattern rifle musket, thus bringing the American System to England.

In addition to the machinery several skilled men from the American armouries were imported, notably James Henry Burton who became chief engineer at Enfield. After serving his five year contract there, Burton returned to America to be brought into the Civil War as a Confederate lieutenant-colonel to superintend all their armouries. To set up a new armoury in Georgia, Burton returned to England to obtain the machinery and tools necessary from Greenwood & Batley of Leeds but, although designed and built, the equipment was not delivered. After the Confederate defeat Burton returned to Greenwood & Batley, which became a principal manufacturer of complete sets of machinery and gauges for arms production, supplying Birmingham Small Arms Company, the Austrian armoury at Steyr, the Belgian Government factory at Liège, the Imperial Russian armouries at Tula, Sestrovelsk and Izhersk, the French armouries at St Etienne, Tulle and Chatellerault, and arms factories in Sweden, Italy, Japan, Hungary and Brazil. Their basic marketing device was to advocate setting up a factory to produce a minimum of 1000 rifles and bayonets per week, for which they would provide all machinery, gauges and power plant, and erect and start up the factory for about £135,000 f.o.b. in 1891. This English company learned the value of the 'American System', but despite its example the ideas were slow to spread to other manufacturers. In the now United States, by contrast, these production methods were adopted quickly in the making of clocks, watches, sewing machines, bicycles, typewriters and agricultural machinery. The principal reason for this difference, and for the consequent change of fortune between the American and English machine tool industries, was the shortage of skilled labour in the USA. Immigrants with skills could rapidly become masters, and for those with imagination and courage the opportunities of a good life farming in the West competed with the prospect of factory work in the East. This created the high level of wages offered and the need to build skill into the machine tools so that they could be operated by whatever unskilled labour was available. Output being so directly related to the efficiency of the operation of machinery, wages were paid under contract and direct 'piecework' systems, marking the beginning of 'payment by results' and the material success of all classes, well outstripping that in Europe.

The 'American System', or mass production, in different industries called for many machines designed to operate at higher speeds and perform multiple operations, and for new processes to meet the needs of the product. Grinding had remained a simple process of creating an edge on a cutting tool or external finishing of cylindrical objects, using wheels of similar form to that in the oldest illustration from the Utrecht Psalter of 850 AD and work guidance by hand. New products such as the sewing machine could only be made commercially by mass production methods, and the success of these products called for greater refinement in the accuracy of their components; grinding therefore had to become a precision operation for both cylindrical and surface finishing. The first attempts to do this involved mounting a grinding wheel on the cross slide of a lathe and driving it from the same overhead shaft which powered the spindle. Several refinements in the design provided covered guideways, specially weighted carriages to avoid chatter, and a reversible, variable speed drive to the carriage.

Brown & Sharpe, the great machine tool and gauge making firm of the USA produced such a grinding lathe in about 1865 for the needles, bars and spindles of the Wilcox & Gibbs sewing machine they were manufacturing. This machine required such careful control that in 1868 Joseph Brown designed his Universal Grinding Machine, the forerunner of all precision grinding machines which in turn made possible the production of accurate gauges, measuring instruments and cutting tools of all types. The success of this machine depended very largely on the use of grinding wheels which were made with emery or corundum bonded by a material to give it strength while allowing the abrasive particles to cut. In England in 1842, Henry Barclay had experimented in producing a vitrified wheel, followed by Ransome in 1857 with a soda bond which was much more successful and Hart produced a similar wheel in the USA in 1872. Sven Pulson, also in the USA, managed to make an effective clay wheel in 1873 on the lines of Barclay's attempt 30 years earlier. Feldspar was used in the formula when F. B. Norton patented the process in 1877, marking the beginning of progressive development of artificial grinding wheels. Small arms manufacturing also created the milling machine and, although its actual inventor is unclear, it appeared in 1818 in the arms workshop of Simeon North. According to a drawing by Edward G. Parkhurst, it consisted of a spindle carrying a cone pulley mounted between two heavy bearings with a toothed cutter on the spindle extension and a carriage, on which the work was mounted, traversed by hand underneath the cutter at 90° to its axis.

Between 1819 and 1826, according to the report of an official investigating commission into the work being done by John Hall in the rifle section of Harpers Ferry Armoury, there existed plain and profile milling machines designed by Hall to cut the straight and curved surfaces of his rifle components and eliminate the handwork of filing to templates. Another machine of doubtful origin is thought to have been built *c*. 1827; the next, made in the Gay, Silver & Co. workshops *c*. 1835, incorporated an improved cutter spindle support and vertical adjustment to the headstock. Frederick Howe was trained in the Gay Silver shop and when he joined Robbins & Lawrence in 1847 he built a production milling machine based on patterns from the Springfield Armoury which resembled the basic structure of the lathe in bed, headstock and tailstock, with the cutter supported and driven between the head and tail centres. The work was traversed across the bed by hand-operated cross slide and cut applied by raising the cross slide. A development of this design with greater rigidity, made in 1852, was sold to the Enfield small arms factory in England. This design was also the one modified by Pratt & Whitney at Phoenix Iron Works into the famous Lincoln Miller supplied all over the world by 1872. The next stage in the development of the milling machine occurred in 1861, when the problem of the slow and expensive method of filing the helical grooves from rod to make twist drills was brought to the attention of Joseph R. Brown by Frederick Howe. Brown's solution was the Universal Milling Machine which, with its knee and column construction and geared dividing head on the swivelling table, was to become the most flexible and widely used machine tool, second only to the lathe, and the basis of the Brown & Sharpe Manufacturing Company's machine tool business (see Figure 7.9). An inclined arbor in the dividing head allowed the production of tapering spirals or straight grooves for cutting such requirements as reamer teeth and was based on Brown's work in gear cutting. For heavier work the Brown & Sharpe Universal Milling Machine was made in 1876, with an over arm to give increased support to the cutter. The pioneer

Figure 7.9: Joseph R. Brown's universal milling machine of 1861.

milling machines of the early clockmakers used a rotary file type of cutter and, as larger gears were required, similar machines were made to match the profile of the teeth to be cut by hand chiselling and filing, following the idea of Vaucanson in 1760. These methods were inadequate for the new milling machines because the small teeth allowed only a small depth of cut and presented difficulties of sharpening. Brown solved this problem by patenting in 1864 a formed cutter which could be face ground without changing its profile. With these cutters the milling operation was fully established, other cutter development following closely: notched in 1869, inserted teeth in 1872, face milling with inserted teeth in 1884.

Safety bicycles, increasingly popular in Britain and on the continent of Europe during the 1880s (see Chapter 8), were imported into the USA and were immediately recognized as candidates for mass production. Special machines were devised for hub forming, boring and threading by the Garvin Machine Co. of New York. A rim drilling machine was built by Rudolph & Kremmel in 1892 which could drill 600 wheel rims in a ten-hour day. Spokes were made of wire and threaded by rolling or cutting. The cutting machine allowed a boy to thread 4000–4500 spokes a day. Thread rolling was carried out by machines made on the pattern of the Blake & Johnson, *c.* 1849. By 1897 bicycles were being produced at the rate of two million a year in the USA before the market collapsed when the era of the automobile arrived.

The need for many small parts in these mass-produced goods encouraged the use of sub-contractors who produced specialized components on machines designed for the purpose in small workshops. Ball bearings were gradually introduced into bicycle manufacture and cycle bearing companies were formed producing balls either by rolling hot metal or turning from bar on special machines similar to the Hoffman used in England. All balls were finished by grinding, and the cups and cones were turned, casehardened and ground on a Brown & Sharpe machine or similar. The vast number of screws required for the percussion locks of 30,000 pistols had caused Stephen Fitch to design and build the first turret lathe in 1845. This had a horizontal axis and carried eight tools which could be fed into the work in turn to perform eight successive operations without stopping to change tools. This idea was so time-saving that it was rapidly taken up by other manufacturers after 1850, although the vertical axis turret became most favoured by Robbins & Lawrence, who also produced the box tool and hollow mill used in operations on the turret lathe. Automatic screwing machines followed with the invention by Christopher Miner Spencer of the brain wheel control of tool slides by adjustable cams and roller followers. His machine used the collet chuck and closing mechanism patented by Edward G. Parkhurst in 1871, which was similar to Whitworth's collet chuck, and the production machine patented in 1873 had two 'brain wheels', one to control collet opening and closing, allowing the bar stock to be moved and locked into working position, and the other to control tool slides. Spencer set up the Hartford Machine Screw Company to take advantage of the great market in screws but failed to protect the 'brain wheel' in his patent, which allowed many copies to be made. Spencer subsequently designed a three-spindle version to make screws from wire using a radial arrangement of tools on the Fitch pattern. From this was developed a four-spindle model by Hakewessel, used initally by the National Acme Screw Co. of America to manufacture screws but from about 1906, in association with George O. Gridley, manufactured as the Acme–Gridley automatic lathe. Other automatic lathes followed; New Britain Gridley, Conematic, and Fay from different companies.

Many of the mass-produced goods also required gears and, more importantly, the new manufacturing machinery also depended on the ability to transfer motion and change speeds between shafts accurately and without vibration through gears made in large quantity. The cast gears used extensively in early machines and continuing in the textile industry were no longer satisfactory for high-speed accurate machining by the new machine tools. The transition between clockmaking and engineering gear production occurs in James Fox's machine of *c.* 1833, which used formed tooth cutters and a screw micrometer division of the index cylinder to machine a large gear blank on a vertical spindle. John George Bodmer also patented a gear cutting machine in 1839 capable of cutting internal gears, spur and bevel in addition to external spur, worm wheels and racks, but his formed metal cutting tools were very difficult to sharpen. Whitworth's gear cutters of 1834 were the first machines with involute cutters, geared indexing and cutters driven by a flat belt through a worm and wheel. By 1851 this type of machine was made with a self-acting in-feed and had the appearance of a heavy lathe with the headstock carrying a vertical spindle on which the involute cutter was mounted, the saddle carrying the wheel to be cut on a horizontal axis shaft with an indexing wheel. Joseph R. Brown's precision gear cutting machine of 1855 made use of the involute form backed off gear cutter invented in 1854 and used with the Universal Milling Machine (see p. 408). A Troughton & Simms dividing engine was used to make the index plate divisions. Other large gear cutting machines were developed for engine and power transmissions such as the Gleason template system machines for cutting large spur and bevel gears. The describing generating method of producing the gear tooth profile was first applied practically by Herrman in 1877 and used a single point reciprocating tool similar to the shaper action, all the motions to generate the gear tooth being given to the blank by various attachments to make spur or bevel gears with epicycloidal or involute teeth. A similar type was produced by C. Dengg & Co. of Vienna, in a patent of 1879 for epicycloidal teeth on bevel gears. Pratt & Whitney developed an epicycloidal machine in 1880 employing a circular cutter controlled by a disc rolling on the outside of a ring representing the pitch circle of the gear. Smith & Coventry of England exhibited a describing generating machine at the Paris Exhibition of 1900, and an example of the original Robey-Smith machine designed by J. Buck in 1895, which has two cutters to work on opposite sides of the same tooth, can be seen in the Science Museum, London. The chainless bicycle stimulated the development by Hugo Bilgram of Philadelphia in 1884, of a machine to produce the bevel gears required. It used a single point cutter and obtained the rolling action by swinging the carrier of the blank around the axis in line with the apex of the pitch cone by means of two flexible steel bands passing round the control cone on the blank spindle. The gear shaper invented by E. R. Fellows in 1897 is the most significant and widely used. It employs the moulding generating principle, with a fully formed gear as the generator, hardened and with the cutting edges backed off. The machine works as a vertical slotter, with the cutting tool and gear blank rotating in mesh. The application of a worm sectioned to provide cutting edges was first used by Ramsden in 1768 for his instruments but the method was first used for production in 1835 by Joseph Whitworth, using a worm cutter as a 'hob' to make spiral gears. Several other patents were obtained for hobbing machines but the first fully geared example to be used was that made by George B. Grant, *c.* 1887.

By the end of the nineteenth century the general machine tools were established in their operating principles and many refinements had been made in machine spindle and carriage drives by gearing systems such as the gearbox invented in 1891 by W. P. Norton and incorporated in Hendey-Norton lathes. The features of interchangeable manufacture were well known and the mass production machinery ready for the explosion in manufacturing which was to take place in the twentieth century with the proliferation of the motor car and aeroplane.

TWENTIETH-CENTURY ORGANIZATION OF PRODUCTION

The key to further expansion lay in increasing the speed of cutting by machine tools and in organizing production to balance the machining and assembly of components so as to achieve the maximum rate of production of the completed product. Frederick W. Taylor was influential in both aspects. He carried out experiments on the shape, cutting angles, lubrication and materials of cutting tools to establish optimum efficiency and made dramatic improvements in output and tool life, culminating in his development of High Speed Steel containing tungsten. He demonstrated it at the Paris Exhibition of 1900, cutting mild steel at 36.5m (120 feet) per minute with the cutting tool red hot, and this revelation was so fundamental that all types of machine tool were to need redesign to increase rigidity, speed range and power to take advantage of the new tool material. The introduction of Stellite, a chromium, cobalt, tungsten alloy, in 1917, carbides by Krupps of Essen in 1926, and ceramic tools were to have similar effects later. A comparison of cutting speeds possible with these tool materials for grey cast iron shows: HSS 22.8m (75ft)/min.; Stellite 45.7m (150ft)/min.; carbide 122m (400ft)/min.; ceramic 183m (600ft)/min.

Taylor also introduced a systematic examination of all parts of the manufacturing process through time study of each operation and the application of piece-rate bonus schemes which became known as the Taylor System of Scientific Management and established the principles of workshop management on production lines.

The first four-stroke gas engine was started by Otto in 1862 and his liquid fuelled engine with electric ignition in 1885, at about the same time as Karl Benz used his engine for the first motor in 1886 (see p. 306ff). Besides the engine the motor car presented many problems in its manufacture. The need for a light but rigid chassis, and the design of suspension, steering and transmission were initially solved in a crude way but were rapidly refined in two main streams of development, the European 'bespoke' type, represented first of all by the Mercedes of 1901 and later by the Rolls Royce Silver Ghost, and the 'motoring for all' concept in the USA exemplified by the Model T Ford. The former was produced by machine tools and methods which would be regarded as archaic by American standards, the latter with the benefit of much experience in interchangeability and 'scientific management'. Britain's motor industry was established around 1894 in Coventry and Birmingham, where cycle, sewing machine and textile manufacturers already existed with one major machine-tool maker serving these industries. With the coming of the motor industry this firm grew to become one of the most famous machine tool manufacturers in the world, Alfred Herbert Ltd. Frederick Lanchester was the first British producer of a motor car by interchangeable methods and designed machines, methods, jigs and tools for the purpose. One of his most important machine tools was the worm hobbing machine of 1896, and he also developed hardened steel roller bearings ground to a tolerance of 0.0002in (0.0051mm) for use in his gear box and back axle. In his system of gauges he adopted the unilateral system of tolerancing in preference to the American system of bilateral dimensioning, which allows variation above and below the nominal size.

Production grinding as a precision operation was established in locomotive work, but was even more essential to the automobile industry. The crankshaft journal grinding machine produced by Charles Norton in 1903, the camshaft grinding machine by Norton and Landis of 1911, the planetary grinding machine for internal bores made by James Heald in 1905 and the piston ring face machine, also by Heald in 1904, enabled engine production to meet the rapidly growing demand for cars propelled by the internal combustion engine. At the London Motor Show of 1905, 434 types of car, British and imported, were available and between 1904 and 1914, UK car registrations rose from 7430 to 26,238.

The early American car builders, White, Cadillac and Olds, initially built their cars one at a time, but when Ransome E. Olds moved to Detroit in 1899 he introduced the assembly line and used components manufactured elsewhere by subcontract. Engines from Leland and Faulconer and transmissions from Dodge were used to make the Merry Oldsmobile at the rate of 5000 cars per year in 1904. But it was Henry Ford in 1913 who made the production line move. His 'Tin Lizzie', the 1909 Model T, was mechanically simple, durable and cheap at less than $600, because this was the only way to achieve the output required. Using a rope and windlass, the chassis was pulled along the 250ft (76m) line and parts added to the chassis at assembly stages. The $12\frac{1}{2}$ hours at first required was reduced to $1\frac{1}{2}$ hours for the complete assembly. Standardization of models and component parts was a feature of the production organization, which by 1916 had raised output to 535,000 cars. By 1920, Ford was building half the cars in the world, and fifteen million Model Ts were made before the line was discontinued in 1927. Many other motor manufacturers were to emulate Ford, but the large investment in plant required caused bankruptcies. Such high-volume production posed new problems for machining in making parts in sufficient quantities to match $1\frac{1}{2}$-hour assembly times and machine shops grew and subcontractors prospered. E. P. Bullard was one machine-tool maker who met this challenge with his Mult-A-Matic vertical turret lathe, which machined flywheels in one minute. The Cincinnati Milling Machine Co. produced a version of Heim's centreless grinder, patented in 1915, using a regulating wheel and narrow work rest with slanted top in 1922; the design of Han Reynold in 1906 for a centreless grinder was used in the French & Stephenson machine in England. With these machines valve stems were finish ground at 350 per hour and by 1935 the Cincinnati Co. had introduced the cam type regulating wheel to through grind parts fed from a magazine for gudgeon pins, king pins etc. Between the wars other machine tools were improved to take advantage of various new processes such as gear hardening, which required precision gear grinding machines using formed wheels, or shaving, using a helical cutter with serrated teeth. The broaching machine, patented by John La Pointe in 1898, became a vital machine in workshops to produce, in one pass, bores or flat surfaces by the progressive cutting action of increasing sizes of successive teeth. His pull broach was manufactured in 1902 and hydraulic power was employed in 1921 by the Oilgear Company machine.

The early twentieth century also saw the introduction of the individual electric motor drive for machines which was eventually to eliminate the 'forest of belts' typical of all early production factories.

An event of considerable consequence for metrology and the improvement of workshop accuracy and inspection was the development by Carl Edward Johansson in Sweden of precision gauge blocks to be used in calibrating bench gauges. These were made in a set of 52 to give, in combination, any division between 1 and 201mm. He fixed the temperature at 20°C for the standard sizes and a face width of the gauges at 9mm for the proper and most accurate use of the blocks. The system was patented in England in 1901 and brought with it many accessories which could be used in conjunction with the blocks to provide a whole system of gauges of high accuracy. This reduced the need for skilled men to carry out measurements using micrometers by enabling operators quickly to judge the accuracy of their own work. 'Go' and 'No Go' gauges had been used earlier, but Johansson was the first to introduce adjustable anvils in a cast-iron frame with both 'Go' and 'No Go' gaps set in one inspection instrument with the accuracy of his gauge blocks, or slip gauges as they became called. To check a diameter the operator tried the gauge which should pass the 'Go' anvil but not pass the 'No Go' end. The difference in the anvil dimensions is the tolerance allowed on the dimension to which the part may be made. Angle gauges, plug gauges and micrometers were also produced, and the blocks were used in setting up sine bars and as checks on parts using comparators. W. E. Hoke in the USA made gauges with a similar purpose in about 1920. Johansson's gauge blocks became the dimensional standards for the whole world by 1926 and he is referred to as the 'Master of Measurement'. The use of gap gauges also became common for the inspection of screw threads based on the idea of William Taylor patented in 1905 which specified the 'Go' end as full thread form but had 'No Go' gauge ends to check specific elements such as the core diameter. Taylor was also interested in measuring surface roughness, but it was Richard Reason of Taylor, Taylor & Hobson who in 1937 produced the first Talysurf which used electrical amplification of the movement of a stylus over the surface being tested to give a measure of its roughness.

The two most significant general purpose production machine tools of the first half of the twentieth century were the heavy production grinding machine by Norton in 1900 and the Bridgeport turret milling machine. The former was designed with the size and power to plunge grind cylindrical parts using wide wheels to eliminate finish turning in the lathe, and its crankshaft version of 1905 revolutionized the engine industry. The Bridgeport machine consists of a multi-purpose head for drilling and milling with its built-in motor drive mounted on an arm which can be swung over the work table. This machine was originally produced by Bannow & Wahlstron in 1938 and has been the most ubiquitous machine tool since.

In 1921 the machine which was to have the greatest impact in the toolroom was produced by the Société Genevoise of Switzerland. The jig borer (Figure 7.10) was developed from the pointing machine of 1912 used in the watchmaking industry for accurately locating holes in watch plates. Scaled up to engineering standards, it gave 0.0001in (0.00254mm) standard in locating holes and was invaluable in the manufacture of the gauges and jigs increasingly called for in the mass production industries where tasks were being made shorter and less skilled. The Hydroptic machine, with hydraulic work table and optical scale, was introduced in 1934. To further maintain this level of accuracy after heat treatment, the jig grinder was developed in the USA by Richard F. Moore in 1940.

Another Swiss machine tool which continues to be of importance is the type of automatic lathe also born in the watch industry, introduced by Schweizer in 1872, but now widely used in many industries. It has a quite different operating action in its moving headstock which provides the feed of the work into various tools, giving a more versatile and higher operational speed than the American auto. The general purpose centre and turret lathes continued to be improved to give greater accuracy and flexibility of operation and by the late 1920s had hardened and ground beds, thrust bearings using balls and rollers and special turret tools, such as the self-opening diehead of 1902 made by Vernon, and the

Figure 7.10: The jig borer, produced by the Société Genevoise of Switzerland in 1921.

collapsing tap devised by C. M. Lloyd in 1916. Major improvements were made in spindle speed capacity and rigidity to match the continued increase in capability of cutting tool material. Many types were produced from the tungsten carbide of 1926 to the tungsten, titanium carbide developed in 1938 by P. M. McKenna. They were all very brittle, and cutting tools were made with steel shanks and tips of carbide brazed to them. Modern tools use specially designed tips to be carried in a holder fitted to the tool shank. In 1914–15, Guest in England and Alder in the USA had devised systems of selecting grinding wheels in terms of grit, grade and bond, but the carbide tools were too hard to be ground by ordinary abrasive wheels and it was the Norton company which produced a small diamond bonded wheel capable of dealing with them in 1934.

To obtain rapid, accurate speed control the infinitely variable gear drive patented by G. J. Abbott of London in 1924, using a link chain to engage an expanding or contracting 'V' pulley, became widely adopted to provide stepless speed changes. The pre-selector gear box was patented by F. A. Schell in Germany in 1929 and Alfred Herbert & Lloyd of Coventry in 1932, and Alfred Herbert used it in their Preoptive headstock from 1934.

Increasing accuracy of machining called for an equally accurate measuring device in the hands of skilled machinists, and the early screw calliper invented by Jean Laurent Palmer in 1848 was developed by Brown & Sharpe with enclosed screw and named the Micrometer calliper in 1877. Ciceri Smith of Edinburgh invented a direct reading micrometer in 1893 and Brown & Sharpe produced their version of this in 1928. Johansson of Sweden and Carl Zeiss of Germany made micrometers in 1916 and 1918 respectively. Moore & Wright were the first English company to make this instrument, which became the symbol of the skilled craftsman; their micrometer of 1930 employed a new thread grinder made by the Coventry Gauge and Tool Co. using a diamond cut ribbed wheel which gave a pitch accuracy within 0.0001in (0.00254mm), the best in the world at this time.

In the high-volume flowline production world of automobile manufacture the trend was to incorporate machine tools into a conveyor-based transfer machine, each cutting station being designed to carry out appropriate operations on the faces of the workpiece presented to it by the location arrangement on the moving platen. Each station was designed to carry out its work in a similar time period. The first machine of this type was installed at the Morris Engine factory in 1923 and was 55m (181 feet) long with 53 operations.

WELDING, ELECTRO-FORMING AND LASERS

Joining metals by heat and filler was practised in bronze statuary *c.* 3000 BC and the first welded iron joint appears in a headrest from the tomb of Tutankhamun, *c.* 1350 BC. The heat required to achieve local melting of iron and steel had to await the production of oxygen by Priestley in 1774, acetylene by Davy in 1800 and Linde's method of extracting oxygen from liquid air in 1893 before the oxy-acetylene welding torch could be devised by Fouché and Picard in 1903 to give an adjustable flame of 3250° C. A patent for welding by electric arc, filed in 1849 by Staite and Auguste de Meitens, devised a method using carbon electrodes, but it is the use of the consumable bare steel rod by the Russian N. G. Slavianoff in 1888 which makes this device recognized as the beginning of metal arc welding. Considerable skill was required to strike and maintain the arc with a bare metal electrode, and coatings have been developed to improve arc stability and provide a protective layer for the weld. Kjellberg in 1907 introduced the first flux coating and in 1909 an asbestos yarn-covered electrode was used. Many different materials have been employed as coatings, including cellulose, mineral silicates and carbonates, to suit welding in different conditions.

Manual metal arc is the most used welding process, but Elihu Thompson's invention in 1886 of the use of electric power to achieve welded joints by resistance heating is very significant, as it took welding from the maintenance function into production. This butt-welding machine was the forerunner of the resistance spot welder between 1900 and 1905 and the seam welder. The submerged arc process, developed for the shipbuilding industry in the USA and USSR in the mid-1930s, employs a powder flux to completely cover the weld pool and the end of the electrode wire. It allows high welding current and low electrode usage in automatic processes. Gas shielding of the arc and weld metal from atmospheric contamination had also been considered from 1919 by Roberts and van Nuys and in the 1930s interest centred on the inert gases, but it was not until 1940 that the Northrop Aircraft Company experimented with tungsten electrodes and helium gas which was used successfully on thin gauge stainless steel. Further advances using other gases followed to give techniques able to cope with the welding of aluminium alloys and other difficult metals. With these developments electric welding became the key process in the rapid fabrication of locomotives, ships and automobiles (linked with the contour sawing of steel made possible by the band saw developed by Leighton A. Wilkie in 1933) and played a huge part in the wartime production effort required from 1939.

During and after the Second World War many developments took place in standard electric welding processes such as submerged arc and inert gas tungsten arc welding. The latter has had the addition of an insulated water-cooled nozzle to form a chamber around the electrode to produce an arc plasma in the form of a flame when the arc is struck from the electrode to the nozzle. This is a non-transferred arc which is used for metal spraying with the addition of powdered metal to the plasma. If the arc is struck between the electrode and the work it is constricted in passing through the nozzle orifice and this is the transferred arc used for cutting and welding. It is used at low currents for sheet metal and at about 400amp for welding thick metal using the keyhole technique. It requires an additional inert gas shield when used for welding. The inert gas metal arc process was introduced in 1948 and uses an electrode of metal which is consumed in the process as a filler. Many other gases have been used as shields, notably argon and CO_2, and it is particularly useful and economic in welding aluminium using small-diameter wire and direct current with the electrode as positive.

Friction welding, introduced in the Soviet Union in 1953, uses the heat developed by contrarotating parts held in contact at high speed until the temperature is high enough for welding when rotation is stopped and the ends forced together.

A number of processes using electrical methods of forming workpieces have been devised which have proved useful in particular applications. Electrolytic machining of metals was introduced by Gusseff in 1930 employing the principles of electrolysis and shaped cathode, so that passage of an electrical current through a suitable electrolyte caused the anode to be eroded to match. The workpiece was placed at the anode.

Chemical milling developed from the ideas of decorative etching produced by Niepce in 1822 and used by Richard Brooman in his patent of 1865 for forming tapered rods. The success of the process depends on the selection of etching chemicals and the inert materials used for masking. In 1936 Griswold in New York patented the technique which has been widely used in the production of printed circuit boards for electronic communications and control equipment. In 1953, M. C. Sanz led a team in the USA to develop the process for manufacturing parts used in aircraft structures and it was used in creating structural parts for the Space Laboratory and Apollo programme (see Chapter 13). In 1961 the process was used to photo-etch transistors on silicon chips.

During the Second World War in 1943 electrical discharge machining, commonly called spark erosion was introduced principally to form cutting and shaping tools made of very hard alloys and carbides which were difficult to produce by any other methods. Modern machines give three-axis cutting by use of the travelling wire technique.

Ultrasonic methods were first developed in 1935 by Sokolov to detect flaws in metal, and this technique has continued as a non-destructive way of checking welded joints. In 1957 the use of this level of oscillation about 20kc/s was used in drilling round and shaped holes, the reciprocating tool being fed with an abrasive slurry which does the actual cutting.

The laser (Light Amplification by Stimulated Emission of Radiation) was invented in 1958 by Townes and Schawlow, although their priority has been disputed. Theodore H. Maiman made the first practical laser in 1960 using a cylindrical ruby crystal. The coherent light produced by the helium-neon laser is used in the interferometer for many applications where movement in very small displacements is to be measured. The laser is also used in metrology as an alignment tool, the narrow beam being transmitted to a photo diode at the end of the surface being checked. At points along the surface being checked, a zone plate, equivalent to a weak lens, is positioned so that if any deviation is present the image on the photo diode will be changed and indicate misalignment. The laser has found many applications in such areas as eye surgery, holography, military range finders and sights, surveying and cutting and welding. A laser beam was used in 1962 to bore a hole in diamond as a demonstration of its power and is now used to make wire drawing dies. The first computer-controlled cloth cutter using a laser was employed by the clothing firm John Collier in 1974. High precision welding can be performed by laser and it completes joints very quickly, but is more difficult to focus and control than the electron beam. This technique, introduced in 1957 to weld nuclear fuel elements, can only be used in a vacuum, but has become increasingly important in making welded joints in micro-electronic devices and on the special alloy material used for jet engine turbine blades. In 1986, CO_2 lasers became available with 600–3000W power and these work on most metals, composites and plastics. Industrial machines with two laser beams to cut three-dimensional sections also exist. At the other end of the scale a laser micro-welder was designed and built by International Research and Development in 1974 to be used by hand in attaching such things as thermo-couples and strain gauges to research and production equipment. The 'Diffracto gage' laser strain measurement system accurate to 1–5μm was announced by the Canadian Company in 1973.

WARTIME ADVANCES

The Second World War was a greater trial of strength in terms of technology and materials than any previous conflict; as well as producing many inventions it accelerated the adoption of new ideas in manufacturing techniques and the organization of production. Because the work force had to grow tremendously it was essential to employ unskilled women on repetitive but precision work. Although they proved to be far more dextrous than men and able to keep up a higher and continuous rate of output on the sort of light capstan lathe operation involved in such areas as fuse work, and made some remarkable achievements in aircraft production and in unpleasant conditions such as welding, intensive training methods and simplified operations were needed. The principles of flow production and interchangeability were applied fully and the few skilled men available were used only to set up machines, maintain cutting tools, jigs and fixtures and, in some cases, operate statistical quality control. These control methods were designed on the basis of probability theory to ensure that all output was good by periodic measurement and recording of dimensions to observe trends in accuracy and correct faults before scrap was produced. Considerable use was also made of 'Go' and 'No Go' gap gauges and large-scale displays to indicate satisfactory dimensions without measuring. Many different types of device were used, such as the Sigma comparator of 1939, which employed a mechanical movement in conjunction with electrical contacts to illuminate lights: red for scrap, green for pass and yellow for rectification. Fluid gauges using liquid, such as the Prestwich, and the air gauges designed in 1928 by Marcel Menneson of Solex in France were the basis of high magnification comparators made by Sigma and by Mercer in England and the Sheffield Corporation in the USA and used in almost every engineering factory. The English versions used pressure difference, as in the Solex, but the Sheffield product measured air flow as it was restricted by the component. All these comparators depended on their setting up in conjunction with a master gauge or component with the tolerance range shown by indicators. The simple rack and pinion dial gauge originally made by B. C. Ames in the USA in 1890 was primarily produced as competition for the micrometer of Brown & Sharpe and L. S. Starrett, although its measuring accuracy was not comparable with that of the screw micrometer. More precisely made dial gauges, conforming to the British Standard set in 1940 and used with slip gauges to set the dial, became important and cheap inspection instruments widely used to this day. In addition to the main production shop's need for inspection instrumentation, the toolroom inspection department and metrology laboratory required accurate measuring machines and instruments to check gauges, jigs and fixtures and, as so many of these were manufactured in Germany by Zeiss, it was necessary for copies to be made by British firms. One vital instrument was the optical dividing head used for essential work in the aircraft and armament industries, and improved versions were made by Cooke, Troughton & Simms and George Watts in conjunction with the Coventry Gauge & Tool Co. by 1939.

American industry supplied war materials to the French and British from 1939 on a payment basis, but the Lend Lease Act proposed by President Roosevelt and passed in 1941 produced a massive turn-round of their own manufacturing. All automobile production was stopped and car plants turned over to aircraft production in large new factory buildings. These were quickly erected, but the biggest problem was the need for machine tools to fill them. This was solved by two-shift working, 12 hours for six days then two days off, in the machine tool works and by 1942 targets were achieved so effectively that at the end of the war there were 300,000 surplus machine tools in the USA. Apart from the organizational success in achieving high-volume production, probably the most significant technological developments of wartime affecting machine-tool design were the extensive employment of servo-mechanisms in gun control and radar and electronics in communications and control. These were to be the basis of modern numerically controlled machine tools.

CONTROL REVOLUTION AND ELECTRONIC METROLOGY

Alan Turing's work on calculating machines enabled him to design the machine to decipher the German 'Enigma' machine codes and he built an electric computer with memory in 1944. In 1948, John Parsons in the USA was transferring the ideas of punched-card control of accounting into the control of machine tools by demonstrating with a universal precision milling machine that all its movements and speed changes could be controlled by mathematical computation. A government contract for more development was given in 1949 to MIT Servo-Mechanisms Laboratory, which produced the first numerically controlled (NC) machine tool in 1954. The system was based on a Cincinnati vertical spindle contour milling machine modified to servo control of table, cross slide and spindle through hydraulic units, gears and leadscrews giving a motion of 0.0005in (0.0127mm) for each electrical pulse received from the control unit. This was provided by a valve electronic unit and Flexowriter using eight-column paper tape to represent an eight-digit number on each line, with holes as units and unpunched positions as zeros. Feedback, the basis of automation, was provided by synchronous motors geared to the slide motions which provided electrical signals to the control for comparison with the input signal to verify the slide position. This machine, and the report on its practical possibilities, was the beginning of NC machine development taken up by many companies. Giddings & Lewis produced Numericord, a contouring machine for aircraft frames which employed magnetic tape to control five axes of motion simultaneously. This system required a computer to interpolate the decimal calculations of points on the profile entered from paper tape, which was then time co-ordinated and recorded on the magnetic tape to give contour control. Other systems were limited to 'point to point' control of the tool position and employed simple electrical relays and limit switches.

The next step in development came with automatic tool change to increase the range of operations performed, the first example being the Barnes drilling machine in 1957 which employed four spindles that could be used in turn. The full concept of a single spindle which could take a number of tools was that of Wallace E. Brainard, who in 1958 designed the automatic tool changer for the Kearney & Trecker Milwaukee Matic machining centre, the first numerically controlled multi-function machine capable of automatic tool changing. It was a horizontal spindle machine with work mounted on an indexing table so that four sides could be machined. A rotary drum on the side of the machine held 30 tools coded on the shank so that they could be selected at random. Tools were collected by a double-ended arm which gripped the tool to be changed, indexed 180° and inserted a new tool in 8.5 seconds. This was the first NC machine imported to Britain, where similar efforts were being made to establish numerical methods of control. Ferranti, in conjunction with Serck, produced a control system to drive a numerically controlled drilling machine to produce the large number of accurately positioned holes required in the end plates of heat exchangers in 1956.

Ferranti also produced in 1959 the first co-ordinate measuring machine, which employed linear diffraction gratings to measure distances on two planes by electronic counting of fringe movement. Many similar optical grating systems have been developed and fitted to machine slides and spindles using photo cells and electronic counters to give accurate positional readings. Electrical gratings such as the EMI Inductosyn use a similar principle, alternating currents in the fixed grating inducing currents in the slider which can be used to indicate position. The latest use of the diffraction grating is in the hand-held electronic micrometer developed by Moore & Wright in 1973 from designs made by Patscentre International. It employs fine pitch gratings of 0.004mm, produced photographically, and a special micro circuit to count the optical fringe movements registered by the photo cell to allow the micrometer to read to 0.001mm or 0.0001in for the inch version.

Post-war research and development in all areas of manufacturing has become more and more costly as the equipment required to make advances has become increasingly complex, and in consequence this has become a company-based operation with large R & D departments continuously making improvements in products which are progressively more anonymous and difficult to assign to individuals. In metrology it was still possible to identify the work of Reason at Taylor-Hobson on the measurement of spherical and cylindrical parts and the problem of 'lobing', whereby a dimension may be judged true by diametrical testing but is 'out of round'. A machine was produced to his design to cope with the measurement of 'roundness' in 1949, the Talyrond, although it was six years before similar machines were made for sale. The machine employed a large vertical column carrying a stylus and electrical transducer which could rotate about the table holding the piece being tested and a circular graph rotating in phase with the spindle to show the variations in the surface of the specimen. Modern developments have enabled the original valve amplification to be up-dated on both Talysurf (see p. 415) and Talyrond, and computers are now used to provide digital sampling and give arithmetical results of surface texture and roundness.

The development of electronics in the Talysurf and Talyrond influenced the company to produce an electronic level to complement their optical alignment telelescope and autocollimator in checking the straightness of surfaces. A pendulum design was finally produced whereby a weight suspended on thin wires would act as an armature for an electro-magnetic transducer. The Talyvel, as it is called, competes with the photoelectric collimator as the most accurate instrument available today for surface contour measurement.

The general machine tools also underwent the control revolution in arrangements for programming slide operations, tool indexing and speed changes by the use of electrical relays, limit switches and hydraulic servo-mechanisms. Many manufacturers produced 'add on' devices for standard lathes, and by the mid-sixties most lathe manufacturers had modified their main lines to provide plug board control of spindle speed and feed rates to cross slide and capstan slide. The 'plug board' consists of a grid of holes, each vertical line corresponding to a step in the sequence of operations, and each horizontal line to a command signal input. When each vertical line is energized all 'plugged' functions will operate simultaneously, and when a timed or external signal, such as that from a limit switch, has indicated completion of these functions the next column is energized. The controller continues stepping line by line until the full cycle is completed. The ultimate development of this type of control appears in the purpose-built lathe by Alfred Herbert named the 3M Robot in 1977. This provided spindle speed control by pneumatic valves and hydraulic feed to front, back, profile and turret slides. Similar 'plug board' controls were introduced on milling machines such as the Parkson Miller-matic of 1964, which had a separate free-standing console containing the electrical circuitry.

At the same time some manufacturers were adopting more sophisticated and expensive numerical control methods such as that for the Société Genevoise Hydroptic 6A first made in 1958, which also incorporated design features of production machines added to their jig borer accuracy. One-inch eight-column paper tape is used to control all the functions of the machine, and accurate table settings are achieved by synchronous motors and photoelectric microscopes. The Dowding 'Atlantic' two-axis co-ordinate drilling machine of 1964 used the Ferranti control system with eight-hole paper tape input, transistorized circuitry and feedback of positional information by the use of diffraction gratings, ruled to 1000 lines per inch, and optical block scanners on the co-ordinate positioning table.

In the 1970s even standard machine tools with the usual hand controls were being fitted with linear and radial gratings to slides and leadscrews to be read by photo cells and translated by electronic processors to provide amplified digital read-outs for simplified control.

FLEXIBLE MANUFACTURING SYSTEMS

In large production factories Britain long persisted in a 1950s design of manufacturing systems, organization and job structure originally arranged to suit high-volume, low variety markets in Henry Ford style, but technological change had entered an accelerating spiral and market demands required greater flexibility. The dedicated transfer machine, with its high cost and rigidity of operation, could only be justified commercially by large-scale production of a constant product design, and this type of machine has been replaced by conveyor linked systems of standard machine tools and machining heads which can be easily dismantled and rearranged to accommodate design changes. Increasing use of bowl feeders and parts dispensers with devices to manipulate workholding jigs between machining sections established automation at the stage when overall control by computer, integrating all the individual machine programme sequence control, became a possibility.

By 1969, D. T. N. Williamson had completed the first flexible manufacturing system (FMS) in the Molins System 24, which provided a range of machines capable of performing multiple operations on parts with different configurations. Each machine in the process had a single horizontal spindle and a tool changing system controlled by a computer program which also controlled depth of cut, feed rate, cutting speed on each cycle and the transference of workpieces between stations. Virtually any part within the physical capacity of the machines could be completely machined by this system, which also featured swarf collection by conveyor and sophisticated oil cleaning and pumping for the hydraulic servo mechanisms. The system, patented in 1970, is a notable British 'first' in manufacturing organization.

The extension of the range of control became possible with the introduction of the concept of computer aided design (CAD) in the USA. This is a method of drawing ideas of shape on a video display and recording them. In the 1960s experiments were made in Britain on interactive graphics using mainframe computers, and since the appearance of microcomputers in the early seventies the techniques have grown rapidly. Pictures are changed by electronic pen, keyboard or digitizers which convert analogue information to digital to allow its manipulation and presentation in the computer. A database of design principles, information on materials and so on is required and is time consuming to prepare. The software for the Clarks Ltd shoemaking CAD took four years to design, but allows presentation and modification of shoe style, colouring and flattening to provide the patterns for leather cutting. CAD can make tapes for numerically controlled machine tools by drawing cutting paths around a two-dimensional drawing which is the basis of computer aided manufacture (CAM). With the control tapes for CAM to provide the instructions for component machining by the numerically controlled machine tool, and the development of mechanical handling and inspection machinery which is also controlled numerically, the target of completely automatic manufacture is achievable. The mechanical handling devices required are the flexible manipulator, the robot and transporters, automatic guided vehicles (AGV).

The name 'robot' first appeared in the play *R.U.R.* by Karel Čapek, performed in London in 1923, for human-like machines created by Dr Rossum to perform tasks. The modern industrial robot is not humanoid in form, but some of the threatening myths attached to fictional monsters, and the Luddite fear of machines which destroy jobs, have had an effect on their acceptance. The simple robot in industry consists of an arm and gripper which operates in only three axes, as the simple 'pick and place' device used to transfer parts between work stations. To this may be added a rotational movement to give increased scope. The articulated arm with dual jointed wrist and elbow movement to give six axes of freedom is the most generally accepted 'robot' with central control by computer, a manipulator at the end of its arm and a sensing device. The most usual sensors are touch, pressure, heat, and vision through photo cells or TV camera. George Devol in the USA invented a programmable robot and made a patent application in 1954. Joseph Engelberger, the founder of Unimation, the biggest manufacturer of robots, met Devol in 1956 and began manufacturing his 'Unimate'; although it was found to infringe a British patent for an industrial robot filed in 1957 by Cyril Walter Kenward, a cash settlement cleared the way for further development in the USA. The first British robot was designed and produced by Douglas Hall, and Hall Automation began in 1974 with a paint-spraying robot called RAMP (random access multi programme). Most of the early applications of robots were in paint spraying, welding and other operations with hazardous environmental conditions, and automobile manufacturers became the largest users. Further improvement in hydraulic drives and the alternative stepping motor, as used in the Swedish ASEA robot, combined with precise feedback control made possible by the incorporation of micro-processor systems, created the accuracy and consistency of placement required for the use of robots in precise assembly work. Modern robots are now capable of returning to the same spot repeatedly within 0.005cm (0.002in) at a speed of 5m (200in) per second and will work continuously for long periods. Visual sensing, first introduced by Hegginbotham and Pugh in laboratory conditions early in 1970, has become the immediate goal of development for applications in assembly work from conveyor supplied parts and in inspection between stages in 'cell' manufacture. A visual sensor was produced at Oxford University in 1984 incorporating a solid state camera to operate while the robot is welding seams using a laser. General vision systems using multiple microprocessors to interpret video picture elements to allow recognition by robots are becoming available to industry and have already been used in medical research.

Increased flexibility in use is demonstrated by the Cobra system produced by Flexible Laser System of Scunthorpe in 1983 which uses a 400W CO_2 laser made by Ferranti Incorporated in an ASEA robot which can cut, trim, weld and drill holes in a variety of materials.

The value of the AGV was first demonstrated in the FIAT body shell production layout in 1979 through their need to achieve flexibility in meeting market fluctuations economically. Assembly and welding is fully automated and can produce different models in any sequence. Robot welders operate within each gate and the AGVs, electrically powered and under the guidance of the central computer, carry body shells on coded pallets to them. The use of AGVs to achieve out-of-sequence operations helps to make more economic use of workshop floor space and get away from the operational sequence organization of flow production (first seen by Pero Tafur, c. 1439, in the equipping and victualling of ships in Venice) by achieving the flexibility of batch production through the use of manufacturing cells for similar products using group technology and FMS principles. The AGV is also a key factor in the creation of computerized warehousing for stock material, finished products, packing and dispatch using underfloor cables carrying instructions.

Three-dimensional co-ordinate measuring machines with computer control of the sensing probe are now in use, with the necessary feedback system to provide information to the main control computer on faulty operations and they could be incorporated in overall manufacturing systems with robot interface between the machining centre and measuring machine. One of these measuring machines, built by the Mitutoyo Company of Japan, can be seen in operation at the Science Museum, London.

THE AUTOMATIC FACTORY

With the addition of the business systems, computer control of the whole process of manufacturing from market forecast to dispatch could be achieved. Computer integrated manufacturing (CIM) would incorporate: forecasting of finished product demand; customer order servicing; engineering and production data source; master production schedule – planning with alternative strategies; purchasing and receiving – quotations, orders, quality and storage; inventory management – quantities and timing to meet plan; order release – authorize production; manufacturing activity – use of capacity at start, minimize 'work in progress' and manufacturing lead time; plant monitoring – control of order through shop, co-ordination of inspection; materials handling; plant maintenance – labour utilization, preventive maintenance; stores control- – materials location; cost control – production information for accounting.

One of the most advanced examples of total automation in the UK, and an operating CIM, is the JCB Transmissions factory in Wrexham producing gearboxes for earth movers. Computers control routine operations and machines, while the master control computer, programmed daily, is in charge of the overall management of the factory. There is manual overview and manual intervention if necessary. The basis is the high-value low-volume FMS for the gearbox cases, the assembly and inspection system, and the automated warehouse, all linked by AGVs. The FMS consists of seven machining centres with large tool magazines containing 80 different tools and capable of machining thirty-two variations of components. Measuring and washing machines are also controlled by the computer which is programmed weekly. Six AGVs are used to carry work between these stations and general distribution work. The AGVs are programmed to follow induction loops in the factory floor with an on-board microprocessor for driving to the loops. Two assembly lines using five manned stations have additional small AGVs to carry parts between them. The plant became operational in 1986.

In the same year, Yamazaki Machinery UK built a new machine tool factory in Worcester at a cost of £35 million to produce machining centres, which are the building bricks of FMS. The factory is temperature controlled and all machining, storage, fabrication and assembly is under CNC. Precision grinding and jig boring is accurate to less than $5\mu m$, parts are fixtured and palletized by robots and transferred by automatic stacker cranes, prismatic machining is carried out by machining centres and lathes allowing multiple operations in one fixture. Sheet metal work is similarly controlled using laser cutting and automatic folding, bending and welding machines. The software control system for the CIM, which took 40 man-years to develop, runs on an IBM S/38 computer for central control of production and scheduling and three DEC Micro Vax FMS-CPU operating the on-line system.

Many assembly operations, such as those in the electronics industry, are highly organized. With controlled positioning of the chassis or board for fitting, the required component is presented in a dispenser, the location for the component indicated by a light beam, and a picture of the component presented to the operator, all under computer control, so that all the operator has to do is grip the component, place it in position and operate some fixing device. This seems only a small step for a robot to take – perhaps it represents a giant step out of the factory for man.

PART THREE

TRANSPORT

8

ROADS, BRIDGES AND VEHICLES

IAN McNEIL

ROAD CONSTRUCTION

Before men began building roads, there were merely tracks, sometimes deliberately cleared but more often worn by repeated passage of man and his domesticated animals from one place to another. In sandy country tracks started by such traffic tended to be enhanced by the subsequent rush of rainwater so that hollow ways were created as much as twenty feet deep, as in parts of Surrey. In marshy lands, like the Fens of eastern England or the Somerset Levels, timber trackways were laid down showing concentrated labour by the relatively small populations of the time and concerted action. Carbon-14 dating has established that these originated as early as the years around 600 BC. The making and use of such trackways presupposes the establishment of permanent settlements – encampments, and later villages and towns. (The walled city of Jericho, built around 8000 BC, is generally accepted as being the first major urban centre.) The routes taken by these early tracks generally followed the easiest possible path, avoiding excessive gradients, dense woodlands, soft marshland and crops, where these had been sown. They also tended to follow the higher slopes of the hillsides where possible, so as to avoid the possibility of attack by an enemy from above. Many of Britain's roads today follow the routes of the trackways of primitive man, which accounts for the somewhat erratic courses that they pursue.

The application of the wheel to transport, said to have originated at Erech in Sumeria about 3500 BC when tripartite or board wheels were first added to a sledge, allowed much greater loads than could be carried either by men or on the backs of pack animals, and this called for more substantial tracks to support them. These were the first true roads. The Romans, the first real road-builders, appreciated the need for adequate drainage, a proper foundation and a

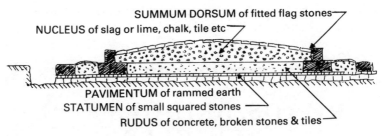

Figure 8.1: Section of a typical Roman road.

cambered, impermeable surface (see Figure 8.1). They constructed 372 major roads, a total of some 53,000 Roman miles (1 mile = 1000 passus: 1 passus = 1.5m (5ft)). These military roads, connecting all the principal towns of the Roman empire from Africa to Scotland and from Spain to Greece generally followed the straightest possible routes and were largely paved with slabs of stone or lava. A map of the Roman roads of Britain shows a remarkable similarity to that of the modern motorway network. The legendary straightness of Roman roads derives, it is said, not from the prowess of their surveyors nor from their desire for the shortest route irrespective of gradient, but from a design weakness in their vehicles. There was no means of steering the front wheels of their four-wheeled wagons, used for baggage, fodder and trade. They could only be inched sideways to change course by means of crowbars. The theory is plausible but incapable of proof.

With the retreat of the Romans and their armies from Britain between AD 410 and 436 little was done to maintain existing roads or build new ones. It was not until after the Norman invasion over six hundred years later that the military value of a good road system was again appreciated. It was, however, the Church which had the greatest influence and, until the dissolution of the monasteries by Henry VIII following his adoption of the position of the Supreme Head of the Church in England in 1534, it was the Church that bore the brunt of road maintenance, including the provision of inns or other accommodation where travellers could rest overnight and obtain refreshment.

In 1555 an Act of Parliament was passed for 'for amending the High-ways being now both very noisom and tedious to travel in and dangerous to all Passengers and Carriages'. Under this Act every parish was bound to appoint two of their number to act as Surveyors and Orderers. At first four and later six days were fixed upon which 'two able men with a team and tools convenient to work 8 hours upon every one of those days' from 'everyone having a term or Ploughland' while every cottager was 'bound to work himself or find a sufficient labourer to work for him'. The tendency was to elect as waywardens the least popular and most incompetent men of the parish and for them to attend only to the roads that they themselves habitually used, to the neglect of the rest. The worst standards of repair were in parishes where no stone was available except by the pillaging of ancient monuments, either buildings or previous roads. Probably the greatest failing of the parish system was that it was too localized: it fragmented the responsibility for road maintenance into tiny divisions at a period when few people travelled outside a radius of ten or fifteen miles of their home in their whole lifetimes. Thus any conception of a parish road forming a part of the national network of inter-city communications was non-existent. No priority was given to the upkeep of trunk roads passing through the parish.

The first Turnpike Act of 1663, 'for repairing the High-ways within the Counties of Hertford, Cambridge and Huntingdon', established a new principle and a new system. By this time both people and goods were travelling longer distances and more frequently, churning up the old roads, which were mostly ungraded and unpaved, more than ever. Yet it was not for another century that many investors had money available to put into turnpikes, for passage along which a charge was made. By 1750 there were only 169, twenty years later 530, but over 1100 by the 1830s, just at the time when the age of the railway was being born to sound the death knell of road transport – until the introduction of the internal combustion engine some fifty years later. By a Turnpike Act, the trustees, originally the justices of the locality concerned, were empowered to charge a toll according to a scale laid down in the Act, the income from such charges to be devoted to the upkeep of the carriageway. In 1706–7 a precedent was established to allow bodies consisting of persons other than the justices to become trustees and to petition Parliament for an Act. The system was hardly a success, since the government made no decisions as to the highways that were of most importance, so that turnpikes grew up and spread haphazardly without any overall plan, the stretches of main road that were included being so purely fortuitously. In the 1830s nearly 4000 trusts were responsible for 35,400km (22,000 miles) of road out of a total estimated at 169,000km (105,000 miles). Worse still was the fact that no systematic or scientific approach was made until about 1756 to the construction and maintenance of highways, whether turnpikes or not. The techniques of the Romans, expensive but effective, had long been forgotten. It was John Metcalfe of Knaresborough, blind since smallpox at the age of six, followed by Thomas Telford and John Loudon McAdam, who were to change all this.

In 1765, 'Blind Jack' Metcalfe contracted to build a turnpike road three miles long between Harrogate and Boroughbridge in Yorkshire. By 1792 he had constructed over 290km (180 miles) of roads. He had an uncanny knack of choosing the best routes over undulating territory, travelling the country for surveys quite alone and unaided by companion or assistant. Most important, however, was his recognition that a well-drained, dry foundation was the first essential, laying a smooth surface of stones over bundles of ling or heather set on a foundation of prepared subsoil, the roadway being raised above ditches for drainage, dug at the sides. It is doubtful if he had heard of Robert Phillips, who in 1737 had read a paper to the Royal Society in which he described a construction on similar principles. The blind Metcalfe, it should be mentioned, was so independent that he served in a volunteer regiment during the Scottish rebellion of 1745. No record exists of his having shot any of his own side.

Thomas Telford was trained as a stonemason, becoming Surveyor of Public Works for the County of Salop (Shropshire) in 1787. In this capacity he was much concerned with turnpike roads within the county. In 1803 he was appointed to direct the construction and improvement of roads in the Highlands of Scotland with a view to improving that country's economy, for the previous road works of General George Wade had been for purely military purposes. By 1820, Telford had built 1500km (920 miles) of new roads which involved 1117 bridges.

If they knew nothing of the principles of good road construction, the authorities were at least aware of some of the major causes of destruction of their roads. The pressure on the road surface was principally a function of the load and the area of tyre in contact with it. Wheel width was of prime importance, narrow wheels, heavily loaded, causing deep ruts, a problem that was to receive much attention in parliamentary Acts from 1750 onwards. In 1753 wagons with wheels less than nine inches wide were prohibited unless drawn by oxen or at least five horses. Teams of eight, twelve or even more beasts were more common. Daniel Defoe, in his *Tour Through the Whole Island of Great Britain* (1724–6), describes a 'tug' carrying a tree to the dockyard at Chatham which was pulled by no less than twenty-two oxen. In spite of this the timber was left to weather throughout a winter when the road surface was churned into a morass. In 1765 someone had the bright thought that, if the front wheels of a wagon were set closer together than the rear wheels, and both pairs were sufficiently wide, the would in effect act as a roller on a good width of the roadway. Freedom from tolls was accordingly granted to such wagons for a period of five years.

In fact France was the first nation to develop a systematic approach to roadbuilding, largely through the work of Pierre J. M. Trésaguet, starting about 1764. The government had established a school a few years earlier, l 'Ecole des Ponts et Chaussées, to train men for the work of overseeing the construction of roads and bridges. Trésaguet laid a single layer of 15–18cm (6–7in) stones on edge on a base of cambered subsoil, covered them with an equal layer of stones of about 5cm (2 in) diameter and topped this with 8cm (3in) of 'pieces broken to about the size of a small walnut' (see Figure 8.2). A network of many hundreds of miles of Trésaguet's construction was built in France and this was so successful that it extended into many other European countries. Telford's method of construction was remarkably similar, but the subsoil base that he used was flat instead of convex in section. It had at the foundation a similar layer of flat

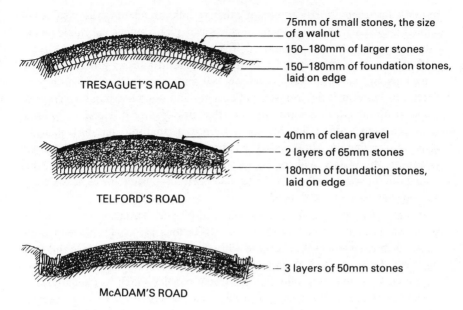

Figure 8.2: Sections of typical roads by Tresaguet, Telford and McAdam.

15–18cm (6–7in) stones placed on edge and tightly packed. Many miles of roads in Britain today follow the tracks of Thomas Telford's ideal alignments and gradients.

Telford's methods were expensive in first cost, in fact too expensive for many of the turnpike trusts, who preferred to adopt the construction of John Loudon McAdam which lasted well enough with a minimum of maintenance.

McAdam's principle was that 'it is the native soil which really supports the weight of the traffic; while it is presented in a dry state it will carry any weight without sinking'. This was certainly true of the loads that were transported by road in McAdam's day, although with the modern juggernaut truck, there might be some doubts, unless the weight was distributed between a sufficient number of axles. McAdam, an Ayrshire man, made a fortune in America and then returned to his native country to become a road trustee and later, in 1815, Surveyor-General of the Bristol roads. His concept of putting 'broken stone upon a road, which shall unite by its own angles, so as to form a solid hard surface' was a sound one, the size of the stones being closely controlled and the way being drained by side ditches. Ideally a 15cm (6in) layer was laid, a further 15cm (6in) being added some weeks later, after traffic had consolidated the first layer. To ensure that the stones were all of the same size his surveyors were equipped with a 6-ounce (170g) weight and scales or alternatively a 2-inch (5cm) ring gauge. It is related that he evolved the latter after inspecting a length of newly laid road one day and finding that a whole stretch of it was of stones of a much bigger size. Up until then the rule was that only stones that could be put into a man's mouth were to be used. The workman responsible for the stretch was admonished, but responded to the rebuke with a broad grin, revealing a capacious and entirely toothless mouth. No finer material was laid on top, as McAdam found that the iron tyres of coaches and wagons ground off chips of stone to fill in the voids in the mass as they passed over it. He was a prolific writer on road building and his books were translated into several languages, thus popularizing the McAdam system abroad. During his lifetime and after his death, his three sons William, James and John Loudon junior continued his road-building work, and by 1823 they had been responsible for 85 trusts comprising over 2000 miles of road.

It was not until 1824 that the problem of properly surfacing the streets of towns was tackled. On the advice of Thomas Telford, a 30cm (12in) deep foundation of broken stones was covered with granite setts about 30cm (12in) long by 15cm (6in) wide and 25cm (10in) deep to withstand the density of traffic. These were far more durable than the cheap cobbles which had been used in the principal streets, at least on those where paving was considered necessary at all. Raised pavements with kerbstone edges had been introduced in Westminster and the City of London in the 1760s. An alternative to granite was the creosoted wood block road surface, the timber for which was mainly imported from Russia from 1838 onwards. A few of these wooden-surfaced city streets survived in Britain into the 1950s.

Trunk roads changed little until the introduction of Dunlop's pneumatic tyre (see p. 449), which became increasingly popular for motor cars from the end of the nineteenth century. Pneumatic rubber tyres had a suction effect that drew up the fragmented stone particles that had been compacted between the stones by the passing of traffic, and dust became a problem. Macadam surfaces made of tar (distilled from wood or coal) or asphalt (a by-product of petrolem) were thus shortly introduced to suppress the problem.

Motorways

A motorway is defined today as 'a limited access road with grade separation, completely fenced in, normally with hard shoulders . . . for the exclusive use of prescribed classes of motor vehicles'. The motorway concept is remarkable in that it originated so early: only some twenty years after Benz and Daimler produced the first motor cars. In 1906 a road, virtually of motorway standards, was proposed from London to Brighton though nothing came of the proposal. The same result came from the plan to build 'the Northern and Western Motorway' from London to Liverpool in 1923.

The first road 'solely for the use of motor vehicles' dates from 1914 when a 65km (40.5 miles) long single-carriageway road with limited access was built in the USA on Long Island. A 10km (6 miles) experimental dual carriageway with no access along its length was built in 1921 to the west of Berlin. Construction of motorways or autostrade increased greatly with Italian activity from the 1920s. A road from Milan to Varese, some 80km (50 miles), was opened in 1925 and later extended a further 50km (31 miles) to Lake Como and Lake Maggiore. In the next ten years seven other motorways, amounting to a total of some 400km (248.5 miles) were built throughout Italy. The German government's plan for a network of motorways (autobahn) started in 1923 and was pursued energetically after the National Socialists came to power in 1933. These were dual carriageways with a central reservation having an anti-dazzle hedge. By 1939, 3300km (2050 miles) of autobahns were completed and in use, 2000km (1243 miles) under construction and a further 3300km (2050 miles) planned.

The first motorway in Britain was the 13km (8 miles) Preston bypass, opened in 1958. By 1980, 2286km (1420 miles) of motorway were in use in England alone. Since then the 203km (126 miles) London Orbital motorway has been completed. Motorways in Britain take up some 30 acres per mile (76,000sq. metres per kilometre), usually of agricultural land.

Modern roads have either flexible or rigid 'pavements', this term being used to describe the form of overall construction. Flexible pavements deflect under heavy loads and flex slightly to accommodate movements due to settlement. Macadam and asphalt are typical 'black-tops' of this type. Rigid pavements are constructed of concrete.

The wearing course, which is 15–40mm (0.6in–1.6in) thick, of a flexible pavement is the top surface and is designed to give an even-running, durable and skid-resistant finish strong enough to resist the tractive forces from the traffic, and weather-resistant. Below this comes the base course of 40–75mm (1.6–3in) thickness and, with the upper wearing course, forming the surfacing. It distributes the traffic loading from the surface to the roadbase. The thickness of the roadbase depends on the intensity of the traffic for which the road was designed and the material of which it is made. It is at least 60mm thick. The sub-base is laid directly on the finished earthwork, the top layer of which is termed the formation.

Concrete roads have thinner sub-bases than flexible roads, the concrete slabs providing the structural layer as well as the wearing surface. The slabs may or may not be reinforced, depending on the traffic loading and intensity for which the road is intended. Transverse joints filled with a sealing compound allow for expansion and contraction.

The normal design life of rigid pavements is 40 years compared with 20 years for flexible pavements.

EARLY ROAD TRANSPORT

First the ox and then the horse were domesticated to become useful as draft animals, as were the mule, the ass and the donkey. Light two-wheeled chariots were used in warfare by many Middle Eastern nations and from them the tradition was handed down to the ancient Greeks and to the Romans. Both had two wheeled chariots as well as four-wheelers, used more for goods transport than for passengers. Chariot racing was popular as a sport before 770 BC.

The Roman two-wheeled *carpentum* was a vehicle used by lower officials, including members of the priesthood. The *carraca dormitoris*, with four wheels, was equipped with beds. The *carpentum mortuarium* was a two-wheeled hearse. The decline in the state of the roads in Britain, after the departure of the Romans, was accompanied by a decline in the use of wheeled vehicles, coastal shipping and river transport being used instead wherever possible. Men, even royalty, preferred to travel on horseback throughout this period: a litter might be used by the aged or infirm or by women. This process continued throughout the Middle Ages up until the second half of the sixteenth century, when carriages and coaches began to be imported from the Continent. Initially carriage travel was mainly confined to towns, only baggage being sent by coach or wagon between towns. The poor, which included the great majority of the population, went on foot if they travelled at all.

When wheeled passenger transport vehicles began to be imported from Europe, British manufacturers were not slow to copy them and here we see the beginnings of an indigenous coach- and carriage-building trade. The word 'coach' is derived from Kocs in Hungary, reputedly the place where they were first built. The first coach known to have been built in Britain was by Walter Rippon for the Earl of Rutland in 1555, while the Earl of Arundel imported one from Germany which was much copied. It is said that Queen Elizabeth I preferred to send to the Netherlands for her carriages though, in ceremonial processions, she also used a litter.

A hundred years later the coaching business had grown to include the use of hackney carriages for hire in towns and regular stagecoach services between them. In 1635 the post became a royal monopoly, 'His Majesty's Mail', and was carried by postboys – or occasionally girls – on horseback and very slowly until after a Post Office Act of 1765, when it was made an offence to convey the mail at less than 6 miles an hour. John Palmer's first mail coaches started operating between London and Bath, a stagecoach journey of seventeen hours, in 1784, covering the distance in one hour less and with greatly increased regularity. The following year saw the introduction of regular mail-coach services all over the kingdom, including such remote destinations as Holyhead and Milford Haven. Four inside passengers were all that were carried. Besant's 'Coach for the Mails', with his improved wheel, was patented in 1786. In this, a collar was formed on the arm of the axle and a plate made to bolt on the inner side of the nave of the wheel so that, unless the plate wore through completely, it was impossible for the wheel to come off, a great improvement over the conventional arrangement in which there was a washer and linch pin passing transversely through a hole in the end of the axle. Furthermore both the axle bearing and the face-plate were lubricated by means of an oil-cup and a groove along the axle. The improvements in speed, punctuality and service brought about by the introduction of the mail-coaches was largely dependent on the improved roads of Telford, Metcalfe and McAdam.

There were many developments in coach design from the seventeenth century onwards which individually amounted to little but which collectively transformed the primitive vehicle which was little more than a box on wheels into the comparative speed and luxury of the mail coach, the landau or the brougham. Springs replaced the earlier suspension from leather straps. Dished wheels were introduced and the use of one-piece iron tyres followed soon after instead of short strips of iron nailed to the felloes of the wheels. Attempts were made to add some form of braking superior to the conventional iron wedge or skid attached by a chain to the perch (or chassis member) and slipped under the front of the nearside rear wheel, at some risk to the coachman, when approaching a descent. Lever brakes, operating a friction pad, usually of leather, to work against the wheel tyre, were not introduced until early in the nineteenth century. Goods vehicles changed little, but from 1780 onwards many new and improved passenger vehicles appeared, a number of them of continental origin.

Of the many types of carriages, coaches, carts and wagons that originated between 1750 and 1900, the most important are listed with their principal characteristics in Tables 8.1–3.

POWERED ROAD TRANSPORT: INITIAL EXPERIMENTS

Until 1760, the only alternative to the roads was transport by river but, in that year, James Brindley started to build his first canal, to carry coal from the Duke of Bridgewater's mines at Worsley through to Manchester. At first greeted with scepticism, the canal was so successful that the price of coal in Manchester was halved. The reduction in price led naturally to a great increase in trade. Soon canal schemes were started up and down the country and the moneyed classes hurried to invest in the wonders of the new inland waterway transport. The 'canal boom' reached its peak about 1793, in which year the Grand Junction Canal, from Oxford to Brentford, was started, the last of the network of canals that now threaded the country. A horse could draw a number of barges, each carrying a far greater load than a wagon, so freight transport moved wherever possible away from the turnpike roads and on to the canals (see Chapter 9). The boom was to be relatively short-lived, however, for the railway age was just over

Table 8.1: Public vehicles.

Name	Approx. date	Usual no. of horses	No. of wheels	Inside passengers
Brouette	1670	1 man	2	1
Stage Coach	1650 on	4	4	up to 14 inc. on top
Stage Wagon	1700 on	10 or 12	4	
Coffin Cab	1750 on	1	2	1
Sedan Cart	1784	1	2	1 or 2
Mail Coach	1784	4	4	4 or 6
Omnibus	1829	3	4	22
American Cab	1830	1	2	4
Hansom Cab	1834	1	2	2
Growler	1840s to 1930s	1	4	4
Knifeboard Omnibus	1850s	2	4	12
Garden Seat Omnibus	1880	4	4	12
Gurney Cab	1880s	1	2	4
Touring Coach	1890s	4	4	up to 22

Table 8.2: Larger carriages and coaches — private and coachman-driven.

Name	Approx. date	Usual no. of horses	No. of wheels	Inside passengers
Vis-à-vis	1768	2	4	2
Berlina	Mid 18th C	2 or 4	4	2 or 4
Landau		1, 2 or 4	4	2 or 4
Landaulet	1790s	1 or 2		2
Barouche	Late 18th C	2 or 4	4	2 or 4
Travelling Chariot	1795–1825	2 or 4	4	2 or 4
Britschka	c.1818	4 or 6	4	2 or 1 (sleeping)
Droshky	Early 1820s	1, 2 or 3	4	4 or 2
Dress Chariot	1830s	2	4	2 (or 4)
Brougham	Late 1830s	1	4	2
Clarence	1842	1	4	4
State Coach	1840s	4 or 6	4	4
Sociable	1860s	2	4	4
Victoria	1860	1 or 2	4	2
Private Omnibus	c.1870	1, 2 or 4	4	6 to 8

the horizon and was to have an even greater effect on the prosperity of both road and canal transport. It may be said to have started properly with the Rainhill Trials in 1829 (see p. 561).

The middle years of the eighteenth century saw the beginning of serious attempts to produce a thermodynamic substitute for the faithful horse, and these were to increase as the cost of fodder soared during the Napoleonic Wars. The steam engine (see Chapters 5 and 11) was at that time the only possibility

Table 8.3: Privately-driven carriages.

Name	Approx. date	Usual no. of horses	No. of wheels	Inside passengers
Continental Chaise	From *c.*1760	1	2	2
Crane-necked Phaeton	*c.*1780	2 or 4	4	2
High Cocking Cart	1790s	1 or 2	2	2
Surrey	1800	1 or 2	4	4
Curricle	1800s	2	2	2
Dogcart	1800	1 or 2	2 or 4	4 or 6
Mail Phaeton	Late 1800s	2	4	4
Cabriolet	1810s	1	2	2
Whiskey	*c.*1812	1	2	2
Dennet Gig	*c.*1814	1	2	2
Tilbury	*c.*1820	1	2	2
Stanhope Phaeton	*c.*1830	1 or 2	4	4
Char-a-banc	1840s		4	
Wagonette	1840s	1 or 2	4	6 or 8
Lawton Gig	*c.*1860	1	2	2
Brake	1860s	2 or 4	4	12 or more
Manchester Market Cart	*c.*1870	1	2	4
T-Cart Phaeton	1880	1	4	3
Spider Phaeton	1880	1	4	3
Village Phaeton	1890s	2 or 1	4	6
Float	1890s	1	2	1
Ralli Car or Cart	*c.*1898	1	2	2 or 4
Governess Cart	*c.*1900	1	2	2
Buggy	End 19th C	1	2	2

available. Nicolas Cugnot, a French military engineer, produced in 1769 a vast, top-heavy, steam-driven contraption (Figure 8.3) that lumbered ponderously round a few of the streets of Paris. Its lack of speed has, been criticized, perhaps unjustifiably, for Cugnot had only intended it as a gun-carriage tractor. However, the military authorities showed little interest and it soon ended up as a museum piece. James Watt's engineer, William Murdock, tried again in 1785 and produced a model of another three-wheeler of a much lighter design. His experiments were not appreciated and his master dissuaded him from wasting too much time on such a 'trivial pursuit', at pain of losing his job. William Symington, the Scottish engineer, designed a steam road carriage in 1796 and built a model which he demonstrated to prospective financial backers. Richard Trevithick persisted longer. After he had tested models, he drove his first full-sized carriage under steam in Cornwall in 1801. The relief of Trevithick and his cousin, Andrew Vivian, on righting it after it had overturned was so great that, when they stopped at an inn 'to comfort their Hearts with a Roast Goose and proper drinks', they were 'forgetfull of the Engine, its Water boiled away, the Iron became red hot, and nothing that was combustible remained of the Engine or the house' where it was sheltered. A second carriage (Figure 8.4) was built the following year and ran for some time in London, but created little interest, only

Figure 8.3: Nicolas Cugnot's steam Traction Engine built in 1769 to draw artillery.

antagonism. Trevithick's steam transport experiments were thenceforth confined to rails. In 1804 a locomotive of his design drew a load of some 10 tonnes and 70 men the 15.5km (9.75 miles) from Penydaran to Abercynon, South Wales, in just over four hours. It was Trevithick's use of high pressure steam, rather than Watt's self-imposed restriction to atmospheric pressure working, that allowed the development of locomotion.

BICYCLES

The idea of using human muscular effort for the propulsion of light road vehicles seems to originate as early as Roman times but, apart from spasmodic experiments, was not seriously developed until the late eighteenth and early nineteenth centuries. The hobby-horse (or dandy-horse) era took off in 1817 when Baron Karl Drais von Sauerbronn of Mannheim demonstrated his 'Draisine' in the Luxembourg Gardens in Paris. It had a wooden frame, a steerable front wheel, and an armrest behind the handlebars which gave the rider a greater purchase in thrusting his feet against the ground. In spite of the inefficiency of this awkward method of propulsion, a journey from Beaune to Dijon, a distance of 37km (just over 23 miles) could be made by Draisine in $2\frac{1}{2}$ hours, an average speed of 15kph or nearly $9\frac{1}{2}$mph. The popularity of the vehicle as a fashionable novelty, rather than a practical means of transport, led to its rapid adoption in other countries as well as France, particularly England (see Figure 8.5), Germany and the USA.

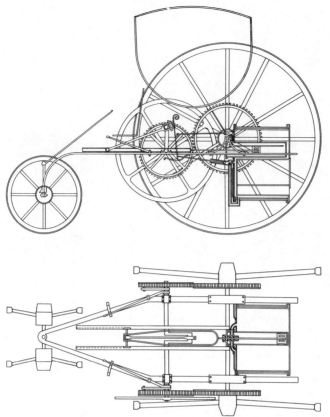

Figure 8.4: Richard Trevithick's steam carriage of 1802. Little public interest was shown and Trevithick turned his attention to rail locomotion.

The first man to appreciate the gyroscopic effect of two wheels mounted in line and the possibility that it gave of lifting the feet from off the ground was Kirkpatrick Macmillan, a blacksmith of Courthill, Dumfriesshire. In 1839 he invented and built the world's first pedal-operated bicycle (Figure 8.6). A curved wooden backbone carried the rider and a saddle, the front wheel being steerable. The iron-tyred wooden wheels were carried on brass bearings, the drive to the rear wheel, which had a pair of cranks keyed to the axle driven by connecting rods, being from two swinging levers, pivoted ahead of the rider and ending in pedals for the rider's feet. Macmillan often rode his machine on the 22.5km (14 miles) stretch between Courthill and Dumfries and the 64km (40 miles) to Glasgow. In spite of the blacksmith's success and that of a number of copyists, the machine did not catch on and experiment in the following twenty years was largely confined to tricycles and quadricycles.

Figure 8.5: An early 'hobby horse' of 1818.

Figure 8.6: Kirkpatrick Macmillan's bicycle of 1839. For the first time, the rider's feet were off the ground.

Figure 8.7: Pierre Michaux's Velocipede of about 1863 was a hobby horse with pedals fixed to the front wheel.

Soon after 1860 the boneshaker made its appearance, basically a variant of the velocipede of Pierre Michaux of Paris. The frame or backbone was of wrought iron and the drive was by pedals keyed directly to the axle of the steerable front wheel (Figure 8.7). The machine incorporated a brake, a lever on the frame pressing a shoe against the periphery of the rear wheel tyre. By 1865 Michaux was building over 400 machines a year which sold for around £8. The first bicycle journal in the world was started at Grenoble in 1869, and the first cycle show took place in Paris in the same year. Wire spoked wheels, tubular frames, solid rubber tyres, mudguards, freewheels and change-speed gears were among the different features on the machines exhibited. French superiority in the bicycle industry was to last until the outbreak of the Franco-Prussian War in 1870.

It was Coventry that was to become the centre of the industry in England, with an early example of both exporting and diversification. The Coventry Sewing Machine Company had received an order for 400 bicycles for France, an export frustrated by the 1870 war. They decided to explore the market in Britain, with some success, and managed to dispose of the stock made for the French order. In the following ten years the diameter of the front wheel, already

Figure 8.8: Starley & Hillman's 'Ariel' of 1870.

larger than the rear wheel, was gradually increased, the rear wheel becoming smaller while the saddle of successive models moved forward until the rider was poised directly above the front axle. In this position he could use his leg power and body weight most effectively. Thus was developed the ordinary bicycle as it was then called, more frequently referred to today as the penny-farthing. In this, the radius of the front wheel was about equal to the length of the rider's leg. The problems of mounting a 183cm (6ft) diameter wheel required considerable skill, as did staying there – and dismounting with a degree of decorum. Figure 8.8 illustrates an example of this variety of bicycle.

In the 1870s and 1880s development was to evolve the safety bicycle (Figure 8.9), forerunner and prototype of the machine we know today, with equal-sized front and rear wheels, the rear one being driven by a chain and sprocket. The diamond frame of tubular steel became standard after 1900 and the pneumatic rubber tyres of J. B. Dunlop after 1893 (see p. 449).

Figure 8.9: Humber Safety Bicycle, 1890.

MOTOR CYCLES

A little-noticed event happened in December 1868, when a small single-cylinder Perraux steam engine and boiler was fitter to a Michaux velocipede. Thus the first truly successful bicycle became the basis for the first motor cycle, starting a separate line of development that was ultimately to lead to the age of the motor car.

This was by no means the first time that such a machine had been suggested: it was a favourite subject for cartoonists, such as Leech, Rowlandson and Cruikshank from about 1820. It led to no commercial development but was undoubtedly the prototype of the motor cycle. Perhaps its most successful follower was the steam tricycle made by L. D. Copeland of Philadelphia in 1884, of which some 200 were built. Progress had to wait for the invention of the internal combustion engine, particularly that of Gottlieb Daimler, who produced a two-wheeler of remarkably primitive aspect driven by an air-cooled four-stroke petrol engine in 1885. His interest in motor cycles soon waned and he devoted himself, as did Karl Benz, to cars (see p. 450).

In 1895 Count Albert de Dion and Georges Bouton, pioneers of steam-driven vehicles for twelve years, produced their first light-weight, high-speed petrol engine and soon fitted it to a tricycle. It ran at 1500rpm, compared to Daimler's 800rpm, weighed 18kg (40lb) and generated $\frac{1}{2}$hp. A $1\frac{3}{4}$hp de Dion Bouton tricycle was first in its class in the Paris–Marseilles–Paris road trial of 1896, averaging 22.5kph over the 1126km course. They licensed many foreign firms to manufacture their engines, as did Daimler. By 1900 over fifty companies in England were building motor bicycles or tricycles. Most of the early models kept strictly to the diamond-shaped frame of the 'safety' pedal bicycle. Coventry was again the centre of the industry which really took off with the repeal of the 'Red Flag' Act in 1896, a maximum speed of 12mph (19kph) then becoming permissible.

At first manufacturers tended to concentrate their efforts on the production of tricycles and quadricycles so that motor-cyclists could obtain the freedom of movement offered by the motor car but with a lower initial expenditure and at reduced running cost. These multi-wheeled cycles had the sociable advantage over the motor bicycle that they could carry a passenger. However, development along these lines virtually ceased with the invention of the sidecar by W. G. Graham in 1903.

The first really practical two-wheeler was produced by Michel and Eugène Werner, two Paris journalists, in 1897. It was little more than an engine-assisted bicycle, with the engine mounted on the handlebars and driving the front wheel. This arrangement was somewhat top-heavy and in 1901 the Werners as well as Messrs Phelon and Moore and others started to build machines with the engine mounted low in the frame and driving the rear wheel by belt or chain, thus establishing the classic layout of motor bicycle used still today. The motor bicycle movement had begun, with popular competitions and trials. Among these was the Tourist Trophy, or TT, races held annually on the Isle of Man, as racing on the public roads of mainland Great Britain was illegal, as it remains today (except for the occasional event licensed by local bye-laws). The first TT was held in 1907 when the winner was a Norton, and this gave a notable boost to the British motor cycle industry. An American 'Indian' won in 1911 with similar benefit to the United States manufacturers. By 1914 the motor-cycle speed record stood at 150kph (93.5mph).

Further development was curtailed by the outbreak of the First World War, the production of machines, except for war purposes, being prohibited after 1915. Many motor-cyclists of the period found employment for their talents as the first army dispatch riders and rendered sterling service throughout the war.

The TT races were re-established in 1920, British motor cycles remaining dominant until 1951 when, for a decade, the Italian makers – Moto-Guzzi, Gilera and M. V. Augusta – took the lead with the German firm NSU. From the early 1960s the Japanese Honda became supreme, by which time the British motor cycle industry was but a shadow of its former glory when, as in 1937, a 1000cc Brough Superior with a JAP engine held the world speed record at 273kph (170mph). In 1978, at Bonneville Flats, Utah, a two-litre Kawasaki Special reasserted the supremacy of the Japanese industry by taking the world speed record at 512.6kph (318.5mph) – a far cry from Ernest and Pierre Michaux's alcohol-fuelled steam cycle of 1868.

The period after the Second World War introduced a general need for fuel economy which did much to promote the moped and the autocycle. The first of the former, in the early 1950s, was the 50cc NSU from Germany, followed shortly by the Italian Vespa and Lambretta machines, with small wheels, 125cc engines and a front apron affording some protection against the weather for the rider. The design of the moped originated in the folding scooters such as the Corgi used by paratroops during the war.

Another development of the same period, in vogue for some years, was the bubble-car. Heinkel and Messerschmitt were both prominent makes of these machines, the designs of which owed much to the cockpits of the aircraft which these companies had been building during the war. The passenger sat unsociably behind the driver; more sociable – and more bubble-shaped – was the BMW machine in which side-by-side seating was provided. Still in Germany, a possible resurgence of the three-wheeled bubble-car was revealed by the Volkswagen Research Division in a design study of early 1986. This vehicle was said to have a maximum speed of 193kph (120mph) and a fuel consumption of only 21.2km/l 59.9 miles per gallon (49.9 US) with a 1.4l engine. Its road-holding at speed was said to be superior to that of the same company's Golf saloon.

THE MOTOR CAR

It is unfortunate, perhaps even unfair, that so many inventions are ascribed to those who brought them to full commercial fruition rather than to those who were their true originators. Just as James Watt was the great improver of Thomas Newcomen's rather inefficient steam engine, those whose names are usually associated with the invention of the motor car had their forerunners, while the same applies to many of its component parts. Rudolph Ackermann, a London printseller, for instance, patented the form of steering known since then by his name. In fact, it was invented for windmill carriages just over a hundred years earlier in 1714 by Du Quet, whose name is virtually unknown. Likewise, John Boyd Dunlop is credited with inventing the pneumatic tyre in 1888 (he fitted them to the rear wheels of his small son's pedal tricycle) and was the first to bring it into general use on any appreciable scale, with the formation of the Pneumatic Tyre Company, later to become the Dunlop Company. In fact R. W. Thompson's 'aerial tyres' were also pneumatic and have been in use for horse-drawn carriages from 1845. While Gottlieb Daimler and Karl Benz can claim to be the first to exploit petrol-engined motor cars from 1886, Siegfried Marcus had some success in running a petrol-driven four-wheeler vehicle in Vienna in 1875 (see Figure 8.10). Unfortunately the Viennese police objected to the noise it made and, as it was to Marcus no more than an engineering exercise, he did not persist in its development. In fact, both Daimler and Benz were

Figure 8.10: Model of Siegfried Marcus' Motor Car of 1875.

almost preceded by Edward Butler, a London engineer, who patented a three-wheeled 'petrol cycle' in 1884, but did not have it running until 1888. It differed from Benz's 'Patent Motor Wagon' (see p. 38, Figure 8; p. 451) in having two steerable wheels at the front and a single driving wheel at the rear. Two horizontal water-cooled cylinders drove overhung cranks on the axle of this wheel drawing petrol from a spray carburettor which was fired by an electric spark. Numerous trials and demonstration runs were made but the lack of enthusiasm of his financial backers and the restrictive road laws in Britain at the time defeated Butler. He broke up his machine and sold the brass and copper for scrap in 1896 – ironically the very year that the laws relating to road vehicles were relaxed.

Thus it is to Daimler and Benz that the credit must be given for building the first motor cars to go into production and to become both a technical and a commercial success. Although they lived only some 96km (60 miles) apart, neither man knew anything of the other's early experiments; ultimately they were to combine forces as Daimler Benz, later Mercedes Benz. Gottfried Daimler, of Cannstatt near Stuttgart, invented the light, high-speed petrol engine in 1884 and in the following year built his 'quadricycle', a curious, wooden-framed motor cycle with a pair of small side wheels mounted on outriggers to prevent it from overturning, such as are sometimes used on a child's fairy cycle today. Also in 1885, Karl Benz built his first model, a three-wheeled

'Patent Motorwagen', which ran in the streets of Mannheim in 1886; its engine was of 984cc and produced 0.9hp (673W) at 400rpm (see p. 38). In the same year Daimler fitted a 469cc engine to a Victoria carriage, working at up to 700rpm and generating 1.1kW (1.5hp). With Wilhelm Maybach, whose spray carburettor contributed so much to the success of his engine, Daimler started building cars in 1899.

Benz produced four cars for sale but found difficulty in attracting buyers. His French agent, Emile Roger, had more success, while Daimler granted licences to both Peugeot and Panhard–Levassor in France to build his latest two-cylinder 1.65hp engines in 1889. Further French interest was shown by the entry of Michelin into the pneumatic tyre market in 1894–5, de Dion Bouton, who adopted Robert Bosch's electric ignition in 1897, and Renault, who built the first car with shaft drive in 1898. The first American car was built by the Duryea brothers in 1893, while Fiat started the industry in Italy with a car driven by a twin horizontally-opposed cylinder engine in 1899.

A significant feature of the early development of the motor car and the internal combustion engine is the relatively small contribution made by British engineers, especially in comparison with their major role in the history of the steam engine and its application to the railway locomotive (see Chapter 11). The plain reason was parliamentary obduracy, which confirmed public apathy by imposing draconian laws on the operation of motorized road vehicles. The 'Red Flag' Act, passed in 1865, required that 'at least three persons shall be employed to drive or conduct a locomotive' on the public roads, a locomotive meaning anything from an agricultural or showman's steam tractor to a three-wheeled, petrol engine car. One such person 'shall precede each locomotive on foot by not less than 60 yards carrying a red flag constantly on display', the law declared. A speed limit of 4mph (6.5kph) was also stipulated and, though the red flag was no longer a requirement after 1878, the pedestrian advance warning was still legally necessary until the Act was finally repealed in 1896, an event commemorated annually today in the London to Brighton Veteran Car Run. The lead of the French is shown by the fact that in the previous year a race from Paris to Bordeaux and back (1178km) (732 miles) had been won by Emile Levassor in a Panhard-Levassor at an average speed of 24kph (15mph). The speed limit in Britain was still only 12mph (19kph).

Naturally there was much opposition to the British laws. Notable was Sir David Salomans of Southborough, who had built an electrically-propelled tricycle in 1874, was one of the promoters of the Self-Propelled Traffic Association in 1895 and organized the first 'motor show' in the country, held in Tunbridge Wells in October of that year. Flouting the Red Flag Act, Sir David drove to the show from Southborough in a Peugeot with an engine of Daimler manufacture to join the other three cars that were on display.

The year after the Act was repealed, Frederick and George Lanchester built the first all-British car, with pneumatic tyres, Ackermann steering and drive, through an epicyclic gearbox, to a rear axle with a differential gear. Lanchester was also the first to develop disc brakes in 1901. An important entrant into the field was Henry Royce, a Manchester crane-maker who, with the Hon. Charles S. Rolls, built three 7.5kW (10hp), two-cylinder cars in 1905. Two years later the 30–37.5kW (40–50hp) 'Silver Ghost' Rolls Royce appeared, shortly to be accepted as the finest car in the world, eagerly sought after by potentates, princes and premiers of every race, colour and creed, a coveted position that the marque has held to this day.

Across the Atlantic inventors and manufacturers were just as active as those in Europe. In August 1859 Edwin L. Drake had struck oil in Pennsylvania at a depth of $69\frac{1}{2}$ feet (21m), starting an industry which, in fifteen years, reached a production of 10 million barrels, each of 360lb (163kg), per day. Initially most of this was used for lamp oil but it was to form a useful source of motor fuel in the future. By 1982, 250,000 million 36-gallon (43.2 US, 163.7l) barrels of oil had been extracted by the world's oil industry.

Henry Ford ran his first 'horseless carriage' through the streets of Detroit in 1896. The Ford Motor Company built six hundred cars at the first mass production car plant at Dearborn, Michigan, in 1903, and built the first of his famous Model T series in 1908, producing 19,000 in that year and 1.25 million by 1920. By 1927 over 15 million had been built, a record in production only to be surpassed by the Volkswagen Beetle of Ferdinand Porsche forty-five years later. The adoption of Ford's mass production techniques in Europe started with the Citroën Type A in 1919.

An important adjunct to the car industry was in the electrical components, in which Bosch, Delco and Lucas predominated, all producing complete electrical systems including headlights by 1911 when Austro-Daimler also offered electric starting as an optional extra, a year before Cadillac installed it as standard. Duesenberg had mechanical four-wheel brakes in 1910 and the first hydraulic brakes on all four wheels originated from the same company in 1920. Triumph led the way with them in Britain five years later. The first entry of the American industry into Europe was in 1924 when General Motors took over the German Opel works. The same year they were producing 100 of their Tree-Frog model a day. Paralleling Ford's oft-quoted dictum about the model T, 'You can have it in any colour you like as long as it's black', the Opel Tree-Frog was only available in green. The first series-produced front-wheel drive car in the world, the Alvis 12/75, appeared in 1928.

The 1930s saw the introduction of the fully synchromesh gearbox, patented by the German component company ZF and first fitted by Alvis to their Speed 20 of 1933. Sir William Morris, later Lord Nuffield, set his sights on producing a British 'people's car' for under £100 and launched the Morris Cowley in 1931. His competitor, Herbert Austin, came out with the inexpensive 'Baby'

Austin 7 about the same time. Ford, which had started building cars at Trafford Park in Manchester in 1911, was, by 1931, producing 500 vehicles a day from its new plant at Dagenham, Essex. They started production of their £100 Ford Popular in 1933.

The 'Baby' Austin was to be the start of the Japanese motor car industry, for in 1935 Datsun started to manufacture it under licence. This firm was to become Nissan-Datsun under the ownership of the Nippon Steel Company. Today with their competitors – Honda, Toyota, Subaru, Mazda, Mitsubishi, Suzuki and Daihatsu – the Japanese motor car industry is the largest in the world, producing over 7 million vehicles a year. Volume production owes much to the use of industrial robots for, of some 10,000 in the world, over 4000 are at work in Japan and over two-thirds of these are in the automobile industry. This achievement is all the more remarkable in the light of the fact that the first Japanese-designed car was built by Toyota as recently as 1955, while motorcycle manufacturer Honda did not produce a car or commercial vehicle until 1962.

Also in 1955, in France, Citroën introduced their revolutionary design the GS 19, with a hydraulic system so comprehensive and so complex that many dealers were reluctant to handle it, predicting that it would not sell, while service garages were equally hesitant to dabble in the unknown mysteries of hydraulics. The car had hydro-pneumatic suspension that could be adjusted to accommodate different loads and road conditions, hydraulically assisted braking, steering and gear change as well as a glass fibre roof and a single-spoke steering wheel. The prophets of gloom were wrong. It did sell and the same principles have been applied to subsequent models. Opposing the trend towards the increased use of hydraulics, the Dutch manufacturers DAF developed a continuously variable mechanical automatic transmission, the 'Variomatic' belt drive, launched in 1958. The following year, BMC, as it then was, ushered in a new era of cheaper motoring with Alec Issigonis's transverse-engined Mini with front-wheel drive, in which four passengers could be accommodated within a body no more than 3.09m long.

AUTOMOTIVE ENGINES

As related elsewhere (see Chapter 5), the steam engine was the first thermodynamic form of power, used originally for mine pumping by Savery and Newcomen and producing rotary motion when Watt added his sun-and-planet gearing in 1783. Even before this the minds of many ingenious men had turned towards propelling ships and carriages, the two forms of transport then in use, by engines. Jonathan Hulls published his proposals for marine propulsion in 1737, Nicolas Cugnot built his steam-driven gun carriage in 1770. It was not until the advent of 'strong steam', i.e. steam at high pressure, that the steam engine became light enough and small enough to be useful as a mobile engine, Richard Trevithick, Sir Goldsworthy Gurney, Thomas Hancock, William Murdock and William Symington were among the most notable pioneers of steam-driven road transport (see p. 458).

The introduction of the gas engine, first by Etienne Lenoir in 1860, and then the development of the four-stroke cycle by Nikolaus Otto in 1876, came at a time when road transport was in a decline in opposition to the success of the railways which were sweeping many lands. William D. Priestman built the first successful oil engine in 1886 and Herbert Ackroyd Stuart developed it further with hot-bulb ignition. Rudolf Diesel had studied under Professor Carl von Linde who had shown him a cigar lighter based on the principle of the Malayan fire piston, in which rapid compression of a gas generated sufficient heat to cause it to explode or ignite. Thus was born the idea of compression-ignition, successfully translated into a practical vehicle engine at the Maschinenfabrik Augsburg AG in 1893, the prototype of the diesel engine, one of the two principal types used in road vehicles today.

The petrol engine, like the oil engine, developed from the original gas engine. It differs from gas and oil engines in that liquid fuel is vaporized before it reaches the combustion chamber and that it is light enough to vaporize at atmospheric pressure and temperature. The first inventor was Samuel Moray who experimented with such an engine, using a mixture of alcohol and turpentine, at Orford, New Hampshire, USA, between 1826 and 1829. Otto's partner, Eugen Langen, had in fact heard of the use of petroleum residue by Siegfried Marcus, and went so far as to try to propel a tramcar on the same principle in Liège by a gas engine.

It was, however, the Germans, Daimler and Benz, who produced the first successful petrol engines, both incorporating their engines into practical road vehicles. Daimler was greatly assisted by Wilhelm Maybach who developed the surface carburettor and subsequently the spray type, efficient atomization of the liquid fuel being an essential feature of the petrol engine.

The four-stroke cycle (suction, compression, power and exhaust) could be considered wasteful, having only one power stroke for every two revolutions of the crankshaft. On the other hand, the stroke that preceded the ignition stroke was devoted to compression of the explosive mixture, the lack of which had made earlier engines, such as those of Lenoir, inefficient. In 1880 Sir Dugald Clerk devised a two-stroke cycle in which the cylinder is exhausted at the end of the expansion stroke and a new charge admitted at the start of the compression stroke, a low pressure pump being required for this. Increasing the pressure, which Clerk did some ten years later, in effect initiated the concept of supercharging. Two-stroke and four-stroke engines have since co-existed, each filling a useful place in the scheme and neither proving to have an overall superiority.

Just as the increased cost of fodder and the rising price of horses during the Napoleonic War gave a fillip to the development of the steam carriage, so did the oil crisis of 1973 encourage designers to save petrol and to try to develop engines which would run on alternative fuels such as methyl, ethyl and hydrogen. None reverted to the roof-mounted balloon gas tanks that became a feature of motoring during the Second World War and little success was achieved. However, as the world's petroleum resources are undoubtedly finite, a reversion to these experiments may well come about towards the end of the century. 'Lean burn' engines which use a lower octane petrol and some 10 – 15 per cent less of it have been developed so far by Austin Rover and Ford in Britain. They are based on an American system using a three-way catalytic converter and fuel injection system. The further development of fuel cells is likely, as well as the production of efficient light-weight batteries for electrically-propelled town runabouts.

Experiments to improve on the conventional internal combustion engine and its efficiency have been a constant preoccupation of engineers in the automobile industry since the days of Daimler, Benz and Diesel. There have also been spasmodic attempts to design a petrol or oil engine that does not conform to the orthodox internal combustion engine configuration. The most notable of these are the Rover gas turbine and the Wankel engine, the latter having made some impact on the commercial market.

The Rover gas turbine motor car, the first in the world, made its appearance in 1950. Based on a standard Rover 75 converted to a two-seater open body, the car was powered by a twin-shaft Rover T8 gas turbine developing 200hp at 40,000 compressor rpm driving the rear axle through a 6:1 helical reduction gear. It was started by an electric motor of 12 volts which rotated the compressor shaft and also supplied the ignition plug in the combustion chamber, the trembler coil and the batteries, but which cut out when the turbine was running. The fuel was paraffin. In 1952 it reached a speed of nearly 245kph (152mph) a world record for this type of car. The 1.2 tonne car could reach 160kph (100mph) from a standing start in 13.2 seconds. The car was not fitted with a heat exchanger which resulted in an unacceptably high fuel consumption of only 4mpg (1.4km/l).

Professor Felix Wankel started work on a preliminary design for a rotary combustion engine in 1954 and the first experimental model was run in 1957, although his ideas had originated much earlier (see p. 325ff.). The first production car to which it was fitted was a German NSU Spyder which, with a 500cc Wankel engine located at the rear, achieved a top speed of 150kph (93mph) and an acceleration of 0–96kph (0–60mph) in 14 seconds. The engine gave 41kW (55bhp) at 6000rpm with a compression ratio of 8.5:1. The touring fuel consumption of the lightweight car was 12.4km/l (35mpg, 29 US) and, in town driving, it registered only 9.2km/l (26mpg, 21.6 US). Though the Wankel engine has since been licensed to such manufacturers as Mazda in Japan and there seem to be advantages in terms of weight and size, the relatively high fuel consumption figures oppose the general adoption of this radical and ingenious substitute for the orthodox design of internal combustion petrol engine.

TRAMS AND TROLLEYBUSES

Tramways had long been in use in collieries and were the origin of railways (see Chapter 11). They also spawned the street tramways which became popular as public transport for passengers in the nineteenth and early twentieth centuries. They originated in the USA where, due to the soil conditions and the lack of plentiful stone for road building, conditions were particularly favourable for their employment. The first horse-drawn tram was built in 1831 by John Stephenson for the New York & Harlem Street Railway, opened in 1832. Its coach-built body had three compartments and was slung on leather straps on an unsprung underframe running on four wheels, with the driver's seat on the roof. It carried thirty passengers. The idea of the street tramway gradually spread to other American towns and then overseas, reaching Paris in 1855 and Birkenhead, on Merseyside, in 1860.

An American merchant, appropriately named G. F. Train, was responsible for the Birkenhead tramway and, later the same year, laid a line between Marble Arch and Notting Hill Gate in London, opened in 1861. He used the 5 inch (12.5cm) flat rail with a $\frac{7}{8}$ inch (22mm) raised flange at one side that had originated in Philadelphia in 1851. It weighed 50lb per yard (24.8kg/m). Train laid similar lines from Westminster Bridge to Kennington and between Buckingham Palace and Victoria Station but, having failed to obtain parliamentary sanction, the concession was withdrawn by the city authority and no further tramways were laid in London until 1869 when the North Metropolitan tramline started up. Horse-drawn trams are still to be seen in operation today in the summer season in Douglas, Isle of Man.

Not unnaturally steam enjoyed a period of favour as a form of traction, its foremost proponents being Messrs Kitson & Co of Leeds, though Manning, Wardle & Co of the same city were before them, having built a car for export to Pernambuco, Brazil, as early as 1866. The first steam tram in England, patented by J. Grantham, ran between Victoria and Vauxhall in London, starting in 1873. Early steam trams had the engine and boiler mounted within the passenger cars, but in some of the later designs the source of dirt, smoke and heat was set at a distance from the passengers, engine and boiler being mounted on a separate trailer. Kitsons, probably the most successful builders of steam trams – they built over 300 of them between 1879 and 1901 – started building to the design of W. R. Rowland for Copenhagen in 1876 and developed their own designs by experiment from 1878. It is notable that six steam tramcars exported to New Zealand in 1879 were all still working in 1937. The last steam tramway in England ran between Wolverton and Stony Stratford until 1926.

Electric traction was the optimum solution for tramways but in most cases had to wait until a suitable source of supply was available. In some instances, as at Bristol, the tramway company built its own power supply station. Leeds led in Britain, having electric trams in 1891, although Werner von Siemens had built them for Berlin where they were running in 1882. A roof-mounted collector was used in most cases, taking electric current from overhead wires strung from poles, although, in some installations, a collector-shoe running through a narrow slot between the rails into a wider electrified groove performed the same function. Rail gauges used varied widely, 3ft 6in, 4ft and 4ft $8\frac{1}{2}$in (106.7cm, 121.9cm, 143.5cm) all being common. Varying gauges were often found in neighbouring towns, some almost contiguous, a point certainly not in favour of the tramways.

The main disadvantage of street tramways was that halts could only be made by the car pulling up in the middle of the roadway with a consequent danger to ascending and descending passengers from being struck by other vehicles in spite of preventive legislation. An alternative was the trolleybus, which could collect current from the same overhead wires and yet, running on wheels as opposed to rails, could draw up close to the kerb. Developed in the USA before the First World War and having pneumatic tyres, the trolleybus became popular municipally between the wars. It still had, however, the other disadvantage, that one trolleybus could not overtake another on the same line. As a result, their heyday was a short one. London withdrew its fleet in 1960 but they survived in Bradford from 1911 until 1972. Glasgow retained its tramcars until September 1962. They are, of course, still used in towns where they are particularly suitable such as Moscow, where the streets are wide and there is relatively little other conventional traffic.

BUSES

Until late in the nineteenth century, the steam engine was virtually the only form of motive power available for 'locomotion' apart from human and animal power. The improvements in the roads, especially those made by Telford and McAdam (see p. 434–5), were conducive to efforts to replace the horse with a steam engine, as was the considerable increase in the cost of fodder for horses that was brought about by the war with Napoleon. Trevithick's use of high pressure steam was a major contribution towards reducing the weight as well as the size of the steam engine so as to make it an acceptable form of traction for road vehicles.

The period 1820 to 1840 was a particularly active one, with a host of steam carriage experiments of which, perhaps, those of Walter Hancock and of Sir Goldsworthy Gurney were the most notable. Some were for private carriages but others, carrying between 12 and 24 passengers, particularly in the London area, plied for hire and competed with the horse-drawn 'omnibus' that George Shillibeer had first put into service in July 1829. Drawn by three horses harnessed abreast, it ran between Paddington Green and the Bank. Later 'buses', particularly from the period of the Great Exhibition of 1851, had seats longitudinally along the roof, the passengers, seated back to back, being separated by a 'knifeboard' backrest. Horse buses were not finally replaced by internal combustion engined buses in London until 1911.

ELECTRICALLY OPERATED VEHICLES

The need for fuel economy revived interest in the subject of electric vehicles and their viability. Around 1900 there was a short period when they became popular as 'town carriages', largely on account of their quietness and smoothness of operation compared with vehicles driven by the internal combustion engines of the time. The Krieger Electric Brougham, made by the Compagnie des Voitures Electriques was typical. The front wheels of this heavy machine were each driven independently by a pair of motors (compound wound 4-pole) supplied by twenty 193 ampere-hour cells giving a range of some 80km (50 miles) when fully charged and a maximum speed of 29–32kph (18–20mph). The lead-acid batteries must have made up much of the car's 2-tonne weight, a failing of many more recently developed electric vehicles which has restricted their use to such applications as milk floats, runabouts for lazy golfers and motorized transport for the elderly and disabled. It is noteworthy that the Krieger vehicle had a regenerative braking system which allowed the motors to become generators when coasting. A similar characteristic applied to certain designs of bus with hydrostatic transmission which were developed in the 1960s.

At the beginning of the twentieth century, electric vehicles competed seriously with the internal combustion engine. Four out of ten cars sold in the United States were electric, another four were steam-driven while only two had petrol engines. The first road vehicle to travel at over 100kph (62mph) was Camille Jenatzy's battery-powered 'La Jamais Contente' in 1899. The London Electrobus Company, formed in 1906, planned to have 300 electric buses running within a year. Unfortunately the technology was not sufficiently advanced and passengers on their first bus in 1907 objected strongly to the acid fumes from the batteries to which they were subjected. The company was wound up three years later.

Recent attempts at economy have included Sir Clive Sinclair's C5 electric three-wheeled car which seemed to have little more in its favour than the fact that it could be driven without holding a licence. Production, never notable, soon ceased for it would seem that Sir Clive or his marketing team misread the signs. Today, only some 40,000 of the 21 million or so vehicles in Britain are electric, but there is a chance of a real resurgence with the introduction of Alcan's new and experimental aluminum/air battery.

The Alcan battery would have an alkaline or saline electrolyte, with the aluminium anode as the replaceable element. Another alternative is the sodium/sulphur battery developed by Chloride Silent Power. Here again is the promise of a battery with three times the output and only half the weight of a lead-acid battery which would greatly increase the chances of electric drive as a replacement for the internal combustion engine. Non-vehicular applications are also conceived by the makers for the future.

ROAD TRANSPORT ANCILLARIES

The automobile age has spawned a host of ancillaries, of greater or lesser importance, to the carriageways and their accompanying vehicles. It is estimated that if William Symington (see p. 441) had built a full-scale carriage and run it on the roads, he would have had to stop every 8 or 10 miles (12–16km) to replenish his boiler water as the engine was non-condensing. In all probability supplies would not have been readily available, muddy ditch-water being hardly conducive to good boiler operation and long life. This is an early example of the need for the now ubiquitous service station. When in 1888 Frau Benz, unknown to her husband, took a day-trip with her two teenage sons in one of his experimental cars, they stopped to buy petrol from an apothecary, the only retailer who then stocked it, and later at a cobbler's to have a new piece of leather fixed on to the wooden brake block. Naturally, at both places, they topped up the radiator with water.

The petrol pump was invented as early as 1885 by Sylvanus F. Bowser of Fort Wayne, Indiana, but he did not have the motor car in mind: the cask in which he stored kerosene was tainting the nearby buttercask in Gumper's General Store. Not for a further twenty years did he apply his idea to the petrol used in automobiles. Bowser's pump measured out a gallon (US, 3.785l) at each stroke. In 1925 he produced a pump which registered the amount drawn off on a dial gauge and in 1932 he incorporated an automatic price indicator. Hand-operated Bowsers were installed at filling stations in the UK in 1920 and the following year an enterprising proprietor in Manchester put in an automatic Bowser pump at his garage.

The first known service station or garage dates from 1896 when A. Borol in Bordeaux advertised his motor workshop and the fact that he would send out petrol, oil and spares to motorists in need of them. He was also the agent for Peugeot. A service station was advertised in the local press in Brighton in 1897.

The advertisement also appeared in the *Autocar* magazine. The Automobile Gasoline Company opened the first bulk storage filling station at St Louis, Missouri, in 1905. Two years later Standard Oil of California set the trend in canopied forecourts when they opened the first at Seattle, Washington. In the UK, the first was at Aldermaston, Berkshire, in 1920, with a hand pump supplying National Benzole, an air compressor, water tap, fire extinguisher and a telephone. The station was manned by Automobile Association patrolmen and was for the use of members only.

('Garage' is, of course, a word of French origin, meaning a place of shelter, and is contemporary with chauffeur, literally a man employed to warm or heat up the engine before starting. This was a lengthy but essential process with hot-tube ignition such as preceded electrical systems. A bunsen-like spirit flame was applied to the outer end of the tube, the other end of which ran into the combustion chamber.)

Road and street lighting are very different from medieval days when an Act called for the citizens of the City of London to light their streets, using containers shaped like frying pans and filled with fat or tallow. Mention must be made here of traffic lights which are allied to railway signals but specifically an adjunct of the road. The first in Great Britain were lit by red and green gaslights, a pair of coloured filters being carried in the extension of a pair of semaphore arms. They were erected in 1868 in Bridge Street, New Palace Yard, Westminster, but did not immediately find universal favour for the gas exploded, killing an unfortunate police constable. The inventor was a Mr J. P. Knight, as one might expect a railway signalling engineer.

Cleveland, Ohio, was the first city to have electrically powered red and green lights, in 1914. Amber was added to these colours for the traffic lights installed in New York in 1918, which were manually operated. In 1925 traffic lights were installed at the junction of St James's Street and Piccadilly in London, manually controlled by a policeman in a central 'signal-box'. The next year saw time-controlled lights for the first time, in Wolverhampton. In 1932 the first vehicle-operated traffic lights in Britain started operation at the junction of Gracechurch Street and Cornhill, in the City of London.

Painting white lines on the road surface originated in Wayne County, Michigan, in 1911 and was introduced into Britain in 1914 when they were used at Ashford, Kent. The first traffic signs came before the motor car, for they were put up by the Bicycle Union in 1879. The first for motorists were warnings on Birdlip Hill, Gloucestershire, erected by the Royal Automobile Club in 1901. The safety of pedestrians involved traffic islands, which started in Liverpool in 1862: the first in London was privately installed in St James's Street by Colonel Pierpoint so that he could cross safely to his club. Further consideration for pedestrians was shown by the introduction of pedestrian crossings, the first in Britain being in Parliament Square in 1926 and soon after round Piccadilly Circus. Lighted globes were used in the Belisha beacons first in Kensington in 1934, while the present zebra crossings date from 1951.

An equally necessary but, to the motorist, less attractive device is the parking meter first heard of in the 1930s. Carl C. Magee of Oklahoma City applied for a US patent for such a meter in December 1932, but the patent was not granted until 1936. In the meantime Professor H. G. Thuesen and Gerald A. Hale of Oklahoma State University had also made a similar application which was successful and resulted in the first motorist, the Revd C. H. North, having the doubtful record of being booked for outstaying his welcome. The first parking meters in Britain were installed, appropriately enough, outside the United States Embassy in Grosvenor Square, London, in July 1958.

ROAD MAPS

Road books, showing the main towns, features and crossroads and the distances between them had existed as early as Roman times, the Itinerarium Antoninus being compiled about AD 200. The Romans probably measured distances in paces, not the most reliable method. Leonardo da Vinci illustrated in his notebook a form of 'perambulator' or odometer, in which a pebble fell into a box every time the wheel revolved but, like so many of the inventions that he sketched, it is doubtful if it was ever made. The frontispiece of John Ogilvy's *Britannia*, published in 1675, shows a waywiser being trundled along a roadway with a more sophisticated form of gearing to count the revolutions of its wheel. Waywisers, for measuring road distances, became quite common by the mid-eighteenth century and, with more people travelling, there was a growing demand for road books. When John Cary first became involved with the mail service in 1794, he was to be paid 9d a mile by the Post Office to measure the routes travelled, but he also took care to retain the right to publish the results of his inspectors in a road book, which he did in 1798. Publishers of earlier road books were quick to produce new editions using Cary's measurements: Cary sued them – and won.

Today we can rely on precise and accurate maps of the whole of Great Britain produced by the Ordnance Survey by the process known as trigonometrical surveying. The first set of such maps on a scale of an inch to a mile was begun in the late eighteenth century shortly after the Ordnance Survey was established in 1791 by the Duke of Richmond as Master-General of Ordnance. It was concluded in the following century. The initial purpose of the first triangulation was, in fact, to establish the distance between London and Paris.

All maps were originally printed by engraving processes as they were successively developed; today the techniques of computer aided design and photo-litho printing, as applied to cartography, allow a high degree of accuracy and frequent updating.

PART THREE: TRANSPORT

BRIDGES

The building of bridges is almost as old as the making of roads, or at least tracks, occurring at about the same time as man's first settlement in agricultural communities and the very beginnings of permanent villages in the Neolithic Age, some 10,000 years ago or more. At first, bridges were of very limited span: simple beams of single logs laid across gaps or streams and suited only to pedestrian traffic. The branch of a vine or similar creeper, on which a man could swing across a gap, could hardly be classed as a suspension bridge, but it was surely the antecedent of the first single rope spans. As more and more animals became more and more domesticated – the dog, the ox, the sheep, the donkey and the ass – it became increasingly necessary to widen the single log beam bridge and to give it a reasonably flat upper surface. Where long enough logs were not available to span the gap to be bridged it was but a small advance to build a pier and double the span or several piers and hence to multiply it. In the fifth century BC Herodotus describes a multi-span bridge at Babylon having a hundred stone piers joined by timber beams averaging over 1.5m (5ft) in length, the whole bridge being 200m (660ft) long. In other cases, where the piers consisted of wooden piles, the technique was possibly adopted from the Bronze Age lake dwellers of Switzerland who developed it about 2500 BC, or from other northern peoples who built similar structures towards the retreat of the last Ice Age, such as those at Star Carr in Yorkshire.

To a limited extent, stone slabs were used in place of logs to make clam bridges if single span or clapper bridges if of several spans. Some have survived in Britain until today, particularly in the Cotswolds, and on Dartmoor and Exmoor. Existing since prehistoric days, they have long outlasted their timber equivalents, but they are few and stone is too brittle to be used successfully except for the smallest spans. It first appears as early as 3500 BC in Mesopotamia. The Egyptians understood the principle of the arch about the same period but appear to have used it more in buildings than in bridges.

The Greeks, although they did learn the art of the arch in the fifth century BC, preferred to use the beam in the form of post and lintel construction. The Romans, great engineers, needed bridges for the vast network of roads which reached to the farthest corners of their great empire, both for trade and for military purposes. Though they were familiar with the stone and brick arch, they, too, had a preference for timber construction in bridges and used piers and lintels wherever possible. Typically, Julius Caesar in 55 BC is recorded as having built a timber bridge 550m (1800ft) long over the River Rhine in ten days. It had fifty spans. Possibly it was the speed of construction that formed the attraction of this form of construction and materials to the Romans.

Roman engineering was advanced enough to involve the use of coffer dams for the construction of bridge piers over rivers. These consisted of two concentric rings of wooden piles, the space between which was packed with brushwood and clay. The biggest arch bridge of Roman construction that we know of was a viaduct at Narni on the Via Flaminia which had four spans making a total of 42m (138ft). The arches, like all those of Roman origin, were semi-circular in shape, making for a rather high rise at the crown of the bridge, though no worse than the later pointed Gothic arches of mediaeval days. Much effort and ingenuity was to be expended by later engineers in attempts to produce a flatter arch form.

With the eclipse of the Romans in the fifth century AD, the lands once ruled by the might of Imperial Rome fell into a state of semi-anarchy governed by a myriad of small tribes with little need for and still less interest in bridge-building beyond occasional repair of a relic of Roman days. Not for another five centuries did the Christian Church take on the responsibility for the upkeep of some of the existing roads and bridges and, occasionally, the erection of new ones.

The span and strength of the beam were both greatly extended by the invention of the truss, a framework of relatively light members generally arranged in triangles. The Renaissance architect Andreas Palladio is said to have constructed the first bridge of truss type in 1570, with a span of some 30m (100ft) to cross the River Cismon in the southern Dolomites, south-east of Bolzano. He used timber. The same material was much used in America with the spread of the railroads to the West in the nineteenth century when Ithiel Town, Thomas Pratt and his father, Caleb, and William Howe, as well as James Warren in Britain, all evolved different designs using wood. Also in Britain, 1847 saw the introduction of an iron truss by Squire Whipple; two years earlier, Robert Stephenson had ingeniously combined a truss with an arch to form the bowstring girder bridge. The truss, taking the place of the string of the bow, tied the ends of the arch together so as to obviate the need for heavy abutments to absorb the thrust. Such a construction was used for the High Level Bridge over the River Tyne, completed in 1845, whose six 38m (125ft) spans carry the railway on the upper deck with a roadway running below it, a further innovation in its day.

The truss bridge perhaps reached its zenith in the 2-mile-long, 85-span bridge designed by Sir Thomas Bouch and completed in 1878 across the River Tay – one of history's most dramatic failures. It was largely his miscalculation of the effect of wind pressure that resulted in the loss of the central spans and a train with seventy-five passengers. One of the central spans was 245ft (74.5m) long and weighed 194 tonnes. Malleable iron was the material used.

In cantilever bridges, the projecting portion of a beam is fixed in the pier or abutment at one end only – possibly opposed by a similar structure arising from the opposite side of the gap. The design appears to have originated in China in pre-Christian times. The Forth railway bridge, completed in 1889, is a prime example, and the first major bridge in Europe to be built of steel. It has two

Table 8.4: The world's longest span bridges.

Construction	Main span m	ft	Completed	Location
Suspension				
Akashi-Kaikyo	1780	5840	1988	From Honshu to Shikoku, Japan Double Deck for Road & Rail
Humber	1410	4626	1978	Humber Estuary, UK
Arch				
New River Gorge	518.2	1700	1976	West Virginia, USA
Bayonne	510.5	1675	1931	New Jersey Bayonne–Port Richmond, USA
Cantilever				
Quebec	548.6	1800	1918	
Forth	521	1710	1889	Firth of Forth, UK
Box Girder				
Hooghly River	457	1500		Near Calcutta, India
Sainte-Nazaire	404	1325	1975	Brittany, France, over R. Loire
Concrete (except arches)				
Brotonne	320	1050		R. Seine, France
Pasco-Kennewick	299	981		Washington, USA

equal main spans of 521m (1710ft) that is, three double cantilever frames with short suspended trusses of 106m (350ft) each between them. Some 45,000 tonnes of Siemens-Martin steel were used in the construction.

An earlier bridge, the first of the modern cantilevers, was built in 1867 by Heinrich Gerber across the River Main at Hassfurt in Germany, at first giving the name of a Gerber to a cantilever bridge. Its central span was 130m (425ft). Charles Shaler Smith followed it with the Kentucky River Viaduct with three spans of 114m (375ft) in 1876.

The box girder bridge first came into prominence with Robert Stephenson's Britannia Bridge over the Menai Straits, completed in 1850, in which the railway ran through the box girder. It consisted of a twin tubular structure with tubes of rectangular section and supported on masonry towers at each end of their 140m (460ft) span. Each tube was fabricated from wrought iron plates and sections on shore and floated out across the water on pontoons whence it was lifted 30m (100ft) on to the piers by hydraulic jacks. In 1970 the bridge was severely damaged by fire and was rebuilt with a continuous beam, the railway running below with motor traffic over the top. The spans of Stephenson's bridge were so linked that they effectively formed cantilevers. His design, a novelty in its day, is perhaps not always recognized as a box girder today when girder construction combined with suspension cables and towers has become the accepted form for long spans. Some of the world's longest spans are given in Table 8.4.

For lesser spans, the arch bridge has remained a favourite, some 70 per cent of the 155,000 road bridges in England today being brick arches. It was also the design used for the first iron bridge in the world. The castings of which it is built were made by Abraham Darby III in 1779, at Coalbrookdale in Shropshire, and the bridge still stands today, though it is no longer in use for vehicular traffic. Its span is just over 30m (100ft) with a rise of 15m (50ft) at the centre. The main structural members are five semi-circular ribs made up of castings about 21m (70ft) long, a remarkable size and weight for the time. The structure weighs 386 tonnes.

The arch bridge has become something of a standard both in steel and in reinforced concrete in motorway networks in many countries. Standardization of design has done much to cut costs, an important factor when bridge construction accounts for some 25–30 per cent of total motorway expenditure.

Suspension bridges have, until recent years, been much the prerogative of American engineers, notably John Roebling and his son, Washington A. Roebling, the first to use spun cables, first at Pittsburgh and across the Niagara Gorge in 1855. John Roebling did not live to see the completion of his greatest achievement, the 470m (1539ft) span of the Brooklyn Bridge 40.5m (135ft) high over the East River in New York, opened in 1883 after fourteen years in the building. The Roeblings triumphed, but at some cost personally. The father died in 1869, three weeks after a boat crushed his right foot. His son, in 1872, was paralysed by 'caisson disease' and, deprived of the use of his voice, he directed the work, with his wife's help, until its completion. The wire cables used were made of steel for the first time.

The Golden Gate bridge in San Francisco, completed in 1937, was for many years the longest in the world with 1280m (4200ft) span and with towers 227m (746ft) high. In contrast, the suspension span at the Tacoma Narrows over the Puget Sound was only 853m (2800ft). Aerodynamically unstable, this bridge was also inadequately stiffened and, in 1940, only four months after it was opened, it oscillated in a gale at such a destructive amplitude that the main span fractured and fell to the water 63m (208ft) below. At the time, so great was the degree of twist, that one side of the roadway was 8.5m (28ft) above the other.

The failure of this bridge, filmed at the time by an amateur cameraman who narrowly escaped being on the bridge when it failed, more than anything brought the attention of bridge engineers to focus on the stiffness of a structure and its aerodynamic stability and, as far as the box girder design was concerned, stressed the failures that occurred during building in South Wales and in Australia. It may appear surprising that James Finley had studied the stiffening of girders in suspension bridges (with chain cables) as early as 1801, a quarter of a century before Thomas Telford built his Menai suspension road bridge.

Today great attention is paid to both stiffness and aerodynamic characteristics, particularly of long suspension bridges, as in the 1006m (3300ft) span of the Forth road bridge (1964), the 987m (3240ft) Severn bridge (1966) and the 1410m (4626ft) of that over the Humber (1978). Inclined suspenders, alternately opposed, help to avoid motion of the deck – which in the case of the Severn bridge is only 3m (10ft) deep and in the Humber bridge only 4.5m (15ft). When completed, the Akashi–Kaikyo bridge, due in 1988, between the Japanese islands of Honshu and Shikoku will be the world's longest span at 1780m (5840ft). Theory has shown that with present materials, techniques and knowledge, a bridge of some three times this length could be built today.

Natural cement was used by the Romans and was rediscovered by a builder on the Isle of Sheppey in 1796, whence it became known as Sheppey stone. A similar discovery was made by Canvass White near Syracuse, New York, in 1818. Artificial cement was patented in 1824–5 by a Leeds bricklayer, Joseph Aspdin, and was tested and promoted by Col. Charles W. Pasley of the Royal Engineers School at Chatham. Aspdin named it Portland cement after its outward resemblance to Portland stone. However, it was of little value for bridge building as, except in compression, cement and concrete have little strength unless reinforced against tension and bending with steel. The originator of reinforced concrete was Joseph Monier, a Paris gardener, who devised it for making flower tubs in 1867. He later patented its use for railway sleepers, floors, bridges, arches and other structures. One of the main problems with reinforced concrete is that it continues to harden for up to a year after casting; this process is accompanied by some shrinkage which can involve cracking, particularly if the structure is only hinged at the abutments in the case of bridges. A third hinge is needed at the centre of the span to overcome this. The Swiss engineer Robert Maillart was one of the first to realize this and hence the potential of this material for bridge building. He built a concrete arch over the River Inn in Switzerland in 1901. Joseph Melan of the Technical University in Vienna was another pioneer of reinforced concrete which he used in his longest span of 100m (427ft) over the River Ammer in Bavaria in 1929.

Prestressing of concrete has greatly improved its properties for bridge building. A Frenchman, Eugène Freyssinet, conceived the idea as early as 1904, but was unable to put it to practical use until high tensile steel was developed and became readily available and this was not until after the Second World War. High tensile steel wires running through ducts in the lower flanges of the arches were pulled very tight to exert the prestressing tension, after which the ducts were filled with liquid cement to prevent corrosion of the steel. Between 1948 and 1950, Freysinnet built five identical arch bridges over the River Marne in France, each having a single span of 74m (242ft).

Since the Second World War prestressed concrete has played an important role in bridge building, first in rebuilding the great number of bridges destroyed in the hostilities. Later it has been invaluable in the construction of the numerous bridges and flyovers in the networks of motorways, autobahns and autostradas that span the islands and continents on both sides of the Atlantic and the Pacific.

A glossary of terms used in bridge building

Abutment a construction that takes the thrust of an arch or vault or supports the end of a bridge

Arch a curved structure, usually in the vertical plane, that spans an opening or gap

Beam a long, straight piece of wood, metal or concrete used as a horizontal structural member

Caisson a watertight compartment or chamber, open at the bottom and containing air under pressure, used for carrying out work under water

Caisson disease Divers working at higher than normal atmospheric pressure breathe in increased nitrogen, which is dissolved in the bloodstream. When they return to normal pressure, the nitrogen forms bubbles which, if collected in the capillary vessels, causes cramps ('the bends'). If it collects in the joints, damage to nerve endings can cause paralysis, temporary or permanent

Cantilever a beam or girder fixed at one end only; a part of a beam or structure which projects beyond its support

Centring a temporary structure, usually of timber, which serves to support an arch under construction

Coffer dam a watertight structure enclosing an area below water level, pumped dry so that construction work can take place

Corbel a projection of timber or stone etc., jutting out from a wall to support a weight

Falsework a temporary framework used during building

Girder a substantial beam, usually made of iron or steel

Keystone the central stone forming the top of an arch or a dome or vault (also called a quoin or headstone)

Lintel a horizontal beam as used over a door or a window

Pier a vertical member or pillar that bears a load

Pile a column of timber, iron, steel or concrete, driven into the ground or river-bed to provide a foundation for a structure

Soffit the underside of a structure such as an arch, beam or stair

Spandrel the space between the shoulder of an arch and the surrounding rectangular moulding or framework, or between the shoulders of adjoining arches and the moulding above

Springer part of an arch where the curve begins; the lowest stone of this

Starling an arrangement, usually of piles, that surrounds a bridge pier to protect it from erosion caused by scouring, debris etc.

Truss a structural framework of wood or metal, usually arranged in a formation of triangles, forming a load-bearing structure

Voussoir a wedge-shaped stone or brick used in the construction of an arch or vault.

TUNNELS

Tunnelling, for canals, roads and railways, is a relatively modern technique where any sort of mechanization is involved, although though irrigation tunnels, called qanats, are found in Iran and have existed for many hundreds of years. These tunnels, dug by the Persians and Armenians to bring water to the towns, are ten, twenty or more miles long and date from the eighth century BC in some parts. A number of vertical shafts were dug and then joined by horizontal tunnels. Similarly, manual labour was used in digging the canal tunnels in the systems created in Britain as well as on the continent of Europe from 1750 to 1850, even earlier in France. A total of some 69km (43 miles) of Britain's canals are tunnelled, the longest tunnel being at Standedge on the Huddersfield Narrow Canal, this being 4951m (5415 yards) long. Mining tunnels date at least from Roman times and existed all over the empire wherever there were minerals to be exploited.

Modern tunnelling, using more or less complicated machinery, originated in 1818 when Marc Isambard Brunel designed a machine copying the principle of working of the shipworm, *Teredo navalis*. The head of this mollusc is protected by two jagged concave triangular shells, between which a proboscis protrudes, like the centre pin of a carpenter's bit. This equipment enables the shipworm to grind the hardest oak into a nourishing paste, while the petrified excreta, having passed through the animal, forms a smooth lining to the tunnel behind it. In Brunel's machine, thirty-three miners working in cells in the box-shaped iron casing which was advanced by screw jacks, the tunnel being lined with bricks as the face moved forwards. The boring beneath the Thames between Rotherhithe and Wapping was the world's first underwater tunnel, completed in 1841. It is 1093m (1200yds) long. The twin tunnels, originally used as roads, are still in use today as a railway.

A second tunnel between Rotherhithe and Wapping was built in 1865 by Peter Barlow and James Henry Greathead, using a similar shield advanced hydraulically. Cast iron segments were used to line the bore, rather than the brickwork used by Brunel and in most of the earlier canal tunnels. Greathead's pioneering work formed the basis for most of the subsequent London Underground Railway.

Rotating cutting heads, hydraulically-operated arms for lifting and holding the cast iron (later steel) lining segments while they were bolted up together, and conveyors to remove the excavated spoil from the face were added subsequently to make the tunnelling machine almost entirely automatic, as it is today. Different companies, too, have gone into the manufacture of specialist machines for tunnelling through chalk, rock or sand in addition to the clay of Brunel's and Greathead's early work.

The first British tunnel of any length was for a railway under the River Severn. Built between 1880 and 1893, it was over 6.9km (4.3 miles) long, 3.7km (2.3 miles) of which were under the water, and 76 million bricks were used in the tunnel lining.

An alternative to boring a tunnel through the solid ground for a river crossing is the immersed tube tunnel in which tubular sections are built up on land, towed out and sunk into position before the ends are removed. Rotherhithe, again, was the site of the first experiment in this method by Robert Vazie and Richard Trevithick between 1802 and 1808. They built 319m (1046ft) of 1.5m (5ft) high by 0.9m (3ft) wide brick tunnel across the River Thames, but ran out of money before the work could be completed. However, they did establish the practicability of the system which has been much used since, notably for the John F. Kennedy tunnel under the River Scheldt at Antwerp. Five prefabricated reinforced concrete sections averaging 102.2m (335.3ft) long make up the 512m (1680ft) of the prefabricated part of the two three-lane roadways and the two-track railway tunnel, completed in 1969.

The sections of an immersed tube tunnel are usually laid in a trench dug in the river bed. Hydraulic jacks are sometimes used to draw them together before they are connected to each other. The immersed tube has been likened to the underwater equivalent of the cut-and cover method of construction on land such as was used for some of the first underground railway lines in London as in other cities.

Compressed air plays an important role in the tunnelling engineer's equipment. In hard rock tunnelling it often provides the power for driving face-cutting tools or the percussive tools used for drilling holes in which to place explosive charges for blasting. In drilling the Mont Cenis tunnel, from 1860 onwards, a 170cu.m/min (6000cfm) air supply at 6 atmospheres (87 psi, 6kg/cm^2) was obtained from water descending 50m (164ft) through cast-iron pipes, the entrained air being separated out through valves at the bottom of the fall. In this case ten hydraulic air compressors were used to obtain the supply, these being based on the principle of the Trompe or Trombe, originating in Italy in the sixteenth century, then much used for providing air supply to brass or iron smelting furnaces. A similar arrangement was used in drilling the St Gotthard tunnel in 1873. Hydraulic rather than pneumatic power was used when work commenced on the Ahrlberg tunnel in 1880. Thomas Doane's mechanical air compressor, patented in 1866, was used on the Hoosac tunnel and started the seeds of an industry in which the Americans have long been among the leaders, together with Sweden where rock drilling is very much a part of life.

When, in the early 1960s, the Mont Blanc road tunnel was being constructed between Aosta and Chamonix, air-operated drills were used mounted on a three-deck gantry, the drills being supported on hydraulically-operated booms, while the pumps which supplied pressure hydraulic fluid were driven by compressed air motors. Compressed air motors are often preferred to electric motors in mining and similar situations where there is a risk of fire or explosion.

Where compressed air drills are used, the exhaust air is often used as an air supply for the operators as well as to ventilate and cool the tunnel while work is proceeding. Pressurizing a tunnel which is being dug under water also performs the useful function of keeping out the water, should there be a leak in the lining. The maximum pressure that can be safely used is about 50 psi (3.5kg/cm^2)

equivalent to 33.5m (110ft) of water, without damaging effects to those working in the tunnel by 'the bends' (compression sickness or caisson disease).

Today, the world's longest traffic tunnel, i.e. not for water or sewage, is the 54km (33.5 mile) railway tunnel that runs between the islands of Honshu and Hokkaido in Japan. This even exceeds the probable length of the proposed Channel Tunnel. This project is hardly new. It started as early as 1802 when Albert Mathieu, a French mining engineer, first put forward plans for a tunnel under the English Channel between France and England. In spite of the interest of Napoleon Bonaparte this came to nothing, as did the plans of another Frenchman, a civil engineer named Aimé Thome de Gamond who devoted himself to the matter for thirty-four years from 1833. At the 1851 Great Exhibition in the Crystal Palace in Hyde Park he tried to raise capital for the venture but abandoned it on account of its cost, at that time estimated as £33,600,000. The debate has persisted through the years, many more schemes being put forward. A Labour government approved the project in the 1970s, but cancelled the contract after a tunnelling machine had been positioned below ground on the English side. In 1986, the governments of Great Britain and France signed a treaty for the building of a 50km (31 mile) double rail tunnel, financed by money from the private sector. Initial estimated costs were something over £2000 million. Much has been made of the difficulty of adequately ventilating a road tunnel of the usual noxious fumes, largely carbon monoxide, emitted by the internal combustion engine, so the idea of an electrified railway, carrying road vehicles has prevailed.

EARTHMOVING AND ROADBUILDING MACHINERY

Man has performed prodigious feats of construction without the help of any machinery. The pyramids of Ancient Egypt were built entirely with slave labour. Silbury Hill, in Wiltshire, England, is some 40m (131ft) high and covers an area of 2ha (5 acres). Archaeologists estimate (using radio-carbon dating methods) that it was built about 2600 BC with nothing but picks made from the antlers of deer, shovels and scrapers from the shoulder blades of oxen, and woven baskets to remove or shift the spoil, very nearly 14,000 tonnes of which had to be moved.

Throughout most of the railway age, cuttings, embankments and tunnels were entirely man-made, sometimes with horses to remove the spoil. A railway navvy was expected to shift 20 tonnes of earth a day in a 14-hour shift – and did so. The mechanical digger or steam shovel originated in the United States in the 1830s, William Otis exporting his second and third machines to Russia and a fourth to England in 1840. The American invention made little impact on British contractors, although it was accepted from the start in its country of origin. Probably the first British contractor to use a mechanical digger was

Thomas Brassey, who had contracted in 1852 to build the Grand Trunk Railway of Canada, from Quebec to Lake Huron. To make up for the shortage of local labour, 3000 navvies were imported from England but cholera and frostbite took a high toll and many returned home in spite of the high wages that Brassey was paying. In about 1854, Brassey resorted to the American invention which enabled him to complete the line by 1859.

Trunk road and motorway construction today involve vast investment in mechanical plant, yet the speeds of construction differ remarkably little from those of the railway era. The wage bill is, however, greatly reduced. Diesel-powered internal combustion engines have superseded the steam engine. The variety of machines employed is extensive, from a number of different types of excavator to quite small back-hoes and front-end loaders and the serviceable dump-truck. Both caterpillar-tracked and wheel-mounted versions of most machines are in use, depending on the nature of the terrain. Like the steam shovel, most types of earth-moving machinery have originated in the United States where, historically, there has long been a shortage of labour. This is a field in which the USA excels in both production and usage as well as generally building the largest and most powerful machines in the world, although these are often used for other purposes, such as opencast mining, rather than road-building. The world's largest land machine, for instance, 'Big Muskie', weighing some 12,000 tonnes, is a dragline excavator with six 5000hp motor generator sets and a bucket of 168 cubic metres (220 cubic yards) capacity. The biggest tractor shovel is the Clark 675 with twin turbo-charged diesel engines, each of nearly 450kW (600hp). Its 18 cubic metre (24 cubic yard) bucket reaches a height of 6.5m (21.3ft). The dump truck record, the Terrex Titan, will carry 350 short tons (324 tonnes) and is powered by a 3300hp engine. Its ten wheels have 3.35m (11ft) diameter tyres. The Marion Power Shovel Company make the biggest bulldozer, weighing 136 tons (126 tonnes) and having an engine of 984kW (1320hp).

In the last 25–30 years a new earth-moving device has come into service. The bucket wheel excavator is a product of Europe rather than America, Orenstein & Koppel of Lübeck in West Germany being the foremost makers. At the front of this track-mounted machine is a long boom carrying a rotating wheel which has buckets spaced around its periphery. Along the boom runs a conveyor belt, while a further boom and conveyor, able to be independently swung, carries the spoil back from the central tractor mounting. The largest bucket wheel excavator O & K make is reputed to be able to shift 198,900 cubic metres (260,600 cubic yards) a day with a 21.5m (71ft) diameter wheel, 5 cubic metre (6.5 cubic yard) capacity buckets and a drive of 2.5MW (3350hp). This machine is generally too powerful for road construction.

Early earth-moving machines were mostly employed on railway works and hence were rail-mounted. Attempts were sometimes made to spread the load of broad steel wheels by ingeniously linked boards pivoted from their peripheries, but the real answer to the problem of moving across soft ground was the caterpillar track. As early as 1770 Richard Edgeworth patented 'a portable railway or artificial road, to move along with any carriage to which it is applied'. Sir George Cayley also, in 1825, took out a patent for 'an endless chain formed of jointed links which fold themselves round the periphery of two wheels'. These patents were, perhaps, in advance of their time and it was not until the early twentieth century that tracked vehicles moved out of the experimental stage. In America, Benjamin Holt of the Holt Tractor Company of California started commercial production of tracked steam-powered tractors in 1904. This firm later became the Caterpillar Tractor Company. In England, Ruston Hornsby developed the patent of David Roberts for tracks of separate links joined by lubricated pins. A tractor thus shod was demonstrated to officials of the War Office in 1907 but, largely due to the indifference of that authority, Rustons sold their patents to Holt. Tracked vehicles were used as artillery tractors first but it was not until 1916 that Winston Churchill, then First Lord of the Admiralty, ordered the first batch of 'landships', better known by their code name 'tanks', which replaced the earlier 'water carriers'. They were not an immediate success but later came to be an important military weapon (see p. 993ff.). After the war the use of tracks became common in earth-moving machines, while the internal combustion engine became the standard power unit.

Wire ropes and pulleys are still used for raising and lowering long booms such as are used in drag-line and bucket-wheel excavators. However, for relatively short lifts, as in scraper, grader and bulldozer blades or the buckets of dump trucks, front-end loaders and back-hoes, hydraulic systems have largely superseded rope drive. The first application was by La Plante Choate on the Caterpillar Angledozer about 1935 and then on the same company's scrapers. John Allen, Whitlock and J. C. Bamford in Britain were not far behind. This use of hydraulics came about largely through the development in the 1930s of nitrile rubbers. Their compatibility with mineral oils allowed the use of the latter in place of water, which had previously been used as the hydraulic medium, with its attendant problems of corrosion and excessive seal wear in the cylinders. There being relative motion between the main structure and the blade or bucket of the machine, flexible hoses were necessary, to which the new material was admirably suited. Working pressures have gradually increased so that, today $210 kg/cm^2$ (3000psi) is considered low and pressures of $350 kg/cm^2$ (5000psi) are by no means uncommon in mobile equipment.

Many other types of specialized machinery are used in road-building and surfacing. The steam-roller has been replaced by the diesel-engined road roller while surfaces of tarmac can now be torn up by a single machine that burns the old tar off the aggregate and returns the latter for re-use. Similar developments have taken place in equipment for mixing and laying concrete as well as in ripping it or breaking it up when it has passed beyond its useful life.

FURTHER READING

Bird, A. *The history of carriages* (Longman, London, 1969)
Caunter, C. F. *Cycles. History and development. Part one* (Her Majesty's Stationery Office for the Science Museum, 1955)
—— *Motor cycles – a technical history* (Her Majesty's Stationery Office for the Science Museum, 1982)
Charlesworth, G. *A history of British motorways* (Thomas Telford, London, 1976)
Damase, J. *Carriages* (Weidenfeld & Nicolson, London, 1968)
Georgano, G. N. (Ed.) *The complete encyclopaedia of motor cars 1855–1982* (Ebury Press, London, 1982)
Institute of Mechanical Engineers, *Engineering heritage, volume I – the automobile over 75 years* (1963)
Johnson, E. H. K. *The dawn of motoring* (Mercedes-Benz UK Ltd, 1986)
Pannell, J. P. M. *An illustrated history of civil engineering* (Thames & Hudson, London, 1964)
Smith, D. J. *Discovering horse-drawn carriages* (Shire Publications, Princes Risborough, 1980)
Steinmann, D. B. and Watson, S. R. *Bridges and their builders* (Dover, 1957)
Stephens, J. H. *The Guinness book of structures* (Guinness Superlatives, London, 1976)
Tarr, L. *The history of carriages*, translated by Elisabeth Hoch (Vision Press, London, 1969 and Corvina Press, Budapest, 1969)
West, G. *Innovation and the rise of the tunnelling industry* (Cambridge University Press, Cambridge, 1988)

9

INLAND WATERWAYS

JOHN BOYES

Water creates both a barrier and a means of communication. The origins of man's realization that the barrier could be transformed into a means of communication for transport of both men and materials on inland lakes and rivers are lost in the mists of antiquity; unquestionably the concept of using a river's current for this purpose must have arisen independently in many parts of the world. The sight of a log floating down a stream would soon have inspired an observer to develop the idea. Inflated skins, rafts, hollowed-out trees and fabricated rudimentary boats, the use of poles, oars and primitive sails all had to be tried. But centuries of trial and error, success and failure must have passed before the sophisticated navigational systems of the pre-Roman era gave rise to the pictorial and documentary sources that provide the contemporary information from that period. It was from those formative beginnings using the existing unimproved rivers that the complex integrated waterway networks of the last three hundred years have evolved.

THE ANCIENT WORLD

The presence of rivers in ancient civilizations, such as the Tigris and Euphrates in Arabia, the Nile in Egypt, the Ganges in India and the Yellow River (Huang He) in China, led to the development of appropriate vessels which provided the means for the generation of trade in a linear direction between riparian towns and communities or between similar settlements on the banks of a lake. Basic civil engineering work within the competence of the locality would have been undertaken to improve the efficiency of navigation, but the great breakthrough came when it was realized that artificial channels on a grander scale than irrigation canals could be dug to link rivers together, to take navigable water across watersheds between one river basin and another and to provide water communication to places remote from the natural river courses.

In pre-Roman times canals to by-pass the cataracts of the Nile had been built by about 2000 BC. Cargo handling facilities were provided to load and unload monoliths such as the massive obelisks erected in Egypt. The most famous waterway project of the classical world, the Corinth Canal, was not to be realized until modern times. It seems that China had one of the earliest, if not the earliest, true navigational canals (though it was probably conceived as an irrigation channel) in the Hung Kou Canal, later known as the Pien Canal. It was possibly built in the time of Confucius in the late sixth or early fifth century BC and linked the Yellow and Huai rivers. Though running through fairly flat country it was some 418km (260 miles) long. About 215 BC it was necessary to provide a freight canal, initially for war materials, across mountainous country to link tributaries of the Yangtse, the Hsiang flowing north and the Li flowing south, with the western rivers in south-east China. This was the Ling Chhu or Magic Canal and its importance lies in the fact that it was the earliest contour transport canal, winding round the mountainside for about 32km (20 miles). In order to maintain the levels techniques were used which were not developed elsewhere in the world until centuries later – the construction of massive embankments and dressed stone retaining walls.

About AD 600 an improved and extended canal was built on a new alignment but roughly parallel in part to the old Hung Kou and this steadily developed with its branches into what became known as the Grand Canal with a total length of nearly 1770km (1110 miles) from Tianjin to Hangzhow. It included a fairly low summit level of 42m (138ft) above sea level. It was intensively used and it is recorded that in 838 a train of 40 boats, many lashed two or three side by side, was drawn by two water buffaloes. Periods of decay and siltation were followed by improvements, for example through the Yangtse gorges, and it is still in use. There is good evidence that the Chinese developed an early form of pound lock – a chamber the length of a boat with gates at each end so that boats could be raised or lowered in the chamber by the operation of sluices.

Although such developments had taken place over many centuries from earliest times there was a long interval in the mediaeval period before a fresh wave of innovation heralded the genesis of a new waterway age more or less simultaneously in many countries.

THE BRITISH ISLES

The long tradition of waterway transport, and indeed the use of rivers for other purposes, has been of fundamental importance in the economic development of Britain. Most, if not all, rivers of any size have been used in part for transport for millennia but it was the Romans who seem to have brought engineering skills to improve the natural watercourses or to dig new lengths to serve their military needs. Just how far they extended in this field is a matter of archaeological debate. Certainly they were responsible for the earliest canal still in use in

England today – the Fossdyke Navigation in Lincolnshire providing a connection between the Trent at Torksey and the Witham at Lincoln. Less certain is whether the Caerdyke from Cambridgeshire into Lincolnshire was solely a drainage channel or a navigation and that again is a source of debate.

In the Dark Ages, between 400 and 1000, there is virtually no evidence of waterway improvement: one of the few references to the use of rivers is the stranding of the Danish fleet in the Lea probably north of Waltham Abbey by the excavation of additional channels in the tenth century. It is also the Lea which provides an early documentary reference to river improvement in an order issued in 1190 authorizing the Abbot of Waltham, in what is now Essex, 'to turn aside the course of the water of the Lea in . . . Waltham as he wishes without harming anyone and for the advantage of navigation'. The Lea again illustrates the importance of river transport in mediaeval times when an Act of Parliament was passed in 1424 establishing a commission to deal with the problem of shoals in the bed of the river which were impeding the navigation. This was the first time an Act had been passed to improve a single river.

The first true canal to be built in England since the Romans was a short one designed to facilitate the passage of small sea-going vessels to Exeter opened in 1566. Lock gates were installed and although mitre gates are shown in a drawing of 1630 there is no certainty that they were originally used. The Exeter Canal was further enlarged and improved from 1698 to 1701.

Although the Exeter Canal had a pound lock separating the non-tidal canal from the tidal range of the sea the first pound lock to be built on an English inland navigation was not constructed until 1577. This was at Waltham Abbey, on one of the several channels of the Lea, and was 70ft by 22ft (21.3 × 6.7m). Owing to local disputes and general decay it survived only until 1592. Elsewhere on the rivers the differences in levels were overcome by flashlocks – contrivances which penned back the water; when boats required a passage downstream the stakes forming the pen would be removed and the resulting cascade of water carried the boats downstream and over the shallows below the lock. Boats working upstream had to be winched through as the fall slackened and then the vertical stakes would be replaced to pen back the upper level again. Such flash locks existed until the beginning of the present century on the Thames and other rivers.

The seventeenth century saw the growth of progressively articulate opinion on the value of river improvement and authority for some engineering work was obtained. By the end of the Commonwealth (1660) there were some 1100km (685 miles) of river navigation which rose to 1870km (1160 miles) in the next half-century with improvements on, for example, the Great Ouse from Bedford, the Thames, the Mersey and Irwell and the Aire and Calder, while others were actively being canvassed.

It was in the eighteenth century that the real waterway network expansion occurred, spurred on by growing industrial enterprise. At the beginning of the century the Suffolk Stour received its Act in 1706. A Board of Trustees for the Lea was established in 1739, following disputes between the bargemen and the New River Company established to supply water to London from springs in the Lea valley and who were augmenting their supply from the river itself. Soon after this came the first true canal to be built in the British Isles since the Exeter Canal – the Newry Canal of 1742 linking Newry with Lough Neagh in Northern Ireland. Proposals had been made at the end of the previous century but no action had then been taken. Now it was built, 29km (18 miles) long with 14 locks, and foreshadowed the building of many canals independent of river navigations. The engineer was William Steers, who had also engineered the Mersey and Irwell Navigation and had built docks in Liverpool. His pupil, Henry Berry, built the first modern canal in England, the St Helens Canal, opened in 1757.

In 1759 the Duke of Bridgewater, who had been on the European Grand Tour and had seen canals on the Continent, obtained an Act to build a canal from his pits at Worsley to carry coal to Manchester. He was fortunate in obtaining the services of James Brindley, a millwright who had great natural engineering ability, who was engaged to re-survey and build the canal. Following a second Act the canal was completed to Manchester in 1761, crossing the Irwell on an aqueduct. This was the first British boat aqueduct and Brindley's massive masonry structure of three arches to carry the canal survived until the construction of the Manchester Ship Canal, opened in 1894, when the original aqueduct was replaced by a swinging aqueduct which remains unique. The face of one of Brindley's arches has been re-erected by the side of the road leading from Barton to Eccles.

The fame of the Bridgewater Canal is that at the Worsley end the canal continued underground to the coal faces in the pits and at one time there were some 74km (46 miles) of waterways underground on three levels linked by inclined planes allowing the boats to be transferred between levels. The boats were taken virtually to the coal face where the coal was loaded into demountable containers. When full the boats were propelled manually to the open air at Worsley basin where they were assembled into gangs of two or three and towed to Manchester. The full containers were then removed and replaced by empty ones for the return journey to Worsley – a very early example of containerization. The underground ceased operation in 1887.

The immediate success of the Bridgewater Canal stimulated a fever to construct other canals and Brindley's genius was in demand. He was the engineer for the Trent and Mersey, the Staffordshire and Worcestershire, the Coventry, the Oxford and the Birmingham Canals, and as all these were under construction at more or less the same time supervision had to be delegated. Brindley died in 1772, the year in which the 128km (76 miles) Staffordshire and Worcestershire was completed linking his Trent and Mersey Canal at Great

Heywood with the river Severn at Stourport and with the Birmingham system at Aldersley. The Trent and Mersey, 150km (93.5 miles) was not completed until 1777, but Brindley's standards left their mark on the dimensions of the Midland canal system as we know it today. Among his major works were the Harecastle tunnel on the Trent and Mersey which took eleven years to build (Telford completed a second tunnel in two years in 1827); and the Trent aqueduct near Great Heywood carrying the Staffordshire and Worcestershire across the Trent. Brindley's efforts were the stimulus for industrialists and others elsewhere in the country to embark on new undertakings or to improve further the existing navigations.

The national system had grown slowly from the figure of 1866km (1160 miles) in 1720 to 2250km (1400 miles) in 1760, but by 1780 it had reached over 3200km and was to exceed 4800km (3000 miles) in 1800. It took another 50 years to reach its maximum extent of just over 6437km (4000 miles) before the non-economic waterways started to close down in the face of railway competition and railway purchase.

Included in these developments were the Brindley-surveyed Coventry and Oxford Canals which were to provide the first link of the Midland system with London. Brindley had adopted the contour system of construction so that his canals looped tortuously round hills and valleys, particularly on the Oxford Canal. Nevertheless the four great rivers, the Trent, Severn, Mersey and Thames were linked by 1790.

Meanwhile, further north, a new canal and to much greater dimensions than Brindley had favoured was projected with John Smeaton as engineer – the Forth and Clyde Canal. It had been considered in the Restoration period and a further survey was made in 1726 but it was after William Pitt (later Earl of Chatham) suggested a canal built at public expense that Smeaton made his survey in 1764. He later revised his survey and in 1768 the Act was passed. Work started soon after, but by 1778 funds which had been privately subscribed, contrary to Pitt's view, were running low and the government in 1784 agreed to lend £50,000. The canal was completed in 1790, Smeaton having resigned in 1773. This was one of the canals where a regular passenger service operated. It started in 1809, included a night service from 1824, and ceased about 1848.

Another major east–west link proposed in the 1760s was the Leeds and Liverpool Canal. Surveyed by John Longbotham, it was controlled by two committees, one in Liverpool and the other in Bradford. Brindley was asked to arbitrate over the line which should be taken in Lancashire and later he was appointed chief engineer but withdrew. Longbotham was then appointed engineer. It received its Act in 1770 and by 1774 the famous five-rise staircase at Bingley was completed, remaining today one of the visual attractions of British waterways. The second major engineering work was the Foulridge tunnel under the Pennines at summit level, completed in 1796; while the third great feature was the 1.2km (0.75 mile) embankment across the Calder valley at Burnley completed in 1799. However the full 204km (127 miles) canal was not completed until 1816 – 46 years after it was started – because of the work involved in the 23-lock flight at Wigan.

The problems arising from the meandering course of the Oxford Canal and the increasing traffic between Birmingham and London necessitated a more direct line. In 1791–2 a line was surveyed from Brentford on the Thames to Braunston near Rugby on the Oxford Canal for 70 tonne boats. William Praed was chairman and William Jessop engineer. Work started as soon as the Act was obtained in 1793 and this involved the construction of two tunnels. Braunston tunnel was opened in 1796, but difficulties occurred at Blisworth because of flooding. Apart from the tunnel, the canal was complete by 1800 so goods were transported by a horse tramway over Blisworth Hill until the tunnel was completed in 1805. Another problem was the crossing of the Ouse at Wolverton where boats at first locked down to river level and then up again on the other side. This system was replaced in 1805 by an aqueduct which collapsed in 1808. Replaced by an iron aqueduct, it reopened in 1811. Traffic soon became extremely heavy. Known as the Grand Junction Canal it acquired smaller canals including both the Old Grand Union Canal and the Leicestershire and Northamptonshire Canal which met at a flight of ten locks forming a bottleneck at Foxton. In 1900 the Grand Junction's owners decided to overcome this problem by constructing an inclined plane carrying two tanks or caissons tranversely on the slope. The tanks were hauled by cables, the haulage power being provided by a stationary steam engine to the top of the plane, but the cost of maintaining the boiler and engine in steam rendered it uneconomic and by 1917 it had been taken out of use.

Following the completion of the Grand Junction to Paddington basin in London, a canal was promoted to link west London with east London and the lower Thames. This was completed in 1820 as the Regents Canal and terminated in a dock at the eastern end with a lock into the Thames. Over the years this dock was enlarged. The last length of canal to be built in the south of England was opened on 1 April 1968 as a link between the Limehouse Cut of the Lee Navigation and the dock to enable the Lee traffic to enter the Thames via the Regents Canal Dock lock and so to abandon the old and difficult Limehouse Cut Lock. (Although the waterway is in the Lea valley and in part was and is the River Lea its legal title is the Lee Navigation.)

On 1 January 1929, after some years of negotiations, the Grand Junction amalgamated with three canals in Warwickshire and the Regents Canal to form the Grand Union. The Warwick and Birmingham and the Warwick and Napton were both built as part of the improved line between London and Birmingham, and the Birmingham and Warwick Junction was added in 1844 to cope with the increased traffic.

PART THREE: TRANSPORT

Although the Severn was connected to the 'national' network of canals it was a treacherous river to navigate in its lower reaches, so in 1792 a proposal was made to construct a ship canal for 300 tonne vessels from Gloucester to Berkeley Pill. An Act was obtained in 1793. Work started the following year but by 1800 only 29km (5.5 miles) had been built. Work ceased for want of money until 1819 and again in 1820–3. It was then restarted and the 26.5km (16.5 miles) were finally completed in 1827 and capable of taking 600 tonne vessels. Instead of continuing south to Berkeley Pill, the canal terminated at Sharpness Point where a much larger harbour and entrance lock were opened in 1874. The Gloucester and Sharpness Canal Company also constructed extensive docks at Gloucester. Later, by leasing the Worcester and Birmingham Canal, the company obtained a through-water route to Birmingham.

An east–west link in southern England had been canvassed since the sixteenth century, but it was not until 1794 that an Act was passed to link the Kennet at Newbury with the Avon at Bath. The Kennet and Avon Canal, completed in 1810, included the Caen Hill flight of 29 locks at Devizes, two notable aqueducts by John Rennie, who was the canal engineer, the Dundas at Limpley Stoke and the Avoncliff, and two interesting pumping stations for supplying water to the canal, one at Claverton powered by a waterwheel and one at Crofton powered by steam.

While many smaller canals were constructed, some of which included pioneering structures such as the boat lifts on the Grand Western in Somerset and the inclined planes on the Bude, Chard and Shropshire Coal Canals, the last major canal of the inland canal system was the Birmingham and Liverpool Junction Canal. Becoming in 1846 the Shropshire Union Canal, it ran from Autherley on the Staffordshire and Worcestershire north-westwards to Nantwich on the Ellesmere and Chester Canal. Proposed by Thomas Telford and receiving its Act in 1826, it was completed in 1835. Unlike the Brindley contour canals, it was straight with deep cuttings and high embankments. The Ellesmere and Chester had incorporated the Ellesmere Canal, under whose 1793 Act the great iron aqueduct with masonry piers was later built across the Dee at Pontcysyllte and opened in 1805. Also on this line was the Chirk aqueduct across the Ceiriog.

In the north of Scotland the Great Glen forms a natural line of fairly low-lying country relative to the high lands on each side and as early as 1726 the line was viewed as a possible ship canal route from coast to coast. It was again surveyed in 1773–4, in 1793, and by Telford in 1801–2. Wartime considerations demonstrated the advisibility of such a link, as it would enable vessels to avoid the attentions of French privateers and also provide a safer passage than the rigours of the Pentland Firth. Acts were passed in 1803 and 1804. Telford was appointed engineer, William Jessop consultant, and the two worked in harmony until the latter's death in 1814. The canal was a series of cuts linking several lochs, including Loch Ness, with locks sometimes in groups as at Neptune's Staircase, the flight of eight locks at Bonavie, and sometimes singly. It was not completed until 1822 but, although it proved its worth at times it was less valuable as a major ship canal than had been hoped. An earlier, though much shorter, ship canal in Scotland was the Crinan Canal, 14km (9 miles) and 15 locks across the Mull of Kintyre, constructed between 1793 and 1815.

In Ireland the Shannon, the longest river in the British Isles, which had always been navigable in parts, began to be improved in the mid-eighteenth century, while at the beginning of the nineteenth century links with Dublin gave it added importance. After a number of abortive starts in the mid-eighteenth century work restarted in the 1770s on the Grand Canal from Dublin and 24km (15 miles) had been opened by 1779. This canal, while designed to reach the Shannon at Shannon Harbour, which it did in 1803, also had the Barrow branch running from near Robertstown southward past Monasterevin to Athy, reaching there in 1790 where it joined the Barrow Navigation. A further branch left westward at Monasterevin for Mountmellick in 1850. The original terminus in Dublin was at St James's harbour west of the city centre, but with the completion to Athy it was felt that there should be a connection with the river Liffey close to St James's Harbour. Instead it was built on a semi-circular route round the south of the city to join the Liffey on the east of Dublin. At this point a dock system, the Ringsend Docks, was built by William Jessop, and both the docks and the new line with seven locks were opened on 23 April 1796. An extension beyond the Shannon to Ballinasloe was opened in 1828. The Grand Canal was unusual in that a series of hotels was built alongside the canal for the benefit of passengers travelling by boat. Excellent surviving examples are at Portobello, Dublin, opened in 1807, and Robertstown, opened 1801, though neither still operates as a hotel.

A rival canal running from Dublin to a more northerly point on the Shannon was promoted in the 1780s. This was the Royal Canal and work began in 1789 at the Dublin end. It leaves the Liffey on the north bank almost opposite the Ringsend Docks. Construction progress was slow and in 1802 John Rennie was called in to advise but it was not completed to the Shannon at Tarmonberry, 145.5km (90.5 miles) from Dublin, until 1817. It included a very deep cutting to the west of Dublin known as the Deep Sinking and a high embankment over the Ryewater. In 1845 it was bought by the Midland Great Western Railway so that a railway could be built alongside. Over the years traffic declined and it was officially closed in 1961.

Towards the end of the nineteenth century the largest of the British canal projects was conceived. Manchester, at the hub of the cotton industry, was suffering because of the Liverpool port charges and the rail monopoly between Liverpool and Manchester. Suggestions were made in 1876 that a ship canal should be built from the Mersey to Manchester and the idea was accepted. Despite the opposition of the authorities in Liverpool and the railway companies, an Act was obtained in 1885 but the Manchester Ship Canal Company had to repurchase the old Bridgewater Canal from the railways and they had difficulty in raising the capital. The target was reached with a month to spare. Work commenced in 1887 and at times 17,000 men, with a great deal of novel mechanical equipment, were employed on cutting the canal. Construction included the entrance locks at Eastham and four other sets of locks. Brindley's original aqueduct was replaced by the Barton Swing Aqueduct. At the Manchester end a dock system was established which became very busy with cotton and grain traffic. Through sailings started on 1 January 1894 and the canal was formally opened on 21 May 1894. Additional docks nearer the Liverpool end were opened in 1922 and 1933 and a new Queen Elizabeth II dock for oil tankers at Eastham, just below the entrance to the Ship Canal, in 1954. With the decline of the cotton industry and the changing pattern of sea transport the Manchester Ship Canal also became moribund.

Sir Edward Leader Williams, who before becoming engineer to the Manchester Ship Canal had been engineer to the Weaver Navigation, had conceived the idea of constructing a lift between the Weaver and the Trent and Mersey Canal at Anderton to overcome a difference in levels of 15m (50ft) and the design by Edwin Clark was the progenitor of several lifts on the Continent. Completed in 1875, it was built as a hydraulically counterbalancing lift with two tanks supported on pistons. The cylinders in which the pistons worked were interconnected and the flow of hydraulic fluid could be started and stopped by opening and closing a valve in the connecting pipe, so that the descent of one tank controlled the ascent of the other. In 1882 one cylinder burst and both cylinders were renewed but by 1905 corrosion was affecting both cylinders again and it was decided to convert the lift to two independent tanks, each counterbalanced by a series of weights suspended on cables carried over 2m diameter pulleys mounted on the summit of the structure.

During the first half of the twentieth century the carriage of bulk materials, coal, timber and grain, steadily passed to rail and then road transport. For example, the Lee traffic, which had consisted mainly of these three types of cargo, fell from over 610,000 tonnes in 1911 to an occasional barge in the 1960s. In 1948 most of the canals in the United Kingdom were nationalized, coming ultimately under the control of the British Waterways Board. Commercial traffic soon fell to negligible proportions on the narrow canals, and was gradually replaced by a thriving leisure industry aided by an energetic campaign for canal preservation, (see p. 516).

FRANCE

In mediaeval France water carriage was mainly served by the great rivers such as the Loire, Seine, Rhône, Somme, Saône and Dordogne, and although the

Romans had suggested a link between the Moselle and the Saône the first major French canal was not started until the seventeenth century. Even river navigation was to a much lower standard than would be acceptable later. Traffic was mainly downstream, either using rafts of timber as on the Sarre, where they could be floated down to water-powered sawmills on the banks, or on purpose-constructed boats, as on the Dordogne and elsewhere, which at the end of a single voyage could be disassembled and their timber sold. The demand for timber from Holland for naval construction led to a considerable traffic in rafts down the Sarre, Moselle and Rhine. Even the busiest river in France, the Loire, was only navigable from All Saints to Pentecost (November to April) each year.

At the beginning of the sixteenth century the return of Francis I to France accompanied by Leonardo da Vinci turned ideas to linking rivers to provide a through route between the Mediterranean and the Atlantic, avoiding the long haul round the Iberian Peninsula. Two proposals were actively considered, neither of which came to fruition at the time but which were later to be built as canals; one linking the Saône and the Loire and the other Garonne and the Mediterranean.

By the end of the sixteenth century the growing demands of Paris for food, plentifully available in the Loire valley, stimulated Henry IV and his adviser the Duc de Sully into considering a link between the Seine and the Loire. Surveys were made and a scheme prepared by Hugues Cosnier was approved on 11 March 1604 whereby a canal would be built from Briare on the Loire to Montargis on the Loing. The Loing was navigable and joined the Seine to reach Paris. Work started in 1605 but was abandoned following the assassination of the king in 1610. By then 42km (26 miles) had been constructed together with 35 locks, which included a seven-rise staircase at Rogny. Recommendation for its completion was made in 1628 but it was not until 1638 that authorization to resume was granted by Louis XIII to Guillaume Boutheroue and Jacques Guyon, Sieur du Chesnoy. The canal was completed in 1642 and became the second summit level canal, and the first modern one, in Europe. The seven-rise staircase of locks at Rogny, a remarkable civil engineering development at the time, was superseded in 1887 by six individual locks but the remains of the staircase are still to be seen, a tribute to the builders. The Canal de Briare is still in use with a recent modern feature – the longest canal aqueduct in the world across the Loire at Briare. Completed in 1896 and designed by Gustave Eiffel, it is 662m (2172ft) long consisting of an iron trough mounted on 15 piers. With facilities for draining it into the Loire, it links the Canal de Briare with the early nineteenth-century Canal Lateral à la Loire and obviated a difficult level crossing over the Loire.

In the south of the country the valley between the Massif Central to the north and the foothills of the Pyrenees to the south suggested the site for a waterway link between the Mediterranean and the Atlantic and various surveys were made. Nicholas Bachelier, instructed by Francis I in the sixteenth century, proposed a canal from the Garonne at Toulouse to the Aude and into the Mediterranean and this was the first survey of many. One of the problems was the supply of water to the summit level. This was solved by the 50-year-old Pierre Paul Riquet, a native of Beziers, who, with no previous experience of canal engineering, realized how the waters of the Montagne Noire could be tapped and used as feeders for the summit level. Ultimately, after further delay but with the recommendations of the Archbishop of Toulouse and of Colbert, Minister of Finance, Riquet was granted a patent on 18 November 1666 to construct the canal in eight years by January 1675: he was then 62. Though work was pushed forward with great pertinacity – there were at the peak period 12,000 working on construction – many new difficulties, both physical and financial, conspired to delay completion. Riquet died on 1 October 1680, prevented by a few weeks from seeing his final triumph, for the official opening of the Canal du Midi took place in the presence of many dignitaries on 15 May 1681.

Riquet's solution to the summit-level water supply was to divert water from the small rivers flowing down from the Montagne Noire into a large reservoir, the Bassin de St Ferreol, and then down the mountain in a feeder into a holding basin at Naurouze. The dam for the St Ferreol reservoir was the first for water for a navigable canal and was the biggest civil engineering undertaking on the project. This was supplemented in 1791 by a second, smaller, reservoir, the Bassin de Lampy. From Naurouze water was available to supply the locks both to the east and the west. That this solution was effective has been proved by its continued use for 300 years. A second important contribution made by Riquet was the world's first canal tunnel, the Malpas tunnel 165m (541ft) in length. One end is still unlined and the winds of three centuries have eroded the soft friable rock so that it looks like the entrance to a cavern. Riquet's third triumph was the construction of an eight-lock staircase at Fonserannes on the opposite bank of the Orb from Beziers. This flight was bypassed in 1985 by a water slope similar to the one constructed on the Canal Lateral à la Garonne at Montech.

Originally the canal crossed the Orb at Beziers on the level, but in 1856 a new line leading from the seventh lock on the Fonserannes staircase was constructed leading to the present aqueduct across the river. This was built with the architectural grandeur commensurate with Riquet's original conception. A further problem on the canal which received a unique solution was the level crossing of the river Libron, west of Agde. Here the bed of the canal is actually below sea level, and as the Libron is liable to bring down large quantities of silt in flash floods which could not be cleared by scouring. Riquet devised a flat-topped barge which could be sunk in the bed of the canal so that the flood water passed over its deck without diminishing the flow velocity and allowing the silt to settle out of suspension into the lower canal bed. This has been superseded by a structure diverting the Libron's flow alternately into one of two channels separated on the canal by a chamber the length of a barge like a lock chamber with gates at each end. Each of the channels has a deck which can be lowered over the canal so that the flood water passes over this. As the barge passes through the chamber the flood water can be directed first through one channel and then the second so that its flow is not impeded. This ensures that the canal can still be used even though the Libron is in flood.

The 240km (149 mile) long Canal du Midi provided half the distance of the link between the Mediterranean and the Atlantic, but it did not overcome the difficulties of navigation on the Garonne between Toulouse and Bordeaux. Bypassing at least part of this section was first proposed by Riquet, but authorization had to wait until 1838. It was then agreed that a 194km (120.5 mile) lateral canal should be built, leaving only the 54km (33.5 miles) between Castets-en-Dorthe and Bordeaux with the navigation in the tidal Garonne. The Canal Latéral à la Garonne was opened from Toulouse to Agen in 1850 and completed in 1856. It includes the 539m (1768ft) long masonry aqueduct over the Garonne near Agen. Between Toulouse and Noissec near the short branch canal to Montauban are the five locks at Montech which were bypassed in 1974 by the experimental water slope. This is one of the most significant developments in inland waterway engineering since the invention of the pound lock.

The water slope, designed by Professor Jean Aubert, Inspecteur-Général des Ponts et Chaussées, consists of an inclined rectilinear concrete flume linking the upper and lower pounds of the canal. A wedge of water in which the barge floats is pushed up the slope by a moveable dam, rather like a giant bulldozer blade, called a shield. At the top of the slope the wedge of water makes a level with the surface in the upper pound and the gate which holds back the upper pound falls under the pressure of the wedge. To propel the shield at Montech two 1000hp diesel-electric locomotives are used. The controls are linked so that one unit is a slave to the other manned unit. The units are mounted on rubber tyres to give better adhesion up and down the slope, but there is also a raised continuous plinth against which horizontal guide wheels mounted beneath the unit, run to keep the unit on the set track. The shield has neoprene rollers mounted on the two sides and the bottom to squeegee against the contour of the flume, ensuring minimal loss of water from the wedge. The slope of the flume is three per cent and its effective length of 443m (1453ft) gives a change in level between the upper and lower pounds of 13.3m (43.6ft) (equivalent to five locks). In front of the shield there is a cradle with two arms which adapt to the shape of the bow or stern of the barge. The speed of travel is 4.5kph (2.8mph). The experiment was a success and a similar slope was constructed and is in operation at the Fonserannes staircase at Beziers on the Canal du Midi. The German authorities also carried out a feasibility study with a view to its use at Scharnebeck on the Elbe Seiten Canal but in the end decided on a lift instead.

In the north progress on the Canal du Midi was watched with interest and immediately after its completion the Duc d'Orléans, Louis XIV's brother, obtained a concession to build the Canal d'Orléans to link Orleans on the Loire with Montargis on the Loing. This was completed by 1692, but the capacity of the Loing between Montargis and the Seine was insufficient to cope with the traffic generated by this and the Briare canal. Accordingly, in 1724, the Canal du Loing was completed parallel to the river and passing through Nemours to join the Seine near Fountainbleau.

The great expansion in the north-west of France came later and was intimately linked with the Belgian waterway system. An early development was the opening in April 1734 of the Canal de Picardie, later to become part of the Canal de St Quentin, linking the Somme and the Oise near Chauny. Otherwise the position at the beginning of the eighteenth century was that the Sambre was navigable north-east from Maubeuge and coal was imported from Liège via the Meuse and the Sambre. The Escaut was only navigable in France between Valenciennes and Condé by 75 tonne boats, but from Condé into Belgium 150 tonne boats could pass. The Scarpe was navigable in theory from Arras, but in practice Douai was the limit. The boatmen of Arras attempted to organize themselves into a guild in 1704 to obtain privileges; but their boats were very small and the hauliers had to be in the water. The Deule was navigable above Lille on the Haute Deule and below on the Basse Deule but there was no navigable link between them, although in 1693 a link between the Scarpe at Douai and the Haute Deule at Lille had been opened.

The far-sighted Sebastien de Vauban realized the need for progress in communications in the area and advocated in 1705 an improved Scarpe between Mortagne at its junction with the Escaut and Douai; a link between Valenciennes and Gravelines; a link between Douai and Bouchain by the Sensée; the completion of navigation through Lille, and of the connections between the Lys and the Aa. It was over a century before all these waterways were completed. Of those mentioned the Neuffossée Canal linking the Aa at St Omer with the Lys at Aire was completed in 1774 with a five-rise staircase lock at Arques, just south of St Omer.

The traffic steadily grew on the Neuffossée Canal which formed a useful link with the coast but the staircase lock at Arques, or as it is usually known, Fontinettes, created a bottleneck. By the 1860s boats could wait for up to a week to pass through and it became the practice to allow boats to pass up and down on alternate days, thus permitting the locks to be used more intensively. Ultimately the authorities decided to install a parallel flight of locks, but this decision coincided with the opening of the English Anderton lift in 1875 in Cheshire between the Weaver and the Trent and Mersey Canal. Edwin Clark was commissioned to design a similar structure on a much larger scale at Fontinettes, which adequately dealt with the traffic problem. Each of the two tanks was capable of taking a 300 tonne Freycinet-type péniche. The lift worked on the same principle of counterbalancing as at Anderton by putting extra water in the tank at the upper level and so forcing the piston of the lower tank to rise, final balancing pressures being obtained by a hydraulic accumulator. The lift was opened in 1888 and continued in operation into the 1960s. Then the enlargement of the canal to cope with the increased size of barges raised questions as to the future of the lift. A new large-capacity lift was considered, but instead a large shaft lock almost on the site of the old five-lock staircase was constructed and the lift was retained as an industrial monument.

Further south a very old proposal, the Canal du Charollais, now the Canal du Centre was to link the Loire south of Nevers at Digoin with the Saône to form part of a north–south link across France. This scheme was revived and work started in 1782, opening to navigation in 1793. Work also started in 1793 on the Canal de Bourgogne to link the Saône and the Yonne, again to provide a north–south connection.

The French Revolution put a temporary end to most major works and it was not until Napoleon, with his vital awareness of the importance of canals, came to power that projects likely to come to fruition were again considered.

In the west in Brittany, in order to stimulate industry and agriculture, a start had already been made in 1580 in making the River Vilaine navigable inland to Rennes, and in 1627 the States of Brittany approved the canalization of the Aulne between Brest and Carhaix, but the work was not pursued. In 1783 a network of canals with Rennes as the centre was proposed on the grounds that they would provide communication which would avoid the English blockade. Such canals were those linking Brest and Nantes, Pontivy and Lorient, and Rennes and St Malo. The plans lay dormant until revived by Napoleon in 1810. On 7 September 1811 stonelaying at the first lock on the Nantes–Brest Canal took place at Chateaulin but work was suspended between 1814 and 1822. It was opened in stages as lengths were finished, but it was not completed throughout until 1836. Unfortunately it was severed later by the construction of the great dam and reservoir at Guerledan.

Another canal which was early considered was a link between the Loire at Decize and the Yonne at Auxerre. This, the Nivernais Canal, was started in 1784 but again the Revolution temporarily put an end to construction, delaying its opening until 1834. It is a delightful canal and forms yet another north–south link on the French canal system. One of its feeders, the Rigole d'Yonne, crosses the Yonne valley at Montreuillon over a 33m (109ft) high and 152m (502ft) long aqueduct constructed entirely in masonry and opened in 1842.

East–west links were still unsatisfactory, so a major waterway, the Marne–Rhine Canal providing communication between Paris and Strasbourg and also giving access to the industrial areas of Alsace and Lorraine, was started in 1846. It passes through difficult country and is unusual in having two summit levels. The 310km (192 mile) waterway was completed in 1853. At the same time a second canal, the Aisne–Marne Canal, connected it with the industrial areas of north-east France, a connection which improved after 1881 by the construction of the Oise–Aisne Canal. Also linked with the Marne–Rhine Canal and consequential on the Franco–Prussian war of 1870–1 was the Canal de l'Est running north on the canalized River Meuse, with an interesting short tunnel at Revin built to cut off a great loop of the river; the canal's southern portion forms a link to Corre at the start of the navigation down the Saône to Lyons. Another modern development can be seen on the Marne–Rhine Canal. At Arzviller, a few kilometres east of Strasbourg, the canal rises through 17 locks up the Zorn valley and at the top of the flight enters a tunnel where, until recently, electric mules towed the barges through. There were considerable delays in passing the locks, so in the 1960s two new lengths of canal were built, one at the level of the canal below the lowest lock and the other at the level above the highest lock. Linking the two levels is a steep inclined plane where the barge is carried up or down transversely in a tank. The difference in level is 44.5m (146ft) and the slope is 41°. Only one tank is used but provision was made for the installation of a second should the need arise. The incline was opened for traffic on 27 January 1969. Further modernization took place on the western side of this summit in 1964 with the replacement of six locks with a single deep lock at Rechicourt.

Also branching off the Marne–Rhine Canal is the Canal des Houillères de la Sarre, or the Sarre Coal Canal, suggested in 1838 when the Marne–Rhine was still under consideration. The project ran into considerable opposition but was sanctioned in 1861 when it was to be financed jointly by Prussia and France as it would benefit both countries. It was constructed between 1862 and 1866 and the first barge from Gondrexange on the Marne–Rhine Canal reached Sarreguemines on 15 May 1866. Over the years it has been gradually deepened and modernized. With the work in hand by Germany in canalizing the Saar (Sarre) to modern standards from the Moselle near Trier to Saarbrücken, this canal may prove a valuable link between the Moselle and the Strasbourg area.

In the east and south the construction of the Rhône–Rhine Canal from Strasbourg via Mulhouse, with a further link to the Rhine at Basle and a short branch canal to Colmar, continuing via Besançon to the Saône north of Chalon and thence to the Rhône at Lyons, has over the years completed another vital link for north–south traffic. The first section from St Symphorien on the Saône to Dôle on the Doubs was completed in 1802. The remainder of the section between Strasbourg and the Rhône were opened as completed until the first vessel to make the full journey passed through the lock at Strasbourg on 31 December 1832.

The Rhône itself has been largely tamed in this century with the construction of dams for power stations and the associated deep locks for barge traffic, and with the necessary improvements to the Rhône–Rhine Canal it will soon be possible for the large Rhine barges and pushtows to travel from Rotterdam to the Mediterranean at Marseilles where a new link, the Liaison Rhône–Fos, was opened in 1983 taking 4500 tonne pushtows.

The Moselle has been improved with new freight handling facilities at Thionville allowing 3200 tonne pushtows to run directly to the Rhine at Koblenz, and the modernization of the Moselle at Liverdun now allows the Marne–Rhine traffic to use the canalized river instead of crossing by an aqueduct. Improvements south of Dunkirk now allow the Europa barges to serve the port and connect with the Escaut to both Belgium and central France.

Finally, the river navigation of the Seine has been transformed. Although navigation from the sea to Paris and beyond has existed for centuries, until the nineteenth century it was difficult and costly partly because of the current passing bridges and partly because of low water in summer. In addition many mills had been built, both on the banks and jutting out into the river, and these caused obstructions. Despite works carried out elsewhere in France, canalization of the Seine did not begin until after 1840 when five locks and dams were built giving a depth of 1.6m (5.2ft). Twenty years later this was improved to 2m and three more locks were added. After 1910 a major reconstruction took place and the number of locks was reduced to the present total of six. Further improvement was completed with enlarged locks in 1964. An example of this development can be seen at Andresy, just below the confluence of the Oise and the Seine, where there are two small locks built in 1880, a dam and large lock 160m by 12m (525ft by 39ft) with a depth over sill of 5m (16.4ft) opened in 1959, and a new lock of 185m by 24m (607ft by 79ft) opened in 1974. This is fitted with a cylindrical sector horizontally rotating gate at the upstream end of the lock. The lock is capable of taking a 5000 tonne pushtow. All the locks below the port of Gennevilliers below Paris have now been enlarged to these dimensions. This has also allowed the development of inland ports as at Boneuil and Gennevilliers and more recently, post-1976, at Limay on the site of a former quarry. Now sea-going coasters of up to 200 tonnes as well as the usual self-propelled barges of up to 1350 tonnes can operate through to Paris.

THE LOW COUNTRIES

The generally flat terrain of Belgium and the Netherlands, much of it reclaimed from the sea, has been particularly favourable to the development of inland navigation, and in the north it has been considerably influenced by the extensive and complicated delta of the Rhine.

Further south, in the area between Amsterdam and Rotterdam, it is true to say that the rivers and canals determine the Dutch landscape. In Roman times there were three main channels of the Rhine into the sea. One flowed to the north and entered the Ijsselmeer (Zuider Zee). The middle one, now known in its remaining stretches as the Oude Rijn (Old Rhine), entered the sea to the west of Leiden. The third, the southern branch, is the modern outlet via Rotterdam and the Hook of Holland. The Romans wished to connect the middle and southern outlets in a north–south direction and, because of the difficulty of cutting an effective channel through the inland peat, they constructed a canal lateral to the coastal sand dunes from the Old Rhine west of Leiden to the Rhine near Maassluis. This was known as the Vliet. The southern portion has disappeared, but the part between Leiden and Rijswijk still survives. Further east a more complicated route from Muiden on the Ijsselmeer using the Vecht to Utrecht joined the Lek near Gein close to where the present Merwede Canal enters the Lek. A canal was built through the centre of Utrecht in 1150 and still survives as the Oude Gracht (Old Canal). With local city and provincial power politics able to demand and purchase rights it is not surprising that yet another north–south route was developed soon after 1200. The Gouwe was constructed from near Alphen a/d Rijn on the Old Rhine southwards through Gouda to the Hollandse Ijssel and thence to the Lek. Finally in 1220–6 a lock called a *spoje* was installed at Spaarndam and could have been a pound lock. At this time neither Amsterdam nor Rotterdam existed.

Utrecht responded by allowing its outer moat to be used for navigation, although commercial interests were against this on account of loss of trade within the city. New ports were established along the Rhine at Schiedam in 1260 and later at Delfshaven and Rotterdam with their own connections with the interior. In order to permit larger vessels to pass through without their masts fouling the superstructures of the locks, mitre gates were installed at a new lock at Spaarndam in 1567 and at a similar lock at Gouda in 1578.

In the seventeenth century new canals were built specially for passenger traffic conveyed in horse-drawn boats known as *trekschuiten*. Between 1630 and 1660 over 650km (404 miles) of these canals were built linking city with city, though not necessarily to form a continuous waterway. Nevertheless they constituted the first internal passenger network in the world, with the boats running to a timetable and providing on many routes a frequent service which was a boon to travellers. Some routes persisted well into the railway age.

Meanwhile Amsterdam was growing and needed improved transport connections with the sea. An early route was for vessels to sail east across the Zuider Zee to Kampen and then up the Ijssel past Zwolle and Deventer to join the Neder Rijn (Lower Rhine) near Arnhem. An alternative route was to sail along the coast of the Zuider Zee to Muiden and then up the Vecht to Utrecht and the Kromme (Crooked) Rijn to Wijk bij Duurstede. In time the Crooked

Rhine silted up and a new canal was made from Utrecht to Vreeswijk known as the Vaartse Rijn or Rhine Navigation leading into the Lek by a lock built in 1284. This very early lock was 54.5m (178.8ft) long and 4.75m (15.6ft) wide.

One of the early new canals for *trekschuiten* was the Weespertrekvaart, a canal between the Amstel at Amsterdam and the Vecht near Weesp constructed in 1638, which allowed vessels to reach the Rhine without traversing the dangerous passage across the Zuider Zee. By the beginning of the nineteenth century this had become neglected. Between 1815 and 1822 William I restored it and it was then known as the Keulse Vaart (Way to Cologne). The Lek also was silting by this time, and so a further canal, the Zederik Canal, was built from Vianen opposite Vreeswijk to Gorinchem on the Waal (otherwise the Merwede) and completed in 1825. In 1890 this canal and the Keulse Vaart south of Utrecht were improved to take 2000 tonne craft, and together they became known as the Merwede Canal. The density of traffic generated posed further problems. In 1921 more than 300 ships passed the lock at Vreeswijk on each of 29 days, and by 1930 waiting time to pass the lock could be up to twelve days. Because of these delays it was decided to construct a new Amsterdam–Rhine Canal and this was started in 1937. It involved upgrading the Merwede Canal from Amsterdam to Utrecht and then, bypassing Utrecht to the west, it followed a new alignment to Wijk bij Duurstede where it crossed the Lek on the level and continued to Tiel on the Waal. A branch, the Lek Canal from south of Utrecht to Vreeswijk, was completed in 1937 to benefit traffic travelling west to Rotterdam. The Amsterdam–Rhine Canal was opened in 1953; by the 1970s it had proved so successful that it was becoming congested and further enlargement had to be made. This was completed in 1981. The problems of crossing the Lek with pushtows were the subject of intensive studies at the Delft Hydraulics Laboratory. At both sides of the river egg-shaped basins were formed and these created whirlpool eddies keeping the main current in its bed. In this way the creation of shoals was avoided and very little dredging is required. The new massive lock structures are dramatic examples of modern canal architecture.

By the late eighteenth century large ships were finding it difficult to reach Amsterdam owing to the silting of the Zuider Zee so a proposal to link Amsterdam with Den Helder, which had permanent deep water, was approved; construction started on the Noordhollandsche Canal in 1819 and its 82km (51 miles) were completed in 1825. It was not particularly successful and in 1876 it was superseded by the North Sea Canal giving a more direct route to the sea at Ijmuiden.

The network of small waterways in north Holland has offered few problems to its intensive use as an integrated transport system, but with the increased size of barges against the old dimensions of bridges and locks it was necessary to improve conditions. In 1935 work started on building a new canal across Groningen province and Friesland to link Groningen and Lemmar on the Ijsselmeer and this has been enlarged for 1350 tonne barges and 400 tonne coasting vessels.

It is known as the Van Starkenborgh Canal in Groningen and the Prinses Margriet Canal in Friesland. The Eemskanaal constructed in 1866–76 from Groningen eastward to Delfzijl on the Ems estuary has also been enlarged. A westward branch from Suawoude near Leeuwarden, the Van Harinxma Canal to Harlingen facing Terschelling, has also been built. In this group, too, is the Winschoter Diep running eastward from Groningen to the German border. These major waterways supplement the many minor ones.

Following the Treaty of Vienna in 1815, what is now Belgium, the Netherlands and Luxembourg were united under William I of the Netherlands. He recognized the importance of water transport and one proposal of significance was to link the Meuse and the Moselle so that the long, tedious journey via the Rhine could be avoided. This was to be accomplished by linking the Ourthe, which (as a tributary of the Meuse meeting it at Liège) was navigable to Barvaux, at that time the limit of navigation, across Luxembourg to join the Sure near Diekirch and thence via Echternach into the Moselle almost opposite its confluence with the Saar. Work started in 1826 and some excavation, including the beginning of the Bernistap tunnel, was carried out before the Belgian Revolution occurred in 1830. Work then ceased; Belgium, now a separate country, was at the time financially unable to continue the work and it was never resumed.

Rather more successful was the Ghent–Terneuzen Canal which superseded an earlier small canal giving Ghent access to the sea via the Schelde in 1827. Terneuzen on the Schelde became Dutch after 1830 while Ghent remained in Belgium, but the canal served Ghent well, being reconstructed in the 1870s and again in 1910; in 1968 its capacity was increased to take ships of 60,000 tonnes.

The major development in recent canal construction has been the Schelde–Rhine Canal to allow Antwerp traffic to gain the Rhine. A new canal has been built from Antwerp to bypass the Western Schelde and enter the Eastern Schelde at Kreekrak locks. The route then crosses the Eastern Schelde and across Tholen to the Volkerak which has been dammed off from the open sea as part of the great Dutch Delta Plan for prevention of the kind of flood disaster which occurred in 1953. To reach the Hollandsche Diep and thence the Rhine, it is necessary to pass the Volkerak locks. Over the locks is now a major road junction, yet the whole area was open sea in the 1960s. A curtain of air bubbles rising from the bottom of the locks creates a turbulence which inhibits the mixing of the fresh and salt water, thus preventing salt water from entering the new fresh-water areas created by the Delta Plan and so preserving the ecology and fertility of the new land areas.

Further south in Belgium changes in the coastline have affected the northwestern area. Before the fifteenth century Bruges had ready access to the sea through its outport at Damme, but a storm in 1404–5 caused the Zwijn, which led to Damme from the sea, to silt up. A new channel was built to Sluis and then in 1622 a canal to Ostend was constructed and later extended to Nieuport and Dunkirk. Eastwards a canal was built in 1613–23 from Bruges to Ghent and though this has been improved Bruges has never regained her former commercial importance.

Much more important were the canal developments in connection with the coalfields in the south of the country and the trade through to France. The Mons–Condé Canal was completed in 1818, crossing into France to link with the Escaut (Schelde), and later in 1826 the Antoing–Pommeroeil Canal enabled barges to travel from Mons to south of Tournai on the Escaut entirely within Belgian territory.

Also within this construction period came the building of the Brussels–Charleroi Canal for small barges in 1832. This was enlarged to 300 tonne standard during the long period 1854–1914 and the number of locks was considerably reduced. In the period after World War Two it was reconstructed to 1350 tonne standard and this included the major engineering work of the Ronquieres inclined plane. In this reconstruction the summit level was lowered to lengthen it from 11 to 28km (6.8 to 17.4 miles) three locks replaced the eleven between Charleroi and the summit, and seven locks and the inclined plane replaced the previous twenty-seven on the Brussels side. The canal was also realigned. The inclined plane is 1432m (4698ft) long and barges are carried in two independently operated tanks or caissons whose weight can vary between 5000 and 5700 tonnes according to the amount of water in the tank. Each tank is attached by cables to a counterweight of 5200 tonnes which also travels up and down the incline, and can take four 300 tonne barges or one 1350 tonne barge. There have been considerable savings in time between Charleroi and Brussels since the incline was opened in 1968.

This canal links near Charleroi with the Canal de Mons or Canal du Centre, construction of which began in 1880. This was an extension of the Mons–Condé Canal and in its course there is a change in level near La Louvière of 66.19m (217.2ft) in 8km (5 miles). Construction was under way at the same time as the La Fontinettes lift was being built on the Neuffossée Canal in France, so it was decided that this escarpment should be surmounted by four lifts and Edwin Clark was engaged as designer. All were to be of the hydraulic type, but of steel girder construction instead of brick and masonry. The first, at Houdeng-Goegnies, was opened by King Leopold II on 4 June 1888. The other three, at Houdeng-Aimeries, Bracquegnies and Thieu, were not opened until 1917. Modernization of this canal will replace the four lifts by a single lift of 73m (239.5ft) at Strepy-Thieu to take 1350 tonne barges.

The Maas (Meuse) in the east of Belgium has always been a difficult river because of its variable flow, and canals were proposed to pass the hazardous stretches, particularly between Maastricht and 's Hertogenbosch. The Zuid-Willemsvaart Canal was built between 1823 and 1826 and was enlarged in

1864–70. Further improvements took place by the construction of the Juliana Canal, 35.6km (22.1 miles) between 1915 and 1935 replacing the longer and smaller capacity Zuid-Willemsvaart. A link between the Maas and the Waal was built between 1920 and 1927 near Nijmegen and the Albert Canal was built between 1930 and 1940 to link the Meuse near Liège with the Schelde at Antwerp.

In the west of Belgium the Baudouin Canal has been built between Bruges and Zeebrugge, and at Ghent a *ringvaart*, or perimeter canal, circling the city was completed in 1969. Work on upgrading the line from the Escaut to Brussels docks will enable it to take 9000 tonne pushtows.

GERMANY

Although Germany as a political unit was a nineteenth-century creation out of a number of independent states, internal transport by water has been a major factor in the territory's economic development over the centuries. It is only, however, within the last two hundred years that the present waterway network has evolved, linking the great rivers of the Rhine, Danube, Elbe, Oder and their tributaries. The Rhine–Main–Danube Canal nearing completion in the late 1980s (see p. 499) was conceived centuries ago when Charlemagne saw the potential value of a link between north-west and south-east Europe. The construction of the Fossa Carolina, of which traces still remain in the vicinity of Weissenburg in Bavaria, was started in AD 793 using demobilized soldiers and prisoners of war, but it was left unfinished and the fulfilment of the vision of boats sailing from the North Sea to the Black Sea had to wait for another thousand years.

Meanwhile the need for improved transport from the salt producing areas near Lüneburg led to the construction of the Stecknitz Canal linking the Elbe and the Baltic at Lübeck. Transport had previously been by a devious route partially overland and partially by water using the rivers Ilnenau and Delvenau, an almost impossible route because of the interposition of mill dams. There is a ridge between the Elbe and Lübeck and crossing this involved a rise on each side to the summit. The Stecknitz Canal, completed at the end of the fourteenth century, thus became Europe's first summit level canal with the first boat passing through in July 1398. It was still a tedious journey as flash locks were used and boats could take several weeks to travel its 100km (62 miles). With various alterations to its line and reconstructions to give greater capacity, the link is continued today as the Elbe–Trave Canal. Preserved as an industrial monument is the Palmschleuse, a circular lock to the north of Lauenburg dating from 1734.

On the other side of the country the growing importance of Berlin made it imperative that transport links should be improved between the Spree in that city and the Oder. In the early fifteenth century locks had been built on the Havel and Spree to provide a connection between Berlin and Hamburg and this was followed in 1540 by a decision to build a canal to the Oder. Lack of money delayed the start but eventually it was completed in 1620 after the start of the Thirty Years War (1618–48). It was short-lived: having already started decaying during the war it ceased to exist soon after its end. Known as the Finow Canal, it was reconstructed later and now on its third realignment carries a large tonnage between Germany and Poland.

Further south a second link between Berlin and the Oder was provided by the Oder–Spree Canal, constructed soon after the Thirty Years War, running eastwards from Berlin to join the Oder just south of Frankfurt-on-Oder at Briekow. It was completed in 1669 and was important for the westbound traffic from the Silesian coalfield. It was also the third European summit level canal. The needs of Berlin were still not satisfied and two further canals were built in the eighteenth century, thus creating in Berlin a focal point of the German waterway system: the Plaue Canal, now known as the Elbe–Havel Canal, built in 1743–6 linking the Elbe north of Magdeburg to the Havel at the Plauer See near Berlin, and in the east an improved Finow Canal built in 1744–51 to reinstate the link between Berlin and the northern Oder leading to Stettin.

The aftermath of the Thirty Years War also led to suggestions for improvements to the River Ruhr to provide transport from the coal measures being exploited in the valley. The nobility of Kleve, part of the domains of the Elector of Brandenburg, persuaded the Elector to allow the canalization of the Ruhr in order to supplement the shortage of supplies of coal in the Duchy of Kleve. Nothing, however, was done until 1768 when Frederick II gave permission for a Dutch financial company to transport coal down to the dock on the Rhine at Ruhrort. This dock had been developed in 1716 and its subsequent importance, generated by the coal traffic together with the rise of Duisburg, led ultimately to the present-day inland port of Duisburg-Ruhrort on the Rhine – the largest inland waterway port in Europe.

At first the canalization of the Ruhr was ill-conceived for, when in 1772 coal began to be shipped downstream from Witten, the river was impeded by fixed dams at each of which the coal, and other cargoes, had to be transhipped to barges lying in the pound below. A few years later the merchants suggested building locks and, after disputes with millowners and those enjoying fishing rights on the river, these were constructed in timber. In 1837–42 the locks were reconstructed in stone and the capacity of the barges steadily increased from 30 tonnes to 150–180 tonnes. Following competition from the railways in the 1860s and 1870s, traffic had virtually ceased by 1890.

However, it was in the nineteenth century that the great surge forward came in German waterway construction, accompanied by the use of a more sophisticated technology which has continued to the present day. This was presaged by the creation of a navigation on the Ems in 1835 as the Haneken Canal and on the Lippe in 1840 from the Rhine to Lippstadt, and each of these canals led to the building of larger canals to serve the Westphalian industrial development.

Towards the end of the nineteenth century authorization was given for the Nord-Ostsee Canal, or Kiel Canal, linking the Baltic at Kiel with the North Sea at Brunsbüttel. This is a major ship canal for large sea-going vessels and designed to eliminate the long haul round the north of Denmark. Work started in 1887 and the canal was opened in 1895. Ships at that time still carried tall masts, so the bridges were necessarily at a high level which caused structural design problems, as the terrain through which the canal passes is generally flat. The most interesting construction was the transporter bridge at Rendsburg which, uniquely among such bridges, also carries a main railway line over the upper girders. It also remains one of the very few transporter bridges still in use. At present the only alternative to the high bridges are the vehicle ferries which have to cross through the very high density of canal traffic. Since opening the canal has been widened and deepened and such work is still continuing. The present width is being increased from just over 100m (328ft) to 160m (525ft) over its whole length of 99m (61.5 miles).

In 1886 the 269km (167 miles) Dortmund–Ems Canal was authorized to link the Ruhr industrial area with Emden and the North Sea. The canal was completed in 1899 and it, too, contained technical innovations. It was promoted to provide an all-German waterway for the Ruhr basin traffic, thus avoiding the Rhine passage through the Netherlands. The northern end near Emden incorporated the earlier Haneken Canal whose capacity was improved. At the southern end Dortmund is situated at a higher level than the main line of the canal. There is a scarp at Henrichenburg about 15km (9.3 miles) west of Dortmund with a difference in level of 13.5m (44.3ft). With the success of the Anderton and Fontinettes lifts in England and France in mind, the authorities decided to install a lift but on a different principle. Construction was by Haniel and Lueg of Dusseldorf and started in 1894. The method of operation consisted of mounting the tank on five floats known as *schwimmers* positioned in wells. The wells were 9.2m (30.2ft) in diameter and 30m (98.4ft) deep and were axially in line below the tank. The floats were 8.3m (27.2ft) diameter and 12m (39.4ft) high. The tank, 70m (229.7ft) long and 8.6m (28.2ft) wide, with the floats comprised a single unit whose rise and fall was constrained by vertical iron guides in the superstructure. The whole system is very carefully balanced so that if excess pressure is applied, by increasing the volume of water in the tank, the system will sink and if the pressure is decreased it will rise. It should be emphasized that the *schwimmer* is a true float and not a piston. As traffic increased it was decided to supplement the lift with a shaft lock which would also act as reserve in case of lift failure. The shaft lock was built between 1908 and 1917, completion being delayed by the First World War. Because the upper pound to Dortmund has no major feeder, water has to be pumped back from the lower pound to compensate for that passing through the lock, and in order to minimize the loss of water ten side ponds were constructed, five on each side on five different levels, into which water from the lock can be directed during emptying and the water used again to fill the lock.

Meanwhile, the canal below the Henrichenburg lift was linked to the Rhine at Duisburg–Ruhrort by the Rhine–Herne Canal in 1914 and in the same year the improved navigation, the Wessel–Datteln Canal, was opened on the line of the old Lippe Navigation. Thus two important links were completed between the Rhine, the industrial area north of the Ruhr and the North Sea.

Further north an east–west link between the Rhine and the Elbe was proposed in the late nineteenth century but authorization for construction, including an intermediate connection at Minden with the Weser, was not given until 1905. The plan included a lock-free length of 211km (131 miles). By 1916 the waterway, the Mittelland Canal, was open to Hanover but it did not reach the Elbe until 1938. Three major engineering projects were successfully undertaken on this canal. At Minden a masonry aqueduct 375m (1230.3ft) long crosses the Weser valley and at the western end of the aqueduct is a shaft lock which, uniquely, has enclosed side ponds and connects the canal and the river. There are also two locks at the eastern end of the aqueduct linking the two waterways. The third major work is the lift at Rothensee near Magdeburg, now in East Germany. This took several years to build and works on the same principle as the Henrichenburg lift but is balanced on two floats. It was intended as a link with the Elbe and the main line of the canal was to cross the river on another great aqueduct. Because of the Second World War and the changed political situation, with Germany divided into two states, the aqueduct was never constructed. Vessels have to descend the lift, cross the Elbe on the level and then rise by locks to the level of the Elbe–Havel Canal.

Yet another east–west link was completed in the north in 1935 with the construction of the Küsten Canal between the Ems at Dörpen via Oldenburg to the river Hunte which in turn flows into the Weser at Elsfleth. This is a major waterway taking 1000 tonne craft. It also connects with smaller canals such as the Elisabeth–Fehn Canal with locks 27.6m (90.6ft) long and 5.4m (17.7ft) wide as against 105m (344.5ft) long and 12m (39.4ft) wide on the Küsten Canal.

In the east traffic increased on the Hohenzollern Canal, formerly the Finow Canal, and now the Havel–Oder Canal in East Germany, causing delays at the four locks at Niederfinow, north-east of Berlin. In 1926 work started on replacing the locks with a single counterweighted lift and this was completed in 1934 to become the highest lift (36m) in the world at the time. It is in constant use today for traffic to and from Poland.

Increased traffic on the Dortmund–Ems Canal was also giving rise to problems with the lift and shaft lock at Henrichenburg, so in 1958 construction of a new lift was started using the same principle of floats as in the original structure but supporting the tank on two instead of five. The tank was also designed to take 1350 tonne barges. This was brought into use in 1962, although the old lift continued in operation until 1970 when it was scheduled for scrap. Later it was reprieved, to be preserved as an industrial monument. Traffic still continues to be heavy and plans for another shaft lock have been proposed. Another problem affects this area. The honeycomb of mine workings has caused so much land subsidence that continuous sheet piling has to be carried out to maintain the water level in the Dortmund to Henrichenburg section; fields which were once level with the canal are now several metres below it.

The problems of variable flow in the Elbe, which carried traffic from Hamburg to the Mittelland Canal at Magdeburg, together with the establishment of the political boundary between East and West Germany running to the west of the Elbe, led to the construction after the Second World War of another major waterway, the Elbe Seiten Canal, or Elbe Lateral Canal, which leaves the Elbe near Lauenburg and follows a roughly southward course parallel to the East/West German boundary to the Mittelland Canal near Wolfsburg. Construction began on the 115km (71.5 miles) canal in 1968 and was completed in 1976. It was designed with only two changes of level – one at Scharnebeck, where boats are transferred from one level to the other by a lift with two counterweighted tanks, and the other at Uelzen by an unusual lock. The Scharnebeck lift has a difference of level of 38m (124.7ft), exceeding the Niederfinow lift, and can take a Europa barge of 1350 tonnes in a tank 100m (328ft) long and 12m (39.4ft) wide. Some idea of its size can be gained from the facts that the weight of the tank and water is about 5700 tonnes and each tank is balanced by eight counterweight sets consisting of 224 laminated concrete discs each weighing approximately 26.5 tonnes. The eight sets are housed in four concrete towers. It takes three minutes to overcome the difference in level. The Uelzen lock is unusual in that most of the structure is above ground and it compasses a change in level of 23m (75.5ft). As with other deep German locks there are side ponds to lessen the amount of water passing from the upper to the lower level during each locking operation. Elaborate pumping equipment is also installed to transfer water to the upper level.

Charlemagne's vision of a trans-Europe canal linking the North and Black Seas was revived by Ludwig I, King of Bavaria, when he ordered a survey in 1828. In 1836 work started on a link between Bamberg on a tributary of the Main and Kelheim on the Danube and was completed in 1845. Unfortunately it was of limited capacity and the rivers at each end, though nominally navigable, were often dangerous, but it proved the feasibility of the link. In 1921 the idea was again revived and two different lines were proposed, the Rhine–Main–

Danube and the Rhine–Neckar–Danube (the Neckar Navigation had already been greatly improved). The Rhine–Main–Danube was favoured and became one of the great canal projects of the world. On completion, after many vicissitudes both political and financial, it will provide a 3500km (2175 miles) long unbroken waterway across Europe.

The Main was canalized for 292km (181.4 miles) from Aschaffenburg to Bamberg in 1962, following which a new canal section to Nuremberg of 70km (43.5 miles) was opened on 23 September 1972. There will be a 100km (62.1 miles) link between Nuremberg and Kelheim and the Danube has been improved between Kelheim and Regensberg. Barrages have been built above and below Passau at the Kachlet dam and the Jochenstein dam. There will be 16 locks between Bamberg and Kelheim, eleven rising from the Main to the summit level and five falling to the Danube. On the river sections of the canal the opportunity has been taken of building hydro-electric power stations adjacent to the locks to help to pay for the canal construction costs.

THE RHINE

Navigation on the Rhine has been undertaken from time immemorial and over the centuries its relentless flow has been largely tamed and vessels have been designed to thrust themselves upstream against the powerful current. The iniquitous toll system levied by the riparian owners has been abolished, allowing a greater freedom of navigation and thus assisting trade. Above its complicated deltaic system (see p. 490) in its middle reaches the Rhine has benefited from the trade brought to it by the Ruhr, Moselle, Neckar and Main, as well as by the various canals which have been connected to it.

Its potential as a through route has already been mentioned in connection with the Rhine–Main–Danube Canal (see p. 498), but it was also anticipated on a different line in the seventeenth century. Then the Dutch traders wished to reach their eastern markets without being hampered by the Spanish navy, the Dunkirk corsairs and the Algerian pirates round Gibraltar. The natural Rhine–Rhône links were in the hands of Spain or its allies, so in 1635 a plan was made to reach the Rhône by a navigation through friendly Switzerland. The proposal was to follow the Rhine above Basle and then by the Aare from Koblenz near Waldshut through Olten and Solothurn, then through the Bielersee and by what is now the Zihl Canal to Lake Neuchâtel. A canal would then be built from Yverdon to near Morges on Lake Leman (Geneva) and so to the Rhône. The Canal d'Entreroches between Yverdon and Lake Leman was started in 1638. It reached Entreroches in 1640, a distance of 17km (10.6 miles). A further 8km (5 miles) was built to Cossonay, opened in 1648, but it never reached Morges. It had crossed the summit level and if it had been completed would have been the second summit level canal in Europe, instead of the

Canal de Briare (see p. 483). It was improved and enlarged in 1775 but later abandoned.

In modern times the use of the Rhine's current for hydro-electric power generation has enabled substantial locks to be built which permit pushtows to reach the ports of Strasbourg and Basle. The first vessel direct from Britain to Basle arrived in 1936. The Swiss Rhine fleet consists of 484 vessels and the tonnage passing through the port of Basle in 1981 was $8\frac{1}{2}$ million tonnes. The navigable Rhine continues for 14km (8.7 miles) above Basle to Rheinfelden, a section which includes a new lock built by the Swiss at Birsfelden.

ITALY

In the Leonardo da Vinci Museum of Science and Technology in Milan there is a display of enlargements of Leonardo's navigation drawings with models constructed according to those drawings. This is one indication of the early interest in inland navigation in Italy; another is that in Milan itself there is a lock gate incorporating a vertically pivoted paddle door of the kind designed by Leonardo. For here in Milan was one of Europe's early canals. A watercourse for supply and irrigation had been built at the end of the twelfth century from the River Ticino a little below its exit from Lake Maggiore, and in 1269 it was enlarged and made navigable. It was then known as the Naviglio Grande and was used to convey marble for the erection of Milan Cathedral. Where it entered Milan a lock was built to allow the boats to gain access to the city moat. This lock was the Conca di Viarenna, the lock chamber of which has been retained as a historic monument dating back to the end of the fourteenth century. Some 50 years later another canal parallel to the Ticino was built from the Naviglio Grande through Bereguardo to Pavia, about 6km (3.7 miles) above the junction of the Ticino and the Po. Soon afterwards the Martesana Canal was built to link the Adda at Trezzo sull'Adda with Milan.

The Po is the most important and the greatest river in Italy. It rises to the west of Turin but does not become navigable as a continuous waterway until some 75km (46.6 miles) below Turin and then only for small craft. However, in the lower reaches south of Venice a complicated system of waterways developed in the nineteenth and twentieth centuries linking the Po and the Venetian lagoon near Chioggia. The greatest development in northern Italy has taken place in the years since the Second World War. The Po has been improved so as to take 1350 tonne barges and pushtows consisting of a pusher and one barge. A new inland port has been constructed at Cremona and a new canal of 63km (39.1 miles) is being excavated from there to Milan, although financial restrictions have delayed completion. The Po divides at Pontelagoscuro, the southern branch running past Ferrara to Porto Garibaldi while the northern branch on its way to the Adriatic has a canal leaving from its north bank at Volta

Grimana leading to the Venetian lagoon. This is the Po–Brondolo Canal and en route it crosses the Pontelongo Canal running to the west. Another major 1350 tonne canal, the Fissero–Tartaro–Canalbianco Canal, will bypass part of the lower Po and link the river Mincio with the Po–Brondolo Canal. A third 1350 tonne canal to Padua will run roughly parallel with the old Brenta Canal.

GREECE

One of the earliest canals ever conceived was one to provide a waterway across the Isthmus of Corinth between the Aegean and Ionian Seas. Both Greek and Roman leaders were mindful of the glory that would follow the creation of the Corinth Canal, but it was only Nero who attempted the operation. There had been, and indeed it continued after Nero's efforts, a plan by which ships could be taken across the Isthmus in wheeled cradles running in grooves cut in the rock, so forming a guided trackway. Nero died before the work was completed. Several centuries passed before the work was restarted, initially under the persuasion of Colonal Istvan Türr who obtained a concession from the Greek king in 1882. The company went bankrupt in 1889 and construction was then taken over by a Greek company which completed the cutting in 1893. Much of the work was excavation through solid rock to a depth of 300 to 450m (984 to 1476ft). The canal, slightly more than 6km (3.7 miles) long, obviated a sea passage of some 320km (199 miles) round the south of Greece. It was closed by the Germans in 1944, restored and reopened after the war, and damaged again in an earthquake in 1953, since when it has been reopened.

SWEDEN

Belief in the feasibility of a navigation across Sweden from the Kattegat to the Baltic, to avoid the long haul round southern Sweden and to give security from attack by enemy forces, dates back many centuries. Although vessels had regularly used the Göta river their passage was impeded by the falls at Lilla Edet and more so by the falls and escarpment at Trollhättan. The first Swedish lock was built at Lilla Edet in 1607; although it was burnt in 1611 it was subsequently reinstated. At the same time a new line, the Karls Grav, was started from Vänersborg on Lake Vänern through Vassbotton to join the Göta river above the Trollhätten falls, but it was not completed until 1752. It had one lock at Brinkebergskulle. This remains Sweden's oldest canal. Meanwhile, in 1718, Christopher Polhem, the famous Swedish engineer, was commissioned by King Charles XII to build a canal from Gothenburg to the Baltic. Unfortunately the king was killed in battle with the Norwegians in the same year and the project languished. A further start on making locks at Trollhätten was made in 1755 but following a landslide that work, too, was abandoned. However, in 1793 a new company took over the construction and on 14 August 1800 the route, requiring eight locks at Trollhätten, was completed to Lake Vänern. In 1832 the canal was completed from Lake Vänern through Lake Vättern to Söderkoping and the Baltic. The Trollhätten flight was then the restrictive factor on the size of vessels which could traverse the full length of the canal and so a new flight was started there in 1838 and completed in 1844. The 1800 and 1844 flights, although on a different alignment, were used concurrently until 1916. By the beginning of the twentieth century it was again necessary to provide larger locks. In 1905 the government assumed control of the canal and in 1910 work on a third different alignment was begun. Completed in 1916, it has locks 87m (285.4ft) by 12.5m (41ft) with 4m (13ft) draught, later increased to 4.6m (15ft). There are three locks in a staircase, then a short pound, and finally a single lock to raise vessels to the upper level. In the 1970s considerable improvements were made to the fairways on the river and the Göta Canal today carries a steady freight traffic as well as providing a regular passenger service.

Also from Lake Vänern leading towards the Norwegian border is Sweden's most beautiful waterway, the Dalsland Canal. Constructed between 1865 and 1869, it has 25 locks in its 255km (158.4 miles) length: only about 8km are artificial cuts, the remainder of the route using natural lakes. At Häverud (where there is a canal museum) the canal climbs one side of the valley down which a stream cascades. Half-way up the valley the canal crosses the gorge in an iron trough and passes along a ledge cut in the rock face before leading to the next lake. Once used by the iron working industry, it now carries little if any commercial traffic.

Another early canal is the Strömsholms Canal from the western end of Lake Mälaren north-west of Stockholm. It was built between 1775 and 1795 and its 108km (67.1 miles) were used for the transport of iron ore. It bypasses the Kölback river and has 26 locks between Lake Barken and Malaren. It is now used entirely by pleasure craft.

THE SOVIET UNION

The Soviet Union includes in its extensive geographical area several river systems which are among the largest in the world, and many have long navigable stretches. Several of the rivers are more than 1600km (1000 miles) long and many have extensive tributaries, in three cases giving drainage areas of over 25 million km^2 each. But the Soviet Union faces the same problem as Canada, the fact that the rivers are frozen over for five to seven months of the year. Most rivers run from a central watershed either north to the White Sea or the Baltic or south to the Caspian or the Black Sea and as their headwaters are reasonably close, and there are tributaries flowing east or west to the main rivers, reasonable links could be made. One of the earliest was inspired by Peter the Great and work started on a Neva (Lake Ladoga)–Volga link in 1703 with a short canal near Vyshniy-Volochok, roughly half-way between Moscow and St Petersburg. An improved line followed which became successful. In the south also at the beginning of the eighteenth century, a start was made on linking the Volga and the Don so that traffic could go to either the Black Sea or the Caspian Sea.

At the end of the eighteenth century new schemes were being proposed including an improved Neva–Volga link much further north, the Mariinski route via Lakes Onega and Ladoga which was opened in 1810. Another route linking the Black Sea and the Baltic was constructed using the Dnieper, the Berezina river and the western Dvina to join the Baltic at Riga. This involved the construction in 1804 of a short canal near Lepel about half-way between Minsk and Vitebsk. Two other canals completed this phase of canal building. The Tikhvine Canal, opened in 1811, is the link between the waterway system to the west from Dereva on a tributary of the Syas river, via the Syas itself and Lake Ladoga, and to the east via the upper reaches of the Mologa river which ultimately flows into the Volga at Nijni-Novgorod (Gorki) and thence to the Caspian Sea. The second, opened in 1828, links the Sheksna river with Lake Kubinskoe and thence by a roundabout river route to Archangel on the White Sea.

Later in the nineteenth century the Moskva was canalized for 160km (99 miles) from Moscow to the Oka, giving improved communication between the city and the Caspian. By the mid-nineteenth century there were more than 80,500km (50,000 miles) of navigable rivers and canals in Russia. Further improvements in the north-west came in 1885 with the opening of a dredged and embanked channel down the River Neva from St Petersburg to Kronstadt.

After the establishment of the Soviet regime in the Soviet Union several waterway routes were developed, but the most important new construction was the Volga–Don Canal. First conceived in the nineteenth century, it was started in 1938 but because of the war was not completed until 1952. It links Volgograd (previously Tsaritsyn and then Stalingrad) with the Don at Kalach, a distance of 101km (62.8 miles).

Since the Second World War the Dnieper has been improved for 3000 tonne vessels and the six locks provided include those bypassing Dneiperpetrovsk (previously Ekaterinoslav). Developments are also taking place in Siberia, mainly in association with the provision of hydro-electric power: the outstanding example is the Krasnoyarsk dam on the Yenisei river. There a double inclined plane has been constructed to lift boats over the dam, one inclined plane leading from the upper level above the power station on to the crest of the dam and a second plane lowering the boats into the tail water below the station. The two planes are not in a direct line, so the caisson or tank, travelling on rails, is turned through 142° on a turntable before the second plane can be traversed.

Opened in 1976, it carried smaller vessels and passenger boats until the first 1500 tonne petrol barge crossed it in 1983.

EASTERN EUROPE

Like the Rhine, the Danube has been one of the main arteries of trade through Europe and improvements have been made intermittently since Roman times when a towpath was cut in the Carpathian gorges to assist the passage of boats through the very difficult sections and small lateral canals were constructed. A proposal by Charlemagne to link the Danube and the Rhine in 793 came to nothing although some digging was done. A thousand years later the construction of the Rhine–Main–Danube Canal is bringing the idea to fruition (see p. 499). In what is now Yugoslavia a link between the river Theiss (Tisa) and the Danube near Bezdan was begun in 1795 and completed in 1802; it was extended in 1856 when a lock said to be the first built in concrete in Europe was included in the extension. From this, the Franzen, or King Peter I, Canal, a branch, the Franz Joseph or Mali Canal, was built between 1855 and 1872.

In 1829 the Danube Steam Navigation Company was formed and operated eventually over the whole length of the river. Difficulties arose at the outfall into the Black Sea as Russia controlled that part of the river and proposals were made in 1837 to build a canal between Cernavodi and Constanta, a distance of 64km (39.8 miles). Water could not be guaranteed at the summit level and the project was abandoned. The concept was revived in 1949 and work began on the link. It stopped in 1953, but was restarted in 1975 under the control of Romania. The canal, taking 5000 tonne vessels and 3000 tonne pushtows and having two locks, was formally opened in 1984. The Romanian government proposes to build a 74km (46 miles) canal from the Danube to Bucharest.

The problem of the Iron Gates gorge has been solved by the construction of a hydro-electric scheme with a bypass canal and this has enabled the river flow to be controlled. With the Danube capable of handling a greater volume of traffic in greater units, proposals are under consideration for a link with the Oder and the Elbe to the north, while to the south the Yugoslavian authorities are actively considering a canal connecting the Danube and the Adriatic which would involve the construction of several inclined planes and tunnels.

In Poland the Elblaski Canal, part of which is river navigation, joins the Vistula at Torun. Built during the period 1844–60, it is 163km (101 miles) long and includes four inclined planes, at Buczyniec, Katy, Olesnica and Jelenie, over which barges are carried dry in cradles. The power to initiate movement is derived from a waterwheel on each plane. Between 1881 and 1885 five of the original seven locks on the canal were replaced by a fifth inclined plane at Calvny. The five planes are still working and are maintained as a tourist attraction.

In the south the Oder navigation is extended into the Silesian coalfield by the Gliwice Canal, originally built between 1788 and 1806 as the Klodnitz Canal and improved in 1930–8, when it was known as the Adolf Hitler Canal, it was given its more appropriate name after the war. Gliwice is an important inland port and handles a large tonnage, some of which goes south to Czechoslovakia and more goes north along the Oder for export. Further north the Bydgoszcz Canal (Bromberg Canal) forms a link together with river navigations between the Oder and the Vistula. The Vistula and the Oder are being improved to take 3500 tonne pushtows and in the mid-1980s a new canal, the Slaski Canal (Silesian Canal) was started. Poland will also receive great benefit from the proposed link between the Danube and the Oder.

SPAIN AND PORTUGAL

In northern Spain the Imperial Canal in Aragon was built as an irrigation canal parallel with the Ebro between Tudela and Saragossa in 1528, but between 1770 and 1790 it was enlarged, the first six barges arriving at Saragossa in 1784. The Castile Canal was intended to link the Douro and the Castilian plain with the coast at Santander, but its navigation was limited to a main line and a branch in the plain. Although started in 1753 it was delayed owing to wars and the first barge did not reach Valladolid until 1835 and Medina de Rioseco until 1849, where a plaque in the church of St Mary commemorates the event. In Portugal major improvement works on the Douro have taken place, including lock construction.

THE SUEZ CANAL

For long the idea of a waterway from the eastern end of the Mediterranean to the Gulf of Suez at the head of the Red Sea had been a mirage, but it became a reality when Ferdinand de Lesseps, who had been French consul in Cairo in the 1830s and had visualized the success that could be achieved by its completion, found nearly 25 years later that Prince Said, whom he had known had unexpectedly become King of Egypt. From him de Lesseps received authority to build a 164km (102 mile) canal linking the two seas. Because of his devotion to railways and the fear that the Suez Canal might be detrimental to the Egyptian Railway engineered by himself, Robert Stephenson vehemently opposed the scheme and his influence led to the virtual British boycott of the proposal. It therefore became a prestigious French undertaking, opened on 17 November 1869 by the Empress Eugènie of France. Britain's attitude changed when the vital importance of the canal to the sea route to India was realized and in 1874 the British government purchased King Said's share in the Canal. Freedom of navigation to all vessels was granted in 1888. After his triumph in creating the Suez Canal de Lesseps went on to plan the Panama Canal (see p. 514). In 1956 the Egyptian Government under President Nasser nationalized the canal but navigation continued. In 1967, as a consequence of the war between Israel and Egypt, the canal was closed. After clearance of minefields, wrecks and other obstructions it was reopened in 1975 and Egypt engaged on a massive reconstruction scheme which was completed in 1980. On average, 55 ships a day pass through the canal.

JAPAN

An interesting canal was built in Japan between 1885 and 1890 to link the navigable lake of Biwa via the Yodo river to Osaka Bay. The Biwako Canal (Biwa River Canal) in its 21km (13 miles) had three tunnels and two inclined planes. The boats were carried on cradles and at each plane there were parallel tracks. The lower incline at Fushima was operated by water power but the upper one near Kyoto was electrically powered by means of Pelton waterwheels (see p. 244). The canal ceased to operate about 1914 but the upper incline is still in existence.

CANADA

For the exploration and early exploitation of trade in Canada the rivers were invaluable as guidelines through the otherwise trackless plains and forests, but as the country became settled and defence contingencies were added to the trading pattern it was clear that improvements to the waterways would ultimately have to take place. Progress along the St Lawrence was hampered by rapids and falls as well as by floods. In 1779 an army captain, William Twiss RE, was instructed to build a canal past some rapids at Côteau du Lac between Cornwall and Montreal. By the end of the navigation season in 1780 a canal 275km (171 miles) long and 2m (7ft) wide had been completed for traffic, although it was still in an unfinished state. Three locks were built, making the Côteau du Lac Canal the first locked canal in the New World, and it was finally completed in 1783. Two other short canals were also built. The original Côteau Canal was superseded by a single lock canal in 1817 and that survived until 1845.

Just above Montreal the Lachine rapids on the St Lawrence formed a barrier to through traffic. In 1680 there had been a proposal to provide a bypass canal, but this suggestion was not considered to be of overriding importance as there was a good portage road between Montreal and Lachine and loads were not heavy. But the war of 1812 and the subsequent opening of the Erie Canal (see p. 511) fuelled the Montreal merchants' anxieties to remove impediments to through waterway traffic. In 1819 they formed a company for building a canal to ensure their control over the trade. Eventually the Lachine Canal was opened to traffic in 1824 with seven locks. This was the first modern canal in Canada.

Between 1843 and 1848 the number of locks was reduced to five and their size increased. It was further improved in the 1880s but the construction in the 1950s of the St Lambert and Côte Ste Catherine locks on the St Lawrence Seaway spelled the end of the Lachine Canal; it was closed to traffic in 1970 but maintained for its leisure use and historical significance.

The 1812 war with the United States had also highlighted the problem of finding an alternative military route between Montreal and Lake Ontario which would not be exposed to attack from across the international boundary of the St Lawrence river. The solution lay in using the Ottawa river and building a canal across from that river to Lake Ontario, but the Ottawa had rapids which required bypassing and three canals were projected for this purpose. Despite military considerations the canals, the Carillon, the Chute à Blondeau and the Grenville were not officially opened until 1834.

Meanwhile a canal had been authorized from Hull on the Ottawa to Kingston, the military base on Lake Ontario and in 1826 Colonel John By, who had worked on the Côteau Canal in 1802, arrived in Quebec to be responsible for building the Rideau Canal. His task was to build a 210km (130 mile) navigation through virgin forest via rivers and lakes virtually unknown, and to do it as quickly as possible. Its construction included the magnificent flight of eight locks rising from the Ottawa river in what is now Ottawa, but which was originally called Bytown in the Colonel's honour. His other achievement was the Jones Fall dam, 18m (60ft) high and 106.5m (350ft) long at its crest. This change in level in the canal is accommodated by three locks. Opened in May 1832, the Rideau Canal is one of the very attractive waterways of the world.

Always the obvious direct route between the sea and the Great Lakes, the St Lawrence dominated the Canadian waterway scene, but between Lakes Ontario and Erie rose the natural barrier of Niagara Falls. From the earliest days it was recognized that a canal with locks would be necessary to pass the Niagara escarpment of 100m (327ft). Although a report at the end of the seventeenth century to Louis XIV suggested a canal, it was realized that the expense would be vast. The question was posed again in 1793, but it was left to William Hamilton Merritt to make the first proper survey in 1818 of a possible route and the impetus was given by the construction of the Erie Canal in the United States (see p. 510) which could have an injurious effect on the Canadian carrying trade. In 1823 an Act was passed and after considerable difficulties the first Welland Canal was opened on 20 November 1829. There were 40 locks, 30m by 6.7m (100ft by 22ft) with 2m (7ft) of water over the lock sills. The waterway used part of the Welland river as its route. It was reconstructed in 1845 with the number of locks reduced to twenty-seven but increased in size to 45.5m by 7.5m (150ft by 24ft 6in). Between 1883 and 1885 it was again reconstructed as the third Welland Canal with locks 85m by 14m (280ft by 46ft) and with 4.2m

(14ft) of water over the sills. Only one of the locks was eliminated in this reconstruction. At the beginning of the twentieth century it was decided to build a completely new canal to much larger dimensions. Although the earlier route from Port Colborne on Lake Erie was retained above the escarpment, a new line through the lower ground was planned to terminate at a new entrance on Lake Ontario 8km (5 miles) to the east of Port Dalhousie, the old entrance to the canal. The number of locks was reduced from twenty-six to seven with a special eighth guard lock at Port Colborne. The locks are 262m by 24m (859ft by 80ft) with 9m (30ft) of water over the sills. This gave a usable length of 233m (765ft) but the Port Colborne lock was 365m (1200ft). Three of the locks formed a massive staircase. Work started in 1913 but because of the First World War was not completed until 1931. Formal opening was on 6 August 1932. Throughout construction the old canal was fully maintained. In 1967 work started on a 13.4km (8.3 miles) channel to move the canal 2.4km (1.5 miles) east of Welland city centre where the canal divided the city in half. Providing a much easier flow for the canal traffic, it was opened on 28 March 1973.

Further west the narrow 112km (70 miles) long St Mary's river links Lake Superior with Lake Huron and there is a variable fall of about 6m (20ft) between the levels of the two lakes. Where the fall occurs stand the twin cities of Sault Ste Marie, generally known as 'Soo'. A canal, though only for canoes, was built here in 1797–8 by the North-west Fur Company with a lock 11.6m (38ft) long. It was destroyed during the 1812 war but later rebuilt. As the river is the international boundary there were difficulties in obtaining authorization for the construction of locks, but eventually this was granted in 1852. The first vessel used the locks in 1855. Although built by the United States they were also used by Canada. Improvements were made in 1871 and 1881. In 1887 Canada undertook its own lock on the north side within its own territory but it was not completed until 1895. Since then the United States has reconstructed its locks and there are now five parallel locks, the largest of which is 42m by 24m (138ft by 80ft).

The Trent–Severn waterway was suggested as far back as 1833 and was based on an old Indian trail. It consists of a number of lakes and rivers linked by short lengths of canal to provide a route between Trenton on Lake Ontario and Port Severn in Georgian Bay on Lake Huron. It was built in stages mainly to serve local needs, but ultimately with the idea of providing a through route for grain traffic. The first boat did not traverse the whole length until July 1920 and by that time it was evident that the canal was not a viable through route for commercial traffic. It does, however, include three interesting features. At Peterborough there is one of two hydraulic lifts of similar design to those at La Fontinettes and La Louvière in Europe (see pp. 487, 493); the other is on the other side of the summit level at Kirkfield. Peterborough has the maximum rise of any lift of this type in the world, with a difference in level of 19.8m (65ft) and was completed in 1904. Kirkfield, with a rise of 15.2m (50ft) was completed in 1907. Peterborough is in masonry and concrete, whereas Kirkfield is a steel girder structure. The third feature is an inclined plane or marine railway installed at Big Chute where the land falls 17.6m (58ft) in about 183m (600ft). This was constructed for small craft when it was realized that there would be no commercial traffic using the through route. It was opened in 1919 and reconstructed to a larger capacity, and also to carry boats horizontally, in 1978. There was a second marine railway at Swift Rapids, but that has been replaced by a lock. Like the Rideau, the Trent–Severn now carries a great volume of pleasure traffic. At the Lake Ontario end at Trenton it connects with a short canal, the Murray, across the Isthmus of Murray designed to make open lake navigation round the isthmus unnecessary.

In the east the Shubenacadie Canal in Nova Scotia is of interest in that Thomas Telford advised on its construction. It crosses from the Bay of Fundy via the Shubenacadie river to Halifax Harbour, 70km (44 miles) with 15 locks. It was started in 1826 and opened in 1861. Also in Nova Scotia a Baie Verte Canal was proposed and later the proposal was changed to that of a ship railway – the Chignecto Marine Transport Railway – but although much of the track was laid across the isthmus and part of the ship lift was built, finances ran out and the project was abandoned.

The greatest Canadian waterway development was the St Lawrence Seaway carried out jointly with the United States. Although agitation for improvements to the river came from the USA after the First World War, and an agreement had been reached in 1932 whereby the two countries would jointly construct the works, the idea was rejected by the USA. A further agreement in 1941 was again rejected by Congress. In 1952, Canada announced that it would undertake the work alone, but on 15 July 1953 the US Federal Power Commission issued a licence to develop the USA share of the project. This was challenged right through to the Supreme Court, which finally rejected the appeal against the licence in 1954. At the same time the President approved legislation to create a St Lawrence Seaway Development Corporation. Work then started to extend deep-sea facilities into the heart of North America. The major work was completed by April 1959, and now only seven large locks (two in the USA), instead of 22, are required to take sea-going vessels inland to Lake Ontario and thus they are able to travel the whole distance of 3768km (2342 miles) inland to Duluth on Lake Superior. The Seaway was formally opened at a joint ceremony on 26 June 1959 by Queen Elizabeth II and President Eisenhower.

THE UNITED STATES

As more and more settlements became established in North America the need for extended communications became evident, and as early as 1661 Augustine

Herman proposed a canal linking Chesapeake and Delaware Bays to save the long distances and dangers involved in making the Atlantic sea passage between them. A more serious proposal came from Thomas Gilpin in 1764, but construction had to wait until after Independence. Even though the Chesapeake and Delaware Company was incorporated in 1799, work did not begin until 1802, only to be abandoned in 1805 and not resumed until 1824. Completed in 1829 and still important, the canal was the first stage of the long intracoastal waterway that was to develop in the next 150 years. It was enlarged in 1935 to a channel 76m (250ft) wide and 8m (27ft) deep and again in 1954 to 137m (450ft) wide and 11.5m (38ft) deep.

George Washington had seen the need for canal communication to the interior and he had advocated a link between the eastern seaboard and the Ohio. After Independence, in 1784 he supported a canal connecting Lake Champlain and the Hudson river; the canalization of the Potomac, again towards a link with the Ohio; and what was to become the Schuylkill and Susquehanna in Pennsylvania. In 1785 the James River Navigation was begun in Richmond, Virginia, to improve the river as far inland as Buchanan. By 1789 part of the navigation had been opened, including probably the first locks in the USA, and by 1816 the 322km (200 miles) navigation was completely open. It was later to be bypassed by the James River and Kanawha Canal, which received its charter in 1835 and reached Buchanan in 1851. It was intended to cross the Allegheny Mountains by a series of locks and a 14.5km (9 miles) tunnel to join the Kanawha river, a tributary of the Ohio, but the mountain section never materialized.

A third early canal was the Dismal Swamp Canal running southward from Chesapeake Bay to Albemarle Sound. This is another link in the intracoastal waterway, completed in 1805 and subsequently, as with the other early canals, improved. Another canal opened on the coastal route, the Chesapeake and Raritan, was completed in 1834 but does not now form part of the intracoastal waterway as it closed in 1932. On the other hand, the Cape Cod Canal, opened much later, still remains.

In 1817 a rather remote vision took practical form and was to influence canal thinking in the whole of North America. This was the decision of New York State to go ahead with the building of the Erie Canal. As early as 1792 the New York State Legislature had authorized the Western Inland Lock Navigation Company to establish a route from the Hudson to Lake Ontario by improving the Mohawk river and linking it to Lake Oneida and the Oneida river. At the same time the Northern Inland Navigation Company was authorized to improve the Hudson to Lake Champlain. In 1796 the Western Company constructed a one-mile canal at Little Falls on the Mohawk but no further progress was made. It was recognized that public funds would be required, but the legislature was not yet convinced. Congress would not support it, but in 1817 New York State had a new Governor, DeWitt Clinton, elected on the programme of building the new canal; on 4 July the first sod was cut at Rome on the Mohawk. The canal was to be 580km (363 miles) long, 12m (40ft) wide and 1.2m (4ft) deep, and the schedule was tight, for it was to be completed in eight years. By 1820 the middle section was complete and it was navigable from Utica to Montezuma. There was disagreement between Buffalo on Lake Erie and Black Rock on the Niagara river as to which should be the western terminus; Buffalo won, with Albany on the Hudson as the eastern terminus. By October 1825, within the stipulated time, the canal was navigable throughout. In that time 83 locks, including five double locks at Lockport on the Niagara escarpment, together with 18 aqueducts across rivers and large streams, had been built. By 1833 an average of one boat every 17 minutes was traversing the canal and transport costs between Buffalo and New York had fallen from $100 to $40 a ton.

The Erie Canal's success soon meant that it was too small for the amount of traffic it generated, so from 1835 to 1862 work proceeded on doubling the width and depth. It had been intended that this work would be completed in five years, but funds were suddenly stopped and it was not until 1854 that money again became available. In 1884 the locks were lengthened to 67m (220ft) and it was proposed that the depth should again be increased, but this last proposal was dropped.

At the beginning of the twentieth century, after rejecting the idea of a ship canal, it was decided to build a barge canal to replace the existing waterway. This was constructed parallel to the old canal between 1905 and 1918 and opened on 15 May 1918. The Oswego Canal, 38km (24 miles) long, joins the Erie Canal to Lake Ontario from Syracuse. The Erie Barge Canal, 560km (348 miles) long, has 35 locks and is the main line of the New York State Barge Canal System. This includes a number of branch canals completed between 1823 and 1857 totalling another 190km (117 miles). It is noteworthy that although the Federal government has financed many harbour and river systems within the USA, not one dollar of public money has been spent on the Erie complex and yet this has been of tremendous economic value to the government. Although freight traffic has fallen off, pleasure use has increased and proposals are being examined to provide year-round availability.

Further south, in the Pennsylvania region, there had been developments expanding the Schuylkill and Susquehanna and building new canals to carry coal from Pennsylvania to New York and other eastern seaboard communities. In 1823 a new waterway was surveyed and in 1829 the Delaware and Hudson was complete over its 173km (108 miles) from Homesdale across the Delaware river to Kingston on the Hudson. It had 108 locks taking 20 tonne boats, and was deepened in 1842-4 and rebuilt ten years later for 130 tonne boats. The level crossings of the Delaware and Lackawaxen rivers were handicaps, so these crossings were replaced by aqueducts almost at the same time the Orb crossing on the Canal du Midi was replaced (see p. 00). Also carrying coal to New York, but by a more southerly route, was the Morris Canal running from Easton on the Delaware to Newark. Conceived in 1824, it was opened in 1832 and then extended to Jersey City in 1836. It was remarkable for having 23 inclined planes, water-powered by turbines, over which boats were carried dry in cradles running on rails. The Morris Canal joined the Lehigh Canal at Easton and in order to pass boats from one canal to the other the Morris was widened following financial reorganization in 1845. With railroad competition it gradually declined and it was closed in 1924.

The Lehigh Canal had started as a river improvement. Arks were loaded with coal and taken down the river to Philadelphia, where the coal was discharged and the boats broken up. There were no return trips. The rapids were passed by what was termed a 'bear trap' lock. This was in effect an inclined plane or water chute down which the boats were launched in a 'flash' of water. The slope was eased off at the bottom to prevent the boats hitting the water too violently and breaking up. The Lehigh Canal was opened in 1823. It was reconstructed in a conventional manner in 1829 and after a very successful period gradually closed down owing to damage from floods, until it finally ceased operation in 1942.

The growing industrial importance of Philadelphia and its relationship with Pittsburgh and the Ohio required improved transport and so in 1825 a canal was authorized between the two cities. The problem was the 600m (2000ft) range of the Allegheny Mountains between them, and it was decided that the route would be partially rail worked. The first section, planned as a horse wagonway and later converted to a locomotive-hauled railroad, ran westwards from Philadelphia to Columbia. From there a canal was built along the Susquehanna valley to Hollidaysburg, and what became known as the Allegheny Portage Railroad began there. This consisted of a series of five inclined planes over which the canal boats were carried to the summit of the ridge at 700m (2300ft) and then the line descended by five more planes to Johnstown on the western side of the range. The cradles on the planes were counterbalanced with steam assistance. From Johnstown a 60-lock canal ran to Pittsburgh. The 630km (394 mile) link was completed between the two cities in eight years and the first boat passed through in 1834. Unhappily the route did not survive long, for it was bought by the Pennsylvania Railroad who closed the western canal in 1863 and most of the remainder by the end of the century. Most of the other canals between the Ohio and the eastern seaboard had also succumbed to the thrust of the railways.

All the foregoing were canals associated with industry but further south there was an attempt to use canals for the benefit of agriculture. South Carolina was sparsely populated but it had a major river system formed by the Santee, Saluda and Wateree rivers. The first canalization, proposed about 1785, led to general improvements between Columbia and Charleston and by 1820 no less than ten small lateral canals had been constructed which carried a great deal of cotton until by the 1850s the railways were taking over. There are still remains of some of these canals and one, the Landsford Canal by the Wateree river, is being restored as a State Park.

The great developments leading into the modern waterway scene were on the Mississippi and the Ohio. The Ohio is formed by the confluence of the Allegheny and the Monongahela rivers at what is now Pittsburgh, and from the beginning of the nineteenth century the industrialists of the town had been trying to get improvements to the river, which was notorious for the treacherous rocks and shoals in its bed. These problems were exacerbated in summer because of periods of drought. With an improved river, traffic would be able to pass down the rivers to New Orleans and thence world-wide. Various proposals were offered for improvements during the century but nothing was done until a powerful figure, General Moorhead, who had rescued the Monongahela Navigation Company, building locks and dams with his own capital, visualized similar improvements to the Ohio but with the injection of public money. This the Federal government would not supply for a long time. Eventually in 1870 work started on the Davis Island Lock and this was the forerunner of the complete transformation of the Ohio. The lock, 183m by 33.5m (600ft by 110ft) incorporated an entirely new feature which was to be adopted in various parts of the world on very wide locks. This was a horizontally travelling gate moving into a recess in the lock wall, with its weight carried on wheels running on rails laid transversely to the axis of the lock. It thus replaced the mitre gates which for the large capacity locks were becoming very heavy and cumbersome. The creation of the lock, which would eventually take pushtows, and the dam alongside provided the control which enabled a harbour to be formed at Pittsburgh, materially assisting the growth of the city.

The Ohio connected with the Mississippi and thus with the Missouri and in all this river system there were roughly 20,000km (12,000 miles) of navigable waterways although, being river navigation, a vast amount of training works had to be undertaken to make them safe. Once steamboats were introduced in about 1812, traffic on all these rivers steadily increased until the onslaught of the railways caused a decline after the 1880s. This decline lasted until 1918, but thereafter traffic again increased following the realization of the economic advantages of water transport. This has been emphasized by the spectacular growth of pushtowing and the capacities involved on the Mississippi and its associated rivers.

One of the tributaries of the Mississippi is the Tennessee which flows north to join the Ohio just east of Cairo, Illinois. Flowing south through Alabama is the Tombigbee river which enters the Gulf of Mexico at Mobile. The Tombigbee is navigable and it had as one of its tributaries the Coosa river, which had also been made navigable by the end of the nineteenth century. The Tennessee had by the end of the Second World War been improved to a (2.7m) deep navigation for 1040km (648 miles) from Knoxville, Tennessee, to the Ohio. It was then considered advisable to link the Tennessee and the Tombigbee rivers by a 375km (234 miles) canal. This was authorized in 1946, but work did not begin until 1971. It was finally completed with locks 183m by 33.5m (600ft by 110ft), the present standard size for major waterways, to accommodate pushtows, and was opened on 18 January 1985, the first commercial tow having passed through a day earlier.

THE PANAMA CANAL

The idea for a canal or other means to transfer vessels across Central America had long been canvassed and various schemes on different sites proposed, but it was not until de Lesseps had called a congress in 1879 that a decision was taken to build a canal across the Isthmus of Panama. A French-sponsored company, with capital raised mainly in France, undertook the work, but by 1888 it was bankrupt. The United States, for political and military reasons, decided to take over the scheme and in 1894 a new company was formed, though there was still argument over the site. Eventually a new State of Panama was created, which granted a concession to the United States of a Canal Zone. Work restarted in 1908 and, amid appalling difficulties including medical ones, it was completed with two flights of locks at Miraflores and Gatun and a single lock at Pedro Miguel through which ocean vessels were hauled by power units on the banks. It was formally opened on 15 August 1914 and remains one of the major ship canals in the world.

CANAL AND RIVER CRAFT

Virtually every waterway system had its own designs of craft built largely for the convenience of the nature of its traffic, the dimensions of the waterway and the peculiarities of the system. Collectively, but usually inaccurately, the vessels have been generally known as barges but locally as well as technically they had distinctive names. Their capacity has ranged from the shallow draft 25 tonne capacity boats on the Chelmer and Blackwater Navigation to the 2000 tonne Rhine barges. On British canals with narrow locks and limited dimensions in tunnels and bridges the boats with a beam of approximately 2m (7ft) were known as narrow boats to distinguish them from the wide boats used on the broader waterways.

On the earlier waterways the boats would be hauled, often by gangs of men; later by mules or horses. Where the canal passed through a tunnel in which there was no towpath, recourse had to be had to legging. This was carried out by professional leggers who worked in pairs lying on their backs on a plank which extended across the beam of the boat. They would then 'walk' along the sides of the tunnel, thus propelling the boat. The alternative was to shaft or pole the boat through the tunnel by pushing with a pole against the tunnel wall.

Even on confined waters and rivers it was possible with skilful handling to use sailing barges, or, to quote their more specific names, keels, wherries, sloops or *trekschuiten*. They continued to be used, and indeed still are, even after the development of a steam propulsion which goes back to the end of the eighteenth century, for example Fulton's experiments at Worsley, near Manchester, in 1797. Diesel-type engines were introduced at the end of the nineteenth century and by the First World War they were in regular use.

Apart from unit propulsion with the engine mounted integral with the vessel itself installations were adopted where mechanical haulage was provided by traction units running either on rails or on roads along the bank of the waterway, although these never found favour in Britain. Towards the end of the nineteenth century the newly available electricity was adopted in a similar way to an electric tramway system, with the traction unit taking its power from an overhead line as was installed on the Brussels to Charleroi Canal in the 1890s. The tractor was attached to the barge by a cable and would haul the barge until it met another approaching from the opposite direction. Cables would be exchanged between the tractors which would then reverse direction while the barges continued as before. Elsewhere a similar shuttle service was performed by diesel tractors. Non-powered boats could thus be hauled without difficulty.

Yet another system was to lay a cable or chain along the bed of the canal and a tug, by means of winches mounted on its deck, would haul itself along and at the same time tow a string of barges behind. This system was particularly useful through tunnels but it was also used elsewhere, as on the River Neckar in Germany.

With the large present-day barge the almost universal mode of propulsion is by a powerful diesel engine on the vessel itself or by a diesel-powered tug which will either tow or push a group of barges. The formations of barges which are pushed are known as pushtows and the constituent barges are so tightly assembled that the pushtow becomes in essence a large unified vessel. The advantage is that the expensive power unit is not left idle when the barge is unloading or discharging. The barges themselves are, of course, non-powered and types have been developed which can be carried on board ship across the ocean and then linked together to proceed to their destinations along inland waterways.

THE CONTEMPORARY SCENE

The use of waterways in all countries has in general shown a steady progression through four phases: first, the employment of natural water resources – rivers, lakes, estuaries etc – with little or no adaptation; second, improvement to remove dangers from their courses, possibly to bypass rapids and rocky areas and to even the flow of the current by dams and weirs; third, the construction of artificial canals to link river valleys and to provide local transport; fourth, the enlargement of existing rivers and canals to accommodate mass national and international transport of goods.

The first three phases arose from the need, in the absence of other satisfactory modes of transport, to provide a reasonably integrated system by which goods could be moved within commercially acceptable time limits. The fourth phase, while including that object, also involves the growing realization of the economic saving in energy costs offered by water transport. This saving is partly demonstrated by the relative energy requirements (with pipeline costs as a baseline) per tonne/kilometre for various modes of transport: pipeline 1, water transport 2, rail 3, road, 10, air 100. An example of the side benefits in canal construction and operation is illustrated in the advance incorporation of hydro-electric power generation plants on the line of the new Rhine–Main–Danube canal (see p. 494), where the sale of power is partly financing canal construction and will reduce operating costs.

The fourth, current, phase is therefore not a mere expansion of an existing and otherwise outmoded means of transport but a stage in the evolution of a co-ordinated system which can operate efficiently within modern economic constraints. In addition, moving bulk freights to waterways relieves congestion in urban areas, and many countries are realizing that improvement of an existing water route is cheaper than building a new motorway – and that it is less costly to maintain. Moreover, those navigations in current use serve a multi-purpose function of water supply, drainage and environmental benefits, as well as transport.

Intensive land use in Britain has militated against developments which have been seen elsewhere in improvements to waterways. Even here, however, where the decline in commercial traffic almost led to the complete abandonment of the system, the environmental benefits have been recognized. At the critical moment there began an upsurge in interest in leisure activity and the recognition of the need for greater facilities. The realization that canals could offer a multi-purpose and desirable source of leisure enjoyment was stimulated by L. T. C. Rolt's account of his experiences on the waterways in his book *Narrow Boat*, first published in 1944. In turn this enthusiasm led to the formation of the Inland Waterways Association and the burgeoning voluntary involvement in the restoration of redundant and abandoned waterways.

The restoration movement has now spread to many other countries, and especially in the United States. Even in those countries where there is a major commercial activity and new waterway construction those canals which no longer carry intensive freight traffic, for example the Canal du Midi and the

Nivernais Canal in France, are supporting an important growth in the leisure industry. Other minor and branch canals such as the Canal du Berry, also in France, are undergoing restoration and maintenance. There is now virtually world-wide recognition of the tourist potential and ecological value of waterways which for so long have been regarded solely as of importance in the industrial and commercial economy of a country.

GLOSSARY

Balance beam the heavy beam projecting landwards from a lock gate which, in providing the leverage for opening the gate, balances the weight of the gate. Found especially on British lock gates

Cylindrical sector rotating gate a modern type of lock gate which rotates horizontally through 90° from the vertical position where it closes the lock to a horizontal position in the bed of the canal to allow boats to pass over it

Flash a body of water suddenly released to provide greater depth to allow navigation over shoals and shallows in a river

Flash lock an arrangement on rivers whereby the water could be held or penned back by a removable barrier across the flow. When a boat travelling downstream wished to pass the barrier was removed and a flash of water rapidly flowed through the opening carrying the boat over shallows below the barrier. For vessels travelling upstream a wait was necessary until the flow had diminished and then the boats would be manually hauled or winched to the upper level. Also known as staunches

Inclined plane a device by which boats can be carried from one level to another either on cradles or in tanks travelling up and down a slope

Lift a device on the same principles as an inclined plane but with the tanks rising and falling vertically

Lock a means by which boats can move from one level of a navigation to a different level

Paddle door a gate either pivoted or capable of rising and falling for opening and closing the orifice through which water enters or leaves a lock

Péniche the standard 300 tonne French canal barge. A law passed in 1879, under the direction of Charles de Freycinet, included minimum dimensions for the French waterways: depth 2m (7ft), locks 38.5m by 5.2m (127ft by 7ft), bridge clearance 3.7m (12ft). Barges which conformed to these constraints were said to be of Freycinet type

Pound lock a chamber, generally in brick or masonry, with gates at each end in which the water is impounded and through which boats pass from one level to another. The change is effected by the opening of paddles, paddle doors or sluices to allow water to enter or leave the chamber when the gates are closed. The gates are opened when the water level in the lock is the same as that either below or above the lock. The gates can be single, as on some narrow canals, or double so that the gates meet at an angle to withstand the water pressure. In this case they are called mitre gates. Also used are guillotine gates, which are raised and lowered vertically; radial gates, which turn on a horizontal axis to lie under the water below the draught of the vessel; or sliding horizontal gates which retract into recesses constructed in the side of the lock

Shaft lock a modern type of very deep lock with a fixed structure across the lock at the lower end. This structure houses a guillotine gate which rises from the canal into the structure and thus exposes a tunnel through which the boats pass to the lower level of the canal. Normal gates are provided at the upper level but gates would be impracticable at the lower level owing to the height of water impounded in the chamber

Staunch, see **Flash lock**

Summit level where a canal passes across a ridge from one valley to another the highest part of the canal is called the summit level. Into this generally run the feeders which supply the principal water for the locks on both sides of the ridge

Tide lock a lock provided where a canal enters the sea or a tidal estuary to allow vessels to pass at different states of the tide

Transporter bridge a type of bridge, generally in flat country, a number of which were built at the end of the nineteenth and beginning of the twentieth century, sufficiently high to allow tall-masted vessels to pass under the structure. A frame ran on rails at the upper level of the bridge and from this was suspended a cradle on which road traffic crossing the canal could be carried. The cradle passed to and fro between piers on the banks and thus shipping was not impeded by the passage of road traffic. Examples can be seen in England at Middlesbrough across the Tees and at Rendsburg in Germany across the Kiel or Nord-Ostsee Canal

FURTHER READING

The literature on world canals and waterways as a complete survey of the subject is regrettably limited. For a modern overview of the past and present scene the only comprehensive work is Charles Hadfield, *World canals. Inland navigation past and present* (David & Charles, Newton Abbot, 1986). A much shorter account is given in Charles Hadfield, *The canal age* (David & Charles, Newton Abbot, 1988). An earlier work dealing primarily with the engineering features is L. F. Vernon-Harcourt, *Rivers and canals*. Two volumes, 1896. Again on special engineering structures is David Tew, *Canal inclines and lifts* (Alan Sutton, Gloucester, 1987).

For individual countries both the quantity and the quality is very variable. The British Isles is admirably covered in Charles Hadfield's series *The canals of the British Isles* (David & Charles, Newton Abbot, 1967–72). Canada is fully dealt with in Dr Robert Legget *The canals of Canada* (David & Charles, Newton Abbot, 1976). Professor Joseph Needham's monumental work *Science and civilisation in China* Vol. IV.3 (Cambridge University Press, Cambridge, 1971) contains a detailed account of the improvement of rivers and construction of canals in China. Roger Calvert *Inland waterways of Europe* describes European waterways with a greater emphasis on cruising rather than on detailed history.

10

PORTS AND SHIPPING

A.W.H. PEARSALL

OAR AND SAIL

Antiquity to the Middle Ages

Man's first attempts to travel by water, most likely on rivers and lakes, were presumably by sitting on floating logs; in time it was discovered that a log could be hollowed out, by cutting or burning, and that a raft could be constructed by tying branches together to make a floating platform. With the ability to work materials came the evolution of a recognizable boat, with a supporting keel and sides made up of wooden sections. Other available materials included wicker, reeds and animal hides, from which coracles – circular, oblong or square – could be constructed. All these craft could be propelled: by paddling with the hands or with a short, flattened piece of wood; by rowing, with two longer wooden oars, wider at one end; or by sculling, working a single oar over the rear end of the vessel; and all could be to some extent steered by the action of paddle or oar against the water. Such simple boats have continued in use to the present day in unindustrialized societies: the coracle was common in inshore waters around the British Isles within living memory.

The archaeological record demonstrates that by the first millenium BC larger vessels were made up of the elements familiar in wooden boats today. The foundation was a sturdy timber keel, running the length of the hull, into which bow and stern post were scarfed – joined by bevelling the two members and keying one into the other – with strakes, or side planks, running from bow to stern and secured, successively, with thongs, trenails (hardwood dowels or pins), and iron nails. As hulls grew too long to allow the use of single timbers for the strakes, 'U'-shaped frames were added as transverse shapers and strengtheners, and linking their tops with beams as further reinforcement gave thwarts, on which rowers could sit, or as support for decking. The two standard forms of outer planking also evolved at this early stage: carvel, where the strakes are shaped to fit edge to edge, and clinker, where each overlaps the one below it. A variation on the framed type of construction was found in ancient Egypt, where a shallow, lightly built hull was reinforced by overhead arched trusses, a method used for river craft into modern times.

Vessels were small, less than 30m (98ft) long, and virtually double-ended; many were solely oar-driven, although sea-going ships might have a square sail set on a single mast; steering was by means of a large oar over one of the stern quarters. In the classical era the two main maritime divisions were already established, between merchant and fighting ships. The trader had a greater beam:length ratio, and was probably decked to protect the cargo. War galleys were longer and narrower, for speed, and given manoeuvrability by numerous oars, usually handled by slaves. The largest Greek warship, the trireme, is thought to have had its oars arranged in three tiers, but this layout remains a matter of dispute. Classical war galleys were equipped with a pointed ram at the bow, used to hole enemy ships, and carried foot soldiers and archers. This type of combat ship was used in the Mediterranean into the sixteenth century – Christian and Turkish galleys fought at the battle of Lepanto in 1571 – and its general configuration was echoed by the Viking longship of the Dark Ages.

In fact ship design changed only in detail in the centuries up to the mediaeval era. Individual advances are difficult to trace, but local variations in hull form to suit local conditions began to evolve, and increased trading in bulk commodities demanded bigger ships which could carry larger cargoes over greater distances. Mediaeval merchant vessels included the hulk, with a rounded hull, bow and stern, and the cog (Figure 10.1), also rounded at bow and stern, but with a flatter bottom more suited to the tidal and shallow coastal waters of northern Europe. The sternpost rudder, directly controlled by the horizontal tiller, was introduced in the mid-thirteenth century. Warships were fitted with fore- and aftercastles, superstructures at bow and stern from which fighting men operated, and these were also adopted in merchant ships.

With increased size, oars were no longer a practicable means of propulsion. The square sail, set on a transverse yard but without a boom to stiffen its lower edge, was probably still the commonest sailing rig, although it has been suggested that it was used in the manner of a fore-and-aft sail as much as for simply running before the wind. To suit wind conditions, sail area could be reduced by tying tucks in the canvas with reefing points, or increased by lacing on additional strips of canvas, later called bonnets. The triangular fore-and-aft lateen sail, familiar in the Mediterranean continuously from antiquity to the present day, was adopted in Europe for the carrack, the foremost trading vessel of the thirteenth and fourteenth centuries. This was square rigged on foremast and mainmast, with the lateen sail on the mizen (stern) mast, and with a cog hull.

Figure 10.1: Model of a mediaeval 'cog'.

Fifteenth to eighteenth centuries

Quite suddenly, and by a chain of development which it has not yet been possible to trace fully, the three-masted square-rigged ship succeeded the carrack in the mid-fifteenth century (Figure 10.2). The additional masts and increased sail area made them more manoeuvrable than their single-masted predecessors and better able to sail to windward, square sterns provided improved accommodation, and they became progressively larger. They offered the technical conditions for oceanic voyages, and opened up the great age of discovery.

The original form of the three-masted ship never changed fundamentally, although it was greatly modified over the course of time. It carried a fore- and mainmast set with square sails only, the fore usually slightly shorter than the main, and a much shorter mizen with one square sail (later more) and a fore-and-aft sail set aft of the mizen and later known as a spanker. Forward there was a bowsprit, inclined at about 30° and projecting beyond the stem, carrying headsails. Originally the headsails were square, set on a yard under the bowsprit and on a small mast rigged in a perilous-looking way at its end. Towards the end of the seventeenth century these were replaced by triangular fore-and-aft sails set on stays running from bowsprit to foremast, which did their work far

Figure 10.2: A Flemish ship of *c.*1560.

more effectively; later, similar sails were set between the masts. Other improvements came with the increasing size of ships, and with the need to carry more sails for greater speed; individual sails became larger and increased from one or two to three or four on each mast by 1800. The masts themselves, becoming loftier and more square in proportion, soon outran the average tree and had to be assembled from two or three sections, one above the other; often, too, the largest sections had to be 'made' from smaller timbers. Sails were cut better and set more tautly to obtain the best from the wind: blocks, containing pulleys to increase manual capacity when hauling on ropes, appeared at an early stage, and multiple blocks with metal sheaves and bearing made running rigging easier to work.

Changes in hull design also improved the qualities of ships. The basic layout, with a deep waist between high castles, remained, but the forecastle was soon much reduced in height. The aftercastle, later known as the poop or quarterdeck, remained somewhat more pronounced and provided accommodation for officers and passengers, if any. The sheer, or fore-and-aft curve of the deck, was also gradually reduced, which greatly reduced the wind resistance of these ships and made them more weatherly. To accommodate the mizen mast, tiller

Figure 10.3: A Dutch 'fluyt' of the early seventeenth century.

steering had given way in the fifteenth century to the whipstaff, a vertical lever which operated a yoke attached to the rudder; this in turn was succeeded in about 1710 by the wheel and tackles, making ships much more manageable. For example, in 1700 flag officers of the British navy were still making sure that the 'great ships' were safely in harbour by the end of September; but in 1759, Admiral Hawke was able to maintain the blockade of Brest far into November and then fight a decisive battle close inshore in Quiberon Bay in a rising gale.

During the period from 1500 to about 1850 the average merchantman was designed along similar lines to the contemporary warship, although at 100–200 tonnes it would have been much smaller. However, much uncertainty still exists about the detailed layout of early merchant vessels, at least until the emergence of the Dutch fluyt at the end of the sixteenth century (Figure 10.3). This was a capacious three-masted vessel, rounded at the stern, cheap to build and run, without armament and so needing only a small crew. A sound bulk carrier, it was the basis of Dutch commercial supremacy and was copied by other seafaring nations. The largest merchant ships were the East Indiamen (Figure 10.4), of all nations, usually 500–600 tonnes in the seventeenth and eighteenth centuries, but rising to 1200 tonnes in the 1790s. These tonnages compare with the 500 and upwards of a small naval frigate, rising to more than 2000 for a first rate ship of the line.

Such big ships were approaching the limits for wooden construction, and schemes were being evolved for strengthening them. Gabriel Snodgrass

Figure 10.4: A merchant ship of the late eighteenth century.

introduced a straighter side to the hull and a flat upper deck, much improving the strength of the whole, and Sir Robert Seppings, master shipwright at the Chatham naval dockyard, developed a system of diagonal bracing to increase the strength of wooden hulls. In actual hull form there were few developments, although Sir Anthony Deane in England, Frederick af Chapman in Sweden and various French naval architects made advances in the theoretical principles of sailing performance. (French ships had a good reputation for speed, but were often found to be weak in construction.) The Baltimore clipper (in fact a schooner), which emerged in the early nineteenth century, represented the most important development to date in hull form: its long, low profile, with sharply raked stem and overhanging stern counter, resulting in the minimum area possible in contact with the water, made these revolutionary vessels the fastest then afloat.

Another important innovation of the eighteenth century was coppering, or sheathing ships' bottoms with copper plates (earlier experiments with lead having proved unsatisfactory) to protect them from boring by shipworms and fouling by barnacles and weed. Once it was recognized that copper nails must be used, to avoid the galvanic reaction that occurred between the copper plating and the iron bolts at first used to attach it, coppering much improved ships' speed, as well as reducing the need for maintenance, and it was widely adopted after 1780.

The nineteenth century

From the early 1800s the very considerable increase in the volume and extent of trade, especially to more distant parts of the world, created economic conditions that encouraged both more and still larger ships and new commercial methods. By 1815, although the majority of merchant ships continued to be built along conventional lines, some significant changes began. Several trades demanded greater speed, sometimes, as in the tea trade, to reach markets quickly, in other less reputable activities, such as the slave trade, to elude pursuers. Many more passengers were seeking conveyance, and regularity of performance began to be important. Although steam would not become a serious competitor on most ocean voyages until around 1850, economical working became more necessary as the expanding markets stimulated increased competition alongside their many opportunities.

The square-rigged three-master was initially supplanted for coastal trading, by the brig, similarly rigged but with only two masts, but the sharp and fine Baltimore hull form, schooner rigged (fore-and-aft sails on two or more masts, with square topsails on the foremast), steadily replaced both ship and brig. For longer voyages the same fast and economical hull was applied to larger and larger square-rigged vessels, for their size slowly but steadily increased. These fast square-riggers were the true clippers, famous in the China tea trade from the 1840s. They were also run by some of the Atlantic packet lines, which carried mail under government contract and therefore needed to offer speedy, reliable and regular service; here, too, the threat posed by steam made speed crucial. The discovery of gold in California in 1848, and in Australia in 1852, created a demand for passages at any price and, as steamships could not yet undertake such long voyages, the clippers, mostly built in the USA of softwood, prospered. Even after the boom years clippers, now mostly British built, continued until about 1870 in the China trade and 1890 in the passenger and wool trades to Australasia, and their example encouraged smarter sailing in general.

After 1860 iron rapidly replaced wooden construction for deep-sea vessels. The larger sizes already in existence were too much for wood – the American clippers quickly wore out – and by the early 1870s ships of over 100m long were being built. Concomitant improvements in rigging were needed: sails had reached their maximum practicable size and were subdivided, and were fitted with patent reefing systems such as Cunningham's, which enabled the area of sail to be reduced without sending so many men aloft. Iron or steel masts and yards, and wire rigging, replaced wood and hemp, and sometimes sailing ships had steam-powered donkey engines to drive their winches, all of which allowed crews to be reduced.

These improvements helped sail to remain competitive, even after the opening of the Suez Canal in 1869 (see p. 505) enabled steamships to take over

Figure 10.5: An iron clipper of the 1870s – the *Piako*.

more and more of the main trades between the Orient and Europe. In the 1870s four-masted ships and barques (three-masters, square-rigged on fore- and mainmast, fore-and-aft rigged on mizen, see Figure 10.4) were extensively built, largely for bulk cargoes such as grain and nitrates. Later a few five-masters were built. Big sailing ships of 1500 to over 3000 tonnes remained profitable for bulk trades and were built up to the 1890s, a few even up to 1905, but after that date their decline was rapid. A few survived in the Australian grain trade and the Baltic timber trade between the wars.

The other characteristic type of the last days of commercial sail was the schooner (see p. 524), still commonly built of wood until 1900. Small coastal schooners remained very active until 1914, while in North America some very large craft of this rig were built for coal and timber carrying, many with four or more masts, one with seven.

Since the 1970s evidence of an increasing energy crisis has directed attention once more to the use of wind power for shipping. Taking advantage of aeronautical discoveries, controllable sails of aerofoil type have been fitted to several otherwise conventional cargo vessels, but the economic benefits have yet to be measured.

STEAMSHIPS

Paddle steamers

Paddle-wheels, operated by men or animals, are said to have been used by the Chinese to power small craft at an early date, and there were vessels driven by treadmill in European waters during the seventeenth century. Early attempts to propel boats by means of steam-driven paddle-wheels were those of Denis Papin (1707); the Marquis de Joffroy d'Abbans (1776), who in 1783 used a horizontal double-acting steam engine developed independently of Boulton and Watt; and the American John Fitch (from 1786), whose experiments included the use of steam-driven oars. Financial backing, however, was not forthcoming until William Symington's *Charlotte Dundas*, driven by a stern paddle-wheel between two hulls, became the first commercial steamship, operating on the Forth and Clyde Canal from 1802. Other pioneers included Robert Fulton, whose *Clermont*, with a pair of side paddle-wheels, ran commercially on New York's rivers from 1807, and Henry Bell, with the first coastal steamer, the *Comet* of 1812, which ran off the west coast of Scotland. By about 1825 there was an extensive system of steamship services around the British Isles and to Europe, as well as on American rivers. The majority of these vessels were side-wheel paddle steamers with low-pressure non-condensing engines, conforming (with suitable strengthening for engines and boilers) to normal wooden ship layout.

The first transatlantic steamship crossing has been claimed for the American *Savannah* as early as 1819, but she was a hybrid vessel and completed most of the voyage under sail. Among early eastward crossings under steam, the twenty-five-day voyage of the *Royal William* in 1833 is notable. However, transatlantic steam navigation as a commercial reality dates from 1837, with both the *Sirius* and Isambard Kingdom Brunel's first marine essay, the *Great Western*, designed to extend the Great Western Railway from Bristol to New York.

The British government, recognizing the political and commercial value of fast, regular communication with distant regions, was quick to support the establishment of four subsequently famous shipping companies: the Royal Mail Steam Packet Co (1839) to the West Indies and South America; the Peninsular Steam Navigation Co. (1840, later the Peninsular and Oriental) to India and later Australia and the Far East; the Cunard company (1840) to North America; and the Pacific Steam Navigation Co. (1839) for service along the west coast of South America. These routes usually employed ships of 1500–2000 tonnes with large engines, usually of side-lever type, by now fitted with jet or surface condensers which conserved fresh water.

PART THREE: TRANSPORT

Figure 10.6: An early paddle steamer – the *Clarence*.

Iron ships and screw propulsion

The Staffordshire ironmaster John Wilkinson and others had experimented with iron hulls before the end of the eighteenth century, using the methods of wooden ship construction, but the idea was long regarded with scepticism. By the 1840s, however, even sailing ships were beginning to outgrow the capabilities of wood; it was becoming evident, too, that the strength of iron would allow the building of hulls that were not only larger but which could be clear of structural members, improving cargo capacity and handling as well as the flexibility of ship layout. The screw propeller seems to have been invented almost simultaneously in 1836, by Francis Pettit Smith in England and John Ericsson in Sweden; in 1845 the British Admiralty conducted trials between two identical frigates, the *Alecto*, with a pair of paddle-wheels, and the screw-driven *Rattler*, in which the latter was clearly superior.

Brunel's second ship, the *Great Britain* of 1845, was a very large (over 3000 tonnes) iron screw ship of advanced design, with the first watertight bulkheads. The minor damage she incurred on running aground in Dundrum Bay, Ireland, in 1847 demonstrated the strength of her hull construction. His *Great Eastern* of 1858, designed to run to the Far East, was even more far-sighted, a vast ship of nearly 19,000 tonnes and more than 200m long, with both a screw and a pair of paddle-wheels to provide sufficient power. Although both ships were technically successful, they were too large for the commercial needs of their time and were not emulated.

From about 1860 to the present day, merchant ships and warships followed different lines of development and must be considered separately.

MERCHANT SHIPPING

Apart from a short period when composite construction (iron frame with wooden planking) was popular, after 1860 only iron ships were built for ocean work, to be replaced from 1880 by the steel made widely available by the Siemens-Martin process (see p. 171). While hull structure followed in essentials that of wooden ships, with keel and ribs to which the plates were bolted, later riveted, iron and steel sections were far less bulky than wood. It was also now possible to install watertight bulkheads and double bottoms (both pioneered by Brunel) and to adapt the hull for a variety of specialized requirements, whether for cargo, for a particular route or, as became increasingly the case, to minimize harbour and other dues calculated on tonnage. In the bigger ships made feasible by iron, multiple decks could be provided for more and better passenger accommodation (a trend that had begun with the largest wooden Atlantic packets), and steamers with three, four and eventually more decks were built.

With cargo vessels the introduction of the screw propeller rapidly brought steam to the forefront. The original bulk carrier steamships, built by Charles Mark Palmer and others for the coal trade between the north-east of England and London, were one-deck ships and the first to use water ballast tanks to reduce turnaround time in port, an invention attributed to John McIntyre. Very quickly these ships extended their activities further afield, becoming the predecessors of the tramp steamers. For regular trades more complex ships appeared with several separate holds and one or two lower decks, allowing cargo to be stowed according to its type or destination.

The vast majority of these steamships were driven by screw engines. Paddle steamers were retained until the early 1860s by the ocean contract packets, and until well into the twentieth century by river, coastal and cross-channel passenger steamers, especially where high speed was required and where the diameter of a large single screw was too great for the available draught of water. On such routes, too, the availability of fuel was not a major consideration, but on longer voyages it virtually governed the extension of commercial steam navigation, other than the subsidized mail services. A world-wide network of coaling stations was established to replenish the bunkers of steamships on ocean voyages – the lack of possible sites for such facilities significantly delayed the introduction of steamship services to Australasia – and the supply of these ports provided in itself an extensive trade for the preferred Newcastle and South Wales coal as well as from other sources. The development of more economical engines and boilers progressed steadily to answer this problem.

Attempts were made in the 1850s to use steam twice, but it was not until the mid-1860s that the famous Holt engine and other designs brought the compound engine into practical use combined with improved boiler construction to allow higher pressures. Holt's Ocean Steamship Company's service to China in 1866, combined with the opening of the Suez Canal in 1869, marked the death-knell of sail on the routes to India and the Far East in favour of compound steamships. From this period, too, the tramp steamer emerged, delivering and picking up freight in an *ad hoc* manner from port to port and making serious inroads into the bulk cargoes carried by large sailing ships. With the development of the triple expansion engine (see p. 286) about 1880, the last bastions of sailing ships as regular passenger and high quality cargo carriers were threatened – the routes to Australia and New Zealand. Both countries, as well as Argentina and the Mid-west of the United States, also provided more bulk cargoes for the tramp steamers, such as grain to feed the industrial populations of western Europe, and refrigeration developed to the point where large consignments of frozen meat in fast cargo ships became a prime business.

In the same period, the 1880s, engines for fast ships were produced, running at higher revolutions and much more compact, and in the 1890s quadruple expansion engines came into use. Twin screws had been tried earlier, with limited success, but also in the 1890s they were widely adopted and, to provide increased power for larger and faster ships, triple and quadruple screws were introduced. The great expansion in trade, and in commercial methods, allowed larger cargoes to be readily handled and opened the way for the economic use of the much bigger ships already shown to be feasible by the *Great Eastern*. In the 1870s the long ship was pioneered by the Belfast shipbuilders Harland & Wolff, with a length: beam ratio of 9 or 10:1, and from the 75m (248ft) of the 1860s the length of the largest ships rose to more than 120m (394ft); an especially rapid enlargement during the 1890s brought lengths up to 200m (656ft) or more, with beam increasing rather more in proportion to around 7 or 8:1. By 1914 great liners of 240m (787ft) were in service. Until the 1890s virtually all the passenger accommodation was in the hull, but as facilities grew more and more luxurious it was sited on the hull in superstructure. As more power needed more boilers, the great 'funnel race' began, the number of funnels being regarded as an index of quality.

Further important changes in the engine room accompanied these advances to produce the merchant ship as it was to remain until the 1970s. The water-tube boiler was developed in various forms and, after much trouble, eventually became a reliable steam producer, especially when combined with the Parsons steam turbine (see p. 290). The early turbines were directly coupled to the propeller, but by 1914 gears had been developed to a state where they could be inserted into the shaft to give slower and more economical propeller speeds. In some ships low-pressure turbines were fitted together with reciprocating engines.

From the 1880s, Russia used indigenous petroleum fuel for ships' boilers, and by 1914 copious supplies of oil were being produced by the countries around the Persian Gulf. Its advantages as a fuel, and its ease of handling and cleanliness compared with coal, ensured its popularity; and after 1918, in more stringent economic conditions, the savings it offered, in time and in the need for numerous stokers, made oil the most used fuel for steamships.

It was also the fuel of a new and serious competitor to steam – the internal combustion or diesel engine (see p. 303ff.), first used in Russian vessels around the Caspian Sea. Much development work by the Danish firm of Burmeister & Wain made it suitable for ocean-going vessels; the *Selandia* of 1912 was the first deep-sea ship fitted with diesel engines, which made rapid progress especially among Scandinavian owners, while its application to submarines was crucial.

Electric power was also adopted for ships, but with less widespread success. Current was generated by either steam or diesel engines and fed to motors which were small enough to occupy otherwise little-used space aft. While the installations were technically successful, electric drive was found to be uneconomic except in specialized cases where exact control was important, such as ferries or cable ships.

The coal-fired steam engine did not remain untouched in this period of diversity, for many improvements were made in both reciprocating and turbine types, such as the use of higher steam pressures, poppet valves and other refinements such as mechanical stokers with pulverized fuel.

Types of ship did not, in general change greatly, although there were many modifications of detail. The main novelty was, of course, the appearance of the oil tanker in large numbers, the first having been built in 1886. From the earliest days these vessels had their engines aft, a feature which became increasingly common on small coastal steamers. For safety reasons tankers had to be closely subdivided into numerous compartments, and they were equipped with expansion tanks and pumping systems to handle the cargo. The new Isherwood longitudinal system of construction, in which much of the strength is in stringers running fore-and-aft as well as in the more usual frames, was found especially advantageous for tankers. The inter-war period also saw a considerable increase in the number of refrigerated ships, not only for meat and dairy products but also for fruit.

Other new methods of building were introduced, many to be of great utility in the emergency building programmes of the Second World War, such as the American Liberty ships. These included welding, instead of the time-honoured riveting of plates and sections, and large-scale prefabrication in shops away from the slip. Another change was in the use of lighter-weight materials for superstructures.

In the post-war years surprisingly few radical changes in ship design were evident, apart from the general adoption of a more streamlined appearance pioneered by more enterprising owners in the 1930s. Crews now needed better quarters, entailing more superstructure, while advanced navigational and radio equipment developed during the war years was generally fitted. The diesel engine steadily improved in efficiency, although gas turbines seemed to have little application in merchant as compared to fighting navies. Thanks to steel hatch covers and improved methods of construction, ships could now be given larger hatches than ever before, with consequent improvements in cargo handling.

The growth of air travel, and increasingly stringent safety requirements, steadily drove the great passenger liners out of operation or into the cruise business, for which new and luxurious, but smaller, vessels were built. From the 1960s onwards the cargo ship, refrigerated or otherwise, steadily gave way to the container ships, very large and fast, with holds designed to carry cargo of all kinds in internationally standardized rigid containers which ports of all major nations have been adapted to handle. The old tramp ship has been replaced by the much larger bulk carrier working to specialized terminals. Oil tankers have grown to vast size: the very large crude carrier (VLCC) attains 382,500 tonnes and ultra large crude carriers (ULCCs) of up to 408,000 tonnes have been built. All these new types adopted engines aft and, progressively, diesel engines, especially after the increase in oil prices of the 1970s.

WARSHIPS

The sailing man-of-war had developed slowly; a seaman of the seventeenth-century Dutch Wars would not have been too unfamiliar with a naval ship of the 1840s. The only essential difference from the contemporary sailing merchant vessel was in armament: numerous guns were mounted along the naval ship's sides, on two or three decks for a line of battle ship, only one on frigates or sloops, and also on forecastle and quarter-deck in larger types. Naval ships were rated according to the number of guns they carried: three-deckers 90 or more, two-deckers (the majority of ships of the line) 70 to 80, frigates 26 to 60, sloops 20 or fewer.

The introduction of steam and iron had a revolutionary effect on warship design. Steam-driven paddle-wheels were fitted to naval vessels from the late 1820s to supplement their sail power. Because the wheels and boxes took up most of a ship's sides, pivot guns (used earlier to some extent in smaller ships) were more often mounted, as well as larger guns fore and aft; many of the guns of this period fired explosive shells, as proposed by the French General

Paixhans. Paddle-wheels, however, occupied much valuable space and were exceedingly vulnerable to gunfire; once the superiority of screw propulsion had been established a very rapid transformation of navies took place between 1850 and 1860. For frigates and ships of the line it was the practice to fit boilers and engines into existing ships, often lengthened, or to complete ships on the stocks in a similar way. For smaller craft, iron or slender wooden hulls following the clipper form (see p. 525) were adopted, in conjunction with a few pivot guns.

The experience of the Crimean War (1854–6), together with further advances in gun-making, initiated another radical change. The need to engage powerful fortifications, already to the fore in the 1830s and 1840s, increased and produced the idea – said to originate with the French Emperor Napoleon III – of vessels covered with an armour of iron plates. Several such floating batteries were built by both Great Britain and France, slow but heavily armed, and it was a short step to seagoing craft on the same principle. The French *Gloire* of 1859 was a wooden ship with armour attached, but the British *Warrior* of 1860 was built of iron throughout, and of great speed for her day. The ship of the line, even with screw engines, was outclassed, and the naval world was plunged into a period of great uncertainty and rapid change.

The early ironclads were armed, like their predecessors, with broadside guns, although these were larger and fewer. Additionally, they were now rifled, for greater accuracy, and in most countries breech-loading, too, although in Britain difficulties with the Armstrong type led to a temporary reversion to the old muzzle-loaders. British guns increased in size from the 4.8 tonnes, the largest used during the Crimean War, successively to 7, 9, 12, 25 and 39 tonnes by 1870. Such weights were impossible for the hand-worked wooden gun-carriages of the past and much more elaborate iron carriages with an increasing element of power were developed. The type which eventually proved most satisfactory was the turret, introduced about the time of the Crimean War simultaneously by Captain Cowper Coles RN and Ericsson, the Swedish engineer. It was the only really sound method for mounting very large guns where steam or hydraulic power were to be used for training, elevating and loading.

All through this period great controversy raged about the design of warships, with little practical experience to provide guidance. It was feared that all warships would have to be armoured, but this was clearly too costly to provide enough ships for the duties required, and the smaller types continued along established lines with the corvette now as the largest class below the ironclads followed by sloops, gun-vessels and gunboats in descending order of size. While their guns were adapted to new standards, all these were really sailing vessels with engines, usually fitted with a hoisting screw which was lifted out of the water to reduce drag when the ship was under sail. This compromise was not the result of pure conservatism: all naval ships were liable to have to make ocean passages either where coal was unobtainable or so long that not enough could be carried to fuel the engines, or both.

With the ironclads, the main argument was between the proponents of the gun turret and those who preferred broadside guns. The turret made its mark in the American Civil War with small, slow, low-freeboard ships called monitors after the pioneer vessel which fought a Confederate ship in Hampton Roads in 1862, and the type was used by many nations although only for coastal defence. Some ocean-going ships were built with turrets, but more had broadside guns or the variant central battery type, in which the guns could also fire ahead or astern, using suitable recesses in the ship's side and pivot mounting. This became increasingly unwieldy as guns grew larger and the numbers that could be mounted therefore tended to decline. The ram became popular as a weapon after Hampton Roads and an incident at the Battle of Lissa in 1866, resulting in ships much shorter than the early ironclads in order to achieve better manoeuvrability; a few freak ships were designed with the sole purpose of ramming. In the event, however, several accidental collisions proved more disastrous than they need have been.

Eventually, in about 1880, after years during which every fleet of the world was composed of a heterogeneous collection of individual ships, a consensus emerged. The battleship had two turrets, one fore and one aft, mounting one or two guns in each, usually 30cm (12in) or thereabouts, with a variety of smaller guns along the sides, and a belt of armour along the waterline, armoured deck over the engines and boilers, and armoured bases to the turrets and over the magazines.

These smaller guns resulted from developments in the smaller classes of ships. With the evolution and increasing economy of the marine engine (see p. 530), it became feasible to abandon sails and, about the same time as the battleship's form was becoming stabilized, the former frigates and corvettes were replaced by cruisers, fully-powered fast ships with a considerable battery of lighter guns, the larger often with some armour and guns up to 25cm (10in). Such ships took over many of the so-called 'police' tasks overseas and also acted as scouts for the main fleets. Below them came small, fast steamers carrying a totally new weapon – the torpedo boats. These craft, and the powerful charges of their torpedoes, seemed to threaten the existence of the great battleships, and one response was to fit both battleships and cruisers with many small guns of the new quick-firing type, whose projectiles were quite sufficient to damage or sink the lightly built torpedo craft. Another was to build larger but still fast vessels to chase and sink the torpedo boats before they could come within range of the big ships. The first essay in this direction, the torpedo-gunboat, was too slow and clumsy; it was succeeded by the torpedo boat destroyer, in practice a larger version of the torpedo boat armed with quick-firing guns. In fact they absorbed both attacking and destroying roles and eventually developed into powerful sea-going warships on their own account.

By 1900, therefore, naval fleets consisted of battleships, large armoured cruisers which scouted ahead, and supported light cruisers which did more of the scouting, and torpedo boat destroyers, which either attacked the enemy battle fleet or countered the efforts of enemy torpedo boats.

In the decade up to 1914 warships, like merchant ships, increased greatly in size, a process accelerated by the great Anglo-German naval race of the period. The appearance of the steam turbine (see p. 288ff.) offered higher speeds with more economy and easier working, while gunners realized that a ship with more guns of one calibre was more effective than one carrying several sizes, such as had developed in the early 1900s. As a result of the reforms instituted in the British navy by Lord Fisher, the *Dreadnought* of 1906 was the first of the class of all-big-gun battleships called by her name, although other countries were about to build similar vessels. At first they mounted 30cm (12in) guns, but the 33.75cm (13.5in) and then the 37.5cm (15in) appeared before 1914. Accompanying the dreadnoughts came Fisher's other innovation, the battle cruiser, less well armoured than a battleship and with fewer heavy guns, but much faster and therefore, because of the extra power needed, much larger. Battle cruisers superseded the earlier armoured cruisers. The turbine was also adopted for a new type of fast light cruiser for scouting and supporting the destroyers, which by 1914 had reached 800–1000 tonnes.

The First World War inevitably speeded up the development of detailed improvements to various types, especially in the case of British light cruisers and destroyers. Submarines, both British and German, similarly made great progress and indeed became almost the decisive weapon, with profound effects on both surface tactics and the protection of trade (see p. 537). The underwater mine, which had been used during the Crimean War and, greatly improved, to much effect in the Russo-Japanese War of 1904–5, was very extensively employed. Mines were laid, often by ships converted from other roles, for both offensive and deterrent purposes and, after early attempts to clear them by aircraft flying off ships, a new specialized type of vessel emerged, the minesweeper. These initially operated in pairs, towing between them a cutting line to sever the moorings of the mines, which were destroyed by gunfire when they reached the surface. A new anti-submarine weapon, the depth charge, was designed to explode at a predetermined depth. Oil proved its great convenience as a fuel and was generally adopted.

Between 1918 and 1939 the lessons of war were applied to new construction. At first few battleships were built, as a result of the disarmament treaties; those that were carried stronger and more concentrated armour, special anti-torpedo protection and 40cm (16in) guns. The naval powers in general added aircraft carriers to their fleets, many converted from the battleships abandoned by treaty. Aircraft carriers had to be large, and many problems of layout had to be overcome, but by the 1930s new ships were embodying the experience gained. Cruisers took a fresh course following the Washington Treaty limits of 10,000 tons (10,160 tonnes) and 8in (20cm) guns, but many countries found these too large and returned to the more economical 15cm (6in) gun ships, usually now with turrets rather than single open-mounted guns. The final wartime designs had produced a very effective destroyer which was the basis of most post-war designs, although France and Japan built some much larger vessels. Submarines displayed similar progress.

In the 1930s rearmament brought some new designs, and the progress of aircraft led to an interest in ships with powerful arrays of dual purpose guns of about 12.5cm (5in) to fight both surface and air targets; these became the secondary guns of new and rebuilt battleships and the main armament of light cruisers and destroyers. Two important innovations were being developed: radiolocation, or radar, as it was later called, and the anti-submarine finding aid asdic, using the general idea of the echo-sounder, nowadays known as sonar.

The Second World War was fought, especially in the larger classes, with the ships recently built and under construction when it started, although there were very large building programmes for small types. Aircraft were shown to be an even more potent threat to the bigger vessels than had been expected, and more strengthening of anti-aircraft armament, both heavy and the light 20mm and 40mm Oerlikon and Bofors guns, took place, sometimes at the expense of the main armament. By the same token, the aircraft carrier emerged as the new arbiter of sea power. While destroyers and submarines tended to maintain pre-war types, with up-to-date modifications, the importance of the anti-submarine campaign for the Allies led to the introduction of two new types of warship, although they were given old names. The corvette, a small ship of about 1000 tonnes, and then the rather larger frigate were both built in large numbers and carried progressively newer weapons. The course of the war also necessitated the construction of new and ingenious landing craft for transporting troops and their equipment from sea-going ships on to beaches. Mine warfare was diversified by the introduction of magnetic and acoustically detonated mines, requiring new techniques to deal with them.

After the war the development of the atomic bomb, combined with general economic conditions, was recognized as rendering big ships vulnerable and prohibitively expensive, while guided missiles were coming forward to replace guns. Battleships soon almost disappeared, while cruisers of the old type survived little longer. Most new construction was of destroyer-sized vessels, armed largely for anti-submarine work. Aircraft carriers remained of primary importance: still of necessity large ships, they were improved by the introduction of the angled deck to provide more space on the flight-deck and then the steam catapult to assist aircraft take-off. Helicopters assumed an important position as part of ships' armament, being used for anti-submarine operations as well as all sorts of other tasks. New types of torpedo have been produced, electrically powered and wire-guided. Mine warfare remains a serious threat; minesweepers and minehunters are usually built of glass fibre plastic to foil varied methods of detonation. Fleet auxiliaries have increased in importance, refuelling and restoring ships at sea. The gas turbine has proved an advantageous power plant for the faster ship, a major facility being that a complete unit can be replaced, rather than having to be repaired *in situ*.

Modern warships, therefore, consist of aircraft carriers, destroyers and frigates (the distinction between them not being very clear), patrol vessels, minehunters and minesweepers, submarines (nuclear-powered and conventional – see below), as well as small gunboats, torpedo boats and missile carriers. Armament includes one or two automatic guns for each vessel, ship-to-ship and ship-to-air missiles, some light guns, and depth charges with various means of discharging them, all backed up by extensive control systems based on radar, radio communication and sonar, together with devices for confusing hostile radar and missile guidance systems.

SUBMARINES

Late mediaeval scholars speculated on the possibility of navigating beneath the surface of the sea, and early attempts to build a practicable craft included those of William Bourne in 1578 and Cornelius van Drebbel in 1620. The first successes were by Americans: David Bushnell's *Turtle* attacked HMS *Eagle* off New York in 1776 and Robert Fulton's *Nautilus* sank a schooner off Brest in a demonstration in 1802. Both boats were driven by screw propellers operated by hand. In 1850 a German craft Bauer's *Der Brandtaucher*, was actually used in war and in the following year was the occasion of the first recorded escape from below the surface. Several other such craft were built in Russia, France and Spain, as well as by both sides in the American Civil War, when in 1864 the submarine claimed its first victim when a hand-driven 'David' sank the USS *Housatonic*, although at the cost of its own loss with all its crew of nine.

All these efforts suffered from problems of controlling the craft, as well as the lack of a suitable power source and a feasible underwater weapon. In the 1860s and 1870s iron or steel provided a material that could be made watertight and the torpedo appeared, but trials of steam, oxygen, compressed air and paraffin engines all failed. In 1886 the French *Gymnote*, designed by Dupuy de Lôme and Gustave Zède, provided a basis for development, using electric motors driven by batteries. The work of J. P. Holland and Simon Lake in the USA produced worthwhile prototypes, and other vessels were built by Laurenti in Italy, Peral in Spain, and the Krupp firm in Germany. All used electricity under water and steam or petrol engines on the surface, until about 1909 when diesel engines became general.

Submarines came into full use during the First World War and, although a number of experimental types were built, moderate-sized boats of simple design were found most satisfactory, the German U-boats (*Unterseeboote*) achieving a high level of combat effectiveness, as they did again in the Second World War. Towards the end of that war the Germans were improving their submarines by fitting the schnorkel, a type of air tube to enable them to remain below water for long periods, as well as developing a submarine of great underwater speed driven by hydrogen peroxide. While the latter proved unsuccessful, all nations were soon using air tubes, called snorts in English, and streamlining hulls to achieve greater underwater speeds.

All these vessels were essentially surface craft which were able to dive and to cruise under water for limited periods. The development in the USA of a nuclear-powered propulsion unit enabled the USS *Nautilus* of 1952 to initiate the era of true submarine craft, capable of travelling submerged, at unprecedented speeds of over 55ph (30 knots), for virtually unlimited periods of time, aided by inertial navigation systems. Some of these vessels are intended to hunt and sink other submarines, while others carry ballistic missiles of strategic use.

HOVERCRAFT AND HYDROFOILS

The ideal of obtaining high speeds at sea by reducing or eliminating the water drag has attracted inventors since the seventeenth century, but its achievement depended upon suitable materials. Just before 1914, an Austrian, and Hans Dinesen, a Swede, built craft depending upon a thin air cushion to hold them above water, but neither succeeded, and it was not until the work of the Englishman Christopher Cockerell in the 1950s that a practical craft was devised. By having a deeper air cushion and using air taken off the aero-type engines, Cockerell's craft could go over waves and equally rough ground, particularly after an improved skirt devised by C. H. Latimer-Needham was provided. A useful craft was built in 1959, and was followed by larger ones for service across the English Channel, carrying passengers and cars, in 1968. These craft are propelled by aero engines and airscrews; another type, the Denny Hovermarine, uses rigid side walls and water propellers, but is consequently more restricted in its movements.

The other approach to the problem uses the principles of the high-speed hydroplanes with stepped hulls and of aerofoils. Before the First World War, Enrico Forlanini and Alexander Graham Bell worked with some success on a hydrofoil which as its speed increased raised itself on to its foils, and skimmed the surface, while between the wars Baron Hanno von Scherkel reached the stage of having a Rhine passenger vessel built. He continued his work after the war, and in 1956 a regular service began between mainland Italy and Sicily, followed by others elsewhere. Much of von Scherkel's knowledge fell into Soviet hands, and many hydrofoils operate in the USSR. The US Navy took a considerable interest in such craft, and both Grumman and Boeing produced patrol vessels, and also passenger derivatives employed since the end of the 1970s on routes across the English Channel.

LIFEBOATS AND LIFESAVING

In Britain, local boats went to the assistance of ships in offshore waters on an *ad hoc* basis until the end of the eighteenth century. Lionel Lukin patented an 'unsinkable' boat for the purpose in 1785, and in 1798, William Greathead won a competition to design a specialized lifeboat. Craft of this type, strong, stable and capacious rowing boats fitted with fenders and protection round their sides, and with buoyancy boxes, were provided by public subscription on a local basis until 1824; in that year the Royal National Lifeboat Institution (as it was later named), formed by Sir William Hillary, began to co-ordinate lifeboat provision and operation throughout Britain. In the late nineteenth century new boats were designed to be self-righting if capsized, using high buoyancy boxes fitted fore and aft. By this time too, most lifeboats could sail as well as being rowed. Although some steam lifeboats were built, they suffered from the fact that steam could not be raised quickly, and the pulling and sailing boats were, in many cases, towed towards the wreck by a tug or other vessel if available. The petrol engine offered good prospects; the first was fitted about 1910 to an otherwise normal boat, and it proved so useful that motors were steadily fitted to existing craft and to all new ones. Different types of boat were built for certain types of station, some being slipway launch, others having to take the boat down a beach, while the largest boats lay afloat.

Since 1945 many changes have been introduced, such as more covered accommodation, modern navigational equipment and more powerful diesel engines. Some boats were built possessing great stability, but not self-righting; after some casualties, all boats have been provided with self-righting equipment. From 1965 onwards, several new types have been developed with much higher speed by abandoning, where possible, slipway launch and the consequent difficulties of design, in favour of being permanently afloat. Recently, a new design has been produced of a fast boat capable of being launched from a slipway. The other important advance in recent years is the use of inflatable or semi-inflatable boats, which can work satisfactorily close inshore, yet have high speed, and these craft have proved invaluable.

Other important components of the rescue services are the Coastguard, which since early in the nineteenth century has provided rocket and breeches buoy equipment to use from the shore, and also navy and air force helicopters, which have vastly improved the possibilities of saving life. All these are now co-ordinated by the Coastguard.

The ship's lifeboat originated with the small boats carried (or towed) by all ships for general purposes. Passenger vessels like the East Indiamen usually had a more liberal supply, but, with the need for legislation to protect the interests particularly of emigrants, came a requirement for passenger ships to carry lifeboats. Scandals in emigrant ships and wrecks of steamers carrying large numbers of passengers produced progressively more stringent rules for ships seeking to have a passenger certificate to have lifeboats, lifebelts and other equipment. The culmination came with the Titanic disaster of 1912, after which ocean-going passenger vessels had to be able to accommodate all their licensed number of passengers in their boats. Coastal ships had only to provide for a proportion, but with enough liferafts to cover the total number. The liferafts usually doubled as deck seats. The circular kapok lifebelt was introduced during this period, while individual life-jackets, also using kapok, were provided for seamen and the RNLI lifeboats and later for passengers too. Passenger ships were required to hold regular exercises for both crew and passengers.

Since 1912 much progress has been made. The actual lowering of the boats from their davits (the small cranes from which they are kept suspended ready for service) is obviously crucial; better types such as the Welin and the gravity type davit kept the boat clear of the ship's side, and these were power worked. Recently a chute method has been introduced. The boats themselves, originally wooden, have been replaced by steel and now often glass reinforced plastic. Some modern ones are covered. Ships must have at least two motor boats; the others can be hand propelled, with the object of keeping the survivors active. Life-jackets are now inflatable, and have small lights and sometimes small radio transmitters. The old raft-seats are now replaced or supplemented by inflatable liferafts, which save both weight and space.

ANCHORS AND CABLES

Anchors and fixed moorings are essential to the safe operation of ships. Originally a heavy stone, or even a basket of stones or perhaps a suitably shaped log, would be secured to a rope and put over the side at the appropriate time. As ships became larger, these expedients were replaced by wood or iron devices shaped like a hook with a spade-like extremity so that it might dig into the bottom. By the sixteenth century, a consensus had been reached that the best results were obtained from a shank (main stem) with two arms, straight or curved in a 'U' or 'V' shape, with wider ends (flukes), and a stock at right angles to the arms to ensure that the anchor would lie in a manner in which it could grip. The stock was wooden, the rest was forged iron, although later stocks were made of an iron bar. Such anchors involved much skilled work in their manufacture, but it was impossible to avoid imperfections in the forging, and despite weights of up to 4 or 5 tonnes, failures with disastrous consequences were not unknown. The anchor cable, the largest of ropes, could go up to 60cm (2ft) in diameter.

In the early years of the nineteenth century, investigation not only led to improved manufacture, but also new types of anchor, while chain cable, pioneered by Lieutenant Samuel Brown RN, proved to be stronger and more manageable than hemp and came into general favour. The new types of anchors, such as Trotman's or Rodgers', usually embodied some method of pivoting the arms to allow them to dig into the bottom when strain came on. In this, they led the way to the stockless anchor, which heralded a complete change in the actual working of anchors.

In sailing days, the anchors hung from catheads forward. Letting go was not too difficult, but to heave in the cable was an 'all hands' task, using the capstan, situated amidships, and a messenger rope running round a bollard forward. The cable, too unwieldy to do other than run from hawse-hole to cable locker, was attached to the messenger by rope nippers and so slowly drawn in. When the anchor came aweigh, it had to be transferred to the cathead by tackles and more hauling. These problems remained in early steamships, and although the development of steam capstans or winches removed much of the heaving work, the difficulties of housing the anchor remained. The cathead was replaced by an anchor bed, but the anchor still had to be hoisted and swung. The stockless anchor could, however, be drawn right into the hawsehole, and there secured. Small vessels had a windlass driven by hand levers with pawls to hold it, the smaller cable being less difficult to work.

Anchors are also used for laying permanent moorings for ships. Sometimes these are chains, picked up as required, but often they have buoys to which the ship's cable is attached. In restricted waters, a ship would lie to moorings fore and aft, but elsewhere she could swing to one buoy ahead. Special designs of anchor are used for these purposes, and for vessels like lightships which are permanently moored in one position, although even stone blocks have been employed in this manner. Screw piles have also been extensively used for moorings. Here, the pile is a tube with a pointed end to drive into the bottom and two turns of a screw thread to take hold. They have also been used in building lighthouses. An important development in recent years has been the use of buoys of special design not only to moor very large tankers, but also to discharge their cargo via a pipeline and flexible pipes to the buoy.

LIGHTS AND BUOYS

Navigational marks – lighthouses, lightships, buoys and daymarks – not only warn of dangers but provide the mariner with points of reference for his further voyaging. Other prominent buildings or natural features can be used for the same purpose, and coastal views and outlines of hills are usually to be found on charts or in sailing directions. Until the late eighteenth century, in fact, specialized marks were rare, and such views often show also church towers, to mention the most obvious. The relative scarcity of proper lights incidentally casts a different angle on the many stories of false lights used by wreckers, for a deep sea seaman would be far more likely to avoid a light than to steer for it.

As in other matters, the first lights were established by the ancients, usually in the form of towers upon the top of which a coal or wood fire was kept burning, the Pharos of Alexandria being the best known, while Roman towers survive at Corunna and Dover. Similar methods continued throughout the Middle Ages, the fire usually being in an iron brazier. Most of these lights were on headlands or offshore islands, but their distribution was purely fortuitous, sometimes being the work of a public-spirited local notable, at other times perhaps due to a religious foundation or to a nearby port's corporation. Passing ships could be charged light dues, so that the profit motive could also bring lights into being. The efficiency of all, however, could be very variable. In England, the Corporation of Trinity House was given powers in the sixteenth century to supervise matters nautical, including the provision of lights.

From the eighteenth century a steady improvement began. In the same way that it became possible to build breakwaters to withstand the sea, so lighthouses could be built on outlying rocks – the Eddystone being the classic example, with the first house swept away, the second burned down, and the third, with its interlocked stonework, still extant, though now moved ashore. Better illuminants were found – oil, paraffin, gas – and elaborate systems of mirrors and lenses were brought into use to give better direction and power to the light. The provision of lights was taken over by official bodies devoted to that purpose, resulting in many more and far more reliable lights, placed where they were most needed. In England, Trinity House took on this role, while in Scotland and Ireland, the Commissioners of Northern Lights and the Commissioners of Irish Lights were set up, although harbour authorities also provided more local lights. In Europe and America, government departments usually provided the lights.

Lightships were used to provide a permanent light where it was not possible to build a lighthouse, the vessel being moored with heavy cable and anchors. Originally the light would be an oil lamp, and the vessel a dumb craft, unable to move independently, but with the improvement of lighting equipment, lightships came to be provided with generating power to work light and optic, and in some countries, are also self-propelled. The tendency in recent years, however, has been to adopt remote control, and the ships are often replaced by large buoys or small unmanned light floats.

Buoys have been used to mark sandbanks and the margins of channels since at least the sixteenth century. They were constructed similarly to casks, or sometimes of a skin over a framework, or might simply be a piece of timber, usually with a small pole and mark on top to make them visible. During the nineteenth century they began to be built of iron and later steel, and their shape could more easily be varied to differentiate one from another. In the late nineteenth century, systems of buoyage were introduced to ensure that one shape and colour had the same meaning everywhere. Lights were increasingly fitted, too, usually using gas, frequently operating continuously, while others had sound signals by whistle, siren or bell. The principal buoy shapes were those known as can, spherical or conical, spar and cagework, with these upper parts mounted on standard 'hulls'.

Daymarks, or 'beacons' comprise a variety of poles or small towers with distinguishing marks but no light, mostly in small ports and creeks, although a few are substantial erections such as Walton Tower on the Naze estuary in Essex.

The main types of optic used in lights are the catoptric, or reflecting, the dioptric, or refracting, and the catadioptric, which combines the two. Optics are often of considerable size and beauty, and of high efficiency in collecting and concentrating the light from the light source, whether oil or gas as in the past or electricity to-day. Occasionally, one light has provided two beams, where a special danger requires a specific indication. Lights are often shielded so that they only cover specified arcs, and sometimes one of these can be coloured, again to mark a particular danger. For ease of identification, it has long been the practice for adjacent lights to show quite different indications and lights can be flashing (light less than dark) or occulting (light more than dark). Flashes may be in twos, threes or more, while the interval between the signal can be varied.

Modern developments include smaller and much more concentrated light sources, often made from plastic materials rather than glass, and an increasing degree of remote control or automatic operation, while, instead of the former ship as a buoy tender, helicopters are used for servicing lights, especially for relieving crews.

Most of the important lights also carry a fog signal, usually a reed horn, actuated by compressed air, and occasionally independent fog signals are found.

NAVIGATION

The early seafarers could only find their way at sea by hard experience in small vessels totally dependent on natural forces, noting winds, currents and tides and using them to best advantage; the instincts developed in such a hard school remain invaluable to seamen and particularly to fishermen. Most early voyagers found their way by hugging the coast, but gradually, by using sun and stars as well as winds and currents, it became possible to sail across the sea in a generally correct direction. As the celestial system became understood, more use could be made of it, for example, in establishing latitude by observation of the angle of elevation of the sun at noon.

The compass was known at least from the thirteenth century, and by the sixteenth the navigator could also use the lead and line to ascertain depth, the log to estimate speed, and a quadrant, cross-staff or back-staff (Davis's quadrant) to obtain his latitude, while declination tables appeared in the late fifteenth century to assist in the latter. When the sun was not visible, however, no alternative was available, and the navigator could only hope that his estimates of course and speed made correct allowances for currents and leeway. Moreover, the basis for good navigation must be good charts, but these, too suffered from serious defects.

The seventeenth and early eighteenth centuries witnessed a slow improvement, rounded off by the invention of Hadley's reflecting octant in 1731, the basis of the later sextant and a much more accurate instrument for the measurement of angles. One great deficiency remained, the inability to find the longitude at sea. This was overcome to some extent in practice by using latitude sailing, reaching the latitude of the objective well to east or west of it and then sailing along that parallel until the destination was reached. In 1714 the British government offered a reward for the inventor of a reliable method of establishing longitude at sea, and much effort was expended to this end. Eventually, the prize was won by John Harrison, who developed over many years a very accurate clock, or chronometer, so that the difference in time between the meridian of the departure point could be measured against the meridian where the ship actually was. These chronometers were used successfully by Captain Cook, and the method was generally adopted.

About the same time, the improvement of astronomy which had result at least in part from the establishment of the Royal Observatory at Greenwich in 1685, with the object of improving navigation, was enabling the production of tables relating to the position and movement of stars (the Nautical Almanac). Using this knowledge it became possible to take sights with a sextant and then calculate (laboriously) the longitude. To do this required more accurate instruments, and these were now available, made of metal (usually brass) to much higher standards and graduated by precision graduating machines. Similar progress was being made in producing more accurate charts, while sailing directions were increasingly published. Thus navigation was at last becoming a precise science, and the end of the eighteenth and the first half of the nineteenth centuries saw a new era of exploration and surveying, in which the Royal Navy took a very prominent part.

Knowledge of the depth of water is obviously of primary importance to a navigator. Until well into the twentieth century, sounding could only be done in most ships by the hand lead, a long narrow piece of lead, with a recess in the bottom 'armed' with wax to obtain a specimen of the sea bed, attached to a lead line marked in fathoms. The lead was cast ahead in the hope that it would reach the bottom before the ship overran it, and indeed for deep sea soundings the ship would have to be stopped for the laborious running out and heaving in of many fathoms of line. A recording machine was devised by Massey early in the nineteenth century, but this did not avoid the same troubles. In the 1870s Lord Kelvin produced a deep sea machine which could be power worked and which used a special tube filled with a chemical which changed colour in relation to depth. Only the invention of the echo sounder in the 1920s, which timed the return of a noise from the bottom, brought real improvement, and the great advantage that it could maintain a continuous record.

Measuring the speed of a ship through the water was, from the mid-sixteenth century, done by the hand log, by timing the running out of a regularly knotted rope for a fixed period, the number of knots passing being termed the speed, hence the 'knot' being the unit of speed at sea. It was also possible to use the Dutchman's log, that is, to time an object passing down the ship's side between bow and stern. Both methods were obviously approximate, and contributed to the general inexactitude of navigation. In 1802, Massey produced a speed-recording machine similar in principle to his depth recorder, driven by a rotating float towed astern, and this system is still in use in the modern patent log. More recent methods include a recorder driven by a small propeller attached to the hull, while the Chernikeef type uses water pressure on a diaphragm in the hull.

Another of the old inexactitudes was compass error, which was little understood until the work of Matthew Flinders early in the nineteenth century on the magnetism of a ship's hull as well as on deviation and variation. An Admiralty Committee in the 1830s extended this work, and also designed a compass card which was steadier than previous ones. The introduction of iron ships obviously accentuated the effect of hulls upon a compass and necessitated more precautions. Lord Kelvin devised another card in the 1870s, and generally speaking, with regular correction and swinging of the ship, the magnetic compass was reliable and needed little attention. Early in the twentieth century, however, it was found that a gyroscope, left spinning freely, adopts a constant axis and could serve as the basis for a compass, which would give 'true' readings direct, while it also could have repeaters driven off the master, and consequently the gyro compass achieved considerable popularity.

Since 1945 navigation has changed radically. The wartime refinement of radar and the Decca navigator carried further the use of radio techniques begun before the war with directional position finding. Radar gives the navigator a view of the sea and land in his vicinity, together with the shipping, and can provide accurate bearings and distances, with allowance too in many sets for one's own ship's movements, though its use has also produced problems of interpretation. The Decca navigator uses the former DF principle of two or three directional bearings from shore stations, but these are continuously available, and can be plotted direct on to special charts to give an accurate position at a moment's notice. Both these developments are concerned principally with waters near land – though radar can be used anywhere – but the methods of inertial and satellite navigation introduced largely for nuclear submarines in the 1960s, provide ships so fitted with instantaneous and reliable positions when far from land.

CHARTS AND SAILING DIRECTIONS

The earliest charts and their associated texts which give detailed advice to the mariner, sailing directions, are known from classical times, but the earliest surviving 'sea-maps' are the so-called *portolani*, drawn with ink on skins, in mediaeval times. The methods of producing these maps are not certainly known, and, although at a glance many show recognizable outlines of coasts, they are inadequate in detail. There were errors in longitude, while the idea of projection was not understood, nor was the size of the earth known, so that all charts well into the eighteenth century – and even after – embodied errors. In the late sixteenth century Dutch cartographers led, producing printed engraved charts which had wide usage; they were generally known as Waggoners, a corruption of the name of Lucas Waghenaer who first produced them in 1584. The secrecy which many countries maintained with relation to their possessions was to some extent dispelled by these charts, but they were copied extensively and for too long after better work was available. In England, Captain Greenvile Collins produced some more detailed charts of many localities in the 1680s which embodied improved methods of survey. While the Dutch charts certainly included some sailing directions, most prudent mariners also kept their own remarks about the pilotage of ports they had visited.

Commercially produced charts held the field for many years. The French navy set up a Hydrographic Office in 1723, the English East India Company appointed Alexander Dalrymple as its hydrographer in 1779, but the British Admiralty did not follow these enlightened leads until 1795, although naval officers had been surveying regularly long before. Dalrymple's appointment to the Admiralty post was not followed by an extensive output of charts for various reasons, and it was not until 1810 onwards that the Admiralty chart began to establish its reputation. However, from 1815 onwards, the Royal Navy embarked upon a serious effort to chart much of the world, using recent scientific progress to raise the standard very considerably. The long and arduous work of officers such as Smyth, Fitzroy, Owen, King and Denham, together with efficient production and distribution, made these charts more widely available. Meanwhile, Horsbrugh, Dalrymple's successor, produced a valuable series of sailing directions for the eastern seas. From the 1860s, however, the Admiralty also began its own series of sailing directions or Pilots as they are commonly called, and the resurvey, and revision of both charts and pilots was placed on a regular basis.

The British example was followed by many other countries, and for many years most nations have undertaken their own survey work. The work of the surveyors has been much assisted by subsequent developments of new techniques and of instruments such as radar and echo sounders and even aerial survey. Special charts have been found necessary for use with navigational systems such as Decca, and standardized formats have been adopted.

PORTS AND HARBOURS

The importance of good harbours to economic development needs little emphasis. Not only must ships be able to load and discharge in safety, but for centuries they also needed shelter on their voyages or safe anchorages where they could wait for winds. Until relatively recently, however, man depended on the gifts of geography for these important elements of his society.

The earliest craft could readily be beached. When ships became larger, and this was no longer feasible, sheltered places had to be sought where ships could lie peacefully alongside the shore or bank to load or unload. Such sites were mostly on rivers, often, indeed, some way up and not uncommonly at the lowest bridge. Structures in timber, and as early as Roman times in stone, brick or concrete, might be built as quays over which cargo could be worked. Quite extensive works were created in harbours such as Ostia, but much of this knowledge was lost in the Dark Ages, and harbour works remained rudimentary for many centuries. In many cases ships had to lie in the river and load or unload from barges. In such ports, the tide was of much assistance to ships in working up and down the river. London, Antwerp, Rotterdam and Hamburg may be given as examples of this type of port, and countless others may be found. Away from the sandy and low-lying shores of northern Europe, however, reliance was usually placed upon a natural harbour in a bay or arm of the sea, where a point sheltered from wind and sea could be found and the necessary works built. For naval purposes, where less cargo work was done, such harbours were especially popular, as access was easier than with a river. Portsmouth, Brest and Toulon typify such places, while Venice and Marseilles are more commercially oriented examples. Beach cargo working, for coasters, remained in use until the early twentieth century. For shelter during a voyage, or as a safe place in which to await fair winds, the small merchant ship of the sailing era might lie in the lee of an island (the Texel off the coast of Holland, for example) or headland; offshore sandbanks often enclosed sheltered anchorages, such as the Downs or Yarmouth Roads off the east coast of England.

Occasional efforts to build small breakwaters against the full force of the sea usually ended, sooner or later, in failure, with the notable exception of the mole at Genoa in north-west Italy: the Molo Nuovo, built in 1643 survives today.

Only during the eighteenth century did engineers develop the ability both to build and to maintain structures against the battering of the sea – Plymouth Breakwater, Whitehaven, Ramsgate and Kingstown Harbours, for example, and these provided amply for shelter as well as improved port facilities. The great increase in trade beginning in the mid-eighteenth century began to make existing ports congested, while the slow increase in the size of ships became much more marked in the first half of the nineteenth century, and rendered the old practice of lying on the bottom at low tide more and more undesirable. Thus efforts were directed towards the provision of wet docks, in which ships could lie quietly in impounded water behind locks and where cargo could be better protected and stored. A few such docks already existed, such as the Howland Dock on the Thames, and some in the naval dockyards and at Liverpool, but from 1800 onwards many more were built, generally by companies set up for the purpose or by municipal authorities, and dock or harbouur management entered a new era far removed from the small-scale efforts of the past. In ports where the tide was less of a problem, similar improvements were nevertheless made to provide open docks with storage and working areas behind the quays. All these works required heavy excavation and building of robust walls, and thus a high level of engineering skill and organization.

The advances in port engineering and organization were timely, for the radical changes in ship design, size and purpose throughout the nineteenth century were to set port authorities a hard task to keep abreast. Longer, wider and deeper locks were needed, and the docks to which they led had equally to be enlarged, while the coming of railways required the provision of additional facilities. Steamships, for economic reasons, had to work their cargo far more quickly than in the past, so that improved equipment and methods were needed. To the former warehouses close to the quay were added transit sheds to organize and collect the cargo before loading or after discharge, large storage areas for bulk cargoes such as timber were necessary, while cranes, steam at first and then hydraulic and electric, together with grabs for coal and suction elevators for grain, had to be provided, as well as refrigerated warehouses for meat and later fruit, and tanks for various liquids. The scale of these requirements also increased as individual ships carried more cargo. Thus all ports, to keep up to date, expanded on to new sites; many, of course, dropped out.

An era of specialization followed. Many new ports, or greatly expanded ports, were built to meet special needs – railway companies for packet ports or coal shipment ports, mining companies to ship their products – while in some parts of the world, such as India or Africa, completely new ports could be built for areas hitherto ill-served by nature, examples of which are Madras, Port Harcourt and Mombasa (Kilindini). Even in the late Victorian era, passenger traffic was quite ill-served, tenders often having to be used at such important places as London and Liverpool. As the number of passengers in one vessel increased with the size of ships, improvement was needed, and preferably outside the dock so that the ships could leave without reference to tide. Floating landing stages were therefore provided, with railway station and road approach, waiting rooms, customs premises and space for circulation. In more recent years, these quays and other berths have also needed ramps to allow cars and lorries to drive on and off ro-ro ships, with parking areas and buildings to match.

Perhaps the first of the specialized cargoes to require separate treatment was oil, which, owing to the risks of fire, was always dealt with at special facilities, usually long jetties with the refinery at the shore end well away from other habitation. This tendency has spread with the increasing use of very large bulk carriers for other cargoes. Coal, iron ore, grain and sugar are now received, as many have long been shipped from special jetties, often adjacent to the works which will use the cargo. The jetties are fitted with modern unloading gear- – grabs, cranes and conveyor systems – and are sometimes of considerable length to reach deep water. Owing to their often exposed positions and the size of the ships using them, the construction of these jetties presents many problems. While containers, the other modern cargo system, are usually handled over quays in less remote situations, they need very large special cranes and much stacking ground, as well as a deep-water berth.

Cargo handling

The handling of cargo from shore to ship and then again from ship to shore, sometimes with further intermediate handlings such as to a lighter, was always recognized as unsatisfactory owing to its liability to breakage and pilfering, but only for some bulk cargoes was it possible to avoid it. Commodities such as coal or stone could be tipped on board, and, after the introduction of the grab crane, lifted out, while grain and other cargoes of similar characteristics could be handled via pipes and suction pumps. Smaller but more valuable and fragile loads could only be barrowed on the quay, then lifted in small batches by the ship's gear, in sailing days often by blocks and tackles from the yards, later by derricks rigged on the masts, and then by dock cranes, all using slings and nets. Weights were limited to a tonne or so, although this figure rose as derricks or cranes achieved greater capacity. Beyond this weight, special heavy cranes had to be obtained. By the 1950s, however, both cranes and derricks could handle up to 5 tonnes, and many ships had also a heavy derrick capable of lifting, in some cases, 80 or 100 tonnes, while port authorities also had floating cranes available of even greater capacity.

Efforts to avoid handling go back many years, and took two forms, the container and the roll on, roll off vessel. The ro-ro originated with the train ferry:

railway wagons could be shunted on to a special vessel and off at the other side, using ramps or lifts to overcome tidal variations. The first such ferry was across the River Forth in 1849, and it was followed by others, especially in Denmark, while rather similar vessels were introduced for road traffic across rivers, all very short crossings. Longer distance train ferries with seagoing vessels began with the Great Belt crossing in Denmark in 1883, and this was followed by others across the Baltic, in the Great Lakes area and between Japanese islands, and elsewhere.

By 1939 the need for similar vessels to convey motorcars was recognized and a few were in service, often on the same routes. During the war, the situation was altered by the construction of large numbers of landing ships and craft for what in essence was a similar purpose, though these relied on bow ramps or doors rather than the stern loading usual in the existing vessels. After the war, converted landing ships or craft were utilized to open new routes, and further new vessels were built for fast cross-channel routes. It was not until the 1960s that a fresh wave of development began based on the conveyance of lorries. New ships were needed with greater headroom and more deck space, as well as loading facilities at both ends. A vast reorientation of trade towards short sea crossings resulted, as well as the decline of many ports. In the late 1970s many deep sea vessels were being provided with ro-ro facilities.

The second radical development was the container. This also began with the railways, which, after using small containers to hasten the transfer of baggage from express to cross-channel steamer before 1914, introduced larger ones between the wars to provide a door-to-door service in reply to the competition of road hauliers. These containers were also increasingly used for cargo to Ireland. In the 1950s, a new type of container was developed in America of rectangular shape, allowing them to be easily stacked and to fit closely into suitably fitted ships, and also to be quickly coupled and uncoupled to the crane which transferred them between ship and shore, often direct to lorry or train. These ISO (International Standards Organization) containers have revolutionized the deep sea cargo trade as ro-ro has short sea crossings. Very large and fast ships can carry several hundred such containers, and are able to do the work of several older ships, so much so that the former shipping companies have been compelled to form consortiums to operate the relatively few vessels needed.

Other new methods have not made so much impact. Lash and Bacat (Barge Aboard CATamaran) ships are variations on the idea of loading a barge inland and taking it in a seagoing vessel overseas to be distributed by waterway.

Dredgers

One of the most important aspects of maintaining a port is to keep a constant depth of water. In areas of large tides, silting can be a serious problem, sandbanks can change very considerably and suddenly, and an increase in the size of ships may require the channel to be deepened. For all these purposes, various types of dredger have evolved. In the seventeenth and eighteenth centuries, ballast was dug from river bottoms by a spade-like arrangement and lifted to the surface, and this could be used for dredging if necessary. Such an arrangement could be adapted to steam power, and then evolved into a chain of buckets on a strong rigid frame, the spoil being emptied into chutes and thence into a barge, usually a hopper barge, which could convey it well out to sea and deposit it through bottom doors. Bucket dredgers of this type were the mainstay of the dredging plant of a port. For quaysides and areas which a bucket dredger could not work, a vessel, usually itself a hopper, fitted with grab cranes was used. Suction dredgers were also employed, drawing up silt through a pipe trailing on the bottom. In recent years, the efficiency of this type has greatly improved, and they are now generally used in place of the other types. Other machines such as rockbreakers and piledrivers are also found, usually in harbour construction, while the suction principle is also widely used to obtain gravel and sand for non-nautical purposes.

SHIPBUILDING AND DOCKYARDS

Shipbuilding in the days of timber ships was a more opportunistic business than it subsequently became with iron and steel. Until the end of wooden shipbuilding, it was a usual practice to build small vessels up to 30m (100ft) long or thereabouts on a convenient site where the ground sloped down to a river, offering a suitable site for the building ways on a firm foundation. Here, using local timber as often as not, the keel could be laid, the frames erected and the hull planked employing only very simple lifting appliances, and with timber scaffolding round the hull, reached by ramps, to allow the shipwrights to work above ground. The hull would be held upright by cradles at various points standing on the ways, and wedged by shores. These would be removed shortly before launch, when the ways would be greased, and the hull on the cradles would slowly slide down to the water. Nearby would be saw pits and perhaps sheds for shaping and preparing the timber, with a steaming kiln to bend it. If no more orders came, the site could be abandoned and revert to nature.

Obviously there were also many shipyards which were well-established businesses in important ports, but the essentials were the same, with associated activities nearby, such are ropewalks, rigger's shops, mastmakers and blacksmiths (for much iron was used). There would also be dry docks, in which ships could be either built or more often, repaired. Such a dock would be excavated in the river bank, lined with timber piles and provided with timber mitre gates closing to form a 'V' against the river outside. The ship would be brought in at high tide, and carefully placed with its keel on a row of blocks placed in the centre of the bottom of the dock. She would then be supported on either side by shores against the dock side, before the dock would be drained by sluices as the tide ebbed. From the eighteenth century onwards, docks began to be lined with brick or stone, and the sides were often stepped to facilitate fixing the shores. Later still, pumping machinery was provided to empty the dock, while instead of gates the entrance was often fitted with a caisson, a floating structure which fitted the entrance closely, but which could be floated to one side or into a special recess to allow entrance or exit of ships.

If no dock was available, ships under repair had to be careened, or hove down to a steep angle using special blocks and tackles to the mastheads. In tidal waters, however, gridirons or hards could be used, the ship standing upright, and work being done between tides.

The building and maintenance of royal fleets was everywhere entrusted very largely to state dockyards, which, incorporating as they did all the various necessary trades in the one establishment, became the earliest large-scale industrial undertakings, employing more than 1000 men even in the late seventeenth century. The largest warships were so much bigger than any merchant ship as to be beyond the capacity of private shipbuilders, and this was a powerful influence in the rise of dockyards; in addition, the existing ships had to be cared for, an peace as in war, while large stocks of all kinds of stores had to be kept up against the eventuality of hostilities. In many cases, the dockyards embodied the most advanced techniques of their day, such as the early use of stone for docks, large-scale manufacture of rope, and early in the nineteenth century mass-production methods in Marc Isambard Brunel's block-making machinery (see p. 405); in matters of detail there is ample evidence of careful thought in layout. At Sheerness in the 1820s Rennie had the opportunity of creating a new yard to replace the old ramshackle one. Most of his buildings were iron-framed and fire-proof, there were tunnels to bring masts from the masthouse to the storage ponds and many other sound ideas, including steam pumping engines.

Navies had also to provide for their other needs. Naval victualling required the preparation of much preserved food, and this was found, again for reasons of scale, to be best done by the service. So one finds breweries, bakeries and slaughterhouses and meat preservation and packing houses under naval control, together with storehouses. Guns, ammunition and powder were often provided by other government establishments, in co-operation with the army, while naval hospitals were early examples of large-scale care for the sick.

In the nineteenth century all shipbuilding establishments changed dramatically, as, in many cases, did their geographical locations. Steam engines, iron ships and larger ships called for great extensions to dockyards and reorganization of private yards, with foundries, plate shops, machine shops, steam hammers and all the equipment for handling iron and later steel. An important additional trade, usually by specialist firms, was the production of armour plate.

Although the size of merchant ships increased to become comparable to that of warships, the state dockyards continued to build some of the ships required for navies, although the proportion was smaller than in days of wood and sail, and they also added submarines to their range.

DIVING

Men have used various devices to work or remain under water since Greek divers destroyed a boom during the siege of Syracuse in 415 BC although in no case could they stay down very long. Such means were simply to hold one's breath or to have an air bag or a crude helmet, or even an air pipe from a bladder on the surface. By the sixteenth century the diving bell, where pressure keeps out the sea water, enabled wrecks to be examined and even some salvage work such as that of Phips in the Bahamas in 1687. In the eighteenth century, air was piped down to diving bells, while diving suits were used by Lethbridge and Freminet, and Borelli pumped air down to his suit. These improvements permitted work to be done under water to a depth of about 15m (50ft), and this was obviously important for harbour works. In 1819, Augustus Siebe patented his 'open' diving suit, with pumped air, followed in 1837 by his 'closed' suit, the real pioneer of modern diving dress. With its aid, much salvage, survey and construction work became possible. At the same time, however, other possibilities were being explored, such as a diving chamber, one of which reached 74.5m (245ft) in 1865, and a semi-independent suit using compressed air, both being French. In connection with this activity, underwater apparatus of various kinds was developed and information was gathered about the effects of depth and of compressed air and other factors on the divers themselves. Following the First World War, gold was recovered from both the *Laurentic* and the *Egypt* in very deep water. Divers on the *Egypt* descended to 119m (393ft) using steel articulated suits, and this idea was extended to deep diving research craft such as Beebe's Bathysphere in the 1930s and Piccard's Bathyscaphe in the 1950s.

In the 1920s had come the first moves towards another style, of quite independent diving, with foot fins, goggles, and compressed air carried by the diver, or later helium. During the Second World War Jacques Cousteau perfected his Aqualung along these lines. Together with advances made by the navies of the world, the basis was laid for the great progress in underwater activity since, not only in general exploration and oil-rig working, but in the recovery by the aid of much new equipment of articles from wrecks and the fragments of crashed aircraft, begun with the Comet disaster of 1954.

Working under compressed air at great depths can produce bubbles of nitrogen in the bloodstream, inducing compression sickness ('the bends') which can prove fatal. It is averted by bringing divers to the surface at a slow and carefully controlled rate.

FURTHER READING

For the development of the ship, the several volumes of the British National Maritime Museum series *The ship* may be recommended. Published by Her Majesty's Stationery Office for the Museum.

For the history of navigation, the best general work is Commander W. E. May, *A history of marine navigation* (G. T. Foulis, Yeovil, 1973).

For lighthouses, see Douglas B. Hague and Rosemary Christie, *Lighthouses: their architecture, history and archaeology* (Gomer Press, Llandyssul, 1975).

For ports, James Bird's *The major seaports of the United Kingdom* (Hutchinson, London, 1963) offers many insights.

For dockyards, Jonathan Coad's *Historic architecture of the Royal Navy* (Gollancz, London, 1982) is invaluable.

As a general compendium, Peter Kemp's *Oxford companion to ships and the sea* (Oxford University Press, Oxford, 1976) will prove most informative.

11

RAIL

P.J.G. RANSOM

RAILWAYS BEFORE LOCOMOTIVES

Rail transport as we know it today is descended from railways used underground in the mines of central Europe in the sixteenth century. These had rails of wood and along them small wagons were pushed by hand. German miners may have introduced the concept to Britain late in that century. The first line in Britain on the surface was completed in 1603–4 by Huntingdon Beaumont: it ran from a coal mine near Wollaton for two miles to a point close to Nottingham where the coal could be sold. A similar line was built in Shropshire the following year, and Beaumont built three more in Northumberland.

These were imitated by other coal owners, gradually at first because of the unstable political state of Britain in the seventeenth century, but with increasing frequency during the eighteenth. Some of these wagonways carried coal from pit to point of sale, but far more delivered it to a wharf for onward carriage by barge or ship. Such a layout lent itself to the construction of lines on gentle continuous gradients down which laden wagons ran by force of gravity; horses hauled them empty back uphill. For their period, the engineering of these lines was advanced. The wagons were guided by flanged wheels and braked by brake blocks bearing on the treads (a practice as yet unknown on road vehicles); cast-iron wheels started to come into use, probably, during the 1730s. The wooden track was well built and carefully maintained, cuttings, embankments and bridges enabled gradients to be easy, and, in Northumberland and Durham, lines terminated at high level on staithes so that coal could be fed by gravity into ships waiting below. The most notable surviving relic of the period is Causey Arch, Co. Durham, the 32m (105ft) stone span of which once carried a wagonway across a ravine. The first wooden railway in the USA was built in 1795, and by 1800 the total length of wooden wagonways in Britain, made up from many short lines, was probably approaching 480km (300 miles). By then another fundamental development had taken place.

To simplify the replacement of worn rails, the practice had grown up of making rails in two parts, one on top of the other, so that only the top part needed to be replaced. In 1767 at Coalbrookdale, Shropshire, hard-wearing cast iron was for the first time substituted for wood as the material from which the top part of the rails was made. Track of this type was installed on the line built to extend the Trent & Mersey Canal's Caldon branch to limestone quarries at Caldon Low, Staffordshire. Between the end of the canal at Froghall and the quarries lay an ascent of over 200m (670ft) in 5.6km (3.5 miles), practicable for a horse-and-gravity railway but scarcely so for a canal. The whole line was authorized by Act of Parliament in 1776, the railway section being the first to be so authorized.

The obvious step forward of making the rails entirely of cast iron seems to have occurred in South Wales, where ironworks were using iron edge rails, deeper than they were wide, in the early 1790s. William Jessop, foremost canal engineer of the day, employed iron rails of T-section in 1793-4 on the Leicester Navigation's part-rail, part-canal Forest Line. This was during the great period of canal construction in Britain (see Chapter 9), and many short horse railways, and some not so short, were built in connection with canals. The terms tramroad or tramway (the derivation is uncertain) came to be used to describe them. See Figure 11.1 for an illustration of an iron-railed horse railway.

The form of track used on most of them, however, proved eventually to be a diversion, a blind alley, from the proper path of progress. Iron rails of L-section were used, in combination with unflanged wheels running loose on their axles; rails were laid with the flanges located along their inner edges, probably so that stones and other obstructions on the running surface could be cleared away easily. Light narrow gauge track of this type had been developed by John Curr for use in coal mines and was first used above ground in 1788; subsequently tramroads, or 'plateways', using comparatively heavy L-section rails laid on sleepers made from stone blocks were extensively promoted by Jessop's associate Benjamin Outram, and came into general use. Only in north-east England, with its long tradition of flanged wheels, were edge rails still used extensively: there, iron first replaced wood as the rail material in 1797.

In 1799, Jessop recommended that an iron railway should be substituted throughout the length of a proposed canal for which it appeared, on investigation, that water supplies would be insufficient. The consequence was the Surrey Iron Railway, a plateway opened in 1803 between Wandsworth and Croydon: it was the first public railway independent of canals, that is to say, it was built and run by a company incorporated by Act of Parliament for the purpose, and it was available for use by all comers. During the next four decades many other public horse railways were built; on one of them, the Oystermouth Railway running westwards out of Swansea, the innovation was made in 1807 of carrying passengers (in a horse-drawn coach) – the first fare-paying passengers

Figure 11.1: A horse railway built in 1815 to carry coal to Belvoir Castle, Leicestershire, from a canal wharf was still in use a century later. Cast-iron fish-belly edge rails can be seen.

carried by rail. Eventually, the routes of horse railways in Britain totalled about 2400km (1500 miles).

In the USA, construction of the Baltimore & Ohio Railroad as a horse railway commenced in 1828. Its track was made up of wooden rails with longitudinal wrought-iron straps fastened along them to form the running surface: the first railway of this type in the USA had been completed in 1826. The first public railway in Austria was the horse railway built from Linz to Budweis between 1827 and 1832.

Horse railway engineering works included tunnels, viaducts and inclined planes. Self-acting inclined planes had originated on the Continent and were used on British wooden railways from the mid-eighteenth century onwards, in locations where gradients too steep for ordinary horse-and-gravity operation were unavoidable. Laden wagons about to descend were connected by a rope, which led round a winding drum at the head of the incline to empty wagons at its foot so that the former, during their descent, could haul the latter up. But where it could not be guaranteed that the weight of descending wagons would regularly exceed that of those ascending, an external source of power was needed. By the end of the eighteenth century, the stationary steam engine had been developed far enough to provide it.

One of the earliest such installations was made by the Lancaster Canal Co. in 1803–4. The deep valley of the River Ribble separated two completed sections of the canal, and a tramroad was built to link them which descended by inclined planes into the valley, and crossed the river by a low-level bridge; it was intended as a temporary expedient pending construction of a high-level aqueduct which in fact was never built. The incline at Preston, on the north side of the river, was powered by a steam engine. Cable haulage on inclined planes, both powered and self-acting, then became a common feature of railways built during the ensuing forty years. In some cases gradients were steep, in others not – the Cromford & High Peak Railway, for instance, was built between 1825 and 1831 with inclined planes ranging in gradient between 1 in 7 and 1 in 60. Some railways achieved powered transport over considerable distances by means of successive inclined planes: the preserved Bowes Railway, Tyne & Wear, is of this type.

This method of operation was still further developed in north-east England by Benjamin Thompson, who achieved power traction over level or gently undulating lines by positioning stationary engines at intervals. When an engine was drawing a train of wagons towards it by means of the 'head rope', there was attached to the rear of the train the 'tail rope' from the previous engine, running off an out-of-gear winding drum. When the wagons arrived at the engine, the tail rope was detached to become the head rope for a train in the opposite direction. This system was installed continuously over five miles of the Brunton & Shields Railroad, completed in 1826, where trains were conveyed by rope haulage at overall speeds of nearly 9.6kph (6mph) including stoppages for rope changing.

THE FIRST STEAM LOCOMOTIVES

By 1826, however, steam locomotives – though not, as yet, particularly efficient ones – were already in commercial service. They had originated with the work of Richard Trevithick who, in about 1797, had conceived the brilliantly simple notion that James Watt's separate condenser patent for stationary steam engines (see Chapter 5) could be avoided by using steam at a high enough pressure for the exhaust to be led to the atmosphere, so that the condenser could be eliminated altogether. A side-effect of this development was that boiler and engine could then be made compact enough to become the power plant of a self-propelled vehicle. The first attempt to build a locomotive to Trevithick's ideas was made by the Coalbrookdale Company in 1802, but little is known about it and it seems unlikely that it worked satisfactorily.

It was otherwise with the next locomotive, built under Trevithick's supervision at Penydarren Ironworks, South Wales. On 20 February 1804, without difficulty, it conveyed 10 tonnes of iron, five wagons and seventy men riding on them for more than 14km (9 miles) along the Penydarren or Merthyr Tramroad, a plateway completed in 1802. Probably this locomotive had a cylindrical boiler with, set into it, a single horizontal cylinder, and cylindrical furnace and return fire tube: and was carried on four wheels driven through gears. Steam was exhausted up the chimney to draw the fire. Unfortunately the track was not substantial enough for a locomotive and too many brittle cast-iron rails were broken: so the locomotive was altered for use as a stationary engine. A similar fate met another Trevithick-type locomotive built at Gateshead the following year which had been intended for Christopher Blackett's wooden-railed Wylam Wagonway in Northumberland. In 1808, Trevithick demonstrated a locomotive on a circular railway in London; this locomotive had a vertical cylinder and the rear wheels were driven directly by connecting rods.

By 1811 the Napoleonic wars had inflated the price of horse fodder so much that J. C. Brandling was considering use of a steam locomotive on his Middleton Railway, an edge railway carrying coal from a colliery into Leeds. John Blenkinsop, who was Brandling's colliery manager, believed that a locomotive light enough to avoid damaging the track would be too light to haul a useful load by adhesion, so he invented a rack railway in which a pinion on the locomotive would mesh with teeth cast along the side of one of the running rails. The locomotive, designed and built in Leeds by Matthew Murray, was a development of Trevithick's 1808 locomotive with two vertical cylinders, cranks at right angles and the rack pinion driven through gears. The first of these locomotives came into use in June 1812: it was the first steam locomotive to go into regular service, for it worked well, hauling 30 coal wagons at 4.8kph (3mph), and was soon joined by others.

Blackett was prompted to attempt locomotive haulage again. He had already re-laid his wagonway as a plateway; his manager William Hedley carried out experiments, first with a model and then with a full-size test carriage driven by men turning handles, and concluded that a locomotive could work by adhesion alone if its two pairs of wheels were coupled by gears. The first locomotive which was then built at Wylam was a failure, its boiler producing insufficient steam, but the next, completed in the spring of 1814, was a success, except that it damaged the track. To spread the load it was converted to run on two four-wheeled bogies, which it and subsequent similar locomotives at Wylam did until about 1830 when the line was re-laid as an edge railway and the locomotives put back on four wheels. These locomotives ran until the 1860s and two survived to be preserved, as *Puffing Billy* and *Wylam Dilly* in the Science Museum, London, and the Royal Museum of Scotland, Edinburgh, respectively.

George Stephenson

One of those who came to watch the early trials of the Wylam locomotives was George Stephenson, a self-taught mechanical genius who was at this period chief enginewright at Killingworth colliery near Newcastle upon Tyne. A wagonway with iron edge rails connected the colliery with the river and for this in 1813 Stephenson was instructed by his employer, Sir Thomas Liddell, to build a locomotive. Its design was probably based on the Middleton type of locomotive, but drive was by adhesion, not rack. So when this locomotive, called *Blucher*, was completed in July 1814 it pioneered what ever since has been the conventional arrangement: it was the first locomotive both to be driven by adhesion, and to have flanged wheels running on edge rails.

For the next decade and a half, other engineers seemed to lose interest in locomotives and Stephenson almost alone and step by step introduced improvements — geared drive gave way to wheels coupled by chain, for instance, and eventually to coupling rods. He built more locomotives, and in about 1819 he laid out the Hetton Colliery Railway, from Hetton-le-Hole to Sunderland, as the first railway designed to have no animal traction. Its 13km (8 mile) route included two sections to be worked by locomotives, while the rest comprised powered and self-acting inclined planes.

An essential preliminary to widespread adoption of locomotives was, as Stephenson well realized, improvements to the track. Although edge rails were stronger than plate rails, particularly when they were 'fish-bellied', cast-iron rails could be no more than about a metre long, and track laid with them was a continuous succession of joints. Rails rolled from malleable iron provided the answer, for not only could they be made as much as 3m (15ft) long, but they were not brittle. In 1821 an Act of Parliament was obtained for the Stockton & Darlington Railway: Stephenson persuaded the promoters to use both malleable iron rails and locomotives and became the railway's engineer. To build locomotives, George Stephenson, his son Robert and other partners set up Robert Stephenson & Co. in 1823.

In 1825, George Stephenson surveyed a railway between Liverpool and Manchester, but a Bill for this line was rejected by Parliament. Later the same year, however, the Stockton & Darlington Railway was triumphantly opened, Stephenson's latest locomotive *Locomotion* hauling a train made up of the company's sole passenger coach and 21 wagons fitted with seats. In service, the S & DR used not only locomotives but also horse haulage, gravity operation and powered and self-acting inclined planes. The locomotives, indeed, proved in the early years unreliable, for they did not steam well enough for a continuous journey as long as 30km (20 miles).

Liverpool & Manchester

In these circumstances the directors of the 50km (31 mile) Liverpool & Manchester Railway were uncertain what form of motive power to use. They had successfully obtained an Act of Parliament in 1826 for a route surveyed by Charles Vignoles, and two years later construction, under George Stephenson, was well advanced. An independent report recommended cable haulage. The L & M directors then decided to hold a prize competition to find a locomotive which would be 'a decided improvement on any hitherto constructed'. This competition, the Rainhill Trials, was held in October 1829 on a completed section of the railway.

Of the four steam locomotives which came to Rainhill, only one, the *Rocket*, proved able to do all that the organizers of the trials stipulated, and indeed far more (see Figure 7, p. 36). *Rocket* was built by Robert Stephenson & Co., and entered by a partnership of George and Robert Stephenson and Henry Booth. Robert at this date was developing locomotive design fairly rapidly, but it was Booth who suggested that the boiler of their trials locomotive should have not a single large-diameter fire tube (as was then usual) but a multiplicity of small-diameter copper tubes through which hot gases from the fire would pass. In this way the heating surface would be much increased, and so would production of steam. This multi-tubular boiler proved highly successful – *Rocket* achieved 48kph (30mph) against the stipulated 16kph (10mph) of the trials – and became the conventional form of locomotive boiler. A water-jacketed firebox, which in *Rocket's* case had been added at the rear of the barrel, was soon incorporated into the structure of the boiler proper, and a smokebox, with chimney mounted upon it, added at the front.

To the English, Booth and the Stephensons are the originators of the multi-tubular firetube boiler: the French credit Marc Seguin. Seguin had obtained two locomotives in 1828 from Robert Stephenson & Co. for the Lyons & St Etienne Railway, then under construction. They proved even less reliable than the locomotives on the Stockton & Darlington, probably because curves on the St Etienne line were much more frequent and demands on the locomotives therefore greater. Seguin designed a boiler of return-tube type, in which the return part was multi-tubular. After experiments with a stationary boiler of this pattern, Seguin built a locomotive incorporating it in 1829, but whether this or *Rocket* was completed first is uncertain. Seguin and Booth were, however, ignorant of one another's activities at this period. Seguin's locomotive would not have steamed as well as Stephenson's for, unlike *Rocket*, in which exhaust steam was turned up the chimney, it relied on fans to blow the fire. Evidently

Seguin quickly discovered their inadequacy, for he too was soon using exhaust steam to draw the fire.

Americans, too, were at this stage closely observing railway development in Britain. In 1829, Horatio Allen obtained the locomotive *Stourbridge Lion* from Foster, Rastrick & Co. of Stourbridge. (Her sister locomotive *Agenoria* is now the oldest locomotive in Britain's National Railway Museum.) She was intended for a railway belonging to the Delaware & Hudson Canal Co., but on trial she caused excessive damage to light track laid with wooden rails and iron straps, and was not put into regular service.

The mainstream of locomotive development continued to be that of the Stephensons in Britain. *Rocket's* cylinders were originally inclined at about 45°, which enabled her to be carried on springs, but she was still cumbersome at speed. The cylinders were soon lowered so as to be nearer horizontal. Then, in 1830, Robert Stephenson designed the locomotive *Planet* for the L&MR with cylinders positioned horizontally beneath the smokebox, and drive from the connecting rods to a crank axle. In its layout, *Planet* was the ancestor of all later conventional steam locomotives.

THE RAILWAY AT WORK

The Liverpool & Manchester Railway was opened on 15 September 1830. It had been promoted to carry freight between the growing port of Liverpool and the growing industrial town of Manchester, but it was in fact as a passenger carrier that it saw great and near-instant success. Britain at this period was crisscrossed with a highly developed network of horse-drawn coach services – over the previous fifty years improvements to the roads, and in administration of coaches, had halved journey times (see Chapter 8). Coaching was an effective and admired means of transport for passengers and mails. Yet however good the administration, however smooth the roads, coaches could never exceed the speed of galloping horses, which in practice, allowing for halts to change horses every ten miles or so, meant overall speeds of about 16kph (10mph).

Rocket demonstrated at Rainhill for the first time the superiority of the steam railway over all other known forms of land transport. This was confirmed as soon as the Liverpool & Manchester was opened. A journey of four and a half hours by coach was reduced to one of two hours or less by train. Furthermore, the fare of ten shillings for the best coach accommodation was reduced to seven shillings for first class rail. Train speeds were subsequently increased, and fares reduced. By 1832 there was but one road coach left between Liverpool and Manchester, where formerly there had been 26. This effect was to be repeated wherever main line railways were built in Britain.

The actual vehicles in which the L&M passengers were carried cannot have seemed too unfamiliar. In the absence of a precedent the first class railway coaches were made up of three road coach bodies of the period mounted upon a single four-wheeled railway wagon underframe. When, later, railway coach bodies were built as a unit, the practice of dividing the accommodation into separate compartments continued. Those wealthy enough to travel by road in their own carriages took advantage of railway speed: private carriages, with their occupants, were carried on flat wagons wherever the railway coincided with a traveller's intended route, or part of it. For those accustomed to travel outside on road coaches, second class railway coaches, simple but roofed, were provided, third class coaches, for the poor, were little more than open boxes on wheels, but to occupants whose only other option was travel at walking pace, whether on their own feet or by road carrier's wagon, the speed alone must have seemed miraculous. Third class passengers were to be entitled by law from 1844 to be carried in carriages provided with seats and protected from the weather, in trains which ran at least once a day over each line.

Railway coaches were coupled together by chains; dumb buffers, that is, extensions of the wooden frames, were soon made slightly resilient by leather padding. Spring buffers and screw couplings minimized the jerks of starting and stopping: they were invented by Henry Booth and introduced in the early 1830s. For goods traffic, which began nearly three months after passenger traffic, the L&MR provided four-wheeled open and flat wagons; for cattle and pigs, open wagons with slatted sides, and for sheep double-deck vans. Coal was carried in open wagons, and also in small containers, the practice of canals in the vicinity.

The track of the Liverpool & Manchester Railway was made of malleable iron rails, weighing 35lb/yd (17.4kg/m) and 15ft (4.5m) long, carried in cast-iron chairs mounted on stone blocks or, in places where subsidence was possible, on oak sleepers. As for the gauge of the track, wagonways of north-east England had always been laid to gauges between four and five feet, convenient for a wagon to be hauled by a single horse, and George Stephenson adopted the gauge of 4ft 8in. (142cm) for the Stockton & Darlington Railway and then the Liverpool & Manchester. Later it was found necessary to add half an inch, to produce the British standard gauge of 4ft $8\frac{1}{2}$in. (1.435m). The rails soon proved too weak for locomotives and from 1833 rails weighing 50lb/yd (24.8kg/m) were substituted. The track was double, each track being used for trains in one direction: this was an early and important advance on the Stockton & Darlington, which was single track with passing sidings, an arrangement which would scarcely have been practicable with the much higher speeds of the L&M. Policemen were stationed at intervals of a mile or so along the route and indicated to drivers whether the road was clear or obstructed, using hand signals, or swinging lamps at night. From 1833, flagpoles were installed at junctions and crossings, from which flags were displayed to warn of obstructions, and the following year the first fixed signals were erected. Vertically pivoted boards were positioned at right angles to the track to indicate danger or parallel to it for 'clear'. The practice arose of allowing one train to follow another only after a fixed minimum interval of time had elapsed – the time interval system.

Apart from its western extremity, where tunnels operated by cable haulage led down to the docks and up to the passenger terminus, the route was for the most part easily graded and without sharp curves. To achieve this, there were substantial cuttings, embankments and viaducts. In these and other civil engineering works such as a skew bridge at Rainhill, and construction of that part of the line which lay across the Chat Moss peat bog on a foundation of hurdles and branches of trees, the builders of the railway used established techniques developed for canal and road construction.

Locomotive development continued swiftly. The Stephensons had shown the way, but other engineers were quick to develop their own ideas and establish locomotive building works. Notable among them was Edward Bury, whose locomotive *Liverpool*, built in 1830, anticipated *Planet* in the use of near-horizontal cylinders driving a crank axle, but did not at first have a multitubular boiler: not until one was fitted did she perform satisfactorily. The frames of *Liverpool* and subsequent Bury locomotives were positioned 'inside', or between the wheels, and made from iron bar, giving them a deceptively insubstantial appearance. *Planet*, by contrast, had substantial outside frames of the 'sandwich' type – timber strengthened on both sides by iron plates, and additional inside frames which supported the crank axle. Sandwich frames became for many years a feature of Stephenson locomotives.

The original *Planet* had her four wheels uncoupled: she was a 2–2–0. (The usual Whyte notation used to describe locomotive wheel arrangements gives the number of leading wheels, followed by the number of driving wheels, coupled together in groups where this applies, followed in turn by the number of trailing wheels.) In 1831, Robert Stephenson & Co. built two modified *Planet*-type locomotives, for the L&MR's goods trains, as 0–4–0s, with the two pairs of wheels coupled by rods. Then, because four-wheeled locomotives were damaging the L&MR's light track, they developed the 2–2–0 *Planet* into the 2–2–2 *Patentee*. Locomotives with this wheel arrangement were to be familiar on British passenger trains for the rest of the nineteenth century. The 0–4–0 type was similarly developed into the 0–4–2 and the 0–6–0, which became the standard British goods locomotive.

The first locomotives with horizontal cylinders positioned outside the frames were 2–2–0s built in 1834 by George Forrester of Liverpool for the L&MR and for the Dublin & Kingstown Railway, and subsequently for the London & Greenwich Railway. The latter two, opened in 1834 and 1836–8 respectively, were the first steam railways to reach the capitals of Ireland and England. Neither was very long, and to that extent they were typical of many railways opened in the early and mid-1830s. Some of these lines adopted the steam locomotive technology of the L&MR, and branch lines, built by separate companies, had linked the L&MR with Warrington to the south (1831) and Wigan to the north (1832), subsequently extended to Preston. Others still continued in the earlier horse-railway tradition. Such was the Festiniog Railway, opened as a horse-and-gravity tramroad with an exceptionally narrow gauge of less than two feet (61cm) as late as 1836.

THE FIRST TRUNK LINES

Far more important than the short railways were the great trunk railways incorporated and built during the 1830s. Even before the Liverpool & Manchester Railway had been built there had been unrealized schemes for extensive long-distance horse railways: now the L&MR had demonstrated that long-distance steam railways were not merely practicable but highly desirable. None the less, it was only after considerable opposition in Parliament (from landowners, and those whose business lay with road or canal transport) that Acts were passed in 1833, on the same day, for the London & Birmingham Railway 180km (112 miles) long and the Grand Junction Railway 125km (78 miles). The latter was indeed to provide a grand junction line, from Birmingham to Warrington, whence Liverpool and Manchester were already accessible by rail, so that it would join three of the most important provincial towns in England. It was built, for the most part, under the supervision of the engineer Joseph Locke; the contractor for construction of its viaduct at Penkridge, Staffordshire, was Thomas Brassey, at the start of a career which was to make him the most famous of all Victorian railway contractors.

Robert Stephenson, engineer of the London & Birmingham, had a harder task than Locke, for his route lay across the grain of the land: deep cuttings through the ridges had to be excavated at Tring and Roade, and tunnels at Watford and (famous for its length of over 2km (1.25 miles)) Kilsby. By these means the ruling gradient on the line was kept to 1 in 330, for as long before as 1818, George Stephenson had carried out experiments which had shown that a gradient as easy as 1 in 100 would reduce the load which could be hauled by a locomotive by half compared with a level line. Where the line left its London terminus, Euston, there was almost a mile of 1 in 70: on this inclined plane. stationary engines and cable haulage were at first used.

Other long-distance lines were also being authorized. The London & Southampton Railway received its Act of Parliament in 1834, and much of it was built by Locke and Brassey. The Great Western Railway, from London to Bristol, was authorized in 1835, and the engineer Isambard Kingdom Brunel successfully convinced his directors that the track should in due course be laid to the broad gauge of 7ft $0\frac{1}{4}$in (2.14m). In the same year the Newcastle & Carlisle Railway, which had been authorized, pre-Rainhill, in 1829 as the longest

British horse railway at 101km (63 miles) opened its first section, with locomotives. An injunction was served on it to prevent their use, and the company had to obtain a further Act to permit them.

In the following year, 1836, Acts of Parliament were passed for a whole series of railways to form a cross-country route between Gloucester, Birmingham, Derby, Leeds, York and Newcastle; and also for the Manchester & Leeds Railway and (to extend the Great Western route) the Bristol & Exeter. In 1837 a second cross-Pennine route was authorized, the Sheffield, Ashton-under-Lyne & Manchester Railway, and so was the Lancaster & Preston Junction Railway, to extend the line of the London & Birmingham and Grand Junction Railways further northwards.

Also in 1837 came the first fruits of this period of railway promotion and construction, with the opening of the Grand Junction Railway on 4 July. Part of it was laid with a form of track developed by Locke – double-headed rails, of 84lb/yd (41.7kg/m), held in chairs mounted on wooden sleepers. It was the prototype of the track, using bullhead rails, which was to be conventional in Britain for a century. Over Locke's track, Stephenson-built 2-2-2s hauled trains of coaches similar to those on the L&MR, and took four and a half hours for the 156km (97 miles) from Manchester to Birmingham: that is to say, an average speed of 34.75kph (21.6mph) including stops.

The first section of the London & Birmingham was opened later the same month, but extreme difficulties with the engineering works delayed completion. At one stage Robert Stephenson had no less than thirteen steam engines at work attempting to pump dry the quicksands encountered during construction of Kilsby tunnel. They were, eventually, successful, and the London & Birmingham was fully open throughout in September 1838. Intending passengers approached its station at Euston by a magnificent Doric arch, and at Birmingham Curzon Street the station building had a massive Ionic portico: the start of a world-wide custom that city stations were to receive the grandest architectural treatment of their day (see Chapter 18).

Opening of the L&B meant that there was through railway communication not only between London, Liverpool and Manchester, but also, within a month, as far north as Preston. One immediate consequence was that the travelling post office, in which letters were sorted on the move, became a permanent institution between London and Preston. The system had first been tried experimentally earlier in the year on the GJR.

Allied with this was the development of mail-bag exchange apparatus, with which bags could be dropped from, and picked up by, a train travelling at speed. It was first installed at Boxmoor on the London & Birmingham in 1837. Edward Bury was appointed locomotive superintendent of the L&BR, and supplied it with locomotives to his design.

The London & Southampton Railway was completed in 1840, having first changed its name to London & South Western Railway. The London & Croydon Railway, authorized in 1835 and in effect a branch of the London & Greenwich, was opened in 1839, and the route extended to Brighton by the London & Brighton Railway opened in 1841. The Great Western opened its first section, at the London end, in 1838, and reached Bristol in 1841: Box tunnel, by which the line pierced a spur of the Cotswolds on a gradient of 1 in 100, had taken five years to make. By 1841 the Bristol & Exeter was already open to Bridgwater, and reached Exeter in 1844. The broad gauge Bristol & Gloucester Railway was opened the same year, and at Gloucester for the first time broad gauge met standard gauge, for the Birmingham & Gloucester had been opened in 1840. Most of the other railways authorized in 1836 to make a south-west to north-east route were completed between 1839 and 1844; the Manchester & Leeds was opened in 1840 and the Sheffield, Ashton-under-Lyne & Manchester in 1845. The first trunk railway in Scotland was the Edinburgh & Glasgow authorized in 1838 and opened in 1842.

RAILWAY PROMOTION

The early 1840s were a period of consolidation, when the railways authorized during the latter part of the previous decade were built and opened. Then, as the early main lines started to demonstrate how successful railways could be, not only as public utilities but also in their return to investors, interest in railway promotion re-awoke and soon, as speculators replaced investors, inflated rapidly into a vast financial bubble, the railway mania of 1845. Speculators would have covered Britain with a dense network of lines, but many of these railways were paper schemes which vanished with the collapse of the mania in 1846. Nevertheless there were sound substantial projects among them, and from this period there date many railways which were to form important components of the British railway network. From a little over 2570km (1600 miles) in 1842, the distance covered had grown to 9790km (6084 miles) in 1850, and by 1860 the basic railway network of lowland Britain was in existence. There remained to be built lines through the upland areas of Wales and Scotland, branch lines in country districts elsewhere, suburban lines, and main lines which would provide alternative routes to some of those already in existence and would be built, in accordance with the spirit of Victorian Britain, by competing companies. In this way most of the remainder of the railway system was built up over the ensuing sixty years.

Brunel's broad gauge had enabled the Great Western to run trains which were faster, steadier and more comfortable than those of the standard gauge railways: the disadvantage was the need to transfer passengers, freight and frightened livestock between trains wherever there was a break of gauge. It first became apparent at Gloucester and the supposed chaos of transfer at that point was probably exaggerated by protagonists of the standard gauge companies established in the Midlands and the North of England, which did not wish to see the Great Western authorized to extend into their territory. The outcome was the government-appointed Gauge Commission of 1845 which heard evidence from both sides, competitive trials of broad and standard gauge trains (suggested to the commission by Brunel) in which the broad gauge trains proved vastly superior, and the Gauge of Railways Act 1846, which nevertheless made the 4ft 8½in gauge the legal standard, to which all new passenger railways in Britain must be built unless they extended existing broad gauge lines. The greater extent of standard gauge railways, 3600km (2236 miles) in 1844 compared with 375km (223 miles) of broad, had carried the day. Most broad gauge lines were eventually either made mixed gauge, by the addition of an extra rail, or narrowed; the last broad gauge sections were eventually converted in 1892. In the mid-1840s there was nevertheless a general feeling among engineers that, purely from an engineering viewpoint, a broader gauge than 4ft 8½in was desirable, even though Brunel's 7ft was excessive; and for Ireland, where the length of railways completed was still comparatively small, the Gauge of Railways Act set the standard gauge at 5ft 3in (1.60m), existing railways being altered to suit.

MAIN LINE MOTIVE POWER AND OPERATION

Locomotive developments, 1840–50

Perhaps the most important general consequence of Brunel's broad gauge was the inducement it provided for designers of standard gauge locomotives to improve the breed. During the early 1840s the principal passenger trains on the Great Western were being hauled by 2–2–2s with 2.13m (7ft) diameter driving wheels; sixty-two of these had been built to a design which locomotive superintendent Daniel Gooch had developed from *North Star*, the GWR's first successful locomotive which Robert Stephenson & Co had built in 1837. During the gauge trials, one of these averaged nearly 87kph (54mph) from Didcot to London with an 80-tonne train. An enlarged and improved version, the 4–2–2 Iron Duke class of 1846, had 2.4m (8ft) diameter driving wheels and a boiler which provided 431.8m³ (4647.4ft²) of heating surface. Such locomotives ran until the end of the broad gauge.

The broad gauge allowed ample space in which to arrange the components of a locomotive. The standard gauge, it seemed, did not, particularly in view of the opinion of the period that, for stability, the centre of gravity must be kept low. In an attempt to increase boiler size and power, Robert Stephenson introduced the long boiler type of locomotive in 1841: the wheelbase was short, the boiler long, with the firebox overhanging the rear axle. Such locomotives not unnaturally proved unsteady at speed, particularly when built with outside cylinders. During the gauge trials, one of them overturned. It was to a long-boiler locomotive, however, that R. Stephenson & Co. in 1842 first fitted link motion valve gear in place of the gab gear used previously.

T. R. Crampton believed that a locomotive would ride steadily if its centre of gravity lay as near as possible to the horizontal line of the drawbar, and in 1842 patented two types of locomotive to achieve this. In one of them, the type which is generally associated with his name, the driving axle was situated to the rear of the firebox: this not only enabled a large boiler to be positioned low down, but meant that the cylinders, though outside, could be brought back to a position about mid-way along the locomotive and minimize any tendency for it to nose from side to side. Locomotives of this type were built from 1846 onwards; the London & North Western Railway (which had been formed from the London & Birmingham, Grand Junction and Liverpool & Manchester Railways) was an early user.

In 1848 an exceptionally large Crampton locomotive, the *Liverpool*, was built for the LNWR with 2.4m (8ft) diameter driving wheels and a total heating surface of 212.75m^2 (2290ft^2) *Liverpool* had three pairs of carrying wheels and a total wheelbase of 5.6m (18ft 6in). She was on view at the Great Exhibition of 1851 where the public could compare her with the latest GWR broad gauge 4-2-2, *Lord of the Isles*. But whereas the Great Western locomotive was an example of a type already proven and highly successful, *Liverpool* damaged the track and was not perpetuated. She remained something of a freak, though Cramptons of lesser size did enjoy some popularity for several years on British railways and considerably more abroad. A much greater freak was the single locomotive built to Crampton's other patented type, in which the boiler was located below the driving axle. This locomotive, *Cornwall*, was built by the LNWR at the same time that it built the first of its rear-driving-wheel Cramptons; her driving wheels were 2.6m (8ft 6in) in diameter and their axle passed through a recess in the top of the boiler. In 1858, by which date concern for a low centre of gravity had largely passed, *Cornwall* was rebuilt with a new boiler mounted in the normal location.

Before that, indeed, during a period when many strange and curious designs for standard gauge locomotives were attempted, some engineers had already shown the way ahead: that it was wholly possible to design standard gauge locomotives which were simple and well-proportioned, rode well and had ample power for the work to be done. Notable among them was David Joy. As chief draughtsman for E. B. Wilson's Railway Foundry, Leeds, Joy designed the *Jenny Lind* class 2-2-2 of which the first example was built for the London & Brighton Railway in 1847 (see Figure 11.2). The boiler was long and low, but the inside frames on which cylinders and driving axle were mounted stopped short at the firebox; leading and trailing wheels were mounted on outside

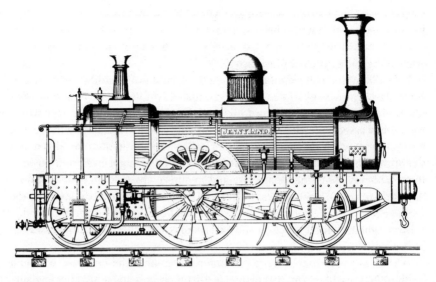

Figure 11.2: *Jenny Lind*, simple and well-proportioned, was the first of a highly successful class of passenger locomotives designed by David Joy and built from 1847 onwards.

frames, which meant that the firebox could be wide and the total heating surface was over $75m^2$ ($800ft^2$): this, and a steam pressure of 8.3 bar (in a period when around 6 bar was usual) produced a machine of adequate power which, in the words of her designer, 'at high speeds, always rolled softly, and did not jump and kick at a curve'.

Many comparable designs of 2–2–2s were introduced, and 2–4–0s developed from them. There had been isolated instances in the 1830s and 1840s of locomotives with four wheels coupled, but they began to come into general use in the 1840s. The same applied to tank locomotives, and (to a lesser extent) locomotives with pivoted bogies on which two pairs of carrying wheels were mounted.

Atmospheric traction

During the 1840s there was a period when it seemed to many people that steam locomotives might be wholly superseded by atmospheric traction. In this system, patented by Clegg and the Samuda brothers in 1839, a large diameter iron tube was laid between the rails, with a continuous longitudinal slot along its top. This was sealed by a leather flap. A piston within the pipe was connected by a bracket through the slot to the leading vehicle of a train; the flap was opened ahead of the bracket and re-sealed behind it. When the pipe ahead of the train was exhausted by a stationary steam pump, atmospheric pressure propelled the train forwards. On trial, it did so remarkably effectively. The system was installed on a one-mile extension of the Dublin & Kingstown Railway, with an average gradient of 1 in 128, and trains ran at 48kph (30mph) in service and considerably faster on test. It was then installed on part of the London & Croydon Railway, and Brunel began to install it on the steeply graded South Devon Railway. But in practice there were numerous problems – not least that of keeping the long leather flap airtight. The Croydon and Devon installations were never fully operational, the Irish one was discontinued in 1855. There was one feature of the Croydon line, however, which was a true forerunner of future developments: because of the difficulty of installing points and crossings on a line with an atmospheric tube, the single Croydon track equipped for atmospheric traction was, near Norwood, carried up and over the steam tracks by the first flyover.

Signalling

The London & Croydon, or rather its engineer C. H. Gregory, was responsible at this period for another development of lasting importance. This was the semaphore signal. By the early 1840s railways such as the Grand Junction and the Great Western were using disc signals, pivoted discs of various shapes which faced the driver of an oncoming train to indicate danger and were turned edgeways to indicate clear. Oil lamps displayed various colours to give equivalent signals by night. Gregory's signal was an adaptation of the visual semaphore telegraph apparatus of the period (see p. 711ff.). Two arms, one for each direction of traffic, were mounted at the top of the post and operated by handles at the foot: an arm when horizontal indicated 'danger', one lowered to 45° 'caution', and a vertical arm (which was hidden within a slot in the post) 'clear'. The first such signal was installed at New Cross, London, in 1841; they were installed on the London & Brighton Railway shortly afterwards and eventually came into general use. In 1843, at Bricklayers Arms Junction, Gregory introduced a simple form of interlocking to prevent signals for conflicting routes showing clear simultaneously, achieved by actuating the signals by foot-operated stirrups which were linked together. A lever frame operated the points.

Gradually, levers to work signals and points were brought together in lever frames housed in signal boxes, and the signals positioned above the box. The first signalling installation in which points and signals were fully interlocked was designed by Austin Chambers and made at Kentish Town Junction, North London Railway, in 1859. Eventually, signal arms for different directions of travel came to be mounted separately; two indications only were given, 'stop' (horizontal) and 'line clear' (lowered to 45°), and signal posts were positioned at the place which, when 'stop' was shown, trains must not pass.

Of greater importance even than effective signals for railway operating was the electric telegraph, which fortuitously was brought to practical form by William F. Cooke and Charles Wheatstone (see p. 714) at precisely the period that the first trunk railways were being built. Cooke carried out successful experiments alongside the London & Birmingham Railway between Euston and Camden Town in 1837 and, although the L&B company did not take up the telegraph at that time, the following year at the instance of Brunel the Great Western decided to do so. The electric telegraph was completed between Paddington and West Drayton in 1839 and subsequently extended, and it was used to pass messages of all sorts, from requests for horses to be available for passengers on arrival at Paddington, to notifying the presence of a suspicious character on a train which led to his arrest for murder.

The electric telegraph was first used to control train operations in 1840 on the London & Blackwall Railway, opened that year. This line was worked by reciprocating cable haulage: when coaches standing at the stations had been connected to the cable, the stationary engine driver was telegraphed that it was safe to start winding. The following year, 1841, the electric telegraph was installed through Clayton tunnel, London & Brighton Railway, between signalmen at each end with instructions not to let a train enter until the previous train had emerged. This was the first instance of the block system, of trains separated by an interval of space rather than time. Other installations were soon made through tunnels and on cable-worked inclines. The Norwich & Yarmouth Railway, opened in 1844, was the first railway to be worked throughout by the block system, using the electric telegraph on principles developed by Cooke, and because of this it had been possible to build it as a single line with passing loops. That was an economy, but in general the installation of a full electric telegraph system on an existing line was expensive, and it was many years before the block system wholly superseded time interval working.

EARLY RAILWAY DEVELOPMENT IN THE UNITED STATES

In the Britain of the 1830s there was already a comprehensive transport network for passengers and goods by road, canal and coastal shipping built up over the previous eighty years, but steam railways showed sufficient advantages to be superimposed on this. In the USA there was no such network, despite construction of some canals and long-distance roads in the eastern states and, particularly, rapid increase of steamboats on rivers and coastal waterways since their introduction in 1807 by Robert Fulton. Distances were longer and the demand for improved transport was even greater, especially for better means of crossing the Appalachian Mountains to link the coastal states with the interior. So although the earliest lines were short, as in Britain, and often connected with canals, much longer lines were soon being built. The first wholly steam-operated railway in the USA was the South Carolina Railroad, which opened its first short section late in 1830; by 1833 it had 220km (136 miles) of track in operation. The Baltimore & Ohio was 122km (76 miles) long by 1835 and entirely operated by steam also; in that year there were some 1600km (1000 miles) of railway in the USA, and by 1850 this had grown to nearly 14,500km (9000 miles).

Therefore, while British railways had substantial and expensive engineering works designed apparently to last for all time, a different engineering tradition rapidly developed in the USA. (So did a distinct railway terminology, but for simplicity British terms will be used here. A list of some British terms with their American equivalents appears on pp. 607–8). In the USA there was a strong inducement to build railways cheaply and quickly, with sharp curves, steep gradients and timber trestle bridges – improvements could came later, and be paid for out of revenue. Iron was scarce, timber plentiful: so track at first continued to consist of iron straps along wooden rails. Iron flat-bottom rails were first imported from England in 1830, but it was not until 1844 that they were first rolled in the USA, and subsequently American track was typically made up of light flat-bottom rails spiked to innumerable sleepers spaced far more closely together than elsewhere. Various gauges were used from 4ft $8\frac{1}{2}$in. up to 6ft (1.43–1.83m); the broad gauge lines were eventually converted to standard.

Light, hastily laid track prompted consequential developments in locomotive design. Robert Stephenson is said to have recommended the use of bogies to a party of visiting American engineers as early as 1830 or thereabouts, though the *John Bull* supplied by Robert Stephenson & Co to the Camden & Amboy Railroad, New Jersey, in 1831 was a 0–4–0 derived from *Planet*. Having seen her, Matthias Baldwin built his first locomotive: the Baldwin Locomotive Works were in due course to build in enormous quantity. After a couple of years a separate two-wheeled sub-frame or pilot was added to *John Bull* at the front, to reduce the number of derailments to which she was subject. *John Bull* is now preserved by the Smithsonian Institution, and is the oldest surviving workable steam locomotive.

The first bogie locomotive in the USA, a 4–2–0, was built by John B. Jervis in 1832, and the first locomotives of this type built in Britain were completed in 1833 for export to the USA. Such locomotives were well adapted to American conditions and became popular; at the end of the decade demand for greater power led to the enlargement of the type into the 4–4–0 with a front bogie and driving wheels before and behind the firebox. With bar frames derived originally from the designs for Edward Bury (but subsequently little used in the UK) this became the classic American locomotive of the nineteenth century.

Bogies were also adopted for both goods and passenger rolling stock; this meant that vehicles could be double the length of their British counterparts, and the interior layout of American passenger coaches was derived, apart from some very early examples, not from road coach practice as in Britain, but from the river steamboats with which Americans were already familiar. Instead of separate compartments, American coaches had open interiors with seats either side of a central passageway, and were entered from end balconies.

Most American railways were single track with passing loops. Development of the electric telegraph was invaluable in enabling a general dispatcher to issue instructions to agents at stations to deviate from the timetable if necessary. The telegraph was first used for this purpose in 1851.

During the 1850s the American railway system continued to expand rapidly, with the emphasis on states west of the Appalachians. Where railways were built into regions still to be settled, grants of public lands were made to the railway companies which they could sell, once their lines were open, to pay off construction costs and, from subsequent settlement, provide traffic. The pattern thus emerged, throughout much of the USA, that railways came first and settlement followed.

By 1860 there were over 48,000km (30,000 miles) of railway in the USA and still construction continued, with scarcely a hiccup caused by the Civil War. Now began perhaps the greatest of all railway epics – construction of the first transcontinental route to link the East Coast with the West. The Central Pacific Railroad, run by hard-headed local merchants, built eastwards from Sacramento, California, up and over the Sierra Nevada, and immigrant Chinese proved their worth as navvies a decade before there was any sort of railway in China itself. The Union Pacific, headed by shady Eastern financiers spurred on by land grants per mile of track and aided by the tracklaying abilities of Irish navvies, built rapidly west from the railhead on the Missouri. When the two lines met at Promontory Point, Utah, in 1869 they had built a railway 2864km (1780 miles) long.

CONTINENTAL EUROPE

While trunk railways were being built in Britain and the USA, they started to spread to European countries also. The first steam railway in Germany, running five miles from Nuremberg to Furth, was opened in 1835. A Stephenson 2-2-2 hauled the first train. By the end of the decade long-distance lines were being built and by 1850 some 4800km (3000 miles) were open. In Belgium a state railway system was planned from the start, with the first section opened in 1835, and the first steam railway in Austria was opened in 1837.

Construction of main line railways in France was delayed by ideological arguments about whether funding and control should be by the state or by private enterprise. Work on the first long-distance railway, from Paris to Rouen, began in 1841. It was financed in part by the London & South Western Railway (which saw it as an extension of its own line), engineered by Locke and built by Brassey, and was opened in 1843. Meanwhile legislation of 1842 produced a plan for a national system radiating from Paris, and by 1848 about 2000km (1250 miles) of railway had been built.

Steam railways came later to some countries – Switzerland, for instance, where the first, 24km (15 mile) long, line from Zurich to Baden was opened in 1847. In Imperial Russia the first 22.5km (14 mile) line from St Petersburg to Tsarskoye Selo was opened in 1837, but it was another ten years before the first section of the St Petersburg–Moscow railway was open, using the broad gauge of 1.525m (5ft) which became the Russian standard. Further afield, the first steam railway in India was opened in 1854, and so was the first in Australia: both used broad gauges, of 5ft 6in (1.675m) and 5ft 3in (1.60m) respectively. From these starts, railways quickly spread. In Australia, although Victoria and South Australia used the 5ft 3in gauge, New South Wales built to 4ft 8½in (1.425m).

Locomotives and rolling stock on the continent of Europe were for the most part at first derived from British practice, although both the Crampton and long-boiler types saw much more prolonged use than in Britain – the former particularly in France, the latter, for slow-moving freight trains, in many countries. Outside cylinders tended to be preferred by continental designers, who indeed tended to place more fittings of all sorts on the outsides of their locomotives than their British counterparts. British designs generally had a simpler and, to many eyes, more handsome appearance. Notable continental contributions to locomotive engineering were the Belgian Egide Walschaert's valve gear, first used in 1848, and the injector for boiler feed water invented by the Frenchman Henri Giffard in 1859. American influence was noticeable in Austria, where the mountain-crossing exploits of American railroads such as the Baltimore & Ohio had been noted, and in Russia.

NARROW GAUGE

By the early 1860s traffic on the 1ft 11½in (60cm) gauge Festiniog Railway had increased to the point at which its engineer C. E. Spooner considered it necessary to replace horse haulage by steam locomotives. It was generally considered impossible to build locomotives for so narrow a gauge, but eventually two small 0-4-0 tank locomotives, with tenders for coal, were supplied by George England in 1863; they proved successful enough for passenger trains to be introduced in 1865. Traffic continued to increase and the former horse tramroad was in due course fully re-equipped as a steam railway. In 1869 it received its first double-ended locomotive carried on bogies to Robert Fairlie's patent of 1864, which proved considerably more powerful than two ordinary locomotives, and in 1872 two bogie coaches, the first in Britain, were put into service.

A railway of less than 2ft gauge, and abounding in sharp curves, which nevertheless carried heavy traffic efficiently and economically through a mountainous region, was a spectacle which had a marked effect on the thinking of railway engineers. This was emphasized by the absence of other railways of less than standard gauge, a consequence of the Gauge of Railways Act passed in 1846. It was realized that there was considerable scope for railways of narrow gauge, particularly in mountainous areas, where they could be constructed cheaply compared with standard gauge lines; not only were the engineering works smaller, but sharper curves meant there need be fewer of them; and the locomotives and rolling stock were small and cheap to buy and run in proportion.

Some narrow gauge lines were built in Britain – the Talyllyn Railway, opened in 1866, was the first narrow gauge railway laid out for steam traction from the start – but the effect in Britain, where there were still memories of the gauge controversy of the 1840s, proved less marked than in many other countries. An extensive network of 3ft (91.5cm) gauge railway was built in the Rocky Mountains in the USA; many railways with a gauge of one metre or less were built in India; in Australia, Queensland opened its first railway on the 3ft 6in. (1.06m) gauge as early as 1865 and Western Australia followed suit. Light railways of narrow gauge were built in many European countries as feeders to the standard gauge network whether the terrain was mountainous or not. France, Austria, Germany, Holland, all made use of them; and the metre gauge was much used in Switzerland, particularly in the Alps.

After about 1870 it became rare for a developing country, in which railways were being built for the first time, to use the full standard gauge; and in some instances short existing standard or broad gauge lines were narrowed. This was the case in New Zealand, where the 3ft 6in gauge was adopted early in the 1870s. The same thing happened in South Africa, where in 1872 the Cape government expropriated the few existing standard gauge lines and narrowed them to match the 3ft 6in gauge system built to open up the country. Most other southern or central African countries built to either the 3ft 6in or the metre gauge.

JAPAN AND CHINA

The introduction of railways to Japan is of particular interest. The Americans had completed a railway across the isthmus of Panama in 1855 and among its passengers, in 1860, was a party of samurai, members of a mission sent by Japan to the USA to ratify a treaty of friendship. These were the first Japanese to encounter rail travel, apart from a few shipwrecked mariners rescued by American ships and taken to the USA before return to Japan. The Japanese were amazed by the experience of rail travel, not merely by the speed, but even by the presence of several people travelling together in the same coach. For Japan had still to emerge from 200 years of isolation during which the government not only minimized contact with the outside world almost to non-existence, but also restricted travel within Japan itself as much as possible. There were no passenger vehicles: travellers had the options of an extremely uncomfortable palanquin, a packhorse or their own feet – most, whether from choice or economic necessity, chose to walk – and they were required to present a permit at barriers built across the highways at intervals in order to be allowed to proceed at all.

Attendants of the samurai on the American mission recorded extremely detailed observations of the Panama railway, from the locomotive, its construction and working to details of the track and signalling. In 1869 the Japanese government changed and isolation ended: the new rulers decided to construct railways. The superintendent appointed to the government railway bureau in 1872 was Inoue Masaru who during the 1860s had been studying civil engineering at London University (illegally, by then current Japanese thinking). From his studies in Britain at this period he concluded that a gauge of 1.06m would be most suitable for his mountainous homeland, and this was the gauge adopted.

The first railway in Japan, from Tokyo to Yokohama, 27km (17 miles) was opened in 1872. Japan was still totally unindustrialized: Britain provided not merely capital and equipment, but engineering and operating staff. But from the first the Japanese intended (with British encouragement) to build and operate their own railways and railway equipment themselves. Within 20 years they were doing so. The first locomotive built in Japan, a 2–4–2 tank locomotive, was completed in 1893. It incorporated many parts made in Britain, and the locomotive superintendent under whose guidance it was constructed was R. F. Trevithick, Richard Trevithick's grandson. By 1907 there were 10,760km (6690 miles) of railway in Japan, with over 2000 locomotives. Locomotive manufacturing progressed so rapidly that from 1915 imports were almost entirely discontinued.

China's entry into the railway age was even later and much slower. A first, short, line built by the British in the mid-1870s was dismantled later in the decade after a Chinese worker was run down and killed by the train, and it was not until the late 1890s that railway-building started in earnest, mostly on the standard gauge.

PULLMAN AND WAGONS-LIT

By the 1850s in the USA the marvel of rail travel was beginning to wear thin, and the discomforts, particularly during the night, were all too evident. At the end of the decade, several people including George Mortimer Pullman were experimenting with simple sleeping cars, and in 1865 Pullman's first purpose-built sleeping car, the *Pioneer*, entered service on the Chicago & Alton Railroad. By day it appeared a normal, though luxurious, passenger coach of the usual American pattern; by night, the seats converted into longitudinal sleeping berths, and upper berths, also longitudinal, hinged down from above the windows. Partitions separated berths from one another, heavy curtains separated them from the central gangway. Vehicles such as this proved popular and their use spread rapidly. Simple dining cars first went into service in 1863, and in 1868 a de luxe Pullman dining car was running on the Chicago & Alton Railroad.

In the late 1860s the Belgian Georges Nagelmackers visited the USA and was impressed by what Pullman was achieving with his sleeping cars. He returned home in 1869, determined to emulate it in Europe by forming a company to operate sleeping cars internationally, across frontiers between countries as well as between different railway administrations. By the end of 1872 he had sleeping cars on trial over three routes: Ostend–Berlin, Paris–Cologne and Vienna–Munich. They were four-wheeled vehicles, with three compartments each for four passengers, convertible for day or night use and with a side corridor. Nagelmackers's activities got off to a difficult start, however: finance, persuading railways to co-operate, and attracting the passengers all presented problems.

James Allport, general manager of the Midland Railway in England, made an extensive tour of American railways in 1872 and was as impressed as Nagelmackers had been by Pullman's sleeping cars. The Midland, old-established as a provincial railway based on Derby, had opened its own route to London, St Pancras, in 1868, and was even then building it own route through the Pennines to Carlisle. When complete this would enable the Midland, in co-operation with two Scottish companies, to compete for Anglo-Scottish traffic with the existing West Coast Route from Euston and the East Coast Route from Kings Cross. On both of these, six-wheeled sleeping cars were introduced in 1873.

But the Midland route to Scotland was long, and by British standards steeply graded. Some extra attraction was needed. Allport was already having improved coaches built, and in 1873 the Midland arranged with G. M. Pullman to build and send over some Pullman sleeping cars (they were shipped in parts and assembled at Derby) for operation on the Midland Railway. The first, the *Midland*, entered service early in 1874: it was much more luxurious than anything seen in Britain previously. It was also, of course, carried on bogies, the first standard gauge bogie coach in Britain. Several others followed, some of them 'parlour' or drawing room cars, luxury vehicles for daytime use. Pullman sleeping cars provided a great inducement to passengers to use the Midland route to Scotland when it was completed in 1876.

They were not, however, the only inducement. Ever since the Liverpool & Manchester, most British railways had offered three, and sometimes four, classes of passenger accommodation. On the Midland, Allport, pursuing his policy of improvement, had abolished second class in 1875. Second class coaches were re-graded third class and the old third class ones replaced as soon as possible: all third class carriages were to have upholstered seats. At the same period, the Midland started to build bogie coaches – with compartment bodies of British pattern – for its long-distance trains. It also continued to assemble Pullman cars at Derby for use on other railways as well as its own, and on occasion to modify them. In 1878 a parlour car was rebuilt to include a kitchen and became the first dining car to run in Britain the following year, on the Great Northern Railway between Leeds and London. Another car was modified in 1881 to become the first railway vehicle, anywhere, equipped with electric light.

On the Continent, Nagelmackers founded the Compagnie Internationale des Wagons-Lit in 1876 to take over his operation, and despite earlier difficulties it had 53 sleeping cars and 22 contracts to operate them. In 1881 the company first operated a dining car, and in 1883 the Orient Express, it passenger coaches being Wagons-Lit company sleeping cars and dining car, was established between Paris, Vienna and Giurgevo (now Giurgiu), Romania, whence ferry, rail and steamer services carried passengers onward to Constantinople. Six-wheeled coaches were used at first, but after a few months the company's first bogie sleeping cars and dining car were available. From 1889 the journey was made throughout by rail.

LARGER LOCOMOTIVES

Although the 4–4–0 was the typical American locomotive during the middle of the nineteenth century, by the 1860s and 1870s larger locomotives were needed for increasingly heavy trains. Traffic was increasing, but long single-track routes with few passing loops resulted in a tendency for trains to become longer and heavier rather than more frequent. Also, rail which was still light necessitated construction of locomotives with many wheels to spread axle loadings along the track. Three enlarged versions of the 4–4–0 developed. The ten-wheeler or 4–6–0, first built in 1847, was initially used, like the 4–4–0, for both passenger and freight trains. The 2–6–0 was introduced in the 1860s and, more powerful size-for-size than a 4–4–0, came to be used increasingly for freight trains. Where something more powerful still was needed, 2–8–0s were used – the first was built for the Lehigh Valley Railroad, Pennsylvania, about 1866. The 4–6–0 was largely displaced from freight trains, but locomotives of this wheel arrangement were used increasingly for heavy passenger trains. Speeds increased as the standard of track improved.

For special purposes much larger locomotives were built. The Central Pacific, on the east side of the Sierra Nevada, had a climb of 850m (2800ft) in 40km (25 miles), and for this a 4–8–0 was introduced in 1882. The following year an even larger version, the 4–10–0 *El Gobernador* was built, the largest locomotive of her day in the world, She remained an isolated example, but the 4–8–0 became the first of a type used successfully on this line for many years.

Brakes had originally been applied throughout a train by brakesmen moving from vehicle to vehicle. In 1868, George Westinghouse introduced continuous air brakes, operated throughout the train by compressed air supplied by the locomotive. They had the disadvantage that if a coupling broke and the train split, so did the air pipe, with the effect that the brakes ceased to operate on the disconnected part of the train. Westinghouse then developed the idea of an automatic air brake, patented in 1871, which would operate automatically on all vehicles in a train if it became divided. Air reservoirs on each vehicle were charged with compressed air from the locomotive through the train pipe. Reduced air pressure in the train pipe, deliberate or not, caused a valve to open and admit compressed air from the reservoir to the brake cylinder.

In Britain, trains tended to be lighter and more frequent than in the USA, and track, at least until towards the end of the century, was heavier and more substantial. W. Bridges Adams had invented the fishplate for joining rails in 1847, bringing to an end the practice of connecting rails by butting them together in joint chairs. Steel rails were first laid in 1857, and by the 1870s they were common on main lines. Locomotives, therefore, tended to be smaller and less powerful than in the USA, but could have high axle loadings. 'Single driver' locomotives, that is, with a single pair of driving wheels, continued to be popular for fast, light trains, for designers feared that coupling driving wheels together reduced a locomotive's ability to run fast. However the typical British locomotive for passenger trains in the last three decades of the century was the 4–4–0 with inside cylinders and frames and a front bogie; the first were built for the North British Railway in 1871. The inside-cylinder 0–6–0 continued to be the usual locomotive for freight trains.

British trains, then and now, were and are smaller in height and width than their standard gauge counterparts elsewhere: the price of having pioneered railways is that bridges and tunnels on early lines which seemed, when built, to be of ample size have later proved restrictive and so the British loading gauge, that is the dimensions to which trains can be built or loaded without risk of fouling lineside structures, is smaller than that of other countries.

During the 1870s British railways experimented, amid considerable controversy, with various types of continuous brakes – some operated by chains, or atmospheric pressure (vacuum brakes) rather than compressed air, some non-automatic, some automatic. In 1889 there was a serious accident near Armagh on the Great Northern Railway of Ireland: an excursion train fitted with non-automatic vacuum brakes stalled on a gradient and was divided, after which the rear portion ran away and collided with a following train. Eighty people were killed. Later the same year the Regulation of Railways Act 1889 made automatic continuous brakes compulsory on passenger trains. However, while some railway companies adopted Westinghouse brakes, more of them used vacuum brakes – a less effective system, since it is dependent on atmospheric pressure, though simpler, for the partial vacuum in the train pipe and cylinders could be created by a steam ejector on the locomotive rather than a pump. The desirable aim of having all goods wagons fitted with continuous brakes has eluded British railways down to the present day. Most European railways adopted air brakes on both passenger and freight stock.

The effect of the Armagh accident would have been mitigated if the block system had been in full operation on the section of line concerned, which was still being worked on the old time interval system. The 1889 Act not only made continuous brakes compulsory, but also the block system, and interlocking of points and signals. In the USA, block signalling came into use in the 1860s, and interlocking in the 1870s.

Another important safety device which came into use in Britain at this period affected single lines. These had been worked by timetable and telegraph, or (from about 1860) by train staff and ticket, or both. The train staff was a wood or metal staff, unique to its section of line, handed to a driver as authority to take his train on to it. Where two trains were to follow one another, the driver of the first was shown the staff and given a ticket authorizing him to proceed. This arrangement lacked flexibility, and in 1879 Edward Tyer introduced his electric train tablet apparatus: several metal tablets, each of which could authorize a driver to enter a section, were housed in instruments at each end of the section which were electrically linked so that only one tablet could be removed from them at any time. The tablet was a driver's authority to proceed, and the system enabled several trains to follow one another through a section. Electric train staff instruments introduced soon afterwards performed the same function but used train staffs of similar form to those used for staff and ticket working.

COMPOUNDS

Compounding, that is to say expansion of steam successively in one cylinder after another, was applied to stationary engines in the 1760s and marine engines in the 1850s, after which, because of the resulting economy, it became popular in this usage of the steam engine (see p. 278). Application to locomotives was delayed until the late 1870s, presumably not only because of the difficulty of adapting it to the cramped and specialized layout of the locomotive, but also because exhaust steam was needed at a fairly high pressure to draw the fire, instead of being condensed as was usual in marine practice.

The first compound locomotive was designed by the Frenchman Anatole Mallet and built in 1876; it was a 0-4-2 tank locomotive with two cylinders and ran on the Bayonne–Biarritz Railway. From this start various systems of compounding were developed by French, German, English and American engineers, the principal variations being in the number and layout of cylinders employed. Some systems were successful, some less so: compounding became a subject of controversy as long as steam locomotives were being developed in the Western world, opinions being divided about whether its inherent economy was not outweighed by the extra complication and expense in manufacture, maintenance and operation.

Possibly the least successful but best publicized compounds were those designed for the London & North Western Railway by F. W. Webb from 1882 onwards. On many of these the two high pressure cylinders drove the front pair of driving wheels and the low pressure cylinder the rear pair, front and rear not being coupled together in the interest of free running. Valve gear for the low pressure cylinder was operated by a slip eccentric, with the ludicrous result, when the locomotive had backed on to a train, that if the front driving wheels slipped on starting, instead of moving the locomotive forward to set the slip eccentric, the rear wheels would revolve in reverse. Once these locomotives did get away, they could run well.

Two-cylinder compounds based more closely on Mallet's original ideas were more successful, and were built by August von Borries, locomotive superintendent of the Prussian State Railways, from 1880, and in England by T. W. Worsdell from 1884 for the Great Eastern Railway and subsequently for the North Eastern. A 4-4-0 locomotive of this type was eventually rebuilt with three cylinders, a single high-pressure cylinder inside and two low pressure ones outside, and became the prototype of similar locomotives built by S. W. Johnson for the Midland Railway from 1901 onwards which proved highly successful over many years. In the United States, Samuel Vauclain's system of compounding used four cylinders, a pair of high and low pressure cylinders being positioned on each side of the locomotive with the piston rods of the cylinders in each pair connected to a common crosshead. In France, Alfred de Glehn and Gaston du Bousquet developed successful four-cylinder compounds, with the two high pressure cylinders driving one axle and the low pressure ones another.

Mallet himself had meanwhile been developing his ideas in a divergent direction. In 1884 he patented a four-cylinder compound system in which the two high pressure cylinders drove one set of wheels at the rear of the locomotive and the low pressure another at the front: these wheels were carried on a separate set of frames, articulated to the locomotive's main frames. Locomotives to this patent were first built by Decauville Aîné, a firm which since the middle of the previous decade had been developing light and portable railways of 40, 50 and 60cm gauge for agricultural and industrial use. In 1888, Decauville demonstrated a 0-4-4-0 Mallet tank locomotive on a temporary 60cm gauge passenger line at Lâon, and followed this by a similar but more extensive line serving the Paris international exhibition of 1889 which used ten Mallets. Thenceforward the term Mallet came to be associated with articulation rather than compounding. The type became popular on narrow gauge railways on the continent of Europe, and was also used on main lines. The first Mallet in America was a 0–6–6–0 built to the standard gauge for the Baltimore & Ohio in 1904. In the USA increasing freight train loads had already led to construction of locomotives as large as 2–10–2s, and Mallet's articulation principle, which had first been used on very small locomotives, was soon seized on in the USA as a means of building some extra-large ones.

SPECIALIZED RAILWAYS

By the early 1860s, traffic in London streets suffered much from congestion, and it was largely to help relieve it that the Metropolitan Railway was built from Paddington to Farringdon Street: the first underground railway, its tunnels were built beneath streets by 'cut-and-cover': that is, the course of the railway was excavated and subsequently covered over for the street to be reinstated. It was opened in 1863, worked by steam locomotives which, to minimize exhaust nuisance so far as possible, condensed their used steam, and in due course it was extended. By 1868 it had reached South Kensington, and the Metropolitan District Railway was opened from there to Westminster. Both railways were subsequently extended further. New York and other American cities solved similar problems by constructing elevated railways on continuous viaducts above the streets.

New York had had horse-drawn passenger tramways through its streets since 1832, and Paris since 1855. The first street tramway in Britain was built in 1860 in Birkenhead by the American G. F. Train, and the first successful tramway in London was opened in 1870. Horse trams became a familiar part of the steet scene in many British cities; in the late 1870s steam trams came into use (generally, small, enclosed steam locomotives were used to haul tramcars), and from the 1880s cable haulage, which had first been applied to street tramways in 1873 in San Francisco, was also used in Britain.

For climbing steep gradients, J. B. Fell patented a central-rail system in 1863. The additional rail, of double-headed cross-section, was laid on its side, and additional powered wheels positioned horizontally beneath the locomotive were pressed against either side of the central rail by a combination of levers and bevel wheels. The system worked satisfactorily over a gradient of 1 in 12 on a test line in Derbyshire, and was then installed on a temporary line over the Mont Cenis pass in the Alps, pending completion of the tunnel beneath. Its best-known application, however, was to the Rimutaka Incline in New Zealand, 4.8km (3 miles) of 1 in 15 on the line from Wellington to Masterton, where it lasted until an 8km (5 miles) diversionary tunnel was completed in 1955.

The first mountain rack railway, the Mount Washington Cog Railway, was built by Sylvester Marsh in New Hampshire, USA; the first section was opened in 1868 and the summit of the mountain was reached the following year. Marsh had revived Blenkinsop's early rack railway idea, but positioned his ladder-type rack centrally between the running rails to climb gradients as steep as 1 in 5. The Swiss engineer Niklaus Riggenbach had been working independently on a similar type of rack railway, and opened his first line up the Rigi mountain near Lucerne in 1871, reaching the top in 1873.

Roman Abt, one of Riggenbach's engineers, developed an improved type of rack in the early 1880s in which the rack teeth were machined in a steel bar; two bars were positioned side by side with the teeth of one bar staggered in relation to those of the other. The system was first used on a line in the Harz Mountains in Germany, opened in 1886. In this application, only parts of the line were rack-equipped, the remainder being easily enough graded to be worked by adhesion. Such rack-and-adhesion lines were built in mountainous areas elsewhere, both in the Alps, such as the Brunig line in Switzerland, and further afield, such as the Nilgiri line in India, and locomotives able to work by adhesion or rack were built for them. The Abt system became the most widely used rack system.

The only rack railway in Britain, the Snowdon Mountain Railway, was completed in 1896 using this system and Swiss-built locomotives. Unusually, guard rails of inverted 'L'-shape are positioned either side of the rack, the consequence of an accident on the opening day when a locomotive's rack pinions mounted the rack so that it ran away, became derailed and fell over a precipice. Subsequently, if the pinions started to lift, grippers beneath the locomotive would make contact with the guard rails and prevent the pinions from becoming disengaged from the rack.

The Pilatus Railway, opened near Lucerne in 1889, used the Locher rack system: a rack rail positioned horizontally has teeth both sides, and horizontal pinions mesh with it. This enabled the line to be built with a ruling gradient as steep as 1 in 2: it remains unique. Such sharp gradients have otherwise been the prerogative of funiculars, short railways independent of main systems, operated by cable haulage over gradients too steep for rack. Such lines were built in Britain as cliff railways from the 1870s onwards, and on the continent of Europe from the same period.

EARLY ELECTRIFICATION

Attempts were made to drive railway vehicles by electricity in the 1830s and in 1842, Robert Davidson successfully operated a battery-electric locomotive on the Edinburgh & Glasgow Railway, hauling about six tonnes at 6.5kph (4mph). The first electric railway was operated by Werner von Siemens at an exhibition in Berlin in 1879. It was of metre gauge and a third rail between the running rails was used to supply electricity, generated by a dynamo at 150 volts, to a small four-wheeled electric locomotive hauling a passenger-carrying train. The railway was subsequently operated in Dusseldorf, Brussels and, in 1881, at the Crystal Palace in London. The same year Siemens started to operate a 2.5km ($1\frac{1}{2}$ miles) electric railway on the outskirts of Berlin, on which passengers were carried in horse tramcars to which electric motors had been fitted.

In 1883, Magnus Volk built his little electric railway along the sea front at Brighton, Sussex, which in modified form still runs, and the first section of the Giant's Causeway electric tramway was opened in Ireland. In the same year Leo Daft in the USA built an electric locomotive: current came from a central rail and this 8.9kW (12hp) locomotive, called *Ampere*, could haul ten tonnes. It had an electro-magnetic brake. Two years later Daft built an improved electric locomotive for the New York Elevated Railroad. From the mid-1880s electric street tramways were built in North American cities on which the cars collected current by trolley poles from overhead wires. The first electric tramway in Britain was opened in Blackpool in 1885: cars collected current from conductors in a conduit beneath the road surface. Overhead current collection was substituted later.

The first electric underground railway was the City & South London Railway, a deep-level tube railway opened in 1890. Its sixteen locomotives each had two 37.3kW (50hp) motors with armatures mounted on the axles. Current at 500 volts DC was supplied from a conductor rail. The same system of electricity supply was used on the Liverpool Overhead Railway opened in 1893, the only elevated railway built in Britain. Here too were seen for the first time in Britain electric trains consisting of motor coaches and trailer cars, without separate locomotives, and electrically operated semaphore signals which returned automatically to danger as a train passed.

In 1894–5 the Baltimore & Ohio Railroad electrified 11km (7 miles) of lines across the city of Baltimore, part overhead and part underground, at 675 volts DC with electricity drawn from a rigid overhead conductor. In 1899 the Burgdorf–Thun Railway in Switzerland was electrified using three-phase alternating current for the first time, at 750 volts, collected from double overhead wires, and in 1903 the 30.5km (19 miles) line from St Georges de Commiers to La Mure in France was electrified using direct current with a voltage as high as 2400: locomotives of 373kW (500hp) hauled trains weighing over 100 tonnes up gradients as steep as 1 in 38. The same year, on an 8km (5 mile) test line near Berlin electrified at 10,000 volts three-phase AC, speeds as fast as 209kph (130mph) were being reached.

MONORAILS

It was during the period before the First World War that monorails first caught public attention, although as early as 1824, H. R. Palmer had proposed a monorail in which divided vehicles would hang down either side of a single rail supported on trestles. An experimental horse-powered line was built, without lasting effect. About 1880, however, the French engineer C. F. M. T. Lartigue built about 190km (120 miles) of similar lines in North Africa along which wagons which hung down either side of the rail like panniers were hauled by horses or mules. The system was demonstrated in London in 1886, but with a steam locomotive designed by Mallet: it had a pair of vertical boilers and two grooved wheels. This led to construction of the 15km (9 miles) Lartigue monorail Listowel & Ballybunion Railway in Ireland, cheap to build, opened in 1888, steam worked, and successful in operation until 1924.

The managing director of the Lartigue Railway Construction Company was F. B. Behr who developed a high-speed electrically powered monorail, demonstrated in Belgium at 132kph (82mph) in 1897. He promoted a company which was authorized in 1901 to build such a line between Liverpool and Manchester, upon which the cars were to travel at 175kph (109mph): but the Board of Trade was hesitant over their braking abilities and capital for construction could not be raised. It was at this period, however, that the Wuppertal Schwebebahn was built in Germany, an overhead monorail with electrically powered cars suspended from it. The first section was opened to traffic in 1901; the line eventually extended to some 13km (8 miles) and continues to operate successfully to the present day.

In 1903, Louis Brennan patented a monorail system in which the cars would run on a single rail and use gyroscopes to maintain their stability; this was built and demonstrated in 1909 with a petrol-electric vehicle, but despite the attraction of cheap construction was never put into commercial use. There have been many subsequent proposals for monorails, some of which have been built. The Bennie monorail, in which a suspended car was driven by an airscrew, was demonstrated in the 1920s; the most notable of many systems proposed since the Second World War has been the Alweg system in which the track is a hollow concrete beam supported on concrete pylons; not strictly a monorail, for a narrow track on the top of the beam is used for the carrying wheels of the vehicles, while additional wheels bearing on the edges of the beam cater for side thrust. With the exception of the Wuppertal line, monorails and the like have seen little use in public service, and have always had greater success in exciting public imagination.

THE PEAK YEARS

In the period 1910–14, after eighty years of continuous development, railways were the foremost means of land transport. The only challenge to their supremacy came from the electric tramways that were beginning to bite into the inner suburban traffic of large cities, and they were simply a specialized form of railway. Neither the motor road vehicle nor the aeroplane had yet made a serious impact. The developed western countries – Europe and the USA – had largely completed their main line networks, although construction of lesser lines continued. Elsewhere there were few parts of the world where railways were unknown, but long-distance lines were still being built and in some countries, notably China, the railway system was still in its infancy (see p. 577).

The United States

The railway network of the United States extended at this period to about 404,000 route kilometres (253,000 miles), owned by some 2500 companies of which about 1000 operated the system's trains. There were about 60,000 locomotives. The USA was if anything somewhat overprovided with railways, and although fortunes had been made, that huge network had not been achieved without a great many company bankruptcies also. Approximately nine-tenths of the system was single track. Automatic couplings between wagons had been obligatory since 1893.

Britain

In Great Britain, there were about 33,800 route kilometres (21,000 miles) of railways; about half the system was double track, and services were by comparison frequent with many trains of moderate weight. Track was laid with bullhead rail weighing from 30 to more than 50kg/m (100lb/yd); most other countries used flat-bottom rail. The total included about 885km (550 miles) of light railways, some of standard gauge, some narrow, built under the provisions of the Light Railways Act 1896. This was a belated attempt to encourage improved rural transport by reducing both costs of promotion and standards of construction. It was largely ineffective, as comparable figures for some European countries, below, will indicate.

Continental Europe

France had a total of 50,700 route kilometres (31,500 miles) of railways of which about 38,600km (24,000 miles) were classified as *d'intérêt général*, the most reasonable interpretation of this term being main or trunk lines providing long-distance transport. Most of the system was operated by five big companies with concessions of limited duration from the state, the land on which the railways were built being considered public property. There were also nearly 11,300km (7000 miles) of railways *d'intérêt local* providing local transport within particular districts. Some 8500km (5300 miles) of these, and a small part of the railways *d'intérêt général*, were of narrow gauge – usually one metre, though 60cm was also used, largely at the instance of Decauville Aîné. In Germany, which then included Alsace-Lorraine, railways had been under the control of the imperial government since 1871, and were mostly owned by the individual German states. Main-line routes totalled about 60,000km (37,300 miles); there were a further 10,800km (6700 miles) or so of light railways, locally-owned and mostly of narrow gauge.

Of smaller European countries, Belgium had a main-line system of nearly 4700km (2900 miles), mostly state-owned, with a further 3900km of light railways, mostly of one metre gauge and with some sections electrified. The Swiss railway system extended to 3500km (2172 miles) of standard gauge, mostly owned by the Swiss Federal Railways, and 1000km (644 miles) of metre gauge, mostly private. A total of 760 route kilometres (472 miles) was already electrified. This included most of the metre gauge lines, but it also included important main lines – the federal railways' Simplon line with its 22.5km (14 mile) tunnel completed in 1907 and the company-owned Berne–Lötschberg–Simplon Railway opened to connect with it through the 14.5km (9 miles) Lötschberg Tunnel in 1913. This, the last of the great European main lines to be completed, was the final link in a chain of railways connecting France with Italy via Berne and avoiding the German-owned lines in Alsace-Lorraine. The Simplon and Lötschberg lines had been built for electric traction: the Simplon was electrified at 3000 volts 3-phase AC, but the Lötschberg line, following successful experiments elsewhere, was electrified at 15,000 volts single-phase AC. It had a ruling gradient of 1 in 37, up which 1864.25kW (2500hp) locomotives could haul trains weighing 310 tonnes.

Further afield, imperial Russia, where the great railway-building period had been the last decade of the nineteenth century, had a network of some 70,800 route kilometres (44,000 miles) in use. The first trans-Siberian line had been completed in 1904. A traveller could leave London, Charing Cross at nine a.m. on a Monday and Ostend the same day, arriving in Moscow on the Wednesday and Vladivostock at 2.50 p.m. on the Friday of the following week.

The British Empire

In Canada, although the Canadian Pacific's transcontinental line had been completed as long before as 1881, the railway network was still expanding fast: in 1912 there were 43,000 route kilometres (26,700 miles) in use and another 14,000km (8860 miles) under construction. India, which then also included what are now Pakistan, Bangladesh and Burma, had 27,000km (17,000 miles) of 5ft 6in (1.67m) gauge, 22,000km (13,800 miles) of metre gauge and 4300km (2,700 miles) of 2ft 6in (76cm) and 2ft (61cm) gauges, to give a total of about

53,300km (33,500 miles). The broad and metre gauges formed separate but interwoven networks. Narrower gauges were used for feeder lines and for lines in mountainous regions such as the 2ft gauge Darjeeling Himalayan Railway. In Australia on the other hand, where some 27,000 route kilometres (17,000 miles) of railways were operated by the states or the Commonwealth government, gauge varied according to the choice made by each state. The principal gauges in use were 5ft 3in (1.60m), 4ft 8½in (1.43m) and 3ft 6in. (1.06m). The mileage of 3ft 6in gauge exceeded the other two put together, but the 4ft 8½in gauge had been chosen for the trans-Australian railway then under construction by the Commonwealth government: it would connect with the 3ft 6in gauge lines of Western Australia at one end and the 5ft 3in gauge lines of South Australia at the other. It was opened in 1917. New Zealand had 4500km (2800 miles) of 3ft 6in gauge open; in South Africa there were about 13,600km (8500 miles) of 3ft 6in gauge and in Rhodesia a further 3800km (2350 miles), although the Cape-to-Cairo railway of Cecil Rhodes was and would remain a dream.

South America

Railways based on British practice, because they had been built by British capital, were common and extensive in South America. One such (with some mixture of American practice also) was the Central Railway of Peru, completed from Callao to Oroya in 1893 and subsequently extended. It gave access to the silver-mining area in the high Andes which in 1817 had attracted Richard Trevithick in search of the fortune that steam engines had failed to produce for him in England. Trevithick came with stationary mine-pumping engines, but proposed construction of railways also. That they were not built at the time is scarcely surprising: when the Central of Peru was eventually built its course, much of it climbing by successive zig-zags, took it over a summit no less than 4817.7m (15,806ft) above sea level. No other railway reached such an altitude.

Early twentieth-century locomotives

In the USA during the peak years, the high proportion of single track was associated, as both cause and effect, with the practice of running infrequent but heavy trains: compound Mallets as large as 2–8–8–2s were in use. Firing such large machines with coal was not easy: those of the Southern Pacific Railway burned oil. This had the added advantage that they could run cab-first, with the tender coupled behind the smokebox. In 1913 a triplex Mallet type locomotive was built for the Erie Railroad, with a third set of eight driving wheels beneath the tender, so that there were two high pressure and four low pressure cylinders. This locomotive proved able to haul a train weighing 16,300 tonnes at 24kph (15mph), the train being more than 2.5km (1½ miles) long. Its main task, however, and the task of the others like it built in 1916, was to help freight trains up steep gradients.

For fast passenger trains, the old 4–4–0 had evolved into the 4–4–2 'Atlantic' type during the early 1890s: with this arrangement, the firebox could be much enlarged by making it wider than the frames and positioning it above the trailing wheels. Reverting to the six driving wheels of the 4–6–0 produced a 4–6–2 in which a large, wide firebox could be allied with a large long boiler: the remaining components of the locomotive then fitted into place with a symmetry that was to make the 4–6–2 'Pacific' the classic express locomotive of the twentieth century. The first Pacifics were built by Baldwin in the USA in 1901 for the New Zealand Railways; from the following year the type was in service in America.

In Britain, lighter and more frequent trains meant that locomotives were generally smaller than those in the USA. However, eight-coupled 0–8–0 locomotives were in use for heavy freight trains on main lines, and a few railways had started to operate 2–8–0s. The 4–4–2 had been introduced to Britain by H. A. Ivatt of the Great Northern Railway in 1898, with limited success until four years later he produced a large-boiler, wide-firebox version. The first 4–6–0s in Britain were goods locomotives built for the Highland Railway in 1894; the first for express passenger work were built by the North Eastern Railway in 1899–1900. Within a few years the heaviest British expresses were usually being hauled by 4–4–2s or 4–6–0s. The only 4–6–2 so far built in Britain, by the Great Western Railway in 1908, had been a failure. The Great Western, however, was going through a period of rejuvenation which had commenced after the conversion of the last broad gauge lines in 1892. Locomotive superintendent G. J. Churchward was steadily improving locomotive design and experimenting to that end. In 1903–5 he obtained three De Glehn compound Atlantics from France for comparative trials with his own locomotives. He eventually adopted neither compounding nor Atlantics, but from the details of their design much was learnt and the four-cylinder simple (i.e., non-compound) 4–6–0 became the standard GWR express locomotive.

Continental railways employed locomotives which, though not as large as those of the USA, were generally larger than those in use in Britain. Belgium, France, Germany, for instance, all had Pacifics in service, and the Northern Railway of France was trying the 4–6–4 wheel arrangement. It was from Germany that there came perhaps the most important development in locomotive design since the multi-tubular boiler: the superheater. There had been many attempts to design superheaters for locomotive boilers, and none was particularly satisfactory until the Prussian engineer Wilhelm Schmidt designed the firetube type, in which steam en route from boiler to cylinders is passed through U-tubes or elements which are themselves housed within large diameter fire-tubes within the boiler barrel. This brings the steam close to the hottest gases leaving the firebox. Superheaters of this type were first used on the Belgian State Railways, and introduced to Great Britain in 1906 on two goods locomotives built by the Lancashire & Yorkshire Railway. In 1908, D. E. Marsh equipped his new 4–4–2 tank locomotives, designed to haul London, Brighton & South Coast Railway expresses, with Schmidt superheaters and these, in comparative trials with London & North Western 4–4–0s, showed marked economy in coal and water consumption. In 1910 the LNWR introduced its own superheated express locomotives and superheating quickly became standard practice worldwide.

In 1909, Beyer Peacock of Manchester used the patent of inventor H. W. Garratt for the first time when they built two articulated locomotives for a 2ft (61cm) gauge line in Tasmania. In the Beyer Garratt design the boiler supplies steam to two separate engine units, between which it is supported on pivots. One of the engine units carries the water tank, the other the coal supply. The next Beyer Garratt locomotive was built in 1911 for the 2ft gauge Darjeeling Himalayan Railway. Like the Mallet, the Beyer Garratt started small, but was eventually to be built in very large sizes. The breakthrough was to come in 1921 when a 2–6–0+0–6–2 Garratt was built for South African Railways: hauling the same load as a heavier Mallet, it proved faster and more economical. It also rode much better. Beyer Garratts were subsequently widely used (except in North America where Mallets were preferred) and especially in Africa.

Electrification

Several railways had been experimenting with petrol–electric railcars from about 1906 onwards. These self-propelled passenger vehicles used electric transmission to overcome the internal combustion engine's inflexibility. Railways which tried them included the London, Brighton & South Coast, the Delaware & Hudson, the Victorian Railways in Australia (see Figure 11.3) and the New Zealand Railways. The first, experimental, diesel-engined locomotive, of 746kW (1000hp), was built for the Prussian State Railways in 1912. Its mechanical transmission was inadequate for the task it had to perform, but the following year the first diesel railcar, with electric transmission, went into service in Sweden.

By 1912 there were numerous electrified railways in operation. The principal reasons for electrification included ease of ventilating long tunnels, low cost of hydro-electric power in some countries, high power for steeply graded lines and good acceleration for lines with dense suburban traffic and suffering, perhaps, from electric tramway competition. Electrified lines now also included 116km (72 miles) of the New York, New Haven & Hartford Railroad, a 117.5km (73 mile) section of the Prussian State Railways and suburban lines of British main

Figure 11.3: First electrification in Australia was a section of Victoria Railways' 5ft 3in gauge Melbourne suburban lines electrified at 1500 volts DC in 1919. Australian Information Service.

line companies in the London, and Liverpool areas. The first section of the Paris Métro had been opened in 1900, the first New York subway line in 1904. The Metropolitan and District underground railways in London had been electrified between 1901 and 1905, and several more deep level tube railways had been made beneath London.

Signalling

When the District Railway was electrified it was equipped with automatic signalling of a type already in use in the USA: track circuits actuated electropneumatically operated signals. Those in the open were semaphores; in the tunnels, coloured spectacle glasses moved in front of gas lamps. The Metropolitan then used a similar system, but lit the tunnel signals electrically. From these the modern colour light signal evolved. In 1905 the District Railway became the first to install an illuminated track diagram in a signal box. In 1906 the Great Western Railway started to install automatic train control, using a system in which a ramp between the rails made contact with a shoe beneath the locomotive to apply brakes when it approached a signal at danger.

Ancillary buildings

Railway buildings at this period – passenger and goods stations, signal boxes, locomotive and carriage sheds, engineering works and administrative offices – ranged, in enormous quantity, from the minuscule to the magnificent and reflected the architectural tastes and techniques of their time. Civil engineering works such as the Forth Bridge, or the Victoria Falls bridge across the Zambesi River, ranked among the greatest then built, and likewise reflected available methods and materials. Reinforced concrete was coming into use for railway structures, and also as a replacement for wood as sleeper material.

THE AFTERMATH OF WAR

The First World War greatly encouraged the use of the internal combustion engine. In a largely static war, armies of both sides were supplied by temporary railways of 60cm gauge, worked by both steam and internal combustion locomotives: the British army alone took delivery of over 1200 small petrol and petrol–electric locomotives. It was the first large-scale use of internal combustion rail vehicles.

However the main railway systems in Europe, where not actually damaged in the fighting, were much over worked, and suffered from lack of maintenance. The war did far more to encourage the development and use of motor road vehicles, and of aircraft, than it did of railways. Afterwards, in the 1920s, the motor bus, the motor car and the motor lorry became serious competitors for rail traffic (see Chapter 8), as in the 1930s did the aeroplane (see Chapter 12). Simultaneously, railways were hard-hit by industrial depression. In Europe and North America, there was an increasing tendency for those railways unable to meet competition to be closed. Slow-speed narrow gauge lines were particularly hard hit. In France, during the 1920s, plans for expansion of rural light railway systems were arrested in mid-progress, and motor-bus routes substituted. In Britain railway routes reached a peak of 32,893km (20,443 miles) in 1927, and then started to decline as the extent of lines closed exceeded that of new lines opened. In the USA routes had already passed a peak of 408,745km (254,037 miles) in 1916. Railways were no longer the automatic choice for transport over distances greater than a few miles, nor were they any longer at the forefront of technological progress.

They could, however, fight back, although since railway undertakings are large and cumbersome organizations they appear to have a built-in resistance to change, and it was sometimes long after circumstances had altered for the worse that a reaction eventually appeared. What was needed was to improve and accelerate services, and provide them more economically; to do this, traditional techniques, such as steam locomotion, were improved, and new technologies were adopted, particularly for motive power and signalling. During the 1920s and 1930s colour light signalling was coming into use on main lines in both Britain and the USA, and, in signal boxes, route setting panels were first adopted: thumb switches set up a route on an illuminated diagram, interlocking being done electrically. Centralized traffic control developed in the USA, enabling train despatchers directly to set points and signals many miles away instead of telegraphing train orders.

Electrification of railways continued, notably on the Gotthard line (the original Swiss north–south line over the Alps) in 1921, on the Pennsylvania Railroad between New York and Washington in 1935, and on the Southern Railway – one of the four big companies into which most British railways had been grouped in 1923 – where London suburban electrification was extended to the main lines to Brighton, Hastings via Eastbourne, and Portsmouth in the 1930s. Direct current was supplied at 660 volts (later increased to 750 volts) by a conductor rail, a system originally chosen for compatibility with London underground lines with which the suburban lines were linked. Southern electrification was primarily for passenger trains and electric haulage was initially little used for freight.

INTERNAL COMBUSTION

In the USA petrol–electric railcars at first proved insufficiently robust to be reliable in the hands of men accustomed only to steam. But in 1923 the Electro-Motive Company completed its first such vehicle. This company had been set up by Hal Hamilton, locomotive fireman turned road vehicle salesman, to develop a robust, reliable, petrol–electric railcar which would be built under contract using bought-in components. He was successful: over the next seven years his company produced some 500 railcars, steadily improving designs and establishing a reputation for good service. Its successor would in due course be instrumental in seeing the diesel engine adopted for railway locomotives. In the introduction of internal combustion to railways, Hal Hamilton occupies a position similar to that of George Stephenson in relation to steam power: he did not invent it, but he more than anyone else made it work and got it adopted.

The most important and far reaching of the new technologies was that of the diesel engine, with its much greater thermal efficiency than the steam engine; the problem, initially, was to exploit so inflexible a power unit as railway motive power. Mechanical transmission proved practicable only for the lowest powers; for units powerful enough to be of real value hydraulic or electric transmission had to be adopted. One of the earliest enthusiasts for diesel traction was Lenin, at whose instigation funds were allocated in 1922 for development of diesel locomotives. Two were built in Germany, one with electric transmission and the other mechanical, for the Soviet railways. Both appear to have run successfully.

In 1923 the Canadian National put into service some diesel–electric railcars with engines made by William Beardmore & Co. in Scotland. In the same year the first American diesel–electric locomotive was built by American Locomotive Co. using a 223.7kW (300hp) Ingersoll-Rand engine and General Electric electrical equipment. This had the wheel arrangement B-B, indicating two bogies, each with four wheels coupled by rods. With the electric and diesel locomotive came a new style of notation for wheel arrangements: 'A' for one powered axle, 'B' for two with their wheels coupled by rods, 'C' for three likewise and so on. Addition of 'o' indicates a group of individually powered axles, i.e. 'Bo', 'Co', and numbers such as '1', '2', '3' indicate unpowered axles. Descriptions such as 1–D–1 and Co-Co became typical. The notation derived from a system used in continental Europe to describe steam locomotive wheel arrangements.

The 1923 American diesel–electric locomotive was intended as a shunting locomotive and as such, on trial, shunted the yards of several US railways. It quickly demonstrated qualities of the sort that were to characterize diesel locomotives *vis-à-vis* steam: it could work almost round the clock without delays for lighting up or ash-disposal, it was easy to drive and economical to run, but it cost twice as much as a steam locomotive. For the last reason, and because diesel engines were still heavy and slow-running, American railways did not immediately take to diesel shunting locomotives, but the building consortium sold two in 1925, two more in 1926 and from then on their numbers gradually increased.

In England in 1932–3 the London Midland & Scottish Railway fitted the frames from a 0–6–0 tank locomotive with a 298kW (400hp) diesel engine driving the wheels by hydraulic transmission. This was part of an attempt to reduce the costs of shunting and it was successful, for from that time onwards the LMS built or purchased diesel shunting locomotives, usually with electric transmission, in increasing quantities.

These had not, however, been the LMS's first essay with diesel traction, for in 1928 it had rebuilt a four-coach suburban electric train as a diesel electric, using 373kW (500hp) Beardmore engine. The first diesel railcar to go into regular, rather than experimental, service in the British Isles was a 3ft (91.5cm) gauge diesel mechanical railcar built for the County Donegal Railways in 1931; the line already had several years' experience of petrol-engined railcars. In 1934 the Great Western Railway initiated diesel railcar workings for both main and branch line services, using railcars built by AEC Ltd, with engines and transmissions similar to those used in London buses. Power from twin six-cylinder engines mounted on opposite sides of the underframe was transmitted through fluid couplings and epicyclic gearboxes.

Diesel railcars were also coming rapidly into use in France at this period: when the Société Nationale des Chemins de Fer Français (SNCF) was formed in 1938, following nationalization of the main line companies, it owned some 650 railcars. The most striking application of diesel power in the early 1930s, however, was in Germany. By 1933 a two-car articulated diesel-electric train, the *Flying Hamburger*, was averaging 124kph (77mph) for the 286km (178 miles) between Berlin and Hamburg, the fastest regular train service in the world. Three-car sets were later introduced, some of them with hydraulic transmission.

Meanwhile in the USA in 1929 the motorcar manufacturers General Motors Corporation purchased an internal combustion engine maker, the Winton Engine Co., and followed this in 1930 by acquiring Winton's chief customer: the Electro-Motive Corporation, which then became a division of General Motors. A lightweight 447.5kW (600hp) diesel engine was successfully developed by General Motors; with electric transmission, this was used to power a three-coach streamlined high-speed train built by the Edward G. Budd Company for the Chicago Burlington & Quincy Railroad. Budd used newly-developed techniques to weld the stainless steel from which the lightweight coaches were built. This train was the outstandingly successful *Pioneer Zephyr*, completed in 1934. It demonstrated its potential soon after completion by running the 1633km (1015 miles) from Denver to Chicago non-stop at an average speed of nearly 125.5kph (78mph), and ushered in an era of high-speed lightweight diesel trains in the USA.

Furthermore, in 1935–6 General Motors fitted diesel engines into a few locomotives built under contract; then in 1937 it set up its own manufacturing plant for diesel electric locomotives. The first were 1800hp A1A-A1A locomotives for passenger trains. For the first time the driver's cab was set back behind a projecting nose, to give protection in case of accident on ungated level crossings. A version lacking a driver's cab, called a 'booster unit', was available for multiple operation when greater power was needed. By 1937 it was already possible to travel from coast to coast in diesel-powered trains. In 1939, General Motors introduced a freight locomotive in which four units produced a total of 4027kW (5400hp), and in 1941 the first 'road switcher', or enlarged shunting locomotive able also to haul local passenger and freight trains.

LATE STEAM

The threat of competition from other forms of motive power and other forms of transport spurred steam locomotive engineers to develop steam traction to its peak. In 1935 the London & North Eastern Railway put on a streamlined steam train, the *Silver Jubilee* between London and Newcastle upon Tyne (Figure 11.4). Chief mechanical engineer Nigel Gresely had considered high-speed diesel units such as were running in Germany, and rejected them in favour of steam, for he had built his first Pacific in 1922 and developed the type steadily since. For the *Silver Jubilee* he produced the streamlined version, called A4 class, and it was one of this class, no. 4468 *Mallard*, which in 1938 achieved the speed record for steam of 202.7kph (126mph) which remains unbroken. In 1937 the LNER and the LMS had put on streamlined expresses between

Figure 11.4: The London & North Eastern Railway's answer to road and air competition was the *Silver Jubilee* streamlined express introduced between Newcastle upon Tyne and London in 1935.
National Railway Museum.

London and Edinburgh, and London and Glasgow, respectively; but the outbreak of war in 1939 halted further development.

In the USA streamlining was first applied to a steam locomotive in 1934: to a 4–6–4 of the New York Central Railroad. Streamlining of trains at this period really had as much to do with increasing the glamour of the train as with reducing its wind resistance: certainly it was effective in doing the former, and many other railways introduced luxury streamlined steam trains. Notable were the Chicago, Milwaukee, St Paul & Pacific – its *Hiawatha* trains hauled by streamlined 4–4–2s designed for maximum speeds of 193kph (120mph) ran in direct competition with Burlington *Zephyrs* – and the Southern Pacific: its 4–8–4 hauled *Daylight* trains between Los Angeles and San Francisco, multi-coloured with bands of orange, red, black and silver along their entire length, were arguably the most glamorous, the most spectacular, of all.

Streamlining was the outward aspect of locomotives of advanced design. Many improvements were made in locomotive design at this period – for example, in the USA bar frames and separate cylinders developed into cast steel engine beds, and lightweight alloy steel connecting and coupling rods reduced hammer-blow damage to track. For freight, locomotives of Mallet type, but with simple expansion, were built in increasingly large sizes, the largest eventually being the Union Pacific's 'Big Boy' 4–8–8–4s of 1941 which could haul trains weighing more than 5000 tonnes on lines where the ruling gradient was 1 in 122. In France, André Chapelon studied, re-designed and enlarged locomotive steam and exhaust passages, and markedly improved performance. One of his earliest successes was the Kylchap double blastpipe, designed in conjunction with a Finnish engineer named Kylala: the first of the A4s to be fitted with this device was *Mallard*.

DIESEL TAKE-OVER IN THE UNITED STATES

By the time the USA entered the Second World War in 1941, many American railways had some diesel locomotives in service. The builders of steam locomotives, who were starting to produce diesel units, were restricted by wartime regulations to building diesel locomotives only for shunting – but General Motors was allowed to build them for freight train haulage in limited quantities. Wartime conditions meant that all locomotives, steam and diesel, were used to the maximum, and confirmed that diesels could haul greater loads faster and more economically than equivalent steam locomotives. After the war, the Budd Co. introduced diesel railcars using General Motors engines driving through torque converters. General Motors was able to keep the prices of its diesel locomotives down by offering a standard range built by mass production, and diesel power then took over from steam with quite remarkable speed, so that by 1960 there were very few steam locomotives left at work. The diesel locomotive in the USA eclipsed the highly developed steam locomotive as rapidly and effectively as the steam railway in England, a century before, had eclipsed the highly developed coach services on the roads. Even electrified lines in the USA were not immune, and where they represented interruptions in long diesel runs they were converted to diesel too.

After the war luxury passenger trains were revived, but only briefly, for American passenger trains, whether diesel or not, were soon subject to intense and increasing competition from widespread private car ownership on the one hand and internal airline services on the other. Eventually, the likelihood of apparent total collapse of long-distance railway passenger traffic was only prevented by the establishment of Amtrak, the state-supported National Railroad Passenger Corporation, in 1970. This operates a limited but nationwide network of long-distance passenger trains (475 stations served by 38,400km (23,860 miles) of routes in 1985), for the most part over the tracks of railway companies which continue to operate their own freight services. Most of the 735km (346 mile) North East Corridor route (Boston–New York–Washington) has however been owned by Amtrak itself since 1976, and is electrified at 12.5kV AC 25 Hz: electrification has been on the increase again.

POST-WAR EUROPE

The situation in Europe after the Second World War was different. For one thing, there was no indigenous source of oil. For another, many continental railway systems came out of the war badly damaged and needed effectively to be rebuilt. Electricity became the preferred source of power, with diesel power used on routes where limited traffic did not justify electrification, or as a temporary expedient pending electrification. The railway systems of France and the Netherlands, for instance, had to be largely rebuilt after the war and the opportunity was taken to electrify substantial mileages. On the Continent, prosperity leading to widespread private car ownership came later than in the USA, and by then the railways were better placed to meet it. Nor are distances between cities so great, so air transport was less of a threat, and was furthermore counteracted by the establishment of the Trans-Europ Express network in 1957. This was set up jointly by the state railways of Belgium, France, West Germany, the Netherlands, Italy, Luxembourg and Switzerland to operate fast daytime diesel and electric expresses between the principal cities of those countries. Austria, Denmark and Spain were later added. Through running was made possible between lines electrified on different systems, by construction of electric trains able to operate from as many as four different types of supply, and on to the Spanish broad gauge system by adjusting wheelsets at the frontier. Thirty years later, the system had evolved into the Eurocity Network set up in 1987, with 64 express train services linking 200 cities in 13 countries.

BRITISH RAILWAYS

The railways of Britain, after the Second World War, had the worst of both worlds. They were neither so badly damaged as to need total reconstruction, like some continental systems, nor were they relatively unscathed, like those of the USA. They were badly run down and had to struggle along as best they could during a period of post-war austerity. The position was exacerbated by the period of re-organization which followed nationalization of the four large companies, and most of the surviving small ones, into British Railways in 1948. BR did start to install an automatic warning system (as automatic train control was re-named) throughout main lines, using magnetic induction instead of physical contact between ramps and locomotives, and it developed a standard coach with a steel-framed body. But despite work done by the old companies to develop diesel power, a range of standard steam locomotives (low-cost and, by steam standards, easily maintained) was introduced in 1950–1.

In 1954, British Railways did put its first diesel multiple unit trains into service. Eventually in 1955 a comprehensive modernization plan envisaged electrifying some lines and introducing diesel locomotives and railcars on others, but by the time it could start to take effect at the end of the decade it was already almost too late: increasing prosperity had made car ownership widespread, and air services were well established. The 1963 report, *The Reshaping of British Railways*, prepared under the chairmanship of Dr R. Beeching, proposed that railways should be used for transport where 'they offer the best available means, and . . . should cease to do things for which they are ill suited'. The consequence was frenzied modernization of equipment, accompanied by drastic closures of branch lines, secondary main lines, and wayside stations on lines which were retained. One effect was that steam locomotives ceased to be used in normal service in 1968. Comparable dates for some other countries are: the Netherlands, 1958; Irish Republic, 1964; France and Japan, 1975. The comparatively late date for France indicates, rather than any tendency to be old-fashioned, that the changeover from steam to electric and diesel started earlier and was carried out in a more ordered manner. As early as 1954 a French electric locomotive had achieved a world speed record of 330kph (205 mph).

The late start made in Britain with large-scale electrification did at least mean that advantage could be taken of the latest technology. Many countries had standardized their electrification systems between the wars – Switzerland, for instance, on 15,000 volts single-phase AC at $16\frac{2}{3}$ cycles, France and Holland on 1500 volts DC. Britain, too, had chosen the latter, but little electrification had been carried out to it. Meanwhile, following earlier trial schemes in Hungary and Germany, the SNCF had electrified a 79km (49 miles) section between Aix-les-Bains and La-Roche-sur-Furon at 20,000 volts single-phase, 50 cycles AC. The voltage was later increased to 25,000. It was confirmed that although locomotives for this system were more expensive than for the DC system, the cost of the overhead and sub-stations was very much less. The system was then used in northern France, and adopted for Britain in 1956, in time to be used on most lines electrified subsequently other than those extending the third-rail system in the south of England. The line from London Euston to Birmingham, Liverpool, Manchester and Glasgow was electrified throughout at 25kV by 1974. In 1958, British Railways had decided to install long-welded rails on main lines throughout the system.

The earliest main line diesel locomotives provided under the modernization plan tended to be underpowered, multiple operation on the American pattern being generally impracticable in Britain because of cramped track layouts. The position was ameliorated by introduction in 1961 of Deltic Co-Co diesel-electrics of 2460kW (3300hp) which powered the principal London–Edinburgh expresses for the next twenty years, and more so by general introduction of 2050kW (2750hp) locomotives from 1962. Speeds of 160kph (100mph) were common by the early 1970s; in 1976 the first High Speed Trains went into service. With a pair of power cars each containing a 1678kW (2250hp) diesel engine driving an alternator to feed four traction motors, these trains were designed to reach 200kph (125mph) in regular service; in 1986 one reached 238kph (148mph) on test between Darlington and York.

With techniques such as these, British Rail has been able to contain the challenge of road and air for passenger travel. Experience with freight has been less happy. Probably too much traffic had already been lost to the roads by the early 1960s. At that time many trains were still made up of four-wheeled short-wheelbase wagons, loose coupled and without continuous brakes, which were worked from marshalling yard to marshalling yard. It was calculated that the average transit time was $1\frac{1}{2}$–2 days for an average journey length of 108km (67 miles); nor was the arrival time predictable. Measures to improve the situation have included the introduction of many block trains to carry commodities such as coal, oil and stone in bulk, freightliner trains to carry intermodal freight containers to scheduled services, and the TOPS (Total Operations Processing System) computerized system of freight information and transit control which came into operation in 1975 to monitor continuously the movement of goods rolling stock and locomotives. This and construction of high-capacity air-braked long-wheelbase wagons able to run at speeds up to 120kph (75mph), and so to offer same-day or overnight transits throughout the country, has enabled wagonload traffic to make something of a comeback.

FREIGHT CONTAINERS AND BULK FREIGHT

Freight containers are older than steam railways – containers were used, for example, on the Derby Canal tramroad to transfer coal between plateway wagon and canal boat and, probably, road vehicle. Containers interchangeable between rail and road were in use in the 1920s in both Europe and the USA, and were well established in the 1950s on British and Irish railways. Recent emphasis on containers, however, has come from international agreements, reached in the late 1960s, for standard dimensions, strength and lifting fittings of stackable containers to be carried by ship, rail and road. As such they have come into widespread use, one notable example being carriage of goods from Japan to western Europe via the Trans-Siberian Railway.

An alternative, in countries where both the rail gauge and the loading gauge of railways are large enough, has been carriage of road vehicle trailers by rail. Such Piggyback services carry trailers over the long-haul portions of their journeys in North America, western Europe, and Australia. Operation in Australia has been aided by extension of the standard gauge rail system: the Trans-Australian railway (now operated by Australian National Railways) has been linked with the New South Wales system, and the standard gauge extended, by building new lines or dual-gauging old ones, to Melbourne, Perth and Alice Springs. The completion of standard gauge throughout between Sydney and Perth (over a route where formerly there were three breaks of gauge) was followed by the successful introduction, as late as 1970, of the *Indian-Pacific* luxury passenger train covering the 3960km (2461 mile) route in about 66 hours. Nevertheless the most remarkable railway operation in Australia is that of the isolated Hammersley Iron Ore Railway in the northern part of Western Australia. Its 386km

(240 mile) route was built during 1964–72 to carry iron ore from mine to port, and over it ore is carried in mile-long trains of up to 210 wagons. Average net trainload during 1982–3 was about 17,650 tonnes. The heaviest regular trainload in Britain is small by comparison, but still no mean figure: 3350 tonnes of limestone carried in 43 wagons.

Bulk freight operations such as these exemplify the continuing and indeed increasing use of railways in developed countries for specialized functions. Another of these is rapid transit: swift passenger transport within large cities and conurbations. The term embraces traditional underground and tube railways, the upgraded tramway systems of certain cities (mostly on the Continent) which, rather than abandon trams when street congestion made them untenable, put the city-centre sections underground, and recently-built systems combining the best features of both, including reserved-track surface operation where possible. Of these systems there are many, for the principle is as attractive in western countries such as the USA, to reduce traffic jams caused by excessive numbers of private cars, as it is in Communist countries for moving large numbers of carless people. In Britain, the Tyne & Wear Metro, of which the first section was opened in 1980, uses two-car train sets based on continental tramcar design, operated over routes which are partly newly constructed (and much of that in new tunnels beneath central Newcastle upon Tyne) and partly over former British Rail routes.

Rapid transit lends itself to automated control of trains, to a greater or lesser degree. As long ago as 1927 the Post Office brought into use its own tube railway, to carry mails beneath London, with automatic operation. The track circuits, instead of operating signals, disconnect current from the section to the rear of each train. This railway, however, carries no passengers. On the Victoria Line, built between 1962 and 1971 and when opened the first new tube railway for passengers in central London for more than 60 years, trains are driven automatically once the train attendant has pressed the start buttons (two of them, because one might be pressed accidentally). On the Bay Area Rapid Transit system (BART) serving San Francisco, a cental computer supervises local automatic control. This 120km (75 mile), 5ft 6in (1.06m) gauge system was built between 1964 and 1974; electricity is supplied by conductor rail at 1000 volts DC. To lure habitual motorists out of their cars, trains run at speeds up to 128kph (80mph) and average 67.5kph (42mph) including stops, approximately twice the average speed of London underground trains. On the Lille Métro, of which the first section was opened in 1983, trains are driven entirely automatically, without any staff on board: television is used to enable the control centre to survey events. Trains run on rubber-tyred wheels, with additional horizontal wheels bearing on guide rails which also act as conductor rails. Rubber-tyred trains have been in use on part of the Paris Métro since 1957; rubber tyres for railway use were developed by the Michelin company during the 1930s.

On London's Docklands Light Railway, the cost of the complex guidance system needed for rubber-tyred trains was not considered justified, but its 750 volts DC trains are driven entirely automatically. On-train computers control the driving of trains, and are continually monitored by a central computer comparing train positions. The railway was opened in 1987, re-utilizing in part the route and structures of the London & Blackwall Railway of 1840 which pioneered use of the electric telegraph.

Between Birmingham Airport, Birmingham International Station and the National Exhibition Centre a 603m (660yd) line uses the maglev (magnetic levitation) system: controlled magnetic fields replace wheels, springs and dampers. The system was developed by British Rail's Research & Development Division during the 1970s. Vehicles operate singly and are propelled by linear induction motors.

NEW HIGH-SPEED LINES

The latest systems are diverging increasingly from the traditional railway, with steel flanged wheels running on steel rails. The traditional form has however been used for the two noted systems built new for high-speed long-distance passenger trains: the Japanese Shinkansen and the French Ligne à Grande Vitesse (LGV).

By the 1930s the Japanese had developed a dense railway network on the 1.06m (3ft 6in) gauge and were planning a new 1.43m (4ft 8½in) gauge line to relieve the most heavily used route between Tokyo and Osaka. This was eventually constructed by Japanese National Railways as the Tokaido Shinkansen line: much of it was built on viaducts or through tunnels to achieve an alignment suitable for service speeds up to 209kph (130mph). The line had a route length of 515km (320 miles) was electrified at 25kV and opened for regular traffic in 1964. The fastest trains were scheduled to take 3 hours 10 minutes from Tokyo to Osaka, including two intermediate stops. It was not merely the speed of the 'bullet' trains which was phenomenal, however, but also the frequency – trains at intervals of fifteen minutes – and the number of passengers. Each train could, and often did, carry 1400 passengers – in a country where only a century earlier the concept of more than one person riding in a vehicle caused surprise, and travel was at walking pace. A network of Shinkansen lines has since been built and by 1988 routes had reached a total of 1859km (1162 miles). Eventually it is intended to extend it through the Seikan Tunnel linking the main island of Honshu with the northern island of Hokkaido. This tunnel – test trains were run through it early in 1988 – is at 53.85km (33.5 miles) the world's longest under-sea tunnel. Its track is laid to 1.06m gauge pending extension of the Shinkansen network.

Figure 11.5: French *Train à Grande Vitesse* runs at speeds up to 270km/hr (168mph) on the high speed line built between 1976 and 1983 to link Paris with Lyon and south-east France.
French Railways—Vignal.

The French had a comparable problem to the Japanese, excessive traffic on the SNCF's busiest main line, from Paris to Dijon and Lyon. Absence of a gauge problem, however, meant that expensive construction of new sections and stations within cities could be avoided: standard gauge trains à grande vitesse (TGVs) (see Figure 11.5) depart from the Gare de Lyon in Paris to run over the traditional route for 30km (18 miles) before diverging on to the new LGV line. They regain the old route at the approach to Lyon, or sooner according to their final destination, for TGVs continue over ordinary railways as far afield as Lausanne, Geneva and Toulon. The total route length of new line is 390km (242 miles) it shortens the distance between Paris and Lyon from 512 to 425km, (318 miles to 264) and it is laid out for trains to run at 270kph (168 mph). Construction started in 1976. To minimize environmental disturbance, a strip of land alongside was used for a new long-distance telephone cable, and the route is also paralleled by new roads. Electric traction meant that the ruling gradient could be made as steep as 1 in $28\frac{1}{2}$. Track is laid with rail of approximately 60.3kg/m (122lb/yd) and points are designed so that wheels and flanges are supported continuously and trains can take even the diverging route at speeds as high as 220kph (136mph). The first section was opened in 1981 and the whole line in 1983.

The trains themselves consist of eight coaches articulated together plus a power car at each end; two sets can be worked together in multiple. Power car noses are streamlined, their shape established by wind tunnel and computer experiments. The twelve traction motors are spread along the train and, to minimize weight directly on the bogies, are slung beneath coach bodies, with sliding propeller-shaft transmission. The trains operate on 25kV 50Hz AC on the new line to produce 6450kW (8650hp); on old lines they operate at reduced power on 1500v. DC. Some are equipped also to operate additionally on 15kV $16\tfrac{2}{3}$Hz AC so that they can run over the Swiss Federal Railways to Lausanne. On the LGV, cab signalling indicates maximum permitted speed, above which the brakes are applied automatically, and trains are in radio contact with the control office in Paris. On 26 February 1981, during pre-opening trial runs, one of these trains reached 380kph (236 mph).

A northward extension of the service occurred in 1984 with the establishment of a through TGV service between Lyon and Lille, using the Ceinture line around Paris to run without booked stops between Lyon and Longeau (Amiens). The same year purpose-built TGVs went into service carrying mails previously conveyed by air. A second LGV in the form of a 'Y' from Paris to Le Mans and Tours is under construction and the SNCF envisage one from Paris to Lille, where it would split into lines to Brussels, north Germany, Amsterdam, and London through the Channel Tunnel, for construction of which Anglo-French agreement was reached in 1985.

SURVIVING STEAM

While specialized roles have been evolving for railways in developed countries, railways elsewhere – in Asia, Africa, South America for instance – have continued to play their traditional role as general carriers. New railways have been built in places as far apart as East Africa, where the 1850km (1150 mile) Tanzania–Zambia Railway was opened in 1975, and western Canada, where the British Columbia Railway's route length was extended from 696km (433 miles) to 2845km (1768 miles) during the 1960s and 1970s.

The most remarkable example, however, is China. There, according to G. Sowler writing in the Newsletter of the Friends of the National Railway Museum (May 1985), there were in 1949, at the end of twelve years of war and civil war, a mere 11,000km (7000 miles) of railways in usable condition. By 1980 rebuilding and much new construction had produced a system of 53,000km (33,000 miles). The first railway bridge across the Yangtse River, across which railways in the north had previously been linked only by train ferry with those in the south, was completed in 1956.

China is one of the few countries still to use steam locomotives to any great extent, and the only one in which they have continued to be built in any quantity. In the early 1980s the locomotive works at Datong, established only in the 1950s, was producing 250 to 300 new steam locomotives a year. Most of these were QJ or 'March Forward' class 2–10–2s. The other two countries still using large numbers of steam locomotives in everyday service are India and South Africa. South Africa Transport Services' Rail Network had in 1983 a total of 1647 steam locomotives, which may be compared with 2056 electric locomotives, 1549 diesel, and 1375 motor coaches for electric multiple unit trains. Experiments to improve steam locomotive efficiency have continued, including equipping locomotives to burn coal on the gas producer system. With this system, which originated in Argentina, a minimal amount of air, mixed with steam, is admitted below the grate and combines with the coal to make gases which are then burned efficiently in air admitted through vents in the firebox sides above the grate.

RAILWAY PRESERVATION

In Western Europe, North America and Australasia active steam locomotives continue to be things of the present, rather than the past, because of the widespread growth of railway preservation in these areas. During the nineteenth century and the early part of the twentieth, historic locomotives and rolling stock at the end of their working lives were occasionally preserved, by railway administrations and museums, instead of being cut up. In 1927 the Stephenson Locomotive Society, a voluntary body, purchased the 0–4–2 *Gladstone* for preservation on withdrawal by the Southern Railway, and in 1938 the LNER brought out Great Northern Railway 4–2–2 no. 1, preserved since withdrawal over 30 years earlier, and ran her in steam with contemporary coaches, largely as a publicity stunt to publicize, by comparison, the advances made with the latest streamlined expresses.

After the Second World War, Ellis D. Atwood rescued locomotives, track and equipment from closed two-foot gauge railways in the state of Maine, USA, and used them to build and operate the 8.8km ($5\frac{1}{2}$ mile) Edaville Railroad round his cranberry farm. In 1950 when the Talyllyn Railway was about to close, the Talyllyn Railway Preservation Society was formed to preserve and operate it (much of the equipment was then some 80 years old) on a voluntary basis. Subsequently the Festiniog Railway Society was formed to revive the Festiniog Railway, closed since 1946 but not dismantled: in 1954, Alan Pegler bought a controlling shareholding in the railway company (later he gave this to charitable trust) and, with voluntary support from the society, the railway was reopened by stages between 1955 and 1982. Meanwhile, in 1963, Pegler had purchased privately the famous Gresley Pacific locomotive *Flying Scotman* on withdrawal by British Railways, with an agreement to run her over BR tracks.

From these beginnings, steam railway preservation has spread far and wide, from California to New Zealand. Steam railways, steam excursions over ordinary railways, and railway museums have become popular and important tourist attractions. Running them has become a widespread and absorbing activity for volunteers and professionals alike. The skills and the equipment needed to maintain and operate steam railways, which would otherwise have been lost, have survived. Working replicas of important early locomotives have been built.

It is appropriate that Britain, which gave steam railways to the world between 1830 and 1870 but later lost its lead in the progress of railway technology, should a century later have given the world the concept of voluntary railway preservation. It is a happy paradox that, in summer during the late 1980s, the only through train service that allowed a day visit from London to the best-known British inland tourist resort, Stratford-upon-Avon, was the Sunday luncheon steam excursion; and the only public transport at all on Sundays between Fort William and Mallaig was the steam train.

GLOSSARY

Adhesion frictional grip of driving wheel upon rail
Bogie short-wheelbase four- or six-wheeled pivoted truck supporting end of locomotive, carriage or wagon
Rail, double-headed rail in which, in cross-section, a vertical web links bulbous upper and lower heads of equal size
Rail, bullhead rail similar to double-headed (q.v.) but with the upper head larger than the lower
Rail, flat-bottom rail of inverted 'T' cross-section with a bulbous upper head
Rail, fish-bellied rail of cast or malleable iron with lower edge convex for strength, between points of support; so-called from its appearance
Skew bridge bridge with line of arch not at right-angle to abutment
Tank locomotive steam locomotive in which supplies of water and, usually, fuel are carried upon the locomotive itself instead of in a separate tender
Torque converter transmission comprising centrifugal pump in circuit with turbine, or similar device to give infinitely variable gear ratios
Valve gear, gab Steam engine valve gear in which each eccentric rod terminates in a 'V'-shaped gab, set at right angles to it to engage a pin on the valve rod, The appropriate rods are engaged and disengaged for forwards or reverse running
Valve gear, link motion steam engine valve gear in which the ends of the eccentric rods for each cylinder are attached to a link, sliding about a die block connected to the valve rod. As well as selection of forward or reverse running, this enables the point in the piston stroke at which steam is cut off from entering the cylinder to be varied, so that economies can be made in steam consumption

SOME BRITISH RAILWAY TERMS WITH THEIR NORTH AMERICAN EQUIVALENTS

British	North American
Bogie	Truck
Chimney	Smokestack
Coupling	Coupler
Engine driver	Engineer
Fireman	Stoker
Footplate	Deck
Guard	Conductor

Petrol-electric	Gas-electric
Railway	Railroad
Shunting	Switching
Sleeper	Tie
Tram	Streetcar
Wagon	Car

ACKNOWLEDGEMENT

I am most grateful to Saelao Aoki and the staff of the Seimeikai Foundation, Tokyo, and to Caroline McCarthy of French Railways, London for their assistance in providing information for this chapter.

FURTHER READING

Ahrons, E. L. *The British steam railway locomotive 1825–1925* (The Locomotive Publishing Co. Ltd., 1927)

Allen, G. F. (Ed.), *Jane's world railways* (Jane's Publishing Co. Ltd., London annual publication)

Bruce, A. W. *The steam locomotive in America* (W. W. Norton & Co. Inc., New York, 1952)

Ellis, C. H. E. *British railway history*, vol. I (1830–1876) 1954, vol. II (1877–1947) (George Allen & Unwin Ltd., London, 1959)

Haut, F. J. G. *The history of the electric locomotive* (George Allen & Unwin Ltd., London, 1969)

Lewis, M. J. T. *Early wooden railways* (Routledge & Kegan Paul, London, Ltd., 1970)

Nock, O. S. et al. (Eds.) *Encyclopaedia of railways* (Octopus, London, 1977)

Ransom, P. J. G. *The archaeology of railways* (World's Work Ltd., Tadworth, 1981)

Ransome-Wallis, P. et al. (Eds.) *The concise encyclopaedia of world railway locomotives*, (Hutchinson, London, 1959)

Reck, F. M. *On time – the history of the electro-motive division of General Motors Corporation* (General Motors Corporation, La Grange (Illinois), 1948)

Snell, J. B. *Early railways* (Weidenfeld & Nicolson, London, 1963). Covers period down to 1914, worldwide

Stover, J. F. *American railroads* (University of Chicago Press, Chicago, 1961)

Whitehouse, P. B., Snell, J. B., and Hollingsworth, J. B. *Steam for pleasure* (Routledge & Kegan Paul, London, 1978). World guide to surviving steam railway operations

12

AERONAUTICS

J.A. BAGLEY

EARLY ATTEMPTS AT FLIGHT

Man first tried to fly by imitating the birds, and a large number of individuals strapped themselves to artificial wings, often covered with feathers. Those who attempted to elevate themselves from ground level by vigorous flapping invariably failed; those who launched themselves from the tops of towers usually fell more or less vertically, but occasionally managed to glide a short distance before hitting the ground, often catastrophically. The history of aviation up to the late eighteenth century is littered with these pioneers, and with optimistic attempts to build artificial birds, powered by clockwork springs or gunpowder; otherwise it consists largely of theoretical speculations founded on totally inadequate premises. Although fascinating from many points of view, these activities contributed nothing to later technology.

BALLOONS

Man first successfully became airborne in the last quarter of the eighteenth century, suspended beneath a large fabric bag filled with heated air. Although there is no apparent reason why this technology should not have appeared centuries earlier – and there are indeed strong hints that it may have done so in the pre-Columbian civilizations of South America – it is a matter of historical record that the brothers Etienne and Joseph Montgolfier initiated the development of balloons with their experiments in 1782–3 at Annonay in France. The Montgolfiers were sons of a fairly prosperous paper manufacturer, and Joseph was a somewhat dilettante student of the physical sciences. His activities were clearly inspired by the publication of Joseph Black's account of the discovery of gases lighter than common air; Montgolfier conceived the idea that the smoke rising from a fire must contain a gas with this characteristic, and that it should be possible to contain enough of this gas in a fabric bag to enable the bag to rise.

His successful public demonstration of the principle on 5 June 1783 with a spherical hot-air ballon of about 10m (33ft) diameter, which rose to a height of about 2000m (6500ft) and landed 10 minutes later 2km (1.25 miles) away, had a seminal importance out of all proportion to its modest achievement.

When the news of this demonstration reached Paris, the scientific community there, better acquainted than Joseph Montgolfier with the current state of scientific research into the properties of gases, assumed that Montgolfier's gas was Cavendish's 'inflammable air' (later named hydrogen). This could most conveniently be made by pouring sulphuric acid on to iron filings, and Jacques Alexandre César Charles set out to repeat Montgolfier's experiment using hydrogen. Joseph Black had already attempted to demonstrate this on a small scale, but had failed to find a suitably impervious fabric to contain hydrogen; Charles successfully used fine silk coated with a solution of rubber (which is reputed to have been based on a secret process for manufacturing contraceptive sheaths for the French Court). Charles was also successful in making hydrogen gas on an unprecedented scale to inflate his 3m (10ft) diameter balloon for a public demonstration in Paris on 27 August 1783. The balloon was liberated in the presence of a great crowd and rapidly disappeared into the clouds. Three-quarters of an hour later it landed in the main street of Gonesse, a small village some 20km (12.5 miles) from Paris, whose frightened population attacked the strange monster with pitchforks. It expired with a frightful hissing and a poisonous stench which convinced the peasants of its infernal origins.

These demonstrations of model balloons led very quickly to manned flights – the Montgolfiers narrowly won the race when their hot-air balloon carried Pilâtre de Rozier and the Marquis d'Arlandes across Paris on 21 November 1783. On 1 December, Charles and Marie Noel Robert (one of the two Robert brothers who constructed the balloon) flew for two hours in a hydrogen-filled balloon. They landed 45km (28 miles) from Paris, and Charles then took off again by himself, ascending to about 3000m (9800ft) where he suffered a sudden sharp pain in the right ear – now a common affliction of airline passengers in a rapidly-climbing aircraft.

These spectacular achievements led immediately and directly to a sudden surge of enthusiasm for ballooning in France, followed quickly in the rest of Europe; this proved however to be surprisingly short-lived and ballooning virtually ceased in Europe for several years.

Technically, Charles's hydrogen balloon was almost perfect, and there was little further development in manufacture or operation for almost a hundred years. Its silk fabric, coated with a varnish of rubber dissolved in turpentine, was sufficiently impervious to hydrogen to permit flights of several hours' duration. A net of rope meshes covering the top half of the balloon carried a wicker basket for the passengers and a quantity of ballast to be dumped as the lifting hydrogen slowly leaked away, or was expanded by the heat of the sun. To release the hydrogen on landing, or to counteract a rising air current, a spring-loaded flap valve at the top of the envelope was operated by a cord extending down to the basket.

The Montgolfiers' hot-air balloons were much larger – air heated to about 100°C has a lifting capacity of 27kg/100m³ (17lb/1000ft³), whereas hydrogen provides 96kg/100m³ (60lb/1000ft³) – so they were much more difficult to handle, especially during filling and launch. The idea of a special gas contained in dense smoke was soon found to be erroneous, but Montgolfier's preference for making a very smoky fire by adding wool, old shoes or more noxious ingredients to his fire to make a denser smoke was well-founded: the particles of greasy soot effectively clogged the pores of the fabric and retained the hot air which was really responsible for lifting his balloons.

Hot-air balloons continued to be made and used throughout the nineteenth century, but hydrogen balloons were dominant for 150 years. The only significant technical developments in hydrogen balloons after Charles's first demonstration were the addition of a ripping panel which could be easily opened by the pilot to facilitate landing in high winds, usually attributed to the American John Wise in 1839; and a trail rope, which was a heavy rope, dropped from the basket so that the end rested on the ground. When the balloon rose, the effective weight of the trail rope increased, so slowing down the rise and stabilizing the flight path. This invention is usually credited to Charles Green (about 1830), but had certainly been proposed earlier, perhaps first by Thomas Baldwin in 1785.

The material of the balloon remained unchanged for two centuries – silk was preferred for strength and lightness, but cheaper cotton was often used. The fabric was coated with a suitable varnish, frequently involving a rubber solution. The British army introduced balloons made of gold-beater's skin in 1883. This material had been used for many years to make model balloons, but as it was available only in pieces up to 1m × 0.3m (3ft × 1ft), it was not easy to make a full-size balloon. Gold-beater's skin is a natural material, derived from the intestines of cattle, and the British army is reputed to have consumed almost the entire output of the Chicago meat-packing industry for several years around 1900.

Hydrogen was usually generated by adding sulphuric acid to iron filings and collecting the gas over water in a large-scale version of the school laboratory apparatus. However, for filling military observation balloons during the war against Austria in 1798, the French chemist Charles Coutelle developed a portable furnace in which hydrogen was produced by passing steam over red-hot iron, and variations of this process were often used for generating hydrogen for military balloons thereafter, because the gas produced was acid-free and the apparatus was easier to carry than large quantities of sulphuric acid.

Throughout the nineteenth century, ballooning was largely an activity for showmen, with a limited number of ascents for scientific (mainly meteorological) purposes. Most of these flights were made using coal-gas, which was conveniently available in most towns from about 1820. Charles Green was the first to use coal-gas in London in 1821; although the Academy of Lyons had suggested it in 1784. Coal gas is heaver than hydrogen, so a somewhat larger balloon was needed, but the lower cost and the convenience of inflating from a permanent supply were great advantages.

Hot-air balloons were used occasionally during the nineteenth and early twentieth centuries, but their very large size compared to gas balloons made them difficult to handle, and it was only when sheer size was a desirable feature required by an entrepreneur that they were noticed. However, it was often realized that if a continuous supply of heat could be contrived, hot-air balloons could be useful in remote places where coal gas or hydrogen was not conveniently available. Experiments on these lines were regularly made, using alcohol or petroleum fuels, from the latter part of the nineteenth century onwards.

Another interesting but abortive proposal was to combine the constant lift of a hydrogen balloon with the variable lift of a hot-air balloon carrying a fuel supply. This concept was first tried by Pilâtre de Rozier, the pilot of the first Montgolfier balloon, in 1785. He made a combination balloon which had a conventional hydrogen balloon on top of a cylindrical bag heated by a brazier, fuelled with straw blocks. Its only flight, on 15 June 1785, was an attempt to cross the English Channel from Boulogne; this ended in disaster when the balloon caught fire after a few minutes' flight. Subsequent attempts to produce combination balloons were equally unsuccessful, though not so spectacularly fatal.

Although the use of balloons for military observation purposes was pioneered by the French in 1794, it was neglected thereafter. During the Franco-Prussian War of 1870 attention was focused on the successful use of free balloons to allow individuals and dispatches to leave Paris during the seige; tethered observation balloons were employed on both sides but were not significantly useful. However, in the aftermath of the war, experiments were resumed in Britain, France, Germany and elsewhere into the use of captive balloons for observation work. The major technical advances arising from this work were the development by the British army of steel cylinders to hold compressed gas and special wagons to carry them; and portable winches to raise and lower the balloon quickly. Electrolysis of water was also introduced as an alternative method of generating acid-free hydrogen on a large scale.

It was soon realized that in strong winds, a spherical balloon was very unstable when tethered, especially near the ground. Two German officers, Major von Parseval and Captain von Sigsfeld, developed (about 1896) an elongated balloon with large inflated tail-fins which flew at an angle to the wind, like a kite, and was much more stable. Such kite-balloons were then taken up by most other countries; a version invented in 1916 by the Frenchman Albert Caquot supplanted the Parseval design during the First World War. Apart from their use at the front for observing the results of artillery bombardment, Caquot balloons were used also as an anti-aircraft screen around London – the so-called 'balloon barrage' – and this use was repeated on a large scale in the Second World War.

After a number of high-altitude balloon flights in which the pilots were killed, a sealed pressurized cabin was first used by the Swiss scientist Auguste Piccard in 1931 for a flight to 15,780m (51,775ft) to investigate cosmic radiation. The spherical aluminium cabin or gondola was designed and made in Belgium, but closely resembled a published design of 1906 by Horace Short which was never built. Piccard's balloon was made of rubberized cotton, and dispensed with the conventional net – the gondola was supported from a band sewn into the envelope.

A number of similar flights were made in the 1930s in several countries. The last pre-war flight, by the Americans Stevens and Anderson, used helium as the lifting gas, and reached 22,066m (72,395ft).

Helium was identified as a minor constituent of natural gas in some American oilfields in 1907, and its potentialities for inflating balloons and airships were realized on the outbreak of the First World War. Although heavier than hydrogen, it is completely non-flammable. Several plants were installed to separate helium from natural gas in the United States in 1917–18; the American military services prohibited its export and from 1923 it replaced hydrogen entirely in US military airships, but was rarely used in balloons because of the cost.

The resumption of high-altitude balloon flights for cosmic ray research after the Second World War utilized the recently developed lightweight polythene films for the envelope. Initially these were not thought safe enough for manned flight, so from 1947 a series of unmanned flights was made using helium-filled balloons developed by the American Otto Winzen and funded by the US navy and air force. Eventually a number of manned flights were made in the USA with polythene balloons between 1956 and 1961.

A by-product of these high-altitude flights was the revival of hot-air ballooning. In 1960, the American Paul E. Yost introduced the 'Vulcoon', with an envelope made of nylon fabric laminated with an internal mylar plastic film. Commercially available propane gas cylinders (normally used for fuelling portable cooking stoves) fed a burner system which heated the air in the balloon. The single pilot sat on a sling, although small wicker or aluminium baskets were soon introduced. Although nominally developed as a military project by Yost's company Raven Industries, these new hot-air balloons were primarily produced for sport-flying. After a somewhat hesitant start, the revival of hot-air ballooning spread world-wide during the 1970s, and improved designs of envelopes and burners made it possible to build a large variety of sizes and shapes, including many novelties to advertise commercial products.

The new fabric technology was also applied to gas balloons, and long-distance flights lasting several days became possible. The first successful balloon flight across the Atlantic was made by the Americans Ben L. Abruzzo, Max L. Anderson and Larry Newman in August 1978 in a helium-filled balloon.

AIRSHIPS

Almost as soon as the free balloon had been invented, attempts were made to control its direction of flight, by using hand-operated paddles (tried by Lunardi in 1784), airscrews (Blanchard, 1784) or even by letting out a jet of hot air from the side of the envelope (Joseph Montgolfier, 1784). It was soon realized that it would be an advantage to reduce the cross-section area of the balloon, so elongated and pointed shapes were proposed. The French General J. B. M. Meusnier produced a detailed design for such a dirigible balloon or airship in 1785; it had the form of an ellipsoid, 79m (260ft) long with a capacity of 60,000ft^3 (1700m^3), and was intended to be driven by manually-powered airscrews.

Meusnier's most significant invention in this design was the ballonet – an air-bag inside the envelope, into which air was pushed by a bellows to maintain internal pressure inside the airship. Varying the amount of air in the ballonet compensated for changes in the volume of hydrogen as the altitude of the ship changed, while maintaining the external shape of the envelope. Although Meusnier's airship was never built, it embodied many of the features of later designs.

As long as manual effort was the only available power source to propel an airship, little progress was possible. The idea of using a pair of horses working a treadmill was proposed in a design by Dr E. C. Génet in 1825, but the first satisfactory power source was a 2.25kW (3hp) steam engine employed by the French engineer Henri Giffard in 1852. This unit, with its coke-fired boiler, was slung some 12m (40ft) below the cigar-shaped envelope, which was 44m (144ft) long and contained 2500m^3 (88,000ft^3) of hydrogen. Giffard's airship was calculated to have a maximum speed of 10kph (6mph) in still air, so in practice it was impossible to fly to a pre-determined destination except down-wind.

An Austrian engineer, Paul Haenlein, was the first to build an airship with an internal combustion engine; in 1872 he constructed a four-cylinder gas engine of Lenoir pattern, but it was too heavy for his balloon which was never flown. He was followed by two French army engineers, Charles Renard and Arthur Krebs, who in 1884 flew their airship *La France* powered by an electric motor of 6.5kW (8.5hp) supplied with current from batteries. This is usually regarded as the world's first successful airship, since it was able to return to its starting point on five of its seven flights; however, it was capable of no more than 24kph (15mph) in still air so was only able to fly in the lightest breezes.

With the development of petrol-engined road vehicles in 1886 (see Chapter 8), it seemed that a suitable power unit for airships had appeared. The German engineer Karl Woelfert, in conjunction with Gottlieb Daimler, produced an airship with a 1.5kW (2hp) single-cylinder Daimler engine which was test-flown in 1888. This was too small to be practicable; Woelfert's next airship with a 4.5kW (6hp) Daimler engine probably never flew because it lacked sufficient lift; and a larger airship built for trials by the Prussian army in 1897 caught fire and killed the inventor on its only flight in Berlin. In the same year, at the same place, a unique all-metal airship designed by the Austrian David Schwarz and fitted with a Daimler engine was wrecked on its first trial flight.

In France, the Brazilian amateur enthusiast Alberto Santos-Dumont fitted a 1.5kW (2hp) De Dion-Bouton engine in the first of a series of small airships which he flew fairly successfully. In 1901, in his No. 6 airship, he just succeeded in flying from St Cloud to the Eiffel Tower and back inside the time limit of 30 minutes and thus won a prize of 100,000 francs offered by Henri Deutsch de la Mearthe. The enormous publicity surrounding this flight gave Santos-Dumont's airships a rather unwarranted reputation, for in truth they were hardly capable of outflying Renard and Krebs' *La France* of 1884.

Inspired by Santos-Dumont's activities, the French Lebaudy brothers, owners of a large sugar refinery, commissioned their chief engineer Henri Julliot to build a much larger airship. This machine *Lebaudy I* was 57m (187ft) long with a capacity of 2250m^3 (80,000ft^3); a 30kW (40hp) Daimler engine driving two 3m (9ft) diameter airscrews gave it a speed of about 40kph (25mph) and it was the first really practical airship when it flew in November 1902.

During the next ten years a considerable number of essentially similar airships were made in several European countries, and these were further developed during the First World War. Most were non-rigid airships, colloquially known as Blimps, which had envelopes made of rubberized cotton or linen fabric, whose shape was maintained by having the gas at slightly greater than atmospheric pressure: this pressure was generated by forcing air into internal ballonets by scoops behind the propellers. The car (often called a gondola) containing crew and engines was slung beneath the envelope, usually hanging from support patches sewn into the fabric (see Figure 12.1). The so-called semi-rigid airships (which included the Lebaudy types) had a rigid keel of wood or a metal framework, attached directly to the bottom of the envelope; the car was slung from this.

Probably the largest of the non-rigid airships built during this period were the British *North Sea* class of 1917, with two 250hp (166kW) engines and a crew of ten; these were 80m (262ft) long with a capacity of 10,000m^3 (360,000ft^3). During and after the Second World War much larger airships of this general type, filled with helium, were developed by the US navy, eventually

Figure 12.1: A 'Coastal' class airship of the Royal Naval Air Service setting off for an anti-submarine patrol during the First World War, 1914–18. This is a typical non-rigid airship, inflated by hydrogen, and powered by two 112kW (150hp) Sunbeam engines.

culminating in the Goodyear ZPG-3W of 1958 with a capacity of 42,500m^3 (1.5 million ft^3), powered by two 1120kW (1500hp) engines. These carried early-warning radar systems for fleet protection, with large aerials inside the envelope.

Small non-rigid airships continue to be built in the 1980s, using modern synthetic materials for the gas-bag and cabins; the basic configuration, however, remains essentially similar to the successful Blimps of the First World War.

The 'semi-rigid' design was particularly developed in Italy, the largest being the *Roma* of 34,000m^3 (1.2 million ft^3), launched in 1919. Its six 500hp engines gave a maximum speed of almost 70mph. This was sold to the United States and its destruction by fire in 1922 caused the US military authorities to use only helium for inflating their airships thereafter.

The largest and most spectacular airships were the 'rigids', which had a wooden or metal framework structure, covered externally with a fabric skin, and containing a number of internal gas bags to provide lift. Accommodation for the crew and passengers was either in a cabin attached directly to the main hull, or in a separate gondola slung below it. Engines were slung from the hull, usually in discrete pods distributed along the length of the ship.

The ill-fated Schwarz metal airship was technically a rigid airship, for it had an internal structure of metal tubes supporting the external skin, but the accepted originator of the classical rigid airship was Count Ferdinand von Zeppelin. His first design, LZ.1, was developed with the assistance of Professor Müller-Breslau, a structural engineer. A framework of internally-based ring girders joined by longitudinal members was made in aluminium – the whole structure being a 128m (420ft) long cylinder of 11.75m (38.5ft) diameter with tapered ends. Two gondolas were slung beneath the hull, each containing a 10.5kW (14hp) Daimler engine. Seventeen individual gas bags made of cotton with a rubber lining were installed between the frames, and the outside of the hull was covered in a varnished cotton fabric. LZ.1 was launched in 1900, and flew only three times because it had inadequate controls. A second ship, LZ.2, built in 1905, introduced triangular section girders made of the new high-strength duraluminium alloys. This had two 63kW (85hp) motors and a more satisfactory control system, but crashed on its second flight. However, LZ.3 of 1906 proved sufficiently successful to spawn a long line of airships, and successful passenger-carrying services were operated from 1911 with LZ.10 and three other ships. The main uses of Zeppelins were to be for military purposes, and notably for the inauguration of night-bombing attacks on French and British targets, in which they were able to avoid the opposition of guns and aeroplanes by flying at altitudes above 6000m (20,000ft).

There were other rigid airships, notably the wooden-framed German Schutte-Lanz designs of 1911–18. The British government sponsored a series of designs culminating in the passenger ships R100 and R101 (Figure 12.2) of 1929, and the US Navy purchased the *Akron* and *Macon* built by Goodyear (with considerable Zeppelin input) in 1931–3; but the Zeppelin company continued to dominate the field. Their LZ.127 *Graf Zeppelin*, built 1928, was 236m (775ft) long with a volume of 106,000m³ (3.7 million ft³), and operated passenger services on a regular schedule across the South Atlantic for several years. However, the spectacular losses of the British R.101 in 1931 and the LZ.129 *Hindenburg* in 1937, following a series of earlier accidents, brought about a cessation of work on rigid airships. Although new designs were proposed in the 1970s using modern materials and various novel design principles, it seems unlikely that the rigid airship will reappear.

HEAVIER-THAN-AIR FLYING MACHINES: THE PIONEERS

An entirely new approach to achieving dynamic flight with heavier-than-air apparatus was initiated by Sir George Cayley, a scholarly Yorkshire landowner with wide practical interests who remained fascinated by flying throughout his life. Although at various times he worked on model helicopters and clockwork-

Figure 12.2: The British airship R101 leaving the passenger access tower at Cardington during test flying, October 1929. Water-ballast is being discharged near the nose, and the four 'forward' engines are stationary while the ship is manouvered by the 'reverse' engine.

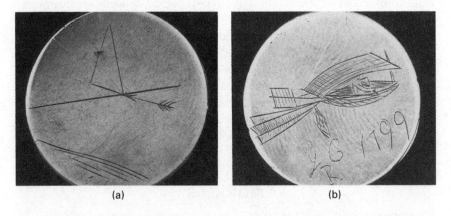

(a)   (b)

Figure 12.3: Sir George Cayley scratched this sketch (a) on a small silver disc in 1799 to illustrate his concept that the aerodynamic force on a wing could be resolved into lift and drag. On the other side (b) he shows an aircraft with the wing, a boat-shaped nacelle for the pilot, a controlling tail unit and a pair of propulsive flappers.

powered airships, his main contribution was to formulate clearly the basic principle of dynamic flight using a rigid wing surface to provide lift and a separate propulsion unit to provide forward motion (see Figure 12.3). In a classic paper published in three numbers of *Nicholson's Journal of Natural Philosophy* (November 1809–March 1810), he summarized the basic principle in the words:

> The whole problem is confined within these limits, viz. To make a surface support a given weight by the application of power to the resistance of air.

In 1804 he had measured the lift produced by a flat plate moving through the air at a small angle of incidence, using a whirling arm driven by a falling weight, similar to that used by John Smeaton in 1752 for comparing various designs of windmill (see Chapter 4). With the data thus obtained he was able to design and make model gliders incorporating a kite-like wing and a stabilizing tail unit. He also found that the stability of his gliders in the lateral plane was improved by bending the wing to incorporate a dihedral angle. There is some evidence that he built what would now be called a hang-glider in about 1810 – a full-size winged machine supporting a man who launched it by running forward (probably downhill) and controlled its flight by moving his body. Towards the end of his life he made two gliders which certainly carried a live load – a 'boy-carrier' in 1849 and a 'man-carrier' in 1853. Unfortunately the evidence for their construction and operation is little better than anecdotal and although it is clear that a controlling tail unit was fitted at least to the later machine, it is not clear whether the occupant could really control the flight path to any significant extent.

Although Cayley experimented with hot-air engines, clockwork springs and even a gunpowder motor as potential power units, he never solved the problem of propelling his aircraft and was consequently unable to exploit his vision that

> an uninterrupted navigable ocean, that comes to the threshold of every man's door, ought not to be neglected as a source of human gratification and advantage (1816).

STEAM POWER

Most of Cayley's work was never published, and although the significance of his paper of 1809–10 was later well recognized, it had less influence on his contemporaries than he had hoped. However, he did directly inspire William Samuel Henson, whose widely publicized patent of 1843 for an Aeriel Steam Carriage fixed the idea of Cayley's classical aeroplane shape in the minds of later workers. Henson postulated a high-wing monoplane with cambered wing section, externally braced with streamline wires to reduce drag, with a separate fuselage containing accommodation for passengers and crew, and housing a steam engine which drove two airscrews behind the wing. A large tail unit comprising horizontal and vertical rudders was intended to steer the machine.

The most impractical feature of Henson's machine was the intention to launch it down an inclined rail. The idea was to use the force of gravity to accelerate the aircraft, so the steam engine needed only to be sufficiently powerful to overcome drag in forward flight. Clearly Henson recognized that the weight of the power plant – including boilers, fuel and water – would be critical, and must be kept as small as possible.

Henson's project was never built, though there is evidence that contracts were placed for construction of the airframe and engine. The company formed to exploit it suffered much ridicule as a result of its excessively optimistic publicity for passenger-carrying services to India and China, and the scheme foundered.

Henson then combined with John Stringfellow to test a large model of about 6.6m (20ft) wingspan – something that might prudently have been done earlier. However, no real success was obtained, partly because the trials (on a hillside near Chard in Somerset) were conducted at night to maintain secrecy and avoid ridicule. These tests were made in 1847; in the next four years or so Stringfellow continued the work alone, building a number of models and light steam engines. Limited success was obtained, but any real demonstration of free flight was prevented by Stringfellow's lack of a suitably large building in which to fly his models. In 1851, frustrated by lack of funds to procure such a building, he postulated 'an Aerial tent of canvass or calico rendered impervious to air and to be filled and kept up by a blowing machine so that no timber would be required to support it'. This imaginative prevision of the inflated structures of the late twentieth century remained as no more than an idea.

Stringfellow briefly resumed his aeronautical work around 1866, stimulated by the formation of the Aeronautical Society of Great Britain under the presidency of the Duke of Argyll; at the Society's Exhibition at the Crystal Palace in 1868, he exhibited a steam powered triplane which seems to have been rather less successful than his model of 1848. He also exhibited one of his earlier small steam engines, and received a prize for it as the lightest practical power unit entered.

Steam power being virtually all that was available in the nineteenth century, it is not surprising that it was employed by later inventors, who made little further progress towards successful mechanical flight. Clément Ader, a famous French electrical engineer, built his *Eole* between 1882 and 1890, with a 15m (49ft) wingspan patterned on the model of a bat. Driven by a steam engine of about 13.5kW (18hp) and piloted by Ader himself, it made a single straight-line 'flight' of about 50m (160ft) just clear of the ground on 9 October 1890. Inspired by this, Ader was funded by the French government in 1892 to make a new machine which appeared in 1897. This *Avion III* (No. 2 having been abandoned before completion) had two steam engines driving curious feathered propellers, and huge bat-like wings of complex construction. It was twice tested in October 1897, but failed totally to fly and was blown off its track and damaged on the second attempt, after which the War Ministry refused further support. The machine still exists in a French museum and is the oldest full-size 'aircraft' to survive.

Equally abortive was the enormous steam-powered test-rig built by Sir Hiram Maxim in Kent in 1893–4. After extensive tests of aerofoils and other components on a whirling arm and in a wind tunnel, Maxim built a carriage propelled by two very lightweight steam engines of 135kW (180hp) each, running on a level railway track some 550m (1800ft) long. On this carriage were mounted wing surfaces extending to some 400m^2 (4000ft^2), which were assembled in various configurations. The lift and drag developed were measured, and the carriage was restrained from rising more than a few centimetres by outrigged wheels running beneath a set of guard rails alongside the main track. Although testing continued for a couple of years, nothing came of it. Maxim did not attempt to build a real flying machine until 1910, and that was totally unsuccessful. He ascribed his failure variously to lack of money, inability to find a sufficiently large open space to continue trials, and to the need to develop a better power-plant. The last reason is certainly valid, but the real reason for his procrastination seems to have been a fear that he might be subject to ridicule if he failed.

GLIDERS

The first man to make repeated flights with a heavier-than-air machine was the German Otto Lilienthal. His original intention was to develop an ornithopter, but lacking a suitable power source he developed a series of fixed-wing gliders between 1891 and 1896. These machines were launched from a variety of eminences, including a special constructed earth mound some 15m (50ft) high. He supported himself from his forearms, placed through sleeves under the wing roots, with his lower body hanging below the wing, launched himself by running downhill into the wind, and controlled the flight by swinging his body to manipulate the centre of gravity. His gliders were fairly crudely made, with a single-surface fabric wing supported by wooden spars which could be folded up for easier portability. The machines themselves were not technically significant, but the publicity given to his numerous flights was a spur to other workers, particularly as for the first time photographic illustrations of a man flying successfully were given wide circulation. The publicity given to Lilienthal's flying was enhanced when he was killed by a crash in August 1896.

Lilienthal's activities inspired the American railway engineer Octave Chanute to make a number of hang-gliders in 1896. Chanute was an enthusiast who had collected a great deal of information on earlier theories and experiments, which he summarized in a seminal work, *Progress in Flying Machines*, published in 1894; his practical experiments were therefore founded on a firm basis of knowledge and engineering experience. The most significant machine was a biplane hang-glider with a cruciform stabilizing tail unit, which was successfully and regularly flown on the shores of Lake Michigan by Chanute's assistant A. M. Herring. Herring attempted to develop a power unit for a version of this machine, and claimed to have made two short hops (with an engine driven from a compressed gas supply sufficient for 15 seconds' running) in October 1898.

THE WRIGHT BROTHERS

Wilbur and Orville Wright of Dayton, Ohio, were directly inspired by reading about Lilienthal's gliding activities. They were the first seriously to consider the problem of controlling a flying machine. Earlier experimenters had, explicitly or implicitly, considered the flying machine as a stable vehicle to be steered in the desired direction. The Wrights recognized that a machine moving in three dimensions in a turbulent atmosphere would need more dynamic control by its pilot.

Their solution was a biplane wing structure with a wire-braced truss structure. The central part was rigid, but the outer parts could be twisted under the pilot's control. The tip section on one side was inclined at a greater angle to the airflow, thus lifting that wing; the other tip was twisted in opposition to lower it; and the wing thereby banked. This inclined the lift vector and thus turned the aircraft. At first they used a fixed vertical fin at the rear of the aircraft like the flights of an arrow to stabilize the aircraft; then converted this into a rudder acting simultaneously with the wing twist; and finally separated the two controls to allow the pilot freedom to sideslip if he wished.

For control in pitch, the Wrights used an auxiliary aerofoil ahead of the wings, directly operated by the pilot. (They followed their predecessors in assuming that control about the pitching axis – vertical motion of the aircraft – was essentially independent of the control about the rolling and yawing axes – horizontal motion. The theoretical explanation of this fact was given by the Welsh mathematician G. H. Bryan in 1904, but it is doubtful that any of the pioneer aviators were aware of this until several years later, by which time it was recognized as an empirical truth.)

The Wrights started experimenting with gliders in 1899, and applied for a patent in March 1903 on their configuration and on the basic concept of twisting (or warping) the wing for control in roll, which was granted in 1906. They also built their first powered machine, using a light four-cylinder petrol engine which they designed and constructed themselves (having failed to find a commercial supplier able to meet their requirements). The engine drove a pair of

Figure 12.4: Orville Wright made the world's first powered flight with a heavier-than-air machine on 17 December 1903 near the Kill Devil Hill in North Carolina.

rear-mounted wooden propellers: not the least of their practical achievements was to design more efficient propellers than any previously used, by a combination of inspired theoretical assumptions and experimental research.

The first 'Flyer', as they called it, was flown successfully on 17 December 1903 (Figure 12.4). The four flights made that day are now recognized as the first controlled powered man-carrying flights in history – although the longest lasted less than a minute. The first Flyer was wrecked by a gust of wind shortly after its fourth landing, and the events of that day are important only because the Wrights went on to perfect their design in 1904 and 1905.

Their second and third Flyers retained the same configuration, but by altering the location and relative sizes of the control surfaces they improved the controllability. The first machine was so unstable in pitch that it was almost uncontrollable, but by October 1905, after about 120 experimental flights (mainly very short), they were able to fly for over half an hour under perfect control. They still eschewed stability, believing that they could only maintain adequate control in turbulent air by deliberately aiming for an unstable aircraft; in this they were undoubtedly influenced by their experience as bicycle riders and manufacturers.

The Wrights' successful development of a practical flying machine was, surprisingly, largely unpublicized, and they ceased flying after mid-October 1905 to avoid risking premature disclosure which might invalidate their patent application. After the patent was granted, they spent two years negotiating with various governments and commercial syndicates for the exploitation of their invention, and only resumed flying in 1908. They practised for a week in secret in May to renew their piloting skills, and then started flying in public – first in France in August 1908 and then near Washington, USA, in September. These flights demonstrated their total pre-eminence in the art of flying, and firmly established their claim to priority.

EUROPEAN PIONEERS

By this time other reasonably practical aeroplanes were in existence, or imminent. The development of reasonably light petrol engines for motorcars and boats, especially in France, provided the power source which the nineteenth-century aviators had lacked. Spurred by news or rumours of the Wrights' activities, a number of Europeans had been experimenting since 1904. Their machines employed a variety of configurations, but the first one to achieve a realistic flight was the biplane built by the Voisin brothers in France for Henri Farman. In November 1907 he flew this for over a minute, and on 13 January 1908 won a prize of 50,000 francs for an officially observed closed circuit flight of just over 1km in $1\frac{1}{2}$ minutes. Farman's machine was superficially not dissimilar to the Wrights' configuration, with biplane wings and forward elevator, but the large tail surfaces were mainly fixed stabilizing surfaces with a limited rudder. The machine was deliberately designed to be as stable as possible, and there was no roll control at all. During the next few months the Voisins built several similar machines for other pilots, and Farman modified his aircraft: in these, vertical fin surfaces (called side-curtains) were added between the wing struts, and the stabilizing tail surfaces somewhat reduced. By September 1908 the modified Voisin-Farman was capable of long flights of up to 45 minutes' duration, and after fitting ailerons to obtain roll-control (as a substitute for the Wrights' wing-warping), Farman made the world's first cross-country flight from Bouy to Reims (27km (18 miles)) in 20 minutes, on 30 October 1908.

Subsequent developments effectively brought the Wright and Voisin-Farman configurations closer together, the Wrights moving towards more stable designs and other designers following Farman towards more effective controls. By mid-1909 the so-called box-kite design (which retained this nickname although the side-curtains which had suggested the soubriquet were soon discarded) had become dominant – biplane wings with ailerons on the outer trailing edges, a forward elevator, and a rear stabilizing tail unit incorporating elevator and rudder surfaces, powered by a petrol engine of around 37.3kW (50hp) driving a directly-coupled pusher propeller. The 'fuselage' of these machines was a box-girder, usually totally uncovered, and the basic structural materials were wood, fabric and steel wire.

Figure 12.5: Louis Blériot standing on the seat of his Type XI aircraft, in which he flew the English Channel on 25 July 1909. The 3-cylinder Anzani engine and wooden Chauvière propeller are well shown in this picture, but the top bracing wires are less clear.

Among the wide variety of configurations tried by other would-be aviators, another practical shape emerged: the tractor monoplane. The first successful machines of this type were the Antoinettes designed by Frenchman Léon Levavasseur, but the most famous example of the breed was the machine in which Louis Blériot made the first flight across the English Channel on 25 July 1909 (Figure 12.5). This had a monoplane wing of wood and fabric, wire-braced to a steel frame structure above the fuselage. The fuselage itself was a simple uncovered wooden box structure with internal wire bracing, carrying a vertical rudder and a fixed horizontal tailplane at the rear end. The outer third of the tailplane was hinged to operate as the elevator for control in pitch. On the front of the fuselage was a sturdy well-sprung undercarriage with wire wheels, and the 18.6kW (25hp) three-cylinder Anzani engine driving a two-bladed wooden airscrew.

This was Blériot's famous 'Type XI', designed by Raymond Saulnier, which was built in very large numbers for several years, with increasing engine power and stronger structure. It inspired numerous imitations, but monoplanes in general acquired a poor reputation as a result of various accidents which led to an official (but temporary) ban by the British War Office in September 1912, and somewhat similar lack of faith on the part of some military organizations. Insofar as there was any common basis for these accidents, they seem to have arisen from a lack of knowledge about the distribution of loads on the wing, and a tendency to overstress the structure by excessively tightening the wing bracing.

The third standard type of aircraft was the tractor biplane, which eventually supplanted the pusher 'box-kite' and numerically dominated the world's skies up to about 1940. The first of these was the otherwise insignificant Goupy II biplane, built in the Blériot factory in March 1909; more significant was the Breguet I of July 1909, because the manufacturer continued in business for many years and his later machines introduced metal structure for the wings (covered in fabric) and fuselage (covered with aluminium sheet).

The classic type of tractor biplane was surprisingly late in emerging, the first important example being A. V. Roe's Type D, first flown at Brooklands in April 1911, of which half a dozen were built. Roe had previously built a number of triplanes, all with a tractor engine installation – the first two had a fixed triplane tail unit and were controlled in pitch by varying the incidence of the wings. The next triplanes retained the triplane tail unit, but this was now hinged to act as an elevator control and the wings were fixed rigidly to the fuselage. Roe's final triplane and his Type D biplane, however, had a simple monoplane tail unit with fixed tailplane and hinged flap-type elevator. There was no vertical fin, simply a rudder.

In December 1911, the Royal Aircraft Factory at Farnborough produced a tractor biplane designed by Geoffrey de Havilland which they designated B.E.1 (for 'Blériot Experimental No. 1' – Blériot being regarded as the pioneer of the tractor form of aeroplane). This had a water-cooled 45kW (60hp) Wolseley V-8 engine, shortly replaced by an air-cooled 60hp Renault of the same configuration. B.E.1 became the progenitor of a long line of aircraft, and the vehicle for many experiments by the Factory's research staff.

MILITARY AND COMMERCIAL APPLICATIONS

The most significant development was the search for an aeroplane which was inherently stable, so that the military pilot could concentrate on his observation of the ground and the troops upon it, but was adequately controllable. The solution was found in a practical application of G. H. Bryan's theoretical work, together with aerodynamic measurements of scale models in a wind tunnel and on a whirling arm. The principal vehicle for these experiments was an aircraft designated R.E.1 (for 'Reconnaissance Experimental 1') in which the wings were rigged with a dihedral angle of about 3°, and large ailerons replaced the wing warping used on early B.E.s. A successful conclusion to the experiment in March 1914 led to large-scale production of variants of the B.E. by British

Figure 12.6: The classic biplane configuration was adopted by the Royal Aircraft Establishment to produce a stable aeroplane for military reconnaissance. This is a B.E.2E, first delivered to the Royal Flying Corps in 1916.

industry; unfortunately wartime experience from August 1914 onwards demonstrated that the stable aeroplane was unnecessarily stable, and lacked the manoeuvrability needed for aerial combat (Figure 12.6).

By the time the First World War broke out in August 1914, the aeroplane could be considered as a technically sound machine, although it lacked any clearly defined operational use in either civil or military roles. Probably rather more than 4000 aeroplanes had been built world-wide, the vast majority being single-engined machines of up to 75kW (100hp) carrying one or two persons at up to 130kph (80mph). World records at the end of 1913 were all held by French aircraft: 1021.2km (634.54 miles) distance in a closed circuit by A. Seguin on a Farman pusher biplane; 6120m (20,079ft) altitude by G. Legagneux on a Nieuport tractor monoplane; and 203.8kph (126.67mph) by M. Prévost on a Deperdussin monoplane with a streamlined monocoque fuselage made of overlapping thin strips of wood.

Practical machines had been developed in several countries which could take-off and land on water, and a few experimental amphibian machines had appeared. A flying boat built by Benoist was used for a few weeks in 1914 to run the world's first scheduled service for passengers on a 35km (22 mile) route from Tampa in Florida; and the Curtiss company was testing a large twin-engined flying-boat in the USA for a projected transatlantic flight when the war broke out.

The largest aeroplanes in the world at this time were the four-engined *Bolshoi* and *Ilya Murametz* machines built by Igor Sikorsky in Russia. The latter, which flew in January 1914, was powered by four 75kW (100hp) Mercedes motors, and carried up to 16 passengers in an enclosed cabin. It was essentially of conventional configuration, with biplane wings on which the four engines were mounted driving tractor propellers, and a conventional tail unit at the rear of the fuselage.

Technical innovations in aeroplane design during the First World War were surprisingly few, but the scale of aircraft manufacture increased enormously: around 200,000 machines were built in those four years. By the end of the war, engines of 300kW (400hp) were available in quantity, so the average speed and weight of aircraft had increased considerably. Multi-engined machines were built in several countries, and aerial bombardment became an accepted form of warfare, while effective fighter aircraft armed with machine guns were developed as a counter to both bombers and reconnaissance aircraft.

The vast majority of aircraft were still mainly built of wood, but metal construction was introduced on a significant scale, especially in Germany. The Fokker company produced welded steel tube fuselage structures on a production scale, but few other manufacturers followed suit because of the problem of adequately inspecting the welds. More significant was the development of the metal cantilever wing by the Junkers company. The founder, Professor Hugo Junkers, had been granted a patent in 1910 for a thick wing which needed no external bracing, but this was not successfully built until December 1915 when the J.1 appeared, with an all-metal structure mainly in steel. Two years later the J.4 biplane appeared, using the high-strength aluminium (dural) alloys which had been developed for airship construction.

A cantilever wing made it possible to set the fuselage on top of the wing, and subsequent Junkers all-metal designs were mainly of this low-wing monoplane configuration. The most significant design of this type was the F.13, developed in 1919 as a civil transport aircraft carrying four passengers in an enclosed cabin. The two pilots originally occupied an open cockpit, but later models were fully enclosed. The F.13 can reasonably be regarded as the world's first airliner, and over 300 were built, continuing to serve in various countries for about 18 years.

The same basic structure, with corrugated dural skins covering a skeleton of dural spars, ribs and fuselage frames, was used by the Junkers company for a range of transport aircraft up to the three-engined Ju 52/3m, built in very large numbers, including the four-engined G-38 of which only two were built, whose wing was thick enough to house six of the 34 passengers.

Figure 12.7: The first passenger services from London to Paris in 1919 were flown by converted bombers like this Handley Page 0/7. Ground facilities for the passengers were limited, but curtains were provided in the cabin.

THE INTER-WAR YEARS

With the end of the war, a large number of aircraft, aircrew and trained mechanics became available, and a small proportion of these assets were employed on civil flying, mainly passenger carrying. These services started with converted wartime bombers (Figure 12.7); although these were soon superseded by aircraft designed for the purpose, for some years there was little change in basic configuration (except for the Junkers machines mentioned above). A typical British airliner design introduced in 1927 was the Armstrong Whitworth Argosy, a biplane with three 300kW (400hp) Jaguar engines, carrying 20 passengers in a rectangular fuselage with steel-tube frame covered in fabric. The wing frames were also of steel construction. Another typical transport aircraft of the period was the Fokker F.VII/3m, built in both the Netherlands and the USA. This had Fokker's welded steel tube fuselage, fabric covered, coupled to an all-wooden cantilever wing mounted above the fuselage. Three engines of about 224kW (300hp) were fitted, and up to 10 passengers were carried.

The almost universal preference for three engines in transport aircraft at this period was an insurance against engine failure – few twin-engined aircraft were capable of maintaining height on the remaining engine if one failed. Typical cruising speeds of these aircraft were about 160kph (100mph), flight times were

Figure 12.8: The Supermarine S.6 which won the 1929 Schneider Trophy contest for Great Britain. The Rolls-Royce engine produced 1415kW (1900hp), and virtually the entire surface of the aeroplane was utilized as a radiator to dissipate the equivalent heat.

around three hours, and operating heights below 1000m (3000ft). Flight at night or in clouds was rarely risked.

Military aircraft of this period were, in general, no more technically advanced than their civilian contemporaries, but there were advances in the specialized area of high-speed contest flying, exemplified by the international competition for the Schneider Trophy for seaplanes. Remarkable speeds were achieved by streamlined machines with small frontal area and large engine powers – boosted by exotic fuel mixtures. Consecutive Schneider Trophy contests were won in 1925 by the Curtiss R3C-2 at 374.3kph (232.6mph), in 1926 by the Italian Macchi M.39 at 396.7kph (246.5mph), in 1927 by the British Supermarine S.5 at 453.3kph (281.7mph), and in 1929 by the Supermarine S.6 at 528.9kph (328.6mph) (Figure 12.8). In 1931 the series ended when the Supermarine S.6B achieved 547.3kph (340.1mph) in an uncontested event.

These aircraft were in no sense practical vehicles, but they pointed the way to a new generation of military and civil aircraft which appeared in the 1930s. The modern airliner was born in the USA, when the Boeing 247 first flew in February 1933. This was an all-metal aircraft, in which the fuselage and wing skins contributed to the torsional stiffness of the structure: this stressed-skin principle had been foreshadowed in the wooden monocoque fuselages adopted by a few manufacturers, but henceforward became universal in all fast aircraft. The

Figure 12.9: The 21-seat Douglas DC-3 of 1936 was the first truly effective civil airliner capable of providing a regular, reliable transport service and generating a profit for its operator from passenger revenue alone.

247 had two engines, each powerful enough to maintain height if one failed; aerodynamic research had demonstrated how these engines could be cowled to reduce drag while maintaining adequate air cooling. The undercarriage was retracted by electrically driven screw-jacks to reduce drag further, and wing flaps were fitted to improve the wing lift at low speeds. Much research was undertaken at this period to improve the lifting efficiency of wings, allowing a progressive reduction in the wing area (and therefore wing weight and drag) needed to support the aircraft's weight. This improvement in wing loading (i.e., weight divided by wing area) over the previous generation of airliners, coupled with improved streamlining, produced higher cruising speeds – around 250kph (155mph). The Boeing 247 was rapidly followed by other types of similar configuration, notably in the USA by the 14-seat Douglas DC-2 and 21-seat DC-3 (Figure 12.9). With these aircraft, the American airlines were able to provide regular, reliable and safe passenger services. Safety and reliability were enhanced by effective de-icing measures, blind-flying instruments, air-to-ground radio communications, meteorological forecasting services and the provision of hard-surfaced runways at airports. The pattern was followed in Europe, and eventually world-wide.

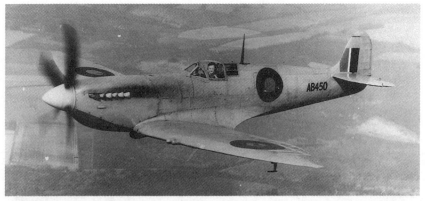

Figure 12.10: A Supermarine Spitfire VII of 1942, typical of the high-performance fighter of its time, with all-metal structure and thin cantilever wings. The engine is a 1120kW (1500hp) Rolls-Royce Merlin 61 with two-speed supercharger and the cabin was pressurized to allow operation up to 12,000m (40,000ft) altitude.

THE SECOND WORLD WAR

The new technologies were applied also to military aircraft, albeit somewhat hesitantly. New designs of fabric-covered biplane fighters with fixed undercarriages were still being introduced into Royal Air Force squadron service as late as February 1937, but by the outbreak of the Second World War in 1939 the front line fighter and bomber units of all the major air forces had been equipped with cantilever monoplanes with enclosed cockpits and retractable undercarriages. Engines of around 750kW (1000hp), supercharged to maintain power at altitudes around 6000m (20,000ft) became the norm. The Supermarine Spitfire may be taken as typical: in 1939 the original ('Mark 1') version with a Merlin engine of 738kW (990hp) driving a two-bladed wooden airscrew had a maximum speed of 588kph at 6150m (353mph at 20,000ft). Wartime development of the basic design produced the final operational version, Mark 22, with a 1530kW (2050hp) Griffon engine and five-bladed airscrew; this achieved 735kph at 8000m (457mph at 25,000ft) (see Figure 12.10).

Progress in bomber design was rather less impressive, but the twin-engined Vickers Wellington 1 bomber of 1939 – with two 750kW (1000hp) engines, it carried 2050kg (4500lb) of bombs at a speed of around 315kph at 3075m (195mph at 10,000ft) – can be compared with the Avro Lancaster 3 at the end of the war – with four 1200kW (1600hp) engines, it carried up to 6350kg (14,000lb) of bombs about 55kph (30mph) faster at 6150m (20,000 ft). The most efficient bomber of the war was the De Havilland Mosquito, which was quite outside the main conventional design trend. The De Havilland company had concentrated on civil aircraft production up to 1939, and their bomber was based on the pre-war DH 88 Comet built for long-distance air racing. The Mosquito structure was built entirely of wood, and the aircraft had a crew of only two, dispensing with the heavy defensive armament of conventional bombers. With two 1275kW (1700hp) Rolls-Royce Merlin engines it carried up to 2250kg (5000lb) of bombs at some 500kph (310mph). The same basic airframe was used in numerous other roles – as a photo-reconnaissance aircraft operating up to 13,000m (38,000ft) with a pressurized cabin, as a radar-equipped night fighter, as a low-altitude strike aircraft and even as a high-speed transport.

Air transport played a substantial part in the military operations of the Second World War. Using variants of civil airliners and conversions of bomber designs, large numbers of passengers and significant quantities of freight were carried on regular services throughout the world, including for the first time regular passenger services across the North and South Atlantic. In addition, direct intervention in land battles was provided by troops dropped by parachute or carried in large towed gliders.

Other technical developments during the war which had a significant and lasting effect on the operational use of aircraft included the introduction of various radar navigation systems and of reliable automatic pilots. An oxygen supply was routinely provided to aircrew flying above 4500m (15,000ft), and pressurized cabins were introduced on a limited scale for operations above 10,000m (35,000ft). Another very important influence on the future development of aircraft operations was the proliferation of hard-surfaced runways throughout the world, largely replacing grass airfields. This allowed landing and take-off speeds to rise, allowing an increase in wing-loading and thus in cruise performance.

DEMISE OF THE FLYING-BOAT

Improvements to runways also caused the virtual elimination during the 1950s of large flying-boats from both civil and military operations. Before and during the war, these aircraft were widely used for anti-submarine warfare and for long-range passenger transport: the first regular airline services across the Atlantic and Pacific Oceans were provided by four-engined flying-boats made by the Boeing and Sikorsky companies. A good deal of research work went into improved hull shapes for a post-war generation of large flying-boats of high performance, including such aircraft as the Saunders-Roe ten-engined boat intended to carry 85 passengers for 8000km (5000 miles), and the Martin P6M Seamaster four-jet military reconnaissance machine. None of these came into service because land-based machines could meet similar design targets more efficiently. The last four-engined passenger flying-boats were withdrawn from scheduled service on major routes in 1959.

PART THREE: TRANSPORT

EXPANSION OF CIVIL AVIATION

A new generation of passenger transport aircraft was initiated on 31 December 1938 with the first flight of the Boeing 307 Stratoliner, a four-engined airliner with a pressurized passenger cabin. Only a few Stratoliners were built, but it set the pattern for the Lockheed Constellation (January 1943) and Douglas DC-6 (February 1946). These two American manufacturers went into large-scale production at the end of the war, and variants of these types operated almost all the major long-distance air services until the 1960s. A typical version, the Constellation 749A, with four 1850kW (2500hp) engines carried fifty passengers at 525kph at 6100m (325mph at 20,000ft) for a range of 4000km (2500 miles). With this type of machine, the number of passengers carried increased exponentially: the major world airlines carried 3 million passengers in 1939, 27 million in 1949 and 98 million in 1959. By 1980 piston-engined airliners had virtually disappeared from service: the reliability of the gas turbine in service had caused the engine manufacturers to stop making piston engines of more than 1000hp, and all large and medium-sized transport aircraft were powered with turbojet or turboprop engines.

INTRODUCTION OF JET PROPULSION

The most significant technical development during the Second World War was the appearance of jet-propelled aircraft, powered by gas turbines. In a sense, all aircraft propulsion can be classified as 'jet propulsion', for a conventional airscrew produces a jet of air which has been accelerated, and the propulsive impulse is equal to the increment in the momentum of this air jet; but the term is commonly restricted to the case where the air jet is accelerated inside the engine unit. The incentive to develop this system arose from the reduced efficiency of conventional propellers at high speeds: when the tip speed approaches sonic velocity, the drag increases sharply owing to the formation of shock waves. This phenomenon effectively constrained propeller-driven aircraft to speeds substantially less than the speed of sound, even in a dive.

In practice, a propulsive jet can be produced in two ways – from a rocket engine or a gas turbine. Rocket engines inevitably imply high effective fuel consumption, because the oxygen for combustion has to be carried in some form as well as the fuel itself; consequently rocket-propelled aircraft have been limited to very specific and unorthodox roles. The most noteworthy example was the German Messerschmitt Me 163 interceptor fighter developed in 1944 to counter American bomber formations. Other applications for rocket power have been mainly to boost the take-off of conventionally-powered aircraft, or for supplementary short-duration boost at high altitudes.

The gas turbine engine, in which the air passing through the engine is continuously compressed, heated and expelled through a turbine which drives the compressor, was postulated as a desirable alternative to a reciprocating engine soon after the steam turbine had been invented (see Chapter 5), but its practical development awaited the availability of suitable materials able to withstand high stress at high temperatures, and the incentive of a perceived requirement for high-speed flight. Under this stimulus, development began almost simultaneously around 1936 in several countries. The names of Frank Whittle in Britain and Hans von Ohain in Germany will always be pre-eminent as the most successful pioneers in the field, but several others were close behind. Von Ohain was the first to achieve a flight engine, and this powered the Heinkel He 178 on its first flight in August 1939. Whittle followed in May 1941, and his engines developed in Britain were also built in the USA and laid the foundation of the gas turbine engine industry in those countries. But the first production jet aircraft was the German Messerschmitt Me 262 which was powered by the Jumo 004 gas turbine developed by a team at the Junkers company. The Jumo 004 had a multi-stage axial flow compressor rather than the centrifugal compressors used by Whittle and von Ohain; this type of compressor has dominated subsequent gas turbine development.

The Me 262 was capable of a top speed around 870kph (525mph) and could operate at higher altitudes (up to 12,000m (40,000ft)) than its piston-engined contemporaries; and the four-engined Arado Ar 234 bomber with similar performance was potentially almost impossible to intercept, but never came into large-scale service before the end of the war.

After 1945 the major air forces of the world rapidly adopted the gas turbine for new fighter and bomber designs. The problem of excessive drag caused by shock waves at near-sonic speeds which had previously affected propeller blades now recurred on the aircraft wings. In particular there were difficulties with control, because deflection of a control surface had unexpected effects on the airflow over the wing. The combination of increased drag and control problems gave rise to the popular misconception of a 'sound barrier', preventing flight at higher speeds. In this climate, the American rocket-powered Bell XS-1 experimental aircraft made the first flight at a speed greater than the speed of sound on 14 October 1947, but this was really something of a freak machine. The real solution was found in the adoption of swept wings, and in substantial programmes of theoretical and experimental research to understand the detailed phenomena of transonic airflow. The first successful swept-winged aircraft powered by a gas turbine was the North American F-86 Sabre, a single-seat fighter first flown in October 1947 and later built in large numbers and in several different versions. An F-86 prototype was flown at supersonic speed (in a dive) in April 1948; aircraft of similar configuration were subsequently developed in several countries.

The useful measure of an aircraft's speed now became the Mach number – the flight velocity as a proportion of the speed of sound at the flight altitude.

As engines of higher thrust were progressively developed, Mach 1 in level flight was achieved by the North American F-100 in 1953, Mach 2 by the Lockheed F-104 in 1958, and Mach 3 by the Lockheed A-11 in 1963.

A wide variety of wing shapes has been employed by these high-speed military aircraft. The drag of a wing at high speed depends primarily on its thickness/chord ratio and on the sweepback of the maximum thickness locus; the strength and stiffness of the wing depends on the same basic variables, and the aircraft designer also has to solve practical problems such as retracting the undercarriage and accommodating the fuel. Therefore the F-104 used a very thin unswept wing; the English Electric Lightning used a wing swept back by 60°; and the Dassault Mirage used a delta wing with leading edge similarly swept but trailing edge unswept, thus reducing thickness/chord ratio by increasing the wing chord. All have comparable performance.

These high-speed aircraft also share common features of powered control surfaces, actuated by hydraulic pressure in response to the pilot's control movements, but monitored by electronic control systems to give the appropriate movement to produce the desired aircraft response over a wide range of flight speeds and altitudes. In the 1980s the increasing reliability of electronics made it feasible to dispense with a physical connection between the pilot's control mechanism – the traditional 'control column' – and the actuators, so that electric signalling or even signalling by modulated light signals in an optic fibre (colloquially 'fly-by-wire' or 'fly-by-light') became feasible. Experimental installations have been made in several countries, and the first production aircraft with a 'fly-by-wire' control system, the Airbus A.320 airliner, was put into service in 1988.

Increasing flight speed has also necessitated changes in materials for the airframe construction: air friction at high speeds produces kinetic heating of the structure. Aluminium alloys of various formulations have been developed with adequate strength for speeds up to about Mach 2, but stainless steel and titanium are needed above that speed.

JET AIRLINERS

It seemed at first that the gas turbine's fuel consumption would be too high for jet-propelled aircraft to be used for civil airline services, so the turboprop engine was developed, mainly in Britain, for this purpose (see Chapter 5). The relative mechanical simplicity of the turbine engine promised greater reliability than the high-powered piston engine, while the application of a propeller improved the propulsive efficiency of the system and hence reduced the fuel consumption – at the cost of limiting flight speeds to around 600kph (400mph). The first proposals for a gas turbine driving a propeller – a turboprop engine – were made as early as 1929 by A. A. Griffith at the Royal Aircraft Establishment in Britain, but after a limited experiment to check his theories on the design of an axial compressor, the work was shelved until 1936 when a complete axial-flow compressor was built and tested. Full-scale tests of a turboprop in flight were made in September 1945 with modified Rolls-Royce Derwent engines in a Gloster Meteor. Rolls-Royce then developed their Dart engine, four of which powered the Vickers Viscount airliner on its first flight in July 1948. After some initial hesitation, the Viscount was taken up by several airlines, mainly for routes between major European cities, and it entered service in April 1953. Over 400 Viscounts were built, and its success encouraged other manufacturers to build turboprop airliners, but on most major routes they were replaced by faster turbojet aircraft. Turboprop airliners were still being built in the late 1980s, mainly for use on secondary routes where relatively small passenger loads are expected, and there is little economic advantage in high speeds.

The introduction of the turbojet engine into airline service was spurred by the realization that the higher flight speeds which it made possible could increase the productivity of a jet airliner sufficiently to outweigh the high fuel consumption. The first jet-propelled airliner, the De Havilland Comet 1, first flew in July 1949, and went into service with British Overseas Airways in May 1952; unfortunately it had to be withdrawn from service after two years as a consequence of structural deficiencies, and Comet services were not resumed until October 1958 (Figure 12.11). By this time the first American jet airliner, the Boeing 707, was ready to enter service. With a 35° sweptback wing it was significantly faster than the Comet, cruising at about Mach 0.82, approximately 900kph at 10,000m (600mph at 30,000ft), and therefore offered a more economic vehicle to the operator. The Boeing 707 configuration was followed by other manufacturers in America, Britain and the Soviet Union, but the 707 retained its initial commercial advantage and was built in far larger numbers – almost 1000 – than its competitors.

The first Comets and Boeing 707s had barely sufficient range for the main North Atlantic routes, but these routes provided the major application for the early jet transports. Extension of services to shorter-range routes led to the development of new designs specifically developed for these markets – the Boeing company produced consecutively the Boeing 720, the three-engined 727 and the twin-engined 737, all of essentially the same configuration but of decreasing size and passenger capacity.

Continual increase in air traffic brought a potential market for much larger airliners. The later 707s carried about 180 passengers in tourist-style seating, so the target for the new designs was set around 400 passengers. Boeing announced its new design, the 747, in April 1966; it first flew in February 1969, and it entered service in January 1970. The basic configuration remained

Figure 12.11: The prototype De Havilland Comet, the world's first jet-propelled airliner, during a test flight in 1949.

essentially similar, with a 38° sweptback wing. Four engines of around 22,000kg (50,000lb) thrust were fitted and these were of a new generation of 'turbofans'. Essentially, the turbofan engine can be considered as a conventional gas turbine with an additional turbine driving a large multi-blade fan as the first stage of the compressor. Only part of the air from this first stage passes through the rest of the engine: the greater part is 'bypassed' in an external duct surrounding the core engine. The combined exhaust stream therefore has a larger mass flow but a lower velocity than that from a conventional turbojet engine, and therefore provides better propulsive efficiency and a lower fuel consumption. Conceptually, the turbofan engine is akin to the turboprop engine, but uses a 'propeller' with many more blades working inside a duct rather than in the open.

The large passenger capacity of the Boeing 747 proved economically attractive, and the pattern was followed by other manufacturers with a generation of so-called jumbos during the next ten years. A somewhat fortuitous by-product of the high propulsive efficiency of the turbofan engine was a relatively lower noise level, associated with the lower jet velocity. This provoked public pressure to reduce aircraft noise generally, so that turbofan engines largely replaced turbojets in civil use in the 1980s.

SUPERSONIC COMMERCIAL AIRCRAFT

Throughout the development of the airline industry there has always been a pressure to increase speed: apart from any influence on the productivity of the vehicle, it seems clear that passengers have preferred to travel as quickly as possible. By the late 1950s, preliminary design studies were set in hand in several countries to investigate the possibilities of a transport aircraft operating at supersonic speeds – several times the speed of any civil aircraft then in service. In the early 1960s the British and French governments agreed to provide funds for an Anglo-French airliner designed to be operated at Mach 2, the US government supported the design of a larger machine to fly at Mach 3, and the Soviet Union also started development of a Mach 2 design.

The Anglo-French design eventually appeared as the prototype Concorde, which flew in March 1969; the Russian Tupolev Tu-144 flew three months earlier, but the American project never materialized. Concorde and Tu-144 were remarkably similar in general configuration, with highly swept delta wings of complex shape, a long slender fuselage and four under-wing engines. To obtain sufficient thrust for take-off and transonic acceleration, the engines utilized after-burning, a process hitherto restricted to military operations in which additional fuel is injected into the exhaust stream of the engine.

The economics of operating a supersonic airliner are very marginal, and only two airlines, British Airways and Air France, put the Concorde into operation, in January 1976. The Russian Tu-144 was operated only experimentally and never entered full service.

Concorde, flying regular transatlantic services at around 2150kph at 15,000m (1350mph at 50,000ft), is faster than almost all military aircraft now flying, but in many ways its technology can be considered conventional. It is built of aluminium alloys, with limited use of titanium and steel. The aerodynamic controls are hydraulically operated, with multiple redundancies to provide adequate reliability in service; in addition, fuel is moved during flight to adjust the aircraft's centre of gravity to match the variation of the aerodynamic lift distribution as speed increases through Mach 1.

JET-SUPPORTED FLIGHT

The development of increasingly powerful jet engines brought in the possibility of building aircraft with a thrust/weight ratio greater than one, which would be potentially capable of vertical flight on jet thrust alone. A precursor was a rocket-propelled fighter, Bachem Ba 349 Natter, produced in Germany in 1945. This was launched vertically using rocket power, but it was envisaged as an expendable missile with the pilot returning by parachute from his mission.

The first successful demonstration of jet-supported flight was provided in Britain by the Rolls-Royce Thrust Measuring Rig – better known by its nickname *Flying Bedstead* – in August 1954. The rig comprised two Rolls-Royce Nene engines mounted horizontally in a tubular framework on four wheels, with the jets deflected downwards through cascades of vanes in the nozzles. The pilot sat on top of the rig, and it was stabilized and controlled by four small air-jets on outriggers. Essentially, this rig was capable only of vertical flight; the next development was to mount four lightweight Rolls-Royce RB 108 engines vertically in a delta-winged aircraft, the Short SC.1, with a fifth RB 108 mounted horizontally for forward propulsion. Jet reaction controls were provided to stabilize and control the aircraft in the hover, and conventional aerodynamic controls for use in forward flight. The first of two Short SC.1 machines first flew in 1957, and achieved a transition from hovering to forward flight in 1960.

A number of other aircraft were built or conceived to this formula, notably the German Dornier 31 experimental transport aircraft in 1967, but the need to carry separate jet engines for horizontal and vertical flight made them inherently inefficient.

The alternative was to use the same engine for both regimes of flight. In its simplest form this was possible by sitting a more-or-less conventional aircraft on its tail for take-off, and flying a curved trajectory into forward flight. Several experimental machines of this type were flown in the USA and France – the Convair XFY-1 and Lockheed XFV-1 in 1954, Ryan X-13 in 1956 and SNECMA Coleoptere in 1959 – but none were put into production.

A third basic concept using progressive deflection of the exhaust jet from vertical to horizontal has proved to be technically the most successful. In 1960 the British Hawker Aircraft company and Bristol Aero Engines developed the Hawker P.1127 single-seat fighter with a Bristol Pegasus engine (Figure 12.12). The single engine, mounted in a conventional swept-winged aircraft, is provided with four nozzles containing cascades of vanes, which can be rotated through about 100 degrees. The forward nozzles take compressed air from the front fan stage of the engine, and the rear nozzles carry the exhaust. Control jets at the extremities of the airframe provide stability and control in hovering flight and conventional aerodynamic controls take over in horizontal wing-borne flight. Developed from the original P.1127, the later Kestrel and Harrier aircraft were built in substantial quantities and entered service in several countries in the 1970s.

The term VTOL (an acronym of Vertical Take-Off and Landing) is often applied to this class of aircraft. It is misleading: helicopters are excluded, although they are equally capable of vertical take-off and landing; and jet-supported aircraft usually take-off from a running start, either horizontally or from a 'ski-jump' ramp, because this generates additional lift allowing the carriage of a greater load. The label STOVL (Short Take-off and Vertical Landing) is more appropriate.

HELICOPTERS AND ROTARY WINGS

The basic idea of generating lift from a rotating wing can be traced back certainly to the early fourteenth century: an illuminated manuscript of this period gives the earliest known illustration of a child's toy consisting of a four-bladed

Figure 12.12: The Hawker P.1127 supported by the thrust from the four deflected jets produced by its Bristol Pegasus engine.
Hawker Aircraft photograph.

'airscrew' mounted on a vertical spindle. This is rotated by pulling a string wrapped round the spindle, whereupon it flies into the air. Essentially the same principle is involved in a toy with two contra-rotating screws operated by a spring-bow mechanism, which was described to the Paris Academy of Sciences by Launoy and Bienvenu in 1784 during that period of acute interest in all matters aeronautical. A similar device was described by Cayley (see p. 617) in his paper of 1809. Versions of this toy powered by clock springs were made by the Swiss Jakob Degen in 1816, and by several later inventors.

In 1842 an English engineer, W. H. Phillips, flew a model which was driven by jets of steam (generated by the heat from a form of chemical firework) from the rotor-tips, and in 1863 the Vicomte Ponton d'Amecourt built a model helicopter powered by a conventional reciprocating steam engine which was probably too heavy to fly. The Italian engineer Enrico Forlanini succeeded in flying a similar model for 20 seconds in 1877 by the expedient of pre-heating the boiler before attaching it to the model.

The first serious attempts to build a full-size piloted helicopter were made in France in 1907, by Louis Breguet and Paul Cornu, who were both aiming for a

50,000-franc prize offered for the first one-kilometre flight. Breguet's machine had four biplane rotors, driven by a 18kW (24hp) Antoinette engine. In September 1907 it hovered for a minute, but was stablized by four assistants holding the machine's four corners. Cornu's machine also had an Antoinette engine, driving two sets of two-bladed rotors; when tested in November it successfully hovered just above the ground for several minutes. Two months later the prize was won by Farman's flight on a fixed-wing Voisin (see p. 624), and Breguet and Cornu both abandoned their efforts; Breguet turned his attention to fixed-wing machines.

For the next thirty years, a considerable number of inventors attempted to make a practical helicopter, with no significant success. As an alternative to the power-driven rotor, the Spaniard Juan de la Cierva produced an aircraft which he called the autogiro. This was a machine on which lift was produced from a rotor which was rotated after the manner of a windmill as the aircraft was pulled through the air by a conventional engine and propellor. In 1923, Cierva built a machine with hinged rotors, free to flap up and down as they rotated, which provided a simple solution to the unbalanced lifting force generated when the rotor was in forward flight. (On one side of the rotor disc, the blade is moving forward relative to the aircraft; on the other side it moves backwards. Adding the forward motion of the aircraft, one blade is moving much faster than the opposite one, and therefore generates more lift unless the incidence is reduced on the advancing blade.)

Cierva was subsequently funded by the British Air Ministry to produce a number of designs which were built in some numbers in Britain, France, Germany and the USA, and the Cierva C.30 model was used for civil and military purposes on a fairly small scale from 1935. Later models of autogiro have been made, mainly as single-seat sports machines, but this type of aircraft has never found a realistic practical use.

One of those who took a licence to build Cierva machines was the German Focke-Wulf company, whose founder, Heinrich Focke, produced the first practical helicopter in 1936, using the hinged rotor system. To counter the torque effect of a power-driven rotor, Focke employed two counter-rotating rotors on outriggers from a basically conventional fuselage. The engine in the nose drove a conventional propeller for forward flight, and also drove the two rotors through bevel gears and inclined shafts. The practicality of Focke's Fw61 helicopter was demonstrated by a 110km (68 mile) cross-country flight in 1938, and by public demonstrations inside the Berlin Sporthalle.

Focke produced a larger development of the Fw61 in 1940, and another German engineer, Anton Flettner, produced a practical helicopter which was built in small numbers for the German Navy, but wartime priorities lay elsewhere and neither design found significant practical application.

Parallel work in other countries produced working prototypes by 1940; by far

Figure 12.13: A Sikorsky S–61N helicopter carrying two dozen passengers: the classic helicopter configuration with one main rotor and a small tail rotor to balance the torque.

the most significant was Igor Sikorsky's VS-300 in the USA, which passed through a variety of configurations to its final successful form in December 1941. A single three-bladed rotor with hinged blades was used to generate lift and, by tilting the rotor disc, forward propulsion, and a small rotor with horizontal axis facing sideways served to counter the torque and acted as a rudder. Sikorsky's configuration has continued to dominate helicopter design to the present day. Although a variety of counter-rotating twin-rotor designs have also been brought into large-scale production, probably 80–90 per cent of all the helicopters built have employed a single rotor and an anti-torque tail rotor (see Figure 12.13).

Helicopters were normally powered by conventional aircraft piston engines (with modified lubrication systems and additional cooling fans) until the 1950s, but then gas turbines were introduced, first by the French company Aérospatiale in 1955. These engines were designed to have a turbine stage driving the rotor through a gearbox – colloquially known as a turboshaft engine. The main advantage was the lower vibration level and greater reliability of the gas turbine and, as in other aircraft applications, the gas turbine has supplanted the piston engine for all but the lowest-powered helicopters.

The helicopter rotor of the 1940s and 1950s was a complicated piece of machinery: the blades were hinged to allow a limited degree of motion at their roots about all three axes, and a variety of springs and friction dampers were employed to allow the blade angles to be varied appropriately. Subsequent development has replaced mechanical hinges and bearings by elastomeric hinges, employing plastic materials with tailored properties of strength and flexibility. Aérospatiale introduced their Starflex rotor hub, in which the three conventional hinges are replaced by a single ball joint of rubber and steel sandwich construction, into their AS350 Ecureil in 1977. At the same time, glass fibre and carbon-fibre composites have been introduced for construction of the rotor blades.

The ability to land and take-off vertically enables the helicopter to perform a wide variety of duties, but its mechanical complexity compared to a fixed-wing aircraft leads to higher costs, so the major users have always been the military. Apart from being used as troop transports and for casualty evacuation, they are used widely for observation work and for ground attack against infantry and armour. Civilian uses include passenger carrying into confined spaces (oil-rig platforms, city-centre roof-top terminals, hospitals, and the like), pipe-line patrols, and specialized crane-type operations. Typically, the Boeing-Vertol 234 carries up to 44 passengers for up to 800km at 250kph (500 miles at 155mph), and the largest production helicopter in the world is the Russian Mil Mi 26 with 20,000kg (44,000lb) payload.

CONVERTIBLE AND HYBRID AIRCRAFT

The helicopter rotor is speed-limited in the same way as the aircraft propeller by the onset of shock waves as the speed increases, and in practice it is difficult to design an efficient helicopter for a speed of more than about 280kph (175mph), which clearly limits its productivity. There has therefore always been a potential requirement for an aircraft using helicopter rotors for vertical take-off, which is capable of conversion in flight to a configuration where the lift is provided by fixed wings. A variety of solutions has been advanced, ranging from conventional helicopters with auxiliary wings and propulsion engines, such as the 40-passenger Fairey Rotodyne of 1957, which set a world helicopter speed record of 307kph (190.7mph) in 1959, to a variety of designs with tilting engines, with or without tilting wings. Typical of the tilt-wing designs was the Canadian CL-84 of 1965: this demonstrated a speed range from hover to over 500kph (310mph). Although many different experimental 'hybrid' aircraft have been built and flown, no design has reached full production status.

RECREATIONAL AIRCRAFT AND GLIDERS

While the greater part of the world's aircraft production has been devoted to vehicles with a military or commercial utility, there has always been a significant interest in purely recreational aircraft. For the most part, these have been technically similar to the more functional machines, but from time to time there have been interesting technical novelties in this field.

Unpowered gliders were used by pioneers like the Wrights as a stepping-stone towards powered flight, but they have also been used as sports aircraft since the 1920s. The origins of the movement lie in the restrictions placed on Germany by the Treaty of Versailles, which effectively forbade any development of powered aircraft in that country. Flying enthusiasts therefore turned to the construction of aircraft without power, which were initially launched from the tops of hills. It was soon discovered that it was possible to maintain height in the up-currents along the windward face of the hill, and then that there were often thermals, rising air currents caused by solar heating of the ground, which could be exploited by skilled pilots in suitable machines. This provided the incentive to produce aircraft with high lift/drag ratios, which developed into a distinctive family of aircraft with very smooth streamlined shapes, wings of long span, and light weight. From the 1930s onwards, such aircraft were produced mainly of wood, and then from the 1950s of glass-fibre and other plastics.

In the 1980s a typical high-performance glider (the German-built Schempp-Hirth Nimbus 3) using both carbon-fibre and glass-fibre in its construction, had a wing of 23m (75ft) span, empty weight of 392kg (864lb) and achieved an optimum glide ratio of 55:1: that is, in still air it travels 55m horizontally for a height loss of 1m.

Gliders used for sporting competitions are single-seaters, and two-seaters are used for training. During the period 1940-50 large transport gliders were developed (mainly in Germany, Great Britain and the USA) for troop-carrying purposes, more clearly related to conventional transport aeroplanes than to sports gliders. The largest types were the Messerschmitt Me 321 with steel tubular construction covered in wood and fabric, carrying up to 20 tonnes, and the all-wooden British Hamilcar designed to carry a light tank or similar load up to 8 tonnes. Such gliders were rendered obsolete by the post-war development of transport helicopters.

There have been repeated attempts to produce very simple and very cheap aeroplanes for recreational flying since the earliest period, Santos-Dumont's *Demoiselle* of 1909 perhaps qualifying as the first such. For the most part, these aircraft have been simply scaled-down versions of conventional aeroplanes, with limited weight and engine power, but occasionally some technical novelty has appeared in this field.

One such was Henri Mignet's *Pou-de-Ciel* ('Flying Flea') of 1933. Mignet was a self-taught aircraft designer with unorthodox ideas, who had already built several single-seat light aircraft. His *Pou* was a wooden tandem monoplane with no control in roll, stabilized by having considerable dihedral angle on both wings, and steered by a large rudder. The rear wing was, in effect, a fixed stabilizing tailplane and control in pitch was effected by altering the angle of incidence of the front wing. In spite of its unusual control system, the *Pou-de-Ciel* was a reasonably successful flying machine; but it suffered a number of fatal crashes which led to official bans in France and Britain. Later the prohibitions were relaxed and the aircraft re-designed to make it safe, but the wave of popular enthusiasm for the model had been killed.

A more significant novelty appeared in 1948 when the American engineer Francis Melvin Rogallo obtained a patent for his flexible-winged glider. In its simplest form, this comprised a triangular wing, with solid booms along the centre line and the leading edges. A single sheet of fabric (of nylon, Terylene or similar artificial fibre) stretched between the booms was free to billow into two half-cones under aerodynamic pressure in flight, very much like the sail of a yacht.

Rogallo's initial invention was tested by the US services in 1963 as a potential lightweight transport vehicle with a 155kW (210hp) engine, but found its real application as a sports vehicle. Initially used as a hang-glider and controlled by the pilot shifting his weight beneath it, it first became popular in 1969 as an adjunct to water-skiing. Further development of the basic idea in various countries added increased wing-span, double-surfaced wings, and a large variety of different control systems including conventional ailerons, wing warping, drag-flaps; and power units which ranged from 4.5kW (6hp) chain-saw motors to specially-developed engines of around 37kW (50hp).

MAN-POWERED FLIGHT

Since the earliest times, mankind has dreamed of flying by human power alone. These ambitions were discouraged by Giovanni Borelli whose posthumous work *De motu animalium* of 1680 showed how inadequate man's muscular powers were compared to those of birds, but by no means all would-be flyers were deterred. Eventually, with increasing aerodynamic knowledge the prospects for ornithopters were seen to be poor, and attention turned to conventional aeroplane shapes with pedal-driven propellers.

In 1935 two German engineers, Helmut Haessler and Franz Villinger, built a light-weight machine with a single propeller, belt-driven from bicycle pedals operated by the pilot; this machine was flown for about two years, and straight-line flights of up to 700m (2300ft) were obtained. The Haessler-Villinger aircraft had an empty weight of 34kg (75lb) and a span of 13.5m (44.3ft); the Italian Bossi-Bonomi Pedialante of 1936 weighed about 100kg (220lb) and had a span of 17.7m (58ft), with a pair of propellers driven by shafts from the pilot's pedals. Broadly similar performance was claimed, with flights up to one kilometre.

A revival of interest in the 1960s was largely due to prizes offered by Henry Kremer and administered by the Royal Aeronautical Society. Initial attempts by various groups produced performances little better than the pre-war German and Italian designs, but eventually the first Kremer prize, for a flight round a prescribed course of about 2km (1 mile), was won by the American Bryan Allen in 1977, flying an unorthodox aircraft of tail-first configuration designed by Paul McCready. In a very similar machine, Gossamer Albatross, the same pilot won a second Kremer prize on 12 June 1979 with a flight across the English Channel – 35.8km (22.25 miles) in 2 hours 49 minutes. To produce this performance, the average power generated by the pilot was 0.34hp (250 watts). Gossamer Albatross had a wing span of 28.5m (94ft) and empty weight of only 32kg (70lb), made possible largely by the use of modern materials including carbon-fibre tubes for the main structure and Mylar film for the covering.

FURTHER READING

Bilstein, R. E. *Flight in America 1900–1983* (Johns Hopkins University Press, Baltimore, 1984)
Cavallo, T. *The history and practice of aerostation* (1785)
Chant, C. *Aviation: an illustrated history* (Orbis, London, 1978)
Cierva, J. de la *Wings of to-morrow; the story of the autogiro* (New York, 1931)
Clement, P.-L. *Montgolfières* (1982)
Gablehouse, C. *Helicopters and autogiros: a chronicle of rotating-wing aircraft* (Muller, London, 1968)
Gibbs-Smith, C. H. *Sir George Cayley's aeronautics, 1796–1855* (HMSO, London, 1962)
—— *Aviation: an historical survey from its origins to the end of World War II* (HMSO, London, 1986)
Gregory, H. F. *The helicopter* (London, 1948)
Jablonski, E. *Seawings, an illustrated history of flying boats* (Hale, London, 1974)
Jane, F. T. *All the world's aircraft* ed. by C. G. Grey and later by others (Jane's, London, 1916–20, 1922 to date)
Kármán, T. von *Aerodynamics: selected topics in the light of their historical development* (Cornell University Press, New York, 1954)
Nayler, J. L. and Ower, E. *Aviation: its technical development* (P. Owen, London, 1965)
Reay, D. A. *The history of man-powered flight* (Pergamon Press, Oxford, 1977)
Robinson, D. H. *Giants in the sky: a history of the rigid airship* (G. T. Foulis, London, 1973)
Rolt, L. T. C. *The aeronauts: a history of ballooning, 1783–1903* (Longman, London, 1966)
Tissandier, G. *Histoire des ballons* (1887)
Valentine, E. S. and Tomlinson, F. L. *Travels in space: a history of aerial navigation* (1902)
Whittle, Sir F. *Jet* (London, 1953)
Wise, J. *A system of aeronautics* (1850)
Wright, W. and O. *The papers of Wilbur and Orville Wright* ed. by M. W. McFarland, 2 vols. (New York, 1953)

13

SPACEFLIGHT

JOHN GRIFFITHS

BLACK POWDER ROCKETS

The origin of the rocket is shrouded in mystery, but it was probably developed in the East. The Sung dynasty Chinese (AD 960–1279) had the technology to make rockets, but there is no definite evidence that they did so.

By 1300 black powder and rockets were known in Europe and Arabia, and it seems likely that their secrets reached Europe along trade routes from the Orient. The earliest European recipe for black powder was given by the Franciscan friar Roger Bacon around the year 1265. He evidently considered it dangerous knowledge; the recipe was in code. The first European picture of a rocket is in Konrad Kyeser's *Bellifortis* of *c*.1400, and seems to have been copied from an Arabic manuscript.

Although occasionally used as a weapon it was as a peaceful firework that the rocket became best known to Europeans, but in sixteenth-century India it was not uncommon for armies to carry thousands of rockets into battle. Some two hundred years later British and French troops striving to dominate India experienced the Indian rocket weapons at first hand. Some captured in 1799 wyere taken to Woolwich Arsenal, near London, and they may be seen in the Rotunda Museum at Woolwich.

The Indian rockets may have inspired the British inventor Sir William Congreve, who turned the black powder rocket into a modern 'weapons system' during the Napoleonic Wars. Congreve realized that rockets could have the firepower of heavy artillery but, unlike guns, were easy to move around. He developed a wide range of rockets and launchers, together with the tactics to use them in warfare. First used against Napoleon's invasion fleet at Boulogne in 1806, Congreve's rockets were later copied by many European armies. However, his rockets had one major failing as a weapon: they were not very accurate. Also, like earlier rockets, they had cumbersome stabilizing stocks. Other inventors such as William Hale developed stickless rockets, but the problem of accuracy was never solved. By the 1870s improved artillery had made the rocket a weapon fit only for colonial wars, and by the First World War it was virtually obsolete. If the rocket had failed as a war weapon, it had found civilian uses. Successful life-saving rockets were developed in the 1850s to carry lines to ships, and Congreve rockets were also used for harpooning whales.

SPACEFLIGHT PIONEERS

Meanwhile, nineteenth-century authors such as Jules Verne and H.G. Wells were writing what came to be known as science fiction stories. For centuries there had been tales about travel to other planets. Victorian authors added technology and 'invented' the spaceship, relying on whirlpools, rising dew or flying birds to carry their heroes into space. Early science fiction directly inspired three visionaries to consider the practical problems of spaceflight. Independently, they all realized that the rocket was the key to spaceflight, and that a new, more powerful type of rocket burning liquid fuels would be needed. Their work eventually led to the conquest of space by such rockets.

Konstantin Tsiolkovsky, a Russian schoolteacher, made the earliest theoretical studies of spaceflight in, although his work did not become well known until the 1920s. Robert H. Goddard, an American physicist, launched the first successful liquid-fuel rocket on 16 March 1926 (see Figure 13.1). Although he was secretive about his work, it is clear that his rockets were the first to use many features now vital to modern space technology. Hermann Oberth, a Romanian schoolteacher, published *Die Rakete zu den Planetenraumen* (The rocket into planetary space), which overnight became a popular success. Unlike Tsiolkovsky and Goddard, Oberth was a publicist of spaceflight, and during the 1920s his work led to growing interest in the possibility and the founding of space-travel societies in different countries.

VENGEANCE WEAPON TWO

The growing interest in rocketry attracted the attention of the German army to the possibility of a secret weapon which had the added advantage of not being banned by the Treaty of Versailles. In about 1930 a Captain Dornberger was ordered to develop 'a liquid fuel rocket . . . the range of which should surpass that of any existing gun'. At that time virtually nothing was known about liquid fuel rockets. Goddard's work was a well-kept secret; the only other experts were amateur experimenters, such as members of the German Verein für Raumschiffahrt (Society for Spacetravel). Dornberger gradually built up a team of rocket engineers. One of his first recruits was an enthusiastic and knowledgeable member of the VfR, Werner von Braun.

Figure 13.1: Robert Goddard standing next to his liquid fuel rocket a few days prior to its successful launch on 16 March 1926.

After Hitler's rise to power in 1933, the formation of the Luftwaffe in 1935 gave added impetus to Dornberger's work. The liquid fuel rocket was seen as a power unit for aircraft, and in 1936 the army and the Luftwaffe began work on a large rocket research establishment at Peenemunde on the Baltic coast. In the same year, Dornberger and von Braun's team began design work on 'Aggregate 4', their fourth rocket design. A-4 was to carry a 1 tonne warhead to a target over 250km (155 miles) away. Its size was determined by the largest size of load that could be easily transported by road or through any railway tunnel, and A-4 was to be 14m (46ft) long and 3.5m (11.5ft) across its fins.

Figure 13.2: The launch of a captured German V-2 rocket during the British Operation Backfire programme in 1946.

It took six years' work to get such an advanced rocket off the drawing board, and even then it was by no means ready as a weapon. After the first successful launch on 3 October 1942, it was nearly three years before it became better known as 'Vengeance Weapon 2', successor of the V-1 'flying bomb' (see Figure 13.2 and p. 1005). The V-2 offensive lasted from September 1944 to March 1945; about 1100 missiles landed in England, and 1750 hit targets on the Continent. Its value as a weapon has been hotly discussed; over some seven months, V-2 missiles delivered about as much explosive as the Allied air forces dropped on Germany during a single day. After the war, Dornberger wrote, 'The use of the V-2 may be aptly summed up in the two words "Too Late" '.

As well as the V-2, Germany developed a wide range of smaller guided missiles, few of which entered sevice, and the ME 163 rocket-powered fighter.

POST-WAR RESEARCH

German rocket technology was one of the great prizes of the Second World War and the USA profited from it most of all. Von Braun's team of rocket scientists chose to surrender to the Americans, who also gained the priceless Peenemunde technical archives and much V-2 hardware. The Soviet forces occupied what was left of Peenemunde and the V-2 production plants, and obtained the assistance of a number of V-2 technicians. After learning all they could about V-2 technology, the USSR and the USA exploited their advantage in different ways. By 1949 both had atomic weapons, but these were heavy and bulky and would need an enormous rocket to carry them.

America relied on aircraft to deliver atomic weapons from European bases within striking distance of Soviet territory, and began developing winged cruise missiles capable of carrying atomic warheads. Rocket development was shelved until smaller, lighter warheads were available. Soviet work concentrated on developing the large rockets that would be needed to carry a nuclear warhead across the Atlantic.

Meanwhile, scientists had realized that advances in rocket technology had made it possible to launch a small satellite. The International Geophysical Year scheduled for 1957–8 provided a good reason; both the Soviet Union and the USA announced satellite programmes for the IGY, but little attention was paid to the Soviet announcement. American belief in their technological leadership was shattered when the Soviet Union launched the world's first artificial satellite, Sputnik 1, on 4 October 1957 (Figure 13.3). On 4 November, Sputnik 2 followed. Both satellites were, for the time, large and heavy, showing that the Russians had developed rockets more powerful than any the Americans had at the time. On 6 December the experimental American Vanguard satellite launcher exploded on its launch pad, in full view of the world's press.

The events of those three months had shown that space technology was a powerful propaganda weapon, and in the 'cold war' climate between East and West, an undeclared 'space race' began between the two superpowers.

MANNED SPACEFLIGHT AND THE SPACE RACE

On 12 April 1961 a Russian became the first man in space, when Yuri Gagarin's Vostok spacecraft completed a single orbit of the earth. Their powerful booster rocket had enabled the Russians to score another space first. As a result, President Kennedy announced that the United States would land a man on the moon 'before the decade is out'. Meanwhile, the American Project Mercury succeeded in sending Alan Shepard on a short hop into space, and on 10 February 1962, John Glenn became the first American to orbit the earth.

The Russian Vostok programme produced more space 'firsts': two spacecraft in orbit at once and the first woman in space, Valentina Tereshkova on 16 June 1963. A 'new' Soviet spacecraft was launched in October 1964, Voshkod. On its first flight it carried three cosmonauts, and on the second and final mission in March 1965, Alexei Leonov carried out the first spacewalk. What was not disclosed at the time was that Voshkod was really an adapted Vostok spacecraft. To get three men into orbit, the escape system had been removed to save weight. In the event of a launch failure, the cosmonauts would have had no means of escape.

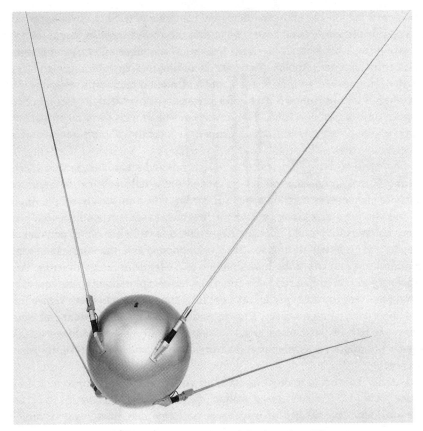

Figure 13.3: A model of Sputnik 1, the first artificial earth satellite.

The next American spacecraft was the two-man project Gemini. Gemini missions in 1965 and 1966 saw the first American spacewalk, and proved that the orbital docking techniques vital to the later Apollo moon missions were practical. But in January 1967 the American space programme suffered its first casualties; astronauts Grissom, White and Chafee were killed in a fire aboard their Apollo spacecraft during a ground test before what was to have been the first manned Apollo mission. The ensuing enquiry uncovered flaws in the spacecraft, costing Apollo a year's delay. A little later the Soviet programme suffered a major setback. Vladimir Komarov was killed on 23 April 1967 when the new Soyuz spacecraft crashed after re-entry into the earth's atmosphere. It appears that the spacecraft was spinning, causing its parachute lines to become tangled.

The Apollo programme prepared for a moon landing with tests of the giant Saturn V (Figure 13.4) booster rocket which was to carry the two part spacecraft into orbit. To save time, the Americans tried a novel approach. Instead of the more usual course of testing the rocket stage by stage, the first Saturn V took off with all its stages live on 9 November 1967. An unmanned test of the complete Apollo spacecraft (including the separate lunar lander) in earth orbit followed on 4 April 1968, and a manned orbital test, Apollo 7, on 11 October. Two unmanned Soviet Zond spacecraft also orbited the moon in September and November 1968. These may have been tests for a manned flight, but Apollo 8 became the first manned spacecraft to orbit the moon in December 1968.

The Apollo 9 astronauts tested the spacecraft and lunar lander in earth orbit during March 1969, and Apollo 10 carried out a rehearsal for the necessary docking manoeuvres in moon orbit during May. Then on 20 July, on the Apollo 11 mission Neil Armstrong became the first man to set foot on the moon. Four more landings followed, punctuated by a near disaster when an oxygen tank on Apollo 13 exploded during its trip to the moon, and the astronauts safely returned to earth using the lunar module as a liferaft to return to earth. With Apollo 17 in December 1972, the American lunar exploration programme came to an end. Government funding was not forthcoming for ambitious future programmes such as moon bases, a space station and a manned landing on Mars. The only part of these plans to survive (although in a modified form) was the Space Shuttle, a re-usable spacecraft originally intended to service the space station.

During the 1970s, manned spaceflight concentrated on long-term stays in space. The USA's sole space station, Skylab, was launched in 1973 and demonstrated the ability of astronauts to carry out emergency repairs in space, but after three crews had visited the station, the programme came to an end. Skylab itself re-entered the earth's atmosphere in 1979, parts falling in Australia. Delays in the Space Shuttle programme prevented an intended attempt to boost the station into a longer-lasting orbit. The first joint American–Soviet spaceflight, the Apollo–Soyuz mission, took place in 1975: after this, there would be no American manned spaceflights until the Space Shuttle.

In contrast, throughout the decade the Russians continued with their Salyut series of civil and military space stations. The first, Salyut-1, was launched in 1971 and the first crew to stay in the station, Dobrovolsky, Patsayev and Volkov, were killed when their Soyuz II spacecraft depressurized during the return to earth. Salyut 6, launched in 1977, stayed in orbit until 1982, and was the first successful space station. Unlike Skylab, Salyuts carried engines to maintain their orbit, but in 1980, a major new development took place, when Salyut 6 was refuelled by a 'Progress' unmanned supply ship. The Russians now had a viable long-duration space station system, capable of being readily resupplied from earth.

Figure 13.4: The launch of a Saturn V vehicle carrying three astronauts to the moon on the Apollo 15 mission, 26 July 1971.
NASA.

In April 1981 the long-awaited first flight of the Space Shuttle took place, and America again had a manned spaceflight capability. The next four years saw the shuttle 'Space Transportation System' demonstrate the capacity to repair satellites in orbit, and return them to earth. The European Space Agency's Spacelab manned laboratory was also flown aboard the Shuttle in December 1983. Scientists on the earth and in space could now collaborate in carrying out experiments in the space environment. In January 1984, President

Reagan announced an American plan to build a permanent manned space station, to become operational in the 1990s. During 1985, Shuttle astronauts also demonstrated that large structures could be built in space, an important step towards the space station. The Russians demonstrated their lead in long term spaceflight during 1984, when three cosmonauts stayed aboard Salyut 7 for 238 days, almost eight months.

As 1985 drew to a close, manned spaceflight was seemingly becoming almost routine and, with four Space Shuttles now operational, NASA had scheduled sixteen missions for 1986. Other nations such as Japan and France were considering developing smaller versions of the Space Shuttle, and the Russians were known to be developing a shuttle of their own (a prototype of which was unveiled in 1989). Then, on 28 January 1986, the space shuttle Challenger exploded shortly after liftoff, killing its crew of seven. Among them was the first American civilian in space, a schoolteacher, Christa McAuliffe. Over two and a half years were spent in correcting faults in the Shuttle boosters before the next launch in 1988. The American space programme was inevitably delayed, but its ultimate direction, a permanent manned space station, remained unchanged.

The first steps towards such a station were taken by the Russians in February 1986, with the launch of a Salyut-sized space station named Mir (Peace), probably the first module of a permanent space station to be assembled in orbit.

SATELLITE TECHNOLOGY

Sputnik, although it was a small, simple satellite, equipped only with sensors to measure the temperature and density of the upper atmosphere, proved the feasibility of launching a satellite and placing it in earth orbit. On the day that Sputnik 1 was orbited, plans for the idea of the launching of a US satellite were resurrected; a launch was promised in 90 days. On 6 December an attempt was made to launch Vanguard 1. The rocket ignited and lifted off the pad. One second later it lost thrust and fell back to earth, exploding upon impact. The satellite, thrown free, rolled across the ground with its beacon transmitter chirping away. Five days later, Werner von Braun and his team at Huntsville, Alabama, were given approval to launch America's first satellite. Within 85 days, Explorer 1 was in orbit. It was a small satellite less than 1m (3ft) long, 15cm (6in) in diameter, with a mass of barely 5kg (11lb). It carried some simple experiments which included a Geiger-Mueller tube to record cosmic rays. Data recorded by any 'events' were stored on a miniature tape recorder and transmitted to earth on demand. This experiment, in fact, was instrumental in locating the Van Allen radiation belts that surround the earth.

Since those early days thousands of satellites have been launched not only by the two major space powers but also by many other nations, some using their own launch vehicles. The earliest satellites were, in essence, primitive devices capable of only simple measurements and having the capability to relay small amounts of data back to earth. However, as with all technology, advances have been dramatic and today the satellite is a sophisticated and complex machine, in many cases possessing its own on-board computer.

Satellites can be divided into a number of distinct types, depending on their purpose. The main groups are: communication satellites, meteorological satellites, military satellites, remote sensing satellites, scientific and astronomical satellites. The earliest were mainly intended for scientific purposes. However, 1 April 1960 saw the launch of Tiros 1. Launched by a Thor-Able rocket this US spacecraft demonstrated the feasibility of using satellites for global weather observations. Following Tiros 1, a series of more advanced Tiros satellites followed, ending with Tiros 10 on 2 July 1965. The latest Tiros satellites are very sophisticated spacecraft which, as well as earth observation equipment, have advanced search and rescue antennas to provide data for locating and identifying ships in trouble or aircraft that have come down in remote places. Since 1966 the entire earth has been photographed at least once a day on a continuous basis. Data can be used to study cloud movements, sea temperatures and to trace ice field movements in the Arctic and Antarctic. Combined with ground and balloon data they serve the meteorological community by providing as accurate a picture of the global weather as is possible.

Communication satellite technology began with the launch of Echo 1 satellite on 11 August 1960. This large reflective (mylar) balloon was used as a passive communication system, simply reflecting radio waves from one point on the globe to another. The first true communication satellite was Telstar 1, launched on 10 July 1962. Although only an engineering test vehicle, Telstar 1 demonstrated the feasibility of using satellites for transmitting television pictures over large distances without the use of land lines. Telstar had a low orbit (952 × 5632km) (590 × 3492 miles) making it impractical for long-term communication use. For practical purposes the satellite's orbit must be geostationary (35,880km (22,245 miles) above the equator of the earth) and Syncom 3 (launched on 19 August 1964) was the first to achieve this. Syncom 3 broadcast live the opening ceremonies of the Tokyo Olympics.

In 1964 the organization INTELSAT (International Telecommunications Satellite Organization) was formed in which a number of nations contributed to a common satellite system, sharing the resulting revenues. Intelsat 1 (Early Bird) was launched on 3 April 1965 to become the first truly commercial communication satellite. Further Intelsat satellites followed, each with a higher capacity than the last. Intelsat 1 could transmit 240 telephone channels and one television channel. By 1980, Intelsat V satellites were capable of transmitting 12,000 telephone and two television channels. Today, there are hundreds of communication satellites in geostationary orbit regularly transmitting television, telephone and telex data all over the globe.

Many nations have launched satellites purely for military purposes, mainly for surveillance and communication. Little information is released on these satellites.

Two areas where satellites have proved invaluable in scientific research have been in the fields of remote sensing of the earth and astronomy. The most important development in remote sensing, observing the earth from orbit, took place with the launch of the American Landsat 1 satellite on 23 July 1972. From its near polar orbit, Landsat 1 returned over 300,000 detailed images of the earth until its retirement in January 1978. The Landsat 5 satellite (launched on 1 March 1984), had a sophisticated array of cameras sensitive at visible and infra-red wavelengths capable of spatial resolution of the order of 20m (66ft). The images returned by the Landsat spacecraft have been put to extensive use by scientists studying cartography, agriculture, oceanography, geology and many other branches of science which have a direct human resource use.

The earth's atmosphere is a severe hindrance to astronomers. It not only absorbs many wavelengths in the electromagnetic spectrum, but its constant state of movement severely limits the spatial resolution possible with ground-based telescopes. Satellite observatories provide a means of eliminating these problems. Hundreds of purely astronomical satellites have been launched and some of the most successful have been the OAO series, IUE and IRAS. The Orbiting Astronomical Observatories (OAO) programme began in 1959 and in 1966 OAO 1 was launched by an Atlas-Agena rocket from Cape Canaveral. However the most successful satellite was OAO 3, launched on 21 August 1972 and named Copernicus. Its battery of UV and X-ray telescopes was responsible for charting 200 previously unknown X-ray sources. It was also the first satellite to observe the source Cygnus X-1, the most likely candidate for a black hole. The International Ultraviolet Explorer (IUE), launched in 1978, was a joint NASA, ESA and UK project; with its ultraviolet telescope it has observed thousands of astronomical sources, many invisible from earth, and has provided astronomers with data on the chemical composition and structure of stars, nebulae and galaxies. IRAS (Infra-red Astronomical Telescope) was launched on 25 January 1983, and like IUE provided much data, this time at infra-red wavelengths. Its primary instrument, a 60cm (2ft) Cassegrain telescope, was cooled to $-271°C$ in a liquid helium Dewar flask. The helium ran out on 21 November 1983.

PROBES TO THE MOON

While most of the publicity and glory of lunar exploration has been given to the manned Apollo missions, it was the unmanned probes of the early and mid-1960s that paved the way for these missions. The first four American attempts at launching a lunar probe were unsuccessful and on 12 September 1959 the Soviet Union launched the Luna 2 probe, which impacted 800km (500 miles) north of the visual centre of the moon. It thus became the first man-made body to reach a celestial object. Very soon after Luna 2, the Russians again achieved a space 'first' when Luna 3 photographed the invisible face of the moon. After these early days, the pace of lunar probe launches accelerated. The American Ranger series of spacecraft were intended to photograph the lunar surface in advance of the Apollo landings. The first six Ranger missions were failures, but Ranger 7 (launched 28 July 1964) sent back more than 4000 high-resolution photographs before impacting in the Sea of Clouds. Two more later Rangers returned more than 13,000 images between them.

In 1963 the Russians were planning for a lunar soft landing. The first attempts were unsuccessful. Luna 9 finally succeeded in 1966 and the spacecraft returned the historic first pictures from the moon's surface. The rapid sequence of the Russian lunar launches leading up to Luna 9 was a direct response to its American competitor, the Surveyor spacecraft. Surveyor 1 soft-landed on the moon barely four months after Luna 9, returning 11,000 pictures over a six-month period. The Surveyor craft were more sophisticated than the Luna vehicles. Further Surveyor landings examined the surface in regions representative of Apollo landing sites. At the same time as the Surveyor craft were landing on the moon, the Americans were launching Lunar Orbiter spacecraft aimed at returning very high resolution photographs of the lunar surface.

During 1969 while all the American efforts were directed towards the Apollo programme, the Russians were landing more lunar craft in a bold attempt to soft-land and return to earth with lunar soil samples. This ambitious programme failed to pre-empt the Apollo 11 landing, but in 1970 Luna 16 did achieve the goal of returning a sample to earth. The Russians never tried to send men to the moon, concentrating solely on robot explorers. Luna 21 carried a rover vehicle (called Lunokhod) which for four months roamed over 37,000 metres (23 miles) on the surface under command from ground control.

PROBES TO THE PLANETS

Since the early 1960s there has been great activity directed towards sending space probes to either make close encounters with, or land on, the planets. Up to the launch of the Jupiter probe, Pioneer 10 in 1972, the efforts of the USA and USSR were concentrated on the inner planets. The first successful fly-by of Venus took place on 14 December 1962 by the Mariner 2 craft. Two further Mariners encountered Venus; Mariner 5 in 1967 and Mariner 10 in 1973. Mariner 10 took the first pictures of the cloud tops of the dense atmosphere of Venus, after which it went on to fly-by Mercury. The Soviet Union, meanwhile, concentrated on soft landings on Venus. Venera 4, launched on 12 June 1967, released a spherical capsule (1m (3ft) in diameter) when 45,000km (28,000 miles) from the planet. A parachute then permitted a slow descent through the atmosphere. During the 94-minute descent, it recorded temperatures of 370°C and sent back data on the pressure, density and chemical composition of the atmosphere. However it was a later spacecraft, Venera 7, that was the first to survive the journey to the planet's surface, relaying data such as pressures in excess of 90 atmospheres (1320psi) and a temperature of 475°C. Later more sophisticated spacecraft succeeded in sending back pictures of the rocky surface of Venus: Veneras 13 and 16 in 1982 actually returned colour pictures and undertook remote analysis of the soil.

Mars, the red planet, has received its due attention from Russian and American spacecraft. Mariners 6 and 7 returned many close-up pictures of the surface in 1969 and the Russian Mars spacecraft actually tried to soft-land on the surface, but without success. However, this feat was achieved by the Viking 1 and 2 spacecrafts in 1976. The sophisticated Viking craft returned excellent pictures of the surface from their scanning cameras, as well as analysing the Martian soil. No organic compounds were found in the samples analysed but the tests did not totally rule out the possibility of there being some form of primitive life on the planet.

Possibly the most stunning and exciting results of planetary exploration have come from the probes sent to the giant outer planets. The two American Pioneer spacecraft, Pioneers 10 and 11 were launched in 1972 and 1973 respectively on trajectories that would enable both craft to fly-by Jupiter. After a journey of 21 months they sent back the first detailed images of the planet. After its encounter with Jupiter, Pioneer 10 used the Jovian gravitational field to swing out of the solar system while Pioneer 11 was re-positioned on a flight path to intercept Saturn. Again, close-up images of Saturn and its extensive ring system were relayed back to Earth. These two missions were 'scout' missions for the more sophisticated Voyager spacecraft, Voyager 1 and 2, identical craft launched on 5 September and 20 April 1977 respectively. The Voyager spacecraft has a mass of over 800kg (1765lbs) with a high-gain antenna 3.7m (12.2ft) in diameter and a sophisticated array of detectors and high and low resolution TV cameras. The spacecraft functions are carried out by an on-board programmable computer and its communication transmitter is designed to transmit data, over a billion kilometres of space, at the rate of 115,200 bits per second, even though its transmitter has a power of only 23 watts.

On 5 March 1979, Voyager 1 made the closest approach (278,000km (172,300 miles)) to Jupiter, returning fascinating detailed images of the clouds in the Jovian atmosphere. The craft made close-up fly-bys of the four Galilean satellites Io, Ganymede, Callisto and Europa, discovering, rather surprisingly, that Io showed signs of active vulcanism. Voyager 2 made its closest approach to Jupiter on 9 July 1979, returning 15,000 images of the planet and discovering five new satellites. Both craft were then programmed for the 20-month journey to Saturn, which they encountered on 12 November 1980 and 25 August 1981 respectively, again returning many startling images of Saturn and its moons. Whereas Voyager 1 then began to travel out of the solar system plane, Voyager 2 encountered Uranus on 26 January 1986 and Neptune on 25 August 1989.

LAUNCH VEHICLES

The success of the early satellites depended greatly on the development of reliable launch vehicles. At the end of the Second World War the Americans, Russians and British captured German rocket hardware and many of the personnel involved in its development (see p. 649). The Americans launched many V2 rockets from White Sands, New Mexico, in a series of tests involving rocket propulsion and guidance. This Bumper Wac programme, as it was called, was only moderately successful, but it gave the Americans much useful experience in rocket technology. At the same time they were developing a number of short and long range missiles. A direct descendent of the V2 was the Redstone missile. Using liquid oxygen as fuel it had a range of 800km (500 miles). Two versions of the Redstone, the Jupiter A and Jupiter C, evolved during the missile test programme of the 1950s, to become significant in the early phases of the nation's space effort. The Jupiter C was first launched on 20 September 1950 and carried a payload nearly 5000km (3000 miles) down range from Cape Canaveral. It was, in fact, a form of the Jupiter C, re-named Juno I, which successfully launched America's first satellite, Explorer 1, in 1958. Following on from Juno I was the Juno II vehicle. Ten Juno IIs were used in space research from 1958 to 1961, but it will best be remembered as the ancestor of the giant Saturn launch vehicle.

The Saturn 1 launch was commissioned in 1958 to provide capacity for large payloads. It was composed of a central Jupiter tank surrounded by eight Redstone tanks. The Saturn's 100 per cent launch record between 1961 and 1965 provided the platform for the late Apollo lunar missions. The vehicle used in the Apollo missions was the Saturn V. This was a four-stage launcher whose first stage was powered by five F1 engines generating, in total 3400 tonnes of thrust.

The USA has developed other launch vehicles primarily for unmanned satellite and space probe missions. The Scout is a four-stage solid fuel vehicle first launched in 1961 and still in service in the 1980s. Other liquid fuel vehicles include the Titan, developed from the ICBM of the same name, and the Delta, developed from the Thor missile.

The USA's most recent vehicle is the Space Transportation System (STS), commonly called the Space Shuttle. It was designed to provide a relatively cheap vehicle with most of its components being re-usable, in contrast to the usual expendable launch vehicle. Conceived in 1969, the STS's eventual design consisted of a winged Orbiter craft with three main engines using liquid oxygen and hydrogen fuel, supplied by an external tank plus two solid fuel rocket boosters. This hybrid rocket combination has a capability of carrying 29.5 tonnes into low earth orbit. The external tank is lost during launch, but the solid boosters are re-used and the Orbiter re-enters the atmosphere after the mission, and lands like a conventional aircraft on a runway. Orbiter's main engines were a major breakthrough in rocket technology. They were designed to be re-usable and throttleable as well as having the largest specific impulse (thrust per kilogram of propellant) of any previous engine. Up to the end of 1985 over 20 Shuttle missions had been performed since the first launch on 12 April 1981.

Whereas the development history of American launch vehicles can be easily traced, the same cannot be said for the other space power, the Soviet Union. The early Sputniks were launched by a vehicle devised from the SS-6 Sapwood ICBM, with a gross lift-off weight of 267,000kg (589,000lb). The A1, A2 and A3 vehicles followed, each with a greater launch capability than the last. The A2e was a basic SS-6 plus an additional stage on the A2 and was first flown in 1961 for the Venus fly-by missions. It was also used for the interplanetary, lunar and other deep space probes such as Prognoz. Later vehicles have designations C1, D1, F1, F2 and G, although few details have been released of these. What is remarkable about Soviet rocketry is that many of the vehicles used in the 1980s date back to the late 1950s and early 1960s, and it seems that little real development took place after these earliest designs were developed.

Other nations and consortiums of nations have developed their own launch vehicles. Europe has the Ariane vehicle; a four-stage launcher capable of placing a 1700kg (3750lb) satellite into low earth orbit. The UK's only launch vehicle, Black Arrow, successfully launched the X3 (Prospero) satellite in 1971, but the project was cancelled soon after the launch. China, Japan and India have also built and flown their own launch vehicles.

FURTHER READING

Baker, D. *The rocket* (New Cavendish Books, London, 1978)
—— *The history of rocketry and space travel* (Nelson, London, 1981)
Von Braun, W. and Ordway, F. *The history of manned spaceflight* (New Cavendish Books, London, 1981)
Von Braun, W., Ordway, F. and Dooling, D. *Space travel – a history* (Harper and Row, London, 1985)
Gatland, K. *Space technology* (Salamander, London, 1981)
Hart, D. *The encyclopaedia of soviet spacecraft* (Bison Books, Greenwich CT, 1988)
Spaceflight directory 1988 edited by R. Turnill (Jane's Publishing, London, 1988)
Lewis, R. *Illustrated encyclopaedia of space exploration* (Salamander, London, 1983)
Ordway, F. and Sharpe, M. *The rocket team* (Heinemann, London, 1979)

PART FOUR

COMMUNICATION AND CALCULATION

14

LANGUAGE, WRITING, PRINTING AND GRAPHIC ARTS

LANCE DAY

LANGUAGE

Man is distinguished from the animal world in no more striking way than in his ability to communicate by means of language. All human activity, except the simplest and most primitive, requires co-operation, and that is not possible without language. Animals do of course communicate by sound, but their cries are inarticulate and general, whereas it is of the essence of human language that the sounds uttered are articulate, that is, they have a structure that can be divided into words having precise and specific meanings.

Language is a system of vocal sounds that are symbolic, that is, they are arbitrarily related to the thing they stand for. Animal gestures and cries, on the other hand, are non-symbolic. An animal seeks to prevent the approach of another by frightening sounds and gestures, even brute force, while man, although not always forswearing these methods, uses sound to symbolize his intention: 'Keep out'. By writing, he has a means of communicating his intention without the necessity of his continued presence. From such basic needs, language has been developed into a highly sophisticated instrument for the expression of the most complex thoughts and the most subtle and uplifting aesthetic emotions.

The origins of language lie hidden in the remote past and are a matter of plausible speculation. The language of present-day primitive peoples offers few clues, since their spoken word is still relatively sophisticated compared with man's earliest attempts at speech. Studying the way babies learn to speak is also hardly relevant, because their speech is derived from adults who have already learned a highly developed language. It is most likely that human language had its beginnings around a million years ago, at about the same time as the earliest toolmaking and distinctively human forms of co-operation. Several theories have been proposed to explain how language arose, some with such curious names as the 'bow-wow', the 'pooh-pooh' and 'yo-he-ho' theories.

The first suggests that primitive language imitated the sounds of nature, particularly animal cries, the second that it arose from the instinctive emotional cries of pain or joy. As the sounds thus emitted are inarticulate, neither explains how man arrived at articulate sounds. The 'yo-he-ho' is better in this respect. Proposed by Noiré in the nineteenth century, it suggests that language arose from involuntary sounds uttered by men engaged in some joint muscular effort as heaving a tree trunk. While heaving they trapped the breath in the lungs, closing the vocal cords, and on relaxing the cords opened and the air was expelled in a grunt or some other sound. This would have given rise to the basic consonant-vowel structure of language and has the additional advantage of associating the origin of language with human co-operative activity. The need to communicate during such activity is most likely to have provided the motivation to derive and develop this method of communication. But whether there was a single point of origin, or development at scattered times and places, and what forms primitive language took must remain virtually guesswork, until we reach the era of written records.

WRITING

Writing is, strictly, derived from speech, but the symbols used in writing are derived from pictures. The use of pictures to convey messages or keep records has been widely used among primitive peoples from the New Stone Age onwards and finds a place in civilized society today, as with the figures of a man and a woman to denote the appropriate toilets. Writing evolved by simplifying the pictures and then conventionalizing them until they were scarcely recognizable as pictures. Next, the signs were made to stand for linguistic components, first words, then syllables and finally individual sounds or groups of sounds (phonemes). The form taken by the signs depended to some extent on the nature of the materials used. If the signs were scratched into clay or incised into wood, the writing tended to be angular and avoid curves. The earliest writing of this, or indeed of any kind, appears to have been that developed by the Bronze Age Sumerians in Southern Mesopotamia between 4000 and 3000BC. At first they drew pictures in clay with the tip of a stylus made from a reed. It was then found easier to press the head of the stylus into the clay, producing a wedge-shaped mark about 8mm long. Signs with different meanings were formed by various combinations of these wedge shapes. The script is known as cuneiform, from the Latin *cuneus*, a wedge. Later the Babylonians and Assyrians supplanted the Sumerians, but took over and developed their writing and cuneiform came to be the almost universal script in the Near East.

In ancient Egypt, on the other hand, scribes painted marks usually on papyrus, made of strips of the stem of a sedge plant glued together, by means of a kind of brush pen and ink. The original pictorial writing goes back to around 3000BC and is known as hieroglyphic, from the Greek 'Holy carved'. Hieroglyphics were used throughout the Egyptian world for religious purposes, but a more conventionalized form was developed for everyday use, known as the hieratic, or priestly, script. By 700BC this had been further simplified to demotic, or popular, script. In other times and places, this stage had been reached; Chinese characters, ideograms, are similarly simplified and conventionalized derivations of picture writing. The really crucial development, the alphabet, derived from Egyptian hieroglyphics. The Egyptians, like the Babylonians and Assyrians, developed a syllabic system with signs for consonants, singly or in groups, and omitting vowels. There were 24 signs for single consonants. When the western Semitic peoples, almost certainly the Phoenicians, developed a script, they took over these signs. The result was not so much a syllabic system but an alphabetic one with the vowels omitted. Around 900BC the Greeks assimilated this alphabet and took the final step of adding signs for the vowels. At last we have arrived at a real alphabet, or system of signs standing for phonemes, so that all sounds of speech can be represented by a mere 25 or thereabouts easily memorized signs. Universal literacy was hardly possible before the Greek achievement, from which other alphabets have stemmed, including the Latin, used in Western Europe and cultures derived therefrom, and the Cyrillic used in Russia and some other countries in Eastern Europe.

To write the new Greek characters required a finer tool than the old brush pen. The Romans took over the reed pen, writing on papyrus, from the Egyptians, but for casual jottings, a handier implement was available – a stylus for making incisions on a wax-coated tablet. In Europe during the Dark Ages the practice of writing inherited from the ancient world was kept alive in the monasteries, a part of which, the scriptorium, was reserved for the scribes who laboriously wrote out the learned and practical works required by the community. It was during the early mediaeval period that quill pen and vellum (originally fine, smooth calfskin: from Old French *velin*, calf) replaced the Roman materials. A number of illustrations in contemporary manuscripts show scribes at work at their desks, surrounded by their writing materials and implements. The quill pen remained in use until well into the nineteenth century. It was made from one of the primary feathers taken from the wing of a bird, usually a goose, sometimes a raven or swan. This pen (Latin *penna*, a feather) was flexible and could be sharpened repeatedly in different ways to produce various kinds of script, yet was not sharp enough to penetrate the writing surface. It was also in plentiful supply. Parchment, also made from animal skins, proved to be a durable, if rather expensive material. Paper was cheaper. It had been invented by the Chinese in AD105, although recent studies suggest an even earlier origin in the second century BC. The art of making it drifted westwards through Islam until it reached Western Europe, by contact with the Moors in Spain in the twelfth century, but its use was not widespread until the fifteenth century, when a massive increase in productive capacity made possible the practice of printing.

The third essential ingredient in writing is ink and the word itself provides a clue to the earliest kinds of ink. The word is derived from the Latin *encaustum*, 'something burnt in', for the ink used throughout the mediaeval period consisted of iron salt and oak galls, which literally burned itself into the writing surface.

With these simple means, for over a millennium, the learned and literary works of Greece and Rome were recorded and communicated, along with the theological, philosophical and scientific works of the Middle Ages. They sustained the intellectual tradition of antiquity and the Middle Ages that the modern world is heir to.

The quill pen, with all its virtues, had many drawbacks: it was fragile and wore out quickly, it could retain enough ink to write only a few words at a time. However, it long reigned supreme in spite of attempts to substitute metal nibs, certainly as early as the sixteenth century. They suffered from the crippling defect of inflexibility, making it impossible to achieve the subtle differences of thickness that are such a feature of early script. The other difficulty was that the metal was rapidly corroded by the ink. The steel pen was to solve these problems. It was on general sale around 1829, but was still uncommon ten years later, being found too scratchy and stiff. By 1849, however, it was in widespread use, thanks to the improvements effected by the leading manufacturer Joseph Gillott: two slits were pierced in the shoulder of the nib, in addition to the one in the centre, greatly improving its flexibility. The mass production of steel nibs, centred in Birmingham, brought down the price, so they could be simply discarded when corrosion set in. After the Second World War the steel nib declined rapidly and now survives only for a few specialized tasks.

Writing with quill or steel pen was always a disjointed business, even with the addition of a small ink reservoir behind the nib. The answer was the fountain pen, which made its first appearance in a work on mathematical instruments by Bion, the English translation of which was published in 1723. But it remained primitive and unreliable for over a century, and it was not until the second half of the nineteenth century that improvements in design and construction made it satisfactory. The patent by John Joseph Parker of 1832 for 'improvements in fountain pens' makes the first mention of a self-filling pen and ushered in a whole spate of patents seeking to perfect the arrangements for filling the barrel of the pen with ink and, more difficult, for feeding the nib with an even and controlled flow. By the end of the century names such as Swan, Parker and Waterman had brought fountain pens into popular use. More modern developments are the interchangeable nib and the ink cartridge – a great help towards the elusive ideal of a leak-proof fountain pen. The search for inks less corrosive than the iron-based one then in use began with Henry Stephens in 1834 and succeeded with the discovery of aniline dyes in 1856 (see p. 201).

The steel pen has been almost entirely, and the fountain pen to some extent, superseded by the ball-point pen. The present form of ball-point derives from the invention by the Hungarian brothers Ladisla and Georg Biró, who applied for a patent in 1938 but moved to the Argentine at the outbreak of war. Development followed, especially in the USA where the Defense Department required a pen that would work at high altitude, with an ink unaffected by climatic changes and yet quick drying. The ball-point came near to meeting these requirements and it swept the board. Its more recent rival, the felt-tip, has complemented and not supplanted the biro.

Both these pens serve also for note-jotting as well as more formal writing. The wax tablets of the Romans were still in use in the sixteenth century; when Hamlet exclaims: 'My tablets, meet it is I set it down', he may well have been referring to them. The schoolboy then struggled with slate and slate pencil, but the discovery of the black lead or graphite mine in Cumberland in the sixteenth century led to the use of this material, first in holders, then in slender columns encased in wood — the pencil.

THE INVENTION OF PRINTING

A massive effort was put into the hand-copying of texts in the Middle Ages, in monasteries and, later, secular scriptoria as well, but it was a laborious and expensive operation, and each time a text was copied there was a fresh chance of error. No two copies were exactly alike. Printing as a means of making many identical copies arose in the late Middle Ages. Playing cards were printed from wood-blocks in which the areas to appear white were cut away, leaving the ink-bearing portions standing out in relief. Pictures with captions were printed from these wood-blocks, the earliest surviving examples being of the Virgin with four saints of 1418 and the St Christopher of 1423, in the John Rylands Library in Manchester. From around 1430 sheets such as these were collected and bound into books, known as block-books or xylographica. For around half a century block-books, produced mainly in Germany, Switzerland and the Netherlands went some way towards satisfying a demand for multiple identical copies. But the number of copies that could be printed from a block of wood was limited and, far more serious, when the print run was finished, the wood blocks had to be discarded and fresh blocks made for a different text.

The invention of printing from movable type was immeasurably superior and led to the early demise of the block-book. Here, the text is assembled from individual type, one for each character. After printing off, the type is distributed ready for use in a fresh work. Movable type was in use in Korea in the thirteenth century and in Turkey some time later, but there is no evidence that its use influenced Johann Gensfleisch zum Gutenberg of Mainz, who is generally accepted as the inventor in Europe of printing with movable type around 1450 – an invention that has exerted a more profound and widespread influence on mankind than any other. All useful inventions have been conceived to satisfy a blatant, or latent, need. With printing, a number of factors converged around the mid-fifteenth century to make it highly likely that someone would invent it. Literacy, confined to the clerics in the early Middle Ages, had by then spread to the laity. The development of education, cultural traditions and prosperity led not only to a desire for literature of all kinds but, in the case of the richer classes, to a taste for collecting fine manuscripts. A brisk trade grew up to satisfy this demand and a way was sought to widen the trade by multiplying texts.

While vellum was the only available medium, printing would have been too expensive and the supply was hardly adequate for large-scale work. Fortunately the textile industries of Western Europe were booming, producing a plentiful supply of waste rag, an admirable material from which to make paper. This made it sensible to look for a way to make multiple copies. Gutenberg appears to have begun his search while a political exile in Strasburg towards 1440. He returned to Mainz between 1444 and 1448 and two years later had perfected his invention to the point of commercial exploitation. In this he was none too successful, for financial difficulties led to foreclosure on his equipment in 1455 and it passed to Peter Schöffer, the son-in-law of the lawyer who had lent Gutenberg the money to set up his business. One book only can be confidently attributed to Gutenberg's workshop – the great 42-line Bible, begun in 1452 and published before August 1456 – the first book to be printed from movable type and a magnificent piece of printing by any later standards.

Gutenberg's achievement lay in the successful combination of existing techniques rather than the devising of new ones – not the last such case in the history of technology. Of these various components, paper has already been mentioned. The others were the press, the type and the ink. The wooden press, with variations in detail, had long been used for wine pressing, in bookbinding and for other purposes, and was only slightly modified for printing. For ink, Gutenberg substituted an oil-based one for the water-based inks of the scribes; it was taken over from the material devised for painting by the Van Eycks. The type was formed by metal-casting techniques which were well established, although the design of mould, of variable size to suit the different sizes of the characters, was ingenious. However, Gutenberg's contribution in respect of the composition of type metal should not be overlooked. He seems to have spent years searching for an alloy with certain properties, such as low melting point for convenience of casting yet strength enough to resist the wear of thousands of impressions. No existing alloy possessed the required combination of properties and the type metal arrived at by Gutenberg, doubtless after much trial and error, appears to have consisted mainly of lead and tin with some antimony. Gutenburg thereby achieved a remarkable technological, if not commercial, success, for the earliest products do not show the usual signs of primitive, halting work but are fine examples, equalled later but hardly surpassed. Also, the fact that the equipment and the process survived virtually unchanged except for minor details for three and a half centuries cannot be entirely due to printers' conservatism. It is a striking testimonial to the soundness of Gutenberg's invention.

THE GROWTH OF PRINTING

Printing spread rapidly, first to German centres of trade and banking such as Cologne (1464), Nuremberg (1470) and Augsburg (1472), rather than seats of learning. By the early 1480s German printers had taken the invention into most of the countries of western and central Europe. Italy was first, with a press at Subiaco as early as 1465, while Paris received its first press in 1470. In Italy the press was given a fresh direction by the new Renaissance learning which it would not have received in essentially mediaeval Germany, for it was Italy that gave us the two faces on which nearly all western types have been based to this day: roman and italic. It was here too that the title-page, pagination and the pocket or handy edition emerged. Rome and Venice followed hard on the heels of Subiaco; Venice was to be the seat of the illustrious Aldine Press, with its dolphin and anchor mark, set up by Aldus Manutius about 1490. He conceived the idea of disseminating scholarly editions of Greek and Roman authors to as wide a public as possible by printing in compact type in small handy volumes no fewer than 1000 copies, whereas the usual print run was a few hundred – 500 at most. Such editions were renowned throughout Europe and did much to spread the knowledge of classical authors that was of the essence of the Renaissance. The printed book was shaking itself free of the manuscript whose appearance it had so faithfully mirrored, and was presenting a distinctive image of its own.

In England printing was introduced in 1476 not by a German but by a native, William Caxton. Having succeeded in commerce as a member of the Mercers' Company, he diversified in late middle age into the manuscript trade and learned the art of printing in Cologne. He set up his press at the Sign of the Red Pale at Westminster, the seat of government and court, rather than the commercial centre of London, in order more profitably to serve the clientele who would be most likely to buy the kinds of books he was most interested in printing – courtly romances and devotional works.

The printed book also proved instrumental in the spread of scientific knowledge. Apart from the encyclopaedic works of natural knowledge such as Pliny's

Historia naturalis, first printed in 1469 and the earliest scientific work of all to be printed, astronomy was the subject most treated, as it was the area of science more advanced than any other. Early in the next century technological books began to appear – on mining, the assay of metals, glassmaking, distilling. In mid-century appeared books that were to initiate profound changes in the way man thought of his place in the universe – above all Copernicus' *De revolutionibus orbium coelestium* printed in Nuremberg in 1543. The book was joined by the periodical, as scientific work gathered pace from the mid-seventeenth century, in order to report progress more quickly and conveniently. The earliest of these periodicals that is still current is the *Philosophical Transactions* of the Royal Society, which began to appear in 1665.

The book was not of course a new medium: the printer took it over from the scribe. But he did invent a new medium – the printed sheet, to advertise or give information. The earliest surviving piece of printing, and also the earliest piece to come from Caxton's press, is of this kind. It was an indulgence, an announcement that a pardon was available to those who undertook certain good works. There flowered a whole variety of ephemeral printing that was to be the lot of the jobbing printer.

During the sixteenth century the printers began to satisfy a growing demand for up-to-date information and entertainment in a more suitable form than the book. From hand-written news reports circulated by hand was born the newsbook and then the news-sheet. At first spasmodic – the earliest English example provided an account of the Battle of Flodden in 1513 – they became regular and emerged as the first true newspapers with the publication of *Aviso-Relation oder Zeitung*, edited in Prague but printed in Wolfenbüttel, and the *Relation* of the Strasburg printer Johann Carolus, both in 1609. 'Corantos', as they were called began to spread and the first appeared in England from the press of Thomas Archer in 1621. These were in the form of newsbooks, but around mid-century the handier news-sheet emerged, among the earliest being the *London Gazette* of 1665, still with us today. At the same time, the periodical began to serve a desire for more leisurely background reading; the aim was well expressed in the cumbersome title of one example, started in Leipzig in 1688: 'Entertaining and serious, rational and unsophisticated ideas of all kinds of useful and agreeable books and subjects'. By the turn of the century the newspaper and magazine were established institutions and made rapid strides. With the introduction of the daily post from Dover to London in 1691, bringing in foreign news each day, the daily paper became a practical proposition. The *Daily Courant* in London in 1702 was the first. Meanwhile the weekly journal developed into the organ for enlightenment and entertainment, above all in the hands of Richard Steele and Joseph Addison with their *Tatler* (1709) and *Spectator* (1711), which exerted an immense influence on eighteenth-century reading. One of the most successful of these periodicals was the

Gentleman's Magazine (1731–1907), which reached a circulation of 15,000 in 1745 and gave the name 'magazine' to this type of production.

The enterprise shown by the printers, publishers and booksellers in developing this remarkably wide variety of reading matter was in striking contrast to the conservatism of their technology. The earliest illustration of a press is a rather crude woodcut of 1499, the 'Dance of Death' print from Lyons. The equally well-known Badius prints of 1507 and 1520 provide a better idea of an early press room. Yet the first detailed account in any language did not appear until 1683–4 in Joseph Moxon's *Mechanick exercises on the whole art of printing*. Clearly little had changed and Moxon's work served as the basis of later handbooks throughout the age of hand printing. A few components were made of iron, but wood remained the principal material, above all for the frame. That restricted the total load that could be imposed by the platen upon the bed of the press and that meant that only half a sheet could be printed at one pull. Production is said to have attained 250 sheets an hour, but that was doubtless an upper limit rather than an everyday average.

Towards the end of the sixteenth century, the composition of the type-metal was changed slightly to give greater toughness to withstand the longer print runs that were becoming the rule. As to the paper, once mediaeval Europe had made slight modifications to the materials and techniques received from the East, there was little change until the Industrial Revolution. The raw material was rag, usually linen, and this was cut up into small pieces, boiled, allowed to ferment to assist disintegration and finally subjected to the action of wooden stampers. That completed the separation into fibres which, mixed with water to the consistency of thin porridge, formed the 'stock' from which the paper was made. The papermaker dipped into the pulp his mould, consisting of a rectangular wooden frame with a mesh of closely spaced brass wires and heavier wires at right angles more widely spaced. The size of the sheet was determined by a second frame, or deckle, placed over the first. When he had scooped up an even layer of pulp on the mould, he gave it a shake to help interlock the fibres and release some of the water. He had made a sheet of paper. The rest of the process, then as now, consisted of getting rid of the rest of the water. The mould wires gave rise to the characteristic wire and chain marks of hand-made paper, although Whatman introduced 'wove' paper in 1757, using a mould with woven wire to produce a smooth surface.

Stamping was mechanized and water-powered during the thirteenth century and this stage of the process was greatly accelerated around 1650 by the invention of the hollander, a vessel in which a vertical spindle with knife blades rotated to lacerate the rags.

Book illustration had settled into a rut. During the early period woodcuts were used, the raised portions receiving the ink with the type, so that text and illustration could be printed together. But towards the end of the sixteenth century, the copper engraving began to supplant the wood block, on account of the finer detail that could be reproduced. In this intaglio process, the ink was received by lines incised into the metal and a sheet of paper in contact with the plate was forced through an engraving press, rather like a mangle. The illustrations thus had to be printed separately from the text and were on sheets interspersed throughout the text or gathered together at the end. Only in recent years has a reconciliation between text and illustration been effected by modern offset litho printing.

TECHNOLOGICAL INNOVATION IN THE NINETEENTH CENTURY

Change, when at last it came, began in the country where the Industrial Revolution had already begun to change so much: England. It revolutionized not only the look and feel of the printed book but the reading habits of the industrialized nations. The move to mechanize paper production began with the making of a model papermaking machine by Nicolas Louis Robert in 1797 at a mill in Essonnes, south of Paris. After applying for a patent the following year, Robert fell out with his backer and owner of the mill, Didot, although the latter did contact John Gamble in England to see if capital could be raised there for a full-sized machine – all this against the background of the Napoleonic Wars. Gamble and the London stationers the brothers Fourdrinier called in the engineer Bryan Donkin and in 1807 they patented a much superior version of Robert's invention. Pulp passed over a 'breast box' in a continuous stream on to a wove-wire screen. The fibrous sheet was passed between rollers to remove some of the water. The paper was then transferred to a felt blanket and a second pair of rollers to remove more water and finally wound on to a drum while still quite wet. Steam-heated drying cylinders were introduced by Thomas Crompton in 1820 and at last it was possible to produce a continuous sheet or web of dry paper. Later, detailed modifications were made to improve quality and widen the range of papers available, but the modern machine, although much larger and faster, operates on the same principle as the early Fourdrinier examples. Even those early machines were able to raise production from the 25–45kg (60–100lb) a day of the old manual process to 450kg (1000lb) a day with a corresponding reduction in cost to the customer. This development was well timed, as the improvements in the printing press that were about to take place would have been useless unless paper production had kept pace.

The first major change in the press was brought about by Charles Mahon, 3rd Earl Stanhope, who in 1800 introduced the first successful iron press, the famous Stanhope press. The iron frame not only enabled the whole forme (the frame containing the body of type for a complete sheet) to be printed at one pull, thus speeding up production, but larger formes could be handled without straining the frame. While retaining the screw-operated platen (the plate which pressed the type against the paper), a system of levers attached to the bar increased the pressure at the moment of contact. This gave beautifully sharp impressions on the paper and also eased the strain on the pressman. Production rate rose to about 300 sheets an hour. *The Times*, soon to be and long to remain in the vanguard of technical progress, was printed on a battery of Stanhopes during the early 1800s.

A number of improved designs of iron hand-press soon followed, the most celebrated being the American Columbian press, invented by Floyd Clymer in 1813. The screw device was replaced by a system of levers assisted by a counterweight in the form of a gilded cast-iron eagle, which gave the press its characteristic flamboyant touch. Many of the masterpieces of nineteenth-century literature were printed on such presses – most book work until mid-century and much thereafter flowed from them – but they could not meet the needs of the newspaper proprietors.

The first machine press was invented by Friedrich König and Andreas Bauer of Würzburg in 1811. As the German engineering industry was not sufficiently well developed to manufacture their invention they moved to England and their first successful press was patented in 1812. A revolving cylinder carried the paper, which received an impression from a reciprocating bed carrying the type forme. The machine could be operated by hand but steam power could also be harnessed. The real breakthrough came when König and Bauer constructed a twin-cylinder machine for John Walter of *The Times*. On that famous night of 29 November 1814, behind closed doors to avoid disruptive attentions from conservative-minded printers, a newspaper was printed for the first time on a machine press. Production shot up to 1100 sheets an hour, but more was to come. After so many years of stagnation progress now became rapid indeed. The Applegath and Cowper four-cylinder press of 1827, made for *The Times* again, carried the rate up to 4000 sheets an hour. Their type-revolving press twenty years later took it up to 8000. It was the newspaper that led the demand for higher speeds and stimulated yet further development.

The next stage was to replace the reciprocating flat-bed by a continuously rotating cylinder press, carrying the type on a curved forme. The first press of this kind was constructed by R. Hoe & Co. in 1846 for the Philadelphia *Ledger* in the USA. The cylinder of this huge machine was about 2m (6.5ft) in diameter and, with four impression cylinders printing off at four points around the circumference, output attained 8000 sheets an hour. Even larger machines of this kind followed, with ten impression cylinders giving 20,000 sheets an hour. Paris received its first Hoe press in 1848 and London in 1856; *The Times* ordered two of them.

The way forward to greater speed was to print on both sides of the paper at once – a perfecting press – and on to a continuous web of paper instead of separate sheets. The first to be constructed was the Bullock press in Philadelphia in 1863 and five years later *The Times* built one, known as the Walter press, after the then editor John Walter III. This could print 15,000 completed copies, or over 60,000 sheets an hour. The printing press thereby achieved its final form; future developments were in size and greater speed rather than in design principles. In the Bullock press the paper, although fed from a reel, was cut up into sheets before printing; on the Walter press a continuous web of paper was fed through, printed on both sides, separated into sheets and copies delivered alternately to two delivery points.

The mass-circulation newspaper, springing up in the 1880s, called for yet more speed and the Walter concept was refined, using several printing formes to a cylinder, combining several cylinders in a machine (Hoe in 1882) and then these were placed vertically. The first folding device was incorporated in 1870. In 1920 the Goss Company introduced the Unit press with several printing units placed in line and the webs from each gathered together at one point for folding and cutting. From the early years of the twentieth century steam or gas power gave way to electricity and the modern high-speed press had arrived.

An essential element in the quest for speed was the rotary principle, by which the type forme was curved to follow the shape of a cylinder, replacing the flat forme on a flat-bed press. At first type was cast in wedge-shaped moulds and clamped to the cylinders, but this led to problems with dislodged type, so curved stereotypes were introduced. During the eighteenth century attempts had been made to make a mould of a forme of type and cast a printing plate from it to enable copies to be run off while the original type was distributed for use of a fresh piece of work. William Ged, a Scottish goldsmith, experimented with plaster moulds in the 1720s, but opposition from the printers, who feared that the innovation would lead to loss of work, put an end to these experiments. In France, however, contemporary work along the same lines found a readier response, particularly in the making of plates for printing currency during a period of high inflation. But it was Stanhope's second great contribution to printing technology to perfect a method for making printing plates: stereotypes. Later the plaster moulds were found to be heavy and fragile and in 1829 the French printer Claude Genoux substituted papier mâché, known as wet flong. This in turn was replaced by dry flong in 1893, paving the way for a mechanized process, the Standard Autoplate Caster by Henry A. Wise Wood in 1900. Stereotypes for rotary presses were of course cast in type metal in curved formes. Recently modern materials have replaced metal for printing certain kinds of material, such as synthetic rubber and plastics.

Not all printers demanded the high speeds required by the newspaper proprietors. For book printing, let alone the miscellaneous work of the jobbing printer, something more sedate was favoured, something the printer could feel was more like a hand-press. Indeed, the hand-press carried on for much commercial book work well into the nineteenth century, but two types of machine were developed to serve his needs: the cylinder and the platen. William Rutt designed a small hand-operated press based on the König & Bauer in which the impression cylinder came to rest after printing a sheet to allow the flat bed carrying the type to be moved back to the starting position. This showed no saving in time over the hand-press, but the Naypeer of 1824, developed by David Napier and commissioned by T. C. Hansard for book printing, was more successful. By 1830, the machine had incorporated two important advances: gripping fingers on the cylinder to take a sheet of paper through the machine, enabling the sheet to be accurately placed on the cylinder; and a continuously revolving cylinder, rising to allow the bed to return after each impression. Between 1830 and 1860 many patents were taken out for cylinder presses. The best known was the Wharfedale, from the name of the valley in Yorkshire where Otley, its place of origin, is situated. Developed by Dawson and Payne in 1859, it was of the stop-cylinder pattern and of robust construction. Similar machines were made in other countries and the name Wharfedale became the generic term for this veritable workhorse of the industry for many years. The version manufactured by König & Bauer in Würzburg in the early years of the twentieth century could print 1200 sheets an hour.

Even that was too elaborate for the jobber and the platen press was evolved to serve his needs. A foot-operated press was patented in London in 1820 by the appropriately named American inventor Daniel Treadwell, but it did not succeed. Other American inventors took up the idea and the Ruggles press of 1851 set the pattern, with a vertical type bed against which the paper-bearing platen closed. This was followed by Gordon's Franklin press in 1856 which became the general form for small jobbing platen presses. When small electric motors to work such machines became available in the early 1900s (see Chapter 6), automatic paper-feed was added and the field was dominated by the Heidelberg and Thompson platens until this kind of press began to be superseded by the small office offset litho machine in the 1950s.

The Fourdrinier papermaking machine, the power-operated rotary, cylinder, and platen presses, and the stereotype, made possible the mass dissemination of literature of all kinds. There was the improving literature promoted by such bodies as the Society for the Diffusion of Useful Knowledge founded in 1826 with such notable publications as the *Penny Magazine* and *Penny Encyclopaedia* founded in 1832 and 1835 by the great popular educator Charles Knight. They made possible a wide readership for Dickens's bestsellers; they also produced a flood of cheap trash. The Industrial Revolution had helped to bring about a desire for both educational and entertaining reading matter among the artisan and new rising middle classes, but it was the technical advances that led the revolution in reading habits of the masses.

The copper engraving had by 1800 done yeoman service for centuries, but copper is a soft metal and ill-suited to the wear imposed by long print runs. Steel engraving, begun in the USA in 1810 for bank notes, spread to Europe and soon superseded copper, thus helping to bring down the price of illustrated books and place them within reach of the masses. The wood engraving was also a popular medium, brought to perfection by Thomas Bewick. Here, the wood block was cut across the grain instead of with it, as for earlier wood cuts, and finely engraved illustrations could be made cheaply and quickly, retaining quality after long print runs. They made possible cheap illustrated magazines like the *Penny Magazine*, the *Illustrated London News* (1842) and *Punch* (1841).

For higher quality work, such as art reproductions, the mezzotint and aquatint gave good results. Mezzotints are a kind of engraving in which the metal plate is roughened with a 'rocker' with a serrated edge. The areas corresponding to the lighter parts of the picture are then smoothed with a scraper to a varying degree. Aquatints are produced by etching the metal plate, that is, eroding it with acid, through a porous ground. The lightest portions are then etched and given a varnish coating to protect them from further action of the acid. Darker areas are then formed by successive etching and varnishing. By these means, subtle gradations of tone can be produced. Mezzotints were used mainly for reproducing paintings and portraits, aquatints for topographical scenes and descriptive illustations.

Both methods were supplanted by lithography, invented by happy accident by Johann Alois Senefelder of Munich when he found that after marking the flat surface of a piece of limestone with a greasy crayon, dampening the surface and passing an inked roller over it, the ink covered the greasy marks but was rejected by the rest. If paper was pressed on to the surface, it took up the image in reverse. Senefelder began to use lithography ('stone printing') in 1800. He tried unsuccessfully to keep the process to himself, publishing no account of it until 1818. The following year Ackermann brought out an English translation, *A Complete Course of Lithography*, and from 1830 lithography had virtually triumphed for printed illustrations, except for mezzotints for paintings, steel engravings for cheap books and wood engravings for both popular periodicals and books.

A slab of limestone had certain disadvantages; it was cumbersome and broke if handled clumsily. Metal plates were sometimes used instead – zinc from around 1830 and aluminium became a practical proposition in 1904. The versatility of the process was greatly increased by the offset principle, first used to print on to tinplate. The hard unyielding printing surface was replaced by a resilient rubber surface by transferring or 'offsetting' the image from one to the other and then printing on to the tin surface. Robert Barclay patented this process in 1875 but printing on to paper had to await the development of a suitable rotary press for offset litho printing, by Harris in the USA and George Mann & Co. in Britain in the early years of this century.

For much of the nineteenth century, the great mass of the printed illustrations produced required the services of armies of artists, whether for 'artistic' work or for purely factual representation for instruction purposes or news media. The invention and development of photography from the late 1820s (see p. 731ff.) provided a method of factual representation, but the large-scale mechanical reproduction of photographs was not easy: it was not until the 1880s that a photo-process engraved plate, that could be printed in a press like a hand-engraved one, became a possibility. The *Daily Telegraph* in 1882 was the first newspaper to use photographic line blocks, known as zincographs because sensitized zinc plates were used, while the *Daily Graphic* was the first to publish such illustrations in bulk. So far only the lines in a photograph could be printed. The breakthrough to reproducing the whole range of tones was achieved with the invention of the half-tone process. The original was copied by a process camera in which a cross-line screen was interposed in front of the negative, producing a photograph consisting of thousands of dots which, viewed from a distance, gave the illusion of tone variation. From the 'dotted' negative, a 'dotted' sensitized printing plate was prepared, which could be printed off with the rest of the text. The use of half-tone blocks in newspapers dates from 1877 by the Jaffé brothers in Vienna, taken up by Horgan in the USA with the *New York Daily Graphic* in 1880. The London *Daily Graphic* was using them in 1894 and the daily newspaper illustrated by a new breed, the press photographer, was in being. Harmsworth's *Daily Mirror*, starting in 1903, the first newspaper to be sold at a halfpenny, epitomized the new mass journalism, made possible by rotary press, papermaking machine, stereotype and half-tone block.

Another way of mass-printing photographs has been photogravure, an engraving or intaglio process. Artists had for centuries graved in copper plates lines that held the ink, which was transferred by pressure to the paper. In 1879, Karl Klič exhibited plates derived from photographs. The ink was held in a myriad of minute cells in the surface of the plate, of varying depth according to the tones of the original. Klič next developed a gravure cylinder for use in a rotary press. The cylinder dipped into free-flowing ink which was then wiped off by a 'doctor' blade, leaving the ink in the cells.

With all this plethora of progress in nineteenth century printing technology, one process remained obstinately at the primitive, hand-worked stage – typesetting. The making of the type was successfully mechanized as early as 1838 with the Bruce typecaster, developed by David Bruce in the USA, using a pivotal action of the mould to and from the mouth of a hot-metal pump. It was exhibited in London at the Great Exhibition of 1851 and its use became widespread in Britain soon afterwards. The Wicks caster of 1878, producing complete sets (founts) of type could turn out some 60,000 types an hour. But every letter to be printed still had to be taken out of the type case by the compositor, placed in his hand-held composing stick and transferred to the galley in batches to be made ready for printing. After printing, each type had to be returned to its place in the type case. A veritable army of compositors was needed to satisfy the voracious appetite of the rotary presses. Many inventions were devised to mechanize this process, but none was really satisfactory. The first approach was to assemble and distribute the types themselves mechanically. The first record of such a machine is of that patented by the London-based New Yorker Dr William Church in 1822, but it seems never to have been constructed. Many later forms were tried and some gave useful service, but none was at the same time reliable and mechanized the whole process. One of the most successful was the Kastenbein, used by *The Times* for many years. Here, the type was assembled mechanically for printing and melted down after use; a fresh fount of type was then cast by a Wicks caster. The problem was solved by following a different route – assembling moulds or matrices of the required characters, casting the type and returning the matrices to their original positions ready for use again. It was a Württemberg watchmaker and immigrant to the USA, Ottmar Mergenthaler, who achieved the breakthrough with the Linotype machine, which produced casts of slugs, or complete lines of type. The first model was completed in 1885 and was first used in the office of the *New York Tribune* the following year. An improved model came out in 1890 and hundreds were constructed for sale in the USA, Great Britain and later in Europe. By 1900, 21 London dailies were being set on the Linotype.

At much the same time, in 1887, the Monotype machine was developed by the American Tolbert Lanston. Its keyboard unit produced a strip of paper tape with holes punched in patterns corresponding to the characters required and this controlled the assembly of matrices in the caster, from which types cast singly were ejected and fed to an assembly point until a complete line had been formed. It became a commercial proposition in 1894 and an advanced model was installed by Cassell & Co. in London in 1900. Some hand setting carried on until the First World War, but from the 1920s Monotype and Linotype reigned supreme – the former mainly for book work, the latter for newspapers.

For the first half of this century hot-metal type setting and relief (letterpress) printing produced the vast majority of the public's reading matter. For long runs of illustrated magazines gravure printing was more suitable. The *Illustrated London News* was the first in which pictures and text were printed simultaneously by gravure, from 1913. This process has still not entirely given way to offset lithography for the 'popular illustrateds'.

For half a century printing seemed quite settled in its hot-metal, letterpress ways and few would have predicted that in a relatively few years a revolution in technology would sweep aside these time-honoured processes and almost do away with metal type. The new photocomposing process was envisaged in patents even in the nineteenth century and inventors were seeking ways of dispensing with metal types in the 1920s, but it was not until after the Second World War that commercially practical machines became available. Photocomposing began to be used more widely after improvements in lithographic printing were made and quicker, cheaper methods were found of making letterpress printing blocks from photographs. In the 1950s some hot-metal machines were adapted to become photosetters by replacing the matrix with a photographic negative. Meanwhile, a new range of specially designed machines, such as the Lumitype of French origin, was being developed, in which the keyboard produced a punched paper tape with coded instructions for operating a photographic printer.

Computers were introduced in order to decide about line length, justifying the right-hand margin, spacing between words and so on, decisions previously taken by the compositor. Two kinds of photosetter emerged, one where the keyboard generated a paper or magnetic tape that controlled the operation of the photo-matrix and one where the operator controlled this direct. A beam of light created the images of the characters on the photographic film or paper. The next stage used a system of electronics to build up the images of the characters in a cathode-ray tube, as in the Linotron 505. Beyond that, the character image could be stored in the form of magnetic impulses in a computer (digital storage), cutting out the need for a photographic matrix. Here, we have reached the stage where a reporter with a keyboard unit can generate impulses through a telephone line and the computer-controlled photosetter can produce the copy from which, via a process camera, a printing plate can be made for printing by offset lithography. This method dispenses not only with metal type, but with the compositor himself. The printing industry has traditionally viewed technical advances with suspicion, understandably when they cause job losses, and the new computerized photosetters have certainly met their share of opposition. But, properly managed, the advance has been beneficial and hot-metal technology belongs to the past.

COLOUR PRINTING

The nineteenth century also saw the flowering of colour printing. Until that time, nearly all the colour that appeared in books was applied by hand, but then several processes were developed for colour printing on a commercial scale. They all, like the Baxter prints and chromolithographs, entailed repeated printing in exact register of each colour that constituted the final picture, a laborious and expensive procedure. Technical progress in this field depended on a proper understanding of the theoretical background of the nature of colours and how they could be produced by mixing the three primary colours, red, blue, and yellow. Some progress was made in applying the three-colour theory to photography, and with that and the invention of the half-tone screen, the way was clear to convince the eye that it sees the range of colours in the original when what it really sees is a collection of dots of yellow, magenta (red) and cyan (blue) ink (known as *process* colours) of varying intensities. A fourth colour was added later – black, to give greater depth of tone. The first example of three-colour printing was exhibited by F. E. Ives at Philadelphia in 1885, and it became a practical proposition from 1892. It could be applied to any of the three printing processes. Four plates are made by photographing the coloured original four times through a half-tone screen, using a different coloured filter each time – blue, green, red and yellow for printing successively in yellow, magenta, cyan and black ink. More recently, the orthodox process camera began to be replaced by more sophisticated equipment, that is, by the photoscanner. It was in 1937 that Alexander Murray working for Eastman Kodak built the first scanner, in which a scanning head picked up light from a coloured transparency and a photomultiplier converted this into electronic signals. From these, various kinds of output can be derived, according to the printing process. Laser light is now used to expose the output film.

OFFICE PRINTING

A kind of typesetting has been carried out for over a century by non-printing operatives – on the typewriter. Ideas for 'writing machines' go back to the eighteenth century and many designs were conceived, including some that resulted in embossed marks on paper to enable the blind to read by touch. The first commercially successful typewriter was evolved by the American inventors Christopher Latham Sholes and Carlos Glidden to type letters on to paper. They persuaded the firm of gunmakers Remington to manufacture it and the job was given to their sewing machine division. Hence the first practical typewriter, when it appeared in 1873, bore a resemblance to a sewing machine. Development was at first slow, but during the 1880s many different designs appeared. Most, like the Sholes and Glidden, used type bars to carry the type, striking down, up or sideways on to the paper. Others carried the types on a wheel or a cylinder, but it was the Underwood No. 1 of 1897 with side-strike type bars, enabling the typist to see what she was typing, that proved to be the standard arrangement for many years to come. Sholes and Glidden first arranged the keys alphabetically, but found the type bars jammed. Sholes's mathematician brother-in-law was asked to rearrange the keyboard so that the movements of the type bars did not conflict, bearing in mind the combinations of letters occurring in the English language and their frequency of occurrence. The result was the 'qwerty' arrangement used ever since in spite of spasmodic attempts to reach a more logical layout.

The New York YMCA bought six Remingtons in 1881 to begin classes for young ladies – eight turned up for the first lesson. Five years later there were an estimated 60,000 female typists in the USA. The typewriter, used jointly with shorthand, not only revolutionized office work; it brought about a social revolution providing millions of women with the possibility of paid employment where none existed before. Sholes said: 'I feel that I have done something for the women who have always had to work so hard. This will more easily enable them to earn a living.'

The first of two major improvements to the typewriter was the application of electric drive. The first electric typewriter was made by Blickensderfer as early as 1902 (he had invented the portable in 1887). Other designs followed, but it was not generally in use in offices until the late 1950s. Then came a radical departure: the IBM 72 of 1961 abandoned the type bars and moving carriage and substituted a moving 'golf ball' typehead. Another improvement was the introduction of automatic control, first mechanically with punched paper tape and then electronically by the computer. The former became common in the 1960s but was soon replaced by computer control, and the typewriter becomes the word-processor, with no difference in principle between it and a computer-assisted composing machine.

The single copy has rarely sufficed in offices and various processes were developed to make additional copies. James Watt invented a copying machine in 1780, whereby letters written in a special ink could be transferred to dampened tissue paper by pressure, and in 1806, Ralph Wedgwood patented his Stylographic Writer using 'carbonic' or 'carbonated' paper – the first recorded appearance of carbon paper. Ink transfer methods were in use until the present century, but for greater numbers of copies the stencil duplicator provided the answer. The first British patent was taken out in 1874 by Eugenio de Zuccato, employing varnished and then wax-coated paper. In 1881 the name of Gestetner enters the scene with his cyclostyle or wheeled pen for preparing the stencil. The stencil could be easily printed off on the single-cylinder duplicating machine invented by Harry W. Loowe in 1897 or the double-cylinder machine of Gestetner developed in 1900–2.

The stencil and duplicator were a form of screen printing. In this process, ink is forced through the fine mesh of a silk screen, partly blanked off by a stencil plate. It was used in China and Japan long before printing arrived in Western Europe. In nineteenth-century France, it was adopted by a manufacturer in Lyons to print textiles. The process found other applications and in the 1930s in Great Britain and the USA, it was used to print on to a wide variety of materials, such as glass, wood and plastics, with various shapes. It was originally an entirely hand-process, but now screens are prepared by photosensitization and printing carried out on automatic presses. Much packaging, like the ubiquitous plastic bag, receives its striking and colourful designs by means of screen printing.

The stencil and duplicator gave good service for many years and are still in use, but they have been in a large measure replaced by two processes. One is the small offset litho machine, introduced in 1927 and marketed by Rotaprint. The other was the electrostatic copying machine which emerged in the 1950s under the name of xerography (Greek: 'dry writing') and heralded the photocopying revolution. Here, an image of the document is reflected or projected on to an electrically-charged selenium surface. Light from the non-text areas of the document removes the charge, leaving the surface charged where the image of the text falls on it. Carbon powder with an opposite electric charge is cascaded on to the selenium, adhering to the charged portions. This powder is then transferred to paper, also electrically charged, and is fused on to the paper by heating, leaving a permanent image of the original text.

OPTICAL CHARACTER RECOGNITION

So far, all the typesetting methods we have considered have required human agency for the conversion of, say, an author's manuscript or typescript into type. Now, even that has been rendered unnecessary by the process known as optical character recognition (OCR). Here, the original copy is scanned optically and the light signals converted into electronic impulses that operate a computer-controlled photosetter. However, there are strict limitations over the nature of the copy the scanner can 'read' and a human agency is likely to be needed for most copy for the foreseeable future.

At the receiving end, too, the long reign of printed page and book appears threatened. For many purposes information is held in remote computer databases and may be called up when required on a computer terminal or microcomputer. The vast collection of references to scientific literature hitherto available only in the form of published abstract journals is now stored in this way; library catalogues are consulted on a VDU (visual display unit) and the public can access useful information on their television screens at home using broadcast Viewdata services. But all this is for reference to specific information. For most reading purposes, ranging from serious study to light entertainment, the printed page and the book seem set to survive.

FURTHER READING

Adler, M. H. *The writing machine* (George Allen & Unwin, London, 1973)
Barber, C. L. *The story of language*, rev. edn (Pan Books, London, 1979)
Dudek, L. *Literature and the press* (Ryerson Press, Toronto, 1960)
Kubler, G. A. *A new history of stereotyping* (privately published, New York, 1941)

Moran, J. *The composition of reading matter* (Wace, London, 1964)
—— *Printing presses* (Faber & Faber, London, 1973)
Proudfoot, W. B. *The origin of stencil duplicating* (Hutchinson, London, 1972)
Steinberg, S. H. *Five hundred years of printing* 3rd edn (Penguin Books, London, 1974)
Wakeman, G. V. *Victorian book illustration* (David & Charles, Newton Abbot, 1973)
Whalley, J. *Writing implements and accessories* (David & Charles, Newton Abbot, 1975)

15

INFORMATION: TIMEKEEPING, COMPUTING, TELECOMMUNICATIONS AND AUDIOVISUAL TECHNOLOGIES

HERBERT OHLMAN

INTRODUCTION

Information technologies (which we will take to include communications) may be regarded as extensions of human sensory-motor capabilities: particularly, the senses of sight and hearing; and memory, speech and manipulative skills, such as writing.

For seeing, up until a few hundred years ago there were few means of enhancing human vision. However, in prehistory man had discovered means to represent his own activities and those things important in his life, such as game animals. The earliest representations have been found on the walls of caves: in Spain in 1879, and in France beginning in 1896. The most important find was in the Lascaux caves in France in 1940. Whereas the Spanish paintings in the Altamira caves have been dated between 8000 and 3000 BC, those in Lascaux are the earliest human creations ever found, dating from 15,000 to 13,000 BC. Artists covered and recovered these walls, not with primitive scratched drawings, but with paintings sophisticated in conception and masterful in execution; their longevity is a tribute to their choice of materials. This evidence indicates that the mastery of materials and techniques needed to record accurately observed experiences, as well as the urge to make such records, occurred much earlier than other marks of civilization.

Also, in prehistory the heavens were mapped, and the movements of sun, moon and planets tracked with only the naked eye. However, it was only with the scientific revolution in Europe, and particularly in the seminal seventeenth century that the inventions of the telescope by Galileo, and the microscope by Leeuwenhoek and Hooke, extended the range of human vision outwards to the macrocosm, and inwards into the microcosm.

For hearing, artificial aids also may be traced into prehistory. Human cries and shouts to signal the finding of game or warn of danger were soon supplemented by beating sticks on hollow logs (the drum), and blowing on hollow shells (the trumpet). The cries of primitive man eventually led to speech, and the noisemakers to music. The Greeks were among the earliest to understand the mathematical basis of music, and developed scales and means of recording melodies. Also, they built amphitheatres for the performance of music and drama, concentrating and reinforcing sound towards their audiences. Early aural inventions were the ear trumpet to help the hard of hearing and, its inverse, the megaphone, to direct the human voice towards an assembled crowd.

Human memory was vastly extended by the inventions of counting and writing. With the coming of writing, the human race no longer had to depend upon oral-history transmission from generation to generation. The first counting was probably done on the fingers (our words digit and digital indicate this history), and with piles of small, natural objects such as pebbles (our words calculate and calculus). To count beyond ten, toes and other bodily members were added, and then the tally stick was brought into use. The recording of numerical quantities goes back to the Babylonians who devised cuneiform 'writing' on clay tablets before 3000 BC. Even older are positional notation and the use of zero, probably developed first in northern India, and passed on to the West via the Arabic culture which flourished from AD 600 to 1200. The earliest mechanical aid, the abacus, goes back at least to 450 BC, and was in use in some form in the East thousands of years earlier. It was the first digital computer, enabling users to add and subtract quantities using beads as counters.

THE EVOLUTION OF INFORMATION TECHNOLOGIES

The scheme of organization depicted in Table 15.1 will help in understanding the evolution of information technologies. The table is divided horizontally into three parts:

1. Theoretical disciplines and discoveries which underlie invention, such as philosophy, mathematics, logic and psychology; and the pure sciences, particularly physics and chemistry;

2. Applied sciences, engineering and invention; particularly those based upon the phenomena of optics, electricity and magnetism.

PART FOUR: COMMUNICATION AND CALCULATION

Table 15.1: The evolution of information technologies.

		SCIENTIFIC BASIS	Up to 1700		1700–1799		1800–1849	
THEORIES, DISCOVERIES, PURE SCIENCES		PHILOSOPHY MATHEMATICS LOGIC	1600 Napier	logarithms			1847 Boole	algebra
		PHYSICS CHEMISTRY PSYCHOPHYSIOLOGY	1600 Galileo 1420 Brunelleschi	pendulum perspective	1747 Franklin 1725 Schuttz	current electricity photo chem.	1820 Oerstead 1821 Faraday 1821 Henry 1824 Roget	electromag. electromag. electromag. persit. of vision
		MECHANICS ACOUSTICS	c.450 B.C. 725 1335 1642 Pascal 1656 Huygens c.1675	abacus mech. clock wt.-driven clock mechanical calculator pendulum clock pocket watch	1762 Harrison c.1790	marine chronometer wristwatch	1805 Jacquard 1822 Babbage 1833 Babbage	punch cards Diff. Engine Analyt. Engine
		OPTICS ELECTRO-OPTICS	10th C. Alhazen 1608–9 Lippershey Galileo 1660 Huygens 17th C. Locuwenhock Hooke	camera obscura telescope magic lantern microscope	1790 Chappe	semaphore	1826 Niepce 1826 Paris 1832–33 1834 Horner 1835 Fox Talbot 1838 Wheatstone 1839 Daguerre 1839 Dancer 1849 Brewster	photography thanmatrope stroboscope zoetrope negative/positive sterescope photography micro-photography
APPLIED SCIENCES, ENGINEERING INVENTION		ELECTRICITY MAGNETISM ELECTRONICS					1800 Volta 1809 von Soemmerling 1837 Cooke & Wheatstone 1837 Morse 1840 Bain 1843 Bain 1847 Bakewell	battery telegraph telegraph telegraph electric clock facsimile telegraph facsimile
		ELECTRO-MAGNETICS						
SERVICE INNOVATIONS		MARKETING MANAGEMENT SOFTWARE						

Table 15.1 (cont.).

1850–1899		1900–1940		1941–1960		1961–1975		1976–1985	
		1936 Turing	universal comp. mach stored-prog. concept						
		1944 von Neumann		1948 Shannon	information theory				
1855 Maxwell	light primaries	1924 Appleton & Barnett	ionospheric reflection	1945 Clarke	commun. satellites				
1860 Maxwell	electromag.								
1883 Edison	Edison effect								
1895 Roentgen	X-rays								
1878 Edison	phonograph	1931 Zuse	binary calc'r	1957 USSR	sputnik satellite				
1895 Berliner	disk phonog.								
1859 Scott-Archer	wet-collodion process	1926–27	sound-motion-pictures	1948 Gabor	holography	1962 Land	Polacolor		
				1948 Land	Polaroid camera				
1878 Muybridge	motion photograph	1930s Mannes & Godowsky	colour photog.	1950 Samatu	Filmorex				
1882 Marey	mot. pic. camera	1938 Carlson	electro-photog.	1960 Mairnani Townes	laser	1970s Philips	videodisc	1983 Philips, Sony	Compact Disk
1889 Eastman	roll film								
1892 Edison	kinetoscope								
1895 Lumière bros.	mot. pic. system								
1876 Bell; Gray	telephone	1904 Fleming	diode	1942–46 Univ. Pa.	ENIAC computer	1962	MOS integ. cct.	1976 SwTPC	micro-computer
1880 Hollerith	punched-card machines	1904 de Forest	Audion	1943 Turing	Colossus comp.	1963 Philips	compact cass.	1977 Apple; Commodore	micro comp.
1884 Nipkow	scanning disk	1923 Zworykin	iconoscope	1944–45 M.I.T.	Whirlwind	1963	electronic calc.	1978 IBM	floppy disks
1897 Braun	cath.-ray tube	1925 Jenkins; Baird	mechanical television	1946 Turing	ACE	1971	digital watch	1979 Sony	Walkman cassette player
1898 Poulson	magnetic recording	1927 Farnsworth	electronic television	1947 Eckert & Manchly	UNIVAC	1973 Scelbi	micro-computer	1981 Osborne	portable microcomp.
		1928 Kleinschmidt	tele-typewriter	1948 Goldmark	LP record	1975 Altair	micro-computer	1981 IBM	PC micro-computer
		1929 Morrison	quartz clock	1948 Libby	atomic clock				
		1931 Blumlein	stereo-phonics	1948 Bardeen	transistor				
		1937 Reeves	PCM	1949 Forrester	mag. core store				
		1937 Stibitz	Complex Computer	1956 IBM	mag. disk store				
		1939–44 Aitken	Mark 1 computer	1956 Poniatoff	videotape recording				
				1959 Kilby	integ. cct				
1887 Hertz	e/m propog.	1906 Fessenden	radio broadcasting	1954 Townes 1960	maser commun. satellites	1963 Hughes Syncom	geo-stationary comm. sat.		
1890 Branley	coherer	1933 Jansky	radio astronomy						
1896 Marconi	wireless telegraphy	1933 Armstrong	FM radio						
		1935 Watson-Watt	radar						
1856 Field	transatlantic cable	1920	sched. radio broadcast	1953 NTSC	colour TV standard	1964 Kemany & Kurtz	BASIC	1979 Bricklian & Frankston	Visicalc
1860	pony Express	1928 GE	first TV broadcast	1954 Backus 1959	FORTRAN COBOL	1969 Wirth 1974–79	Pascal Ada	1980	teletex
		1933 Germany	first telex						

3. Service innovations, particularly social inventions such as trade, finance, and advertising.

Vertically, the chart is divided into broad time periods. Because most information-related inventions are relatively recent, we have put all early developments under the heading 'up to 1800', and divided the nineteenth and twentieth centuries into six segments of ever-decreasing duration.

There is also a three-fold classification of discovery and invention: first, by broad discipline or skill; second, by specific scientific or technological area; and third, by earliest reported discovery or invention date. Logically, there should be a cause–effect relationship from theory/discovery to invention to dissemination; however, in many cases an invention preceded theoretical understanding of how it works, as well as practical applications. There are other examples where imagination, expressed through writings and drawings of a fantasy or science-fiction nature, preceded either.

The day of the fabled lone inventor working for decades in his basement or garage, and by an unlikely combination of skill and luck getting his invention successfully to market, is probably gone for ever. In truth, most successful inventors of the nineteenth and early twentieth century had partners or assistants, but it was not until the organizational genius of Edison, with his 'invention factory', that we see the beginning of the end of cottage-industry invention. Today, with few exceptions – mostly in microcomputer software – inventions are the result of corporate efforts. Where individuals receive credit, it usually goes to a team of from three to ten. In this article and on the chart, more than 100 of the most important discoveries, inventions and service innovations are listed; hundreds of others could not be included in a limited survey. But how to choose which inventions are the most important ones is a contentious question. We all have our favourites and what may seem a frivolous luxury to one may be thought a necessity to another. However, it is possible to suggest objective criteria with which to rank inventions. We will not consider the entire range of inventions, but only those related to information technologies (including communications in the older, generic sense, where it includes transportation):

1. Suppose we ask, 'How much of our life is affected by a certain invention?' By this criterion, the most important would be the generation and distribution of electricity, and particularly the electric light; television, to which the average family devotes several hours every day; and the motor car and road system, which has given personal transport dominance over older, public forms such as the railway.

2. Another pertinent question would be, 'How much of our income is spent using an invention?' Here the motor car would no doubt come out on top; typically, it accounts for more than 10 per cent of disposable personal income.

3. Still another measure might be, 'What percentage of the population own the invention?' Here, the radio receiver might come out on top, probably followed by television, and then the motor car.

4. From a less personal point of view, we could ask, 'What would be the effect on society if we had to do without the invention?' Here, electricity, the telephone and motorized transport would be the most significant. However, newer inventions, such as the computer and the aeroplane, although not used by most people directly, greatly affect their lives.

5. Another way, as shown in Table 15.2, is in terms of how many people are employed world-wide in industries which have developed from electrical or electronic inventions. To the ten largest (category III), we could add clock and watch production; photography and motion pictures; data (nonvoice) communications; and the transportation industries, particularly automotive and aerospace.

THE TIMING OF INVENTIONS

An invention must be considered in the context of its time. No matter how great its ingenuity or ultimate worth, it cannot succeed without certain prerequisites:

1. The necessary materials, elements and components must be available; for example, transistors only became possible after we understood the solid state of materials, and integrated circuits only after transistors became common and cheap.

2. There must be entrepreneurs or venture capitalists ready to fund the development, early production and dissemination of the product; there is always a high risk of failure no matter how much of a breakthrough the invention seems to offer, and conservatism of vested interests must be overcome.

3. There must be an unmet, even though unexpressed, need for the invention. And there must be opinion leaders ready to take a chance on accepting a new idea, even though neighbours may scoff at their foolishness – who could have been brave enough to sign a subscription for the first telephone?

It is interesting that none of these criteria relate to the theory underlying an invention; inventions often precede their theoretical explanation. For example, for several centuries (from Franklin to Marconi) electricity was the basis of many inventions, but the force behind it – the electron – was only discovered by J. J. Thompson at the beginning of this century. On the other hand, Maxwell's elegant theory of electromagnetism preceded Hertz's demonstration of wireless communications.

Table 15.2: The expansion of electronics and its effect on industry.

Date	Inventor	Category I Small (approx. 1–10,000)	Category II Medium (approx. 10,000–100,000)	Category III Large (over 100,000)
1831	Faraday			Electricity generation
1831	Faraday	Transformers		
1837	Cooke *et al.*	Relays		
1837	Davenport		Electric motors	
1839	Dancer	Microfilming		
1839	Grove	Fuel cells		
1843	Bain	Facsimile repro.		
1857	Wray	Mercury arc lamps		
1860	Plante	Accumulators		
1868	Léclanché		Dry batteries	
1876	Fitzgerald	Paper capacitors		
1876	Bell			Telephones
1877	Edison	Carbon microphones	Phonograph	
1877	Siemens	Loudspeakers		
1878	Swan/Edison *et al.*			Electric lamps
1885	Bradley		Carbon composition resistors	
1887	Berliner		Gramophones	
1895	Rontgen	X-rays		
1896	Marconi			Radio
1897	Braun	Cathode ray oscillographs		
1898	Poulsen		Magnetic recorders	
1901	Cooper-Hewitt	Fluorescent lamps		
1906	de Forest		Three-electrode tubes	
1908	Cahill	Electric organs		
1910	Claude	Neon lamps		
1915	Longevin	Sonar		
1918	Northrup	Induction heating		
1919	Zworykin & Farnsworth			Television
1924	Appleton, Briet *et al.*			Radar
1926	Round	Screened grid tubes		
1927	Fox Movietone News		Sound on film recording	
1928	Tellegen-Holst	Pentode tubes		
1932	Knoll, Ruska	Transmission electron microscope		
1933	Jansky	Radio astronomy		
1934	Dreyer	Liquid crystals		
1935	Knoll *et al.*	Scanning electron microscope		
1937	Carlson		Xerography	
1938	Frisch & Meitner		Nuclear fission	
1939	Hann, Varian Bros	Klystrons		
1939	Aitken and IBM			Computers
1940	Centralab		Thick-film circuits	

Table 15.2 (cont.)

		Category I	Category II	Category III
1943	Kompfner et al.	Travelling wave tubes		
1943	Eisler		Printed wiring	
1948	Gabor	Holography		
1948	Bardeen et al.			Transistors
1950s	MIT, Bell	Modems		
1953	Townes et al.	MASER		
1953	Mallina et al.	Wire wrapping connections		
1954	Chapin et al.		Solar batteries	
1958	Ampex	Video tape-recorders		
1958	Various	Explorer, Vanguard, Pioneer 1, etc. (satellites)		
1958	Schalow	LASER		
1959	Bell Labs.	Tantalum thin-film circuits		
1959	Hoeni		Planar process ICs	
1960	Bell Labs.		Electronic telephone exchanges	
1960	Various			Digital & linear integrated circuits
1960	Allen & Gibbons	LEDs		
1961	Digital Equip.		Minicomputers	
1963	Bell		Electronic calculators	
1963	Rown-Sittig	Surface acoustic wave devices		
1964	Baran	Packet-switching communications		
1964	Lepselter	Beam lead connections		
1964	Various	Telemedicine		
1965	Various	Intelstat Satellites		
1966	Kao and Hockham	Optical fibres		
1967	Dolby		Audio noise-reduction system	
1969	Bobeck et al.	Magnetic bubbles		
1970	Boyle Smith	Charge coupled devices		
1972	Intel			Microcomputers
1972	Magnavox		Video games	
1976	Various	Marisat satellites		
1977	Sinclair	Pocket TV		
1978	Bell	Light-powered telephone		
1978	Philips	Laser optical recorder		
1978	Intel	One megabit bubble memory		
1979	Sony	CCD colour TV camera		
1980	NEC Toshiba		256 K RAM	
1980	STC	Fibre optic submarine cable		
1982	AERE	Fission track autoradiography		

Courtesy of Mr G. W. A. Dummer and Pergamon Press

Every invention of the first rank – in the context of this chapter, clocks, calculators, computers, photography, the phonograph, motion pictures, the telephone, radio and television, and satellites – has spread rapidly from its origin, and become used by the majority of the population instead of just being the plaything of a few devotees.

Also, people do not actually 'own' most of these inventions. The only ones which are really self-contained are clocks, watches and calculators. Cameras, radios, television, the telephone and satellites are useless without elaborate support services. For example, the telephone instrument, which is all that we see and that most people are conscious of, is only the terminal apparatus of what is perhaps the most complex and capital-intensive industry ever known. Similarly, whereas we own motor cars, to use them we must have a highly developed and maintained infrastructure of highways, petrol stations, and repair shops.

TIMEKEEPING

Mankind first developed a sense of time from observations of nature. For the short term, he observed the movement of heavenly bodies – the sun and moon held particular importance, heralding the seasons and the months. For the long term, birth and death events – of themselves and their livestock – marked the passage of time. One of the earliest inventions was the astrolabe, which astronomers used to track stars and planets. The first such instrument may have been made in the second century by the greatest astronomer of ancient times, the Greek Hipparchus, and was brought to perfection by the Arabs.

Early artificial means of timekeeping to provide an estimate of the hour were all analogue in nature, whether passive, like the sundial, or dynamic, like the sandglass or water-clock, which measured time by rate of flow. The sundial probably began with a stick thrust into the ground; the position of its shadow corresponded to the hour of the day. Very elaborate sundials have been constructed which compensate for the sun's relative position during the year, and ingenious pocket versions have been popular from time to time. However, all such instruments are worthless if the day is cloudy, and after the sun goes down.

Therefore, particularly for stargazers, independent means of time estimation were important. The simple sandglass, in which sand is made to run through a small opening, was adequate only for short durations. However, running water can power a mechanism indefinitely. The greatest water-clock ever made was a building-sized astronomical device, constructed by Su Sung in China in 1094, to simulate the movements of sun, moon and the principal stars. Chinese philosophers thought that because water flows with perfect evenness, it is the best basis for timekeeping. However, in this belief they were wrong; the secret of accurate timekeeping is to generate and count regular beats, which is a digital rather than an analogue process. This may be surprising, and not just to the ancients in China, because except for intra-atomic events time can be regarded as a continuous phenomenon.

The mechanical clock

Three elements necessary to the invention of the mechanical clock were brought together early in the fourteenth century:

1. A weight-driven mechanism to provide constant force; unlike water-clocks, such a mechanism is relatively independent of temperature.
2. Energy (gravity pulling the weight) transmitted to a train of wheels, or better gears (Hero of Alexandria, one of the earliest known inventors, developed the principle of the gear wheel in the second century).
3. Back-and-forth motion, such as provided by a controlling device or regulator whose oscillations (beats) 'keep track of time', with an escapement to count the beats by blocking and releasing the wheel train (early mechanical clocks did not separate these functions).

In summary, the essence of clockwork is the conversion of stored energy into regular movements in equal intervals; in modern terms, analogue-to-digital conversion takes place.

The earliest mechanical clock on record was said to have been built in China in 725 by I Hsing and Liang Ling-tsan. In Europe, the earliest weight-driven clocks were built in the fourteenth century: they did not indicate time visually, but struck a bell every hour. One was built for Wells Cathedral in Somerset in 1335. The first public clock that struck the hours was built in Milan in the same year, and late in that century clocks appeared for domestic use. Each beat of these clocks was several seconds long, so their accuracy was poor. The first astronomical clock, built by Giovanni de'Dondi between 1348 and 1362, was well in advance of its time, showing the motion of the sun, moon and planets and providing a perpetual calendar.

Spring-driven clocks appeared in the fifteenth century, but were even less accurate because a spring could not provide a constant driving force. The fusee wheel (see p. 392), a grooved, conical spindle around which a cord was wrapped to compensate for this irregularity, was first used in 1430 for Philip the Good's Burgundy clock. By the middle of the fifteenth century, reasonably accurate spring-driven clocks became available, which could be made much smaller. About 1500, Peter Henlein, a German locksmith, built what is said to be the first portable clock.

In 1584, Joost Burgi invented the cross-beat escapement, which increased accuracy almost a hundred times. However, this was soon made obsolete by the crucial invention of the pendulum at the beginning of the seventeenth century. Galileo had observed that a free-swinging weight gave a reliable and consistent period, dependent only on its length (as long as the swing was small). A pendulum 990mm (39in) long gave a period of one second, and would lose only one second a day if this length was kept within an accuracy of 0.025mm; thus, a perfect pendulum would have an accuracy of 1 part in 100,000.

Galileo did not actually build such a pendulum-driven clock; Christiaan Huygens in the Netherlands seems to have been the first, in 1656. Huygens's clock kept time to an accuracy of 10–15 seconds per day, a vast improvement on the verge-and-balance system which varied more than 15 minutes a day; in 1670, the minute hand was introduced. In 1675, Huygens invented the balance wheel which allowed further miniaturization, and led to the development of the pocket watch.

During the seventeenth and eighteenth centuries, great attention was devoted by maritime nations to improving the accuracy of timekeeping. Although ships at sea could determine their latitude by 'shooting' the sun or stars, to determine longitude they also had to know the time difference between a fixed meridian and their position at sea. Therefore, navigation in the east–west direction was very uncertain (perhaps this was why Columbus thought he had discovered India). A prize was offered by the British Admiralty for a clock which, carried aboard a rolling and pitching ship, could still keep time to an accuracy sufficient for the determination of longitude. The prize was won by John Harrison in 1762 with his No. 4 chronometer which made two beats every second.

Modern clocks and watches

About 1790, Jacquet-Droz and Leschot invented a wrist-watch in Geneva, but it was not until Adrien Philippe of Paris brought his 1842 invention of the stem-winding mechanism to Patek in Geneva in 1844 that a practical wrist-watch became possible. Today, because of its convenience and as an attractive adornment, the wrist-watch has all but supplanted the pocket watch.

A clock based on a tuning fork was invented by Louis Breguet in France in 1866, but although this further miniaturized the regulation function, a watch using this principle was not made until 1954 by a Swiss, Hetzel. Alexander Bain invented the first battery-driven electric clock in 1840, although a battery was not incorporated into a clock until 1906. In 1918, the domestic electric clock using a synchronous motor was introduced; this depended on the mains frequency for its accuracy, and stopped if the current was interrupted.

In 1929, Warren Morrison in the United States used the natural period of vibration of a quartz crystal to give new accuracy to timekeeping, better than 0.0001 second per day. A ring of quartz was piezoelectrically driven to oscillate at 100,000 vibrations per second, and the resulting electrical signal frequency-divided a number of times to drive a human-readable display.

The atomic clock was developed by William F. Libby in the United States, the first model being constructed at the National Bureau of Standards in 1948. The atomic clock was the first artificial timekeeper which was better than the stability of the earth's rotation. The functioning of present-day atomic clocks depends upon the natural resonance of caesium atoms. Since 1972, the international time standard has been the atomic second, defined as the duration of precisely 9,192,631,770 periods of the radiation corresponding to the transition between two hyperfine levels of the ground state of the caesium-133 atom. This is an accuracy of 1 part in 2×10^{-11} per month, more than ten times that of the best quartz clock. A master clock which controls up to fifty slave clocks was marketed in 1963 by Patek Philippe; it is synchronized with an atomic clock by radio signals. Since 1960, the international standard for length has also been based on atomic resonance: one metre is defined as 1,650,763.73 wavelengths in vacuum of radiation corresponding to the orange-red line of the krypton-86 atom.

In 1961, an electronic clock was patented by P. Vogel et Cie of Switzerland, followed by an electronic watch patent a year later. However, it was not until 1971 that the first electronic digital watch was marketed by Time Computer Corporation in the United States. This watch, called Pulsar, cost $2000, and used light-emitting diodes (LED) to show the time; because they use so much current, the wearer had to push a button to read the display. Most people are used to interpreting the two hands of an analogue timepiece; the digital watch provides a direct, numerical readout. Present digital-watch technology is quartz-based, and instead of gear-train and hands, uses an integrated circuit to calculate the time, showing numbers representing hours, minutes and seconds continuously on a low-current-drain, liquid-crystal display (LCD). Digital timepieces may also provide indications of the month, date and day of the week, alarms, corresponding time in other zones, and even a calculator function.

Inexpensive digital watches were offered in kit form in the mid-1970s by the British inventor Sir Clive Sinclair, but few could be made to work. The first acceptable consumer products came from Timex Corporation and Texas Instruments in the US, and Seiko and other companies in Japan; they offered an unbeatable relation between price and performance, and put the mass-market, mechanical watch industry in jeopardy. However, even in the most expensive digital watches, the LCD display is inelegant; therefore, many watchmakers are now using digital electronics to control the hands of conventional-looking watches.

COUNTING, CALCULATING AND COMPUTING

The first mechanical calculator (indeed, the world's first digital computer) was the abacus, whose exact origins are lost in prehistory; however, counting devices were in use in the Far East by 450 BC. An abacus is a small wooden frame into which wires are fastened; each wire represents a numerical place value (units, tens, hundreds, etc.), and is strung with a number of beads, each of which represents a digit with that place value. By manipulating the beads, addition and subtraction may be performed at a rate which in skilled hands is astonishing. In 1946, a Japanese clerk using a soroban (abacus) easily won a speed contest against an American soldier using an electric desk calculator.

Many early civilizations developed analogue or digital systems for measurement or calculation. The megaliths of Stonehenge and other stone circles were erected about 2000 BC to predict important astronomical events. Polynesians in outrigger canoes fitted with sails navigated thousands of kilometres of the Pacific Ocean using direct observation of natural phenomena, sometimes supplemented with 'maps' made of knotted cords attached to a wooden frame. In the sixteenth century, the Incas of Peru developed the *quipu*, an intricate pattern of knotted cords, for calculation.

Logarithms and the slide-rule

Michael Stifel, a mathematician of Jena, coined the term exponent in 1553; and as early as 1544, he had discovered the properties of positive and negative exponents by noting the connection between two sequences, one of which is an arithmetical progression and the other geometrical. About 1600, the Scottish mathematician John Napier invented logarithms; and the Swiss watchmaker Joost Burgi independently invented antilogarithms. The beauty of this system of calculation was that problems of multiplication were reduced to those of addition, and division reduced to subtraction. There was a ready market for the tables designed to calculate the products of sines which Napier published after twenty years of labour, because every advance in science and technology demanded greater precision. In 1624, Napier's friend and collaborator Henry Briggs published the first tables of logarithms to the base ten (decimal logarithms), which became the basis of all surveying, astronomy and navigation until recently.

Napier also developed a calculating system using his 'natural' logarithms, which came to called Napier's bones. In 1622, the English mathematician William Oughtred invented the slide-rule, an analogue calculator based on logarithmic scales. The slide-rule was much quicker than using tables, but the accuracy of practical-sized instruments was limited to three or four figures. The slide-rule in its case swinging from the belt was a trademark of students and practitioners of engineering until the mid-1970s, when the inexpensive, pocket-sized, scientific calculator made it obsolete.

Mechanical calculating machines

The invention of the first calculating (more accurately, adding) machine which did not require the memory skills of the abacus operator is attributed to the French mathematican and philosopher Blaise Pascal in 1642. His priority has been disputed: a letter written to Johannes Kepler by Wilhelm Schickhardt of Tübingen in 1624 describes a similar device, which he called a calculating clock. Pascal's calculating machine used gear wheels, each of which had the numbers 0 to 9 engraved on its face (see Figure 15.1). There was a wheel for units, one for tens, one for hundreds, and so forth; gear wheels were used for carrying. In 1671, Gottfried Wilhelm Leibniz conceived a calculating machine which could multiply numbers and such a device was built in 1694, although it proved unreliable.

The eighteenth century saw no radical innovations in calculating devices; however, a great deal was accomplished in the related field of horology, where greater timekeeping accuracy stimulated more precise machining of gears and small mechanisms. This refinement in mechanical technology would stimulate Charles Babbage to conceive the first true digital computer early in the nineteenth century.

In 1805, Joseph-Marie Jacquard in France invented the perforated cards which, linked together, were used to control weaving patterns in the Jacquard loom (see p. 823). Jacquard's system was the direct ancestor of punched cards, the key element of the tabulating machines which were the principal product of IBM during the first half of the twentieth century, and the main form of external storage for digital computers until the 1960s.

Most of the concepts which underly the modern digital computer were first enunciated by Charles Babbage, an English mathematician. Based on the division of human labour which Gaspard Riche, Baron de Prony, had used to calculate logarithms and trigonometric functions by the method of differences, he conceived of a machine to accomplish the task without human intervention. In 1822 he built a prototype of what he called the Difference Engine (see Figure 5, p. 32), which calculated the first thirty terms of a polynomial. Based on this promising demonstration, the government funded a three-year contract to build a working machine which could both calculate and print tables. The Difference Engine was never completed, although the government spent £17,000 and Babbage at least as much out of his own pocket. Apparently, this failure was not because the machinists of his day could not construct gears and levers to the required precision, but because of inadequate project management.

In 1833, Babbage had a new and much more ambitious conception, the Analytical Engine, whose attempted realization occupied much of the rest of his life. Although a mechanical solution, the only one available in his day, could never

PART FOUR: COMMUNICATION AND CALCULATION

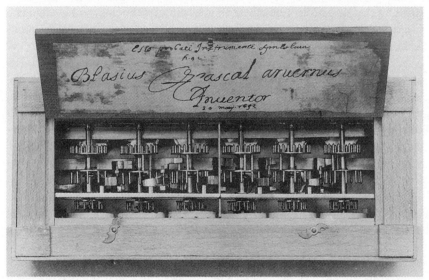

Figure 15.1: Model of Blaise Pascal's 1642 calculating machine.

have accomplished the task, the Analytical Engine was the ancestor of all current digital computers. The machine was to have four parts: a store, made of columns of wheels in which up to a thousand 50-digit numbers could be placed; the mill, where arithmetic operations would be performed by gear rotations; a transfer system composed of gears and levers; and a mechanism for number input and output. For the latter, he planned to adapt Jacquard's punched cards. Babbage also anticipated many other of the most important features of today's computers: automatic operation, external data memories, conditional operations and programing.

Between 1836 and 1840, Babbage devised some fifty user programs for the Analytical Engine, using two strings of Jacquard cards, one of which was for the arithmetic operations and the other to specify variables; the strings of cards could be looped, and could move forwards or backwards independently. In 1840, Babbage gave seminars in Italy; as a result, Luigi Menabrea wrote a paper on the Analytical Engine which, translated into English by Ada Augusta, Countess of Lovelace, is the best description we have of Babbage's ideas. (Her name has been given to Ada, the most powerful, general-purpose computer language so far developed.) In 1846, Babbage stopped working on the Analytical Engine, and began to design a new Difference Engine. Georg and Edvard Scheutz built one on the basis of a published description (which included no drawings), and showed it to Babbage in 1855; a copy of the Scheutz engine was used to calculate life expectancy tables, but proved unreliable. Babbage then tried to simplify the design of the Analytical Engine; models of the mill and printing mechanism were under construction at the time of his death in 1871.

In 1847 the English mathematician George Boole published *The Mathematical Analysis of Logic*, which showed that logic could be expressed using binary symbols. As in the binary number system, Boolean algebra has only two symbols, 0 and 1; however, instead of using them as binary digits for arithmetic computation, they are interpreted as the outcomes of logical operations, 0 denoting 'false' and 1 'true'. (In 1938, Claude Shannon of the Bell Telephone Laboratories realized that the Boolean 'true' and 'false' could be represented by the 'on' and 'off' states of bistable electronic components, and this became the basis of computer logic design.)

Herman Hollerith devised a suite of punch-card machines in 1880 to assist in processing the United States census; but the cards stored only data rather than programs. In 1896 he formed the Tabulating Machine Company, which became International Business Machines in 1957 and is now called just IBM.

Digital computers

Alan M. Turing, working at the University of Manchester, published in 1936 a paper, 'Can a Machine Think?', which has been a challenge to computer philosophers ever since. In it he proposed a theoretical 'universal machine', whose purpose was nothing less than to determine whether any mathematical statement could be proved true or not. Later, Turing became the first mathematician to be employed in the British cryptanalysis service where he developed relay-based machines to break the Enigma-enciphered transmissions of the German navy. The resulting device, Colossus, became operational in 1943; it used 1500 valves (called vacuum tubes in the US). While not a stored-program machine, Colossus may have been the world's first true automatic digital computer.

Credit for this invention has become a matter of dispute, much of the development work being hidden because of the Second World War. As early as 1931, Konrad Zuse in Germany had constructed a mechanical calculator using binary components. This 'Z1' was followed by a Z2 in 1939, but it was not until 1949 that he built the Z3 which used electromechanical relays.

In the late 1930s, Dr John V. Atanasoff sought a device to perform mathematical calculations for twenty degree candidates at Iowa State College. Rejecting mechanical solutions, he outlined ideas for memory and logic elements and built the first electronic digital computer with the assistance of Clifford Berry. It was called ABC – Atanasoff-Berry-Computer. This very early vacuum tube system seems to have had no influence on subsequent developments, but in 1974 a Federal judge ruled Dr Atanasoff the true inventor of the concepts required for a working digital computer.

In 1937, George R. Stibitz at the Bell Telephone Laboratories conceived a binary calculator using telephone relays; with S. B. Williams and other colleagues, their Complex Computer was completed in 1939, and performed useful work at Bell Labs until 1949. Between 1939 and 1944, Harold H. Aitken, of Harvard University, with engineers of the International Business Machines Corporation, built the electromechanical Automatic Sequence Controlled Calculator, or Mark I.

Between 1942 and 1946, the valve-based Electronic Numerator, Integrator and Computer (ENIAC) was built at the Moore School of the University of Pennsylvania to compute ballistic tables for the US Army Ordnance Corps. This was followed by the Electronic Discrete Variable Automatic Computer (EDVAC), built between 1947 and 1950. Also starting about 1947, J. Presper Eckert and John Mauchly of the Moore School conceived the Universal Automatic Computer (UNIVAC), which was built for the US Census Bureau and put into operation in 1951; eventually 48 UNIVAC I computers were made. Their Eckert-Mauchly Corporation was absorbed into Remington Rand and Sperry Rand's Univac became the main rival to IBM for several decades.

John von Neumann, a Hungarian immigrant at the Institute for Advanced Study in Princeton, became a consultant to the atomic bomb project at Los Alamos. In 1944, Herman H. Goldstein, working for the Ordnance Corps, told von Neumann about the Moore School's work on ENIAC, and he began an active collaboration with them. A report written by von Neumann with Goldstein and Arthur W. Burks of the University of Michigan, first described the design of an electronic stored-program digital computer.

The Whirlwind computer was constructed between 1944 and 1945 at MIT's Lincoln Laboratory, under the direction of Jay W. Forrester. It was built for the

US Army Air Force to simulate flight, and was the first system to use data communications and interactive graphics.

In 1946, Alan Turing joined the UK's National Physical Laboratory (NPL), and immediately set about designing a computer. He presented NPL's executive committee with what some claim as the first detailed design for a stored-program electronic digital computer (others claim von Neumann). Turing's proposal gave a cost estimate for construction of £11,200, and envisaged a plan for national computer development. Approval was given and funding obtained for the Automatic Computing Engine (ACE), but by 1948 not a single component had been assembled, and Turing resigned. Only a Pilot ACE was built, but it had a complex, 32-bit instruction format; 800 vacuum tubes; and mercury-delay-line storage of 128 32-bit words. A few years earlier, M. V. Wilkes and his colleagues at Cambridge built the Electronic Delay Storage Computer (EDSAC), using ultrasonic storage; and F. C. Williams and his group at the University of Manchester built their Mark 1 based on Williams's cathode-ray tube (CRT) storage.

In 1948, John Bardeen, Walter Brattain and William Shockley of the Bell Telephone Laboratories published the results of their work on solid-state electronic devices, which began just after the war in 1945. This paper described the invention of the point-contact transistor, which has almost completely displaced the valve as the active component in electronic circuits. For this work, the three shared a Nobel Prize in 1956. Also in 1948, Claude Shannon of Bell Labs developed his mathematical theory of communication (also called information theory), the foundation of our understanding of digital transmission. In 1949, Forrester invented magnetic-core storage for the Whirlwind, which became the standard internal memory for large-scale digital computers until semiconductor memory was introduced in the mid-1960s.

Computer operation and programing

During the 1950s, medium to large-scale systems, known as mainframes, were introduced for business use by IBM, Univac and others; they cost from hundreds of thousands to millions of dollars; required special housing and air-conditioning; demanded a sizeable staff; and called for a high degree of security. From the beginning they were operated as a closed shop; that is, users never got near the actual hardware. Punched cards or paper tape were the primary input media; users entered their programs and data by keypunching. A deck of cards or roll of paper tape had to be physically delivered to the mainframe location. Hours, sometimes days, later – if there were many big jobs in the queue – users would get their output, usually in the form of fan-folded paper printouts. Frequently, finding some slight mistake, they would have to resubmit the whole job again.

The earliest computers had to be programmed in machine language, which meant that both instructions and data had to be input in binary form, and the location of each address in memory had to be specified (absolute addressing). For humans this procedure is extremely tedious and prone to error. A big improvement was assembly language, which assigned alphabetical codes to each instruction, allowed the use of decimal numbers, and permitted relative addressing. However, the relationship between the lines of code and the machine's memory was still one-to-one. Many users wanted to be able to specify their instructions to the computer in a language closer to algebra, and in 1954, John Backus of IBM published the first version of FORTRAN (Formula Translator), the first 'higher-level' computer language. In 1959, under the urging of the US Department of Defense, a standardized higher-level language for business use was developed, known as COBOL (Common Business-Oriented Language). By the mid-1960s, 90 per cent of all scientific and engineering programs were being written in FORTRAN, and most business data-processing in COBOL.

Even though personnel responsible for running mainframes believed that they must maintain closed-shop service for security reasons, an important advance in user convenience was made in the 1960s – remote job entry (RJE). RJE was still batch-processing and users had to wait several hours for their results. However, instead of having to carry decks of cards and pick up print-outs at the mainframe, they could go to a nearby RJE station where a card reader and high-speed printer were connected by dedicated line to the central computer.

Most mainframe computers sold to businesses were not intended to be used interactively by anyone except their operators, who sat at a large console composed of a video display unit (VDU) and keyboard. Operators monitored the job-stream, and mounted magnetic tapes on which user programs and data were stored when the VDU alerted them. Computing was still a scarce resource, and users were charged for each task required to fulfil their job requests. The introduction of interactive access was largely user-driven; users put pressure on data processing departments to provide faster and easier access than was possible under batch processing. Also, relatively low-cost terminals (a few thousand dollars per station) had to become available to make this economically feasible. Special applications requiring real-time processing pioneered this development in the late 1950s and early 1960s – particularly air defence (the US Air Force's SAGE system) and airline reservations systems.

In 1959, Christopher Strachey in the UK proposed the idea of time-sharing mainframe computers to make better use of their capabilities for small jobs; each such job would be processed for only a second or two, but users, working interactively, would not experience noticeable delays because their interaction time is at least 1000 times slower than the computer's. In 1961, MIT demonstrated a model time-sharing system, and by the mid-1960s access to large, central computers by means of low-cost terminals connected to the public switched telephone service had become a commercial service. Terminals were usually Teletypes which, although designed for message communications, were mass-produced and inexpensive; so, rather than seeing results on a CRT screen, users would get their output printed on a roll of paper.

Within large organizations, interactive VDU terminals became available to regular users. They resembled stripped-down operators' consoles, and became known as dumb terminals because they were completely dependent on the mainframe for their operation, having no memory of their own. Communication with the mainframe was on a polled basis via high-speed, dedicated links, and transfer of information was in screen-sized blocks.

Paralleling the evolution of mainframes towards open access, rapid technological developments were taking place to provide more computing power in smaller packages for less money. In 1963, the Digital Equipment Corporation (DEC) marketed the first minicomputer, the PDP-5, which sold for $27,000; two years later the more powerful and smaller PDP-8 was made available for only $18,000. This was the beginning of a new industry oriented towards more sophisticated customers who did not need or want the hand-holding service which was a hallmark of IBM. Such users, particularly scientists and engineers, preferred an open system into which they could put their own experimental devices, and develop their own applications programs.

The micro breakthrough

This trend culminated in the development of the integrated circuit (IC), popularly known as the microchip, or just 'chip', which enables digital-circuit designers to put essential mainframe functions into one or more tiny chips of silicon. The first patent for a microelectronic (or integrated) circuit was granted to J. S. Kilby of the United States in 1959. By the early 1960s, monolithic operational amplifiers were introduced by Texas Instruments and Westinghouse. In 1962, Steven R. Hofstein and Frederick P. Heiman at the RCA Electronic Research Laboratory made the first MOS (metal-oxide semiconductor) chip. Then Fairchild Semiconductors came out with the 702 linear IC in 1964, the result of a collaboration between Robert Widlar and David Talbert; soon the 709 was put on the market, one of the early success stories in the greatest technological revolution of the twentieth century. Widlar stressed active components in preference to passive ones. In 1964, Bryant (Buck) Rogers at Fairchild invented the dual-in-line package (DIP), which has come to be the universal packaging form for ICs.

In the software area, John G. Kemeny and Thomas E. Kurtz at Dartmouth College developed a higher-level computer language for student use called

BASIC; from its beginning in 1964, BASIC was designed for interactive use via 'time-sharing' on minicomputers. BASIC has become the standard higher-level language for microcomputers, often being embodied in a ROM (read-only memory) chip, which is also known as firmware.

As with the early development of the large-scale digital computer, invention priorities are often disputed, but, according to one source, the first electronic calculator was introduced in 1963 by the Bell Punch Company of the UK. This contained discrete components rather than ICs, and was produced under licence in the US and Japan. The first American company to make electronic calculators was Universal Data Machines, which used MOS chips from Texas Instruments and assembled the devices with cheap labour in Vietnam and South America. In 1965, Texas Instruments began work on their own experimental four-function pocket calculator to be based on a single IC, and obtained a patent on it in 1967.

Westinghouse, GT&E Laboratories, RCA and Sylvania all announced CMOS (complementary metal-oxide semiconductor chips) in 1968. They are more expensive, but faster and draw less current than MOS technology; CMOS chips have made portable, battery-operated microcomputers possible.

In 1969, Niklaus Wirth of Switzerland developed a higher-level language he called Pascal, in honour of the inventor of the first mechanical calculator. Pascal was originally intended as a teaching language to instil good programming style, but has become increasingly popular for general programming use, particularly on microcomputers where it is now the main rival to BASIC. However, Pascal is usually compiled (the entire program is checked before it can be run), whereas BASIC is interpretive (each line is checked as it is input).

The idea of creating a simple, low-cost computer from MOS chips – a microcomputer-on-a-chip – seems to have occurred first to enthusiastic amateurs. In 1973, Scelbi Computer Consulting offered the Scelbi-8H as a kit for $565. It used the Intel 8008, the first 8-bit microprocessor, and had 1 kilobyte of memory. In January 1975, *Popular Electronics* magazine featured on its cover, the Altair 8800, proclaimed as the 'world's first minicomputer [sic] kit to rival commercial models'. Altair was the product of MITS, a tiny Albuquerque, New Mexico, firm, which sold it in kit form with an Intel 8080 microprocessor and a tiny 256 bytes of memory for $395. Users had to enter data and instructions bit-by-bit via sixteen toggle switches, and read the results of their computations in binary form on a corresponding array of sixteen pairs of panel lamps, but these early machines were received enthusiastically by dedicated 'hackers'. (To put this price in context, mainframes cost a million dollars or more, and minicomputers tens to hundreds of thousands of dollars.)

The microcomputer market developed rapidly, even though no software was available. Users had to develop their own applications using machine language which was (and remains) binary. However, in June 1976, Southwest Technical

Products Company offered the SwTPC M6800 with an editor and assembler, and shortly afterwards with a BASIC interpreter. By the end of 1976, the floodgates had opened; it was perhaps a fitting testament to this seminal year that Keuffel and Esser stopped production of slide-rules – donating the last one to the Smithsonian Institution in Washington.

Microcomputers remained largely an amateur market ('hacker' came to be the preferred term because users were constantly tinkering with their innards) until the Apple II and Commodore PET came on to the market in 1977. For a couple of thousand dollars, customers were offered an alphanumeric keyboard for data entry and control; a colour display (their television set); a beeper which could be programmed to play simple tunes; a pair of paddles to play video games; a high-level language (BASIC) in ROM, and an inexpensive means of storing and retrieving data and programs (their audiocassette recorder).

However, Apple and its rivals did not have much effect outside home and school environments until the floppy disk was miniaturized and reduced in price to a few hundred dollars in the late 1970s. A floppy diskette could store up to 100 kbytes and speeded access to data and programs by a factor of twenty. Thus, the entire contents of RAM (random-access memory), which disappeared when the power was turned off, could now be saved in a few seconds and loaded back quickly when needed again. Even so, the microcomputer remained a home or consumer product, largely for games, until useful applications programs were developed. In 1979, the VisiCalc program was developed by Dan Bricklinn and Bob Frankston specifically for the Apple II micro. VisiCalc enabled data to be entered on large (virtual) speadsheets; in that form, the data could be manipulated arithmetically *en masse*, with its results printed out in tabular form automatically. The business community now found that the technology of microcomputers had something unique to offer, and bought Apples in increasing numbers simply to use VisiCalc.

In 1981, Adam Osborne, an Englishman who had emigrated to the US, introduced the Osborne 1, the first complete, portable microcomputer. Selling for less than $2000, it included a full-sized keyboard, CRT display, dual floppy drives, and was 'bundled' with some of the best software available at the time: the CP/M operating system; WordStar for word processing; SuperCalc for financial analysis; and interpretive and compiled BASIC systems.

During all these developments, owners and operators of mainframes had taken little notice. Microcomputers were regarded as amazing but rather expensive toys. Mainframe services had been organized for computing specialists, and users had to employ these specialists – systems analysts and programmers – as intermediaries. In contrast, with only a few hours' instruction, a non-specialist could learn a user-friendly microcomputer software package directly relevant to his work. It soon became apparent that, because microcomputers could be purchased without being scrutinized by a committee and sent out for competitive bids – the standard procedure for items costing hundreds of thousands of dollars or more – an employee could, and would, get one to put on his desk. At last the mainframe masters had to take notice.

Almost from the beginning of the stored-programme digital computer, IBM has dominated the market to a remarkable extent. However, in no sense did they lead the introduction of micros, but entered the market only when it showed signs of maturity. Nevertheless, the IBM PC quickly took a dominant position in the market from its introduction in 1981.

Since this 'legitimatization' of micros, software of great versatility and power has been introduced for the IBM PC and 'clones', as its imitators are known. Notable among these are integrated packages, such as Lotus 1–2–3, which was marketed in 1982. Lotus combined database, spreadsheet and graphics in a single program. As a result of their convenience, versatility and low cost, micros are already far more numerous than mainframes and minis, and will account for more than half the revenue of the entire computer market in the last half of the 1980s.

Communications between mainframe and micro

Because IBM dominates the mainframe market, many of its standards have become *ad hoc* world standards. However, most of these are inappropriate or meaningless in the microcomputer world. Mainframes use large 'word' sizes; that is, each such entity they handle is usually 32 bits (even 64 bits when high-precision computation is required); in contrast, early micros used 8-bit bytes, and more recent ones – led by IBM's own PC – 16-bit bytes. A few micros, particularly those emphasizing graphics, such as Apple's Macintosh, now use 32-bit microprocessors.

Also, IBM developed its own codes to interpret the bit patterns as numbers or letters. Their Binary-Coded Decimal (BCD) system – developed in response to the needs of business to encode each decimal digit separately – was extended to handle larger character sets by EBCDIC (Extended Binary Coded Decimal Interchange Code). With BCD, only four bits were used, providing a maximum of sixteen codes; the four codes not required for the digits 0–9 were used for the decimal point and plus and minus symbols. However, no letters could be represented directly. A 6-bit code was also developed which provided 64 codes, enough to handle upper-case letters, decimal digits and some special characters. However, as text processing developed, it became apparent that more than 64 codes would be needed. EBCDIC, IBM's 8-bit system, provides 128 or 256 codes, allowing not just upper- and lower-case letters, but many special symbols and codes for control (nonprinting) purposes.

At about the same time, other manufacturers got together and under the aegis of the American Standards Association developed ASCII (American Standard Code for Information Interchange) which has since become a world standard under the auspices of the International Standards Organization (ISO). ASCII is also an 8-bit system with 256 possible codes, but assignments are entirely different from EBCDIC. However, even IBM, when it entered the microcomputer market, chose ASCII. Therefore, when any microcomputer- – IBM or another make – communicates with an IBM mainframe, code conversion is required at one end of the link or the other.

Another source of incompatibility is the type of communications used between mainframes and their terminals. In the mid-1970s IBM developed a system to handle remote dumb terminals based on using an 'intelligent' device between communications lines and terminals: the cluster controller. A mainframe could poll remote terminal clusters to see if they had a request, but remote terminals could never initiate a dialogue. Also, communication links were established to run at fairly high speeds – 9600 bits per second is common – because they had to poll every terminal on the system in turn (even if an operator was absent). The system IBM developed for this is known as Binary Synchronous Communications (BSC), and requires a synchronous modem (modulator–demodulator) at each remote cluster, as well as at the mainframe.

However, when micros began to offer communications capabilities, manufacturers chose much less expensive asynchronous modems and protocols and, taking advantage of the built-in 'intelligence' of personal computers (PCs) as terminals, let the micro initiate the conversation. Still another source of incompatibility is that micros were first developed for home and hobbyist markets. They provided graphics (often colour) and audio capabilities at the time of their greatest market growth, which were essential for games, the most popular application at that time. Most mainframes, on the other hand, did not offer graphics and sound capability at remote terminals, only text.

As more and more businesses encouraged the spread of micros into their day-to-day operations, they found that their most valuable data were often locked within mainframes. Therefore, unless someone laboriously re-entered these data from existing printouts, another way to transfer (download) from mainframe to micros had to be found. Also, the rapid development of micro technology led to smaller and smaller systems, and eventually battery-operated laptop portables. The pioneering systems were the Epson HX-20 and the Radio Shack TRS-80 Model 100, which were introduced in 1982 and 1983. The laptops provided essential microcomputer applications, including word processing and telecommunications, in a briefcase-size package for less than $1000. Users could now upload data gathered in the field to mainframes for processing.

Laptop micros can communicate with remote systems wherever there is a telephone by means of a type of modem which does not have to be physically connected to the public switched telephone network (PSTN) – the acoustic coupler. Rubber cups which fit over the telephone handset provide a temporary connection. The purpose of any modem is to convert pulses, which cannot be passed through the PSTN, to audible tones of standard modem frequencies, which can. However, this type of communication is slow – usually 300 bits per second (bps) for acoustic couplers, and 1200 or in some cases 2400 bps for directly connected modems – and sensitive to noise and others types of interference which increase error rates. At 300 bps, it would take about twenty minutes to transfer all the data on a standard double-sided IBM diskette (360 kilobytes – equivalent to about 100 typewritten pages).

The next step is to employ synchronous modems, which are used when terminals are attached to mainframes. If an IBM PC or compatible microcomputer is to be used, it must have a synchronous adapter card and 3270 emulation software. This hardware and software adds about $1000 to the cost of each PC, and provides terminal emulation, but not compatible file transfer between mainframe and micros. File transfer capability requires a coaxial communications card in the PC, which is connected to a controller. No modem is required at the PC, because the controller allows many devices to share a high-speed modem. Usually there are from 8 to 32 ports on a controller, which can support interactive terminals, remote-job-entry terminals, remote printers – and now PCs. However, special programs are required in the PCs, and often software must be added at the mainframe to support file transfer.

Another problem in attaining true emulation is that the keyboard of an IBM PC is quite different from IBM 3278 terminals, which have a large number of special function keys. Operators who are familiar with terminal keyboard layouts find it difficult to learn PC equivalents, and particularly frustrating if they have to switch back and forth. With a PC emulating a 3270-type terminal, for instance, two keys may have to be pressed simultaneously, or even a sequence of keys struck, where one keystroke would do on the terminal keyboard. Still another problem is that, as has been mentioned, IBM mainframes and terminals communicate with a character code known as EBCDIC, whereas PCs use ASCII. Therefore, EBCDIC-ASCII conversion is required at one end or the other (it is usually done at the mainframe). The best emulation systems permit not just whole files, but selected records or even fields to be downloaded for processing by PC software.

THE TELEGRAPH

In the early 1600s, the Jesuit Flamianus Strada, impressed with William Gilbert's research on magnetism, suggested that two men at a distance might communicate by the use of magnetic needles pointing towards letters on a dial; George Louis Lesage, a physicist of Geneva, did some experiments along these lines in 1774, but electricity was still too poorly understood to provide a practical means of communication.

Visual telegraphy

A French abbé, Claude Chappe, with his three brothers invented and put into use the first practical system of visual (or aerial) telegraphy, the semaphore. They had first tried, and failed, with an electrical solution. In 1790, they succeeded in sending messages over half a kilometre, using pendulums suspended from two posts. However, public demonstrations were greeted with hostility by a revolutionary populace who thought the system would enable royalists to communicate with their king. Chappe, who had spent 40,000 livres of his own on the scheme, sought police protection for his towers.

Chappe presented his invention to the French Assembly on 22 May 1792, and on 1 April 1793 the invention was favourably reported to the National Convention as a war aid. However, some vital questions were raised: would it work in fog? could secret codes be used? The Convention ordered an inquiry, for which it appropriated 6000 livres, and after a favourable committee report, Citizen Chappe was given the title of Telegraphy Engineer with the pay of a lieutenant, and ordered to construct towers from Paris to Lille. The 14km (8.7 miles) distance between towers was too far for best results, and Chappe's business management was poor; however, the line was completed in August 1794 (Figure 15.2(a)). Many messages were sent during construction, but the first to be recorded after completion was on 15 August, relaying the capture of Quesnoy from the Austrians; this was made known in Paris only one hour after troops entered the town.

To operate a semaphore, the station agent manipulated levers from ground level, whose movements were faithfully followed by wooden arms placed three metres (10ft) above, on top of a stone tower. The central member, the regulator, was 4m (13ft) long, and arms of 1.8m (6ft) were pivoted from both ends. Regulator and arms could be placed in 196 recognizable positions in addition to those where the regulator was vertical or horizontal, because Chappe had decided that the only meaningful signals were when the regulator was in an inclined position. For communication at night, a lantern was placed at each end of the arms and at each of the pivots. There were 98 positions with the regulator inclined to the left for inter-operator communications, and another 98 with the regulator inclined to the right for dispatches. Chappe formulated a code of 9999 words with the assistance of L. Delauney, and the perfected list was published with each word represented by a number. These pairs were printed in a 92-page code book each page of which carried 92 entries, giving the somewhat reduced total of 8464 word-number equivalents.

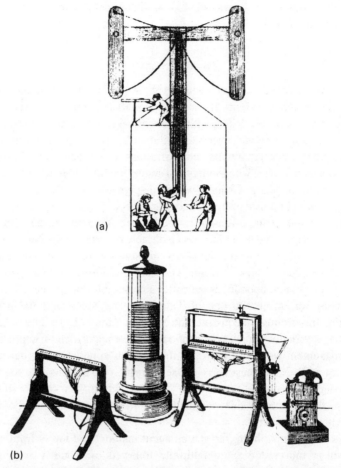

Figure 15.2: Three stages in the evolution of the telegraph: (a) Chappe's 1794 semaphore; (b) von Soemmering's 1809 electrolytic telegraph.

A station agent telegraphed the page number with his first signal, and the number of the word on the page with his second. Phrase and place-name vocabularies, each again of 92 by 92 items, were developed; to distinguish them, an initial signal indicated the code; the second, the page; and the third, the item on the page.

No public excitement over the semaphore system occurred until 1 September 1794, when Condé was recaptured. The Convention then ordered the extension of the line from Lille to Ostend, and a second line to be built to Strasbourg, which was completed in 1798 with 46 towers at a cost of 176,000 francs. Despite the practical success of these telegraph lines, they did not bring in revenue, but rather entailed great expense; maintenance and service in the eighth year of use cost 434,000 francs. Napoleon, losing enthusiasm for the

Figure 15.2: (c) Cooke and Wheatstone's 1837 5-needle telegraph.

venture, reduced the appropriation to 150,000 francs a year, and cancelled a planned Paris-Lyon line.

Claude Chappe thought that his system might benefit business, as well as give an advantage in war. During the Restoration, interest in the telegraph increased and Chappe's dream was at last realized – but he committed suicide in 1805, despondent over the slow progress and suffering from bladder trouble. However, his brothers continued to perfect the system; Abraham worked at overcoming the fog problem, anticipating British use of hydrogen fires during the Second World War. Semaphore systems remained in use well into the nineteenth century, even after the electric telegraph was developed (see p. 714).

Figuier, writing in the 1860s, claimed that aerial telegraphers could send a dispatch in two minutes from Lille to Paris (a distance of 240km (150 miles), requiring 22 stations); in three minutes from Calais (270km (168 miles) and 33 stations); in eight minutes from Brest (600km (373 miles) and 54 stations); and in twenty minutes from Toulon (more than 1000km (620 miles) and 100 stations). The equivalent rates of speed would be, respectively, 7200, 5400, 4500 and 3000kph; even the worst case is three hundred times better than the then fastest system – a relay of horses and riders, such as the famed Pony Express in the USA.

The electric telegraph

The idea of the electric telegraph preceded its practical possibility by many decades. Chappe's semaphore had demonstrated the need for rapid communication of messages, but until electricity was better understood, little progress was possible. There were attempts to do the job with static electricity, but the high voltage and very low current characteristic of this form of electricity did not allow reliable transmission beyond a few metres.

The inventor of the earliest electric telegraph did not wait upon the epoch-making discoveries of Oersted, Faraday and their contemporaries (see Chapter 6). In Germany, S. T. von Soemmerring observed that electric current passed through an acid solution caused bubbles to appear (electrolytic decomposition of water into its elements, hydrogen and oxygen). He invented a telegraph system using this principle in 1809, which used 26 parallel wires to transmit letters of the alphabet a distance of up to two miles (Figure 15.2(b)). He even designed an ingenious alarm to alert the receiving operator (this was really the first relay, although not electromechanical). However, the expense of so many conductors made the system economically impracticable.

André-Marie Ampère, in 1820, invented the galvanometer which enabled electricity to be measured, and suggested using a galvanometer needle for telegraphy. W. F. Cooke and Charles Wheatstone invented a five-needle telegraph, which was patented in 1837 (Figure 15.3(c)). Although still a parallel data-communication system like Soemmerring's, this reduced the number of wires to only six, by representing alphabetic characters with a 2-out-of-5 code. By 1839, they had set up a 13-mile telegraph for the British railways using this system. It was not long before the cost of multiple conductors stimulated reduction of the number of needles to two, and finally to one. So the needle telegraph became the first serial data-communications system, with codes defined by sequences of needle deflections to make up each character.

Samuel F. B. Morse, already a noted American painter, became interested in the possibilities of electrical communication in the 1830s. He built a telegraph system in 1835: the sender used notched metal strips to encode the alphabet, and the receiver was an electromagnetically-driven pendulum with a pencil attached which wrote the coded signals on a moving roll of paper. Morse's knowledge of electricity was rudimentary, but by 1837 he had patented the telegraph, replaced the type-bar sender with what we now call a telegraph key, and simplified the receiver to a pen which put marks (dots and dashes) on paper tape. Early telegraphs used two wires, but it was soon found that one would do, the earth acting as the return path.

With some government support, Morse started a telegraph service between Washington and Baltimore in 1844. When the line was extended to New Jersey, it attracted customers in the financial community who appreciated the commercial value of instantaneous communications. The visual receiver was replaced by a sounder, because operators could transcribe aurally transmitted codes to paper faster than the marks on paper tape. This early human-factor discovery shows the importance of inventor-user interaction in bringing an idea to practical fulfilment. The so-called Morse code was actually the work of Morse's assistant, Alfred Vail.

The telegraph was one invention which did not have be marketed; the public was eager and ready to pay for good communications. By 1852, more than 18,000 miles of wire covered the eastern third of the USA, with service provided by many small companies. In 1856, Western Union was created, and extended the telegraph to the west coast in 1861 with government support; a message cost senders $1 per word.

There had been several attempts to develop submarine cables for telegraphy, but the first, under the English Channel, ruptured in 1850. Improvements in cable-making, and the development of specially equipped cable-laying ships, enabled construction of a reliable link from Dover to Calais in 1851. A much more audacious undertaking, a transatlantic cable, was proposed by an American, Cyrus W. Field, in 1856, and was successfully laid by HMS *Agamemnon* between 1857 and 1858. However, a few months afterwards, operator error put 2000 volts across the cable which rendered it unusable.

It was not until after the American Civil War, in 1865, that Field attempted another link; this time, the cable was spun out from a single ship, Brunel's *Great Eastern*. This sail-and-steam leviathan carried almost 5000km (3,100 miles) of cable weighing over 5000 tonnes. However, after laying 2000km (1250 miles) between Ireland and Newfoundland, a defective length snapped; after ten days' unsuccessful grappling, the spot was marked and the attempt abandoned. Another cable was manufactured by the Telegraph Construction and Maintenance Company of the UK, and laid by the *Great Eastern* in 1866: the commercial communications link so long sought between Europe and the United States was at last established.

The first telecommunications network – organizational rather than physical– was that of the Associated Press, begun in the 1840s by a group of New York newspapers to share telegraph expenses. Many technical improvements were made in both the United States and Europe, including the paper-tape perforator of Wheatstone (1855), and the multiplex telegraph of Emile Baudot (1874). However, the most radical departure, which led the way to the modern era of telegraphy, was the invention of the teletypewriter (or teleprinter) by E. E. Kleinschmidt in the US in 1928. This allowed operators to compose messages using a typewriter-like keyboard to punch paper tape, which was then torn off and fed into a tape reader for transmission. At the receiving end, the message was printed out on paper strips (later, directly on a roll of paper). Thus, the days of the telegraph operator, sending with a lightning touch on the key, and listening to the clicks of the sounder were ended. However, Morse is still used in radiotelegraphy, where it can get a message through static and difficult transmission conditions when electromechanical and electronic alternatives are unworkable.

Telex

The sending and delivery of a telegram has always been a hybrid type of service, part electronic and therefore very rapid, part manual and therefore slow and labour-intensive. The sender had to go to a telegraph office and print out his message in capital letters on a special form; the clerk had to count the words and compute the amount due from a tariff schedule; then the telegraph operator had to key in the message. At the other end, the reverse took place; the message would be printed out on a paper strip, cut into segments and pasted on a form, and then had to be delivered by hand to the recipient.

By the time most customers had their own telephones (the 1930s in the US and Canada; later elsewhere), the US telegraph carrier Western Union would accept outgoing telegrams over the phone, and read incoming telegrams to recipients who were phone subscribers. However, the unique feature of the telegram among public telecommunications services has always been the delivery of a written message, providing legal proof to sender and receiver. Most customers would insist on physical delivery even though the message had been read to them over the telephone; in such cases, telegraph officers would resort to the postal system.

For business users, the whole system seemed archaic. In the US they turned increasingly to the telephone whenever written proof was not essential. In Europe, with its many languages and complex cross-border tariffs, the telephone did not have the same ease of use or economy. Therefore, when the first telex network was put into service in Germany in 1933, a great unmet demand was released. From its start, with only nineteen subscribers in Berlin and Hamburg, telex service had an explosive growth. In Germany subscriptions grew at more than 100 per cent annually up to the Second World War; similar growth, but at lower rates, was experienced in other European countries. By the early 1980s, the number of telex subscribers world-wide exceeded 1.5 million, more than half of them in Europe.

Today, telex is a world-wide, switched public teleprinter service. Therefore it is like the telephone, in that subscribers are loaned terminal equipment, can dial up other subscribers themselves, can receive messages, and are charged on the basis of time and distance. However, unlike the telephone, messages must be keyed in and received on teleprinters. Up to the 1970s, the usual practice was to keypunch a series of messages on paper tape, dial up the recipient's teleprinter, and put the strip of tape in the sender's teleprinter; this store-and-forward process has been greatly enhanced by the use of computers in telex officers, and by microprocessors and electronic memory capability in terminal equipment. Also, in store-and-forward operation, the same message may be sent to many recipients (a type of message broadcasting), and re-dialling done automatically when receiving terminals are busy.

One of the greatest advantages of telex is that teleprinter equipment and service available to subscribers is uniform throughout most of the world. Unfortunately, a faster service, Teletype, was introduced in the US side-by-side with Western Union's telex, and until the FCC forced AT&T to relinquish their public switched message service, conversion was difficult. However, since Western Union took it over, computers do this conversion without customers having to worry about it. The international telex system uses the five-level Baudot code, which requires a shift from letters to numerals and back again, and there is no provision for indicating letter-case. The speed of transmission is very slow: 50 bits per second, equivalent to 66 words per minute (a word is defined to be six characters, including space). In international practice, the telex system still has these limitations, but within certain countries faster and more flexible equipment is allowed. In the early 1980s, a completely new public switched message service was introduced called teletex. It operates at 2400bps, 40 times faster than telex, and allows upper-and-lower-case letters, formatting codes and a host of other features: teletex can best be described as communicating word processing. However, new equipment is required in both central offices and subscribers' premises; total cost is higher for low-volume users; and in the mid-1980s teletex service was only available in a few countries. Therefore, it is unlikely to displace telex in international use until the 1990s.

THE TELEPHONE

In addition to an understanding of electromagnetism, a knowledge of acoustics and human hearing was essential to the development of the telephone. However, with the exception of Bell, most of the early experimenters were physical scientists, engineers or dedicated tinkerers, rather than biologists, physicians or psychologists.

Acoustic transmission

The science of acoustics developed out of the theory and practice of music, and the refinement of musical instruments. An understanding of the laws of harmony started with Pythagoras in ancient Greece. Shorthand symbols had been used to record Greek and oriental speech (ekphonetics), and between the fifth and seventh centuries AD a system was developed to show melodic movement using symbols called neumes. The staff was created in the ninth century – a single coloured line. Guido d'Arezzo suggested using three and four lines, the latter becoming the standard for recording Gregorian chants. Today's five-line staff appeared in the eleventh century, but did not come into general use until the seventeenth century; some composers would like to have six or even eight lines.

In 1796 a German student, G. Huth, musing on ways to overcome the disadvantages of Chappe's semaphore (see p. 711), such as excessive operating cost and interruption of service by fog and rain, hit upon the ideas of using large speaking-tubes (megaphones). Specially selected and trained operators would listen to the message from one tower, and turning their megaphones in the direction of the next, repeat the message immediately. Huth wrote that 'this . . . difference might deserve a different name. What could be more appropriate . . . than the word derived also from the Greek: Telephone or Fernsprecher?'

Actually, Huth's idea was less practical than the semaphore, because visual signals are usable over much further distances than voices or even whistles or drum beats. The speaking voice, even using a megaphone or electronic amplification, is audible only up to a few hundred metres; a human whistle, screams and yodelling carry further. Mechanically aided sound-generation systems, such as noisemakers and musical instruments are also useful; African drums can carry messages for many kilometres. Today, aerosol-powered shriek-alarms can carry at least a kilometre, and police and ambulance sirens can be heard over a radius of several kilometres even in noisy cities.

Sounds also can be sent through hollow tubes, and through denser matter (liquids or solids) they can be transmitted much further and at higher speeds than in air. In 1838, Romershausen proposed a telephone which would transmit speech through narrow tubes. Modern hydrophones listen in on the cries of dolphins and whales, and detect other underwater sounds over a range of frequencies far beyond human hearing.

Electrical speech transmission

However, although such means could carry intelligible messages, there is less privacy, greater interference from noise, and greater annoyance to the public at large than in visual signalling. What was needed to make sound transmission a practicable means of communication, just as with the telegraph before it, was a better understanding of the nature and application of electricity. In 1837, about the time that Cooke and Wheatstone were building the first commercial telegraph system, Professor Page in Salem, Massachusetts, discovered that an iron rod which was suddenly magnetized or demagnetized would emit sounds; this came to be called Page's effect, but no practical applications emerged.

In 1854, Charles Bourseul in France predicted the coming of speech transmission, and outlined a method for its realization that was correct except in one critical feature: he expected that it would be achieved using make/break circuity (like the relay and telegraphy, which is digital) rather than with circuits capable of handling continuous current (which is analogue). Philip Reis in Germany constructed a telephone along the lines suggest by Bourseul in 1861, but was able to transmit only tones rather than speech. In 1868, Royal E. House, who had invented a printing telegraph in 1846, patented what he called an electrophonetic telegraph, but did not realize its capabilities to transmit speech. Ironically, we are today on the threshold of ISDN (Integrated Services Digital Network) which will transmit all signals – speech, video, text and data – digitally.

The breakthrough came in 1876 when Alexander Graham Bell and Elisha Gray almost simultaneously – and independently – filed patents for successful speaking telephones. Although, after long patent litigation, Bell was awarded priority, Gray's system was technically superior. Bell was an early example of the brain drain from the Old World to the New. His parents had emigrated from Scotland to Canada after having lost two sons to tuberculosis, and Bell later settled in Boston where he became a teacher of the deaf. From this background, rather than as a physical scientist or engineer, Bell approached the problem: as a teacher trying to find better ways to teach deaf students to speak. For this purpose he had invented a special notation for recording speech. Bell's earliest working models used no external source of electricity, and transmitter and receiver were almost identical. This system – the voice-operated telephone – is still used over short distances. Switched telephone systems require external power.

On 7 March 1876 the famous command, 'Mr Watson, come here. I want you!' was uttered by Bell to his assistant, who instead of only hearing Bell's voice from the other room, heard it over the primitive induction device with which they were experimenting. This, the first transmission of a human voice over wire, happened only three days after the first telephone patent had been issued to Bell. He had filed for this patent on 14 February, only three hours before Gray had filed a caveat that he was working on a similar device. Bell's patent was probably the most valuable ever granted, giving birth to what has become the world's largest private service organization, the Bell Telephone Company.

Bell Telephone was formed only a year afterwards, the first two telephones being leased to a Boston banker. Also in 1877, Bell, who had married the deaf daughter of his first backer, Thomas Sanders, went to England on his honeymoon; mixing business with pleasure, he gave many demonstrations, and presented Queen Victoria with a pair of telephones. However, although the British Post Office's chief engineer recommended obtaining rights to manufacture and use the new technology, the Post Office claimed that this was not necessary to protect its monopoly – because the telephone was just another form of telegraph.

In 1883, Edison discovered what eventually became the basis for signal amplification in both wire and wireless communication – he observed that when he added another element (electrode) to a light bulb, that current could flow across the evacuated space between filament and electrode. This came to be called the Edison effect, but was only a laboratory curiosity until 1904, when Fleming invented the two-element rectifying diode: shortly afterwards, in 1906, Lee de Forest added a third element which enabled small voltage changes to control the flow of large currents, giving birth to the triode and the age of radio.

THE GRAMOPHONE

All sounds, whether voice, music or simply noise, depend upon physical vibration of some medium – if our ears had the ability to detect extremely low frequencies and faint sounds, we might even hear the ripples in a calm lake and not just the roar of ocean waves. Unlike the extremely limited ability of our eyes to sense the immense range of the electromagnetic spectrum directly (less than one per cent), we can hear about ten per cent of the sound spectrum which is inherently limited because physical molecules must vibrate to emit and transmit sound.

It has been reported that a visitor entering an ancient Chinese temple is greeted with the Chinese equivalent of 'close the door' upon entering and, when he does, 'thank you'. The mechanism for this automatic doorman is said to be a pointed object which moves over a serrated strip when the door is opened, and then moves backwards when it is closed, the Chinese for 'thank you' being the reverse of 'close the door'. If true, this invention is similar to noisemakers, where one stick is rubbed over another which has a regular pattern of notches cut into it; however, the ability to cut notches so that a semblance of the human voice is reproduced strains credulity. Practical systems for the automatic recording and reproduction of sound have become available only recently, and voice synthesis and recognition are still in their infancy.

Mechanical sound recording and reproduction

Techniques of recording vibrations occurred before means were found to reproduce them. In 1857, Leon Scott de Martinville, designed a device for tracing soundwaves on a cylinder covered with lampblack, which he called a 'phonautograph'; however, he could not find a way to replay the sound. In 1877, Charles Cros in France outlined a scheme to convert Scott's waveform trace into a physical groove by photoetching, and play the sound back by a point attached to a diaphragm which followed the groove. He dubbed this device the 'paleophone', but lacked funds to put the idea into practice.

The first machine actually built which could both record and reproduce sound was Edison's 1878 cylinder phonograph, for which he filed the first detailed patent application in England. This had been preceded by an 1877 patent for a means to record Morse code by making indentations on paper stretched over a disc or cylinder, and another which outlined a method for transmitting the human voice over telegraph wires.

The seminal 1877 Edison patent contained in its specifications not only details of his original tinfoil-on-cylinder model, but also claimed disc recording, wax materials, electromagnetic (rather than acoustic) recording and reproduction, mass production of recordings and amplification. Unfortunately, Edison failed in his efforts to make the equivalent claims in the US. As a consequence, a tangle of litigation marred the early decades of the phonograph, just as it did the contemporary invention of the telephone. In fact Bell, with his brother and Charles Tainter, patented a rival system, the graphophone, in the mid-1880s which used removable waxed-paper cylinders.

The US patent, dated 19 February 1878, was titled 'Phonograph or Speaking Machine'. The drawing showed a spiral-grooved cylinder wrapped with tinfoil; on one side was a short horn (the mouthpiece) containing a diaphragm to which a rounded needle (the stylus) was fastened at right angles; opposite, a larger horn with an upward-facing opening was similarly outfitted (the speaker). Thus, when Edison spoke his famous 'Mary had a little lamb' into the mouthpiece, he was startled to hear his voice reproduced on the other side a fraction of a second later. Bell's first telephone (see above) was similar in its simplicity, being able to act either as sender or receiver.

In 1888, Edison also patented a wax-cylinder machine, but one year earlier, Emile Berliner had patented his gramophone; although depicted in the drawing as a cylinder machine, by the time he built his first model a year later, it was a disc machine. Another innovation introduced by Berliner was lateral recording of sounds in spiral grooves. Scott's tracings had been lateral, but Edison and his followers had employed vertical (hill-and-dale) recording. By 1895, Berliner had combined his flat disc, Scott's lateral recording method and Bell and Tainter's wax coating, and formed the basis of the modern mass-production recording industry. Edison never exploited the alternatives of his 1878 British patent, sticking with cylinders almost until the First World War; reluctantly, in 1912 he brought out a disc phonograph. He was right about some of the advantages of the cylinder: it provides identical grooves over the entire surface, rather than spiralling to a smaller and smaller radius; and hill-and-dale recording can handle a much greater dynamic range without shortening recording time.

The original application of the graphophone and phonograph was not in entertainment but in business. In the late 1880s, the North American Phonograph Company was formed to rent out both types of cylinder machines to record dictation (for a similar machine see Figure 15.3); however, the graphophones proved unreliable, and the confusion of offering two incompatible machines to customers led to failure of the company. Rather, it was as an entertainment medium that the phonograph and gramophone scored their early

Figure 15.3: Cylinder phonograph for recording and transcribing dictation; the governor regulated the speed of the weight-driven 'motor'.

successes. In 1888, Gianni Bettini in New York used a phonograph to record the singing voice. However, until the commercial pressing industry began around 1900, multiple copies could only be made by using a bank of machines, or copying from one machine to another.

Magnetic and optical recording and reproduction

Edison in 1878 and Tainter in 1885 had proposed magnetic recording, but it was not until 1898 that Valdemar Poulsen invented a system for recording on a steel wire: the telegraphone. One of the biggest advantages of a magnetic-recording medium was its erasability, so that it could be used again and again, particularly important in business applications such as dictating. Poulsen also suggested magnetically coated tapes and discs in his patents, but exploitation of these media had to wait until plastics were discovered and a coating industry in place. In 1920 an Austrian, Pfleumer, developed a plastic magnetic tape, but it was not until 1934 that magnetic recording tape was first manufactured by BASF in Germany. After the war, largely due to the enthusiasm of Bing Crosby, tape became the preferred medium for pre-recording radio programmes. Tape became standard in recording studios soon afterwards.

Another means of sound recording is optical, and it would not have been surprising if Edison, one of the inventors of the motion picture, had hit upon this means of adding sound. In fact Edison's laboratory had been able to synchronize film and discs in 1889, and Joseph Poliakoff, a Russian, patented a photoelectric system in 1900. Augustus Rosenberg patented a stereophonic talking film system in 1911. However, none of these systems were commercialized. In 1926, Warner Brothers tried to stimulate lagging interest in its films by making synchronized-sound motion picture – 'talkies'. They chose to use a sound-on-disc system, and contracts were awarded to Western Electric and the Victor Talking Machine Company. The first picture to be released under this system was Don Juan, but the big success – which spelled the end of the silent film era – was The Jazz Singer (1927), starring Al Jolson. Warner Brothers then installed disc-recording equipment in their studios, naming the system Vitaphone. Also in 1927, the first commercially successful sound-on-film system was marketed by Fox Movietone News. This used a variable intensity method of modulating a light beam (the Aeo-light) to expose a track on the side of the film negative.

Stereophonic sound and the long-playing record

In 1931, Alan D. Blumlein at EMI in the UK and the Bell Telephone Laboratories in the US almost simultaneously brought out the first complete stereophonic sound systems. Blumlein produced a reel of test film, but it was not until 1942 that Walt Disney's feature-length cartoon *Fantasia* brought stereo to the public. In 1948, RCA and others developed a system for coating a strip of magnetic oxide on 35mm film, which provided a higher dynamic range and permitted re-recording. This became the standard method for providing sound on amateur films – 16mm, and later Super-8.

Also in 1948, Peter Goldmark, working at CBS Laboratories in the US, invented the long-playing record (LP). This was a radical improvement of disc recording. Previously, discs had been made of a rigid, composite material and coated with shellac – if one was dropped, it cracked or broke – and even a 12-inch-diameter disc could hold only four minutes of sound. The LP was thinner, lighter and was made out of flexible vinyl plastic, which provided much lower surface noise. Also, LPs could play almost thirty minutes of high-fidelity music on a side, thanks to the invention of microgroove recording (100 grooves per cm instead of 36), and the slowing down of the rotational speed ($33\frac{1}{3}$rpm instead of 78). Almost at the same time, RCA brought out a rival microgroove system using a smaller, 45rpm disc. However, its playing time was about the same as the 12-inch 78s – only long enough for singles of hit songs. Even though the two incompatible systems confused the public, LP became a success and in a few years 78rpm production was stopped forever.

In 1958, the 45–45 cutting system for stereo recording, originally invented by Blumlein, became a world standard. This puts both channels in a single microgroove. The introduction of stereo recordings required the audiophile to replace much of his equipment; he had to have a new, more compliant stylus, two matched amplifiers, and two loudspeakers. Despite this additional complexity and expense, stereo hi-fi caught on quickly with the public.

Magnetic tape recording and reproduction

Also about this time, stereophonic reel-to-reel recorders using $\frac{1}{4}$in-wide magnetic tape became popular, because owners could record selections off the air, and tape could play for hours without interruption. However, a certain amount of manual dexterity was required to thread the tape through the magnetic heads and drive mechanism (similar to the skill needed to thread the film through an amateur movie projector). In the early 1960s, RCA tried to market a cartridge tape, but it was not until 1963 that Philips in the Netherlands hit upon the right production and marketing strategies to make their compact cassette a worldwide standard for audio recording. It took only a moment to snap a cassette into a recorder, and both were miniaturized to such an extent that using batteries, they could be carried about. One innovative marketing technique of Philips was to offer manufacturing details and working drawings of their design free of charge to all others willing to adhere to the standard. This was unheard-of in a world where most firms tried to ensure that potential competitors were licensed under their patents or locked out of the market. However, the great advantage to customers of being certain that any cassette would work in any recorder led to the explosive growth of the cassette industry.

A compact cassette is only $10 \times 6.4 \times 1.2$cm ($3.9 \times 2.5 \times 0.5$in), yet, because its tape speed is one-half or one-quarter that of $\frac{1}{4}$in reel-to-reel, it can play up to an hour on each side. Great improvements in the formulation of the tape coating have been made, the original iron oxide being supplemented by chromium dioxide and, recently, metal; and in 1967, Ray Dolby in the US brought out his patented noise-reduction system. All these developments have meant that cassette audio reproduction, in terms of extended frequency and dynamic range, high signal-to-noise ratio and low distortion, approaches the best of the older reel-to-reel systems. The compact cassette is a two-hub system, a miniaturized version of reel-to reel; the tape never leaves the enclosure, but is exposed only as it passes by the recording or playing heads.

A rival system, the Lear Jet Pack endless-loop tape cartridge, a one-hub system, makes ingenious use of the principle of the Moebius strip; both surfaces of the tape may be recorded and played without rewinding. Also, instead of the four tracks of the compact cassette (which provide two stereo or four monaural channels, but require turning the cassette over to use the other side), the Lear Jet cartridge had eight tracks providing four parallel stereo channels. It had its best market in automobiles during the 1960s and early 1970s, but the greater fidelity and reliability of the cassette players, and the advantages of a single system for both home and car use, proved decisive. Then in 1979, Sony introduced its Walkman, play-only cassette machines which were half the weight and a fifth the size of older cassette recorders. The idea was to bring cassette listening through earphones to stroller and jogger; not only would the earphones blot out noise and distraction, but the sound would not disturb others. The Walkman was an immediate success, and much imitated; by the mid-1980s, high-quality players were available only slightly larger than the cassettes themselves.

Even this technology is now being challenged by another, even more revolutionary audio system, the Compact Disc or CD (see p. 755).

RADIO AND RADAR

Radio waves

The notion of transmitting information without the use of wires must have seemed to our ancestors only a hundred years ago to be magic. In England, during the 1860s, James Clerk Maxwell developed analogies between Faraday's conception of electric and magnetic lines of force, and incompressible fluid flow. In a brilliant series of papers published in 1861 and 1862, he presented a fully developed theory of electromagnetism, based on Faraday's field concept. In his theory, Maxwell was able to explain all existing electromagnetic phenomena, and to predict the transmission of electromagnetic waves.

Heinrich Hertz in Germany proved the existence of these waves in 1887, using what was probably the first antenna, two metallic plates each attached to a 30cm (12in) rod. The two rods were placed in line and balls attached to their nearer ends, separated by a 7mm ($\frac{1}{4}$in) gap. When a spark crossed the gap, a similar spark was generated in the microscopic gap of a 70cm (27.5in) diameter loop of wire at the other side of the room.

In the 1890s, Edouard Branly in France observed an effect of Hertzian waves on metal filings in a tube. The normally nonconducting filings became conductive only when the waves were being generated. Branly called the tube of filings a coherer; it was what is called today a rectifier, and became the first practical detector of electromagnetic waves. In 1893, J. J. Thompson in the UK published a book intended to be a sequel to Maxwell's. In it he made the first suggestion of propagating electromagnetic waves through hollow tubes. It is interesting that Hertz's original experiment and Thompson's idea both dealt with what has come to be called microwave radiation, the basis of radar, satellite telecommunications, diathermy and the microwave oven.

Wilhelm Röntgen, a German, experimenting with a tube invented by W. Crookes in 1878, discovered in 1895 what he called X-ray (X standing for the unknown, as in algebra). A current passed through the Crookes tube caused a piece of paper coated with barium platinocyanide to fluoresce. Röntgen soon started using photographic plates in his experiments, and in 1895 the first X-ray of a human hand was made, clearly showing the underlying bone structure; a year later, the first X-ray of a human skull was made.

Thus, before the turn of the century, the means to generate and detect a large portion of the electromagnetic spectrum had been found, even if not well understood (Röntgen eventually died from the effects of his X-rays). Microwaves were centimetres long, light waves a fraction of a micrometre, and X-rays one ten-thousandth of a micrometre – a range of about eight orders of magnitude (10^8). The waves we call radio are longer than microwaves, ranging over another seven orders of magnitude, from ultra-high frequency (UHF) which are about 30cm (1ft) in wavelength, to extremely low frequency (ELF) waves thousands of kilometres long.

Wireless telegraphy

In 1896 the first patent for wireless telegraphy, that is the transmission of messages without wires, was granted to Guglielmo Marconi in the United Kingdom. By the following year the Wireless Telegraph Company had been formed to exploit the invention, and in 1899 the Marconi Wireless Company of America was set up. American Marconi, as it was called for short, soon began the manufacture of wireless equipment for commercial and military markets. In 1904, Alexander Fleming patented a two-electrode valve in the United Kingdom. The invention was based on his observations of the 1884 Edison effect, in which a one-way flow of electricity (rectification of alternating current) occurred when a metal plate was sealed inside a carbon-filament lamp.

Although Marconi's original invention was designed for fixed (point-to-point) and ship-to-shore message communication, the idea of wireless as a one-way medium to transmit speech to many people – broadcasting – was quick to follow. On Christman Eve 1906, Reginald Fessenden made the first documented broadcast of speech and music from Brant Rock, Massachusetts. His transmitter was a 1kW, 50Hz alternator built by the General Electric Company in Schenectady, NY, under the direction of E. F. W. Alexanderson. The signal was received clearly in many locations and even on ships at sea.

Lee de Forest made some experimental broadcasts from New York in 1907 and from Paris in 1910, but his 500W arc transmitters were inherently noisy. However in 1906 one of his associates, General Henry H. C. Dunwoody, patented a solid-state detector using the newly invented material carborundum. These crystal detectors soon became the heart of early radio receivers. Used with headphones which drew very little power, crystal sets had the great advantage of not requiring any external source of electricity. In 1906, de Forest applied for a patent for a 3-electrode tube which he named the Audion. It was first though of as a device to amplify weak currents, but de Forest soon extended the patent application to include detection. The additional electrode greatly extended applications of what came to be called, generically, thermionic valves; unfortunately, like many other inventions of commercial importance, the triode patent led to extensive and bitter litigation between de Forest and Fleming.

Early radio receivers were mostly constructed by amateurs who could make their own components. They required large outdoor antennas (aerials) and the use of headphones; in operation, locating and retaining the signal of a broadcasting station (tuning in) demanded skill and patience. In 1917, Lucien Lévy in France patented a superheterodyne circuit, which made tuning much easier, reduced power requirements and simplified receiver construction.

Radio broadcasting

Marconi's Wireless Telegraph Company first broadcast daily concerts from Chelmsford, Essex, between 23 February and 6 March 1920. The transmitter had a power of 15kW, and used an antenna strung between two 137m (450ft) masts; programmes were picked up as far away as 2400km (1500 miles). On 15 June 1920 a concert by the celebrated Dame Nellie Melba was broadcast on 2800 metres (equivalent to 107kHz), with a power of 15kW; and on 14 November 1922 the British Broadcasting Company began a daily programme. Educational broadcasting also was pioneered in the UK, beginning on 4 April 1924. Fifty years later, the British Broadcasting Corporation (BBC) was providing more than 3100 broadcasts a year to schools, and 500 for further education.

Broadcasting has a unique advantage over other communication media – its messages are carried without any physical link between sender and receivers. Unlike newspapers, magazines and the post, no rails, roads or vehicles are needed to carry messages, resulting in large savings of time and expense. Unlike the telegraph or telephone, no wires need be laid, so radio messages can reach the remotest and most inaccessible locations, including ship at sea and aircraft.

In the United States, the first organization to receive a radio licence specifically for broadcasting was KDKA, a Westinghouse station in Pittsburgh. Their first broadcast, on 2 November 1920, announced the results of the Harding–Cox election. The first radio advertisement was bought by American Telephone and Telegraph and broadcast over WEAF, New York, on 28 August 1922; the following year, the first radio network was created, between New York and Boston.

Without government involvement, the spread of radio broadcasting in the United States was unconstrained during the 1920s. Whereas in 1921 there were only two stations officially on the air, by 1925 there were more than 500 (not including thousands of very low-power, unauthorized do-it-yourself stations). The first frequency assignment in the US was on 830kHz; by 1925 the spectrum from 550 to 1,350kHz was allocated exclusively to AM (amplitude modulation) broadcasting, which is also called medium wave (MW). In 1922, George Front in the US installed a radio in his Model T Ford; and in 1927 the first product designed for automobile use, the Philco Transitone, was introduced. Today probably more radio listening is done in cars than in all other locations.

Radio stations added more and more power to reach larger audiences, with some stations exceeding a million watts before government regulation brought some order to the scene. For example, broadcasts of one Cleveland station electrified nearby fences and provided free lighting for many households in the neighbourhood; and some people who had tooth fillings even claimed that they could hear (but could not turn off) the broadcasts – the filling acting as a rectifier, and the programmes being heard by bone conduction, apparently. Stations even interfered with each other on opposite sides of the continent, and frequency and power regulation became essential. In any case, advances in technology made unlimited power unnecessary to reach vast audiences: radio stations became linked into networks via leased telephone lines.

In 1933, Edwin Armstrong in the US patented his FM (frequency modulation) radio system and in 1938, General Electric installed the first FM broadcasting station. In contrast to AM, FM is noise-free and uses a much greater band width, so it became the preferred method for broadcasting high-fidelity music and eventually stereophonic sound; on the other hand, because the channels assigned to FM are in the very high frequency (VHF) band, signal reception is limited to line-of-sight. Whereas listeners with ordinary AM sets often picked up stations from overseas, particularly at night when their signals bounced off an expanded ionosphere, FM waves penetrated and were lost in space.

Many other improvements were made in radio broadcasting and receiver construction, but not until after the Second World War did components became small enough to make truly portable radios. The invention of the transistor in 1948 made such miniaturization possible, and the first transistor radio appeared in 1955. During the next decades, prices dropped from hundred of dollars to about $10 for the simplest sets, and 'transistors', as they are called for short, have become ubiquitous, perhaps the product owned by more individuals than any other in the world.

Radar

In 1924, Sir Edward Appleton and M. F. Barnett in the UK found that they could bounce radio waves off the ionosphere, the existence of which had been hypothesized by Oliver Heaviside in the UK and Arthur Kennelly in the US in 1901. A year later, Gregory Breit and Merle A. Tuve were the first to use a pulse technique in similar experiments. During the next ten years, the US and several European countries experimented with radio waves to detect aircraft and ships at sea. In 1933, A. H. Taylor, L. C. Young and L. A. Hyland in the US applied for a patent for a radio detection system, as did Henri Guton in France the next year. However, for the next decade, research became shrouded in wartime secrecy.

In 1935 in the UK, a committee headed by Sir Henry Tizard was set up to investigate military applications of radio waves, and in that year Sir Robert Watson-Watt with six assistants developed the first practical radar (Radio Detection and Ranging) equipment for the detection of aircraft. Both the Allies and the Axis powers had radar during the Second World War, but its most dramatic role was during the Battle of Britain in 1940, when early-warning radar gave the Royal Air Force a critical tactical advantage. Since the war, it has become essential not only for a host of military purposes, but also for civilian air traffic control.

Radar operates by sending out pulses of radio energy, a tiny portion of which is reflected by objects in its path and picked up by highly directional antennas. The range of such objects can be calculated automatically from the difference in time between transmitted and received pulses. Microwave frequencies work best; early-warning radars use 10m wavelengths, positional radars 1.5m, and high-resolution systems wavelengths down to 10cm.

In 1933, Karl G. Jansky of the Bell Telephone Laboratories discovered that radio waves were being generated by stars. He used a 30m (100ft) long rotatable short-wave antenna. This began a new era in astronomy – radio astronomy – which has greatly expanded our knowledge of star creation and life processes, and has detected many stars invisible to light telescopes. Immense dishes are used for this purpose, the largest being one in Puerto Rico, with a diameter of 3km ($1\frac{3}{4}$ miles), which is capable of detecting phenomena out to 15,000 million light-years away.

PHOTOGRAPHY

In 1725, Johann Heinrich Schultz, a German physician and professor, was trying to make a phosphorescent material by mixing chalk with nitric acid containing dissolved silver. He observed that sunlight turned the mixture black, and realized that he had a photosensitive material (even though he did not use this term). He made photographic impressions of words and shapes cut out of paper and placed against a bottle of the solution, but although he published his discovery in 1727, he never made the image permanent – one shake of the solution and it was gone forever.

Tiphaigne de la Roche, in his sceince-fiction novel *Giphantie* (1760), prophesied the 'sun picture': a canvas could be coated with 'a most subtle matter, very viscous and proper to harden and dry', which when held before an object to be painted would retain its image. He even had the notion of present-day snapshot photography: 'This impression . . . is made the first instant . . . [and the canvas] is immediately carried away into some dark place an hour after the subtle matter dries, and you have a picture . . . [which] cannot be . . . damaged by time.'

Optics

Alhazen of Basra, a tenth-century Arab mathematician and scientist, observed the inverted image of a sunny landscape on the wall of a darkened tent. He realized that the critical element was a small hole in the tent material, and made use of the discovery to observe eclipses of the sun. This was the first recorded instance of what came to be called the camera obscura (dark room).

It might be thought that lenses, and especially the convergent (or convex) lens and its focusing properties, would have been discovered far back in antiquity. Glass vessels have come down to us from civilizations which flourished several thousand years BC, such as the Egyptians. However, the earliest recorded use of glass globes (probably filled with water) to kindle fires is that of Lactantius, about AD 300. Lenses may be made out of many substances as long as they are clear, that is, do not absorb much light: Arctic explorers have astonished their Eskimo hosts by using blocks of clear ice cut in the form of lenses to start fires. More recently, cast ice lenses have been used in place of glass lenses in cameras, and provide a good, if soft-focus, image.

All the lenses so far described are simple convex ones. Only with the development of optical science, and in particular the compound lens, have we been able to approach the theoretical limits of optical magnification. It was found that by using different curvatures and types of glass, and in particular by setting several lenses a distance apart in blackened tubes, magnification could be increased without worsening distortion. Almost all modern optical instruments except simple magnifiers and spectacle lenses use such compound lenses. Apparently, Johannes (or Hans) Lippershey in the Netherlands made a crude telescope using two lenses in 1608, and Galileo constructed his own upon hearing about it the following year. These devices were actually opera glasses, using a convex objective and concave eyepiece, but with such instruments Galileo made observations which revolutionized astronomy. Dutch spectacle-makers are also supposed to have made the first microscopes, about 1590. Anton von Leeuwenhoek brought the single-lens magnifier to such perfection that he was able to observe single-celled organisms for the first time. However, the use of two lenses for microscopy seems not to have been accomplished until the mid-1600s, when Robert Hooke in England built the first compound microscope.

Another precursor of photography was psychological; men began to see the natural world as it really is, and not filtered through the mind of Aristotle. In Italy and the Low Countries, artists studied the interplay of light with objects, but up until the fifteenth century, drawings of three-dimensional objects looked flat. About 1420, Filippo Brunelleschi discovered perspective. Brunelleschi assumed that objects are perceived as a cone of rays proceeding from the eye; he imagined a plane just in front of and at right angles to the eye on which he could 'project' an image. The camera obscura, which essentially does this projection optically for the artist, did not come into use as a drawing aid until late in the sixteenth century. The Dutch mathematician Rainer Gemma-Frisius published an illustration of a camera obscura used to register the solar eclipse of 24 January 1544, and Girolamo Cardano of Milan reported that a convex lens placed in an opening of a darkened room would produce a bright, sharp image. Up until that time, the clearest images had been produced by a hole of small diameter, which gives a dim, somewhat soft, but wide-angle view. Attempts to brighten the image by widening the aperture blurred the image severely; only a convex lens – probably invented in Europe around the thirteenth century – can provide images which are both sharp and bright.

The development of optics flourished in the sixteenth century with Kepler and Galileo, and the artist's camera obscura was soon reduced in size to a table, and then to a box. Johann Zahn invented the reflex camera obscura in 1685, using a mirror angled at 45° to reflect and invert the image on to a horizontal sheet of frosted glass. It was a great convenience to artists to be able to observe and trace an image which was right side up and in a horizontal plane.

In 1796, Tom Wedgwood, a son of the potter Josiah Wedgwood, was experimenting with silver salts to record images of botanical and insect specimens. However, he had no means to fix the images, which grew blacker at each examination until soon they were gone. If he had learned to fix them with ammonia or salt, he might have been the inventor of photography, but this achievement had to wait until the following century.

The invention of photography

Nicéphore Niepce made the world's first true photograph of a scene in 1826, in St-Loup-de-Varennes, France. (However, a monument put up by the town claims 1822.) His original interest was in improving lithography, which had been invented in 1796 by Alois Senefelder (see p. 678). In 1814 Niepce substituted a tin sheet for the heavy limestone from which lithography took its name. He wanted light to do the work of transferring an existing engraving, made transparent with oil, to the sheet. He tried a varnish made of bitumen of Judaea dissolved in animal oil, coated on to sheet of glass, copper or pewter. He made a sandwich of engraving, sensitized plate, and a sheet of glass, which he exposed to the sun for two to four hours, which hardened the exposed varnish. Then, in a darkroom, he immersed the plate in an acid bath to dissolve the varnish protected by the lines of the engraving; the remaining hardened varnish formed the image.

He called this process, as well as the images he made in a camera obscura, heliography. The earliest known such image is a view of neighbouring rooftops made by exposing a pewter plate in a camera obscura for eight hours: because of this extremely lengthy exposure of the first true photograph, the sun can be seen lighting both sides of the rooftops. Niepce soon changed his material to silver-plated copper, and gave the plates extra sensitivity by iodine fuming. He also tried glass plates, viewing them as transparencies, so it is surprising that he did not realize he could use these as negatives to make multiple prints – the basis of most modern photography.

This invention was not made until 1835, by Henry Fox Talbot in England. In a camera obscura, he exposed a sheet of paper which had been sensitized with silver chloride. The first such negative he made was of a window in his home, Lacock Abbey in Wiltshire. Unlike the plates of Niepce, Fox Talbot's process required an exposure of only thirty minutes. Sir John Herschel, who named the invention 'photography' and also coined the terms 'negative' and 'positive', suggested removing the unexposed silver chloride to fix the image permanently, using hyposulphite of soda. With its name shortened to hypo, this is still the standard fixing solution. In 1840, Talbot discovered (as had Daguerre two years before) that an exposure of only a few minutes could provide a latent image, which could be developed by chemical means afterwards. Fox Talbot used gallic acid as a developer, and after drying the paper negative, made it transparent by a wax coating. Then he contact printed it on to another sheet of silver-chloride paper to make a positive; he called this process calotypy.

The first person to make a commercial success of photography was Louis-Jacques-Mande Daguerre in France in 1839. He based his invention on Niepce's process, but brought out the latent image with mercury vapour and fixed it permanently with a hot solution of table salt. The resulting image was a positive which could be viewed only by reflection at certain angles, but the quality of the images was superb and many have lasted to the present time. As an artist-showman, Daguerre had created his Diorama in 1822 in Paris, with 14×22m (46×72ft) painted canvases; therefore, he was well placed to exploit his invention commercially. However, subjects for portraiture found it torture to sit under a baking sun for 10 to 20 minutes. Soon, the $6\frac{1}{2}$ by $8\frac{1}{2}$ inch (16.5×21.5cm) plate was reduced to quarter-size, bringing exposure time down to a few minutes. Then, by the use of large aperture achromatic lenses, and by enhancing the mercury vapour with bromine and chlorine, exposure time was reduced to thirty seconds, and portraits for everyone became a reality. The craze for daguerreotypes, as they were called, spread rapidly from Europe to the United States, and dominated the profession for decades.

A Scottish painter, David Octavius Hill, and a chemist-photographer, Robert Adamson, made 1500 superb pictures between 1843 and 1848 using Talbot's calotype paper-negative process. In 1848, Niepce de Saint-Victor developed a system using a glass plate coated with a solution of iodized albumen, and sensitized with silver nitrate. Then L. D. Blanquart-Evrard coated paper with albumen to make positives, eliminating much of the paper's inherent roughness. Blanquart-Evrard established the first mass-production photographic printing plant in Lille, France. He employed forty girls, each of whom performed one specialized operation. In 1851 he published the first album of original photographs, and the following year the first book illustrated with them.

Frederick Scott Archer, a British sculptor and photographer, invented the wet collodion process in 1851, displacing the calotype and daguerreotype as the most popular photographic medium until the dry-plate process was introduced by R. L. Maddox in the 1870s. Scott Archer made collodion by dissolving guncotton, ether and alcohol, blending it with a solution of silver and iron iodides. This was applied to a glass plate, which was then immersed in a silver nitrate solution and exposed wet in the camera; the plate had to be developed while the collodion remained moist, giving the photographer only ten minutes after the exposure to develop the plate. This was physical development: silver was deposited to make the latent image visible, and then it was fixed in sodium cyanide.

The great advantages of the wet collodion process were that exposure required only two to three seconds, and its tonal range was greater than that of other existing processes. However, a photographer in the field had to carry more than 50kg of equipment with him, including cameras, distilled water, chemicals, plates, trays, and even a darkroom. Nevertheless, the process provided the first portable photographic system, which made possible the magnificent portraits and landscapes of such pioneering photographers as Bisson Frères (the Alps, 1860); Roger Fenton (the Crimean War, 1855); William Notman (the Canadian north, early 1860s); Nadar (portraits; first aerial photographs, 1862); Julia Margaret Cameron (portraits, 1860s to 1870s); and Mathew Brady (American Civil War, 1860s).

Microphotography

The processes of Daguerre and Scott Archer exhibited very high resolution, that is, the ability to capture fine detail. In 1839, J. B. Dancer, an optical instrument maker from Liverpool and Manchester, made the first microphotograph. (Microphotography is the copying of objects on a very reduced scale, in contrast to photomicrography, which is the recording of the enlarged image of an object under a microscope.) He made his microphotographs in a daguerreotype camera fitted with a $1\frac{1}{2}$-inch (3.8cm) microscope lens, and attained the remarkable reduction of 160:1 (linear). Dancer's invention gave birth to the document recording segment of the photographic industry, now called microrecording or micrographics. Early uses were more trivial. By 1859, a fashion began which featured $\frac{1}{16}$ to $\frac{1}{8}$-inch (1.5–3mm) microphotographs incorporated into jewellery and domestic items. The tiny transparencies were glued in contract with a jewel lens called a Stanhope. It is said that certain immodest subjects exhibited in this way brought the art of photography into disrepute.

Sir David Brewster was the first to suggest using microphotographs to communicate dispatches in wartime. However, this was not done until 1870–1, when Dagron, who had established a studio in Paris for the manufacture of miniature photographic novelties, found his business ruined by the Franco-Prussian War. With Paris itself under siege, carrier pigeons sent out in balloons provided the only means of communication over enemy lines. However, a pigeon could only carry a tiny piece of paper. Dagron escaped to Tours in a balloon with his equipment and an assistant. There he devised a system in which all private and public letters were printed on large sheets of paper, each of which could hold 300,000 characters – the equivalent of about 100 typewritten pages today. These sheets were photographically reduced on to thin collodion films, each 50 × 70mm (2 × $2\frac{3}{4}$in); twenty of these could be rolled into a quill. Flown by pigeon to Paris, they were projected on to a large screen and transcribed by a small army of copyists. A remarkable 100,000 dispatches were sent in this way, forty copies of each being made to ensure that at least one got through.

After the war the process languished until an inexpensive material became available early in the twentieth century, the 35mm motion-picture film. Special cameras and emulsions were developed for this purpose, and microfilms of everything from rare books to fragile newspapers and bank cheques are now archived on this strong, long-lived medium with great savings in storage and postage. The idea of the pigeon post was revived in the US during the Second World War: starting in 1942, V-mail service, as it came to be known, was provided between servicemen and their families. Letters written on special blanks were photographed on to 16mm microfilm, which was sent overseas by airmail; the received images were enlarged on to rolls of photographic paper, cut into individual letters, and folded for mailing to recipients.

Photographic emulsions had been supported by metal plates, then by paper and glass; these materials were in single-sheet form and each had to be handled individually. In 1889, George Eastman, frustrated with the difficulties of the photographer's lot, invented roll film, and with it simple cameras preloaded with 100 exposures. Amateurs would take 'snapshots' and send camera and film to Eastman's company where they would be developed and printed – 'You push the button and we do the rest'. With Eastman, and his company Kodak, photography became an industry with a mass market.

Stereoscopy

No matter how realistic early photographic images seemed, they lacked two essentials of nature: colour and depth. The illusion of depth was made possible with the invention of the stereoscope by Sir Charles Wheatstone in 1838. He created a device to present a pair of perspective drawings, one as seen by the left eye alone and the other by the right eye. However, this created little interest until photography made the recording of stereoscopic views quick and easy; Fox Talbot made calotypes of still-life for some of the first stereoscopes. In 1849, Sir David Brewster invented a stereoscope with two magnifying lenses 7cm (2.75in) apart, corresponding to the distance between the two eyes. Not finding an English optician to manufacture and market the device, he took it to Jules Duboscq in Paris. At the Great Exhibition of 1851, the Duboscq and Soleil stereoscope was exhibited using a collection of daguerreotype stereo images to great acclaim. However, daguerreotypes were not the best medium for stereoscopy because their shiny surfaces could only be viewed at certain angles. Brewster also invented a stereoscopic camera in 1849, but it was not manufactured until 1853; until then, it was necessary to use two cameras side by side, or one camera which could be moved sideways.

The stereoscope enjoyed great success; by 1860, hundreds of thousands of stereographs of subjects from all over the world were available. Oliver Wendell Holmes was an enthusiastic user, and designed a hand-held stereoscope which became standard. Interest in stereoscopic photography has waxed and waned ever since, but the need to use special cameras, viewers and projectors has kept it from becoming as popular as conventional photography for all but the dedicated amateur.

Colour photography

The mixing of a few primary-colour pigments to make the vast array of colours we see was known to painters for centuries, but colour photography had to wait until James Clerk Maxwell in 1855 showed that there were also light primaries. In 1861 he projected an image of a multi-coloured ribbon using three superimposed colour images. In 1868, Louis Ducos du Hauron in Paris suggested techniques for making colour photographs on paper, and these became the basis for all colour photography for the next 60 years. In 1911–12, Fischer proposed an integral tripack, superimposing three emulsion layers, and in 1919,

Christenssen suggested dying the layers to separate the colours optically, but the technology available at that time was not up to the task. In the early 1930s two young American musicians, Leopold Mannes and Leopold Godowsky, invented the first practical process based on the integral tripack; they sold it to the Eastman Kodak Company, which placed it on the market under the trade name Kodachrome in 1935. It was first used as 16mm motion picture film, then extended to 8mm, and only later made in 35mm for colour transparencies. Agfacolor, a process based on additive colour combination rather than the subtractive process of Kodachrome, was marketed for 16mm motion pictures by Agfa in Germany in 1936. The professional motion picture industry used Technicolor, a complex, multi-emulsion system using dichroic mirrors, which had originally been developed as a two-colour system in 1918. The first three-colour Technicolor production was Disney's animated cartoon 'Flowers and Trees' (1932); it was not until 1942 that Technicolor introduced a monopack process, reducing the number of negatives from three to one.

Electrophotography

The photographic process had always involved the use of chemicals, which usually had to be in solution or vapour form; other imaging processes such as television, which were based on electronics, could not provide a fixed image. In 1938 an American patent attorney named Chester Carlson made a crude electrophotograph on his kitchen table. He named his process xerography, because it used a dry powder which clung to those parts of a semiconductive plate which had been electrified. He had great difficulty in trying to bring the invention to market; twenty companies refused to consider it, and he turned to the Battelle Development Corporation for help. In 1947 they finally sold the rights to a small photographic materials company, the Haloid Corporation. After about ten years of development, the first automatic system for making dry images on ordinary paper was marketed, the Xerox 914; the company became the Xerox Corporation, and today rivals Kodak, its Rochester, New York, neighbour which had turned down Carlson, in the multinational industry of image technology.

Instant photography

The next breakthrough in photography appeared just after the Second World War. Edwin Land, already well known for his invention of Polaroid in the 1930s (the first economical material for polarizing light), put the world's first instant camera on the market late in 1948. The Polaroid 95 produced sepia-toned prints directly, a finished print being delivered from the camera in just one minute. Then, in 1962, Polaroid brought out Polacolor, which made instant photography in colour a reality. Thus, in little more than a century, the art and science of the image had so progressed that what originally required an artist-scientist with a portable chemical laboratory, could now be accomplished by a child with a hand-held, self-contained box capable of capturing the fleeting moment and producing a permanent record on the spot.

Lasers and Holography

Light is only a tiny portion – less than 1 per cent – of an electromagnetic spectrum which ranges over more than twenty orders of magnitude, from waves many kilometres long to those of atomic dimensions. Components analogous to optical lenses, mirrors, diffraction gratings, prisms, filters and sensitive materials have been constructed to control and record these waves.

Holography is a type of 'lensless' photography invented by Denis Gabor in the UK in 1948. It is a process based upon the phenomenon of light interference, allowing both the amplitude and the phase of the propagating waves to be recorded (hence the name holography: 'recording of the whole'). By this means, true three-dimensional images can be shown; it is possible for an observer literally to walk around the image, seeing it as it would be in reality.

Holography remained an interesting, but impractical invention until powerful sources of monochromatic, coherent (meaning that the waves are all in step) radiation were developed. Einstein had postulated the possibility of stimulated emission of radiation in 1917, but it was not until the 1950s that this idea was embodied in a practical device. Fabrikant in the USSR filed for a patent in 1951, but failed in his efforts to use caesium for optical amplification. J. Weber in the US described such possibilities in 1953, and N. G. Basov and A. M. Prokorov of the USSR made detailed proposals for a beam maser (Microwave Amplification by Stimulated Emission of Radiation) the following year. However, Charles H. Townes, a professor of physics at Columbia University, working with his graduate student James P. Gordon and postdoctoral researcher Herbert Zeiger, built the first successful maser in 1954.

Townes's group, working with Arthur L. Schawlow of Bell Telephone Laboratories, thought out a scheme for an optical maser. Gordon Gould, another Columbia graduate student, devised the term laser (Light Amplication by Stimulated Emission of Radiation); he went to work for a defence contractor, TRW, which, proposing his laser ideas to the receptive US Department of Defense's Advanced Research project Agency, received a million dollars for its development – three times as much as they had requested.

All these researchers were concentrating on gaseous systems, hoping to be the first to demonstrate a practical laser. However, it was Theodore H.

Maiman, working with ruby masers at the Hughes Research Laboratories, who accomplished the feat. Maiman decided to try this gem material because it was rugged and did not have to be used at very low temperatures, even though Schawlow had calculated that ruby could not work. He put a tiny rod of ruby inside the spiral of a photographic flash lamp, and so achieved the first laser emission in 1960. Maiman's submission of his results was rejected by the editor of *Physical Review Letters*, so that his first published report appeared in the British journal *Nature*. A few months later, Townes and Schawlow demonstrated their helium-neon laser, and Townes shared a Nobel prize in 1964 with Basov and Prokhorov for the maser–laser principle. Maiman, who had built the first true laser, did not share in the prize.

Holography and the laser exhibit very well the dual nature of light: holography can only be understood by considering light as waves, but lasers are best conceived in terms of light as tiny energy packets – photons. Because holography is a means of recording wave phenomena, not only images made by light may be so captured, but waves from other parts of the electromagnetic spectrum – particularly microwaves; as well as those generated mechanically (by vibration) – particularly water and sound waves.

In 1962 several semiconductor lasers were announced in the US, and both the technology and the application of coherent light sources has advanced rapidly ever since. Such diverse systems as the proposed US Strategic Defense Initiative ('star wars'), thermonuclear fusion power, and telecommunications through fibre optics depend upon lasers. However, the first consumer products to employ lasers are videodisc and compact disc using digital recording techniques, the former being designed for motion-picture and television images and the latter for music (see pp. 753–6).

Motion Pictures

The motion picture, as we are familiar with it today, is the culmination of a number of inventions, but, even more than still photography, it depends on human physiology and psychology – for it is an optical illusion.

The notion of projecting an image on to a screen in a darkened room before an audience is implied in the term 'camera obscura'. However, up to the middle of the seventeenth century, no one seems to have thought of it in terms of theatre. In 1654, P. Kircher in Germany described a small projector for transparencies and about 1660, Christian Huygens constructed a magic lantern, but such projectors were not manufactured on a large scale until 1845, when they quickly spread throughout Europe. The magic lantern can be thought of as an inside-out camera obscura: instead of projecting a sunlit outdoor scene into a darkened room, it projects scenes drawn on glass slides and placed between a source of light and a lens on to a light-coloured surface. Artificial illumination for the magic lantern progressed rapidly to oil lamps, and then to limelight and electric arc. Light was concentrated by putting a concave mirror behind the light source, and additional (condenser) lenses were added to collimate the light beam before it reached the slide.

Several developments which foreshadowed the invention of the motion picture occurred in the nineteenth century, such as the use of multiple lanterns, which allowed one view to dissolve into another, and, between 1850 and 1890, increasingly elaborate mechanisms built into the slides themselves to provide a simulation of motion. Another development which was necessary before motion pictures could become a reality was a flexible transparent film, so that a rapid succession of images could be pulled past the light source. George Eastman's celluloid film was placed on the market in 1889. The third essential element was an understanding of visual psychophysiology, and how it could be fooled. Of course, magicians had always depended on the dictum that 'the quickness of the hand deceives the eye', but it was not until 1824 that P. M. Roget made a scientific examination of what has come to be called persistence of vision: the eye is incapable of resolving motion below a threshold of about $\frac{1}{20}$ second.

In 1826, J. A. Paris, an English physician, introduced the thaumatrope, which had different images on each side of a disc; when spun, it gave the illusion of superimposition of the images. In 1833 a Belgian, Plateaux, constructed what he called the phenakistoscope (Figure 15.4); and an Austrian, Simon von Stampfer, independently invented a similar device called a stroboscope. A cardboard disc had a series of drawings around its periphery, each depicting a successive stage of an object in motion. The rotating images were viewed in a mirror through slots cut in the periphery, and gave the illusion of a moving image.

A similar instrument which was based on putting the pictures inside a cylinder, the zoetrope ('wheel of life'), was described by Horner in 1834, but did not come into widespread use until Desvignes popularized it in 1860. In 1877, Emile Reynaud patented his praxinoscope, whose images could be projected on to a screen; Reynaud also devised a system using paper strips to extend the number of drawings beyond the dozen possible around the periphery of a wheel or cylinder. The use of photographs instead of drawings, and even the provision of sequential stereo views was accomplished by William Thomas Shaw in 1861, using a viewing device which had a rotating shutter to keep the eye from seeing the transitions from image to image.

The photographic recording and display of objects in motion was accomplished first by Edweard Muybridge, an English photographer who had emigrated to America. In 1872 he was hired by the ex-governor of California, Leland Stanford, to settle a bet that Stanford's trotting horse Occident had all four feet off the ground at once when in a gallop. However, the wet-plate process was too slow to give incontrovertible proof, and it was not until 1878 that

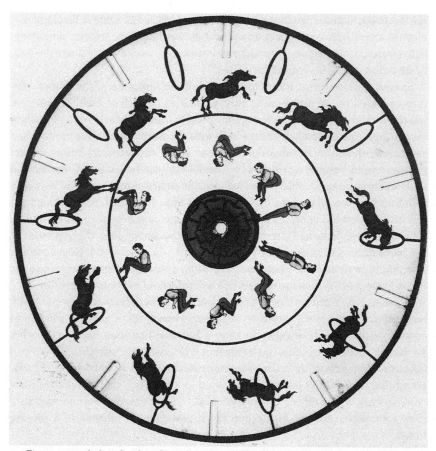

Figure 15.4: A disc for the 1833 phenakistoscope; two different, complete picture cycles are incorporated.

Muybridge was able to put together a system which settled the bet (in Stanford's favour). Muybridge used twelve cameras with drop shutters, and specially treated wet plates permitting exposures of only 1/2000 second.

In 1881, Muybridge met Etienne Jules Marey, a physiologist, and the painter Meissonier in France, where he showed some of his sequential photographic slides in motion by fastening them to a large disc; another counter-rotating disc with slots and an arc light, condenser and lens served to throw the images briefly on a screen. Muybridge claimed that he had first used this device, which he named a zoopraxiscope, in 1879 at Stanford's house. This was the first motion picture projector. On his return to America in 1883, with encouragement from the painter Thomas Eakins, he greatly improved his recording technique at the University of Pennsylvania. He used batteries of twelve cameras with special high-speed lenses and shutters (which he patented as the electro-

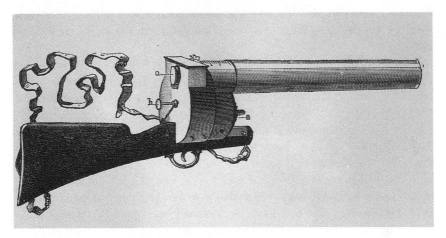

Figure 15.5: Marey's 1882 photographic revolver.

expositor); after this, he designed a portable version. However, instead of pursuing the invention of what was destined to become a vast industry, his interest stayed with the recording and analysis of animal and human locomotion.

The next important steps were made by Marey, who was interested in studying birds in flight. In 1882 he designed a portable multi-exposure camera in the form of a gun, with which he was able to take twelve exposures in a second, each of 1/720 second duration (Figure 15.5); a later device took ten exposures of 1/5000 second each. Taking this photographic gun to Naples, he 'shot' seagulls against the sky; the result was a series of silhouettes from which he modelled life-size images in wax which he viewed in a large zoetrope. In 1888 he used a roll of sensitive paper and built what could be considered the first motion picture camera, but this was limited to about forty exposures.

Many other inventors were at work on motion picture devices before 1890, such as Donisthorpe (1876), Friese-Greene and his associates (1884 on) and Le Prince (1888). In 1892, William Dickson, a Scotsman working with Edison, built a camera which could take 46 pictures a second on Eastman celluloid film, and a corresponding viewing device, the kinetoscope. Edison credited the zoetrope and zoopraxiscope with giving him the idea, but claimed that the kinetoscope made significant improvements: pictures were taken from a single viewpoint; they were taken more rapidly, and the inter-image interval was reduced to below $\frac{1}{7}$ second to give smooth motion; and (most important from the practical standpoint) celluloid film provided an image carrier of indefinite length. (Early kinetoscope films were in the form of an endless loop about 20m long (66ft), which was moved continuously past a viewing lens, transitions between images being hidden by a revolving shutter.) Edison exhibited the device in 1893–4, but thought of it as a mere novelty and failed to patent it in

England. However, his choice of film width (35mm) and perforation design have become standard for normal-width motion pictures down to the present time.

Robert Paul, finding the kinetoscope unpatented in Britain, copied the device and sold several to Charles Pathé in France. Edison, realizing his mistake, refused to allow Paul to use his films, stimulating Paul to build his own camera. Paul also constructed a projector with a Maltese-cross intermittent-motion system (the French astronomer, P. J. C. Janssen had used a revolver camera with a Maltese cross to photograph the transit of Venus in 1874). However, credit for the first successful motion-picture projection system is usually given to the brothers August and Louis Lumière of Lyons, who in 1895 designed a combined camera and projector to which they gave the name cinématographe. They used the same size film as Edison, but with only one round perforation at the margin of each picture, instead of the four square ones used by Edison. They also were the first to give an exhibition for which the public had to pay admission, in 1895.

In 1896, Lumière films were projected in London by the magician Trewey, who was soon giving regular cinema shows; the films were only about 15m long (49ft), so frequent reel changes were necessary. In 1897, Pathé separated the cinématographe into two distinct parts: camera and projector. The Lumières were quick to capitalize on this new entertainment medium, buying up film manufacturers in France and creating production units in the United States and in Russia. Edison, reluctantly abandoning his kinetoscope, was persuaded by his distributors to buy the rights to a projector so his films could be shown before an audience, and not just remain a curiosity – the 'peep-show'. Edison then threw his great organizational talents into the battle, and his company made 1700 movies. He also built the first studio, a rambling affair which could be rotated to follow the sun; it was called the Black Maria, because it was covered with tar paper to increase contrast.

Even though Edison was also the inventor of the phonograph, and had been able to synchronize film and disc in his laboratories in 1889, the system was too awkward for commercial application. A sound-on-film system was patented by Eugène Augustin Lauste in France in 1906 and demonstrated in 1910, and a more advanced system shown by Lee de Forest of the US in 1923, but they did not have any commercial success. What many critics consider the golden era of the motion picture, the silent era, had a chance to flourish for quarter of a century.

Another desired innovation was colour, and attempts were made from as early as 1898. Some films were hand-coloured using stencils, and the French tried tinting certain scenes for dramatic effect. However, although a number of two-colour processes were devised in the early part of the twentieth century, the first commercially viable full-colour film was not made until 1932 (see p. 736).

FACSIMILE AND TELEVISION

While the practice of transmitting pictures over the air is relatively recent, the basic principles behind facsimile and television can be observed in nature, and have been used by man in his art for thousands of years. Many insects have compound eyes; that is, instead of just one lens to focus a scene on the optic nerve, their eyes are composed of thousands. This is the mosaic principle: a multiple of similar, tiny elements composes a single, meaningful whole. The great advantage of using fragments of minerals, glass or shells in art is that they are small, light and multi-coloured, yet can be used to cover immense, multiply curved surfaces with intricate designs, more brilliant than any painted surface, yet far more durable. Hence, mosaics were a natural choice in decorating early Christian churches.

A mosaic design is another example of optical illusion; the eye is tricked into seeing the whole when it cannot distinguish individual elements. In today's communications terminology, digital elements are used to represent an apparently continuous pattern. The same principle was applied to painting in the nineteenth century, particularly by Georges Seurat who used tens of thousands of different-coloured paint dots to depict outdoor scenes. The breaking up of a whole into tiny segments is also the basis of half-tone printing processes (see p. 679), and the underlying mechanism of video cameras.

Another method of dissecting a whole into simpler elements is scanning, which transforms a spatial pattern into a temporal one. This is the technique we use in reading the printed page, and underlies facsimile and television transmission. The eyes move across a line of type, and 'fly back' to the beginning of the next line; similarly, the electron beam in a television receiving tube sweeps across it 625 times every $\frac{1}{50}$ second to compose images, a sequence of which, due to persistence of vision, we interpret as a moving image, the close simulation of reality. However, the invention of television had to wait until the principles underlying motion pictures were better understood.

Facsimile

In contrast, facsimile ideas were being considered soon after the invention of photography and even before electric telegraphy; the basic inventions for the transmission of documents and still pictures were made 30 years before the telephone. In 1817 the element selenium was discovered by Berzelius in Sweden. In 1843, Alexander Bain, the Scottish watchmaker who invented the electric clock (see p. 696), patented an electrochemical recording telegraph, and in 1847, Frederick Bakewell devised the cylindrical mechanism on which images for transmission are placed for scanning, which also provided synchronization. Cylindrical scanning became the basis of most facsimile machines up to the

1980s. During 1861–2, Giovanni Casseli in Italy devised the pantelegraph, which was used by the French PTT to transmit images between Paris and Marseilles during the latter part of the decade. Willoughby Smith of the Telegraph Construction Company in England discovered in 1866 that light would lower the electrical resistance of a selenium rod. This, the photoelectric effect, is the basis of all facsimile and television systems. In 1877, Constantin Senlecq in France published a paper on his telectroscope, the first device to use selenium for scanning.

Facsimile seems to have had a slow but steady evolution from that time, but during the early part of the twentieth century the major use was in transmitting photographs for newspaper publication, a process called wirephoto. It was not until after the Second World War that facsimile devices became available in offices. These desktop transceivers, called telefax, required six minutes to transmit a page over the public switched telephone system; in contrast, large centrally located systems known as bureaufax work at higher speeds between post offices, and use separate transmission and reception devices. In the 1970s the Japanese turned their attention to facsimile, after long frustration with attempting to use telex with the large character sets of their language. They now have a virtual monopoly on the manufacture of facsimile machines, which are capable of unattended transmission and reception at speeds of less than a minute per document.

Television

It was not long after the invention of the telephone in 1876 that imaginative writers and artists were dreaming of the next step, which they called by such names as 'telephonoscope' or 'electric vision'. The French artist Albert Robia was especially entranced, and drew scenes of families watching a war 'live' in their living-rooms, people taking courses without going to school, and housewives window-shopping from an armchair.

In 1884, eight years after the invention of the telephone, a German, Paul Nipkow, patented his ingenious disc, which had 24 holes evenly spaced along a spiral. While the disc spun at 600rpm a lens focused the light samples on a selenium photocell. The output of this cell was a varying electric current; to reassemble the now dissected image, Nipkow proposed using an identical, synchronously rotating disc and a Faraday-effect (magneto-optical) light modulator. Nipkow did not actually build this system (no suitable technology was available at that time), but the Nipkow disc was to become the basis of early mechanical television systems.

In 1897 in Strasbourg, Ferdinand Braun constructed the first cathode-ray oscilloscope, the basis of all present television receivers. In 1907, Boris Rosing at the Technological Institute in St Petersburg proposed using Braun's tube to receive images, and A. A. Campbell-Swinton in a letter to *Nature* in 1908, 'Distant Electric Vision', proposed using cathode-ray tubes (CRT), as they became known, for both transmission and reception. Vladimir K. Zworykin, who had emigrated to the USA from Russia, joined the Westinghouse company in 1919, bringing with him ideas about using CRT devices for television. However, it was not until 1923, after having left and returned to Westinghouse, that he constructed the first practical storage camera tube, the iconoscope. The image was projected on to a mosaic of photosensitive elements inside the tube. This mosaic was then scanned by a beam of electrons which, by a process called secondary emission, released charged elements to form the picture signal.

During the period 1923-4, Charles Francis Jenkins, who had contributed to the evolution of motion-picture projectors in the 1890s, experimented with the Nipkow disc. About the same time, but independently, John Logie Baird in Scotland was engaged in similar experiments. Baird, and Jenkins shortly after, made demonstrations of their systems in 1925, as did Ernst F. W. Alexanderson at the General Electric (GE) Company's headquarters in Schenectady. In 1927, Philo T. Farnsworth, an independent inventor, demonstrated the first complete electronic television system, based on his invention of the image-dissector tube. However, Zworykin, now with the Radio Corporation of America (RCA), contested Farnsworth's patent, and only after long litigation did each receive basic patents on their systems.

These experiments came to preliminary fruition in 1927, when the presidential candidate, Herbert Hoover, appeared in an experimental AT&T telecast. GE put the first home TV set on the market in 1928, and the same year broadcast the first dramatic production, the sound going out over WGY while the picture was transmitted from experimental station W2XAD. This was followed by a science-fiction drama, giving its audience a missile-eye's view of an attack on New York City. In 1929 the British Broadcasting Corporation (BBC) began an experimental, low-definition TV broadcast; and in the US, the Bell Telephone Laboratories transmitted a television image in colour between Washington and New York, using three separate channels to transmit the light-primaries, red, green and blue.

In 1932, Zworykin, now at RCA, supervised the installation of television equipment in their flagship NBC studios in the Empire State Building, and in 1935, RCA's president, David Sarnoff, announced that they would spend a million dollars for TV programme demonstrations. Also in 1935, a station in Berlin began low-definition broadcasting, and in 1936 the BBC began the world's first high-definition TV broadcasting service. In 1937 a mobile TV unit roved the streets of New York City for live pickups. The pre-war development of television technology culminated in the conception of the shadow-mask colour tube in 1938 by W. Flechsig in Germany, and the invention of the first large-screen television projector, Eidophor, by Professor Fischer at the Swiss Federal Institute of Technology in 1939.

The year 1939 also saw the opening of the New York World's Fair, at which Sarnoff premiered commercial television. RCA also demonstrated a complete home TV set; a peculiarity of this set was that viewers did not look directly into the CRT, but saw the tiny images in a mirror – whether this was done because of concern about irradiating the public, or for technical reasons is not known. A few of these sets were sold to enthusiasts; others, more technically minded, built their own to receive the experimental broadcasts, but the war delayed regular television broadcasting.

Television broadcasting

After the war ended in 1945, the US Federal Communications Commission (FCC) resumed television licensing, displacing FM radio's allocation in the frequency spectrum. By mid-1946, 24 new licenses had been issued. The Columbia Broadcasting System (CBS), long RCA's main rival, demonstrated a colour system invented by Peter Goldmark. Although giving a brilliant image, it was incompatible with the existing black-and-white sets which RCA had just placed on the market. Therefore, RCA wanted to delay approval, and in 1947 the FCC postponed any colour decision. It was not until 1953 that the FCC adopted a compatible colour system, called NTSC (National Television Systems Committee). In 1956 an improved system called SECAM (Sequentiel Couleur à Memoire) was devised in France; and in 1962, PAL (Phase Alternance Line) appeared in Germany. All these systems are incompatible, and in consequence no world-wide television system is possible without conversion until a new, high-definition, all-digital replacement is introduced, probably in the 1990s.

In the US, 1947–8 saw the beginning of television entertainment as well as news, pitting NBC and CBS against each other for the flood of advertising revenue. In 1949 the first TV set to appear in Sears, Roebuck's mail-order catalogue was offered for $149.95.

By 1956–7, 40 million (85 per cent) of all US homes had TV sets; there were 500 TV stations; and families were spending five hours a day before the 'tube'. Advertisers flocked to take advantage of the immense persuasive power of the medium, even experimenting with such Orwellian devices as subliminal perception.

In 1962, the first transatlantic television transmission was achieved, using the communications satellite Telstar I (see below).

COMMUNICATIONS SATELLITES

The basic principle under which satellites function is due to Kepler, who first elucidated the laws under which planets orbit the sun (celestial mechanics).

However, the notion of an artificial satellite was apparently first conceived in an 1870 science-fiction story, *The Brick Moon* by Edward Everett Hale. The story featured the use of a visible, 60m (200ft) diameter body orbiting the earth to assist navigators. In 1870, Asaph Hall discovered two natural satellites revolving around Mars, and told Hale that from what he could see, one might be such a 'brick moon'.

Jules Verne's novel *From the Earth to the Moon* (1865) was remarkably prophetic, not in the mechanism used – a train-like projectile shot out of a huge cannon – but in his choice of geographical locations. Verne picked Florida for his launch site (a hundred years later, the first manned moon rocket left from Cape Canaveral), and the cannon was designed in Baltimore (home of the Glenn L. Martin company, which built the Viking rocket, an early improvement to the Second World War V2). However, it was not until 1924 that Hermann Oberth described the potential advantages of artificial satellites in detail, including space stations and huge space mirrors to control weather, in his book *Wege zur Raumschiffahrt* ('Ways to Spaceflight').

After the Second World War serious work on artificial satellites accelerated. In 1945, Arthur C. Clarke published his prophetic article on the advantages of a geostationary orbit for satellites to be used for radio and television communications. Wernher von Braun published in 1952 a popular article advocating the use of a manned space station for military purposes. Other, more peace-oriented proponents wanted small, instrument-carrying satellites for scientific purposes (see Chapter 13).

The first satellite designed specifically for communications was launched in 1960. Dubbed ECHO 1, it was nothing but a huge reflecting sphere for relaying voice and television signals. However, it was soon followed by the first active repeater satellite, Courier 1B, and the communications satellite era began in earnest. In 1962, Teistar I became the first communication satellite which could relay not only on data and voice, but also TV. Although NASA, a US government agency, provided the launch, a private corporation, AT&T, owned the satellite; a few months later, Congress authorized the creation of COMSAT (Communications Satellite Corporation) as a private corporation.

In 1963, the first satellites to be placed in geosynchronous orbit – with all the advantages that Clarke had foreseen – were launched. These, the Syncom series, were built by the Hughes Aircraft Corporation. Because their on-board power was low, and because the Syncoms had to be placed in an orbit 37,000 km (23,000 miles) from the earth's surface, ground stations had to employ 30m (100ft) diameter parabolic antennas (called 'dishes' from their appearance), which cost $3–5 million each. In 1959, Clarke, again foreseeing the next step, published a popular article advocating satellites of great power, whose signals could be received by low-cost antennas connected to ordinary TV sets; in the late 1980s, several such direct-broadcast satellites (DBS) have been put into orbit.

From 1960 thousands of satellites were launched by the US and the USSR for a variety of military and civilian purposes; and by them for other countries. The West, under the multinational consortium Intelsat, preferred geostationary orbits for its communications satellites, three of which are enough to provide world-wide coverage. The East, under Intersputnik, preferred close-in, near-polar orbits, although this means that many satellites are required to provide 24-hour coverage; also, they must be continuously tracked as they flash from horizon to horizon. However, much less power is needed aboard, and smaller earth-station antennas can be employed.

The launch monopoly held by the two superpowers has been challenged since 1979 by the European Space Agency (ESA), using their Ariane rocket, and other countries including China and India have the capability. Also, these countries and Japan have built their own satellites. In the 1980s, the US became more and more dependent on NASA's manned space shuttles for satellite placement, but a launch disaster in 1986 gave new impetus to conventional rockets.

Back to 'wire'?

By the mid-1980s, Intelsat had so many satellite circuits available over the north Atlantic for telephone, data and television transmission that it had surplus capacity. Furthermore, a rival technology, transatlantic fibre-optic cables, was threatening satellites with severe competition. Laser beams, travelling in two pairs of glass fibres, are capable of carrying the equivalent of 37,800 simultaneous telephone conversations. Even greater capacity is in the offing, AT&T's Bell Laboratories having achieved a record of 20 Gbits (1 gigabit = 10^9 bits) per second over one hair-thin fibre using optical multiplexing, the equivalent of 300,000 conversations.

The idea of beaming TV programmes directly into homes by means of very high-power direct broadcast satellites (DBS) is being challenged by co-axial and hybrid fibre-coax distribution systems. Thus, we may come full circle, back to the original concepts of telegraphy – a 'wired world', rather than one principally dependent upon space technology.

INFORMATION STORAGE TODAY

In principle, any physical phenomenon can be used for recording information, but until the scientific revolution information always remained readable by the naked eye. Only after optical (photographic), electromechanical (telegraphic and phonographic) and magnetic recording methods were developed did we extend our storage capacities to the submicroscopic level.

Also, with the exception of telegraphy, information was recorded in analogue form. The alternative, digital recording, is much simpler in principle: finger-counting and tally sticks were among the earliest digital techniques. The development of the electric telegraph, which used a pulse-code system from the beginning, led to theoretical studies of digital systems and analogue-digital conversion technologies.

Digital systems for data storage are only two centuries old. However, the inventions of Jacquard, Babbage and Hollerith (see pp. 699, 701) were mechanical, and even when they became electromechanical they were relatively slow in operation. The acceleration of science and technological development during the Second World War provided practical electronic, magnetic and optical data storage systems.

Pulse-code modulation and information theory

In 1926, Paul M. Rainey of the United States was granted a patent on a pulse coding system, but nothing seems to have come of it until Alec H. Reeves, an Englishman working in the Paris laboratories of the International Telephone and Telegraph Company (ITT) reinvented it in 1937; the fundamental PCM (Pulse Code Modulation) patents were granted to him in 1938 and 1942. However, no available electronic components were capable of implementing PCM at a reasonable cost. During the Second World War a team headed by Harold S. Black at the Bell Telephone Laboratories designed the first practical PCM system for the US Army Signal Corps to safeguard telephone conversations. Civilian applications had to wait until relatively recently, when semiconductor technology – particularly the integrated circuit (see p. 705) – made them economically attractive. The first significant PCM application was in digitizing voice communications for telephony. For PCM, soundwave amplitude is sampled 8000 times every second and each sample encoded to a 7-bit accuracy (1 part in 128), with an extra bit added for signalling purposes. Thus 64,000 pulses per second (64 kbits per second in today's units) are necessary for intelligible speech transmission.

In 1948, Claude E. Shannon of the Bell Telephone Laboratories published a seminal paper, 'The Mathematical Theory of Communication', which for the first time provided a sound theoretical basis for understanding information and communication phenomena of all kinds. He outlined a 'universal' communication system: a *source*, producing messages; a *transmitter*, encoding them into suitable signals for transmission; a *channel*, through which the signals travel; a *receiver*, decoding the signals; and a message *destination*. Sender and receiver are usually human beings, the other elements being mechanical, optical, electronic or other artificial means. Another element, a source of *noise*, is always present in real communications channels. This general model applies from the simplest to the most complex communication systems.

Another part of Shannon's theory gives a precise measure of information, based on the concept of uncertainty (entropy in thermodynamics; information has been called negentropy by some information theorists). A bit (a neologism formed from '*bi*nary' and 'digi*t*') can represent the uncertainty between 'yes' and 'no' when both are equally likely; the message content, the storage capacity, the transmission rate, and the channel capacity can all be measured in terms of bits. For example, a telephone channel which has a band-width of 3000Hz has a channel capacity of about 60,000 bits per second, and commercial television, with a 6MHz band-width, requires a channel capacity approaching 100 Mbits per second – more than a thousand times greater.

The work of Reeves, Shannon and many others will culminate in a totally digital switched public telecommunications system known as ISDN (Integrated Services Digital Network), which will replace the existing analogue networks which have been the only public carriers for more than a century. ISDN will transmit voice, data, text and images using multiple 64kbit-per-second channels.

Image and data storage and retrieval systems

Starting in the 1950s, many attempts have been made to use electromechanical – essentially phonographic – techniques to store and play back images, such as Phonovid, a still-picture TV system (Westinghouse, mid-1960s); and Teldec, a full-motion, black-and-white TV system (AEG-Telefunken and Decca, 1970). They failed: Phonovid required an expensive scan converter; and Teldec, even though attaining a storage density 100 times greater than that of LP records, could only play 7-to 15-minute programmes. About 1970, EG&G Inc. introduced their Dataplatter, an adaptation of 7in (18cm) diameter, 45rpm phonograph technology which could store up to 5 megabits (Mb) of data. The reasons for the failure of this innovative technology in the marketplace are unclear – perhaps it was too far ahead of its time, coming five years before the microcomputer.

Optical means have been used almost from the invention of photography, but only for images, not for digital data. However, in 1945, Vannevar Bush published a seminal article, 'As We May Think', in a popular American magazine, the *Atlantic Monthly*. Bush had developed the Differential Analyzer in 1931, one of the earliest analogue computers, for solving differential equations. During the war he became director of the US Office of Scientific Research and Development. Looking forward to a conversion of the massive wartime scientific effort to peacetime uses, he pointed out how primitive were our methods of accessing the research literature. After predicting the inventions of dry photography, ultramicrofiche and the voice-operated typewriter, he conceived an automatic system for the mass storage and rapid retrieval of documentary information, which he called 'memex'. Memex was based on an extrapolation of imaging technology, but the selection principle was associative indexing, which only now is coming into the realm of the possible as a technique of 'artificial intelligence'.

It was not long after Bush's detailed prophecy that automatic retrieval of photographically reduced documentary information was embodied in working hardware. The Filmorex system, invented by Jacques Samain in France early in the 1950s, used 70 × 45mm ($2\frac{3}{4}$ × $1\frac{3}{4}$in) film-cards (or microfiche, which French word has been adopted into English). Each Filmorex fiche contained two moderately reduced page images, together with a grid of tiny squares on which was encoded up to 500 bits of index information. When the fiche passed by an array of photocells in the Filmorex selector, code-combinations matching the search request would trip a release mechanism and divert the selected fiche to a bin for viewing. A similar system, Minicard, was developed in the 1950s by Kodak's Recordak Division. This was based on tiny fiche cut from 16mm microfilm, each of which could store up to six pages at a 60:1 reduction, together with more than 1500 bits of data. Roll microfilm was used for optical data storage in FOSDIC (Film Optical Scanning Device for Input to Computers) to speed up the processing of the vast amount of data collected for the 1960 US census; data were reduced images of punched cards, compressed vertically by photographing them through an anamorphic lens.

Videotape

In 1956, Alexander M. Poniatoff demonstrated the first videotape recording (VTR) device; and in 1958 his Ampex Corporation installed the first VTR in American television studios. This invention radically changed TV programme production and dissemination. The first VTR machines used 2 inch-wide (5cm) tape and were very expensive, but programmes could be recorded and played back immediately, unlike film; this capability was of incalculable value for television production. The public could no longer tell whether it was watching a programme 'live' or 'canned'. Attempts had been made to store television on tape before, but these had been brute force systems based on tape speeds hundreds of times greater than used for audio – with concomitant mechanical problems. Poniatoff saw that the required megahertz band-width could be achieved without these penalties by using magnetic heads rotating at right angles to the linear tape motion. Ampex also realized that VTR technology could store high-resolution still images. In the 1960s they marketed Videofile, which could store facsimiles of 250,000 pages (about 5Gbits: 1 gigabit=10^9

bits) on a reel of videotape. These systems cost several million dollars and only a few were purchased by government agencies and contractors, such as NASA for space programme records; Bell Telephone Laboratories for maintaining Nike-X data; and the Southern Pacific Railway for waybills.

The Ampex professional system was the standard for commercial television studios for several decades. Much less expensive recorders using one inch (2.5cm) wide tape attracted educators and training establishments which wanted to experiment with this alternative to film production; these VTRs use helical recording (tracks at a slant instead of almost vertical). Many helical recorders were in use during the 1960s, and Philips soon marketed a videocassette recorder (VCR), which was much easier to operate for the inexperienced. Philips one-hub cassettes were not very reliable, and image quality was inferior. Sony brought out a two-hub system using $\frac{3}{4}$ inch-wide (19mm) tape, called U-Matic. These recorders sold for $2000 or more, but were very reliable, versatile and provided a high-quality colour picture, and U-Matic became the standard for educational and industrial use.

During the 1970s, Phillips, Sony and JVC (Japan Victor Corporation) brought out rival two-hub consumer systems using $\frac{1}{2}$-inch (12.5mm) tape. By the early 1980s the most popular was JVC's VHS (video home system); Sony's Betamax had only 10 per cent of the market, and the Philips V2000 was scrapped. By 1985, inexpensive VHS recorders capable of recording up to four hours of TV on one cassette, and with built-in timers for unattended recording were dominant.

In the early 1970s, mass memory systems based on lasers and holography (see p. 737) were developed. W. J. Hannan of RCA invented a home system, Selectavision, which used clear vinyl tape for the storage and playback of colour television programmes; images were stored holographically. Precision Instruments Company developed an optical terabit memory (1 terabit=10^{12} bits), the Unicon, based on burning 1μm holes in metal-coated polyester tape with a laser. However, such optical media, like electromechanical ones, had a major disadvantage–data could be recorded only once, and could not be erased. Erasable and reusable magnetic systems were preferred for active mass storage, with optical systems relegated to archival support, usually for images.

Computer data storage

The earliest form of magnetic mass-data storage was magnetic tape, which had been originally developed for the analogue recording of sound waves (see p. 724). However, even with multi-track digital recording, which allowed the bits encoding each character to be stored in parallel, tape had to be scanned sequentially. Computers also needed rapid random access to small amounts of data. Magnetic cores strung on a grid of wires gave the highest speeds, but were very expensive on a per-bit basis (one bit per core); therefore, magnetic drums were introduced for back-up mass memory. In 1956, IBM introduced RAMAC (Random Access Method of Accounting and Control for Automatic Computers), the first magnetic disk system. Up to 100 rigid disks were mounted on a common spindle which spun at high speed; data were written and read (computerese for recorded and reproduced) by magnetic-sensing heads flying in and out only microinches above the disk surfaces.

In 1962, IBM marketed the first cartridge magnetic disks, which could be mounted on a disk drive for access, but removed for storage – a similar system to that employed for magnetic tapes. About 1970, IBM introduced a non-rigid magnetic disk which has come to be called the floppy disk, or simply floppy. This has become the preferred medium for off line storage of data and programs for small computers. The original floppies were 8in (20.3cm) in diameter and could store 100,000 bytes (100 kbytes) on a side; the encoding system, data density and file format were set by IBM, becoming a *de facto* standard. However, when microcomputers became popular in the late 1970s, a $5\frac{1}{4}$in (13cm) version was introduced, soon dubbed the minifloppy. Unfortunately, formats were never standardized, so that by the early 1980s there were a hundred different ones: single-density, double-density, even quad-density; and both single-sided and double-sided. Despite this diversity, the $5\frac{1}{4}$in floppy has become the universal external storage medium for microcomputers, providing from 200 kbytes to more than 1 megabyte (1 Mb) per disk. In the early 1980s, further miniaturization took place: a $3\frac{1}{2}$in (9cm) diameter disk, sometimes called a microfloppy. These disks can store 1.5Mb and are more reliable because they are enclosed in a rigid case.

Also, in the late 1970s, rigid disks of 8in (20cm), and then $5\frac{1}{4}$in (13cm) diameter were introduced to provide much greater mass-storage capacity for microcomputers. They are called hard disks to distinguish them from floppies, and are usually not removable, being housed in a dust-free casing. They spin at much higher speeds than floppies, and their plated-metal surfaces are capable of much higher recording densities, providing users with from 20 to more than 100 Mbytes of on-line data access.

Optical Media

The dominance of magnetic media for the storage of all types of information is now being challenged by optical media, particularly those based on the use of lasers for recording and reproduction. The first videodisc system was developed by Philips in the Netherlands in the 1970s for the storage and playback of still and motion pictures; all information is stored digitally in tracks as microscopic pits ('micropits'), and read out by means of reflected laser beam. Thomson CSF, a French company, developed a similar format, but data is recorded as microscopic surface blisters, and read out transmissively. Recording techniques are based on pulse-width modulation, and both use discs the same size as LP records, 12in (30.5cm) in diameter.

The Philips system was originally intended for home use; consumers would buy videodiscs of their favourite movies, just as they bought pop songs or other music on gramophone discs. However, instead of about five minutes of music (on 45rpm records) or thirty minutes (on $33\frac{1}{3}$rpm LPs), videodisc enthusiasts would get up to an hour of sound, colour motion pictures or TV programmes. A number of manufacturers have marketed laser videodisc players, including Philips; Magnavox, a US company; and Japanese companies such as Pioneer and Sony.

However, videodisc marketing not only had to contend with three incompatible broadcast television standards (see p. 746), but also with the well-entrenched videocassette industry which so dominates the home video market that it threatens to replace motion-picture theatre distribution as the primary way in which the public sees movies. Therefore, although multinational companies such as General Motors and IBM adopted videodisc technology for industrial training, it was a failure in the consumer market. The worst drubbing was taken by RCA; by the time they withdrew their contact-stylus, capacitive-sensing system, they had spent about $100 million on its development and marketing.

For education and training, a single videodisc can store up to 54,000 individually addressable, colour, still images; or a mixture of stills and motion sequences. Archival videodisc systems are an ideal distribution medium for large colour slide collections, and are becoming a useful tool for museums. For example, the Smithsonian Institution's Air and Space Museum in Washington, DC, the world's most visited, has put its archives of Wernher von Braun, the rocket pioneer, on a single videodisc for sale to the public. They are also making their photographic collections (almost a million items) available in this format.

Even more important than the vast storage capacity of videodiscs is the possibility of student interaction using a built-in microprocessor or external microcomputer. Interactive videodisc is the most promising method so far devised to implement the long-sought promises of CAL (computer-assisted learning), alternatively known as CAI (computer-assisted instruction) in the United States. Apart from text-oriented 'drill-and-practice' lessons, CAL has been too expensive in cost per student-hour; however, by making available the distribution of 50,000 images at a selling price of $20, videodisc-based CAL has the potential of becoming the least expensive form of audiovisual instruction ever devised. The main obstacle is the high cost of developing well-designed and tested instructional programmes – it may take a professional team of educators and technologists 100 hours or more to develop a one-hour programme.

Optical Data Recording

Another use for laser-optical technology is the storage of digital data, a natural extension because all types of information (pictures, sound, text, data) can be recorded by laser-based technology. More than a gigabyte can be stored on one side of a 12in (30.5cm) DOR (digital optical recording) disk, ten times more than on a similar-sized magnetic disk; however, random access times and transfer rates are much slower. For very large memory requirements, multidisk systems are available, such as Philips's Megadoc, which uses a jukebox mechanism holding up to 64 optical disks. Within a quarter of a minute, users can access one record (equivalent to an A4-sized page) out of a store of 15 million such records. Megadoc could provide the all-electronic library, which, unlike present automated bibliographic systems, would provide immediate access to facsimiles of original documents.

At the other extreme is the unit record, particularly important in transaction processing systems. Such systems are characterized by the storage of only a small amount of information, which must be updated frequently, in each record. Drexler Technology in the United States has developed the LaserCard, which can store up to 400 pages of digitized text (2 Mbytes) on a credit-card-sized piece of polycarbonate. This medium can store ten thousand times more data than magnetic-strip plastic cards, and lasts twenty times longer. Promising applications include individual, pocket-sized medical histories; maintenance manuals and other reference publications; and software for microcomputers.

Philips and Sony, working together, developed a laser-read optical format for audio recording, which was introduced in 1979 as the Digital Audio Disk (DAD). Without the incompatibility problems of television, Philips and Sony were able to persuade other manufacturers to adopt it. Stereophonic music recordings using this medium were introduced to the public about 1983, under the name Compact Disc (CD); they are 12cm ($4\frac{3}{4}$ inches) in diameter, and 1.2mm thick. Although their dynamic range is 96 decibels, 30 times greater than on LP records, CDs provide virtually noise-free listening during quiet passages. Also, because the recording is read by reflected laser beam, they are virtually indestructible.

Sound to be recorded for CD use is sampled 48,000 times a second, each sample being encoded with 16-bit resolution. With the addition of error-detecting bits, the digital band-width approaches 1 Mbit per second, fifteen times greater than PCM-based telephony. A compact disc can provide uninterrupted play for up to an hour; any track can be selected and played automatically; and sequences of selections can be pre-programmed by the user. Soundwaves are encoded by PCM, the basic information carrier being pits burned by laser into the master, each of which is only $0.16 \mu m$ high and $0.6 \mu m$

wide. Such dimensions are those of light wavelengths, which range from 0.4 μm (violet) to 0.7 μm (red) – hence the iridescent colour patterns that CDs exhibit, characteristic of lightwave interference. This is the greatest density of artificial information storage yet achieved in a consumer product, and the error-correction capability of the code used is such that a 2mm hole drilled in the record will not affect playback.

In 1985 the same technology was made available for the storage of digital data. Virtually the same players can be used as for music reproduction, using a switch to bypass the digital-to-analogue conversion required by audio amplifiers and loudspeakers. Each CD-ROM (for Compact Disc Read-Only Memory) can store up to 500 Mbytes of data. This is 100 times greater than state-of-the-art floppy-disk technology. However, as has been true of many new media, finding the right place for CD-ROM technology is a problem. What is the best application for a read-only (soon to be write-once, read-only) storage system which has a data storage capacity equivalent to 1000 printed books?

One market in which many CD-ROM-based services are interested is the publication of bibliographical databases. Such databases were originally available only in printed form, but as publishers accumulated vast amounts of machine-readable text they realized that a market existed for on-line access to frequently updated versions of the same material. Customers use computer terminals or microcomputers connected to large, central computers via telecommunications, so remote database searching is expensive and demands considerable training. CD-ROM promoters believe this new technology could be more cost-effective: users can interact with the medium without entailing constantly accumulating communications costs, and the difficulties of 'logon' procedures which require accurate keying of long sequences of characters are eliminated.

Present CD-ROM storage capacity is too small for retrospective searching applications, which may require access to data files going back 10–20 years, and the discs cannot be updated. However, as CD-ROM technology advances these problems will be mastered, so that by the early 1990s publishers will at last have a convenient, compact and economical alternative to print – heralding the era of electronic publishing.

FURTHER READING

Introduction

Dummer, G.W.A. *Electronic inventions and discoveries: electronics from its earliest beginnings to the present day*, 3rd edn (Pergamon Press, Oxford, 1983)

Grogan, D. *Science and technology: an introduction to the literature*, 4th edn (Clive Bingley, 1982)

Ryder, J.D. and Fink, D.G. *Engineers and electronics: a century of social progress* (IEEE Press, New York, 1984)

Timekeeping

Landes, D.S. *Revolution in time* (Harvard University Press, Cambridge, 1983)

Counting, calculating and computing

'The computer issue', *Science*, vol. 228, no. 4698 (26 April 1985)
Shurkin, J. *Engines of the mind: a history of the computer* (Norton, 1985)

The telegraph

McCloy S.T. *French inventions of the eighteenth century* (University of Kentucky Press, Lexington, 1952)

The telephone

Shiers, G. (Ed.) *The telephone: an historical anthology* (Arno Press, New York, 1977)

The gramophone

Hope, A. 'A century of recorded sound' and 'From rubber disc to magnetic tape', *New Scientist* (22/29 December 1977, pp. 797–9; 12 January 1878, pp. 96–7)

Radio and radar

Barnouw, E. *A history of broadcasting in the United States*, 3 vols. (Oxford University Press, New York, 1966, 1968, 1970)
Briggs, A. *A history of broadcasting in the United Kingdom*, 3 vols. (Oxford University Press, London, 1961–5)
Susskind, C. 'Who invented radar?', *Endeavour*, New Series, vol. 9, no. 2 (1985) pp. 92–6

Photography

Pollack, P. *The picture history of photography*, 2nd edn (Harry N. Abrams, Inc., New York, 1970)
Desilets, A. et al., *La photo de A à Z* (Les éditions de l'homme, Quebec, 1978)
Okoshi, T. *Three dimensional imaging techniques* (Academic Press, New York, 1976)

Facsimile and television

Barnouw, E. *Tube of plenty: the evolution of American television* (Oxford University Press, New York, 1975)

Satellites

Clarke, A.C. *The making of a moon: the story of the earth satellite program*, 2nd edn (Harper & Brothers, New York, 1958)

McDougall, W.A. *The heavens and the earth: a political history of the space age* (Basic Books, New York, 1985)

Information storage today

Information takeover (videocassette with printed Fact File, *New Scientist/Quest Video*, 1985)

PART FIVE

TECHNOLOGY AND SOCIETY

16

AGRICULTURE: THE PRODUCTION AND PRESERVATION OF FOOD AND DRINK

ANDREW PATTERSON

INTRODUCTION

From the time that the first steps were taken to obtain food by the cultivation of the land, the proper practice of that technology has been of concern to the whole of the community that depended upon it. The need to record the methods of the experts so that their knowledge might be utilized by those less proficient, or might be passed to future generations, was felt by many early authors. As a result various literary sources have been left to us, from which we are able to gain glimpses and sometimes great insights into the agricultural practices of the past. The state of preservation of this literature, and the range of its content, has affected the quality of knowledge available. The most complete material in terms of subject matter and timespan comes from the Chinese. Classical Mediterranean writing has provided quite detailed information on technology, although little of social or economic interest. Middle Eastern sources have provided tantalizing glimpses of laws of property and husbandry, and also volumes of account books and records of palace or state stores. In contrast, the mediaeval period has left us with only snatches of technical and economic information, and our knowledge of European agriculture is very scanty until the eighteenth century. Only in very recent times has any consideration or study been given to those areas of the world outside the European influence, and outside the knowledge left by the written word.

In addition, archaeology has provided extensive data with the retrieval of the actual equipment used, and with the discovery of pictorial representations of equipment and practices. In the past twenty years it has also occupied itself with the collection and interpretation of economic data, such as the plant and animal remains which sophisticated retrieval techniques have increasingly yielded from excavations. What is perhaps most surprisingly illustrated from the prehistoric evidence, is how quickly the basic tool kit was established, and how sophisticated it was in terms of the tasks for which it was designed. Huge steps have been taken in the development of machinery in the intervening ten or so millennia, but machinery merely allows greater acreages and bigger yields to be handled more easily and more swiftly: the basic processes of cultivation and harvest, in both crop and animal husbandry, have really changed very little.

Until very recently the search for agricultural origins was pursued by Europeans who brought their own particular regional bias to the interpretation of the data before them, and indeed this bias influenced the type of data they sought in the first place. Since the origins of European religion are to be found in the Middle East, it is perhaps not surprising that the origins of civilization were sought here also. Civilization in this context was seen in terms of a socially stratified society, believing in the concept of a divine being, and with the ability to write of this mythology for the benefit of existing and future generations. To the early nineteenth-century antiquarian only a sedentary population, living in an urban setting, was capable of the sophisticated thought needed for such ideas, and only an efficient agricultural system was capable of producing sufficient food to allow the priests, metal workers or merchants to devote their time and energies to their trades and professions. Within this setting those who produced the food would be totally involved with the process, and would have no input into the other activities.

These beliefs were held well into this century, and it is only since the Second World War that we have achieved a better understanding of the strategies of the various subsistence economies that still exist on this planet. In doing so we have perhaps been able to guess more accurately the economies of prehistoric societies, and the shifts in those societies that resulted in the profound change from hunter gatherer to farmer.

Because of the background to this study, the bulk of archaeological research has been directed to the Middle East, the Mediterranean littoral and to Europe. It is therefore not surprising that most of the information available to us is from this region, and that the oldest dates that can be applied to artefacts and technologies should also reflect this bias; but there are *still* vast areas of the world that have yet to be explored above the ground, and which may contain remains below that could alter our view of the origins and spread of agricultural technology, or which might push its date even further back than can at present be contemplated.

HUNTER GATHERER TO FARMER

The studies of hunter gatherer societies that have been conducted over the past thirty years indicate that these groups spend embarrassingly little time on food procurement, that they utilize only a small percentage of the plant and animal resources available to them, and that therefore the population density within the territories that they occupy is well below the density that the area could sustain. There are not many hunter gatherer groups available to study, and those that do exist inhabit marginal areas which have been ignored or discarded by those societies whose economies are based on a settled agriculture. If these marginal areas can be so effectively utilized today, it is not unreasonable to suppose that the prehistoric hunter gatherer, living in more favourable climates, enjoyed a comparable standard of living with perhaps even less energy expenditure.

The relative importance of plants and animals within the diet of these prehistoric groups is difficult to determine. Studies of modern groups show a dominance either of animals, as in the Ache tribe of East Paraguay, or of plant material, such as characterizes the diet of the Kung of the Kalahari in southern Africa. The latter have been frequently held as the example to prove that our ancestors were predominantly vegetarian; in fact this dietary balance is much more likely to be an adaptation to an environment from which animals have been eliminated by over-exploitation in comparatively recent times.

In attempting to understand the transition to an agriculturally based economy, it has been necessary to study the subsistence strategies at either side of it, and it is perhaps ironic that in this process we should have found a better awareness and respect for the abilities of those hunter gatherer groups who are still so successfully exploiting their harsh environments. Although the projection of ethnographic studies back in time must be viewed with caution, it is worth noting how superbly efficient are the modern groups within their own environment. Western society tends to view their technology, culture and subsistence as crude, and yet few of its own members would be able to survive in these environments. To achieve this success it is essential that their understanding of plants and animals should have attained a high level. It is not unnatural that this should have been so, since these peoples are as much affected by as affecting the environment in which they live.

Before the invention of radio carbon dating it was very difficult to establish relative chronologies and particularly difficult to make comparisons between artefacts found on different continents. However, as soon as this new technology was applied to the organic material derived from different sites throughout the world, it became quite apparent that the long-held belief of the origins of agriculture in the Middle East, and its subsequent spread from there to all parts of the world, was difficult to sustain. There now exists ample evidence to support the concept of the independent origins of both animal and plant domestication in many unconnected sites in the world, and it is likely that an economy based on plant husbandry was established well before one which incorporated a full animal husbandry.

Cultivation

In terms of the numbers of people who base their diet on a particular commodity, the three major plant species for human consumption are wheat, rice and maize. The bulk of the evidence for the first has again been found in the Middle East, in that area generally referred to as the Fertile Crescent, which extends from the Euphrates and Tigris river basins, through south-eastern Turkey, and also along the eastern Mediterranean. Arguments have been put forward for a wider area of origin extending along the north coast of Africa, and also into eastern Europe. Whatever the geographic source, the timing of the shift from the collection of wild species to the cultivation of seed and the subsequent selection which led to domestication seems to lie about 10,000 years ago.

The ancestors of modern bread wheat are varied but can be traced back to primitive einkorn, and perhaps more importantly to emmer wheat. The latter, a plant with a great resistance to fungal attack and a very high protein content, is to be found in the archaeological context in most areas of early settlement agriculture, ranging from the Fertile Crescent, down the Nile valley, up into Anatolia and the Balkans and into continental and northern Europe.

Grains of domesticated wheat, dating from between 7500 and 6500 BC, have been discovered at various sites in the Fertile Crescent. Tell Mureybit in Syria has yielded samples from the earlier date, those discovered at Jarmo in Iraq to 6750 BC. Hacilar in Turkey has realized a date of 7000 BC, and there is evidence for the spread of domestic forms into Europe by the fourth millennium, and into the Nile valley a thousand years earlier. It appears in the Indus valley by the third millennium, and the earliest evidence from China dates to the middle of that millennium.

Early domesticated barley has also been found at Mureybit, and in Jericho by about 7000 BC, Greece by 6200 BC, and Anatolia by about 6000 BC. The presence of barley is of particular significance, because it will tolerate poorer soils than wheat, but in the region in question the poorer soils could only be effectively used if irrigation techniques were also employed. Although the presence of barley does not necessarily indicate the use of irrigation, evidence for this technology can frequently be demonstrated on these sites.

The wild ancestors of maize are difficult to identify, and there are a number of contenders for the title. Some Mexican material retrieved in the 1960s can be dated to 5000 BC, and are definitely domestic. Their cultivation moved slowly southwards, reaching Central America by at least the middle of the third millennium, and possibly considerably earlier.

The early history of rice is less clear. It has been retrieved from a site in Thailand at levels dated to 3500 BC, and in China in levels dating to between 2750 and 3280 BC. It has also been dated to 2500 BC in India.

Animal husbandry

It is perhaps not surprising that the successful hunter groups, with their detailed knowledge of animal behaviour, should eventually begin to manipulate both this behaviour and their own, in order to exploit animal resources more efficiently. Past assessment of this transition has always assumed that this change in strategy was merely to facilitate the procurement of meat. However it has recently been suggested that it was the products realized from live animals which were the true incentive for such a dramatic alteration in life-style.

Milk and wool are the most obvious products and in addition the occasional letting of blood is attested from ethnographic evidence. However the most significant of these secondary products to subsequent development is the power of the animal itself. Probably in the beginning only realized by their use as pack animals, the idea was later conceived to harness the animals to haul loads. The earliest evidence suggests that this draught potential was first utilized for the transport of goods and people, but then later the idea was transferred to mechanical tasks such as ploughing. Later still, with the invention of gearing, animal power could be exploited for industrial purposes (see Chapter 4). The concept of these secondary products provides some explanation for the very limited number of animal species that have become established as domestic animals, since there would be a very much tighter specification of requirements than would have been the case if the procurement of meat had been the sole purpose of the exercise.

If it is accepted that it was these secondary products that were the stimulus to animal domestication, then there is the need to review also the stages of change from hunter gatherer to farmer. The most logical progression would appear to be for a hunter of wild game to begin to control the movements of the wild herd to his own advantage, and within a limited territory of his choice. This type of economy is to be found today in Finland, where the Lapps control and exploit herds of reindeer. The logical progression from this is to the pastoralist with a fully domesticated herd, and then to a settlement economy making use of the milk and wool obtained from this herd. This progression also allows for the plant gathering element of the economy to move to a settled regime, planting and harvesting crops, and exploiting the animals for both their strength and their manure.

However, modern pastoralist economies are very dependent on settlement agriculture for supplies of industrial goods and feeding stuffs for their animals, and also as a market for their surplus products. Indeed in many societies the pastoralists themselves are landholders and therefore control most of the steps within the whole economy. It seems much more likely that a system of settlement agriculture gave rise to a nomadic tier within its economy, than that the process should have occurred in the other direction.

Fishing

The cultivation and harvest of domestic plants implies the existence of settled communities if only because the storage of produce has little point unless it is easily accessible later in the year. Early interpretation of archaeological data assumed that the opposite was also true, and that permanent settlement was only possible with an established agriculture. While this might be the norm, there is sufficient ethnographic evidence to show that other means of subsistence are capable of supplying adequate food for year-round occupation of a site. For example, the Kwakiutl and Haida tribes inhabiting Nootka Sound in North America were but two of the tribes who exploited the salmon runs to such effect that they were able to establish large permanent settlements with a very rich culture. The fish were consumed fresh or gutted and then either sun-dried or smoked for storage. In addition to the salmon, an oil-rich fish known as candle fish was also exploited. They were pressed to extract the oil from the flesh, which was then eaten. Oil extraction, smoking and salting were the most usual methods for the storage of fish, whatever their size, before the invention of refrigeration. Smoking seems to have been practised from the very earliest of times, and strong evidence for its use is to be found at the site of Lapenski Vir, which was situated on the Iron Gates gorge on the River Danube. This site would appear to have been seasonally occupied about 4500 BC, and existed because of the quantity of catfish which could be caught during the summer months. The site has realized some of the earliest fishing hooks to have been retrieved from an archaeological site. Harpoons have a much longer ancestry, being in use in Europe at least 13,000 years ago, and there is evidence for the use of nets in the Middle East as early as 11,000 years ago.

It has been suggested that the hunter gatherer economy is not only an efficient strategy, but that in prehistoric times the resources to be exploited were much more abundant than today. There is therefore the need to question why it should be that the economy should shift to one that required a much larger energy expenditure to achieve subsistence requirements, and in which the plant and animal exploitation was limited to a very small number of species. Such a change would not only have limited the cuisine, but also made the economy much more susceptible to environmental and other changes, which is at first glance a retrogressive step in terms of species survival. This last point is highlighted in modern Africa, where the hunter gatherers might be experiencing leaner times, but where the short-term climatic conditions have had a devastating effect on both the pastoralist and agricultural communities, who have a negligible safety margin within their territory.

It has been argued that it was the development of the social complexity of the human species that allowed the transition to be made. Alternatively, population pressures may have forced the change. Whether this pressure was increased by an actual rise in population numbers, or whether changes in climate reduced the biomass available to an existing population, once the transition had been made it was very difficult, though not impossible, to revert. Once the change had been made, the group could persist as an unstratified society only so long as the availability of land matched the population being supported by it. However, when the stage was reached when marginal land had to be exploited, and to do so a sophisticated technology such as irrigation was required, it would appear that social stratification began to appear within that society. The division may have been wealth, but it was later to be one of specialization, since as communities expanded, so certain groups within those communities became increasingly divorced from the production of food.

ARABLE FARMING

Irrigation

In order to raise crops, there is a need to manipulate water supplies whenever land in less favourable climates is settled. By its nature water is an awkward commodity to transport artificially, and therefore early crop husbandry was confined to those areas with a regular and dependable rainfall, or to areas very close to river systems. Some of these rivers flooded seasonally, and in the case of the Nile or Euphrates this fact was exploited to take advantage of the water dispersed in this way, and also of the fertile silt that was left behind as the waters receded.

In areas as flat as the large river basins, the need to trap this seasonal flood water was realized, and lagoons were created from which the water could gradually be released as and when it was required. There is evidence that this practice had been established by the second millennium in Egypt. The system was dependent on the understanding and political control of extensive distances of the river system, and also required an administrative system, not only to organize the building in the first place, but also to control the release and fair distribution of the water. These were prerequisites for any irrigation scheme, and imply the establishment of an already structured society before the expansion of agriculture into less favoured territory could take place.

Irrigation takes many forms, each determined by the nature of the land, or the source of water being exploited. In hilly regions the construction of terraces will slow down the normal passage of water en route to the river basin. The size of a terrace field will be determined by the steepness of slope on which it is built, but even in gentle terrain it has been found that an area of about one sixth of an acre is the optimum if critical levels of fall are to be maintained. Extensive terrace systems appear to have been established in China by the Han period, where they are associated with wet rice farming. Similar systems existed throughout Asia, either fed from natural streams, or by artificial canal networks. For scale and age the Chinese canals surpass any others in the world (see p. 474–5), but there were also extensive projects in widely differing geographical regions. An extensive system of canals was begun in antiquity and completed about AD 500 to connect the Euphrates and Tigris river systems, and irrigate the lands that lay between them. In South America, also, there is evidence of early irrigation practices, systems being established in Peru, for example, by the second century.

Irrigation is not simply the process of releasing controlled volumes of water on to land, but carries major difficulties because of the high mineral content of the water. Salination is one of the problems, and it is possible that the decline of Sumerian power in what is now southern Iraq, which began in about 2400 BC, was caused by the decline in wealth brought about by a faltering agriculture. Babylon became dominant in the area about 1700 BC, but its power may also have been undermined as the problem moved gradually northwards along the river system.

Ground water is also a source for irrigation schemes, and its retrieval may be seen in the qanat system in Iran, to which the Assyrian King Sargon made reference as early as the seventh century BC. It consists of deep, slightly inclining tunnels, connecting an underground highland water source with fields lying many miles below. Such systems still exist in Pakistan, Iran, Iraq, Syria and southern regions of the USSR.

Soil preparation

To achieve the optimum growth from a seed it must be planted at a specific depth in soil which has been prepared in such a way that it contains a fine structure of particles, enclosing sufficient air and water for the needs of the young seedling. In addition the ground should be free of unwanted plants which may compete with the desired crop, and also free from plants or debris from the previous crop, particularly if the crop is sown more than once in succession, since there is then the danger of disease transfer. The desired conditions may be achieved in a number of ways, perhaps the simplest of which is by turning the soil with a spade to break up and invert the fresh ground, and then pounding the clods in some way, so as to produce the fine tilth required.

Spade cultivation is still carried out in many parts of the world, and can be a very efficient form of production. It is ideally suited to the cultivation of small plots of land, whether these be cottage gardens, or terrace strips whose size is dictated by the steepness of the slope on which they lie. However, where larger tracts of land are available speed can only be achieved by using large numbers of people, and the depth of cultivation is controlled by the strength of the human frame.

With the extra power available from animals, combined with a tool formed to utilize it, there was a significant benefit to be derived not only from the increase in acreage which might be cultivated, but also from the speed with which this acreage could be tackled. In its simplest form the tool need be no more than a suitably shaped branch. All that was required was a pointed part, the share, which would enter the ground and disturb the soil as it passed through it, a handle so that the share could be guided, and lastly a beam to which the draught animal could be attached. The required shape could be obtained by the careful selection of timber, or by joining several pieces of wood together. Unfortunately, since wood is perishable, the evidence which exists for the earliest ploughs is very limited, and its date of origin is therefore difficult to determine. The earliest known representation of a plough was found at the site of Uruk in Iraq, and dates to the third millennium. The actual examples preserved by and found in the peat bogs of Europe date to about two and a half thousand years ago. These are not the earliest known evidence for cultivation implements since a piece found in Satrup Moor in Jutland dates to about six thousand years ago. This seems likely to have been drawn with human muscle power, and cannot therefore be classified as a plough.

These very early examples are not true ploughs, but are known as ards. They lack the piece known as a mouldboard, which projects from the side of the implement and serves to lift and turn the soil as the plough passes through. The ard is ideally suited to light sandy soils through which it can be passed frequently to produce the correct tilth without causing excessive water loss. It is therefore still an extremely useful tool under arid conditions, and can be found in many areas of the Mediterranean and Middle East. For heavier soils the mouldboard is an essential requirement, and the exploitation of these soils was delayed until its development shortly before the Roman period.

The development of the ard and plough were of significance because of the extension in cultivated acreages that they made possible. Not only were substantial increases in the carrying capacity of the land achieved but, perhaps more significantly, it allowed for the production of food by a smaller percentage of the population.

While cereal production had been practised for a considerable time without the use of the plough, once it had been invented, and once the package of cereal and plough had become established, the diffusion of cereals was almost without exception accompanied by the diffusion of the plough and animal traction. Thus it would appear that both cereals and the plough arrived together in India about 4500 BC, and in moving further east arrived in China about 4000 BC. The plough also moved westwards from the Middle East, if indeed that is where it originated, but although apparently traceable to the common form, its subsequent development was to take a very different track from that of the eastward moving counterpart.

In China the problem of soil friction against the wooden plough parts was solved with the introduction of metal facings. One of the most significant parts to be protected was the mouldboard, which by the early Han period, that is by about 200 BC, was frequently made with metal alone. With this material they were able to produce it in a curved form, thus achieving a great reduction in the friction and therefore the draught required to pull the plough through the soil. In Europe stones embedded into the wood of the plough were the original solution to the problem of wear, and this practice continued well into the mediaeval period. The mouldboard developed in Europe was made with an unshaped piece of wood, and although it achieved the desired effect, it did so at huge cost in the increase in friction and therefore the animal power required to pull it. It would appear that the introduction of metal facings on to the woodwork of the mouldboard did not occur in Europe until the eighteenth century. There is no proof of connection, but it is part of the curious coincidence that the Jesuit and other European contacts with China should occur at the same time as numerous developments were appearing in agricultural technology in Europe during the Age of Enlightenment.

The early European ploughs frequently shown in mediaeval illustrations as clumsy wheeled implements pulled by up to eight oxen certainly existed, but lighter ploughs which could be pulled by small draught teams were also much in evidence, particularly on lighter soils. Their rectangular construction so apparent in the illustrations of the time, left a lot to be desired, but little attempt was made to improve on it until the eighteenth century. In 1730 there appeared in England a plough that has become known as the Rotherham plough, but also carries the name of the Dutch plough, perhaps as an indication of an earlier ancestry. This plough had a triangular framed construction which made it not only lighter but also very much stronger. It was this basic design that the Scottish engineer James Small was to use for his calculations on plough design, and particularly on the design for the mouldboard in 'A Treatise on Ploughs and Wheeled Carriages' (1784).

James Small was one of the earliest people to apply scientific methods to plough design and it was on the basis of his calculations that the form of the curved mouldboard became established. The efficiency and design of the curved mouldboard, particularly the low draught requirements that it allowed, were to reduce the number of animals required for traction, and therefore the amount of land that was needed to feed them. The very discovery of this device in China is perhaps the reason why the Chinese were to develop a dry land farming that was so little dependent on animal power, and was therefore able to support a much larger population within a given area of land.

Despite the advances represented by the Rotherham plough and those that followed it, the wooden plough was difficult and slow to produce and impossible to reproduce accurately in every detail. In 1808, Robert Ransome patented the first of his all-metal ploughs. He also introduced the concept of easily interchangeable parts, which greatly simplified the task for the ploughman who was faced with breakages or misalignments. Ransome was to follow this patent with many others of significance, not the least of which was the discovery that if, in the casting of a metal share, one area of the casting was allowed to cool more quickly than another, then a difference in hardness would be induced between these two parts. When in use the parts in question would wear at different rates and by the careful design of the casting, the share could be made to be self sharpening, thus removing one of the more tedious tasks that faced the ploughman during the course of a working day.

Changes in the finer detail have occurred but the working parts of the plough have changed little from these nineteenth-century developments. Even for the modern tractor plough the parts that actually work the soil, as against those parts that support them, have changed so little that the names are generally interchangeable between the old and new ploughs. What has changed with the coming of the tractor (see p. 788), and the plough that has been specifically designed for use with it, is not only the increase in area that can be covered in a given time, but also a great increase in the accuracy with which this ploughing is carried out. Hydraulic lift mechanisms on the tractor can be set in such a way that the depth ploughed is very tightly controlled despite changes in the topography of the land or in the hardness of the soil.

Despite its long association with cultivation, the plough has passed through a number of phases in its popularity. It has been argued that the slow process of turning the soil is unnecessary and that it is frequently a threat to scarce soil moisture and structure. The argument has perhaps more validity today when the sophistication of herbicide sprays can eliminate the majority of weeds whose destruction is frequently cited as the main function of the plough.

The design of the mouldboard is such that it will always turn the soil over in one direction. Changes in design, whether for the horse-drawn or the tractor-drawn plough, have been brought about in the attempt to increase the productivity of the implement. The longer the working parts are in the soil, and turning the soil, the more work is achieved. Ideally, therefore, the greater the ratio of the field's length to its width, the less time will be wasted in turning the plough at the end of one bout ready for the trip back down the field.

With the mouldboard fixed to one side of the plough this efficiency was achieved by ploughing in long strips, beginning at what would be the centre of the strip, and gradually working away from that centre, until the time taken to cross from one bout to another became too great, and a new strip was begun, lined up so that it would eventually join with the first. This method, when repeated in exactly the same place each year, resulted in alternate peaks and troughs of land. The practice was encouraged on heavy clay lands since it greatly improved the drainage, and therefore made them more productive. The pattern that was created, known as ridge and furrow, is generally associated with mediaeval practice, but in fact the method was employed well into the nineteenth century, by which time other means of drainage were becoming available.

Soil drainage

Originally designed to turn the soil before seed bed preparation, the plough has several other forms. Improvements in drainage can be achieved with a conventional plough (see above), but also by creating a ditch with a plough which has two mirror-image mouldboards attached side by side. In the early nineteenth century a highly specialized plough was developed specifically for draining clay soil. Clay, being plastic, can be moulded, and if a solid cylinder is passed through it, a hollow tube will be formed which will retain its shape for a considerable period, acting as a drain for excess soil water.

The mole plough, as it was christened, gained particular significance because John Fowler, an engineer from Leeds in Yorkshire, began to experiment with steam traction engines and winches to provide the power for this work. By the 1850s he had developed a two-engine system which was to provide a comparatively cheap and fast method of drainage, with the result that hundreds of acres of cold wet land in Britain and elsewhere were brought into productive use. Extending the idea further, he devised a system whereby the mole would drag behind it a string of clay pipes, thus creating an instant and durable drainage system. Although the materials may have changed, plastic tubes replacing clay pipes, much of modern drainage is achieved in the same way. Fowler took the process a stage further by creating a conventional reversible plough with six furrows a side, which could then be pulled backwards and forwards across the field, turning the soil at each pass. Although these machines never ploughed huge acreages, part of the significance of Fowler's achievement was that his were the first successful attempts to apply mechanical power to field operations.

SOWING

To maximize the yields from a given area of land, it is necessary to sow the seeds as evenly as possible, so that each developing plant will have its own source of nutrient without competition from its neighbours. It is also necessary that the seeds are placed at a particular depth, so that the young seedlings can establish themselves before emerging from the soil, but not so deep that the nutrient reserves in the seed are used up before the new leaves are able to start photosynthesizing. An even depth of seed will ensure an even growth throughout the crop, and this in turn will result in the crops reaching maturity simultaneously. Any variation from these optimum spacings will result in a fall in yield.

The easiest way to spread seed on to a prepared area of land is to broadcast it. This method involves taking handfuls of seed and hurling them out in an attempt to cover a given area as evenly as possible. After broadcasting the land will be harrowed so as to cover the seed, and therefore make it less available to birds and other pests. A major disadvantage of broadcasting is that it leaves seeds randomly distributed, and it is therefore impossible to weed between the plants. Whatever the skill of the broadcaster, the sowing will only be an attempt at the optimum and will never achieve it, and almost since seeds have been sown, methods have been sought for a mechanical means to regulate depth and spacing.

The earliest known information concerning the design of a seed drill is to be found in the pictorial representations carved on Sumerian seals dating to the third millennium BC. These implements resembled the ard used in the area, except that attached to the handle was a funnel and tube passing down behind those parts which moved the soil. The seed was dropped into the funnel, and was then deposited at an even depth and in as straight a line as the ploughman was able to achieve. A similar device still exists in India, and indeed it was noticed and commented upon by European travellers as early as the eighteenth century. Despite various attempts to design a machine in Europe, it was not until the late nineteenth century that the drill became a standard part of farm equipment. This does not mean that the drilling of seed in rows was not practised, but rather that it was achieved by other means. One method was to broadcast seed immediately after ploughing, and then to cultivate the ground. In this way the seed tended to roll down the steep sides of the furrow slice and thus formed a row along the furrow bottom. This method may have created straight rows, and may even have placed the seed at an even depth, but it was still an expensive practice as far as the seed itself was concerned.

An alternative method was to use a stick to make holes at a set spacing and at a set depth, and then to drop seeds into them by hand. Again this would have produced the desired result, but this time the cost was in terms of labour. By the third century BC the concept of sowing crops in rows was already well established in China. This was accomplished either by sprinkling seed along a ridge, or else the seeds could be individually planted by hand. The latter was used in dry land corn farming and also for wet land rice production. A further refinement to individual planting is that of transplanting, which although generally associated with rice can also be applied to other crops. The earliest references to this occur in literature dating to the second century AD. The process is not only extremely efficient in terms of land productivity, but also increases the yield from individual plants. It is utilized in highly intensive enterprises such as market gardening and horticulture.

FERTILIZERS

Since animal and plant exploitation have always been closely linked, it is reasonable to suppose that the earliest agriculturalists noticed, if they did not understand, the boost to plant growth that could be derived from animal manure. Similarly the early farmers who were dependent on seasonal river flooding to provide soil moisture, would also have been aware of the fertilizing properties of the silt contained in that water. Human manure was also known for its fertilizing properties, and while its use tends to be associated with Far Eastern farming, its value was well recognized in Europe, and it continued to be used as a fertilizer well into the twentieth century, passing under the title of 'night soil'.

Another practice with a long ancestry and wide distribution is that of green manuring, whereby growing plants are ploughed back into the soil, and then allowed to decompose. The plants ploughed in may be the weeds established after harvest, or they may have been specifically sown for the purpose. The Romans and Greeks favoured the use of legumes, recognizing their high nitrogen content centuries before its existence was established, or the reasons for its presence in these plants was understood.

Most of what society has viewed as waste has found its way on to the land at one time or another. Animal and human manure have been joined by blood, bone, rags and sundry other materials which were thought to be of benefit to the soil. There was certainly an amount of artificially manufactured material that joined this, though the references to it are limited. In a curious premonition of future developments, the German chemist J. R. Glauber suggested in 1648 that the methods used to produce the chemicals necessary for the manufacture of gunpowder during the Thirty Years War should be put to use for the benefit of agriculture.

For the most part the benefit of a particular material has been observed and acted on, but scientific experiment was not part of the process, though the use of trial plots was mentioned in Blith's *The English Improver*, published in 1652. Just how little the processes involved in plant nutrition were understood is well illustrated in Jethro Tull's *The New Horse-Hoeing Husbandry*, first published in 1731, which advocated frequent hoeing between the crop plants in the mistaken belief that it would release nutrients from the soil. It was for this reason that he advocated drill husbandry, rather than the reason for which it was subsequently adopted, which was as a method of weed control.

By 1807 the chemist who had been appointed by the Bath and West Society was offering soil analysis to farmers, utilizing a system devised by Humphry Davy. In 1835 the first imports of guano arrived in Britain, and at about the same time the first imports of nitrate of soda arrived from Chile. In the early years of the nineteenth century coprolites found in Suffolk were being used in the manufacture of superphosphates. The German chemist Justus von Liebig, published his discovery of a method for producing an artificial phosphate manure in 1840. John Bennet Lawes is said to have discovered the same method in 1839, but did not patent it until 1842. However with the profits derived from this process he bought land and a house at Rothamsted in Hertfordshire, where he was joined in 1843 by Joseph Gilbert. Together they set up field experiments and laboratories and produced a standard of experimentation which established Rothamsted as one of the major research institutes engaged in soil research, a reputation it still holds today.

Sulphate and nitrate salts of ammonia were generated as by-products of the manufacture of coal gas, and these were utilized as a rich source of nitrogen fertilizer. However as early as 1898 the British Association was warning that mass starvation would be caused by a lack of nitrogen fertilizers brought about by the depletion of natural nitrate resources. At about the same time it was discovered that atmospheric nitrogen and hydrogen could be made to react to form ammonia, and in 1913 the chemical idea conceived by Fritz Haber was being used, in the industrial process devised by Carl Bosch, to produce the gas in quantity (see p. 223–4). The ammonia was used both for the manufacture of explosives and also for fertilizer, thus finally realizing the suggestion made by Glauber three centuries before.

Liquid fertilizers, in the form of town sewage or farmyard effluent, have long been used and recognized as a source of fertilizer, and less obvious by-products of manufacturing industry have been used on the land, such as the ammonia solutions derived from gasworks. However the impurities contained in such by-products, and the general bulk and difficulty in handling, precluded the extensive use of liquid fertilizer until the 1950s. The production of highly concentrated liquids such as urea and ammonium nitrate, and a better understanding of the effects of application rates, have now made it an efficient method in which to make the chemicals available to the crop plant.

The growing understanding of the plant nutrition cycle that was occurring in the mid-nineteenth century was matched by a similar increase in the study and understanding of the chemical and physical properties of the soil itself. Some of the earliest work took place in the United States, as a result of which the first soil map was produced in 1860, based on a survey of Maryland. The careful study of soil and the matching of chemical application to the requirements revealed by this study, is now an essential part of the process of maximizing production from the land.

PEST CONTROL

It would have been obvious to both the early farmers and the contemporary gatherers of wild plants, that if something ate or damaged the plants which they themselves valued, then the returns they could expect from their labours would be reduced, whether the unwelcome intruder had been an elephant or an aphid. The larger pests could partly be controlled by fencing, and evidence for this appears quite early in the archaeological record. Insect damage, and that caused by fungal parasites, is less easy to control, but would have been no less apparent.

In the natural world there is a wide diversity of plants and animals living in a state of approximate stability one with the other. This diversity is achieved by the often very limited environment that any one species will tolerate, and this will apply whether the relationship of species is of mutual benefit, of rivalry or is parasitic. Agriculture reduces this diversity, and therefore reduces the land's ability to withstand environmental changes without significant effect on the human economy. A plant species normally found scattered over a wide area is more difficult to harvest than one which is cultivated in a defined location. Unfortunately, it is also much easier for other species whose existence is detrimental to that crop, to spread within it when it is sown to the farmer's best advantage.

There are a number of ways in which plant pests can be eliminated, or at least their damage limited. Chemical control is the most commonly used in western agriculture, and although there is evidence even in classical times for its use, its true significance is very recent. The first method of control used by the early farmer was the attempt to limit the availability of the crop to its pests, whilst trying to maximize it to their own use. This could be achieved by practising a swidden, or slash and burn, regime, so that a given area of land would be exploited on a monocultural basis for only a short period and was then left to revert to its natural state. In this way the crop pests were not given long enough to build up their population to an extent that was potentially dangerous to the farmer. However, where the human population was too large to allow this continual movement from one piece of virgin land to the next, a different approach was necessary.

Chemical control of pests was initially limited to substances that could be derived from natural sources, such as plants themselves. Certainly used in classical times, it was not until the mid-eighteenth century that experimentation led to an understanding of the principles involved. Early workers in France, recognizing that the tobacco plant was toxic to aphids, initially used the ground-up plant as an insecticide. Later it was recognized that the chemical nicotine contained in the plant was the responsible agent, and this knowledge led to the identification of a number of other useful plant derivatives which could be used in the same way.

Of the chemicals used for insect control one of the most significant was dichlorodiphenyltrichloroethane, isolated in 1939 by the German chemist Paul Müller. Though DDT has now passed out of use because of its more recently discovered and damaging effects within the food chain, its use in the control of insects which act as carriers of disease in domestic and agricultural animals, but more particularly in the control of the carriers of human diseases such as malaria, had a profound effect in many regions of the world. This has not only been to improve the health of those populations living in affected areas, but has also allowed the settlement and exploitation of land previously unusable because of the insect pests within it.

One of the earliest documented successes against fungal attack also occurred in France, when in the 1840s a mixture of sulphur and lime was found to be an effective agent against a powdery mildew which attacked grape vines. Thirty years later lime was to appear with copper sulphate in a mixture which effectively controlled the fungus which caused potato blight. This particular combination, known as Bordeaux mixture, had originally been used to spray on plants at the edge of vineyards in the hope that the violent colour would put off prospective pilferers. Its fungicidal properties were recognized when it was noticed that the crops at the edge were healthier than those deeper into the vineyard which had been affected by downy mildew.

Although chemical pesticides are economically the most important, biological control can sometimes be more effective. In 1762, Indian mynah birds were introduced into Mauritius in a successful attempt to control the red locust, the first of many examples in which a pest in one country has been controlled by the introduction of a potential predator from another. One of the major successes of this method was the introduction of a South American caterpillar into Australia to control the prickly pear cactus. It has been claimed that up to 25 million acres of previously unusable land was reclaimed in this way. While there are great advantages in this method the dangers of a poorly researched project are all too apparent, and it is inevitable that there have also been disasters. Another type of biological manipulation that has been used in recent years has been the introduction of sexually sterile insects into an insect population in order to reduce, if not completely check, its breeding. There is a political advantage to be gained from the application of pest control schemes which have the appearance of being natural, but in fact any attempt to eradicate a particular plant or animal species will alter the ecological balance and therefore carries with it the dangers of unexpected and often spectacular side effects.

WEED CONTROL

One of the essential differences between an agricultural economy and one based on the collection of wild plants is that the former is dependent on the successful harvest of a very limited number of plant species. Any unwanted plants that grow within the crop are a threat to the nutrient supply of that crop, and by the nature of things weeds were often more suited to the newly prepared ground than were the crops themselves. The practice of fallowing, that is, leaving land free from crop exploitation every alternate or third year, was partly because of the need to re-establish fertility after a crop, but it was also used to attack any build-up of weeds. Fallowing did not mean that land was left unattended, but was in fact a period when the land was continuously ploughed and cultivated so as to create an ideal environment for the weeds to germinate and develop slightly, before being destroyed by the next cultivation. In this way the number of weed seeds remaining in the ground when the crop was sown would be much reduced.

Fallowing will serve to limit the build-up of a weed population, but it will also limit the time in which a piece of land is in productive use. An alternative would be to choose a crop which will smother the seedling weeds, while also being of value to the farmer, and alternating this crop with those which were necessary for survival, but which created conditions suitable for the establishment of weed populations. For example the thick, spreading top of turnip will cover anything which germinates at the same time, and if the crop is then fed to livestock by fencing them on to a small area each day, the tight pressure of hooves will further clean the land. By introducing the turnip into a rotation, a check on weed plant intruders can also be controlled mechanically, by weeding, and this very labour-intensive method was all that was available to most farmers in the past, and is still utilized within many present economies.

Chemicals will also kill plants, but to be of any use to the farmer they must kill only those plants which are unwanted. Early experiments on weed control made use of the fact that the major crop plants were grass derivatives, and therefore had long narrow leaves, while the major weeds were broad-leaved. If a chemical mixed in water is poured on to the former it tends to run straight off, but on the latter it will form into droplets on the hollows and angles of the leaf, and therefore remain in contact long enough to do damage. Experimental work in France at the end of the nineteenth century identified iron and copper sulphates and nitrates as suitable chemicals for use with cereals.

Since the end of the Second World War the research and application of herbicides has relied less on mechanical selectivity and more on the differences in chemical and particularly hormonal activities which occur in different plants at different stages of their development. The earlier complicated chemicals, which have become known by their abbreviations such as 2,4-D or 2,4,5-T, have now given way to a whole cocktail of chemicals which are available to farmers. These chemicals have become vital to the achievement of high yields from continuously cropped land, and with a reliance on single species enterprises. A new move in chemical control is to utilize natural amino acids found in plants to exaggerate processes that take place naturally. By the careful selection of the correct amino acid for a specific weed, it is hoped that a new generation of herbicide will be developed which will be environmentally safer than those in present use.

CROP ROTATION

A single crop grown repeatedly on the same area of land allows the expensive back-up facilities of capital and machinery to be spread over a number of years, but does raise problems of loss of fertility or build-up of pests. This single-minded approach can be tempered by the use of a sequence of enterprises which are mutually beneficial.

A rotation is devised to suit a particular set of circumstances and requirements. At its simplest it is to be seen in the slash and burn technique. As simple, but more productive, is the system known as alternate husbandry, in which an area of land is used for a short time for arable production, and then sown to grass and grazed for a number of years, the animal manure making a significant contribution to the fertility of the land. There is increasing evidence for the practice of crop rotations even in the very earliest settlement areas, but the first references to alternate husbandry appear in the agricultural writings of Columella in the first century AD. It was later practised by the Cistercians in France, where as early as 1400 they were advocating five years of grass, followed by two years of corn and then reseeding with grass.

More complicated rotations were created to resolve more difficult problems and in Europe all the traditional constituents of rotations were known and being used at least as early as the Roman period. Legumes, turnips and corn were all documented if not widely grown. In mediaeval East Anglia extremely complicated rotations of fourteen-year duration have been recorded. It therefore seems likely that formal rotations were in use over a much wider area, and at a much earlier period than the seventeenth-century English writers would suggest. By 1880 leguminous crops were an established part of the farming regime throughout Europe, and they represented a significant proportion of the total arable acreage. This situation is significantly different from that of a hundred years earlier and is an indication of the recognition of the nitrogen fixing properties of this crop. Over the same period the reduction in the acreage left fallow is significant and highlights the benefits derived from the change of rotation made possible by the additional use of legumes, not only in terms of increased yield per acre, but also in terms of the increase in the numbers of productive acres.

HARVESTING

Reaping and binding

By the careful preparation of the land, and the attention paid to the plant from germination to maturity, the farmer will come to harvest, which is perhaps the most critical stage of all. One of the most common finds in an archaeological context is that of microlithic flint blades which are thought to have been used in the construction of sickles. These thin, sharp pieces of flint were mounted into a handle and fixed with pitch or similar material. They occur with early Ubaid pottery at Eridu, in Iraq, and have been dated to the sixth millennium BC. In these cases the handle and blade holder is made of clay, the blades being made up from a series of small flakes of carefully prepared flint. The presence of these flints, or the complete sickles themselves, is frequently cited as one of the vital pieces of evidence for the transition from the hunter gatherer economy to an agricultural one, together with pottery and grinding stones forming the traditional agricultural package, but in fact each could also have been of value to the gatherer. Under the microscope these flint blades frequently show a particular gloss which is certainly of plant origin, but may as easily have been caused by the cutting of grass or reeds for roofing materials as by the cutting of corn stems, and therefore their presence on a prehistoric site should not be seen as automatic proof of a sedentary farming existence.

The wild cereal species are very brittle just below the ear, and recent experiments have shown that it might be easier to pick this corn by hand, rather than cutting it with the primitive sickle; ethnographic evidence suggests that in light sandy soil it may have been harvested merely by uprooting. Less fragile strains were to be developed as domestication and selection occurred, and a cutting blade became more necessary for the harvest. The flint sickle gave way to one made of bronze, and then of iron. Examples of this implement in use can be seen in the Egyptian wall paintings, and later still in Roman mosaics. Innumerable examples have also been discovered in archaeological sites.

Mosaics and actual examples also provide evidence for the existence of the scythe by at least the Roman period, but there is no indication that it was used on corn. The scythe consists of a blade of about a metre (3ft) in length, attached to a long handle which allows for its use in a more upright posture than is possible with the sickle. It is potentially a more effective tool for harvesting corn than the sickle but, in Europe at least, their widespread use seems to have been restricted to the cutting of grass until the nineteenth century. However in China by the Han period, between 200 BC and AD 220, an implement was in use which should perhaps be called a scythe, although its form differed considerably from the European equivalent. Resembling an upturned walking stick, it has remained an essential harvest tool to the present day.

The earliest application of the scythe to corn in Europe is of unclear date. To gain the full advantage from the long blade it is necessary to attach a basket or cradle to the handle. This cradle will collect all the corn cut in one sweep, and allow it to be deposited out of the way of the next sweep, and conveniently placed for the workers following, whose job it was to bind the corn into convenient sheaves. The earliest representation of a scythe and cradle is in a thirteenth-century illustration, and it is also to be seen in Brueghel's *Spring*, dated to 1565.

The sickle was slower to use than the scythe, but it was much cheaper and could be afforded by the harvest labourer. The more expensive scythe was generally purchased by the farmer, and it has been suggested that this situation changed the status of the harvester from one in which he was essentially a contractor, to one in which he became purely a hireling, and that this produced a resistance to its introduction in certain parts of Britain. Even today, in those parts of the world where the corn is cut by hand tool rather than by machine, there is great variety in the distribution of the two types of implement.

Speed is a necessity of the harvest more than any other aspect of the farming year. A mechanical means to bring in the corn has therefore been a matter of interest for centuries. The earliest references to a successful device can be found in the first century AD writings of Pliny, and also those of Palladius about three centuries later. Both authors referred to a harvester which was supposed to have been used in northern France at the time in which they were both writing. The machine consisted of a two-wheeled barrow, along the front edge of which was a metal comb. When it was pushed into the standing corn, the stalks passed between the teeth, and the ears of corn were plucked off and fell back into the container. Until quite recently this description was thought to have been fanciful, or at least describing an oddity without very much practical use. However in recent years four stone relief carvings have been found or identified which carry a clear representation of the device that the writers had been describing.

No further evidence for this machine or any developments from it exists. However, centuries later on the other side of the world something based on a similar line was to appear. Whether John Wrathall Bull read the classics or had a knowledge of the Roman texts is unclear, but in 1843 he submitted a model of a harvesting machine to the Adelaide Corn Exchange Committee in the hope of winning a prize being offered for a harvester suited to Australian needs. Sad to relate, neither Bull nor any of the other competitors won the prize, but a miller and farmer by the name of John Ridley saw the model and built himself a full-scale example based on the same principles. Having proved to himself that the idea worked, he took out patents and began production.

This basic design was continuously improved, and in 1885 the stripper-harvester was produced (see Figure 16.1). This machine was capable of harvesting, threshing and winnowing the corn in one operation. It has been suggested that this Australian invention had such an effect on binder sales in the Australian, Argentine and American markets that the two major manufacturers, International Harvester and Massey Harris, were forced into combine harvester production. By 1909 the first self-propelled example had been built, using an internal combustion engine, though it was not to become a production reality until twenty years later.

PART FIVE: TECHNOLOGY AND SOCIETY

Figure 16.1: Australian stripper harvester, *c.* 1900. The forerunner of the modern harvester-thresher.

In Europe the search for a harvesting machine was also being conducted in the early years of the nineteenth century, and the Society of Arts offered a prize for a number of years. Eventually the Revd Patrick Bell designed a machine, which was built by his local blacksmith and successfully used to harvest corn in the summer of 1828. Bell decided that his machine was so important that the restrictions in production that would occur if it was patented should be avoided. Unfortunately this meant that those imitations that were built were of poor and uncontrolled quality, and the idea lay dormant for a number of years owing to the bad reputation they gained.

In the USA, Cyrus McCormick designed a reaper, which he patented in 1834 and put into limited production about the same time (Figure 16.2). While enjoying considerable success in America, it was some time before it was to appear in Europe. The incentive was the Great Exhibition held in London in 1851, and more particularly the trials organized that summer by the Royal Agricultural Society of England. Resistance to the machine was not pure conservatism, but also had a technical and economic basis. Not least of the problems was the ridge and furrow drainage system to be found on most of the heavy lands of Britain, which made it difficult, if not impossible, to operate the reaper. New drainage techniques being developed at about the same time were eventually to remove this particular obstacle, but the heavy investment needed created its own delay.

These early reaping machines merely cut the corn and laid it on the ground

Figure 16.2: The McCormick reaper, patented in 1834. This machine employed the essential features of the modern reaper: vibrating sickle-edged blade, fingers to hold the grain and reel, divider and platform to receive it.

for it to be tied into sheaves by hand, as had occurred with the sickle and scythe. Several manufacturers had managed to develop machines which would reap the corn, and also tie it into sheaves. The early machines used wire to bind the bundles, but in 1880 a farmer-inventor called Appleby produced the first successful twine binder, using a knotting mechanism that is still found on modern balers. Appleby's invention rested on the accumulation of the previous experiences of a number of people, not least that of William Deering, who was the first to manufacture the machine. The cutting of corn now became a one-man operation, but the stooks had still to be stacked in the field to dry, and later to be carted to the barn for storage. During the winter months the grain had to be threshed from the straw, and then cleaned or winnowed to remove chaff, weed seeds and any other light material which might have become mixed up with it.

Threshing and winnowing

Because of the very weak stem that is a characteristic of the primitive cereals and the wild grass from which they derived, the seeds are easily dispersed. However the awns that surround the individual seed are very difficult to remove, and this would have presented problems to the early gatherers of the wild form, or cultivators of the primitive domestic varieties. As natural and human induced selection has caused the gradual evolution of these cereals, the features of most use to the farmer have been favoured. Thus the stem nodes have become stronger, allowing a considerable amount of rough treatment to the plant during the course of harvest, without consequent loss of seed. At the same time species have been developed in which the awns are more readily removed or, in some cases, completely absent. These benefits have been achieved at some cost, since the very high disease resistance of the primitive ancestors such as emmer has been lost, and the protein level of modern grain is also considerably lower.

Threshing is the process whereby the individual seeds are separated from the stem of the plant, and also from the awns which surround them. Winnowing is carried out afterwards to produce a clean sample. Threshing requires a repeated and vigorous pounding of the corn so as to separate the various parts of the plant, but without damaging the seed itself. This can be accomplished by spreading the harvested crop on to a clean piece of ground and driving livestock over it so that their hooves crush the grain stems, or a device such as the norag or threshing sledge can be used. Alternatively, human effort can be utilized and the corn beaten with sticks. These may be just straight sticks, or may be specially developed flails which allow a greater force to be applied from a more upright stance. Either way the work is arduous and tedious.

Wild cereal stands are still to be found in certain areas of the Middle East and Turkey, where they are occasionally harvested. This corn has firstly to be scorched or parched to make the awns more brittle and easily separated, and the same practice would have had to have been carried out by the prehistoric food processor. The changes that occurred as the early domestic species developed made this process unnecessary, but apart from this the methods employed for threshing the grain have remained remarkably similar throughout the world and over several millennia. Occasionally, and in certain parts of the world, reference may be found to some mechanical device to ease the process, but this has always been applied on the threshing floor. Where the climate will allow, this is situated in the open, but in northern Europe the slow winter threshing was conducted in an area specially set aside in the storage barn.

The first practical threshing machine was invented in 1786 by the Scot Andrew Meikle (Figure 16.3). It consisted of a revolving drum along the edge of which were set four slats. The drum turned very close to a rounded mesh cage, the concave, whose curvature matched very closely the diameter of the drum. The combined effects of the slats and the pressure induced between the drum and concave, forced the seeds away from the other parts of the plant. The design of the Scottish machines required rollers to feed the corn gradually into the threshing chamber, with the result that the straw was severely damaged. In 1848, John Goucher in England patented the rasper bar cylinder which removed the need for these rollers, leaving the straw in a condition suitable for threshing, and also caused less damage to the seed. The peg tooth drum was invented in the USA in 1823. The principle employed was similar to the bars mounted on a drum, but teeth were also introduced to improve the threshing efficiency. The concept was very popular in America, and was used also in the newly developing combine harvesters (see pp. 785–6), but in Europe, where a

Figure 16.3: Ground plan of Meikle's thresher from Robert Forsyths 'Principals and Practice of Agriculture,' *c.* 1805.

higher premium was placed on straw, the rasper bar predominated, and is still to be seen on modern conventional combine harvesters.

By 1830 the threshing machine had been readily adopted into northern Britain, and was also making an appearance in the south. However the southern labourer depended on threshing during the winter months to provide work when it was impossible to get on to the land, and a series of riots resulted in the destruction of a large number of machines and a check in its rate of adoption. In the north the industrial manufacturing towns were able to offer higher wages than the farmer, and farm labour was more difficult to recruit. The machine therefore provided a much needed solution, as it did in other areas of the world where labour was always in short supply, such as America and Australia. By the 1850s resistance to the thresher had been overcome, and its overall adoption was speeded by the development of machines which could be driven by portable steam engines, allowing several farmers to share the capital investment.

Combine harvesters

The ideal harvest machine was of course one which was able to reap, thresh and clean the corn all in one operation, and this was achieved surprisingly quickly after the appearance of the first reaper. Hiram Moore's earliest machine first 'cut, threshed, separated, cleaned and sacked' a field of wheat in 1838 in Michigan, USA, but the success was not immediately followed through, partly because of wrangles over patent rights, and also because the machine only worked efficiently on crops that were ripe and dry. When a further example of Moore's machine was tried out in California in 1854, it harvested 600 acres (243 hectares) of wheat in that one season. The 1838 machine cut a width of fifteen feet (4.6m) at each pass, and the developments that occurred in California followed this pattern. The farmer George Berry, in the Sacramento Valley, developed a machine with a forty foot (12.2m) cut, which was operated in the late 1880s, and about the same time another of his machines, powered by a straw burning steam engine, was the first self-propelled harvester to be produced. These, and later examples produced by manufacturers such as Best & Holt, could only have been developed for the great prairies of America, and apart from the much more compact Australian machine of the 1840s (see p. 781), there were no developments elsewhere until the early part of the twentieth century.

By the late 1920s there were half a dozen American manufacturers producing machines capable of cutting between ten and eighteen feet (3 – 5.5m) of corn at one pass. The first of these crossed the Atlantic in 1927, being shown at the Empire Exhibition in London. The following year two machines were tried out in Britain, and trials were also conducted in continental Europe. Two European manufacturers recognized the potential, Clayton & Shuttleworth beginning production in Britain in 1929, with sales directed at the export market, and Claas in Germany testing a machine in 1931 which was aimed at domestic sales. Claas sold some 1400 of these machines over the next ten years, but generally acceptance in Europe was very slow. There were, for example, only 120 in Britain by the outbreak of the Second World War. The depressed state of European agriculture was part of the reason, but there was also a resistance from the corn merchants, suspicious of the quality of the corn that resulted from the drying process, which with the combine, became a necessary addition to the European harvest.

In 1934 the Allis Chalmers company of America introduced their 'All-Crop' harvester. With a cutting width of five feet (1.5m), it would be pulled by a low powered tractor, was ideal for the small farmer, and most important of all, it cost less than a binder to buy. Within four years it dominated the US market and ensured that the binder was a machine of the past in that market, outstripping binder sales in the US by 1936. Since American manufacturers were so powerful in world markets, this ensured replacement of the binder in overseas markets as well. The war years had a profound effect on the spread of this machine, and of farm mechanization generally. On the Continent progress was virtually at a standstill. In Britain the massive ploughing-up campaign gave a stimulus to the spread of machinery in much the same way as it had during the First World War. But British manufacturers were geared to the production of tanks and other military equipment, and most of the machinery arriving on farms was therefore from North America. In America itself, the demands for grain and other agricultural produce from Britain, Russia and China provided a stimulus for further mechanization, perhaps most graphically demonstrated by the 'harvest brigade' formed with the new self-propelled combines developed by Massey Harris. In 1944, 500 machines started harvesting in Texas in March, and completed over one million acres by September of that year.

Since the war manufacturers have produced ever more sophisticated machines capable of handling larger acreages and heavier crops, but the general principle of the machine has remained the same. However, since the late 1970s many new designs have been tried in the attempt to accommodate the staggering increases in crop yield that have been achieved in the past fifteen years.

FARM TRACTION

The establishment of an economy based on plant domestication before one based on animals does not preclude the domestication and exploitation of a single species even earlier. French evidence suggests that horses may have been used from very early times, either as beasts of burden or for riding, on the basis of a characteristic form of dental wear noticed on a Palaeolithic horse jawbone. This raises the possibility that European hunter groups were aware of the potential of horse power well before any attempts were made at plant cultivation. This in itself was an important step, but much more significant was to be the utilization of this power for draught. Once again it is the Middle Eastern sculptures and frescos that provide the earliest evidence for the use of draught animals. A horse can carry only a limited load on its back, but when used in harness to pull a wagon it is capable of moving many times this weight. In the earliest representations it is the now extinct relative of the horse, the onager, that is most usually portrayed, and then always in association with the chariot rather than the more humble plough. As a power source for cultivation the ox predates the horse.

The method of harnessing for the ox is by means of a yoke attached to the horns, and this place of attachment is still to be found in many parts of the world today. In Europe this practice gave way to a design of yoke which fitted across the withers of the animal. The horse's place in the early history of traction has traditionally been limited to the chariot and subsequently as an animal of prestige or war rather than of labour. Various arguments have been presented to explain this, ranging from the economic ones that the horse was more expensive to keep, and realized no useful meat at the end of its working life, to the technical one that it was not until the invention of the horse collar that the horse's greater speed and power could be utilized. The collar possibly originated in China, appearing in Europe by about 1000 AD. Mediaeval manuscripts show the horse being used with the harrow, which could only operate effectively at the greater speed possible with this animal. It is also occasionally shown in harness with the ox, but hardly ever on its own. Pictorial and documentary sources both testify to European plough teams of eight oxen yoked to the plough in four pairs, and it has always been supposed that these great teams were necessary because of the poor design and heavy draught of the ploughs being used. More recent work would suggest that the large teams were more to ensure that each individual animal was lightly loaded, and could therefore work day in, day out, over a very extended period. It is likely that the horse was used more in agriculture than surviving records might imply, even in mediaeval times, but it was nevertheless a gradual process of replacement that led to the horse predominating in the extended European scene, and some areas hung on to the use of the ox until animal power was replaced by the tractor. In each part of the world a suitable animal was chosen from the native stock: the water buffalo has possibly the widest distribution after cattle and horses, but the yak, camel and llama were all chosen because of their suitability to their native environment. South and Central America is exceptional in that it is the only area in which city states were developed on the strength of an agricultural system that depended on human muscle alone. In this region animals were used as beasts of burden, but there is no evidence for their ever being used for traction.

Animal traction in agriculture is generally associated with the direct traction of implements on the land, but the invention of effective gearing systems so important to early industrial development, also had its counterpart on the farm. Horse wheels had provided power for machinery at least since Roman times, and in association with water power they were an important means for driving the fixed equipment of the homestead, such as the corn mill. The development of the steam engine was to see the displacement of much of this equipment, but for many smaller farmers animal gears provided a cheap and reliable power source.

Steam engines by their very construction were of great weight, and fear was expressed that such weights might cause damage to the structure of the land. The early pioneers of the internal combustion engine were hampered by the attitudes of the steam age as well as by the materials and manufacturing technologies available to them. The early machines, both European and North American, were therefore built on similar lines to the steam engines they were to replace, and frequently weighed in excess of four tonnes.

It is therefore not surprising that it was in the drier prairies of North America, in which there was the need to cultivate huge areas with limited man and horse power, that the agricultural motor was to be most readily applied. In Great Britain the pattern of development was being repeated, but one individual in particular recognized the need for a more compact power source. Dan Albone, a bicycle maker in Bedfordshire, produced his first machine in 1902. This motor weighed only $1\frac{1}{2}$ tonnes, and yet was capable of pulling a two-furrow

Figure 16.4: Ivel tractor pulling two mowers, *c.* 1905.

plough, and later models were capable of three furrows, or of pulling two binders in tandem. Almost immediately Albone's Ivel tractor (Figure 16.4) was being exported to countries on every continent. Albone himself died in 1906 and without his guidance the company produced no new ideas of significance; the initiative passed back to the Americans, or to those companies which were designing and producing for the American market.

Up to the First World War the tractors being produced continued to be of the size determined by their ancestry. A very large number of companies were engaged in establishing a position in the market-place, but it was the Ford company that was to have the greatest impact on future developments. Henry Ford brought to the problem two important contributions. The first was technical, in that his tractor was built without a chassis, but with the engine, gearbox and transmission casings forming the structural strength. The idea was not new, having first appeared on the Wallis Cub in 1913. However, when Ford's mass production manufacturing techniques were applied to the concept, he was able to market a machine which was not only much lighter than its rivals, but also significantly cheaper.

As Ford was perfecting his new Fordson tractor, the British government was implementing a huge ploughing-up campaign to produce the corn necessary to replace that which was being denied to Britain by the successes of the German U-boat blockade. They therefore agreed to purchase the first 6000 Fordsons. The basic structural design of this machine is to be found on modern examples.

To it have gradually been added a series of new features and also some which had been invented much earlier, but had never been followed up. For example, the idea of the power-take-off shaft, which had first appeared on the British Scott tractor in 1904 and on the French Gougis tractor in 1906, was reintroduced by the International Harvester company as an optional extra on their model 8–16 Junior in 1917, and as standard equipment on their 15–30 tractor which appeared in 1921. The power-take-off shaft allowed the power of the tractor's engine to be transferred to the machine it was towing. Previous to this the moving parts of a machine such as a binder were driven via gearing from its ground wheels. This system was inefficient because of the slip of these wheels as they rolled on the ground, but it also meant that when a blockage occurred the machine had to be stopped, and the power that might have cleared the blockage was therefore not being provided. With the power-take-off, when the tractor was stopped, the tractor's engine power could still be used to drive the equipment concerned.

Most early tractors were merely used as horse replacements to provide the power to drag a particular implement through or across the ground. However as early as 1917 the Emerson-Brantingham tractor had a mechanical lift feature which allowed the direct mounting of specially designed implements. The idea reappeared in 1929 on the John Deere GP tractor in a mechanical form, and on the John Deere model 'A' in hydraulic form in 1934. At the same time Harry Ferguson in England was developing a hydraulic system which has become a standard feature on modern tractors. Ferguson's system consisted of three arms to support the rear-mounted implement, the bottom two of which were raised or lowered by hydraulic pumps, and the top one was connected to valves which controlled the flow of oil to and from these pumps.

Equipped with this system the implement could be carried by the tractor, greatly increasing its manoeuvrability compared to the trailed arrangement of its predecessors. Additionally the top link on the tractor could be set up in such a way that changes in soil depth or hardness would act on the hydraulic valves so that the depth of the implement would alter and compensate for the changes acting on it. Tractor hydraulics have now become extremely sophisticated, and are used to make running adjustments on both the tractor itself and the machinery it is towing. Additionally hydraulic pumps are used on equipment such as combine harvesters as a substitute for transmissions. The advantage of the system is that, unlike a gearbox, where the range of power selection choices is determined by the number of gears, the hydraulics system offers a limitless range of speed.

Pneumatic tyres were issued as standard equipment on the 1935 Allis Chalmers Model 'U', and quickly became a standard fitting on most makes of tractor in America; their introduction into Europe was delayed by the shortage of rubber during the war. Pneumatic tyres greatly increased the grip of the tractors' wheels on the soil, allowed greater speeds of operation to be achieved with more efficient fuel consumption, and also allowed the tractor to be taken on to the public highway without the time-consuming need to protect the road from damage by the iron wheels. Their introduction did much to check the development of tracked crawler tractors, which had appeared very early on the scene in response to concern about soil damage. Modern low pressure tyres aim to compensate for the ever increasing size of modern machinery. This trend itself is in part to maximize capital investment in plant, but it is also an attempt to reduce the number of passes across a field that are needed to carry out all the year's processes. The larger the tractor, the larger the implement it can carry or pull, and the greater the acreage covered in one trip down the field. There is a limit to this concept, and while it might have a validity in many areas of the world, the heavy reliance on petrochemical inputs makes it unsuitable in others. Animal or human power will always be a vital input into an agricultural system existing in a situation where the cost or supply of fuel outstrips the resources available. Indeed, many traditional systems maximize available resources in a way that is inconceivable within Western technology. The Indian sacred cow scavenges for its feed at little cost to the community in which it lives, and yet it provides that community with the calves which will grow into the draught oxen vital to its existence.

DAIRY FARMING

In terms of returns gained from a given level of input, milking whether of cow, sheep, goat or horse, is many times more efficient than the production of meat. As a form of subsistence economy it has the added benefit that in really hard times the slaughter of a percentage of the herd can provide a satisfactory safety net. If it is not carried to extremes, the herd can be brought back to a previous population level reasonably quickly after the return of the good times.

The earliest evidence for milking comes from a fresco from Ubaid in Iraq, dated to 2500 BC. Later representations from the same area, and also from Egypt, suggest that milk from sheep and goats was also used. Curiously shaped pottery sherds discovered at Ghassul in the Jordan valley, in levels dated to the middle of the fourth millennium, have been identified as butter churns on the basis of their resemblance to the churns still used today by Arab nomads. This early processing of milk is important, since it demonstrates the ability to convert a highly perishable commodity into one which has some storage potential. Any society which changes its economy from one based on gathering and hunting, and which may therefore store little beyond its most immediate needs, will very quickly be faced with the need to store the products of a bountiful season in order to see it through less favourable times.

The pastoralist whose diet is dependent on the milk products from the flock or herd is perhaps the most precariously balanced exploiter of marginal areas, since lactation is so dependent on adequate food and water supply. Left to its own devices, it is likely that a human population density will be established which reflects the carrying capacity of the land and the territory in which the shepherd or cattle herder moves. There may be occasional short-term weather changes which cause distress to a community, but in general these effects would be local, and can be survived because the stocking rate would originally be well below the level supportable within that area. Unfortunately it is only in more recent times that western agencies have recognized the significance of the low stocking rates in terms of a safety net, and that the placing of wells, and therefore water, into areas that had previously been without, admittedly increases the carrying capacity of the land, but only by making the fall-back area unavailable in bad times.

The actual process of removing milk from the animal changed little from the earliest times until well into the twentieth century. Selective breeding has greatly increased the yield from any one animal, but while farm labour remained relatively cheap, there was little incentive to look for alternatives to hand milking. Although some experiments on vacuum milking machines had been carried out in Britain and America in the 1860s, it was not until Murchland's patent in 1889 that a practical machine was developed (Figure 16.5). However, the continuous pressure of this machine caused damage to the sensitive udder tissues and it needed Dr Shield's pulsating mechanism to realize the full potential of the idea. Even then hand milking was a continued practice in much of Europe until the middle of the present century.

The very perishable nature of milk limited the ways in which it could be exploited in increasingly urban societies. Town dairies allowed a certain volume to reach a broader market, and the farms on the edges of smaller towns were able to supply them with fresh products. Processing into cheese or butter increased the ability of the product to travel, but it was not until the coming of the railways that the major markets could be reached. Not long afterwards the development of refrigeration (see p. 796) again changed the market structures dramatically, since now countries as far away as Australia and New Zealand were able to supply Europe with the perishable dairy products they were so well suited to produce. These changes in transport provided the stimulus for the development of more sophisticated mechanical means for milk separation and churning which were to occur in the late nineteenth century. Improvements in road transport also extended the ability of farmers to get liquid milk to the railway, or at least to a central collecting point, and as this occurred so the production of butter and cheese on the farm gave way to the factory-based system of today. These outside influences, together with the increasing decline in labour willing to work long and often antisocial hours, has in turn created an increasingly mechanized and automated cowshed.

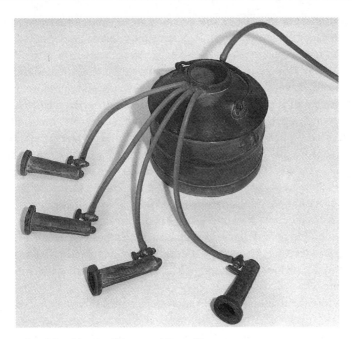

Figure 16.5: Murchland milking machine, 1889.

The early milking machines consisted of a bucket attached to a pump which created the vacuum in it. From the bucket ran a length of rubber piping to which the 'cluster' was attached. This cluster consisted of four carefully designed cups which acted to simulate the calf's sucking action so as to stimulate the lactating animal to release her milk. The vacuum is not intended to actually suck the milk from the animal, but rather to carry the milk away once it has been released. For the cowshed the vacuum is supplied by a pipeline running along the shed in such a way that the bucket could be connected at different points along its length. As each cow was milked the bucket was carried to the dairy where the milk was cooled and placed in churns ready for transport. In modern machines the cluster still exists, but the milk now passes down pipelines directly to storage tanks where it is cooled, and from which it is directly collected.

Milk in its liquid form is the most perishable of agricultural products, and for any economy that is dependent on it for a large part of its diet, there must exist the ability to process it into a storable form. There are many ways in which liquid milk may be treated to prolong its usefulness. In 1862, Louis Pasteur, realizing that bacteria were the causative agents in the spoiling of milk, demonstrated that if it was heated to 70°C, the bacteria it contained would be killed, and the milk would therefore keep several days longer. Later developments of pasteurization were to improve the fitness of liquid milk for human consumption, since heating was also shown to destroy the bacteria responsible for tuberculosis and brucellosis. Both diseases reduce the efficiency of cattle, but more significantly can also be transmitted to humans by the consumption of milk.

By removing a large percentage of its water content, and therefore increasing the relative proportions of fat, milk can be turned into butter, which has slightly better keeping properties. Butter-making was a laborious and time-consuming task, and offered plenty of scope for contamination. The first stage of the process is to separate the fat content of the fresh milk from the heavier water and other solids. To achieve this, milk is traditionally allowed to stand for several hours in shallow bowls so that the cream will form on its surface. The cream is then either skimmed off using special ladles, or else the heavier liquid is drained off from below. It is then placed in a churn and agitated continuously for a considerable period, during which time the fats will coagulate into butter, and the liquid fraction, known as buttermilk, can be drained off. The length of time required to achieve this result will again be dependent on factors such as temperature.

Even before the manufacture moved from farmyard to factory, methods were employed to mechanize the arduous work of churning and the horse wheel, already in use in other contexts, was adopted on some larger farms. However it was not until the 1870s that mechanical means were applied to the cream separation process. The early work on these developments occurred in Germany and Scandinavia and were quickly adopted in the major milk producing areas of the world. The principle commonly applied was that of the centrifuge. Milk placed in a fast-spinning chamber will separate in such a way that the heavier solids pass into an outer chamber and are drained off, while the lighter cream will be collected in the inner chamber. Not only is this method much faster, and therefore removes the need to leave standing milk open to contamination for long periods, but it also separates the milk more efficiently, realizing a higher percentage of the cream contained within it. Developed in Scandinavia, hand-powered or belt driven separators had become standard features of larger farm dairies by the late 1880s. The de Laval was exhibited at the Royal Show in England in 1879, and although there were numerous others involved in the development, it is this name that is most closely associated with early cream separators.

By a different process it is possible to make soft cheeses, and by applying physical pressure to a particular type of soft cheese, so as to squeeze even more water from it, it is possible to make the hard cheeses which can be stored for appreciable periods.

When milk is allowed to stand and the cream content rises to the surface, a slow fermentation will take place which will raise its acidity. If rennet is added to the milk at the correct temperature and acidity, then a solid curd will be formed containing the fats, and from which will be expelled a large percentage of water. The liquid fraction, known as the whey, is drained off, and the curd will be cut to allow more water to drain from it. In each of the stages a slightly different approach will alter the physical nature of the curd, and therefore its final texture and flavour as a cheese.

In the final stages the curd will be milled into very fine pieces, before being placed in a mould and then pressed under heavy weights for a number of days. This will allow further water to be squeezed out, and will result in a firm or hard cheese, depending on the accumulation of processes up to this point. Finally the cheeses will be left under controlled conditions to mature, during which time they will have to be frequently turned to ensure an even treatment throughout their mass.

Each of the stages described is fraught with difficulties. Environmental conditions such as temperature and humidity will affect the fermentation process, and therefore the acidity of the curd. Bacterial action may be intended at certain stages, and for particular cheeses, but the presence of the wrong strains of bacteria, or the introduction of desirable bacteria at the wrong time, can have disastrous consequences on the final results. The milking of animals in the open, the use of wooden receptacles, and the difficulty of keeping animals or equipment clean under such conditions, leave ample opportunity for contamination to occur. The problem was often compounded by the fact that the milk was left uncooled, thus allowing the bacterial population to multiply.

In Europe there is little evidence for the provision of buildings specifically set aside for milk processing until the sixteenth century. Up until that time both the processing and the maturing is likely to have been carried out within the domestic or the livestock areas of the holding, with all the associated problems of contamination that this implies. By the early nineteenth century the need to sterilize equipment had been recognized, and although it may still have been common practice to milk cows in the field until the latter half of that century, the new farmsteads being created at this time would invariably be equipped with a separate area for the stalling and milking of cows.

By the mid-1800s temperature control was also recognized to be an advantage, and the realization brought about changes in dairy design to create a cool environment in which to process and store the dairy products. In the USA the availability of commercial ice was utilized to the same end, and by 1843 ice was being used to facilitate the transport of this perishable commodity. The idea was quickly extended and by 1851 refrigerated rail cars were being used to transport butter over great distances.

Because of the high risks of contamination, and also because of the bulky and awkward nature of the raw material, the processing of milk has to be carried out as quickly, and therefore as close to the lactating animal, as possible. As transport systems have improved, and as the urban market has increased in size, so the processing has moved further and further away from the farmstead. The availability of refrigeration to farm scale operations has allowed the raw material to be cooled to a temperature at which chemical changes are minimized, and this factor, linked to improvements in road and rail transport, has made possible the centralization of processing into factories that can take advantage of economies of scale. Essentially the principles of butter- and cheese-making have not altered, but a better understanding of the processes involved and an ability to control environmental factors between the animal being milked and the shop counter has greatly reduced the risks, and permitted the presentation of a consistent and high quality product.

POULTRY FARMING

Poultry are another species of great significance in the history of animal domestication, although the most widely distributed representative, the domestic fowl, would appear to have arrived rather late on the scene. Earliest evidence from the Indus valley can be dated to about 2500 BC, although claims are made for an earlier Chinese origin. It had reached the Middle East at least by the first millennium BC, and the Mediterranean by the sixth century. The ancient Egyptians incubated poultry eggs by placing them in dung heaps, where the heat would remain constant whatever the outside temperature, a method recorded by the Roman writer Pliny, but beyond this example it would appear that the rearing of chickens remained a natural process until the nineteenth century. The constant temperature was the essential feature of the dung heap, and until a device could be invented to control heat application, an artificial means of incubating was not feasible. In 1883, Hearson invented a device which made possible the thermostatic control of temperature, and with this it was possible to regulate the temperature in a water jacket surrounding a container for the eggs. While this might have greatly facilitated the incubation process, and in doing so improved the percentage of chickens obtained from a batch of eggs, the poultry industry continued to be based on free range, labour-intensive methods for both egg and meat production. Although developments did occur earlier, it is only since the Second World War that full automation has allowed the economies of scale that have made poultry the cheapest meat available to the consumer. Egg production has been mechanized in the same way, and although the economic benefits are obvious, the poultry industry more than any other represents to a number of people a disquieting attitude by western society to the welfare of farm animals.

FOOD PRESERVATION

Refrigeration

The storage of any food materials raises problems because it is food to much else other than man. Animals, bacteria, fungi, insects and plants can all cause losses to stored material. As with much of human discovery, it is likely that methods of storage were discovered by accident and exploited from necessity. Thus it is possible to imagine a hunter in the Arctic regions being forced by unfortunate circumstance to sample the meat from a frozen carcase discovered by chance. It would not be such a major step to freeze meat deliberately in times of abundance to avert a repetition of the incident. Evidence for the intentional freezing of food in prehistoric times has been found in several excavations in the Ukrainian Republic of the USSR. At a site near the town of Mezhirich, in a habitation area dated to about 15,000 years ago, storage pits cut into the permafrost have been excavated.

For affluent groups living outside these areas of permafrost, but aware of such storage methods from trading contacts, the idea of moving ice to permit long-term storage may not have been such an absurd idea. In more recent times the distances involved may have been quite considerable. There is certainly literary evidence in classical times of the use of snow and ice to preserve food, and the practice was still to be found in the nineteenth century, when ice houses were a feature of the wealthier estates. Ice was certainly used in areas in which it was comparatively accessible or could be easily transported. As early as the sixteenth century some of the coastal Mediterranean towns could boast of a year-round supply of ice. By the 1660s the first ice houses were being built in Britain and other areas with similar supply problems. In America the potential market for ice was readily exploited for both home and overseas sales. In 1833 a shipment of ice was successfully transferred from Boston to Calcutta, and this was to be the start of a massive trade. One company alone required storage facilities for nearly 20,500 tonnes of ice. Ice-making machines were to appear by the middle of the nineteenth century, but the trade in ice continued well beyond this time.

The invention of refrigeration brought the concept into practical application and was to have far reaching effects not only on the diets of the western world, but also on their agricultural economies. In 1875 a small quantity of chilled beef was successfully transported from New York to Britain. Two years later the SS *Paraguay* sailed from Buenos Aires bound for France with a consignment of frozen meat. The success of this voyage led to a flood of meat from North America, over 32,600 tonnes being transferred in 1880 alone. Australian exports began the same year, though the response to the idea was much more slowly taken up, only 400 carcases of mutton and some 51 tonnes of beef reaching the UK in that year. Initially natural ice refrigeration was used, but in 1879 the first refrigeration machine was fitted into a ship, and by 1891 this had become the normal method used. The labour-extensive range farming possible in these new territories allowed frozen products to be sold in Britain at a price below that achievable by the home producers. This factor, together with the development of steamships, which made possible the swift transfer of grain from the prairies of America to the mills of Europe, was to have a devastating effect on British agriculture in particular. Recovery did not occur until the First World War, when this trade was hindered by submarine blockade.

Once frozen, most products can be stored for appreciable periods, but the early freezing procedures were slow, and the volume that could be handled therefore limited. In 1924, Clarence Birdseye developed a method for quick freezing which allowed for greater volumes of produce to be handled speedily. Although he sold his company within a few years, the name was retained and became for a time synonymous with the idea of convenience food. In 1930 frozen peas were first introduced, and in America it was possible to buy precooked frozen foods as early as 1939.

Canning

Australia and America both raised huge herds of cattle and sheep, and refrigeration was not the only attempted solution to the lack of a home market. In 1809 the Frenchman Nicolas Appert invented a method for the long-term storage of perishable foods, whereby the foods were sterilized by boiling and then sealed into glass jars. An English merchant developed the idea by using metal containers, and in 1811 the first canning factory was established in Britain. The Australians began to export canned meat in 1847, selling mostly to the British Admiralty. The armed forces of various countries seem to have been the major market for these early products, but by 1867 the Australian shipments were available for general sale in Europe, and by 1871, South American producers were also contributing to these supplies.

Salting

Salting was also an important method of meat preservation, and although its use continues to the present day, in the West it is most usually found as a means of producing a distinctive type of meat, such as bacon or ham, rather than purely as a means of achieving long-term storage. The hieroglyphic sign for salt does not appear in the early scripts of the Middle East, and its use is therefore likely to have developed later. Its importance is reflected in the word 'salary' which derives from the Roman word for salt. People do exist whose diet is totally devoid of added salt, but it is unusual, particularly if the food preparation involves any degree of boiling. As a preserving agent it had great economic importance, and was used extensively before it was superseded by refrigeration. Initially the latter increased rather than decreased the volume of meat being salted, since the optimum temperatures for the process precluded summer curing. With the advent of refrigeration machinery it became a year-round occupation.

Margarine

Long-term storage of food has become a major stage in the journey from the farm gate to the table and chemical rather than biological processes have become increasingly important. The production and sale of margarine is an example of this trend. In 1869 the French chemist Mège-Mouriez patented a process which he had devised to utilize the hard suet fat from cattle, by processing it and adding to it other animal wastes to produce a soft butter substitute. It was sometimes blended with butter, but increasingly vegetable oils were used to alter the taste and texture of the product. In 1923 the first margarine made purely from vegetable oils was put on the market by Van den Bergh & Jurgens. Present-day anxieties over animal fats have benefited the sales potential of a product which met with considerable market resistance in its early development.

Cereal storage

Plant products are also susceptible to damage, and even cereal grain, which gives such an appearance of long-term stability, can be very easily spoilt. It also provides appetizing food to rats, mice and many other lesser pests. In prehistory there is strong evidence for the storage of grain in pits in the ground. These are to be found in widely varying geographical situations, and the distribution of their use indicates that the idea was independently discovered by several totally separate communities. It is essential that the ground is well drained and that there is a low water table. The pits may be lined, though this is not always the case, and once filled they are sealed with a plug of clay. The seeds in contact with the moist walls will germinate, and may begin to rot, but both processes use up oxygen and give off carbon dioxide: this breakdown continues until the carbon dioxide concentration reaches such a level that it kills off the bacteria, and any other pest which might be contained in the pit. The contents of the pit will then remain in extremely good condition until such time as the seal is broken. Recent experiments have shown that corn stored in this way will be of sufficiently good quality for human consumption, and also that a very large percentage of the seed will subsequently germinate, and the method is therefore of value in the storage of seed corn from one year to the next.

During the Roman period in Europe, and even earlier in China, purpose designed buildings were being used for the storage of very large quantities of grain. As early as 700 BC grain storage organized by the state was being carried out in China. The motives behind the action were obviously to store in good times the food that would be required in bad; the incentive was as much political as humanitarian, since the idea was to stabilize prices, and therefore to ease public unrest. The effective running of the concept required strong political control, and it is perhaps significant that it was not until about AD 600 that a successful scheme was brought into being, and then run for an extended period. Biblical and other sources attest to the recognized need to store grain against bad times, and the efficient and powerful state system that was required to put such ambitious policies into practice. To work properly, and to ensure that the system was not abused by those charged with its administration, there was the need for rigorous book-keeping, and this in itself may have been the stimulus for several societies to develop their own scripts, and ultimately to establish the written word as a means of communication.

The storage of grain above ground leaves it highly exposed to wastage from birds and rodents and was the cause of huge losses in antiquity, as it continues to be today. To an extent building design can limit these losses, but it can do little to diminish comparable losses from insect and fungal attack. Perhaps one of the most urgent problems still to be resolved in Africa is not just the production of food, but also its safe storage. European mediaeval records give a consistently and depressingly low return on yields, suggesting figures for seed harvested which are as low as only three or four times the volume sown, but it is more likely that the amount is in fact the quantity used, and the low figures therefore reflect not only poor yields, but also losses subsequent to harvest. In contrast the Chinese literature of the same period, and even records as early as 200 BC, make claims to a thirty-fold return on grain sown. In part this will reflect the higher yields possible because of very high labour input, but it is probably no coincidence that insecticide treatment, of both seed corn and corn stored for subsequent consumption, is a subject considered by the Chinese chronicler but not the European counterpart.

In order for cereals to store well it is essential that the moisture content in the grain should be below 12 per cent of total weight. This was originally achieved by harvesting and drying the grain in stooks in the field, a method which was simple and efficient in the right climate or season, but prone to disaster when conditions were less favourable. Corn drying could be achieved by spreading the grain in the sun, or artificially in specially designed buildings, as was sometimes practised by the Romans. Up to the twentieth century these methods were all that was available to the farmer, and the population as a whole lived with the spectre of lean times which could be inflicted by the vagaries of climate. The arrival of the combine harvester, particularly in the European context, made an artificial means of drying an essential process of harvest. Efficient drying systems, linked to the capabilities of the modern harvester, have allowed the recovery of crops in conditions which would have been the cause of major shortages before the war. This capacity, together with improvements in genetic and chemical technologies of crop growth, has allowed western agriculture to claim the ability to feed itself and provide a reserve for poor years.

Sugar

The use of sugar as a preserving agent has a wide and long history, and can be applied directly as in jams and preserves, or used as a base for fermentation processes which will produce alcohol, itself a useful preservative. Sugar can be derived from a whole range of plants, but in its commercial form was initially obtained from sugar cane. Grown in India as early as 1000 BC, its cultivation was taken westwards by the Arabs. It was introduced into southern Europe in the eighth century AD, and first introduced into the New World in 1493 when Christopher Columbus took cane to the West Indies. This area became the major world producer, being able to compete successfully against others because the labour required for its production was supplied by slaves from West Africa and elsewhere. In the 1740s the German chemist Andreas Marggraf developed a method for the extraction of sugar from sugar beet, and although the effect was far from immediate, the process was eventually to make Europe a major sugar producer, with profound consequences to the Caribbean economy.

Alcohol is a product of the fermentation of the natural sugars that are to be found in plant materials. As such it is likely that it was encountered accidentally by prehistoric man who, attracted by its flavour and effect, later developed its use in the production of a range of beverages. Grapes, cereals and many other plants have been utilized to produce alcoholic drinks, the strengths of which would depend on the method of production. Although distillation may have occurred accidentally because of the kind of receptacle used in some other process, it is unlikely that it was knowingly and deliberately carried out until well into the historical period.

Whatever the motives for producing alcoholic beverages, the end result was not only palatable, but it stored extremely well and could also be nutritious. William Cobbett, writing in nineteenth-century Britain, condemned the introduction of tea, since it replaced beer in the diet of the labourer, and therefore deprived those who could least afford it of the nutritious benefit of that beverage. Others would have viewed alcohol less charitably, but its production in its many forms has occupied considerable acreages, and been of concern to farmer and brewer for thousands of years. More recent developments have seen the potential of an industry producting fuel alcohol from agriculturally produced raw materials in the attempt to replace expensive or diminishing supplies of oil. It is likely that agriculture will be increasingly seen as a source of industrial raw materials.

CONCLUSION

The term 'revolution' has been applied to three separate transition points in agricultural history. The change from a hunter gatherer economy to one based on a settled agriculture is traditionally called the Neolithic Revolution, but the change was hardly sudden, and it is in fact difficult if not impossible to identify any critical date at about which the change occurred. The second point to be identified was shortly after the Industrial Revolution and, to give the farming industry comparable status, the term Agricultural Revolution was applied to the period in the later nineteenth century when mechanization of both farmyard and field operations was becoming normal practice. But if a revolution did occur it was considerably before this time, and it did not involve machinery. For example, in the United Kingdom between 1701 and 1801 the population increased from about 9.4 million to 15.7 million. Although a certain amount of food imports did occur over this period, these people were in effect fed by British farmers. This achievement was accomplished by changes in rotations, selective breeding of plants and animals, and increasing awareness of the biological and chemical principles involved in agriculture, and the application of techniques based on this awareness. It was also a period when scientific experimentation was being applied, and in this the period is not dissimilar to the latest revolution, which has occurred since the Second World War. It is in the laboratory that the chemicals for fertilizers and sprays are developed, and here also breeding experiments have produced the multiple increases in crop yields. On the experimental stations animal breeding, and particularly veterinary techniques such as artificial insemination and embryo transplants, have been developed. The skill of the farmer is still of the utmost importance in the application of these advances, since only with good husbandry can they be fully realized. The skill and inherent knowledge of the farmer are of particular importance in those economies that can least afford the benefits of the laboratory that are dependent on oil for their implementation. It is likely that the laboratory will produce further startling increases in production potential, but research of this sort can only be afforded by those who produce more than sufficient to meet their needs. These nations have produced the technology which has the potential to feed the world's population, but the technology to achieve the necessary distribution economically has yet to be realized.

17

TEXTILES AND CLOTHING

RICHARD HILLS

INTRODUCTION

Whether we are attracted to the recent idea of man as a 'naked ape' or to the more picturesque scene of Adam and Eve in the Garden of Eden suddenly becoming aware of their nakedness, it is evident that the human race, alone among animals, has needed some form of clothing to act as protection from both the heat of the summer sun and the cold of winter. While birds are endowed with bright feathers to differentiate and attract the opposite sex, humans very quickly started to improve their appearance by decorating their apparel, so that the tale of textile invention has become linked with fabrics which show colour and beauty as well as mere functionalism. It is this story of luxury linked with necessity that adds another dimension to textile history when compared with the development of some other technologies.

Animal skins probably provided the earliest clothing worn by man, at first untreated and then in the form of leather. Strictly, leather is the middle layer of the skin, without the epidermis on the outside, to which the hair or wool is attached, or the inner layer of fat and flesh below. Both these must be scraped off. Smoking may have been one early method of curing leather to prevent it developing an unpleasant smell, but a dried skin becomes hard unless it is treated with fats or oils. The most important curing method uses tannin and the skins are soaked in pits before being impregnated with grease and oil. Leather used to play a more important role in the domestic economy than it does today. Not only were clothes, gloves and shoes made from it, but so were buckets, bottles for liquids, leather hosepipes, as well as book bindings. Leather began to be used for upholstery and wall coverings in the seventeenth century and of course there were also leather balls for games. Because no two skins are alike, and will vary in size, shape and thickness, tanning and curing leather has remained very much a hand craft and has been difficult to mechanize.

TEXTILE FIBRES

Textiles are created from a mass of fibres processed in a variety of ways. The average length (staple length) of the individual fibres determines how they are to be spun. Cotton fibres have the shortest staple length of those commonly used, at a minimum of 3.8cm (1½in), while linen is around 25cm (10in). Man-made fibres which imitate natural ones are manufactured in continuous lengths and later cut up into staple lengths similar to their natural counterparts. In spinning, the fibres must be locked together by being twisted around each other to form a yarn. A yarn is a single strand of twisted fibres which, by itself, would unwind or coil up into snarls. More turns, or twist, give a harder, stronger yarn than one with fewer. For sewing or knitting by hand, the yarn is doubled with at least one other to make a thread which will not unravel. In the cotton industry, the thickness, or count, of the yarn was determined by length, how many lengths of 840 yards (768m) weighed one pound (454g), so the finer the yarn, the higher the count.

From these basic fibres is made the wide range of fabrics and garments available today. A hat, for example, may be made of felt, a type of fabric which may have been one of the earliest made by man. An overcoat will, most likely, have been woven from wool clipped off sheep. The heavy cloth will need fulling or milling to make it compact and, in addition, may have been raised and sheared to give it a smooth finish. The linings of the sleeves, back and pockets will be totally different cloth, today mostly based on man-made fibres. A raincoat will have cloth with cotton fibres that will then have been waterproofed. Such garments are often called 'mackintoshes' after the person who first perfected a waterproofing process (see p. 849). Underneath these outer garments, suits for either men or women may have been made from wool treated in one way to give worsted cloth (named for a small village to the north of Norwich) or in another to give woollen tweed. Denim trousers are made from a tough, tightly woven cloth, but the same cotton fibres could have been spun and woven in a different way for making into softer shirts or knitting into underwear. Again, cotton fibres may be subjected to different finishing processes, of which mercerization, to give a sheen, was one of the earliest (see p. 849). Weaving elastic webbing is a specialized skill, and so is making lace. Shoelaces may well have been produced on a braiding machine, while a narrow fabric loom can weave ribbons or the labels sewn inside a garment to show who produced it and how it should be washed. Women's stockings and tights (whether made of silk or nylon) have to be shaped specially to fit the curves of their legs. On the feet may be worn shoes made from leather. This short review has not covered furnishing or industrial fabrics, which again come in an almost endless variety of fibres, shapes and types.

Silk

Silk from the domestic worm has the longest staple length of any of the natural textile fibres; it is also the toughest and is extremely resilient, for it can be extended between twenty and twenty-five per cent of its length before it breaks. Another outstanding property of silk is that it reflects light and so can be made to look luxurious, while it is also ideal for wearing next to the skin.

The silkworm was domesticated in China in the third millennium BC, but, although cloth was exported to Europe and was highly prized by Roman emperors, cultivation of pure silk did not reach Europe until the sixth century AD. The silk grub makes a cocoon around itself for protection while the chrysalis is being transformed into a moth and produces two protein substances within its body. One of these, fibroin, forms the main core of the fibre, while the second sericin (the so-called 'silk gum'), is laid on top of the fibroin. The fibroin is secreted from two glands and, when the silkworm is ready to spin its thread, it extrudes a little of this solution, fixes it on to a support and then stretches it out by drawing back its head. In this way, the grub gradually winds around itself a cocoon of a double filament which may be up to 1000m in length.

When the moth is about to emerge from the cocoon, it exudes another substance which melts one end so that it can creep out through the hole. This of course destroys the continuity of the filament and renders that cocoon useless for winding into pure silk. During the early 1860s, Samuel C. Lister persevered in finding ways of mechanically processing this waste silk and so created a new industry. Normally, the chrysalis is killed, for, by immersing the cocoon in hot water, the sericin melts and the filament can be pulled off. The beginning and the end of the cocoon are too fine for practical use and these go to the waste silk processors.

If the cocoon is brushed while wet, the loose exterior can be pulled off and soon a single continuous filament will adhere to the brush. This is still too fine to be used on its own, so five, six or seven cocoons are immersed in hot water and unreeled side by side. The remaining sericin hardens as it dries and helps to stick the filaments together. This combined filament is still too fine for knitting or weaving, so it is twisted or 'thrown' together with as many others as are necessary to make up the correct weight of thread. Eventually the thread, or the finished fabric, is boiled and the sericin discharged. It is only then that the original double filaments exuded by the silkworm are separated and the true nature of silk, with its sheen and shimmer, begins to appear.

The throwing of silk (from the Old English *thraw*, to whirl or spin) was one of the earliest processes of the textile industry to be mechanized, because the staple length of the fibres was so long. A fibre length of hundreds of yards meant that there were few breakages after the initial unreeling, so not only could the final thread be made very fine, but the operative, and later the machine, needed no skill or 'feel': in fact a silk winding machine can be left running unattended. In any other textile industry constant attention is essential, to deal with breaks in the yarn being spun or woven.

Linen

In the Bible, pure white garments worn in heaven were considered to have been made from linen, which is known to have been used for textiles as early as the seventh millennium BC. It is washable and can be made into very fine, thin cloth so that it was popular with the ancient Egyptians. Linen comes from the bast fibres of flax. The flax plant has a single erect stem up to about a metre tall and branching only at the top, where it bears a number of blue flowers. The plants used to be pulled up by hand, including the root, just before the seeds ripened, but today this can be done by a machine. After drying, the seeds are removed and crushed to make linseed oil and cattle-cake. The stem of flax is made up of a thick woody core surrounded by an outer bark. Between them, parallel to the core, lie the linen fibres.

Flax is the most difficult of all fibre sources to prepare for spinning. To separate the bast fibres from the woody core and bark may demand up to seven operations. After removal of the seeds, the flax must be retted. Today this is done with modern chemicals in steam-heated vats, but the traditional method was to submerge bundles of flax in stagnant or slow-running water for two to three weeks until the bark and core began to decay. The stalks were taken out and dried in the sun. After this, the stalk had to be broken, usually by pounding with a mallet. This started to break up the bark and core and loosen the fibres. Today this is done be passing the stalks between fluted rollers. Scutching follows, when the stalks are beaten by flat wooden blades. Water-powered scutching mills were introduced into Northern Ireland during the early part of the eighteenth century. The final stage in the preparation is hackling, where the fibres are drawn through iron spikes set in a board to remove the final pieces of bark or core. Flax fibres can be any staple length up to 25cm (10in) and are very fine, which is the reason they can be spun and woven into light cloth.

Wool

Sheep were among the earliest animals to be domesticated, probably well before the fourth millennium BC. The fleece grows from two types of follicles, primaries and secondaries. The primaries produce the coarser fibres of the outer coat (hair or kemp), which have been mostly bred out of modern sheep. The secondaries grow the more numerous, finer and shorter fibres of the undercoat and are almost always the wool fibres in the strict sense. Their natural crimp and the scaly outer covering of the cuticle enable wool fibres to interlock easily in spinning. Not only does this give maximum stability to woven cloth, but the fibres do not need to be tightly twisted together, therefore airlocks can be formed which give good insulation. Woollen garments can be made flexible and are crease resistant as well as being water-repellent.

In antiquity, the fleece was pulled off the sheep as it moulted, with the result that a great deal was lost. The Romans certainly had shears and clipped their flocks. The staple length varied according to the breed of the sheep. Breeds of long-haired sheep might yield wool 40–25cm (16–10in) long, but short-haired varieties would average 15cm (6in) or less. The fleece would be washed to remove some of the natural grease, or lanolin, and dirt such as burs picked out. In terms of textile technology, discussion of wool includes fibres from goats (angora, cashmere) and other animals such as the camel, alpaca, llama and vicuña.

Felt

Advantage was taken of the scaly nature of wool to make another early form of fabric, felt. The thin outer covering of the fibres, the cuticle, is composed of overlapping scales which point in one direction. This means that they can slide past each other when moved one way, but will interlock if pulled in opposite directions. It is this characteristic which enables wool to be felted or matted. By using heat, moisture, pressure and vibration, the feltmaker causes the wool fibres to lock together to produce non-woven fabrics which range from the soft and bulky to those which are so solid (such as that for hammers in pianos) that they have to be cut with a saw. A mass of loose hairs or fur, which has been damped with hot water, can be made into a lump of felt by rubbing and squashing it. This can either be flat (for example modern carpet underlays) or it can be shaped (for example a hat or a shoe). It was for these latter purposes that felt was made in antiquity and is still used today. Making hats by planking, that is, rolling the fur on a board by hand, unrolling it, immersing it in hot water and re-rolling it again, was carried on until well after 1970 in the hatting industry. Felt can be a cheap fabric to produce because it consists essentially of only a single process. Recently, there have been attempts to copy it with non-woven fabrics using modern man-made fibres.

Cotton

Cotton was the only fibre from the seed of a plant which had any commercial value in antiquity. It came either from a tree or from a perennial plant. The fibres grow on the seeds within the seed pod or boll. When this splits open in the summer, the seeds and cotton wool are picked out, traditionally by hand; in developed countries this is now done by machine. Ginning, separating the cotton wool from the seeds, was usually done in the cotton fields themselves. The staple length of cotton varied according to the place where it was grown. The average fibre length was 2.5–3.5cm (1–1⅓in) but in the nineteenth century a good quality West Indian cotton might grow to 4.5cm (1.8in). Colour and character also varied according to origin, for Egyptian cotton was yellower and West Indian was a silky white. Of all the fibres, cotton needed the least preparation, for after ginning it was ready for the spinning processes. Cotton lacks the harshness of linen and is softer to the touch. It is readily absorbent and this is a quality relevant in hot climates, where the cotton plant flourishes. It was used in these places for clothing long before the birth of Christ. It can be washed easily and has another characteristic, for it burns without smell. For this reason, it was used in the colder northern parts for candle and lamp wicks long before it was made into garments there.

EARLY TEXTILE PROCESSES

Primitive spinning techniques

The origins of textile production are lost in the mists of antiquity. Possibly prehistoric man wanted to secure his stone axe-head more firmly to its shaft, so he twisted some strips of leather together to make a sort of rope. Possibly one of the women, sitting by a fire in a cave, was playing with fur from the skin of an animal they had killed and pulled it out and twisted it into a sort of yarn. A couple of yarns twisted together into a thread could be used for sewing up skins to make garments. The yarn, or thread, could be stored by winding it round a stick, and soon the stick itself was being used to put the twist into the fibres. Put a weight on the stick to act as a flywheel and to give it more momentum and there is the first spindle and whorl; the earliest spinning device which is still used today in some parts of the world.

Up to 1800, textile production formed a vital part of the economy of most households. Much later than this, Queen Victoria was photographed working at her spinning wheel. The term 'spinster' was originally used for a single woman because she was expected to spin and weave the household textiles she would need in her married home. Learning to spin rough yarn and to weave coarse cloth was fairly easy, but great skill was needed to produce the finer qualities. The raw materials were important, for fine textiles cannot be produced without fine fibres, but the actual spinning implements in a peasant society are so simple that the quality of their construction is unlikely to be reflected in the final product, which is determined solely by the skill of the individual spinner.

While it is possible to spin directly with the raw materials, the fibres of wool, flax and cotton really need some basic process to begin disentangling them and laying them parallel to each other. The remains of the bark in flax, the seeds or dirt in cotton and grit or burs in wool would be removed at this stage. How this was done in antiquity is not clear, for carding seems to have been a mediaeval invention. Wool was certainly combed in antiquity, and it is possible that the other fibres were combed in some way too. A flat piece of iron, measuring 25–35cm (10–14in) long by 10cm (4in) wide, had teeth about 22cm (8½in) long cut in one end. One comb would be mounted in the top of an upright stake, the wool put on it and another comb pulled through the wool. Short fibres could be pulled out and discarded when spinning the finest yarns, while the long ones were straightened. In the mediaeval period, and later, the combs were kept hot and the wool was well greased.

The prepared fibres were wound up and loosely tied to the top of a distaff. Distaffs were basically short sticks, 20–30cm (8–12in) long. Sometimes they were forked to hold the fibres better and sometimes they were elaborately carved. In antiquity, they were held in the left hand, but later longer ones were used which could be stuck in a belt round the spinner's waist, leaving both hands free for manipulating the fibres. The fibres can be drawn out (drafted) from each other while they are in a loose mass, but, if they are twisted, they will stick together. If sufficient twist is put in, the fibres will be locked together so firmly that they can no longer slide past each other and so a strong yarn is made. Spinning consists in drawing out the requisite number of fibres to form the correct thickness or weight of yarn and then twisting them so they are locked firmly together.

All spinning methods, until the recent invention of 'break' or 'open end' spinning (see p. 842) have relied on two simple principles for twisting and winding on. The first principle is that, if a rod or spindle is rotated, any length of yarn tied to it and held in the same line as it is pointing will be rotated too and so will be twisted. The second principle is that, if the yarn is held at right angles to the spindle, it will be wound on as the spindle is rotated. Some people have claimed that the earliest spinning aid, the spindle and whorl, is a machine because there are two distant parts: the spindle itself, a round piece of wood tapering unequally to a point at both ends, while the whorl is a circular weight with a hole through it which fits over the longer of the spindle's tapers until it is secured near the bottom. Both are caused to rotate and the whorl acts as a flywheel; and both have to be proportioned to the type of fibres that are to be spun.

To spin any yarn, the necessary number of fibres to form the correct thickness, or count, must be drawn from the mass of wool on the distaff and then twisted to give them coherence. A short length of yarn is attached to the top of the spindle and the spindle and whorl are set spinning, dangling from the right hand. The weight of the whorl keeps the spindle rotating for a considerable time while the fibres are being drawn out, or drafted, by the fingers of both hands. The fingers of the right hand allow a little twist from the spinning spindle to reach the drafting zone, which causes the fibres to stick together. The spindle hung from about waist height so that barely a metre could be spun before that length had to be wound on to the spindle and a new one started. To wind on, the yarn had to be unhooked from the spindle tip; the spindle was grasped in one hand and turned while the yarn was guided on from the side by the other hand. By this painfully slow process must have been spun all the linen, wool and cotton yarn used not only by our prehistoric ancestors, but also by the ancient Egyptians, the Greeks, Romans, Saxons and in fact, anybody living before about 1300 AD. Yet the spindle and whorl had two important advantages over the later spinning wheels. They were portable and could be used as women walked to and from market or while guarding flocks. Then, by resting the bottom tip on a stone bearing to relieve the weight on the yarn, very fine threads could be spun which could not be matched on spinning wheels (see p. 813).

Primitive weaving

Weaving consists of interlacing two sets of threads, the warp and the weft. The warp, which is prepared first, consists of a given number of parallel threads, usually all the same length. These threads are most likely to have been passed through some devices both to keep them at the correct distance apart (the reed) and to enable selected ones to be drawn apart from the others (the heddles). The weft is inserted across the warp threads at right angles to them in a single traverse (a shoot or pick) at a time. How it crosses the warp threads will determine the pattern. It will need to be beaten up against the previous pick to form a closely woven cloth and, at the edges, the warp threads may be selected in a special order to form a strong selvedge to stop the sides unravelling.

Textiles discovered in archaeological excavations give some idea of the sort of looms used in antiquity. The simplest are those used for weaving narrow fabrics such as ribbons or belts. One method was to make a heddle-frame which consisted of thin vertical strips of wood or bone each with a hole in the middle. The strips were set slightly apart in a frame, so that slots were left between them. The warp threads were passed alternately, one through a slot and the next through a hole. One end of the warp could be tied to the weaver's belt and the other to anything convenient which could take some tension. By depressing the heddle-frame, the weft could be passed above the warp threads passing through the holes but below those in the slots which were now at the top. This gave one opening, or shed. By raising the heddle-frame, the weft could be passed back; this time over the warp threads in the slots and under those in the holes, which gave the other shed. The heddle-frame could be used to beat up the weft too, and the finished fabric had a plain weave. More refined versions of the heddle-frame looms, made up into small boxes, may be seen in many museums in the north-east of the USA, for they were easy to operate and could help to while away profitably the long evenings in the remote villages of the New World. Ribbons or tapes with fringes could also be woven on them.

Another method of weaving narrow fabrics was to use tablets. Flat plates of bone, either triangular or square, had a hole drilled at each corner and a warp thread passed through. By twisting the tablet, a different thread, or threads, could be brought to the top position and the weft passed through the shed between them. The width of the fabric was determined by the number of tablets and the delicacy of the pattern was created by the skill of the weaver in knowing which combination of tablets to turn to give the next shed. Tablet weaving was suitable for making belts or webbing, and both tablet and heddle-frame weaving were used for making the starting borders for the most important loom used in antiquity, the warp-weighted vertical loom.

In many archaeological excavations in Europe, evidence has been found of the vertical loom in the form of loom weights, but so far no actual looms have survived. However, reconstructions have been made based on those still in use in the north of Scandinavia. The warp-weighted vertical loom had two wooden uprights, 2m (7ft) high or more, which did not stand vertically, but were made to lean against a wall at a convenient angle. They were joined across the top by a cloth-beam which could probably revolve in holes in the uprights and might be as much as 3m (10ft) long, determining the maximum width of the cloth. From each upright at about breast height there projected a short bracket, the end of which was usually forked to support one end of the heddle rod. Lower down, a shed rod spanned the gap between the uprights.

First the warp had to be prepared, which was done by weaving a starting border. The warp of this border would form the width of the cloth and its weft became the warp of the cloth by pulling out the weft at one side of the border into long loops. Usually this was done on a frame, so that all the loops had equal lengths. The ends of the loops were cut and this gave the alternate threads for the main warp. The border was fixed to the cloth-beam at the top of the loom. All the odd-numbered warp threads were tied in groups to loom weights and hung over the shed rod. All the even-numbered warp threads were tied to another set of loom weights and hung down perpendicularly behind the shed rod. The angle between the two sets of warp threads gave one shed through which the weft was passed and beaten up by either the fingers or a sword stick. The back threads were tied by loops which passed either side of the front ones to the heddle rod, so that, when the other shed was wanted, the heddle rod would be pulled forward and placed in the forks of the short brackets in front of the loom. This brought the back, even-numbered, warp threads to the front of the odd-numbered ones, and so the weft could be shot through the second shed for plain weaving. It was possible to weave a simple pattern of, say, a two-by-two twill on these looms, but nothing more complicated.

Another type of vertical loom used in antiquity was the two-beam vertical loom. This had wooden uprights, about 2m high, standing vertically. At the top and near the bottom were two horizontal beams. One of these beams, usually the top, could be adjusted to tension the warp. To warp up this loom, the yarn was tied to a bar at the back, passed over the top beam, down the front of the loom, under the bottom beam and up to the bar, where its direction was reversed. This gave a pair of yarns at the front for forming the shed. Carpets are still knotted by hand on a similar sort of loom in eastern countries, but for weaving cloth it was limited because the shed could not be opened very wide so there was difficulty in passing the weft through.

There is much uncertainty about the origins of the horizontal loom in the West. There is no mention of it in Roman literature and there are no pictorial representations earlier than the Middle Ages. It was probably used in China first and spread to Asia Minor during the third to fourth centuries AD. Its advantage lay in having plenty of space where multiple heddles or a draw-cord device could be arranged for weaving figured textiles. Yet textiles with very complex patterns can be woven on the simplest looms by the weaver himself selecting which warp threads are to be raised and which lowered. A sword stick is inserted to keep that shed open while the weft is passed through. Bedouin Arabs still weave strips of carpet in this way. Because it takes a very long time to select each pick, not only is it exceedingly painstaking, but it is almost impossible to repeat the pattern exactly.

THE MIDDLE AGES

Carding

At some time during the Middle Ages, a preparatory process appeared for disentangling the mass of raw wool fibres. It was called carding from *cardus*, the Latin name for a type of thistle or teazle which has small hooks covering its seed heads. The teazle heads were mounted side by side on wooden frames. The carder had a pair, one in each hand, so that he could separate the fibres from each other by drawing one card across the other (see Figure 17.1). When the fibres had been sufficiently disentangled, they were transferred to one card from which they could all be taken off and made into a rough roll or sliver. Usually this was spun into the finished yarn, but often it was drawn out into a coarse, loose length, lightly twisted, called a roving, which was spun subsequently into the finished yarn. It is certain that by the 1730s, the natural teazle was being replaced by bent metal wires, mounted in rows on letter strips which were nailed to wooden boards (see p. 830). This gave not only a large carding area but also one that was covered more evenly with hooks or points which were stronger than the teazle heads.

TEXTILES AND CLOTHING

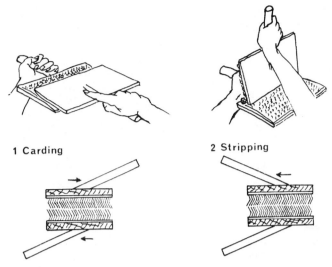

Figure 17.1: The principles of carding by hand. Drawing by Richard Hills.

Spinning wheels

Winding silk may have led to the development of the charka in India some time before AD 750. A spindle was mounted horizontally on a post and the whorl converted into a pulley driven by a band passing round a very much larger wheel which was turned by hand. This type of spinning wheel seems to have reached the British Isles during the fourteenth century when it had become much stronger and was mounted on legs so the spinner stood to work it. It had various names, such as the Great wheel or Jersey wheel (see Figure 17.2). With the right hand, the spinner turned the large wheel which rotated the spindle very rapidly. The left hand held the wool and was drawn away from the spindle, paying out the fibres while the right hand turned the wheel to put in twist. The yarn was held at an angle to the spindle so that, after a few turns, the yarn wrapped itself round the spindle until it reached the end. With continued turning of the spindle, the last coil of yarn slipped over the end and twisted the fibres. When the left hand had been drawn away as far as the arm could reach, twisting continued until the necessary strength had been achieved. To wind the yarn on, the spindle had to be reversed, or backed off, to unwind the coils at the tip. Then the left hand brought the yarn to a right angle with the spindle and guided it to form a nicely shaped cop or conical ball of thread, as the right hand was turning the wheel in the original direction again. Thus spinning with the great wheel was a multi-stage process, consisting of spinning, reversing, winding on and spinning again. Flax, wool and cotton could all be spun on this wheel and, while the art could be learnt quickly, great skill was required to produce fine yarn with the same consistency throughout its length.

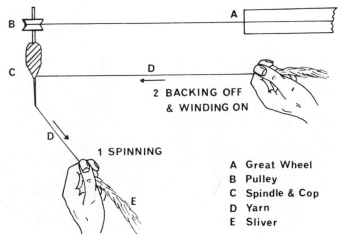

Figure 17.2: Stages of spinning on the Great wheel. Drawing by Richard Hills

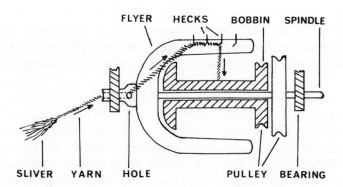

Figure 17.3: Spinning with flyer and bobbin on the Saxony or flax wheel. Drawing by Richard Hills.

Another type of spinning wheel, often called the Saxony wheel, flax wheel or flyer wheel, appeared soon after AD 1400. A large wheel drove the spindle to which was attached the flyer, a crosspiece with arms (see Figure 17.3). The wool passed through a hole in the end of the spindle so that it was twisted as the spindle was rotated by the whorl pulley. On the shank of the spindle was a bobbin with a pulley driven by a second band from the large wheel. The arms of the flyer guided the wool from the hole in the tip of the spindle to a position at the side of the bobbin. As the bobbin and spindle rotated at different speeds, because their pulleys were of different diameters, the wool was pulled through the hole and wound on to the bobbin. In this way, twisting and winding on were combined into a single, continuous process. Soon a crank and foot pedal were fitted to drive the large wheel, leaving both hands free to pull out and control the fibres. This wheel was more suitable for spinning the longer fibres of flax and wool.

It is not known whether it was the silk or the other textile industries which pioneered the flyer. In the other industries, the yarn was twisted as it was wound on to the bobbin. In the silk industry, the reverse happened. After the initial reeling of silk from the cocoons, the necessary number of filaments to make the final thread were unreeled and wound together on a bobbin. This was placed vertically in a frame with a spindle and flyer. There was a reel above which was turned to pull the silk off the bobbin while the spindle and flyer were rotated to put in twist. The weight of the bobbin resting on its ledge was sufficient to give enough drag to prevent it rotating at the same speed as the spindle and so the silk was twisted as it was pulled off the bobbin. Waterpowered silk throwing mills had appeared in Italy by 1500; Leonardo da Vinci drew a picture of a flyer and bobbin which had an automatic traverse to distribute the yarn evenly.

Weaving

The Middle Ages also saw great strides in the improvement of weaving techniques. In the thirteenth century the city of Winchester in Hampshire had two guilds of weavers, one for the vertical loom and one for the horizontal. Gradually the vertical loom disappeared from Britain because weaving on it was too slow. For example, a reed could not be used on the vertical loom. A reed was made from thin vertical strips of cane, spaced with slots between them rather like a comb but bound with strips of wood along the top and bottom. Pairs of warp threads passed through the slots so that not only was the warp kept evenly spaced to the correct width but the reed could be used to beat up the weft. Weaving was speeded up and the quality of the cloth improved.

Behind the reed on a horizontal loom could be placed heddle frames which controlled the lifting of the warp. The frames were raised by foot pedals, which was not feasible on the vertical loom, and this left both hands free to throw and catch the shuttle. On a horizontal loom it was possible to fit in many more heddles and still give them sufficient lift to open a reasonable shed, which made it possible to weave more complex patterns. Still finer patterns could be woven by using a draw boy system (Figure 17.4). While a complex pattern could be selected by hand, a method of repeating this across the width and along the length of the cloth had to be devised so that the same pattern would appear time and again. To repeat a pattern across the width of the cloth, the appropriate warp threads could all be joined to a vertical leash, or lingo, so that by lifting this, all the others would be lifted too. If the pattern is based on 200 ends, then the leashes of numbers 1, 201, 401, 601, etc. will be joined together.

PART FIVE: TECHNOLOGY AND SOCIETY

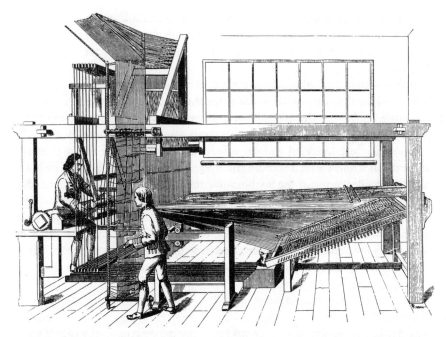

Figure 17.4: Draw boy loom. The weaver is sitting on the left while, in this case, the drawboy is standing beside the loom.

To repeat the pattern along the length, the identical warp threads must to lifted at the appropriate pick. All the vertical leashes for that pick were pulled to one side and, on the draw boy loom, a piece of string looped round them. The same was done for the next pick and similarly to the end of the pattern. A boy (hence the term draw boy loom) sat above the warp in the top of the loom. When he had sorted out the vertical leashes in the first loop of string, he wrapped them round a forked piece of wood which he twisted to raise the warp threads to form the shed. While he was sorting out leashes in the next loop, the weaver threw his shuttle and beat up the weft. It was possible to combine the draw boy system with a set of heddles which made weaving a little quicker. Draw boy looms were still being used to weave the finest cloth of gold at Isfahan in Iran at least until the fall of the Shah in 1975.

Fulling

Woollen cloth must be washed to remove grease and dirt, and this can be done at the same time as it is fulled. The tension in the warp threads during weaving will prevent the cloth being woven very tightly so, to make it more compact and to thicken it, advantage is taken of the felting characteristic of wool. The primitive way of fulling was to soak the cloth in a solution of fuller's earth (an alkali) and pummel it with the feet. This was the origin of the *ceilidh* in the Scottish Highlands and Western Isles, when the women used to sit round in a circle with the cloth on the floor in the middle. While they pummelled it with their feet, it was a convenient time for gossip. It is also the origin of the place names Wauk (or Walk) Mill in England and Pandy in Wales.

A mechanized way of fulling was available from the eleventh century on the Continent. The first record of fulling mills in England was in 1185 when two are mentioned, one in the Cotswolds and one in Yorkshire. A waterwheel turned a shaft with cams or tappets on it which lifted a pair of hammers alternately. The hammer heads pounded the cloth placed in a trough. The angle and slope of the heads ensured that the cloth rotated, to treat it equally throughout its length. The fulling mill was the first application of power to any textile process. In England it caused a shift of the industry out of the towns, with their well-established and restrictive guilds, into the hilly country where streams abounded and where flocks of sheep flourished and from the lowland east to the upland west. It also gave English textile producers a competitive edge over their rivals in Flanders, so it became more profitable to spin and weave cloth locally rather than to export the raw wool. The waterpowered fulling mill helped to give rise to the famous broadcloth industry in England. Two weavers worked at the broadcloth looms because they were too wide for one person to throw the shuttle through and catch it at the other side of the cloth. This cloth was finished by fulling, drying, stretching, raising, shearing and setting.

Some fulling stocks remained in use to within the last decade. However, in 1833, John Dyer of Trowbridge patented the milling machine. The cloth is sewn up into a continuous length and the selvedges may be sewn together so that it becomes a long tube. The bulk of the cloth sits in the bottom of the machine where fuller's earth and water can be fed in. A pair of rollers pulls the cloth out of the liquid and squashes it into a tapering spout. Some air will be trapped inside the sewn-up cloth and this will form a bubble in front of the rollers so that the cloth is opened up and then squeezed in a fresh place as it passes through.

Fulling made the cloth more compact at the expense of shrinking it. To bring it back to a standard size, it was stretched and dried on a tenter frame which consisted of two unequal pairs of bars, two the length and two the width of the cloth. Each bar had tenter hooks nailed regularly along it. One end bar and the upper side bar were fixed. The ends of the cloth were hooked on to the end bars and stretched out by pulling on the movable one. The top edge was hooked on to the fixed long bar and the bottom on to the lower one which could be pulled down in to a slot and pegged. In this way the cloth could be stretched out taut to its finished size as it was dried (hence the term 'on tenter hooks').

Hanging cloth out on tenter frames would take too long for modern production. Today a stenter is used. The tenter hooks are mounted on grippers which form a pair of continuous chains. The grippers close automatically on either side of the cloth as it is fed in. Guides push the two chains of grippers away from each other until they have stretched the cloth to the correct width. They then take the cloth through a drying section heated by steam pipes. At the far end, the grippers are released and pass back under the stenter while the cloth is rolled up.

A blanket has a raised surface or nap on which the wool fibres have been pulled out of the weave deliberately. At first raising was done by hand by pulling teazles mounted on wooden frames across the surface of the cloth. Gig mills, in which the teazles were mounted on rollers turned by hand or by water power, had made their appearance by the fifteenth century. Although today wire hooks, similar to card clothing, are used, teazles were long preferred for the highest quality woollen cloths and were used until the 1970s in some mills.

To give a smooth surface to cloth such as the old broadcloth, the nap was then sheared off. Hand-operated shears of enormous size cut the loose fibres while the cloth was laid over a curved table-top. It required great skill and delicacy to achieve a smooth finish. Various attempts were made to mechanize the process by placing several pairs of shears in a frame and operating them by cranks, but this was not successful. A rotary shearing machine was first patented in England in 1794 by the American Samuel Dore, but again it was not successful. In 1815, J. Lewis of Brimscombe near Stroud, Gloucestershire, patented a rotary shearing machine with cutters that were forerunners of the cylinder type of grass mowing machine used today. The cloth was moved underneath the blades which could be of the same width, so only one operation was needed for each side. By 1830 hand shearing was extinct.

THE SEVENTEENTH AND EARLY EIGHTEENTH CENTURIES

Knitting

Non-woven fabric can be made by taking a yarn and looping it into a series of interlocking stitches so that there is no separate warp and weft. Crocheting with a single hooked needle probably preceded knitting with a pair of pointed needles. Knitting may have been a Northern European, possibly Scandinavian, invention, but archaeological evidence has so far thrown little light on this subject. Certainly we know that knitting with needles had appeared in Britain by the fifteenth century. This looped fabric has much greater elasticity than a woven one, so that it moulds much more readily to the shape of the human body and is therefore particularly useful for closely fitting garments such as caps, socks and stockings, vests and pullovers. One inherent problem is that a dropped loop or broken thread can result in the whole garment unravelling. The common way of hand knitting has never been mechanized. Skilled knitters can follow very complicated and beautiful patterns such as the multicoloured Fair Isle designs.

Another method of knitting, peg knitting, formerly very popular, has almost died out except as a child's toy in the form of headless nails driven into the end of a cotton reel. Loops are formed round the nails or pegs and then the yarn is wound round. The first loops are drawn over the yarn, over the tops of the pegs and then dropped into the centre hole or slot where the rest of the fabric hangs, leaving the next set of loops on the pegs. Sometimes the loops or stitches are lifted by a hooked needle.

These rows of pegs may have given the Revd William Lee of Claverton, Nottinghamshire, the inspiration for the knitting machine he invented in 1589. It was a remarkably complex machine, needing at least seven or eight operations to make one row of stitches, and it is surprising that it succeeded as well as it did. A row of needles is placed horizontally and the last row of stitches is looped round their shanks with the fabric hanging below them. The needles are made from springy steel drawn out into a fine point. The point is bent back to form a hook that almost touches the shank of the needle. Normally these bearded needles are left open, but they can be closed by pressing the point into a small hollow ground in the shank. The knitter lays a shoot of yarn across the shanks of the needles in front of the first loops. Between each needle is a specially shaped piece of flat metal called a sinker. By using a foot pedal, a slurcock is drawn across the frame and causes the sinkers to push the yarn into loops over the needles. Next the sinkers push these loops forward under the points of the needles into the hooks at their heads. A presser bar is brought down to close the needle beards. The sinkers are raised so that a different part of them can be brought forward to push the first set of loops over the tops of the needle points and over the ends of the needle hooks as the presser bar is released. The arch of the sinkers draws the loops left in the needle hooks, and now part of the knitting, back on to the shanks ready for the next shoot of yarn. Finally the sinkers are raised again.

Although complex, these bearded needles could be made finer than later types, so framework knitting has continued to be a commercial proposition up to the present day. Flat strips only could be knitted on these frames, but the width could be altered by casting on or off stitches. Therefore, stockings made on these frames had to be sewn up by hand. The other disadvantage of the original machine was that it could knit only plain stitch, which did not have the necessary flexibility for stockings or socks. In 1758, Jedediah Strutt of Derby developed the Derby rib frame, which could be attached to the front of an ordinary frame. On it, vertical needles could be inserted between the horizontal ones to reverse every other stitch and so create a ribbed fabric. The fortune he made with this invention helped him finance Arkwright's spinning machine (see p. 827). The idea was extended further so that loops could be moved from one needle to another, thus creating a pattern of holes in the fabric. In this way, a sort of lace could be made, or openwork shawls. Later a way was found of knitting three or four pieces side by side at the same time, and by 1800 the stocking frame had been developed to the limits of its versatility, being able to knit forty distinct types of fabric. To operate these frames was hard physical labour, so they were normally worked by men. Although very complicated, even the early ones soon achieved a speed of over 600 loops per minute, six times faster than knitting by hand, while later, silk stockings were being produced at the rate of 1500 loops per minute.

Braiding

On May Day, it was the custom to set up a maypole on the village green. Children were separated into two groups to dance round it in opposite directions, each child holding a ribbon attached to the top of the pole, and they wove in and out of those coming towards them. Eventually the pole was wrapped up in a braid. In commercial braiding, the children are replaced by bobbins with thread wound on them. The maypole can be replaced by a rope or by the centre core of an electric cable which is covered with a smooth outer sheathing of braid. A shoelace may be a type of braid with no centre core. Sometimes the bobbins do not make a complete circle but are turned back at one point. This gives a flat braid for ribbons and red tape or candle wicks. The Japanese developed the technique of braiding to a fine art for making multicoloured ceremonial cords.

In 1748, a 'maypole' braiding machine was developed in Manchester. The tracks of the bobbins were pairs of semicircles and the bobbins had to be directed on their proper courses by deflectors. These machines could be turned by hand or by power. On later versions the tracks were lozenge-shaped and the bobbins were mounted on elongated slippers which could pass across the acute junction crossovers without any deflectors. Braid is still made on this type of machine today.

Early mechanized weaving

A loom on which more than one ribbon could be woven at once may have been invented by Anton Moller in Danzig in 1586. There was great opposition to it, both there and a little later in Holland when it appeared in that country. It arrived in England from the Low Countries, hence the name Dutch loom, and was being used in London by 1616 and Lancashire by 1680. At first only four to six ribbons could be woven at the same time, but this was improved to twenty by 1621 and eventually fifty and more. The shuttles were longer than the width of the ribbon and on the early looms they were knocked through the shed by pegs fitted on to a sliding bar. Later the shuttles incorporated a rack, and pinions on the shuttle race, or slay, drove the shuttles through their sheds. In Switzerland, in 1730, means were devised for driving these looms by power, which was prohibited. An attempt to set up a waterpowered manufactory in Manchester in 1765 by a Mr Gartside also failed because one person was still needed to tend each loom in case any threads broke and to change the shuttle when the weft ran out. In 1787, Thomas Gorton ordered a 10.5kw (14hp) steam engine from Boulton and Watt (see Chapter 5) to drive his smallware factory at Cuckney, Nottinghamshire, which had 56 weaving frames as well as a room full of spinning machinery. In 1791, Richard Gorton patented a power loom which was weaving fabrics of fairly narrow width.

Narrow fabric looms on these principles are still used today, particularly when fitted with a Jacquard harness (see p. 823) for weaving nametapes or labels. Ribbons with a simple pattern are being produced on looms where the shuttles have been replaced by needles. The needle pokes the weft through the warp and another catches the loop to complete the stitch, doubling the weft. The speed of production is of course very many times faster than on the old looms.

The flying shuttle

It is quite possible that John Kay of Bury, Lancashire, saw narrow fabric looms at work and from them derived his idea of the flying shuttle which he patented in 1733. Before this invention, the weaver had to pass the shuttle through the shed by hand, so the width of the cloth was limited to the distance he could reach. When he had passed the shuttle through, he had to change the position of his hands to beat up the weft with the reed. For broadcloth, two men were necessary to throw the shuttle to each other. The reed was mounted in a frame called the slay. Kay's shuttle ran along a ledge or race on the slay in front of the reed into a box at either end. It had wheels to lessen the friction. The weaver's right hand held a stick, the picking peg, to which were attached cords to pickers sliding inside the boxes. A jerk of the picking peg flicked the shuttle out of one box and sent it across the loom into the one on the other side. The left hand held the centre of the slay and beat up the weft with the reed once the shuttle had passed through. Not only did the weaver not have to change the position of his hands after each pick, which increased the speed of weaving, but also one person could weave much wider pieces.

At first the flying shuttle was unpopular because workers were afraid that they would lose their jobs since it speeded up production. Also, it probably needed many minor modifications to make it work satisfactorily. For example, the race must be set at a slight acute angle to the reed or the shuttle will fall off, and the shuttle must be shaped to that angle. The wheels are positioned so that the shuttle will run in a gentle circle, again to help keep it close to the reed. Then the points on the shuttle must be carved so that when it is held by them, the centre of gravity lies towards the back. It was probably not until after 1747, when Kay had to flee to France to escape the hostility aroused by his invention, that he made the final, and possibly crucial, modification. On earlier shuttles, the bobbin, or pirn, of weft was free to rotate on its spindle and the weft was pulled off from the side through a hole in the middle of the shuttle. With the faster speed of the flying shuttle and its sudden deceleration in the box, the pirn would have continued to rotate, causing a snarl in the weft. This fault was overcome by placing the pirn firmly on the spindle or skewer in the shuttle and drawing the weft off over one end and out through a hole at that end of the shuttle. The normal manner of winding the yarn on to the pirn was to traverse it from side to side along the face of the bobbin so that parallel layers were built up. Kay's new method was to wind the yarn in short traverses at one end of the pirn to begin with, and to build these up into a cone. Then he continued winding on this short conical surface until the entire length of the pirn had been filled and he had made a cop. This method of winding was soon copied on the great spinning wheel (see p. 813) because then this cop could be put straight into the shuttle without rewinding.

The flying shuttle became much more popular in Lancashire after 1760, when John Kay's son Robert invented the drop box. Instead of a single box for the shuttle at the end of the slay, he introduced a tiered box holding at first two but later up to four shuttles. This could slide up and down, so that any shuttle could be lined up with the race. Shuttles with either different colours or different types of weft could be put on the shelves and the weaver could select which one he wanted by pressing a lever on the reed with his left hand. This invention helped to speed up the weaving of cloths like checks with their different colours which formed a large part of the Lancashire output. John Kay also made another contribution to textile technology when, around 1730, he substituted strips of thin metal instead of split cane in the reed.

Pattern weaving

Methods for speeding the selection of warp threads to make patterns were developed on the Continent, particularly in France, presumably through the demands of the luxurious royal court. Two things had to be developed for pattern weaving. First was a simple memory system in which the pattern could be stored so that it could be reproduced pick by pick. Then a method of lifting the appropriate warp threads and leaving the others down had to be devised to link up with the memory. Ideally, each warp thread ought to be selected individually.

In 1725, Basile Bouchon invented a very ingenious apparatus consisting of a combination of hooks and needles, the lifting part, working in conjunction with a paper pierced with holes, the memory. The holes in the paper allowed some needles to penetrate and lift the hooks, which they controlled. In turn these hooks lifted the warp threads to form the design. The spaces on the paper pushed back the remaining needles and so the hook linked with them, and their corresponding threads remained down. The advantage of this mechanism over the earlier draw boy system was that the pattern could be changed merely by replacing the paper. But the system was limited, partly because of the small number of needles it could accommodate and probably through the weakness of the paper.

In 1728 another Frenchman, Falcon, invented a new arrangement with the needles placed in rows which permitted many more cords (the leashes, or lingos) to be used. The hooks were picked up on a series of crossbars called the griffe. He replaced the paper with a series of cards, each card representing one pick. The mechanism was moved to the side of the loom but still had to be worked by a second person. In 1745, Jacques de Vaucanson put the selecting mechanism on top of his unsuccessful power loom, but it was not until 1801 that Joseph-Marie Jacquard took out a patent for an improved figuring mechanism. The best features of earlier inventors were combined into a machine which needed only slight modifications to make it work properly. Jacquard took Falcon's cards punched with holes like Bouchon's paper, pressed against the needles by the pierced cylinder invented by Régnier, to select the pattern. He placed his machine on top of the loom, like Vaucanson, and arranged it so it could be operated by the weaver himself through a foot pedal. Although Jacquard was awarded a gold medal in 1804, his invention failed in the way it pressed the card against the needles. This was improved by Skola in 1819, and it was his Jacquard machine which finally solved the problem of repeating complex patterns in weaving (Figure 17.5). His apparatus was easily adapted for fitting on to power looms.

THE INDUSTRIAL REVOLUTION

Spinning machines

All the new inventions which emerged during the eighteenth century for producing different types of fabrics more quickly precipitated a crisis: because all flax, wool and cotton still had to be spun on the single thread mediaeval spinning wheels, the supply of yarn became inadequate. It is difficult to estimate how much yarn a weaver needed, or how much a spinner could produce, because spinning and weaving the finer counts took much more skill and time. Different writers give various estimates of the number of spinners required to keep one weaver at work, ranging from four to eight or even ten to twelve. If, as a rough average, we assume that the output of about eight spinners was needed

Figure 17.5: The Jacquard loom. The weaver is working the Jacquard foot pedal with his right leg while his wife is winding the pirns for the shuttle.

per weaver, then the weaver's difficulties in securing regular supplies will be appreciated.

Some accounts of the period around 1760 say that weavers had to walk many miles to find spinners, and had to pay them at a higher rate or give them presents in order to have the yarn ready in time for the cloth to be woven within the promised period. Often the spinning was of poor quality, the yarn being lumpy and irregularly twisted. In such circumstances, it is not surprising that many people were trying to invent a satisfactory spinning machine.

Many people must have looked at the silk throwing machines and wondered if the spinning of cotton or wool could be mechanized in the same way. Thomas Cotchett tried to build a water-powered mill with Italian silk throwing machines at Derby in 1702. He failed, and it was left to the half-brothers Thomas and John Lombe to establish the first successful power-driven English textile mill in 1717 at Derby beside Cotchett's earlier site. Spindles and flyers with reels above them were placed in three tiers around the periphery of a large circular frame. The waterwheel drove large horizontal wheels inside the frames which turned the spindles by rubbing against their rims. This twisted the silk which was wound on to the reels, also turned by the waterwheel. Similar machines were introduced to a mill at Stockport in 1738, so people in the Lancashire area would have heard about them.

The earliest claimants to making a machine for spinning cotton are Lewis Paul and John Wyatt who took out two patents in Paul's name in 1738 and

1758. The variety of ideas included in their first patent suggests a machine still in the experimental stage. Little work was done after 1743 and Paul died soon after the second patent had been granted. Paul and Wyatt built a circular machine somewhat on the lines of the silk throwing machines. They tried rollers for drawing out the rovings and wound the yarn on to bobbins past flyers. They failed, probably because the rollers they used were too large in diameter. They attempted to use the principle of spindle drawing, where the cotton was stretched to draw it out between the rollers and the spindle. Where this was done with a flyer, the cotton would have been drawn out unevenly and irregularly twisted, with many breaks. Various mills were established with these machines but none lasted long. Their machine was tantalizingly near to success, and might well have been a commercial proposition for doubling a thread, but it was based upon the wrong principles for spinning cotton.

The spinning jenny

It was some time before anyone else embarked on a machine for spinning after the failure of Paul and Wyatt. One day in about 1763, a spinning wheel in James Hargreaves's home near Blackburn, Lancashire, was knocked over accidentally. As it lay on its side, still rotating, it inspired him to try making a spinning machine. To understand the principles of the spinning jenny (Figure 17.6), it is necessary first to understand one method of spinning by hand. When spinning with the great wheel, it was possible for the spinner to grasp firmly the sliver or roving, leaving three of four inches of unspun fibres between the fingers and the spindle. Her fingers, still tightly clamped, were pulled away from the spindle, drawing out (drafting) the fibres. If the spindle was rotated gently at the same time, the fibres cohered as they were pulled out. The roving could be drawn out into a fine yarn because the twist ran into the thinnest places first, binding them together, while the thicker parts were more loosely twisted and so could continue to be drawn out. If the spinner did not twist enough, the yarn disintegrated, while if she twisted too much, it locked together before it was drawn out finely enough. When it was fine enough, she had to put in more twist to hold it all firmly together, and then she could back off and wind on in the usual way.

To spin like this required a very sensitive touch and would seem impossible to mechanize, yet it was this method which Hargreaves used on his jenny. He placed the spindles at the far end, slightly inclined from the vertical towards the spinner. They were driven by bands from a large wheel turned by the right hand. Instead of the spinner's fingers, he had a sliding wooden clamp, or clove, to hold the eight rovings he spun on his first machine. The left hand drew back the clove as the yarn was being spun and pushed it forward to wind on. A length of roving was released through the clove as it was drawn away from the spindles. The clove was closed again and the roving drawn out while the spindles were

Figure 17.6: A replica of the Hargreaves spinning jenny.

twisted gently. When the clove was nearly at the end of the slides, and the roving drawn out fine enough, the spindles were rotated quickly to put in twist to lock the fibres together. After this, the spindles were reversed to back off and finally turned once more in the spinning direction to wind on as the clove was pushed back in again. The yarn was guided on to the spindles by a faller wire operated by one foot.

Only a soft, lightly twisted yarn could be spun on the jenny because, when the yarn was wound on, a short length was left between the tips of the spindles and the clove. When the next length of roving was pulled through the clove, the twist from the yarn would run into the roving and, if it had been twisted too much, would lock the fibres in the roving and prevent it being drawn out. Therefore the yarn spun on the jenny was really only suitable for weft and not warp. By 1770 the number of spindles on the jenny had been increased to sixteen, but was soon raised to 80 or even 120. One person working an early jenny could supply one weaver. The jenny was cheap to produce, being mainly of wood, and comparatively small, so it was used principally in the spinners' homes.

James Hargreaves was born in Oswaldtwistle in Lancashire; son of a poor hand loom weaver, in about 1720. After he had developed his machine, he was

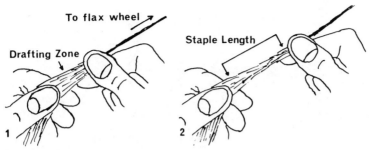

Figure 17.7: Principles of drawing out the fibres by hand. Drawing by Richard Hills.

forced to leave owing to the wrath of the local people who thought it would create unemployment. He moved to Nottingham and set up a factory producing yarn for the hosiery industry. Although he patented his machine after his arrival in Nottingham, he was unable to enforce the patent rights. He died in 1778, a disappointed man, for, although he himself never made a fortune, others did with his jenny and, in the end, he was forced to install Arkwright's waterframes in his factory because the jenny did not produce satisfactory yarn for the hosiery industry.

The waterframe

In 1769, Richard Arkwright of Preston, Lancashire, patented a spinning machine which had rollers for drawing out the cotton and flyers and bobbins for spinning. Arkwright may have been familiar with the flax spinning wheel (see p. 814) from his childhood and, from its continuous action, realized that it was the obvious type to try to mechanize if some mechanical action could be found to copy the spinner's fingers. The importance of the flax wheel lies not only in its continuous spinning principle but also in one of the ways in which spinners drew out the fibres on it. While the fingers of one hand held the sliver, the fingers of the other held the spun yarn and pulled out the right number of fibres from the sliver. Then the spun yarn was released, allowing the twist to run into the newly drawn out fibres. This distance the fingers could be pulled apart, the drafting zone, was determined by the staple length of the fibres (Figure 17.7).

In his prototype machine, Arkwright placed four spindles with their flyers and bobbins vertically near the bottom. The friction of the spindles turning caused the bobbins to rotate as well, so they were slightly retarded by light string brakes. At the top, he placed the vital drawing rollers which have barely changed up to the present day (Figure 17.8). There were four sets, each consisting of pairs of rollers. The top roller of each pair was covered with leather, and was kept firmly in contact with its counterpart below by a weight. The bottom rollers were made from metal and wood, and had flutes cut

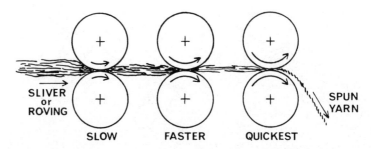

Figure 17.8: Principle of the cotton drawing rollers patented by Richard Arkwright in 1769.
Drawing by Richard Hills.

along them. These lower rollers were connected together by gearing so that one shaft could turn them all. The back pair of rollers, where the thick cotton roving entered, rotated slowest and each succeeding pair turned faster. In this way, the roving was drawn out until the correct number of fibres was passed out between the nip of the front rollers, where the twist from the spindle locked them firmly together and made the finished yarn.

Arkwright did not have the skill to make this machine himself, so he employed John Kay, a Warrington clockmaker, to help him. Earlier Kay had helped Thomas Highs with a spinning machine and, according to one version of the story, Kay showed Highs's machine to Arkwright, who copied it and took all the credit for the invention. However, Arkwright added two vital contributions. First, he realized that the successive pairs of rollers must be set the correct distance apart to correspond with the staple length of the fibres being spun. The drafting zone must be a little longer than the staple length. If the rollers were set close enough for two pairs to grip the same fibre, this fibre could not be drawn past its neighbours and would break. If the rollers were set too far apart, so that the fibres floated between them and were not gripped by either, the drawing would be uneven and lumpy and a broken yarn could result. Ideally, all the fibres should be the same length and the rollers set a little further apart than this length.

In order to hold the fibres tightly in the nip of the roller, Arkwright made his second contribution. He hung weights on the top rollers so that they pressed firmly against their lower counterparts. Paul and Wyatt (see p. 824) had not weighted their rollers and, in addition, did not seem to have realized the vital importance of the staple length. Without the weights, the twist will run back between the rollers and lock the fibres together in the drafting zone so they cannot be drawn out. By January 1768, Arkwright had returned to Preston and built a spinning machine in the old Grammar School. He could have made more and sold them locally, like Hargreaves, or he could have allowed his machines to be built on a small scale and used in cottages, for they could have been turned quite easily by people in their own homes. Instead, Arkwright's

Figure 17.9: An early Arkwright waterframe, possibly dating from 1775.

business sense told him that his machine had much greater potential because it could be set up in factories and driven by power.

In 1769, Arkwright and his two partners, John Smalley and David Thornley, were granted a patent and they moved to Nottingham where they established a factory powered by a horse. Then the partnership was enlarged with the addition of Jedidiah Strutt and Samuel Need, a Nottingham banker. A water-powered mill was built at Cromford, Derbyshire, in 1771 which was the first spinning mill as we know them today and here the final development of the spinning machine and of the preparatory machinery (see below) was carried out. Because the Cromford mill was driven by a waterwheel, Arkwright's spinning machine became known as the waterframe (Figure 17.9). The drag of the bobbin meant that only a hard-twisted yarn of medium counts could be produced on the waterframe. This was suitable for the warp of coarser types of cloth, so at last it was possible to weave in England cloth made entirely from cotton: previously most warps had been linen. Yarn was spun on the waterframe more cheaply than by the spinner at home and soon there was an enormous demand for it as both home sales and exports boomed.

Carding machines

Having produced a successful spinning machine, Arkwright needed to develop a series of preparatory machines in order to start factory production. On the waterframe, he fitted a cam motion to raise and lower the bobbins so that they were filled evenly in place of the earlier hecks (hooks) on the flyers. For the waterframes to be able to spin a continuous yarn, each spindle had to be supplied with a continuous length of roving which, in turn, had to be made from a continuous length of carded sliver. In other words, having mechanized one part of the spinning sequence, it was necessary to mechanize all the others to maintain the momentum. This is typical of the sudden upsurge called the Industrial Revolution, during which one invention stimulated another in quick succession. Once the balance of the old domestic industry had been upset, invention followed invention until textile production was completely industrialized.

Over the years there had been few improvements to carding. In 1750, William Pennington had patented a machine for making the holes in the leather backing through which the wires were inserted and, soon after that, Robert Kay developed a machine for cutting, bending and inserting the wires or points through the leather. Larger cards, sometimes called stock cards, were tried. The lower card was firmly fixed to a table and the upper one might be suspended by a series of pulleys and weights to help ease the load. Paul and Wyatt had found that it was necessary to mechanize carding to keep their spinning machines supplied, but they did not really progress beyond an enlarged hand or stock card. A more significant advance was made by D. Bourn in 1748 when he patented a carding machine on which the fibres were passed from one cylinder to another. The surfaces of the cylinders were covered with card clothing and, while it was easy to take the fibres off one set of points by another, Bourn did not solve the problem of taking the fibres off the final set which had to be done by hand with a needle stick.

Other people tried to make carding engines, but it was Arkwright who solved the problem and patented the solution in 1775. Descriptions and pictures of ten different machines for preparing cotton were given in the patent, but some were of little significance except that they show that Arkwright was trying to mechanize all the preparatory processes. His opening and cleaning machines were never constructed and, at Cromford, women beat the cotton with sticks to open it and then picked out the dirt by hand. The cotton had to be fed into the carding engine, which Arkwright did by laying a length of cloth on a table and spreading cotton evenly on the top. Then both were rolled up and put at the back of the carding engine. The fibres were fed in as the cloth was unrolled and left on the floor. Later the cloth was shown to be unnecessary.

To give a continuous sliver from the carding engine, the whole of the surface of the cylinders needed to be covered with points. Arkwright prepared a long narrow strip of card fillet which he wound in a spiral round the cylinder so there were virtually no gaps between the points. The type of carding engine which soon became the standard pattern had a main cylinder of large diameter. The cotton was pulled out between the points on its surface and much smaller cylinders set round the upper circumference. Later these cylinders were replaced by flat cards to give a better coverage. A second cylinder, the doffer, somewhat smaller than the main one, rotated at a different speed to take the carded cotton off the main one. Arkwright solved the problem of removing the cotton from the doffer with his crank and comb. Across the front of the doffer was placed a strip of metal called the comb, because the bottom edge was cut into fine teeth. It was connected to a crank which raised and lowered it so that the teeth pushed the cotton off the wire points of the doffer. The cotton came off in a broad gossamer web which was collected into a single roll, passed through a pair of rollers and allowed to drop into a can placed in front.

The drawing frame

The sliver coming from the carding engine had many irregularities and was too thick to be fed straight into the rollers on the waterframes. Therefore it had to be made thinner by being passed through drawing frames. Four or six slivers were passed through a set of rollers and were drawn out in the same ratio as the number fed in, so that one sliver of the same thickness appeared at the other end but four or six times as long. Because the thickness of the original slivers was averaged out, this sliver would be more even than those fed in and also the bent fibres were straightened out. After passing the slivers through two or even three drawing frames, Arkwright had a good quality sliver, but it was still too thick for his waterframes. He could reduce it by passing it through more rollers, but the thinner sliver had no cohesion and fell to pieces when handled.

Arkwright's patent of 1775 shows that he tried three different ways to overcome this problem, but, although he produced a reasonably satisfactory solution, a more adequate machine was not developed until after his death. His solution was to pass the sliver through a set of rollers and allow it to fall into a rotating can so the sliver was twisted as it entered and coiled against the side by centrifugal force. The sliver was twisted enough to keep it together but not enough to prevent it being drawn out. These rovings had to be wound on bobbins by children to fit the waterframe.

Through the success of the first mill at Cromford, Arkwright built others in the area and later helped to finance mills in other parts of the country. One, for example, was at New Lanark in Scotland, which he founded in 1784 with David Dale as his partner. Robert Owen became the manager at New Lanark in 1799 and carried out his social experiments there in the following years. Arkwright died in 1792, a very wealthy man, one of the few textile inventors to do so. His example was copied by many others and textile mills began to spring up all over the country, ushering in the first phase of the Industrial Revolution.

The spinning mule

To explain the last great spinning invention of the Industrial Revolution period, it is necessary to go back to 1779 when Samuel Crompton revealed his 'spinning mule' to the public in Bolton, Lancashire. Crompton, a weaver by trade, wanted to produce a better yarn than he could on the jenny (see p. 825) and started work on his mule in about 1772. He combined the principles of the jenny and the waterframe by placing his spindles on a carriage which moved away from the drawing rollers mounted on a frame at the back. To commence spinning, the rollers were put in gear to draw and pay out the cotton at the same time as the plain spindles were rotated to put in twist while the carriage was moving out. No yarn was wound on at this stage, so the cotton was suspended between the nip of the rollers and the tips of the spindles. The cotton was only lightly twisted to give it enough strength to stop it breaking. The rollers stopped paying out the cotton before the carriage reached the full extent of the draw, so the yarn was stretched on the last part of the draw, helping to even it out. Then twist was put in to lock the fibres together. Hence the mule combined the ease of drawing out the cotton in Arkwright's rollers with the gentleness and quality found through the stretching action and spinning on bare spindles. Very soon Crompton was spinning finer yarns than any hand spinner.

With the spinning sequence finished, the yarn had to be wound on. First the spindles had to be rotated backwards to clear the yarn from their tips and then, with the help of a faller wire, the yarn was guided on to the spindles in the shape of a cop, as the carriage was pushed in. When the carriage was fully in, the faller wire was raised, so that the yarn wound itself in a loose spiral up to the tip of the spindle and the mule was ready for the next draw.

Crompton built his first mule, which had 48 spindles, at Hall-i'-th'-Wood, Bolton, but soon aroused the suspicions of the local people because he was producing such fine quality yarn more cheaply than anyone else. He was pestered to such an extent by people trying to see his machine that he finally agreed to make it public in return for a subscription to be raised among the local manufacturers. In view of the fortunes made from his invention by others, the sum of £70 which was agreed was a ludicrous amount, but he never received even this, since only £60 was collected. His invention was immediately taken up by others and, from then on, he had nothing to do with further development of the mule. He did receive a grant of £5000 from Parliament in 1812, but this fell far short of his expectations and he died in 1827 in straitened circumstances.

Other inventions

Power sources

The three great spinning inventions, by themselves, would not have been able to alter the textile industry so dramatically if there had not been parallel improvements in other sections to enable these to increase production too. The enormous expansion of power-driven mills would have soon outstripped the number of waterpowered sites, and continued growth was possible only through an alternative source of power. Probably the first use of steam to drive a textile mill was in 1783 at Arkwright's mill on Shudehill in Manchester. Here the stream was too small to drive the mill, so an atmospheric steam engine pumped the water which had turned a waterwheel back to the upper mill pond so that it could be used over and over again; an early example of a pumped-storage scheme.

In 1785, James Watt installed one of his first rotative steam engines to drive directly George Robinson's spinning mill near Papplewick, a little to the north of Nottingham. This 7.5kw (10hp) engine was a much more efficient way of using steam power and was the forerunner of many thousands of mill engines. Some of the later ones generated 2250kw (3000hp) to drive spinning mills built in the early 1900s. Like all other areas of industrial production, the textile industry moved from waterpower through steam to electricity as they were successively introduced (see Chapters 4, 5, and 6).

Sources of cotton

The production of all-cotton cloth rose dramatically in the 1780s after the successful development of the Arkwright spinning system and the introduction of the spinning mule. Soon the value of cotton goods sent for export had exceeded that of the old-established woollen industry. The demand for raw cotton was met by increasing the acreage of plantations, first in the West Indies and then in the southern states of North America. Between 1790 and 1810, the output of raw cotton from the United States rose from 680,000kg to 38.5 million kg ($1\frac{1}{2}$ million to 85 million lb) per annum. When the Civil War came in 1861, the American slave plantations were satisfying five-sixths of an ever-increasing world demand; Britain took 450 million kg (1000 million lb) per annum, the

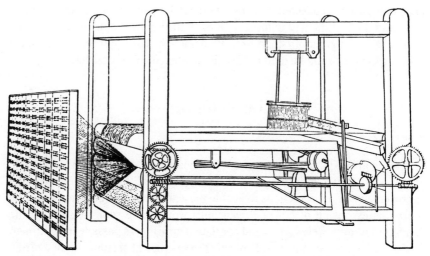

Figure 17.10: Edmund Cartwright's second power loom of 1786.

rest of Europe about two-thirds of that amount and American rather less than one-third.

Such a volume of cotton wool could not have left the cotton fields without Eli Whitney's cotton gin. In 1793 he invented a machine which used small circular saw blades to strip the cotton off the seeds more effectively than by hand. More important, this machine could gin a new range of short staple cotton which could be grown far inland in parts of America where other types would not survive. The cotton arrived in England much cleaner and so could be selected more carefully and classified more uniformly, which led to better spinning.

Power looms

In 1784 the Revd Dr Edmund Cartwright happened to fall into conversation at Matlock with some gentlemen from Manchester. One of them remarked that, as soon as Arkwright's patent had expired, so much cotton would be spun that there would not be enough people to weave it. Cartwright replied that Arkwright must set his wits to work to invent a weaving mill, but the Manchester gentlemen unanimously agreed that the thing was impracticable. Cartwright decided to make a powerloom himself and took out a patent in 1785. He described it as having 'the warp placed vertically, the reed fell with the weight of at least half a hundredweight, and the springs which threw the shuttle were strong enough to throw a Congreve rocket'. It needed two men to work it. He patented a more conventional loom in the following year (Figure 17.10) and set up a factory at Doncaster which failed. Slow progress was made by other people so that, by 1804, there were about 2400 powerlooms at work in Britain (see p. 844).

Finishing

For centuries, the traditional way of bleaching cotton cloth had been by treating it with buttermilk and leaving it out on the grass of bleachfields to expose it to sunlight. The quality varied and the process would have been too slow to cope with the increased production. In 1785 the French chemist C. L. Berthollet had found that a strong bleaching solution was made by passing chlorine through potash. He explained the process to James Watt who was visiting Paris. Watt told his friend Charles Tennant, a bleacher near Glasgow, who improved the process in 1797 by using milk of lime instead of potash. Then, in 1799, he made a dry bleaching powder from chlorine and solid slaked lime which was much more convenient to use. Without this bleaching powder, the cotton industry could not have achieved its enormous expansion (see p. 833).

There were also improvements in mordants and dyes for colouring and printing the fabrics. Printing was revolutionized for, in the place of a hand-block about one foot square with a different one for each colour, a calico-printing machine with rotary cylinders was invented in 1786 by a Scotsman named Bell, and was first used at Masney near Preston. Although the copper cylinders were difficult to make and engrave, for they were rolled from copper sheet and the soldered seams often burst, these machines were able to increase the output of the cheaper printed calico cloth. Rollers became the normal way of printing virtually all cloth until silk screens started to be introduced in the 1930s. Today, roller printing has all but disappeared and has been replaced by either flat or rotary screens. A very small amount of hand-block printing survives for very special fabrics.

THE NINETEENTH AND TWENTIETH CENTURIES

Preparatory machines

Little progress was made with opening machinery before 1800. Then in 1801, Bowden tried to make a machine to copy the hand method of using sticks to beat or batt the cotton on a table of tightly stretched cords through which the dirt fell. For coarser cottons, the devil or willow began to come into general use a little before 1810. It was probably invented by the Strutts of Belper and consisted of a large spiked drum which revolved in a casing also set with spikes. The lumps of cotton from the tightly-packed bales were carried round by the drum and torn apart by the spikes. This opened up the cotton and allowed some of the dirt to fall out. Machines based on the same principle have remained in use up to the present both for opening up the bales and for tearing apart weaving waste or rags for reprocessing.

Further cleaning and opening stages were still needed. Soon after 1800, Snodgrass first used a scutcher at Johnston, near Paisley in Scotland, which he derived from the corn threshing machine. Beaters separated the cotton from the remains of seeds and dirt. While the heavier refuse fell out and dropped below a wire drum, the cotton was carried forward by a current of air and at the end of the scutcher was rolled up into a broad lap. This lap might be put through a second scutcher to clean it further before it was weighed to check it for consistency and placed at the back of the carding engine. Scutchers and openers have been designed in various forms, such as the Porcupine or Creighton openers, but the basic principles remain the same today.

Carding

While carding engines have also remained basically the same, various changes have taken place to make them more efficient. The cotton may be passed through two carding engines, an opener and a finisher. For carding cotton, the small wire-covered rollers above the main cylinder were replaced by flats to give a better coverage. To save having to stop and clean these flats twice a day, in 1834, J. Smith connected them up to form an endless chain which moved slowly over the upper part of the main cylinder. At one end were fitted cleaning brushes. More recently, the wire type of card clothing has been replaced round the large cylinders by metallic clothing. This is like a long length of a band-saw wound closely round and round. It permits far higher carding speeds which are too fast for Arkwright's crank and comb (see p. 83), so the cotton is taken off by a small cylinder also covered with metallic clothing.

Carding engines with rollers have remained popular for wool and cotton waste. The rollers are set in pairs, one of which teases out the fibres and the other, the clearer, strips them off the first roller and returns them to the main cylinder. In the woollen industry, the carding, or rather scribbling machines as they are called, have become massive, with three or four main cylinders in succession. There may be a couple of sections with the wool being transferred between them by a 'Scotch feed' to ensure an even distribution of the fibres. Such machines may be up to 23m (75ft) long.

Condenser carding and spinning

Just as carding engines have been specially designed to suit particular types of fibres, so preparation methods differ for spinning them. In 1822, J. Goulding of the USA, invented a method of dividing the carded web longitudinally into individual strips by covering the doffer cylinder alternately with card fillet and plain rings. The strips had to be lightly twisted into slubbings to make the fibres cohere before being wound on to bobbins. This system does not seem to have been used much in Britain until after 1850, when the twist was inserted by passing the strips between leather belts or aprons which were moved from side to side as they rotated to take the fibres off the doffer. In 1861 the tape condenser was invented by C. Martin of Verviers in Belgium, in which the fibres were taken off the whole surface of the doffer and then split up into strips.

This condenser strip was spun on mules specially adapted for the purpose. The strips were wound side by side on long bobbins which were placed at the back of the mules. Rollers paid out the fibres but did not draw them out: the drawing, or drafting, was done by stretching. As the carriage moved out, the spindles rotated slowly to put in just sufficient twist to make the fibres stick together. Then the rollers stopped paying out the fibres, well before the carriage had reached the end of its run, so that the fibres were drawn out during the last part of the movement of the carriage. When the carriage had reached its fullest extent, the spindles were speeded up to put in the correct amount of twist. This became the usual way of spinning woollen yarns and a few mules are still kept running for very high quality products. A similar system was adapted for spinning the very short fibres of cotton waste, but today ring frames (see p. 841) have replaced the mules. The yarn from this industry was used for weaving flannelette sheets, dusters and dishcloths.

The self-acting mule

By 1800, two systems for cotton spinning had appeared. In one, the sliver from the carding engine was fed into stretcher mules, or at this period in the woollen industry into slubbing billies, in which it was converted into a thinner roving suitable for spinning on the proper mules. A great many people made all sorts of improvements to the mule so that by 1790, the carriages contained over 150 spindles. Great strength was required to work such a mule manually and it was a highly skilled job to wind the cotton on to the cop evenly and with the correct tension. Not only was there the problem of moving the carriage in at the right speed to compensate for the varying diameters of the cone-shaped cop itself, but the spindle tapered, so the amount of yarn wound on for each revolution decreased as the cop was built up. The addition of a counter faller helped the mule spinner to keep an even tension, but, even with this, building up a properly shaped cop which would unwind without snarls in the shuttle required much practice.

It was obviously desirable to work the mule by power so that one spinner could manage more spindles. In 1790, William Kelly of New Lanark was able to drive the spinning sequences by power but failed with winding on. Mules on which the carriage was drawn out, and the rollers and spindles turned, by power soon became standard, but the spinner still had to wind on by hand. He worked a pair of machines, so that while one was drawing out, he was winding on the other. Piecers, often small children, were employed to join the broken ends and

PART FIVE: TECHNOLOGY AND SOCIETY

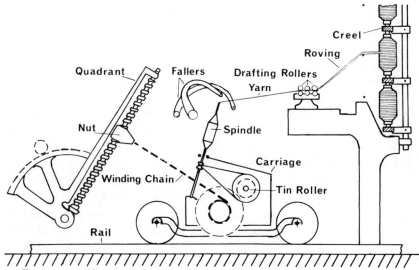

Figure 17.11: The principle parts of the self-acting spinning mule. Drawing by Richard Hills.

to keep the mules clean. By 1800 the size of these semi-powered mules had increased to around 400 spindles.

During a strike of mule spinners in 1825, some millowners approached Richard Roberts in Manchester to see if he could find a way of making the mule self-acting. He finally solved the problem of winding on in 1830 with his quadrant. He realized that there was always the same length of yarn to be wound on after each draw, so he made the movement of the carriage turn the spindles. The number of turns given to the spindles was varied by moving one end of the winding chain up or down the quadrant lever. The quadrant was further refined with self-adjusting mechanisms worked by the counter faller wire to compensate for the taper of the spindle. Later, nosing motions were fitted for spinning very fine counts. Roberts also introduced the camshaft, which changed the belt drives and brought the various clutches in and out of gear. See Figure 17.11 for a diagram of this machine.

The self-acting mule developed into one of the most intricate machines ever devised. Some mules spun the finest yarns ever produced and other variants spun the coarsest. There were mules for very short staple condenser waste cotton yarns and others for the long fibres of worsteds. Eventually, some medium counts cotton mules were about 40m (130ft) long with over 1000 spindles. These were the basis of the prosperity of the Lancashire industry, for they could spin a higher quality yarn than anything else. The last of these mules ceased operation in 1974 and few are left except in museums, because they are too slow for modern production and have to be attended by highly skilled operators.

The roving frame

Where waterframes were used for spinning, Arkwright's revolving can frames began to be replaced with roving frames soon after 1800. On these, the sliver was drawn out through a set of rollers and a flyer put in the twist before the roving was wound on to a bobbin. Because the friction of the bobbin caused the fragile roving to break, both the bobbin and the spindle were driven through gearing. The roving had to be wound on to the bobbin at the same speed all the time even though the diameter, and hence the circumference, was increasing as it was filled. As the rollers paid out at the same rate all the time, the speed of the bobbin, but not the flyer, had to decrease in the same ratio as the increasing diameter, or the rovings would have become thinner and thinner as the bobbin was filled.

Arkwright had drawn a picture in his 1775 patent of a belt being moved along a coned pulley to give a speed difference. A better solution, which appeared soon after 1800, was the use of a pair of coned pulleys, facing in opposite directions, which saved having to retension the belt. But even this system had problems through the belt slipping owing to the power needed to drive all the bobbins. This was solved by the introduction of the differential drive which meant that the belt had to provide only the difference between the flyer and bobbin speeds. A. Arnold patented this in America in 1822 and H. Houldsworth followed in England in 1826. It was found that with double-flanged bobbins, the rovings tended to stick against the flanges and break as they were unwound. Therefore, around 1840, the rovings were wound on to tubes in parallel layers, but each layer was shorter than the last, giving a double-coned effect tapering towards the ends.

As quality improved, roving frames gradually replaced the stretcher mules, and they are still used today to prepare the cotton. Flyer frames based on the same principles were developed for preparing the longer fibres of flax and wool for worsted. Such has been the improvement in these machines, through better engineering combined with improvements in carding, that whereas there would have been at least three in a Victorian mill, a 'slubber' supplying an 'intermediate' supplying a roving frame, today there is only one.

Combing

To prepare worsted yarns for spinning combing has always been necessary. Not only are the wool fibres untangled and straightened, but the short ones are deliberately removed. Those fibres that are left are more uniform in length, so they can be spun into a finer yarn with a more silky appearance. In 1792, Edmund Cartwright (see p. 834) patented a combing machine which was nicknamed 'Big Ben', because its action resembled the punching movements of a local prizefighter. It found a certain amount of favour, but, in spite of attempts by many other people, there was no really satisfactory combing machine in Britain until the 1850s.

In 1846, J. Heilmann, of Mulhouse in France, took out a British patent for a machine to comb cotton. Only a very basic idea can be given of this complex machine. The fibres were first made into a narrow lap (about 30cm (12in) wide) which was fed into rollers turned intermittently. They presented a short section of the lap to nippers and then stopped. The section beyond the nippers was combed to take out the short fibres. Those remaining were released by the nippers and further rollers joined them on to those that had been already combed. Then the first rollers turned once more to present another batch of fibres. Combing cotton enabled spinners to produce finer, smoother and more lustrous yarns. Later the Heilmann was replaced by the Nasmyth comber. The waste, or short fibres, from these machines was used in the condenser spinning industry (see p. 836).

Meanwhile, G. E. Donisthorpe of Leeds had obtained combing patents in 1842 and 1843. He joined forces with S. C. Lister and more patents were taken out up to 1852. However, Heilmann successfully claimed that these infringed his patent and Lister was able to exploit his own patents only by buying up Heilmann's rights. The Lister comb was a circular machine, suitable for wool, and worked on a different principle except that, at one stage, the fibres were held by nippers. James Noble patented another circular combing machine in 1853, again for wool. Three years later, Isaac Holden patented yet another circular machine which used a 'square motion' comb for removing the short fibres. Each machine had its own characteristics which rendered it suitable for particular types of wool. The short fibres, or noils, are an important raw material for the woollen industry. The combed sliver containing the long fibres is wound into a ball called a top ready for drawing and spinning.

Gill boxes and Casablanca spinning systems

For drafting long fibres, the rollers have to be spaced apart a distance corresponding to the longest fibre length. Therefore there will be a large gap between them where the fibres are unsupported, and if there is much variation in length, the drafting will be unequal. In 1834, P. Fairbairn patented an arrangement consisting of a revolving tube placed between the sets of rollers through which the fibres passed to be given a false twist. The better solution, however, lay in the development of a much earlier patent obtained by A. Thompson of London in 1801 for hackling flax. The flax passed through a series of vertical pins mounted on a continuous belt which was slowly rotated as the fibres were drawn through. The longer fibres were pulled through the pins and straightened by the drawing rollers at the front while the shorter ones were held steady by the pins and presented later to the rollers. Many variations of the gill boxes were developed for spinning worsted, linen, jute and other long fibres.

In 1912, Fernando Casablanca patented in Spain his high draft system for fibres of medium length. This consists of two aprons round the middle pair of rollers which reach almost to the front ones. The aprons lightly press the fibres together in the drafting zone and yet allow the quicker rotating front rollers to pull fibres out of the aprons quite easily. This enables the slivers or rovings to be reduced in thickness more quickly and evenly. The Saco-Lowell-Uster Versamatic Automatic Draft Control, introduced to the drawing frame in 1953, checked the evenness of the sliver and adjusted the speed of the rollers to compensate for any variations. This, coupled with the Casablanca systems, has enabled the sliver to be first made and then drawn out more evenly so that the intermediate roving frames are no longer necessary. In fact, the Japanese have manufactured a ring frame capable of spinning directly from the drawn sliver.

The throstle and ring frames

Development continued of the principle of spinning used on the waterframe. Soon after 1800, the waterframe, which had been made principally from wood with brass gears driving groups of four spindles and rollers, was replaced by the throstle. This had cast-iron framing and cast-iron gears which were so strong that only one set was needed at the end of the machine to drive all the rollers and spindles. The flyers, bobbins and rollers remained unaltered. While Arkwright himself was unsuccessful in spinning worsted, his waterframes were adapted for this in the 1790s at mills such as Dolphinholme near Lancaster or Davison & Hawksley's near Nottingham. Matthew Murray in Leeds was also using them to spin flax in the 1790s.

Around 1830, a new type of spinning machine began to make its appearance in America but did not really begin to penetrate the English mills until after 1870. On the ring frame, the drawing rollers remained the same but the flyer was replaced by a tiny traveller which slides round a ring. The bobbin was replaced by a tube held by friction on the spindle. The rotation of the spindle and tube twisted the fibres as they left the nip of the rollers and the winding on was caused by the yarn passing through the traveller at the side of the tube. The weight of the traveller and its friction on the ring caused it to drag behind the speed of the spindle and so the yarn was wound on. The rail holding the rings moved up and down to guide the cotton on to the tube in a cop build.

At first, only coarse counts could be spun on ring frames. Gradually, however, through better preparation of the roving, and higher precision and better design in the manufacture of the frames themselves, the ring frame became able to spin both finer and coarser and softer yarns and so slowly ousted the mule.

With the adaptation of various systems of fibre control between the rollers, such as gill boxes or the Casablanca systems (see above), the ring frame has been adapted to spin a greater range of fibres and also to do in one operation the work of two or three earlier roving and spinning machines. In the 1960s it largely replaced the mules in all sections of the textile industry thanks to its greater simplicity and productivity.

Open end spinning

The ring frame itself began to be replaced during the 1970s by open end or break spinning. The problem with the older ways of spinning has been that the package containing the spun yarn has had to be rotated to put in the twist. The weight increasing as it filled, prevented high speeds and limited its size. During the 1950s the Shirley Institute in Manchester experimented with some ways of break spinning, but it was not until 1960 that the first commercial open end spinning machine was marketed by the Czechoslovakians. A sliver of fibres is fed to an opening device, usually a spiked roller or drum, and drawn apart so that the fibres move forward more or less individually. They are carried by an air stream and collected on the inner surface of a cup-shaped rotor with a hole in the end which revolves at 30,000–60,000rpm. During each revolution a thin layer of fibres is deposited in a collecting groove on the inner surface of the cup. Layers of fibres are laid one upon another until a strand is built up. This is peeled from the groove and twisted by the rotor as it is drawn out of the hole in the end. From here the strand, or yarn is wound on to a bobbin. Open end spinning has a very high rate of production and has been developed to give a fine quality yarn with some of the characteristics of that spun on the mule.

In 1964 the Repco self-twist spinning machine was patented. A pair of rovings is drawn out and passed through aprons which are moved from side to side to give a false twist. The twist lies alternately in opposite directions and would soon untwist itself but, while one roving moves straight forward, the other goes round guides so that, when it meets its pair, they are twisted in opposite directions. As the fibres try to untwist themselves, they combine into a single loosely twisted yarn which is suitable for knitting into woollen pullovers and similar garments. Once again, the rate of production is very high.

Weaving

Sizing

Sizing was essential with cotton warps to give them sufficient strength and tenacity. Without it, the threads would not have stood up to the friction in the loom. Two threads usually passed between each slot or 'dent' in the reed and, every time the heddles changed the shed, these threads rubbed past each other. Then the motion of the reed backwards and forwards and the shuttle passing across caused constant friction which would have soon worn through the warp threads unless they were properly sized.

For handlooms, sizing seems to have been carried out in two stages. The yarn was measured off into the correct length, removed from the warping frame or drum and coiled into a ball which was boiled for some hours in size before being dried. Once in the loom, more size had to be applied which was put on short stretches of the warp and allowed to dry before it could be woven. Sizing may have taken up to one-third of the weaver's time and labour and compelled him to stop weaving for long periods.

For the powerloom to become fully economic, the warps had to be prepared before they were 'entered' into the loom so that weaving could be continuous. Ironically, the solution was found by William Radcliffe of Stockport who was seeking to improve handlooms. He took out a patent in 1804 (in the name of one of his workmen, Thomas Johnson, so that it could not easily be traced) for a machine on which the warps could be dressed with size, dried and wound on a beam ready to be put into the back of the loom. He used a current of air with fans to dry the yarn, but later these were replaced by steam-heated drying cylinders. This proved to be the key to continuous weaving on the powerloom and the same process is still used today.

The powerloom

While the actual mechanics of a powerloom are quite simple, the refinements necessary to make it work properly are difficult to achieve, as Edmund Cartwright found to his cost. For example, it is necessary to prevent the shuttle rebounding out of its box. The handloom weaver does this by tensioning the picker peg and releasing it when he feels the shuttle hit it. On the powerloom, the same result has to be attained by leather buffer straps and side springs in the shuttle box. Should the shuttle become stuck in the warp, the loom will beat up the reed and force the shuttle into the warp threads, breaking them, unless it is stopped instantly. Sometimes a loose reed is fitted which can hinge back if it hits an obstruction. In either case, a brake to stop the loom is essential. Should the weft break or run out, then there must be a weft-stop motion to put on the brake, and ideally there should be a warp-stop motion for when a warp thread breaks. Also there was the problem of winding the cloth on precisely the same amount each pick so that the beating up was even. This was solved by the sand roller, which had a rough surface to grip the cloth.

Many people contributed to the development of the powerloom. For example, Austin of Glasgow used cams in 1789 to work the picking stick to throw the shuttle and some of his looms were built in Scotland in 1800. Robert

Millar of Dumbartonshire in 1796 patented an improved protector which stopped the loom altogether when the shuttle failed to enter its box. Between 1803 and 1813, Messrs Marsland & Horrocks of Stockport, Cheshire, patented various improvements including the crank for operating the reed. During the 1820s, Richard Roberts continued to make improvements and constructed his powerloom principally out of cast iron. Therefore the numbers of powerlooms increased steadily from around 2400 in 1804 to 14,150 in 1820, 55,500 in 1829 and 100,000 in 1833.

On most of these early looms, only plain cloth could be woven, such as cotton calico or sheeting. Richard Roberts was also partly responsible for adapting the powerloom to weave simple patterns. Raising the different sets of heddles was contrived by using wheels with projections which caused the appropriate heddle to be lifted. By 1838, eight changes of pattern could be worked in this way, but a really satisfactory system was not evolved for another twenty years, when the Blackburn dobby appeared. On a dobby, pegs, inserted into small bars made up into a continuous chain, lift levers to raise the appropriate heddle. Where no peg has been inserted, that heddle remains down. Dobbies can control quite complex patterns on up to 18 heddles or shafts. In about 1846, Squire Diggle of Bury mechanized the drop box so that different types or colours of wefts could be used and later this too was controlled by the dobby. In the 1830s power-driven Jacquard looms began to be introduced from the silk industry to weaving worsted and the Jacquard has remained in use ever since for weaving complex patterns, today sometimes controlled by a computer.

Automatic looms

On the traditional Lancashire loom, the weaver had to stop weaving to fit a full pirn in the shuttle each time it ran out of weft. Between 1889 and 1894 an Englishman, J. H. Northrop, took out various patents for a loom in which the weft package in the shuttle was changed automatically when one was empty. The first Northrop loom was made by the Draper Corporation in the USA in 1894. Each time the shuttle went into one of the boxes, a feeler was poked in through a slot in its side. As the weft was unwound, the feeler penetrated further until it triggered a mechanism which pushed another full pirn into the shuttle and ejected the empty one through a hole in the bottom. The full pirns were stored in a rotary holder which was replenished manually. Self-threading shuttles had to be designed, which helped to lower the incidence of weaver's pneumosilicosis caused by breathing in dust as they sucked the weft through the hole in the end of the old shuttles. Warp-stop motions were also fitted to Northrop looms so that all the weaver had to do was to repair breakages of warp or weft and refill the pirn-holder.

Although Northrop looms were introduced to Britain around 1900, they never became very popular because they could weave with only one type of weft and the British textile industry specialized in a variety of fancy fabrics. While in 1909 the USA had 200,000 Northrop looms compared to 8000 in Britain, 95 per cent of all American looms were automatic by 1930, compared with only 5 per cent in Britain. The Northrop became the most popular of the automatic looms and many are still in use. To reduce further the number of weavers needed to supervise them, in 1957 the Leesona Corporation's Unifil winder was introduced to the European market. Here the pirns are wound on the loom itself, so that, instead of 2–300 being necessary for each loom on the old system, the Unifil needed only twelve, and only half a dozen of these were full of yarn. The empty ones were stripped of any remaining weft and returned to the top for rewinding.

The days of looms with shuttles are numbered because, for many years people have been experimenting with ways of inserting the weft without wasting the energy that a shuttle requires to accelerate it through the warp, only to be lost as the shuttle strikes its box. Sulzer Brothers of Switzerland took out their first patent in 1929 for a loom in which a small gripper or bullet 8.75cm (3.5in) long is shot across the loom from one side. It pulls a shoot of weft off a cone, passes through the shed, releases the weft and is returned underneath to the starting position. A number of bullets enable high weaving speeds to be maintained, often at the rate of over 300 picks per minute compared with the 120 of a Lancashire loom. The ends of the weft are tucked into the warp to form a neat selvedge. A single-coloured-weft machine was available commercially in 1950, two colours in 1955 and four in 1959. By 1969, over 20,000 of these looms had been sold, even though the precision engineering necessary in their construction made them very expensive.

In the meantime, a successful rapier loom had been developed by the Draper Corporation in 1945. A sword stick, or rapier, takes the weft from one side of the loom to the middle, where another grips it and pulls it through the rest of the way. Various manufacturers offer these looms today. Next people concentrated on abolishing bullets or rapiers altogether by using air or water jets instead to shoot the weft across. Max Paabo produced his air-jet Maxbo loom in 1954, which runs more quietly and quickly than ordinary looms. In the following year the Czechoslovak engineer Vladimir Svaty introduced a water-jet loom at the Brussels exhibition. The water-jet loom is suitable only for certain types of synthetic fibres which are not affected by moisture.

VELVET, TOWELLING AND CARPET WEAVING

Fabrics woven with a pile or loops have had a long history. In the Middle East, it is still possible to see carpets being woven in the traditional way on vertical looms. Lengths of coloured wool are knotted by women or children to the warp threads to form the pile which is trimmed with scissors to make it even. In the traditional way of weaving velvet, there were two sets of warp threads. One made the backing or ground and was woven with weft in the usual way. But after a couple or so shoots of the weft, a rod was placed over the ground warp and under the pile warp, which was wound on a separate beam with a little friction to prevent it unwinding. The pile warp was lowered over the rod and was bound in by the next shoots of weft. After a few rods had been placed in position, the first one was taken out by cutting the loops of the pile with a sharp knife run along a fine slot along the top of the rod. Some velvet is still made in this laborious way for special royal occasions such as a coronation. Today, most velvet is woven as a double cloth. The weft threads binding the two warps together are cut to form the pile which of course separates the double cloth into two pieces of velvet.

Shortly before 1850, one of the Christy Millers from the hatting family in Stockport, saw terry towelling in Turkey. After some experiments he devised a way of weaving it on powerlooms. As with velvet, two warps are used. The reed can be set in two positions to beat up the weft. With the reed in the back position, two shoots of weft are inserted. After the third shoot, the reed is put in the forward position so that it pushes the three shoots of weft along the warp. The weft slides over the ground warp, which is held tight, but the freer pile warp moves with the weft and is pushed up into loops. After the next three picks, the loops are pushed down so that the pile is formed on both sides. Towelling, or velvet if the loops are cut, can be woven with the all loops on one side if the number of picks is even. It is said that Queen Victoria saw some of this towelling at an exhibition and decided to use it in the nursery, hence the name Royal Turkish Towels.

Weaving carpets by power started in the 1850s on enormous and complex machines. In one system, each pick, or row of tufts, had its own set of bobbins and each tuft its own bobbin. Each bobbin was wound with weft of the appropriate colour to form the design and a length was pulled off and knotted before the next set of bobbins was presented. E. G. Bigelow of Massachusetts invented a loom which was improved for weaving both Brussels and Wilton carpets in the 1850s and 1860s. Then cheaper carpets began to be made by having a design printed on to the weft threads wrapped round a drum. Eventually in 1876, H. Skinner of New York developed a machine for making Axminster carpets. Today, the design is sometimes printed on to cheaper carpets after they have been woven as plain ones.

KNITTING MACHINES

William Lee, on his original knitting machine (see p. 819), used only one yarn, which is termed weft knitting. A quicker method was warp knitting which seems to have originated in Nottinghamshire about 1775. A beam, similar to that used in the loom with warp threads wound round it, was placed at the back of the knitting machine and the threads passed through eyelets or guides. Needles looped the threads into themselves and across to the others. The fabric that was formed was not so elastic as that made on the stocking frame but was more akin to a woven fabric and so more suitable for underwear. In 1791, William Dawson, a Leicester framework knitter, applied a notched wheel or cam for selecting patterns which led to warp knitting machines being driven by power.

A circular knitting machine patented by Marc Isambard Brunel in 1816 used the bearded needle and presser. For some years it was little used because it produced only a tubular fabric. In 1847, M. Townsend of Nottingham added a mechanism which fashioned ribbed knitting on circular machines. They became popular for knitting stockings which, although not fully fashioned, had sufficient stretch to fit the leg. A. Paget of Loughborough made a circular knitting machine in 1855 which could be driven by power. Such machines became more popular when many knitting heads were placed around the circumference, which increased production because more yarns could be used at the same time.

The most important nineteenth-century development in knitting was the patenting in 1856 of the latch needle by Townsend. The hook of the needle was closed by a hinged strip of metal, the latch. Normally the needle worked in a vertical or inclined position, so that the latch hung down with the previous loop on the shank below it. The principle of knitting was similar to the Lee machine except that the equivalent of the sinkers was fixed. Weft yarn was placed in the hook of the needle. The needle was drawn down between the fixed plates which replaced the sinkers, and so the new loop was formed. At the same time, the shank of the needle slid through the old loop which closed the latch and so passed over the hook, incorporating the new loop on its way. As the needle moved up again, the latch either fell open or was pushed down by the loop in the hook which moved on to the shank and so the needle was ready to receive weft again. The hinge of the latch could not be made as narrow as the bearded needle, so really fine work continued to be knitted on Lee's machines. The latch needle so simplified knitting that soon both circular and flat-bed domestic machines appeared on which socks, gloves, pullovers and the like could be knitted at home.

Another Leicestershire inventor, William Cotton, patented in 1864 a way of making fully-fashioned work that proved to be adaptable to knitting a dozen or more stockings simultaneously. This was, of course, a power-driven machine. In the USA in 1887, a machine was invented for sewing up the seams of knitted garments to prevent them unravelling. Development of knitting has continued ever since so that today computer-controlled knitting machines work at the rate of six million stitches a minute.

LACE MACHINES

John Heathcote was born at Duffield near Derby in 1783 and was apprenticed to a hosiery manufacturer and framesmith. He found that a patent he had obtained in 1804 for making lace had been anticipated, so he turned his attention to the principles of making pillow or bobbin lace which he mastered in 1809. He analysed the complex array of bobbins that the lacemaker used and came to the conclusion that fundamentally there were two sets of threads. He worked out their respective movements and made one set into a warp. The other he wound on bobbins. His patent machine of 1809 made a type of net curtain, Brussels lace, without patterns. To obtain the correct movement of both warp threads and bobbins required complicated mechanisms which Heathcote was finally able to construct. He was highly praised for the excellence of his machine, but his mill at Loughborough, Leicestershire, with £10,000 worth of machines, was destroyed by the Luddites in 1816, so he moved to Tiverton in Devon where production continued until recently.

Another framesmith, John Levers, brought out a variation of Heathcote's machine in 1813, which was later adapted for making patterned lace because it used only a single row of bobbins instead of two. It was for this type of machine that, in about 1833, Hooton Deverill adapted the Jacquard mechanism to select the patterns. The Jacquard needles pushed the warp threads sideways to form the holes in the lace while the bobbins were moved round them to bind them together. Soon patterns as complex as any made by hand could be produced on these machines and lace became cheap enough for most people to be able to hang it in their windows as curtains or to use it for trimming clothing.

SEWING MACHINES

The invention of sewing machines had an enormous impact on the textile industry, stimulating demand for cloth because making up garments became so much quicker. They were one of the first mass-produced consumer durables. In the same way as with knitting machines, it was not possible to reproduce the stitches of the hand sewer. Two threads formed a lock-stitch, using a needle with the eye at its point. The sewing machine is essentially an American invention, for Elias Howe of Massachusetts patented a machine in 1846, although it could only sew lock-stitch in straight lines. It was improved when Isaac M. Singer of Boston, Massachusetts, produced the first practical domestic sewing machine in 1851. This initiated a big sewing-machine industry, covering new types of stitches, new types of needles, presser bars, shuttles and all sorts of other mechanisms. The adaptation of sewing machines to other industries followed quickly; for example, to the shoe industry in 1861, and they have played a prominent role in the clothing and making-up industries.

MAN-MADE FINISHES AND FIBRES

Charles Macintosh, a Glasgow chemical manufacturer, found that naphtha was a solvent for rubber, so he could make a varnish that could be brushed on to cloth. The naphtha evaporated, leaving the rubber as a layer impermeable to air and water. If two pieces of this cloth were pressed together with the rubber inside, a satisfactory waterproof cloth was obtained which had no tacky surface. The first factory for producing macintosh cloth was opened in Manchester in 1824. The process was improved and speeded up by the introduction from the United States of calendering, passing the cloth between heated rollers, in 1849.

The rubber industry did not really begin to grow until after the introduction of vulcanizing by Charles Goodyear, another American, in 1841. Up to that time, although very many rubber articles such as shoes were made, they were liable to become tacky with heat or solid when cold. Vulcanized rubber was stable, so that its use increased dramatically in Britain by something like 400 per cent between 1840 and 1858, when the vulcanizing patent expired. Heavy waterproof garments were replaced with lighter cloth and there was a growing demand for rubber shoes and other forms of footwear. Rubber was used increasingly in engineering industries for springs, valves, conveyor belting and so on.

It was soon discovered that textile fibres could be treated with chemicals to enhance their appearance or give special characteristics. In about 1840, John Mercer of Lancashire had discovered that caustic soda would make cotton fibres more elastic and easily dyed and also, when kept in tension, give them a more lustrous effect. Mercerized cotton was not exploited commercially until about 1890 when it became one of the earliest man-made finishes. Mercer also found that cotton could be dissolved in a solution of copper oxide in ammonia, a process later exploited in the manufacture of artificial silk. Meanwhile in 1857 at Krefeld on the Rhine, it was discovered that the preliminary treatment of silk with certain metallic salts before dyeing, weighting, made the product both more lustrous and much heavier which improved the feel and the appearance. These were the forerunners of the crease and shrink resistant and the non-iron fabrics of today.

Since the 1930s the range of available textiles has been widened by the proliferation of man-made fibres. One of the first people in this field was Joseph Swan, while working on the development of the incandescent electric light in 1879 (see p. 366). At first he carbonized mercerized cotton threads for the filaments of his lamps, then made them by extruding nitrocellulose dissolved in acetic acid. There was something about these silky threads which caught his attention and suggested their possibilities as a textile. He made some of special fineness, the first artificial silk, which his wife crocheted into doilies.

In 1892, C. F. Cross and E. J. Bevan produced fibres by treating cellulose first with caustic soda and then with carbon disulphide to make xanthate. This was dissolved in a solution of dilute caustic soda to produce a viscous liquid known as viscose. After being aged, it was extruded through fine holes in a spinneret and coagulated in a dilute solution to produce continuous filaments of regenerated cellulose which became known as artificial silk. In 1900 the Topham box was invented, a device for collecting the extruded filaments on the inside of a cylindrical container rotating at high speed. In 1904, Courtaulds purchased the British rights for viscose rayon so that by 1913, Britain had emerged as the largest producer, ahead of the USA. Production at this period concentrated on continuous filament yarns to rival silk. Shortly before the First World War, other ways of making rayon had been found, for example with cuprammonium viscose.

The introduction of one type of artificial fibre encouraged chemists to look for more. Nylon 66, a polyamide made from adipic acid and hexamethylene diamine, was discovered by W. H. Carothers in 1937. The molten polymer was extruded through spinnerets and the cooled filaments stretched by some 400 per cent to orient the fibres. J. T. Dickinson and J. R. Whinfield made yet another synthetic fibre in 1941 from the polyester derived from terephthalic acid and ethylene gycol. Owing to restrictions imposed in Britain during the Second World War, this fibre was developed initially by the Du Pont Company in the United States, where it was marketed under the name Dacron. When ICI was able to manufacture it in Britain, it acquired the brand name Terylene. New types based on polyacrylonitrile (acrylics) appeared after the war. For example, Du Pont produced Orlon in 1945 and the Chemstrand Corporation Acrilon in 1952. These later fibres were designed to imitate wool or cotton. Although manufactured in long lengths, they had to be given the characteristics of their natural counterparts by crimping and heat-setting. They were then cut to appropriate 'staple' lengths so that they could be processed on the normal carding and spinning machines. While the early man-made fibres were based on cellulose, some of those developed after the Second World War were derived from chemical by-products of the oil industry. Man-made fibres have not only made today's clothing cheaper; modern fabrics keep their shape better, are hard-wearing and easier to wash than those of earlier generations.

CLOTHING MANUFACTURE

Apart from the individual handwork still carried out in the bespoke tailoring and couture businesses for the few who can afford it, most clothing today is made by machinery and mass-production methods. The latter are largely dependent on the statistics of measurements to evolve a system of standard sizes which originated first from measurements taken by the Medical Department of the US Army during the First World War. Similar figures for women's sizes did not become generally available until the 1950s.

Before the various parts are cut out to the required pattern, a roll of the cloth to be used is mounted on a Spreading Machine which traverses the bed of the machine to and fro, automatically counting the number of folds laid. The first portable cutting machine, that of Joseph Bloch, dates from 1888 and had a circular knife which was not ideal for cutting round curves. In the same year, in the USA, G. P. Eastman produced a reciprocating cutting machine which used a straight blade.

SEWING

Hand-sewing was carried out from about 40,000 years ago in Palaeolithic times with bone or ivory needles. The manufacture of iron sewing needles originated in Germany in the mid-fourteenth century, at first with a closed hook instead of an eye which was introduced in the Low Countries in the same century. When the process was fully developed, the needles were made in pairs, the cut wires being put inside two heated rings and rolled backwards and forwards to straighten them. Then followed the dangerous process of pointing them using a revolving stone. The eyes were then punched in pairs in a simple drop-hammer. The excess metal was then trimmed off by grinding, the pairs split before finishing by hardening and tempering, scouring and polishing, the last process taking about six days.

The sewing machine is probably the most universal and important machine in clothing manufacture, being used in both industrial and domestic applications. The first working machine was built by Barthelemy Thimmonier, a working tailor of St Etienne in France in 1830. It was a chain-stitch machine, made largely of wood. He set up a factory with 80 machines to manufacture army uniforms, which was destroyed by rioting rival tailors. Although his machine was improved and patented in England and the USA, Thimmonier died impoverished in 1857.

Several inventors were trying to develop a lock-stitch sewing machine in the USA, including Elias Howe who came to England where he sold the patent rights to a London maker of corsets and umbrellas, William Thomas, for £250.

In 1851 Isaac Merritt Singer patented a lock-stitch machine worked by a treadle, and with a toothed wheel to advance the fabric, which proved to be at least ten times as fast as hand stitching. It was his partner, a lawyer, Edward Clark, who introduced door-to-door salesmen and hire-purchase in 1856 ($50 cash down or $100, with a cash deposit of $5 and $3 a month thereafter). The industrial Singer sewing machine was followed by the domestic model in 1858.

All the early models were hand- or foot-operated. Electric motors were added to sewing machines from 1889 onwards though they found little application to domestic machines until the 1930s and factory machines tended for some years to be driven from overhead shafting.

FASTENINGS

Fastenings are an essential component for the majority of garments and one in which great advances have been made in recent years. Toggles, of wood, bone or antler, were first used and engaged loops of leather thong. The toggle was later refined to form the button, originally made from shell, bone or wood.

The button, in fact, was originally used more as an ornament than as a fastening, the earliest known being found at Mohenjo-daro in the Indus valley. It is made of a curved shell and about 5000 years old. There was a gradual increase in the use of buttons throughout the Middle Ages, particularly for decorative purposes – in fact until the eighteenth century when production became mechanized in the factory with a consequent reduction in price, bringing the product within the popular rather than the luxury range. Mother-of-pearl, glass and precious and base metals were added to the range of materials used. Cloth buttons, formed on a disc of brass with a shank added, became common in the early twentieth century.

The hook-and-eye fastening dates from Roman times when it was used for fastening soldiers' leather breast armour. It came into use again on the continent of Europe in the sixteenth century, the first British hooks and eyes coming from the firm of Newey's in 1791.

The press stud stems from the invention by Louis Hannart in 1863 of an 'Improved clasp or fastener for gloves and other wearing apparel, for umbrellas, travelling bags . . .'. From the late nineteenth century manufacture took place in Germany reaching a peak in the early years of the twentieth century. In 1885 H. Bauer patented a spring and stud fastener while M. D. Shipman patented a similar design in the USA the following year.

Production of existing fasteners, particularly the press stud, was greatly threatened by the invention of the zip fastener, originally patented by Whitcomb L. Judson of Chicago, USA, in 1896 and at first only used on boots and shoes. It was not a success, tending to jam or to spring open. It was also expensive, being largely made by hand. Eventually the Automatic Hook and Eye Company of Hoboken, New Jersey, took on Dr Gideon Sundback who improved the design until, in 1918, a contractor making flying suits for the US Navy placed an order for 10,000 fasteners and in 1923 B. F. Goodrich & Co put zips in the galoshes that they manufactured. Success was assured from then on.

A recent competitor to the zip fastener is Velcro, the 'touch-and-close' fastener. Invented by George de Mestral, first patented in 1954, but not made in the UK until the 1960s, UK patents were granted to Selectus Ltd of Biddulph, Stoke-on-Trent. Two separate strips of synthetic material, usually nylon, one incorporating a pile of loops and the other of hooks, are set in tapes. Both tapes are finished by heat treatment in the form of an electric current passing through the hooks and loops, which sets and hardens them. There are numerous variations of the arrangement but all, to a greater or lesser extent, resemble the working of the burdock or teazle found in nature.

WATERPROOF CLOTHING AND ELASTIC

The Amazonian Indians are said to have used rubber to waterproof clothing as early as the thirteenth century. When this material, the rubber sap exported from Brazil, became available in Britain, Charles Macintosh, the Scottish chemist, set up a factory in Manchester where he dissolved the rubber with naphtha. (see p. 849).

Linseed oil was also used, painted on the fabric, to make oilskin. In the early 1840s Charles Goodyear of New York first vulcanized rubber treating it with sulphur which converted it to an elastic material, less subject to temperature changes. Plastic waterproofs date from the 1940s. PVC treatment of the fabric avoids temperature sensitivity.

Macintosh's partner, Thomas Hancock, was involved in the application of rubber to the production of elastic fabrics as were Messrs Rattier & Guibel of St Denis, Paris, producing elastic webbing cloth in 1830. This quickly took the place of spiral brass wire for shoulder straps and for ladies' corsets. A patent for elastic-sided boots was granted to James Downie in 1837.

The umbrella originated in China some 3000 years ago where it was considered a symbol of rank; in the form of feathers or leaves covering a frame of cane, it was used mainly as a protection against the sun. This was its main purpose until the early years of the eighteenth century when the French covered them with oiled cloth to make them waterproof. Earlier attempts had been made in Italy where leather was used as early as 1608.

Jonas Hanway is credited with being the first Englishman to use an umbrella. This was in London in 1756, although ladies had used sunshades earlier in the century and umbrellas were kept at coffee houses to escort patrons to their carriages or sedan chairs. For a man to carry an umbrella was considered effeminate and was extremely unpopular with the hackney coachmen. Their use grew, however, and by 1790 it was said that umbrella-making was a great trade in London. In 1818 the Duke of Wellington banned his troops from carrying them into battle. The curved steel frame was invented by Samuel Fox of Deepcar, Sheffield, in 1874. Hans Haupt of Berlin invented the telescopic 'brolly' in 1930.

CLEANING

Until the mid-nineteenth century washing with soap and water was the only way of cleaning clothes and, with woollen fabrics in particular, it was more common to dye them a darker colour to conceal the dirt. In about 1850 Jean-Baptiste Jolly-Bellin, a Paris tailor, spilt some camphene, a kind of turpentine, on an article of his wife's and found that the area stained by the spirit was cleaner than the rest. He opened up a business for 'Nettoyage à Sec', the first dry-cleaning business. The garments had to be unstitched before brushing with camphene and then sewn together again.

In 1866 the firm of Pullars of Perth set up a business with a postal service throughout Britain, for dry cleaning using benzene or petroleum, both highly inflammable but not requiring the garments to be dismembered. Carbon tetrachloride was used in Leipzig by Ludwig Anthelm in 1897, but was found to corrode the equipment and was dangerous to breathe. It was replaced by trichlorethylene in Britain in 1918 and in the USA in 1930. Perchlorethylene, non-inflammable and giving off no noxious fumes, is now used.

FOOTWEAR

Footwear was the product of individual boot- and shoemakers until Marc Isambard Brunel set up his army boot factory in Battersea after seeing the victors of Corunna return home unshod, their feet bandaged and festering in rags. The factory employed 24 ex-servicemen, most of whom were disabled, to operate 16 separate processes including punching, tacking, welting, cutting, trimming and nail-driving, all by machinery. The factory could produce 400 pairs of boots a day. Others had tried to mechanize the process before with little success, but from then on the machine-made industry was to grow. However, with the ending of hostilities, Brunel made little money from the venture. A pair of 'water boots' was sold for half a guinea, superior shoes for 12 shillings and Wellington boots for 20 shillings a pair. Brunel boots shod Wellington's guards at the Battle of Waterloo.

18

BUILDING AND ARCHITECTURE

DOREEN YARWOOD

PRIMITIVE BUILDING

For an undeveloped rural community, the overriding factor in building was the availability of materials. Transportation was always difficult; roads were few, vehicles simple and the easiest means was by water. Consequently the materials used in building were, apart from very important structures, those obtained locally. Most often these comprised earths, wood and brushwood, reeds, turf and thatch and flint. Stone, the heaviest of building materials to transport and needing the greatest degree of skill to prepare, was reserved for structures of supreme importance (see pp. 859–61).

Unbaked earths

Earth simply dug out of the ground has been widely utilized from earliest times until the later nineteenth century, when its use still survived for simple, rural dwellings. Various ways of making earth into a suitable building material have been developed over centuries according to the type of earth available and the climate of the area concerned.

Construction using earth is generally of two main types. The more common way is to make the substance plastic by the addition of water, then to press it into a mould. The other is to ram dry earth to give cohesion.

Clay earth is the most suitable material for the plastic form. It was mixed with water to give a stiff mud consistency then, as was discovered empirically, the addition of various substances would improve its strength and durability. Lime was added to help the setting quality and the addition of straw and gravel or sand gave better cohesion; other types of aggregate might be included: pebbles or fragments of slate, stone or brick. Buildings constructed of such plastic earth blocks require to be based upon a plinth of more permanent materials – stone, brick, flint or rubble – up to a depth of about 45–60cm (one-and-a-half to two feet) in order to keep the house dry and prevent invasion by vermin. Earth walls also need to be plaster-faced and to be colour-washed annually to make them waterproof and to improve insulation.

In countries with a hot climate a sun-dried brick was made from such plastic clay mixes. It is known now generally as adobe, which is the Spanish term for this material, a word derived from *adobar*, to pickle or to stew. Adobe was the earliest building material in Ancient Egypt and Mesopotamia and has been used until recent times in Spain, south-east Europe, the south-west region of the USA and in Central America. The clay, of a mud-like consistency, was mixed with straw or grass and sand or loam, then moulded into brick form and left to dry in the hot sun for a week or so.

The rammed earth method is known in Britain as pisé (from the French verb *piser*, to pound). It is a building material traditional to many parts of the world – Europe, Africa, Asia and America. In Europe it was employed by the Romans but chiefly in areas with a dry climate, which is necessary for its successful use. The clay earth was contained between shutterboards, then beaten or rammed against a hard surface by a wooden rammer until it cohered into a block. Stabilized earth is a modern equivalent but this is mixed with cement and so is stronger and easier to handle.

Wood

One of the earliest means of walling was by making panels of wattle and daub. Vertical wooden stakes or branches were interwoven with reeds or brushwood. These wattles were then plastered on one or both sides with mud (the daub) and often reinforced with turf or moss and left to dry in the air.

Where there was ample afforestation, timber was used for building in the form of whole or split logs. This method has been traditional over the centuries in the coniferous areas of Scandinavia, southern Germany and the Carpathian Mountains. In hardwood forest regions such as England, the timber-framed method of construction was developed during the Middle Ages (see pp. 862–4).

Flint

In districts lacking easily available stone, flint has been utilized from early times as an extremely durable substitute. It is a pure form of silica found in chalk deposits where it consists of very hard, irregularly-shaped nodules. These may be used whole, embedded in mortar, in conjunction with stone or rubble. Flints are easy to split and, during the Middle Ages, a craft developed whereby such split flints were used decoratively in patterns with mortar and stone. The shiny blue-black faces of the split flints formed an ornamental contrast against the cream-coloured stone. The craft of flint splitting is known as knapping. The decorative form of using such knapped flints is called flushwork.

Roofing materials

Walls constructed of earth blocks, timber or wattle and daub were suitable only to support lightweight roofing. Again, the materials used for this were those most readily available. These included thatch made from reed, straw or heather, or turves or moss. For more important structures a tiling made from wood was usual. Such wood tiles are known as shingles. In England shingles were of oak, each tile measuring about five inches by ten inches (12.5 × 25cm) and laid with an overlap; they were fashioned to be thicker at the lower edge.

TRABEATED CONSTRUCTION

The word 'trabeated' derives from the Latin *trabs*, a beam and denotes a form of construction in which a horizontal member is supported by vertical ones. A more descriptive and so popular term nowadays is post and lintel.

This method of building has been in use continuously since earliest times in many parts of the world, but its effective employment depends upon the availability of strong, permanent materials to provide posts which will safely support the lintels. Cultures of the ancient world which built in this way included notably the Egyptian, the Minoan and the classical under Greece and Rome. In Mesopotamia, where neither stone nor timber was in good supply and the major permanent building material was brick, the constructional form was based upon the arch, since brick is not suited to a horizontal spanning of an opening (see pp. 872–4). In modern times the material in use is steel (see pp. 896–9).

The ancient Egyptians were important pioneers in building in trabeated form, using stone, marble and granite. There were two chief types: the columned hall or courtyard, where rows of columns supported continuous lintels, and the pylon entrance, where massive, sloping masonry blocks were constructed to flank an entrance and these supported equally vast lintels above.

The earliest peoples in Europe to develop a high art and architectural form were the Cretans between about 2000 and 1400 BC. The means and knowledge for them to do this derived from Egypt and were transmitted from the Bronze Age centres in the interior of Asia Minor to the coastal area round Troy and from there to the island of Crete. The Cretans built in trabeated manner using columnar supports which characteristically and unusually tapered downwards towards the base, like those which were excavated in the Palace of Minos at Knossos. The Cretans also constructed small rectilinear buildings which are also of post and lintel type, the vertical walls supporting the horizontal ceilings;

typical of modest building through the ages, both lintels and posts have become continuous surfaces to make slabs.

It was the Greeks who transformed the trabeated type of architecture into an art form of superb beauty and technical perfection. Like the Egyptians they preferred this structural method to that based upon the arch which they regarded as being less stable and liable to collapse. The Greeks developed a system of orders, a word referring to a grouped number of constituent parts. Each order possessed specific relative proportions between its parts and certain distinguishing features in ornamentation and mouldings peculiar to itself. The size of a building did not affect these proportions, which remained constant, so the differing scale did not impair the aesthetic or functional aspect.

Each order comprised a column (often with a base) and at its head, a capital. This capital was a decorative feature but also a functional one in that its upper surface area was greater than that of the column, so that it better supported the lintel above. The lintel, known as an entablature, consisted of three horizontal sections. The upper one, the cornice, was made up of mouldings which projected forwards, so directing the rain away from the sides of the building. The central section was a decoratively sculptured frieze of which the famous and disputed sections of the Parthenon (the Elgin Marbles) are on display in the British Museum, London. The lower section was called the architrave.

Until the Hellenistic period (300 BC onwards), the Greeks used the post and lintel system of construction for almost all their building needs. They built colonnaded temples, civic structures, theatres, town fortifications and, above all, stoas. The stoa was a roofed promenade colonnaded at the front and sides and walled at the rear, where doorways led into shops and business premises. Long single- or double-storeyed stoas were characteristic architectural features of ancient Greek cities. Situated in the market place (the agora), they provided a meeting area for social and business purposes sheltered from the sun and rain. These sophisticated colonnades had evolved from the early discovery that a lintel stone supported on two columns could form an opening; the colonnade was an extrapolation of this concept.

The beauty of Greek trabeated architecture derives not from diversity of form, for this is limited and very simple, but from its subtle and detailed attention to line and proportion. Greek builders developed a series of refinements of line, mass and curve which were meticulously and mathematically calculated to give vitality and plasticity to a building as well as to make it appear correctly delineated. The true vertical and horizontal line of a white marble feature, especially when silhouetted against a brilliantly blue sky, appears concave to the human eye. To offset this illusion in their major buildings, the Greeks made all planes, whether horizontal, vertical or sloping (as in pediments), very slightly convex. The convexity was so subtle that, unlike in some eighteenth- and nineteenth-century European versions of such classical architecture, the line appeared to be straight. In the Parthenon, the finest of all Greek buildings, entablatures, pediments, capitals, columns and base platforms are all so treated. The curves are not arcs of circles but parabolas. To give an indication of the subtlety of the correctional refinement, the stylobate – the stepped base upon which the Parthenon stands – is 69.5m (228ft) in length on the long sides of its rectangular form. In the centre of such a side the convex curve rises to only 11cm ($4\frac{1}{4}$ in).

The Romans followed upon the order system established by the Greeks, expanding and developing the theme and incorporating arcuated into trabeated construction. After the Renaissance revival of classical architecture in fifteenth-century Italy such trabeated forms were to be seen in buildings all over the western world through the centuries until after the First World War but in a more limited scope than in ancient Greece, for the arcuated form of classical structure, developed by the Romans, was found to be more suited to the northern climate. Post and lintel construction was also used for simple bridge building in timber, slate or stone as in the clapper bridges of Dartmoor where flat slabs are laid across vertical piers.

In modern building the use of steel and reinforced concrete lends itself to post and lintel construction. This can be seen in the steel-framed skyscraper where it is employed in grid pattern (see p. 897). A similar grid form of trabeated building was also utilized in timber-framed structures of the fifteenth to seventeenth centuries (see p. 862). With modern materials, much greater flexibility of design is possible when using a columnar structure supporting beams or a roof than was possible in the days of ancient Greece. Then, using stone or marble for building, it was necessary to erect columns at restricted and specified intervals to act as safe supports to the entablature lintels. With the use of alloy steels, reinforced concrete or plastics reinforced with glass fibre, slender vertical supports may be distanced much further apart, so giving better natural illumination and visibility to the interior. This is in marked contrast to the afforested columns of the Temple of Karnak at Luxor, for example.

MASONRY

Masonry is the craft of building with stone and similar materials – marble, granite, slate – of cutting and carving, dressing and laying.

The prepared blocks or rough-hewn pieces have, in most periods, been bound with mortar but the skilled craft of walling by dry-stone means has been practised since ancient times. In England, dry-stone walling to divide up areas of land for animal pasture has been traditional over the centuries in areas where stone is plentiful; particularly, for example, in the Pennine and Lake District regions of England where sheep need to be securely contained. Among the more primitive cultures, the Mycenaeans in the Peloponnese region of Greece were noted for their dry-stone walling in which they incorporated huge blocks of stone. These were so heavy that they remained in position, the interstices being filled with smaller stones. The citadels of Tiryns and Mycenae (fourteenth century BC) still show extensive evidence of this type of work. The builders of classical Greece (fifth century BC) coined the term cyclopean for this masonry. They viewed with awe construction using such immense blocks and, believing it to be superhuman, attributed it to the mythical Cyclopes.

Until about the sixteenth century AD in Europe, most stone walling was of rubble construction, that is, using smallish pieces of irregularly-shaped stone, the interstices filled with mud and/or mortar. There were various types of rubble construction and, as time passed, these became more regular in pattern. The haphazard arrangement of stones is known as random rubble. If the pieces were laid in rows, or courses, it is coursed rubble, if they have been roughly squared, it is squared coursed rubble.

The craft of cutting stones into precisely squared pieces and finishing it smoothly is known as ashlar. Even in later times this was always costly, so it was only in use for important buildings – cathedrals, abbeys, civic structures etc. More commonly ashlar blocks were reserved for dressings round windows and doorways, for gables, lintels, arches, pediments, buttresses and quoins. In good quality ashlar the stones have horizontal and vertical faces at true right angles and are laid in horizontal courses with tightly fitting vertical joints. A smooth finish was also necessary for carved stonework, for ornament and classical columns. A high quality freestone is most suitable for ashlar work as, for example, the oolitic limestones and some sandstones. In England, Portland stone, as used by Christopher Wren at St Paul's Cathedral in London, was most effective.

Other important building stones include granite, marble and slate. Most forms of granite are igneous rock, formed when molten rock cooled. They are extremely hard and durable. They are crystallized and will take a high polish but are so hard to cut that decorative carving was rarely attempted before the introduction of power tools. In Britain, the city of Aberdeen is an example of this. The local material was widely utilized when much of the city was laid out in granite in the Greek Revival style of the early years of the nineteenth century but, owing to the intractability of the material, the classical mouldings are bare of the usual egg and dart and acanthus decoration of the style. For the same reason the moulding edges are as sharp now as when they were cut 150 years ago.

Much of the building of the ancient Greeks was of marble, particularly in Greece itself where Pentelic and Hymettian white marble from the mountains behind Athens and Parian from the island of Paros were available in quantity. From Eleusis came the grey Eleusian marble used by builders for decorative contrast. In areas of Greek dominance such as southern Italy, where limestone and conglomerate had to be used, the Greeks coated it with a stucco made from powdered marble; the stucco has now worn off and the limestone is exposed, giving a brownish hue quite different from the brilliant white of the Athenian marbles.

The Greeks rarely used mortar but fitted the blocks of stone and marble with great accuracy, using metal dowels and cramps to hold them together. The drums of the columns were so finely fitted that the joints were barely visible. When erecting fluted columns (those carved with narrow vertical channels in them), the Greeks carved the flutes in the top and bottom drums of the column then lined these up to complete the flutes when the column had been erected.

Slate is a hard, fine-textured, non-porous rock, formed originally by great heat and/or pressure, from clay, shale or volcanic ash. It may be sawn for use as masonry in the same way as marble or stone Because of its fissile character, which enables it to be split along its natural laminae, it has more commonly been utilized as a roofing material or for paving and shelving. It can also be employed for fittings such as chimneypieces.

Slate, granite and marble have all been widely used over the centuries as a cladding material, that is, for facing a building which is made from a less attractive substance, In modern times, with the availability of improved equipment for quarrying and sawing, slate has been increasingly used in this way for expensive structures and the material gives an attractive and durable finish. In Italy, where stone is only available in limited quantity yet there is a wealth of marbles of infinite variety in colour and markings, the Italians have for centuries built in brick and clad in marble. This has been the tradition whether the building was a mediaeval cathedral or modern power station.

ADVANCED TIMBER CONSTRUCTION

Apart from the earths, wood is the oldest building material. It has been used as much as, possibly more, than any other in large areas of Europe where afforestation was extensive. Several types of wooden structure were developed over the centuries. These varied from region to region, partly according to the climate and partly to the type of timber available, whether hardwood, as in England where it was chiefly oak, or softwood pines and firs, as in Scandinavia and many mountain areas of Europe such as the Alps or the Carpathian regions of Hungary and Romania.

Cruck building

A common early form of construction was by the use of crucks or, as they came to be called, blades. The origins of this type of building are obscure. Some authorities believe it stemmed from the boat-building traditions of Scandinavia, others that it has affinities with the pointed arch of Gothic design (see p. 874).

More probably the form evolved from practical needs and limitation of knowledge as did the similarly-shaped early corbelled stone structures of areas as far apart as Mycenae in Greece or the Celtic oratories of western Ireland.

In early cruck construction the basic timbers, the crucks, were massive whole or split tree trunks, the lower edges of which were fixed into the ground in pairs at intervals in the long sides of a rectangular building – the interval was generally some 3.5–5m (12–16ft) – and the narrower tops were then bent over to be fastened together in the centre so forming an apex to the roof. A long horizontal timber, the ridge piece, was extended along this whole roof, the crucks being affixed to it. Parallel to the ridge piece were other slenderer poles and these were crossed at right angles by more timbers. The interstices were filled with branches and mud and covered on the exterior by moss, thatch or turf.

The great drawback to the single-storeyed cruck building was its lack of headroom, as it had no vertical walls, so the design was further developed into two-storeyed buildings as the crucks were taken up higher. The next stage was to extend horizontal tie beams across the interior and beyond the crucks to attach to horizontal timbers, at right angles to the tie beams, which would carry the roof and act as the summit of the now vertical walls. Gradually cruck structures became more advanced and practical. The lower ends of the crucks were set into low stone walls or timber sills instead of directly into the ground. The crucks were shaped before use, a pair forming a symmetrical arch. More complex roof designs followed (see pp. 865-6).

Timber-framing

In regions of northern Europe, especially where hardwoods were available in quantity, as in England, a more sophisticated method was evolved in which split timbers were attached together to form basic framework for a building, which would have vertical walls and a gabled roof (see Figure 18.1 for an example of this type of structure). The panel spaces between such framing were then filled with other materials – wattle and daub, laths and plaster or brickwork. This type of construction is also known as half-timbering, because the tree trunks were halved or cleft, not complete logs. The trees were stripped of their bark and soaked for a year or more in water to season the wood before use.

Many designs of timber-framed structures were developed over the centuries. In England there were two fundamental types: the box-frame and the post and truss. The first of these was built with horizontal and vertical timbers crossing at right angles and fastened together to make box shapes. In the latter designs, vertical posts supported a truss, that is, a group of strong timbers formed into a framework within the triangle of the roof gable.

The simplest type of truss is a triangle formed from three beams and, in building, grew out of the use of the gabled roof. The Greeks of Hellenistic

Figure 18.1: Timber-framed inn. The Feathers Hotel, Ludlow, Shropshire, late sixteenth century.
Photograph Doreen Yarwood.

times invented the truss, finding the triangle to be a rigid form. The earliest surviving description of a truss is in one of the volumes of *De Architectura* by Vitruvius, the Roman architect and engineer of the first century BC. In timber-frame construction there is a horizontal timber (a tie beam) and two sloping timbers fastened together at the apex. The weight of any roof creates a thrust which attempts to push the upper part of the walls outwards which is counteracted by the horizontal beam tying the sloping timbers together. In later centuries the theory of this single rigid triangular truss was elaborated, as in the classical trussed bridge design in which combinations of triangles were used. Later still, in the nineteenth century, with the availability of cheap, strong iron and steel, large, complex trussed roofs were constructed to cover railway sheds and stations (see p. 893). In modern times, vast and complicated trusses have been created to enclose great auditoriums (see p. 879).

Whatever the design of timber-framing there was a similarity of basic parts and methods of construction. In order to exclude the damp, the framework was usually erected upon a low wall of more impermeable building material, such as stone or brick. Upon this footing were laid horizontally baulks of timber called sills or plates. Strong vertical posts (the studs) were then mortised into this sill and their upper ends tenoned into another horizontal plate.

In most types of timber-framed buildings the upper storeys projected over the ones below. This overhang is called a jetty. When a building was jettied on all sides, additional support was provided for the structure by internal cross members known as dragon beams which extended diagonally from the centre to the corners of the upper floors. Jettying had several purposes. Of importance on limited town sites, it gave a greater area of floor space to the rooms of the upper storeys. The lower storeys were protected by those above from rain and snow at a time when there were no down pipes or gutters. Jettying also gave a more balanced stability to the upper floors themselves.

Oak timber-framed buildings did not require preservative sealing. The black and white effect which many surviving structures show is due to a nineteenth-century custom of painting them with a coal derivative tar with the plaster panels whitened for contrast. Many owners are now stripping their timber-work back to its original silvery hue.

Timber-framed buildings were largely prefabricated and they may, without too much difficulty, be taken down and re-erected as has been done in a number of instances, for example, on open-air museum sites. Originally the larger timbers were cut and shaped on site but smaller ones were made in a carpenter's workshop and brought to the building. The infilling was put in floor by floor, so completing the structure as the building was erected. A form of decoration known as pargeting, using abstract patterns or representational figures, was sometimes impressed into the wet plaster between the timbering.

Stave churches

A different type of timber construction may be seen in the stave (mast) churches of Norway. In these the roof was not supported on the walls; each part of the structure was self-supporting. The building was based upon a skeleton framework of vertical poles (the masts) which were slender tree trunks. The lower ends of these were sunk into horizontal sills and the upper ends supported the roof. There were cross beams near the top of the poles. The walls of such churches are stave screens, self-contained wooden panels – rather like modern woven fence units – which rested upon the timber sills and were attached to the poles but did not take any weight or thrust. In advanced designs of churches of this type, there is an inner ring of 'masts', so giving a nave and aisles to the interior.

Stave church design is based upon timber shipbuilding in which each section is a self-sufficient unit. The interiors are tall and dark, bearing a resemblance to inverted timber ships. They were constructed from about the eleventh century until the Reformation, and during the nineteenth century about 500 churches were extant. Their number is now greatly depleted and most survivors have been extensively restored. A superb example is St Andrew's Church of Borgund near Laerdal; it was built *c.*, 1150 and is about 15m (50ft) in height, erected in six storeys. Another example near by is the church of the tiny community at Urnes.

Log construction

From the sixteenth century onwards timber building in Scandinavia followed the more usual pattern of that in eastern Europe which was to build with whole logs or logs split only down the centre, constructing with the curved side of the log on the exterior of the building. Many examples of structures of all types, erected over hundreds of years until the early twentieth century, may be seen in the open-air museums where they have been re-erected to save them from total loss. Of particular interest is the Norsk Folke museum at Bygdøy Park near Oslo, the Aarhus Museum (Den Gamle By) in Denmark and the Village Museum in Bucharest. The English equivalent, displaying timber-framed building, is the Weald and Downland Museum at Singleton near Chichester.

The timber-trussed roof

This is an open interior roof system composed of a rigid construction of timbers (the truss) which were pinned and tenoned together to provide a stable combination which could resist all thrusts. Such roofs, constructed in Europe chiefly between the eleventh and sixteenth centuries, represent the timber equivalent of the stone vault (see pp. 874–6).

The earliest designs were based on cruck construction, but gradually a more sophisticated system was evolved empirically which would provide better headroom and not obstruct the interior visibility. The medieval roof was generally gabled at each end and, at least in northern Europe, was fairly steeply pitched. A long baulk of timber extended along each side of the building on top of the walls: this was the wall plate. Another such baulk was the ridge pole or ridge piece, extending along the roof apex. Between these two were other beams running parallel to them, fixed at intervals down the slope of the roof pitch: these were purlins. Crossing these beams at right angles were the rafters which extended from wall plate to ridge piece. At regular intervals were heavier rafters, called principal rafters; in between were the thinner common rafters (see Figure 18.2).

The simplest and earliest form of timber-trussed roof was the tie beam design. These tie beams were heavy baulks of timber spanning the interior at intervals and pinned or tenoned into the wall plates in order to offset the outward thrust of the roof upon the walls. Upon the beams were generally set one or two vertical posts extending from them up to the ridge piece. If there is one central post, it is called a king post, if two, they are queen posts. Horizontal beams set higher up the roof are called collar beams. Connecting timbers were incorporated to make the structure rigid. If straight they are called struts, if curved, braces. A coupled roof is one constructed without tie or collar beams but only struts or braces. It gives better internal visibility but is less strong. Vertical posts extending from collar beam to the ridge piece are called crown posts.

In the late fourteenth century came the final and more complex design: the hammerbeam roof. Hammer beams are like truncated tie beams. Extending at wall plate level, they are supported from stone corbels embedded in the walls by arch-braced wall posts. These shorter hammer beams gave better visibility than the continuous tie beams. Vertical posts (hammer posts) stand upon or are tenoned into the ends of the hammer beams and are attached to the collar beams above where they join the purlins. Hammerbeam roofs are richly decorated by carving, gilding and painting: many, especially the later examples, for instance at Hampton Court Palace, are particularly ornate. The finest hammerbeam roof is Hugh Herland's earlier structure of 1394 covering the interior of Westminster Hall in the Palace of Westminster in London.

BRICK AND TILE

Brickwork

Brick has been used as a building material for thousands of years. It has many useful qualities: it is cheap and easy to make, the raw material is available almost anywhere, it can be manufactured in different sizes, colours, qualities and tex-

Figure 18.2: Timber-trussed roof with braced collar beams. Manor house, Solar, c. 1475–85.
Drawing by Doreen Yarwood.

tures and it is durable, withstands the weather and is a good insulating material. It is also convenient to use in conjunction with other materials, partly as bonding courses for added strength (as in Roman walling) and partly as a colour contrast, a custom practised by many architects, for example, Sir Christopher Wren, who used Portland stone as dressings for his brick buildings.

Early bricks, in areas of the world where the climate was sunny, were sun-dried. Harder bricks were obtained by being exposed to heat from fire. Such burnt bricks were being made in the Near East by 3000 BC. Later references show such use in many parts of the world. The fifth century BC Greek historian Herodotus tells us that such bricks were made to build the city walls of Babylon. The Chinese used them in the third century BC to construct parts of the Great Wall.

The Romans (as did the Etruscans before them) made extensive use of brick in all forms of construction. Such bricks were sun-dried until the first century BC, after which kiln-burnt bricks became more general. It was the proud boast of the first Roman emperor, Augustus, that he found Rome a city of bricks and left it a city of marble, but this only referred to the facing of the buildings, which were still largely of brick construction underneath.

Roman builders used brick for arched openings, as bonding courses in walling, for relieving arches in vaults (for example the Pantheon in Rome) and for the colossal spans of the great basilicas and baths where they utilized brick compartments in vaulting in conjunction with concrete. This wide use of brickwork was extended further in the eastern area of Roman influence which later became the Byzantine Empire. Roman building bricks were hard, large and thin, resembling large tiles, and were laid with thick courses of mortar between. Floor bricks were were also rectangular but smaller. Hypocaust pier bricks were square.

After the collapse of the western part of the Roman Empire in the fifth century AD, brickmaking died out in much of western Europe, especially in Britain where it was not revived until the thirteenth century. Throughout the Middle Ages, however, brick was a vitally important material along the coastal regions of the North and Baltic Seas and bricks were being made from the early Middle Ages along this coastal belt which extended for 1200 miles from Bruges in Belgium to Novgorod in the USSR, the area made wealthy under the administration of the Hanseatic League. Because it lacked other permanent building materials brick was developed as a constructional and stylistic material for all types of building from cathedrals and castles to guild halls and manor houses. Brick is not an easy material to make into decorative forms, but a simplified version of both Gothic and the later classical architecture of Europe was created over the centuries by skilled craftsmen of the region and resulted in a remarkable unity of style over so many countries in such a large area.

Bonding

In early brickwork, bricks were laid irregularly and, in walling, were often mixed with flint and/or stone. As the work became more skilled bricks were laid in a form of bonding, a uniform arrangement which ensures strength and is also decorative. Many different types of bonding have been developed over the centuries. These vary from area to area but it has also been found that some bonds are stronger than others and these are often more costly, so a bond is chosen for a specific commission according to the quality and appearance required whether it be an important building or a garden wall.

In England the most usual bond to be employed in medieval brickwork was that termed English bond, which consists of alternate courses of headers and stretchers. A header is a brick laid so that only its end appears on the wall face; a stretcher is laid lengthwise to the face. This type of bond was in general use in Tudor structures but, by the 1630s and 1640s, when brick building in England was becoming much more common, partly due to the depletion of timber stocks, Flemish bond gradually replaced it in popularity. This is laid with alternate stretchers and headers on each course; an early example is Kew Palace, built 1631.

The great age of English brick building was the century beginning in 1660 when high quality bricks were manufactured from a variety of clays. Limited technical advances were made and the cost of brickmaking decreased, leading to a growth of brick building. Unfortunately in the later eighteenth century, this situation attracted the attention of government: a brick tax was imposed in 1784 and soon was twice increased, resulting in a serious curtailment of brick building of smaller structures in rural areas.

The brick tax was repealed in 1850. Contemporaneously the building needs following the tremendous population explosion and those of urbanization stemming from the Industrial Revolution led to an immense acceleration in brickmaking. Partly this boom was to serve the housing building industry but it was also for the construction of railway viaducts, tunnels, cuttings and stations, as well as factories and warehouses. Technical advances and improvements in brickmaking rendered the material the cheapest and the most ready to hand. Bricks were made for all purposes in different qualities by different processes. A wide range of colours was available and Victorian architects readily adopted the fashion for polychromy in their designs.

Brickmaking

Three chief stages of production are involved in the making of bricks from clay by heating or burning them. This has been so since the earliest bricks were made though, with the passage of time, the methods for carrying out the processes have become more complex and mechanized. The raw material has to be dug out of the ground, it then needs to be prepared and moulded into shape and, afterwards, burnt.

After the clay has been dug, it has to be puddled. This is a preparation process which includes removing pebbles and extraneous matter from the clay, then mixing and squeezing it with the addition of water and sand to give an even consistency. Often more than one clay is used in the mix so that the properties of one will counteract or improve the properties of others. Before the seventeenth century puddling was done by men treading the substance barefoot. After this the pug-mill gradually replaced the human labour. The mill was cylindrical; it contained a central shaft with blades attached which acted in the same manner as a butter churn. Early mills were powered by a horse or donkey on a treadmill (see Chapter 4). Later the pug-mill contained rotating wheels and, according to date, was powered by water, steam or electricity.

The early way to mould the bricks into the shape required was by human hands. During the Middle Ages moulds were used and later these were made with detachable base so comprising a stock mould and stock board.

Clamps were constructed in early times for brick burning. The clamp consisted of stacked bricks intermingled with brushwood which was the fuel. The clamp was lit from the exterior and allowed to burn itself out. This method caused bricks to be unevenly fired, the ones on the interior of the clamp being burnt more than those on the exterior and so, having lost more water, they were smaller and darker in colour. It was this which gave to fifteenth- and sixteenth-century brickwork an interesting variety of shape and tone. On a site with a suitable clay sub-soil, the whole brickmaking process might at this time be carried out where the building was to be constructed. Gradually the kiln system superseded the clamps and, after about 1700, coal and coke replaced wood as fuel.

At this time advances in brickmaking slowly took place but it was the mid-nineteenth century before these resulted in mass-production systems. Then many hand processes became mechanized, also delay in drying bricks before firing was eliminated by mechanical drying. The introduction of the Hoffman Kiln of 1858 was a great advance as its multi-chamber system enabled the firing process to operate continuously with brick stacks at different stages of firing or cooling instead of waiting for a fortnight for each setting of bricks to dry as before. The development of the Fletton process in the 1890s was a further great advance. Fletton is an area near Peterborough in Cambridgeshire, where a certain shale-clay is found in quantity. This material has a low water content which makes it suitable to be pressed into a brick mould without previous drying. Finally, brickmaking machinery replaced remaining hand processes for making bricks about 1885, vastly increasing the speed at which bricks could be turned out.

Tiles

Also fired in a kiln, tiles are less porous, harder and smoother than bricks. They are traditionally used for roofing and also for wall and floor covering, where they might be decorated and glazed. In England, in the later seventeenth century, the craft of tile-hanging was developed. Also known as weather-tiling because of its use as protection from rain and snow, it is ornamental, too, and varied designs were adopted. The tiles were hung on wood battens plugged into the walls or were pegged or nailed into mortar courses. In general, tile-hanging was practised more in the south of the country and slate-hanging in the north.

Another use for tiles was found in the second half of the eighteenth century; these were called brick tiles, alternatively, mathematical or weather tiles. These were generally made to the same linear dimensions as bricks and their purpose was to be applied to walls as a tiled skin to cover the original material of the wall behind which might be timber and plaster. An older, simpler building might thus acquire the appearance of a more fashionable brick one. These tiles were employed by many of the fashionable architects of the day and especially in the south-east of England; the Royal Crescent on Brighton sea-front (*c.* 1800) is a notable example where black glazed tiles have been used.

Terracotta and Coade stone

Terracotta is also a fired earthenware material which is harder and less porous than brick. It is used chiefly for decoration on buildings where it is moulded into ornamental or sculptural forms. Its use was introduced into England in the early sixteenth century by Italian craftsmen as can be seen, for example, in the roundels on the gateways of Hampton Court Palace by Giovanni da Maiano. Architects of the nineteenth-century Gothic Revival utilized terracotta extensively as a decorative medium. A well-known instance of this is Alfred Waterhouse's Natural History Museum in London (1873–9), where the architect faced the façade in two shades of the material – buff and grey – and decorated it with a wealth of animal and plant sculptures.

The mix for terracotta includes grog, that is, previously fired earthenware ground to a powder and this was also the basis for Coade stone, an extremely hard and durable material which was employed as an artificial stone by most of the leading architects of the day, including Robert Adam, Sir William Chambers, John Nash and Sir John Soane.

The manufactory of Coade stone was established by Mr and Mrs Coade in 1769 in Lambeth and it was run by their daughter Eleanor. The formula for the material was kept very secret but modern analysis has shown it to be a combination of china clay, grog from pulverized stoneware and various fluxes. It was a particulary stable, smooth substance eminently suited to making larger-than-life-size statuary and ornament and it has, in many instances, endured city atmospheric pollution better than actual stone, This may be seen, among other famous buildings, at the Royal Naval College in Greenwich, Buckingham Palace and St George's Chapel, Windsor. Coade stone, under its later owners, continued to be used until, in late Victorian times, the fashion for polychromy ousted it from favour.

THE ARCH

An arch is a structure composed of wedge-shaped blocks, which are called voussoirs, and designed to span a void. The blocks support each other by mutual pressure and therefore capable of withstanding considerable weight and thrust. A wooden framework, centering, was set up underneath the arch to support it until all voussoirs were in position.

Parts of an arch

Often the central voussoir at the top of an arch is larger and made into an architectural feature: this is then called a keystone. The point or line where the arch springs up from its supporting wall or column is known as the springing line. The lowest voussoir on each side of the arch is thus known as a springer. The support – wall, pier or column – is the abutment. The span of an arch is the distance between abutments and the height in the centre is the rise. The outer curve of the line of voussoirs is called the extrados, the inner is the intrados; the under surface is the soffit. These parts and the different forms of arch are shown in Figure 18.3.

History of building with arches

Knowledge of arcuated construction, that is, building structurally with arches rather than in the post and lintel manner (see p. 857), was gained by most of the early cultures but in some of these, notably the Egyptian and the Greek, plentiful supplies of stone and marble were available so trabeated structures were possible. Moreover, the builders held the opinion strongly that the arch was a less stable form of structure. Indeed, the Greeks had a proverb which said 'the arch never sleeps', referring to the truism that, if the arch support were to be weakened, the ever-present outward thrust of the curve of the arch would cause collapse. So, in both Egypt and Greece, arcuated structures were confined to small and utilitarian buildings.

In Mesopotamia, however, in the area between the rivers Tigris and Euphrates, adequate supplies of neither timber nor stone were available for building and sunbaked bricks made from the local clay constituted the chief building material. Bricks are too small to span a lintel opening, so an arcuated structural system was developed for both arched openings and vaulted roofing. The true arch, that is, one built with radiating self-supporting voussoirs, was known here

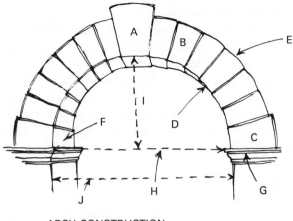

Figure 18.3: Arch construction.
Drawing by Doreen Yarwood.

as early as 2400 BC. Sun-dried bricks are not as strong as kiln-baked ones so the Mesopotamian structural spans of this time were not large and, due to the softness of the material, remains are far less in quantity than are the stone structures of contemporaneous Egypt.

In Italy the Etruscans introduced the concept of arcuated principles at an early date, building both arched openings and vaults. A number of their fine arched city gateways survive; most of these date from a later time, the fourth to third century BC as, for example, that at Volterra. These structures were in stone, as were a number of the corbelled vaults but, as in Mesopotamia, Etruscan arcuated work was on a fairly small scale and it was the Romans, particularly in the days of the Empire, who fully exploited the arcuated possibilities for arcades and vaults. The facing and structure was of brick, forming compartments, which were then filled with their exceptionally hard concrete.

The arches so far discussed, as well as those built in the revived classical tradition of the Renaissance and the Baroque eras, were round arches. These might be semicircular, wherein the height of the arch above the springing line is equal to the radius on the line or they might be segmental or stilted; in the former case the arch is wider and lower, in the latter it is taller and narrower. A further variant is the horseshoe arch of Islamic origin in which the curve of the arch is carried on below the semicircle.

Romanesque architecture of the early Middle Ages (Norman, in England) also employed the round arch. Indeed, much of the Romanesque building, especially in Spain, Italy and southern France where remains of Imperial Rome had survived in quantity, was strongly derivative of Roman classical work.

For most of the medieval period the pointed arch was in use and this is characteristic of what is termed Gothic architecture. There were many variations on the pointed form from the early, narrow lancet shape, through the equilateral to the late, wide obtuse and to the four-centred types in the ogee and Tudor forms.

The pointed arch is fundamental to Gothic architecture in northern Europe. It was developed solely to provide a flexible form suited to the building of the stone vaults which were made necessary by the hazard of fire which destroyed many Romanesque timber roofs, causing them to have to be rebuilt time and time again.

It was the French who introduced the pointed arch into Gothic architecture – aptly, they call it the *arc brisé*, the broken arch – in the Abbey Church of S. Denis, now in a suburb of Paris, in 1135. The pointed arch was not new. It had long been employed in the Middle East and, in Europe, was used in areas under Arab influence such as Moorish Spain, Sicily and Provence. The new style originated in northern France in the area of the Île de France, because here there was a more stable economy at the time.

The reason for the introduction of the pointed arch was that the round arch presents great problems in vaulting a church. This is because nave, choir, transepts and aisles often have different heights and widths. Vaulting is made in bays and the semicircular arch lends itself to a square bay. The bay is decided by the positioning of the supporting piers or columns. In Romanesque building the bay was traversed at roof level by diagonals crossing one another from pier to pier. As the diagonals were longer than the four ribs connecting the four sides or faces of the bay, it was impossible for all of these ribs to be semicircular in form. Either the vault had to be domical, so giving an undulating ridge rib, or the side arches stilted. The pointed arch, which could be varied infinitely in proportion of width to height, was far more flexible for constructing stone or brick coverings to nave and aisles of different spans and roof levels.

THE VAULT

In building structures the term 'vault' is used to describe a covered form of roofing based upon the principle of the arch. Traditionally, over the centuries, such vaults were constructed, as were the arches, from wedge-shaped blocks of stone or brick supported upon a wall, pier or column. During the nineteenth century plaster vaults were often erected reinforced by an iron structure. In modern architecture reinforced concrete, steel and glass fibre are among the materials used for this purpose. As in the building of an arch, a wooden supporting framework – centering – was erected underneath the vault to be constructed and removed when the vault structure had been completed.

It was the Romans who first fully exploited the structure of the vault, although knowledge of such building principles had been known to most of the earlier cultures. Using their strong mix of concrete together with brick as a structural and facing material, Roman builders covered very great spans with such vaulting, particularly in their great basilicas and thermal establishments. The vaults were massive but of the simplest forms. Three types were used, all based on the round arch shape. Commonest was the half-cylindrical, known variously as the barrel, wagon or tunnel vault. This was a continuous semicircular or obtuse arch, extrapolated as in a railway tunnel. Where two such barrel vaults of equal diameter met at right angles as in a cruciform building, this created a square plan groined vault, the groin being the angles of intersection. The Romans also built domes which were generally hemispherical as in the Pantheon.

In the basilicas and buildings constructed for the bathing establishments, the vaults were very thick. They were supported either on the also massive walling or on heavy columns or piers. The square piers in the Basilica of Maxentius in Rome (AD 308), for instance, measure 4.25m (14 ft) across. The majority of important Roman vaults were coffered, that is, they were decorated by carved octagonal, square or diamond-shaped sunken panels, partly for ornamentation, but also because coffers reduced the weight of the vault and so its lateral pressure upon the walls.

Romanesque builders, basing much of their church construction upon the Roman basilican model, continued to erect round-sectioned barrel and groin vaults but, since the knowledge of Roman concrete had been lost they dared to cover only narrow spans. In a few instances, notably the remarkable early example of Durham Cathedral where the choir was vaulted in 1104 (later rebuilt) and the nave in 1135, Romanesque builders constructed rib vaults, but these were rare, being unsuited to vaulting over square bays.

With the introduction in France of the pointed arch into vaulting design, builders were able to break away from the constraints imposed by the square bay. They adopted rectangular ones which, with the use of the pointed arches, gave greater flexibility. They also constructed rib vaults in which the stone ribs were built first, supported by centering, then the infilling between the ribs – the web – which consisted of pieces of stone or of bricks, was inserted. This web was of much thinner material than the Roman or Romanesque groined vaulting had been, so rendering the whole roofing of greatly reduced weight.

The early rib vaults were quadripartite in design, that is, the bay roofing was divided into four compartments by two diagonal ribs crossing one another in the centre. On the Continent, especially in France, the quadripartite design continued to be built for a long period but in England, often, the quadripartite vaulting bay was divided transversely into two parts so giving six compartments: a sexpartite vault. In most English cathedrals and churches a ridge rib extended longitudinally along the whole interior at the roof apex level and the rib intersections were covered by decoratively carved bosses. In France the ridge rib was rarely used.

From the early fourteenth to the mid-sixteenth century vaulting designs gradually became more elaborate and complex. This was especially so in England where the last phase of Gothic architecture, the Perpendicular, lingered late and had no parallel on the Continent. The first development was to insert tiercerons, intermediate ribs extending from the vault springing to the ridge rib. Exeter Cathedral is a fine example of this. Soon the lierne vault evolved (Figure 18.4 shows a modern replica). Lierne ribs, named from the French *lier*, to tie, might extend in any direction from the structural ribs and might join any other rib. Their use was entirely decorative and later examples are extremely complex as is evidenced in the choir vault of York Minster.

The final phase of English Gothic vaulting was the fan vault which evolved from the desire for a structure which would accommodate ribs of different curves as they sprang from the vaulting shaft of the capital. The radiating ribs of the fan are of equal length and the bounding line is in the shape of a semicircle. The whole group of ribs is made into an inverted cone, the radiating ribs so creating a panelled and cusped all-over surface pattern. King's College Chapel, Cambridge (Figure 18.5), Sherborne Abbey Church and Bath Abbey Church are supreme examples illustrating the use of this type of vault over a long period.

An essential concomitant to the building of arches, vaults and domes was the development of the buttress. This architectural feature is intended to withstand a lateral force and it is arch building which particularly requires this means to counteract the thrust which it generates. There are two chief forms of buttressing used for this purpose. One is that pertaining to an arcade, for example, arches extending sideways to divide a church nave from its aisles, where each counteracts the lateral thrust from its neighbour. The other type is where a heavy buttress is built up against the outside wall of a building to reinforce the walls at points believed to be most vulnerable.

The Romans, who developed building with the arch and vault, constructed strong, heavy walling, reinforced at intervals with wall buttresses to offset the lateral pressure of their barrel and groined vaults. Romanesque builders did the same. Gothic vaulting aroused greater problems. The buildings were higher, the vaults high and wide and, soon, the walls were being built thinner and being

Figure 18.4: Lierne high vault. Church of St Mary, Wellingborough. Architect: Sir Ninian Comper, 1906–30.
Photograph Doreen Yarwood.

pierced by larger windows. Two types of buttress were developed. The tower or pier buttress consisted of a reinforcing buttress, thicker at the base than at the top and descending in stages with sets-off from a lofty top pinnacle to a projection at ground level of some eight to ten feet (2.5 to 3m). The lateral pressure is felt at the top of the buttress but the sheer weight of the massive buttress converts the thrust downwards to ground level.

As the Gothic period advanced, cathedral and abbey churches in Europe were being constructed higher and wider and with ever larger window openings. The usual cruciform church with nave and choir roofs built much higher than the sloping aisle roofs meant that it was impossible to construct a pier buttress between the clerestory windows to offset the thrust of the nave or choir

Figure 18.5: Fan vault. King's College Chapel, Cambridge, 1446–1515. Drawing by Doreen Yarwood.

high vault, so the flying buttress was developed. It was discovered empirically that the maximum thrust was to be felt at a point just below the springing line of the vault on the interior wall so it was at this point on the external wall face that a bridging arch was built to slope outwards and slightly downwards to join a pinnacled wall buttress. This type of construction served a dual purpose; the counterthrust at the given point on the exterior wall conveyed the vault pressure away from the building and down to the ground; also, by means of the heavy pinnacle above, it helped to offset the vault thrust. In very complex church designs, where five aisles had been built, a double flying buttress system had to be constructed with arches spanning each of the two flanking aisles. The French architects developed the flying buttress and it is in France that the eastern arm chevet of a cathedral displays the elegant and dramatic forest of arches and pinnacles which are so impressive. In England, with lower vaults and roofs, flying buttresses were needed less and are most evident in the later buildings of the fifteenth and sixteenth centuries as, for example, at the Henry VII Chapel of Westminster Abbey in London.

With the introduction of Renaissance architecture and the return of classicism the domed covering became more common than the vault. In modern times, however, the development of new materials has made possible the erection of thin shell vaults of half-cylindrical or barrel form which may be extended over long spans without requiring piers which obstruct the interior vista. These are particularly suitable for industrial, sports or entertainment purposes. Reinforced concrete, lightweight metals such as aluminium and plastics reinforced with glass fibre are the most usual materials for this.

THE DOME

A dome is a convex rounded roof covering the whole or a part of a building witha base on the horizontal plane which is circular, elliptical or polygonal. In vertical section the dome may be hemispherical, partly elliptical, saucer-shaped or formed like a bulb (the so-called onion domes to be seen in eastern Europe). The English word dome derives from the Latin *domus*, a house, and the Italian *duomo*, cathedral, or House of God. It has been traditional in Italy to build domes over churches and so 'dome' had been adopted for this meaning in England. The more usual Italian term is cupola, derived from the Latin *cupula*, diminutive of *cupa*, flask, barrel.

While the Romans extensively developed the theme of the arch and the vault they largely neglected the possibilities of domical coverings. They were the first to build domes in western Europe but these were most often supported on buildings with circular walling as in the most famous example, the Pantheon in Rome. Begun in AD 120, this has a very large dome – 43m (142 ft) in diameter – which is hemispherical in form. Its continuous thrust is contained by the 6m (20 ft) thick cylindrical walling into which it is embedded and further support is provided by stepped rings of masonry encircling it above wall top level. This results in an uninterrupted view of the dome from inside the building, but on the exterior only the upper part is visible.

The Romans also covered square interiors with groined vaults, as in their bathing establishments and it is thought that they knew how to build a domical vault over a square space. This is a domical form made of four webs separated by groins which fuse to a point; it is not a true dome and its design lacks flexibility of proportion. Building a dome over circular walling presented fewer structural problems but limited the spatial capabilities of interior architectural design.

In Renaissance and, even more so, Baroque architecture, the dome became a fundamental characteristic of design and a variety of different forms of construction were attempted and built. To achieve, in particular, the Baroque spatial effects of lighting and of solid stone structures ornamented by painted and sculptured curved surfaces appearing to float upwards, it was necessary to support the dome upon free-standing piers instead of the more constricting walling. The basis for this type of construction was the Byzantine development of the dome, evolved in Constantinople, capital city of the Byzantine Empire, a construction which, in its turn, had derived from earlier buildings in Mesopotamia, Anatolia, Persia and Syria.

The essential problem in building a dome of circular section upon a drum, also of circular section, which can contain windows to light the interior, over a square is to transform the square base into a circle upon which the drum can be supported. This may be achieved two different ways: by the use of the squinch or the pendentive. Both methods derive from Asia; the former can be seen in extant remains as early as the fourth century in Persia, the latter in Syria and Asia Minor. In squinch construction an arch, or series of arches, is built across the upper angles of the square to transform the square base to an octagonal one and so more easily support a circular base. This method was extensively employed in the Byzantine period and later in both east and west, for example, in the cathedral at Le Puy in France.

The pendentive is a more ambitious and imaginative, as well as more successful, solution. It was widely used in the best Byzantine building, from S. Sophia in Constantinople (built 532-8 by the architects Anthemios of Tralles and Isodorus of Miletus) onwards, and has been the basis of dome construction in the west from the Italian Renaissance to the present century. A pendentive is a spherical triangle. In using this form of construction the triangular spaces between the square section of the plan of the piers and the circular section of the base of the dome are built as if they are parts of a lower, larger dome, so their section is like that of an arch carried across the diagonal of the square space to be covered. The lower dome possesses a horizontal section which is concentric with the plan of the intended dome. As the lower dome is too large to fill the square space, it is cut off in vertical planes formed by the four sides of the square. When the four remaining parts of the lower dome have been built high enough to form a complete circle within the square section, this circle provides the basis for supporting the actual dome. If this design is not further developed it is known as a sail vault because it gives the appearance of a sail filled with wind anchored at the four corners.

If the dome set upon pendentives is further constructed, the drum of the dome is set upon the circular section of the lower dome and the drum carries the upper dome. Externally most large domes carry a lantern. Internally the drum is articulated with a classical order and windows while the pendentives and dome are decorated by paintings and/or mosaics.

Renaissance domes

It was accepted by the classical architects that a hemispherical dome provided the most satisfactory exterior silhouette from the aesthetic viewpoint. This, however, created problems. These were partly of a structural nature, because a hemispherical dome exerts a particularly strong thrust. They were also of a visual character because, while externally the hemisphere provides a pleasing shape and an imposing landmark, internally it is too tall and dark and its decorative covering of paint or mosaic cannot readily be seen. The solution, adopted by many architects, was to build two domes, a taller exterior one and a lower interior one.

The first great Renaissance dome was the one built by Filippo Brunelleschi to cover the unfinished medieval Cathedral of Santa Maria del Fiore in Florence. To this commission Brunelleschi had to bring his considerable knowledge of mathematics and of the structure of Roman vaults. The practical problem for this early date (1420) was considerable: how to construct a permanent dome to span the 42m (138 ft) diameter space. Brunelleschi dared not build a hemispherical dome on the existing drum, which had no external abutment, so he compromised and proceeded step by step. His dome is a pointed one, constructed on Gothic principles with ribs supporting a lighter infilling, and is taller than a hemisphere to offset the thrust. He then built an inner dome of slightly less height which is decoratively painted on the interior.

Michelangelo faced not dissimilar problems with St Peter's Basilica in Rome. His exterior design was for a hemispherical dome but when this was completed in 1590, after his death, by Giacomo Della Porta, it too had acquired a slightly pointed form. Michelangelo raised his great dome, also of 42m (138 ft) diameter, almost as large as that of the Pantheon, upon a classical drum and supported both drum and dome on four pendentives rising from four massive piers set at the church crossing. Unlike the Pantheon, however, there was no supporting enclosing circular wall, so Michelangelo encircled the dome with iron chains, which have had to be replaced ten times since the original construction in order to contain the thrust.

At St Paul's Cathedral in London (completed 1710) Sir Christopher Wren supported his smaller drum and dome 34m (112 ft) diameter upon eight piers. He also designed a novel triple structure to give stability to his hemispherical outer dome and flatter interior one. His outer dome is constructed of wood covered with lead. Inside this is a pyramidal brick cone which supports the lantern, while the inner decoratively painted masonry dome is seen from the interior. Wren's colonnaded drum with its continuous entablature supports the dome better than Michelangelo's drum at St Peter's but, even so, encircling iron chains were used, now replaced by two stainless steel girdles.

Modern domes

The traditional domical covering has largely been abandoned in modern building but the technical development in materials – of plastics, steel and reinforced concrete – have led to a wide variety of curved shell coverings to large structures, particularly for auditoriums, exhibition halls and sports stadiums. Such shells, made of reinforced concrete, may be used to span a space as great as Brunelleschi's Florence Cathedral but need to be only a few inches in thickness. For example, in 1957 in Italy, Pier Luigi Nervi, who had been experimenting since 1935 with reinforced concrete, employed it to cover his Palazzo dello Sport in Rome. In this building great 'Y'-shaped structures transmit the weight and thrust of the dome to the ground.

The geodesic dome, invented by R. Buckminster Fuller and built for the United States display pavilion at the Montreal Expo 67, was almost spherical and had a maximum span of 76m (250ft). It was constructed of short struts of light metal (this is usually aluminium) and was covered by a plastic skin. The short structural elements were interconnected in a geodesic pattern, that is, by the shortest possible lines. Such domes, ideal for use as factory buildings, will withstand high winds and are very lightweight and may be quickly erected.

CANTILEVER CONSTRUCTION

In this type of structure the building element – block, beam or girder – is fixed securely at one end by a downward force behind a fulcrum and carries a load at the other, free, end; or the load might be distributed evenly along its whole length. When such a cantilever beam bends under its load, its surface becomes convex, unlike a similar beam secured at both ends which would curve in a concave manner under load.

An early example of cantilever building is in the use of a corbel. The single corbel is in the form of a block of stone (or other permanent material) part of which is fixed into a wall. The projecting part of the block may then be used to support a vertical strut or curved brace as in a timber-trussed roof (see p. 886). A development of this was in the corbelled dome or squinch (see p. 880). In this type of construction each successive layer of building stones projects further from the wall in steps; it is the weight of the stones above and behind (the fulcrum in this case) which maintains the stability of the corbelling. Such corbelled domes were being built as long ago as 2500 BC in Sardinia and this method of building was used in the famous tholos tombs at Mycenae in Greece: the so-called Treasury of Atreus there dates from about 1330 BC.

The major use of large cantilever spans has been in bridge building. In the simpler of such structures the metal girders extend from the piers or towers, where they are firmly fixed, to meet and join and so produce a continuous deck for the roadway. Such a constructional method is often used where a fast-flowing river or soft subsoil makes the erection of piers difficult. In these cases a cantilever beam structure can be extended from each bank of the river to join in the centre.

In a more complex design there is an anchor span between the bank and the tower, a cantilever arm extending beyond the lower and a suspended span joining this cantilever arm to another such arm which is projecting from the next tower. These metal girders form a truss framework for the bridge structure. Probably the best-known example in Britain of such a cantilever construction is the railway bridge built over the Firth of Forth near Edinburgh in Scotland between 1882 and 1890.

Cantilever construction is also widely used in staircases and balconies. The beautiful, so-called geometrical staircases of eighteenth- and nineteenth-century-date, often built on circular or elliptical plan, were constructed with one end of each step built securely into the wall of the well and the other resting upon the step below. Iron balconies of the same period were supported upon cantilever brackets similarly affixed into the wall at one end.

The cantilever is extensively employed in present-day structures because of its suitability for use with reinforced concrete and steel materials (see p. 900). In a steel-framed building the metal beams may be allowed to project, to carry the glass curtain-walled façades. Concrete balconies are carried on slabs extended outwards from the interior ones.

ROOFING

A roof is the supporting framework and the covering of a building which shelters the occupants from the weather.

In areas where the need is for protection from rain, wind and snow, as in northern Europe, the pitched or sloping roof was evolved. In general, the further north the region the steeper the pitch, so that snow may more easily be thrown off. There have developed a variety of different designs of sloping roof of which there are six principal types: gable, hipped, lean-to, mansard, gambrel and helm.

Roof shapes

The simplest of the sloping roofs is the gable pattern, also known as a saddleback roof. This is the type seen in the classical temples of ancient Greece where the roof slopes down on each side from the ridge and terminates at the ends in a gable or, as in classical architecture, a pediment. In the warm Mediterranean climate of Greece, only a low pitch was necessary. Before the end of the seventh century BC the pitch had to be steeper because the roof covering was of thatch or clay but, from this time, with the adoption of roofing tiles, a shallower pitch became possible. It is notable that the Greek pediment is a lower, more elegant shape because of this roof pitch than the Roman and, particularly, the Renaissance designs of northern Europe.

The hipped roof, which had also evolved by ancient classical times, is different from the gable in that the ends slope downwards to the cornice instead of finishing in gables. A lean-to is a single sloping roof built against a wall which rises above it. In Europe a gambrel roof ends in a small gable at the ridge though in America, the term is applied to a different type of roof which has a double pitch. This is more like the mansard roof where, in the double slope, the lower one is of a steeper and longer pitch than the upper; the roof is named after the seventeenth-century French architect François Mansart.

The helm roof is a church steeple design. The square tower rises to four sloping faces meeting in a point at the top and gabled at the bottom.

Flat roofs have been constructed since the most primitive of times. These were suited primarily to countries which had a hot dry climate and were the usual design for all kinds of structures from palaces to simple huts. In ancient times they were extensively built in Mesopotamia, the Middle East, Crete, Mycenae and classical Greece. The flat roof continued to be traditional in such areas over the centuries but was less suited to northern climes because of a tendency to leak in wet weather. In hot climates the flat roof was not only trouble-free in this respect but could provide extra sleeping space on hot nights. With the use of modern building materials and modern methods of construction and sealing, the flat roof has been widely adopted in all areas during the twentieth century, particularly for industrial and commercial buildings.

Roofing materials

Since early times a wide variety of materials have been in use for covering the exterior surfaces of roofs. These include thatch and wood shingles, metals such as lead or copper and slates or tiles. The Romans roofed with slate, which is particularly easy to split along its natural laminae, and it has been a common material for this purpose ever since. Sandstone and limestone of a fissile character have also been utilized for this purpose.

Tiles, like bricks, are a ceramic material fired in a kiln; they are fired to be harder and smoother than bricks. Roman roof tiling generally incorporated two types of tile, the *tegula*, the flat tile and, covering the joints between these, the rounded *imbrex* tile. In later tiling in Britain the flat tile was most common but the pantile, the large curving 'S'-shaped tile, which was common in Europe especially the Mediterranean region, was imported from Flanders in the 1630s and has been manufactured in Britain since the early eighteenth century.

Twentieth-century development of reinforced concrete as a building material (see pp. 891–3) has led in modern times to striking innovations in roof shapes. The parabolic or paraboloid form has been experimented with and constructed by engineers such as Pier Luigi Nervi in Italy. Such designs were based upon the form of the parabola which may be defined as 'the path taken by an object which, when expelled or thrown, travels originally at a uniform speed to fall eventually to earth under the influence of gravity'. The parabolic curve was first defined and named by Apollonius of Perga (modern Turkey) about 225 BC. A paraboloid roof is one some of whose plane sections are parabolas. A variation upon this is the hyperbolic paraboloid roof which is constructed in a double curve of hyperbolic form, one of which is inverted.

PLASTER

During the Middle Ages plaster, for both exterior and interior wall covering, was made from lime, sand and water mixed with various other ingredients which, trial and error over a long period of time had shown, would help to bind the mixture and prevent cracking. These ingredients included feathers, hair, blood, dung, straw and hay.

Plaster of Paris, the best quality plaster obtainable, was introduced into England about 1255–60. It was made by burning gypsum (calcium sulphate) and mixing it with water; this gave a finer, harder material. It was called plaster of Paris in England because at that time in France the chief source of supply of gypsum was in the Montmartre district of Paris. Plaster of Paris was reserved, in England, for fine finishes in important buildings because, as an imported material, it was expensive. However, before long, gypsum deposits were discovered and worked in England, chiefly on the Isle of Purbeck, Dorset, and in the valleys of the River Trent in Nottinghamshire and the Nidd in Yorkshire.

The beautiful decorated plaster ceilings and walls of English interiors of the sixteenth to eighteenth centuries were created by craftsmen who learnt their trade from Italian stuccoists. Such Renaissance plasterwork had evolved in fifteenth-century Italy where the craftsmen had long experimented with the medium and had developed a type of plaster which had been used by the Romans. This was malleable and fine, yet set slowly, giving time for the design to be worked; when set it was very hard. The Italians called it *stucco duro*. It contained lime and some gypsum but also powdered marble. Elizabethan stuccoists in England also experimented with this plaster and empirically found that the addition of certain other ingredients would improve the mix. Such ingredients, which varied from area to area, included beeswax, eggs, ale and milk.

During the eighteenth century, when decorative plasterwork became more delicate and lower in relief, a number of patent stucco compositions were marketed. The makers of such plasters claimed that the materials were finer, harder and easier to work. Mr David Wark patented one such plaster and Monsieur Liardet (a Swiss clergyman) another. The firm of Adam Brothers purchased both these patents and in 1776 obtained an Act of Parliament which authorized them to be the sole makers and vendors of the material. They called it 'Adam's new invented patent stucco'.

By 1800 improvements were being made to stucco to make it more robust for exterior work. The fashion was to build in brick, then face this material with stucco and paint it to imitate a costlier stone building. John Nash, the designer of the Regent's Park Terraces, developed such a stucco made from powdered limestone sand, brickdust and lead oxide. Later cement was added to plaster for exterior work. By 1840 several patent gypsum plasters were being sold: Keene's Cement was one such.

In the twentieth century, soon after the First World War, a plasterboard was being made for interior use, chiefly for ceilings. There was an acute shortage of plasterers immediately after the war and this board became very popular. It was made in panels which consisted of a layer of gypsum plaster sandwiched between sheets of strong paper. As the plasterers returned to work the use of this plasterboard declined. The Second World War brought an even more acute shortage of both plaster and craftsmen. Plasterboard reappeared on the market and this time was a greatly improved product so leading to its widescale use. The new material could be applied easily and directly to brick or concrete surfaces, joints were almost invisible and improved thermal insulation was achieved by adding a backing of aluminium foil.

CONCRETE AND CEMENT

Concrete is one of the oldest materials made by man. It is composed of four major ingredients: sand, stone, cement and water. The first two of these are comparatively inert and it is the last two which are the active ingredients. When all four are mixed together a wet plastic substance results and this can be applied and moulded as desired. The cement and the water cause a chemical reaction which gradually transforms the mix into a rock-hard conglomerate. The world concrete derives from the Latin *concretus*, grown or run together.

The stone content may be in the form of pebbles or flints, crushed rock or broken-up materials of many kinds: clinker, cinder, slag, bricks, old concrete. Especially suitable are granite, hard sandstone and rounded pebbles found in natural gravel beds.

In early times the word cement referred to a mixture of broken stone held loosely together by a binding material of lime, clay, burnt gypsum and sand. The word derives from the Latin *caementum*, rough stones, rubble, building material. This 'almost concrete' substance preceded the making of true concrete in which the source of the cement was primarily quicklime (calcium oxide), usually obtained by burning pulverized limestone. This process was called calcining.

It is not precisely known when concrete was first made as the early development of cementing agents produced a mortar which was insufficiently strong to bind the whole mass durably together and, being friable, these materials have not survived. The oldest example of concrete discovered to date was made about 5600 BC. Concrete was certainly made and used in ancient Egypt: wall paintings exist which depict these processes and fragments have been found in a number of places where it acted as an infill mortar in stone walling. The Greeks also made a lime or burnt gypsum mortar for walling with sun-dried bricks.

It was the Romans who developed the use of concrete in a structural manner, not merely as an agent for binding and covering. It appears likely that their first use of the material was in emulation of Greek builders and examples of their work have been found which date from about 300 BC. Towards the end of the second century BC the Romans experimented with a volcanic earth which they found in the region of the town of Pozzuoli near Mount Vesuvius. Because the earth was like a reddish-coloured sand, the Romans took it for sand and mixed it with lime to make concrete. This concrete was of exceptional strength and hardness because the pozzolana, as it is called after its town of origin, is in reality a volcanic ash which contains alumina and silica and this combines chemically with the lime to produce a durable concrete. It is probable that the Romans were unaware of the reason why their new concrete was so superior, but doubtless their engineers and architects quickly comprehended its value. They experimented with the material and discovered that it would set as well under water as in air. They mixed it with an aggregate of broken stone and brick and built structures of concrete and brick faced with marble.

By the first century AD the Romans were experts at handling concrete and were taking full advantage of its possibilities as a structural medium. It enabled them to span and roof spaces of great magnitude, achievements which were not equalled until the introduction of steel construction in the nineteenth century. They built aqueducts, breakwaters, bridges and massive foundation rafts, and vaulted their great building interiors. They experimented with the reinforcement of concrete with bronze strips and dowels in order to improve its tensile strength, but these experiments were not very successful because the higher rate of thermal expansion of the metal caused cracking in hot weather.

The Romans built walls and vaults of great thickness in this hard, durable pozzolana concrete which would resist any stress, but these walls were so heavy that the engineers needed to develop a system of lightening the structure by the introduction of relieving arches and earthenware jars into the wall. They also mixed a quantity of light porous volcanic rock into the aggregate. Concrete was utilized for all possible needs, from floors upon which mosaic was laid to the construction of Hadrian's Wall.

Pozzolana concrete was widely used for building in Europe for some 800 years. It is due in no small measure to the extent of building in this material that so much of the work survives. The Romans had developed the use of concrete as an immensely important structural medium yet, after the collapse of the western half of the empire in the fifth century AD, the greater part of this knowledge and experience was lost for about 1300 years. The making of concrete did not die out, but its use was mainly confined to infilling in walling and as a foundation material. The Normans built concrete foundations to many castles and medieval masons did the same in ecclesiastical architecture. The thirteenth-century Salisbury Cathedral is a notable example.

It was in the mid-eighteenth century that builders began to search for a means to make a cement which would set quickly and be strong and durable. Pozzolana and lime mixtures had been imported from Italy by English builders since the sixteenth century but the secret of Roman concrete had not been understood. The English engineer John Smeaton was successful in his researches to find a satisfactory cement for the new Eddystone lighthouse situated on a rocky reef in the English Channel some 22.5km (14 miles) off Plymouth. The two previous lighthouses here had been constructed of wood and Smeaton was commissioned in 1756 to build a stronger one of stone. His problem was to make a hydraulic cement (one which would set under water) as Roman pozzolana cement had done; this was essential to cement the stone blocks together in order to withstand the onslaught of the sea in such an exposed position. Smeaton found that the best results could be obtained by using lime obtained from a limestone containing an appreciable quantity of clay. The limestone he used came from Wales and he added to this some Italian pozzolana. The lighthouse, built in 1759, lasted until replaced in 1876. Smeaton published his researches, *A Narrative of the Eddystone Lighthouse*, in 1791.

Further experimentation was carried out in a number of European countries in the succeeding 60 years. James Parker, an Englishman, patented his Roman cement in 1796. This was made in Kent at Northfleet by calcining stones found on the beach on the Isle of Sheppey. The stones, limestone with a high clay content, were burnt at a high temperature then powdered. In France the engineer Louis J. Vicat began his researches into French limestone in 1812. His publication of these show that he discovered that the best hydraulic lime derived from limestone containing clay, so producing alumina and silica. He experimented with the use of different clays, adding these to slaked lime and burning the mixture. Other investigators followed in experimentation into making artificial cements from clay and chalk. One of these was the Englishman James Frost who patented his British cement, made at Swanscombe in Kent, in 1822.

Joseph Aspdin, an English bricklayer, patented a better, more reliable product in 1824; he called it Portland cement. This cement was also made from a mixture of chalk and clay but was calcined at a higher temperature before being ground. Aspdin used the name Portland because he thought that it resembled Portland stone in colour. His cement works were in Wakefield in Yorkshire but his son set up another works at Rotherhithe on the Thames.

Modern Portland cement is quite different from Aspdin's, having resulted from many later improvements. A major one of these was achieved in 1845 by Isaac Charles Johnson, manager of the White & Sons Cement Works at Swanscombe, in Kent. In his process the raw materials were burnt at a much higher temperature until the mass was almost vitrified. The resulting clinker was then finely ground to produce a cement far superior to Aspdin's original product.

Between 1850 and 1900 concrete was gradually employed more widely for building purposes, but greater use of the material was inhibited by the high cost of making Portland cement. Production methods were slowly adapted to reduce this cost. One of these was to replace in the 1880s the millstones used for grinding the clinker by iron mills. Later, these in their turn were replaced by tube mills filled with pebbles and, in a further advance, steel balls replaced the pebbles.

Early nineteenth-century kiln design was also a handicap in reducing the costs of making the cement. Parker had made his Roman cement in a bottle or dome kiln of the type which had traditionally been used to burn lime. Isaac Johnson introduced a horizontal chamber kiln which was an improvement and this was bettered by a shaft kiln which enabled the raw materials to be burned continuously in a vertical shaft. However, the cost of manufacture was most sharply reduced by the introduction of the rotary kiln which is the basis of modern design. An early example of rotary kiln was introduced in 1887 at Arlesey in Hertfordshire to the designs of Frederick Ransome. It was 8m (26 ft) long, small indeed compared to a modern kiln which would probably exceed 90m (300 ft) in length.

Reinforced concrete

Plain concrete is strong in compression and so resists the weight of a load placed upon it, but weak in tensile strength: it has a poor resistance to forces attempting to pull it apart. As a result if such concrete is used to manufacture beams, arches or curved roof coverings, the outward and downward thrusts will cause the material to fail. When the Romans used concrete to construct their immense vaults and domes they built massively and the concrete was poured into brick compartments which provided the necessary strength. Similarly arch voussoirs were of brick.

If concrete is reinforced by metal bars, wires or cables embedded in it, the elastic strength of the metal will absorb all tensile and shearing stresses and so complement the high compression resistance of the concrete to provide an immensely strong constructional material; Roman experimentation with bronze for this purpose had little success (see p. 887) but similar trials in the late eighteenth century using iron were very satisfactory because, over the normal range of temperatures, the iron and the concrete displayed the same coefficients of thermal expansion and contraction. In modern construction steel has replaced iron. Using the steel as reinforcement economizes in the utilization of this expensive material. The concrete, which shrinks around the reinforcement as it sets, grips the metal firmly, so protecting it from rusting and weathering. It also reduces the fire hazard, for the steel, which would warp and bend if it became sufficiently hot, is protected within its concrete casing.

The theory of reinforcing concrete with iron to increase its tensile strength was advanced in late eighteenth-century France. Rondelet, Soufflot's collaborator in building the Church of Geneviève in Paris (the Panthéon), experimented by using a metal-reinforced mortar and rubble aggregate. In England, Sir Marc Isambard Brunel constructed an experimental arch in building his Thames tunnel of 1832. He incorporated strips of hoop iron and wood into a brick and cement arch.

Experimentation with metal reinforcement was carried out in several countries during the nineteenth century. In England William B. Wilkinson took out a patent in 1854 for embedding wire ropes or hoop iron into concrete blocks for building fireproof commercial and industrial premises. It was, however, the French who were the chief pioneers in this field. Joseph Monier experimented with large garden tubs reinforced with wire mesh, then developed a system (1877) of reinforcing concrete beams. He also took out patents for reinforcing bridges and stairways. François Coignet exhibited a technique of reinforcing with iron bars at the Paris Exposition and François Hennébique developed hooked connections for reinforcing bars in 1892.

By the late nineteenth century reinforced concrete construction was developing speedily in America, Germany and Austria. As early as 1862 a patent had been taken out for filling a cast-iron column with concrete. In 1878 steel rods were being inserted into concrete columns. Soon after 1900 a stronger column was being produced by inserting spiral steel hoops. Reinforced concrete was being widely used for constructing bridges: an early example was built in Copenhagen in 1879. Longer spans were being handled in the USA and France; the bridge of 1898 at Châtellerault in France is an example.

In Britain reinforced concrete was being developed more slowly and plain concrete was more widely used during the nineteenth century. Pre-cast concrete had been available since about 1845 for paving and balustrades. In 1875, William Lascelles patented a system for building houses from pre-cast concrete slabs fixed to a wooden framework; panels for walls and ceilings could be cast with relief decoration. Plain concrete was being used for building factories and bridges. The Glenfinnan Viaduct in the Scottish Highlands, which carried the railway from Fort William to Mallaig, is a survivor of this type of design. It was built in 1897 by Robert McAlpine and has 21 massive concrete arches. J. F. Bentley also used plain concrete in the massive Roman manner in the domes of his Westminster Cathedral in London. In the early years of the twentieth century Britain slowly adopted reinforced concrete construction, a process speeded up by London offices being opened for business by the two French pioneers, Coignet and Hennébique. Most of the early building was industrial. The first reinforced concrete frame building was the Royal Liver Building begun in Liverpool in 1908.

Pre-stressed concrete

The next advance in the development of concrete as an important structural material came just after the First World War. This was in the field of pre-stressed concrete which greatly increased the material's potential. The purpose of pre-stressing is to make more efficient and economic use of materials. It makes possible the creation of slender, elegant forms which are at the same time immensely strong. The theory is that the metal reinforcement should be stretched before the concrete is poured into position and the pull maintained until the concrete is hard. The reinforcement is in the form of wire cables stretched through ducts and these are placed so as to create preliminary compressive stresses in the concrete in order to avoid cracking under conditions of load.

To be successful pre-stressing requires high quality concrete and high-tensile steel. In the late nineteenth-century experimentation with the process, results were not very satisfactory owing to the quality of the materials. The pioneer of the process was the French engineer Eugène Freyssinet. As an army engineer he had experimented with concrete bridge building and, after 1918, he formed his own company to continue his development. In 1934 he achieved fame when he saved the Ocean Terminal at Le Havre from sinking into the mud under the sea by building new foundations of pre-stressed concrete.

After 1945 steel was in limited supply and architects turned more and more to reinforced and pre-stressed concrete. Improved technical understanding had shown this to be a material with almost unlimited structural possibilities. It was not costly and the needs of an individual project could be exactly calculated in terms of reinforcement and strength of concrete. It could be cast into any form and this gave a greater freedom of architectural expression.

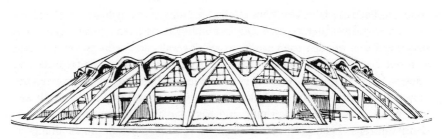

Figure 18.6: Concrete. Palazzetto dello Sport, Rome. Architect and engineer: Annibale Vitellozzi and Pier Luigi Nervi, 1957.
Drawing by Doreen Yarwood.

A leader in the use of this material was the Italian architect and engineer Pier Luigi Nervi. From his early work on Florence Stadium (1927–30) and his later imaginative hangars at Orvieto and Ortebello, where he developed the use of pre-cast concrete elements, he went on to create original and beautiful structures of infinite delicacy and variety. The Palazzetto dello Sport, built for the 1960 Rome Olympics, was typical (Figure 18.6). This was a 'big top' but in concrete not canvas, a structure 60m (194 ft) in diameter anchored by 36 concrete ties. In the 1940s Nervi felt constrained by the limitations imposed by the use of timber formwork to shape the concrete and he went on to experiment with his own version of ferro-concrete in which he erected metal-framed skeleton structures to support the cladding, a method with which much of his subsequent work was identified. The great roof of the 1949 Turin Exhibition Hall is a major instance of this.

Nervi had been pressed to use his ingenuity in reinforced concrete by a shortage of steel in post-war Italy. Similarly, necessary economies in a number of countries led to the development of building methods which saved time and cost. Pre-cast cladding in high-rise construction was one answer and Le Corbusier's experiment in urban living, the Unité d'Habitation in Marseilles was a pacesetter. This structure and others inspired by it gave rise in England, in 1954, to the term brutalism. This referred to the use of concrete in its most overt, naked form, handled in strong masses. It was an English derivation from the French *béton brut*, meaning concrete left in its natural state, unfinished, displaying the timber graining impressed upon it after the formwork has been removed. Formwork, also called shuttering, is the timber or metal form into which concrete is poured and which is removed when the concrete has set. Other means of producing a rough but even texture on the concrete includes bush-hammering after the concrete has set by use of a hammer with grooved head employed to mark the surface. Typical brutalist work is the Queen Elizabeth concert hall on London's South Bank (1967) and, of the same date, the Tricorn Centre in Portsmouth.

During the 1950s and 1960s vast numbers of reinforced, pre-stressed buildings were erected all over the world, but by the 1970s disillusion was apparent. Concrete is an inexpensive building material but it weathers badly Much work has gone into the problem and better quality structures with better surface concrete are now being erected. When concrete is faced, as it so often is in Mediterranean countries, by mosaic, faïence or marble, its structural advantages are undeniable. In great engineering projects such as the Thames Barrier Scheme (opened 1984) reinforced pre-stressed concrete has no rival.

IRON AND GLASS

Ironwork

The nineteenth century was very much the age of iron, just as in the twentieth century steel is such a vital constructional material, but the story of iron as a decorative and structural material began much earlier.

In Europe iron had been regarded as a durable, utilitarian material since the early Middle Ages but its use was limited by the lack of power and technical knowledge which hampered both quantity and quality of production. At this time the iron in use was wrought, that is, hammered to beat out impurities and make it into the desired shape. It was manufactured into utensils, weapons and agricultural implements. With the later, improved blast furnace design, cast iron could be produced and, by 1600, adequate supplies were available for the needs of the time (see Chapter 2).

By the later eighteenth century, with the development of steam power, the adoption of coked coal furnaces and the evolution of the engineering industry, the Industrial Revolution in Britain had reached a point where large-scale iron structures had become possible and the material was gradually adopted for wider use. The famous Ironbridge at Coalbrookdale spanning the River Severn was a landmark. Constructed from iron elements made in Abraham Darby's foundry there, the bridge, designed by Thomas Farnolls Pritchard, was completed in 1779.

In Britain, from the last years of the eighteenth century, iron was being utilized for a wide range of industrial building, for factories, warehouses, textile mills, bridges. None of these was of note architecturally; they were strongly built and functional engineering structures. The iron was used for roof trusses and covering, for wall panels, for column supports and for window frames, as well as for the heavy machinery. But in the same period iron was also being employed for architectural projects in both a structural and a decorative manner. For example, in 1794 Sir John Soane covered the 7m (23ft) diameter oculus over his Consols Office in the Bank of England with an iron and glass lantern; Thomas Hopper fan vaulted his great conservatory at Carlton House in

London in 1812 with iron; in 1818, John Nash built his Chinese-style staircase in the Royal Pavilion in Brighton entirely from iron. Increasingly iron was being used for supporting columns, galleries and roofing in churches, Rickman's Liverpool churches of 1812–15, St Michael, Toxteth Road, and St George, Everton, among them. Decoratively, iron was extensively employed at this time for ornamental galleries, balconies and railings.

For many years until the early 1860s iron was used for structural features, especially roofing, in buildings because it was believed to be more fireproof than timber, as indeed it was. When the danger to iron structures under intense heat became apparent its use lessened and it was finally replaced by cased steel. Early instances of roof replacement with iron include Soufflot's covering of 1779–81 over the staircase hall leading up to the Grand Galerie of the Louvre and Louis' roof of 1786–90 over the Théâtre Français, both in Paris.

By the 1840s more extensive use of iron roofing was being taken up in many areas. Barry covered his new Palace of Westminster with iron roofing and, in 1842, Montferrand completed his cast-iron dome for St. Isaac's Cathedral in St Petersburg. Before long American architects were following his lead as in T. U. Walter's vast and spectacular dome over the Capitol in Washington (1855–65).

At this time also iron was becoming increasingly important for heavy structural work. In New York in 1848, James Bogardus erected his first four-storeyed factory with iron piers and lintels and went on to more ambitious iron urban buildings. Others emulated his example for factories, department stores and apartment blocks. In Britain in 1851 at Balmoral Castle the Prince Consort ordered a prefabricated iron ballroom. At Saltaire, in Yorkshire much of Sir Titus Salt's magnificent new textile mill (1854) in the town which bears his name was of iron construction. In 1889 in Paris was erected the tallest of the nineteenth-century structures in iron; Gustave Eiffel's 300m (984 ft) tower to commemorate the centenary of the start of the French Revolution.

The nineteenth century was also the great age of railway and bridge building and for both of these iron was the chief structural material. In addition to the track and the vast railway sheds with their trussed iron roofs, iron was extensively used in the architecture of railways, in columns, brackets, trusses and roofing. It was also a decorative medium, being employed for window frames and balconies, railings and canopies. In the Victorian Gothic period, when great London terminals such as St Pancras, King's Cross and Paddington were constructed, famous architect/engineer partnerships were energetically building stations, hotels, viaducts and sheds. The great engineers of the day included Isambard Kingdom Brunel, Thomas Telford and Robert Stephenson, the architects Francis Thompson, Sir George Gilbert Scott and Matthew Digby Wyatt.

Great bridges were also being constructed to carry the railway lines across rivers, canals and other railway routes. Iron was utilized for such structures in girders, piers, cantilevers and chains. Famous among surviving bridges are Brunel's Clifton Bridge over the Avon gorge near Bristol (1837–64) and his Royal Albert Bridge over the River Tamar which divides Devon from Cornwall (completed 1859) and Robert Stephenson's tubular Britannia Bridge over the Menai Straits which separate North Wales from the Isle of Anglesey (completed 1850) and his High Level Bridge over the River Tyne in Newcastle.

Ferro-vitreous construction

By the 1820s, as a result of technical advances in the making of both iron and glass, the idea of combining the use of these two materials for specific types of building was experimented with widely. Their employment together resulted in some practical as well as aesthetically attractive interiors. An early use of these materials was for the construction of conservatories and glasshouses but for some time wood was used to frame the glass panels, this being gradually replaced by iron. The Palm House in Kew Gardens is a magnificent surviving example of this type of construction. Designed by Decimus Burton and erected by the Dublin engineer Richard Turner in 1844–7, this remarkable structure comprises over 4180m^2 (45,000 sq ft) of greenish glass. Inside a decorative iron spiral staircase gives access to an upper gallery from where tall plants may be closely viewed. Sir Joseph Paxton was responsible for the Great Conservatory at Chatsworth House (begun 1836, now demolished), a project which he followed by his entry for the competition held for a building to house the Great Exhibition of 1851 in Hyde Park. Aptly dubbed (in *Punch*) the 'Crystal Palace' this (then) unusual structure was a prefabricated glasshouse of vast dimensions: 563m (1848ft) long, 124m (408ft) wide and over 30m (100ft) high. It contained 3300 iron columns, 2150 girders, 183,000m (600,000ft) of timber and 83,600m^2 (900,000 sq ft) of sheet glass (see prefabrication p. 899).

From the 1830s to the 1880s iron and glass were used together to construct large, naturally illuminated, elegant interiors. These materials employed together were regarded as more fireproof than wood and glass though, in the second half of the century, it was being realized that an unclad iron skeleton to a building could collapse dangerously when fire raised the temperature to a certain level. This proved only too true on the night of 1 December 1936 when the Crystal Palace was in a short space of time almost totally destroyed assisted, in this case, by the contents, much of which were inflammable.

By the 1840s the roofs of many large interiors were being covered by ferrovitreous construction. Bunning's Coal Exchange in London (1846–9, now demolished) was a superb example. In 1854–5, Sydney Smirke filled in the court of his brother's British Museum with the domed reading room. It was soon discovered that this type of roofing was ideal for railway station sheds as

Figure 18.7: Iron and glass. State Department Store GUM. Red Square, Moscow. Architect: A. N. Pomerantsev, 1889–93.
Drawing by Doreen Yarwood.

evidenced at, for example, Lewis Cubitt's King's Cross in London (1851–2) and Duquesney's Gare de l'Est in Paris (1847–52).

Iron and glass continued to complement each other as the century advanced although, as both quality and quantity of steel production improved, steel increasingly replaced iron for constructional purposes. In many European cities great undercover shopping arcades and galleries were built where people could stroll, sit at café tables or window-gaze. A number of these survive: Mengoni's Galleria Vittorio Emanuele II in Milan is an impressive example. Another is the remarkable department store in Moscow's Red Square called GUM (Figure 18.7). Built in 1889–93, the interior of this great building comprises three parallel barrel-vaulted galleries, each about 300m (1,000 ft) in length, with balconies and walkways at different heights, all serving shops. There are iron connecting walkways from one section to another. The interior is like the nave and aisles of a church though the aisles are nearly as wide and high as the nave. There are hundreds of shops and stalls, all covered by an iron and glass domed roof.

Steel-framed high-rise building

The skyscraper was conceived and named in America where, by the 1880s conditions were ripe for this type or architectural development. In the big cities, notably New York and Chicago, steeply rising land values provided the incentive to build high and the structural means to do so had become available. However, the desire to build high, at first for commercial and office needs, had been frustrated for several decades by the twin problems of load-bearing walls and of transporting the users of a building from floor to floor. The second of these difficulties was solved when Elisha Graves Otis adapted the traditional and age-old goods hoist for passenger use. In 1852 he devised a safety mechanism which would hold the lift in place in the shaft by means of spring-controlled pawls if, by chance, the controlling rope gave way. In 1854, Otis personally demonstrated his device and by 1857 the first elevator was installed in a New York department store.

The next essential development to enable buildings to rise above about ten storeys was the steel-framed structure. As early as 1849, William Johnston was erecting his seven-storey Jayne Building in Philadelphia in an architectural style which though eclectic, displayed the vertical design format for the façade which was later to become characteristic of high-rise structures. However, until the early 1880s buildings were still being erected with traditionally load-bearing walls up to ten storeys in height. To rise still higher would require the walls to be impractically thick at base in order to carry the load above. It was the emergence of the load-bearing metal framework, structurally independent of the external walling, which made the true skyscraper possible. An early landmark in this development was the Home Insurance Building in Chicago built in 1883–5 by William Le Baron Jenney. In this he devised an iron and steel framework of columns, lintels and girders. The building was quickly followed by the fully developed steel skeleton construction of the Tacoma Building by Holabird and Roche in the same city, where the walls were merely cladding.

Despite these revolutionary structural ideas which made tall skyscrapers possible, architectural design continued to be largely eclectic, the wall cladding being dressed with classical columns and entablatures or, as in Cass Gilbert's 52-storey Woolworth Building of 1913 in New York, in Gothic detail (Figure 18.8). The leader in designing an architectural style suited to take advantage of the new structural method was Louis H. Sullivan, who treated his elevations to accentuate the height and with continuous pilasters to stress the steel frame beneath rather than to hide it as his predecessors had done. His Wainwright building in St Louis (1890–1) is characteristic. A few years later, in 1894–5, came his masterpiece, the Guaranty Building in Buffalo, based on almost free-standing piers which anticipated Le Corbusier's later use of *pilotis*. The elevations are sheathed in terracotta and rise to a decorative, non-eclectic cornice.

Since the 1920s American skyscrapers have risen even higher from the decorative 380m (1247ft) Empire State Building to the 443m (1453ft) Sears Tower in Chicago. Europe was slower to follow the American lead until after 1945 but the international style soon took over there also. Even in eastern Europe, where in the early 1960s the Soviet form of neoclassical architecture was superimposed upon high-rise buildings leading to innumerable skyscraper 'wedding cakes', after 1965 internationalism prevailed too.

Figure 18.8: Steel-framed skyscraper. The Woolworth Building, New York. Architect: Cass Gilbert, 1913.
Drawing by Doreen Yarwood.

The steel elements of framed building may be joined together by bolting, riveting or welding, Carbon steels may corrode through oxidization and above temperatures of about 370°C may lose their strength so, for both fireproofing and protection from the atmosphere, steel members are generally encased in concrete. The steel skeleton of a modern skyscraper is constructed as a grid made up from vertical columns and horizontal girders or beams. Foundations and floors are made of reinforced concrete.

MODERN INDUSTRIAL CONSTRUCTION

The process of replacing craftsmanship in building by the mass-production of materials and of decorative and structural features began on a small scale in Britain in the eighteenth century; the Industrial Revolution brought steam power which was to revolutionize the machine tool and engineering industries.

The idea of making the parts of a building in a workshop then assembling them on the building site was, however, an earlier one. Medieval timber-framed structures had been partly constructed in this way (see p. 864) and Leonardo da Vinci in the late fifteenth century suggested an extension of this idea. In North America prefabrication methods supplied the need for rapid provision of homes for the men taking part in the California Gold Rush of 1849.

All of these instances were largely concerned with building in wood. The famous ferro-vitreous prototype of prefabrication was Paxton's Crystal Palace of 1851 (see p. 895), erected in less than five months in Hyde Park then dismantled the following year and re-erected in Sydenham, South London. Originally the parts were standardized and made in quantity for assembly on the site.

The growth of such constructional methods advanced slowly during the later nineteenth century, then received an impetus with the shortage of buildings which existed immediately after the First World War. This was when standardized steel window frames, structural steel framing and pre-cast concrete panels for walling and roofing were being manufactured at speed. It was the need to build quickly in Europe after the devastation of towns during the Second World War which led to wide-scale prefabrication. Temporary houses were constructed in factories and delivered by lorry for assembly on site. In Britain four lorryloads were required for the building of one house. One of these loads carried a complete kitchen and bathroom unit containing the necessary plumbing. These houses, known familiarly as 'pre-fabs', lasted for many years longer than their intended lifespan. The idea of utilizing steel and plastic in this way to mass-fabricate complete bathroom units had first been put forward in the 1930s in America by Buckminster Fuller, but at that time the cost was too high. The mass-production level required after the war in so many European countries from Britain to the USSR made the project cost-effective.

An essential concomitant of prefabrication is standardization of manufacture of separate building parts and, by this time, a system to establish an overall three-dimensional unit of measurement was required. The system developed, known as modular design, ensures the accurate fitting of all building parts of whatever material and wherever or by whomsoever manufactured. In Britain the Modular Society was founded in 1953 with members drawn from all concerned with the building industry, from architects to clients and craftsmen.

A further development, in which prefabrication, modular co-ordination and functional planning are incorporated, is called systems building. In such projects a complete building is planned for manufacture, not just separate parts. Its specifications, such as heating, lighting, ventilation, capacity, load-bearing requirements, building materials, site characteristics etc. are studied and computerized. This type of system has been utilized for housing and school buildings.

During the twentieth century many technical advances have taken place in the manufacture and use of materials which have made new building methods possible. As early as 1911, Walter Gropius in Germany was experimenting with glass cladding in his Fagus Factory built at Alfeld-an-der-Leine. This was a revolutionary design which heralded the introduction of the curtain wall of steel and glass, hanging in front of a steel-framed structure and separated from it, introduced in 1918 by Willis Polk in San Francisco. After 1945 architects all over Europe were emulating the American structures which had been built in great numbers in the intervening years.

With the development of reinforcement and pre-stressing, concrete has become the ubiquitous twentieth-century building material. Since 1950 large construction firms have adopted on-site pre-casting methods, economizing greatly in transport and handling. Other technical advances have included float and solar glass, laminated wood and particle board products. The potential application of plastics to structural needs had not yet been fully developed, although experimental work and manufacture points to their satisfactory potential in combination with other materials. A notable instance of this is the reinforcement of certain plastics, for example, polyester resins, with glass fibres. Apart from load-bearing units, different plastics are being utilized in ever increasing quantity and variety in every aspect of building as in, for instance, cisterns, piping and roofing panels.

FURTHER READING

Brunskill, R. W. *Traditional buildings of Britain* (Gollanz, London, 1982)
Brunskill, R. and Clifton-Taylor, A. *English brickwork* (Ward Lock, London, 1977)
Clifton-Taylor, A. *The pattern of English building* (Faber & Faber, London, 1972)
Clifton-Taylor, A. and Ireson, A. S. *English stone building* (Gollanz, London, 1983)
Condit, C. W. *American building: materials and techniques from the beginning of the colonial settlements to the present* (University of Chicago Press, Chicago, 1968)

Copplestone, T. (Ed.) *World architecture* (Hamlyn, London, 1985)
Fintel, M. (Ed.) *Handbook of concrete engineering* (Van Nostrand Reinhold, 1974)
Gloag, J. and Bridgewater, D. *History of cast iron in architecture* (George Allen and Unwin, London, 1948)
Hitchcock, H. Russell *Architecture: nineteenth and twentieth centuries*, Pelican History of Art series (Penguin, London, 1982)
Lloyd, N. *History of the English house* (The Architectural Press, London, 1975)
Raeburn, M. (Ed.) *Architecture of the western world* (Orbis, London, 1980)
Whiffen, M. and Koeper, F. *American architecture 1607–1976* (Routledge & Kegan Paul, London, 1981)
Yarwood, D. *Architecture of Italy* (Chatto and Windus, London, 1970)
—— *Architecture of Britain* (Batsford, London, 1980)
—— *Architecture of Europe* (Chancellor Press, London, 1983)
—— *Encyclopaedia of architecture* (Batsford, London, 1985)
—— *A Chronology of Western Architecture* (Batsford, London, 1987)

19

THE DOMESTIC INTERIOR: TECHNOLOGY AND THE HOME

DOREEN YARWOOD

SURFACE, COVERINGS AND DECORATION

Walls

Before the late thirteenth century the inside surfaces of the walls of a well-to-do home were usually thinly plastered over the stone or timber-framed structure (see Chapter 18). For decoration this plaster was then painted in patterns or scenes. For warmth in the winter months, hangings of wool or richer fabrics were affixed: in wealthy households tapestries replaced hangings.

The practice of lining interior walls with wood dates from the thirteenth century. Such a lining was not panelling but made of boards fixed vertically, tongued and grooved to give a flat face to the interior of the room. The preferred wood was wainscot, a medieval term for a high quality oak grown along the Baltic coastal plain from Holland to Russia. The term is believed to derive from 'wain', an early form of wagon for which the wood was also used. Before long wainscoting came to refer to any type of wood wall lining.

The next stage was panel-and-frame construction, a system of joined framed panelling which was developed in Flanders and Germany and introduced into England in the fifteenth century: it was an important technical advance. The thin wood panels were tapered on all four sides to fit into grooves made in the framework of thicker horizontal and vertical members which were mortised and tenoned together, then fastened with wood pegs. This method of wood covering, made by a craftsman joiner, was more satisfactory than the earlier one as it allowed space for contraction and expansion in the atmosphere and so maintained its form. Early panels were decorated by painting. From about 1485 linenfold panelling, carved to represent folded material, was introduced and this was replaced by the mid-sixteenth century with a more varied form of carved decoration or one made by inlays of coloured woods.

Paper wall hangings – wallpaper – were first made as a cheaper substitute for the patterned tapestries and embossed leather hangings which lined the walls of wealthy homes. These papers were in small squares or rectangles, often applied to the wood panelling. With the development of printing (see Chapter 14) wood blocks were made to print designs, at first in black on white paper, with the colour being painted in afterwards by hand. Later, several blocks were used to print in different colours. The earliest example found in England of such patterned papers is dated 1509: it is printed on the back of a bill proclaiming the accession of Henry VIII.

In France in the 1620s the idea of flock wallpaper was introduced in order to create a cheaper imitation of the costly Italian cut velvet hangings. The wallpaper was printed in a pattern with wood blocks which applied gum instead of coloured ink. While this was wet finely chopped coloured wool (and later silk) was blown on to the surface using a small pair of bellows. Also in France, in the 1760s, wallpaper was first made in long strips by pasting sheets 12 by 16 in (30.5 × 40.5cm) together before printing; this gave a roll 12 in wide by 32 ft long (30.5 × 975cm).

Imported hand-painted Chinese papers were available, at high cost, to European households from the early seventeenth century. These were prized and very popular as they blended with the then fashionable oriental porcelain and lacquered furniture which was also being imported. European imitations of these Chinese papers followed later.

Paper in continuous rolls was produced in France from 1799, but mechanization of wallpaper printing had to wait until the nineteenth century. Patents were being taken out by the 1830s for machines; these were not very satisfactory, but after about 1845 improved designs were able to produce papers which gradually replaced hand-printed ones. Before this time, the costly flock and Chinese papers had been protected from damp and dirt by being mounted on wood frames backed by fabric. The cheaper machine-printed wallpaper rolls could be pasted to the wall and replaced when they became dirty.

Ceilings and walls

During the Middle Ages the principal rooms in secular buildings were roofed with open timber trusses, (see Chapter 18).

By the later fifteenth century it became more usual to ceil interiors. Such flat ceilings were supported by carved, moulded beams crossing one another at right angles, the interstices being infilled by boards or plaster. From the mid-sixteenth century onwards ceilings and friezes were decoratively plastered all over, the ornamental forms changing over the centuries and reflecting the evolution of style from strapwork to Renaissance, baroque and rococo designs. During the later seventeenth and eighteenth centuries walls were also often decoratively plastered.

Mediaeval plaster was made from lime, sand and water mixed with various other ingredients which, it had been empirically discovered, helped to bind the mixture and prevent cracking: these included dung, blood, feathers, straw and animal hair. The best and finest plaster of the early thirteenth century was plaster of Paris, so called because it was made by using gypsum (calcium sulphate), the chief source of which was then in Montmartre in Paris. Before long other sources of supply were discovered in Europe, but the name survived.

In the late sixteenth century the ceilings of important Western European interiors became ornately decorative thanks to the introduction into northern Europe of the *stucco duro* developed by the craftsmen of fifteenth-century Renaissance Italy. These craftsmen had experimented with a plaster which had been used by the Romans (see Chapter 18); this contained lime, some gypsum but also powdered marble. It was malleable and fine yet set slowly – giving time to work the design – and finally very hard.

During the neo-classical period of the later eighteenth century, when plasterwork designs were especially delicate and detailed, a number of patent plasters were marketed which were finer and set harder. By the end of the century, with advancing industrialization, plaster mouldings and ornament were beginning to be manufactured by mass production methods so that the builder could apply these ready-made to his ceilings, friezes and walls.

Plasterboard came into use in the early 1920s as a result of a shortage of plasterers after the First World War. It was manufactured in panels which consisted of a layer of gypsum plaster sandwiched between sheets of strong paper. The post-Second World War product was much improved. It could be applied directly to brick and concrete surfaces, its joins were imperceptible and a backing of aluminium foil was added to improve thermal insulation. Plasterboard was later superseded by plastic tiling.

Floors

Until the later nineteenth century floor surfaces on the ground storey were covered by stone flags, bricks or tiles, mosaic, polished wood boards or blocks; upstairs floors were usually of wood. Bricks and tiles were in use from very early times in Europe, the tiles being fired to be harder and smoother than bricks. Roman floor bricks were rectangular, smaller and thicker than their walling bricks and were fired to be more like modern tiles than bricks. Encaustic tiles were widely used for both floors and walls in the Middle Ages, and their use revived in the nineteenth century; they were made of earthenware into which coloured clays had been pressed to make a design before glazing and firing.

The craft of lining floors and walls with mosaic is much the same today as it was in the days of ancient Rome. The mosaicist designs his picture then re-creates it in tiny cubes of ceramic, stone, marble or glass which are laid upon a mortar base. The top surface of the cubes is painted in colours and, if earthenware, glazed. The cracks between the cubes are filled with a grouting compound.

Wood floors were most commonly of planks, very wide in earlier times and of hardwood, generally oak. Quality wood floors might be of parquet, that is, covered by thin pieces of hardwood laid in patterns, most commonly herringbone.

Linoleum was the first of the successful easy-care floor coverings which could be washed or polished: it represented a great breakthrough in alleviating the drudgery of floor scrubbing. The Englishman Frederick Walton set up the first factory to make linoleum in 1864 at Staines in Surrey. A base of burlap was coated with a cement which he made from linseed oil, gum and resin with the addition of colour pigments as desired. Linoleum was made in rolls; it continued to be manufactured until replaced soon after 1950 by synthetic plastic coverings (see pp. 906 and 947).

The making of carpets and rugs by flat weaving and, later, by knotting, dates from the early centuries AD in the Middle East, Central Asia and the Far East. The Moors introduced the craft into Spain where beautiful carpets were made by the thirteenth century.

In England the ladies of a household were stitching small carpets from about 1575. These were of cross-stitch or embroidered needlework designs worked on a frame on a canvas ground. Both these and the imported carpets from Persia and Turkey were considered to be too valuable to place on the floor and were used as wall or furniture decoration.

The French tapestry workshops such as Gobelins had been established during the seventeenth century. From these developed the carpet workshops of Savonnerie and Aubusson which made woven carpets of, at first, oriental design but which later reflected European art forms. After 1685 many of the Huguenot carpet weavers fled to England to escape religious persecution and set up their craft under royal patronage. During the following century manufacturers such as Wilton, Axminster and Kidderminster became renowned for their carpet production.

The steam-powered carpet loom, devised in the USA in 1839 by Erastus B. Bigelow, was a great advance. At first the machine would produce only flat-weave carpets but soon it was adapted to make a Brussels type of pile. By mid-century the Jacquard attachment was incorporated into machines for patterned carpet weaving (see Chapter 17). Before long the use of the steam loom was extended to the Wilton type of cut pile, the loops of the pile being raised by one set of wires while another set with knife-blade attachment was incorporated to cut the pile. Later the power loom was able to produce cheaper 'tapestry' carpets and, using two looms, chenille carpets.

Hand-made tufted carpets, based upon the idea of candlewick bedspreads, were pioneered in the USA soon after 1900. In the 1920s sewing machines were being adapted to make tufted fabrics but it was about 1950 before a suitable machine was designed to produce tufted carpets in volume. In modern tufted carpet production the tufts are held in position by a latex coating and the carpet is backed by a polyurethane layer. From about the 1950s also synthetic fibres were beginning to challenge wool as the prime carpet material.

Facing materials

A patent for plywood was taken out as early as 1840. Early plywood was cut by saw and the thin sheets glued together, each layer being placed so that the grain ran at right angles to that of the previous one. The development of efficient knife-cutting machines, which have the advantage of being able to cut round a log and so produce a continuous layer, also the introduction after 1945 of synthetic resins, have made possible cheaper, stronger plywood. Particle board is another modern product, made from chips and shreds of wood compressed with synthetic resin.

In 1913 a method was found in the USA to make a laminated plastic sheet and by the 1930s decorative plastic laminates were being manufactured, although it was not until after 1945 that these were developed as a tough, attractively-patterned, easy-care finish suited especially for bathroom and kitchen surfaces.

The best-known name in these veneers, which are applied to all kinds of core materials, such as wood or particle board, is Formica laminate. This veneer, 1.5mm (0.059 in) in thickness, consists of seven layers of paper impregnated with phenolic resin, on top of which is a patterned sheet of paper impregnated with melamine resin and a further overlay sheet of alpha cellulose paper similarly treated. The whole laminate 'sandwich' is bonded chemically under controlled heat and pressure into a single veneer.

FURNISHINGS AND FURNITURE

Before the sixteenth century the home was furnished with few fabric coverings and these were mainly made at home or locally by the simple textile processes of the day; similarly, the fabric most often used was that produced nationally – in Britain, wool. The richer materials of silk, velvet and brocade were costly imports. In Britain only the larger homes had window glass, the window openings generally being fitted with wooden shutters which were closed at night and to exclude very cold air. By the mid-sixteenth century window hangings were becoming fashionable, and during the following century they were being made from rich, patterned fabrics and fitted with pelmets or frills. Seating furniture was covered by similar fabrics or by home-stitched tapestry work which was edged with fringe or tassels.

The mechanization of all the textile processes, which was initiated early during the Industrial Revolution, especially in England, brought cheaper furnishing materials within reach of many homes. Such fabrics were used for covering many kinds of furniture and for draping bedsteads, windows, doorways and, in the nineteenth century, even chimney-pieces. The introduction of the sewing machine, which became an essential item in many households by the second half of the century, encouraged home furnishings. In the same period colour schemes were greatly influenced by the availability of aniline dyes which made strong, vivid colours popular for fabrics as well as wallpapers.

Upholstered furniture appeared soon after 1600; before this cushions were used to soften hard seats while bedsteads possessed a wood-board or rope-mesh foundation upon which were piled one or two feather beds. Seventeenth-century upholstered chair seats comprised simply the cushion encased in the covering material and were padded with feathers, wool, horsehair, down or rags.

Springs were available in the eighteenth century but they were used for carriages, not furniture. In 1826, Samuel Pratt patented a spiral spring intended to make a swinging seat to be used on board ship to counteract seasickness. Soon he appreciated its possibilities for furniture and the spiral spring began to be incorporated into chair and sofa design. Such seating furniture became very bulky because of having to accommodate the springs inside a great thickness of horsehair and feather stuffing. The French *confortable* easy chair was typical. In 1865 the woven wire mattress for beds, supported on spiral springs, was patented.

In more modern times new materials have replaced the sprung and stuffed upholstery with a simpler, more comfortable, cheaper and more hygienic upholstered form. This is based upon plastic foam material used for cushions and seat backs; it is supported on rubber webbing or cable springing. Mattresses vary in softness and resilience according to taste and utilize combinations of sprung, padded and suspension systems.

Furniture-making

The most far-reaching changes in the technology and consequent style of cabinet-making were brought about by the late medieval introduction of panel-and-frame construction (which was applied to furniture as well as wall panelling, see p. 902), the development of veneering in the seventeenth century, the progress of mechanization in the nineteenth century and the introduction of new materials and techniques in more modern times.

Until about 1660 oak was the chief furniture wood. It was used both solid and panelled and was decorated by carving, painting and inlay of coloured woods or other materials such as ivory. In the craft of inlay these materials are deeply recessed into the solid piece of furniture; it differs from marquetry, which is applied as a thin veneer.

The age of oak was followed by the age of walnut, a wood which lent itself to a more delicate style and one which, being more costly and available in smaller quantity, was especially suited to veneering. This is a cabinet-making technique where thin sheets of decorative high quality word are glued to the flush surface of a carcase of cheaper wood. The art of veneer is old: it was practised in ancient Egypt and by the Romans. The idea of so decorating furniture was revived with Italian Renaissance inlay (*intarsio*). The craft was then developed in France and the Low Countries and in the later seventeenth century the Huguenot refugees brought it to England. It was during the nineteenth century that some poor quality workmanship gave rise to the word 'veneer' becoming synonymous with the veiling of a shoddy article with something finer.

Veneering is a highly skilled craft and, by cutting different parts of a tree, beautiful patterns may be obtained. Walnut is particularly suited for this, the 'oyster' designs coming from small branches cut laterally and the 'burr' veneers from malformations of the tree. Other woods, such as maple, olive, laburnum and kingwood, were also used.

By the early eighteenth century the technique had so advanced that it was possible to saw fine woods to a thickness of 1.5mm (0.059 in) and glue large sections to the carcase. Veneering then gave place to marquetry. In this, different woods of varied colour and pattern are cut into thin layers and made up to fit into a design which is then married into the ground veneer and the whole marquetry panel is then applied to the piece of furniture. Floral, arabesque and geometrical patterns were used. Particularly elaborate designs also incorporated metals, ivory, mother-of-pearl and tortoiseshell; the French cabinet-maker André Charles Boulle made this form of marquetry internationally known.

From the 1720s mahogany – the most famous of all furniture woods – was increasingly imported into Europe and America from the West Indies. This is a wood superbly suited to carving and caused veneered furniture to go out of fashion, although it was revived in the neo-classical work of the later eighteenth century by such designers as Adam, Sheraton and Hepplewhite.

Lacquered furniture was imported into Europe from the Far East from the mid-seventeenth century. This richly decorative process originated in China and was then perfected in Japan. The sap of the sumac tree was used to lay a dark, glossy surface on the furniture, which was then painted and gilded with oriental designs. Such pieces possessed great *cachet* in Europe, blending as they did with the equally fashionable oriental porcelain also being imported at the time.

Imported lacquered furniture was very expensive and in short supply. Also this type of sumac tree was not grown in Europe, so European imitation pieces were made and sent to the orient to be lacquered then returned. By the early eighteenth century the Dutch, and later the French and English, had perfected a method of imitating oriental lacquer which they (naturally) termed 'japaning'. In japanning, as it came to be called, a coating of whiting and size was applied to the piece of furniture, then it was further coated with several applications of varnish and the pattern gilded. Relief designs were built up with a paste made from gum arabic and whiting.

Another form of furniture decoration, particularly fashionable in the years 1690–1730, was gesso (the Italian word for gypsum). A paste was made up from whiting and parchment size and applied in several coats until it was thick enough to incise, or even carve, a design upon the piece of furniture, which was then finally gilded all over. Gilded gesso work was especially suited to the baroque styles of the time.

Papier mâché was a material particularly beloved by furniture makers of the years 1830–60; the process is an ancient one, originating in the orient, its European use stemming, as its name suggests, from France, whence it was introduced into England in the 1670s. The material is a prepared paper pulp, pressed and baked to produce a very hard, durable substance. A fine quality papier mâché was produced in England in the eighteenth century from a method patented in 1772 by Henry Clay of Birmingham. In this process the sheets of paper were soaked in a mixture made up from resin, flour and glue, then were applied sheet by sheet to a moulded core, after which they were baked, sanded smooth and finally decorated and japanned. In the nineteenth century the Clay method was still in use but there was also a different process whereby the pulped paper was pressed between dies and then received many coats of varnish to produce a very hard material which could be turned, planed or filed. The article of furniture was then painted, gilded and, often, inlaid with mother-of-pearl and given a final coat of varnish.

Although machines to carry out many wood-working processes had been invented by 1800, the furniture trade was slow to adopt mechanical methods until the unprecedented increase in population triggered off a heavy demand for new furniture. By the 1870s the process of mechanization had accelerated and designs were being made specifically for such production means. Machines suited to furniture-making, powered first by steam and later by electricity, were gradually made available to saw, plane, bore, groove, carve, mould and mortise.

In the twentieth century all stages of manufacture of a piece based on the work of a talented furniture designer can now be carried out by machinery so providing good furniture for sale at reasonable prices. Mechanization has not completely replaced handcraft, but a hand-made piece will be very expensive. Much of the best furniture now available is the product of both methods. All modern technological means of reducing costs and speeding production are employed, but the use of good materials and careful hand-finishing ensures a quality article.

While machinery has modernized the processes of furniture-making, new materials, and different methods of handling existing ones, have been developed during the twentieth century, particularly since 1950. One example is metal. Japanned iron tubing had been widely employed during the nineteenth century for making bedsteads and rocking chairs but the introduction of tubular steel (later chromium-plated) led to its more extensive use, for instance for the cantilevered chair first introduced by Marcel Breuer in 1925, and its reappearance in a variety of guises since. For lightness steel is often replaced by aluminium.

A second instance, and one more generally employed in domestic furniture, is lamination, a method evolved by some Scandinavian designers in the 1930s. The new bonding resins developed for military purposes during the Second World War, together with the availability of modern techniques of moulding and bending plywoods by electrical means of heating without danger of damage and splitting, have encouraged the use of lamination in furniture design. These curved shapes are especially suited to seating furniture.

The production of a range of plastics since the 1950s has introduced a new, tough, inexpensive material to furniture-making. In particular, polypropylene moulded by injection methods has made possible all kinds of unbreakable, stain-proof furniture especially useful in kitchens, bathrooms and nurseries.

Wall mirrors and chimney glasses were very important features in interior decorative schemes of the principal rooms of a home especially during the seventeenth and eighteenth centuries. This was partly for aesthetic reasons but also, importantly, because of their reflective contribution to the inadequate level of artificial illumination of the time. Many mirror frames incorporated clusters of candle-holders.

After the introduction of the making of plate glass (see p. 197) in England in the 1620s, larger mirrors of higher quality were being manufactured. Before this mirror glasses had been small and of poor reflective quality. The term 'plate glass' seems to derive from the early glass of this type which was a luxury product reserved for coach windows and mirrors: these were called 'looking glass plates'. The method of giving reflective properties to the glass was a lengthy one of coating with tinfoil amalgamated with mercury. The experimental discoveries by Baron Justus von Liebig, the German chemist, in 1835 of a new chemical method of depositing metallic silver upon glass led to a greatly improved process of silvering; this incorporated a shellac coating and a backing of red lead.

HEATING AND LIGHTING

Ignition

It is almost impossible, even for the most imaginative person living at the present time, to appreciate the problems over hundreds of years of making a flame in order to light a fire to provide warmth and means of cooking and, most important, to make a light to push back the enveloping darkness. Yet these difficulties were the lot of everyone living in an age up to about 1850.

People at home tried to keep at least one fire, usually in the kitchen, perpetually burning. At night it was damped down, the ashes drawn over the embers, then covered with a curfew for safety's sake. In the morning the curfew was removed and fresh life was blown into the embers with the assistance of a pair of bellows. A curfew is a metal cover with a handle; the name is a corruption of the French word *couvre-feu*. During the Middle Ages a bell was rung in the streets at a prescribed hour each evening to warn people to cover their fires. The term has survived in the military sense.

From primitive times until the later eighteenth century there were two chief ways of creating a flame, by wood friction and by tinder box: in the damp climate of northern Europe the tinder box was the usual method. The metal box, in later times fitted with candle and holder, contained a piece of hard stone (the flint), one of metal, which was an iron and sulphur compound, and the tinder, which was of dry, flammable material such as charred linen, dried fungi, feathers or moss. The flint was used to strike the metal (it was later called a 'strike-a-light') to pare off a tiny fragment and so create heat. The fragment, with luck, fell upon the tinder, ignited it and, by blowing on it, a flame could be induced. It was not an easy process, especially on a dark winter morning; it needed experience and patience.

Tinder pistols were developed in the seventeenth century from the flint-lock pistol. They were expensive. Early designs contained tinder and, when the pistol was triggered, a spark ignited a charge of gunpowder the flame from which lit the tinder. By the eighteenth century many designs incorporated a flint and steel and after 1800 more sophisticated pistols also held sulphur matches, extra tinder and a candle in its holder.

For most people the tinder box continued to be the usual way to obtain a flame until matches became fairly cheap after 1850. For the well-to-do, however, a number of expensive, ingenious, and often dangerous, ways of obtaining a light were invented. From about 1775 to 1835 these were based upon the interaction of one chemical upon another. They were termed generally instantaneous lights. There was the development of the phosphorus match which was marketed in various forms. One was the phosphoric taper (1780, wax-tipped with phosphorus) which was carried on the person sealed in a little glass tube; when the tube was carefully broken it would ignite on contact with the air. The phosphorus box (1786) was first made in Paris as *le briquet phosphorique*. It contained sulphur matches, a bottle coated internally with phosphorus and a cork. Friction was required for ignition.

More dangerous still was the chlorate match (also known as an acid-dip match), the head of which was tipped with a mixture of chlorate of potash, sugar and gum arabic. For ignition the head had to be dipped in vitriol (sulphuric acid). The equipment for this type of ignition was enclosed in a small container called an instantaneous light box. Henry Bell's partly mechanized version of 1824 was little safer, especially when operated by an awakened sleeper on a dark morning.

A number of patents followed in the early nineteenth century. There were John Walker's friction lights of 1827, then Samuel Jones's Promethean match of 1828 and his lucifer of 1831. Then came the yellow phosphorus-tipped matches which ignited more readily, sometimes too readily, causing accidents. Yellow phosphorus also caused the terrible condition of 'phossy jaw' suffered by match factory workers when it entered their bodies via defective teeth.

Professor Anton von Schrotter's discovery of amorphous red phosphorus in 1845 led to the development of the safety match, credited to John Lundstrom of Sweden in 1855. In this he divided the chemical constituents between the match head and the striking surface on the box, so markedly reducing the chances of spontaneous combustion.

Lighting

A level of artificial illumination which enables people to continue their domestic activities as well during the hours of darkness as during those of daylight is a very recent boon, applicable only to the twentieth century. Before the 1830s the only artificial light came from candles and oil lamps and the level of illumination was exceedingly low. Constant attention had to be given to trimming wicks and the burning fat dripped and produced a smoky, smelly atmosphere.

The cheapest form of domestic illumination was the home-made rushlight produced by repeatedly dipping a dried, peeled rush into tallow (derived from animal fat) which had been melted in a greasepan.

Candles were also generally home-made from tallow poured into moulds into which the wicks had first been inserted. Beeswax candles burned brighter, needed less attention and smoked less, but were costly and subject to tax. With the expansion of the fishing industry in the late eighteenth century sperm whale oil began to be used to make candles and, when in the following century, paraf-

Figure 19.1: Colza oil-burning Argand reading lamp. The burner is fed by oil flowing under gravity from the reservoir above the level of the wick. The oil level in the reservoir is controlled by a float valve which closes the filling hole.

fin wax was produced from petroleum, a blend of the two materials gave a greatly improved flame. The later stearine candle owed much to the work on the chemical nature of fats by the French chemist Michel Eugène Chevreul; it was a firmer candle and gave a brighter light without the accompanying acrid odour.

The use of simple oil lamps made of earthenware or metal dates from very early times. These were containers for the fuel, which derived from various sources – olive oil, animal or vegetable oils, rape seed (colza), oil from kale, whale oil and paraffin – and for a floating wick; this needed frequent adjustment and trimming.

The first significant contribution towards improving the illumination of the oil lamp came in 1784 with the Argand design (Figure 19.1); up to this time the illumination level was only one candlepower per wick. Ami Argand, the Swiss chemist, devised a tubular wick which was a hollow cylinder enclosed in a glass chimney. The resulting current of air so improved combustion that the light from a single wick could be increased up to ten times. This lamp was then further improved by replacing the existing gravity feed fuel container by one with a mechanical pump. The Frenchman Carcel produced one such design in 1800 and this was followed by that of another Frenchman, Franchot's simpler Moderator lamp in 1836. Twentieth-century paraffin oil lamps give a much brighter light because the fuel is vaporized before it reaches the flame, so less energy is wasted.

Twenty-six miles of gas mains had been laid in London by 1816 and soon gas illumination of the streets and public buildings was acclaimed a great success. Domestic consumers were less satisfied. The mid-century standard batswing or fishtail burner flickered and smoked; it contained sulphur compounds which emitted an unpleasant smell and the fumes damaged furnishings and dirtied the interior decoration.

The refinement of the gas mantle revolutionized gas illumination. It gave a much better light and did away with much of the dirt and smell. Attempts to make a satisfactory mantle had begun by 1840 but until the invention of the Bunsen burner in 1855 it had not proved possible to produce a gas flame of high enough temperature. In the Bunsen burner air was mixed with the gas before ignition, so giving a hotter flame. The incandescent gas mantle was devised by the Austrian Carl Auer von Welsbach in 1886. In it he was making use of the principle that when the temperature of a substance is raised sufficiently it begins to glow and so emits more of its energy in the form of light. His mantle was made of knitted cotton impregnated with solutions of rare earth oxides (he found that the best mixture proved to be 99 per cent thorium oxide and one per cent cerium oxide). When the cotton mantle had burned away, the skeleton of the material retained its form. The first Welsbach mantle was manufactured in 1887 and improved versions followed including the inverted mantle of 1903.

The arc lamp, its principle based upon the voltaic pile, was the earliest form of electrical illumination. This could not be developed until the invention of Gramme's ring dynamo (1871) (see p. 360) but soon afterwards arc lighting was being installed for use in the streets, factories and railway stations. But the arc lamp was totally unsuitable for domestic illumination; the light was too dazzling, it was too costly, it was difficult to adjust (see p. 362ff.).

Electric lighting in the home had to await the incandescent filament lamp. The concept had been understood from before W. E. Staite's first demonstration in Sunderland in 1847, but there were three difficult problems which had to be solved before a satisfactory lamp could be manufactured. The principle was the same as for the incandescent gas mantle, but for the electric counterpart it was very difficult to produce a filament sufficiently durable to survive the high temperatures; the technology available made it almost impossible to evacuate sufficient air from the lamp bulb; and it was even more of a problem to provide a satisfactory seal for the wires carrying the electric current where they passed into the bulb.

A successful carbon filament lamp was produced almost simultaneously in the USA and in England, by Thomas Alva Edison in America (1879) and Joseph Wilson Swan in England (1878 or 9) (see p. 365ff.). Both were manufactured, at first in competition; later the firms were merged to become Edison and Swan United Electrical Company.

The carbon filament lamp dominated the market until about 1910. Meanwhile, better metal filaments were being devised: osmium (1902), tantalum and then tungsten (1906). The tungsten filament with its high melting point of 3410°C was very durable, but the high temperatures blackened the inside of the bulb. The research carried out at the General Electric laboratories in the USA under the American chemist Irving Langmuir led, in 1913, to the development of the argon-filled incandescent lamp with coiled-coil filament. Today's electric lamp gives about four times as much light as a carbon filament lamp for the same consumption of electricity. Later improvements include interior frosting of the bulb and the introduction of a fuse to avoid explosion of the lamp if a filament burns out.

The fluorescent lamp, introduced in the 1930s, was developed from the mercury vapour discharge lamp of the type in use for street lighting. It has become popular in the home in more modern times especially for bathroom and kitchen use. The lamps are coated on the inside with several different phosphorescent compounds (phosphors). The ultra-violet light which is produced by the discharge of electricity through the mercury vapour in the tube is absorbed by the phosphors which emit light. The colour of such lamps can be altered by using different types of phosphors. Fluorescent lamps use little electrical power and the tubes last much longer than the equivalent filament lamps.

Heating

Until the beginning of the seventeenth century most rooms were heated by wood burning on the open hearth. The hearth surface was of brick or stone and the logs were supported on this by andirons. Alternative fuels were peat, turves or heather. Supplementary heating was provided by charcoal burned in portable metal braziers.

Soon after 1600, depletion of Britain's forest stock, used in profligate quantity without adequate replanting for hundreds of years for shipbuilding, charcoal-burning, glass-making and general building, became so serious that various forms of legislation were passed by Parliament restricting from this time onwards indiscriminate felling of certain woods. Gradually coal began to replace timber as the normal domestic fuel, though this was a slow process and many households, particularly in rural areas, continued to burn permitted and dead woods until the early nineteenth century.

This changeover to coal as a domestic heating fuel marked the end of the traditional open hearth design. It is difficult to kindle and burn coal on a flat hearth, as a draught is needed under the fire; also andirons, which satisfactorily support logs, are not suited to containing lumps of coal. A metal basket was devised to hold the fuel which had bars in front and a slatted base below; the basket was attached at the sides to andirons similar to those previously in use and this raised the basket on legs above the floor of the hearth. Such grates, in general use by the later seventeenth century, were called dog grates. The name derives from the alternative term for andiron which was firedog, referring to the shape of the leg which resembled the hind leg of a dog. The next stage was the basket grate in which the fireback was also incorporated into the design. The beautiful eighteenth-century grates of polished steel and brass, such as those which Robert Adam designed, were of this type.

As time passed less wasteful though much less beautiful fireplaces were produced. The chimney draught was more carefully controlled by reduction of the fire-basket area, as in the hob grate, for example. A smoke canopy was fitted, also a chimney register; this was a metal plate which could be slid across the flue opening to adjust the air intake. With the contraction of the fireplace opening, the chimney flues became narrower also, leading to the iniquitous practice of employing small boys to climb inside the flues to sweep them as these were no longer wide enough for men to enter: this practice only ended in 1875.

The modern open fire burns smokeless fuel; convection heating and provision of hot water may be incorporated. Gas or electric lighting devices are usually fitted. The free-standing solid fuel stove is an alternative. Nineteenth-century versions of these were ornate designs in cast iron; their modern counterparts are usually finished with coloured glazed enamel.

Paraffin heating stoves became available in the 1860s. Their use was chiefly in rural areas where gas (and later electricity) were very late in coming. Nineteenth-century examples, made of ornamental cast iron, were primitive and odorous; they were markedly inefficient. The Valor Oil Stove of the 1920s was a great improvement; in this type a brass vessel contained the paraffin. The circular wick in its holder stood upon this removable container.

After the invention of the Bunsen burner many attempts were made to design a gas fire on the radiant principle in which a gas flame would heat a material to incandescence. The difficulty was to find such a material: fire-bricks, woven wire, pumice balls were all experimented with. Success was finally achieved with asbestos tufts fixed into firebrick and gas fires began to be sold in considerable numbers from 1882. Fire-clay radiants were then introduced, the familiar columnar ones in 1905. In the early twentieth century the idea of convector gas radiators was revived; the Embassy design of 1920 was a notable example. The present-day version of such heaters became available from the 1950s.

The use of electricity for heating lagged behind its adoption for lighting. In Britain the earliest heaters put on sale were the 1894 designs of Crompton &

Co. They consisted of cast-iron radiant panels in which the heating wires were embedded in enamel. They were not very successful because of breakages due to the differing thermal expansions of the iron and the enamel. A different type of heater was then devised by Dowsing of Crompton, Manchester, in which he used two or four large sausage-shaped carbon filament bulbs. Despite a reflector backing, these gave out little heat.

It was A. L. Marsh's work on the resistance wire made from nickel-chrome alloy (1906) and, later, the element devised by C. R. Belling in which the wire was coiled round fire-clay formers, which revolutionized electric fire design. Despite this, it was some time (after 1930) before electric heating competed with gas and solid fuel in the home, partly because many households were not yet wired for electricity and partly due to its higher cost.

A scheme to exploit off-peak electricity, in heaters containing concrete blocks which would retain heat when switched off in the intervening hours, was begun in the 1930s; its success was then minimal. In the 1960s the more efficient storage heaters, which then contained bricks and a fan to maintain a flow of air through the heater, brought a better response but it is the modern, much improved slim-line heater which has been most successful.

It is well known that the ancient Romans employed a method of central heating by hypocaust; all over Europe remains of such systems have been excavated, showing their use for all types of buildings including homes. This is a heating method whereby hot air from a basement furnace is passed under the floor and through wall flues to heat all the rooms of a dwelling. The hypocaust was the underfloor chamber, the term deriving from two Greek words meaning 'the place heated from below'.

With the collapse of the Roman Empire in the west the hypocaust idea in this part of Europe fell into disuse and it was many centuries before central heating systems were once again devised. Several methods were developed in the eighteenth and nineteenth centuries to use steam for this purpose in factories and other large buildings. Domestic central heating in Britain is very much a feature of post-1945 building, although in Europe and North America, because of much colder winter weather, schemes were given more urgent priority.

WASHING, BATHING AND TOILET FACILITIES

The marked reluctance to wash or bathe the person more often than was absolutely necessary which pertained in Europe in the centuries before about 1850 stemmed chiefly from the problems of heating sufficient water and its disposal and the unreliability of the domestic water supply. Similarly, the inadequacy of sewage disposal facilities made the evacuation of the waste products of the human body, at best, an uncomfortable, unhygienic and malodorous proceeding and, at worst, a dangerous hazard to health and even life.

Under the administration of the Roman Empire there had been adequate supplies of fresh water for washing and for sanitation purposes and heating of water by the hypocaust method. The majority of town dwellers in European countries ruled by Imperial Rome lived in flats (*cenacula*) in apartment blocks (*insulae*). These flats were mostly overcrowded and lacked amenities, so it was customary for people to use the public baths (*thermae*, from the Greek *thermos* = hot). These great bathing establishments provided free, or at least very cheap, facilities for the population to attend daily to bathe, relax, chat, carry out business, receive massage and medical treatment, eat and drink, take part in sports or be entertained as they so desired. More well-to-do citizens living in a house (*domus*) would have water laid on in their homes, a means of heating and disposal of it and adequate bathing facilities. This would apply even more to inhabitants of the country house (*villa*).

The Romans attached great importance to an adequate water supply. All over Europe they built great aqueducts to bring the water from the mountain and hill regions to the cities on the plains. In the capital city of Rome, for example, eleven great aqueducts carried the water for miles across the Campagna to supply the daily consumption of 1615 million litres (355 million UK gallons, 427 million US gallons). Vitruvius, the first century BC architect and engineer, gives the desirable rate of fall in building an aqueduct as 150mm per 30.4m (6in per 100ft) and large detours were made to avoid sudden descents.

To the Romans, using the latrine was a communal, even social, activity and communities were provided with public (or private in larger houses) sanitary latrines, well equipped with running water and washing facilities. This can be seen in many parts of the Empire as at, for example, Ostia Antica, the port of Rome, where public toilets in the forum baths survive with twenty marble seats, washing facilities and fittings for revolving doors. Remains of rows of toilet seats also survive at the military fort on Hadrian's Wall at Housesteads, beneath which is a channel for perpetually running water.

Washing and bathing

After the collapse of the western part of the Roman Empire, the medieval monasteries maintained a fair standard of bodily cleanliness. Cold water was provided in basins or troughs for daily washing and warmed for less frequent bathing; this had to be carried to wooden bathtubs in ewers.

From about 1530 until the mid-nineteenth century personal cleanliness lapsed to a very low level, mainly due to a lack of domestic facilities. People washed in the kitchen or in their bedrooms. Whatever the class of household, water had to be carried into the house from an outside pump and heated in a cauldron over the kitchen fire. Wealthier people, who employed servants to carry out these tasks, bathed in the privacy of their bedrooms in wooden or metal tubs. The poorer section of the population bathed in front of the kitchen fire.

The water supply was intermittent until about the mid-eighteenth century. The introduction of the ball valve at this time helped, as it meant that the main house tap did not need to be constantly turned on and off. At the same time fashionable furniture designers were making purpose-built pieces for bedroom use such as washbasin stands and toilet tables which were fitted with shaving mirrors and spaces for toilet articles.

From about 1850 onwards technical advances began to give much improved facilities. Running water was supplied, at first to the ground floor, only later to bedrooms. Washbasins with taps were fitted to the walls. With better water supplies many different designs of bath appeared on the market. Chief of these were the full-length lounge bath, the sitz-bath, where the bather washed the bottom half of the body retaining (for warmth) his clothes on the top half, the hip bath for washing legs and feet only and the slipper or boot bath. This metal bath was very sensible. It was shaped like a boot and in it the water kept hot and only the bather's head and shoulders were exposed; the bath could be topped up with hot water as needed. There were also shower baths fitted with curtains and a tank of water above which was operated by a hand-pump.

Heated baths, using charcoal or gas for fuel, were expensive but became popular in well-to-do households where baths were fitted with taps connected to a cold water supply. The usual charcoal heater was a vessel containing the fuel which was put into the bottom of the bath and then ignited. Two breathing tubes were attached to the vessel to enable the air to circulate and the fuel to burn well. When the heater was alight the bath was filled with cold water. The fuel continued to burn, since the tops of the tubes remained above water level.

There were two chief types of gas heater. One was portable; it was connected to a gas lighting bracket and, like the charcoal heater, was placed in the cold water. The other, a more popular design, was fitted to the exterior of one end of the bath. A gas burner was swung out to be lighted and then swung back so that the flames played directly on to the underside of the bath (for example, see Figure 19.2). Such devices were advertised to heat a bathful of water in six minutes. They also carried a warning to turn the gas off before entering the bath.

From the 1880s bathrooms were designed in new houses; these contained a bath, often with shower fitting, a bidet and washbasin. Soon after 1900 the more expensive porcelain-enamel finished bath began to replace the painted cast iron one. In more modern times this has given way to the plastic bath.

Apart from the few heated baths hot water for bathing continued to be a luxury until the satisfactory development of the gas geyser. Until the late nineteenth century water was still heated over the kitchen fire and carried upstairs in jugs. A gas geyser had been invented in 1868 by Benjamin Waddy Maughan,

PART FIVE: TECHNOLOGY AND SOCIETY

Figure 19.2: Gas heated bath by G. Shrewsbury, 1871.
Note swinging bunsen burner at one end, no flue fitted. Advertisements for this type of bath read 'By the use of the above, a hot bath can be obtained in six minutes for less than 2d. Every family should provide itself with this invaluable requisite, its limited cost placing it within the means of all and its simplicity within the management of a child.'

taking the name from the Icelandic word *geysir* = a gusher or hot spring. But for a long time gas geysers were hazardous to operate. There were no pilot jets, the water supply was unreliable and bathrooms were ill ventilated. Even well into the twentieth century, unless the user carefully followed the lighting instructions, an explosion could result. It was in the inter-war years that safety devices were fully incorporated into geyser design, after which the gas geyser became a common item of household equipment.

Electric instantaneous water heaters were marketed from the 1920s and by 1930 immersion heaters were being installed in lagged hot water tanks but for some time these were much less numerous than gas geysers as electric heating was more costly.

Sanitation and toilet facilities

During the Middle Ages monastic houses had their necessarium built near or over a water supply; these facilities, like the Roman ones, were communal. Public latrines were provided in towns; London Bridge had a large latrine which served over 100 houses. In castles and manor houses the garderobe was built into the thickness of the great walls and drained into the moat. The exits to these conveniences can still be seen in the exterior walls of the White Tower at the Tower of London.

Between about 1500 and 1750 sanitary arrangements deteriorated and the close stool and chamber pot largely replaced the garderobe. The close stool, used in well-to-do homes where it would be emptied and cleaned by servants, was a bowl inside a padded, lidded box. Elaborate examples had arms so they were like chairs to sit upon. Pots were emptied into the street. A large house might be equipped with a 'house-of-easement', sited in a courtyard or basement and accommodating two or more people.

In the eighteenth century sewage disposal was still primitive and insanitary. The usual method was a privy or closet outside the back of a house; in poorer quarters one privy served many houses. These privies had wooden seats and were built over small pits. They had to be emptied by buckets, which were then carried through the house to larger communal cesspools and these were emptied regularly by night-soil men. In many cases the pits drained through into the drinking water supply, with serious results. Indoors, chamber pots were used. These were often concealed in pieces of furniture, for instance, in the dining-room sideboard for the relief of gentlemen after the withdrawal of the ladies after a meal and, in the bedroom, in a night table or night-commode.

The invention which finally abolished the malodorous germ-ridden forms of privy was that of the water closet. The original inventor of this was Sir John Harington, a godson of Queen Elizabeth I, who constructed a valve closet in 1596 at Kelston near Bath. It had a pan with a seat and a cistern above. Harington described his invention in his *Metamorphosis of Ajax: A Cloacinean Satire*. ('Jakes' was current slang for a privy.)

Unfortunately, owing to the lack of good and reliable water supplies and drainage systems, 179 years were to pass before the first WC was patented by a London watchmaker, Alexander Cumming. This also was a valve closet with overhead cistern; when a handle was pulled up, the valve opened to empty the contents of the pan into a waste pipe and water entered to flush the pan. This was the principle common to subsequent WCs, but its sliding valve was inefficient and remained unsatisfactory until, three years later in 1778, Joseph Bramah improved the valve and so produced a WC which was the best standard pattern for a hundred years.

Despite the great improvement which Bramah's closet represented, only comparatively few houses were fitted with valve closets and more common, at least until the 1890s, were the cheaper pan closet or long hopper closet, both of which were inadequately flushed and so became very soiled. Over the years foul smells and gases permeated the house from such noisome appliances.

Even more hazardous to health were the sewage disposal methods before the 1860s and, for many years, the introduction of the WC made this situation worse. Closets were tucked into the corners of rooms, even into cupboards with almost no ventilation. The waste pipe emptied into a cesspool as before, but the connecting pipe provided an easy entry into the house for foul gases. Safety measures came only slowly. In 1792 a stink trap was invented to keep out undesirable odours and, by the 1840s, cesspools were made illegal and the WC had to be emptied into the sewers. There were problems here also, as town sewerage systems were less than satisfactory until later in the century.

A widely used alternative in the nineteenth century was the mechanical earth closet invented in 1860 by the Revd Henry Moule, vicar of Fordington in Dorset. His contrivance was a wooden box with seat, below which was a bucket. Above, at the back, was a hopper full of dried earth or ashes. When a handle was pulled, these would be released to cover the contents of the bucket. Such closets were still on sale in 1910, price £1.10s. Portable versions were used in bedrooms.

In the 1870s new designs of WCs were invented which gradually replaced the 100-year old Bramah type. Twyford produced his Washout Closet in which water was always present in the pan, but a strong force of water was needed to flush it. There was also a syphonic closet with two flushes on the market, but it was the washdown system which finally triumphed and is still in use.

CLEANING IN THE HOME

Before twentieth-century technology brought labour-saving appliances into the home, keeping the surfaces and utensils in a house clean and bright was extremely hard work: it was monotonous, unremitting labour. The surfaces themselves – plain and polished wood, tiles, bricks, stone and metal – were kept clean by sweeping, dusting and scrubbing with soap and water. All the equipment to do this work was made from natural, easily obtained materials. Brushes and brooms were made from twigs, bristles or hair, mops of wool, cotton or linen, dusters of soft cloth or feathers; wooden handles were attached as needed.

The cleaning of textiles and carpets, the polishes for furniture and metals, the removal of stains, were achieved, somewhat imperfectly by today's standards, by the use of a wide variety of substances and chemicals which, it had been found empirically, worked to a limited extent. For example, carpets were sprinkled with tea leaves to lay the dust and lend fragrance to the room; they were then swept and/or beaten. Furniture polishes were made from a number of different recipes but were mainly based upon linseed oil, vinegar, turpentine or spirits of wine. Pastes were made from beeswax, curd soap and turpentine. Black lead was used to polish the cast ironwork of fireplaces and fire-irons. Emery paper was rubbed on the blackened fire bars. Household silver was cleaned by the application of hartshorn powder, a substance which was obtained from shavings of a hart's antlers mixed into a thick paste with spirits of wine or water. The paste was applied to the silver, allowed to dry, then polished off with a brush and leather. The application of the paste was a weekly ritual. The silver was also rubbed up daily with plate-rags. These soft cotton rags had been impregnated by being boiled in a solution of new milk and hartshorn powder: the forerunner of a silver-cloth.

Until the mid-nineteenth century almost all the cleaning equipment and agents were made at home. Even soap was not generally purchased until William Hesketh Lever's attractive manufactured soap became available later in the century. Home-made soap came from animal fats rendered down in the kitchen then boiled for hours with home-produced lye which had been obtained by pouring water through wood ashes contained in a lye dropper. The resulting soap was very unattractive. It was greasy, the consistency of putty and smelled unpleasant until flower perfumes had been added.

Mechanical cleaning devices

Knife-cleaners

The last quarter of the nineteenth century saw the introduction of a number of rather primitive and usually unwieldy mechanical aids to cleaning. The knife-cleaning machine was an important household asset. Before the manufacture of stainless steel domestic knives in the 1920s, steel knives required careful regular cleaning to keep them bright and unmarked. The centuries-old method had been to polish them, after washing, on a knife board. This board was covered with leather and the knife blades, with the addition of emery powder, were stropped on this.

The Kent Knife Cleaner, patented 1870 and later, was typical of the rotary knife cleaning machines introduced in the later nineteenth century (Figure 19.3). It consisted of a circular wooden drum mounted on a cast iron stand. Inside the drum was an inner wheel fitted with leather leaves or alternating felt pads and bristle attachments. The washed knives were inserted (up to twelve at a time) into holes around the edge of the drum. Abrasive powder was poured into the machine, a crank handle was turned and the knives were cleaned and polished. Unfortunately the action of the machine was so drastic that the knife blades became progressively thinner and shorter. By 1890 alternative designs of

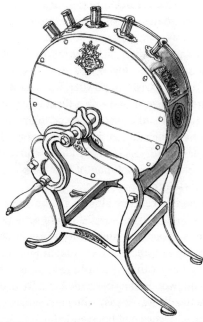

Figure 19.3: Kent knife cleaner, *c.* 1900.
Drawing by Doreen Yarwood.

machine appeared on the market. These were smaller, more convenient and less hard on the knives. Characteristic was the Uneck model which resembled a miniature metal mangle with rubber rollers. It could be clamped to a table and the knives were cleaned and polished by being passed through the rollers.

The sweeper

A machine to sweep floors was patented as early as the beginning of the nineteenth century and by 1865 a number of carpet sweepers were available. These were not very efficient as they consisted of cast-iron boxes running on two rollers and containing roller brushes rotated by a belt pulley attached to the back roller. This belt drive was not powerful enough to counteract the frictional resistance of the brush as it moved over the carpet.

Melville R. Bissell, the owner of a china shop in Grand Rapids, Michigan, was allergic to dust and suffered greatly when unpacking his china from its straw and sawdust wrappings. This encouraged him to design the first really satisfactory and effective carpet sweeper (patented 1876). Though less streamlined than a modern one, it worked on the same principles. The brush was rotated by means of the friction of four rubber-treaded carrying wheels against the drum to which it was fitted. Frictional carpet resistance was minimized by setting the brush spirally in tufts round the drum so that only a few bristles at a time could be in contact with the carpet. By 1880 the Bissell carpet sweeper was becoming a success.

In Britain the Ewbank sweeper (a similar design) was the best-known make. In 1911 it retailed for 10s 6d. One of the early models of the Bissell sweeper possessed 'sidewhiskers', that is, rotary brushes at the corners of the sweeper, so placed to clean the edges of skirting boards and alongside furniture. This feature has been revived in the modern Bissell sweeper.

The suction cleaner

During the second half of the nineteenth century there were many attempts to design a machine to blow away or suck up dirt; several such, not very efficient, types of apparatus were made, including a foot-operated model designed by Sir Hiram Maxim. It was, however, the construction engineer Hubert Cecil Booth who produced the first successful machine in 1901: this was a powered suction cleaner.

Booth's interest in the matter had been aroused the previous year when he was invited to attend a demonstration at St Pancras Station in London of a new machine to clean railway carriages. This blew the dirt from one end of the carriage to the other by means of compressed air. Booth did not believe this to be a good way of cleaning and thought sucking to be better than blowing. In a now well-known domestic experiment he placed a piece of damp fabric over the arm of his easy chair and sucked it. The resulting black ring on the cloth proved the point to his satisfaction.

He patented (and named) his vacuum suction cleaner in 1901 and formed the Vacuum Cleaner Co. to manufacture the machines (now Goblin BVC Ltd). But, although his machine worked very well and was in demand, it suffered the handicap of all machines being designed at that time to relieve drudgery in the home: lack of a small electric motor to power it. Large electric motors had been available since the early 1880s but it was not until the Yugoslav-born American inventor of the prototype AC motor, Nikola Tesla, designed the first small one-sixth hp (0.125kW) motor to drive a Westinghouse fan in 1889 that the means of using electricity to power household appliances became a possibility (see Chapter 6). Even so it was the twentieth century before motors were manufactured to service such appliances.

The first Booth cleaner was 1.25m (4ft) in length. Painted bright red and costing £350, it was powered by a 5hp (3.75kW) piston engine and was transported through the streets on a horse-drawn van. A team of white-coated operators accompanied it to clean, for a fee, carpets, curtains and upholstery: the process was too costly for all but a few. Among the select clientele was Westminster Abbey where, in 1902, the Booth cleaner removed an enormous

PART FIVE: TECHNOLOGY AND SOCIETY

Figure 19.4: Sir Hiram Maxim's 'Little Giant' dust extractor, 1909. Dust is trapped in the canister which contains sawdust as a filter. Suction obtained by working long handle. Canister stands on top of a diaphragm pump. The spring-loaded valve in the lid acts as a safety valve and measure of suction is obtained. Price in 1909 5 guineas.

quantity of dirt from the previously un-vacuumed carpet laid down for the coronation of Edward VII and Queen Alexandria.

Attempts were made on both sides of the Atlantic to produce a more convenient and cheaper cleaner but the stark choice remained: between a large, expensive, heavy, powered machine and a small, cheaper one without power and needing two persons to operate it. Booth's produced one of the former type, the Trolley-Vac, powered by an electric motor and drawn round the house on a trolley. This was still too expensive (35 guineas in 1906) and too heavy to take upstairs. It was plugged into the electric light sockets. There were several models of the second type, many of them American. Nearly all worked by means of one person powering a bellows by hand or foot and another operating the suction device on the end of a long handle. There was also a one-person-operated design which resembled an overgrown bicycle pump (Figure 19.4): it was very hard work and not very efficient.

The first person to succeed in marrying a small electric motor to a convenient suction cleaner was J. Murray Spangler of Ohio who devised in 1907 a prototype which worked. It consisted of a sweeper inside a tin canister and a dustbag supported on a long handle. Spangler lacked the capital to put his invention on the market, so he sold the rights to a firm of saddlers who developed it and manufactured in 1908 the first one-person-operated, electrically-powered domestic vacuum cleaner. The machine was an instant success and became so well-known that the name of the firm became synonymous with the machine's function: if Spangler had not sold out to W. H. Hoover, people might well talk about 'spanglering' the carpet instead of 'hoovering'.

By the 1920s the characteristic Hoover triple action had been incorporated into the design of the machine. This vibrated the carpet to loosen the dirt deep in the pile; it was an agitating action achieved by combining the suction with a revolving brush. In 1926 the action was improved by the addition of revolving bars which helped the brush action by beating the carpet on a cushion of air. This gave rise to the company's well-known slogan, 'It beats as it sweeps as it cleans'.

Hoover's success was followed by many other designs developed by different companies, notably the Magic of 1915, the British Vacuum Cleaner Company's version of 1921, and the Bustler of 1931. Electrolux then pioneered the horizontal canister type in 1917, followed quickly by BVC under the trade name of Goblin. Both upright and canister designs are still made. There are also turbo-powered cleaners and, since 1980, many machines are electronically controlled.

Carpet shampooers and floor polishers

The increased adoption after the Second World War of wall-to-wall carpeting and of modern floor coverings led to the introduction of the upright shampooer/polisher. First developed in 1958 under the presidency of Melville R. Bissell (grandson of the inventor of the Grand Rapids carpet sweeper and founder of the firm) this was quickly followed by the Hoover version. Both designs are easy and efficient to use. They are fitted with interchangeable heads, which enable the machine to shampoo, scrub or polish.

Dishwashers

As with so many labour-saving appliances in the nineteenth century, many of the new ideas came from the USA: American ingenuity was spurred on by an urgent need as there was not available, in this newer culture, the vast army of servants which European homes were able to employ. Patents were issued for dishwashers as early as the 1850s but before 1914, when the first dishwasher powered by an electric motor became available, the mechanism required almost as much labour to operate it as washing up by the traditional means. Water still had to be heated on the kitchen range then carried to the machine in jugs or buckets. Homemade soap then had to be cut up and put in to the water. To wash the dishes one turned a handle to operate paddles or a propeller. Finally the dirty water had to be emptied out manually.

After the introduction of the first electrically powered machine several years had still to pass before other processes were mechanized. Between 1920 and 1940 improvements were made. Machines were plumbed in and increasingly were automated. In Britain the public was slow to enthuse about the dishwasher. Even today, for the small family, it does not rate high on the list of home essentials.

LAUNDERING

Washing

In the years before about 1775–80 the cleaning of fabrics, whether in household or personal use, was a great problem. It was very hard work to heat enough water and to carry it to fill and empty the washing tubs and cauldrons. Soap was very expensive and was, therefore, mostly home-made. Also, many fabrics were unsuited to washing. Velvets, brocades, silks (often fur-lined) were shaken or beaten and marks imperfectly removed by various cleaning fluids, often derived from grapes. The most common cleaning agent was fuller's earth which might be combined with lye.

The traditional washing method, in use from Roman times, was to agitate the fabrics in hot water and soap in immense iron cauldrons; the water might be heated in the kitchen and carried out in ewers or a fire was lit under the cauldron. Particularly dirty fabrics were first buck-washed, that is, steeped in a solution of lye. They were put into wooden tubs which had holes in the base; lye water was poured over them and the material beaten or stamped upon with bare feet to loosen the dirt, with more lye water periodically added. After washing, the fabrics were rinsed in clean water and wrung out by hand. These methods were fairly satisfactory, though wearing, for linens and cottons but had a disastrous effect on woollens which quickly became felted.

Because of the hard work involved and the deleterious effect of the washing methods upon the garments, washdays were infrequent, generally monthly or less often in the sixteenth and seventeenth centuries but a little more often in the eighteenth. Washdays lasted two or more days. Large country houses had special laundry rooms containing water heating equipment and sinks for rinsing.

Better facilities and equipment to help in the washing process gradually became available from the late eighteenth century onwards. The copper was introduced, the word deriving from the fact that early examples were made from this metal; later ones were of cast iron and, by the mid-nineteenth century, had become wash boilers, that is, a fire burned beneath them to heat the water. When gas was piped into homes, gas-heated coppers replaced the solid fuel versions. A tap was fitted to empty the vessel but, until the end of the century, it was still filled manually. The galvanized, zinc-coated copper appeared in the twentieth century and, in the 1930s, enamelled coppers powered by electricity or gas were in use.

Other aids gradually introduced included the wooden dolly stick used to pound the washing up and down in the wash-tub and so force the soapy water through the clothes, the later posser with its conical metal head which had a similar function and, the most characteristic nineteenth-century aid, the wooden washboard; this was faced with zinc-covered corrugations against which the clothes were rubbed to remove the dirt.

During the nineteenth century many and varied attempts were made to design a washing machine. The first patent for a machine to 'wash, press out water and to press linen and wearing apparel' was taken out as early as 1780 by a Mr Rogerson from Warrington, Lancashire, but it is not clear whether the machine was ever made. Other ideas followed, but it was after 1850 before serious attempts were made to manufacture a working machine.

Between 1850 and the early years of the twentieth century, a number of different washing machines were manufactured but they were all powered by human energy. The filling and emptying had to be done by hand and, in most models, the water had to be heated first. Soap was chopped up manually and added to the hot water.

The design of these nineteenth-century machines varied considerably, but they were all based on a mechanical means of imitating the contemporary traditional method of washing, using a dolly stick and washboard. The mechanisms which were devised to achieve the agitation of the water ranged from a machine which could be rocked by the foot like a cradle to one which incorporated a dolly stick attached to the lid or base of the tub; in others the washing was churned round by paddles or pegged wooden rollers (Figure 19.5). Apart from the rocking model, the motive power was provided by a lever operated by hand. Many designs had corrugated interior walls and floors to simulate the washboard action. Most were made of wood and shaped like a tub or box. Metal was used for stands and fittings.

By the 1880s a few machines were being made which could heat the water in the tub; gas jets were used for this, alternatively a coal-fired boiler. The first electrically-powered machines were made in the USA soon after 1900. In early models the existing machine was adapted to the addition of an electric motor and this was often sited beneath the tub – a dangerous proceeding, as many tubs leaked and machines were rarely earthed.

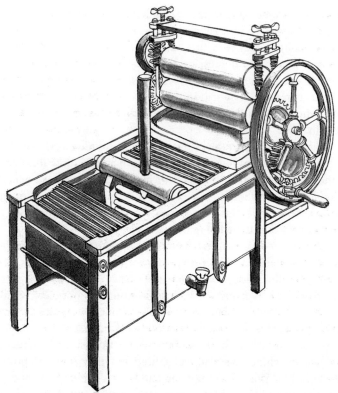

Figure 19.5: Wooden washing machine, late 19th century. Handle is moved backwards and forwards to agitate washing by means of a swinging wooden gate. Rubbing boards fitted on each side. Large mangle with wooden rollers turned by handle on wheel.
Drawing by Doreen Yarwood.

It was the late 1920s before the American washing machine was redesigned to take full advantage of electrical power and to meet the needs of a mass market. An all-metal tub replaced the wooden one and this was then enclosed in an enamelled cabinet. There were two chief methods of creating the washing action. One agitated the water by a revolving disc fitted with fins mounted in the base of the tub and operated by a driving mechanism beneath. The other was a perforated cylinder which was driven to rotate, first in one direction, then in the other.

In Britain washing machines sold in small numbers in the 1920s and 1930s; only after 1945 were they manufactured and sold in quantity. The automatic washing machine with programmed cycles of washing, rinsing and spin-drying became a widespread success in the 1960s. Since then models have steadily become more sophisticated, including a choice of programmes to suit different fabrics and biological control as well as economy. Many models now have microprocessor control.

Drying

For centuries the only method of drying laundry was to stretch out the hand-wrung garments over bushes or frames to dry in the sun or upon airers around the kitchen fire. In large houses of the later seventeenth century onwards part of the laundry building was devoted to drying and finishing. The drying compartment was heated by a solid fuel stove round which were immense drying racks of wood and metal which could be run on wheels towards or away from the heat. The racks could accommodate blankets, sheets and large curtains. The clothes were aired on a creel, or wooden airer, suspended by pulleys from the ceiling. Wooden tables which were used for starching, ironing, finishing and crimping, extended round the walls of the rest of the room.

Many attempts were made from about 1780 onwards to design equipment which would extract water from fabrics by mechanical means. The urgent need for this stemmed from the textile and dyeing industries and most nineteenth-century ideas concentrated on utilizing centrifugal force to do this. An early, successful machine was that built to the order of Angela Burdett Coutts in 1855 to be sent out to the Crimea to handle linen for hospital army casualties. This iron and wood machine measured 1.8m (6ft) square and was 2.1m (7ft) high, cost £150 and was shipped out in parts and reassembled on the spot. It comprised a washing copper and a drying closet. The latter functioned on the centrifugal principle; it could dry 1000 linen articles in less than 25 minutes.

With the introduction of the small electric motor, a domestic spin drier, on centrifugal principles, was developed in the USA in the 1920s. Improved models followed and, after 1945, washing machines were incorporating spin driers. Tumble driers were also developed.

Pressing, mangling, ironing

During the last two millennia a number of ideas have been put into practice to smooth household linens and personal garments after they have been washed. In Europe, until the early sixteenth century, fabrics were not sufficiently fine to benefit from the use of a heated iron, so the equipment was designed chiefly to press and smooth out creases from the material by unheated means. The wooden clothes press – called a *prelum* by the Romans – followed the same idea as a press for wine or olive oil. It could be made as part of a stand or could stand upon a table. Many designs incorporated drawers to hold the finished linen. The press consisted of two heavy, smooth wooden boards between which the folded linen was placed. Pressure was maintained upon the boards with the aid of a wooden turnscrew.

During the sixteenth century the mangling board and roller came into general use. The idea spread from Holland, Denmark and northern Germany; the word mangle derives from the Dutch and Middle High German *mangelen*, which itself stemmed from the ancient Greek word for an 'engine of war': indeed, the later box mangle resembled a mighty weapon. The material was wrapped round the roller (itself about 50cm (1ft 8in) long), which was placed upon a flat table. The mangling board (a flat piece of wood about 66cm (2ft 2in) long and 8cm (3in) wide, with a handle on top) and then passed backwards and forwards over the roller until the fabric was smoothed. This method produced quite a high standard of pressing and was in use until well into the nineteenth century. The idea was exported by Dutch colonists, particularly to North America and South Africa.

The final pressing stage in Europe came with the development of the mangle which took place during the eighteenth and nineteenth centuries. The process was an almost dry one. The heavy, cumbersome box mangle was an eighteenth-century version intended primarily for use in large houses or communal laundries where there was a great quantity of linen to be pressed. The mangle was a large wooden box, 2m (6ft 6in) long and 1m (3ft 3in) each wide and deep, filled with stones. The box was part of a wooden framework standing on the floor. The damp, clean linen was wrapped round wooden rollers which were then placed under the box by means of a lifting device and a crank handle was then turned to make the heavy box trundle backwards and forwards over the rollers. A flywheel helped to overcome the inertia of starting the box in motion and there was a system of stops to prevent the box running off its bed on to the floor.

After about 1850 the box mangle was slowly replaced by a standing design, smaller and easier to operate, with large wooden rollers turned by a handle. By the end of the century the mangle had evolved into a handier design with smaller, rubber rollers; it was then intended to expel water before pressing.

The heated iron, to smooth damp fabric, was introduced in Europe in the late sixteenth century, mainly by the Dutch who were the leaders in making decorative brass irons; cheaper everyday models were soon being produced everywhere by the local blacksmith in iron. The ancestor of such heated smoothing irons was the oriental pan iron, in use in the Far East since early medieval times and made to the traditional pattern ever since. This design of brass or bronze resembled a small saucepan and had a wooden handle: the pan, containing burning charcoal or coal, was rubbed over the fabric.

Until the emergence of the self-heating iron in the mid-nineteenth century there were two chief types of heated iron: the box iron and the sad iron. The box iron (also known as a slug iron and with a variant in the charcoal iron) was hollow and designed to contain a heated cast-iron slug or burning charcoal.

Figure 19.6: 'Cannon' charcoal iron with bellows.

The slug iron had a gate at the rear to be lifted up for the insertion of a red-hot slug, the charcoal iron a top plate which was removed to put in the fuel. Charcoal irons also had top chimneys for the escape of smoke and holes inserted along the sides to aid combustion (Figure 19.6). The sad iron, or flat iron, was made of solid cast iron. Several such irons were in operation at once, two or three heating up on the fire while one was in use. The word 'sad' derives from its medieval meaning of 'solid' or 'heavy'.

Using any of these irons was skilled work. The correct heat for the fabric had to be gauged – usually by spittle or by holding the iron near the cheek. A charcoal iron was liable to scatter smuts on the fabric from its chimney. It needed skill, too, to keep several sad irons at the correct heat so that time was not wasted; a cast-iron heated laundry stove, where irons could be kept hot, was used for this purpose. Surprisingly the detachable wooden handle was not introduced for use with sad irons until the nineteenth century and iron handles became very hot.

Many varieties of special purpose irons were designed to smooth different parts of a garment. There were mushroom, egg and ball irons (named according to shape), long-handled sleeve irons and bustle irons and many different irons to crimp, flute and goffer the material. The tally iron was introduced from Italy in the sixteenth century to dry and form starched ruffs: the word is an English corruption of Italy. Also sometimes termed goffering iron, this consists of a stand fitted with cigar-shaped metal barrels. Iron poking or setting sticks were heated and inserted into the barrels, then the damp, starched fabric of the ruff was drawn over the barrel until stiff and dry. Tally irons continued in use for centuries after ruffs had gone out of fashion, for finishing the ruffled edges of garments. More elaborate crimping devices followed, such as the ridged fluting iron and, in the nineteenth century, the crimping machine which resembled a miniature mangle with ridged brass rollers.

The later nineteenth century saw the rapid development of many forms of self-heating iron, some of which were efficient, others often downright dangerous. Gas-heated irons, from the 1850s, were the first to be satisfactorily operated. Then came the use of other fuels: oil, paraffin, naphtha, methylated spirits, carbide, acetylene and petrol.

The first electric iron appeared in the USA in the early 1880s, but this was not connected to the electric supply and had to be heated frequently on a special stand. A French design based on the arc lamp principle followed. In Britain, Crompton's were producing electric irons from 1891 which, like later kettles and vacuum cleaners, had to be plugged into electric light sockets. The electric iron, however made little progress until the late 1920s because comparatively few households were wired for electric current. Electric irons of the 1920s and 1930s were heavy, weighing about 4kg (9lb) because, owing to fluctuation and unreliability of the power supply, the iron needed to be able to store heat for the times when it had to cut out. Thermostatic control came in the late 1930s; before this the temperature of the iron still had to be judged in the same way that it had been since the sixteenth century. The steam-or-dry iron appeared in the 1950s. Modern versions are now much more controllable and streamlined.

THE KITCHEN AND COOKING

The kitchen

Over the centuries the kitchen has been the heart of a home, whether cottage or castle. This was especially so before the twentieth century when the kitchen was working centre of the home, a place always warm (for the kitchen fire was permanently kept alight), and one where food, hot water and a means of drying clothes, made wet by working all day in the open in all weathers, were always available. In building structure, in architectural and decorative style, in up-to-date materials and finishes the kitchen has, until modern times, lagged behind the rest of the house; it rarely aroused the interest of architects and, since its workers were women and mainly servants, their comfort and interests rated very low. It was only after the First World War, when a new world of work had been opened up to women so that they declined to return to domestic service, that well-to-do and middle-class housewives demanded a more attractive, easily cleaned and run kitchen changing the pattern permanently.

In the Roman kitchen the layout and facilities were simple but adequate. During the earlier years of the Republic the kitchen was situated in the atrium, but in larger homes under the Empire it became a separate room at the back of the house. Slaves did most of the work – stoking the fires, obtaining the fuel, cooking, cleaning and looking after the stores; the lady of the house supervised and planned their work and made up the menus. This was important, for the Roman cuisine was sophisticated and varied. A rainwater cistern supplied piped water for preparing food and washing up at a mortar-covered stone sink; a drain carried the water away. Wood or stone tables were provided, shelves and racks for utensils lined the walls and great storage jars stood upon the stone or brick floor.

During the Middle Ages the culinary standard fell sharply. A millennium had to pass after the departure of the Romans before even the ruling classes could once more enjoy a varied diet and adequate skill in the cooking and presentation of food. Apart from rural cottages and small town houses, cooking for many was communal, particularly in the great monasteries and castles. For fear of fire such kitchens were generally housed in buildings separate from the main structures. They were large with immense fireplaces and ovens set in the walls which, like the floor, were of stone. In the roof of these one-storeyed buildings were louvres from which steam and smoke could escape.

Very slowly, after about 1500, kitchens became better equipped and furnished and were more comfortable. By the seventeenth century arrangements for feeding, even in larger houses, were less communal and kitchens were smaller, less lofty and draughty. With the gradual changeover in building materials from timber to the more permanent brick or stone, the danger of fire was lessened and the kitchen had become part of the main house. Country houses, particularly the larger ones, were self-sufficient in food and general welfare so, apart from the kitchen itself, there were many adjacent rooms for different purposes: as a minimum there was a bakehouse, scullery, pantry, buttery, walk-in larder, dairy, brewery and laundry.

By 1700 storage space in the kitchen had become more generous. Cupboards and shelves lined the walls and the dresser had made its appearance. This had begun life as a flat board or table fixed to the wall upon which the food was prepared (dressed) to make it ready for cooking. Gradually it acquired shelves above and cupboards below for greater storage availability. There were also settles, stools and chairs as well as preparation tables. Larger sash windows gave more light to such kitchens and a trough sink was fitted with water pipes connected to a pump served by the well outside in the yard.

With nineteenth-century industrialization came taps and running water, better artificial illumination from oil lamps and, later, gas burners and, towards the end of the century, the introduction of linoleum for more easy-care floor coverings (see p. 905). Yet the concept of labour-saving in either equipment or materials was of low priority. There was a more than adequate supply of cheap domestic labour to carry out daily scrubbing, steel burnishing and cast iron black-leading. In fact, the nineteenth-century kitchen was often less comfortable and attractive than its eighteenth-century counterpart. It was generally darker because, in towns, houses were built high and narrow to economize on costly land space, and the kitchen was relegated to the basement. This was also considered advantageous as the servants could not then waste their time looking out of the window at passers-by. The nineteenth-century kitchen also became dirtier and more difficult to clean due to the coal-burning kitchener and the lighting by open gas burner.

Only after 1918 did comfort and convenience genuinely begin to be considered. Slowly gas lighting gave place to the cleaner electricity. Hot and cold piped water were laid on. Closed cupboards replaced open shelves and, in 1925, Hygena Cabinets Ltd introduced the fitted kitchen cabinet, an idea which stemmed from America. Gradually the ideas of labour-saving devices and easy-care surfaces were put into practice and the concept, long current in America, of applying time-and-motion studies to save time and energy in kitchen work, were adopted. By the 1950s the 'unit kitchen' was being built into new houses and the kitchen had become a bright and colourful place. Thirty years later the development of plastics (see p. 906) and of microelectronics have revolutionized kitchen design as well as all its functions.

The preparation and storage of food

The form and appearance of utensils used in preparing food barely changed for 2000 years. The needs of the housewife were constant: she required containers to hold powders, solids and liquids, also equipment to grate, sieve, pound, grind, mash, squeeze, press, mix and beat. Because her requirements did not alter and the needs of all human beings are similar, equipment to carry out these processes bore a close resemblance wherever it was to be found in the civilized world. In Western Europe a competent housewife would certainly find herself familiar with the utensils of any kitchen at any time between the days of Ancient Rome and the early nineteenth century.

The natural materials readily to hand were used over the centuries to make these utensils: wood, horn, earthenware and metals such as iron, brass, copper and pewter. By the later eighteenth century development in materials had made kitchen work easier and pleasanter. Advances in the ceramics industry, under Wedgwood and Spode for example, brought inexpensive and attractive earthenware and china into everyone's home whatever their income and, later, the production of aluminium and then stainless steel made equipment which was easier to use and keep clean.

By the eighteenth century also, design of equipment had become more varied and complex but it was the mid-nineteenth century before the manufacture and advertisement of labour-saving gadgets began to get under way. Many of these ideas came from the USA, where there was a perennial insufficiency of servant labour to carry out the boring tasks of preparing food in the kitchen with the use of knife or chopper. Hand-operated mechanical devices were marketed which each claimed to reduce magically such chores as peeling and coring apples, stoning raisins or cherries, slicing bread, cleaning knives or making sausages.

The breakthrough, as with all other labour-saving devices in the home, came with the introduction of the small electric motor. This only became available for the domestic market in suitable form and at reasonable cost soon after the First World War when such motors were adapted into the design of kitchen equipment such as blenders, mixers, beaters and choppers. Although sophisticated electrical equipment of this type was in use in America by the 1920s, electric food mixers, for example, did not come into common use in Britain until after 1945. The Kenwood design appeared in 1947 as a free-standing table model which developed over some years to become, in the words of the manufacturers, a 'complete food preparation appliance'. In later models the Kenwood Chef, with its impressive array of attachments, could extract juice, shred, slice, beat, liquidize, make cream and grind coffee. The introduction of the electronic (thyristor) device made electrical operation more economical. More recently the food processor has taken over all these chores in a simpler, less cumbersome design and using a less costly range of attachments.

Various ways of preserving fresh food – pickling, drying, salting and smoking – were used over the centuries (see Chapter 16). In addition, in the nineteenth century, canned food could be bought in shops. Paradoxically, though, the tin opener only became available some time after the tins of food could be purchased. Early cans were made by hand and the lids soldered on. Helpful instructions were printed on them suggesting that the best method of opening was by hammer and chisel. The domestic can opener was introduced in the USA in the 1860s and was supplied with cans of bully beef. Not surprisingly the opener – of painted cast iron – was in the form of a bull's head with its tail shaping the handle.

The sixth method of preserving food, by chilling, had been practised in a limited way in larger homes over generations by using ice formed naturally. This was then cut up into blocks in winter and stored underground in ice-houses. In the early nineteenth century in wealthy countries with hot summer climates, such as the USA, the ice box was used in homes. It was a wooden chest or storage cupboard lined with zinc: there was an insulating material – asbestos, cork, wool, felt, charcoal, ash, for example – packed between the wood and zinc layers. Ice, imported from northern countries, was delivered daily by the ice man. Ice boxes came much later to England and were still being sold in Harrods in 1929, price £6 19s 6d.

The household mechanical refrigerator, which would actually lower the interior air temperature, did not appear until just before the First World War, some 35 years after its introduction in long-distance, sea-going vessels carrying meat or fish. The Domelre model was on sale in Chicago in 1913 and the Kelvinator came out the following year in Detroit. Steam power was used at first, later an electric motor. Again, with a cooler summer climate and a less wealthy population, Britain lagged behind the USA. The first model sold in Britain was a French compressor design in 1921. By 1923 the Frigidaire Corporation of the USA was manufacturing in Britain and soon the successful Electrolux water-cooled absorption refrigerator followed. The A. B. Lux Company of Sweden had been founded in 1901 and the 'Electro' was only added in 1919. The company purchased in 1922 the results of research carried out by two Swedish students at Stockholm University and, after four years' development of the appliance, brought out their refrigerator. It had a capacity of about one-third of a cubic metre (11.75ft^3) and in Britain cost £48 10s.

This, of course, was only the beginning. After 1945 the refrigerator became increasingly a household necessity. It was soon more efficient, and offered better insulation, automatic de-frosting and door-storage capacity. The home freezer, developed as a miniature version of the bulk food container freezers in commercial use, followed, either in chest form or as part of a refrigerator. Since 1960 it has been regarded as essential in many homes.

Cooking on the open hearth

The Roman hearth was a raised one; the preferred fuel was charcoal which burned in circular holes set into the surface of the hearth. The cooking vessels of earthenware or metal (generally bronze), were supported upon iron tripods over these charcoal fires or stood on an iron gridiron. A hooded wall flue carried the toxic fumes from the burning charcoal. Such hearths can be seen in many sites but particularly in Pompeii in Italy, where the charcoal remains *in situ* in the cavities as do also the gridiron and the cooking vessels just as they had been in use before the eruption of Vesuvius.

From the early Middle Ages until the development of the kitchen range in the late eighteenth century, all cooking of food, with the exception of baking which was carried out in ovens, was done above or in front of the open hearth. By the twelfth century this was built into a wall under a large archway with a wide flue for the escape of smoke. The brick or stone floor of such down hearths was raised slightly and upon this stood the great andirons which supported the burning logs of the fire. As coal gradually replaced wood as a fuel, the firegrate developed for kitchen use in the same way as in other rooms of the house, though its design was less ornamented and more practical.

Cooking by boiling, stewing and simmering was done in vessels suspended over the fire. Such vessels varied in size but were of similar design. They were made of earthenware or metal – iron, bronze, brass, copper – and had rounded bottoms to achieve an even distribution of heat. Some vessels stood on stubby legs and could be set beside the fire in the embers to cook food slowly or keep it hot. All vessels had handles and some were fitted with lids.

From the early Middle Ages onwards ways were experimented with for suspending the vessels over the fire and the chimney crane was developed for this purpose, its final, sophisticated form not being reached until the seventeenth century. The earliest designs consisted merely of iron bars or hooks fixed into the wall of the chimneypiece with vessels hung from these on chains. This was a limited system, as only one or two pots could be suspended at any one time and it was difficult to adjust the position of these relative to the heat of the fire. It was only too easy for the cook to burn herself in replacing one pot with another. Then the ratchet-hanger was made which enabled the cook to raise or lower a pot, but the chimney crane was a great advance and as it developed greater manoeuvrability was achieved. In its final form a vertical iron post was sunk into the floor of the side of the hearth. From this extended at right angles a horizontal bar, making a bracket which could be swung like a gate through an angle of 90°. There was also a mechanism to raise and lower the height of pots, also to move them along the bar towards or away from the heat source. Many of these three-motion cranes, which were usually made by the local blacksmith, were decorative, incorporating ornamental ironwork.

Apart from boiling and simmering food could also be toasted in front of the fire and, over the centuries, many designs of toaster were developed, from the small down-hearth toasters standing on low feet on the hearth to tall adjustable types. Camp ovens, for slow cooking of smaller dishes were set on short legs in the embers of the fire. These were of cast iron and were lidded.

Roasting was carried out on the revolving spit in front of the fire and the large dripping tray was set beneath the spit to catch the fat. Different designs of iron or steel spit were made (also by the local blacksmith). The straight spit was in general use but there was also the pronged spit, designed to grip the meat firmly, and the basket spit to contain a small animal or fowl. The spit (or spits, for several could be operated simultaneously to roast different sizes of joint) rested in ratcheted supports known as cob-irons or cobbards. These could be adjusted to vary the height of the spits.

The motive power for turning the spit continuously developed over the centuries. For a long period – up to the end of the Middle Ages – the power was human. A man or boy turned a handle and he was called a turnspit. He was responsible for seeing that the meat or fish was evenly and fully cooked: the phrase 'done-to-a-turn', derives from this operation. A screen was set up to protect the turnspit from the heat of the fire. In the sixteenth century the dog turnspit began to replace the boy. Short-legged dogs were bred especially for this purpose and worked in pairs, taking it in turns to pad round the inside of a wooden wheel set high up on the side of the chimney breast. The handle of the spit was then replaced by a grooved wheel which was turned via a continuous chain and a pulley by the dog treadmill. Dogs continued to be used as turnspits until well into the nineteenth century, especially in rural areas. A treadmill and a stuffed dog are on display in Abergavenny Museum in South Wales. Of a breed now extinct, the dog is golden, long-haired and long-bodied, with short legs and a sweeping tail.

From the late sixteenth century onwards alternative means of motive power were being developed to turn spits. The mechanical jack, which was in use in most larger homes by about 1650, consisted of a system of weights and gears and was gravity driven. (The word 'jack' was commonly used to apply to a mechanism which would replace the labour provided by a human being.)

An eighteenth-century invention was the smoke jack, powered by the uprush of hot air ascending from the fire. A vane was fixed horizontally into the chimney at the point where it narrowed to ascend the shaft and the heated air caused it to revolve. The vane was connected by a system of gears to the power shaft and thence by chains which could turn the wheels of several spits. It was a wasteful method as it required great heat from the fire and was generally only used in large houses to power a roasting range. One was fitted (and is still on display) in the kitchen of the Royal Pavilion at Brighton.

The nineteenth century brought two other aids to roasting: the hastener and the bottle jack. The hastener was a large metal screen, half-cylindrical in form, standing on legs in front of the fire. In the open side facing the fire hung a small joint and its cooking process was speeded up by the reflected heat from the metal shield. In the other side of the hastener was a door through the opening of which the cook could baste the joint. The bottle jack was a spring-driven small brass cylinder which could be hung on a hook from the mantelpiece or from a hastener. A small roast was suspended from a wheel under this jack which, when wound up caused the joint to turn first one way then back again for a prescribed time.

Ovens were generally built into the wall by the side of an open hearth, so sharing the same chimney flue. They were brick-lined, with a brick or stone floor, and could be closed with a metal door. The fuel, usually brushwood, was placed on the floor of the oven and lit. When the oven was hot enough, the ashes were raked out and the food inserted. Residual heat was later used to dry herbs or firewood.

Cooking with solid fuel by kitchen range

During the nineteenth century the enclosed kitchen range, made of iron and steel and burning coal, gradually replaced the open fire for cooking purposes. This represented the logical progression from the hob grate which, developing from the dog and basket grates, had begun to enclose the coal and so create better draught conditions for burning coal (see p. 916). Over many years it became understood that a further enclosure of the fire would be more economical in use of fuel as well as giving better heat control and making it possible to provide an oven and a boiler for hot water as well.

It was in the second half of the eighteenth century that several inventions were patented which finally led to the first successfully functioning kitchen range. As was only to be expected, Britain being world leader in industrialization at the time, this range was a British one, designed by Thomas Robinson in 1780. It was an iron roasting range comprising an open coal fire with removable bars and, on top, iron hobs on either side. Roasting could be done in front of the range and cooking pots for stewing and boiling were set on the flat hobs. A hinged trivet was fitted to the top bar of the grate; this could be swung forward to take a kettle or pan. At one side of the range was a brick-lined oven and on the other a hot water tank. Robinson's range represented a landmark in the development of means of cooking but it had considerable drawbacks. Most important of these was the oven which, being only heated on one side, tended to burn the food on this side and leave it undercooked on the other. Also, the top of the fire grate was open, emitting smoke and burning fuel extravagantly.

In 1802, George Bodley, an Exeter ironfounder, patented the enclosed kitchen range. In this the fire grate had a closed top, so controlling the heat and directing the smoke up the chimney. The heating of the oven was better controlled than in Robinson's range, as Bodley introduced flues to convey the heat round the far side of the oven. He also left space below the oven for air to circulate.

Neither of these two ranges was very efficient but improvements came gradually. Between 1800 and 1900 many different models were marketed and by 1850 ranges were being built into new houses and replacing open hearths in existing ones. Interestingly, during the whole of the nineteenth century, despite the plethora of designs to choose from, there continued to be two main types: the open range based on Robinson's original and the closed one of the Bodley type. The open range, which generally came to be called the cottage range or Yorkshire range, was more popular in northern areas as it made the kitchen warmer, providing heat and cooking simultaneously. The closed range, usually called a kitchener, was more suited to southern regions as the fire could be enclosed when not needed for cooking.

Late nineteenth-century ranges were large and much more efficient. They were fitted with one or two ovens, four to six hobs, hot water tank, a warming closet for plates and many gadgets. Heat and smoke were carefully controlled and fuel consumption was much more economical. Much of the ancillary equipment used with the older open hearth continued to be employed with the range: the hastener, bottle jack, kettle tilter, trivets, toasters.

Ranges were still being made in the 1920s for homes which had no gas or electric supply but, as the range was nearing the end of its useful life, cooking by solid fuel was given a boost by the introduction of the Aga cooker. This was the brain-child of the Swedish Nobel prize-winner Dr Gustav Dalen, who devised a cooker based on sound scientific heat-conservation principles. Clean, economical, easy to use and needing a minimum of attention, it was made of cast iron and enclosed in an enamelled insulated jacket: Dalen used kieselguhr, a diatomaceous earth, as insulating material. The heat was conducted to the various parts at precisely the correct temperature for the types of cooking needed: boiling, baking, grilling etc. The temperature of the hob plates was maintained by the fitting of insulated hinged covers which were retained in place when the plates were not in use. The Aga was introduced in Britain in 1929. It is still in use today but the modern version is thermostatically controlled and can function on solid fuel, gas or oil.

Cooking by gas

In the early nineteenth century, as the supply of gas was extended (see Chapter 3), experiments took place to try to use the fuel for cooking, but the problems of making it financially viable were formidable. The cost of gas was much greater than coal; there were no meters, so there was difficulty in measuring consumption; more important still, the public were prejudiced against the use of gas, believing it to be dangerously explosive and the fumes and impurities in the gas to be harmful to any food cooked by it.

Despite these problems inventors continued to try to make satisfactory appliances. One of the early workable devices was a griller of 1824, made at the Aetna Ironworks near Liverpool. It was a kind of gridiron made from a gun-barrel, with holes pierced in it, then twisted round to a circular shape. It could be used horizontally with pans placed on top or vertically so that a joint could be roasted in front. Commercial, box-shaped cookers were marketed from the 1830s and in 1847, A. A. Croll read a paper to the Royal Society of Arts in London on the 'Domestic Uses of Gas'. In this he referred to Alexis Soyer, the famous French chef, who took the lead in promoting cooking by gas because he believed it to be clean, efficient and economical in comparison with coal; he had introduced it into the kitchens of the Reform Club in London in 1841.

By 1855 there were many designs of appliance on the market and gas cooking had been clearly demonstrated to be a practical proposition, but it was only used by a small minority of the population. The chief reasons for this were the still entrenched fear of using this type of fuel and its high cost. The incorporation of Bunsen-type atmospheric burners helped to popularize the cookers, but because gas was still more expensive than coal, coupled with the fact that most people already possessed a coal range, acted as a great deterrent.

Two important incentives were then offered by the gas companies which tipped the scales in favour of gas cooking. The first was the introduction in the 1870s of facilities for hiring cookers instead of insisting on purchase. The

Figure 19.7: Parkinson cooker, *c.* 1890.
Made of cast iron with silicate packing for insulation. Water heater at left side.
Plate warming hood over hotplate.
The Gas Council.

second, which came in the 1890s, was the installation of prepayment slot machines, which brought gas within the financial reach of everyone. By 1911 the Gas Light and Coke Company were taking 230 million pennies annually from their machines, an amount which weighed nearly 2000 tonnes.

From the 1890s gas largely took over from coal as a cooking fuel, but the design of the cooker continued to be cumbersome and difficult to clean. It was made of the same materials as the kitchen range – cast iron, which needed black-leading, with taps and fittings of brass and steel, which were hard to keep shiny (see Figure 19.7 for an example of this period). It was only with the dearth of servant labour which occurred after the First World War and the consequent demand of housewives for a more attractive and labour-saving appliance, which they then had to clean themselves, that the manufacturers responded.

Cookers appeared clad in cream enamelled iron sheeting, replaced in the 1930s by coloured vitreous enamel. The design was streamlined and ovens were lined with easy-to-clean materials. Technical advances followed also, the most important of which was the introduction in 1923 of oven thermostatic control – Regulo – developed by Radiation Ltd.

Innovations after the Second World War included the standardization by manufacturers of their thermostat settings, the introduction of the high-level grill, automatic ignition to burners, thermostatic hot-plate burners and lift-off oven doors. The design was further streamlined and the cooker lost its legs, the space at the bottom being utilized for a storage drawer. The Sola Grill appeared in the 1960s. Since 1965 has come conversion to natural gas, automatic oven timing, self-clean oven linings, electric spark ignition for burners, graduated simmer control and the flame-failure safety device.

Cooking by electricity

The possibilities of cooking by electricity were being explored as early as the 1890s. In 1891 an electric cooker was demonstrated at the Electrical Fair at Crystal Palace and two years later a model electric kitchen was on display at the Chicago World's Fair. Meanwhile, in Britain, Crompton, in collaboration with Dowsing and Fox (see p. 916–17), was designing and manufacturing a range of electrical appliances. In 1900 the Crompton Company catalogue was advertising ovens, hot-plates, saucepans, frying pans, kettles, coffee urns and hot cupboards.

Most of the early cookers, like the Crompton designs, consisted of a number of separate appliances. The oven was like a safe, made of metal sheeting on a metal frames. It was heated by elements at top and bottom, controlled by brass switches; the elements consisted of cylindrical ceramic formers coiled with wire. The oven had no hot-plate but there were a number of ancillary appliances – griller, kettle, frying pan and hot-plate – which were all wired separately and plugged into a panel of wall switches; these appliances were put on top of the oven or on the floor beside it.

Like gas cookers, electric cookers were for many reasons slow to attract custom. By 1900 they were a practical proposition but they were, like their gas counterparts, black, ugly and difficult to clean. Also, people distrusted them. The heat was not visible and it was easy to burn oneself. The elements were as yet unreliable; they easily burned out, they took a long time to heat up and so were expensive to use. Most people already had a gas cooker and/or solid fuel kitchener and, lastly, only a minority of homes were yet wired for electric current.

Various attempts were made in the 1920s to convert customers to 'cooking electric'. Electric lighting companies offered to hire cookers. Smaller cookers of more attractive design were produced. C. R. Belling brought out his 'Modernette' made of light steel which had two boiling burners on top, a grill, plate warmer and oven, all in one conveniently designed cooker. But, like all electric cookers of the time, the burners were of open type and if the contents of a saucepan boiled over, the spillage ran directly down on to the heating element, usually causing a short circuit.

It was not until the 1930s that electricity began seriously to compete with gas for cooking, Figure 19.8 shows an example of the first generation of successful

Figure 19.8: Creda electric cooker, 1933.
First model to have thermostatic control of oven. Hot plate has two elements, one of the new spiral tube type, the other a metal plate with coiled wire element beneath to heat grill.

electric cookers. By 1939 over a million households were using electricity compared to nine and a half million opting for gas. But, by this time also, most households were wired for electricity and great improvements had been made in the design and technical performance of the cookers. Appearance and materials had kept pace with gas appliances. Functionally the fast-heating tubular-sheathed radiant rings had replaced the earlier burners and later solid hotplates, which had been slow to heat, and ovens were thermostatically controlled.

Since 1950 electric cookers have marched in step with gas cookers in all modern advances. Importantly, what had always been the great drawback of electric cooking – the excessive time taken for elements to heat up and consequent high expense of using the appliance – has been overcome. Boiling rings heat quickly and split rings lead to economy. There is a wide range of heat control. With the 1970s came the ceramic hob and the cool-top hob, both products of modern technology. In the 1980s the cost of electric current is now only marginally greater than gas and new cookers are so fast, clean and efficient in conservation of energy that running costs are not dissimilar. Also electric cookers tend to be cheaper to purchase than their gas counterparts. These factors have dramatically altered the nation's preferences and over 40 per cent of the population use electric cookers.

The pressure cooker

Denis Papin, the French physicist, invented the first pressure cooker, which he called 'A New Digester or Engine for sofning Bones', in 1679 and he demonstrated it two years later at the Royal Society. His vessel was designed with a tight-fitting lid and had a safety valve to guard against excessive rise of pressure.

The idea of a pressure cooker is that a vessel is designed to seal in and control the steam which normally escapes when cooking is done in a saucepan. As the steam is retained, the pressure rises and so does the temperature at which the water boils. The pressure forces the super-heated steam through the food, so greatly reducing the cooking time needed.

After Papin the idea languished, but it was revived during the nineteenth century when a cast-iron version was marketed: it was called a digester. Cast aluminium was used in the 1930s and pressure cookers, as they were then termed, were fitted with pressure gauges on top. The modern cooker is made of aluminium or stainless steel and has a non-stick finish. Pressure cookers have been in less demand since 1960 owing to the introduction of such alternatives as readily available convenience foods, the auto-timer fitted to cookers, the slow cooker and the microwave oven.

Microwave ovens

Microwave cooking makes use of short radio waves (as short as 12cm (4.75 in)). Such waves were first generated by Britain and the USA for the operation of radar during the Second World War. Americans then adapted the research for cooking and produced a microwave cooker; the method was introduced to Britain in 1959.

Whatever the means of cooking the effect of the heat is to raise the temperature of the food, by increasing the agitation of its constituent molecules. In traditional cooking methods by solid fuel, gas or electricity, the surface of the food is heated from an exterior source and it is only cooked to the centre by such heat after a considerable period of time has passed. In microwave cooking the radio waves penetrate very quickly to the centre of the food and raise its temperature. Microwave cooking will not, therefore, 'brown' food; if this is desired it must be done afterwards by conventional heat.

The interior of a microwave oven has polished reflecting surfaces which concentrate the microwaves on the food. Also, a rotating fan is attached to the roof of the oven in order to spread the heat evenly. As is well known, microwave cooking and de-frosting is extremely fast. As it is only the food which is heated and not the oven, such cooking is extremely clean as well as inexpensive. The output power of most microwave ovens is thyristor-controlled and microprocessor control is gradually being introduced.

PLASTICS IN THE HOME

No material has ever changed the appearance of the home as much as plastics have done and it is not only appearance but their resistance to wear and their easy-care characteristics which have altered everyone's way of domestic life. The word, which derives from the Greek *plastikos*, meaning 'that which may be moulded', describes their quality of being easily shaped to any required form.

The development of plastics took place rapidly in the years since 1920 (see Chapter 3). In the modern home a very wide range of plastics is in use for a great variety of purposes. Three types which particularly serve domestic needs are polyvinylchloride (PVC), polyethylene and polystyrene and each of these may be found in different forms which meet different needs. For example, PVC (often called simply vinyl) may be manufactured to be rigid or flexible, thin or thick, transparent or opaque. The rigid type is suited to the making of structural parts such as piping and guttering; it is also used for making gramophone discs. Vinyl floor tiles or soles for footwear are manufactured from a more flexible version. A softer type is used for a wide range of goods from furnishing fabrics and garments to wallpaper and hosepipes. Transparent PVC provides the wrapping film used for a multitude of purposes in the kitchen.

There are two chief types of polyethylene – more commonly known as polythene. An early type was produced as long ago as the 1930s, when it was manufactured into sheets of fairly heavy wrapping material. High density polythene was developed in the 1950s. This is heavier and more rigid and has proved most suitable for making a range of useful containers such as buckets, large bowls and dustbins.

Polystyrene appears in three chief forms in the home. There is the toughened type which lines refrigerators, the clear but strong and rigid type made into moulded containers for eggs, butter, salads etc. and expanded polystyrene, which is a lightweight foam ideal for a broad range of goods from moulded packaging to ceiling tiles.

A more recent development is the use of glass fibre as a reinforcement for plastic. Polyester is ideal for this purpose and this combination of materials has made the use of plastics possible for heavy, structural parts of a house as well as for articles such as water tanks.

Apart from all these widely used plastics there are others of particular assistance in the kitchen. The development of non-stick cookware has been made possible by the use of polytetrafluorethylene, generally abbreviated to PTFE, which is inert to a wide range of chemicals and is also resistant to sunlight and to moisture. Polypropylene is a rigid thermoplastic which is highly resistant to liquids and solvents, so is used extensively for kitchenware of all kinds as well as garden equipment such as wheelbarrows.

Elsewhere in the house melamine resins are made into unbreakable tableware and phenolic resins into electrical fittings. Carpets are backed with polyurethane foam. Lastly there is the extensive range of synthetic textiles manufactured into garments and furnishings fabrics, dating from the first production of nylon by Du Pont of America in 1938, after eleven years of research costing $27 million, to the later numerous varieties of polyester and acrylic fibres marketed under the plethora of different trade names adopted by their countries of manufacture.

FURTHER READING

Adamson, G. *Machines at home* (Lutterworth Press, Cambridge, 1969)
Conran, T. and C. *Kitchens past and present* (Hygena Ltd., 1976)
Haan, D. de *Antique household gadgets and appliances c. 1860–1930* (Blandford Press, London, 1977)
Harrison, M. *The kitchen in history* (Osprey Publishing, London, 1972)
Hole, C. *English home life 1500–1800* (Batsford, London, 1949)
Joy, E. *The Country Life book of English furniture* (Country Life, London, 1964)
Lindsay, J. S. *Iron and brass implements of the English home* (Tiranti, London, 1964)
McRobert, R. *Ironing today and yesterday* (Forbes Publications, London, 1970)
Megson, B. *English homes and housekeeping 1700–1960* (Routledge & Kegan Paul, London, 1973)
Singer, C. et al. *A history of technology* (7 vols.) (Clarendon Press, Oxford, 1958–78)
Wright, L. *Clean and decent* (Routledge & Kegan Paul, London, 1960)
—— *Home fires burning* (Routledge & Kegan Paul, London, 1964)
Yarwood, D. *The English home* (Batsford, London, 1979)
—— *The British kitchen* (Batsford, London, 1981)
—— *500 years of technology in the home* (Batsford, London, 1983)

20

PUBLIC UTILITIES

R. A. BUCHANAN

INTRODUCTION

The complexity of modern life and, in particular, the organization of large towns and cities on which it is based, can only be sustained by an intricate network of public services. When towns were generally small, such services were simple because the population could depend upon local wells and rivers for their water supply, and food and raw materials were available from the adjacent countryside. But the magnificent Roman system of roads and aqueducts should remind us that even ancient cities required basic services in order to function efficiently. The skills which had created these engineering achievements were virtually forgotten in western Europe for many centuries, during which most towns were small and their services rudimentary. During these centuries, the existence of a river or a fast flowing stream was usually all that was necessary in the way of water supply, providing for drinking, drainage, sewage and garbage disposal, transport, power for industry, and fire-prevention measures. As long as towns remained small and could retain an intimate association with the surrounding countryside, such multiple use of natural watercourses could sustain them indefinitely.

With the onset of rapid industrialization in western Europe at the end of the eighteenth century, however, towns grew prodigiously as centres of trade and industrial production, and with this development the need to provide adequate services became critical. Failure to make adequate provision almost caused a breakdown of British town-life in the 1830s and 1840s, when the ravages of cholera and other diseases induced by overcrowding in insanitary conditions were most severe. But from the middle of the nineteenth century a determined effort was made to improve the conditions of life in the towns of Britain. The appalling squalor reported in dozens of official statements by doctors and administrators, and described so vividly by Friedrich Engels in his account of Manchester in 1844, became a thing of the past, replaced by towns which had a good supply of fresh water, an efficient means of disposing of organic waste and garbage, and all the other services of paving, lighting, transport, shopping and trading facilities which have come to be taken for granted as the normal features of life in 'megalopolis' – the huge urban conglomerations which have become the dominant pattern of existence for most people in the developed nations of the twentieth century.

All the public services which sustain megalopolis have been created by the co-operation of many individual technologies developed over a number of years. These technologies are the subjects of other parts of this volume but the present chapter is concerned with the development of those public services which have made life in large towns and cities possible.

The services considered fall into two categories:

(1) The primary services of water supply (sources, treatment, distribution, drainage and sewage disposal), power supply (gas, electricity, hydraulics and pneumatics) and waste disposal.
(2) The secondary road, postal, telephone and telegraph services.

WATER SUPPLY

Water supply is the most basic of the public service industries. No town can survive for long without a supply of water for drinking, washing and various industrial processes, so that if a convenient source such as a river or a well does not exist it has to be provided, or if natural sources are inadequate they have to be supplemented. Substantial resources of materials, man-power and technological skill have thus been required wherever town life has flourished on a large scale in order to ensure that the need of an urbanized population for a constant supply of water can be fulfilled. The ancient civilizations of Egypt and Mesopotamia established intricate irrigation works, bringing water to their fields and cities. Pergamon in Asia Minor and the Greek city-states constructed elaborate systems of culverts and aqueducts to bring water to their citizens. And the Romans made a fine art of the technology of bringing water long distances in masonry aqueducts to serve the large cities of their Empire, and especially Rome itself. The remains of many of these aqueducts survive as eloquent testaments both to the engineering skill of the men who designed and built them, and to the quality of urban life which they made possible.

Sources

All water becomes available in the course of the hydrological cycle, whereby water is evaporated from the oceans and precipitated on the land surface, whence it can be obtained from rivers and lakes, or recovered from springs and wells if it sinks into the ground to replenish the underground accumulation of water or aquifer. In certain geological conditions where the aquifer is under pressure in a basin of water-bearing strata, the water will rise spontaneously up a shaft bored into it – an artesian well. Normally some form of pumping or water-raising apparatus is required. By far the simplest form of urban water-supply is thus a river, and for many large towns, including London, this has remained the largest source down to the present time. Although direct extraction from the turbid tidal reaches of the River Thames was abandoned in the mid-nineteenth century, the Metropolis Water Act of 1852 prohibiting the extraction of drinking water below Teddington Weir, the bulk of London's water supply still comes from the Thames and the Lea, through the elaborate catchment reservoirs and treatment plant between Staines and Surbiton, and in the Lea Valley. London, of course, has had a long history of water supply difficulties, but most of the successful schemes such as the waterwheel-powered pumps built into an arch of the Old London Bridge by the Dutch engineer Peter Morice in 1582, and the more ambitious scheme of Sir Hugh Myddelton's New River project in 1619, which brought water into the city through an artificial channel from the Hertfordshire countryside, relied upon river catchment in one form or other. Subsequently all rivers with a reliable flow of water in western Europe and many in other parts of the world have been liable to have a proportion of their volume extracted for public supply either immediately or at a considerable distance from the point of extraction. Of large rivers flowing through industrialized countries like the Severn in England and the Rhine in West Germany and the Netherlands it has been recognized that any cup-full of water drawn from them might make several cycles through supply and waste disposal in the course of flowing from their source to the sea.

The device of impounding rivers to provide reservoirs for water supply, river control and irrigation, is very ancient. One of the oldest civil engineering structures in the world is a dam built for such purposes at Sadd el-Kafara in Egypt, 32km (20 miles) south of Cairo, dating from the third millenium BC, and many similar archaeological relics of dams survive. Most early dams were gravity dams – being embankments or masonry walls built straight across a valley and resisting the load of the water by their weight. More sophisticated forms are the buttress dam in which a fairly slim masonry wall is supported by a series of buttresses on its down-stream face, and the arch dam in which a thin curved wall depends upon arch action for its strength as the main load is thrown into the adjoining cliffs, making it possible to build a lighter structure than in other forms. The arch dam has proved particularly suitable for large modern dams in mountainous terrain, where concrete has been used as the main structural material. The spectacular Hoover Dam across the Colorado River in America is an outstanding example of this type, demonstrating the achievement of modern civil engineering in using new materials in an elegant manner to perform on a huge scale one of the basic public services – providing a good water supply.

The systematic use of impounding reservoirs to supply water to urban communities in modern times dates from the early nineteenth century, when Robert Thom created in 1827 what was then the largest artificial lake in Britain to supply Greenock on the Clyde estuary in Scotland. It retained this distinction until the Vyrnwy and Thirlmere projects at the end of the century (see below), but by that time extensive water supply projects had been undertaken in many parts of Britain. The first extensive municipal water supply project was that devised by J. F. La Trobe Bateman for Manchester between 1851, when the first part became operational, and 1877, when the works were complete. These consisted of five major reservoirs on the River Etherow in the Longdendale valley in the Pennines to the east of the city, with their associated aqueducts and auxiliary reservoirs. Bateman used earth embankment dams and had great difficulty in making them stable.

Techniques of dam construction improved substantially in the second half of the nineteenth century, and pioneering work in Europe on the construction of high-wall masonry dams was adopted in Britain by the end of the century in the Manchester Thirlmere extension in the Lake District (1874–1890), the Liverpool Lake Vyrnwy scheme in north Wales (1881–1892) and the complex of reservoirs to supply Birmingham built in the Elan valley in central Wales (1890–1904). In the Manchester Haweswater project in the Lake District (1923), concrete became an important constructional material.

In those places where surface water has not been easily available, thirsty towns and cities have exploited the potential of underground sources. Sometimes natural springs could be directed with little difficulty into water supply systems, but more normally wells have been sunk to reach the water table or to give access to artesian supplies where favourable geological conditions have prevailed. In some dry inland regions such supplies are vital to any permanent settlement. The construction of wells and bore-holes encouraged the use of powerful pumping engines to maintain a continuous flow and thus provided the primary need for the large steam engines which have figured prominently in the history of British water supply. One of the most extensive groups of water works deriving their supply from wells were those built under the direction of Thomas Hawksley to extract water from the New Red Bunter Sandstone for Nottingham, but Hawksley built similar systems for many other British and continental towns. The fine beam pumping engine built at Ryhope in Durham in 1868 under his direction for the Sunderland and South Shields Water Company has been preserved in working order, as has also the elaborately embellished engine and engine-house at Papplewick, north of Nottingham (1884).

Treatment

Before water can be distributed for public consumption, various processes of treatment are required by modern legislation to ensure that it maintains wholesome standards. The first requirement is filtration to remove the coarser forms of pollution, but chemical treatment of dangerous organisms is also now obligatory in most countries. Filtration was used by Thom in his Greenock scheme in 1827, using mechanical and slow sand filtration processes. Two years later, in 1829, James Simpson brought into use his sand-filtration process at Chelsea waterworks in London. Together with Bateman and Hawksley, Simpson was amongst the leading British water-works engineers of the nineteenth century. His pioneering efforts at Chelsea were partly in response to a loud public outcry against the foul condition of the River Thames from which water was being drawn for public supply at that time. Simpson made a study of methods of filtering water supplies and built an experimental arrangement in 1827 in which water was first allowed to settle and then passed through a downward-flow filter consisting of a four-foot bed of carefully graded sand and gravel. It was this method which he then applied to Chelsea waterworks in 1829. Although successful in giving a vastly improved supply of water, the method took a surprisingly long time to be generally adopted, mainly because of scepticism about its beneficial effects, but also because of the high cost and substantial space required for filter beds. The impact of cholera, however, strengthened the case for filtration, and the principle was made obligatory in Britain by the Metropolis Water Act of 1852. This measure, confirmed by subsequent Royal Commissions and Acts of Parliament, served to give a sound legal basis to the provision of wholesome water in Britain, although chemical purity of water supply was not achieved until the introduction of chlorination to kill off undesirable bacteria. Like filtration, this was accepted with some reluctance by the water authorities, and it only became general practice in Britain after the Second World War. The further addition of chemicals such as fluoride to harden children's teeth against decay is a matter of continuing controversy about the medication of water supply, but the principle of the responsible authorities being required to provide pure water to every domestic and industrial consumer is no longer in doubt in the developed countries.

Distribution

The distribution of treated water to its consumers has called on the skills of several generations of civil engineers. In Britain, even the early impounding reservoirs were usually in upland valleys comparatively remote from the towns where the supply was needed, so that extensive culverts and aqueducts were necessary to get it to its destination. The fact that the reservoirs were usually in high land, moreover, meant that the supply could flow largely under the effect of gravitation. As the reservoirs were established farther afield, so the need for siphons, tunnels, and pumping engines increased. For example, the scheme designed by Bateman at Loch Katrine in the Trossachs to supply Glasgow is 48km (30 miles) from that town; Thirlmere is 160km (100 miles) from Manchester; and the Elan valley is 120km (75 miles) from Birmingham. The works by which these distances are covered are rarely obtrusive or easily visible to the casual visitor. In most cases, the level of the reservoirs is such to permit a siphon action to convey the water to the taps of the customers, but in some areas it is necessary to supplement this with water towers, to which the water is pumped in order to allow it to gravitate to its point of use. Powerful steam engines were frequently installed by waterworks in the nineteenth century to perform this function, and several of these have been preserved in various parts of Britain, although the work is now done by more efficient and convenient electrical pumps.

Water may be allowed to flow freely in open culverts, but when it is put under pressure by pumping or by passing it through a siphon, or when it is being conveyed underground, it has to be contained in strong piping. The traditional form of water pipes had been those made of elm trunks bored out through the core and fitted end-to-end, but the tremendous expansion of the British iron industry made available new piping materials which were quickly demonstrated to be extremely versatile when used for water pipes. First came cast-iron pipes, flanged and bolted together, providing excellent water mains for laying underneath busy city streets, just as they also provided ideal gas mains. For larger pipes carrying water overground, wrought-iron pipes were constructed by curving plates and riveting them together in a manner similar to that used in the construction of boilers and the hulls of iron ships. Such wrought-iron pipes were well-suited for the aqueducts carrying water across valleys and other obstacles on the routes from reservoirs to its urban consumers. Subsequently, wrought-iron was replaced by mild steel, which remains in widespread use for conveying water although concrete piping is now used for some functions. The use of iron and steel piping has enabled water to be pumped for considerable distances in arid regions of the world where it is not practicable to use open culverts. For instance, inland cities in the USA like Denver, are supplied with most of their water from sources many miles away: in this case, it is collected on the western slopes of the Rocky Mountains and transferred through tunnels across the continental water-shed to the city which is on the arid eastern side of the mountains.

Water supply has become increasingly a matter of water conservation as an expanding population with rising expectations presses on the finite fresh-water resources of the world. It has always been a scarce commodity in arid regions. The twentieth century attempts to conserve the water of the River Nile, from the first Aswan Dam completed under the supervision of Sir Benjamin Baker in 1902 to the Aswan High Dam finished in 1970, have extended the work of many generations of Egyptian engineers in this respect. In Australia, the driest of all the land masses in the world, most of the areas of human settlement are on the coast, but the development of gold-mining settlements at Coolgardie and Kalgoorlie in the arid deserts of Western Australia was made possible by the opening in 1903 of a 645km (400 mile) pipeline from the Mundaring reservoir near Perth. An even more ambitious project in the south east of Australia, the Snowy Mountain scheme, during the 1950s and 1960s involved converting a large eastwards-flowing river into one flowing westwards into the dry but potentially fertile interior, and providing a large amount of hydro-electric energy in the process. In parts of the world with even scarcer resources of fresh water, the desalination of sea water is the sole means of maintaining essential water supplies. The first large plant of this type was installed in Kuwait in 1949, and despite heavy initial expense on the equipment required to produce the evaporation and condensation of sea water, there is likely to be more use of it in the future as even more-favoured parts of the world begin to find that their available water resources are inadequate to meet expanding needs.

Drainage

A corollary to a good water supply is a good system of drains. In the first place, drains are required to remove surface water, and secondly to allow the removal of sewage.

With the increase in urban building and the paving of streets, natural means of allowing surface water to percolate the sub-soil or to run off in streams are blocked, and it becomes necessary to provide artificial drains for this purpose. As rain rarely falls consistently, the provision of drains has to allow for the periods of maximum rainfall in order to protect settlements from the hazard of serious flooding. In tropical areas this involves 'monsoon ditches' alongside main roads to carry off the very heavy rain accompanying frequent thunderstorms, but in more temperate latitudes the service of drains is usually less conspicuous, although some modern cities have found it necessary to undertake storm-water and river-control works in order to guard against exceptional conditions of prolonged rainfall or flash floods. Towns on tidal estuaries have comparable problems when parts of the built-up area are below peak high-tide level. Normally, potential flood conditions can be controlled by extensive river embankments and dykes, as demonstrated in the Netherlands and Belgium around the mouths of the Rhine. In some places, however, the gradual sinking of the land creates a long-term problem and makes the coincidence of exceptional weather conditions with a particularly high tide a dangerous hazard. Parts of the Netherlands and south east England experienced such a combination in

1953, and the construction of the Thames Flood Barrier at Woolwich was undertaken as a defence against a repetition of this event, which could put at risk large parts of London. The Barrier, completed in 1984, consists of ten rising sector steel gates between hooded piers spanning the 520m (1710ft) of the river at Woolwich Reach. The gates are normally submerged horizontally, allowing shipping to pass freely, but in an emergency they can be raised in about 30 minutes to stand vertically and hold back any potential tidal surge. The system has not yet been put to a serious test, but it is a considerable comfort to Londoners to know that this public service is now prepared.

Sewage disposal

Water supply has been an essential factor in the development of successful methods of disposing of large volumes of organic waste. Modern systems of sewage disposal began with the introduction of water-borne techniques inspired by Edwin Chadwick's public health reforms in Britain in the 1840s, and the insistence of engineers such as Hawksley on providing a continuous rather than an intermittent supply of water wherever it was required. As towns acquired reliable sources of fresh water, so a proportion of this became available to flush water closets and to maintain a continuous flow of sewage waste through a network of well constructed main and tributary sewers. The design and construction of sewers were carefully managed. In the Victorian era good quality brick became the favoured material for the building of the larger sewers, in a pear-shaped cross-section, the narrower end at the bottom to ensure a strong flow of liquid even when the level was low. Such sewers thus had a self-scouring action so long as they remained in use. Most of the early systems of sewers aimed at discharging their effluent into a river downstream from the town being served, but gradually it was recognized that this was not acceptable for inland towns where other settlements depended upon the same river for their water supplies, and even those with access to the sea and tidal estuaries were encouraged to consider the value of recovering minerals from the waste. Thus followed the development of 'sewage farms' for securing the partial purification of water to be released back into the rivers and for exploiting those parts of the waste which could be used for agricultural fertilizers and other purposes. The techniques of sewage treatment involved some form of sedimentation process, with the addition of chemicals as appropriate to speed the bacteriological decomposition and to remove unpleasant odours.

The most serious problem confronting public health experts and the engineers who served them in the middle of the nineteenth century was that of providing a system of drains and sewers for the metropolis of London. This was achieved between 1855 and 1875 by the Metropolitan Board of Works and its engineer Joseph Bazalgette. The principal feature of Bazalgette's system was a main drain along the north bank of the Thames, partly to accommodate which the Victoria Embankment was constructed. This cut across older drains and received the discharge from many tributary sewers before finally discharging itself into the tidal waters of the Thames below Blackwall. A similar network provided for the removal of sewage south of the Thames, to discharge at Crossness. On both sides of the river the gradient was such that it was necessary to assist the flow of sewage by the installation of several powerful pumping engines, several of which survive as industrial monuments although no longer active. Bazalgette received a well-deserved knighthood for his work, which did much to remove the curse of cholera and other fever diseases from London. Similar systems were installed in other large cities in Europe and elsewhere. Indeed, from an engineering point of view it is difficult to improve on the system so long as the principle of water-borne sewage disposal is adopted, so that new work has tended to concentrate on improving the quality of sewage treatment; the replacement of old sewers, which have frequently now given over a hundred years of excellent service in many of our big cities, with the minimum of maintenance; and the adoption of concrete and plastics instead of the traditional constructional materials of brick and iron.

POWER SUPPLY

Gas

Coal-gas had been recognized as a by-product of coal before the eighteenth century, but it was William Murdock who, at the end of that century, first discovered how to apply it as a public service (see also p. 208ff.). Murdock was the agent responsible to Boulton and Watt, the Birmingham manufacturers of steam engines, for the erection of their machines in Cornwall. Murdock's daily work, however, did not prevent him from exercising his inventive and active mind in other directions, and in 1792, while pursuing methods of preparing wood preservatives from materials produced by the distillation of coal, he collected coal gas from his primitive retort and used it to light his house in Redruth. On returning to Birmingham in 1798 he continued his systematic investigation of coal distillation and his coal-gas plant was used to illuminate the Soho Foundry during the celebrations accompanying the Peace of Amiens in 1802. Simultaneously, Phillipe Lebon in France produced elaborate plans for manufacturing coal-gas, incorporating a process to wash it in order to extract useful by-products, and applying it to equipment for lighting and heating, and also for inflating balloons. Lebon was killed in Paris in 1804, and little was done to take advantage of his research, but in the same year Boulton & Watt started seeking orders for Murdock's gas-making equipment and installed it in several large mills beginning with Phillips and Lee of Salford in 1806. The equipment employed was crude, burning gas in small jets known as cockspurs, generating offensive smoke and smells.

Boulton and Watt abandoned the manufacture of gas apparatus in 1814, by which time the quality of their equipment had been overtaken by rivals and particularly by that of Frederic Albert Winsor, an eccentric anglicized German professor who launched the Gas Light and Coke Company in London in 1812, and by Samuel Clegg, who had trained under the chemist Dalton in Manchester and worked with Boulton and Watt before being recruited by the new London company in 1812, just as the demand for gas lighting was increasing rapidly. By the end of 1815, some 26 miles of gas mains had already been laid. Clegg used cast-iron retorts and provided for cooling and washing the gas, and for purifying it in lime. His system was extensively adopted by the gas companies which sprang up in most British towns, often in close competition by the middle of the nineteenth century. Some companies introduced oil-gas to replace coal-gas in the 1820s, using a process based on the distillation of whale oil, but it proved to be an expensive alternative and by the middle of the century most installations had been converted to coal-gas and merged with rival undertakings. Gas was now being very widely used, although suspicion of it had not entirely disappeared and it was excluded from the Great Exhibition in 1851. Gas had been used mainly to illuminate streets and public places, its fumes making the application of gas-lighting in domestic rooms unbearable. But the introduction around 1840 of the atmospheric burner, mixing air with the gas just before combustion, greatly enhanced its utility, and the subsequent applications in the Bunsen burner (1855) and the gas-ring (1867) increased its performance as a heater in boilers and ovens. Gas cooking, however, did not become common until the 1870s and the gas-fire with radiants was introduced in 1880. Dry gas meters also became widespread in this period as a means of registering the amount of fuel supplied to the domestic consumer.

A vigorous international market developed for gas equipment, and British engineers were responsible for many installations in Europe, America, and Australasia. As a measure of rationalization was achieved between competing suppliers, coal-gas came to exercise a virtual monopoly in urban illumination (the paraffin lamp developed for use elsewhere), and the industry became complacent about its prospects until the advent of the electric incandescent filament lamp presented a serious challenge from a totally different technology. Thanks to the incandescent gas-mantle patented by the Austrian Carl von Welsbach in 1885, the gas industry continued to provide an efficient system of illumination into the twentieth century, while it transferred its main emphasis to heating functions, both for industrial processes and for domestic heating and cooking. The distillation of coal in retorts underwent little change until the 1960s, although the traditional horizontal retorts operating a batch-production process were gradually replaced, first by inclined and then by vertical retorts which could maintain continuous production. In some smaller gas works however horizontal retorts remained in operation until the end of coal-gas manufacture. The re-introduction of oil-based processes and then natural gas, obtained as a by-product from North Sea oil fields, brought about a spectacular change in the gas industry in Britain in the 1960s and 1970s. Coal-gas retorts disappeared, and even the large bell-tank gas-holders which had long featured on every urban skyline were reduced in number as the gas was processed at a few new units and conveyed around the country in a network of high-pressure mains. Gas has thus managed to remain an efficient and convenient source of heat energy in all developed nations, even though its long-term viability has come to depend on the availability of oil and natural gas and the economics of recovering them from places which tend to become steadily more inaccessible.

Electricity

The generation of electricity is treated in Chapter 6, but it is appropriate to mention it here as the source of many of the most valuable utilities and services in modern societies. While scientific knowledge of electric power increased rapidly in the nineteenth century, with the discovery of both chemical and mechanical means of producing a continuous current, the electricity supply industry was remarkably slow to take advantage of this accumulating expertise. Not only did the coal-gas industry appear to have established a comfortable monopoly of urban lighting by the second half of the nineteenth century, but the risk capital was not readily available for a source of energy which required heavy installation costs and was then likely to have only spasmodic use. The solution to these problems came in the form of the electric incandescent filament lamp which provided a means of adapting the brilliant light available from arc lamps to a domestic scale; and the development of electric traction in the form of the tramcar and underground train.

The filament lamp was developed by Thomas A. Edison in the USA, leading him to the production of a commercially viable electric light bulb in 1881, and by Joseph Swan in Britain. They joined forces to market their invention in Europe.

Electric traction began with the adaptation of horse-drawn tramways, which had been established in most major cities of the Western world by the 1880s, for electrically-propelled vehicles. Werner von Siemens opened the first public electric tramcar service in Berlin in 1881, but this was essentially an experimental line and many problems regarding the engines, the power transmission system, and the design of the cars had to be overcome before the electric tramcar could offer a really efficient system of transport. This came in 1888, with the opening of the network in Richmond, Virginia, and in 1891, when the first British system began operations in Leeds. Thereafter electric tramcars became vehicles of mass transport – 'the gondolas of the people', in the vivid phrase of Richard Hoggart – and were responsible for a rapid suburban expansion of town life.

To meet this expanding need for electricity, it became feasible at last to establish power stations generating sufficient electricity to supply a substantial district. Edison pioneered such electricity supply systems as early as 1879, when he demonstrated a complete system of generating plant, distribution network, and lighting apparatus, at his Menlo Park laboratories, and he quickly received contracts from municipal authorities to install such systems in the United States. The first complete installation in Britain was that constructed by Sebastian de Ferranti at Deptford for the London Electric Supply Corporation in 1889 (there had been earlier partial schemes at Brighton in 1881, and in London). Whereas Edison supplied direct current from his power stations, Ferranti and most subsequent electrical engineers adopted alternating current at high voltage, which has substantial advantages over direct current for long-distance transmission. The Deptford plant consisted of six steam engines driving 10,000 and 5,000 volt alternators, and amongst other innovations it employed new types of cable and an electricity meter to measure consumption by the customer. In America, George Westinghouse installed hydro-electric generators at the Niagara Falls in 1893, and like Ferranti's system these operated on alternating current.

As with gas works at the beginning of the nineteenth century, by the end of the century every town in Britain was promoting or seeking to promote its own electric power station, and again like the earlier experience of gas works, many towns found themselves with competing power stations. Enterprises tended to merge, however, as they became larger, and the advent of the steam turbine as a more efficient alternative to the reciprocating steam engine promoted this process by encouraging the construction of increasingly larger installations. By the 1920s these were being combined into national grids, and the process of rationalization continued with fewer but larger power stations producing more and more electricity. Oil-burning furnaces replaced coal-burners to produce steam for the turbines in many power stations, and since the 1950s a significant number of nuclear-fuelled power stations have also come on line. However, a combination of economic and political factors has ensured the retention of coal as an important fuel.

The power produced by these large stations is distributed at high voltage over a wide area by overhead wires carried on pylons or by underground cables. It is reduced to lower voltages by local transformers for use by the customers, and the instant availability of electric power has become an assumption on which modern industry and domestic life rely as a matter of course. This universality of electric power has been a powerful factor in increasing the mobility of industry and population.

But perhaps the most conspicuous public service of electricity has been in the provision of street lighting. Gas had done much to make urban streets safer and more wholesome at night by reasonably efficient forms of lighting, which eventually came to include automatic control through a clock and a pilot-light, and gas mantles protected from the elements by a glass hood. But for sheer simplicity and ease of maintenance, gas street lighting could not compete with electricity and it was gradually displaced, while electric lighting also became readily available in areas beyond the reach of town gas mains and in new towns which never had cause to use gas lighting. With the introduction of the gas-discharge lamp, moreover, filament bulbs were replaced by neon lights and by mercury and sodium-vapour discharge lamps, which have become generally used in the streets of human settlements all over the world. The sodium lamp, in particular, has become very popular because of its remarkable cheapness and its efficiency in giving a diffused light – despite its unfortunate yellow colour. Electric power, in short, has become the largest single public service in modern societies.

Hydraulics and pneumatics

Just as the existence of a dependable supply of water made the provision of water-borne systems of sewage-disposal practicable, it served also to promote the use of hydraulic power systems. The possibility of using the pressure in a fluid both to multiply a force in its application and to transmit the force from one part of a machine to another or to another quite different machine had been fulfilled first by the Yorkshire cabinet-maker turned engineer, Joseph Bramah, who took out his patent for an hydraulic press in 1795 (see Chapter 7). Progress was slow for several decades, despite the versatility of Bramah's machines, and the credit for developing hydraulic power into a popular means of power transmission belongs to William Armstrong, a lawyer in Newcastle-upon-Tyne who became one of the most innovative engineers of his generation. He patented his idea for an hydraulic crane in 1846 and established a large enterprise in Newcastle to manufacture it. Hydraulic power was found to be an ideal medium for operating lock gates in harbours, for raising and swinging movable bridges, and for working elevators in warehouses and public buildings. By the end of the nineteenth century, hydraulic power systems complete with pumping engines and miles of cast-iron mains had been established for public use in several of the largest British cities, and in places further afield including Antwerp, Melbourne, Sydney and Buenos Aires. These remained operational until well into the twentieth century, but they could not compete with electricity for efficiency and convenience, once this alternative was available. Although it has declined as a public service, hydraulic power has undergone important development in the twentieth century, being now used in oil-pressure systems for automobile braking and many other power-transmission applications.

Another form of power-transmission which has made a contribution to public services has been compressed air. Pneumatic devices in the shape of bellows have been familiar to craftsmen from antiquity, but little general use was made of them until the twentieth century, when they became very valuable tools in the construction industry and in mining, and in various forms of machine tool. They have also provided an essential part of the modern jet engine in aircraft, and they have become familiar to motorists the world over as a means of inflating pneumatic tyres. They have been particularly valuable in deep-mining processes because they do not cause pollution of the air supply and they minimize the danger of accidents from sparking or short-circuiting. The facility for transmitting power over long distances through flexible hosing has made pneumatic power convenient for many heavy but portable tools such as the drills used in constructional work. With a compressor coupled to a small diesel engine, air-powered drills have become a familiar feature of modern roadworks, As with hydraulic systems, the compressed medium is used to drive a piston or a ram which can then deliver a reciprocating action or which can be readily converted to rotary movement. It is less suited than hydraulic power to large-scale applications and has not been popular for municipal public services, but a system of underground compressed air pipes was installed in Paris in the 1880s, carrying power to pneumatic devices in various parts of the city.

WASTE DISPOSAL

Modern life generates many forms of waste, and important public services are devoted to disposing of them. The water-borne removal of organic waste has already been considered, because this is so closely related to the provision of a constant supply of fresh water. Other forms of waste removal can be more episodic. Thus, industrial and domestic garbage is collected regularly in modern towns and subjected to various forms of processing to recover metals, glass, and other reusable materials. The remainder is usually pulverized and compacted before being disposed of by emptying into holes such as those left by old quarrying operations, or dumped from barges into river estuaries from which it can be dispersed by the movement of the tides. A few of the most progressive public authorities in Sweden and elsewhere have installed the expensive plant necessary to burn such garbage, using it as a fuel to heat water and generate power. There is little doubt that this is the most desirable solution to the problem posed by the disposal of constantly growing masses of rubbish, after everything of commercial value has been extracted from it, but the heavy initial costs are a formidable deterrent, tempting authorities to seek cheaper, if less satisfactory, solutions. It should also be admitted that even the best incinerators of this type are known to produce dangerous fumes, so this method of disposal is not problem-free.

Other forms of waste such as those produced by many modern industrial plants pose even more serious problems of disposal. Some of them, like the waste from nuclear energy installations and certain chemical works, can present considerable health hazards. Normally, the producers are prevented from emitting these effluents into the air and local rivers by legislative controls, but these are not always adequate and in some cases insufficient to prevent serious accidents with which neighbours to nuclear power stations and large chemical works have had to cope. The maintenance of surveillance in these situations is an important public service, and modern technological equipment such as that used for monitoring the presence of radioactive materials and for protecting personnel against dangerous materials, performs a vital part of this service.

ROADS AND POSTAL SERVICES

Amongst those secondary services which, although not basic to the existence of civilized social life, go far towards making it more pleasant and satisfactory, the provision of good roads and a postal system are very important. Towns can function without properly surfaced roads and pavements, but the volume of urban traffic makes such improvements desirable at an early stage of town development, and corresponding improvements in roads to neighbouring districts and places further afield provides immediate benefits in terms of communications, so that the efficiency of a postal system is closely related to the condition of the roads. The pressure to improve urban roads came from local residents anxious to protect their property, and resulted in the paving of market places and the surfacing of roads with cobbles or granite setts. The condition of inter-urban roads was not susceptible to the same imperatives for improvement, although it was occasionally subject to military contingencies such as led to the construction of sound main routes between Paris and the frontiers of France, and to the military roads built by General Wade in the 1720s to pacify the Highlands of Scotland. Elsewhere, they were usually left to the inefficient care of local authorities with little incentive and scant resources to undertake substantial road improvement.

Industrialization brought both an increase in traffic, which caused more rapid deterioration of inadequate roads, and a new determination to improve the roads, stimulated in part by the need for a more efficient postal service. In Britain, this conjunction of means and needs produced a vigorous programme of road improvement in the second half of the eighteenth century, carried out mainly by Turnpike Trusts which sought increasingly effective professional advice from road builders like John Loudon McAdam and his family. The many miles of improved 'macadamized' road which resulted involved a fairly simple technology discussed in Chapter 8. Stage coaches were quick to take advantage of the improvements and which served, amongst other things, to extend and accelerate the mail services. Stages coaches were first used in place of mounted post-boys for British mail services in 1784, and by the 1830s these vehicles were able to average ten miles an hour over long distances, providing prompt deliveries of mail in most parts of the country. Later improvements became necessary in the roads with the advent of motor vehicles and the pneumatic tyre, because the latter tended to reduce the macadamized surfaces to dust with a consequent alarming increase in the rate of decomposition of the roads. The solutions to this problem, first by tar-spraying and then by the development of highly resilient tarmac and asphalt surfaces, supplemented in some places by concrete, restored the road system and were generally adopted both in the towns, where the old-established cobbled streets tended to disappear, and on the inter-city highways.

Postal services have taken advantage of successive improvements in transport technology, readily adopting the railways, often with the provision of special sorting coaches and other equipment, and then accepting the facility for rapid transit over long distances to all parts of the world offered by regular air-line services. Otherwise, however, they do not involve very sophisticated technological equipment, or at least they have not done so until the recent introduction of electronic devices for scanning and sorting letters. The introduction of Rowland Hill's 'penny post' in 1840 brought a great increase in the volume of mail handled by the postal services and the use of adhesive stamps created a need for perforating machines, making it easier to distribute stamps. Franking machines, used to cancel the stamps and to register the time and place of postage were also adopted, and letter boxes were introduced to facilitate the collection of mail, considerable thought being given to the manufacture of a safe receptacle, resulting in the cast-iron pillar box bearing the Royal monogram which, in a succession of forms, has become a traditional feature of the British postal service. More recently, the introduction of air-mail, area-coding, automatic sorting devices, pre-paid stamping machines, and optical character readers (OCR) (see Chapter 15) designed to read key elements in the address and thus accelerate the sorting process, have all helped to cope with an increasing volume of postal communications. Only the last of these involve a high degree of electronic sophistication, and for the most part postal services all over the world remain heavily dependent on human labour. They have, nevertheless, achieved a remarkably efficient standard of handling the mail.

TELEGRAPH AND TELEPHONE SERVICES

Most countries have assumed responsibility for operating their postal services as some form of public corporation, and many countries have extended this measure of state control to their telegraph and telephone services, although virtually all of these began as private enterprise ventures and in some countries they have remained such. The electric telegraph began in Britain in 1837 as the result of an invention by William Cooke and Charles Wheatstone, who succeeded in transmitting an electric signal along a wire and registering its reception with a deflecting needle on a dial (see p. 714). This system was adopted in the following year in Britain by the Great Western Railway in order to operate its signalling system, for which purpose it became indispensable to the railways in their great mid-century boom because it allowed distant signal boxes to keep in contact with each other. Samuel Morse meanwhile promoted his own system with its characteristic code of dots and dashes in the United States, and the network built up by him under the terms of the Morse Bill in 1843 played a vital part in the westward expansion of American settlement. The establishment of a cable across the English Channel in 1851 put the business communities of Paris and London into instantaneous contact with each other and demonstrated the practicability of submarine cables. Several attempts were made to lay a trans-Atlantic cable before the S.S. *Great Eastern* was successful in doing so in 1866, and thereafter other cables were quickly prepared to bring the advanced societies of the world into telegraphic contact. The achievement of this objective was signalled in 1872 when the Mayor of Adelaide in South Australia exchanged telegraph greetings with the Lord Mayor of London.

The telephone was invented in America in 1876 by A. G. Bell, but it made little progress until two years later when D. E. Hughes invented the electromagnetic microphone which, when added to the telephone, amplified the weak voice currents produced in Bell's device and thus made it suitable for long-distance communication (see p. 719). As such, the telephone was eagerly adopted by the American business community, and after a slower start it became a universal tool of business and domestic communication in Europe also. The first British telephone company was formed in 1878, and soon afterwards telephone exchanges were established in all the main towns and cities of the country and inter-city lines were constructed. By 1884, there were 11,000 'telephone stations' (i.e. offices and homes with telephones) in Britain, while the United States at the same time had 148,000. As with the development of the telegraph system, British telephony was pioneered by private companies and then taken over by the General Post Office and conducted virtually as a government monopoly. The telegraph network had been taken over in 1868, and by 1912 most of the telephone network had been similarly acquired. By 1958, the GPO was responsible for about 7 million telephones, amounting to about one for every seven inhabitants of Britain, which compared with 67 million telephones in the USA, giving about one for every two-and-a-half inhabitants. Technologically, the telephone service has continued to improve, with automatic exchanges, trunk dialling schemes, a trans-Atlantic telephone cable carrying 35 channels which was opened in 1956, and now the ever-expanding facility for satellite links through telephone stations in synchronous orbits round the Earth which has brought the whole of the planet within the scope of instantaneous voice communication. This has been a technological revolution of tremendous significance, and has provided an outstanding public service.

CONCLUSION

In several of the public services considered in this chapter, it is important to remember the contribution made in the past 25 years by electronic devices and especially by computers. These ubiquitous innovations of modern technology have made themselves indispensable in many aspects of industrial manufacturing, commerce, transport and communications. It is not surprising, therefore, to find them widely adopted in the public utilities like gas and electricity supply (though less obviously in water undertakings, which still impress mainly by their civil engineering works).

These valuable products of modern high technology should not be taken for granted. A sophisticated technology underpins all the basic services which the citizen of modern megalopolis has come to rely upon with remarkable confidence. Consternation is complete when there is a serious breakdown in the system, so that the lights go out, the elevators stop working and the water dries up in the taps. In such circumstances, an awareness of the complexities and interdependence of modern technologies is at least an aid to peace of mind and, at best, a help towards working out the solution to the problem.

There are, of course, other problems associated with many of the technologies employed in the public services in addition to the possibility of outright failure. Air and noise pollution, for instance, are insidious facts of modern town life; acid rain derives from our thoughtless consumption of carbon fuels; the urban environment tends to become spiritually and culturally restrictive. But without ignoring these problems, it is salutary to recognize the enormous service of technology to the community in the shape of the public services, and, while trying to eliminate harmful by-products and side-effects, to seek their extension to communities which at present do not enjoy their benefits. It is important, in short, to envisage a goal whereby the sort of public services at present enjoyed by citizens of developed nations are made available to everybody.

21

WEAPONS AND ARMOUR

CHARLES MESSENGER

PREHISTORIC WEAPONS

Among the earliest and most widespread of man's weapons is the spear. Originally it was merely a wooden pole with one end sharpened with a stone or piece of bone, but once palaeolithic man had discovered fire, some 500,000 years ago, charring was also used to harden and sharpen the tip. The next stage was to insert pieces of stone or bone in order to reinforce the point, and then to fit a stone head. From a very early period there were two types of spear, thrusting and throwing. The throwing spear, or javelin, tended to be lighter and in order to increase its range a device called the spear thrower was introduced. This acted as a lever and was a piece of shaped wood, bone or horn with a hook or recess into which the end of the spear fitted.

Two of man's other original weapons are the club and the axe. The club was originally made of hardwood, the head being larger than the handle. Like the tip of the spear, the head was then reinforced with stone. The original axe was made entirely of stone, simply an almond-shaped head sharpened by flaking. In about 3500 BC the wooden haft or handle was introduced, being attached to the stone head by means of bent wood, horn sockets, lashing and gum. Then, during the neolithic or New Stone Age era, 7000–2000 BC, the art of grinding, polishing and drilling of stone was developed, which radically increased the effectiveness of the axe, both as a tool and as a weapon of war. The head was now often fitted to the handle by means of a circular hole drilled through it, known as the 'eye'.

Perhaps surprisingly, the bow was already in existence around 15000 BC. It was first developed by the Mediterranean civilizations, and was taken up in northern Europe during the ninth millennium. From the start, yew, because of its good tensile characteristics, was the preferred wood, although in colder climates, where yew did not grow, elm and occasionally pine were used. Most bows were man-sized, and by the third millennium the composite bow, strengthened with horn and sinews, was in use in some regions. The string was normally made with plaited leather strips. Stone arrowheads were used, and the arrow itself was straightened by passing it through a hole drilled in bone or horn. In order to obtain arrows of standard size – important in terms of accuracy – they were shaved by means of a hollow tube cut as grooves in a split stone. The other basic weapon was the slingshot, a spherically shaped stone which was projected from a leather sling, which the firer whirled above his head in a circular motion in order to impart increased velocity to the stone.

THE BRONZE AGE AND CLASSICAL ERA

As with much else in the history of technology, the discovery of metal and the coming of the Bronze Age in about 2000 BC had a dramatic impact on weapons. The development of the forced-draught furnace, in particular, enabled the known ores to be smelted and to be fashioned into shapes that could not be achieved, or at least only with much difficulty, by the stone craftsman. Furthermore, damaged metal weapons could be recycled. An added advantage was that much longer cutting and thrusting weapons could be made. Daggers had existed in stone, but, using bronze and copper, the sword could now be made. Both these metals are, however, relatively soft, and in order to make a more durable weapon, which would not bend easily, the metal was strengthened by hammering, then by the addition of lead at the smelting stage (see Chapter 1). Initially, the sword was merely a thrusting weapon, with a strong central rib running down the centre of the blade and smaller side ribs, but gradually a cutting capability was introduced, with double cutting edges. With the discovery of iron, around 1000 BC, weapons became much tougher, but it was a much more difficult metal to work than copper and bronze, and hence for a long time the three coexisted.

The growing effectiveness of weapons in their ability to kill and maim caused increasing attention to be paid to personal protection. The original form of armour, consisting of layers of linen wrapped round the body, was used by the Egyptians in the third millennium. Hide was also used and gradually metal strips were introduced; the Sumerians in Mesopotamia had long cloaks reinforced with metal discs during the first half of the third millennium BC. Two basic types of early armour were scaled and lamellar. The former consisted of a short tunic on which were sewn overlapping bronze scales, while lamellar armour had pliable metal plates, or lames, which were laced together in slightly overlapping horizontal rows. Later, in about the 5th century BC, chain mail was developed by interlinking metal rings, or sometimes wire. Like body armour, helmets were originally made of cloth, but this gave way to leather, metal, or a mixture of the two. Apart from the basic conical style, helmets with cheek pieces to protect the face from sword cuts became popular. Often they were elaborately decorated, including horns and crests, not merely from male vanity but more to make the wearer look imposing and formidable in the eyes of his enemies.

The third major item of personal protection was the shield, which was certainly in common use by the beginning of the second millennium BC. Shields existed in several different shapes – round, rectangular and oval – and were made of leather, leather-covered wood and wickerwork. They also often had overlaid thin strips of metal which were used both decoratively and to provide additional protection.

Before 1000 BC the main centre for both military and political development was bound by the three major rivers of the Middle East, the Nile, Euphrates and Tigris, with the two dominant countries being Mesopotamia and Egypt. From about 3500 BC the dominant weapon in Mesopotamia was the chariot, which gave warfare much greater momentum and punch than hitherto. Originally it was drawn by asses, until the horse arrived from the steppes of Mongolia around 2000 BC. Chariots were used to make frontal charges on the enemy in order to create panic, their crews being equipped with both javelins, to engage at medium range, and spears for hand-to-hand fighting. By 1500 BC, with the development of the spoked wheel, means were found to make the chariot lighter and hence more mobile, thus increasing its effectiveness as a weapon of shock action. Surprisingly, the Egyptians did not use the chariot until about 1600 BC, but it quickly became the basis of their military might. Armed also with the double convex shaped composite light bow, with a range of 275–365m (1200–1600ft), which they used both mounted in their chariots and on foot, they became a formidable force. Indeed, it was the arrow projected by the light composite bow, with its reed shaft and bronze head, which brought about the need to consider personal protection.

It was not until the rise of the Assyrians at the end of the second millennium BC that horse cavalry began to appear, and then in only a secondary role on the battlefield, being used to harry the enemy's flanks, while the chariot remained the decisive weapon. Early cavalry were armed with both bows and spears, but their horses had merely a bridle, with no stirrups.

So formidable was the Assyrian army that opposing forces would not take to the field against it if they could avoid it. Instead they relied on the protection of the fortified city, a concept which had been in existence since the third millennium. An example is the fortress of Megiddo, which was built at the beginning of the nineteenth century BC. The base of its main wall was 2.13m (7ft) and it had 5.5m (18ft) salients and recesses, with a crenellated parapet on top. To counter these strong defences the Assyrians introduced battering rams designed to break down the main gates to the city. They were mounted in wooden towers, which were roofed and protected by metal plates, and were borne on six wheels. Under the roof was a platform used by archers to shoot at the defenders on the walls. Tunnelling and scaling ladders were also used.

By 500 BC, the Greeks had become the major military power and they made two significant contributions to the history of warfare. The first was the phalanx, a close order formation made up of hoplites, infantry equipped with 2.44m (8ft) spears and swords and dressed in horsehair-plumed helmets, breastplates and calf and shin plates known as greaves, with a 0.91m (3ft) diameter round shield held on the left arm. This tightly packed 'mobile fortress' was frequently more than a match for looser and less well disciplined bodies of enemy. The other development engineered by the Greeks was the invention of torsion artillery, in the shape of the catapult. It was the Alexandrian mathematicians who developed the theory of the catapult, showing how there was a direct correlation between the proportions of the various parts and the diameter of the 'straining hole' through which the skeins which controlled the tension passed, and the Greeks who put it into practice. They had two types of catapult (or *ballista*, as the Romans were to call it). The *katapeltes* were used to project arrows, javelins and smaller stones – a 3.63kg (8lb) stone could be projected accurately to a range of 228m (750ft) – while the larger *petrobolos* could hurl stones of up to 25kg (55lb) in weight. The skeins themselves were made of twisted human hair and sinew. A further refinement was the use of fire arrows, either with their heads wrapped in inflammable material and ignited just before firing, or made red hot by heating in coal fires.

Unlike the Greeks, the Romans were not innovators but very practical engineers, who applied the ideas of their predecessors. Perhaps their most outstanding feats of engineering were the numerous aqueducts which are still to be seen today. The Romans have been called 'the greatest entrenching army in history' and it was a constant principle that when legions halted after a day's march they constructed a fortified camp, usually square in shape, with ramparts, palisades and ditches. Apart from the comfort afforded, it also meant that they always had a secure base from which to operate. Roman camps, especially those near rivers, are the foundation of many of today's European towns and cities. As with the Greeks, the main element was the regular infantry of the legions, whose members were armed with a short stabbing sword, javelins and spears. The main shield used as the *scutum*, large and semi-cylindrical rectangular in shape, which when rested on the ground would come up to a man's chest. With this shield they went one stage further than the mobile phalanx of the Greek hoplite by developing the *testudo* or tortoise, especially useful in sieges. While the outside ranks protected the front and flanks with their shields, those on the inside put theirs over their heads in order to provide protection from arrows and missiles fired from above. Cavalry still played a secondary role and, indeed, the Romans tended to rely on mercenaries or 'auxiliaries' to provide it as well as their archers and slingers. One new weapon of war introduced was the elephant. The Greeks had used it as heavy cavalry, but it was the Carthaginians who brought it to the fore at the end of the third century BC, and their celebrated general Hannibal took elephants on his march across the Alps which led to defeat of the Romans at Cannae in 216 BC. The Romans finally gained their revenge at Zama in 202 BC by using trumpets to panic and stampede the beasts.

THE DARK AGES

By 400 AD the Roman Empire was beginning to disintegrate in the face of Vandals, Goths and then Huns from the east. These races relied almost entirely on the horse, and they brought with them the stirrup, which was responsible more than anything else for the gradual elevation of cavalry to the decisive arm on the battlefield. Stirrups gave the rider the necessary stability to withstand the shock of the charge and meant that cavalry could now be committed against the main force of the enemy rather than merely harrying his flanks and being used in the pursuit. This, combined with brilliant horsemanship and the hardiness of their mounts, which could travel up to 160km (100 miles) a day, as well as the use of mounted archers *en masse*, made the Huns especially a devastating military power.

One relic of the Roman Empire in south-eastern Europe became the Byzantine Empire, which successfully withstood the ravages of Huns, Arabs, Turks and Magyars for five centuries before succumbing to the Turks at the Battle of Manzikert in AD 1071. The Byzantines built their armies around the cavalryman, having taken the Hun model to heart. He was dressed in a long chain-mail shirt down to his thighs with steel helmet, gauntlets and shoes and was armed with a heavy broadsword, dagger, short bow and lance. His horse had a saddle and iron stirrups and was sometimes protected by a poitrel or breastplate. There were two types of Byzantine infantry, light, which mainly consisted of archers, and heavy, armed with lance, sword and axe. In battle they held their ground while the cavalry achieved victory through repeated charges at the enemy.

During this period the first chemical weapon made its appearance. This was Greek Fire, which was certainly used by the Byzantine defenders of Constantinople during some of its numerous sieges in the seventh and eighth centuries. It was, in essence, liquid fire and a forerunner of today's napalm. No precise recipe for making it exists, but the main ingredients were sulphur, pitch, nitre and petroleum, which were boiled together. It was effective on both land and sea and appears to have been used in siphons or as a grenade made of glass or pottery. During sieges it was often delivered by means of *ballistae*.

The equivalent to the Byzantine Empire in the west was that of the Franks. Initially they were reliant on infantry, which was only lightly armoured and armed with axes and barbed javelins, used as both throwing and thrusting weapons. The axe, which was rather like the Red Indian tomahawk, was an accurate and lethal throwing weapon, as well as being used in close combat. Not until towards the end of the sixth century did the wealthier of the Franks begin to equip themselves with armour, and they really came into prominence with Charles Martel's victory over the Arabs at Tours in 732. Under his grandson Charlemagne, Frankish military power reached its zenith. By now, like the Byzantine army, the core was the heavy cavalry, armed and protected in much the same way. The weakness lay in their relative indiscipline, and after the death of Charlemagne, Frankish might withered away.

During the ninth and tenth centuries western Europe was also ravaged by the Vikings from Scandinavia. The main Viking weapons were the heavy battleaxe and the sword. Their sword blades were formed by twisting together strips of steel and iron and heating the resultant 'plait' and beating it flat. The different ways in which the two metals reacted to the process resulted in a distinctive decorative pattern being engrained in the blade. Viking shields were wooden and initially round, but these gradually gave way to a long triangular shape with slightly convex top and sides.

THE AGE OF CHIVALRY

In western Europe the Normans took over from the Franks as the dominant military power and this marked the beginning of the golden age of the mounted knight. The Norman knight was equipped in much the same way as the Byzantine and Frankish heavy cavalryman, but he was better protected in that he also had chain mail leggings. His helmet was of a distinctive conical shape with a nose piece, and he carried the later Viking style of shield. His lance was normally 2.40 – 2.75m (8–9ft) long and his two-edged sword was 112cm (44in) in length. These would remain as such for 400 years. He also carried on his saddle a battleaxe or a mace. This differed from the primitive club in that, instead of being entirely of wood, it had a stone or metal head, and it was this weapon which was to force the change from mail to plated armour.

The secret to the success of the mounted knight was his use in combination with archers. King Harold of England's defeat by the Normans at the Battle of Hastings in 1066 is a classic example of this tactic, with the archers being used to disrupt the Saxon infantry, armed with spears and axes, and then the knights charging. The conventional bow was still being used at this stage, but it would not be long before it was rivalled by a new version, the crossbow. Its origins lay in the handbow modified for use in an animal trap: it was left loaded and sprung by means of the victim pressing a cord or rod. It was the Chinese who transferred this concept into a hand-held weapon. The essence lay in a lock which held the bowstring back until released by a trigger. In order to tense the bowstring, the archer anchored the bow with his feet and then drew the bowstring back, either by hand or with the help of an iron hook attached to his waistbelt, and then hooked it in the lock. Later models employed a windlass to draw the bowstring back. In this way he could obtain a far greater tension than on the conventional bow and hence a longer effective range. The arrow itself was inserted in a groove which was cut into the stock. By the eleventh century the crossbow had reached western Europe, but because of its lethality when compared with other weapons of the time the Church considered it barbarous; the second Lateran Conference of 1139 declared it an unsuitable weapon of war for Christian armies and laid down that it should only be used against infidels. This did not deter elements of Christendom from wholeheartedly adopting it, in spite of the punishment of excommunication threatened by the papacy, and Genoese and Gascon crossbowmen became much in demand as mercenaries in many European armies.

The English, however, although they used the crossbow for sporting purposes, had a rival to the crossbow, the longbow. This was reputedly introduced from Wales in the latter part of the thirteenth century and was as tall as a man. It had almost the same penetrative power as the crossbow, but with the added advantage of a much higher rate of fire, because the arrow could be strung more quickly. In mediaeval Britain archery was the major sport and longbowmen soon became very proficient: this was demonstrated at the Battle of Crécy against the French in 1346, where the longbow, because of its rate of fire, dominated the crossbow.

By the thirteenth century the penetrative power of the crossbow and longbow, together with the crushing blows which could be delivered by the mace, meant that chain-mail armour was no longer effective and plate armour began to be introduced. The first item to change was the helmet, which now totally enclosed the knight's head and face, with two slits so that he could see through it. This gave way to the basnet, which was shaped like a skull cap, with a collar attached to protect the neck and a hinged visor. Over his chain-mail shirt (the hauberk) the knight began to wear a coat of plates, his leggings were similarly overlaid with plate armour, and he wore gauntlets made of steel plates and scales. During the fourteenth century complete plate armour was developed, built round the breastplate or cuirass, and it was now that the mediaeval craftsman began to demonstrate a high degree of engineering skill. His object was to give the knight complete protection yet without hindering his movements. This meant not only acquiring the ability to perfect the hinge, but also a sculptor's understanding of human anatomy, and the construction of body armour became as much an art as a science. For the armour to function properly it needed to be made to measure, using a light but strong steel, and a knight was prepared to pay a very high sum for it. By the fifteenth century, Milan and Augsburg had become the two main centres of armour manufacture.

If the knight was now becoming better protected, his horse was not. While he could fight effectively when mounted, once on the ground his armour meant that he was relatively immobile, and to have his horse killed under him made the knight very vulnerable. This led to the development of the bard, or set of horse armour. Initially this consisted of a trapper made of mail or quilted cloth, which protected the body and upper legs of the horse and looked like a horse blanket, but gave way gradually to plate. This included the shaffron, made of articulated plates to protect the head, the peytail to cover the chest, and body plates. The problem was that a complete bard drastically impeded the horse's mobility and only parts of it were usually worn. As it was, in order to carry the knight, a breed of large horse was gradually introduced, whose descendants are reflected in today's heavy draught horses.

This up-armouring of the knight and his horse meant that the shield progressively became less important and it was gradually reduced in size to a small triangular shape. With the heraldic emblem of its owner painted on it, it became a means more of identifying individuals on the battlefield rather than of protection.

The nature of the feudal system which dominated the Middle Ages, in which nobles were granted land in return for their allegiance to the monarch, resulted in a growth of castles so that these lands could be better protected. This in turn meant that sieges became a dominant characteristic of mediaeval warfare. The simple Norman keep gave way to increasingly elaborate designs of castle. In order to prevent the besiegers from sapping up to the walls by excavating trenches, moats were constructed, with drawbridges leading across to the entrances. These themselves had a metal grille, a portcullis, which could be lowered in order to inhibit the use of battering rams on the wooden gates. Outworks were developed, including water defences, with sluices to control the water levels. Towers were built at the outer corners of the walls so that fire from each pair could completely cover the wall between them.

Siege artillery did not increase in effectiveness at the same pace. There was, however, one new development, the trebuchet. The drawback of the ballista was that when the fibres used to propel the missile became wet, they lost their resilience and therefore catapults were seldom suitable for siege warfare in more northerly areas. The origins of the trebuchet are to be found in the counterbalanced well sweep used by the Egyptians in the second millennium BC. This worked on the principle of an unequal lever with a heavy weight attached to the shorter arm. The longer arm had a bucket on the end of a rope and the force created by releasing the weight would jerk the bucket upwards and fling its contents in the same way as a sling. It was the Chinese who first adopted it as a weapon, the *huo-pa'o*; they were certainly using it as such by the eleventh century AD and within a short time it had been introduced into Europe. The trebuchet could provide an effective means of creating a breach in a castle's walls, and a stone of 135kg (300lb) could be thrown with significant force to a range of 275m (900ft). Animal carcasses were also used as projectiles in order to create pestilence. Indeed, the defences of castles were often so formidable that disease or starvation often became the only means through which the besiegers could gain ultimate success.

THE INTRODUCTION OF GUNPOWDER

Gunpowder is generally considered to be the most significant technological development in the history of warfare, yet its origins are obscure, and have often been confused with those of Greek Fire. It would seem that the Chinese used it in fireworks but never as a weapon of war and, unlike other Chinese inventions, the Europeans discovered gunpowder independently. The first recorded recipe for its use is attributed to the English monk and scholar Roger Bacon. He recommended a mixture of six parts of salpetre (potassium nitrate), five of charcoal and five of sulphur, which produced 'a bright flash and thundering noise'. This, however, was to use it simply as an explosive rather than a propellant, for which different proportions are required. In essence, the saltpetre, having a high oxygen content, is the fuel for the combustion of the charcoal and sulphur. The charcoal produces significant amounts of carbon monoxide and dioxide accompanied by much heat, while sulphur, burning at a lower temperature, promotes the combustion of the mixture and acts as a paste to bind the other two substances together. It also ensures even burning, which is essential for a propellant. When deciding the proportions required to make gunpowder for a particular purpose, it must be remembered that charcoal and sulphur accelerate combustion, while saltpetre retards it. On the other hand, charcoal produces great amounts of heat and gas, while sulphur causes quick burning but gives rise to solid residues.

The first recorded use of gunpowder as a propellant in a firearm was at Metz in 1324, and the earliest type of gun was made of iron and cast in the shape of a vase. It was called a *pot-de-fer* and was placed on a flat board supported by trestles. It had a touchhole on top and the propellant was ignited by a glowing tinder at the end of a rod. Both arrows and stone and iron shot could be fired from it. By the latter half of the fourteenth century hand-guns were also in use. The earliest models had pole-like butts attached to the barrel by iron bands and were designed to be held under the arm or resting on the shoulder, with the other hand being used to apply the match to the powder in the touchhole. Both hand-guns and cannon, or bombards as they were more commonly called at the time, often had hooks on the bottom in order to help control the recoil. Another type of artillery weapon in service by the end of the fourteenth century was the *ribauld*, which looked like a set of organ pipes and, mounted on a form of chariot, could fire a number of projectiles sumultaneously – an early form of multi-barrelled rocket launcher.

These early guns were of only limited range and were very inaccurate, as well as often being as dangerous to the firer as to his enemy. Their impact on the battlefield was therefore only to be gradually felt. One early development, however, was that they produced a demand for a new type of mercenary. Because nations did not maintain standing armies and the handling of guns required a particular expertise which part-time soldiers could seldom hope to achieve, a breed of professional musketeers and artillerymen came into being.

THE RENAISSANCE

In spite of the introduction of gunpowder, cavalry remained the dominant arm throughout the Middle Ages, but towards the end of this period there were signs that the scales were beginning to swing back in favour of the foot soldier. The infantryman's weapons had remained the spear, usually over 2.4m (8ft) long, and the sword. The problem with these was that they did not give him enough stand-off distance when engaging charging cavalry. A solution to this came in the fifteenth century with the pike. It was the Swiss who first developed this as a spear, but lengthened to up to 6.75m (22ft). With this they demonstrated that they could not only keep the cavalryman with his lance at bay, but also use it as an offensive weapon, and other nations soon followed suit. The Swiss, too, along with the Germans, also developed the halberd. The word itself comes from the German *Halm*, 'staff', and *Barte* 'axe', and it developed as a multi-purpose weapon at the end of the thirteenth century. It was, in effect, an axehead mounted on a 2m (6.5ft) pole with a spike on top. In the fifteenth century a second spike was added behind the axe blade in order to break open helmets and armour, and this new weapon took the name of halberd, while its predecessor became known as the *vouge*. Pikemen and halberdiers now operated together, the latter wielding their weapons with both hands. In view of its shorter length and many uses, the halberd was a very much more flexible weapon in close combat than the pike, although often it was the concentrated force produced by the mass of pikemen in the advance, the so-called 'push of pikes', which carried the day.

The hand-gun also made a significant advance during the fifteenth century with the invention of the matchlock; for the first time the firer could use both hands to steady the weapon while firing. It consisted of a spring and tumbler system released by a trigger, which would activate the serpentine ('S'-shaped lever) which held the burning match (cord of twisted flax, hemp or cotton soaked in saltpetre, which smouldered rather than burned) and introduce it into the priming pan, thus firing the charge. An added refinement was the addition of a cover over the pan to protect the charge from damp. The matchlock hand-gun was called an arquebus, and it was the Spanish arqebusiers who first proved its advantages during a series of campaigns in the first quarter of the fifteenth century, culminating in the Battle of Pavia against the French in 1525. The arquebus, which marked the knell of the bow, was also able to keep both cavalry and pikemen at bay, provided the fire of the arquebusiers was strictly controlled by the use of volleys. Nevertheless, it suffered a number of disadvantages. For a start it was cumbersome and heavy, and therefore had to be rested on a stand in order to fire it. The rate of fire was also very low, about one round every two minutes, and the arquebus was clearly of little use when it came to hand-to-hand fighting, in which the firer needed the protection of the pikemen and halberdiers.

By the mid-sixteenth century the cavalryman was able to regain some of the power that he had lost to the arquebus through the pistol. The word originally pertained to any short-barrelled hand-gun which could be fired with one hand. Matchlock pistols existed, but these did need two hands to operate, if only in order to adjust the position of the lighted match, and hence were of little use to the cavalryman, who needed one hand to control his horse. The problem was solved with the invention of the wheel-lock in *c.*1500. The principle is much the same as that of the cigarette lighter. When the trigger was pulled it enabled a spring loaded wheel to revolve. This in turn caused a cam to open the pan cover, while the wheel itself made contact with a piece of iron pyrite held in a cock which was lowered into position before firing. This contact produced the spark which ignited the primer in the pan. Armed with this weapon, German cavalry developed a new tactic, the caracole, whereby each line of cavalrymen rode up to the opposing infantry, discharged their pistols and then retired to the rear to reprime and reload. Fifty years later came a further development, that of the snaphance lock or snap-lock. Here the lock, when released forwards by the trigger, came into contact with a slightly convex steel plate mounted on an external arm above the pan, and it was this which produced the spark. Its main advantage over the wheel-lock was improved reliability.

During the latter half of the fifteenth century there were also significant improvements to artillery guns. By the end of the century they could be divided into two categories, heavy ordnance, which was held in reserve until required in siege operations, and light or field artillery which gave direct support to infantry and cavalry on the battlefield. Barrels were now cast in bronze and by the 1450s heavy guns weighing as much as 35 tonnes were in existence. These had the capability of successfully engaging the thickest castle walls, as the Sultan Mohammed II demonstrated at the siege of Constantinople in 1453. Employing no less than 62 heavy guns firing 91kg (200lb) cannonballs, he demolished the triple walls of the city in many places within six weeks. Field artillery began to be made in standardized calibres, firing projectiles ranging from 0.9kg (2lb) to 29kg (64lb) in weight. Carriages were constructed so that the guns could be moved and set up for firing quickly. The earliest type consisted of a solid wooden beam with the front end mounted on an axle fitted with two wheels and the rear end resting on the ground in such a way that the gun, which was mounted on the front end by the means of iron hoops, was in a horizontal position. In order to adjust elevation, and hence range, the lower part of the beam, which later became known as the trail, had either to be dug in or raised by the means of stones or sods of earth. To obtain any accuracy was clearly a time-consuming business, but this was radically speeded up by the beginning of the sixteenth century with the introduction of gun trunnions. These were lugs projecting from either side of the barrel and slotting into iron-clad grooves cut into twin parallel beams, which replaced the pre-trunnion single beam. Transoms and crossbeams held the gun in place and alterations in elevation were achieved by wedges inserted between the breech and one of the transoms. Formulas had by now been worked out for the correct mix of gunpowder ingredients to give the best performance to each type of gun. These improvements meant that, after almost two centuries, artillery had joined cavalry and infantry as a vital element in any army.

It was during this period that the origins of today's armoured fighting vehicle can be discerned. While the concept of the 'war car' originated with the chariot, especially when crewed by archers or javelin throwers, the first to harness the concept to gunpowder was John Zizka, leader of the Hussite movement in Bohemia in the early 1420s. He mounted guns on peasant wagons and, when threatened by the enemy, formed them into a circular laager in much the same way as the American settlers and the Boers in South Africa would do some 450 years later. There are, however, few concrete examples of other armies taking this idea up, although Leonardo da Vinci designed an armoured war car and similar ideas appeared in other military treatises during the sixteenth and early seventeenth centuries.

SEVENTEENTH AND EIGHTEENTH CENTURIES

The improved effectiveness of guns during the sixteenth century caused the designers of armour to look for means of giving protection against bullets. At the same time, full armour was becoming increasingly unpopular because of its discomfort and restrictions on mobility, especially with foot soldiers who often refused to march in it. Consequently, it gave way to 'half armour', which was little more than a cuirass, while full armour was only worn on ceremonial occasions.

Armour was still seen as primarily a defence against cutting and thrusting weapons and, as such, was to be retained by cavalry for some time to come. For the infantry, apart from the problem of discomfort, its final demise was brought about by the disappearance of the pike from the battlefield. The reason for this was the gradual introduction of the bayonet from the second quarter of the seventeenth century. The word itself has two derivations, both French. In the city of Bayonne, in south-west France, a larger dagger called the *bayonnette* was introduced, while the verb *bayoner* means to put a spigot in a cask. The French were the first to adopt it and the initial model was called the 'plug bayonet' because it was literally plugged into the muzzle of the musket (see below). This had the obvious disadvantage that the soldier could not fire his weapon with the bayonet fixed, but it nevertheless made the pike superfluous, thus greatly simplifying the handling of infantry on the battlefield. By the end of the seventeenth century this problem had been overcome with the introduction of the ring bayonet, which had a sleeve that slid on to the barrel. In order to clear the line of fire from the muzzle the blade was offset. While the pike quickly disappeared, the halberd was retained for many years to come by some armies, being carried by senior non-commissioned officers and used to adjust the soldier's aim during volley firing.

Much attention was also paid to making the infantryman's firearm easier to handle and more efficient. By the second half of the seventeenth century, the matchlock had given way to the wheel-lock and snap-lock and these were brought together into the flintlock, which combined the steel and pan cover as one element. It also had an intermediate position for the cock, 'half-cock', which could be used as a safety mechanism when the weapon was loaded. In order to simplify loading, the paper cartridge was developed. Initially, this was nothing more than a screw of paper which contained a measured amount of propellant charge, but soon the ball projectile was also included. Finally, the weapon was considerably lightened, which meant that the rest was no longer required. By the end of the seventeenth century the military smoothbore flintlock had become generally known as the musket (although the term itself was originally derived from the Italian *moschetto*, meaning a sparrow hawk, which was used to describe a small artillery piece in the culverin class). The musket would undergo little change during the eighteenth century, and in the hands of an experienced infantryman it had a rate of fire of two rounds per minute and an effective range of 230m (750ft).

The firepower of the mounted soldier was also enhanced during the seventeenth century by the introduction of the carbine. This was, in effect, a shortened musket with a sliding ring fixed to a bar so that it could be carried on a shoulder sling. By the end of the century, as a result of this innovation, cavalry was divided into two types: the horse, which was armed with sword and pistols and fought mounted, and dragoons, with sword and carbine, who were mounted infantry and did not wear armour.

The increasing effectiveness of firearms meant that, apart from cavalry, the sword's importance as a weapon of war declined. During the sixteenth century priority had begun to be given to thrusting rather than cutting and the art of fencing was developed. This led to lighter and slimmer blades and the introduction of the rapier and smallsword, which became a standard article of civilian dress for any eighteenth century gentleman. While the foot soldier now had the bayonet to take the place of the sword, the early part of the eighteenth century saw the sabre begin to be used by cavalrymen. They still required a cutting weapon, and the long, curved, single-edged blade of the sabre, which came to Europe from Asia, effectively satisfied this need.

Another new infantry weapon which appeared during the seventeenth century was the hand grenade. Originally the term grenade applied also to explosive balls fired from artillery, since both worked on the same principle of a hollow ball filled with black powder (another name for gunpowder) and fitted with a short fuse. By the end of the century European armies had companies of grenadiers within their infantry battalions, easily recognizable by their tall conical caps, since it was found that the standard brimmed or tricorn hat was too easily knocked off in the act of throwing the grenade. By the middle of the eighteenth century, however, the hand grenade had begun to fall out of favour and it would not reappear for another 150 years (see p. 991). A variation on the hand grenade was the grenade projector or launcher. This came in two types, either a firearm with a specially designed barrel or a discharge cup which could be fitted in the muzzle of a musket. Because the fuse had to be lit before the grenade was fired, there was the inherent danger of a premature explosion caused by a misfire and this weapon was never widely adopted. Like the hand grenade, it would have to wait until the twentieth century before being perfected.

The grenade fired from artillery weapons, later called a bomb, was used with the mortar, its name coming from the Latin *mortarium* on account of its early shape, which was like that of the mortar which, with a pestle, was used to grind substances. Short and squat, the mortar could fire very large projectiles at a very high trajectory over walls and earthworks and hence, from the fifteenth century onwards, it became a valuable member of the siege train. It was not until the beginning of the seventeenth century that it began to fire explosive projectiles, and indeed was the only type of artillery which could, because of the time lag in a conventional gun between lighting the fuse, ramming it down the barrel and firing. These early large mortars did have one drawback in that the force of recoil was almost entirely vertical and the base of the mounting therefore had to be wide and static in order to spread the shock. Hence it was not fitted with wheels and lacked the mobility to be a battlefield weapon, being used merely in long sieges and as part of the fixed armament of fortresses. Naval vessels, called bomb ketches, were also introduced in the eighteenth century. Equipped with mortars, they were used for bombarding forts and fleets in harbour.

A new type of gun, as opposed to mortar, made its appearance towards the end of the seventeenth century. This was the howitzer, derived from the Ger-

Table 21.1: European field artillery *c.* 1700.

Calibre	Names	Point blank range* (given in paces. 1 pace = 0.58m or 0.63 yd)	Max. range at 45°
24-pdr	Demi-canon (Spain) Demi-canon (France) 24-pdr (England)	8–900	5000
16-pdr	Culverin (England) Quart-canon (Spain)	450	5000
12-pdr	Quart-canon (France) 12-pdr (England)	400	4500
8-pdr	Demi-culverin (England)	400	4000
6-pdr	Saker (England)	500	3500
4-pdr	La Moyenne (Spain) Courte-Moyenne (France)	300	3500
3-pdr	Minion (England)	450	3200

* Point blank range in those days meant the range achieved with the barrel at the horizontal.

man *Haubitze* (= catapult). This came in between a mortar and a field gun and had a comparatively short barrel and a marked parabolic trajectory. Like the mortar it fired bombs and also suffered from the problems of requiring two 'fires', one to light the fuse and the other to detonate the propellant charge, with the inherent danger of premature explosion. Because of the problem of co-ordinating the burning time of the fuse with the desired range, air-bursts were common (and largely ineffectual, since this was before the invention of shrapnel: see p. 986). Nevertheless, the howitzer, because it was mounted on a gun carriage, did have advantages over the mortar in longer range and greater mobility. It also became more reliable once the double-linstock had been introduced and perfected. The linstock was a pole with slow match attached to fire the charge, and the double-linstock enabled the two 'fires' to be carried out simultaneously.

By 1700 field artillery had become further standardized by type in that there were generally common sizes used by all European armies. Table 22.1 gives an idea of these and their ranges.

Smaller calibres did exist, being known as 'regimental guns', implying that they were an infantry rather than artillery weapon, but these were disappearing. Navies had larger calibres, usually the 36-pdr, but with some 40-, 42-, 48- and even 64-pdrs.

The most common type of ammunition was the solid shot, now usually of cast iron and known as roundshot. To ensure that it was of the right size for a particular calibre of gun it was passed through a ring gauge before loading. Heated shot was sometimes used, normally by shore batteries against ships, with a wet wad placed between the charge and the shot in order to prevent premature detonation or 'cock-off'. At sea, chain-shot, two roundshot linked by a chain, was often used to cut the enemy's rigging. Varieties of grapeshot, normally musket balls in a container, were available to engage infantry and cavalry at close quarters.

Priming became more efficient in the early 1700s with the introduction of a powder-filled tin tube which was inserted in the touch-hole, and bag charges, measured amounts of propellant sealed in canvas bags. The early part of the eighteenth century also saw the first serious attention being paid to the science of ballistics. Before then guns were either laid directly, that is by lining the target up with the line of the barrel, or by using a gunner's quadrant for longer ranges than point blank. This consisted of two arms, one longer than the other, with a metal arc between them marked in $12°$ intervals and a plumb line. The longer arm was placed in the bore and the angle of aim above the horizontal indicated by the plumbline. Each marking on the arc was known as a 'point' and range was calculated in terms of the number of times above point blank range that each point represented, with $45°$ mistakenly believed to be the correct setting to achieve maximum range for the piece. The first development was the drawing up of mathematical rules for firing mortars by the Frenchman de Blondel, who drew largely on Galileo's work on the laws of movement. Little note was taken of de Blondel in military circles, but another Frenchman, de Bélidor, in his *Le Bombardier français* (1731) went several stages further, publishing complete tables of ranges for charges and elevations, and furthermore proved that the charges currently used were excessive, resulting in premature wearing of barrels and carriages and thus decreasing accuracy. Until this time there had been a long-held theory that the ideal weight of charge was two-thirds that of the shot, but as a result of de Bélidor's findings, which showed that the power of a charge was greatest at the end of the explosion, the regulation amount of charge was reduced by half, paving the way for the construction of shorter and lighter guns.

The practitioners of eighteenth century warfare regarded it as a science rather than an art and believed that it could successfully be waged according to precise rules. Nowhere was this better demonstrated than in siege warfare. The seventeenth century is remarkable for the number of distinguished military engineers it produced. Most notable of all was the French Marshal Sébastien le Prestre de Vauban, whose influence on the art of fortification was to outlive him by almost 200 years. While retaining the traditional fortress layout of inner enclosure, rampart, moat and outer rampart, Vauban applied the principles of geometry to ensure that the enemy was subjected to the maximum amount of fire, by both cannon and musket, at any point in the defences. The basis of his design was the obtuse-angled bastion, with the point facing the enemy, thus giving him a very small target, and with the ability to cover the walls adjacent to it with enfilade fire. The core of the fortress was shaped like a pentagon and successive lines of defences forced the enemy to commence his siege operations from a distance as well as providing defence in depth. Vauban was also adept at laying siege to fortresses and his major contribution to this art was the digging of parallel trenches, connected by zig-zags, which enabled the besieger to engage any part of the defence from behind cover.

THE INDUSTRIAL AGE

The Industrial Revolution of the late eighteenth and first half of the nineteenth centuries had as dramatic an effect on military technology as it did on all other aspects of life. One of the most significant changes was brought about by the development of the factory and the mechanization of manufacturing processes. Not only did this mean the beginning of mass production of arms, but also that their parts became interchangeable. Indeed, the first to take advantage of this was a French gunmaker, Le Blanc, who in the early 1780s began to manufacture musket parts, which, after checking with a standard gauge, could be fitted to any of the same type of musket. The concept was taken up in the USA before the turn of the century by Eli Whitney, and by 1820 the principle had been introduced into the US government armament factories at Springfield and Harpers Ferry. In addition, by the end of the eighteenth century the art of boring out the barrels of cannon had reached a level where, for the first time, true standardization of calibre could be guaranteed (see pp. 396–7).

For the infantryman, the most significant development at the onset of the nineteenth century was the rifle. The idea of rifling, that is, scoring spiral grooves inside a gun barrel in order to impart spin to the projectile, and hence achieve greater range and accuracy, was not new. Sporting rifles were in vogue from the beginning of the sixteenth century, but there was little military enthusiasm for them because of the difficulties of manufacture, a consideration which applied to a number of seemingly promising new concepts of the nineteenth century. Not until around 1750 did armies begin to pay serious attention to the rifle. It was first used in Austria and Germany, where it was called a *Jaeger*, or 'hunter', and the troops who used it were light infantry who adopted the same name. Swiss and German immigrants took it to North America and from it the famous Kentucky rifle was produced. The British, too, at the end of the century formed a Corps of Riflemen armed with the Baker rifle. The main problem with these early rifles was how to reconcile the requirement for the ball to be easily rammed home with the need for it to be a tight enough fit for the rifling to have effect. The first solution to this came in the 1820s when a Frenchman, Captain Delvigne, designed a rifle with a smaller chamber than the bore. Using an iron ramrod, the firer tapped the soft lead ball into the chamber in such a way as to expand its shape to fit the shoulders of the rifling. The main breakthrough came, however 25 years later when another Frenchman, Captain Minié, introduced his Minié ball. This had a hollow base and the introduction of the hot propellant gases into it caused the bullet to expand into the rifling. The bullet itself, rather than being spherical, was cylindro-conoidal in shape and marked the beginning of the bullet shape as we know it.

Early rifles still worked on the flintlock ignition principle, but in the 1800s a Scotsman, Alexander Forsyth, produced an idea which would radically reduce the incidence of misfires in firearms. The percussion lock employed a volatile chemical detonating powder, or fulminate, contained in a bottle-shaped magazine which could be rotated so as to allow a quantity of fulminate into the pan. This was then struck by a hammer-shaped cock. The next development followed quickly. This was the percussion cap, wherein the bottle-shaped magazine gave way to individual pellets, copper tubes and finally copper caps of fulminate. The greatest breakthrough came, however, from a Swiss, Samuel Pauly, who, in conjunction with the French gunsmith Prélat, produced in 1812 the first cartridge to contain its own primer. It was made of paper, but with a reusable metal base with a cavity in the centre for the fulminate, and a striker operated by a spring causing ignition. This was called centre-fire and is the most common type of cartridge today. Another current type, rimfire, where the fulminate is contained in the rim of the cartridge base and crushed between the striker and the face of the breech, was also developed at much the same time.

These early cartridges relied on breech-loading, the next highly significant development in firearms. Like rifling, breech-loaders had been in existence for a long time; models were developed as early as the fourteenth century. Various attempts were made through the ages to develop a satisfactory breech-loading system – removable barrels, pivoted and separate reloadable chambers – but the main problem was achieving a reliable gas-tight seal, or obturation. Pauly's work on the cartridge encouraged a former employee, Nikolaus von Dreyse, to make the necessary breakthrough. In 1827 he produced a cartridge with fulminate contained in its base. A few years later he created a bolt-action breech-loader to take this cartridge, with a striking needle operated by a spring. Von Dreyse's needle gun, as it was called, was adopted by the Prussian army in 1840 and made a major contribution to its victories over the Danes in 1864 and the Austrians in 1866. During the next thirty years all armies introduced breech-loaders, many based on the principle of the needle gun, with a sliding and sometimes rotating bolt to assist obturation. A totally gas-tight breech was only finally achieved, however, when in 1866 an Englishman, Colonel E. M. Boxer, developed a brass cartridge case with a percussion cap set into the base. On firing, the heat of the propellant gases expanded the walls of the cartridge against the chamber, thus creating a seal.

Armed with an effective breech-loading rifle, the infantryman could now reload while lying down behind cover and this radically increased his effectiveness, especially in defence. His firepower was also being enhanced in other ways designed to produce a faster rate of fire than even the single-shot breech-loader. The pioneering spirit behind this was the American Samuel Colt, who in 1835 took out a British patent for a pistol with cylindrical revolving chambers, with a nipple at the base, separated by partitions in order to prevent a neighbouring charge from being detonated. The action of the hammer rotated the cylinder and aligned each chamber with a barrel. The bullets still had to be rammed into the chamber, but in 1851 an Englishman, Robert Adams, went a stage further in producing a revolver with special wadded bullets and a self-cocking action. The final development came in 1857 when Smith and Wesson patented the first revolver designed to fire metal cartridges. These were rimfire, but centre-fire cartridges quickly followed.

During the same period parallel efforts were being made to develop a repeating rifle. The first to be produced was by Walter Hunt in 1849. This had a tubular magazine fitted under the barrel and was the forerunner of the famous Volcanic, Henry and, more especially the 1866, 1873 and 1876 Winchester rifle models. In the latter, the rounds were driven into the receiver by a long coil spring, having been inserted into the magazine by means of a trap on the right-hand side of the receiver. From here a carrier took the leading round and aligned it with the chamber. A horizontal sliding breech-block, operated by a lever incorporated in the trigger guard, rammed it home. The main drawback of this system was that the accuracy of the weapon suffered through the constant shifting of the centre of gravity, and the magazine took a long time to load. This was overcome in the 1880s by the vertical magazine fitted below the breech-block.

The other means by which the firepower of the foot soldier was increased was through the development of the machine gun. The first was one built by a Belgian engineer, Fafschamps, in conjunction with the firm Fusnot et Montigny Fils, in the 1850s. It consisted of a cluster of breech-loading barrels, and used cardboard-cased cartridges which were fired by a complex mechanism using long firing pins. Subsequent Montigny models were produced using the same principle. Another derivation was the Meudon or Reffye, more commonly known as the *mitrailleuse*. This had 25 barrels of 13mm (0.5in) calibre which were encased in bronze and mounted on a 4-pdr artillery chassis. It was operated by two worm screws, one of which moved the firing mechanism backwards and forwards, while the other aligned the firing plate with the barrels. Unfortunately, although the French employed it during the Franco-Prussian War of 1870–1, its seeming resemblance to an artillery piece caused them to use it in this way, at its extreme range rather than as an infantry weapon, and it was not as effective as it should have been.

A more successful model was that developed by the American Dr Richard Gatling and the principle is still used in the modern US 20mm (0.79in) Vulcan air defence system and the Minigun. Rather than having a moving firing mechanism, the Gatling gun took up an old idea of revolving barrels mounted in a cluster round a shaft. Cartridges – the Gatling was one of the first machine guns to use metal cases – were fed in by means of a hopper, and the whole system operated by a hand crank which operated an endless screw. The US Army adopted the weapon in 1866 and soon many European armies had also bought it. Later, the hopper was replaced by a drum magazine and then by a straight horizontal ammunition strip. Other well known hand cranked types of the time were the Swedish Nordenfeldt and American Gardner.

The first truly automatic machine gun, whereby all the firer had to do was to continue to press the trigger, was built by another American, Hiram Maxim, in 1884. He took advantage of the recoil of the breech-block on firing to use it to extract the empty cartridge case and eject it, re-lock the firing mechanism and compress the return spring. This would then drive the working parts forward, chamber the next round from the ammunition belt and activate the firing pin. Many armies quickly adopted the Maxim gun. Ten years later came the French Hotchkiss which used a different system, gas operation. Here a portion of the propellant gases was directed into a piston situated under the barrel and this then drove the working parts to the rear, where a return spring brought them forward again.

Progress in the development of artillery during the nineteenth century was less dramatic than with small arms, because the technical challenges were often greater and more complex. At the beginning of the century a new type of ammunition was gradually adopted, case shot or shrapnel, as it became more commonly known after its British inventor General Sir Henry Shrapnel. In its original form it consisted of a hollow shell filled with musket balls and fitted with a bursting charge and fuse. Against troops in the open it could be devastating in its effect and by the end of the century it was the main type of ammunition used in field artillery. During the first decade of the century another British officer, William Congreve, developed a 12-pdr explosive rocket and a battery of these was present at the Battle of Waterloo in 1815. Congreve rockets were, however, too inaccurate to be of much effect and the concept was not pursued.

There was a significant improvement in the ignition system on artillery pieces when the French developed the friction tube in the 1820s to replace the slow match and fuse tube. It consisted of a thin brass or copper tube which was inserted in the vent, and contained highly compressed gunpowder, which was detonated by a saltpetre and antimony mixture contained in an inner tube. Running through this was a copper wire, which when pulled created a violent friction and ignited the mixture.

The next radical change occurred in the middle of the century, when roundshot was replaced by the elongated conical projectile. The problem of roundshot in terms of accuracy was that the centre of gravity seldom coincided with the geometric centre, which meant that trajectories would vary considerably. Hence it was difficult to draw up ballistic tables with any precise accuracy.

However the conical round, in order to achieve stability in flight, needed to rotate about its own axis and this could only be achieved by rifling. The achievement of this was more difficult than with small arms because iron projectiles could not be forced into the rifling in the same way as lead bullets; in addition, windage, allowing propellant gases to flow past the projectile while it is in the bore, hence reducing velocity and accuracy, had to be allowed for if muzzle loading was not to become impossibly difficult. The initial solution was to produce shells with studs on them which would fit into the rifling, and the French first demonstrated the effectiveness of rifled guns of this type at Solferino in 1859. Nevertheless, as with small arms, full advantage could not be taken of rifled guns until breech-loading with reliable obturation had come into being.

The first attempts to achieve a satisfactory breech-loading system began in the early 1840s. The key to it was the obturator and various types were used. Early models included pads of material such as leather or hemp, which would expand but were not easily combustible. These would be forced against the bore when firing. For larger guns metal rings were often used. Another system, known as the Wahrendorff system, used a piston-like obturator with a cylindrical head which fitted snugly into the breech. The shell and its charge were first loaded and the obturator closed and locked into the breech by means of a handle through the rear end of the piston. Indeed, most systems in service during the latter half of the nineteenth century used some form of screw mechanism combined with obturator rings. The main trouble was that they were expensive to produce, complicated, and initially radically increased the weight of the gun. Indeed, in 1870, the British army reverted to muzzle loaders for a short time because of the expense.

More efficient methods of barrel manufacture were, however, evolved. Bronze finally began to give way to steel, for field guns especially – the German firm of Krupp was a pioneer in this – while those of larger calibre were made of cast-iron, which was cheaper than steel. To cope with the greater pressures inside the barrel caused by successful obturation, cast-iron hoops were externally secured to the breech end, but this was later superseded by shrinking concentric tubes of metal on to one another. Studded shells were replaced by ones with copper driving bands, which eliminated windage much more effectively, and all-metal gun carriages were built. Recoil systems, designed to reduce the effect of recoil on the gun carriage, thereby making it a more stable firing platform, came into being by the end of the century, firstly mechanical and then hydraulic.

The greater lethality of guns meant that the traditional wooden-walled warship was vulnerable and from the 1850s onwards armoured plate began to be used in increasing quantities, and the age of the 'ironclad' began. Gradually, too, the traditional broadside gave way to larger guns mounted in revolving turrets: the first new-style naval engagement, between the armour-plated *Merrimac* and the *Monitor* with her revolving turret, took place in 1862 during the American Civil War. This war also proved the effectiveness of rifled guns against fortifications, and one result of this was to cause existing fortresses to be modified through the introduction of retractable armoured turrets.

The mid-nineteenth century saw the introduction of a new range of explosives to augment gunpowder. The groundwork was laid by the discovery of nitroglycerine by the Italian Ascanio Sobrero in 1847. Alfred Nobel, the Swedish chemist who was to found the famous prizes for achievements in science, literature and world peace, developed dynamite by mixing nitroglycerine with kieselguhr, a white powder made of the skeletons of unicellular marine organisms called diatoms. At much the same time, in the 1860s, trinitrotoluene (TNT) was produced by Wilbrand and others by nitration of toluene and its derivatives with a mixture of nitric and sulphuric acids. It was Nobel who first produced fulminate of mercury detonators and who discovered gelatin and the gelignites. In the late 1880s there were three significant discoveries. The first was the adoption of picric acid as a charge for use in projectiles, the other two concerned smokeless powder. Gunpowder had often been called 'black powder' because of the telltale smoke signature it gave on the firing of a weapon. In 1885 the Frenchman Vieille discovered that by partially gelatinizing a mixture of guncotton (wood fibre treated with nitric and sulphuric acids and otherwise known as nitrocellulose) and nitrocotton in alcohol and ether he could produce a smokeless powder which was quickly adopted for the cartridges of the current French army service rifle, the Lebel. Three years later, Nobel produced another smokeless powder, ballistite, a mixture of nitroglycerine and nitrocotton. Furthermore, by mixing guncotton with nitroglycerine and mineral jelly, a variation on this, cordite, was produced. These smokeless propellants, like the magazine rifle, machine gun and breech-loading artillery piece, favoured the defence. Moreover, because they were slower burning, the rate of energy release in the barrel was such that the longer the barrel the greater the thrust at lower internal pressures. Hence barrel walls did not have to be so thick and ranges were significantly increased.

Although not related to weapons technology, two other nineteenth-century inventions had a dramatic impact on warfare. The first of these was the railway (see Chapter 11), which greatly speeded up the strategic movement of armies. Indeed, a major reason for the successes of Prussia during the 1860s was that the kingdom's railways had been laid out in order to be able to mobilize and deploy troops to her frontiers with potential enemies with great speed. The other important invention was the electric telegraph (see Chapter 15) which speeded up the passage of information so that fleets and armies could react very much more quickly to situations. Most of the developments reviewed in this section were blooded on the battlefield during the American Civil War

(1861–5) and it is for this reason that it has often been called the first of the modern wars.

In spite of all these advances one aspect of war remained largely untouched. This concerned the horse, still, in spite of the coming of the railway, a major means of transportation. While the railway speeded up strategic movement, that near or on the battlefield was still heavily reliant on horse-drawn transport. Cavalry, too, still played a major role on the battlefield but, apart from receiving modern carbines, their other weapons did not change. The sword remained, although it had now reverted to a straight blade, the lance enjoyed a new lease of life, deriving from the performance of the Polish lancers during the Napoleonic Wars, and many European cavalry regiments still wore the helmet and cuirass.

THE FIRST WORLD WAR

Artillery and trench warfare

Indications that the new military technology favoured defence over attack had been apparent in the latter stages of the American Civil War, the Russo-Turkish War of 1877–8 and the Russo-Japanese War of 1904–5 in the shape of trench warfare. In 1897 a Polish banker, I. S. Bloch, accurately forecast the effect that modern weapon systems would have on a future major war in Europe. Nevertheless, the countries who went to war in August 1914 all believed that they could achieve a quick victory through manoeuvre and concentration of superior force at the critical point. The early battles, especially on the Western Front, soon dispelled this conviction, when it was found that machine guns and magazine rifles operated from behind cover were many times superior to infantry attacking in the open. For the next three years the main technological effort would be directed towards breaking the resulting stalemate.

For much of the time it was artillery which played the dominant role in the belief that fire alone could create a breach in the enemy's trench system. Initially the Central Powers had an advantage, in that they had put more effort into developing heavy siege artillery in the years leading up to 1914 and this quickly proved itself superior to the fortifications of the day, particularly against the Belgian frontier fortresses, which had been thought impregnable. As the war progressed, calibres increased to as much as 15in (38cm) and, because these heavy guns were difficult to move and emplace, both sides used gun carriages mounted on railway lines. They used high explosives shells with separated charges. Field artillery, however, was still the backbone and in the years preceding the First World War it had undergone a dramatic change.

This was pioneered by the French when they produced their famous 75mm (3in) field gun in 1897. The most significant of its several novel features was a hydro-pneumatic recoil system. This meant that the shock of recoil did not force the carriage itself backwards, the end of the trail having a small spade attached which dug into the ground; the recoil was also used to open the breech after firing. Furthermore, it had fixed ammunition with shell and propellant combined, as in small arms ammunition. All this meant a very much faster rate of fire, and this type of gun became known as quick firing or QF. The Germans, with their 77mm, and the British, with their 18-pdr, quickly followed suit. The British and French had a preponderance of shrapnel ammunition for their field guns, although they did have some high explosive ammunition. The Germans, however, pinned their hopes on the Krupp 'Universal' shell, which was a dual-purpose round, containing both TNT and shrapnel. It could be set for an air burst, with the explosive head landing in the middle of the shrapnel impact area and detonating, or for a ground burst, with the shrapnel being thrown outwards. While perhaps not as effective as a single purpose HE or shrapnel shell, it nevertheless gave the Germans an advantage with the onset of trench warfare, where shrapnel had little effect against men dug in. Consequently, the British and French had to re-order their ammunition production priorities.

As the war progressed artillery techniques became more sophisticated. Until the twentieth century artillery had always employed direct fire techniques, that is, the gun was laid directly on the target. Improvements in gun sights, especially with the introduction of the calibrated dial sight, and in ammunition and recoil systems, meant that guns were very much more accurate and the concept of indirect fire was evolved. This meant that guns which could not see the target were directed by an observer elsewhere who could. It was of course essential that the observer had means of communication with the gun position and this was achieved through the use of the field telephone and cable. A variation on this was developed by the British in early 1915 using aircraft with primitive wireless sets in order to spot and correct the fall of shot. Until the last half of the war artillery had to register its targets before supporting a major attack, which meant that surprise was lost and battery positions given away. To get over this predicted fire was developed, firstly through a detailed topographical survey of the Western Front and later by means of a process called calibration. This was the application of a detailed study of each individual gun's ballistic performance, taking into account adjustments in range caused by meteorological factors such as temperature, barometric pressure and wind. By 1918 predicted fire had become very accurate. For counter-battery work enemy gun positions were pinpointed either by taking bearings of gun flashes or, still better, by sound ranging. This made use of an electronic device developed during the war, which could isolate the sound of a particular gun through a system of microphones and a jet of air playing on an electricity-carrying wire. This could detect the wave made by the shell flying through the air by the increased cold air played on the wire which caused a short break in the electric circuit.

The vulnerability of battery positions and troop concentrations to observation from the air brought about a need for camouflage. The use of netting painted in colours to merge with the landscape, under which guns could be placed, became widespread. The roofs of hutments were similarly camouflaged. Snipers used natural foliage, if they could find it, to disguise themselves, and observation posts were often very ingeniously hidden. One operated in the Ypres Salient from a hollowed-out tree for many months. At sea, too, ships were often painted in zig-zag stripes to make them more difficult to identify from a submarine's periscope.

Trench warfare itself brought about a host of new weapons, some of which were revivals from the past. One of these was the hand grenade. Early models were home-made affairs, often using ration tins filled with gun cotton and nails, with a detonator and a fuse which was lit by a cigarette. In 1915 more reliable types were designed and mass-produced, the two most famous being the British Mills Bomb, with its pineapple-like shape designed to break up into segments on detonation, and the German Potato Masher, a stick grenade with streamer attached in order to steady it during flight. Rifle grenades also reappeared and an example of this was the British Hale grenade, which was fitted with a long steel rod inserted into the rifle barrel and fired using a special blank cartridge. There were even grenade throwers based on the old *ballista* concept.

Mortars, too, came back into fashion in the shape of the trench mortar. The first to appear was the German *Minenwerfer* (= minethrower), which was of 90mm (3.54in) calibre and breech-loading. The bomb had an ingenious fuse which would detonate it no matter which way the bomb landed. Later models were rifled and muzzle loading. The eventual British reply was the Stokes Mortar, muzzle-loading and smoothbore. The bomb had a shotgun cartridge charged with ballistite inserted in its base, and it was dropped down the barrel on to an exposed firing pin which detonated the propellant. The most common calibre was 3in (76mm), but 4in (100mm), 6in (152mm) and even 9.45in (240mm) models were produced.

Sapping and mining also saw a resurgence as each side tunnelled under the other's trenches and set off underground mines. The most spectacular example of this were the nineteen mines, representing almost one million pounds (455,000kg) of ammonal and blasting gelatine, implanted by the British under the Messines-Wytschaete ridge over a two-year period. On 7 June 1917, as a prelude to a highly successful operation to seize the ridge, they were exploded: the noise was heard in England. A successor to the mace, the loaded stick or cosh, was much used by patrols in no man's land, and another reminder of mediaeval warfare came in the shape of the steel helmets adopted by all nations as protection against head wounds. German snipers also used body armour, which looked and was made up in the same way as that of five hundred years before.

The infantryman soon needed a machine gun lighter than the Maxim, Hotchkiss and Vickers (a British derivative of the Maxim), which were found to be somewhat unwieldy in the front-line trenches. The first solution to this, and for its time a very successful one, was the light machine gun developed by the American Colonel I. N. Lewis, which was adopted by the British at the end of 1914. The Lewis gun was gas operated, using a piston in the cylinder under the barrel and a sliding and rotating bolt; it used drum magazines and weighed only 12.7kg (28lbs) compared to the 31kg (68lbs) of the Vickers. It was also, like the Vickers, adapted for aircraft armament. The French answer was the Chauchat, also later used by the Americans, but this had a reputation for mechanical unreliability and was replaced in the French army shortly after the end of the war. The Americans produced the Browning automatic rifle (BAR), which they would continue to use in the Second World War.

The most horrific weapon to be used during the war was undoubtedly gas. The use of burning sulphur to give off noxious gases had been known about in ancient times, and during the latter stages of the American Civil War 'stinkshells' to give off 'offensive gases' were considered. There is evidence that the Japanese used burning rags impregnated with arsenic against the Russians in 1904–5. The use of poisonous gas in war had been banned by the Hague Conventions of 1899 and 1907, but this did not deter the Germans from beginning to experiment with it in late 1914. This work was spearheaded by Dr von Trappen, assisted by Professor Fritz Haber, famous for the Haber process (see pp. 223–4). They looked initially at irritants, especially xylyl bromide, a tear-producing agent, using artillery shells as the means of delivery. The first significant use of it came in the German offensive against the Russians at Lódz at the end of January 1915 when, because of the cold weather, the liquid froze instead of evaporating as it should. The Germans next turned to chlorine and discharged it from cylinders at the onset of the Second Battle of Ypres in April 1915. Chlorine attacks the lungs and the initial results, against French colonial troops, were devastating, creating wholesale panic. The situation was eventually restored, with British and Canadian troops wrapping wet cloths over their faces. Urine was used and provided some protection because the ammonia in it had a neutralizing effect on the gas. Proper gas masks were quickly produced. The first type was a flannelette bag impregnated with phenol, with a celluloid eyepiece. Later in the war came the box respirator, which was more comfortable to wear and operate in.

The British retaliated at Loos in September 1915, but here there were problems in that the wind was not sufficiently strong in all sectors to blow the gas into the German lines. As a result, both sides came to favour the artillery shell over cylinders as a delivery means. New types of gas gradually appeared. In

June 1915 came lacrymatory gas, followed in December by phosgene, which, like chlorine, is a choking agent but more difficult to detect. Then, in 1917, came perhaps the most unpleasant of all, mustard gas (dichlorodiethyl sulphide), a blister agent. Blood agents such as hydrogen cyanide and cyanogen chloride, which attack the enzyme in the red blood cell, were also developed.

Almost as unpleasant as gas was the flamethrower, again not a new idea, having been used by the Byzantines 1100 years previously. The French experimented with it in 1914, but the first significant use was by the Germans against the British at Hooge at the end of July 1915. These *Flammenwerferen* were portable, with the drum containing a mixture of oils strapped to the operator's back. The system was operated by means of compressed air, with the liquid being ignited at the nozzle of the tube. Later the British, with their Livens projector, developed a means of firing cans of oil mixed with cotton waste using high explosive, with a time fuse set to detonate them at target end and scatter burning cotton waste over a wide area.

It was inevitable that the internal combustion engine would have an influence on the conduct of warfare, but, in many ways, its influence was gradual rather than dramatic. The armies did increasingly use mechanical transport, but even by the war's end they were still heavily reliant on the horse. Nevertheless, thought had been given to armed and armoured motor cars in the 1890s, and early examples were a design for an armoured car with two machine guns by the American E. J. Pennington in 1896 and the 'Motor Scout' by the British engineer F. R. Simms in 1899. During the 1900s various European countries produced prototype models and the Italians actually deployed armoured Fiat lorries during their war against the Turks in Libya in 1911, but it was not until the outbreak of war in 1914 that serious attention began to be paid to them. The French and Belgians fielded models based on Renault, Peugeot and Minerva chassis, while the Germans and Austrians had Daimlers, Büssings and Ehrhardts, the last-named with four-wheel steering. The first British armoured cars used were created in France from normal touring models with ship's plate incorporated as armour and were used by the Royal Naval Air Service to guard forward landing grounds, but soon standard designs were produced, foremost of which was the Rolls-Royce armoured car, which fought in every theatre and lived on to see active service in North Africa in the early years of the Second World War. Armoured lorries were also developed to carry QF guns, especially for use against aircraft, and motor-cycle combinations carrying machine guns were used.

Trench warfare understandably limited the employment of wheeled armoured fighting vehicles and brought about the birth of the tank, a British invention. No single individual can claim to have invented it, although an Australian called de Mole did take out a patent in 1912 for a vehicle which was remarkably similar in concept to the early British tanks. The requirement evolved by the end of 1914 was for a machine based on the 'caterpillar' tracked system (tracked traction engines for pulling heavy agricultural machinery had been developed before the war; (see Chapter 16)) which could overcome barbed wire, cross trenches and was armed with machine guns. By mid-summer 1916 the Mark 1 was in production and the first models shipped to France in August. They were originally called landships, and the name 'tank' was adopted as a security measure. The Mark 1, with a crew of eight, weighed 28.5 tonnes and was powered by a 6-cylinder 78kw (105hp) Daimler engine with a top speed of 6kph (3.7mph). The 'male' version had two sponson-mounted 6-pdr naval guns and four Hotchkiss machine guns, while the 'female' had five Vickers machine guns and one Hotchkiss. Besides the driver, it required two men to operate the gears. Enormous heat built up inside, which added to the crew's difficulties, and breakdowns were frequent. Nevertheless, it made its battlefield debut on the Somme on 15 September 1916 and took the Germans by surprise. The British produced further, more mechanically reliable models with the same rhomboid shape. The French, too, produced the heavy St Chamond and Schneider, and the very successful light 7-tonne Renault. The Italians also produced two Fiat heavy tank prototypes. The German answer was the A7V, but because of a shortage of raw materials during the last part of the war, few were built. The first tank against tank action took place between three British Mark IVs and three German tanks on 26 April 1918 at Villers Bretonneux. Other portents for the future were the development of self-propelled artillery guns on tank chassis and command tanks equipped with wireless. There is no doubt that the tank made an important contribution to the eventual Allied victory, but it was too slow and cumbersome to be decisive in its own right.

The Germans recognized very quickly the importance of developing effective anti-tank weapons. They were initially lucky in that they had had in service since 1915 the 'K' bullet, with its tungsten-carbide core, which could penetrate the steel plates used to protect snipers and sentries in the front line, and they found that it would penetrate the early tanks until the British produced thicker armour. The next stage was the anti-tank rifle, and this came into service as the Mauser 'T' 13mm (0.5in) calibre rifle of late 1917. They found that the light trench mortar was also effective, but the greatest threat to the tank became the German 77mm (3in) field gun using a steel pointed armour-piercing shell. In 1918 the British produced the No 44 rifle grenade filled with amatol, but no evidence exists as to how effective this was, while the French had the 37mm (1.45in) Puteaux gun firing solid shot.

Air warfare

The aeroplane introduced a new dimension to war. While initially the prime role of the aircraft was reconnaissance, its use as a bomber began early with

British attempts to bomb the German Zeppelin bases in autumn 1914 because of fears that they would attack Britain. First attempts to arm aircraft were very *ad hoc*, with rifles, pistols, grenades and even darts being used, but proper machine guns were mounted as both sides realized the need to achieve air superiority over the battlefield. The main technical problem was how to achieve a forward-firing machine gun which would not damage the propellor blades. Early solutions were steel deflector plates fixed to the blades and the British concept of the 'pusher' aircraft, with the propellor mounted in the rear, as exemplified by the Vickers FB5 Gunbus. The breakthrough came, however, with the development of the interrupter gear by the Dutch aircraft designer Anthony Fokker. This synchronized the firing of the machine gun with the position of the propellor blades – it only fired when they were at the horizontal. He fitted it on his Fokker fighter and for a time the Germans were dominant in the air until the Allies produced a similar system, the Constantinesco 'cc' interruptor gear. The beginning of strategic bombing came with the Zeppelin raids on England in 1915, and this was followed by attacks by aircraft two years later. The Allies retaliated, although the physical effects were a mere fraction of what they would be 25 years later. Bombs, high explosive in soft metal casings, designed to detonate on impact, increased in size from 20lb (9kg) to 230lb (104kg) as the war progressed and aircraft payloads became larger. It was also the air war which largely brought about the introduction of the tracer round so that pilots could correct their fire, and incendiary ammunition for use against observation balloons. Experiments using aircraft-mounted rockets against balloons were also carried out.

The threat of the aircraft brought about the need for air defence weapons. In the main, these were normally converted field guns, which were mounted on a platform in order to incorporate 360° traverse and an elevation of 80° or 90°. Spring catches had to be inserted in the breech in order to prevent the round from slipping out before the breech-block was closed when loaded at high angles, and recoil systems had to be strengthened. The problem of accurately gauging the speed and height of aircraft was overcome by the 'Central Post Instrument' or Tachymeter devised by a French inventor, Brocq. This consisted of two telescopes, one to track the target's height and the other its course. The observers rotated them with cranks connected to a generator which emitted a current proportional to the speed of tracking, with height and speed being reproduced on dials from which the predicted position of the aircraft could be calculated through using a resistance equivalent to the shell's time of flight. To provide early warning of an aircraft's approach, sound detectors were developed, and by 1918 an aircraft could be detected over five miles (8km) away.

Naval warfare

In 1914 the war at sea was still primarily a matter of bringing the enemy's fleet to battle and destroying it. To this end, the main maritime weapon was the battleship, and typical of this was the Royal Navy's *Queen Elizabeth* with her 15in (380mm) guns. Yet, while the British were successful in keeping the German Grand Fleet in harbour for much of the war, thanks to the Battle of Jutland in May 1916, the battle demonstrated that the Germans had superior sighting equipment and that the quality of the British armoured steel plate was inferior. There were, however, two other naval weapons which really came of age during 1914–18, the torpedo and the mine.

In naval parlance a mine is a weapon designed to explode against a ship, and in this sense there was and is no difference between the mine and torpedo. The first serious attempt to produce an effective mine was by an American, David Bushnell, during the American War of Independence (1775–83). He constructed a one-man submarine which carried a watertight keg containing some 150lb (68kg) of gunpowder. This could be released from inside the submarine and had a handscrew for attaching it to the target ship. Releasing the keg set a clock running in the submarine and this enabled the operator to get his vessel clear before the gunlock mechanism on the charge was actuated. Unfortunately, although his submarine was successful in getting close to British warships on a number of occasions, their bottoms proved too tough for the handscrew to penetrate. He also designed a floating mine, consisting of a gunpowder keg on a log float with an ingenious trigger-operated device, but this met with only very limited success. Robert Fulton, another American and contemporary of Bushnell, also used the submarine concept, in which he tried to interest both the British and French during the Napoleonic Wars. The inventor of the Colt revolver also invented an ingenious system for electrically detonating moored mines and in a trial sank a ship at 6.4km (4 miles) range. This idea was taken up by the Prussians during their war with Denmark 1849–50 and the Russians during the Crimean War (1854–6). At the same time a detonation system for the moored contact mine was invented by the father of Alfred Nobel which would remain in use for the next 100 years. This was made up of a hollow lead horn on the outside of the mine which enclosed a mixture of sugar and potassium chlorate, together with a glass phial filled with sulphuric acid. If a ship struck the mine and bent the horn the phial would break and the resultant chemical reaction produced a flame sufficient to detonate the gunpowder charge.

The American Civil War saw many types of mine used, including Nobel-type chemical fuses and electrically detonated mines. Bushnell's submarine concept was also used in the form of the 'David', a small hand operated semi-submersible craft carrying an explosive charge, later types being steam driven. Two types of torpedo were also used. The first was the 'spar', which was an explosive charge on a pole suspended over the bows of a small vessel, which, on approaching the target was lowered on a wire and detonated through contact with the hull. It was successfully used against the Confederate ironclad *Albemarle* in 1864. The second type was the 'towing' torpedo, which was towed by a boat but at a 45° angle rather than directly astern, for obvious reasons. Just after the American Civil War the first self-propelled torpedo was invented by Robert Whitehead. It was driven by a compressed air engine and had two contra-rotating screws to keep it on course, as well as a hydrostatic pressure system to keep it at the correct depth. It did have a number of teething problems, but its first successful use in war came in January 1878 when the Russians, using two Whitehead torpedoes, sunk a Turkish ship in Batum harbour. The new explosives developed by Nobel helped to perfect the torpedo because of their greater power than gunpowder, and very soon torpedoes were being used in two different ways, fired either from a submarine or by a new type of surface warship, the torpedo boat destroyer.

During 1914–18 the sea mine underwent several further refinements. They could either be anchored to the seabed by a weight, or be free floating. In the latter case, various types were used. There were drifting mines suspended from a float, creeping mines on the end of a chain that dragged along the sea bottom, and the oscillating mine, which used a hydrostatic valve to vary its depth below the water surface. Later in the war an ingenious anti-submarine mine, the antenna mine, was developed by the British. This was moored at a depth of some 15m (50ft), with a 12m (40ft) copper antenna suspended on top of it by means of a float. If this contacted the steel hull of a submarine or surface vessel it created a sea cell through a copper element within the mine and this fired the electrical detonator.

The war quickly showed that it was the submarine rather than the surface launched torpedo which was the greater threat, and this was dramatically brought home by the sinking of three elderly British cruisers within 75 minutes by the German U-9 on 22 September 1914. The torpedo, too, had become more efficient, especially as a result of the adoption of the gyroscope for directional control by the Austrian Ludwig Obry in 1895. Greater range was also achieved by the use of alcohol or paraffin as a fuel. The threat became especially severe to the British in 1917 when the Germans embarked on unrestricted submarine warfare against merchant shipping and ways had to be found to defeat the menace. The adoption of the convoy system was one, and, apart from the antenna mine, two other methods of destroying the submarine under the surface were the paravane charge, which was towed behind a surface vessel in the hope that a submarine might hit it, and the more effective depth charge which made its first appearance in 1914, although it did not achieve any results until two years later. It was, in essence, a mine launched by a thrower and set, through a hydrostatic fuse, to detonate at a certain depth, usually 25m (80ft). The submerged submarine still had to be detected, however, and a solution to this was not found until 1918 with the invention of ASDIC, named after the Allied Submarine Detection Investigation Committee and later called sonar (Sound Navigation and Ranging) by the Americans. This consisted of a radio transmitter/receiver, which transmitted sound impulses. If these hit a solid object they would be reflected in the form of an echo or 'ping'. From the transmission-reception time interval the range to the submarine could be worked out and depth charges set accordingly. Anti-submarine warfare had arrived.

THE SECOND WORLD WAR

Air warfare

There was a widely held belief between the two world wars that warfare in the future would be dominated by the bomber aircraft and the gas bomb. The early First World War attempts at strategic bombing, although they may not have caused much physical damage, did have a significant effect on the morale of civilian populations. Attacks on them and centres of government would, it was argued, quickly lead to surrender. For a number of reasons this did not happen.

The horrific effects of gas had led to its banning as a weapon of war by the 1925 Geneva Protocol. Nevertheless, during the 1930s the Germans did develop a new form of agent, nerve gas, designed to attack the nervous system and more lethal than existing types, in the shape of Tabun and Sarin. Neither side, however, used gas in 1939–45, more for fear of retaliation than because it was banned.

As for the bomber, the argument that it was invulnerable to attack because fighter aircraft would not be able to intercept it before it arrived over its target was proved groundless for two reasons. The first was the development of the monoplane fighter, largely thanks to the Schneider Trophy races, with its much higher performance than biplane types, and with wing mounted multiple machine guns. Not only could it climb and fly very much faster, but it also had the necessary firepower. Even so, the fighter still had to have sufficient warning of the bomber's approach and sound locating devices were not enough.

The answer lay in radar. This is based on the principle that all solid and liquid bodies reflect radio waves, and was originally called 'radiolocation' by the British. The first experiments were carried out in 1924 by two Cambridge scientists, Appleton and Barnett, who investigated the measuring of distance by this means and their methods were used to identify the Heaviside Layer, the electrified conducting region in the upper atmosphere which causes downward refraction of radio waves and is called after the physicist who first forecast its existence. A radio transmitter sending out short pulses of radio energy was then devised and was incorporated in VHF radio telephone links. It was noticed that aircraft could be detected through the reflection of these pulses, and in 1935 Sir Robert Watson Watt began work to apply radiolocation to military purposes. Very soon he was able to detect an aircraft and represent its presence by 'blips' on a cathode ray tube, with the transmitter automatically shutting itself off after it had transmitted a pulse to enable the receiver to identify the reflection. Height was measured by comparing the receiving signal on two vertical aerials placed a known distance apart. In order to be able to locate the aircraft accurately, its bearing had also to be measured, and this was achieved employing the same principle, but using a comparison of the received signal detected by two crossed aerials. The system was now called RDF or radio direction finding, but was later changed to the American acronym radar (Radio Direction and Ranging).

From 1936 the British began to set up a continuous chain of radar stations known as the 'Chain Home' network and this eventually provided complete coverage of the coastline. Proof of its effectiveness came during the Battle of Britain in 1940. The information on German bomber streams detected by the radar stations was passed back to controllers, who were then able to direct fighters into the air and, using radio, guide them on to the approaching bombers. The Germans, too, during 1940–2, built up a similar chain across north-west Europe, which was called the Kammhuber Line after the Luftwaffe general responsible for installing it.

It soon became clear that radar could be applied to many other uses. It was installed in night fighters in order to detect hostile aircraft when airborne. It became a vital weapon in the war against the U-boat, complementing sonar in that it could detect submarines on the surface. Early models merely had an audio display, but later a cathode ray tube was incorporated. There were also airborne submarine detection radars, and a type similar to this was used by Allied bombers as a navigation device: here, different types of terrain produced different echoes, which could be represented on a visual display tube, a system known as 'Home Sweet Home', more commonly abbreviated to H2S. Another system, OBOE, had two ground transmitters. One, the 'cat' sent out a signal to an aircraft which enabled it to fly on a line of constant radius from the transmitter, while the other, the 'mouse', aimed its signal to the target, which was on the line of constant radius from the 'cat'. In this way the bomber could accurately attack the target in conditions of poor visibility. The Germans employed a similar system, called *Knickebein*.

The main problem with radar was that it was an 'active' system, which meant that it could be detected and jammed. Thus, German U-boats were equipped with a Metox receiver, which would warn them when an aircraft using radar was in the vicinity, and German night fighters could locate and attack Allied bombers using radar detection devices. The most effective means of jamming radar was simple in the extreme, consisting of strips of aluminium foil dropped by aircraft. These created a 'snowstorm' effect on visual displays. Originally codenamed WINDOW, it is now known as Chaff and is still much used as a defence against radar-guided missiles.

Radar as a navigation aid only came into service during the second half of the war. At the outbreak, British and German bomber crews relied on dead reckoning which was very inaccurate. Bomb aiming, too, was an imprecise system, still using First World War bombsights which relied on the aircraft flying straight and at constant speed over the target, by no means easy under fire. Only with the introduction of gyroscopes into bombsights was a reliable system introduced. The size of bombs increased as the war went on. From the 500lb (227kg) bomb of 1939, the British in March 1945 used the 22,000lb (9975kg) Grand Slam to destroy the Bielefeld viaduct in Germany. Some bombs were thin cased, relying on blast for their effect, while others had thick cases and streamlined shape and were designed to penetrate and fragment. There were also incendiary bombs, some filled with inflammable liquid, others, usually small, using magnesium alloy. Yet, in spite of the increasing weight of the Allied strategic bombing offensive, the German people never did succumb to it as the pre-war theorists and many airmen believed that they would.

Naval warfare

Radar played a large part in the war at sea, especially in the Battle of the Atlantic against the German U-boats, and other weapons were employed on both sides. By 1939 the torpedo, which could now be air- as well as surface and subsurface launched, was driven by compressed air and diesel. Detonation was either on contact with the target or by magnetic proximity, whereby the increased magnetic field induced by an all-metal ship would activate a pistol in the torpedo. This latter system, however, had several problems during the first years of the war and took time to perfect. Towards the end of 1942, the Germans introduced a new type, the *Fat*, in order to increase the chances of hitting a ship when attacking a convoy. This was electrically driven and could be set to circle to left or right. The next stage was the acoustic torpedo, which homed in on the noise of the ship's propellers. The Germans called this *Falke* (= hawk). U-boats also had listening devices incorporated, with receivers built into the hull, so that ships could be detected when the U-boat was submerged. In order to counter sonar, the Germans developed both a rubber skin over the hull, in order to dampen the echo, and a decoy in the form of a cylinder filled with calcium hydride, which could be fired from a torpedo tube and floated under the water to deceive the sonar while the U-boat made its escape.

On the Allied side, besides radar and sonar, an important detection device was High Frequency Direction Finding (H/F D/F or 'Huff Duff'). This was both shore and ship based and detected U-boat radio transmissions, giving a bearing on them. Two sets operating in conjunction could pinpoint a U-boat's position. A further and highly significant weapon, as it was in many other aspects of the war, was the Allied ability to decipher the top secret German codes employed with the Enigma cipher machine, which often gave prior warning of the intentions of the U-boat 'wolf packs'. Depth charges could be dropped by aircraft, as well as launched from a ship. Amatol, minol and torpex were the explosives used, and these were detonated by a preset hydrostatic pistol. A variant of this was the 'Hedgehog', which launched a pattern of 24 projectiles, each filled with 14.5kg (32lb) of torpex, but would only explode on contact with a solid object. The most significant antisubmarine weapon, developed by the Americans in 1942, was Fido, an aerially launched acoustic homing torpedo.

Mention must be made of a particular branch of science which played an important role in the Battle of the Atlantic. Operational Analysis is an observational, as opposed to experimental, science which uses accumulated data to construct a mathematical model in order to aid problem solving and decision making. The father of the concept was F. W. Lanchester, who first used this technique in 1916 in order to evaluate the most efficient use of aircraft in war. Operational Analysis was used during 1939–45 to find answers to such questions as the best method of searching for a submarine, optimum depth charge settings and the best colour to paint anti-submarine aircraft to minimize visual detection by submarines. Employed also in many other aspects of the war, Operational Analysis continues to play a major part in the formulation of requirements for new weapons systems.

Mines, like torpedoes, became increasingly sophisticated as the war progressed. One type developed by the British just before the war followed the original Bushnell concept of a mine which was attached to the hull of the target ship. The limpet mine was disc-like in shape, containing 1.8–4.5kg (4–10lb) of blasting gelatin, and had a time fuse which could be set from a period of thirty minutes up to five hours. It was attached to the hull by magnets and many successes were scored with it. At the outbreak of war the Germans began using an ingenious parachute mine. The parachute was used to control the speed of descent so that on hitting the water an arming clock was started. As the mine sank, a spring-loaded button in the side operated as a hydrostatic switch, which stopped the clock when it reached a certain depth. This acted as an anti-handling device, for once the mine was raised to this depth, the arming clock would restart and run for half a minute before detonation. As it was, once the mine touched bottom an electrical circuit was completed and the mine came alive, ready to detonate under the influence of a ship's magnetic field. The counter to magnetic mines was to pass electric cables round the ship's hull, connect them to her generators and induce a magnetic field to cancel out that of the ship. This was called degaussing after Karl Friedrich Gauss, who introduced the gauss as the unit of measurement of magnetic fields. Acoustic mines and pressure mines, designed to detonate as a result of the change in water pressure created by a vessel as it passed through it, were also developed. A cross between the mine and the torpedo was the human torpedo, which was pioneered by the Italians and successfully used by them to sink the British battleship *Queen Elizabeth* in Alexandria harbour on Christmas Eve 1941. Called 'chariots' by the British, they were shaped like torpedoes and 'ridden' to the target by two frogmen who, on arrival, merely disconnected the explosive warhead, clamped it to the ship's keel and set a time fuse.

The war at sea also saw the rise of the aircraft carrier over the battleship. Air power proved a key factor in defeating the U-boat, and it also provided a serious threat to the surface warship, as was dramatically brought home by the sinking of the British *Prince of Wales* and *Repulse* by Japanese aircraft off the coast of Malaya on 10 December 1941. It was in the Pacific, however, that this phenomenon was most marked: all the major naval battles between the Americans and Japanese were largely conducted by carrier-based aircraft, with the opposing fleets often not seeing one another.

Land warfare

The dominant feature of land warfare during 1939–45 was the Blitzkrieg, or lightning war, concept first put into practice by the Germans and the key to their dramatic early victories. The ingredients of this strategy were mechanized forces and tactical airpower and the object dislocation of the enemy's command and control processes. This meant the ability to think and act faster and was much helped by the development of robust and reliable battlefield radio communications. Aircraft were used as aerial artillery, armed with rockets, bombs and machine guns and controlled by air–ground radio links. Typical of the types of aircraft used in this role were the Soviet Stormovik series, the British Typhoon and, perhaps the most well known of all, the German Ju87 Stuka dive-bomber.

During the years 1919–39 there had been great progress in tank development. Different countries had divergent views on their form and role. The Germans began the war with light tanks, armed only with machine guns, and medium with 50mm (2in) and 75mm (3in) guns. The Americans had much the same categories, but the British had three types – light, infantry and cruiser. Light tanks, together with armoured cars, were for reconnaissance. Infantry tanks were relatively slow moving, thickly armoured and had either a machine gun or 2-pdr (40mm). They were used, as their name suggests, to support infantry on their feet in the attack. Cruiser tanks, although also armed with the

2-pdr, were lightly armoured and fast moving, and employed in the old cavalry role of shock action. The French and the Russians, on the other hand, had light, medium and heavy tanks, the last-named being designed to break through the enemy's defences. Most tank ammunition was solid shot, using the tungsten carbide core developed by the Germans in 1917.

By 1942 the Americans, Germans and Russians had fixed on the 75–76.2mm (3in) calibre as the main tank gun, while the British lagged behind with the 6-pdr (57 mm), but increasing armour thicknesses meant that existing muzzle velocities were no longer sufficient to impart the required kinetic energy to the armour piercing round. The initial German answer was to increase the barrel length of the 50mm (2in) and 75mm (3in) guns. They also introduced a refinement to the standard shot by surrounding the tungsten carbide core with a soft metal jacket. This Armour Piercing Composite Rigid (APCR) round was lighter than the normal shot and hence had a greater muzzle velocity and better penetration, although its performance did fall off as the range increased. In 1944 the British improved on this with the Armour Piercing Discarding Sabot (APDS) round, which remains today one of the main types of armour defeating ammunition. Here the tungsten carbide slug was encased in a pot, or sabot, which broke up as the round left the muzzle. In this way APDS did not suffer the same performance degradation as APCR. Another solution was the tapered bore, used on light anti-tank guns and armoured cars, in which the soft casing surrounding the slug was squeezed as the round travelled up the barrel. The main drawback was that the barrel became quickly worn.

By the end of 1942 the Germans had begun to introduce a new breed of heavy tank based on the highly successful dual purpose anti-aircraft/anti-tank 88mm (3.5in) gun. Tanks like the Tiger and Panther (this retained the long barrelled 75mm (3in)) were heavily armoured as well, and more than a match for the lighter Allied types; the later Jagdtiger even had a massive 128mm (5in) gun. Lighter Allied tanks, like the American Sherman, British Cromwell and Russian T–34, acclaimed by many as perhaps the best all-round tank of the war, were more manoeuvrable. This competition represented the perennial problem for the tank designer, how to reconcile firepower, mobility and protection; overemphasis on one must be at the expense of the other two. It was the British who produced what was to become the standard concept of the tank of the future, the Centurion, which made its first appearance in 1945, just too late to see action. The Centurion was designed to undertake all the roles of the tank – infantry support, anti-tank and shock action – and this concept remains as today's main battle tank. Nevertheless, led by the British, a large number of tanks were developed for specialized roles – mine clearing, bridging, flamethrowers, swimming tanks, air portable tanks. The Germans, Russians and Americans also produced armoured fighting vehicles whose sole role was anti-tank. The German and Russian models were turretless, and thus cheaper to produce than a normal tank, while the Americans went for types with open-topped turrets, which, while having excellent mobility and firepower, paid the penalty of being too thinly armoured.

Besides tank guns themselves, a wide range of other anti-tank weapons was produced during the war. Anti-tank guns, often using tank guns mounted on wheeled artillery carriages and relying on kinetic energy developed through high muzzle velocities, abounded. For the infantryman, his ability to counter tanks was greatly enhanced by the development of hollow charge ammunition. An American engineer called Monroe had discovered in the 1880s that if a slab of explosive was hollowed out, placed in contact with a piece of steel and detonated, the shape of the hollow would be reproduced in the steel, but serious efforts made to apply the Monroe effect to weapons were not made until 1940. Taking a cylindrical-shaped charge, one end was given a conical hollow which was lined with metal. On detonation, the metal liner collapses and is transformed into gas and molten metal which is then projected at the target as a high-speed jet with great penetrative power. For it to be effective detonation has to take place at an optimum stand-off distance from the target to allow the jet to form, and this is normally achieved by a false nose or a rod projecting from the nose. Hollow charge was first used in demolition charges, but then the British introduced the No 68 anti-tank rifle grenade. It was also used to give field artillery an improved anti-tank capability. Then, during the latter half of the war, came a family of hand-held recoilless hollow charge round projectors, the German *Panzerfaust*, American Bazooka and British FIAT. The principle was also used in anti-tank mines, which became a major threat to tanks.

The emphasis on mechanized warfare meant that infantry and artillery had to be able to keep up with the tanks. The result was a wide range of close support self-propelled artillery, often mounted on existing tank chassis, although some chassis were purpose built. For infantry there were both half and fully tracked armoured personnel carriers. However, while the Allies always emphasized the need for wheeled mechanical transport and were, apart from the Russians, fully motorized, the Germans relied heavily on horse-drawn transport throughout the war.

The infantryman's weapons were generally much lighter and had a faster rate of fire than during 1914–18. This was especially so with mortars and light machine guns. One new category of weapon was the sub-machine gun. This works on the 'blow-back' principle. On firing, the propellant gases both drive the projectile forward down the barrel and the breechblock to the rear, where it comes up against the return spring. When firing single rounds it is retained by a sear and released by pressure on the trigger, while on automatic the spring automatically drives the breechblock forward. It normally uses pistol ammunition. The two pioneers of the sub-machine gun were the German Hugo

Schmeisser and American General John T. Thompson, both of whom began to produce a series of models in the years after 1918.

While the Germans with their *Nebelwerfer* and the Russians with the *Katyusha* brought back the concept of the multiple-barrelled rocket launcher, the rockets themselves being filled with high explosive, more dramatic was the German development of free flight rockets. In great secrecy they began work on these in the late 1930s, and they later became known as *Vergeltungswaffen* (Vengeance- or V-weapons). By 1944 there were two types in existence. The first, the V-1, was in essence a flying bomb, fitted with wings, tailplane and a ramjet motor driven by low grade aviation spirit; when close to the target, the engine shut off. It carried 850kg (1874lb) of explosive in its warhead and had a range of 250 miles, and was launched either from a ground ramp or from a specially modified Heinkel 111 bomber. The V-2 was a rocket bomb, fin stabilized, carrying a one-tonne warhead and 9 tonnes of liquid fuel (see p. 649). It was launched vertically and climbed to a height of 80–100km (50–60 miles), when preprogrammed controls would shut off the fuel supply and it would complete its flight along a ballistic trajectory. The V-weapons appeared in service too late to alter the course of the war, but the German experience in this field was to be the foundation of today's ballistic missiles (see p. 1007).

Most awesome by far of all the weapons to appear during the Second World War was the atomic bomb. Its origins lie in the discovery of radioactivity by the Frenchman Henri Becquerel in 1896 when experimenting with uranium. Three years later the New Zealand physicist Ernest Rutherford established that uranium emitted two types of radiation, α-particles and β-particles. While the β-particle was an electron, the α-particle was much heavier and released an intense energy. Later he established that this was a helium nucleus containing two positive particles, protons, and two neutral particles, neutrons. In 1920 the Danish scientist Niels Bohr completed the theory of the atom by establishing that the electrons revolved around the nucleus. Then, in 1930, the Germans Bothe and Becker showed that when beryllium is bombarded with α-particles a very penetrating neutral particle was emitted: the basis for deriving energy from the nucleus was established. The work on this next step was begun by a Hungarian, Leo Szilard, who settled in the USA in the late 1930s and, separately, by two Germans, Hahn and Strassman. From now on it became a race between the Germans and the Western democracies, whose efforts were centred on the Committee on Uranium set up in the United States in 1939. In 1942 the name was changed to the Manhattan Project, and it was based at Los Alamos in New Mexico where a large team of scientists set about trying to design a bomb which would fit into an aircraft, and which would explode when required to but was otherwise safe. To create it, they initially used uranium 235 but, finding that it was a lengthy and expensive business to isolate it from the more abundant and infinitely more stable uranium 238, they turned to plutonium. Much less plutonium was needed for it to be critical, that is, to be of sufficient size for the splitting of the nucleus, fission, to become a chain reaction, in that the resultant neutrons would have more than a 50 per cent probability of hitting another nucleus and thus continuing the fission process in order to release a significant amount of energy. However, the plutonium 239 isotope being used created plutonium 240 if it absorbed a neutron without fissioning, and the latter isotope was prone to spontaneous fission. This meant that it could produce enough energy to destroy the bomb casing before sufficient destructive energy had been created, since the chain reaction would then cease. They therefore conceived the idea of making a sphere of plutonium 239 of such density as to be sub-critical and then to compress it – what was called implosion – so that there would be explosive fission. In the meantime, the German programme fell behind, especially as a result of the Allied destruction of the heavy water plants in Norway (heavy water containing the heavy hydrogen isotope, vital in the manufacture of fissile material). Not until 16 July 1945 was the first nuclear weapon detonation carried out, at Los Alamos. On 6 August came the dropping of the atomic bomb on Hiroshima, followed three days later, on 9 August, by that on Nagasaki. For peaceful uses of nuclear energy see pp. 212–16.

WEAPONS TODAY

Ever since 1945 the world has lain under the shadow of the nuclear weapon, which has grown ever more powerful and efficient. At the end of the Second World War the USA, being the sole possessor of the nuclear bomb, appeared all-powerful, but in 1949 the Soviet Union successfully exploded an atomic device, and the nuclear weapons race was on, complicated by the fact that Britain, France and, latterly, China, also developed a nuclear capability. In the meantime came the development of the hydrogen bomb.

During the course of the Manhattan Project, theoretical calculations had suggested that the intense energies created by the atomic bomb could be used to trigger far more violent reactions in two isotopes of hydrogen, deuterium and tritium, by converting them into helium. This made use of a concept called fusion, in which, under conditions of extreme temperature and pressure, two atomic nuclei can combine to form larger elements. In the case of very light elements, energy is given off in the process, since the new nucleus has less mass than its two parent nuclei. The implication is that in a fusion, or, as it is more commonly called, thermonuclear device, much more energy is released than in an atomic weapon of equivalent weight. It was, however, not until the Soviet Union exploded its first atomic bomb that the Americans began seriously to develop the hydrogen bomb. The main obstacle was how to create fusion before the explosion destroyed the device. This was overcome by making use of photons, particles of light containing energy but lacking mass, which constitute electromagnetic radiation. They travel several times faster than the expanding nuclear explosion caused by the fission trigger and hence are able to create fusion in the core before it is engulfed and destroyed. The Americans successfully exploded their first thermonuclear device towards the end of 1952, but it was not long before the Russians matched this.

By the end of the 1950s what was known as the 'nuclear triad' for delivering strategic nuclear weapons had come into existence. While the nuclear bomber aircraft was and still is retained as one leg of the triad, both the Americans and Russians made use of German wartime expertise to develop ballistic missiles. The first to be brought into service was the US Corporal in 1952. This was a tactical or battlefield weapon of limited range, little more than an improved V-2, adopting a ballistic trajectory after the fuel had been shut off by radio command. It was, however, the first to be armed with a nuclear warhead. Within a few years inter-continental ballistic missiles (ICBM) had been developed, like the US Titan and Soviet SS-9. These work on the same principle as the V-2, but have booster rockets to propel the missile into space, after which they fall away. Both liquid and, now more usually, solid fuels are used; once the missile has reached the desired height the engine is shut off, and a computer is used to trigger alterations in the trajectory in order to ensure that the missile hits the target. That part carrying the warhead is known as a re-entry vehicle, and modern ICBMs have multiple warheads, enabling them to attack several targets simultaneously. These are known as multiple re-entry vehicles (MIRVs). ICBMs can be land-based, launched from hardened silos, or mounted in submarines, which are known as SSBNs (Surface-to-Surface Ballistic Nuclear), and these are the other two legs of the triad.

The main effects of a nuclear explosion are blast, heat, and immediate and residual radiation. Protection against these is generally called 'nuclear hardening'. Immediate radiation is that generated in the first minute of a nuclear explosion, while residual is spread through neutron-induced activity, or fallout. The general effect on human beings is to destroy body cells, especially bone marrow, brain and muscle, and to attack tissue. Protection against radiation is best achieved through shielding by dense materials, although respirators and nuclear, chemical and biological (NBC) clothing does help. Many armoured fighting vehicles also have collective protection in that they can be sealed and pure air drawn in through filters. A further effect is electromagnetic pulse (EMP) which damages electronic equipment.

Two other categories of nuclear weapon exist today. The first is the theatre nuclear weapon, of which a typical example is the US Pershing 2 missile, which can carry a 200 kiloton (1 kiloton = the destructive force of 1000 tons of TNT) warhead (the atomic bombs dropped on Japan were 12 kiloton) a distance of 1600km (1000 miles). A new type is the cruise missile, designed to fly low enough to pass under radar coverage, and which can be air-, ground- and sea-launched. Tactical nuclear weapons continue to exist in the form of missiles and artillery shells, and there are also nuclear mines, both land and sea, as well as nuclear depth charges. Both the USSR and the USA have been trying to perfect active defence against nuclear missiles, the latest attempt being the US Strategic Defense Initiative (SDI), which is intended to intercept missiles at three stages – launch, mid-flight and re-entry. There are fears that this course could be destabilizing.

A major disadvantage of battlefield nuclear weapons is the amount of collateral damage caused by heat and blast, which can hinder friendly as well as enemy forces. Consequently, in 1958 the USA began to develop a weapon which minimized these two effects in favour of increased radiation. The enhanced radiation/reduced blast (ERRB) or neutron bomb relies on the effect of highly lethal penetrative neutrons produced in the fusion process. It has low yield and is a 'clean' weapon, with minimum fall-out, which would contain radiation casualties to the immediate target area. In 1981 the USA announced that it intended to stockpile the bomb, but Soviet protests that it too was destabilizing, in that it too might make nuclear war more likely, resulted in this plan not being carried out.

Missiles are also used in a large variety of conventional roles – air-to-air, surface-to-air, anti-tank, air-to-surface. Guidance on to the target is all-important. Surface-to-air missiles generally use infra-red homing, detecting and locking on to the infra-red given out by the aircraft's engine exhausts. The counter to this is for the aircraft to drop flares to deflect the missile, but the most modern type of missile is now designed to operate in the ultra-violet spectrum, which makes such countermeasures less effective. At the lower levels, multiple cannon, usually mounted on a tracked chassis and radar-controlled, with a very high rate of fire are used. Crucial to all air defence systems is the incorporation of an Identification Friend-or-Foe (IFF) device, which interrogates the aircraft, usually through the form of a radio signal to which the aircraft's built-in responder will reply if it is friendly. Anti-tank guided missiles, whether launched from the ground, vehicle or helicopter, are wire-guided, a wire connecting the missile to the sighting and control unit. Earlier models were manually guided, which meant that the operator had to line up the missile with the target and steer it along this line, but now by means of an infra-red detector in the sight, the operator merely has to keep this on the target and the detector, using the flare in the missile's tail, will sense if the missile is off course and corrective signals will automatically be sent down the wire.

A variation on the guided missile is the precision guided munition (PGM) or 'smart' munition. This is being increasingly used for air delivered munitions, as well as for artillery, mortars and multiple launch rocket systems. The munitions themselves are radar controlled, heat seeking or can be guided on to the target by means of laser. The munition or sub-munition has a laser sensor which locks on to the reflections of a laser beam projected on to the target by an observer equipped with a laser designator.

In spite of the continued ban on chemical weapons, many nations still possess them and they have been used in war since 1945, the Iran–Iraq conflict being one example. Decapacitating agents, like the drug LSD and the CS and VN gas used to control riots, which are not fatal, have now joined the choking, blood, nerve and blister types. There is also a biological threat. Biological agents are micro-organisms which cause disease or deterioration in material, and take the form of fungi, protozoa, bacteria, rickettsiae and viruses. Formerly they could be distinguished from chemical agents in that they exist naturally, but now they can be artificially produced in laboratories and the boundary is becoming blurred. The main means of protection against both chemical and biological agents are the wearing of NBC clothing and the use of chemical antidotes.

Helmets continue to be worn by all armies, but a more recent innovation is the flak jacket designed to protect the body against shell splinters. It was first worn by US air gunners during the Second World War, but is now standard combat wear in a number of armies. Early models used plates of ballistic plastic, but this made them relatively heavy and cumbersome, and the most recent types incorporate an American-developed fibre material called Kevlar, which is much lighter.

Modern main battle tanks, like the American M-1 Abrams, Soviet T-80, British Challenger and German Leopard 2, weigh 45–60 tonnes and have a 120mm (4.7in) gun. Some are still rifled, but an increasing number are smoothbore. The main advantage of this is that barrel wear is less and hotter charges can be used, resulting in higher muzzle velocities, thus increasing effective range against other tanks, which is now 3000m (9840ft). In order to maintain accuracy, fin-stabilized ammunition is used, the fins being folded and automatically positioning themselves as the projectile leaves the muzzle. Sophisticated fire control systems, incorporating ballistic computers, further enhance accuracy. New types of ceramic armour, specifically designed to increase protection against hollow charge projectiles without increasing the weight, are also being introduced.

Mines, too, are ever more sophisticated. Most have anti-handling devices, and in order to make their detection more difficult many are made of plastic. A new type of anti-tank mine is the off-route mine. This is positioned at the side of a road or track and is detonated by the vehicle passing over a pressure tape or by acoustic sensor. Minelaying is very much faster than the old manual methods. Machines resembling agricultural ploughs are used, and remotely delivered mines, using rocket systems or artillery, are coming into service.

As far as infantry weapons are concerned, the most dramatic change during recent years has been an overall reduction in small arms calibre, from 7.62mm (0.3in) to 5.56mm (0.22in). While this has reduced effective ranges, it has resulted in smaller and lighter weapons, and means that the infantryman can carry more ammunition, since it is also lighter than in the past.

Surveillance devices have undergone notable improvements during the past forty years, especially in the shape of infra-red image intensification, which concentrates the ambient light, and low level television systems used to see at night and in other conditions of poor visibility. Radar is becoming ever more effective and one new development is that of sideways-looking airborne radar, which enables an aircraft to survey enemy territory from the safety of its own lines. High-flying 'spy planes' like the US Lockheed SR–71 Blackbird carry long-focus cameras which can cover an area of $325km^2$ (125 square miles) in one exposure from a height of 40km (15 miles), with sufficient resolution for targets as small as artillery guns to be accurately identified. At a lower level, another class of surveillance device is the remotely piloted vehicle (RPV), which can be directed on photographic flight missions with pictures being instantaneously reproduced on a television screen at the ground control station.

The computer, too, is playing an ever increasing part in war. Apart from being used in fire control systems, it is a highly effective means of recording and collating the increasing mass of information available to a commander and assisting him in problem solving and decision making. A new role for the computer which is being explored in the USA is in robotic weapons and surveillance systems. However, while many aspects of warfare are becoming automated to a high degree, the computer will not, at least in the foreseeable future, replace the human brain, which will remain ultimately responsible for the conduct of war.

Attention is turning to the threat of war in space. Already a comprehensive network of military intelligence and communications satellites exists and, in spite of constant efforts to ensure that space has peaceful uses only, it is very likely that anti-satellites (ASATs) will soon make an appearance. War in space is likely to use a new breed of weapon, directed energy, which is now under active development. Directed energy weapons comprise lasers, which use photons of electro-magnetic radiation, particle beams, made up of sub-atomic particles which have mass and destroy a target through nuclear reaction, sonic and radio frequency weapons, which use sonic waves and electrical fields only to focus and accelerate particles.

The pace of military technological development is ever quicker, and the only two brakes that can be applied to it are economic and humanitarian. Nevertheless, and perhaps increasingly, besides influencing the art of war, military technology is applied to many other fields of human endeavour.

FURTHER READING

Beckett, B. *Weapons of tomorrow* (Orbis, London, 1982)
Dupuy, T. N. *The evolution of weapons and warfare* (Janes, London, 1982)
Fuller, J. F. C. *A military history of the western world* (Minerva, New York, 1967)
Montgomery, B. L. *A history of warfare* (Collins, London, 1968)
Reid, W. *The lore of arms: a concise history of weaponry* (Arrow, London, 1984)
Tarassuk, L. and Blair, C. (Eds.) *The complete encyclopaedia of arms and weapons* (Batsford, London, 1982)

THE CONTRIBUTORS

J. A. Bagley read Mathematical Physics at Birmingham University, and learnt to fly with University Air Squadron. He joined the Royal Aircraft Establishment at Farnborough in 1952, and was engaged in research in the Aerodynamics Department until 1976. His main field of research was concerned with subsonic and transonic flow around swept wings and with the aerodynamics of engine installations, on transport aircraft in particular. In November 1976 he joined the Science Museum, London, as the curator in charge of the Aeronautical Collection. He joined the Royal Aeronautical Society as an Associate Fellow in 1961, and was elected a Fellow in 1976. He is chairman of the Historical Group Committee of the Society.

Brian Bowers is an electrical engineer with a doctorate in the history of technology. He has been a Curator in the Science Museum, London, since 1967, and was previously an Examiner in the British Patent Office. He has been Chairman of the Institution of Electrical Engineers' *History of Technology* and *Archives* Committees, and is Editor of their *History of Technology* series. His publications include a biography of Charles Wheatstone, a children's biography of Michael Faraday, and *A History of Electric Light & Power*. His main current research interests are in the introduction of electricity supply, especially in the home, and the relationships between T.A. Edison and electrical engineers in Britain.

John Boyes has lately retired as a Principal Inspector of Factories in the Health and Safety Executive of the British Civil Service. His professional experience has led to a wide interest in technical, and particularly transport, history. As the main author of *The Canals of Eastern England* (1977) he was awarded the first Rolt Memorial Fellowship at Bath University. He has written and lectured extensively on local history and is a member of the editorial committee of the Victoria County History of Essex. He is a member of the Council of the Newcomen Society and editor of its Bulletin; Past President of the Essex Archaeological and Historical Congress; President of the Lea and Stort Rivers Society; committee member of the wind and watermill section of the Society for the Protection of Ancient Buildings; contributor on international waterways to waterway journals; and lecturer for, *inter alia*, the National Trust and the Inland Waterways Association.

R. A. Buchanan is Reader in the History of Technology at the University of Bath, and Director there of the Centre for the History of Technology, Science and Society. He is the Secretary General of the International Committee for the History of Technology, and a member of the Royal Commission of Historical Monuments in England. He has written several books on technological history and industrial archaeology, including *Industrial Archaeology in Britain* (1972). He has recently become Director of the National Cataloguing Unit for the Archives of Contemporary Scientists, at the University of Bath, which is responsible for finding permanent homes in libraries and archives for the manuscript papers of eminent scientists and engineers.

A. K. Corry saw war service with RAF Bomber Command, then joined the first British 'sandwich course' scheme set up by the Engineering Institutions and Ministry of Education. He worked as a Production Engineer in the automobile industry, subsequently taking up a post as a lecturer at the Great Yarmouth College of Further Education, eventually becoming Head of the Engineering Department. He was appointed Regional Engineer for Industrial Administration Ltd., a management consultancy company in 1961, developing group apprenticeship schemes, and became a Technical Adviser to the Minister of Labour to work on the Industrial Training Act. After a few years with Industry Training Board as a Senior Adviser, he joined the Science Museum, London, in 1972 as an Assistant Keeper. Until recently he was Deputy Keeper in the Department of Mechanical and Civil Engineering with responsibility for National Collections from manufacturing industries. He is Editor of the classic *Tools for the Job* by L.T.C. Rolt (1986).

A. S. Darling began his career as a mechanical engineer. He moved into the metallurgical industry and eventually became head of Metallurgical Research at Johnson Matthey and Co. Since that time he has been involved in high temperature alloy development at the British Ministry of Defence. He is the author of many papers and articles on metallurgical and engineering subjects and has been granted numerous Patents in this area.

Lance Day was educated at London University, where he took degrees in Chemistry and in the History and Philosophy of Science. He joined the Science

Museum Library, London, in 1951 and became Assistant Keeper in charge of Public Services. From 1970 to 1974 he was with the Department of Chemistry, where he was responsible for setting up the Iron and Steel Gallery, opened in 1972. In 1974 he was appointed Keeper of the Department of Electrical Engineering and Communications, where he was responsible for the first permanent exhibition in the Museum's East Block Extension, devoted to printing and papermaking. From 1976 until his retirement in 1987 he was Keeper of the Library. He made a number of major acquisitions in developing the Library as the national reference source for the history of science and technology, and was responsible for setting up a new section for the Museum's collections of pictures and archives. He was for many years Hon. Secretary and Member of Council of the Newcomen Society. He is the author of *Broad Gauge*, a Science Museum publication on the broad gauge and other rail systems.

W. K. V. Gale comes from a family long employed in the iron and steel industry. He started work in an iron foundry, moved to technical journalism, and was metals editor of *The Engineer* from 1960 to 1972. Since 1972 he has been a free-lance writer, specialising in all aspects of iron and steel, historical and modern, and is a regular contributor to technical journals. He is a member of The Institute of Metals, a Fellow of the Royal Historical Society, and a Past President of the Newcomen Society, the Staffordshire Iron and Steel Institute, and the Historical Metallurgy Society. He is honorary adviser on iron and steel to the Ironbridge Gorge and the Black Country Museums. He is the author of a large number of articles in technical journals and society transactions and of several books, including: *The Black Country Iron Industry* (1966) (2nd edition 1979), *The British Iron & Steel Industry* (1967), *Iron and Steel* (1969), *The Iron & Steel Industry: A Dictionary of Terms* (1971) (2nd edition 1973), *Historic Industrial Scenes*: Iron and Steel (1977).

John Griffiths is Curator at the Science Museum, London, having responsibility for the Space Technology and Protective Clothing Collections. He was educated at University College, London, from which he has a doctorate, in Infra-red Astronomy.

Richard Hills read History and Education at Cambridge, where he also began research into fen drainage history. This took him to Imperial College, London, where his thesis was published as *Machines, Mills and Uncountable Costly Necessities* (1967). In 1965 he moved to the University of Manchester Institute of Science and Technology where his doctoral thesis was published as *Power in the Industrial Revolution* (1970). Here he started the North Western Museum of Science and Industry and collected the major exhibits now displayed at Liverpool Road Station. He has also published *A Life of Richard Arkwright* (1973),

Beyer and Peacock: Locomotive Builders to the World (1982), *Papermaking in Britain 1488–1988* (1988) and *Power from Steam* (1989).

Ian McNeil is of the fifth generation in his family to be blacksmiths or engineers. After reading Engineering at Cambridge, his career has covered oil exploration, research into gear transmission and gear pumps, and the world-wide marketing of oil-hydraulic equipment. He is the author of *Joseph Bramah: a Century of Invention* and *The Industrial Archaeology of Hydraulic Power*. He has recently retired from the position of Executive Secretary of the Newcomen Society for the Study of the History in Technology and Engineering, based in the Science Museum, London. He is now, Rolt Research Fellow in the History of Technology at the University of Bath where he is working on a history of bellows.

J. Kenneth Major is an architect whose practice specialises in the repair of old buildings and mills. He holds the qualifications of Bachelor of Architecture, Member of the Royal Institute of British Architects and Fellow of the Society of Antiquaries. He is currently Chairman of Council of The International Molinological Society, and a member of the Committee of the Society for the Protection of Ancient Buildings and of its Wind and Watermill Section. His published works include *Fieldwork in Industrial Archaeology* (1975), *Victorian and Edwardian Windmills and Watermills from Old Photographs* (with Martin Watts 1977), *Animal-Powered Engines* (1978), *Animal-Powered Machines* (1985) and *A Pocketbook of Watermills and Windmills* (1986). His repair of the 36ft-diameter waterwheel, pumps and cascade at Painshill Park in Surrey, UK was completed in December 1988 and received a Civic Trust Commendation.

Charles Messenger was educated at Sandhurst and Oxford, where he read History. He was commissioned into the Royal Tank Regiment in 1961, and served as a regular army officer until 1980. He is now a freelance military historian and defence analyst. During 1982–3 he was a Research Fellow at the Royal United Services Institute, London, and is currently Associate Editor of *Current Military Literature* and military consultant to a number of publishers. Among his many published works are *The Art of Blitzkrieg* (1976), *Bomber Harris and the Strategic Bombing Offensive 1939–45* (1984), *The Commandos 1940–1946* (1985) and *Hitler's Gladiator* (1987), a biography of SS General Dietrich.

Herbert Ohlman has a BS degree in physics from Syracuse University and an MS in computer science from Washington University in St. Louis. He was a staff member of several multinational corporations and non-profit organisations in the United States until 1972, when he joined the World Health Organization. Since 1980, he has been an independent consultant in microcomputing and communications based in Geneva. He is the inventor of permutation indexing, one of the first mechanized indexing systems, and the author of articles in information science and technology, particularly applied to health, education and library systems. He was the founding chairman of the Special Interest Group on Education for Information Science of the American Society for Information Science, and a Chairman of the Special Interest Group on Information Retrieval of the Association for Computing Machinery.

Andrew Patterson was educated at Haileybury, and London University's Wye College of Agriculture. He farmed in West Wales between 1973 and 1979, earning little, but learning much. He joined the Science Museum, London, in January 1980 as Research Assistant to the Agricultural Collection, and was promoted to Assistant Keeper with responsibility for the collection in October 1983. In 1986 he joined the Yorkshire Museum of Farming as Keeper of Social and Agricultural History with additional responsibility for the collections at the Upper Dales Folk Museum at Hawes. He is a Founder Member and Vice-Chairman of the Historic Farm Buildings Group, established to co-ordinate recording, and increase awareness of this fast-altering historic resource.

A. W. H. Pearsall read History at Cambridge University, then saw service in the Royal Navy 1944–7, joined the National Maritime Museum, Greenwich, UK first as Research Assistant, then as Custodian of Manuscripts, and subsequently as Historian. He is Vice-President of the Navy Records Society and member of the Publications Committee, member of the Committee of the Fortress Study Group and of the Association of British Transport Museums, and formerly member of Council of Society of Archivists. His Publications include *North Irish Channel Services* (1962) and *Steam enters the North Sea* in Arne Bang-Andersen *et al. The North Sea* (1985).

P. J. G. Ransom has been studying the history of railways for forty years and writing about it for more than twenty. His books include *The Victorian Railway and How it Evolved* (1989), *Scottish Steam Today* (1989), *The Archaeology of the Transport Revolution 1750–1850* (1984), *The Archaeology of Railways* (1981), and *Railways Revived* (an account of preserved steam railways) (1973). He has been actively involved in railway preservation since the early days of the Talyllyn Railway Preservation Society in the 1950s. He was for some years on Board of the Festiniog Railway Society Ltd., and is currently on the council of the Association of the Railway Preservation Societies.

E. F. C. Somerscales read engineering at the University of London, received his doctorate from Cornell University in 1965, and is currently Associate Professor of Mechanical Engineering at the Rensselaer Polytechnic Institute, Troy,

New York. He is the chairman of the History and Heritage Committee of the American Society of Mechanical Engineers and has written a dozen papers on the history of mechanical engineering. In addition to the history of engineering, his teaching and research interests are in the areas of heat transfer and fluid mechanics, and he is currently involved in research on the corrosion and fouling of heat transfer surfaces. In 1988 he received the Bengough Medal from the Institute of Metals. He has over 30 publications on various aspects of heat transfer, and was the co-editor of *Fouling of Heat Transfer Equipment* (1981).

Doreen Yarwood is a professional author and artist, specialising in the history of architecture and design, including costume and interior decoration, in Europe and the United States over the last 3000 years. Trained in art and design, until 1948 she lectured in a number of institutions and in the Women's Royal Air Force. Subsequently she turned to writing full-time and has published 19 books – including *The Architecture of Europe* (4 vols., 1974 and 1990), *Encyclopedia of Architecture* (1985) and *Science and the Home* (1987). Doreen Yarwood currently divides her time between new publishing projects and her position as an extra-mural lecturer for the universities of Sussex, Surrey and London.

INDEX OF NAMES

Abernathy, James 242
Abruzzo, Ben L. 613
Abt, Roman 584
Ackermann, Rudolph 449, 678
Adam, Robert 871, 916
Adams, Robert 985
Adams, W. Bridges 580
Adamson, D. 286
Adamson, Robert 733
Ader, Clément 620
Aeschylus 69, 388
Agricola (=Georg Bauer) 14, 80, 187, 196, 232, 264
Aitken, Harold H. 41, 702
Albert, Prince Consort 104, 201, 894
Albone, Dan 788–9
Alder 417
Aldus Manutius 671
Alexander the Great 74
Alexanderson, Ernst F.W. 726, 745
Alhazen of Basra 730
Allan 380
Allen, Bryan 647
Allen, Horatio 562
Allen, J.F. 284
Alleyne, Sir John 175
Allport, James 578, 579
Ames, B.C. 421
Amman, Jost 390
Ampère, André-Marie 357, 714
Anderson, John 33–5
Anderson, Max L. 613
Anthelm, Ludwig 854
Anthemios of Tralles 880
Apollonius of Perga 885
Appert, Nicolas 798
Appleby 783
Appleton, Sir Edward 728, 998
Arago, François 383
Archer, Frederick Scott 733

Archer, Thomas 672
Archimedes 18
Argand, Ami 211, 913
Aristophanes 71
Aristotle 16, 58, 187
Arkwright, Sir Richard 266, 827–32, 839
Arlandes, Marquis d' 610
Armengaud (the elder) 242
Armengaud, R. 332
Armstrong, Edwin 728
Armstrong, Neil 654
Armstrong, W.G. 40, 244, 351, 961
Arnold, A. 839
Arsonval, J.A. d' 375
Ashley 198
Askin, Charles 98
Aspdin, Joseph 466, 889
Atanasoff, John V. 702
Atwood, Ellis D. 606
Aubert, Jean 485
Austin (textile engineer) 843
Austin, Herbert 452
Ayrton, W.E. 375, 376

Babbage, Charles 31, 398, 699–701
Bachelier, Nicholas 484
Backus, John 704
Bacon, Roger 648, 975
Baekeland, Leo Hendrik 218, 220
Baeyer, Adolf von 202, 218
Bain, Alexander 696, 743
Baird, John Logie 745
Baker, Sir Benjamin 955
Bakewell, Frederick 743
Baldwin, Matthias 573
Baldwin, Thomas 611
Ball, Major Charles 117
Bamberger 54
Barber, John 329
Barclay, Henry 408

INDEX OF NAMES

Barclay, Robert 678
Bardeen, John 42, 703
Barlow, Peter 468
Barnett, M.F. 728, 998
Barry, Sir Charles 894
Barsanti, Eugenio 305
Bartlett, C.C. 100
Basov, N.G. 737, 738
Bateman, J.F. La Trobe 241, 952, 954
Baudot, Emile 715
Bauer, Andreas 675
Bauer, Georg *see* Agricola
Bauer, H. 852
Bauer, S.W.V. 537
Baumann, K. 297
Bayer, Karl Joseph 112
Bayly, Sir Nicholas 90
Bazalgette, Joseph 956–7
Beau de Rochas, Alphonse 305
Beaumont, Huntingdon 555
Beavan, E.J. 849
Becker, H. 1005
Becquerel, Henri 213, 1005
Bedson, George 175
Beebe, Charles William 553
Beeching, Dr R. 599
Behr, F.B. 586
Bélidor, Bernard Forest de 233–4, 982
Bell (textile engineer) 835
Bell, Alexander Graham 538, 719, 721, 965
Bell, Henry 35, 527, 912
Bell, Hugh 105
Bell, Sir Lowthian 105
Bell, Revd Patrick 782
Belling, C.R. 917, 944
Bennet, Revd Abraham 373
Benson, Dr E.W. 98
Bentham (engineer) 30
Bentley, J.F. 891
Benz, Karl 37, 310, 311, 449, 450–51, 454
Berliner, Emile 721
Berry, A.F. 373
Berry, Clifford 702
Berry, George 786
Berry, Henry 477–8
Bertholet 56
Berthollet, Claude Louis 205, 835
Berthoud, Ferdinand 392
Berzelius, Baron Jöns Jakob 201, 743
Besant 438
Bessemer, Sir Henry 167–70, 171, 177
Besson, Jacques 393
Bethell, John 208
Bettini, Gianni 722
Bewick, Thomas 678
Bienvenu 641
Bigelow, E.G. 846
Bigelow, Erasmus B. 905

Bilgram, Hugo 411
Bion, Nicolas 668
Birdseye, Clarence 798
Biringuccio, Vannoccio 187, 396
Birkeland, Kristian 384
Biró, Georg and Ladisla 669
Bissell, G.H. 211
Bissell, Melville R. 924
Bissell, Melville R., jun. 927
Bisson brothers 733
Black, Harold S. 749
Black, Joseph 609
Blackett, Christopher 559
Blanchard, Jean-Pierre-François 614
Blanquert-Evrard, L.D. 733
Blenkinsop, John 559
Blériot, Louis 625
Blickensderfer 683
Blith, Walter 774
Bloch, Joseph 851
Block, I.S. 989
Blondel, N.-F. de 982
Blumlein, Alan D. 723
Bodley, George 941
Bodmer, John George 411
Bogardus, James 894
Bohr, Niels 213, 1005
Bollée family 259
Bolton, W. von 133, 143
Bond, George 406
Boole, George 701
Booth, Henry 561, 563
Booth, Hubert Cecil 925, 926
Borchers, Wilhelm 123
Borelli, Giovanni 553, 646
Borol, A. 459
Borries, August von 582
Bosch, Carl 775
Bosch, Robert 311
Bothe, W.W.G.F. 1005
Böttger, Johann Friedrich 192–3
Bouch, Sir Thomas 463
Boulle, André Charles 908
Boulton, Matthew 28, 34, 91, 155, 275–6, 325, 403
Bourn, D. 830
Bourne, William 537
Bourseul, Charles 718
Bousquet, Gaston du 582
Boutheroue, Guillaume 483
Bouton, Georges 447
Bowser, Sylvanus F. 459
Boxer, Colonel E.M. 984
Boyd, T.A. 317
Boyden, Uriah A. 243
Boyle, Robert 33, 189
Bradley, Charles S. 107
Brady, Mathew 733

1019

Bramah, Joseph 29, 40, 404, 921, 961
Bramante of Florence 28
Branca, Giovanni de 290
Brandling, J.C. 559
Branly, Edouard 725
Brassey, Thomas 470, 565, 575
Brattain, Walter 42, 703
Braun (instrument-maker) 376
Braun, Ferdinand 744
Braun, Wernher von 649, 650, 651, 656, 747, 754
Braunschweig, Hieronymus 187
Brayton, G.B. 310
Brearley, Harry 174
Breguet, Louis 641–2, 696
Brennan, Louis 586
Breuer, Marcel 910
Brewster, Sir David 733, 735
Bricklinn, Dan 707
Briggs, Henry 698
Brindley, James 439, 477–8
Brocq 995
Brode, John 83
Brooman, Richard 419
Brown, Charles 385
Brown, Joseph 407
Brown, Joseph R. 408, 409, 411
Brown, Samuel (first internal combustion engine) 305
Brown, Lieutenant Samuel (chain cable) 541
Brownrigg, William 129
Bruce, David 679
Brueghel, Jan 232
Brueghel, Pieter (the elder) 248, 780
Brunel, Isambard Kingdom 35, 401, 403, 527, 528, 529, 565, 568, 572, 895
Brunel, Marc Isambard 30, 398, 404, 468, 552, 847, 854, 890
Brunelleschi, Filippo 24, 731, 881
Brunner, John 222
Bryan, G.H. 622, 626
Büchi, A. 315, 316
Buck, J. 411
Buckle, William 403
Bull, John Wrathall 781
Bullard, E.P. 414
Bunning 896
Bunsen, Robert Wilhelm 103, 107, 113, 114, 210, 353
Burgess, G.K. 141
Burgi, Joost 695, 698
Burgin, Emile 360–61
Burks, Arthur W. 702
Burton, Decimus 895
Burton, James Henry 406
Bury, Edward 564, 566, 573
Bush, Vannevar 750–51

Bushnell, David 537, 996
Bussy, Antoine-Alexander-Brutus 113
Butler, Edward 450
By, Colonel John 507

Cameron, Julia Margaret 733
Campbell, Wilfred 298–9
Campbell-Swinton, A.A. 745
Caquot, Albert 613
Carcel 914
Cardano, Girolamo 731
Cardew, Philip 376
Carlson, Chester 736
Carnot, Sadi 343
Caro, Heinrich 202
Carolus, Johann 672
Caron, H. 113
Carothers, Wallace H. 218, 219, 850
Carter, John 84
Cartwright, Revd Edmund 834, 839
Cary, John 461
Casablanca, Fernando 841
Casement, Robert 240
Casseli, Giovanni 744
Castner, Hamilton Y. 105, 224–5
Caxton, William 27, 671, 672
Cayley, Sir George 39, 342, 472, 617–19, 641
Cecil, William 305
Celestine III, Pope 246
Cellini, Benvenuto 68
Chadwick, Edwin 956
Chadwick, James 213
Chambers, Austin 571
Chambers, Sir William 871
Champion, Nehemiah 85–6
Champion, Richard 193
Champion, William 87–9
Chanute, Octave 621–2
Chapelon, André 597–8
Chappe, Abraham 713
Chappe, Claude 711, 713
Chapman, Frederic af 524
Charlemagne, Emperor 79, 494, 956
Charles, Jacques Alexander César 610
Charles XII, King of Sweden 501
Chevenard, P. 126
Chevreul, Michel Eugène 913
Childe, V. Gordon 5
Chopping, Henry 258
Christensen, C. 736
Christie, S.H. 375
Church, William 680
Churchward, G.J. 590
Cierva, Juan de la 642
Clark, Edward 851
Clark, Edwin 482, 486, 493
Clarke, Arthur C. 747

1020

INDEX OF NAMES

Clarke, Sir Clement 83
Claus, C.E. 131
Clay, Henry 909
Clegg, Samuel 40, 209, 570, 958
Clement, Joseph 31
Clement, William 24
Clements, Joseph 398, 399
Clerk, Dugald 307, 315, 317, 454
Clinton, DeWitt 511
Clymer, Floyd 675
Coade family 871
Cobbett, William 801
Cockerell, Christopher 538
Codd 198
Coignet, François 890, 891
Coles, Captain Cowper 533
Collins, Captain Greenvile 546
Colt, Samuel 985, 996
Columbus, Christopher 21, 801
Columella 779
Confucius 2
Congreve, Sir William 648, 986
Cook, Maurice 123
Cooke, William F. 41, 357, 358, 572, 714, 965
Cookworthy, William 193
Coolidge, W.D. 133
Copeland, L.D. 447
Copernicus, Nicolaus 672
Corliss, G.H. 282–3
Cornu, Paul 641–2
Corson, Michael 124
Cort, Henry 157, 162
Cosnier, Hugues 483
Cotchett, Thomas 824
Cotton, William 847
Couchon, Basile 822
Coulomb, Charles Augustin de 373
Cousteau, Jacques 553
Coutelle, Charles 611
Coutts, Angela Burdett 931
Cowles, Alfred and Eugene 106
Crampton, T.R. 569
Craufurd, Commander H.V. 95
Crawford, J.W.C. 219
Croesus, King 27
Croll, A.A. 942
Crompton, R.E.B. 360–61, 367, 372, 944
Crompton, Samuel 832–3
Crompton, Thomas 674
Cronstedt, Axel 97
Crookes, W. 725
Cros, Charles 720
Cross, C.F. 849
Crowther, J.G. 42
Cruickshank, William 352
Ctesibius of Alexandria 16
Cubitt, Lewis 896

Cubitt, Sir William 256, 264
Cugnot, Nicolas 441, 453
Cumming, Alexander 921
Curie, Marie and Pierre 213
Curr, John 556
Curtis, C.G. 293–4

Daft, Leo 584
Dagron 733
Daguerre, Louis-Jacques-Mandé 732
Dahl 124
Daimler, Gottlieb 37, 311, 315, 447, 448, 449, 450–51, 454, 615
Dale, David 832
Dalen, Gustav 942
Dalrymple, Alexander 546
Dalton, John 189, 212–13
Dancer, J.B. 733
Daniell, John Frederic 353
Darby, Abraham 84–5, 153–4, 207, 893
Darby, Abraham, III 156, 465
Davenport, Thomas 378
Davidson, Robert 378, 584
Davy, Sir Humphrey 102, 113, 209, 224, 352, 354–5, 774
Dawson, William 847
De Boer 141
de Dion, Count Albert 447
de Forest, Lee 41, 720, 726–7, 742
de Havilland, Geoffrey 626
Deacon, Henry 205, 222
Deane, Sir Anthony 524
Deck, J.T. 194
Deering, William 783
Defoe, Daniel 91, 434
Degen, Jakob 641
Delauney, L. 711
Delisle, Romé 129
della Porta, Francesco 32
della Porta, Giacomo 881
della Porta, Giambattista 273
della Robbia, Luca 192
Delvigne, Captain Henri-Gustave 983
Democritus 212
Deprez, Marcel 381
Desvignes 739
Devereux, Colonel W.C. 122
Deverill, Hooton 848
Deville, Henri Lucien Sainte-Claire 103, 104, 107, 109, 111, 113
Devol, George 426
Dickinson, J.T. 850
Dickinson, Robert 398
Dickson 219
Dickson, William 741
Diderot, Denis 236
Diesel, Rudolph 37, 311, 313, 322, 454
Diggle, Squire 844

INDEX OF NAMES

Dinesen, Hans 538
Dioscorides 75
Doane, Thomas 469
Dockwra, William 84
Dolby, Ray 724
Dondi, Giovanni di 23, 695
Dondi, Jacopo di 23
Donisthorpe, G.E. 840
Donisthorpe, Wordsworth 741
Donkin, Bryan 398, 674
Dony, Abbé Jean Jacques 92–3
Dore, Samuel 818
Dornberger, Captain W.R. 649, 651
Dow, Herbert 118
Downie, James 853
Downs, J.C. 225
Dowson, J.E. 314
Drais von Sauerbronn, Baron Karl 442
Drake, Edwin L. 211, 452
Drebbel, Cornelius van 537
Dreyse, Nikolaus von 984
Du Quet 449
Duboscq, Jules 735
Ducos du Hauron, Louis 735
Duddell, William du Bois 377
Dudley, Dud 153
Dunlop, John Boyd 217, 446, 449
Dunwoody, General Henry H.C. 726
Dupuy de Lôme 537
Duquesney 896
Duryea brothers 451
Dyer, John 817

Eakins, Thomas 740
Eastman, George 734–5, 739, 851
Eberner, Erasmus 80
Eckert, J. Presper 702
Edgeworth, Richard 472
Edison, Thomas Alva 41, 362, 365, 367–8, 370, 376, 719–21, 722, 741–2, 915, 959
Edwards, Humphrey 279
Eiffel, Gustave 483, 894
Einstein, Albert 213, 737
Elder, John 286
Elizabeth I, Queen 232
Ellington, E.B. 40
Elton, Abraham 84
Engelberger, Joseph 426
Engels, Friedrich 949
England, George 575
Ercker, Lazarus 86, 187
Ericsson, John 342, 528, 533
Euler, Leonhard 236
Evans, Brook 98
Evans, Oliver 30, 238, 277, 282
Ewbank, Thomas 380

Fafschamps 985

Fairbairn, P. 840
Fairbairn, William 241–2
Fairlie, Robert 575
Falcon 823
Faraday, Michael 41, 104, 352, 354–6, 358, 377
Farman, Henri 624
Farnsworth, Philo T. 745
Faure, Camille 353
Fawcett 219
Fellows, E.R. 411
Fenton, James 403
Fenton, Roger 733
Ferguson, Harry 790
Fermi, Enrico 214
Ferranti, Sebastian Ziani de 41, 285, 371–2, 376, 960
Fessenden, Reginald 726
Field, Cyrus W. 715
Field, Joshua 398
Figuier 713
Finley, James 465
Fischer (colour film) 735
Fischer (television projector) 746
Fischer, F. 114
Fitch, John 527
Flechsig, W. 745
Fleitman, Thomas 99
Fleming, J.A. 41, 384, 720, 726, 727
Flettner, Anton 642
Focke, Heinrich 642
Fokker, Anthony 995
Ford, Henry 37, 413–14, 452, 789
Forlanini, Enrico 538, 641
Forrester, George 564
Forrester, Jay W. 702, 703
Forsyth, Alexander 984
Fouché 418
Fourdrinier, Henry and Sealy 674
Fourneyron, Benoît 242
Fowler, John 772
Fox, James 399, 404, 411
Fox, Samuel 853
Fox Talbot, Henry 732, 735
Franchot 914
Francis, J.B. 243
Franklin, Benjamin 350
Frankston, Bob 707
Fredkin, Edward 43
Freminet 553
Freyssinet, Eugène 466, 891
Friedrich August II, Elector of Saxony 193
Friese-Greene, William 741
Frisch, Otto 214
Frishmuth, Colonel 105, 108
Frost, James 889
Fuller, R. Buckminster 882, 900
Fulton, Robert 35, 527, 537, 996

1022

INDEX OF NAMES

Fust, Johann 27

Gabor, Denis 737
Gagarin, Yuri 652
Galen 75
Galileo Galilei 23, 25, 695–6, 730
Galvani, Luigi 94, 351
Gamble, John 674
Gamble, Josias 221
Gamond, Aimé Thome de 470
Garratt, H.W. 591
Gartside 821
Gascoigne 398
Gaskill, H.F. 285
Gatling, Richard 985
Geber 220
Ged, William 676
Geitner, Dr 97
Gemma-Frisius, Rainer 731
Genet, E.C. 614
Genoux, Claude 676
Gerber, Heinrich 464
Gerhard, William 104
Gesner, Abraham 211
Gestetner 683
Gibbons, John 162–3
Gibson 219
Giffard, Henri 575, 614
Gilbert, Sir Alfred 108
Gilbert, Cass 897
Gilbert, Joseph 775
Gilbert, William 41, 350
Gillott, Joseph 668
Gilpin, Thomas 510
Glauber, J.R. 221, 774
Glehn, Alfred de 582
Glenn, John 652
Glidden, Carlos 683
Goddard, Robert H. 649
Godowsky, Leopold 735
Goldmark, Peter 723, 746
Goldschmidt, Hans 123
Goldstein, Herman H. 702
Goliath of Gath 388
Gooch, Daniel 568
Goodyear, Charles 216, 849, 853
Gordon, James P. 737
Gorton, Richard and Thomas 821
Goucher, John 784
Gould, Gordon 737
Goulding, J. 836
Gowland, Professor 75–6
Goyen, Jan van 248
Graebe 202
Graetzel, R. 114
Graham, W.G. 448
Gramme, Z.T. 360, 381
Grant, George B. 412

Grantham, J. 456
Gray, Andrew 238
Gray, Elisha 719
Greathead, James Henry 468
Greathead, William 539
Green, Charles 611, 612
Gregory, C.H. 571
Gresham, Sir Thomas 81
Gresley, Nigel 596
Gridley, George O. 410
Griess, Peter 202
Griffith, A.A. 637
Griswold 419
Gropius, Walter 900
Grove, William 353
Guericke, Otto von 273, 323
Guest 417
Guido d'Arezzo 717
Guinand 197
Gundelfinger, Joachim 81
Günther, W. 244
Gurney, Sir Goldsworthy 458
Gusseff 419
Gutenberg, Johann Gensfleisch zum 27, 670–71
Guton, Henri 729
Guyon, Jacques, Sieur du Chesnoy 483

Haber, Fritz 223–4, 775, 992
Hadfield, R.A. 174
Haenlein, Paul 614
Haessler, Helmut 646
Hahn, Otto 214, 1005
Hair, T.H. 266
Hakewessel 410
Hale, Edward Everett 747
Hale, Gerald A. 461
Hale, William 646
Hall, Asaph 747
Hall, Charles Martin 107–108, 111, 224
Hall, Douglas 426
Hall, John 223, 405–406, 408
Hall, Joseph 161–2
Halladay, Daniel 258
Hamd-allah Mustafi 77
Hamilton, Hal 594
Hammond, Robert 370
Hanaman, Franz 133
Hancock, Thomas 216–17, 853
Hancock, Walter 458
Hannan, W.J. 752
Hannart, Louis 852
Hansard, T.C. 677
Hansgirg 121
Hanway, Jonas 853
Harford family (brass-makers) 91–2
Hargreaves, James 825–7
Harington, Sir John 921

INDEX OF NAMES

Harrison, John 24, 544, 696
Hart 408
Hartmann 376
Haupt, Hans 853
Hawksley, Thomas 952, 956
Haynes, Edward 127
Heald, James 413
Hearson 796
Heathcote, John 848
Heaviside, Oliver 729
Hedley, William 559
Hegginbotham 426
Heilmann, J. 840
Heiman, Frederick P. 705
Heinkel, Ernst 39
Henlein, Peter 695
Henne, Heinrich 242
Hennébique, François 890, 891
Henry VIII, King 81
Henson, William Samuel 39, 619–20
Herland, Hugh 866
Herman, Augustine 509–510
Hero of Alexandria 17, 25, 273, 290, 695
Herodotus 27, 69, 462, 868
Héroult, Paul Louis Toussaint 109–11, 176, 224
Herring, A.M. 622
Herrman 411
Herschel, Sir John 732
Hertz, Heinrich 725
Hetzel 696
Hewes, T.C. 239, 241
Highs, Thomas 828
Hill, David Octavius 733
Hill, Rowland 219, 964
Hillary, Sir William 539
Hipparchus 694
Ho Wei 97
Hochstetter, Daniel 82, 232
Hochstetter, Joachim 81
Hofmann, August Wilhelm von 201, 208
Hoke, W.E. 415
Holabird 897
Holden, Isaac 840
Holland, J.P. 537
Hollerith, Herman 701
Holly, Birdsill 285
Holmes, F.H. 358
Holmes, Oliver Wendell 735
Holt, Benjamin 472
Holtzapffel 393
Holzwarth, Hans 331, 332
Homfray, Samuel 35
Honnecourt, Villard de 231
Hooke, Robert 17, 24, 25, 731
Hooper, Captain Stephen 256
Hoover, W.H. 927
Hopkinson, John 361–2

Hopper, Thomas 894
Hornblower, Jonathan 278
Horner 739
Horning, H.L. 317
Horsbrugh 546
Hostein, Steven R. 705
Houldsworth, H. 839
House, Royal E. 719
Houston, E.J. 376
Howe, Elias 848, 851
Howe, Frederick W. 406, 408
Howe, William 463
Hudson, Henry 48
Hughes, D.E. 965
Hulls, Jonathan 453
Humfrey, Sir William 82–3
Hunt, Captain A.E. 108
Hunt, Walter 985
Hunter, M.A. 141
Huntsman, Benjamin 159–60, 397
Huth, G. 718
Hutton, William 98
Huygens, Christiaan 23, 24, 304–305, 696, 738
Hyatt, John Wesley 217
Hyland, L.A. 729

I Hsing 695
Ince, Francis 371
Isodorus of Miletus 880
Issigonis, Alec 452
Ivatt, H.A. 590
Ives, F. E. 682

Jabir *see* Geber
Jablochkoff, Paul 363
Jacobi, M.H. 380
Jacquard, Joseph-Marie 699, 823
Jacquet-Droz 696
Jansen, Zacharias 25
Jansky, Karl G. 729
Janssen, P.J.C. 742
Jefferson, Thomas 405
Jeffries, Zay 126
Jenatzy, Camille 458
Jenkins, Charles Francis 745
Jenney, William Le Baron 897
Jervis, John B. 573
Jessop, William 479, 480, 481, 556
Joffroy d'Abbans, Marquis de 527
Johansson, Carl Edward 414–15, 417
Johnson, Isaac Charles 889
Johnson, Percival Norton 97–8
Johnson, S.W. 582
Johnson, Thomas 843
Johnston, William 897
Jolly-Bellin, Jean-Baptiste 854
Jones, Samuel 912

INDEX OF NAMES

Joubert 377
Joy, David 569
Judson, Whitcomb L. 852
Julius Caesar 462
Julliot, Henri 615
Junghans, S. 177
Junkers, Hugo 324, 628
Just, Alexander 133

Kahlmeter, Henric 85
Kaiser, Henry 121
Kaovenhofer, Andreas 236
Kaplan, Victor 245
Kapp, Gisbert 361
Karavodine, M. 332
Kay, John (of Bury) 821–2
Kay, John (of Warrington) 828
Kay, Robert 822, 830
Kekulé, August 202
Kellner, Carl 225
Kelly, William 837
Kelvin, Lord 112, 545 see also Thomson, Sir William
Kemeny, John G. 705
Kennelly, Arthur 729
Kenward, Cyril Walter 426
Kepler, Johan 746
Kettering, C.F. 317
Kilby, J.S. 705
Kircher, P. 738
Kjellberg 418
Kleinschmidt, E.E. 715
Klič, Karl 679
Knight, Charles 677
Knight, J.P. 460
Komarov, Vladimir 653
König, Friedrich 675
Krais 205
Kranich, Burckard 81
Krebs, Arthur 614
Kremer, Henry 646
Kroll, W.J. 133, 141, 142, 143
Kügelgen 115, 117
Kung Ya 97
Kurtz, Thomas E. 706
Kyeser, Konrad 648
Kylala 598

La Hire, Phillippe 396
La Pointe, John 414
La Rive, A.A. 95
Lactantius 730
Laithwaite, Eric 384
Lake, Simon 537
Lanchester, Frederick 413, 451–2, 1001
Lanchester, George 451–2
Land, Edwin 736
Lane-Fox, St George 365, 368

Langen, Eugen 306, 454
Langer, Carl 100
Langmuir, Irving 915
Lanston, Tolbert 680
Lartigue, C.F.M.T. 586
Lascelles, William 891
Latimer-Needham, C.H. 538
Launoy 641
Lauste, Eugène Augustin 742
Laval, Gustav de 290, 346
Lavoisier, Antoine Laurent 189, 201
Lawes, John Bennet 224, 775
Lawrence, Richard S. 406
Le Blanc 405, 983
Le Corbusier 892
Le Prince, Louis 741
Leakey, Louis 6
Leakey, Mary 6
Leakey, Richard 10
Leavitt, E.D. 285
Lebaudy brothers 615
Leblanc, Nicholas 221
Lebon, Phillippe 40, 209, 957
Leclanché, Georges 353
Ledru 95
Lee, Edmund 258
Lee, Revd William 819, 846
Leeghwater, Jan 250
Leeuwenhoek, Anton von 25, 730
Legagneux, G. 627
Leibniz, Gottfried Wilhelm 699
Lemale, C. 332
Lenin, Vladimir Ilyich 594
Lenoir, Etienne 305, 454
Leonardo da Vinci 162, 390, 393, 395, 461, 483, 500, 815, 899, 978
Leoni 209
Leonov, Alexei 652
Lesage, George Louis 710
Leschot 696
Lesseps, Ferdinand de 505, 514
Lethbridge 553
Leukippos 212
Leupold, Jacob 235, 277
Levassor, Emile 451
Levavasseur, Léon 625
Lever, William Hesketh 923
Levers, John 848
Lévy, Lucien 727
Lewis, Colonel I.N. 992
Lewis, J. 818
Lewis, William 129
Liang Ling-tsan 695
Liardet 885
Libby, William F. 697
Liebermann 202
Liebig, Baron Justus von 201, 775, 910
Lilienthal, Otto 39, 621

INDEX OF NAMES

Lillie, James 241
Linde, Carl von 454
Lindsay, Sir Coutts 371
Lippershey, Johannes 25, 730
Lister, Samuel C. 805, 840
Lloyd, C.M. 417
Lloyd, Edward 84
Locke, Joseph 565, 575
Lombe, John and Thomas 824
Longbotham, John 478
Louis 894
Louis XV, King of France 193
Lovelace, Ada Augusta, Countess of 701
Lowrey, Grosvenor P. 368
Lucretius 212, 350
Ludwig I, King of Bavaria 498
Lukin, Lionel 539
Lumière, August and Louis 742
Lunardi 614
Lundstrom 912
Lusuerg, Dominicus 25
Lusuerg, Jacob 25
Lysippus of Sicyon 68

McAdam, John Loudon 435–6, 963
McAlphine, Robert 891
McAuliffe, Christa 656
McCormick, Cyrus 782
McCready, Paul 647
Macintosh, Charles 216, 849, 853
McIntyre, John 529
McKenna, P.M. 417
Macmillan, Kirkpatrick 443
McNaught, William 279
Macquer, P.J. 129, 193
Maddox, R.L. 733
Magee, Carl C. 461
Maiano, Giovanni da 871
Maillart, Robert 466
Maiman, Theodore H. 419, 737–8
Mallet, Anatole 581, 582–3
Malouin, P.J. 95
Manby, Aaron 164
Manly, C.M. 320
Mannes, Leopold 735
Mansart, François 884
Mansfield 208
Marconi, Guglielmo 726
Marcus, Siegfried 449, 454
Marey, Etienne Jules 740, 741
Marggraf, Andreas 801
Margraff 92
Maritz 396
Marsh (slotmeter) 209
Marsh, A.L. 125, 917
Marsh, Sylvester 584
Martel, Charles 972
Martin, C. 837

Martin, Emil and Pierre 172
Martius, Carl Alexander 202
Martyn, Sir Richard 83
Masaru, Inoue 577
Masing 124
Massey 545
Mathieu, Albert 470
Matther, William 114
Mattucci, Felice 305
Mauchly, John 702
Maudslay, Henry 29, 30, 395, 397–8, 404
Maughan, Benjamin Waddy 919
Mawson, John 366
Maxim, Hiram 365, 368, 621, 925, 986
Maxwell, James Clerk 725, 735
Maybach, Wilhelm 310, 311, 451, 454
Mège-Mouriez, Hippolyte 190, 799
Meikle, Andrew 256, 267, 784
Meissonier (mining engineer) 104
Meissonier, Jean-Louis-Ernest 740
Meitens, Auguste de 418
Meitner, Lisa 214
Melan, Joseph 466
Mellon, Andrew 108
Mellor, Samuel 113
Menabrea, Luigi 701
Mendeleev, Dimitri 213
Mengoni 896
Menneson, Marcel 421
Mercer, John 849
Merensky, Hans 132
Mergenthaler, Ottmar 680
Merica, Paul 99, 122
Merle, Henri 103
Merritt, William Hamilton 507
Messel, Rudolf 223
Mestral, George de 852
Metcalfe, John 433–4
Meusnier, General J.B.M. 614
Meyer, Lothan 213
Michaux, Pierre 445
Michelangelo 881
Midgley, Thomas 317
Mignet, Henri 645
Millar, Robert 843–4
Minié, Captain 983
Moissan, Henri 112, 135, 141
Mole, de 993
Moller, Anton 820
Mond, Ludwig 100, 222
Monell, Ambrose 101
Monier, Joseph 466, 890
Monroe 1004
Montferrand 894
Montgolfier brothers 287, 609–10, 611, 614
Moore, Hiram 785–6
Moore, Richard F. 415
Moorhead, General 513

INDEX OF NAMES

Moray, Samuel 454
Morewood, Edmund 95
Morice, Peter 234, 951
Morin, Paul 112
Morland, Sir Samuel 3, 33
Morris, Sir William 452
Morrison, Warren 696
Morrison, William Murray 112
Morse, Samuel F.B. 714, 965
Morton, James 203
Morveau, L. 102
Moschel, Wilhelm 115, 116
Moss, S.A. 315
Mosselman, Dominique 93
Moule, Revd Henry 922
Moxon, Joseph 673
Müller, Paul 776
Müller, Wilhelm 242
Muller-Breslau, Professor 617
Mummer, Jacob 84
Murchland 792
Murdock, William 40, 161, 208–209, 227, 403, 441, 957
Murray, Alexander 682
Murray, Matthew 399, 403–404, 559, 841
Mushet, R.F. 172, 174
Muspratt, James 221, 222
Muybridge, Edweard 739–41
Myddelton, Sir Hugh 951

Nadar 733
Nagelmackers, Georges 578, 579
Napier, David 677
Napier, John 698
Napoleon I, Emperor of the French 93, 102, 487
Napoleon III, Emperor of the French 103, 533
Nash, John 871, 886, 894
Nasmyth, James 398, 401, 406
Natrus 251
Need, Samuel 829
Neilson, J.B. 162
Neri, Antonio 187
Nero, Emperor of Rome 501
Nervi, Pier Luigi 882, 885, 892
Netto, Professor 105, 108
Neumann, John von 702
Newcomen, Thomas 3, 33, 155, 207, 273, 275
Newman, Larry 613
Niepce, Nicéphore 419, 731–2
Nipkow, Paul 744
Nobel (the elder) 996
Nobel, Alfred 190, 223, 988
Noble, James 840
Noiré 666
North, Revd C.H. 461

North, Simeon 405, 408
Northrop, J.H. 844
Norton, Charles 413
Norton, W.P. 412
Notman, William 733
Nuffield, Lord 452
Nuys, van 418

Oberth, Hermann 649, 747
Obry, Ludwig 997
Oechehäuser, W. von 324
Oersted, Hans Christian 102, 355, 377
Ogilvy, John 461
Ohain, Hans von 39, 333, 635
Olds, Ransome E. 37, 413
Osborne, Adam 707
Otis, Elisha Graves 897
Otis, William 470
Otto, Nikolaus A. 306–307, 454
Oughtred, William 698
Outram, Benjamin 556
Owen, Robert 832
Owens, Michael 198

Paabo, Max 845
Page, Charles 380, 718
Paget, A. 847
Paixhans, General 533
Palladio, Andreas 463
Palladius 781
Palmer, Charles Mark 529
Palmer, H.R. 586
Palmer, Jean Laurent 417
Palmer, John 438
Papin, Denis 3, 33, 273, 527, 945
Paracelsus 80
Paris, J.A. 739
Parker, James 888, 889
Parker, John Joseph 668
Parkes, Alexander 217
Parkhurst, Edward G. 408, 410
Parry, G. 163
Parseval, Major von 612
Parsons, C.A. 3, 37, 290–91
Parsons, John 422
Pascal, Blaise 33, 699
Pasley, Colonel Charles W. 466
Pasteur, Louis 793
Pathé, Charles 742
Patten, Thomas 90, 91
Paul, Lewis 824–5, 828, 830
Paul, R.J. 375
Paul, Robert 742
Pauly, Samuel 984
Paxton, Sir Joseph 895
Pechiney, Alfred Rangod 103, 109
Pegler, Alan 606
Pennington, E.J. 993

1027

Pennington, William 830
Perkin, William Henry 190, 201–202
Perkins, Jacob 286, 287
Perry, John 375, 376
Peter, Peregrinus 350
Pfleumer 722
Pherecrates 71
Philippe, Adrien 696
Phillips, Robert 434
Phillips, W.H. 641
Phips 553
Picard 418
Piccard, Auguste 553, 613
Pickard, James 276
Picolpasso 192
Pidgeon, M. 120
Piercey, B.J. 127
Pierpoint, Colonel 460
Pilcher 39
Pilkington, Dr 197
Pilkington, Sir Alastair 198
Pistor, Gustav 115, 116, 119
Pixii, Hippolyte 356–7
Planté, Raimond Louis Gaston 353
Plateau, J.A. 739
Pliny the Elder 75, 187, 350, 672, 781, 796
Plumier 393, 394
Polhem, Christopher 237, 395, 501
Poliakoff, Joseph 722
Polk, Willis 900
Polly 251
Polo, Marco 19, 27, 77
Poncelet, J.V. 242
Poniatoff, Alexander M. 751
Ponton d'Amecourt, Vicomte 641
Porsche, Ferdinand 452
Porter, C.T. 283-4
Pouillet, Claude-Servain 374
Poulsen, Valdemar 722
Praed, William 479
Praetorius 123
Pratt, Caleb 463
Pratt, Francis 406
Pratt, Samuel 907
Pratt, Thomas 463
Preece, Sir William 368
Prélat 984
Prévost, M. 627
Priess 123
Priestley, Joseph 216
Priestman, William D. 310, 454
Pritchard, Thomas Farnolls 893
Prokhorov, A.M. 737, 738
Prony, Gaspard Riche, Baron de 699
Pugh 426
Pullman, George Mortimer 577
Pullman, John 369
Pulski, François von 5

Pulson, Sven 408
Pythagoras 717

Radcliffe, William 843
Rainey, Paul M. 749
Ramelli 265
Ramsbottom, John 175
Ramsden, Jesse 394, 411
Ransome (grinding wheel) 408
Ransome, Frederick 889
Ransome, Robert 771
Rateau, Auguste 291, 315, 316
Ravenscroft, George 196–7
Rawcliffe, G.H. 384
Reason, Richard 415, 423
Reeves, Alec H. 749
Regelsburger 123
Régnier 823
Rehe 396
Reis, Philip 718
Renard, Charles 614
Rennie, John 480, 481, 552
Reynaud, Emile 739
Reynold, Han 414
Reynolds, E.T. 285
Reynolds, Richard 161
Ricardo, H.R. 316, 317, 322
Richmann, Georg Wilhelm 350
Rickman 894
Ridley, John 781
Riggenbach, Niklaus 584
Riley, James 98
Rippon, Walter 438
Riquet, Pierre 484, 485
Ritchie, S.J. 99
Rites, F.M. 284
Robert, Marie Noel 610
Robert, Nicolas Louis 674
Roberts (welding) 418
Roberts, David 472
Roberts, Richard 398, 399, 400, 838, 844
Robia, Albert 744
Robinson, George 833
Robinson, Thomas 941
Roche (architect) 897
Roche, Tiphaigne de la 730
Roe, A.V. 626
Roe, Charles 90, 91
Roebling, John 465
Roebling, Washinton A. 465
Roebuck, John 222, 275
Rogallo, Francis Melvin 646
Roger, Emile 451
Rogers, Bryant 705
Rogers, George 96
Rogerson 929
Roget, P.M. 739
Rohn, W. 124, 125

1028

INDEX OF NAMES

Rolls, Hon. Charles S. 452
Rolt, L.T.C. 516
Romershausen 718
Rondelet 890
Röntgen, Wilhelm 725–6
Root, Elisha K. 406
Roper, William 406
Roscoe, Henry 113
Rose, H. 104, 108
Rosenberg, Augustus 723
Rosing, Boris 744
Rowland, W.R. 456
Royce, Henry 452
Rozier, Pilâtre de 610
Ruffle, Richard 258
Ruisdael, Jacob van 248
Rustichello of Pisa 19
Rutherford, Ernest 213, 1005
Rutt, William 677

Sabatier 190
Saint-Victor, Niepce de 733
Salomans, Sir David 451
Salt, Sir Titus 894
Samain, Jacques 751
Samuda brothers 570
Sanders, Thomas 719
Santos-Dumont, Alberto 615, 645
Sanz, M.C. 419
Sargon, King of Assyria 57, 768
Sarnoff, David 745
Sauerbronn, Baron Karl Drais von 442
Saulnier, Raymond 625
Savery, Thomas 3, 33, 273, 275
Schawlow, Arthur L. 419, 737, 738
Scheele, Carl Wilhelm 205
Scherkel, Baron Hanno von 538–9
Scheutz, Edvard and Georg 701
Schickhardt, Wilhelm 699
Schiøler 230
Schliemann, Heinrich 62
Schmeisser, Hugo 1004–1005
Schmidt, Wilhelm 590
Schöffer, Peter 27, 670
Schönbein, Christian Friedrich 217, 223
Schröter 135
Schrotter, Anton von 912
Schuckert 375
Schultz, Christopher 82–3
Schultz, Johann Heinrich 729
Schwarz, David 615
Scott, H. 122
Scott de Martinville, Leon 720, 721
Seaward, James 398
Seger, H. 195
Seguin, A. 627
Seguin, Marc 561
Sejournet 126

Semon, W.L. 219
Senefelder, Johann Alois 678
Senlecq, Constantin 744
Senot 395
Seppings, Sir Robert 524
Seurat, Georges 743
Seward 115, 117
Shannon, Claude E. 701, 703, 749–50
Shaw, William Thomas 739
Shepard, Alan 652
Shield, Dr 792
Shillibeer, George 458
Shipman, M.D. 852
Shockley, William 42, 703
Sholes, Christopher Latham 682
Short, Horace 613
Shrapnel, General Sir Henry 986
Sickels, F.E. 282
Siebe, Augustus 553
Siemens, C.W. 171–2, 176, 198, 359
Siemens, Werner von 457, 584–5, 959
Siemens, Sir William 369
Sigsfeld, Captain von 612
Sikorsky, Igor 628, 643
Silliman, Benjamin, jun. 211
Simms, F.R. 993
Simonson, Vedel 4
Simpson, Gerald 230
Simpson, James 953
Sinclair, Sir Clive 458–9, 697
Singer, Isaac Merritt 848, 851
Skaupy, F. 138
Skinner, H. 846
Skola 823
Slavianoff, N.G. 418
Small, James 770
Smalley, John 829
Smeaton, John 236–7, 255, 258, 265, 397, 478, 888
Smirke, Sydney 896
Smith, Charles Shaler 464
Smith, Ciceri 417
Smith, Francis Pettit 528
Smith, J. 836
Smith, Willoughby 744
Snodgrass (textile engineer) 835
Snodgrass, Gabriel 523–4
Soane, Sir John 871, 894
Sobrero, Ascanio 988
Soddy, Frederick 213
Soemmerring, S.T. von 714
Solomon, King of Israel 52
Solvay, Alfred and Ernest 222
Sonstadt, Edward 113–14, 117
Sorel 95
Sorocold, George 234–5
Soufflot 894
Soyer, Alexis 210, 942

INDEX OF NAMES

Spaichel 390
Spangler, J. Murray 927
Spencer, Christopher Miner 410
Spode, Josiah 194
Spooner, C.E. 575
Sprague, Frank J. 381
Staite, W.E. 418, 914
Stampfer, Simon von 739
Stanford, Leland 739–40
Stanhope, Charles Mahon, 3rd Earl of 674, 676
Stanley, Robert 101
Staudinger, Hermann P. 218
Stearn, C.H. 368
Steers, William 477
Stephens, Henry 669
Stephenson, George 35, 560–61, 563, 565
Stephenson, John 456
Stephenson, Robert 35, 463, 465, 505, 560, 561, 562, 565, 566, 568, 895
Stibitz, George R. 702
Stifel, Michael 698
Stirling, Robert 342
Stock, A. 123
Stolze, F. 332
Stone, Henry D. 406
Strabo 69, 74, 76
Strachey, Christopher 704
Strada, Flaminianus 710
Strassman, F. 214, 1005
Stratingh, Sibrandus 378
Strato 273
Stringfellow, John 29, 620
Strutt, Jedediah 266, 819, 829, 835
Stuart, Herbert Ackroyd 310, 454
Stumpf, J. 287
Sturgeon, William 357, 377
Su Sung 694
Sullivan, Louis H. 897
Sundback, Gideon 852
Svaty, Vladimir 845
Swan, Joseph William 41, 353, 365–7, 368, 849, 915, 959
Swedenborg, Emanuel 86
Symington, William 35, 441, 527
Szilard, Leo 1005

Tacitus 72
Tafur, Pero 427
Tainter, Charles 721, 722
Talbert, David 705
Taylor, A.H. 729
Taylor, Frederick W. 412
Taylor, W.H. 378
Taylor, William 415
Telford, Thomas 434–5, 436, 478, 480, 509
Tennant, Charles 205, 221, 835
Tennant, Smithson 131

Tereshkova, Valentina 652
Tesla, Nikola 383, 925
Thales of Miletus 350
Themistocles 71
Thenard, Louis Jacques 95, 205
Theophilus of Essen 65, 68, 81, 196
Theopompus 74, 76
Thimmonier, Barthelemy 851
Thom, Robert 241, 952
Thomas, P.G. 170–71
Thomas, William 851
Thomsen, Christian Jurgensen 4
Thomson, Alfred 371
Thomson, Elihu 376, 384, 418
Thomson, J.J. 213, 725
Thomson, James 243, 244
Thomson, John L. 100
Thomson, Sir William 368–9, 373 *see also* Kelvin, Lord
Thompson, A. 840
Thompson, Benjamin 558
Thompson, General John T. 1005
Thompson, R.W. 449
Thonnelier 28
Thornley, David 829
Thorp 209
Thuesen, H.G. 461
Thurland, Thomas 82
Thwaite, B.H. 314
Titt, John Wallis 259
Tizard, Sir Henry 729
Todd, T.J. 287
Topham, Fred 368
Torricelli, Evangelista 32, 273
Town, Ithiel 463
Townes, Charles H. 419, 737, 738
Townsend, M. 847
Train, G.F. 456, 583
Trappen, Dr von 992
Treadwell, Daniel 677
Trésaguet, Pierre J.M. 434
Trevithick, R.F. 577
Trevithick, Richard 35, 277, 278, 280, 441–2, 469, 558–9, 589
Trewey 742
Tsiolkovsky, Konstantin 649
Tull, Jethro 774
Turing, Alan M. 422, 701, 703
Turner, Richard 895
Turner family (brass-makers) 91
Türr, Colonel Istvan 501
Tuve, Merle A. 729
Twiss, Captain William 506
Twyford 922
Tyer, Edward 581

Uhlhorn, Diedrich 28
Umbleby 325

Vail, Alfred 715
Van Arkell 141
Vauban, Sebastian de 486, 982–3
Vaucanson, Jacques de 395, 409, 823
Vauclain, Samuel 582
Vazie, Robert 469
Verbruggen, Jan 396
Vermuyden, Cornelius 250
Verne, Jules 747
Vernier, Pierre 25
Verrazzano, Giovanni 48
Versnyder, F.L. 127
Vicat, Louis J. 888–9
Victoria, Queen 201, 808
Vignoles, Charles 561
Villette, Professor 92
Villinger, Franz 646
Vitruvius 16, 17, 19, 230, 864, 918
Vivian, A.C. 123
Vivian, Andrew 441
Voisin brothers 624
Volk, Magnus 584
Volta, Alessandro 41, 94, 351–2

Wade, General George 434, 963
Waghenaer, Lucas 546
Waidner, C.W. 141
Walker, Charles Ludlow 92
Walker, John 912
Wallace, William 367
Wallop, Sir Henry 82
Walschaert, Egide 575
Waltenberg, R.G. 122
Walter, T.U. 894
Walton, Frederick 905
Wankel, Felix 326, 328, 455
Ward, Joshua 221
Wark, David 885
Warren, James 463
Wasbrough, Matthew 276
Washington, George 510
Waterhouse, Alfred 871
Watson, Dr 88
Watson, Bishop Richard 95
Watson-Watt, Sir Robert 219, 729, 999
Watt, James 31, 33, 155, 156, 209, 275–6, 277, 290, 325, 397, 398, 683, 833, 835
Wayland the Smith 128–9
Wayne, Gabriel 84
Webb, F.W. 582
Weber, J. 737
Weber, Wilhelm 374
Webster, James Fern 105
Wedgwood, Josiah 193
Wedgwood, Ralph 683
Wedgwood, Tom 731
Weldon, Walter 205, 222
Wellington, Arthur, 1st Duke of 853

Welsbach, Carl Auer von 123, 132, 209, 914, 958
Werner, Eugène and Michel 448
Westinghouse, George 580, 960
Weston 375
Wetschgi, Emanuel 393
Wheatstone, Sir Charles 41, 357, 358, 375, 377, 380, 384, 572, 714, 715, 735, 965
Wheeler 214
Whinfield, J.R. 219, 850
Whipple, Squire 463
White, Canvass 466
White, William 114
Whitehead, Robert 997
Whitelaw, James 243
Whitney, Amos 406
Whitney, Eli 29–30, 405, 834, 983
Whittle, Frank 39, 333, 336, 635
Whitworth, Joseph 398, 401, 406, 411
Widlar, Robert 705
Wikander 230
Wilbrand 988
Wilkes, M.V. 703
Wilkie, Leighton A. 418
Wilkinson, David 395
Wilkinson, John 155–6, 276, 397, 528
Wilkinson, William B. 890
Williams, Sir Edward Leader 482
Williams, F.C. 703
Williams, S.B. 702
Williams, Thomas 90–91
Williamson, D.T.N. 425
Wilm, Alfred 122
Wilson, Robert 401
Wimshurst, James 351
Winsor, Frederic Albert 40, 209, 958
Winzen, Otto 613
Winzer, Frederick *see* Winsor, Frederic Albert
Wirth, Niklaus 706
Wise, John 611
Woelfert, Karl 615
Wöhler, Friedrich 102–103, 105, 107
Wollaston, William Hyde 131–2
Wood, Henry 341
Wood, Henry A. Wise 676
Woolf, Arthur 279
Woolley, Sir Leonard 57, 69
Woolrich, J.S. 357–8
Worcester, Edward Somerset, 2nd Marquis of 33, 273
Worsdell, T.W. 582
Wren, Christopher 860, 867, 881–2
Wright, Orville and Wilbur 39, 622–4
Wyatt, John 824–5, 828, 830

Yost, Paul E. 613
Young, James 211

Young, L.C. 729

Zahn, Johann 731
Zède, Gustave 537
Zeiger, Herbert 737
Zeiss, Carl 417
Zeppelin, Count Ferdinand von 617
Zisman 220
Zizka, John 978
Zonca 265, 266
Zuccato, Eugenio de 683
Zuse, Konrad 702
Zworykin, Vladimir K. 745
Zyl, van 251

INDEX OF TOPICS

Aaron Manby (ship) 164
abacus 686, 697–8
Aberdeen (ship) 286
Aberdeen: granite construction 860
abrasives 407–8
accumulator 353–4
acoustic coupler 710
acrylic fibres 850
acrylic plastics 219
adhesives 220
adobe 856
advertising, radio 727
adzes
    copper 49
    stone 11
aerofoil sails (ships) 526
aeronautical alloys 116, 122, 143
aeronautics *see* balloons; aircraft
Africa, early technology 6, 9, 10
Aga cooker 941–2
Agamemnon (ship) 715
Agenoria (locomotive) 562
Agricultural Revolution 802
agriculture 761–802
    alternate husbandry 779
    animal husbandry 765
    animal power in 267, 765
    arable farming 767–87
    archaeologists' view 761–2
    canals benefiting 512–13
    crop rotation 21, 779
    crop storage 799–800
    dairy farming 791–6
    farm traction 787–91
    fertilizers 170, 223–4, 774–5
    pest control 775–7
    weed control 777–8
air, compressed 469, 553, 962
air defence missiles 1008

air warfare 994–5, 998–1000
    see also aircraft, military
aircraft 609–47
    cantilever wing 628
    convertible and hybrid 644
    engines: gas turbines 39, 330, 333, 335–8, 634–40; piston 39, 319–20; supercharged 316
    jet-propelled 39, 634–40
    jet-supported 639–40
    man-powered 646–7
    military 628, 630, 632–3, 634, 990, 995, 998
    petrol-engined 622–34
    recreational 644–6
    rocket-propelled 634, 635
    'spy' 1010
    steam-powered 619–21
    supersonic 39–40, 638–9
    wing covering (cellulose acetate) 218
    wing shapes 635, 636
    Wright brothers' 39, 622–4
    see also airliners; airships; biplanes; gliders
aircraft carriers 535–6, 1002
aircraft detectors 995
airliners 629, 630–31, 634
    jet 636–9
    supersonic 39–40, 638–9
airships 614–17
Aisne–Marne Canal 488
Albati *see* nickel silver
alcohol 801
Aldine Press 671
Alecto (ship) 528
Alexandria: Pharos 542
alizarin 202
Alkali Act (1863) (UK) 222
Allegheny Portage Railroad 512
alloy steel 172–5

1033

alloys
  aeronautical 116, 122, 143
  age-hardening 121-3, 124, 126
  brazing 80, 87, 96
  cutlery 97, 98, 174-5
  gilding metal 74
  high-temperature 124-8, 144
  magnetic 137
  single crystal 128
  sintered 135-7
  superconducting 144-5
  see also under names of metals
alphabet, evolution 667
Altair calculator kit 706
alternate husbandry 779
alternating current 370-71, 372, 960
  generators 362
  motors 383
  waveforms 377
alum 199-200
alumina 102
aluminium 102-113, 115
  production: by carbon reduction 106; by electrolysis 105, 107-11, 112-13, 224; by magnesium reduction 106; by sodium reduction 104-106, 109
aluminium alloys 106, 122-3, 418
American Civil War 533, 988-9, 992, 996-7
American System 407
  see also mass production
Amerindians
  copper artefacts 48
  platinum artefacts 129
amino acids, selective herbicides 778
ammeter 375
ammonia manufacture 5, 223-4
ammunition
  anti-tank 994
  chain-shot 981-2
  conical 986-7
  fin-stabilized 1009
  grapeshot 982
  hollow charge 1004
  incendiary 995
  rifle (Minié) ball 983-4
  roundshot 981
  shrapnel shell 986
  tank-gun 1003
  tracer 995
  World War I 990
ampere balance 374
Amsterdam-Rhine Canal 491
Amtrak 598
analysis, chemical 179, 188, 189
Analytical Engine (Babbage) 31, 699-701
Anatolia
  copper artefacts 48, 50

lead artefacts 68
anchors 540-41
Anderton lift 486
anglesite 68
aniline dyes 201-202, 907
animal husbandry 765
  see also dairy farming
animal power 13, 260-69, 614, 765, 787-8, 870, 939-40
annealing, bronze 66
anodizing, titanium and titanium alloys 144
antifreeze 212
Antikythera gearing mechanism 17
Antoing-Pommeroeil Canal 493
Apollo space project 653-4, 661-2
aqua regia 220
aqualung 553
aquatints 678
aqueducts 19, 477, 478, 479, 480, 482, 483, 485, 487, 497, 918
arable farming 767-87
arc lamps 362-5, 914
arc welding 418-19
arch 872-4
  see also arcuated construction
arch bridges 462-3, 464-5
archaeological ages 5-6
Archimedean screw 19
architectural orders 858
Architecture Hydraulique (Bélidor) 233
arcuated construction 859
ard 769
Argentan see nickel silver
Argosy airliner 629
Ariane rocket 662, 748
armour
  early 968-9
  Greek 970
  half 978
  horse 974
  mediaeval 973
  Norman 972
  plate 973
  World War I 991-2
armoured vehicles 978, 993, 1003, 1004
Aron meter 376
arquebus 977
arrows 968
  fire 970
  obsidian-headed 195
  stone-headed 11
arsenical copper 56-8, 61-2
Art du Tourneur, L' (Plumier) 394
artificial intelligence 43
artificial silk 849
artillery 977-8, 980-81
  anti-aircraft 995
  friction tube ignition 986

INDEX OF TOPICS

predicted fire 990
rifled 987, 988
role in World War I 989–91
self-propelled 1004
sound ranging 990–91
ASCC (computer) 41, 702
ASCII codes 709
ashlar 860
asphalt 211
assembly line (cars) 413–14
  see also mass production
Assyrians
  chemical tablets 187
  glass 16
  pottery glazes 191
  pulley 18
  warfare 969
astrolabe 694
astronomical instruments 24
  see also telescopes
Athens: Parthenon 858, 859
atmospheric engine 33, 305
atmospheric rail traction 570–71
atomic clock 697
atomic energy 212–16
atomic number 213
atomic structure 213
atomic theory 189, 212–13
atomic weapons 214, 1005–1008
atomic weights 213
Austin 7 car 452–3
Australopithecus 6
autobahns 437
autogiro 642
automatic guided vehicles (AGVs) 425, 426–8
automation 181, 425, 426–8, 796
automotive engines 453–6
  see also internal combustion engines; steam engines
aviation *see* aircraft
aviation fuel 317
axes
  copper 48, 50–51
  Cretan double 62
  Frankish weapon 972
  stone 10, 967
axle 13, 16
azo dyes 202
Aztecs, use of rubber 216
azurite 52

Baie Verte Canal 509
Bakelite 218
balance beam 18
ball bearings, mass production 410
ball-point pen 669
ballistic missiles 1007

ballistics 982
ballistite 988
'balloon barrage' 613
balloons 609–14
band saw 418
barges 514–15
Barker's Mill 243
barley 12, 764
Baroque architecture 880
barque 526
barrel vault 875, 879
BASIC computer language 706
basket grate 916
basketwork 15
bathing 917, 919
battering rams 969–70
battery (electrical) 41, 352–4
  for vehicles 455, 458, 459
battery (hollow ware) 82
battle cruisers 535
battleships 534, 535, 996
  pocket 324
bauxite 104
  purification 111–12
bayonet 979
beam engine 275, 280–81
bearings
  ball, mass production 410
  roller 413
  sintered bronze 137
bell 65
Bellifortis (Kyeser) 648
bellows 13, 18
'bends' (compression sickness) 553
benzene 201, 208
beryllium and beryllium alloys 123–4
bevel gear 17
bibliographic databases 756
bicycles 37, 217, 442–7
  mass production 409–10
biological pest control 777
biological warfare 1009
biplane 624, 626–8, 629, 632
bitumen 211, 220
Biwako Canal 506
Black Arrow rocket 662
Black Country 156
black powder *see* gunpowder
blacksmithing 17
blade tools
  copper 48, 62
  stone 10
blankets, raising pile 818
blast furnace
  iron 149–51, 153, 154, 157–8, 160, 162–3, 178
  zinc 94
blast furnace gas, engine fuel 314

1035

blasting gelatine 190, 223
bleaching 205, 835
bleaching powder 205, 222, 835
bleu de Deck (glaze) 194
'Blimp' 615
Blitzkrieg concept 1002
block (pulley) 522
    mass production 30, 398, 404–405
block printing 669
    textiles 835
    wallpaper 903
bloomery 149, 151
blowpipe 13
Blücher (locomotive) 560
'blue whiteners' 205
boat, electric 378, 380
Boeing 707 airliner 637
Boeing 747 airliner 637–8
Bombardier Français, Le (Bélidor) 982
bombards 975
    see also cannon
bomber aircraft 632–3
bombs 980, 1000
bone china 194
bone tools 9
boneshaker (bicycle) 443, 445
books
    illustration 673–4, 678–9
    printing 671–2, 676–7
    technological and practical 187, 232, 235–6, 242, 251, 265–6, 672
Boolean algebra 701
boot-making 854
Bordeaux mixture 777
boring machines 155–6, 396–7, 403, 415
boron 102
bottle 198
bottle jack 940
bow 9, 967–8, 969, 972
    see also crossbow
bow-drill 9, 389
bow drives 389, 392
bowstring girder bridge 463
box girder bridge 464
brace, carpenter's 26
braiding 820
braking systems
    coach 439
    motor car 452
    railway 580–81
Brandtaucher (ship) 537
brass 72–6, 79, 80, 81, 82–3, 84, 85–92
    coinage alloy 66
    gears 17, 23
Brayton cycle 329
brazing alloys 80, 87, 96
breakwater 548
breech-loading gun 984, 987

brewing, fuels used in 207
brickmaking 868, 869–70
bricks 873
    flooring 904
    sun-dried 856
brickwork 866–9
bridges 462–7
    arch 463–4, 464–5
    bowstring girder 463
    box girder 464
    cantilever 463–4, 883
    clam 462
    clapper 462, 859
    glossary of terms 466
    iron 156, 463, 465, 893, 895
    reinforced concrete 890–91
    Roman 19, 462–3
    suspension 465–6
    transporter 496, 518
Bridgewater Canal 477, 482
brig 525
Brightray C alloys 125
Bristol: soap industry 204
Britannia (Ogilvy) 461
British plate *see* nickel silver
broaching machines 414
broadcloth 817, 818, 821
Brooklyn Bridge 465
bronze 57–66
    aluminium 106, 109
    Chinese 59, 63–5, 73, 79
    gears 17
    Grecian 62–3, 65–6
    lead in 64–5, 76
    patina 65
    Roman 66, 67
    sintered bearings 137
Bronze Age 57–63, 67, 69, 73
    bridges 462
    tools 388–9
    weapons 388–9, 968
Browning automatic rifle 992
Brussels–Charleroi Canal 493
Brussels lace 848
brutalist architecture 891–2
bubble-car 449
bucket wheel excavator 471
building and architecture 22, 855–901
building materials
    bricks 856, 868, 869–70, 873, 904
    concrete 886–8, 889; pre-cast 891; pre-stressed 466, 891–3; reinforced 466, 859, 885, 889–91
    earth 855–6
    flint 856–7
    iron 893–6
    masonry 859–61
    plastics 900

steel 859, 896–9
tiles 870–71, 885; floor 904
wood 856, 857, 861–6, 905, 906
bulldozer 471
Bunsen burner 210, 914
Bunsen cell 353
buoys
   mooring 541
   navigational markers 542–3
bureaufax 744
buses 457–8
butter churning machines 262, 269
butter-making 794
buttons 852
buttresses 876–9
Bydgoszcz Canal 505
Byzantine Empire 971
   architecture 880

cables
   power transmission 371–2
   ships' 541
   submarine telegraphy 715, 965
CAD (computer aided design) 425
calamine 75, 80, 81, 82, 83
calcium carbide 112
calculating devices 31, 398, 697–701, 702, 707
   electronic 706
   see also Analytical Engine, Babbage, computers, Difference Engine
Calder Hall nuclear power station 215
Caledonian (Great Glen) Canal 480–81
calico-printing machine 835
calipers 391, 417
   gunner's 25
calotype process 732
cam 25
CAM (computer aided manufacture) 425
camera obscura 730, 731
camouflage 991
Campbell Diagram 299
can opener 937
canal and river craft 514–15
Canal de Bourgogne 487
Canal de Briare 483
Canal de l'Est 488
Canal de Mons (Canal du Centre) 487, 493
Canal d'Entreroches 499–500
Canal des Houillères de la Sarre 488
Canal d'Orleans 486
Canal du Berry 517
Canal du Centre (Canal de Mons) 487, 493
Canal du Loing 486
Canal du Midi 484–5, 486, 516
Canal Lateral à la Garonne 485
canals 161, 439, 474–518
   associated railway construction 556

glossary of terms 517–18
irrigation 768
water power 241
see also individual canal entries by location
candle-clock 22, 23
candles 912–13
canned food 798, 937
cannon 403
cannon borers 396–7
cantilever bridges 463–4, 883
cantilever construction 882–3
Cape Cod Canal 510
caracole (cavalry tactic) 977
carbide tools 135–7, 412, 417
carbine 979
carbon filament lamp 366, 368, 386
carbon paper 683
carburettor 310, 311, 454
carding 82, 809, 812
carding machines 830–31, 836–7
cargo handling 548, 549–50
   see also containerization
carnallite 114
Carnot cycle 343
carpenter's tools 14–15, 17, 26
carpentum 438
carpet shampooer 927
carpet sweepers 924–5
carpets 905–906
   weaving 812, 845–6, 905–906
carraca dormitoris 438
carrack 520
carriages 437–8, 440
   steam-powered 35, 441–2
cartography 546
cartridges 979, 984
carving, cave 11
caschrom 13
casein plastics 218
cassiterite 58, 60
cast-iron 18, 85, 149–53, 168
   gears 17
   gun barrels 987
   millwork 238
   rails 161, 556
   water pipes 954
Castile Canal 505
casting
   bronze 61–2, 64–5
   continuous 177, 179–80
   copper 49–51, 62, 67–8
   ploughshares 771
castle 974
Catal Hüyük 50, 68
catalysts 222
   Haber process 224
   hydrochloric acid oxidation 222
   hydrogenation of oils 190

INDEX OF TOPICS

catalysts—*cont*
  petroleum re-forming/platforming 212
  sulphur dioxide oxidation 222–3
catapult 969
caterpillar track 472
cathedrals 22
  treadwheels 263–4
cathode ray tube 377, 744–5
cathodic protection 144
caustic soda, manufacture 225
cavalry 979, 988
  Assyrian 969
  Byzantine 971
  caracole tactic 977
  Hun 971
cave art 11, 686
CD *see* compact disc
ceilings 903
cells, electric 352–4
Celluloid 217
cellulose acetate 217–18
cellulose nitrate 217, 218
cement 466, 887, 888–9
central heating 917
Central Pacific Railroad 574, 579–80
Centurion tank 1003
ceramics 190–95
  tools 412
cereals
  cultivation 764, 769
  harvesting 779–83
  storage 799–801
  yields 800
cerussite 69, 71
Chaff (radar jamming system) 1000
chain cable 541
chain mail 968
chain shot 981–2
Chalcolithic Age 48, 52
chalcopyrite 55, 70
Channel Tunnel 470
charcoal 206
  use in smelting 52, 60, 149, 152, 153, 154, 155
chariots 438, 969
charka (spinning wheel) 813
Charlotte Dundas (ship) 527
charts 546
chauffeur (origin of term) 460
cheese-making 794–5
Chemical Essays (Watson) 88
chemical industry 189–90
Chemical Manipulation (Faraday) 352
chemical milling 419
chemical weapons
  gas 992–3, 998, 1009
  Greek Fire 971
Chesapeake and Delaware Canal 510

Chesapeake and Raritan Canal 510
chimney crane 939
chimneys 22
China, Ancient
  bellows 18
  bronze 59, 63–5, 73, 79
  building and architecture 868
  canals 475
  cantilever bridges 463
  cast iron 18
  cereal storage 799
  clocks 392, 694, 695
  fire 7
  gunpowder 19
  optical glass 197
  paktong 80
  paper 18–19, 667
  plough mouldboard 770
  pottery and porcelain 192
  rockets 648
  scythe 780
  steel 18
  terrace field systems 767
  umbrella 853
  voice reproduction 720
  weapons 974
  writing (ideograms) 15, 667
  zinc 78–9, 80
'china ware' 194
chisel 146
chlorate match 912
chlorine
  manufacture 222
  uses: in bleaching 205; in water treatment 953
  weapon 992
chromium alloys
  high-temperature 125, 126–7
  nichrome 386
chronometer 544, 696
chrysocolla 52
cider mill 260
cinematography
  film 218
  see also motion pictures
Cistercians 232
Citroën cars 452, 453
cladding materials 861, 900
clam bridges 462
clapper bridges 462, 859
clay 190–91
Clean Air Acts (US) 319
cleaning 922–8
Clermont (ship) 35, 527
clipper ships 524, 525
clocks 22–4, 392, 694–7
  atomic 697
  candle 22, 23

electric 696
  gears 17
  pendulum 23–4
  water 16, 22–3, 392, 694
  see also chronometers
close stool 921
clothes press 931–2
clothing manufacture 850–51
coaches 438–9, 440, 562, 963–4
Coade stone 871–2
coal 207–208
  canal transport 477
  heating fuel 915–16
  ships' fuel 529
  smelting and refining fuel 83, 84, 153, 157
coal gas 40, 161, 208–10, 957–9
coal gas balloons 612
coal gas engines 305–307, 310, 314
coal mining 207–208, 555, 559, 560
coal tar 201, 208
Coalbrookdale 154, 155, 156, 157, 207, 465, 556, 559, 893
coastguard service 539–40
cobalt alloys
  high-temperature 127–8
  magnetic 137
cobalt glazes 192
COBOL computer language 704
cochineal 200
code-breaking machines 422, 701–702, 1001
Codice Atlantico (Leonardo da Vinci) 390
coffer dams 19, 462–3
cog 520
cog and rung gears 269
coinage metals
  brass 66, 75, 76
  cupro-nickel 97, 99
  nickel silver 98–9
  paktong 97
  zinc 78–9
coins 27–8, 71
coke 85, 153–4, 157, 207
collodion process 733
Colossus code-breaking machine 701–702
colour motion pictures 723, 742
colour photography 735–6
colour printing 681–2
colour television 745, 746
Columbian printing press 675
combine harvester 781, 786–7
combing (textile fibres) 809
combing machines 839–40
Comet (ship) 35, 537
Comet airliner 637
communications satellites 657, 746–8, 965–6

commutator 357
compact disc (CD) 725, 755–6
comparators 406, 421
compass
  gyro 545
  magnetic 21, 350, 545
Complete Course of Lithography, A (Ackermann) 678
component interchangeability 29–30, 405–406, 771, 983
composite materials 138
compound steam engines 278–9, 284–5, 530, 581–2
compression ignition engines 310, 321–5, 454
  marine 323–5, 531, 532
  see also diesel engines
compression sickness 553
computer aided design (CAD) 425
computer aided manufacture (CAM) 425
computer assisted learning (instruction) 754
computer graphics 702, 709
computer integrated manufacturing 427
computer languages 704, 705–706
computers 41–2, 701–703
  applications: in printing 681; in public services 966; in steel-making 181–2; in war 1010
  data storage and retrieval systems 707, 751, 752–6
  fifth generation 43
  interactive terminals 704
  laptop 709–10
  mainframe 703–705, 707; time-sharing 704–705
  mainframe–micro communication 708–10
  micro- 705–708
  mini- 705
  operation and programming 703–705
  precursor 31
  remote job entry 704
Concorde aircraft 40, 639
concrete 886–8, 889
  pre-cast 891
  pre-stressed 466, 891–3
  reinforced 466, 859, 885, 889–91
concrete bridges 466
concrete roads 437
Constellation airliner 634
contact materials (electrical) 133, 138
container ships 532
containerization 477, 550, 601
controls
  machine tool 421–3
  steel strip thickness 181

cooking 938–47
    with electricity 373, 944–6
    with gas 210, 942–4, 958
    with microwaves 946–7
    pressure 946
    with solid fuel 940–42
Copernicus satellite 658
copper 12–13, 47–57, 68
    arsenical 56–8, 61–2
    cold-forged native 47–9
    melted and cast native 49–51
    mining 51, 67, 79–80, 82
    refining 84, 85
    separation from nickel 100
    ships' hull protection 524
    smelted 51–5
Copper Age 5, 48, 51–2, 55
copper alloys 123, 124
    see also arsenical copper; brass; bronze;
    cupro-nickel; gilding metal; Monel
    metal; paktong
coppersmith's tongs 15
copying machines 683–4, 736
corbelling 882–3
cordite 988
Corinth Canal 501
Corliss engines 282–3, 285
corn mills
    animal-powered 230, 262–3, 265, 266
    automatic 30–31
    drive mechanisms 26
    gears 17
    steam-powered 257
    text-book 238
    water-powered 233
    wind-powered 245–8, 250–51
Cornish engines 277–8
Cornwall (locomotive) 569
Corporal missile 1007
corrosion prevention *see* cathodic
    protection; galvanizing
corrugated iron 95–6, 180
corvette 536
Côteau du Lac Canal 506
cotton 804, 807–808, 833–4
    bleaching 835
    carding 836–7
    cleaning and opening machines 835–6
    combing 840
    mercerized 849
    sizing 842–3
    spinning 824–5, 837–8
cotton mill, horse-driven 266
counting 686, 697–8
County of York (ship) 286
Coventry
    bicycle industry 445
    motor cycle industry 448

crane 19, 549
    treadwheel 264
crank 25–6, 276, 390
    bell 269
cream separator 794
creep (metals) 299
Crete
    bronze artefacts 62
    building and architecture 857–8
Crimean War 533
crimping devices 933
Crinan Canal 481
cristallo glass 196
crop rotation 21, 779
crossbow 25, 972–3
crucible furnace 50, 60
crucible steel 159–60, 397
crucibles, Bronze Age 60–61
cruck building 861–2
cruise missiles 1007–1008
cruisers 534, 536
cryolite 104, 105, 107–108, 109
Crystal Palace 895, 899
crystal radio sets 726
cubit 391
cuneiform script 15, 666
cuprite 52
cupro-nickel alloys 97, 98, 99, 101, 124
curfew 911
Curtis turbines 293–4
cutlery alloys 97, 98, 174–5
cutting tools
    sintered carbide 135–7
    steel 159, 174, 412
    tungsten alloy 412
    water-powered manufacture of 240
    see also drills; lathes; etc.
cyclopean masonry construction 860
cylinder boring machine 397
Cyprus: copper artefacts 73

2,4-D 778
Dacron 220, 850
daguerreotypes 732–3, 735
dairy farming 791–6
Dalsland Canal 502
dams
    canal reservoir 484, 507
    coffer 19, 462–3
    river impoundment 951–2, 955
Daniell cell 353
Danube river, canalization 504
data storage and retrieval systems 707, 751,
    752–6
database, bibliographic 756
Dataplatter system 750
davits 540
Daylight trains 597

1040

daymarks 543
DDT 776-7
De Architectura (Vitruvius) 230, 864
De Magnete (Gilbert) 41, 350
De Motu Animalium (Borelli) 646
De Re Metallica (Agricola) 14, 196, 232, 233, 264-5
De Rerum Natura (Lucretius) 212
De Revolutionibus Orbium Coelestium (Copernicus) 672
Decca navigator 546
decimalization 401, 403
Delaware and Hudson Canal 511-12
delftware 192
demotic script 667
depth charge 535, 997-8, 1001
depth sounding 544-5
derrick 211, 549
desalination plant 955
destroyers 534-5, 536
detergents, synthetic 204-205
detonators 988
diamond drill bits 212
diamond mining 267
Dictionnaire de Chimie (Macquer) 129-31
diesel engines 310, 311-14
    marine 531, 532
    see also under locomotives
Difference Engines (Babbage) 31, 699, 701
Differential Analyzer (Bush) 750
diffraction gratings 423
digital watches 697
'dinanderie' 81
diode valve 41, 720, 726
direct current 370, 372
    generators 356-62
    motors 383
directed energy weapons 1010
directional solidification (alloys) 127
dishwashers 927-8
Dismal Swamp Canal 510
distaff 809
dividers 391
dividing engine 394, 395
diving 553
Dneiper river, canalization 503-504
dobby 844
docks
    dry 551-2
    wet 548
dockyards, naval 552
dog grates 916
dog-powered treadmill 269
dog turnspit 939-40
Dolby noise reduction system 724
dolomite 119, 120, 170, 195
domes 879-82
donkey power 230

Dortmund–Ems Canal 496-7, 498
Doxford engine 325
dragline excavators 471
dragoons 979
drainage mills, wind-powered 249-50
drains
    sewage disposal 956-7
    soil 771, 782
    surface water 955-6
Draisine (bicycle) 442
draw boy loom 816
drawing frame 831-2
dreadnought 535
dredgers 550-51
drilling
    oilfield 211-12
    ultrasonic 419
drilling machines, controls for 422, 424
drills
    bow 9
    compressed air 469
    rocking 389
    tungsten steel 174
drive mechanisms 25
drop-valves 282
dry-cleaning 208, 854
dry-stone construction 859-60
drying (laundry) 931
ducted fan engines *see* turbofan engines
Duke of Argyle (ship) 35
dump truck 471
duplicator, stencil 683
Duralumin (alloy) 122
Dutch loom 820
dyers' weed *see* weld
dyestuffs 199-203, 907
dynamite 190, 223, 988

Eagle (ship) 537
ear trumpet 687
earth closet 922
earth construction 855-6
earthmoving machinery 470-72
Ebbw Vale 158, 163, 181
Echo I satellite 657, 747
echo-sounder (sonar) 536, 545, 998
EDVAC 702
Eddystone lighthouse 542, 888
Edison effect 720, 726
Eemskanaal 492
Egyptians, Ancient
    arch 462
    armour 968
    art (profile convention) 15
    boats 520
    bronze artefacts 59, 62
    building and architecture 18, 856, 857, 872, 887

Egyptians, Ancient—*cont*
  canals 474–5
  chicken-rearing 796
  cleansing agents 203
  copper artefacts 12, 50, 56, 59, 62
  copper smelting and refining 54
  dairy farming 791
  dyeing 199, 200
  fuels 206
  glass 16, 195
  glues 220
  iron artefacts 18
  irrigation 767
  lathe 389
  lead artefacts 69
  machines 17–18
  obelisks 13
  pottery glaze 191
  river impoundment 952
  silver artefacts 70
  standards of length 18, 391
  weapons 969
  welding 417
  writing 666–7
Eiffel Tower 894
Eisernen Wasserräder (Henne) 242
El Gobernador (locomotive) 580
elastic fabrics 853
Elbaski Canal 504
Elbe–Havel Canal 495
Elbe Seiten Canal 498
Elbe–Trave Canal 494
electric arc 352
  see also arc lamps; arc welding
electric arc furnace 106, 176, 182
electric boats and ships 378, 380, 531
electric cells 352–4
electric clock 696
electric cookers 373, 944–6
electric current
  measurement 374–5
  see also alternating current; direct current
electric heating 916–17
electric iron 934
electric lighting 41, 358, 360, 369, 386–7, 914–15, 959, 961
Electric Lighting Act (1882) (UK) 369–70
electric motors 355, 381–5
  in domestic appliances 925, 927–8, 929, 931, 937
  induction 383–4
  linear 384–5
  pole amplitude modulated 384
  precursors *see* electromagnetic engines
  repulsion 384
  synchronous 383
  universal 383
  variable speed control 384

electric power generation 41, 284, 356–62, 369–70, 371, 372, 381, 960
  gas turbines 333, 338–40
  nuclear 215
  steam turbines 295–303, 385
  wind 259–60
electric telegraph 357, 572, 574, 713–15, 964–5
  railway signalling system 714, 965
  wartime role 988–9
electric vehicles 378, 380, 458–9
  trams 47, 959–60
electric water heaters 920
electrical appliances 386
electrical measurement 373–7
electricity 41, 959–61
  current 351–4
  off-peak 372–3, 917
  static 350–51; inverse square law 373
  transmission 370–72
  see also hydro-electric power
electricity supply meters 376
electrification
  iron- and steel-making 176
  railways 382, 584–5, 591–2, 594, 598, 599, 600
electrochemistry 352
electrodynamometer 374
electro-forming 419
electrolysis, applications of 105, 107–11, 112–13, 114–19, 224–5, 352
electromagnetic engines 377–81
electrometer, torsion balance 373
electron, discovery 213
electron–beam welding 420
electronic clocks and watches 697
electronics 41–2, 428
electrophonetic telegraph 719
electroscope, gold leaf 373
electroslag refining (steel) 184
elements 187
elephant (weapon of war) 970–71
Encyclopédie (Diderot) 236
energy meters 376
English Improver, The (Blith) 774
engravings
  copper 674
  photo-process 679
  steel 678
  wood 678
ENIAC 42, 702
Erie canal complex 510–11
Eros (statue) 108–9, 111
Eskimos, use of copper 48
etching 419
ethylene 212
ethylene glycol 212

1042

INDEX OF TOPICS

Etruscans
　brass statuary 75
　building and architecture 868, 873
　wood-turning 389
excavators 471
Exeter Canal 476
Experienced Millwright, The 238
Explorer satellite 656
explosives 190, 223, 988
exponents 698
external combustion engines 341–2

facsimile transmission 743–4
factory system 28–9
　water power in 240–42
fallowing 777–8
fan vault 876
Fantasia (motion picture) 723
farm traction 787–91
fastenings 852–3
feldspar 193
felt 807
ferries, ro-ro 550
ferro-vitreous construction 895–6
Fertile Crescent 13, 764
fertilizers 170, 223–4, 774–5
Festiniog Railway 565, 575, 606
fibre optics 199, 748
fibres, textile 804–808
　see also man-made fibres
fighter aircraft 632, 995, 998
　rocket-propelled 634, 639
　STOVL 640
filament lamp 41, 132–3, 365–9, 386–7,
　914–15, 959
filling stations 459–60
Filmorex system 751
filters, sintered metal 137
filtration beds (water) 953
finery 152, 155
Finow Canal 495, 497
fire 7–9
fire arrows 970
fire drill 8
fire lighting *see* ignition
firearms, mass production 30, 405–407,
　983
fireplaces 916
fishing 9, 766
Fissero–Tartaro–Canalbianco Canal 501
flak jacket 1009
flamethrowers 993
flash locks 476, 494, 517
flax 806
　hackling 806, 840
　spinning 841
flexible manufacturing systems (FMS)
　424–8

flight simulator 703
flint
　building material 856–7
　mines 9
　sickles 779–80
　uses: in fire-making 7–8; in glass 196; in
　　pottery 193
flint glass 197
flintlock guns 979
float glass 198
flood control 955–6
floor coverings 15, 904–906
floor polisher 927
floppy disk 707, 753
flour mill *see* corn mill
flowline production 427
fluorescent lamps 387, 915
fluoridation of water 953
flushwork 857
fluyt 523
Flying Bedstead (aircraft) 639
flying boats 627, 633
flying buttress 877–9
Flying Hamburger (train) 595
Flying Scotsman (locomotive) 606
flywheel 16
foam, furniture 907
fog signals 543
Fontinettes boat lift 487
food mixer 937
food preparation 936–7
　see also cooking
food preservation 796–801, 937–8
footwear 854
Ford cars
　Model A 37
　Model T 413–14, 452
Forth and Clyde Canal 478
fortification 982
　camps 970
　cities 969
FORTRAN computer language 704
Fossa Carolina 494
Fossdyke Navigation 476
fountain pens 668–9
four-stroke cycle 322, 454
frame knitting 819–20
Franco-Prussian War 612, 734, 985
Franks 971–2
freeways *see* motorways
friction welding 419
frigate 536
From the Earth to the Moon (Verne)
　747
frozen food 797–8, 938
fuel cells 455
fulling 816–18
fungicides 777

INDEX OF TOPICS

furnaces
    blast: iron 149–51, 153, 154, 157–8, 160, 162–3, 178; zinc 94
    bloomery 149, 151
    copper-smelting 50, 52, 60
    electric arc 106, 176, 182
    electric induction 162
    finery 152, 155
    glass-making 171, 196, 198
    refractory linings 194–5
    regenerative 93
    reverberatory 83–4, 104, 157, 162
    rocking resistor 119
    vacuum melting 125, 133, 183–4
furnishings 906–907
furniture 907–911
fusee 24, 392, 394

galena 68–9, 71
galley, ship's 520
galvanizing 95–6, 180
galvanometers 374, 377, 714
garages 459–60
garnierite 98
gas bath heater 919
gas cookers 210, 942–4, 958
gas discharge lamps 387, 961
gas engines 305–307, 310, 314–15, 454, 614
gas fires 210, 916, 958
gas geysers 919–20
gas industry 40, 161, 957–9
    see also coal gas; natural gas
gas irons 934
gas lighting 208–209, 914, 958
    mantle 123, 209, 914, 958
gas masks 993
gas meter 209, 958
gas turbines 329–41
    aircraft engines 39, 330, 333, 335–8, 634–40
    blades 126–8, 141
    electric power generation 333, 338–40
    helicopter engines 643
    marine engines 340, 537
    motor cars 341, 455
    railway locomotives 334, 340–41
    supercharged 309, 315–16
gas warfare 992–3, 998, 1009
gasoline *see* petrol
Gatling gun 985–6
Gauge of Railways Act (1846) (UK) 568
gauges 400, 401, 413, 414–15, 421
    strain 420
gear cutting machines 395–6, 408–409, 410–11
gears 16–17
    bevel 17

clock 17
cog and rung 269
cycloidal 269
helical 17
lanthorn and trundle 17, 19
mill 17, 236, 238, 247, 250
mortise 17
sun-and-planet 276
gelignite 988
Gemini space missions 653
Genoa: Mola Nuovo 548
geodesic dome 882
German silver *see* nickel silver
gesso 909
geysers, gas 919–20
Ghent–Terneuzen Canal 492
gilding 68, 80
gilding metal 74
gill boxes 840–41
ginning (cotton) 808, 834
glass 16, 195–9
    cut 197
    float 198
    lead (flint) 197
    mirror 910–11
    optical 197
    plate 197–8, 910
    stained 22, 196
    window 22, 196
    see also ferro-vitreous construction
glass fibre 198–9
glass-fibre-reinforced plastics 947
glassblowing 16, 196
glasshouses 895
glassmaking furnaces 171, 196, 198
glazes 191, 194
Glenfinnan Viaduct 891
gliders 39, 619, 621–2, 645, 646
Gliwice Canal 505
Gloire (ship) 533
glues 220
goffering iron 933
gold leaf electroscope 373
gold mining 79
Golden Gate Bridge 465
Gossamer Albatross (man-powered aircraft) 647
gossan 55
Göta Canal 502
Gothic architecture 874, 876–9
governor
    centrifugal 31
    cut-off control by 283
    inertia 284
    for waterwheel 242
Graf Spee (ship) 324
Gramme ring 360
gramophone 720–24

long-playing record 723–4
Grand Canal (Dublin–Shannon) 481
Grand Canal (Tianjin–Hangzhov) 475
Grand Junction Canal 479
Grand Union Canal 479
granite (building stone) 860
grapeshot 982
graphophone 721
Great Britain (ship) 401, 528
Great Eastern (ship) 403, 528, 715, 965
Great Exhibition (1851) 258, 470, 569, 679, 735, 782, 958
    see also Crystal Palace
Great Glen (Caledonian) Canal 480–81
Great Western (ship) 35, 527
Great Western Railway 565, 567–8, 572, 590, 592, 965
Greek fire 971
Greek mill (Norse mill) 19, 229–30, 269
Greeks, Ancient 18
    alphabet 667
    brass artefacts 73–4, 79
    bronze artefacts 61, 62–3, 65–6
    building and architecture 858, 862–4, 872, 884, 887
    cleansing agents 203
    'mathematical gearing' 17
    measuring instruments 391
    music 686, 717
    Norse mill 19, 229–30, 269
    philosophers 187, 212
    pottery 191
    silver mines 69, 70–71
    tin artefact 60
    warfare 970
    zinc artefacts 75
green manuring 774
grenade launcher 980
grenades 980, 991
    anti-tank 994, 1004
grinding 10, 407
grinding machines 407–408, 413, 414, 415
grindstone crank 25
groin vault 875
Groot Algemeen Moolen-Boek (van Zyl) 251
Groot Volkomen Moolenboek (Natrus and Polly) 251, 255
Grove cell 353
guided missiles 1008
gun carriage 977–8
gun-cotton 223
gun turret 533, 534, 988
gunner's calipers 25
gunpowder ('black powder') 19, 648, 975
    charcoal for 206
    grinding machine for 266
    saltpetre for 221

guns
    anti-tank 994, 1004
    breech-loading 984, 987
    early 975
    hand- 976–7, 985
    machine 985–6, 992; aircraft-mounted 995
    naval 532, 533, 536
    rifles 403, 405, 983–4; anti-tank 994
    tank 1003, 1009
    see also artillery
gutta percha 217
gypsum 885, 904
gyroscopes
    in bombsights 1000
    for torpedo control 997
gyroscopic compass 545

hackling (flax) 806, 840
haematite 52
halberd 976, 979
half-timbered buildings 862
half-tone process 679
halogen lamps 386–7
hammer, power 152, 166, 232, 401
hammer ponds 232
hammerbeam roofs 866
Haneken Canal 496
hang-gliders 621–2, 646
harbours *see* ports and harbours
harrow 21
harvesting 779–87
hastener (roasting aid) 940
hats 807
Havel–Oder Canal 497
heat engines 272, 342–3
heating 385–6, 915–17
heddle-frame weaving 810–11
helical gears 17
helicopters 536–7, 640–44
heliography 732
helium balloons 613
helmets 968–9, 973, 991
helve 166
herbicides, selective 778
Hiawatha trains 597
hieratic script 667
hieroglyphics 15, 666–7
'hipping' (hot isostatic pressing) 140–1
Historia Naturalis (*Natural History*) (Pliny) 75, 187, 672
horse armour 974
horse collar 20, 787
horse shoe 20
horse trams 583
horse wheels 261, 264–5, 788, 794
horses 13–14, 20–21, 260, 264, 265–9, 787–8

horses—*cont*
　heavy 974
　of San Marco 67–8
　see also cavalry
hot-air balloons 609–10, 611, 612, 613–14
hot air engine *see* external combustion engine
hot bulb engine 310–11
hourglass 23
hourglass mill 230
Housatonic (ship) 537
hovercraft 538
howitzer 980–81
'Huff Duff' (detection device) 1001
hulks 520
Huns 971
hunter gatherer societies 762–3, 766–7
hydraulic cement 888–9
hydraulic power 961
　boat lifts 482, 493, 508
　earthmoving machinery 472
　mains systems 40–41
　motor car suspension 452
　tractor implement carrying facility 790
hydrochloric acid, 221, 222
hydro-electric machine 351
hydro-electric power 109, 112–13, 118, 244–5
hydrofoils 538–9
hydrogen balloons 610–11, 612
hydrogen bomb 1006–1007
hydrogen peroxide, bleaching agent 205
hydrophones 718
hypocaust 917

IBM PCs 707, 710
ice boxes 937
ice houses 797
ice-making machines 797
iconoscope 745
ignition (fire) 7–9, 911–12
ilmenite 143
Imperial Canal (Spain) 505
impulse turbine 290, 294
Inchmona (ship) 286
inclined planes 479, 480, 488, 493, 503–504, 506, 509, 512, 517, 557–8
indigo 199, 200
indigotin 200, 202
induction furnace 162
induction motors 383–4
infantry
　Byzantine 971
　mediaeval 976
　modern weapons 1009–1010
　phalanx formation 970
　testudo formation 970
information technology 686–94, 748–53

information theory 703, 749–50
ink 668, 669
　printer's 670
inlaid furniture 908
insecticides 776–7
'instantaneous' lights 912
instrument making 24, 393–4
insulators, electrical 194
intaglio process 674
intarsio (furniture) 908
integrated circuit (IC) 42, 705
integrated ironworks 158, 160
Intelsat satellites 657
inter-continental ballistic missile (ICBM) 1007
internal combustion engines 37–9, 272, 303–29, 454, 615
　see also compression ignition engines: gas engines; hot bulb engine; spark ignition engines; Wankel engine
International Ultraviolet Explorer satellite 658
inventors 2, 690
Iran
　bronze artefacts 59
　copper artefacts 48, 50, 57
　irrigation tunnels 467–8, 768
　lead artefacts 69
　pottery 16
　windmills 245
　zinc refining 77–8
Iraq
　dairy farming 791
　lead artefacts 69
　plough 769
IRAS satellite 658
iridium 131
iron 12, 14, 18, 146–7
　building material 893–6
　by-product of copper refining 54
　case hardening 159
　charcoal for smelting 206
　'malleable' 168, 560, 563
　meteoritic 147
　pig 151
　production: electrification 381; gas engines for 314; use of coal in 207, of coke in 207, of water power in 232
　see also cast iron; corrugated iron; ferro-vitreous construction; galvanizing; wrought iron
iron, smoothing 932–4
Iron Age 6, 54, 388–9
　industrial 160–67
iron alloys
　high-temperature 124
　magnetic 137

see also steel
iron bridges 156, 463, 465, 893, 895
iron ships 21, 525, 528, 533
Ironbridge 156
irrigation 767–8
  see also water-raising devices
irrigation tunnels 467–8, 768
ISDN (Integrated Services Digital Network) 750
isostatic pressing 139–41
isotope abundance analysis 69–70
isotopes 213–14

Jablochkoff candles 363
Jacquard punched-card control systems 400–401, 699, 700, 701, 821, 823, 844, 848, 905
James River and Kanawha Canal 510
James River Navigation 510
japanned furniture 909
Jazz Singer, The (motion picture) 723
Jericho 12, 431
jet engine *see* gas turbines
jettied buildings 864
jewellery
  anodized titanium 144
  brass 76, 80, 87
  copper 48, 61
  glazed beads 16
  iron rings 54
  tin bracelet 60
jig borer 415
jig grinder 415
John Bull (locomotive) 573
Joule cycle 329
Juliana Canal 494
Juno rockets 661
Jupiter probes 660
Jupiter rockets 661

Kaiser Wilhelm II (ship) 287
kaolin 193
Kaplan turbine 244–5
Karls Grav (canal) 501
Kennet and Avon Canal 480
kerosene *see* paraffin oil
Keulse Vaart 491
Kiel Canal 496
kilns
  brickmaking 870
  cement 889
  pottery 15–16, 191, 195
  temperature measurement in 194
kinetoscope 741–2
kitchen 934–6
kitchen equipment 936–7
kitchen ranges 940–42
knapping (flints) 857

knife-cleaners 923–4
knitting 818–20
knitting machines 846–7
knock (detonation) 306, 316–18, 335
kollergang (oil mill) 253, 255
Küsten Canal 497

lace-making machines 848
Lachine Canal 506–507
lacquered furniture 908–909
laminated wood furniture 910
lampshades 368–9
Lancaster bomber 632
lance 972, 988
landing ships 550
Landsat satellites 658
language 665–6
lanthorn and trundle gear 17, 19
Laser-Card system 755
lasers 419–20, 426, 752, 737–8
lathes
  automatic 410, 415
  centre 415–17
  clockmaker's 392
  controls for 424
  early 9, 66, 389–90
  industrial 394–5, 399, 400
  instrument maker's 394
  ornamental 393
  pole 26, 390
  screw cutting 393–4, 395
  turret 410, 414, 415–17
latitude measurement 544
'laughing gas' 352
laundering 928–34
laundry, horse-driven 266
Lea river, canalization 476, 477, 482
lead 68–72
  in Roman plumbing 20, 66, 72
lead–acid battery 353–4
lead tetraethyl 317
lean burn engines 455
leather 803
Leclanché cell 353
Leeds and Liverpool Canal 478–9
leguminous crops 779
Lehigh Canal 512
Lend Lease Act (1941)(US) 421
length measurement
  standards 18, 391, 697
  vernier scale 25
Lenoir cycle 329
Lenoir engine 305
lenses 197, 730
level, electronic 423
lewis 19
Lewis gun 992
Leyden jar 351

1047

Liberty ship 531
lierne vault 876
lifebelt 540
lifeboat 539, 540
lifejacket 540
liferaft 540
lifts
   boat 480, 482, 486, 493, 496, 497, 498, 508–509, 517
   passenger 897
lighting
   candles 912–13
   electric 41, 358, 360, 369, 386–7, 914–15, 959, 961; *see also* arc lamps; filament lamps
   gas 208–209, 914, 958; mantle 123, 209, 914, 958
   oil 211, 913–14
   rushlights 912
lighthouse 358, 542, 889
lightning conductor 350
lights, navigational 543
lightship 542
linear motors 384–5
linen 804, 806
liner 37, 530, 532
Ling Chu (Magic Canal) 475
linoleum 905
Linotype machine 680
linseed 806
linstock ignition (artillery) 981
lithography 678
   offset 678–9, 684
Liverpool (Bury locomotive) 564
Liverpool (Crampton locomotive) 569
Liverpool: Royal Liver Building 891
Liverpool & Manchester Railway 561–2, 563
locks (canal) 517–18
   'bear trap' 512
   flash 476, 494, 517
   pound 475, 476, 517
   shaft 496–7, 518
locks (security) 29
   mass-produced 404
locomotives
   diesel 591, 595, 598, 600
   diesel–electric 591, 594–6, 600
   gas turbine powered 334, 340–41
   petrol–electric 591, 593, 594
   steam 35, 558–62, 564, 568–70, 575, 579–83, 589–90, 596–8, 599, 605–606; American 573; European 575; oil-fired 589; preservation 606–607
   streamlining 596–7
   superheating 590–91
   Whyte notation to describe 564

Locomotion (locomotive) 560
locust, biological control 777
lodestone 21, 350
log construction 856, 865
logarithms 698
logs(ships' speed recorders) 545
London
   Docklands Light Railway 603
   drains and sewers 956–7
   electric lighting 363, 364–5, 368, 370
   hydraulic mains system 41
   Kew Gardens (Palm House) 895
   London Bridge waterworks 234, 235
   St. Paul's Cathedral 860, 881–2
   underground railway 382, 468, 583, 584, 592
   water supply 951
   Westminster Abbey 879
   Westminster Cathedral 891
   Westminster Hall 866
London & Birmingham Railway 566
London Gazettee 672
longbow 973
longitude measurement 544, 696
looms
   automatic 844–5
   carpet 812, 845–6, 905–906
   Dutch 820
   early 22
   flying shuttle 821–2
   horizontal 812, 815–16
   powered 834, 843–5
   shuttle-less 845
   vertical 811–12, 815
Lord Chancellor (micrometer) 398–9
'lost wax' casting process 62, 68
lubricating oil 212
lubrication, forced 284
Lumitype machine 681
Luna moon probes 658–9
lustre ware 192

macadam road surface 436
Mach number 635–6
machine guns 985–6, 992
   aircraft–mounted 995
machine tools 29, 392–404
   controls for 422–4
   for mass production 404–417
mackintosh fabric 216, 849
madder 199, 200, 202
magazines 673, 678, 680
magic lantern 738–9
magnesite 121, 194–5
magnesium 113–21
   as scavenger 99
   production: electrochemical 114–19; by sodium reduction 113, 114;

thermochemical 119–21
magnesium alloys 115, 116, 123
magnetic alloys 137
magnetism 41, 350, 355–6
magneto 357–9
mail coach 438
mail services *see* postal services
maize 764
majolica 192
malachite 52
Mallard (locomotive) 596, 598
malleable iron 168, 560, 563
man-made fibres 218, 219–20, 849–50, 948
 dyes for 203
Manchester Ship Canal 481–2
manganese, as scavenger 99
manganese steel 174
mangle 932
Manhattan engines 285
Manhattan Project 214, 1005
manuring 774
maps
 road 461
 see also cartography
marble (building stone) 860
margarine 190, 799
marine engines
 compression ignition 323–5, 531, 532
 gas turbines 340, 537
 steam 286–7, 398, 403
 steam turbines 530–31, 535
marine railways 509
Mariner Spacecraft 659, 660
Marne–Rhine Canal 487–8
marquetry 908
Mars probes 660
Martesana Canal 500
masers 737
masonry construction 859–61
mass number 213
mass production 404–412
 bricks 870
 building units 900
 clothing 850–51
 firearms 30, 405–407, 983
 photographic printing 733
matches 912
matchlock gun 976–7
Mathematical Analysis of Logic, The (Boole) 701
'mathematical gearing' 17
Mauretania (ship) 37
mauve (dye) 190, 201–202
Mayas, use of rubber 216
measurement 391–2
 electrical 373–7
measuring machines 401

automatic 427
optically controlled 422
mechanical digger 470–71
Mechanick Exercises on the Whole Art of Printing (Moxon) 673
mechanization
 agriculture: dairy farming 792; ploughing 771; reaping and binding 781–3, 786–7; threshing 784–5
 boot-making 854
 brick-making 870
 canal barge haulage 515
 corn mills 238
 furniture-making 909–10
 glass-making 198
 iron and steel industry *see* steam engines, applications in iron-making
 paper-making 673, 674
 railway construction 470
 textile industry: carding 830–31; cleaning and opening 835–6; combing 839–40; drawing 831–2; fulling 817; knitting 819–20; lace-making 848; shearing 818; silk throwing 805–806; sizing 842–3; spinning 823–30, 832–3, 837–8, 840–42; weaving 820–23, 843–6
 typesetting 680
 wallpaper printing 903
Megadoc system 755
megaphone 687, 718
Megiddo (fortified city) 969
Meissen porcelain 193
Memex system 751
mercenary soldiers 970, 976
mercerized cotton 849
merchant iron 166
merchant ships 520, 523, 525–32
Merwede Canal 491
Mesolithic Age 10
Mesopotamia
 building and architecture 462, 856, 872–3
 chariot 969
 fuels 206
 petroleum 210–11
 pottery 16
 silver artefacts 69
metallurgy, textbook *see De Re Metallica*
Metamorphosis of Ajax (Harington) 921
meteorological satellites 657
meters, electricity supply 376
metrology *see* measurement
Metropolis Water Act (1852) (UK) 951, 953
mezzotint 678
microcomputers *see* computers, micro-
microfiche systems 751

1049

INDEX OF TOPICS

microfilms 733, 751
Micrographia (Hooke) 25
micrographics 733
micrometers 398, 417
    electronic 423
microphotography 733–5
microrecording 733
microscopes 24, 196, 255, 730–31
microwave ovens 946–7
microwave radiation 725, 726, 729
military aircraft 628, 630, 632–3, 634, 990, 995, 998
    gliders 645
milk, pasteurization 793–4
milking machines 792–3
Mill at wijk bij Durstede (van Ruisdael) 248
milling machines 408–409, 817
    controls for 422, 424
    turret 415
minehunters 537
minelaying machines 1009
mines
    anti-tank 1004, 1009
    land 1009
    sea 535, 536, 537, 996, 1001–1002
    trench weapon 991
minesweepers 535, 537
Mini (motor car) 452
Minicard system 751
mini-mill (steelworks) 184
mining
    animal-powered machines in 260–61, 264–5, 266–7, 269
    coal 207–208, 555, 559, 560
    copper 51, 67, 79–80, 82
    electrification 381–2
    gold 79
    lead 71
    silver 67, 69–71, 79–80
    textbook *see De Re Metallica*
minting 27–8
mirror 910–11
Misanthrope, The (Brueghel) 248
Mississippi river, improved navigability 513
mitrailleuse (machine gun) 985
Mittelalterliche Hausbuch 391, 392
Mittelland Canal 497
modem 709–710
modular design 900
mole plough 772
molybdenum 133
monastic orders 232
Monel metal 101–102
monorail 586
Monotype machine 680
Mons-Condé Canal 493
Monte Penedo (ship) 324
Montech water slope 485

moon, flight to 654, 658–9
mooring 541
moped 449
mordant dyes 199
mordants 199–200
Morris Canal 512
Morse code 715, 965
mortar (weapon) 980, 991
mortise and tenon joint 14
mortise gears 17
mosaic 905
Moscow: GUM department store 896
Moselle river, canalization 489
Moskva river, canalization 503
Mosquito bomber 632–3
Moteurs Hydrauliques (Armengaud) 242
motion pictures 738–42
    colour 735, 742
    film 739
    sound 722–3
motor cars 37–9, 212, 449–56
    brakes 452
    engines *see* internal combustion engines
    gas turbine powered 341, 455
    ignition systems 133
    mass production 412–14, 417
    radios 728
    tyres 217
motor cycles 447–9
motorways 436–7
mould forming (pottery) 194
moulin pendant 236, 270
music 686, 717
muskets 29–30, 405, 979, 983
mustard gas 993
Mycenae
    bronze artefacts 62, 66
    corbelled domes 883
    masonry 859–60

nails 14
nap raising (blankets) 818
Napalm 121
Napoleonic wars 559, 648
Narrative of the Eddystone Lighthouse (Smeaton) 888
Narrow Boat (Rolt) 516
natron 203, 204
natural gas 210, 959
    in gas engines 314–15
Natural History (Historia naturalis) (Pliny the Elder) 75, 187, 672
Nautical Almanac 544
Nautilus (submarine, 1802) 537
Nautilus (submarine, 1952) 538
naval warfare 996–8, 1000–1002
    see also warships

navigation 21, 543–6, 696, 698
  see also radar
Naviglio Grande (Milan) 500
needle gun 984
needles 851
negative feedback 31
Neolithic Age
  metallurgy 47
  weapons 967
nerve gas 998
Neuffossée Canal 486
neutron, discovery 213
neutron bomb 1008
Neva river, canalization 503
Neva–Volga Canal 503
New Horse-Hoeing Husbandry, The (Tull) 774
Newry Canal 477
newspapers 672, 675–6
nichrome 386, 917
nickel 73, 96–102
nickel alloys 124
  high-temparature 125, 126–7
  magnetic 137
  see also cupro-nickel alloys; Monel metal; nichrome; nickel silver; paktong
nickel silver 97–8
nicotine (insecticide) 776
niello (inlay) 62
Nimonic alloys 126
niobium 144–5
Nipkow disc 744, 745
nitric acid 220, 221
  manufacture 224
nitrogen fertilizers 224, 775
nitrogen fixation
  artificial: ammonia manufacture 5, 223–4; via titanium nitrides 141
  natural *see* leguminous crops
nitroglycerine 223, 988
nitrous oxide 352
Nivernais Canal 487, 517
noisemakers 686
Noordhollandsche Canal 491
norag (threshing sledge) 784
Nord-Ostsee Canal 496
Norman architechture *see* Romanesque architecture
Norman warfare 972–3
Norse Mill (Greek Mill) 19, 229–30, 270
North Sea Canal 491
Novo Teatro di Machine et Edifici (Zonca) 265
nuclear energy 212–16
nuclear fusion 1006–1007
nuclear weapons 214, 1005–1008
numerical control 422–3
nylon 218, 850

oared ships 519–20
obelisks (Egyptian) 13
obsidian 195
obturation 984, 987
octane number 317
octant 544
Oder–Spree Canal 495
odometer 461
off-peak electricity 372–3, 917
offset lithography 678–9, 684
Ohio river, canalization 513
oil (mineral) *see* petroleum
oil lamps 211, 913–14
oil mills
  animal-powered 266
  wind-powered 253–5
oil stove 916
oil tanker 531, 532
oilskin 853
olive crushing mill 263
Oman, copper workings 55, 57
omnibus, horse-drawn 458
On Marvellous Things Heard 74
onager 12, 787
operational analysis 1001
optical character recognition (OCR) 684
optical glass 197
orders, architectural 858
Ordnance Survey maps 461
ore-carriers 177
organic chemistry, birth of 201
Orient Express 579
Orlon 850
oscillograph, mechanical 377
oscilloscope, cathode ray 377
osmium 131, 132
osmium filament lamps 386
Ottawa waterways complex 507
Otto cycle 306
Otto engine 307
Oude Gracht (Old Canal), Utrecht 490
ovens
  brick 940
  camp 939
  microwave 946–7
  see also electric cookers; gas cookers
oxen 787–8
oxy-acetylene welding 272

paddle engine 262, 268
paddle steamers 527, 529, 532–3
Page's effect 718
paintings, cave 11, 686
paktong 96–7
Palaeolithic Age 10
paleophone 720
palladium 131, 132
paltrok (windmill) 251–2

INDEX OF TOPICS

Panama Canal 514
panel-and-frame construction 902–3, 907
Panoplia (Amman) 390
pantelegraph 744
pantiles 885
paper 18–19, 22, 667–8, 670
  carbon 683
  production: hand 673; mechanized 673, 674
  see also papier mâché
paper mills, wind-powered 252–3
papier mâché 909
paraffin 211, 212
paraffin lamps 211, 913–14
paraffin stoves 916
Paraguay (ship) 797
paravane charge 997
pargeting 864
Paris
  electric lighting 363
  Pont Neuf waterworks 234, 235
parish system, road maintenance 432–3
parking meters 461
Parsons turbine 290–91
particle board 906
Pascal computer language 706
passenger liners 37, 530, 532
pasteurization (milk) 793–4
Patentee (locomotive) 564
pedestrian crossings 460–61
peg knitting 819
Pelton wheel 244
pencil 669
pendentive construction 880–81
pendulum 23–4, 695–6
penny-farthing bicycle 446
pens 667, 668–9
percussion cap 984
periodicals 672–3
Permalloy 137
Perpendicular architechture 876
Pershing 2 missile 1007
Persia *see* Iran
perspective 731
Perspex 219
pest control 775–7
petrol 212
  octane number 317
  taxation 37
petrol engine *see* internal combustion engines
petrol pump 459
petroleum
  cracking 212
  port facilities 549
  ship's fuel 531
petroleum industry 210–12, 452
phalanx formation 969

Pharos (Alexandria) 542
phenakistoscope 739
Philadelphia–Pittsburgh Canal 512
Phoenicians
  purple-dyeing 200
  silver mines 70
  writing 667
phonautograph 720
phonograph 720–22
Phonovid system 750
phosgene 993
phosphate fertilizers 224, 775
phosphorus matches 912
'phossy jaw' 912
photocomposition (printing) 681
photocopying machines 684, 736
photoelectric effect 744
photographs, reproduction 679
photography 729–38
  aerial 733
  colour 735–6
  film 218, 734
  instant 736–7
  magnesium light 113, 114
  paper 366
  stereoscopic 735
photogravure process 679, 680
photomicrography 734
piece-rate bonus payments 412
piecework payments 407
Pien Canal 475
pier buttress 877
pigeon post 733
pigments
  in cave paintings 11
  grinding machines for 266
pike 976
Pilatus Railway 584
pile driver 19
piles, screw 541
pilots (sailing directions) 547
Pioneer spacecraft 660
Pioneer Zephyr (train) 596
pipes
  gas 161
  water 72, 954
Pirothecnica (Biringuccio) 396
pisé construction 856
pistols 977, 985
pitch 206, 211, 220
Planet (locomotive) 562, 564
planing machines 399–400, 404
plaster 885–6, 903–904
plaster of Paris 885, 904
plasterboard 886, 904
plastics 217–20
  building materials 900
  domestic applications 947–8

foam 907
furniture 910
laminates 906
plate glass 197–8, 910
platforming (petroleum) 212
platinum 129–32
platinum filament lamp 238
ploughing, electric 381
ploughs 13, 769–72, 788
  mole 772
  wheeled 21, 770
plutonium 214–15
plutonium bomb 1005–1006
plywood 906
pneumatic power 962
  see also air, compressed
pneumatic tyres 37, 217, 436, 446, 449
  tractor 790–91
Po waterways complex 500–501
pocket battleship 324
Polaroid cameras 736–7
pole lathe 26, 390
pollution 318, 966
polyamides 218
polyesters 850
polymers 216–20
polypropylene 910, 947
polystyrene 947
polythene 219, 947
polyvinyl acetate (PVA) 219, 220
polyvinyl chloride (PVC) 219, 853, 947
Pompeii
  hourglass mill 230
  plumbing 20, 72
poppet valve 282
porcelain 193–3, 194
Portland cement 466, 889
ports and harbours 547–51
post and lintel construction see
  trabeated construction
post mills 20, 246–8, 249, 250
Post Office Act (1765) (UK) 438
postal services 438–9, 566, 965
  automatic railway 602
  road book 461
potassium fertilizers 224
pot-de-fer (gun) 975
potentiometer 375
potter's wheel 12, 16
pottery 12, 15–16, 190–95
Pou-de-Ciel (aircraft) 645–6
poultry farming 796
powder metallurgy 128–38, 143
  compaction methods 138–41
power stations see electric power
  generation
pozzolana concrete 887–8
Practical Hints on Mill Building (Abernathy) 242

praxinoscope 739
pre-cast concrete 891
precision guided munition 1008–1009
prefabricated buildings 899–900
press stud 852
presses
  clothes 931–2
  printing see printing presses
  screw: coining 28; paper-making 253
pressure cooker 946
pre-stressed concrete 466, 891–3
prickly pear cactus, biological control 777
Prinses Margriet Canal 492
printed circuit boards 419
printing 27, 28, 669–84
  colour 681–2
  office 682–4
  textiles 835
  transfer (pottery) 194
  wallpaper 903
printing presses
  hand 670, 674–5, 677
  mechanized 675–7
prison treadmill 264
privy 921
producer gas 195, 314
production organization, twentieth-century 412–17
Progress in Flying Machines (Chanute) 622
Propontis (ship) 286
proton, discovery 213
PTFE 947
public utilities 949–66
puddling
  brick-making 869–70
  iron-making 157, 161–2, 165
Puffing Billy (locomotive) 560
pug-mill 870
pulley 18
Pullman cars 577–9
pulse-code modulation 749
pumped-storage scheme 833
pumps
  horse-powered 269
  steam-powered 33, 275, 285, 952, 954
  water-powered 235
  wind-powered 258
punched-card systems 699, 701
  see also Jacquard punched-card control systems
Puritan (ship) 287
Purpura, dye from 200
pushtows 515
PVA 219, 220
PVC 219, 853, 947
pyramids, Egyptian 18
pyrrhotite 98

1053

qantas 467–8, 768
QF (quick-firing) field gun 990
quadrant, surveyor's 25, 544
quadricycle 450
    powered 448
quadruple expansion steam engine 286–7, 530
quality control 420–21
quantum mechanics 213
quartz clock 696
Queen's ware 193
quern 25, 229, 270
quill pen 667, 668
quipu 698

racing cars 315
rack railways 559, 584
radar 219, 536, 545, 728–9, 998–1000, 1010
    jamming 1000
radiation protection 1007
radio 725–8
radio astronomy 729
radio broadcasting 727–8
radio waves 725
radioactivity 213
radiotelegraphy 715
railroads *see* railways
railway buildings 592–3, 894–5
railway carriage, electrically driven 378, 380
'railway mania' 567
railways 35–6, 164, 555–608
    American 572–4, 579–80, 589–90, 593–4, 598
    atmospheric 570–71
    automatic train control 602–603
    brakes 580–81
    British 590, 591, 593, 594, 599–601
    British Empire 588–9, 591
    central-rail system 583
    Chinese 577, 605
    coaches 562–3, 574, 579
    construction mechanization 470
    couplings 563
    electricfication 382, 584–5, 591–2, 594, 598, 599, 600
    European 574–5, 587–8, 590, 593, 594, 598–9, 603–605
    first passenger service 560
    freight transport 563, 600–603
    glossary of terms 607–608
    high-speed 603–605
    horse 555–8
    Japanese 576–7, 603
    maglev system 603
    monorail 586
    narrow gauge 575–6
    preservation 606–607

rack 559, 584
rails: cast iron 161, 556; long-welded 600; malleable iron 560, 563; steel 170, 580; wooden 555, 557
signalling systems 563–4, 571–2, 592, 593; electric telegraph 714, 965
single line working 581
sleeping cars 577–9
South American 589
stationary engines 557, 572
track 563–4, 566, 573, 580; gauge 563, 565, 567–8
trunk lines 565–7
tunnels 468, 470, 565, 603
underground 382, 468, 583, 584, 592
wartime role 988
see also locomotives
Rainhill Trials 561, 562
Rakete zu den Planetenraumen, Die (Oberth) 649
Ramelius (horse) 79
Ramepithecus 6
Ranger spacecraft 659
rapid transit systems 602–603
rapier 980
Ras Ratnasmuchchaya 77
Rateau turbine 291–3, 294
Rattler (ship) 528
rayon 850
reaction turbine *see* Parsons turbine
reaping and binding 779–83
reaping machines 268
    see also combine harvester
rectifiers 372, 383, 725
'Red Flag' Act (1865) (UK) 448, 451
Redstone missile 661
re-forming (petroleum) 212
refrigeration 795, 796–8, 937–8
Regents Canal 479
Regulation of Railways Act (1889) (UK) 580–81
reinforced concrete 466, 859, 885, 889–91
relative atomic masses *see* atomic weights
remote sensing satellites 658
remotely piloted vehicle (RPV) 1010
Renaissance architechture 879, 880, 881–2
reservoirs 951–2
retting (flax) 806
reverberatory furnace 83–4, 104, 157, 162
revolver 985
rheostat 380
Rhine–Herne Canal 497
Rhine–Main–Danube Canal 498–9, 504
Rhine Navigation (Vaartse Rijn) 491
Rhine river, canalization 499–500
rhodium 131
Rhône–Rhine Canal 488–9
rib vault 875–6

INDEX OF TOPICS

ribauld (gun) 975–6
ribbon weaving 811, 820–21
rice 764
Rideau Canal 507
ridge and furrow 772, 782
rifle grenades 991, 1004
    anti-tank 994
rifles 403, 405, 983–4
    anti-tank 994
    Browning automatic 992
    repeating 985
ring frames 841–2
Rio Tinto mines 55, 70, 231
rivets 14
    age-hardening 122
road maps 461
road transport 438–42
    see also bicycles; motor cars; *etc.*
roads 37, 963–4
    construction 431–9; machinery for 470–73
    lining 460
    see also motorways
roasting food 939
robots, industrial 425–7
Rocket (locomotive) 35, 561, 562
rocket launcher (weapon) 1005
rocket-propelled aircraft 634, 635
rockets 648–54, 656, 661–2, 747
    weapons 648–51, 986, 995, 1005
roller bearings 413
roller mill (corn) 257
roller printing (textiles) 835
rolling mills 28, 152, 180, 381
Roman cement 888, 889
Roman mill 19, 20
Romanesque architecture 874, 875, 876
Romans 18
    baths and sanitation 918
    brass and zinc artefacts 75
    bridges 19, 462–3
    building and architecture 19–20, 856, 859, 868, 873, 875, 876, 879–80, 885, 887–8, 889
    canals 475–6, 490
    central heating 917
    cereal storage 799
    cleansing agents 203
    cooking 938
    copper and bronze artefacts 66–8
    corn mills 230
    floor bricks 904
    glass 196
    kitchens 934–5
    lead and silver artefacts 72
    measuring instrucments 391
    mill gears 17
    navigational towers 542
    pottery 191
    road books 461
    road transport 438
    roads 431–2
    treadwheel 262
    warfare 970–71
    water-raising devices 231
    water supply 20
    writing 667
Rome 20
    Pantheon 879
    water supply 20
roofing 883–5
roofing materials 857
    aluminium 111
    glass/iron (ferro-vitreous) 895–6
    iron 95, 894
    Monel metal 101
    zinc 93, 96
'ro-ro' (roll on, roll off) vessels 550
Rothamsted Experimental Station 775
Rotherhithe tunnels 468–9
roundshot 981
roving frame 839
Royal Canal (Dublin-Shannon) 481
Royal Institution 352, 354, 355
Royal Society 3, 188–9, 672
Royal William (ship) 527
RR alloys 122
rubber
    natural 216–17, 849, 853
    synthetic 218–19, 472
rubble construction 860
rudder 21, 520
Ruhr river, canalization 495
rushlights 912
ruthenium 131
rutile 143

sabre 980
saddle 13
safflower 200
saffron 200
sailing directions 547
sailing ships 14, 520–26
St Lawrence Seaway 509
salt tax 204, 221
salted food 798
saltpetre 221
Salyut space missions 654, 656
sand-glass 22, 23
sanitation 918, 921–2
satellites, earth 652, 656–8
    communications 657, 746–8, 965–6
    meteorological 657
    military 1010
Saturn probes 660–61

Saturn rockets 653–4, 661
Savannah (ship) 35, 527
saw mills 231, 251
   textbook 236
scavenging (in two-stroke engine) 308–309, 315
Scharnebeck boat lift 498
Schauplatz der Muehlen-Bau-Kunst (Leupold) 235–6
Schedula Diversarum Artium (Theophilus) 196
Schelde–Rhine Canal 492
Schneider Trophy 630
schooners 524, 526
Schuylkill and Susquehanna Canal 511
science, distinct from technology 2–4
Scotch turbine 243
screen printing 683–4
screw cutting machines 392–4, 395, 403, 410
screw piles 541
screw press
   coining 28
   paper-making 253
screw propulsion 528, 529, 533
screws 391–2
   Archimedean 19
   thread: inspection gauges 415; standardization 401–403
   see also woodscrews
scribbling machines 836
sculpture
   glazed 191
   see also statuary
scutching (flax) 806
scutching machines 835–6
scythe 780–81
seaplanes 630
   see also flying boats
seed drills 773
Seine river, canalization 489
Selandia (ship) 323, 531
Selectavision videotape system 752
selenium 743, 744
semaphore 711–13
semiconductors 42, 383, 384
service stations 459–60
Sèvres porcelain 193
sewage disposal 921, 922, 956–7
sewing 851
sewing machines 407, 848, 851–2
sextant 544
shaduf 18
Shang bronzes 63–4
shaping machine (Nasmyth) 401
shear steel 159
shearing (cloth) 818
shears 15

Sheffield: iron and steel industry 160, 170, 240
'Sheppey stone' 466
sherardizing 96
shields 969, 970
   mediaeval 974
   Viking 972
shift working 151
   wartime 421
shingles (roof tiles) 857, 884
shipbuilding 551–3
ships 519–41
   canal and river craft 514–15
   cathodic protection 144
   engines *see* marine engines
   iron 156, 164
   nuclear-powered 215
   rudder 21, 520
   sailing 14, 520–26
   steam-powered 35, 37, 527–31
   Sutton Hoo burial 21
   see also brig; schooners; warships *etc.*
shrapnel 986, 990
Shropshire Union Canal 480
Shubenacadie Canal 509
shuttle, flying 821–2
sickle 780, 781
siege warfare 974–5, 982–3
Silbury Hill 470
silicon carbide 195
silk 805–806
   throwing 815; machines 805–806, 824; mills 815
   weighting 849
silver 69–72
   composites (contact materials) 138
   mining 67, 69–71, 79–80
Silver Jubilee (locomotive) 596–8
sintered materials 129–37
Sirius (ship) 35
sizing (cotton warp) 842–3
Sketches of the Coal Mines in Northumberland and Durham (Hair) 266
Skylab space station 654
skyscrapers 896–9
slag 151
   as fertilizer 170
slash and burn farming 776, 779
Slaski Canal 505
slate, building material 861, 884
slate industry, water power in 240, 244
slide-rule 698, 707
slide valve 282, 344–5, 404
slip (pottery) 191
slitting mill 152
'smart' munition 1008–1009
smelting 13
   copper 48, 51–5

iron *see* blast furnace; bloomery
lead 71
smock mills 249–50, 253, 255, 259
smoke jack 940
smokeless powders 988
'snail' *see* Archimedean screw
snap-lock pistol 977
Snowdon Mountain Railway 584
soaps 203–205, 923
Society for the Diffusion of Useful Knowledge 677
soda, production: by Leblanc process 221–2; by Solvay process 222
sodium, production 224
sodium chlorate, production 225
sodium discharge lamps 387, 961
soil drainage 772, 782
soil preparation 768–72
sonar *see* echo-sounder
sound recording 720–25, 755–6
sowing 772–3
space, war in 1010
Space Shuttle 654–5, 656, 661–2, 748
spaceflight 648–62
Spacelab project 655
spade 768–9
spark erosion 419
spark ignition engines 305, 309, 311, 316–19
    emissions control 318–19
    Wankel 325–8, 455–6
spectacles 197
Spectator, The 672–3
spear 967, 976
sphalerite 77, 78, 80
spin drier 931
spinning 16, 28, 808–810
    Casablanca systems 841
    gill boxes 840–41
    mechanization 823–30, 837–8
    open end (break) 842
    roving frame 839
    self-twist machine 842
    spindle and whorl 809–810
    spinning jenny 825–7
    spinning mule 832–3, 837–8
    spinning wheels 22, 813–15, 825
    throstle and ring frames 841–2
spit roasting 939
Spitfire fighter 632
spokeshave, stone 11
Spring (Brueghel) 780
springs 124
    furniture 907
Sputnik satellites 652, 656, 662
squinch construction 880
SS- 9 missile 1007
stained glass 22, 196

stainless steel 174–5, 300
    welding 418
standards
    ASCII codes 709
    bronze composition 65–6
    building components 900
    electrical 374, 377
    length measurement 18, 391, 697
    time measurement 697
Stanhope printing press 674–5
static electricity 350–51
    inverse square law 373
statuary
    aluminium ('Eros') 108–109, 111
    brass 75
    bronze 62–3, 67
    copper 62, 67–8
    lead 69
    zinc 74, 93
stave churches 865
steam bicycle 447
steam carriages 441–2, 457–8
steam engines 3, 31–7, 154–6, 272, 273–303, 343–4, 397, 453–4
    applications: in agriculture 772, 788; in aircraft 619–21; in airships 614; in iron-making 154, 155–6, 158, 166; in mining 207–208; in minting 28; in pottery industry 193–4; in textile industry 28–9, 833
    atmospheric 34–5
    automatic variable cut-off 281–3
    compound 278–9, 284–5, 530, 581–2; marine 286–7
    Corliss 282–3, 285
    Cornish 277–8
    double-acting 276–7
    high pressure 277
    high speed 284
    marine 286–7, 398, 403
    medium speed 283–4
    pumping 33, 275, 285, 952, 954
    rotative 325
    speed regulation 30–31, 284
    Uniflow 287–8
    valves 282, 344–5
    see also locomotives, steam
steam shovel 470–71
steam trams 456–7, 583
steam tricycle 447
steam turbines 3, 37, 288–303, 346–7
    cross-compound 298
    Curtis 293–4
    high pressure 298, 300
    high speed 299
    high temperature 300–301
    hybrid 297
    impulse 290, 294

steam turbines—*cont*
  marine engines 530–31, 535
  Parsons (reaction) 290–91
  power generation 295–303, 385
  Rateau 291–3, 294
  reheat cycle 299–300
  supercritical 300
  superposition 298
steamship 35, 37, 527–31
Stecknitz Canal 494
steel 18, 146
  alloy (special) 172–5
  applications: as building material 850, 896–9; in drill bits 212; in gun barrels 987; in ships 529; in water pipes 954; *see also* reinforced concrete
  crucible 159–60, 397
  nickel 98
  plastic–coated 180–81
  pre-painted 181
  production 159–60, 167–85; by basic oxygen (L.D.) process 177, 179; by Bessemer process 167–70, 171, 172, 194; by cementation 159; electrification 381; gas engines for 314; by Siemens–Martin open-hearth process 171–2; by Thomas process 170–71
  shear 159
  stainless 174–5, 300, 418
  Wayland's 128–9
steering mechanisms
  ships' 21, 520, 523
  vehicles' 449
Stellite (alloy) 127, 412
stenter 817–18
stereophonic sound 723–4
stereoscopy 735
stereotypes (printing plates) 676
Stirling engine 342
stirrup 13, 21, 971
stoas 858
stocking-knitting machines 820, 847
Stockton & Darlington Railway 560, 563
Stone Age 4, 6, 9–11
stone bridges 462
stone circles 698
stone tools 9–11
stone weapons 11, 967, 968
storage heaters, electric 386, 917
Stourbridge Lion (locomotive) 562
strain gauges 420
Strategic Defense Initiative 1008
stratified charge engines 306–307, 319
Stratoliner airliner 634
streamlining, locomotive 596–7
street lighting 363, 387, 961
street tramways 456
streets, surfacing for 436

strip farming system 771–2
strip mill 180, 181
stroboscope 739
Strömsholms Canal 502
stucco 860–61, 885, 886, 904
sub-contractors 410
sub-machine gun 1004
submarines 144, 535, 537–8, 997–8, 999, 1000–1001
  one-man 996
suction cleaners 925–7
Suez Canal 505, 525
sugar 801
sulphur dioxide, bleaching agent 205
sulphuric acid 205, 221–3
Sumerians
  armour 968
  bronze artefacts 58
  copper artefacts 57, 62
  cuneiform script 15, 666
  decline 768
  wheel 13, 16, 431
Summa perfectionis (Geber) 220–21
sun-and-planet gear 276
sundial 694
  portable 22
'superalloys' 126, 128
supercharged engines 309, 315–16, 322, 332
superconductors 144–5
superheterodyne tuning circuit 727
superphosphate fertilizers 224, 775
supersonic aircraft 39–40, 638–9
surface roughness gauge 415, 423
surveillance devices 1010
surveying
  hydrographic 546, 547
  trigonometrical 461
surveying instruments 24, 25
Surveyor spacecraft 659
suspension bridges 465–6
Sutton Hoo burial ship 21
swords 968, 979–80, 988
  bronze 66
  Norman 972
  Viking 972
  Wayland's 128–9
swidden *see* slash and burn farming
synchromesh gearbox 452
Syncom satellites 657, 747
systems building 900

2.4.5-T 778
tablet weaving 811
Tacoma Narrows bridge 465
tally iron 939
Talyllyn Railway 575, 606
Talyrond 423

INDEX OF TOPICS

Talysurf 415, 423
Talyvel 423
tankers 531, 532
tanks 472, 933–4, 1002–1004, 1009
tantalum carbide tools 137
tantalum lamp filament 133
tape recorders 722, 724–5
Tatler 672–3
taxes
  brick 869
  petrol 37
  salt 204, 221
Tay bridge disaster 463
Technicolor process 735
technology
  distinct from science 2–4
  place in history 1–2
Teldec system 750
telectroscope 744
telefax 744
telegraphone 722
telegraphy 710–15
  electric 357, 572, 574, 713–15, 964–5;
    railway signalling system 714, 965;
    wartime role 988–9
  visual 711–13
  wireless 726; *see also* radio
telephone 716, 717–20, 965–6
  field 990
teleprinter 715, 717
telescopes 24–5, 196, 730
teletex service 717
Teletype service 717
televison 744–6
  direct broadcast by satellite 748
  use of videotape 751–2
telex 716–17
Telstar satellite 657, 747
Tennessee–Tombigbee Canal 514
tenter frame 817
terra cotta 871
terrace fields 767–8
terry towelling 846
Terylene 220, 850
textile fibres 804–808
textile industry
  factory system 28
  water power in 239, 241
  see also spinning; weaving; *etc.*
textiles, furnishing 906–907
Thailand 59
Thames
  flood barrier 956
  water supply from 951
thatch 857, 884
thaumatrope 739
thermionic valves 41, 727
  see also diode; triode

thermonuclear bombs 1006–1007
threshing 783–5
threshing machines 267, 268
throstle frame 841
tidal mills 20
tidal power 245
tiles 870–71, 885
  floor 904
timber-frame buildings 861–4
time measurement 3–4, 22–4, 392, 694–7
Times, The, printing of 675, 676, 680
Timna (Negev) 52, 54, 57, 67
tin 59–60
tin alloys *see* bronze
tinder box 911
tinder pistol 911
tinplate 180, 181
Tiros meteorological satellites 657
Titan missile 1007
titanium and titanium alloys 141–4
  expolosive compaction 139
TNT 988
toasters, electric 939
tongs 15
'tops and bottoms' process (nickel
  extraction) 100
torcs, arsenical copper 61
torpedo boats 534
torpedoes 534, 537, 997, 1000
  'human' 1002
Tour Through the Whole Island of Great
  Britain (Defoe) 434
towelling 847
tower buttress 877
tower mills 20, 246, 248
  see also smock mills
town gas *see* coal gas
trabeated construction 857–9
tracer ammunition 995
trachite 199
tracked vehicles 472
tractor shovel 471
tractors, agricultural 788–91
traffic islands 460
traffic lights 460
traffic signs 460
tramp steamers 530
trams 583
  electric 381, 584, 959–60
  horse 583
  steam 456–7, 583
tramways 161, 456, 555–8
transfer machines 417
transfer printing (pottery) 194
transformers 371
transistor radios 728
transistors 42, 419, 703
transmission cables 371–2

1059

INDEX OF TOPICS

transplanting 773
transporter bridges 496, 518
trapetum 263
Travels (Marco Polo) 19, 27
treadle 25, 390
treadmills 262
  oblique 262, 265, 269
  prison 264
treadwheels 262, 263–4
  oblique 262, 266, 268–9
Treatise on Mills and Millwork (Fairbairn) 242
trebuchet 974–5
trenails 14
trench mortar 991
trench warfare 983, 989
Trenton–Severn waterway 508–509
tricycles, powered 447–8, 451
triode 41, 720, 726–7
triple expansion steam engine 285, 286, 530
trolleybuses 457
trunnions, gun 978
truss bridges 463
trussed roofs 864, 865–6
TT races 448
tungsten 133–4
tungsten carbide 135, 138, 417
tungsten filament lamps 386–7, 915
tungsten steel 174, 412
tunnels 467–70
  canal 478, 479, 484, 488, 506
  irrigation 467–8, 768
  railway 468, 470, 565, 603
Tupolev Tu-144 aircraft 639
Turbinia (ship) 37
turbines *see* gas turbines; steam turbines; water turbines; wind turbines
turbocharging *see* supercharged engines
turbofan engines 336, 338, 638
turbojet airliners 637
turbojet engines 330, 336
turboprop engines 334, 336, 636–7
turboshaft engines 643
Turkey, bronze artefacts 60
turnpike roads 433, 963
turnspit
  dog-powered 939–40
  electric 377–8
turpentine 211
Turtle (ship) 537
tuyeres, Bronze Age 60
two-stroke engines 307–309, 322, 454
type-casting 676
type metal 670–71, 673
typefaces 671
typesetting 679–80
typewriter 682–3

tyres
  iron 439
  pneumatic 37, 217, 436, 446, 449; tractor 790–91
  solid 217
Tyrian purple 200

U-boats 538, 999, 1000–1001
Uelzen lock 498
ultrasonics 419
U-matic videotape system 752
umbrellas 853
underground railways 382, 468, 583, 584, 592
Uniflow engine 287–8
Union Pacific Railroad 574
UNIVAC 702
upholstered furniture 907
Ur 57, 58, 60, 62, 69
uranium, nuclear fission 214
uranium glazes 194
Uranus probe 661

V-1 flying bomb 1005
V-2 rocket bomb 649–51, 661, 1005
V-mail (microfilm) 734
vacuum 32–3, 273
vacuum cleaners 925–7
valves
  drop 282
  poppet 282
  slide 282, 344–5, 404
  thermionic 41, 727, *see also* diode; triode
  Wankel rotating disc 328
Van Harinxma Canal 492
Van Starkenborgh Canal 492
Vanguard rocket 656
Variomatic belt dirve 452
vat dyes 199
vaulting 874–9
Velcro fastener 852–3
velocipede 445
  steam-powered 447
Velox boiler 332, 333
velvet 847
veneers
  plastic laminates 906
  wood 908
Venera spacecraft 659–60
Venice
  glass 196
  horses of San Marco 67–8
Venus at the Forge (Brueghel) 232
Venus probes 659–60
vernier scale 25
Versailles, water supply to gardens 235
videodiscs 753–4
Videofile videotape system 751–2

videotape 751–2
Viking spacecraft 660
Viking weapons 972
viscose rayon 850
Viscount airliner 637
Vitallium (alloy) 127
Vitaphone recording system 723
Vitruvian mill 19, 20, 230, 233, 270
Vliet (canal) 490
Volga–Don Canal 503
Volkswagen Beetle 452
voltaic pile 41, 94–5, 352
voltmeters 373–4
Von der Wasser-Muehlen und von dem inwendigen Werke der Schneide Muehlen (Kaovenhofer) 236
Vortex turbine 243, 244
Voshkod spacecraft 652
Vostok spacecraft 652
Voyager spacecraft 660–61
vulcanization of rubber 216–17, 849, 853

wagons 434, 440
wainscot 902
Walkman (cassette player) 725
wall coverings 902–903
wallpaper 903
wallower 270
Wankel engine 325–8, 455–6
Warrior (ship) 533
warships 520, 532–7, 980, 987–8
 dockyards for 552
wash boilers 929
washing
 personal 917, 918–20
 textiles 928–31
washing machines 929–31
Washington Monument 106
Wasserräder und Turbinen, Die (Henne) 242
waste disposal 962–3
watches 24, 392, 696, 697
water clock 16, 22–3, 392, 694
water closets 921–2
water conservation 954–5
water desalination 955
water heaters 919–20, 928–9
water power 22, 229–45
 applications: in ironworking 150, 152, 154; in textile industry 817, 824, 829–30
 see also hydro-electric power; tidal mills; tidal power; watermills; waterwheels
water-raising devices 231, 235, 263, 273
water slopes 484, 485–6
water supply 19–20, 72, 234–5, 285, 918, 919, 950–55
water towers 954
water turbines 241, 242–5

watermills 19, 20
 glossary of terms 269–70
 Greek (Norse) 19, 229–30, 270
 Smeaton's designs 236–7
waterproof clothing 849, 853
waterwheels 269–70
 horizontal 19, 230; *see also* Norse mill
 iron 239
 large 240, 241
 suspension 239
 textbooks 233–4, 235–6
 ventilated bucket 241–2
 vertical 230
 for water supply 234–5
wattle and daub construction 856
wattmeter 374
waveforms, visualization 377, 720
waywisers 461
Weald, ironworking in 150, 156, 232
weapons 967–1011
 see also guns; swords; tanks; *etc.*
weaving 28, 810–12, 815–16, 820–23
 basketwork 15
 pattern 822–3, 844
 see also looms
wedge-shaped script *see* cuneiform script
weed control 777–8
Weespertrekvaart 491
Wege zur Raumschiffart (Oberth) 747
weights, standard 18
weld (dyers' weed) 200
welding 417–20
 robot 426
Welland Canal 507
Wellington bomber 632
Wessel–Datteln Canal 497
Wharfedale press 677
wheat 11–12, 764
Wheatstone bridge circuit 375
wheel 13, 16, 431
 coach 438–9
 minimum width 434
 potter's 12, 16
wheel-lock pistol 977
Whirlwind computer 702, 703
whitening agents 205
Widia (sintered alloy) 135, 137
Wimshurst machine 351
wind engines 257–60
wind power 14, 20, 245–60
 see also sailing ships; windmills
wind turbines 259
windlass, horse-powered 264
Windmill by a River (van Goyen) 248
windmills 20, 245–6, 259
 glossary of terms 269–70
 sails 246, 247, 255–8
 winding 257

window glass 22, 196
winnowing 784
Winschoter Diep 492
wire drawing 82, 85, 99
wireless telegraphy 726
  see also radio
wirephoto process 744
woad 199, 200
women in industry 420
wood
  bridges 463
  building material 856, 857, 861–6, 905, 906
  fuel 206, 915
  furniture 908
  gas from 40
  gears 16, 17
  millwork 19, 233, 238
  rails 555, 557
  ships 519–25, 526, 551
  tools 9
  wall covering 902–903
  water pipes 954
wood ash, for soap-making 204, 206
wood turning 389
woodcuts 669, 673–4, 678
woodscrews 14
wool 806–807
  carding machines 836
  combing 809, 839–40
  fulling 816–17
  spinning 842
  see also worsted
Woolf engines 279, 285, 286
World War I
  agriculture 789
  air warfare 994–5
  aircraft construction 218, 628
  airships 615
  artillery and trench warfare 989–94
  balloon barrage 613
  dyestuffs industry 202
  internal combustion engine 593
  naval warfare 996–8
World War II 420–21

agriculture 786–7
air warfare 536, 632–3, 998–1000
atomic bomb 214, 1005–1006
balloon barrage 613
land warfare 1002–1005
naval warfare 536, 1000–1002
plastics and rubber development 219
radar 545, 729
shipbuilding 531
V-1 flying bomb 1005
V-2 rocket bomb 649–51, 661, 1005
V-mail 733
worsted, spinning 841
writing 15, 666–9
wrought iron 147–9, 155, 162, 164–6
  drill bits 212
  gears 17
  water pipes 954
Wylam Dilly (locomotive) 560

X-ray tube 134
X-rays 725–6
xerography 684, 736
xylographica 669
xylyl bromide 992

Y alloy 122
yarn 804
Yngaren (ship) 325
Young Mill-Wright and Miller's Guide (Evans) 238

Zederik Canal 491
Zeppelins 617
zinc 74, 76–9, 80–81
  applications 94–6
  in bronzes 67, 73, 74
  Chinese 79, 80
  production 76, 77–8, 87–8, 92–4
zinc alloys *see* brass; paktong
zincographs 679
zip fasteners 852
zoetrope 739
zoopraxiscope 740
Zuid-Willemsvaart Canal 493–4